Coastal Research Library

Volume 25

Series Editor
Charles W. Finkl
Department of Geosciences
Florida Atlantic University
Boca Raton, FL 33431, USA

The aim of this book series is to disseminate information to the coastal research community. The Series covers all aspects of coastal research including but not limited to relevant aspects of geological sciences, biology (incl. ecology and coastal marine ecosystems), geomorphology (physical geography), climate, littoral oceanography, coastal hydraulics, environmental (resource) management, engineering, and remote sensing. Policy, coastal law, and relevant issues such as conflict resolution and risk management would also be covered by the Series. The scope of the Series is broad and with a unique cross-disciplinary nature. The Series would tend to focus on topics that are of current interest and which carry some import as opposed to traditional titles that are esoteric and non-controversial. Monographs as well as contributed volumes are welcomed.

More information about this series at http://www.springer.com/series/8795

Christopher Makowski • Charles W. Finkl
Editors

Threats to Mangrove Forests

Hazards, Vulnerability, and Management

Springer

Editors
Christopher Makowski
Coastal Education & Research
Foundation (CERF)
Coconut Creek, FL, USA

Charles W. Finkl
Coastal Education & Research
Foundation (CERF)
Asheville, NC, USA

Department of Geosciences
Florida Atlantic University
Boca Raton, FL, USA

ISSN 2211-0577 ISSN 2211-0585 (electronic)
Coastal Research Library
ISBN 978-3-030-10301-9 ISBN 978-3-319-73016-5 (eBook)
https://doi.org/10.1007/978-3-319-73016-5

Printed on acid-free paper

This Springer imprint is published by the registered company Springer International Publishing AG part of Springer Nature.
The registered company address is: Gewerbestrasse 11, 6330 Cham, Switzerland

Preface

Throughout the world, mangrove forests serve as a vital part to the stability and health of many coastal environments. These intertidal conduit forests connect land with sea, creating a natural protection defense along the coast and producing an intrinsic ecosystem of which numerous animals, fish, and vegetation rely on for survival. However, mangroves face copious pressures today, both natural and anthropogenic, which threaten their ability to serve as shoreline buffers, juvenile nurseries, and breeding grounds. This volume in the *Coastal Research Library* (CRL) not only delves into the current threats that mangroves face but also focuses on strategic management designs and conservation practices being applied as researchers look to the future. To this end, the volume is divided into five main parts: Part I, Natural Mangrove Systems; Part II, Threats, Vulnerability, and Impacts; Part III, Pollution and Contamination by Hydrocarbons and Heavy Metals; Part IV, Assessment Techniques, Ecosystem Design, and Management Strategies; and Part V, Conservation, Rehabilitation, and Governance. These general subject-area parts are in turn subdivided into chapters that are either archetypes of various hazards, vulnerabilities, and impacts or vignettes demonstrating how various approaches to mangrove conservation and ecosystem design are being applied, either conceptually or in principle.

The following collection of broad-ranging topics provides insight into the common thread that is shared among many researchers of mangrove environments. For example, the geographical scope of this volume ranges between the tropical and subtropical latitudes and offers a wide array of mangrove coastal zones found on five continents. Yet, despite being found in different parts of the world, these mangrove coasts encounter similar impacts and vulnerabilities, especially when it comes to interference from human action, such as urbanization, industrialization, and commercialization. With many of the world's mangroves already compromised, it is important to assimilate the research techniques and field results reported in the following chapters and to recognize the valuable conservation efforts being applied to protect, rehabilitate, and govern these fragile ecosystems.

Part I (Natural Mangrove Systems) contains four chapters highlighting general examples of highly variable mangrove ecosystems found throughout the world. Chapter 1 (Australian Mangroves: Their Distribution and Protection), by Ian D. Cresswell and Vic Semeniuk, discusses the full biodiversity of Australian mangroves, including species occurrences, biogeographic limits, and community assemblages, in order to facilitate conservation efforts and sustainability in these areas. Chapter 2 (The Dynamics of Expanding Mangroves in New Zealand), by Erik M. Horstman, Carolyn J. Lundquist, Karin R. Bryan, Richard H. Bulmer, Julia C. Mullarney, and Debra J. Stokes, presents the unique ability of New Zealand's mangroves, found at the southern latitudinal extent of their global range, to rapidly expand and how varying political viewpoints of such an expansion can lead to misinformed mangrove removal practices. Chapter 3 (Mangrove Forests of the Persian Gulf and the Gulf of Oman), by Alireza Salehipour Milani, quantifies mangrove forest cover throughout the Middle East by using habitat distribution maps generated by Landsat Thematic Mapper (TM) satellite imagery. These mangrove distribution maps are then used for resource inventory cataloging, change detection analysis, and the selection of specialized aquaculture sites. Chapter 4 (Current Status of Mangrove Wetlands in Sinaloa: A Biological Corridor Along the Eastern Margin of the Gulf of California, México), by Marlenne Manzano-Sarabia, Olivia Millán-Aguilar, Francisco Flores-Cárdenas, Lidia Rodríguez-Arredondo, Mayra I. Grano-Maldonado, and Mario Nieves-Soto, continues in a similar vein by describing the mangrove coverage along this Mexican biological corridor and how diverse drivers of change have shaped the coastal landscape over time. Chapter 5 (Geospatial Assessment of spatial and temporal Changes in Mangrove Vegetation of Pichavaram Region, Tamil Nadu, India), by M. Vani and P. Rama Chandra Prasad, uses satellite imagery to examine general spatial and temporal changes in mangrove vegetation along coastal regions of India. This chapter offers encouraging results by showing net area increase over time while presenting an accurate picture of the extent and condition of this particular mangrove ecosystem.

Part II contains seven chapters and deals with general threats, vulnerabilities, and impacts that mangrove forests are either currently being subjected to or have already succumbed to. Chapter 6 (Natural Threats and Impacts to Mangroves Within the Coastal Fringing Forests of India), by Ashis Kr. Paul, Amrit Kamila, and Ratnadip Ray, reports on the natural threats and impacts of mangroves to India's East Coast. These natural causes of loss to the coastal fringing forests include cyclone landfalls, seismic activity with tsunami events, increased hypersalinity, and current sea-level rise. Chapter 7 (An Assessment of Vulnerability and Adaptation of Coastal Mangroves of West Africa in the Face of Climate Change), by Isaac Boateng, assesses the vulnerability of West African mangroves to current climate change effects, as well as to ongoing anthropogenic activities. For example, the natural impacts from increased storm intensity coupled with anthropogenic influences (e.g., urbanization, population growth, clear-cutting) have led to the degradation of mangrove ecosystems in West Africa. Chapter 8 (Historical Losses of Mangrove Systems in South America from Human-Induced and Natural Impacts), by Daniel Gorman, presents a historical and contemporary perspective on the drivers and degree of mangrove loss in South America. Focusing on those countries with the most up-to-date accessible data (i.e., Brazil, Ecuador, and the Guineas), an emphasis is placed on the current and future threats these degraded forests must face. Chapter 9 (Australian Mangroves: Anthropogenic Impacts by Industry, Agriculture, Ports, and Urbanization), by Vic Semeniuk and Ian D. Cresswell, discusses the impacts of various human activities upon Australian mangrove ecosystems. The unique approach of this chapter shows that mangrove threats can be either of an overt (direct) nature, such as clearing or burial, or of a covert (indirect) nature, such as an alteration to the ecosystem's hydrochemistry or a drastic variation in the coastal area's salinity regime. Chapter 10 (Determining the Influence of Urbanization on Mangrove Zones of Northeastern Brazil: Characterization of Ceará State Coastal Zone Organic Matter Inputs), by Stéphane Jean Louis Mounier, Rozane Valente Marins, and Luiz Drude Lacerda, focuses on the vulnerability of river-mangrove systems in northeastern Brazil to the effects of ongoing urbanization. This was accomplished through an organic matter characterization, which helped determine how the pressure of urbanization is adversely impacting these mangrove areas. Chapter 11 (Vulnerability of Mangrove Forests and Wetland Ecosystems in the Sundarbans Natural World Heritage Site (Bangladesh)), by Shafi Noor Islam, Sandra Reinstädtler, and Albrecht Gnauck, investigates vulnerabilities of the coastal deltaic mangrove forest within the Sundarbans Natural World Heritage Site of Bangladesh. This chapter shows hypersalinity, storm effects, agricultural crop production, urbanization, and water intrusion as being among the threats mainly responsible for mangrove degradation.

Part III contains four chapters and examines specific impacts to mangroves from the pollution and contamination of hydrocarbons and heavy metals. Chapter 12 (The Impact of Oil and Gas Exploration: Invasive Nypa Palm Species and Urbanization on Mangroves in the Niger River Delta, Nigeria), by Aroloye O. Numbere, focuses on oil and gas exploration activities around the Niger Delta, which began in 1956 and is responsible for the greatest damage to mangrove forests in this region. This petrochemical pollution then leads to further degradation of mangroves through increased urbanization and the introduction of outcompeting alien species, such as the nipa palm (*Nypa fruticans*). Chapter 13 (Oil-Related Mangrove Loss

East of Bonny River, Nigeria), by Erich R. Gundlach, investigates one of the largest cumulative losses of mangroves in Africa related to oil spillage and pipeline corridor construction. This tragedy is further exasperated by the fact that some of these affected areas may never truly recover due to high concentrations of surface and subsurface standing oil. Chapter 14 (Sediment Hydrocarbons in Former Mangrove Areas, Southern Ogoniland, Eastern Niger Delta, Nigeria), by David I. Little, Kay Holtzmann, Erich R. Gundlach, and Yakov Galperin, probes another impacted area of the Niger Delta, which contaminated much of the local soil and traditional fishing resources. Methods included quantitative sediment sampling to aid in determining the degree of contamination from hydrocarbon concentrations. Chapter 15 (Heavy Metal Distribution and Accumulation from Natural and Anthropogenic Sources in Tropical Mangroves of India and Bangladesh), by Prabhat Ranjana, Karuna Raob, Alok Kumar, and A.L. Ramanathan, identifies trace metal distribution and accumulation throughout the Indian and Bangladeshi estuarine-mangrove complex. Bioaccumulation in fish, gastropods, and bivalves lends further credence to adverse impacts on mangrove biodiversity in this region of the world.

Part IV contains nine chapters and covers assessment techniques, ecological designs, and management strategies with particular emphasis on creating a balance between human needs and mangrove system requirements. This part of the book is thus a logical follow-up to the previous two sections that primarily identified threats or risks to mangrove environments. Awareness of the problem is the obvious first step toward the goal of mangrove conservation; however, the next phase entails addressing logical approaches and techniques to achieve that goal. The following chapters are examples of innovative ideas that came to life in an effort to protect mangrove forests. Chapter 16 (Ecosystem Design: When Mangrove Ecology Meets Human Needs), by Martin Zimmer, introduces a new five-step conceptual design to reconstruct one of the most productive ecosystems in mangrove forests. One of the novel ideas presented in this chapter is the inclusion of ecological processes driven by fauna and microbiota to achieve a functioning mangrove ecosystem. Chapter 17 (Toward Sustainability and Protection of Threatened Coastal Ecosystems: Management Strategies for a Rare Stone Mangrove in Gorda Beach, Armação dos Búzios, Brazil), by Gerson Cardoso da Silva Junior, Kátia Leite Mansur, and Marcelo Obraczka, reports the rare occurrence of *Avicennia* and white mangroves growing on a rocky substrate, prompting increased management of such an oddity. Management strategies proposed include creation of a comprehensive environmental protection unit, education and environmental awareness, dissemination of knowledge about the importance of mangroves, and social control and mobilization of the expanding population. Chapter 18 (Assessment and Management Strategies of Mangrove Forests Alongside the Mangsalut River Basin (Brunei Darussalam, Island of Borneo)), by Shafi Noor Islam, Nurin Hidayati Hj Abd Rahman, Sandra Reinstädtler, and Nur Azri Bin Aladin, discusses assessment and management techniques in Brunei Darussalam, a small country on the island of Borneo. Those techniques proposed involve an increase in the stewardship over coastal mangrove biodiversity, a more hands-on approach to solving socioeconomic problems that lead to coastal degradation, and an increase in public awareness concerning coastal ecology, environmental training, and management education, particularly related to the importance of the sustainable use of coastal mangroves and other natural resources. Chapter 19 (Mangrove Establishment in an Artificially Constructed Estuarine Channel, Sungei Api Api, Singapore), by Suzanna Ramos and Patrick Hesp, demonstrates the advantages of incorporating mangroves into urban tropical environments. This is achieved through the artificial establishment, and subsequent natural regeneration, of mangroves along a man-made estuarine corridor in Singapore. Chapter 20 (Interaction of Mangroves, Coastal Hydrodynamics, and Morphodynamics Along the Coastal Fringes of the Guianas), by Erik A. Toorman, Edward Anthony, Pieter G.E.F. Augustinus, Antoine Gardel, Nicolas Gratiot, Oudho Homenauth, Nicolas Huybrechts, Jaak Monbaliu, Kene Moseley, and Sieuwnath Naipal, offers different modeling techniques to aid in the management of the South American coastal zone. Through the numerical prediction of ocean currents, wave action, and sediment transport, coastal zone managers will be better equipped to protect mangrove forests along shorelines. Chapter 21 (The Beneficial Effects of Mangrove Forest to Sea Defense Structures), by Henk Jan Verhagen, explains the benefits of using mangroves in tandem with flood defense structures (e.g., dikes, levees) to protect the hinterland against storm surge from typhoons, hurricanes, and tsunamis. Not only do mangroves offer an economic advantage, but they also greatly decrease the wave height in front of the defense structure. Chapter 22 (Remote Sensing of Mangrove Forests: Current Techniques and Existing Databases), by Stuart Hamilton, Gustavo Castellanos, Marco Millones-Mayer, and Mara Chen, examines the major remotely sensed mangrove databases from around the world and summarizes the methodological differences between them. By doing so, recommendations are made to the remote sensing community on how improved techniques can better monitor and manage mangrove forests at risk. Chapter 23 (Urban Mangrove Biology and Ecology: Emergent Patterns and Management Implications), by Benjamin Branoff, presents an urban mangrove management model based on mangrove physiology, function, and growth in response to increasing urbanization. By applying this model to affected regions, the future of mangrove management along urbanizing coasts can be vastly improved. Chapter 24 (High-Throughput Techniques as Support for Knowledge-Based Spatial Conservation Prioritization in Mangrove Ecosystems), by Véronique Helfer and Martin Zimmer, shows the effectiveness of establishing spatial prioritizations when planning mangrove protection measures. This management approach prioritizes the societal, biological, and economic values of mangroves while integrating adjacent ecosystems to maintain spatial connectivity throughout the entire ecosystem. In such a way, the spatial and temporal dynamics of mangrove environments, as well as the neighboring communities, are properly managed.

Part V contains seven chapters related to the overall conservation, rehabilitation, and governance of mangrove forests throughout the world. Each of the following chapters cites a specific example of how the conservation and rehabilitation of mangroves are being governed. As one reads through these examples, it is evident that productive efforts are being applied in various coastal regions to promote mangrove survival and longevity. Chapter 25 (Mangrove Concessions: An Innovative Strategy for Community Mangrove Conservation in Ecuador), by Fausto Vinicio Lopez Rodriguez, shows how mangrove conservation is being implemented in Ecuador through sustainable use practices, which include providing legal security to traditional mangrove users, promoting the participation of local communities to aid in the governance of mangrove areas, and restricting the amount of biodiversity that is inadvertently caught in mangrove ecosystems. Chapter 26 (Conserving Mangroves for Their Blue Carbon: Insights and Prospects for Community-Based Mangrove Management in Southeast Asia), by Dixon T. Gevaña, Leni D. Camacho, and Juan M. Pulhin, describes the blue carbon stock potential of the region's mangrove forests. In this context, the development of community-based payments for ecosystem services (PES) schemes is a way of governing mangrove conservation efforts to offset declining trends. Chapter 27 (Mangrove Habitats in São Tomé and Príncipe (Gulf of Guinea, Africa): Conservation and Management Status), by R. Haroun, A. Herrero Barrencua, and A.D. Abreu, discusses the conservation of mangroves through ecotourism activities while postulating that the restoration of remnant mangrove habitats can help support community development and job opportunities in Western Equatorial Africa. Chapter 28 (The Success of Hydrological Rehabilitation in Mangrove Wetlands Using Box Culverts Across Coastal Roads in Northern Yucatán (Southeastern México)), by Claudia Teutli-Hernández and Jorge A. - Herrera-Silveira, presents various restoration, rehabilitation, and reforestation methods being conducted in the northern Yucatan peninsula. While the success of such measures is still being quantified, there is awareness that rehabilitation of mangroves must occur at several localities. Chapter 29 (Mangroves on the Brazilian Amazon Coast: Uses and Rehabilitation), by Marcus E.B. Fernandes, Francisco Pereira Oliveira, and Indira A.L. Eyzaguirre, reports on the initiative of the local Brazilian community to reforest degraded mangrove sites, which then leads to the rehabilitation of surrounding mangrove ecosystem communities. For example, the establishment of new mangrove forest stands and the return of an important invertebrate species, *Ucides cordatus* (a mangrove crab), marked the success of this rehabilitation. Chapter 30 (Mangrove Restoration and Mitigation After Oil Spills and Development Projects in East Africa and the Middle East), by David I. Little, focuses on the protection, restoration, and integration of mangrove forests into landscape-scale conservation projects. This chapter also describes some intergovernmental and industry initiatives for oil spill prevention, preparedness, and response within coastal mangrove areas. Chapter 31 (Environmental Governance as a Framework for Mangrove Sustainability on the Brazilian Amazon Coast), the last chapter in the book by Indira A.L. Eyzaguirre and Marcus E.B. Fernandes, shows the value of considering mangrove ecosystem resources as a type of conservation unit. Once that designation has been granted, this type of environmental governance helps to generate sustainable development through social responsibility, the accountability of management, and the promulgation of laws that enable the protection of mangrove forests.

This volume is but a glimpse of the global situation confronting mangrove forests today, which includes multiple threats of both a natural and man-made influence. Chapters in this book selectively offer a geographical range of current mangrove research being conducted today in the tropics and subtropical latitudes via studies in New Zealand, the Middle East, Southeast Asia, Brunei Darussalam, Brazil, Bangladesh, East Africa, Nigeria, Australia, India, Singapore, Ecuador, Mexico, and the Guianas. Carrying a common theme of mangroves under stress, these vignettes show that most of the world's mangrove forests have already been adversely impacted by anthropogenic urbanization and the development of industrial and commercial infrastructure (e.g., petroleum exploitation, wood harvesting and clearing of surrounding forests, hypersalinity, sand extraction, agriculture, aquaculture farming, fishery development, land erosion, sediment deposition, recreational uses, etc.). Once the man-made threats are coupled with natural pressures (e.g., sea-level rise, increased storm frequency), there is definitely a need for countermeasure action to be taken in support of coastal mangroves.

As such, the other main theme in this volume focuses on the management strategies and conservation techniques being applied to save mangrove forests throughout the world. Innovative ecosystem designs and sustainability assessment methods are receiving more and more acceptance by local municipalities in order to create hospitable environments for mangroves to once again thrive. This shift in rehabilitation governance allows for conservation efforts that are not only restricted to the scientific community but also extend to the social communities that must live in harmony with mangrove ecosystems. Only through widespread awareness of the beneficial nature mangrove forests offer can local populations and governments enable a truly successful global conservation effort. The road ahead for such an effort is a long one, but this text might serve as a good launching point to reveal how different parts of the world are dealing with struggles related to mangrove survival. Through the dissemination of this volume's knowledge, researchers, managers, and the public alike can begin to work together to combat threats facing mangrove survivability in the future.

Coconut Creek, FL, USA
Asheville, NC, USA

Christopher Makowski
Charles W. Finkl

Contents

The original version of this book was revised. An erratum to this chapter can be found at https://doi.org/10.1007/978-3-319-73016-5_32

Part I

Natural Mangrove Systems

Australian Mangroves: Their Distribution and Protection

1

I. D. Cresswell and V. Semeniuk

Abstract
While coastal systems play a key role in the culture, economy and environment of Australia, perhaps the least understood of all Australian coastal systems are its mangroves and salt marshes. Subsequently their protection and conservation has not always been adequately or comprehensively addressed. At the national level there is no coordination of mangrove conservation with each State or Territory managing mangroves within its jurisdiction under a wide range of laws and policies. While at a broad level the importance of mangrove habitat has become better known, and decisions at local and state level in Australia now take into account the need to conserve and manage these coastal assets, over the past 30 years there has also been increasing development pressure which has resulted in clearing or degradation of many areas of mangroves. A national approach is required for implementing consistent management based not only on species but also on mangrove communities, using an approach of community assemblage combined with physical habitat, i.e., a "community-*cum*-habitat" approach, with assemblages viewed in an inter-related suite of habitats.

Current estimates of overall mangrove vegetation cover in Australia range from ~ 900,000 ha to over 1.1 million ha. Nearly every State and Territory in Australia has some mangrove-lined coastal areas but overall coverage is unevenly distributed because of climate and heterogeneity of habitat, with Queensland and the Northern Territory supporting 85% of Australia's mangroves. The distribution of mangroves along the Australian coast can be seen at the sub-continental level as a relatively simple relationship with latitude and climate but this does not account for the full complexity at finer scales where responses become more related to habitat variability, and/or effects of the hinterland in terms of run-off and seepage, and geochemically diverse soils.

Today, some 19% of mangrove area is in protected areas, however the comprehensiveness, adequacy and representativeness of that protection is unknown. One factor working contrary to their conservation has been the notion that "mangroves" are a single unit, ubiquitous along the coast and therefore their degradation has little impact on overall biodiversity or broader ecosystem services. This reflects that within Australia there has been limited adoption of ecosystem-based management and limited coordination of efforts across jurisdictions, and none for mangrove ecosystems. There is a requirement to understand mangrove distribution in Australia in relation to global trends in species diversity and the unique and heterogeneous suites of habitats within Australia.

To comprehensively capture the full biodiversity of Australian mangroves for conservation, it is necessary to address (1) species occurrences, (2) species at their biogeographic limits, (3) community assemblages incorporating structure and floristics, (4) the community-*cum*-habitat approach, and (5) assemblages in inter-related suites of habitats. Australia offers mangrove ecosystems that are globally unique with regard to floristic assemblages, structural and phenotypic responses to their climate and habitat settings, their community-*cum*-habitat expression, and suites of habitats.

Currently Australia's reserve system is ad hoc for mangrove communities with some areas having further "protection" through listing under international agreements. In terms of mangrove protection it is evident that World Heritage and Ramsar Convention listings provide additional levels of protection in addition to national designations as National Park, Nature Reserve, or any of the many other designations of protected areas. Other global conservation instruments deal with mangrove

I. D. Cresswell (✉)
CSIRO Land and Water, Hobart, TAS, Australia
e-mail: ian.cresswell@csiro.au

V. Semeniuk
V & C Semeniuk Research Group, Warwick, WA, Australia

C. Makowski, C. W. Finkl (eds.), *Threats to Mangrove Forests*, Coastal Research Library 25,
https://doi.org/10.1007/978-3-319-73016-5_1

ecosystems only peripherally with no specific instruments providing additional protection for mangroves than exist within jurisdictional legislation.

By undertaking an adequate inventory of Australia's mangroves and their habitats, it will be possible to establish their global, national and regional (sub-national) significance. This will facilitate both conservation and sustainable development in areas where environmental impact assessment is needed. A national mangrove conservation strategy could deliver adequate inventory and management to support the conservation of Australia's mangrove systems.

There is a need to systematically *conserve mangroves in Australia nationally* and also given their global significance specifically to *conserve Australian mangroves from a global heritage perspective.*

Keywords
Mangroves · Australia · Mangrove distribution · Mangrove diversity · Mangrove determinants · Legislation · Protection

1.1 Introduction

Coastal systems play a key role in the culture, economy and environment of Australia. Most Australians believe in the importance of protecting and maintaining coastal assets, and while many may not necessarily understand the great diversity of coastal habitats, the near-shore environment is unquestionably the greatest used and most highly regarded of all the nation's natural systems. Perhaps the least understood of all Australian coastal systems are its mangroves and saltmarshes. At the national level in Australia, each State or Territory manages mangroves within its jurisdiction under a wide range of laws and policies, however, no coordination of mangrove conservation or management issues exists between the jurisdictions. This chapter describes the status and distribution of mangroves and reviews the current mechanisms for protection of mangroves within Australia, and the efficacy of mangrove conservation. It also presents a national approach for implementing consistent management regimes based not only on species occurrence but also on mangrove communities and a community assemblage combined with physical habitat approach, hereafter referred to as "community-*cum*-habitat" approach, and mangrove assemblages in inter-related suites of habitats. These concepts of species treated in isolation, assemblages (or communities), "community-*cum*-habitat" approach, and mangrove assemblages in inter-related suites of habitats are illustrated in Fig. 1.1.

1.2 Current Status

Current estimates of overall mangrove cover in Australia range from approximately 900,000 ha to over 1.1 million hectares (ABARES 2016; Mangrove Watch 2010). Over the past 200 years there has been clearing of mangroves for development, and also expansion of mangroves in some areas. One estimate of the pre-1750 cover of mangroves was approximately 940,000 ha (Department of the Environment and Energy 2016a). Other estimates are that around 17% of Australia's mangroves have been cleared since European settlement (Department of the Environment and Energy 2016b).

Nearly every State and Territory in Australia has some expression of a mangrove-lined coast (Fig. 1.2) with the exceptions being Tasmania located in a temperate climate, and the land-locked Australian Capital Territory. The overall areal coverage of mangroves is not evenly distributed in the different States and Territories of Australia due to climate setting, the availability of suitable habitat, and the variability of habitat (85% of Australian mangroves occur in Queensland and in the Northern Territory).

In Australia, protection of biodiversity has been consistently measured through three lenses, the "CAR" system, where protection is categorised according to Comprehensiveness, Adequacy and Representativeness of the amount of each ecosystem with protected areas (Commonwealth of Australia 1999). While systematic conservation planning against these criteria has been undertaken for over 20 years in Australia's terrestrial and marine domains, mangrove systems straddle the interface between land and sea and have been either completely overlooked or inadequately dealt with in the planning systems for terrestrial and marine reserve systems. One of the significant issues with previous conservation planning has been that thin linear coastal systems at the very edge of the planning area are inadequately detected, and therefore inadequately protected in either terrestrial or marine protected area systems.

At the coarsest level across all of Australia, approximately 19% of the area of mangroves is in protected areas (Department of the Environment and Energy 2016a), however this does not measure comprehensiveness, adequacy or representativeness of that protection, and certainly, to date, has not addressed protection of mangroves focused on the community-*cum*-habitat approach, or mangrove assemblages in inter-related suites of habitats. Table 1.1 presents the area of mangroves in protected areas according to IUCN category[1]

[1] The IUCN-protected area management categories (IUCN 1994) are a standardised global classification of protected areas according to their management objectives. The categories are recognised by international bodies such as the United Nations and by many national governments as the global standard for defining and recording protected areas, and are incorporated into government legislation.

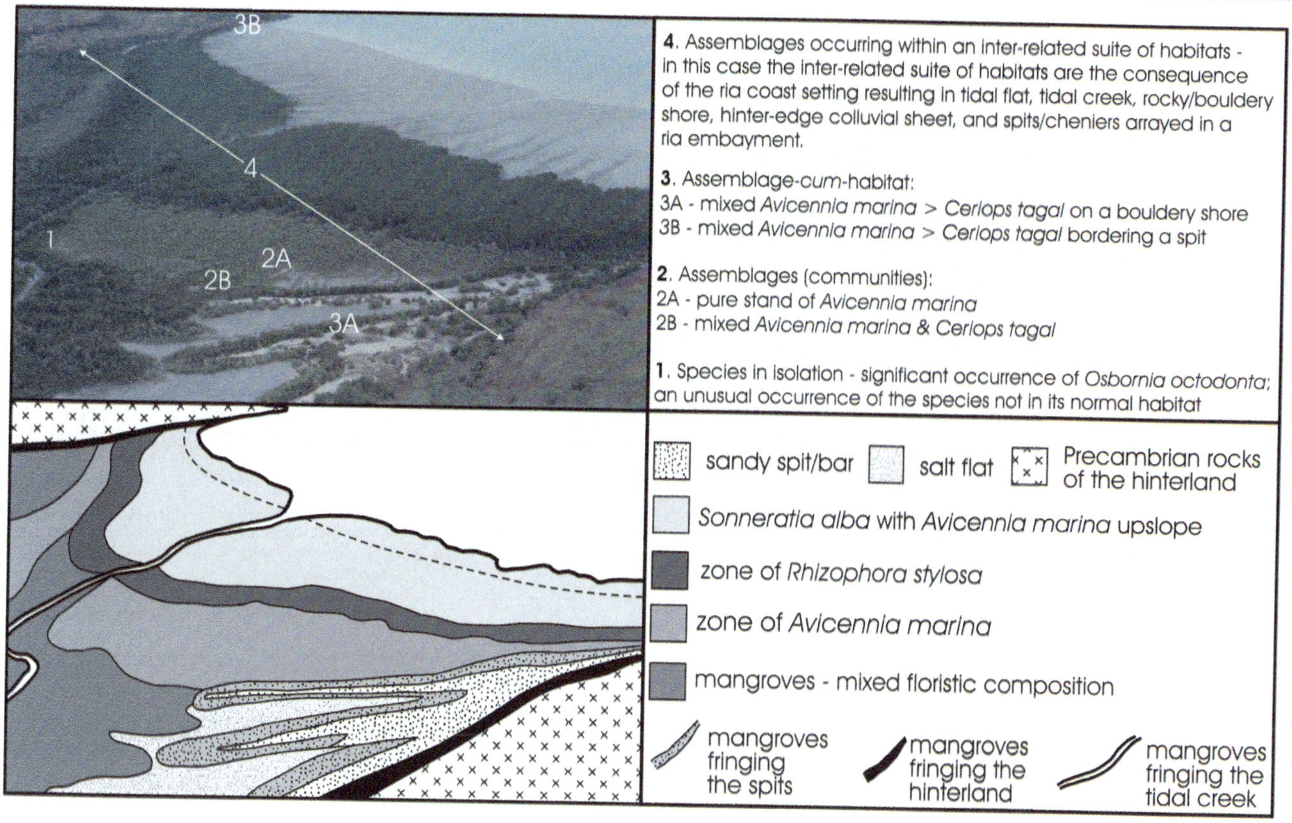

Fig. 1.1 The concept of significance of mangroves based on (1) occurrence of species, (2) occurrences of assemblages (or communities) – these communities may be floristically diagnostic and regionally distinct, (3) assemblages related to a given habitat (community-*cum*-habitat), and (4) assemblages within an inter-related suite of habitats. The example used here is from the ria coast of the Kimberley Region (Adapted from Cresswell and Semeniuk 2011). The concept of distinct mangrove assemblages within an inter-related suite of habitats is covered in the descriptions of mangrove coasts by Semeniuk (1985, 1996), Semeniuk and Wurm (1987), and Cresswell and Semeniuk (2011)

for the whole of Australia. In Australia, forest systems (excluding mangroves) have been through an extensive process to understand pressures and conservation needs, and to establish what are known as Regional Forest Agreements (RFAs) (NFPS 1992). While criticism of that process and the resulting implementation of a CAR reserve system exists, it does have a scientific basis that is open and transparent in its delivery. For mangrove systems, however, no overarching process of documentation and conservation decision-making has been initiated to bring State and Territory planning together in a national framework.

1.3 Previous Work Describing the Distribution of Australian Mangroves

The distribution of mangroves across Australia has been inadequately documented to date, particularly in terms of the major drivers of distribution, namely climate setting and habitat variability. Previous studies have provided some information on the presence/absence of species but little on habitat. Saenger et al. (1977) documented 27 species of mangrove 'trees' in northern tropical Australia, with a further 10 species as associated liane, epiphytic or understorey species, with a descriptive interpretation of mangrove zones noting major species occurrences in each zone. Bunt et al. (1982) recognised 45 mangrove taxa in Australia, Busby and Bridgewater (1986) recognised 38 species, Smith and Duke (1987) recognised 40 species. Duke et al. (1998), using a broad definition that includes ferns, creepers, and herbaceous plants list 39 species occurring in Australia, and Hutchings and Saenger (1987) recognised 'approximately 30 species of trees and shrubs'. Using a broader definition (including climbers, grasses, ferns and palms), Wightman and Andrews (1989) and Wightman et al. (2004) recognised 51 species for the Northern Territory. Semeniuk (1993a), using a much tighter definition that only includes woody trees, recognised 17 species in Western Australia.

Galloway (1982) described 26 broad regions of mangrove distribution in Australia, grouping together 'extensive areas with similar patterns relating to broadly homogenous environments'. Wells (1983, 1985) used an extensive set of vegetation surveys in northern Australia (excluding

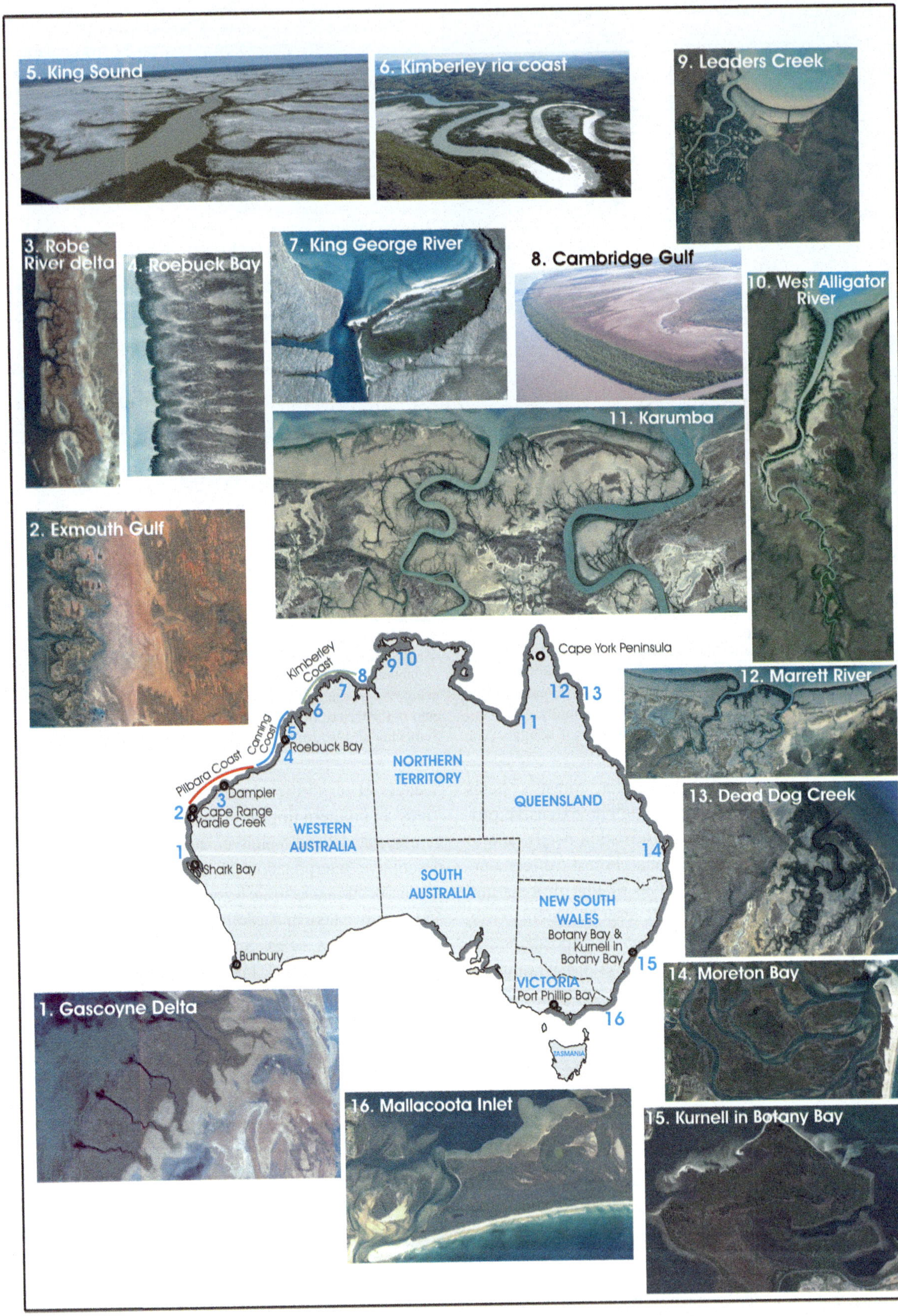

Fig. 1.2 Generalised occurrence of mangroves around the Australian coast (information from the authors, Semeniuk et al. 1978; Semeniuk 1983, 1985; Cresswell and Semeniuk 2011; Duke 2006) with some key sites shown in low-angle oblique and vertical aerial photographs to illustrate diversity of coasts and mangrove settings. The numbered locations of the key sites are shown in bold blue font. The locations are: (1) Gascoyne Delta, showing *Avicennia marina* bordering a salt flat along the edge of the Pleistocene Gascoyne Delta; (2) *Avicennia marina* bordering the arid-zone Yanrey Delta, eastern Exmouth Gulf; (3) *Avicennia marina* bordering the arid-zone Robe River Delta of the Pilbara Coast; (4) *Avicennia marina* bordering the carbonate supratidal flats of Roebuck Plains; (5) the mangrove-fringed tidal-creek network

Table 1.1 Area of Australian mangroves in the main IUCN protected area categories

IUCN protected area category	Area protected nationally (ha)
IUCN Ia	27,000
IUCN II	109,000
IUCN III	7000
IUCN IV	32,000

Queensland) to define five broad regions and species distribution and richness in 2-degree latitudinal blocks. Bridgewater and Cresswell (1999) delineated eleven broad regions for Australia across both saltmarsh and mangrove species. Bridgewater and Cresswell (2003), using specimen records, identified eight major mangrove groups for Australia. Semeniuk (1993a) classified Western Australia's mangrove systems into nine mangrove regions based on species information in combination with geological/geomorphic setting providing a broad-scale setting for the approaches involving community-*cum*-habitat and the mangrove assemblages in inter-related suites of habitats. Duke (2006) described the distribution, abundance and total mangrove area in regional areas for each State and the Northern Territory. Table 1.2 provides a comparison of the major mangrove studies undertaken in Australia, highlighting the differences between studies.

1.4 Distribution of Australian Mangroves in Terms of Climate Setting, Biodiversity, and Habitat Variability

The distribution of mangroves along the Australian coast can be seen at the broadest sub-continental level as a relatively straight-forward relationship with latitude (a surrogate for temperature) and climate (in particular, rainfall). However, this does not account for the full complexity at finer scales down to the local level where the responses become more and more related to habitat variability (Semeniuk 1983, 1985; Semeniuk and Wurm 1987; Cresswell and Semeniuk 2011). Theoretically, given a uniform coast, there would be a gradual decrease in species richness from low latitudes to higher latitudes reflecting climate gradients of temperature and rainfall, and their secondary derivative effects such as evaporation and salinity. However, this is complicated by patchiness in distribution for any given species for reasons of habitat availability, heterogeneity in distribution of habitats at a given locality, patchiness in climate, and the effect of the hinterland in terms of run-off and seepage, and geochemically diverse soils (cf., Semeniuk 1983, 1985, 1993a, b).

There is a patchy distribution of mangrove occurrence, mangrove species, and mangrove communities which can be related to the heterogeneity of the Australian coast (Fig. 1.3). Previous studies by Semeniuk et al. (1978), Semeniuk (1983, 1985), Wells (1983), and Tomlinson (1986) show the distribution of mangroves species along the humid to arid Western Australian coast and the humid Queensland coast, respectively, to be uneven with complicated patterns of presence and absence. While there is an overall decrease in species richness southwards on both coasts the presence or absence of species along the coast is not a simple relationship with latitude. To understand the patterns of mangrove distribution along the Australian coast it is necessary to distinguish the influence of habitat from climate. A primary differentiator is the distribution that is predominantly marine-influenced from that which is predominantly terrestrially-influenced, namely the distribution of those species that occur between mean sea level (MSL) and mean high water springs (MHWS); and the distribution of those species that occur between MHWS and highest astronomical tide (HAT – the highest level of tide), respectively (Semeniuk 1983).

Furthermore, it is necessary to distinguish two tidal zones in the marine-influenced category, namely those mangroves that are inundated daily by marine water (between MSL and mean high water neap MHWN) from those that are inundated only on spring tides (MHWN and MHWS) (Semeniuk 1983, 1985). Those that are inundated daily are less influenced by terrestrial processes such as run-off, rainfall, and influx of terrestrially geochemically related soils, since the daily flushing of the tidal zone surface by tides will ensure that the surface is tidal-marine in nature, and as such the dominant determinants on the mangroves are marine processes and air/sea temperatures. *Aegialitis annulata*, *Aegiceras corniculatum*, *Avicennia marina*, *Bruguiera exaristata*, *B. parviflora*, *Camptostemon schultzii*, *Rhizophora stylosa*

Fig. 1.2 (continued) dissecting the salt flats of King Sound; (6) a meandering mangrove-fringed tidal creek in the ria coast system of the Kimberley Coast; arid zone Yanrey Delta, eastern Exmouth Gulf; (7) mangroves, spits and cheniers at the mouth of King George River, northern Kimberley Coast; (8) zoned mangroves fringing a wide salt flat in northern Cambridge Gulf; (9) small embayment vegetated by mangroves, dissected by a tidal creek, and barred by a spit, Leaders Creek near Darwin; (10) sediment-filled valley tract of West Alligator River, with salt flats and mangrove-fringed channel; (11) complex network of progradational flats, meandering tidal creeks, and mangrove-fringed channels at Karumba, Gulf of Carpentaria; (12) prograded muddy shores and cheniers, dissected by transverse tidal creeks, with channels and foreshore fringed by mangroves in the Marrett River area; (13) sediment-filled small embayment with mangrove-fringed tidal creek and easterly occurring sandy barrier spits at Dead Dog Creek; (14) complex mangrove formation in the protected southern part of Moreton Bay, Queensland; (15) complex mangrove formation protected by sandy spits (barriers) in the Kurnell area of Botany Bay; and (16) flood-tidal delta setting for narrow-fringing mangroves and scattered mangroves in Mallacoota Inlet, Victoria

Table 1.2 Comparison of the major mangrove studies undertaken in Australia

Author	Data type	Mangrove definition	National vs regional	Site locations	Biogeography	Mangrove habitat
Saenger et al. (1977)	Flora (and fauna) records, descriptive interpretation of mangrove zones	Mangroves defined *sensu* MacNae (1968)	National	No site locations	Descriptive interpretation of twelve biogeographic zones	Description of mangrove zonation and mangrove habitat
Lear and Turner (1977)	Descriptive interpretation of data such as flora, and some abiotic factors	Mangroves defined *sensu* MacNae (1968)	National	No site locations, Species noted in 'blocks'	Distribution of number of species occurring latitudinally in 8° latitudinal blocks, and in three coastal sectors	Description of mangrove habitats based on physical, physico-chemical attributes, and biota
Semeniuk et al. (1978)	Site locations and vegetation surveys	Mangroves defined *sensu* MacNae (1968)	Regional – Western Australia	No site locations	Descriptive interpretation of mangrove zones (and biogeography) broadly noting species occurrences within each zone	Description of habitat characteristics, but not site-specific
Semeniuk (1980)	Descriptive, occurrence of species with respect to habitats	Mangroves defined *sensu* MacNae (1968)	Regional – covers only King Sound, WA	Numerous transects to illustrate species occurrence and zoning with respect to various types of habitats	Listing and zoning of species in relation to habitats but not related to regional biogeography	Various habitats illustrated as developed by cliff, sheet, and creek erosion
Bunt et al. (1982)	Vegetation survey	Broader definition of species subjected to tidal inundation growing in association with mangrove species	Regional – North east Queensland	56 site locations	Nine floristically based 'site-groups' identified	Not recorded
Galloway (1982)	Maps and aerial photos to delineate mangrove regions, though not floristics	Not defined	National	Locations for coastal types noted by aerial photographs, but no mangrove sites noted	Mangrove region boundaries based on coastal physiographic patterns	Description of general zonation, based on physical and environmental setting for each region
Wells (1983)	Vegetation survey	Mangroves defined *sensu* MacNae (1968)	National	No site locations	Provides geographic limits of distribution of species	Not recorded
Semeniuk (1983)	Maps, descriptive, occurrence of species with respect to habitat and freshwater seepage	Mangroves defined sensu MacNae (1968)	Regional – covers only WA	Listing of species with respect to sites along a climate gradient and hydrochemical zones	Mangrove species distribution in relation to climate zones and mangrove biogeography	Mangrove habitats and mangrove assemblages and floristics described for different settings and climate zones
Semeniuk (1985)	Maps, descriptive, occurrence of species with respect to heterogeneity of ria habitats	Mangroves defined sensu MacNae (1968)	Regional – north-western to northern Australia	Listing and abundance of species within habitat-determined assemblages with respect to the recurring suite of habitats along climate gradient	Mangrove species distribution in relation to climate zones and mangrove biogeography	Mangrove habitats and mangrove assemblages and floristics therein described for different settings and for different climate zones
Busby and Bridgewater (1986)	Desktop study based on herbarium records	Mangroves defined *sensu* MacNae (1968)	National	No site locations	Species distributions only	Not recorded

(continued)

Table 1.2 (continued)

Author	Data type	Mangrove definition	National vs regional	Site locations	Biogeography	Mangrove habitat
Hutchings and Saenger (1987)	Flora records, desktop study	Mangroves defined *sensu* MacNae (1968)	National	No site locations	Descriptive interpretation of thirteen biogeographic zones	Not recorded
Semeniuk & Wurm (1987)	Maps, descriptive, defining habitats, assemblages, floristics, environmental factors	Mangroves defined *sensu* MacNae (1968)	Regional – covers only Dampier Archipelago, WA	Detailed site descriptions of soils, water, vegetation, and floristics along 28 illustrated transects crossing different habitats	Not dealt with	Suite of mangrove habitats and mangrove assemblages therein described
Semeniuk (1993a)	Descriptive	Mangroves defined *sensu* MacNae (1968)	Regional – covers WA only	No site details	Seven coastal sectors containing nine mangrove regions described	Suite of mangrove habitats described
Suzuki and Saenger (1996)	Vegetation survey	Mangroves defined *sensu* MacNae (1968)	Regional – covers NSW and part Qld	80 sites	Seventeen phytosociological plant communities and their limited distribution	Not recorded
Pedretti and Paling (2001)	Desktop study	Mangrove defined *sensu* MacNae (1968)	Regional – covers WA only	363 sites described	Beard: physiographic regions and geological regions as framework for mangrove biogeography	Provides 'floristic habitat' assemblage for each site
Heap et al. (2001)	Estuary database and remote sensing	Mangroves not defined	National	Each estuary considered a 'site' but only presence/absence of mangrove vegetation not species presence/absence	Not dealt with in respect to mangroves, but subdivide the coastal waters into estuarine classes	Relationship between physical factors and occurrence of mangrove vegetation, but no habitats
Bridgewater and Cresswell (2003)	Desktop study	Mangrove defined *sensu* MacNae (1968)	National	No site locations	Eight 'phytogeographic' groups defined within broader biogeographic framework (IBRA)	Not recorded
Duke (2006)	Literature review	Mangrove defined *sensu* MacNae (1968)	National	No sites listed	Mangrove regions defined uses coastal catchment areas within regional areas	Intertidal position recorded but no habitats recorded
Mount and Bricher (2008)	Desktop study and literature review	Mangroves not defined	National	No sites listed	None recorded	Single habitat mapped for whole of Australia on 'unconsolidated substrates'
Short and Woodroffe (2009)	Literature review	Obligate mangroves	National	No sites listed	Broad coastal sectors described (not defined)	Not recorded

and *Sonneratia alba* are in this category. Those mangroves that are inundated only on spring tides (MHWN and MHWS) are still maintained dominantly by marine processes, but are increasingly influenced by terrestrial processes to landward. *Ceriops australis*, *C. tagal* and *Osbornia octodonta* are in this category. Bridgewater (1989) describes this in a phytosociological context using ordination techniques.

Mangroves residing in the tidal interval of MHWS and HAT occupy habitats that are progressively more influenced by terrestrial processes and terrestrial influences to landward.

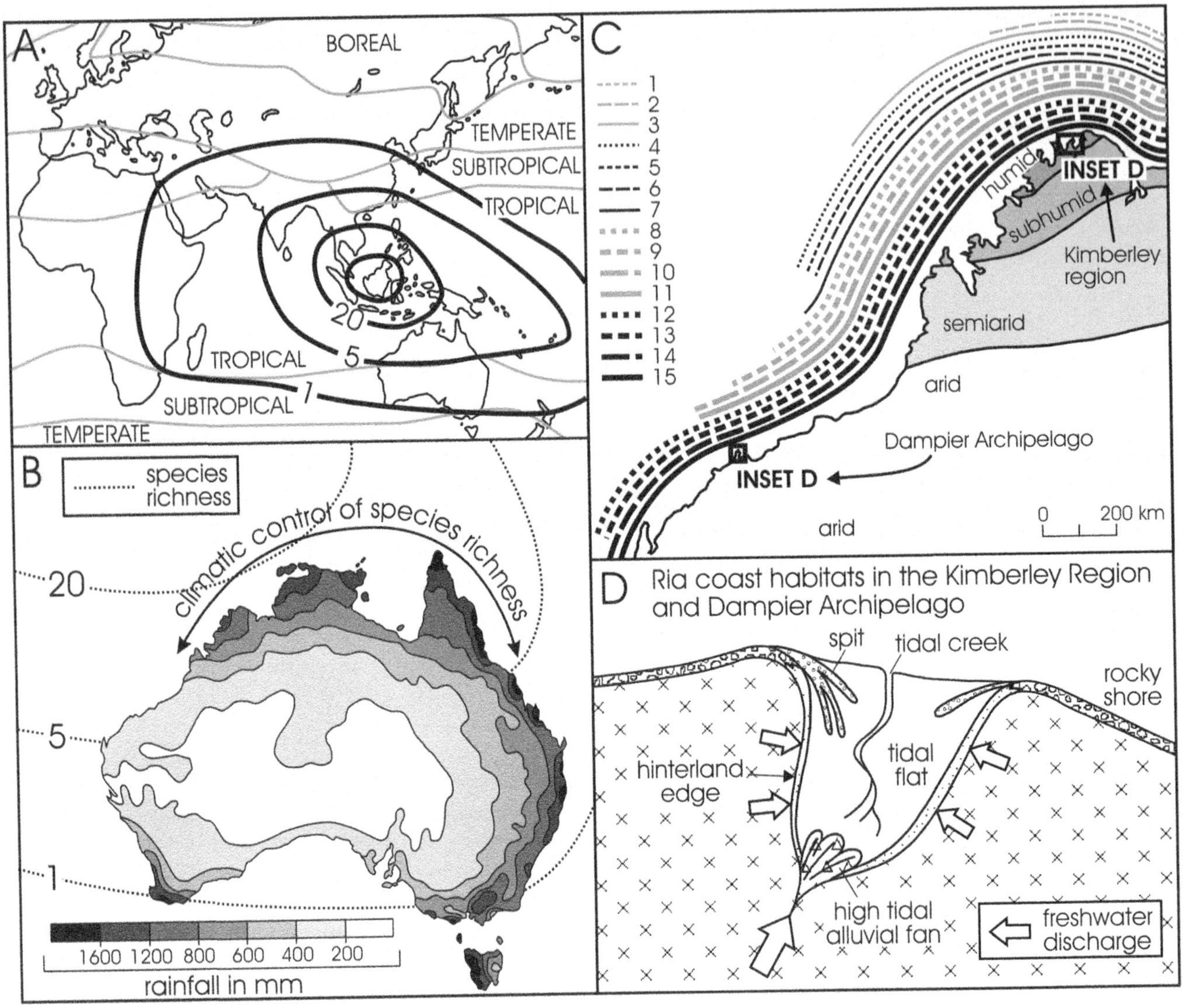

Fig. 1.3 From a subcontinental perspective, Australia resides in the region of the Old World Mangroves with the greatest species richness centred in the Malaysian Archipelago; in this context, the northern perimeter of Australia lies on the contour line involving ~20 mangrove species (Insets A and B) – this would be the regional species pool for the Northern Territory and northern Queensland; thereafter, progressing southwards, there is progressive decrease in the species richness correlated with climate (Inset C)-(Adapted from Semeniuk 2012). For the Western Australian subcontinent and its mangroves, each line parallel to the coast shows the biogeographic distribution of the 15 mangrove species from the Kimberley Region to Exmouth Gulf [Key to numeration of species: (1) *Scyphiphora hydrophylacea*, (2) *Xylocarpus granatum*, (3) *Bruguiera parviflora*, (4) *Camptostemon schultzii*, (5) *Lumnitzera racemosa*, (6) *Xylocarpus moluccensis*, (7) *Sonneratia alba*, (8) *Excoecaria agallocha*, (9) *Osbornia octodonta*, (10) *Aegiceras corniculatum*, (11) *Bruguiera exaristata*, (12) *Ceriops tagal*, (13) *Aegialitis annulata*, (14) *Rhizophora stylosa*, (15) *Avicennia marina*]. Inset D illustrates the range of habitats in ria coast settings that are present in the Kimberley Region and Dampier Archipelago. The species richness and structure of the mangrove vegetation in these six habitats of ria coasts for Darwin (Northern Territory) and the Kimberley region and Dampier Archipelago are described in Semeniuk (1985)

Rainfall in the progressively higher tidal zones, where there is less flushing by tidal inundation, will have an increasing effect diluting the salinity of soilwater and groundwater. The effects of terrestrial processes and terrestrial influences are particularly pronounced where there are freshwater seepages and run-off from the hinterland edge, where geochemically diverse sediments and soils are washed into the edge of the tidal zone, and where mangroves interface with geochemically and hydrochemically diverse environments along the edge of geologically different hinterlands. As such, this upper tidal zone, and particularly along the edge of the hinterland, is where there may be expected to be the greatest diversity of mangroves. In this zone the greatest determinants of diversity are the different hydrological regimes created by freshwater seepage where tidal flat hypersalinity is diluted. This allows the occurrence of mangrove

Fig. 1.4 This oblique aerial photograph illustrates the principle of variability of mangrove composition depending on sampling location and habitat complexity. In this example a shore-normal, elongate small island (with a seaward bouldery shore) acts as an anchor for a sandy tombolo. There are several mangrove habitats present: a rocky shore, a bouldery shore, a shore-normal sandy tombolo (equivalent to a high-tidal spit), a gravelly tidal flat, a muddy tidal flat, hinterland margin, and a sand spit (located in the middle extreme right of the image). The mangrove habitat at approximately MSL varies from right to left, which is reflected in a change in mangrove composition from a *Sonneratia* and *Avicennia* association and an upper slope occurrence of *Rhizophora* on right and left flanks to dominantly an *Avicennia* stand in the central seaward area. *Rhizophora* is most abundant and forms a dense copse on the left side of the tombolo. The mangroves fringing the tombolo are asymmetric in composition, with *Avicennia*, *Ceriops*, and *Bruguiera* forming denser formations on the left flank. There are mangroves fringing the sand spit which also is a reservoir for freshwater, discharging in a series of (six) small "rivulets" that are inhabited by mangrove shrubs

species that normally inhabit the interval of MSL-MHWS into areas of diluted groundwater and soilwater salinity, thereby enabling these species to inhabit tidal levels normally too saline for their survival. A range of other mangrove species also inhabits this high-tidal interval but do not occur in the more marine-influenced tidal zone. In this category of tidal level, and particularly along the hinterland edge, the dominant determinants of mangrove occurrence are air temperature, seawater temperature, rainfall, run-off/seepage (potentially with their attendant geochemically altered signatures) and hinterland margin soils. This zone has a markedly different waterlogging regime from the more marine-influenced tidal zone. Depending on the range of edaphic features present in this zone, mangrove species diversity can be highest, populated by species that normally inhabit lower tidal levels, and species that are unique to this hydrochemical/geochemical environment.

Unfortunately, it is not possible to use data from the literature and State herbaria to generate distribution maps of species diversity according to climate and physiographic zones as the published accounts to date have tended not to identify tidal level or the habitat of mangrove species but rather provide species lists for a given area or region or site. Cresswell and Semeniuk (2011) discuss the problem of mangrove species collections in herbaria, and the quality of the mangrove surveys undertaken to date within the heterogeneous coast of the Kimberley with its heterogeneous, asymmetric, and complex distribution of mangrove species and mangrove communities (Fig. 1.4).

1.5 The Need to Conserve Mangroves

While at a broad level the importance of mangrove habitat has become better known, and decisions at local and State level now take into account the need to conserve and manage natural coastal assets, over the past 30 years there has also been increasing development pressure in the natural coastal environment which has seen many areas of mangrove cleared. One factor working against the conservation of mangroves has been the general impression that "mangroves" are a single unit and ubiquitous along the Australian coastline, therefore any destruction of mangrove habitat has little or no impact on either the overall biodiversity or the broader ecosystem services to which they contribute. However, the impact of the removal of mangrove ecosystems may often be felt far beyond the immediate area where mangrove stands are present.

There is a growing recognition that management needs to take into account the 'ecosystem' as the basic framework on which to build sustainable land management practices. There is no common consensus on how to manage specific elements within an ecosystem, but the general trend is away from species-based management towards managing "systems", their component elements and their inter-relations. A framework of ecosystem-based management of mangroves is required to ensure the true diversity and functioning of mangroves in terms of habitats, vegetation floristics and structure/physiognomy is captured from national to regional (sub-national) to local scales.

As global policies have, by and large, shifted from species-based approaches to an ecosystem-based approach, the emphasis has shifted to the connectivity between species, their habitats and physical environment, and the connection between human interactions with ecosystems. While many countries may have committed to implementing the ecosystem approach and/or ecosystem-based management within their national jurisdiction, management is still, by and large, based on a species-based approach. Within Australia, there is limited adoption of ecosystem-based management with limited coordination of efforts across jurisdictions, and none for mangrove ecosystems.

The United Nations Framework Convention on Climate Change (UNFCCC) has established new approaches to the management of ecosystems in relation to adaptation to climate change. Internationally, one approved afforestation and reforestation baseline and monitoring methodology has been established for degraded mangrove habitats to re-establish multi-species and/or zoned mangrove forests (UNFCCC 2011a). While not applied to Australia, this measure may provide a mechanism to better document the current status of mangrove forests globally and place greater value on maintenance and enhancement of mangrove systems, albeit focused solely on understanding carbon loss and sequestration.

The lack of definition of what is required for ecosystem-based management of mangroves means that managers must rely on species-based management. Greater clarity on what are mangrove habitats is essential for on-ground implementation of ecosystem-based management.

1.6 The Need to Specifically Conserve Australian Mangroves from a Global Perspective

In order to understand the distribution of mangroves in Australia, it is useful to understand the global trends in species diversity. It is generally recognised that two main groupings of species separate the Americas/western Africa species from the eastern Africa/SE Asia/Oceania species. These groupings are referred to as Eastern Hemisphere/Western Hemisphere or Old World/New World or Eastern/Western (Chapman 1976; Tomlinson 1986). The mangroves of Australia belong to the Indo-Malesian group of old world mangroves centred in the Indian-Pacific area (Chapman 1977; Tomlinson 1986), which is the most species-rich region of the world. Duke (2006) describes Australia as having the fourth highest species diversity of mangroves in the world (after the Philippines, Indonesia and Papua New Guinea), with some 41 species. While Giesen et al. (2007) recognise over 250 mangrove species in South East Asia these include many species that are not confined to mangrove ecosystems, but are known to occur there, such as climbers, grasses, ferns and palms.

While southeast Asia is very diverse for mangrove species it is also an area of immense human pressure on coastal systems, with one estimate that nearly two thirds of China's mangrove systems have disappeared over the past 40 years from clearing for development of farming, aquaculture and buildings, and overall the extensive delta systems of Asia contain few natural wetlands (Mao et al. 2006). Other estimates put the area of mangroves globally to be less than half of what it was before 1950 (FAO 2007; Giri et al. 2011), with heavily populated coastal zones leading to clearing and degradation of mangrove systems (Polidoro et al. 2014).

Globally, protected areas cover only a small proportion of the ranges of all mangrove habitats across all five continents where mangroves occur, and are unevenly distributed with many areas of the planet not currently protected (Spalding et al. 2010). Qualitative reviews that document broad-scale habitat destruction or alteration in many parts of the world are clear indications that in many areas both the extent and diversity of mangrove habitats has markedly reduced in the past 50 years. One estimate is that more than 50% of the world's original mangrove forests have disappeared (Valiela et al. 2001) and the annual global rate of mangrove loss continues to be between 1% and 2% (Spalding et al. 2010).

Currently, there is a diverse set of international instruments that may be used for mangrove conservation, all of which have different mechanisms for implementation and which no single nation implements in toto. While synergy is possible between some of these instruments, very little evidence exists that such synergies have led to any specific biodiversity outcomes. There is no plan as to how these different approaches should work together, and no functioning checks and balances are in place to monitor the effectiveness of how all these instruments interact. It has been argued that this complex set of arrangements is doing more harm than good as they lead to a strong perception within governments and more broadly within society that there is adequate legal protection leading to positive outcomes (Johannsdottir et al. 2010).

Another aspect to consider in the development of legislative and management regimes for the conservation of mangroves, taking account of the community-*cum*-habitat approach (or, at the finer scale, a species-*cum*-habitat approach) or mangrove assemblages in inter-related suites of habitats, is that the Australian coastline presents an heterogeneous suite of habitat types at the large scale. For example the range of mangrove habitat types along the vastly different tracts of coast in Western Australia such as the Pilbara Coast, the Canning Coast, and Kimberley Coast results in a plethora of habitats with a consequence of variable communities, variable mangrove vegetation structures, and differential phenotypic responses [cf., Fig. 1.3 of Semeniuk and Cresswell 2013]. See Semeniuk (1983, 1985, 1995, 2008), Semeniuk and Wurm (1987), and Cresswell and Semeniuk (2011) for further information on the complexity of the Australian coastline.

Generally the significance of the vegetation either regionally or nationally in terms of its biodiversity value has been assessed either purely on floristic criteria (i.e., absence or presence of rare, or threatened species) and/or in combination with terrestrial vegetation composition. Mangrove ecosystems have rarely been assessed on the basis of community composition, and never within a community, or as assemblages in inter related suites of habitats. Using the approach of community-*cum*-habitat and/or the mangrove assemblages in inter-related suites of habitats, the structural, physiognomic, and phenotypic attributes of the vegetation can be variably expressed from habitat to habitat. For example, while the mangrove *Avicennia marina* occurs throughout Australia and is considered to be a common species, it occurs in climates that are Tropical humid, Tropical arid, and Temperate, inhabits tidal mudflats, sandy beaches, rocky shores, riverine and creek banks, and tidal pavements (amongst others), and occurs in habitats with groundwaters that are hypersaline, or marine, or brackish. Viewed simplistically from a species occurrence viewpoint, *Avicennia marina* would never be considered as a threatened species because it is widespread. However, it occupies a range of ecological niches and expresses different structural, physiognomic, phenotypic, and reproductive strategies depending on the climate, habitat, and salinity. Therefore in considering conservation of organisms that co-habit with *Avicennia marina* in these different settings, it is necessary to address this full expression across the range of environments that *Avicennia* inhabits. That is, *Avicennia marina* needs to be viewed and assessed and conserved from a community-*cum*-habitat perspective. Semeniuk and Cresswell (2013) addressed this matter in a description/discussion of biodiversity in relation to habitat heterogeneity and indicated that biodiversity assessments need to address the full expression of a species in various habitats for effective conservation.

The next level of complication of identifying significant mangrove systems is the approach involving mangrove assemblages in inter-related suites of habitats because here the assemblages in totality need to be considered. For example the mangrove assemblages inhabiting the diverse habitats along a ria coast present a different suite of ecological patterns to those inhabiting a delta complex or those inhabiting a limestone barrier and barred tidal lagoon complex.

In this regard, Australia offers mangrove ecosystems that are globally unique with regard to floristic assemblages, structural and phenotypic responses to their climate and habitat settings, and in their community-*cum*-habitat expression. Many of the mangrove assemblages and habitat-specific assemblages are globally unique and are inadequately represented in a CAR reserve system.

Therefore in order to comprehensively capture the full biodiversity of Australian mangrove ecosystems, it will be necessary to address each of these five key attributes:

1. Species occurrences,
2. Species occurrences at their biogeographic limits (e.g., *Avicennia marina* in Victoria at its polar limit, or *Rhizophora stylosa* in Western Australia at its southern limit in the State at Yardie Creek; cf. Semeniuk et al. 1978),
3. Community assemblages incorporating structure and floristic composition (cf., Semeniuk and Cresswell 2013),
4. The community-*cum*-habitat approach, and
5. Mangrove assemblages in inter-related suites of habitats.

1.7 The Need to Conserve Mangroves as Species, Communities, and in an Ecosystem Approach

While it is generally understood that mangroves inhabit dynamic physical environments, there is no documentation of that dynamism of change that can result in significant change in habitat, and of course, species presence. In lower-energy regimes mangrove habitats may be relatively stable leading to long periods of "stable" vegetation. However, elsewhere in higher-energy environments, and in response to extreme events and decadal variation, there can be major habitat loss or gain with rapid mangrove response. Mangroves respond to these changes and colonise newly-formed habitats.

This approach was adopted by Semeniuk (1997) in a study of mangroves of the Pilbara Coast in Western Australia wherein the selection of mangrove stands for conservation along the coast was based on (1) their megascale geomorphic setting (e.g., active delta, inactive delta, limestone barrier coasts, archipelago/ria shores, and beach/dune shores) with their range of specific smaller-scale geomorphic units that function as habitats, (2) assemblages and floristics within the habitats, and (3) structural/physiognomic expressions of the various species. This work supplemented the studies by Semeniuk and Wurm (1987) who described the mangroves of the Dampier Archipelago as mangrove assemblages linked to inter-related suites of habitats. The results by Semeniuk (1997) identified 22 tracts of mangrove coast along some 650 km of the Pilbara Coast of Western Australia. The Western Australian government of the day adopted this approach and issued a Policy (Environmental Protection Authority 2001) to deal with any either planned or indiscriminate development of the coast where mangroves were present. It placed emphasis on the importance of mangroves and on the need to 'not destroy them'. However, ultimately, while this Policy provided guidance and policy, it was never translated into legislation.

The development of a consistent framework of coastal geomorphology, coastal processes and climate setting for the whole of Australia is required to achieve a nationally consistent approach to mangrove conservation, such that the importance of a given stand of mangroves can be viewed in a state-wide, national, and indeed, global perspective. An understanding of the various coastal sectors of Australia is needed to determine the significance of any tract of coast within a perspective of the coastal geology, geomorphology, and sediments, and the prevailing coastal processes and climate. To compare and assess mangrove systems around the Australian coast in a robust manner requires identifying and describing coastal sectors to provide the basis upon which to describe the suites of mangrove habitats that occur around the continent. For instance, at the species level, the mangroves of the Sunderbans (Stanley and Hait 2000; Sundarbans National Park 2014) cannot be strictly compared with mangroves of Queensland nor with the mangroves of the ria coast of the Kimberley (Brocx and Semeniuk 2011), but the issue of incomparability becomes more pronounced, and specific sets of mangroves become more evidently globally significant when the mangroves are viewed in their geomorphic/habitat setting.

To this objective, along the coast of Australia and its offshore marine environments, Cresswell and Semeniuk (2017 in prep) have identified at the megascale, 35 coastal sectors reflecting their setting in terms of geology, coastal geomorphology, hinterland geomorphology, coastal processes and climate. Only 19 of these support mangrove vegetation, and these are briefly described below. The classification is described starting at the most northerly breakpoint between sectors along the Australian mainland, Cape York, and then proceeds clockwise around the continent. A summary of each coastal sector containing mangroves is provided in Table 1.3. Figure 1.5 provides a generalised map of the mangrove-inhabited coastal sectors.

Further to this, the coastal sector framework presented here provides a comprehensive and detailed basis to compare the area of mangroves, number and area of protected areas containing mangroves, and percentage of mangrove area inside protected areas for each Coastal Sector (Table 1.4).

1.8 Moving from a Set of *ad hoc* Conservation Reserves to a National Strategy for Australian Mangroves

As discussed earlier Australia's reserve system provides ad hoc protection for some mangroves through inclusion in formal protected areas with some protected areas having further "protection" through designation under an international agreement, such as being part of Ramsar sites (e.g., Kurnell in southern Botany Bay in New South Wales, and Roebuck Bay in north-western Australia), or by being part of World Heritage areas (e.g., the mangroves on the mainland Queensland coast shoreward of the Great Barrier Reef, and mangroves in the Shark Bay World Heritage area). In terms of mangrove protection, while not formally hierarchical, it is evident that World Heritage and Ramsar Convention listing provide additional levels of scrutiny and reporting on top of national designations as National Park, Nature Reserve, or any of the many other designations of protected areas. In addition to inclusion in protected areas, each jurisdiction also has sustainable use laws (often governed through Fisheries departments) and policies that provide levels of protection for mangroves.

While each jurisdiction manages its mangroves through a variety of different protected area types, along with additional levels of recognition through designations under international Conventions, there is no coordination between jurisdictions. In the Northern Territory, the total area of mangroves has been estimated to be approximately 412,000 ha with 12% managed under some form of formal conservation management, which includes 11,000 ha in World Heritage Areas (NRETAS 2007). In Queensland, the total area of mangroves has been estimated as 460,000 ha with some 64,000 ha included in World Heritage Areas (DERM 2012). Where

Table 1.3 Summary of each coastal sector containing mangroves (for details of the Western Australian Coast see Brocx & Semeniuk 2010; and Gentilli 1971 provides information on climate)

	Description	Oceanography	Climate
Barrier Reef Coast	Leeward of the Great Barrier Reef, comprised of J-shaped embayments and estuaries, rocky shores, barrier dunes, local prograded beach ridges, and sweeping beaches, local archipelagos and rias, small open to barred estuaries, deltoid alluvial plains, and locally large lobate deltas, with intervening rocky shores	Open coast is wave dominated, macrotidal	Tropical humid
Fraser Island Coast	Large barriers with protected large estuaries, alternating with large deltas and small, open to barred estuaries	Open coast is wave dominated, mesotidal	Tropical humid
Sydney Basin Coast	Local rias, partly barred and filled estuaries, fully barred and filled estuaries, and small open to barred estuaries, with intervening cliffy rocky shores	Open coast is wave dominated, microtidal	Subtropical humid to temperate humid
Tathra Coast	Long linear sandy barriers, beach ridge barriers, barred lagoons, cliff shores and rocky headlands	Open coast is wave dominated, microtidal	Subtropical humid to temperate subhumid
Gippsland Coast	Long linear sandy barriers, barring lagoons and estuarine lagoons	Open coast is wave dominated, microtidal	Temperate subhumid
Melbourne Coast	Headlands separating J-shaped sandy stretches and sandy barriers with leeward narrow lagoons, and local small estuaries	Open coast is wave dominated, microtidal	Temperate humid
The South Australian Gulfs	Ridges of Precambrian rock, desert dunes, Quaternary limestone, and Tertiary/Quaternary alluvial fans underlying the peninsula, with cliffs, local barriers and their leeward lagoons, local estuaries with/without barriers, headland-and-beach coasts, tidal flats, supratidal flats, coastal dunes	Gulfs are protected, tide dominated, microtidal	Temperate semiarid
Ceduna Coast	Cliffs cut into Tertiary and Quaternary rocks; sediment derived from these reworked into beaches, dunes, and barriers	Open coast is wave dominated, microtidal	Temperate arid
Perth Basin Coast	Sandy beaches, limestone rocky shores, barrier dunes, cuspate forelands, estuaries, and local bedrock outcrop (mangroves only in the southern part)	Open coast is wave dominated, microtidal	Subtropical subhumid in south semiarid in north
Houtman-Abrolhos Coast	Offshore, isolated coral reef islands	Open coast is wave dominated, microtidal	Subtropical arid
Carnarvon Coast	Complex of embayments, lagoons and gulfs formed by tectonically-active barrier ridges	Open coast is wave dominated, microtidal	Subtropical arid in south, Tropical arid in north
Pilbara Coast	Active and inactive deltas, limestone barrier island coasts, beach/dune coasts, eroding embayment coasts, and ria/archipelago coasts	Open coast is wave dominated, mesotidal to macrotidal	Tropical arid
Northwest Shelf	System of emergent limestone islands and sand islands	Wave dominated, mesotidal	Tropical arid
Canning Coast	Carbonate-dominated system of scattered embayments barred by a barrier of limestone or sand dunes	Open coast is wave dominated, macrotidal	Tropical arid in south, Tropical semiarid in north
King Sound	A large estuarine gulf	Tide dominated, macrotidal	Tropical semiarid
Kimberley Coast	Archipelago and ria system resulting from marine inundation of a dissected sandstone and basalt plateau	Tide dominated, macrotidal	Tropical semiarid to subhumid to humid
Cambridge-Bonaparte Coast	System of large estuarine gulfs with intervening coastal plain	Tide dominated, macrotidal	Tropical subhumid
The Top End Coast	Coastal system of intergradational suite of archipelagos, rocky shores, rias, partly filled rias, and coastal plains	Tide dominated, macrotidal	Tropical humid
Gulf of Carpentaria	Extensive low coastal plain bordering a hinterland plateau, numerous rivers, each with delta	Wave dominated, mesotidal	Tropical subhumid to semiarid

this World Heritage designation overlaps with protection under State law, it provides additional oversight and reporting responsibilities.

Though differences exist between protected area types between jurisdictions, with the inclusion of site listing under the World or Ramsar Conventions, the following areas of mangroves can be considered to have the highest level of protection:

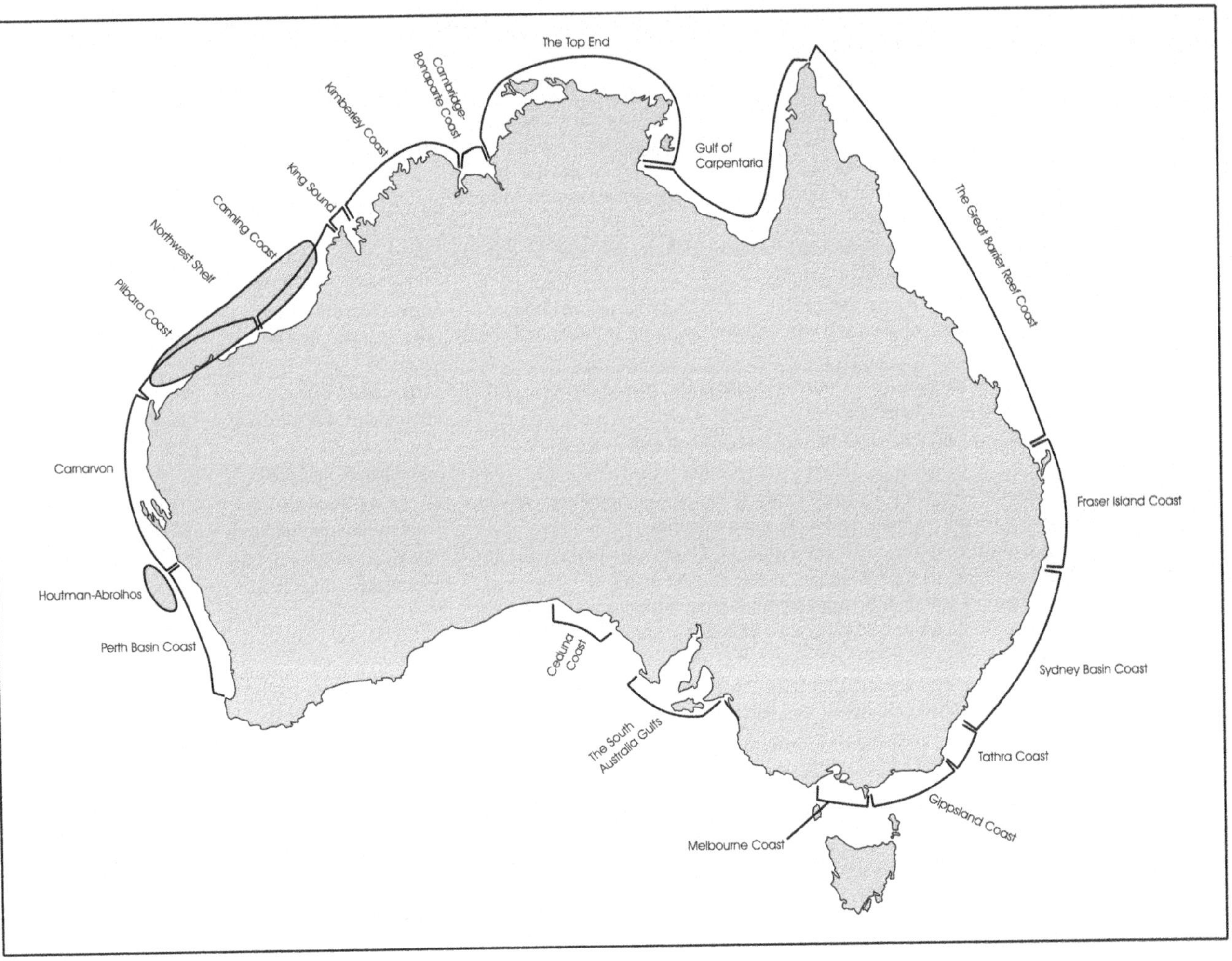

Fig. 1.5 The major coastal sectors of Australia containing mangroves

World Heritage areas (Australian Government 2017a):

1. Mangroves in the Shark Bay World Heritage area in Western Australia;
2. The Ningaloo Coast in Western Australia;
3. Mainland coastal mangroves onshore within the Great Barrier Reef World Heritage area in Queensland;
4. Mangroves on west shore of Fraser Island in Queensland;
5. Wet Tropics of Queensland; and
6. Kakadu National Park in the Northern Territory.

Ramsar sites (Australian Government 2017b; as well as Hale 2008; Hale & Butcher 2009; Kellogg Brown & Root 2010; BMT WBM 2010, 2011a, b; Newall & Lloyd 2012):

1. Kakadu National Park in the Northern Territory;
2. Cobourg Peninsula Ramsar Site in the Northern Territory;
3. Bowling Green Bay in Queensland;
4. Shoalwater and Corio Bays Area Ramsar in Queensland;
5. Moreton Bay in Queensland;
6. Great Sandy Strait Ramsar Site in Queensland;
7. Towra Point / Botany Bay in New South Wales;
8. Myall Lakes in in New South Wales;
9. Corner Inlet Ramsar Site in Victoria;
10. Port Phillip Bay and Bellarine Peninsula in Victoria;
11. Western Port in Victoria;
12. Lavinia Ramsar Site, island off Tasmania;
13. Roebuck Bay in Western Australia;
14. Eighty Mile Beach Ramsar Site in Western Australia;
15. Ord River Floodplain Ramsar Site in Western Australia.

National Heritage Listing (Australian Government 2017c) (noting that while there are mangroves occurring in areas inscribed on the National Heritage List, the area may not have been listed for its mangrove occurrence):

1. West Kimberley Coast in Western Australia;
2. Wet Tropics of Queensland;
3. The Ningaloo Coast in Western Australia;
4. Moreton Bay in Queensland; and
5. Kakadu National Park in the Northern Territory.

Table 1.4 Area of mangroves, number and area of protected areas containing mangroves, % of mangrove area inside protected areas for each Coastal Sector

Coastal sector	Mangrove area (km^2) (GA/Ozcoasts)	Mangrove area (km^2) (NVIS modern)	Mangrove area (km^2) (NVIS 1750)	Mangrove area within PA (km^2) (GA/Ozcoasts)	Mangrove area within PA (km^2) (NVIS modern)	Mangrove area within PA (km^2) (NVIS 1750)	Number protected areas (GA/Ozcoasts)	Number protected areas (NVIS modern)	Number protected areas (NVIS 1750)	% Mangrove within PA (GA/Ozcoasts)	% Mangrove within PA (NVIS modern)	% Mangrove within PA (NVIS 1750)
Great Barrier Reef Coast	1736	2477	2550	363	508	511	108	140	140	21	21	20
Fraser Island Coast	259	250	267	49	55	55	65	68	68	19	22	21
Sydney Basin Coast	119	121	124	32	42	42	65	79	81	27	35	34
Tathra Coast	1	10	11	0	4	2	9	2	10	0	40	18
Gippsland Coast	17	14	18	16	12	14	4	5	5	94	86	78
Melbourne Coast	15	42	63	4	26	28	11	24	25	27	62	44
South Australian Gulfs	72	108	108	17	25	25	7	13	13	24	23	23
Ceduna Coast	35	22	22	6	6	6	2	4	4	17	27	27
Perth Basin Coast	1	0	0	0	0	0	0	0	0	0	–	–
Carnarvon Coast	11	11	14	0	0	0	1	0	0	0	0	0
Pilbara Coast	121	396	440	0	16	18	0	2	4	0	4	4
Canning Coast	17	72	99	0	1	1	0	1	1	0	1	1
King Sound	116	165	184	0	0	0	0	0	0	0	0	0
Kimberley Coast	687	670	778	111	177	210	2	3	3	16	26	27
Cambridge-Bonaparte Coast	703	592	625	264	208	227	24	9	15	38	35	36
The Top End	1437	3018	3020	280	620	620	21	37	37	19	21	21
Gulf of Carpentaria	1012	2110	2461	44	75	122	6	11	11	4	4	5

PA Protected Areas, *NVIS* National Vegetation Information System, *GA* Geoscience Australia

While these sites are inscribed on the National Heritage List, it should be noted that they also may be listed as World Heritage areas, and Ramsar sites.

The level and types of protection between the different States and Territories is complicated and, as such, is beyond the scope of this Chapter. However, to illustrate the range of protection/conservation instruments, Western Australia is used as an example.

In Western Australia, leaving aside inscriptions at World Heritage level or as a Ramsar site, there is protection of mangroves by listing of areas on the National Heritage List (Australia Government 2017c) and by the State Government in various types of Reserves (e.g., Conservation Parks, National Parks, Nature Reserves) (Department of Parks and Wildlife 2017). Furthermore, as noted above, the entire coast of the west Kimberley with its extensive mangrove formations is listed on the National Heritage List. The mangroves of the Leschenault Peninsula, Monte Bello Islands, the Dampier Archipelago, and the Prince Regent River also are captured by the State Government in various types of Reserves.

A whole-of-landscape approach that adequately takes into account mangrove habitat diversity can provide the most sustainable level of coastal development, and indeed remove barriers for development in less important areas. However, to ensure adequate protection and appropriate development, it is necessary to change from the current ad hoc management of mangroves (in terms of protection in protected areas) to ranked assessment of mangrove systems at the local, regional and national level. In this way it would be possible to establish the most appropriate management regime based on an understanding of the importance of any mangrove area, and assessed using agreed criteria. It would also provide a rigorous tool for choosing the best areas for selection of offset areas and investment in compensatory habitats for approved developments.

A management approach that provides a framework to take into account spatial changes in distribution and diversity over time is the most likely to ensure conservation of mangroves and service/benefits they provide to people, and to other ecosystems over the long term.

Implementation of management regimes at an ecosystem level that truly takes into account the effects of environmental drivers needs to be applied at the scales of the drivers (Waltner-Toews et al. 2008). Furthermore, on top of an understanding of the natural environmental drivers for effective management, it is also necessary to understand and manage anthropogenic influences/drivers. Effective management needs to be adaptive to changing human uses and those ecosystem characteristics that are impacted by management measures. In mangroves, an understanding of the major environmental drivers that determine the dynamics of the system (such as prograding or retreating coastlines, or the extent of freshwater seepage and discharges) are essential for long-term management. While regional management of mangroves should be based on the broad regional level drivers that are controlling the distribution of the regional species pool and the existing mangrove habitats, finer-scale management measures may be necessary to accommodate site specific issues, particularly where highly-valued sites of mangrove diversity may exist. This means identifying the diversity of the finer-scale habitats that underpin mangrove assemblages, structural types, physiognomic types, and reproductive strategies of mangroves assemblages and individual mangrove species – an expression of these finer-scale habitats is shown in Fig. 1.3D. Each of the finer-scale habitats may have its own range of tidal level, salinity, seawater-freshwater dynamics and interchanges, sediments/soils, benthic fauna, and microbiota, all of which will influence the maintenance and survivorship of mangrove species and mangrove communities. A diagrammatic representation of some the complex multi-dimensional factors involved in the fine-scale habitats of a ria coastal embayment is shown in Fig. 1.6. From the above discussion, and as evident in Figs. 1.3, 1.4, and 1.6 that show the heterogeneity of habitats and their internal complexity, management outcomes may not be

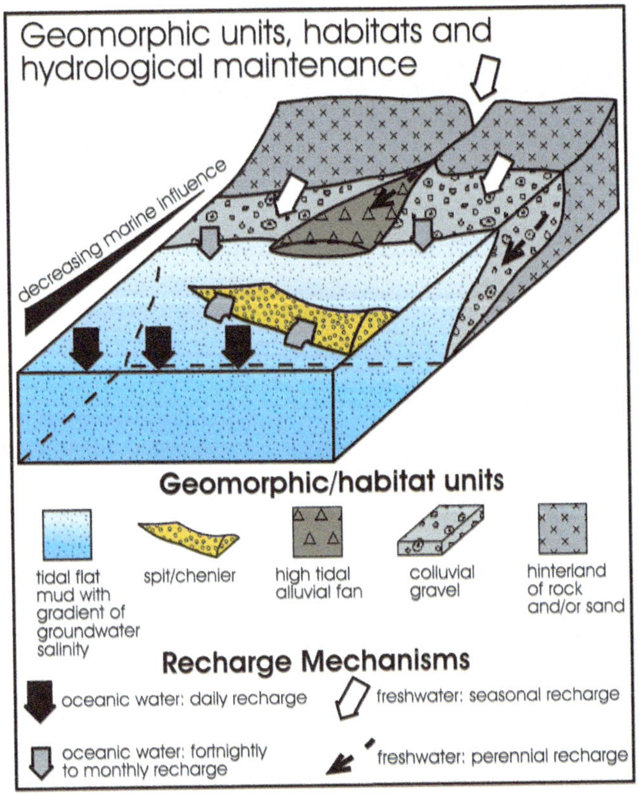

Fig. 1.6 Idealised diagram showing the suite of habitats with diverse substrate types within a ria coast embayment shown in Fig. 1.3D, and the various hydrological/hydro-chemical mechanisms maintaining the mangroves. From MSL to the edge of the tidal flat along the edge of the hinterland, there is a decreasing marine influence

achieved if the measures that are being applied do not take into account the dominant ecological processes and the complexities of the mangrove environment. Under these circumstances, where management rules are poorly matched to the dominant drivers of the ecosystems, then management actions will most likely fail (McDonald et al. 2008).

1.9 The Existing Conservation Regime and Legislation for Australian Mangroves

At the national level in Australia, each State or Territory manages mangroves within their jurisdiction under a wide range of laws and policies, and no coordination of mangrove conservation or management issues exists between the jurisdictions. Legislation introduced in all jurisdictions over the past 40 years has reduced the overall area of mangrove decline caused by development (Rogers et al. 2016). However, in some places, mangroves are severely affected by modified hydrology, exposure, burial, erosion, pollutants, clearing, land-use change, and limited potential for retreat in the face of rising sea levels (Clark and Johnson 2016; Semeniuk and Cresswell 2017, this volume).

There are four main global conservation instruments that have some alignment with mangrove ecosystems in Australia through different mechanisms, some only peripherally, but few directly address mangroves specifically. The Convention on Biological Diversity (CBD) mentions mangroves in many parts of the Convention's work from reporting on national conservation initiatives such as protected areas to sustainable livelihoods in natural systems. However, there are no specific instruments under the CBD that provide additional protection for mangroves than exist within existing jurisdictional legislation. The Convention on Migratory Species (CMS) deals with the conservation of migratory species and several migratory species listed under the CMS utilise mangrove habitats. However, as the CMS does not deal with the habitats themselves, the value for mangrove management and conservation lies solely in information-sharing and knowledge transfer on management of the areas on which migratory species rely. The Convention on Wetlands, most commonly known as the Ramsar Convention, does specifically include mangroves, and includes a number of resolutions concerning the integrated management and sustainable use of mangrove ecosystems, and Australia has listed many sites that contain mangroves. The World Heritage Convention aims to protect the cultural and natural heritage of such outstanding universal value that its conservation is important for current and future generations; Australia has listed a number of sites that contain mangroves, some of which overlap with Ramsar sites.

The Australian Government has limited land management responsibilities and marine management only from 3 nautical miles from the coast. Within State and Territory jurisdictions there are areas of federal land that incorporate the coast, with the Department of Defence being the largest manager of coastal lands that contain mangroves. Notwithstanding the limited direct management nationally there is a plethora of policies and strategies that have been adopted that have relevance to mangrove conservation (see Table 1.5). However, none of these policies or strategies has had a marked effect in establishing a strategic systematic approach to mangrove conservation.

The 2016 Australian State of the Environment (SoE) report reported that in the Northern Territory sea level rise, saltwater intrusion and severe weather events are increasingly severe pressures on coastal environments, particularly floodplains, including suggesting that recent extensive dieback of mangroves is likely linked to climate change (Cresswell and Murphy 2016; Duke et al. 2017; Lovelock et al. 2017). Furthermore, while information on the extent of mangroves exists, there is a lack of monitoring and evaluation of their function, productivity and condition. There is a

Table 1.5 Relevant policies and strategies adopted nationally that include mangroves

Year	Strategy or policy
1992	National Strategy for Ecologically Sustainable Development
1992	Intergovernmental Agreement on the Environment
1995	Commonwealth Coastal Policy
1996	National Strategy for the Conservation of Australia's Biological Diversity
1996	Australia: State of the Environment report (SOE)
1996	State of the Marine Environment report (SOMER)
1997	Wetlands Policy of the Commonwealth Government of Australia
1998	Guidelines for Establishing the National Representative System of Marine Protected Areas
1999	National Local Government Biodiversity Strategy
1999	National Framework for the Management and Monitoring of Australia's Native Vegetation
2001	National Objectives and Targets for Biodiversity Conservation 2001–2005
2001	Coastal Catchments Initiative
2001	Australia: State of the Environment report
2001	Biodiversity Conservation Research: Australia's Priorities
2002	National Framework for NRM Standards and Targets
2003	Framework for a National Cooperative Approach to Integrated Coastal Zone Management
2004	National Biodiversity and Climate Change Action Plan 2004–2007
2005	Directions for the National Reserve System—a Partnership Approach
2005	A National Approach to Biodiversity Decline
2006	National Cooperative Approach to Integrated Coastal Zone Management Framework and Implementation Plan
2006	Australia: State of the Environment report
2008	A National Approach to Addressing Marine Biodiversity Decline 2009

pressing need to quantify carbon sequestration dynamics of mangroves (and saltmarshes) throughout Australia (Clark and Johnson 2016). The SoE report found that mangroves are threatened by climate change and other anthropogenic pressures, and predicted climate change–driven alteration of rainfall patterns around the nation is likely to alter mangrove productivity and growth. In southern Australia, mangroves are threatened by coastal development, and mangroves in north-western Australia are predicted to decline in response to changing salinity regimes associated with drought, but increase southwards in southern Australia because of increasing temperatures and atmospheric carbon dioxide (Alongi 2015; Clark and Johnson 2016).

1.10 Discussion and Conclusions

To maintain the full diversity of mangrove habitats requires a multi-pronged approach from protection in reserves through to active management in high-use industrialised areas of the coast. A national mangrove conservation strategy that relates hierarchically to existing Government biodiversity and coastal management strategies and policies at national and State levels is needed. To facilitate such an approach, a consistent national information-base on mangroves that captures information on each coastal sector, the variations in coastal type within each sector, the variations in mangrove habitat within each coastal type, the ecological drivers in each coastal type, and the variations in species diversity within each habitat type, is required. In this context, for national and for global heritage purposes, it is important not only to conserve species, but also representative communities in their habitat settings, and suites of habitats and their associated assemblages within a diagnostic coastal type and within a range of climate and oceanographic settings.

Through adequate inventory of existing mangroves and their habitats, it should be possible to establish the global, national and regional (sub-national) significance of any mangrove area. This will facilitate sustainable development in areas where Environmental Impact Assessment for development approval is needed. Only through this type of rigorous approach it is then possible to provide compensatory habitat for conservation for all impacted mangrove habitats and to ensure areas are large enough to maintain conservation values. A national approach to mangrove conservation would benefit from a series of analyses at four levels:

1. Identification of those tracts of Australian coast that have international significance from the perspectives and approach of community-*cum*-habitat and mangrove assemblages within inter-related suites of habitats; this level of protection identifies and accords with Australia's responsibility to the global community in preserving globally unique and globally significant stands of mangroves in a manner similar to that preserved in the Sunderbans (Stanley and Hait 2000; MAB 2011; Sundarbans National Park 2014); for an approach to assess International, National, Regional and Local significance see Figure 3 of Brocx & Semeniuk 2007);
2. Identification of those tracts of Australian coast that have national significance from perspectives of community-*cum*-habitat and mangrove assemblages within inter-related suites of habitats;
3. Identification of those tracts of coast within each jurisdiction that have State-wide significance from perspectives of community-*cum*-habitat and mangrove assemblages within inter-related suites of habitats – the works of Semeniuk (1997) and Semeniuk and Wurm (1987) for the Pilbara Coast conform to this and can be used as a model;
4. At a local to regional level, protection of mangroves within an industrialised and urbanised coast, paying due attention to important components of the mangrove system from the perspective of fisheries, coastal protection, education, research, and naturalist activities

This approach will provide a rigorous test of the significance of any mangrove assemblage and from that the degree of protection required may be ascertained. However, protected-area designation may also need to take into account other features of the area, such as the use of the ecosystem by avifauna which may also be part of its designation under international agreements, for example, the use of the Roebuck Bay mangroves and its adjoining low-tidal flats by trans-equatorial avifaunal migrators, and by turtles (Department of Environment and Conservation 2009; Semeniuk and Brocx 2016).

Another outcome from an adequate and consistent inventory will be to analyse the existing Protected Area system to then seek to ensure all major ecosystems with broad buffer zones (that capture underlying physical processes that maintain the ecosystems) are included.

Perhaps though, the biggest hurdle mangrove conservation faces in Australia relates to public awareness of the value of mangrove ecosystems and the roles they play in the provision of ecosystem services. While there has been a distinct increase in the understanding of the role of mangroves in Australia, this is by no means widely understood, and without an appreciation of their value there will be ongoing pressures for their removal.

References

ABARES (2016) Australian forest profiles, Australian Bureau of Agricultural and Resource Economics and Sciences, Canberra

Alongi DM (2015) The impact of climate change on mangrove forests. Curr Clim Chang Rep 1:30–39

Australian Government (2017a) Australia's world heritage list. https://www.environment.gov.au/heritage/places/world-heritage-list

Australian Government (2017b) Australian Ramsar wetlands. http://www.environment.gov.au/water/wetlands/australian-wetlands-database/australian-ramsar-wetlands

Australian Government (2017c) Australia's National Heritage List. http://www.environment.gov.au/heritage/places/national-heritage-list

BMT WBM (2010) Ecological character description for Kakadu National Park Ramsar Site. Prepared for the Australian Government Department of Sustainability, Environment, Water, Population and Communities

BMT WBM (2011a) Ecological character description for Cobourg Peninsula Ramsar Site. Prepared for the Australian Government, Canberra

BMT WBM (2011b) Ecological character description of the corner Inlet Ramsar Site – Final Report. Prepared for the Australian Government Department of Sustainability, Environment, Water, Population and Communities. Canberra

Bridgewater PB (1989) Syntaxonomy of the Australian mangal refined through iterative ordination. Vegetatio 81:159–168

Bridgewater PB, Cresswell ID (1999) Biogeography of mangrove and saltmarsh vegetation: implications for conservation and management in Australia. Mangrove Salt Marshes 3(2):117–125

Bridgewater PB, Cresswell ID (2003) Identifying biogeographic patterns in Australian saltmarsh and mangal systems: a phytogeographic analysis. Phytocoenologia 33(2–3):1–20

Brocx M, Semeniuk V (2007) Geoheritage and geoconservation – history, definition, scope and scale. J R Soc West Aust 90:53–87

Brocx M, Semeniuk V (2010) Coastal geoheritage: a hierarchical approach to classifying coastal types as a basis for identifying diversity and sites of significance in Western Australia. J R Soc West Aust 93:81–113

Brocx M, Semeniuk V (2011) The global geoheritage significance of the Kimberley Coast, Western Australia. J R Soc West Aust 94:57–88

Bunt JS, Williams WT, Duke NC (1982) Mangrove distributions in North-East Australia. J Biogeogr 9:111–120

Busby JR, Bridgewater P (1986) A preliminary atlas of Australian mangroves. AGPS, Canberra. 38pp

Chapman VJ (1976) Mangrove vegetation. J. Cramer, Vaduz. 447p

Chapman VJ (1977) Wet coastal ecosystems. Elsevier Scientific Pub. Co., Amsterdam. 428p

Clark GF, Johnston EL (2016) Coasts: biodiversity: habitat-forming species In: Australia State of the Environment 2016, Australian Government Department of the Environment and Energy, Canberra, https://soe.environment.gov.au/theme/coasts/topic/2016/biodiversity-habitat-forming-species, DOI https://doi.org/10.4226/94/58b659bdc758b. Accessed 20/05/2017

Commonwealth of Australia (1999) Australian guidelines for establishing the National Reserve System. Environment Australia, Canberra. http://www.environment.gov.au/land/nrs/publications/australian-guidelines-establishing-nrs. Accessed 20/05/2017

Cresswell ID, Murphy H (2016) Biodiversity. In: Australia State of the Environment 2016, Australian Government Department of the Environment and Energy, Canberra, https://soe.environment.gov.au/theme/biodiversity, DOI 10.4226/94/58b65ac828812

Cresswell ID, Semeniuk V (2011) Mangroves of the Kimberley Coast: ecological patterns in a tropical ria coast setting. J R Soc West Aust 94(2):213–237

Cresswell ID, Semeniuk V (2017) Australia's major coastal sectors. Coastal & Shelf Science, Estuarine

Department of Parks & Wildlife (2017) List of protected areas of Western Australia. https://en.wikipedia.org/wiki/List_of_protected_areas_of_Western_Australia; https://www.dpaw.wa.gov.au/management/forests/about-our-forests/179-reserves-and-protected-areas; and https://www.wa.gov.au/information-about/recreation/parks-reserves

Department of Environment & Conservation (2009) Roebuck Bay Ramsar Information Sheet (RIS) - 2006–2008 version. Department of Environment and Conservation 17 Dick Perry Avenue, Technology Park, Kensington WA 6983, Australia

Department of the Environment and Energy (2016a) Australia – pre-1750 major vegetation groups – National Vegetation Information System Version 4.2

Department of the Environment and Energy (2016b) Coastal wetlands – mangroves and saltmarshes fact sheet. http://www.environment.gov.au/water/wetlands/publications/factsheet-wetlands-mangroves-saltmarsh. Accessed 20/05/2017

DERM (2012) Queensland Department of Environment and Resource Management. Mangroves. Retrieved 28 Mar 2012, from http://wetlandinfo.derm.qld.gov.au/wetlands/factsfigures/floraandfauna/flora/mangroves.html

Duke NC (2006) Australia's mangroves: the authoritative guide to Australia's mangrove plants. University of Queensland, Brisbane. 200p

Duke NC, Ball MC, Ellison JC (1998) Factors influencing biodiversity and distributional gradients in mangroves. Glob Ecol Biogeogr Lett 7:27–47

Duke NC, Kovacs JM, Griffiths AD, Preece L, Hill DJ, Van Oosterzee P, Mackenzie J, Morning HS, Burrows D (2017) Large-scale dieback of mangroves in Australia's Gulf of Carpentaria: a severe ecosystem response, coincidental with an unusually extreme weather event. Mar Freshw Res 68(10):1816–1829

Environmental Protection Authority (2001) Guidance statement for protection of tropical arid zone mangroves along the Pilbara coastline. Western Australian Government. http://www.epa.wa.gov.au/sites/default/files/Policies_and_Guidance/GS1-TAZ-Mangroves-Pilbara-230401.pdf

FAO (2007) The world's mangroves 1980–2005, FAO Forestry Paper, vol 153. Forest Resource Division, FAO, Rome

Galloway RW (1982) Distribution and physiographic patterns of Australian mangroves. In: Clough BF (ed) Mangrove ecosystems in Australia: structure, function and management. National University Press, Canberra, pp 31–54

Gentilli J (ed) (1971) Climates of Australia and New Zealand, World Survey of Climatology, vol 13. Elsevier, Amsterdam, 405p

Giesen W, Wulfraat S, Zieren M, Scholten L (2007) Mangrove guidebook for SouthEast Asia. FAO and Wetlands International 2006, Regional Office for Asia and the Pacific

Giri C, Ochieng E, Tieszen LL, Zhu Z, Singh A, Loveland T, Masek J, Duke N (2011) Status and distribution of mangrove forests of the world using earth observation satellite data. Glob Ecol Biogeogr 20:154–159. https://doi.org/10.1111/j.1466-8238.2010.00584.x

Hale J (2008) Ecological character description of the Ord River Floodplain Ramsar Site, report to the Department of Environment and Conservation, Perth, Western Australia

Hale J & Butcher R (2009) Ecological character description of the eighty-mile beach Ramsar Site, report to the Department of Environment and Conservation, Perth, Western Australia

Heap AS, Bryce M, Ryan D, Radke LC, Smith C, Smith R, Harris P, Heggie D (2001) Australian estuaries & coastal waterways: a geoscience perspective for improved and integrated resource management. Australian Geological Survey Organisation, Record 2001/07

Hutchings P, Saenger P (1987) Ecology of mangroves. School of Environmental Science and Management Papers

IUCN (1994) International Union for Conservation of Nature. Guidelines for Protected Area Management Categories, Gland

Johannsdottir A, Cresswell ID, Bridgewater P (2010) The current framework for international governance of biodiversity: is it doing more harm than good? Rev Eur Community Int Environ Law 19 (2):139–149

Kellogg Brown & Root (KBR) (2010). Western Port Ramsar Wetland ecological character description. Report for Department of

Sustainability, Environment, Water, Population and Communities, Canberra
Lear R, Turner T (1977) Mangroves of Australia. University of Queensland Press, Brisbane, p 84
Lovelock CE, Feller IC, Reef R, Hickey S, Ball MC (2017) Mangrove dieback during fluctuating sea levels. Sci Rep 7(1):1680
MAB (2011) The Sundarbans biosphere reserve. Retrieved 20 Jan 2012, from http://whc.unesco.org/en/list/798
Macnae W (1968) A general account of the fauna and flora of Mangrove swamps and forests in the Indo-West-Pacific region. Adv Mar Biol 6:73–270. + 1 diag
Mangrove Watch (2010) Mangrove Watch. Retrieved 22 Dec 2011, from http://www.mangrovewatch.org.au/index.php?option=com_content&view=category&layout=blog&id=85&Itemid=3--2-2
Mao L, Zhang Y, Bi H (2006) Modern pollen deposits in coastal mangrove swamps from Northern Hainan Island, China. J Coast Res 22(6):1423–1436
McDonald AD, Little LR, Gray R, Fulton E, Sainsbury KR, Lyne VD (2008) An agent-based modelling approach to evaluation of multiple-use management strategies for coastal marine ecosystems. Math Comput Simul 78:401–411
Mount RE, Bricher PJ (2008) Estuarine, Coastal and Marine (ECM) National Habitat Mapping Project, Project Report, Version 1 February 2008. Spatial Science Group, School of Geography and Environmental Studies, University of Tasmania. Report to the Department of Climate Change and the National Land and Water Resources Audit, Canberra
Newall P R & Lloyd, L N (2012) Lavinia Ramsar Site Ecological character description. Lloyd Environmental report to NRM North. Lloyd Environmental, Syndal, Victoria. 2 March 2012
NFPS (1992) National Forests Policy Statement. Australian Department of Agriculture and Water Resources. http://www.agriculture.gov.au/forestry/policies/forest-policy-statement. Accessed 20.05.2017
NRETAS (2007) Marine and coastal biodiversity of the Northern Territory. Retrieved 2 Dec 2011, from http://www.nt.gov.au/nreta/wildlife/marine/education.html
Pedretti YM, Paling E (2001) WA mangrove assessment project 1999–2000. Murdoch University, report to Environment Australia
Polidoro BA, Carpenter KE, Dahdouh-Guebas F, Ellison JC, Koedam NE, Yong WH (2014) Global patterns of mangrove extinction risk: implications for ecosystem services and biodiversity loss. In: Lockwood JL, Maslo B (eds) Coastal conservation, conservation biology. 19. Cambridge University Press, Cambridge
Rogers K, Boon PI, Branigan S, Duke NC, Field CD, Fitzsimons JA, Kirkman H, Mackenzie JR, Saintilan N (2016) The state of legislation and policy protecting Australia's mangrove and salt marsh and their ecosystem services. Mar Policy 72:139–155
Saenger P, Specht MM, Specht RL, Chapman VJ (1977) Mangal and coastal salt marsh communities in Australasia. Wet coastal ecosystems. Elsevier Scientific Pub. Co, Amsterdam, pp 293–345
Semeniuk V (1980) Mangrove zonation along eroding coastline, King Sound, NW Australia. J Ecol 68(3):789–812
Semeniuk V (1983) Mangrove distribution in northwestern Australia in relationship to regional and local freshwater seepage. Vegetatio 53:11–31
Semeniuk V (1985) Development of mangrove habitats along ria shorelines in north and northwestern tropical Australia. Vegetatio 60:3–23
Semeniuk V (1993a) The mangrove systems of Western Australia – 1993 Presidential Address. J R Soc West Aust 76:99–122
Semeniuk V (1993b) The Pilbara coast: a riverine coastal plain in a tropical arid setting, northwestern Australia. Late Quaternary evolution of coastal and lowland riverine plains of southeast Asia and northern Australia. C. D. Woodroffe. 83: 235–256
Semeniuk V (1995) The Holocene record of climatic, eustatic and tectonic events along the coastal zone of Western Australia – a review. Chapter 21. In: Finkl CW (ed) Holocene cycles: climate, sea level rise, and sedimentation. J Coast Res Spec Issue 17: 247–259
Semeniuk V (1996) Coastal forms and quaternary processes along the arid Pilbara coast of northwestern Australia. Palaeogeogr Palaeoclimatol Palaeoecol 123(1–4):49–84
Semeniuk V (1997) Selection of mangrove stands for conservation Pilbara Region. Report to Western Australian Department Resources Development. Copy of report residing in the Battye Library, Perth, Western Australia
Semeniuk V (2008) Sedimentation, stratigraphy, biostratigraphy, and Holocene history of the Canning Coast, north-western Australia. J R Soc West Aust 91:53–148
Semeniuk V (2012) Predicted response of coastal wetlands to climate changes – a Western Australian model. Hydrobiologia 708 (1):23–43. https://doi.org/10.1007/s10750-012-1159-0
Semeniuk V, Brocx M (2016) Roebuck Plains, north-western Australia – a major Holocene carbonate mud deposit of international geoheritage significance. ProGEO News 1: 3–9
Semeniuk V, Cresswell ID (2013) A proposed revision of diversity measures. Diversity 5:613–626
Semeniuk V, Wurm PAS (1987) Mangroves of the Dampier Archipelago, Western Australia. J R Soc West Aust 69:29–87
Semeniuk V, Kenneally KF, Wilson PG (1978) Mangroves of Western Australia. Western Australian Naturalists Club, Perth Handbook 12. 90pp
Short AD, Woodroffe CD (2009) The coast of Australia. 288p
Smith TJ, Duke NC (1987) Physical determinants of inter-estuary variation in mangrove species richness around the tropical coastline of Australia. J Biogeogr 14:9–19
Spalding M, Kainuma M, Collins L (2010) World atlas of mangroves. Earthscan, London
Stanley DJ, Hait AK (2000) Holocene depositional patterns, neotectonics and Sundarban mangroves in the Western Ganges-Brahmaputra Delta. J Coast Res 16:26–39
Sundarbans National Park (2014) http://whc.unesco.org/en/list/452
Suzuki K, Saenger P (1996) A phytosociological study of mangrove vegetation in Australia with a latitudinal comparison of East Asia. Mangrove Sci 1:9–27
Tomlinson PB (1986) The botany of mangroves
UNFCCC (2011) CDM Executive Board AR-AM0014/Version 01.0.0 Sectoral Scope: 14, EB 61 UNFCCC (2011b). Accessed from http://unfccc.int/files/meetings/awg/application/pdf/cbd_wsi_200511.pdf
Valiela I, Bowen JL, York JK (2001) Mangrove forests: one of the world's threatened major tropical environments. Bioscience 51 (10):807–815
Waltner-Toews D, Kay J, Lister NME (2008) The ecosystem approach: complexity, uncertainty, and managing for sustainability. Columbia University Press, New York
Wells AG (1983) Distribution of mangrove species in Australia. In: Teas HJ (ed) Biology and ecology of mangroves. W. Junk, The Hague, pp 57–76
Wells AG (1985) Grouping of tidal systems in the Northern Territory and Kimberley Region of Western Australia on presence/absence of mangroves species. In: Bardsle KN, Davie JDS, Woodroffe CD (eds) 1985 Coasts and tidal wetlands of the Australian monsoon region: a collection of papers presented at a conference held in Darwin 4-–11 November 1984. North Australia Research Unit, Mangrove Monograph No. 1
Wightman GM, Andrews M (1989) Mangroves of the northern territory. Conservation Commission of the Northern Territory, Palmerston
Wightman G, Danaher K, Dunning M, Beumer J, Michie M (2004) Mangroves: a description of key species groups in the northern planning area. National Oceans Office, Hobart

The Dynamics of Expanding Mangroves in New Zealand

2

Erik M. Horstman, Carolyn J. Lundquist, Karin R. Bryan, Richard H. Bulmer, Julia C. Mullarney, and Debra J. Stokes

Abstract

In contrast to the global trend of mangrove decline, New Zealand mangroves are rapidly expanding, facilitated by elevated sediment inputs in coastal waters as a consequence of large-scale land use changes following European settlement. New Zealand mangroves are at the southern limit of the global mangrove extent, which limits the tree height of *Avicennia marina* var. *australasica*, the only mangrove species present. Mangroves in New Zealand thrive in the sheltered environments of infilling drowned river valleys with abundant supply of fine terrigenous sediments, showing various stages of mangrove succession and expansion dynamics. Bio-physical interactions and carbon dynamics in these expanding temperate mangrove systems show similarities to, but also differ from those in tropical mangrove forests, for instance due to the limited height and complexity of the mangrove communities. Likewise, ecosystem services provided by New Zealand mangroves deviate from those offered by tropical mangroves. In particular, the association of mangrove expansion with the accumulation of (the increased supply of) fine sediments and the consequent change of estuarine ecosystems, has provoked a negative perception of mangrove expansion and subsequently led to mangrove clearance. Over recent decades, a body of knowledge has been developed regarding the planning and decision making relating to mangrove removal, yet there are still effects that are unknown, for example with respect to the post-clearance recovery of the original sandflat ecosystems. In this chapter we discuss the dynamics of New Zealand's expanding mangroves from a range of viewpoints, with the aim of elucidating the possible contributions of expanding mangroves to coastal ecosystem services, now and in the future. This chapter also reviews current policies and practice regarding mangrove removal in New Zealand and addresses the (un)known effects of mangrove clearance. These combined insights may contribute to the development of integrated coastal management strategies that recognise the full potential of expanding mangrove ecosystems.

Keywords

Temperate mangroves · *Avicennia marina* · Mangrove expansion · Bio-physical interactions · Carbon dynamics · Ecosystem services · Mangrove removal · Mangrove management

2.1 Introduction

Mangroves are located throughout estuaries around the shores of northern New Zealand and are part of the island's native vegetation. The presence of mangroves in New Zealand has been dated back to 19 million years BP by association with Miocene deposits (Sutherland 2003). Pollen records show that the mangrove species *Avicennia marina* has been present from around 14,000 years BP (Pocknall et al. 1989), long before New Zealand was discovered.

The early Māori settlers valued the mangroves (mānawa in Māori) for their provisioning function for food and resources.

The original version of this chapter was revised. An erratum to this chapter can be found at https://doi.org/10.1007/978-3-319-73016-5_32

E. M. Horstman (✉) · K. R. Bryan · J. C. Mullarney
Coastal Marine Group, Faculty of Science and Engineering, The University of Waikato, Hamilton, New Zealand
e-mail: erik.horstman@waikato.ac.nz

C. J. Lundquist
National Institute of Water & Atmospheric Research Ltd., Hamilton, New Zealand

Institute of Marine Science, The University of Auckland, Auckland, New Zealand

R. H. Bulmer
National Institute of Water & Atmospheric Research Ltd., Hamilton, New Zealand

D. J. Stokes
Marine Ecology Research Centre, Southern Cross University, Lismore, NSW, Australia

C. Makowski, C. W. Finkl (eds.), *Threats to Mangrove Forests*, Coastal Research Library 25,
https://doi.org/10.1007/978-3-319-73016-5_2

Māori traditionally gathered kaimoana—fish and shellfish—in the mangroves and on adjacent mudflats (Crisp et al. 1990). To date, the Māori belief in the interrelationship of all parts of the environment resonates in the concept of kaitiakitanga—guardianship of the environment—which also applies to estuaries, mangroves included. The fundamental idea behind this concept is that all benefits are reciprocal and that an ecosystem that is sustained and cared for can provide benefits back to humans (Harmsworth and Awatere 2013).

The approach to ecosystem management changed with the arrival of the European settlers, with widespread land use changes such as those associated with agricultural development and construction works initially resulting in a direct loss of mangrove habitat (Crisp et al. 1990; Dingwall 1984). However, following rapid clear felling of the land and consequent erosion, sedimentation in estuaries increased and tidal flats started expanding rapidly, accommodating widespread mangrove expansion since the mid-twentieth century (Lovelock et al. 2007; Lundquist et al. 2014b; Morrisey et al. 2007). These newly established mangroves generally occupy muddy intertidal flats that have expanded into estuaries that were previously dominated by coarser sandy sediments (Ellis et al. 2004; Lovelock et al. 2007; Stokes et al. 2010).

More recently, the rapid expansion of mangroves and the 'muddification' of estuaries has raised concerns about the loss of diversity of estuarine habitats and a concomitant decline of ecosystem functioning (Harty 2009; Thrush et al. 2004), in addition to the more tangible consequences related to recreational, navigational and amenity values of the estuaries (Lundquist et al. 2014b). Therefore, mangrove removal is a common response to mangrove expansion in New Zealand, with little consideration of the ecosystem services that may be impacted, either positively or negatively, due to mangrove removal (Lundquist et al. 2014b; Morrisey et al. 2010; Stokes and Harris 2015).

The New Zealand practice of mangrove removal contrasts strongly with international efforts to restore and expand mangrove area with the goal of enhancing coastal safety by (re-)establishing the attenuating and stabilizing effects of mangroves (Friess et al. 2016a; Lewis 2005; Spalding et al. 2014; UNEP 2014). Moreover, in the face of expected sea-level rise, accelerated sedimentation in mangrove forests could provide coastal zones with a natural resilience that traditional engineered structures are lacking (Alongi 2008; Barbier 2014; Borsje et al. 2011; Bouma et al. 2014; Stagg et al. 2016; Temmerman et al. 2013). Another benefit of mangrove expansion is the potential for carbon burial, combating increasing carbon emissions that are aggravating the severity and impacts of climate change (Bouillon et al. 2008; Bulmer et al. 2016b; Donato et al. 2011; Friess et al. 2016b; Twilley et al. 1992).

In this chapter we address lessons learned from the expanding mangroves in New Zealand, focussing on their (changing) biophysical, ecological and carbon dynamics, as well as local management practices coping with mangrove expansion. By doing so, we hope to provide a useful view on the dynamics and management of coastal areas where mangroves are expanding, either naturally (as in New Zealand) or as a result of successful mangrove restoration schemes.

We start with an introduction to the mangroves of New Zealand, their characteristic geophysical settings and their expansion dynamics (section "Mangroves in New Zealand"), followed by an appraisal of the biophysical interactions in these expanding mangroves based on recent work in contrasting mangrove systems (section "Biophysical Dynamics of Expanding Mangroves"). The resulting morphological development and ecological succession of New Zealand mangroves is discussed from a forest succession perspective, identifying different pathways for the succession of prograding and stable mangroves (section "Succession of Expanding Mangrove Systems"). Both the morphological development and the forest succession are closely interlinked with the carbon dynamics of expanding mangrove systems, to be addressed in section "Carbon Dynamics of Expanding Mangroves". Moving on to the management of expanding mangroves, we first identify the ecosystem services and dis-services related to the expanding mangroves in New Zealand (section "Ecosystem (Dis-) services of Expanding Mangroves"). Perceptions of the consequences associated with mangrove expansion are the main driver of mangrove removals in New Zealand. Historically, only minimal information on effects of mangrove removal was available to inform policy planning and decision making regarding mangrove removal, but recent research has led to more informed mangrove management practices, which will be discussed in section "Mangrove Removal in New Zealand". Analysis of recovery trajectories of multiple decades of mangrove removal practices in New Zealand allows us to compare the ecosystem services provided by both intact and cleared mangrove sites. However, not all consequences of mangrove removal for the biophysical properties and functioning of these intertidal habitats are known and these knowledge gaps will be addressed in section "The Known and Unknown Effects of Mangrove Removal". We conclude with an outlook on the future development of and the (dis-)services provided by New Zealand mangroves and how these could be managed effectively and sustainably.

2.2 Mangroves in New Zealand

2.2.1 Mangrove Distribution

Although mangroves predominantly occur in the tropics, they also can extend into temperate climate regions. Mangroves flourish in areas where mean monthly air temperatures are above 20 °C, with a seasonal range of up to 10 °C (Chapman 1977). Beyond these areas, the poleward extent of mangroves is often limited by the incidence of ground frost and generally coincides with the winter position of the 20 °C seawater isotherm (Duke et al. 1998). The mangroves on the North Island of New Zealand, together with those in southern Australia and eastern South America, are exceptions to the above water temperature limitation. These mangrove communities, at the extremities of the global mangrove distribution, are likely relict populations of a greater poleward mangrove distribution during a warmer period in the past, and may also be facilitated by extensions of irregular ocean currents, warming coastal waters and transporting propagules from the north (Chapman 1976; Duke et al. 1998; Macnae 1966).

Both mangrove area and mangrove diversity generally decline with increasing latitude (Ellison 2002; Giri et al. 2011). Temperate mangroves (*sensu* Morrisey et al. 2010) typically comprise a few species only, as opposed to greater species diversity in mangroves at lower latitudes. Restricted species diversity may limit the range of habitats that are suitable for mangroves at their latitudinal extremes, depending on species' tolerance of abiotic factors such as temperature, day length, aridity, water and soil salinity, tidal range and hydrodynamic exposure (Chapman 1977; Duke 1990; Morrisey et al. 2010; Tomlinson 1986; Watson 1928). According to recent estimates, less than 1% of the total global mangrove cover of ~13.8 million ha is found at latitudes greater than 30° (Giri et al. 2011).

New Zealand's mangrove distribution covers about 26,050 ha (Spalding et al. 2010). Mangroves are found from 34°27′S in the far north of the North Island to 38°05′S at Kawhia Harbour on the west coast and 38°03′S at Ohiwa Harbour on the east coast (Fig. 2.1; Crisp et al. 1990; de Lange and de Lange 1994). Planted mangroves have been observed to establish and survive as far south as 41°13′S on the Hutt River, at the southern end of the North Island (de Lange and de Lange 1994). Frequency and duration of frost events, causing tissue damage in mangroves (Duke 1990; Saenger and Snedaker 1993), form a primary constraint on mangrove expansion at the latitudinal limits of their distribution in New Zealand (Lundquist et al. 2014b). Additional abiotic factors inhibiting the southward expansion of New Zealand mangroves are the prevailing currents and the lack of suitable habitats within the dispersal range of the propagules (de Lange and de Lange 1994). Propagules of the native mangrove remain viable for a few days only, which is too short for wind-induced drift currents to transport them past the rocky shores directly south of the southernmost limit of the present mangrove distribution (de Lange and de Lange 1994; Duke et al. 1998).

2.2.2 Geomorphological Setting of New Zealand Mangroves

Similar to most mangroves around the world, New Zealand mangroves thrive in sheltered coastal environments with a terrigenous sediment supply (Augustinus 1995; Chapman 1977; Thom 1982; Woodroffe 1992). Whereas tropical mangroves also occur in river deltas and at muddy open coasts with extensive intertidal flats, New Zealand mangroves predominantly occur in estuaries, tidal basins, lagoons, tidal creeks and river mouths, as shown in Fig. 2.1 (Chapman 1976; Morrisey et al. 2010; Spalding et al. 2010). The largest mangrove areas are observed in estuaries with large terrigenous sediment supplies such as Rangaunu Harbour (2416 ha) and the Firth of Thames (1100 ha; locations indicated in Fig. 2.1 Lundquist et al. 2014b; Morrisey et al. 2010).

Estuaries in New Zealand are ancestral river valleys and bays that were flooded by rising sea levels during the last post-glacial marine transgression (Hume 2003; Roy et al. 2001). Since the sea level stabilized, around 6500 years BP, the drowned river valleys and bays started infilling. The rate of estuarine infilling depends on the supply of sediment from the river catchment discharging in the estuary and the tidal prism flushing the estuary (Hume and Herdendorf 1992). At the same time, coastal processes caused the formation of sandy barriers and spits across the entrances of some of these estuaries, reducing tidal ranges and currents inside the estuaries (Hume 2003; Roy et al. 2001). The infilling of drowned river valley estuaries, which has accelerated since European settlement in New Zealand (see section "Mangrove Expansion in New Zealand's Anthropocene"), created intertidal flats that were suitable for mangrove establishment.

2.2.3 Mānawa: The New Zealand Mangrove

The mangrove genus *Avicennia* tolerates a wide range of environmental conditions such as air and water temperatures, frequency and severity of frosts, day lengths and salinity (Chapman 1976; Duke 1990; Krauss et al. 2008; Stuart et al. 2007). Hence, *Avicennia* spp. cover the largest latitudinal range of all mangroves, from 32°N on Bermuda to 38°S in Australia and New Zealand (Duke et al. 1998; Krauss et al.

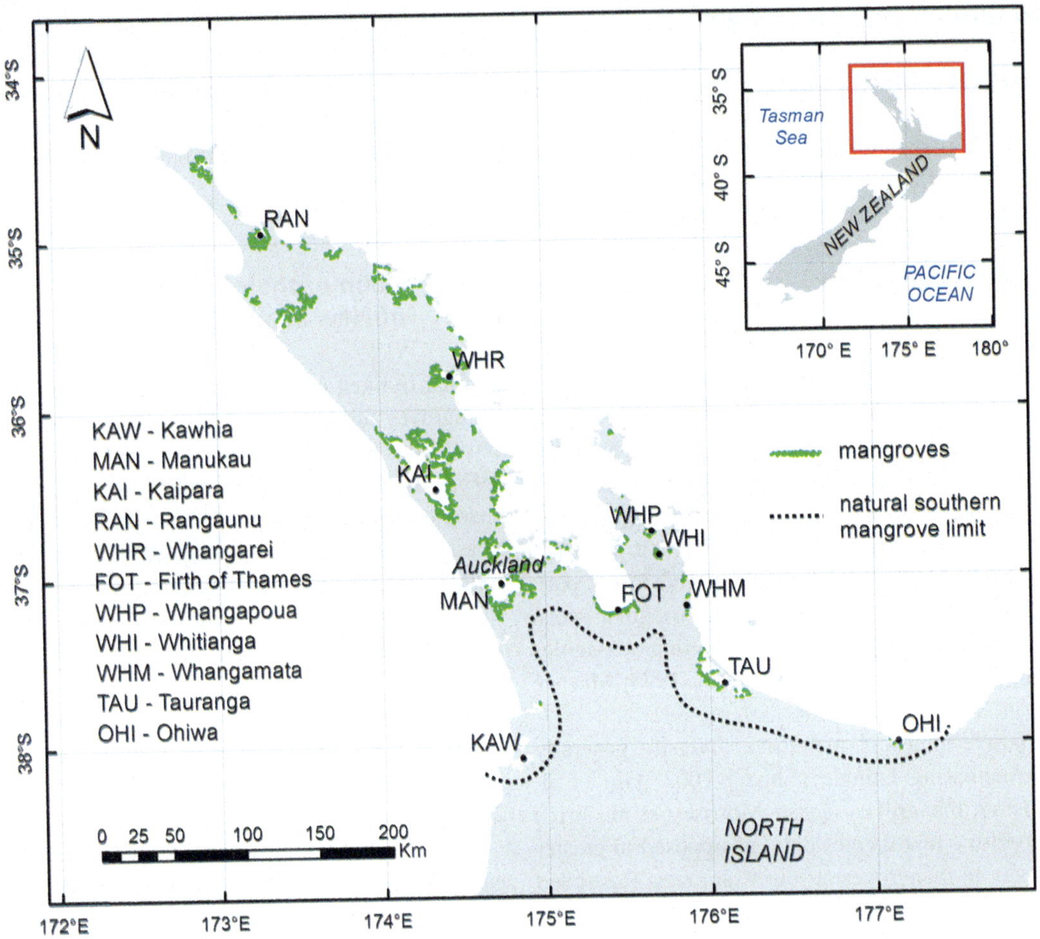

Fig. 2.1 Mangroves in New Zealand are found from the tip of the North Island, down to Kawhia Harbour on the west coast and Ohiwa Harbour on the east coast (mangrove distribution data: Crisp et al. 1990; Spalding et al. 2010). Names of locations are included for mangrove areas that are referred to throughout this chapter

2008; Spalding et al. 2010; Tomlinson 1986). These distributional limits are hypothesized to coincide with a trend towards zero reproductive success, due to the limited summer period available for growth, flowering and fruiting (Duke 1990).

New Zealand mangroves consist of a single species: *Avicennia marina* var. *australasica* (Fig. 2.2), previously named *Avicennia marina* var. *resinifera* (Chapman 1976; Duke 1991; Tomlinson 1986). The same species is also found in south-eastern Australia, New Caledonia and Lord Howe Island (Duke 1991). *Avicennia marina* is also commonly known as the 'grey mangrove' due to the grey shade of its bark. *Avicennia* spp. have pneumatophores (i.e. pencil roots) that emerge from the bed for aeration on waterlogged soils. In New Zealand, these pneumatophores have heights generally ranging between 5 and 25 cm, and reaching up to several decimetres on the deeper banks of creeks and in runnels through the mangroves (pers. obs. by the authors).

Greater temperature stresses towards the latitudinal limits of mangrove distribution are associated with smaller tree heights and reduced net primary productivity (Bouillon et al. 2008; Saenger and Snedaker 1993). Low winter temperatures and frost have been reported to cause a direct loss of stem height due to die back of twigs (Lovelock et al. 2007) and impact on terminal buds, limiting the growth of mangrove tree crowns so that only lower branches can continue growing (Chapman 1976). Such periodic damage due to low temperatures can cause a *stunted* or *dwarfed* growth form, typically featuring a well-developed trunk and a wide, spreading canopy of limited height (Fig. 2.2a); (Chapman 1976; Crisp et al. 1990; de Lange and de Lange 1994; Saintilan et al. 2009). Nevertheless, stunted trees also occur in areas with taller trees, and hence this growth form is not uniquely caused by temperature stresses. De Lange and de Lange (1994) observed that, in New Zealand, stunted growth forms mostly occur on substrates with <50% mud (<20 μm), whereas taller trees can be found on substrates with higher mud contents under the same climatic conditions (Fig. 2.2b, c).

Avicennia marina var. *australasica* can grow to heights of up to about 10 m (Chapman 1976; Duke 1991) and New Zealand mangroves in the far north approach this

Fig. 2.2 *Avicennia marina* var. *australasica* in New Zealand: (**a**) a stunted (or dwarfed) growth form of 0.5–1 m of height in Tauranga Harbour, (**b**) a dense growth form of trees of 2–3 m of height in the Firth of Thames, and (**c**) a tall growth form of trees of up to about 7 m of height in the inner Whitianga Harbour (Photo credits: C. Lundquist and E. Horstman)

height. Although mean tree heights generally reduce with latitude, other controlling factors such as abovementioned substrate type, but also soil drainage and tidal inundation can locally affect tree heights. Hence, stunted trees are also observed in the far north of New Zealand, whereas taller tree growth is observed closer to the southern limits of mangrove distribution as well, with 7–8 m tall trees in e.g. the Auckland region and in the Whangarei and Whitianga Harbours (Chapman 1976; Crisp et al. 1990; pers. obs. by the authors). In well-established mangroves, the taller trees are often observed at the mangrove fringe or close to a creek and tree heights decline in the landward direction (Chapman 1976; Lovelock et al. 2007; Morrisey et al. 2007). In rapidly expanding mangroves, tree age progressively declines towards the forest fringe, which coincides with a decline in height near the seaward fringe where seedlings have only recently established (Morrisey et al. 2007). These tree height gradients become non-uniform on creek-dissected mudflats (May 1999) and in wide stands of mangroves that have extended in a non-linear fashion over time, such as the extensive mangroves at the southern fringe of the Firth of Thames (Lovelock et al. 2010; Swales et al. 2015).

Mangroves are distributed across the intertidal profile from approximately mean sea level (MSL) to the highest spring tide mark (Duke et al. 1998; Tomlinson 1986; Watson 1928). The lower elevation limit around MSL is associated with the inundation duration of the aerial roots under regular tidal conditions and physiological constraints of the mangroves with respect to this variable. Increased inundation frequency and wave activity at lower elevations inhibit the establishment and survival of mangrove seedlings (Balke et al. 2013, 2015; Lovelock et al. 2010). Mangroves can extend to elevations at the upper end of the tidal prism, that are only infrequently flooded by extreme spring tides and episodic storm surges, with the latter substantially (but temporarily) increasing the inundation duration of the mangroves (Swales et al. 2007).

Mangroves in New Zealand can be part of an extensive coastal ecosystem with multiple zones comprising different vegetation types that occupy different intertidal elevations, as schematized in Fig. 2.3 (Chapman 1976; Crisp et al. 1990; Graeme 2013; Wassilieff 2006). The lower parts of muddy or sandy intertidal flats are often colonized by seagrass (*Zostera muelleri*) communities, which can extend from the edge of the mangrove zone down into the shallow subtidal (Chapman 1976; Wassilieff 2006). At higher intertidal elevations, directly in front of the mangroves we generally observe a seaward extension of the mangrove root systems, forming a pneumatophore 'carpet' interspersed with (smaller) more isolated trees and seedlings, named the mangrove *fringe*.

Fig. 2.3 A typical cross-shore profile of an intertidal area with mangroves in an estuary in northern New Zealand (After: Chapman 1976; Crisp et al. 1990; Swales et al. 2015; Wassilieff 2006): the *lower* intertidal is dominated by seagrass (*Zostera muelleri*); mangroves (*Avicennia marina*) thrive in the regularly flooded area above mean sea level; the upper intertidal is home to salt meadows, comprising succulents and herbs such as saltwort (*Sarcocornia quinqueflora*) and remuremu (*Selliera radicans*), and salt marshes, covering rushes such as sea rush (*Juncus kraussii*) and oioi (*Leptocarpus similis*); and at the landward end, coastal scrubs such as flax (*Phormium tenax*) are prevalent

The upper end of the intertidal is the domain of saltmarshes and salt meadows. Saltmarsh vegetation consists of sea rush (*Juncus kraussii*) and oioi (*Leptocarpus similis*) that can form dense grasslands up to 1.5 m of height. Salt meadow vegetation is of lower height and consists of species such as saltwort (*Sarcocornia quinqueflora*) and remuremu (*Selliera radicans*) (Chapman 1976; Wassilieff 2006). The separation between saltmarsh and salt meadow is not very strict, systems can intermingle and pervade into the upper mangroves. At the landward end, intertidal vegetation gives way to coastal scrubs and other land uses.

2.2.4 Mangrove Expansion in New Zealand's Anthropocene

European settlement and subsequent developments initially caused a large-scale loss of mangrove habitat in New Zealand (Crisp et al. 1990; Lundquist et al. 2014b; Morrisey et al. 2007). Human activities such as land reclamations for ports, landfills, airports (e.g. Auckland International Airport), agriculture and industrial and urban development directly impacted the mangrove area. Additionally, construction of causeways, marinas and other engineering works has restricted tidal flows and/or elevated water levels in some estuaries, reducing suitable mangrove habitats (Crisp et al. 1990; Dingwall 1984; Lundquist et al. 2014b).

European settlement was followed by large-scale deforestation of river catchments since the mid-nineteenth century, intensifying land erosion and accelerating infilling of the drowned river valley estuaries (Burns and Ogden 1985; Lovelock et al. 2007; Lundquist et al. 2014b; Morrisey et al. 2010; Nichol et al. 2000; Swales et al. 2007). Sediment supply to the estuaries was enhanced by the typical topography of New Zealand catchments with steep slopes and limited spatial extents, and also by the erratic precipitation pattern with episodic rainstorms. Large supplies of relatively fine terrigenous sediments to the estuaries resulted in increased water turbidity and accelerated sediment accretion, shifting these estuaries to muddier intertidal systems (Morrisey et al. 2010; Nichol et al. 2000; Thrush et al. 2004). Consequently, sedimentation rates increased by an order of magnitude or more (Jones 2008; Nichol et al. 2000; Swales et al. 2009), as illustrated by the estuaries on the Coromandel Peninsula where accretion rates were found to be 50–80 times greater in the second half of the twentieth century than before European settlement (Jones 2008). Although the stabilization of land-use practices has reduced sediment supply from the catchments over the past century, sediment yields and consequent sediment accretion rates are still substantially higher than in pre-European times (Healy 2002; Jones 2008; Swales et al. 2002).

The accelerated estuarine infilling and vertical accretion of tidal flats, providing additional habitat suitable for mangroves, combined with larger nutrient inputs, climate warming and changes in relative sea level (due to accretion/subsidence) have facilitated the expansion of New Zealand mangroves during the past 50–70 years (Burns and Ogden 1985; Harty 2009; Lovelock et al. 2007; Morrisey et al. 2010; Swales et al. 2007, 2015, 2016; Young and Harvey 1996). These processes have resulted in rapid mangrove expansion in the majority of the estuaries on the Coromandel Peninsula (Jones 2008). In the Firth of Thames mangroves have expanded over a width of 1 km over the past 50 years

(Lovelock et al. 2010; Swales et al. 2007, 2016). Historical analyses of aerial imagery of a limited number of Auckland estuaries suggested mangrove expansion rates of up to 20% per year, with an average of 4.1% per year, during the second half of the twentieth century (Swales et al. 2009). Updated analyses for all estuaries in the Auckland region suggest expansion rates of 3.4% per year for the 1940–2014 period. Since 1970, large differences in expansion rates between different mangrove growth forms have been reported, with average expansion rates of 1.1% and 20.7% per year for tall and dwarf mangroves, respectively (Suyadi, University of Auckland, pers. comm.).

The infilling of estuaries as they 'age' causes a shift in the relative proportion of different estuarine habitats, such as a decline of the sandflat area (see section "Ecosystem (Dis-) services of Expanding Mangroves" for related consequences to ecosystem services), in favour of intertidal mudflats and mangrove expansion (Morrisey et al. 2007; Roy et al. 2001; Saintilan 2004). The accelerated infilling of New Zealand estuaries accelerates the shift towards mangrove habitats, and is also associated with increased turbidity levels, which is further enhanced due to sediment resuspension by wave action in these shallow waters (Green et al. 1997; Green and Coco 2007; Roy et al. 2001). These changes come at the risk of reduced habitat diversity and a loss of ecosystem values (Saintilan 2004; Thrush et al. 2004). For example, mangrove cover in the Whangapoua Harbour (on the Coromandel Peninsula) has more than doubled between 1945 and 2006, while seagrass habitat in this estuary has halved over the same period (Jones 2008). Consequences of the rapid succession of New Zealand mangroves are further discussed in section "Ecosystem (Dis-)services of Expanding Mangroves".

2.3 Biophysical Dynamics of Expanding Mangroves

2.3.1 Morphological Change in Expanding Mangroves

Increased sedimentation in New Zealand's drowned river valley estuaries raises the intertidal flat heights and allows mangroves to establish. Sediment loads delivered to mangroves in these estuaries have a terrigenous origin, supplied by catchment run-off, and are primarily composed of fine sediments such as silts and clays (Griffiths and Glasby 1985; Healy 2002). Nutrient addition coupled to the terrigenous sediment input, especially from catchments dominated by agricultural land use, can also promote expansion, although this effect plays a secondary role relative to that of sediment supply (Lovelock et al. 2007; Martin 2007; Melville and Pulkownik 2006). Estimation of precise sediment budgets can be difficult, owing to the highly variable (both temporally and spatially) nature of sediment delivery (Bryce et al. 2003), with large pulses of sediment delivered to estuaries episodically during storm events (Fig. 2.6; Stern et al. 1986) and additional variation over seasonal timescales. Longer-term records (over the past century) of sediment accumulation can be determined using ^{210}Pb geochronology (Nittrouer et al. 1979), and these rates have been shown to be strongly dependent on the sediment supply and geomorphic setting. Within New Zealand, short-term estimates (but integrating over seasonal timescales) of sediment accumulation vary widely and range up to 100 mm year^{-1} (Table 10 in Morrisey et al. 2010).

Following sediment delivery to the system, deposition processes vary between different types of mangrove systems. In more exposed-coast locations, deposition can be controlled by the larger-scale shelf and tidal forcing and baroclinic (river-influenced) flows (Wolanski et al. 2001). These processes often lead to the formation of the classic flat-fronted deltaic shape commonly found in tropical deltaic environments. In the quiescent upper reaches of estuaries, sediment resuspension and transport (asymmetry) are controlled by a balance between tidal currents, water depth, wind-driven currents and wind wave growth (Green et al. 1997; Green and Coco 2007; Hunt et al. 2015, 2017). Mangroves occupy the landward edges of intertidal flats and expand seaward from these estuarine fringes (see section "Succession of Expanding Mangrove Systems" for succession patterns). Deposition is also modulated by estuarine maturity as the tidal prism is smaller for (more mature) infilled estuaries (Friedrichs and Aubrey 1988). In mangrove systems with incised creeks, the presence of the creeks further controls flow routing at low water depths, with sediment delivered to the forest once creek banks are overtopped, and in particular with formation of sheet flow and direct sediment supply from the estuarine waters once water depths exceed the height of the root system (Horstman et al. 2015).

2.3.2 Biophysical Interactions

Mangrove trees introduce considerable complexity to the hydrodynamics within coastal areas. The presence of obstacles (tree trunks, branches, leaves and roots) enhances drag. This vegetative drag removes energy from the system and has been observed to induce large-scale changes in flow directions (Horstman et al. 2013; Kobashi and Mazda 2005; Mullarney et al. 2017b) and also dissipates swell and wind wave energy (Henderson et al. 2017; Horstman et al. 2014; Quartel et al. 2007; Vo-Luong and Massel 2008). Mangroves can also dissipate longer-period waves such as tsunamis (Alongi 2008; Danielsen et al. 2005; Wolanski 2007) or storm surges, provided the forest is of sufficient width (Mullarney and Henderson 2017), see Fig. 2.4. Thus,

Fig. 2.4 Schematic showing wave energy dissipation of swell (dark blue) and longer wavelengths such as storm surge (light blue). Insets showing swell waves propagating into the mangrove forest in the Firth of Thames (Photo credits: E. Horstman)

mangrove forests can act as an 'eco-defence', providing a protective barrier against wave impact and mitigating against coastal erosion, whilst simultaneously offering ecological co-benefits (see section "Ecosystem (Dis-)services of Expanding Mangroves" for these ecosystem services).

Attenuation of hydrodynamic forces in mangroves creates a favourable environment for further sediment deposition (Horstman et al. 2014, 2017; van Santen et al. 2007; Young and Harvey 1996) with spatial patterns of deposition within the forest being controlled by the small-scale interactions between the water and the vegetation elements, as illustrated in Fig. 2.5. In some areas, the enhanced vegetative drag (e.g. from water moving through the pneumatophores) removes energy from the system, creating zones of low flow which prevent sediment resuspension and promote deposition. However, in other cases, the vegetation elements have been shown to generate turbulence in the form of eddies and wakes at small length scales (corresponding to the widths of the obstacles or the spacing between them), which can lead to resuspension and local scour (Bouma et al. 2007; Mullarney et al. 2017a). The shift between these two modes of behaviour is still not well known. In tropical systems, intense turbulence (characterised by dissipation rates of turbulent kinetic energy of up to 0.1 W kg^{-1}, values comparable to those observed near the bed in surf zones) has been observed to occur within pneumatophore canopies, with smaller values observed on the adjacent unvegetated mudflats. Even more intense turbulence was reported during large wave events or in regions with denser vegetation (Norris et al. 2017). In particular, the forest fringe between the unvegetated and vegetated regions, where relatively fast flows first interact with vegetation, is shown to be a particularly dynamic area. This intense turbulence at the fringe has potential to winnow fine sediments, which are then transported and deposited further into the forest, resulting in a gradient in grain size with distance into the forest (Fricke et al. 2017). Likewise, turbulence production and near-bed turbulence intensities have been observed to decrease with vegetation density (see insets in Fig. 2.5), creating a similar gradient in grain sizes of the deposits (Horstman et al. 2017, 2018; van Katwijk et al. 2010).

Results from tropical regions have raised the possibility that mangroves may act as ecosystem engineers, by growing at densities which promote maximal energy dissipation and sediment deposition, thus creating a feedback mechanism which allows for mangrove expansion (Henderson et al. 2017). Furthermore, there are many other small-scale processes, whose influence on deposition in vegetated regions has not, to date, been studied in detail. Examples include the effect of vegetation-generated turbulence on the formation and break up of aggregates of silt and clay particles ('flocs'), which influences the settling velocity of particles (Wolanski 1995), as well as the effects of bioturbation, which changes bed roughness, and biostabilisation, which can increase cohesiveness e.g. through the stabilizing effect of mangrove roots or excretion of extra-cellular polymeric substances (EPS) by microorganisms (cf. Widdows et al. 2004).

The balances between erosion and deposition on the small-scale may be different in mangrove forests in New Zealand, relative to those found in tropical

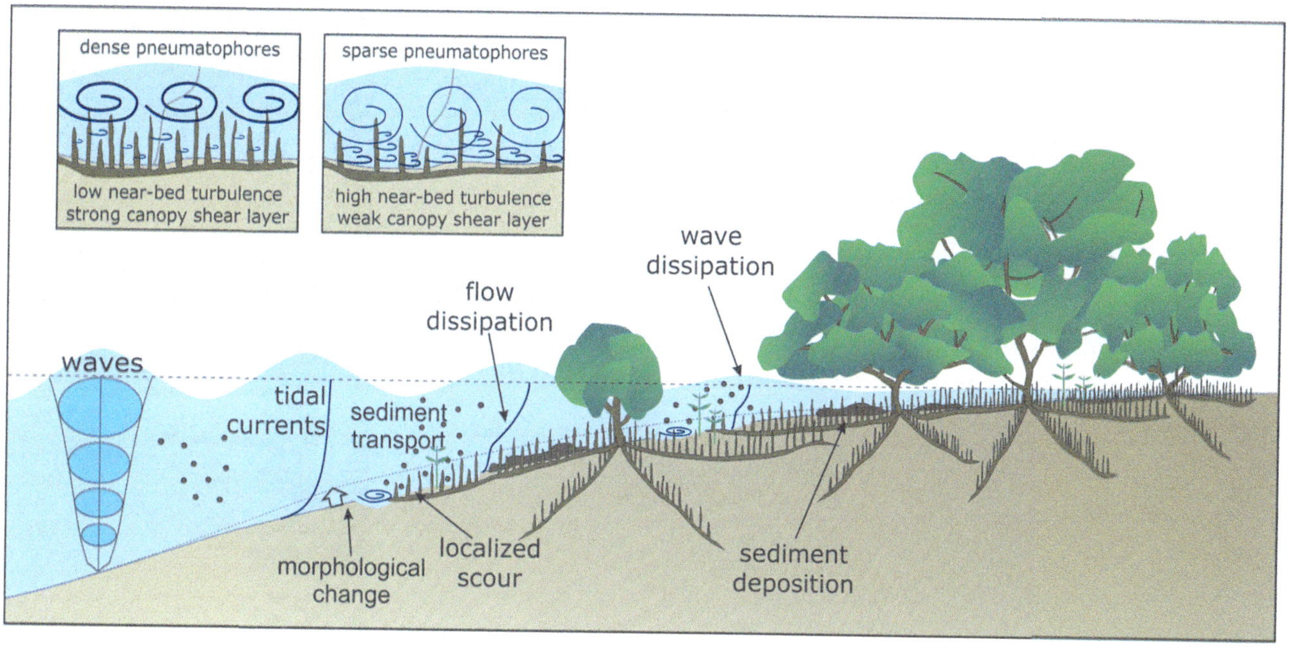

Fig. 2.5 Schematic representing physical processes and biophysical interactions within a typical mangrove forest (insets after Horstman et al. 2018)

environments, owing to substantial differences in tree geometry. New Zealand mangrove trees are smaller (more bush-like, see section "Mānawa: The New Zealand Mangrove") and pneumatophores are thinner, shorter and more flexible. Data obtained in the Whangapoua Harbour (for location see Fig. 2.1) revealed pneumatophores with basal diameters of around 7 mm and mean and maximum heights of 150 and 250 mm, respectively, compared to basal diameters of around 13 mm and heights of up to 820 mm for *Sonneratia* mangroves in the Mekong Delta in Vietnam (Mullarney et al. 2017b). However, similar to observations by Henderson et al. (2017) in tropical mangroves, enhanced deposition is observed in the fringing zone of mangroves in Whangapoua Harbour (Fig. 2.6), where the flow first interacts with the vegetation, facilitating an ongoing seaward expansion of these mangroves (further discussed in section "Succession of Expanding Mangrove Systems").

Although preliminary observations suggest that turbulence generated within the smaller New Zealand pneumatophores is less intense (values of TKE dissipation rates from Whangapoua did not often exceed 10^{-4} W kg^{-1}), the extent of the influence of the other tree elements is not yet clear. In some cases, for the smaller mangroves in New Zealand, the branches and leaves of the trees themselves are submerged during high tides, and therefore, once water levels reach the tree canopy, will likely enhance drag (as observed in similar mangroves in Japan and Vietnam by Chen et al. 2016; Mazda et al. 1997; Quartel et al. 2007) and possibly also generate additional turbulence. Whether these differences suggest that other factors such as sediment supply may play a stronger role in setting sediment deposition patterns than local turbulence conditions is an active area of research (see for example Horstman et al. 2016, 2017).

2.4 Succession of Expanding Mangrove Systems

The structure of any mangrove forest depends on its successional history that evolves in a dynamic equilibrium with the environmental gradients that occur in the coastal zone. Because New Zealand mangroves are monospecific (see section "Mānawa: The New Zealand Mangrove"), their successional history relates to growth forms within *Avicennia* rather than replacement of pioneer species with more mature forest. In addition, mangrove forests exclude understory species which could be a consequence of multiple factors: the high-stress environment making it difficult to cope with the added stress of reduced light (Janzen 1985), competition, or the ability of mangroves to change soil conditions so their own recruits survive better (Bosire et al. 2006). Environmental gradients in the coastal zone relate to resources, regulators and the inundation regime: resources are the factors which are consumed by growth such as nutrients; regulators are physical and biological parameters that cause stress such as salinity, accumulation of mud or predation; and the inundation regime controls the degree to which resources (sediment, nutrients etc.) are delivered and stresses, associated with tidal currents and wave action, are administered within the forest (Twilley and Rivera-Monroy 2009). Of all the mangrove species, *Avicennia* is known to be the most tolerant to

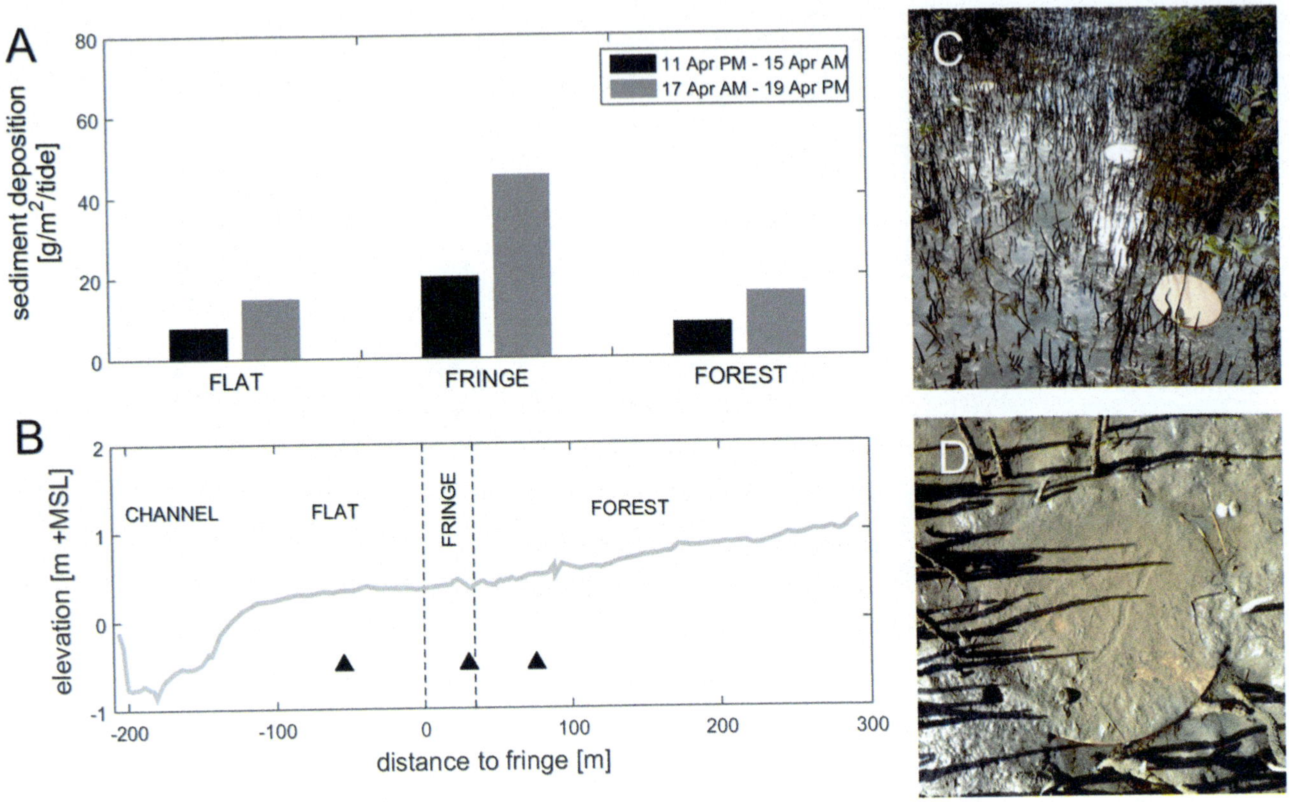

Fig. 2.6 (**a**) Sediment deposition measured in the Whangapoua Harbour (location indicated in Fig. 2.1) along, (**b**) a transect covering tidal flat, fringe and mangrove forest (Horstman et al. 2017). Deposition rates shown for regular conditions (11-15 April 2016) and just after a torrential rain storm (17-19 April 2016). (**c**, **d**) Sediment deposition rates were monitored with sediment traps (Ø 0.33 m) that were deployed in triplicates at the positions marked with ▲ in the cross-shore profile (Photo credits: E. Horstman)

regulating stresses imposed by muddy substrates and salinity (Twilley et al. 1999).

Factors limiting zonation and succession in mangroves are highly complex (Alongi 2009). Within New Zealand, environmental gradients vary greatly around the country due to variations in geomorphic setting, and so mangrove succession patterns are difficult to generalise. According to the original five-class classification by Thom (1982), most New Zealand mangroves can be classified as wave-dominated barrier lagoon type mangroves (typically with autochthonous sediment types), river-dominated type mangroves (typically with allochthonous sediments), or a combination of both. However, due to the rapid infilling of New Zealand lagoons with catchment-derived terrigenous sediments (see section "Mangrove Expansion in New Zealand's Anthropocene"), sediments in these lagoons are more likely to be allochthonous than not. However, even when geomorphic conditions of estuarine mangrove environments are similar, the classic succession states of initial pioneering, rapid growth, maturity, old age and death (e.g. Fromard et al. 1998) do not always occur (Lugo 1997). Disturbance events (sometimes on large scale), restart the succession locally, and therefore succession patterns are sensitive to the external controls on the mangrove structure (Twilley et al. 1999).

Forest development in New Zealand depends on the ability of the fringing mangroves to colonise seaward, which in turn depends on seedling establishment (Lovelock et al. 2010), and on the inundation regime being appropriate for mangrove growth (Swales et al. 2016). It is thought that expansion occurs during 'windows of opportunity' that correspond to relatively quiescent conditions which allow seedlings to establish in stable sediment conditions without being uprooted (Balke et al. 2015). Following seaward expansion of the fringing mangroves, the condition of the landward forest is dependent on the provision of resources and stresses across the (extended) forest fringe (salinity, pH, sulphide levels, nutrients, waterlogging), both of which are controlled by the inundation regime (Lara et al. 2009). If growth is too intense at the fringe, growth and sustenance in the inward forest can suffer due to nutrient depletion by the fringe, but also because of dissipation of energy at the fringe, reducing the tidal range (and sediment and nutrient delivery mechanism) inward. Consequently, the trees are often locally taller at the fringe and diminish in height in the landward direction (section "Mānawa: The New Zealand Mangrove").

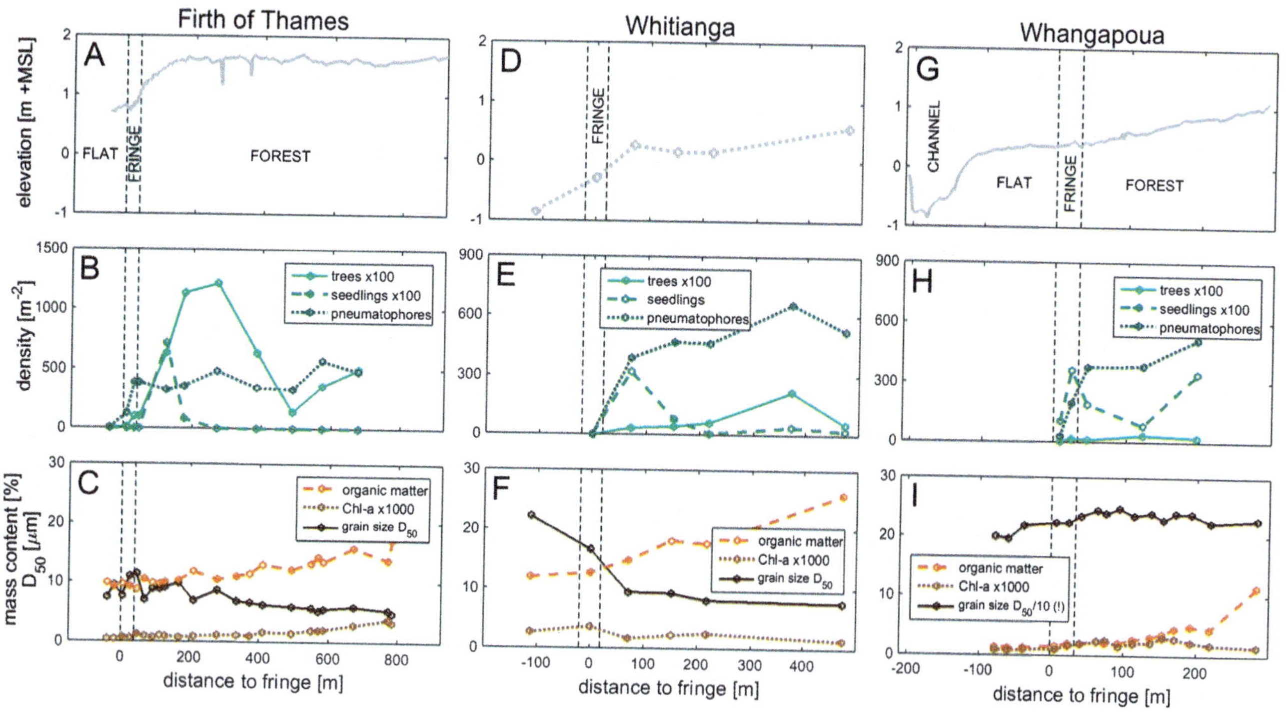

Fig. 2.7 (**a**, **d**, **g**) Elevation profiles, (**b**, **e**, **h**) vegetation distribution, and (**c**, **f**, **i**) sediment characteristics along cross-shore transects in three mangrove sites: (**a–c**) Firth of Thames, (**d–f**) Whitianga and (**g–i**) Whangapoua (see Fig. 2.1 for site locations).

Balke et al. (2015) and Lovelock et al. (2010) both discuss the role of waves on recruitment in mangroves in the Firth of Thames, which is exposed to a large fetch from the north. During La Niña years, the winds align with the orientation of the estuary, increasing the hydrodynamic exposure on the tidal flats and in the mangrove fringe, and seedlings are unable to establish. Conversely, during the protected south-westerlies of El Niño conditions, rapid establishment and expansion has been reported. Consequently, bands of dense trees can be detected as historic 'waves of expansion'. Measurements collected in 2016 show the rapid increase in vegetation density across a very short distance in the Firth of Thames (Fig. 2.7b), with the accompanying flat-topped morphology (Fig. 2.7a) indicating a reduced supply of sediment to the back of the mangroves. The photos in Fig. 2.8a–d provide evidence for the rapid expansion of the mangrove fringe at this site, with very little change inside the dense forest (photo shown in Fig. 2.9b) over the last decade. As described in section "Biophysical Interactions", these dense zones of vegetation at the fringe feed back into the hydrodynamics by dissipating energy and increasing sedimentation at the fringe (see also Fig. 2.6a), with supply to the back forest reduced so that the morphology develops a nearly level surface.

Very few studies exist of forest development at other New Zealand mangrove sites. Yet, our observations suggest that developmental controls might be quite different. Figure 2.8 shows the vegetation dynamics of the Firth of Thames and of two mangrove forests on the Coromandel Peninsula on the northeast coast of New Zealand's North Island over the past 16 years: one in a sandy back-barrier lagoon (Whangapoua Harbour), and one in a more muddy river-dominated estuary (Whitianga Harbour; see Fig. 2.1 for locations). The median sediment particle sizes for these sites are shown in Fig. 2.7. Imagery from GoogleEarth shows the mangroves in the Whangapoua Harbour are changing rapidly (Fig. 2.8i–l), whereas those in Whitianga have remained remarkably stable (Fig. 2.8e–h). Historical aerial photos from Whangapoua (not shown) show no mangroves were present in 1970, corroborating the rapid expansion of these mangroves, whereas mangrove stands in Whitianga were very similar to those observed today. Whitianga was also one of the estuaries visited by Captain Cook in November 1769, when he noted "low flat islands all cover'd with a sort of Mangrove trees" (Beaglehole 1955; p. 196). In the case of the expanding Whangapoua site, mangroves do not expand seaward in waves of establishment, like those in the Firth of Thames, but rather by establishment of seedlings in the gaps between widely dispersed single trees throughout the intertidal. Recruitment is therefore controlled by the conditions in the gaps (Clarke 2004), where light availability is better and growth is higher (Krauss et al. 2008), and

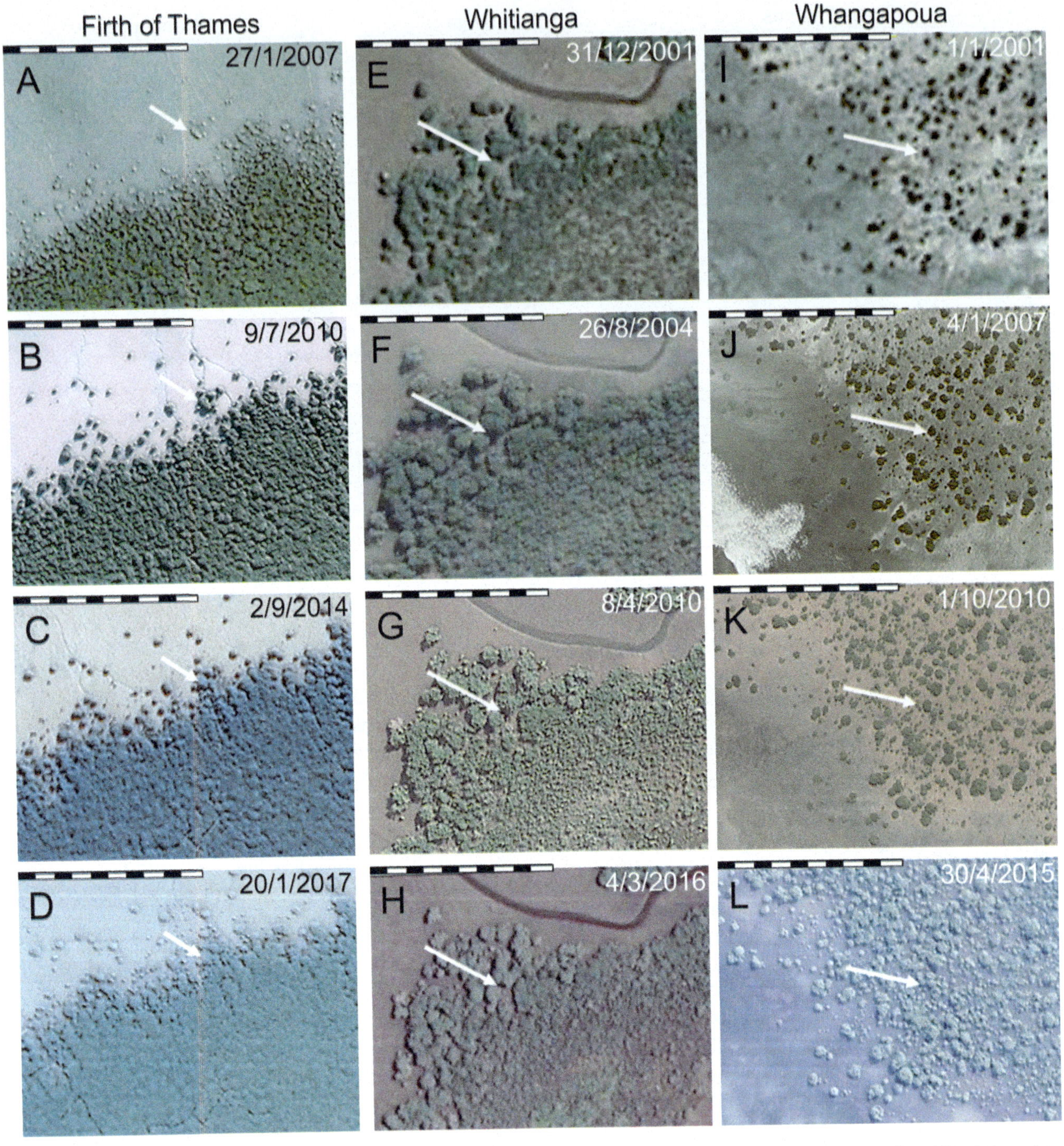

Fig. 2.8 Changes in spatial distribution of the mangrove forest at the three mangrove sites from Fig. 2.7: (**a–d**) Firth of Thames, (**e–h**) Whitianga, and (**i–l**) Whangapoua. The printed scale is 100 m long, and north is upward. Images are extracted from GoogleEarth (map data: panels (**a**, **b**, **f**,**g**, **h**, **l**): Digital Globe; panels (**c**, **d**): CNS/Airbus 2017; and panels (**e**, **i**, **j**, **k**): Thames-Coromandel District Council)

predation might be reduced (Osborne and Smith 1990). The lack of expansion in the riverine site (Whitianga) may reflect the rapid changes in sediment load associated with the common flooding events that regularly inundate the site. Although the lower salinity environment favours enhanced mangrove growth (Krauss et al. 2008), sediment loads can cause fluctuations in the surface which may bury or uproot seedlings (Balke et al. 2013) and shrubs. Ten centimetres of episodic sediment deposition has been reported to kill *Avicennia* trees (Ellison 1999) and a sedimentation rate of 10 cm $year^{-1}$ causes substantially reduced growth (Sidik et al. 2016). Certainly, the longevity of the Whitianga forest

Fig. 2.9 Photos of the forest characteristics along the transects at the study sites from Fig. 2.7: **(a–c)** Firth of Thames, **(d–f)** Whitianga and **(g–i)** Whangapoua. Panels (**a**, **d** and **g**) are taken at the fringe, panels (**b**, **e**, and **h**) are in the middle of the forest and panels (**c**, **f**, and **i**) are toward the back of the mangroves (Photo credits: K. Bryan, E. Horstman, J. Mullarney and D. Sandwell)

has allowed the development of mature stands composed of large trees (Fig. 2.9d).

The above three examples illustrate the large differences in growth form and forest development that can occur over a short geographical distance. The youngest of the three forests discussed here is the Whangapoua site, which is located in a sandy estuary with low sediment supply, and is characterized by short trees (Fig. 2.9g–i) and a linear intertidal profile (Fig. 2.7). Sediment and tidal energy can easily penetrate this sparse canopy to supply nutrients and sediments to the inner forest, accounting for the recruitment and morphology building inward. The Firth of Thames is a medium-aged site, where the sediment loads from the nearby rivers create conditions that are ideal for rapid seaward expansion. The relatively high surface elevation gains (see upper panels in Fig. 2.7 for inter-site comparison), mainly at the seaward fringe of the mangroves, compared to other Indo-Pacific countries (Lovelock et al. 2015), show the importance of sediment loading to expansion in New Zealand mangroves. Together with regulating stresses, the rapid seaward expansion of the intertidal flats accommodates fast expansion of the mangrove vegetation (Fig. 2.8a–d), resulting in a very dense canopy just behind the forest fringe (Fig. 2.9b). Finally, at the oldest forest, in the Whitianga Harbour, we hypothesize sediment loading is too great for rapid development of the fringe; in this case the sparse fringing canopy of very tall, old growth trees (Fig. 2.9d) allows sediments to penetrate inward to the point where the inundation regime is too small and larger mangroves die out to be replaced by inland species (Fig. 2.9f). In both the Firth of Thames and Whitianga, the established vegetation at the fringe is characterised by larger trees (Fig. 2.9a, d), which benefit from the unrestricted supply of water-borne resources and presumably larger trees are also better equipped to withstand the higher stresses due to waves and currents impacting on the mangrove fringe. The smaller shrubs, affected by a limited supply of resources across the forest fringe and more exposed to temperature stresses (section "Mānawa: The New Zealand Mangrove") due to greater distance to the water edge, only become commonplace as the gaps are infilled further into the forest.

2.5 Carbon Dynamics of Expanding Mangroves

Mangrove ecosystems are thought to account for as much as 14% of carbon sequestration by the global ocean (Alongi 2014). Primary production in mangrove ecosystems typically decreases with increasing distance from the equator (Bouillon et al. 2008). However, even at the limits of mangrove distribution, primary production in New Zealand mangroves remains high, estimated at 4–10 tonnes C ha^{-1} year^{-1} based on litter fall rates of 3.24–8.1 tonnes dry weight ha^{-1} year^{-1} (Gladstone-Gallagher et al. 2014; Woodroffe 1982), biomass carbon concentrations (Bulmer et al. 2016a, b), and established relationships between litter fall and wood and root production (Bouillon et al. 2008). As mangroves in New Zealand are estimated to cover 26,050 ha (Morrisey et al. 2010), this rate equates to 104,000–260,000 tonnes of carbon produced each year within New Zealand mangrove ecosystems. High rates of primary production in combination with unique sediment conditions, such as anoxia which slows the breakdown of carbon (Fig. 2.10), result in carbon burial rates in New Zealand mangroves of 0.67 tonnes C ha^{-1} year^{-1} (Pérez et al. 2017). The accumulated carbon stocks within New Zealand mangrove ecosystems are estimated at 117 tonnes C ha^{-1}, including above and belowground biomass, and carbon within the sediment up to 1 m below the surface (Bulmer et al. 2016b).

The high stocks of carbon within New Zealand mangrove sediments also result in high rates of sediment CO_2 efflux (average of 168.5 ± 45.8 mmol m^{-2} day^{-1} (Bulmer et al. 2015)) relative to habitats containing lower organic carbon stocks, such as sandflat ecosystems where sediment CO_2 effluxes are close to zero (Bulmer et al. 2017b). Sediment CO_2 efflux from mangrove ecosystems (Fig. 2.10) is primarily derived from CO_2 released through root/mycorrhizae respiration and microbial respiration associated with the decomposition of organic matter (Bouillon et al. 2008). Other pathways for carbon losses from mangrove ecosystems include the emission of other greenhouse gasses such as CH_4, export of dissolved carbon from the sediment to the water column, and the export of particulate organic carbon from litter and other detritus (Bouillon et al. 2008).

Mangroves in New Zealand are primarily expanding over adjacent unvegetated mudflats, which often used to have a sandier substrate prior to the rapid infilling and muddification of the estuaries (see section "Mangrove Expansion in New Zealand's Anthropocene"). Consequently, mangrove expansion in New Zealand is not only associated with enhanced carbon stocks, but also an increase in sediment CO_2 efflux from the mangrove ecosystems compared to the original sandier ecosystems (Fig. 2.10). However, it is important to note that when carbon losses associated with the CO_2 efflux are subtracted from carbon stocks stored within tree biomass and buried within the sediment, expanding

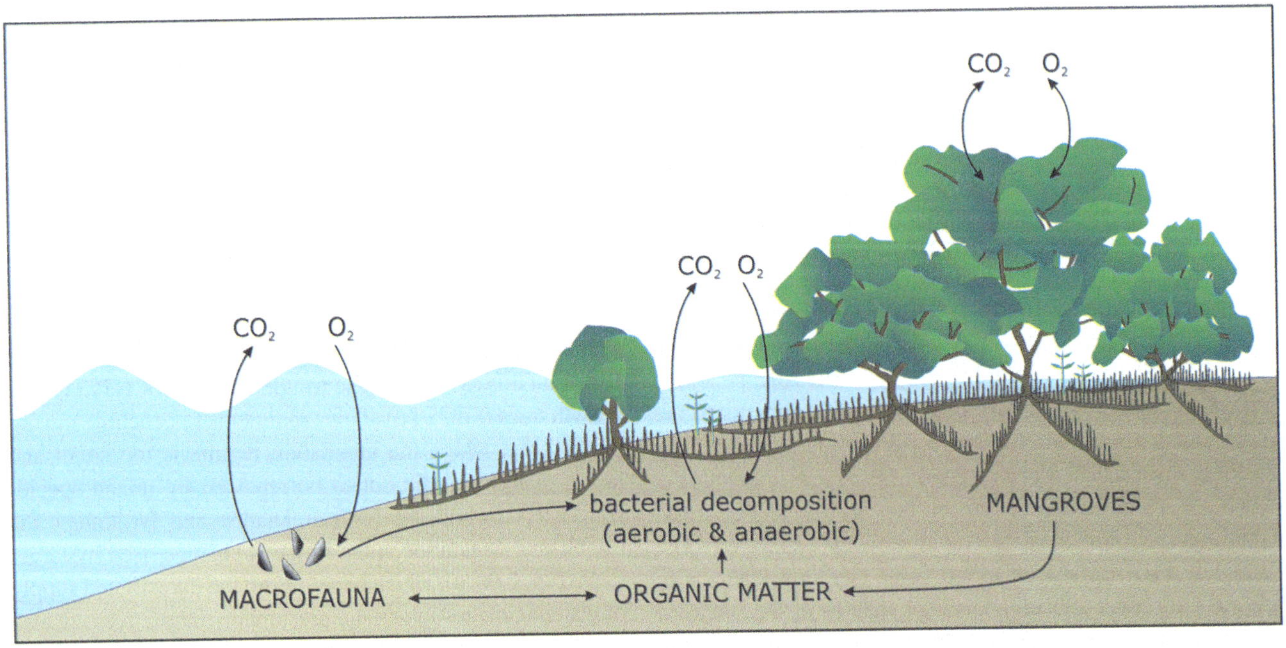

Fig. 2.10 Carbon cycling in a mangrove ecosystem, after Bulmer et al. (2017b)

mangrove ecosystems still provide a net gain in carbon stocks (Pérez et al. 2017). Taking account of these factors, expansion of New Zealand mangroves has been estimated to result in an increase in carbon stocks by 79 tonnes C ha^{-1} (Bulmer et al. 2016b). This gain in carbon stocks following mangrove expansion is not immediate, rather it is estimated to occur over decade to century time scales, based on carbon accrual rates in sediments of *Avicennia marina* forests of 0.67 tonnes C ha^{-1} $year^{-1}$ (Pérez et al. 2017). As mangrove ecosystems in New Zealand have been estimated to be growing in extent by 1068 ha^{-1} $year^{-1}$ (Morrisey et al. 2010), mangrove expansion is estimated to result in an enhanced storage of 716 tonnes carbon per annum.

2.6 Ecosystem (Dis-)services of Expanding Mangroves

The expansion of mangroves in New Zealand initiates change to the ecosystem services delivered by the mangroves and their adjacent estuarine and coastal habitats. The enhanced capacity for carbon storage was described in section "Carbon Dynamics of Expanding Mangroves", and the coastal protection and erosion mitigation benefits of expanding mangroves were addressed in section "Biophysical Interactions". In addition to improving coastal stability, the retention of fine sediments within the mangroves reduces the risk of detrimental levels of siltation across sensitive coastal habitats located in proximity to the mangroves (Fig. 2.11). Moreover, water flushing through the mangroves is 'purified', with nutrients and heavy metals being adsorbed to the sediments and becoming trapped in the mangrove 'sink' (Barbier et al. 2011; Ewel et al. 1998). In this section we discuss the aspects of (mostly) biological ecosystem functioning that may be positively or negatively impacted as a consequence of mangrove expansion.

Globally, mangroves are considered to be an important ecological habitat, through the provision of safe haven, habitat and food for juvenile stages of numerous fish, and a diversity of crustaceans and terrestrial species (Claudino et al. 2015; Gladstone-Gallagher et al. 2016; Nagelkerken et al. 2008). However, studies of mangrove productivity and related ecosystem services are predominantly conducted in tropical settings, or where extensive muddy coastlines or deltas support vast mangrove forests (Lee et al. 2014; Nagelkerken et al. 2008). The local climate and geomorphological setting (cf. section "Mangroves in New Zealand") will influence the overall productivity and ecosystem services that a particular mangrove forest can provide (Ewel et al. 1998; Lee et al. 2014). New Zealand mangroves survive in a temperate climate, are often found in the upper reaches of estuaries and harbours, and are often relatively young, recently colonised forests (see section "Mangroves in New Zealand"). Therefore, the ecosystem services these forests provide may differ from those reported for mature tropical species.

There is no marine or estuarine fauna that is unique to mangroves in New Zealand (Morrisey et al. 2010). However, the habitat still supports numerous invertebrate and vertebrate species (Alfaro 2006; Crisp et al. 1990), including a diversity of terrestrial species (Doyle and Hogg 2015). Similarities exist between temperate and tropical mangrove ecosystems in the broad functional groups that are represented. Most mangroves, in the tropics as well as at higher latitudes, host a range of algal and microbial communities. Epiphytic algae adhere to the mangrove roots, stems and sediments and provide nutrient dense food to grazing snails (Morrisey et al. 2010). Additionally, benthic macrofaunal communities in temperate mangroves consist of a diversity of burrowing

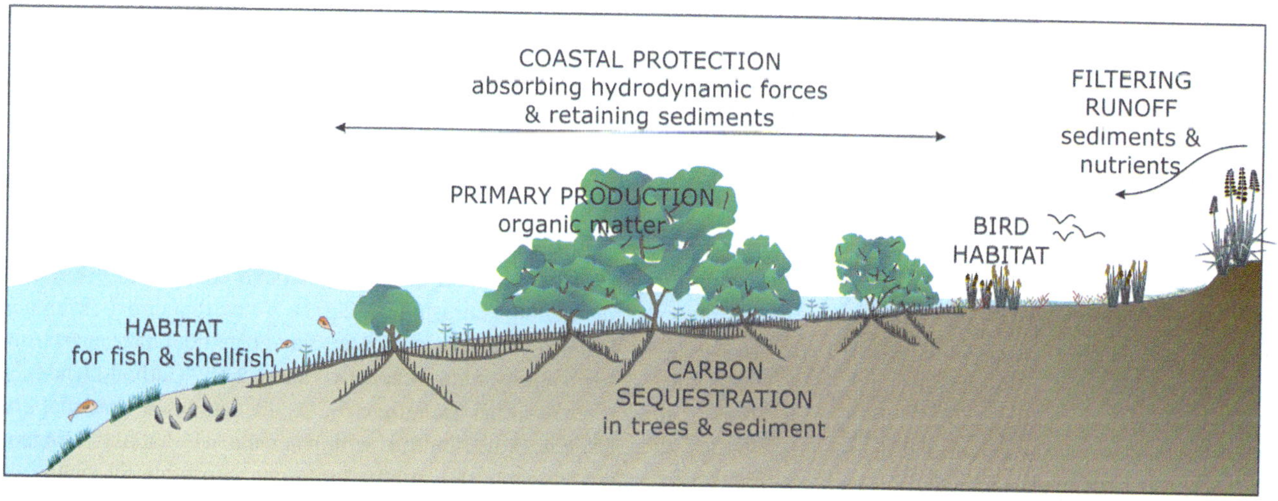

Fig. 2.11 Ecosystem services of New Zealand mangroves

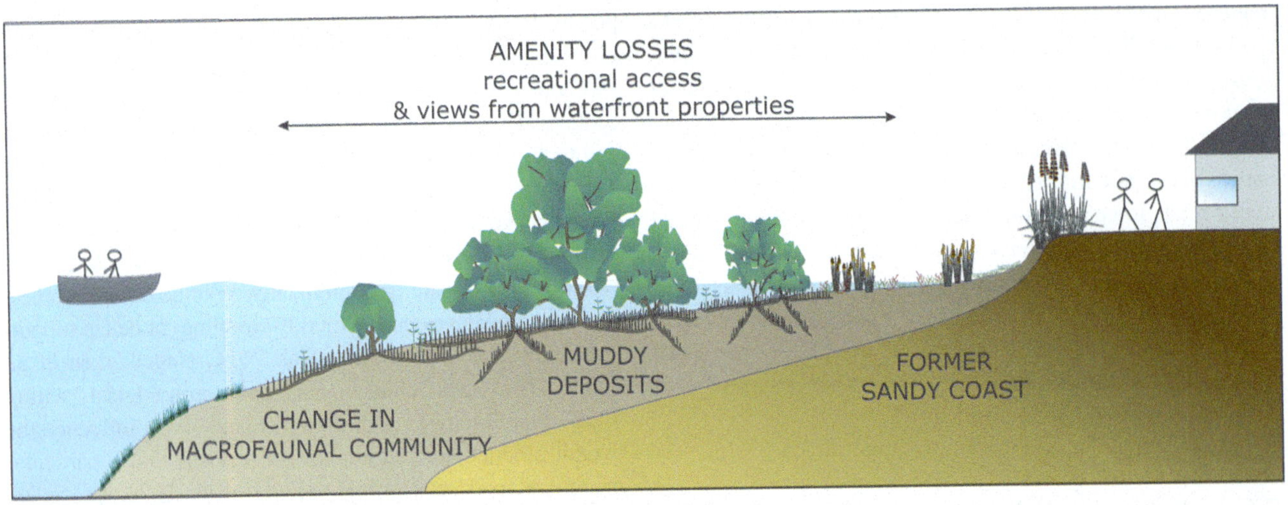

Fig. 2.12 Negative impacts of rapid expansion of New Zealand mangroves

infaunal polychaetes and oligochaetes, gastropods, small crustaceans and crabs (Morrisey et al. 2003).

A review of fish use of temperate mangrove systems suggests that fish are occasionally found in the mangroves, most commonly at the edge of mangrove habitats near deeper channels (Sheaves et al. 2016). Species found to visit mangrove habitats in New Zealand are abundant fish species such as mullet, pilchards, eel, triplefins and parore (Morrisey et al. 2010). Diet studies suggest most fish observed in New Zealand mangroves are feeding from the water column rather than deriving trophic benefits from mangrove ecosystems (Lowe 2013). Migration of larvae and juvenile fish in New Zealand mangroves is limited compared to tropical mangrove systems where multiple tree species provide a complexity of submerged root structures, some of which are deeply or even permanently inundated, allowing for easy fish migration. Many New Zealand mangrove forests, in comparison, are only submerged by the tide for a few hours each tidal cycle and with water depths generally less than 0.3 m (Lundquist et al. 2017). There is, however, site specific variability of tidal inundation depth and exposure throughout the mangrove forests in New Zealand. For example, some fringing mangroves positioned at lower tidal elevations (see Fig. 2.7) can be exposed to inundation depths of over a meter.

The rapid expansion of New Zealand mangroves is also associated with a decline of some coastal ecosystem services, particularly with respect to societal and cultural values associated with the use of coastal habitats. Much of the mangrove expansion has occurred in the past 50 years (see section "Mangrove Expansion in New Zealand's Anthropocene"), within the living memory of New Zealanders who reside at and interact with the coast. As such, this notable change in the coastal landscape is often perceived to be detrimental to the 'health' of the estuary as it used to be (Harty 2009).

Rapid infilling and the associated mangrove expansion in New Zealand estuaries are at the basis of widespread concerns about the impacts on estuarine habitats (Fig. 2.12), such as a loss of seagrass meadows, a decline of wading and roosting grounds for birds, and a reduced variety and abundance of shellfish (Ellis et al. 2004; Harty 2009; Jones 2008; Lundquist et al. 2014b; Saintilan 2004). An increase in muddier, mangrove dominated habitat is typically associated with a change in macrofaunal communities, usually resulting in declines in bivalve abundance and increases in mud tolerant, opportunistic polychaete species (Ellis et al. 2004). The low bivalve abundance and high mud content of the surface sediment within mangrove ecosystems suggest that mangrove expansion will also be associated with reduced fluxes of inorganic nutrients between the sediment and the water column, and a decline in the ecosystem value of this service (Bulmer et al. 2017b). It is important to note that declines in shellfish beds would likely occur regardless of the presence of mangroves, driven by the increased mud content within estuarine habitats due to accelerated sedimentation (Robertson et al. 2015; Thrush et al. 2004).

The broader implications of the increase in muddy, mangrove dominated intertidal areas on estuarine ecosystem functions are difficult to predict. However, the increasing homogeneity of estuarine habitats has been associated with a reduction in ecosystem resilience and an increased likelihood of profound and unpredictable changes to ecosystem functioning (Thrush et al. 2004). A key functional difference between mangrove ecosystems and bare tidal flats, apart from their carbon storage capacity (see section "Carbon Dynamics of Expanding Mangroves"), is the increased potential for nutrient redistribution within mangrove habitat. Biogeochemical processes such as nutrient absorption by mangrove roots or litter fall (Bouillon et al. 2008; Gladstone-Gallagher et al. 2014) are independent of direct sediment-water column exchange. Mangrove ecosystem expansion is also likely to be associated with an increase in the contribution of mangrove derived carbon and nutrients within estuarine food-webs, due to the increased mangrove biomass and associated detritus. These additional pathways may build resilience back into the system which was lost due to an increase in the sediment mud fraction.

2.7 Mangrove Removal in New Zealand

2.7.1 Overview of Recent Mangrove Expansion

Mangrove expansion rates (reviewed in section "Mangrove Expansion in New Zealand's Anthropocene") vary both within and between estuaries. Historically, mangroves were more commonly distributed around creeks incising the upper intertidal flats, often in areas far away from residential areas (Lundquist et al. 2017). However, increased deposition associated with estuarine infilling, and reduced tidal flows due to infrastructure development (e.g. causeways and marinas), have resulted in expansion of intertidal flats. This allows mangroves to establish in locations where they do conflict with other values such as recreational and navigational access and ocean views (Fig. 2.12). Concomitant declines in other estuarine habitats (e.g seagrass meadows, shellfish beds) are caused by increases in sediment supply and deposition, and not, as commonly perceived by the public, by invasion of mangroves into these habitats (Thrush et al. 2004). Regardless, the perceived relationship between mangrove expansion and the decline of other intertidal estuarine habitats feeds the assumption that removal of mangroves will result in the restoration of other estuarine habitats and the values associated with these habitats (Harty 2009).

2.7.2 Mangrove Removal Policies in New Zealand

In response to mangrove expansion and perceived correlation with declines in estuarine ecosystem services, many community groups—estuary care groups—were established to support estuary restoration initiatives. These community groups were often focussing on mangrove removal only, with objectives of restoring historical ecological balances between mangroves and unvegetated sandflat habitats. Consequently, the majority of early mangrove removals (1990s and 2000s) were small community initiatives, using manual removal methods. The first mechanical removal consent was approved in 2010, using a wide-tracking mechanical digger to clear 100 ha of mangroves in the sub-estuaries of Tauranga Harbour (Lundquist et al. 2012). Motivations for mangrove removal include improving recreational and amenity values; restoring ecological habitat such as sandflats, seagrass meadows and shellfish beds; restoring access and navigation; maintenance and use of coastal infrastructure; improving the function of drainage systems; and flood protection.

Mangrove removal policies and consent processes differ between the four regional councils (regional authorities charged with managing natural and physical resources) in whose jurisdiction mangroves are found in northern New Zealand. Overarching national legislation guiding council mangrove removal policies includes the New Zealand Coastal Policy Statement (NZCPS) and the Resource Management Act (RMA). Policy 11 of the NZCPS includes provisions to "*avoid significant adverse effects and avoid, remedy or mitigate other adverse effects of activities on indigenous ecosystems and habitats that are only found in the coastal environment and are particularly vulnerable to modification, including estuaries, lagoons, coastal wetlands, dunelands, intertidal zones, rocky reef systems, eelgrass and saltmarsh.*" Further guidance is provided by section 12 of the RMA which states that "*no person may, in the coastal marine area, destroy, damage or disturb the foreshore or seabed in a manner that has or is likely to have an adverse effect on the foreshore or seabed, or on plants or animals or their habitat, unless expressly allowed by a rule in a regional coastal plan or a resource consent.*"

Most councils allow some provision for non-notified mangrove removal to support public safety, coastal access, and maintenance of infrastructure (e.g. clearing for drainage; preventing tree growth from blocking roads and motorways). Policies for the removal of mangrove seedlings, intended to limit further mangrove expansion, vary across councils; for example, Auckland Council allows manual removal of single stem, non-reproductive seedlings less than 60 cm in height, with formal council notification required if the total area cleared is greater than 30 m^2. Removal of adult, reproductive mangrove trees typically does not comply with most regional coastal policies, therefore requiring application for resource consent which often results in contentious Environment Court proceedings to assess whether adverse impacts are likely to occur, and if the objectives of consents are likely to be achieved. Regular reviews occur of these regional coastal plans, with the latest Auckland regional plan being finalised in 2017. Initial proposed planning conditions for the Auckland region allowed mangrove removal, in response to community desire to remove mangroves; however, these permissive conditions were removed during plan revisions in response to public submissions and a committee hearings process, which reflected on mounting evidence supporting best practice methods and likelihood of success of mangrove removals.

2.7.3 Mangrove Removal Strategies

The two primary methods for mangrove clearance are manual methods typically using chainsaws, and mechanical clearance of aboveground biomass, most commonly using a digger with a shovel/rake attachment (Fig. 2.13). The use of mechanical diggers is associated with greater sediment disturbance than hand clearance, as sediment is compacted in mechanical tracks and the sediment column mixed due to digger/rake activity (Fig. 2.14), with anoxic sediment commonly present on the sediment surface after application (Lundquist et al. 2012, 2017). Cleared aboveground biomass is sometimes removed from the site, but mulching of aboveground biomass and leaving mulch in situ has also occurred. These approaches have been associated with adverse impacts as mulched biomass smothers the sediment and disrupts sediment chemistry, resulting in nutrient release and associated macroalgal blooms (Lundquist et al. 2012). More recent mechanical trials have shown reduced disturbance, by limiting the total area tracked by the digger (Fig. 2.14; Bulmer and Lundquist 2016). Below-ground biomass is generally left in situ, due to both expense and difficulties in removing it without creating substantial sediment disturbance (Alfaro 2010; Lundquist et al. 2012, 2014a; Stokes 2008). Seedling removal is the method of mangrove removals with the least impact, although regular seedling removal is required at most locations to keep them from being recolonised (Lundquist et al. 2017).

2.7.4 Removal 'Success' Rates

The aim of many clearance operations is to restore sandflats that may have been historically present (Harty 2009). However, there is little scientific evidence to suggest that a loss of mud and return to a sandflat system will occur following

Fig. 2.13 Mangrove clearance in New Zealand: (**a**) manual mangrove clearance using chainsaws in Mangere Inlet, Manukau Harbour (Photo credits: C. Lundquist), and (**b**) mechanical clearance using digger with rake attachment at Whangamata (Photo credits: R. Bulmer)

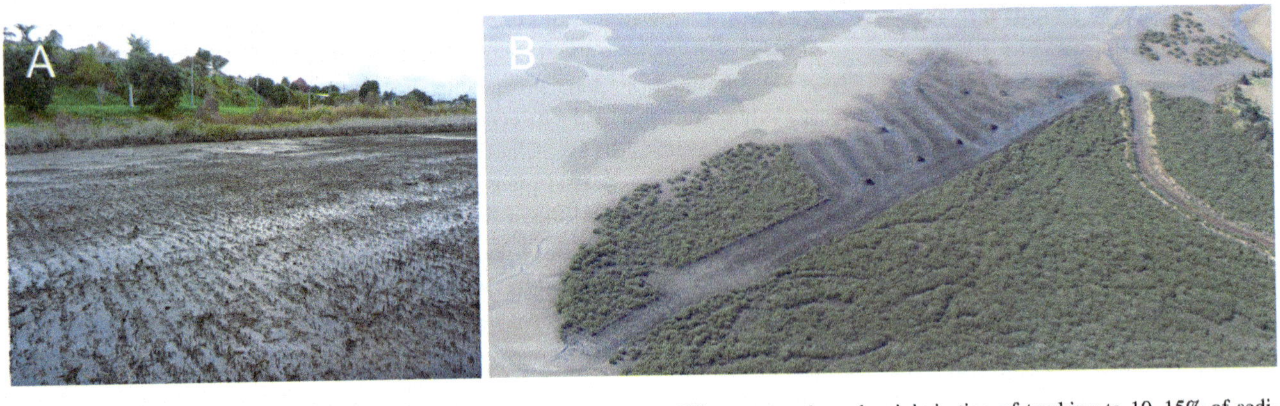

Fig. 2.14 Impacts of mechanical mangrove clearance: (**a**) tracking disturbance from mechanical mangrove clearance at Tauranga (Photo credits: C. Lundquist), and (**b**) reduced impacts of mechanical clearance at Whangamata through minimisation of tracking to 10–15% of sediment surface (Photo credits: Waikato Regional Council)

mangrove clearance (Alfaro 2010; Lundquist et al. 2012, 2014a; Stokes 2008). The conclusions on the impacts of mangrove clearance are further limited due to the lack of baseline measurements prior to mangrove clearance in most published studies (Alfaro 2010). Most early mangrove removals in New Zealand anticipated an immediate return to sandflats, but had limited or no monitoring to demonstrate whether they were successful at achieving their objectives (Alfaro 2010). Furthermore, removals using new methodologies (e.g. mechanical mulching) often lacked adequate trials of these methods to determine if adverse impacts were likely to occur (Lundquist et al. 2012; Stokes et al. 2016). Nonetheless, two large scale removals have been subjected to long-term monitoring, either through associated government funded scientific research, in Tauranga Harbour (Lundquist et al. 2012), or through consent requirements, in Whangamata Harbour (Bulmer et al. 2017a). The Tauranga monitoring assessed removals at three sites in each of two sub-estuaries after removal of 110 ha of mangroves using mechanical mulching with in situ disposal. Monitoring over 5 years showed consistent responses of degraded macrofaunal communities, anoxic sediments, low rates of sediment erosion, and persistence of mulched aboveground biomass that was left in situ, and was further associated with extensive macroalgal blooms (Lundquist et al. 2012, Lundquist unpublished data). The large-scale mechanical removals consented in Whangamata used either burn piles or offsite disposal of aboveground vegetation biomass (Bulmer and Lundquist 2016). A trial and adaptive management process was included in the consent conditions (Bulmer et al. 2017a) to allow comparison and selection of mechanical (or non-mechanical) methods that provided the least risk of adverse impacts and the most rapid recovery trajectory toward sandflat conditions (Bulmer et al. 2017a).

One additional study analysed the status of ~40 mangrove removal sites on the east coast, from Whangarei to Tauranga (Lundquist et al. 2014a). This study showed that most mangrove removals occurred at sites that had already been impacted by changes in sediment characteristics prior to or beyond the observed mangrove expansion. Sites which had adjacent tidal flats with high mud content were unlikely to have sufficiently energetic hydrodynamic conditions to result in rapid erosion of fine sediments and associated decomposition of belowground mangrove biomass. Sandier, more exposed removal sites showed some change toward sandier unvegetated states and macrofaunal communities associated with these sandy sediments, though these sandier removal sites accounted for only a limited proportion of the total areas where mangroves had been removed (Lundquist et al. 2014a, Lundquist unpublished data).

Comparisons between removal methods suggest that mechanical methods are typically associated with larger immediate adverse impacts and longer timeframes before sediment erosion occurs. Comparison of sizes and shapes of removals suggest that removals of alongshore narrow strips of vegetation (<30 m in cross-shore width) on seaward sides of forests were most likely to trend toward sandflat characteristics (Bulmer and Lundquist 2016). Larger or shoreward removals were associated with degraded conditions except at the seaward edges. Cleared narrow strips perpendicular to the shore, a common strategy to create recreational access through mangrove forest, were rapidly recolonised by seeds from adjacent forest, and required regular seedling removal to maintain access (Lundquist et al. 2017).

Generally poor success rates of mangrove removals to date have influenced multiple Environment Court decisions on regulatory policies for mangrove removal, usually resulting in more stringent rather than permissive requirements for removals, and often requiring site-specific assessments for any consents for removing adult mangroves.

2.8 The Known and Unknown Effects of Mangrove Removal

2.8.1 Effects of Mangrove Clearance on Estuarine Morphodynamics

Morphodynamic changes following mangrove clearance are influenced by a range of environmental factors, including the hydrodynamic characteristics of the site, the size of the area cleared, and the pre-existing sediment conditions (further discussed in section "Effects of Mangrove Clearance on Sediment Composition"). A cleared estuarine site positioned in proximity to either a tidal channel or the ocean entrance is more likely to experience surface erosion after mangrove removal, as waves and tidal currents are no longer attenuated (see section "Biophysical Dynamics of Expanding Mangroves"). Strong tidal currents, particularly flood flows, may be responsible for rapid scour after mangrove clearance, or conversely, may introduce fresh terrestrial sediments to the intertidal. Engineering works such as the construction of causeways or revetments, whether undertaken prior or post mangrove clearance, may induce further morphodynamic change. These solid obstructions may result in reduced flushing potential or enhanced erosion by the reflection of incoming waves (Winterwerp et al. 2013).

The capacity of mangroves to trap sediments is well established (Furukawa and Wolanski 1996; Furukawa et al. 1997; Horstman et al. 2015; van Santen et al. 2007). The vertical elevation, or surface elevation, of a mangrove forest can increase by a combination of sediment accretion, root production (and decomposition) and groundwater influx (Krauss et al. 2014). The capacity of mangroves to adapt and survive in a changing environment is contingent upon the rates of change of both mangrove surface elevation and sea-level rise (Lovelock et al. 2015; Webb et al. 2013). However, with the felling of aboveground mangrove vegetation, both the sediment trapping capacity of the site and the production of root material are removed. The resulting combination of sediment flushing, sediment compaction and the structural collapse of the remaining root material will likely produce a significant deflation of the substrate and a loss of surface elevation. Only one study has addressed this process in New Zealand, where surface elevation was reported to fall up to 38 mm year^{-1} in the first 2 years following mangrove clearance (Stokes et al. 2010). Elsewhere, a tropical site in Belize was assessed 1 and 2 years following the clearing of *Rhizophora* mangroves that fringed a small island. Here, the rate of elevation loss was spatially variable, ranging from 2 to 20 mm after 2 years (Hayden and Granek 2015). Similar low rates of elevation loss have been reported where mangroves were destroyed in catastrophic hurricane conditions (Cahoon et al. 2003). In southern Australia, sediment compaction and changed groundwater conditions have been linked to a similar geomorphological change (Rogers et al. 2014).

Regardless of the driving process, the abovementioned observations imply that mangrove clearance results in a rising relative sea level, which then may impact on adjacent saltmarsh habitats or other land use functions (Rogers et al. 2014; Saintilan and Rogers 2013). Exposed shorelines are also likely to become more vulnerable to erosion, especially if the localised rise in relative sea level is exacerbated by predicted global sea-level rise (Hayden and Granek 2015).

2.8.2 Effects of Mangrove Clearance on Sediment Composition

The relationship between mangrove expansion and the higher mud content of intertidal areas is one of the reasons for mangrove removal in New Zealand (sections "Ecosystem (Dis-)services of Expanding Mangroves" and "Mangrove Removal in New Zealand"). Sediment conditions prior to clearance can be characterised by the sediment texture (i.e. grain size distribution), the relative contribution of dissolved and visible organic matter, the amount of structural and fine root material, the density of algae and sediment nutrient loads. Importantly, these sediment conditions may, or may not, change with depth. An aspect of 'recovery' after mangrove removal is the change in these sedimentological characteristics towards a sand-dominated system which is more representative of the conditions that typically preceded the phase of rapid estuarine infilling and mangrove expansion. Benthic faunal community composition is linked to the sediment mud fraction (Robertson et al. 2015), with most bivalve species sensitive to increased levels of silt and clay. As such, mud content is often reported as an index of recovery of benthic fauna as well. Other sediment metrics are typically measured alongside grain size: organic matter content (Fig. 2.15c) and chlorophyll *a* content typically show higher values in a mangrove habitat compared to adjoining bare sandflats (see data presented in Fig. 2.7). Organic content in particular is correlated with the mud content, and so temporal shifts tend to mirror textural changes (e.g. Grellier et al. 2017).

In locations where repeated monitoring has been a requirement for mangrove removal resource consents, changes in sediment composition are reported over 3 years or less. However, the comprehensive assessment of removal sites, undertaken by Lundquist et al. (2014a), reported up to decadal recovery timeframes. Generally, post-clearance mud contents were found to decrease more rapidly at sites that exhibited a higher sand content prior to mangrove clearance, associated with a greater exposure of such sites to tidal currents and wind waves. Conversely, more protected sites showed limited release of the sediment's mud content, even after 5 years.

Stokes and Harris (2015) monitored a non-consented clearing in Whangamata Harbour at 3, 4 and 6 years post-removal, and described a considerable coarsening of the surface sediment between 3 and 4 years post-clearance. The delay in this textural change may be linked to the slow rate of root decomposition. It was speculated that the dense mangrove roots helped to hold the sediments until decomposition reached a point where the hydrodynamics of the site overrode the declining holding capacity of the root matrix. The timeframe sits within model predictions developed by Gladstone-Gallagher et al. (2014), showing that buried *Avicennia marina* var. *australasica* wood and pneumatophores decompose to half their original weight between 317 and 613 days. A broader examination of root biomass across ~40 mangrove removal sites indicated that belowground biomass may persist for as long as 16 years after mangrove removal (Lundquist et al. 2014a, Lundquist unpublished data). In situ decomposition experiments also reported up to decadal delays in decomposition of belowground biomass, associated with low rates of sediment erosion and surface elevation change at most mangrove removal sites in New Zealand (Gladstone-Gallagher et al. 2014; Lundquist et al. 2014a).

Although the rate at which silts and clays are removed from the sediment surface and redistributed is variable after mangrove removal, the abovementioned work demonstrates that generally some textural coarsening is observed after a number of years. It is possible, however, that any textural changes to the surface may mask persistence of muddy

Fig. 2.15 Post mangrove clearance sediment condition in the Waikaraka estuary (a sub-estuary within the Tauranga Harbour): (**a**, **b**) sediment cores collected 2 years after the clearing are still predominantly muddy; (**c**) the remaining root material from the core in panel (**b**) separated into structural roots and fine roots (left to right); and (**d**) the soft sediments of the cleared zone (Photo credits D. Stokes)

material at depth (Fig. 2.15). Quantitative assessments of sediment characteristics below the surface are rare, although these assessments should be undertaken as (i) benthic faunal composition may not shift toward a typical sandflat community if mud remains at depths similar to those where filter feeding bivalves dwell, and (ii) the desired firming up of the substrate may not be realised in the short term.

'Sand armouring' has been observed following mangrove clearance in Whangamata Harbour (Stokes and Harris 2015), where a 5 mm sand cap developed between two monitoring surveys at 3 and 6 years after clearance. The sand cap increased the erosion threshold of the substrate, reducing the sediment's mobility under the typical hydrodynamic conditions of the estuary. This finding implies that continued 'recovery' would require stronger than normal tidal currents or wave forces for any further entrainment and redistribution of sediments to be realised. Sandy silts were observed beneath the surface sand lens, although the grain size coarsened to a silty sand within the upper 20–60 cm. The depth of mangrove muds may have been limited at this site due to the relative immaturity of the mangroves that were cleared. In comparison, a recent assessment of changing soil characteristics following mangrove clearance in Vietnam found that sediment properties evolved to a depth of 35 cm as a result of the clearing activity. The mud and sediment organic carbon content decreased by >65% in the 2 years following mangrove removal, with a variable mud content of <80% from the surface to around 30 cm depth, after which both the cleared and intact mangrove sediment cores presented similar mud content of >90% (Grellier et al. 2017).

Surface sediments may evolve to reflect sandflat conditions, but buried sediment may not mirror the textural conditions at the surface. Nevertheless, sediment core profiles are rarely included in the assessment of mangrove clearance impacts, despite the potential of these sediments at depth to limit faunal recolonization. Further investigation into the temporal changes of the sediments below the surface is required. A cleared site must be exposed to sufficiently strong tidal flows to facilitate the on-going removal and redistribution of muddy sediments. Consequently, recovery may be impeded by low-energy hydrodynamic conditions, and also by continued supplies of terrigenous muddy sediment, particularly where catchments and riparian zones are still poorly managed. Indeed, there may be some locations that will not realise substantial restoration. Monitoring changes in both surface and deeper sediments is therefore a requirement to avoid unjustified conclusions regarding the 'success' of the mangrove removal in terms of the return to a sandier system that could support greater abundance and diversity of intertidal macrofauna (Stokes et al. 2016).

2.8.3 Effects of Mangrove Clearance on Carbon Dynamics

It is estimated that mangrove deforestation generates as much as 10% of carbon emissions from deforestation globally, yet mangrove ecosystems account for only 0.7% of the deforested area (Donato et al. 2011). Not only does mangrove clearance result in a loss of mangrove primary production, estimated at 4–10 tonnes carbon per year for New Zealand (derived from Bouillon et al. 2008; Bulmer et al. 2016a, b; Gladstone-Gallagher et al. 2014; Woodroffe 1982), mangrove clearance also results in a loss of estuarine carbon stocks, estimated at 79 tonnes carbon per hectare of mangrove cleared (Bulmer et al. 2016b). Rather than a rapid loss of carbon immediately following mangrove clearance, this loss is expected to occur over a number of years to decades following clearance due to the slow decomposition of organic matter within New Zealand mangrove sediments (Bulmer et al. 2015, 2017b; Gladstone-Gallagher et al. 2014).

As the majority (>90%) of carbon stocks within mangrove ecosystems are stored belowground in the sediment (Bulmer et al. 2016b), one of the primary pathways for carbon export from mangrove ecosystems following mangrove clearance is via sediment CO_2 efflux (Bouillon et al. 2008). The average rate of sediment CO_2 efflux from mangrove clearance sites throughout New Zealand is 133.9 ± 37.2 mmol m^{-2} day^{-1}, which is comparable to rates observed in cleared mangrove sites in the tropics (Bulmer et al. 2015). The rate of sediment CO_2 efflux from cleared mangroves varies based on a number of factors, such as time since clearance (Grellier et al. 2017; Lovelock et al. 2011), the organic matter content of the sediment column, as well as clearance methodology (Bulmer et al. 2015, 2017b). For example, mechanical mulching and dispersal of aboveground biomass in situ has been associated with sediment CO_2 efflux seven times higher than that of smaller hand clearances where aboveground biomass was removed from the site (Bulmer et al. 2015). Nonetheless, the loss of carbon stocks and carbon emissions associated with mangrove removal is not currently considered in consent applications for mangrove clearance.

2.8.4 Effects of Mangrove Clearance on Other Ecosystem Services

Many of the expected benefits of mangrove clearance rely on a transition (or return) to a previous state of the estuarine system when sandflats were still a common feature. This return should be resulting from the erosion of muddy sediments and erosion or decomposition of belowground biomass post mangrove clearance (see section "Effects of Mangrove Clearance on Sediment Composition"). If clearance sites do return to sandflat characteristics, the associated changes in macrofaunal community structure are likely to support an increase in macrofaunal diversity and abundance and associated ecosystem functioning (Thrush et al. 2004). Nevertheless, regardless of whether mangrove ecosystems are expanding or are being cleared, New Zealand faces a fundamental problem with regard to high levels of terrigenous fine sediment inputs and enhanced sedimentation impacting estuarine habitats and reducing many of the ecosystem services provided (Thrush et al. 2004). As increased mud content and loss of macrofauna is driving many of the declines in ecosystem functioning faced in New Zealand estuaries (Thrush et al. 2004), mangrove clearance appears unlikely to increase provisioning of these ecosystem services (see also section "Ecosystem (Dis-)services of Expanding Mangroves"). Rather, the infilling of estuaries that facilitates mangrove expansion and degrades other estuarine habitats, is a long lasting geomorphological process and clearance sites remain comparable to intact mangrove ecosystems many years after clearance (in regards to sediment conditions, macrofaunal communities and carbon and nutrient cycling; as discussed in section "Effects of Mangrove Clearance on Sediment Composition"). Holistic catchment management approaches that reduce sediment loads into estuaries should thus be an integral part of both mangrove management and restoration of estuarine ecosystems (Lundquist et al. 2017).

Community group perceptions of the effects of mangrove clearance on the restoration of ecosystem services are generally positive, reflecting restoration of ocean views through removal of vegetation, and in some cases, successful restoration of recreational access through clearing of channels through the mangroves that support boat or kayak access. However, these benefits should be balanced with losses of other ecosystem services that benefit humans, such as carbon sequestration (section "Effects of Mangrove Clearance on Carbon Dynamics") and nutrient storage (Bulmer et al. 2017b), flood protection, and buffers against coastal erosion (section "Biophysical Interactions"). Removal of the latter function of mangroves has resulted in visible erosion of neighbouring saltmarsh vegetation at a number of mangrove removal sites (Fig. 2.16).

Initial concerns that mangrove removals might have negative downstream impacts (through sediment erosion and

Fig. 2.16 Erosion of saltmarsh habitat after removal of adjacent mangroves, Tauranga Harbour (Photo credits: C. Lundquist)

subsequent deposition) on neighbouring seagrass meadows and shellfish beds have not eventuated (Lundquist et al. 2012). However, mangrove removals have also not been associated with the return of previously abundant seagrass and shellfish beds, suggesting that further changes in catchment management and reductions in sediment loads are required to benefit these habitats, whose declines were unlikely to be caused by mangrove expansion. Further study is required to assess concerns that mangrove removals could have negative impacts on shorebirds that utilise mangrove habitats for roosting, foraging, or as protection from predators, as limited information collected at Whangamata suggested that there were no impacts of removals on shorebird populations (Wildlands 2014). To further avoid impacts on bird populations, consent conditions typically require mangrove removals to occur at times that are unlikely to disturb roosting or nesting seasons.

2.9 Synthesis: Expanding Mangroves and the Future

The complexity of the interlinked biophysical processes in mangroves, as discussed in section "Biophysical Dynamics of Expanding Mangroves", means that the outlook for New Zealand mangroves with future changes in climate and physical forcings is not well known. Nonetheless, several key drivers of change can be identified: sea-level rise, change in weather patterns and alterations to sediment supply.

Sea-level rise can be difficult to assess in a geologically active area such as New Zealand, as estimations of (relative) vertical elevation changes need to include vertical landmass movements (Swales et al. 2016). Moreover, in other locations, responses of mangroves to sea-level rise have been shown to be highly site specific (Lovelock et al. 2015). Although inundation is increased under higher sea levels, which may imply drowning of mangrove forests (Gilman et al. 2006; Ward et al. 2016), a higher sea level also leads to a greater tidal prism, which can potentially support larger waves and stronger tidal currents with the ability to resuspend more sediment, enhancing inward sediment transport (Green and Coco 2007; Horstman et al. 2015). Additionally, mangrove forests can also accrete through below-ground biomass production (Kirwan and Megonigal 2013). Indeed, forests in regions with ample sediment supply have shown to possess the ability to accrete vertically to keep pace with sea-level rise (Krauss et al. 2014). A similar resilience could be expected for New Zealand mangroves, provided local sediment supplies into the estuaries remain high.

Mangrove root architecture and hence also sediment trapping ability has been found to depend on salinity (Krauss et al. 2014), and therefore changes in rainfall may lead to a changing mangrove distribution, particularly at their upstream extent on riverbanks. Changes in the frequency and intensity of storms may further enhance sediment delivery events (see section "Mangrove Expansion in New Zealand's Anthropocene"), but conversely, may reduce 'windows of opportunity' in which mangroves can expand (see section "Succession of Expanding Mangrove Systems"). Moreover, an excess of sediment supply can lead to burial and death of mangroves (Sidik et al. 2016). Hydrodynamic changes can also affect sediment storage and supply in offshore shelf regions (Liu et al. 2017; Nittrouer et al. 2017), indirectly affecting sediment supply to the mangroves.

Lastly, human interventions provide further unknowns. Activities such as damming reduce sediment supply and may eventually lead to a sediment starved system lacking the resilience to cope with sea-level rise (Allison et al. 2017; Horstman et al. 2015; McBride et al. 2016; Syvitski et al. 2005; Willemsen et al. 2016). Construction of landward boundaries such as dikes and barriers limits tidal fluxes of water and sediments (Willemsen et al. 2016; Winterwerp et al. 2005) and also restricts landward movement of the forest, the latter causing coastal squeeze (Lovelock et al. 2015; Mills et al. 2016). Nevertheless, despite these uncertainties, it is thought that New Zealand's temperate and sediment rich mangroves will be more resilient than tropical mangrove forests (Alongi 2008; Schaeffer-Novelli et al. 2002).

The capacity of New Zealand mangroves to mitigate erosion and to raise surface elevations provides these coastal ecosystems with an intrinsic resilience, potentially enabling them to combat land subsidence and (relative) sea-level rise (McBride et al. 2016). For example, in the Firth of Thames bed elevations inside the mangroves (Fig. 2.7a) have accreted to an elevation of more than a meter above the elevation of the land directly behind the stop bank that separates the mangroves from the subsiding Hauraki Plains (Horstman unpublished data). This mangrove forest has a width of more than a kilometre, and has been observed to reduce tidal water level fluctuations and storm surges, thereby reducing the intrusion of saline (ground)water and enhancing the protection provided by the stop bank. This forest is a typical example of a coastal ecosystem defence, where ecological engineering (by the mangroves) benefits the safety and quality of the coastal zone (Borsje et al. 2011; Bouma et al. 2014; Temmerman et al. 2013).

Regulating ecosystem services, such as the contribution of mangroves to coastal defence (section "Biophysical Dynamics of Expanding Mangroves") and carbon sequestration (section "Carbon Dynamics of Expanding Mangroves"), together with other provisioning ecosystem services (as discussed in section "Ecosystem (Dis-)services of Expanding Mangroves"), are the motivation behind the global surge in mangrove restoration and management efforts (Bosire et al. 2008; Gilman and Ellison 2007; Hashim et al. 2010; Lewis 2005). New Zealand mangroves do not at present suffer from reduced sediment inputs or other physical stressors (apart from the temperature limitation) and therefore these systems provide a useful example illustrating the biophysical interactions and succession dynamics of expanding mangrove ecosystems, as discussed in sections "Biophysical Dynamics of Expanding Mangroves" and "Succession of Expanding Mangrove Systems".

In New Zealand, however, mangrove expansion is associated with a negative connotation, owing to the reduced ecosystem value of temperate mangroves compared to their tropical equivalent, as well as the change of intertidal ecosystems and the loss of amenity values associated with mangrove expansion. The consequent tendency towards mangrove removal, following these negative impacts, has partly been due to un- or misinformed decision making (as discussed in section "Mangrove Removal in New Zealand") and a general lack of knowledge concerning the consequences of mangrove removal. However, with a growing body of knowledge related to mangrove removal, opportunities arise for integrated coastal zone management strategies carefully weighing both the disadvantages and the opportunities related to mangrove expansion. Planning such comprehensive coastal management strategies could benefit from models that allow the simulation of morphological changes in the coastal zone in relation to the establishment and succession of mangroves, also including related impacts on the biophysical interactions in the mangroves and the consequences for coastal stability and safety. A range of such interdisciplinary models exists for salt marshes (e.g. Kirwan and Murray 2007; Temmerman et al. 2005; van de Koppel et al. 2012), but for mangroves the development of these models is still in its infancy (Bryan et al. 2017; Horstman et al. 2015; van Maanen et al. 2015). As shown throughout this chapter, these processes form the basis of an active area of research, which will contribute towards integrated coastal zone management strategies incorporating the services provided by mangroves and honouring the traditional belief in kaitiakitanga—that if an ecosystem is sustained and cared for, it can then provide benefits back to humanity.

Acknowledgements Contributions by EMH, KRB and JCM and collection of the field data presented in sections "Biophysical Dynamics of Expanding Mangroves" and "Succession of Expanding Mangrove Systems" were supported by the Royal Society of New Zealand's Marsden Fund (grant number 14-UOW-011). Contributions by CJL and RHB were supported by MBIE Programme C01X1002, and NIWA Freshwater Programme core funding. Contributions by DJS were supported by Waikato Regional Council and Southern Cross University MERC grants.

References

Alfaro AC (2006) Benthic macro-invertebrate community composition within a mangrove/seagrass estuary in northern New Zealand. Estuar Coast Shelf Sci 66(1–2):97–110. https://doi.org/10.1016/j.ecss.2005.07.024

Alfaro AC (2010) Effects of mangrove removal on benthic communities and sediment characteristics at Mangawhai Harbour, northern New Zealand. ICES J Mar Sci 67(6):1087–1104. https://doi.org/10.1093/icesjms/fsq034

Allison MA, Nittrouer CA, Ogston AS, Mullarney JC, Nguyen TT (2017) Sedimentation and survival of the Mekong delta: a case study of decreased sediment supply and accelerating rates of relative sea level rise. Oceanography 30(3):98–109. https://doi.org/10.5670/oceanog.2017.317

Alongi DM (2008) Mangrove forests: resilience, protection from tsunamis, and responses to global climate change. Estuar Coast Shelf Sci 76(1):1–13. https://doi.org/10.1016/j.ecss.2007.08.024

Alongi DM (2009) Paradigm shifts in mangrove biology. In: GME P, Wolanski E, Cahoon DR, Brinson MM (eds) Coastal wetlands: an integrated ecosystem approach. Elsevier, Amsterdam, pp 615–640

Alongi DM (2014) Carbon cycling and storage in mangrove forests. Annu Rev Mar Sci 6(1):195–219. https://doi.org/10.1146/annurev-marine-010213-135020

Augustinus PGEF (1995) Geomorphology and sedimentology of mangroves. In: GME P (ed) Developments in sedimentology, vol 53. Elsevier, Amsterdam, pp 333–357

Balke T, Bouma TJ, Herman PMJ, Horstman EM, Sudtongkong C, Webb EL (2013) Cross-shore gradients of physical disturbance in mangroves: implications for seedling establishment. Biogeosciences 10(8):5411–5419. https://doi.org/10.5194/bg-10-5411-2013

Balke T, Swales A, Lovelock CE, Herman PMJ, Bouma TJ (2015) Limits to seaward expansion of mangroves: translating physical disturbance mechanisms into seedling survival gradients. J Exp Mar Biol Ecol 467:16–25. https://doi.org/10.1016/j.jembe.2015.02.015

Barbier EB (2014) A global strategy for protecting vulnerable coastal populations. Science 345(6202):1250–1251. https://doi.org/10.1126/science.1254629

Barbier EB, Hacker SD, Kennedy C, Koch EW, Stier AC, Silliman BR (2011) The value of estuarine and coastal ecosystem services. Ecol Monogr 81(2):169–193. https://doi.org/10.1890/10-1510.1

Beaglehole JC (1955) The journals of Captain James cook on his voyages of discovery. Hakluyt Society at the University Press, Cambridge, UK

Borsje BW, van Wesenbeeck BK, Dekker F, Paalvast P, Bouma TJ, van Katwijk MM, de Vries MB (2011) How ecological engineering can serve in coastal protection. Ecol Eng 37(2):113–122. https://doi.org/10.1016/j.ecoleng.2010.11.027

Bosire JO, Dahdouh-Guebas F, Kairo JG, Wartel S, Kazungu J, Koedam N (2006) Success rates of recruited tree species and their contribution to the structural development of reforested mangrove stands. Mar Ecol Prog Ser 325:85–91. https://doi.org/10.3354/meps325085

Bosire JO, Dahdouh-Guebas F, Walton M, Crona BI, Lewis RR III, Field C, Kairo JG, Koedam N (2008) Functionality of restored mangroves: a review. Aquat Bot 89(2):251–259

Bouillon S, Borges AV, Castañeda-Moya E, Diele K, Dittmar T, Duke NC, Kristensen E, Lee SY, Marchand C, Middelburg JJ, Rivera-Monroy VH, Smith TJ III, Twilley RR (2008) Mangrove production and carbon sinks: a revision of global budget estimates. Glob Biogeochem Cycles 22(2):GB2013

Bouma TJ, Van Duren LA, Temmerman S, Claverie T, Blanco-Garcia A, Ysebaert T, Herman PMJ (2007) Spatial flow and sedimentation patterns within patches of epibenthic structures: combining field, flume and modelling experiments. Cont Shelf Res 27:1020–1045. https://doi.org/10.1016/j.csr.2005.12.019

Bouma TJ, van Belzen J, Balke T, Zhu Z, Airoldi L, Blight AJ, Davies AJ, Galvan C, Hawkins SJ, Hoggart SPG, Lara JL, Losada IJ, Maza M, Ondiviela B, Skov MW, Strain EM, Thompson RC, Yang S, Zanuttigh B, Zhang L, Herman PMJ (2014) Identifying knowledge gaps hampering application of intertidal habitats in coastal protection: opportunities & steps to take. Coast Eng 87:147–157. https://doi.org/10.1016/j.coastaleng.2013.11.014

Bryan KR, Nardin W, Mullarney JC, Fagherazzi S (2017) The role of cross-shore tidal dynamics in controlling intertidal sediment exchange in mangroves in Cù Lao Dung, Vietnam. Cont Shelf Res 147:128–143. https://doi.org/10.1016/j.csr.2017.06.014

Bryce S, Larcombe P, Ridd PV (2003) Hydrodynamic and geomorphological controls on suspended sediment transport in mangrove creek systems, a case study: Cocoa Creek, Townsville, Australia. Estuar Coast Shelf Sci 56:415–431. https://doi.org/10.1016/s0272-7714(02)00192-0

Bulmer RH, Lundquist CJ (2016) Whangamata mangrove removal monitoring: summary of 36 month post-removal sampling, NIWA Client Report, Waikato Regional Council. National Institute of Water & Atmospheric Research Ltd., Hamilton

Bulmer RH, Lundquist CJ, Schwendenmann L (2015) Sediment properties and CO_2 efflux from intact and cleared temperate mangrove forests. Biogeosciences 12(20):6169–6180. https://doi.org/10.5194/bg-12-6169-2015

Bulmer RH, Schwendenmann L, Lundquist CJ (2016a) Allometric models for estimating aboveground biomass, carbon and nitrogen stocks in temperate *Avicennia marina* forests. Wetlands:1–8. https://doi.org/10.1007/s13157-016-0793-0

Bulmer RH, Schwendenmann L, Lundquist CJ (2016b) Carbon and nitrogen stocks and below-ground allometry in temperate mangroves. Front Mar Sci 3 (150). doi:https://doi.org/10.3389/fmars.2016.00150

Bulmer RH, Lewis M, O'Donnell E, Lundquist CJ (2017a) Assessing mangrove clearance methods to minimise adverse impacts and maximise the potential to achieve restoration objectives. N Z J Mar Freshw Res 51(1):110–126. https://doi.org/10.1080/00288330.2016.1260605

Bulmer RH, Schwendenmann L, Lohrer A, Lundquist CJ (2017) Sediment carbon and nutrient fluxes from cleared and intact temperate mangrove ecosystems and adjacent sandflats. Sci Total Environ 599–600:1874–1884. https://doi.org/10.1016/j.scitotenv.2017.05.139

Burns BR, Ogden J (1985) The demography of the temperate mangrove [*Avicennia marina* (Forsk.) Vierh.] at its southern limit in New Zealand. Aust J Ecol 10(2):125–133. https://doi.org/10.1111/j.1442-9993.1985.tb00874.x

Cahoon DR, Hensel P, Rybczyk J, McKee KL, Proffitt CE, Perez BC (2003) Mass tree mortality leads to mangrove peat collapse at Bay Islands, Honduras after Hurricane Mitch. J Ecol 91(6):1093–1105. https://doi.org/10.1046/j.1365-2745.2003.00841.x

Chapman VJ (1976) Mangrove vegetation. Cramer, Vaduz

Chapman VJ (1977) Introduction. In: Chapman VJ (ed) Ecosystems of the world: wet coastal ecosystems. Elsevier, Amsterdam, pp 1–29

Chen Y, Li Y, Cai T, Thompson C, Li Y (2016) A comparison of biohydrodynamic interaction within mangrove and saltmarsh boundaries. Earth Surf Process Landf 41(13):1967–1979. https://doi.org/10.1002/esp.3964

Clarke PJ (2004) Effects of experimental canopy gaps on mangrove recruitment: lack of habitat partitioning may explain stand dominance. J Ecol 92(2):203–213. https://doi.org/10.1111/j.0022-0477.2004.00861.x

Claudino MC, Pessanha ALM, Araújo FG, Garcia AM (2015) Trophic connectivity and basal food sources sustaining tropical aquatic consumers along a mangrove to ocean gradient. Estuar Coast Shelf Sci 167(Part A):45–55. https://doi.org/10.1016/j.ecss.2015.07.005

Crisp P, Daniel L, Tortell P (1990) Mangroves in New Zealand – trees in the tide. GP Books, Wellington

Danielsen F, Sørensen MK, Olwig MF, Selvam V, Parish F, Burgess ND, Hiraishi T, Karunagaran VM, Rasmussen MS, Hansen LB, Quarto A, Suryadiputra N (2005) The Asian tsunami: a protective role for coastal vegetation. Science 310(5748):643. https://doi.org/10.1126/science.1118387

de Lange WP, de Lange PJ (1994) An appraisal of factors controlling the latitudinal distribution of mangrove (*Avicannia marina* var. *resinifera*) in New Zealand. J Coast Res 10(3):539–548

Dingwall PR (1984) Overcoming problems in the management of New Zealand mangrove forests. In: Teas HJ (ed) Physiology and management of mangroves. Dr W. Junk Publishers, The Hague, pp 97–106. https://doi.org/10.1007/978-94-009-6572-0_13

Donato DC, Kauffman JB, Murdiyarso D, Kurnianto S, Stidham M, Kanninen M (2011) Mangroves among the most carbon-rich forests in the tropics. Nat Geosci 4(5):293–297. https://doi.org/10.1038/ngeo1123

Doyle EJ, Hogg ID (2015) Assessing the diversity of terrestrial invertebrates in the mangrove forests of the Firth of Thames, New Zealand. Genome 58(5):213–213

Duke NC (1990) Phenological trends with latitude in the mangrove tree avicennia marina. J Ecol 78(1):113–133. https://doi.org/10.2307/2261040

Duke NC (1991) A systematic revision of the mangrove genus *Avicennia* (Avicenniaceae) in Australasia. Aust Syst Bot 4:299–324

Duke NC, Ball MC, Ellison JC (1998) Factors influencing biodiversity and distributional gradients in mangroves. Glob Ecol Biogeogr Lett 7(1):27–47. https://doi.org/10.2307/2997695

Ellis J, Nicholls P, Craggs R, Hofstra D, Hewitt J (2004) Effects of terrigenous sedimentation on mangrove physiology and associated macrobenthic communities. Mar Ecol Prog Ser 270:71–82. https://doi.org/10.3354/meps270071

Ellison JC (1999) Impacts of sediment burial on mangroves. Mar Pollut Bull 37(8–12):420–426. https://doi.org/10.1016/S0025-326X(98)00122-2

Ellison AM (2002) Macroecology of mangroves: large-scale patterns and processes in tropical coastal forests. Trees 16(2):181–194. https://doi.org/10.1007/s00468-001-0133-7

Ewel K, Twilley R, Ong JIN (1998) Different kinds of mangrove forests provide different goods and services. Glob Ecol Biogeogr Lett 7(1):83–94. https://doi.org/10.1111/j.1466-8238.1998.00275.x

Fricke AT, Nittrouer CA, Ogston AS, Vo-Luong HP (2017) Asymmetric progradation of a coastal mangrove forest controlled by combined fluvial and marine influence, Cù Lao Dung, Vietnam. Cont Shelf Res 147:78–90. https://doi.org/10.1016/j.csr.2017.07.012

Friedrichs CT, Aubrey DG (1988) Non-linear tidal distortion in shallow well-mixed estuaries: a synthesis. Estuar Coast Shelf Sci 27(5):521–545. https://doi.org/10.1016/0272-7714(88)90082-0

Friess DA, Lee SY, Primavera JH (2016a) Turning the tide on mangrove loss. Mar Pollut Bull 109(2):673–675. https://doi.org/10.1016/j.marpolbul.2016.06.085

Friess DA, Richards DR, Phang VXH (2016b) Mangrove forests store high densities of carbon across the tropical urban landscape of Singapore. Urban Ecosyst 19(2):795–810. https://doi.org/10.1007/s11252-015-0511-3

Fromard F, Puig H, Mougin E, Marty G, Betoulle JL, Cadamuro L (1998) Structure, above-ground biomass and dynamics of mangrove ecosystems: new data from French Guiana. Oecologia 115(1):39–53. https://doi.org/10.1007/s004420050489

Furukawa K, Wolanski E (1996) Sedimentation in mangrove forests. Mangrove Salt Marshes 1(1):3–10. https://doi.org/10.1023/a:1025973426404

Furukawa K, Wolanski E, Mueller H (1997) Currents and sediment transport in mangrove forests. Estuar Coast Shelf Sci 44(3):301–310

Gilman E, Ellison J (2007) Efficacy of alternative low-cost approaches to mangrove restoration, American Samoa. Estuar Coasts 30(4):641–651. https://doi.org/10.1007/bf02841961

Gilman EL, Ellison J, Jungblut V, van Lavieren H, Wilson L, Areki F, Brighouse G, Bungitak J, Henry M, Kilman M, Matthews E, Sauni I, Teariki-Ruatu N, Tukia S, Yuknavage K (2006) Adapting to Pacific Island mangrove responses to sea level rise and climate change. Clim Res 32(3):161–176. https://doi.org/10.3354/cr032161

Giri C, Ochieng E, Tieszen LL, Zhu Z, Singh A, Loveland T, Masek J, Duke N (2011) Status and distribution of mangrove forests of the world using earth observation satellite data. Glob Ecol Biogeogr 20(1):154–159. https://doi.org/10.1111/j.1466-8238.2010.00584.x

Gladstone-Gallagher RV, Lundquist CJ, Pilditch CA (2014) Mangrove (*Avicennia marina* subsp. australasica) litter production and decomposition in a temperate estuary. N Z J Mar Freshw Res 48(1):24–37. https://doi.org/10.1080/00288330.2013.827124

Gladstone-Gallagher RV, Lohrer AM, Lundquist CJ, Pilditch CA (2016) Effects of detrital subsidies on soft-sediment ecosystem function are transient and source-dependent. PLoS One 11(5):e0154790. https://doi.org/10.1371/journal.pone.0154790

Graeme M (2013) Estuarine vegetation survey: Whangapoua Harbour. Waikato Regional Council, Hamilton

Green MO, Coco G (2007) Sediment transport on an estuarine intertidal flat: measurements and conceptual model of waves, rainfall and exchanges with a tidal creek. Estuar Coast Shelf Sci 72(4):553–569. https://doi.org/10.1016/j.ecss.2006.11.006

Green MO, Black KP, Amos CL (1997) Control of estuarine sediment dynamics by interactions between currents and waves at several scales. Mar Geol 144(1–3):97–116. https://doi.org/10.1016/S0025-3227(97)00065-0

Grellier S, Janeau J-L, Dang Hoai N, Nguyen Thi Kim C, Le Thi PQ, Pham Thi Thu T, Tran-Thi N-T, Marchand C (2017) Changes in soil characteristics and C dynamics after mangrove clearing (Vietnam). Sci Total Environ 593–594:654–663. https://doi.org/10.1016/j.scitotenv.2017.03.204

Griffiths GA, Glasby GP (1985) Input of river-derived sediment to the New Zealand continental shelf: I. Mass. Estuar Coast Shelf Sci 21(6):773–787. https://doi.org/10.1016/0272-7714(85)90072-1

Harmsworth GR, Awatere S (2013) Indigenous Māori knowledge and perspectives of ecosystems. In: Dymond JR (ed) Ecosystem services in New Zealand – conditions and trends. Manaaki Whenua Press, Lincoln, pp 274–286

Harty C (2009) Mangrove planning and management in New Zealand and South East Australia – a reflection on approaches. Ocean Coast Manag 52(5):278–286. https://doi.org/10.1016/j.ocecoaman.2009.03.001

Hashim R, Kamali B, Tamin NM, Zakaria R (2010) An integrated approach to coastal rehabilitation: mangrove restoration in Sungai Haji Dorani, Malaysia. Estuar Coast Shelf Sci 86(1):118–124

Hayden HL, Granek EF (2015) Coastal sediment elevation change following anthropogenic mangrove clearing. Estuar Coast Shelf Sci 165:70–74. https://doi.org/10.1016/j.ecss.2015.09.004

Healy TR (2002) Muddy coasts of mid-latitude oceanic islands on an active plate margin – New Zealand. In: Healy TR, Wang J, Healy J-A (eds) Muddy coasts of the world: processes, deposits and function. Elsevier, Amsterdam, pp 347–374. https://doi.org/10.1016/S1568-2692(02)80088-2

Henderson SM, Norris BK, Mullarney JC, Bryan KR (2017) Wavefrequency flows within a near-bed vegetation canopy. Cont Shelf Res 147:91–101. https://doi.org/10.1016/j.csr.2017.06.003

Horstman EM, Dohmen-Janssen CM, Hulscher SJMH (2013) Flow routing in mangrove forests: a field study in Trang province, Thailand. Cont Shelf Res 71:52–67. https://doi.org/10.1016/j.csr.2013.10.002

Horstman EM, Dohmen-Janssen CM, Narra PMF, Van den Berg NJF, Siemerink M, Hulscher SJMH (2014) Wave attenuation in mangroves: a quantitative approach to field observations. Coast Eng 94:47–62. https://doi.org/10.1016/j.coastaleng.2014.08.005

Horstman EM, Dohmen-Janssen CM, Bouma TJ, Hulscher SJMH (2015) Tidal-scale flow routing and sedimentation in mangrove forests: combining field data and numerical modelling. Geomorphology 228:244–262. https://doi.org/10.1016/j.geomorph.2014.08.011

Horstman EM, Bryan KR, Mullarney JC, Pilditch CA (2016) Model versus nature: hydrodynamics in mangrove pneumatophores. In: Lynett P (ed) Proceedings of 35th International Conference on

Coastal Engineering, vol 35. Coastal Engineering Research Council, Antalya. https://doi.org/10.9753/icce.v35.management.19
Horstman EM, Mullarney JC, Bryan KR, Sandwell DR (2017) Deposition gradients across mangrove fringes. In: Aagaard T, Deigaard R, Fuhrman D (eds) Coastal dynamics 2017. Helsingør, Denmark
Horstman EM, Bryan KR, Mullarney JC, Pilditch CA, Eager CA (2018) Are flow-vegetation interactions well represented by mimics? A case study of mangrove pneumatophores. Adv Water Resour 111:360–371. https://doi.org/10.1016/j.advwatres.2017.11.018
Hume TM (2003) Estuaries and tidal inlets. In: Goff JR, Nichol SL, Rouse HL (eds) The New Zealand coast: Te Tai O Aotearoa. Dunmore Press, Palmerston North, pp 191–213
Hume TM, Herdendorf CE (1992) Factors controlling tidal inlet characteristics on low drift coasts. J Coast Res 8(2):355–375
Hunt S, Bryan KR, Mullarney JC (2015) The influence of wind and waves on the existence of stable intertidal morphology in meso-tidal estuaries. Geomorphology 228:158–174. https://doi.org/10.1016/j.geomorph.2014.09.001
Hunt S, Bryan KR, Mullarney JC (2017) The effect of wind waves on spring-neap variations in sediment transport in two meso-tidal estuarine basins with contrasting fetch. Geomorphology 280:76–88. https://doi.org/10.1016/j.geomorph.2016.12.007
Janzen DH (1985) Mangroves: where's the understory? J Trop Ecol 1 (1):89–92. https://doi.org/10.1017/S0266467400000122
Jones H (2008) Coastal sedimentation: what we know and the information gaps. Waikato Regional Council, Hamilton
Kirwan ML, Megonigal JP (2013) Tidal wetland stability in the face of human impacts and sea-level rise. Nature 504(7478):53–60. https://doi.org/10.1038/nature12856
Kirwan ML, Murray AB (2007) A coupled geomorphic and ecological model of tidal marsh evolution. Proc Natl Acad Sci U S A 104 (15):6118–6122. https://doi.org/10.1073/pnas.0700958104
Kobashi D, Mazda Y (2005) Tidal flow in riverine-type mangroves. Wetl Ecol Manag 13(6):615–619. https://doi.org/10.1007/s11273-004-3481-4
Krauss KW, Lovelock CE, McKee KL, López-Hoffman L, Ewe SML, Sousa WP (2008) Environmental drivers in mangrove establishment and early development: a review. Aquat Bot 89(2):105–127. https://doi.org/10.1016/j.aquabot.2007.12.014
Krauss KW, McKee KL, Lovelock CE, Cahoon DR, Saintilan N, Reef R, Chen L (2014) How mangrove forests adjust to rising sea level. New Phytol 202(1):19–34. https://doi.org/10.1111/nph.12605
Lara RJ, Szlafsztein CF, Cohen MCL, Oxmann J, Schmitt BB, Filho PWMS (2009) Geomorphology and sedimentology of mangroves and salt marshes: the formation of geobotanical units. In: GME P, Wolanski E, Cahoon DR, Brinson MM (eds) Coastal wetlands: an integrated ecosystem approach. Elsevier, Amsterdam, pp 593–613
Lee SY, Primavera JH, Dahdouh-Guebas F, McKee K, Bosire JO, Cannicci S, Diele K, Fromard F, Koedam N, Marchand C, Mendelssohn I, Mukherjee N, Record S (2014) Ecological role and services of tropical mangrove ecosystems: a reassessment. Glob Ecol Biogeogr 23(7):726–743. https://doi.org/10.1111/geb.12155
Lewis RR (2005) Ecological engineering for successful management and restoration of mangrove forests. Ecol Eng 24(4):403–418. https://doi.org/10.1016/j.ecoleng.2004.10.003
Liu JP, DeMaster DJ, Nittrouer CA, Eidam EF, Nguyen TT (2017) A seismic study of the Mekong subaqueous delta: proximal versus distal accumulation. Oceanography 30(3). https://doi.org/10.5670/oceanog.2017.315
Lovelock CE, Feller IC, Ellis J, Schwarz AM, Hancock N, Nichols P, Sorrell B (2007) Mangrove growth in New Zealand estuaries: the role of nutrient enrichment at sites with contrasting rates of sedimentation. Oecologia 153(3):633–641. https://doi.org/10.1007/s00442-007-0750-y
Lovelock CE, Sorrell BK, Hancock N, Hua Q, Swales A (2010) Mangrove forest and soil development on a rapidly accreting shore in New Zealand. Ecosystems 13(3):437–451. https://doi.org/10.1007/s10021-010-9329-2
Lovelock CE, Ruess RW, Feller IC (2011) CO2 efflux from cleared mangrove peat. PLoS ONE 6(6):e21279
Lovelock CE, Cahoon DR, Friess DA, Guntenspergen GR, Krauss KW, Reef R, Rogers K, Saunders ML, Sidik F, Swales A, Saintilan N, Thuyen LX, Triet T (2015) The vulnerability of Indo-Pacific mangrove forests to sea-level rise. Nature 526(7574):559–563. https://doi.org/10.1038/nature15538
Lowe M (2013) Factors affecting the habitat usage of estuarine juvenile fish in northern New Zealand. The University of Auckland, Aucland
Lugo AE (1997) Old-growth mangrove forests in the United States. Conserv Biol 11(1):11–20. https://doi.org/10.1046/j.1523-1739.1997.96012.x
Lundquist CJ, Hailes S, Cartner K, Carter K, Gibbs M (2012) Physical and ecological impacts associated with mangrove removals using in situ mechanical mulching in Tauranga Harbour. NIWA Technical Report No.137. NIWA Technical Report No137
Lundquist CJ, Hailes SF, Carter KR, Burgess TC (2014a) Ecological status of mangrove removal sites in the Auckland region. National Institute of Water & Atmospheric Research Ltd., Hamilton
Lundquist CJ, Morrisey DJ, Gladstone-Gallagher RV, Swales A (2014b) Managing mangrove habitat expansion in New Zealand. In: Mangrove ecosystems of Asia: status, challenges and management strategies. p 415–438. doi:10.1007/978-1-4614-8582-7_19
Lundquist CJ, Carter K, Hailes S, Bulmer RH (2017) Guidelines for managing mangrove expansion in New Zealand. National Institute of Water & Atmospheric Research Ltd., Hamilton
Macnae W (1966) Mangroves in eastern and southern Australia. Aust J Bot 14:67–104
Martin KC (2007) Interactive effects of salinity and nutrients on mangrove physiology: implications for mangrove forest structure and function. Australian National University, Canberra
May JD (1999) Spatial variation in litter production by the mangrove *Avicennia marina* var. australasica in Rangaunu harbour, Northland, New Zealand. N Z J Mar Freshw Res 33(2):163–172. https://doi.org/10.1080/00288330.1999.9516866
Mazda Y, Magi M, Kogo M, Hong PN (1997) Mangroves as a coastal protection from waves in the Tong King delta, Vietnam. Mangrove Salt Marshes 1(2):127–135. https://doi.org/10.1023/a:1009928003700
McBride G, Reeve G, Pritchard M, Lundquist C, Daigneault A, Bell R, Blackett P, Swales A, Wadhwa S, Tait A, Zammit C (2016) The Firth of Thames and Lower Waihou River. Synthesis report RA2, coastal case study. Climate Changes, Impacts and Implications (CCII) for New Zealand to 2100. National Institute of Water & Atmospheric Research Ltd., Hamilton
Melville F, Pulkownik A (2006) Investigation of mangrove macroalgae as bioindicators of estuarine contamination. Mar Pollut Bull 52 (10):1260–1269. https://doi.org/10.1016/j.marpolbul.2006.02.021
Mills M, Leon JX, Saunders MI, Bell J, Liu Y, O'Mara J, Lovelock CE, Mumby PJ, Phinn S, Possingham HP, Tulloch VJD, Mutafoglu K, Morrison T, Callaghan DP, Baldock T, Klein CJ, Hoegh-Guldberg O (2016) Reconciling development and conservation under coastal squeeze from rising sea level. Conserv Lett 9(5):361–368. https://doi.org/10.1111/conl.12213
Morrisey DJ, Beard CM, Morrison MA, Craggs R, Lowe M (2007) The New Zealand mangrove: review of the current state of knowledge. Auckland Regional Council, Auckland
Morrisey DJ, Skilleter GA, Ellis JI, Burns BR, Kemp CE, Burt K (2003) Differences in benthic fauna and sediment among mangrove (Avicennia marina var. australasica) stands of different ages in New Zealand. Estuar Coast Shelf Sci 56(3–4):581–592. https://doi.org/10.1016/S0272-7714(02)00208-1
Morrisey DJ, Swales A, Dittmann S, Morrison MA, Lovelock CE, Beard CM (2010) The ecology and management of temperate

mangroves. In: Gibson RN, RJA A, JDM G (eds) Oceanography and marine biology – an annual review, Oceanography and Marine Biology – An Annual Review, vol 48. CRC Press, New York, pp 43–160. https://doi.org/10.1201/EBK1439821169-c2

Mullarney JC, Henderson SM (2018) The effect of marine vegetation on shorelines. In: Pan-Chang V, Kaihatu J (eds) Advances in coastal hydraulics. World Scientific Publishing Ltd (in press)

Mullarney JC, Henderson SM, Norris BK, Bryan KR, Fricke AT, Sandwell DR, Culling DP (2017a) A question of scale: how turbulence around aerial roots shapes the seabed morphology in mangrove forests of the Mekong Delta. Oceanography 30(3):20–33. https://doi.org/10.5670/oceanog.2017.312

Mullarney JC, Henderson SM, Reyns JAH, Norris BK, Bryan KR (2017) Spatially varying drag within a wave-exposed mangrove forest and on the adjacent tidal flat. Cont Shelf Res 147:102–113. https://doi.org/10.1016/j.csr.2017.06.19

Nagelkerken I, Blaber SJM, Bouillon S, Green P, Haywood M, Kirton LG, Meynecke JO, Pawlik J, Penrose HM, Sasekumar A, Somerfield PJ (2008) The habitat function of mangroves for terrestrial and marine fauna: a review. Aquat Bot 89(2):155–185. https://doi.org/10.1016/j.aquabot.2007.12.007

Nichol SL, Augustinus PC, Gregory MR, Creese R, Horrocks M (2000) Geomorphic and sedimentary evidence of human impact on the New Zealand coastal landscape. Phys Geogr 21(2):109–132. https://doi.org/10.1080/02723646.2000.10642702

Nittrouer CA, Sternberg RW, Carpenter R, Bennett JT (1979) The use of Pb-210 geochronology as a sedimentological tool: application to the Washington continental shelf. Mar Geol 31(3):297–316. https://doi.org/10.1016/0025-3227(79)90039-2

Nittrouer CA, DeMaster DJ, Eidam EF, Nguyen TT, Liu JP, Ogston AS, Phung PV (2017) The Mekong continental shelf: primary sink for deltaic sediment particles and their passengers. Oceanography 30 (3):60–70. https://doi.org/10.5670/oceanog.2017.314

Norris BK, Mullarney JC, Bryan KR, Henderson SM (2017) The effect of pneumatophore density on turbulence: a field study in a Sonneratia-dominated mangrove forest, Vietnam. Cont Shelf Res 147:114–127. https://doi.org/10.1016/j.csr.2017.06.002

Osborne K, Smith TJ (1990) Differential predation on mangrove propagules in open and closed canopy forest habitats. Vegetatio 89 (1):1–6. https://doi.org/10.1007/bf00134429

Pérez A, Machado W, Gutierrez D, Stokes D, Sanders L, Smoak JM, Santos I, Sanders CJ (2017) Changes in soil organic carbon accumulation driven by mangrove expansion and deforestation in a New Zealand estuary. Estuar Coast Shelf Sci. https://doi.org/10.1016/j.ecss.2017.05.009

Pocknall DT, Gregory MR, Greig DA (1989) Palynology of core 80/20 and its implications for understanding Holocene sea level changes in the Firth of Thames, New Zealand. J R Soc N Z 19(2):171–179. https://doi.org/10.1080/03036758.1989.10426446

Quartel S, Kroon A, Augustinus PGEF, Van Santen P, Tri NH (2007) Wave attenuation in coastal mangroves in the Red River Delta, Vietnam. J Asian Earth Sci 29(4):576–584. https://doi.org/10.1016/j.jseaes.2006.05.008

Robertson BP, Gardner JPA, Savage C (2015) Macrobenthic–mud relations strengthen the foundation for benthic index development: a case study from shallow, temperate New Zealand estuaries. Ecol Indic 58:161–174. https://doi.org/10.1016/j.ecolind.2015.05.039

Rogers K, Saintilan N, Copeland C (2014) Managed retreat of saline coastal wetlands: challenges and opportunities identified from the Hunter River Estuary, Australia. Estuar Coasts 37(1):67–78. https://doi.org/10.1007/s12237-013-9664-6

Roy PS, Williams RJ, Jones AR, Yassini I, Gibbs PJ, Coates B, West RJ, Scanes PR, Hudson JP, Nichol S (2001) Structure and function of South-east Australian estuaries. Estuar Coast Shelf Sci 53 (3):351–384. https://doi.org/10.1006/ecss.2001.0796

Saenger P, Snedaker SC (1993) Pantropical trends in mangrove above-ground biomass and annual litterfall. Oecologia 96(3):293–299. https://doi.org/10.1007/bf00317496

Saintilan N (2004) Relationships between estuarine geomorphology, wetland extent and fish landings in New South Wales estuaries. Estuar Coast Shelf Sci 61(4):591–601. https://doi.org/10.1016/j.ecss.2004.07.002

Saintilan N, Rogers K (2013) The significance and vulnerability of Australian saltmarshes: implications for management in a changing climate. Mar Freshw Res 64(1):66–79. https://doi.org/10.1071/MF12212

Saintilan N, Rogers K, McKee KL (2009) Salt marsh-mangrove interactions in Australasia and the Americas. In: Perillo GME, Wolanski E, Cahoon DR, Brinson MM (eds) Coastal wetlands an integrated ecosystem approach. Elsevier, Amsterdam, pp 855–884

Schaeffer-Novelli Y, Cintron-Molero G, Soares MLG (2002) Chapter nine angroves as indicators of sea level change in the muddy coasts of the world. Proc Marine Sci 4:245–262. https://doi.org/10.1016/S1568-2692(02)80083-3

Sheaves M, Johnston R, Baker R (2016) Use of mangroves by fish: new insights from in-forest videos. Mar Ecol Prog Ser 549:167–182. https://doi.org/10.3354/meps11690

Sidik F, Neil D, Lovelock CE (2016) Effect of high sedimentation rates on surface sediment dynamics and mangrove growth in the Porong River, Indonesia. Mar Pollut Bull 107(1):355–363. https://doi.org/10.1016/j.marpolbul.2016.02.048

Spalding M, Kainuma M, Collins L (2010) World atlas of mangroves (version 1.1), A collaborative project of ITTO, ISME, FAO, UNEP-WCMC, UNESCO-MAB, UNU-INWEH and TNC. Earthscan, London. http://data.unep-wcmc.org/datasets/5

Spalding MD, Ruffo S, Lacambra C, Meliane I, Hale LZ, Shepard CC, Beck MW (2014) The role of ecosystems in coastal protection: adapting to climate change and coastal hazards. Ocean Coast Manag 90:50–57. https://doi.org/10.1016/j.ocecoaman.2013.09.007

Stagg CL, Krauss KW, Cahoon DR, Cormier N, Conner WH, Swarzenski CM (2016) Processes contributing to resilience of coastal wetlands to sea-level rise. Ecosystems 19(8):1445–1459. https://doi.org/10.1007/s10021-016-0015-x

Stern MK, Day JW, Teague KG (1986) Seasonality of materials transport through a coastal freshwater marsh: riverine versus tidal forcing. Estuaries 9(4):301–308. https://doi.org/10.2307/1352102

Stokes DJ (2008) Assessment of physical changes after mangrove removal: Whangamata 2008. Report prepared for Environment Waikato TR 2009/13. Hamilton, New Zealand

Stokes DJ, Harris RJ (2015) Sediment properties and surface erodibility following a large-scale mangrove (Avicennia marina) removal. Cont Shelf Res 107:1–10. https://doi.org/10.1016/j.csr.2015.07.011

Stokes DJ, Healy TR, Cooke PJ (2010) Expansion dynamics of monospecific, temperate mangroves and sedimentation in two embayments of a barrier-enclosed lagoon, Tauranga Harbour, New Zealand. J Coast Res 26(1):113–122. https://doi.org/10.2112/08-1043.1

Stokes DJ, Bulmer RH, Lundquist CJ (2016) Addressing the mismatch between restoration objectives and monitoring needs to support mangrove management. Ocean Coast Manag 134:69–78. https://doi.org/10.1016/j.ocecoaman.2016.09.024

Stuart SA, Choat B, Holbrook NM, Ball MC (2007) The role of freezing in setting the latitudinal limits of mangrove forests. New Phytol 173:576–583. https://doi.org/10.1111/j.1469-8137.2006.01938.x

Sutherland JI (2003) Miocene petrified wood and associated borings and termite faecal pellets from Hukatere Peninsula, Kaipara Harbour, North Auckland, New Zealand. J R Soc N Z 33(1):395–414. https://doi.org/10.1080/03014223.2003.9517736

Swales A, Williamson RB, van Dam LF, Stroud MJ, McGlone MS (2002) Reconstruction of urban stormwater contamination of an

estuary using catchment history and sediment profile dating. Estuaries 25(1):43–56
Swales A, Bentley S, Lovelock C, Bell R (2007) Sediment processes and mangrove-habitat expansion on a rapidly-prograding muddy coast, New Zealand. In: Coastal Sediments '07: Proceedings of the Sixth International Conference on Coastal Engineering and Science of Coastal Sediment Processes, New Orleans. American Society of Civil Engineers, p 1441–1454. doi:10.1061/40926(239)111
Swales A, Bell RG, Gorman R, Oldman JW, Altenberger A, Hart C, Claydon L, Wadwha S, Ovenden R (2009) Potential future changes in mangrove-habitat in Auckland's east-coast estuaries. Auckland Regional Council, Auckland
Swales A, Bentley SJ, Lovelock CE (2015) Mangrove-forest evolution in a sediment-rich estuarine system: opportunists or agents of geomorphic change? Earth Surf Process Landf 40(12):1672–1687. https://doi.org/10.1002/esp.3759
Swales A, Denys P, Pickett VI, Lovelock CE (2016) Evaluating deep subsidence in a rapidly-accreting mangrove forest using GPS monitoring of surface-elevation benchmarks and sedimentary records. Mar Geol 380:205–218. https://doi.org/10.1016/j.margeo.2016.04.015
Syvitski JPM, Vörösmarty CJ, Kettner AJ, Green P (2005) Impact of humans on the flux of terrestrial sediment to the global coastal ocean. Science 308(5720):376–380. https://doi.org/10.1126/science.1109454
Temmerman S, Bouma TJ, Govers G, Wang ZB, De Vries MB, Herman PMJ (2005) Impact of vegetation on flow routing and sedimentation patterns: three-dimensional modeling for a tidal marsh. J Geophys Res Earth Surf 110:F04019. https://doi.org/10.1029/2005jf000301
Temmerman S, Meire P, Bouma TJ, Herman PMJ, Ysebaert T, De Vriend HJ (2013) Ecosystem-based coastal defence in the face of global change. Nature 504(7478):79–83. https://doi.org/10.1038/nature12859
Thom BG (1982) Mangrove ecology: a geomorphological perspective. In: Clough BF (ed) Mangrove ecosystems in Australia; structure, function and management. Australian National University Press, Canberra, pp 3–17
Thrush SF, Hewitt JE, Cummings VJ, Ellis JI, Hatton C, Lohrer A, Norkko A (2004) Muddy waters: elevating sediment input to coastal and estuarine habitats. Front Ecol Environ 2(6):299–306. https://doi.org/10.1890/1540-9295(2004)002[0299:MWESIT]2.0.CO;2
Tomlinson PB (1986) The botany of mangroves, Cambridge tropical biology series. Cambridge University Press, Cambridge, UK
Twilley RR, Rivera-Monroy VH (2009) Ecogeomorphic models of nutrient biogeochemistry for mangrove wetlands. In: GME P, Wolanski E, Cahoon DR, Brinson MM (eds) Coastal wetlands: an integrated ecosystem approac. Elsevier, Amsterdam, pp 641–683
Twilley RR, Chen RH, Hargis T (1992) Carbon sinks in mangroves and their implications to carbon budget of tropical coastal ecosystems. Water Air Soil Pollut 64(1):265–288. https://doi.org/10.1007/bf00477106
Twilley RR, Rivera-Monroy VH, Chen R, Botero L (1999) Adapting an ecological mangrove model to simulate trajectories in restoration ecology. Mar Pollut Bull 37(8–12):404–419. https://doi.org/10.1016/S0025-326X(99)00137-X
UNEP (2014) The importance of mangroves to people: a call to action. United Nations Environment Programme World Conservation Monitoring Centre, Cambridge, UK
van de Koppel J, Bouma TJ, Herman PMJ (2012) The influence of local- and landscape-scale processes on spatial self-organization in estuarine ecosystems. J Exp Biol 215(6):962–967
van Katwijk MM, Bos AR, Hermus DCR, Suykerbuyk W (2010) Sediment modification by seagrass beds: muddification and sandification induced by plant cover and environmental conditions. Estuar Coast Shelf Sci 89(2):175–181. https://doi.org/10.1016/j.ecss.2010.06.008
van Maanen B, Coco G, Bryan KR (2015) On the ecogeomorphological feedbacks that control tidal channel network evolution in a sandy mangrove setting. Proc Royal Soc Lond A: Math Phys Eng Sci 471 (2180):1–24. https://doi.org/10.1098/rspa.2015.0115
van Santen P, Augustinus PGEF, Janssen-Stelder BM, Quartel S, Tri NH (2007) Sedimentation in an estuarine mangrove system. J Asian Earth Sci 29(4):566–575. https://doi.org/10.1016/j.jseaes.2006.05.011
Vo-Luong HP, Massel SR (2008) Energy dissipation in non-uniform mangrove forests of arbitrary depth. J Mar Syst 74:603–622. https://doi.org/10.1016/j.jmarsys.2008.05.004
Ward RD, Friess DA, Day RH, MacKenzie RA (2016) Impacts of climate change on mangrove ecosystems: a region by region overview. Ecosyst Health Sustain 2(4):e01211. https://doi.org/10.1002/ehs2.1211
Wassilieff M (2006) Estuaries – plants of the estuary. http://www.TeAra.govt.nz/en/interactive/4623/vegetation-profile. Accessed 1 May 2017
Watson JG (1928) Mangrove forests of the Malay Peninsula. Malayan forest records No. 6. Federated Malay States Government, Kuala Lumpur
Webb EL, Friess DA, Krauss KW, Cahoon DR, Guntenspergen GR, Phelps J (2013) A global standard for monitoring coastal wetland vulnerability to accelerated sea-level rise. Nat Clim Chang 3 (5):458–465. https://doi.org/10.1038/nclimate1756
Widdows J, Blauw A, Heip CHR, Herman PMJ, Lucas CH, Middelburg JJ, Schmidt S, Brinsley MD, Twisk F, Verbeek H (2004) Role of physical and biological processes in sediment dynamics of a tidal flat in Westerschelde Estuary, SW Netherlands. Mar Ecol Prog Ser 274:41–56. https://doi.org/10.3354/meps274041
Wildlands (2014) Monitoring of banded rail and other avifauna before and after mangrove clearance at Whangamata Harbour – Annual report (Mar 2013–2014)
Willemsen PWJM, Horstman EM, Borsje BW, Friess DA, Dohmen-Janssen CM (2016) Sensitivity of the sediment trapping capacity of an estuarine mangrove forest. Geomorphology 273:189–201. https://doi.org/10.1016/j.geomorph.2016.07.038
Winterwerp JC, Borst WG, de Vries MB (2005) Pilot study on the erosion and rehabilitation of a mangrove mud coast. J Coast Res 21(2):223–230. https://doi.org/10.2112/03-832a.1
Winterwerp JC, Erftemeijer PLA, Suryadiputra N, Eijk P, Zhang L (2013) Defining eco-morphodynamic requirements for rehabilitating eroding mangrove-mud coasts. Wetlands 33(3):515–526. https://doi.org/10.1007/s13157-013-0409-x
Wolanski E (1995) Transport of sediment in mangrove swamps. Hydrobiologia 295(1–3):31–42
Wolanski E (2007) Tidal wetlands. In: Wolanski E (ed) Estuarine ecohydrology. Elsevier, Amsterdam, pp 71–89. https://doi.org/10.1016/B978-044453066-0.50005-2
Wolanski E, Moore K, Spagnol S, D'Adamo N, Pattiaratchi C (2001) Rapid, human-induced siltation of the macro-tidal ord river estuary, Western Australia. Estuar Coast Shelf Sci 53(5):717–732. https://doi.org/10.1006/ecss.2001.0799
Woodroffe CD (1982) Litter production and decomposition in the New Zealand mangrove, Avicennia marina var. resinifera. N Z J Mar Freshw Res 16(2):179–188. https://doi.org/10.1080/00288330.1982.9515961
Woodroffe CD (1992) Mangrove sediments and geomorphology. In: Robertson AI, Alongi DM (eds) Tropical mangrove ecosystems. American Geophysical Union, Washington, DC, pp 7–41
Young BM, Harvey EL (1996) A spatial analysis of the relationship between mangrove (*Avicennia marina* var. australasica) physiognomy and sediment accretion in the Hauraki Plains, New Zealand. Estuar Coast Shelf Sci 42(2):231–246. https://doi.org/10.1006/ecss.1996.0017

Mangrove Forests of the Persian Gulf and the Gulf of Oman 3

Alireza Salehipour Milani

Abstract

Mangrove forests are found as isolated units of varying length and width along the coast line of the Persian Gulf and the Gulf of Oman. These forests are scattered within the intertidal and are among emergent plant communities occurring along land-sea margins in the Islamic Republic of Iran, Saudi Arabia, Bahrain, Qatar, United Arab Emirates, Oman and a small part of Pakistan in Govatr Bay. *Avicennia marina* and *Rhizophora* spp. are found in this region. Recently, as the effect of industrialization and development, as well as global environmental changes, concern about status, health, damage, disturbance and reducing mangrove area are a big challenge both for ecologists and engineers in these regions. Hence, producing mangrove map distributions and monitoring their change during the time is necessary. Mangrove habitat maps have been used for three general management applications: resource inventory, change detection, and the selection and inventory of aquaculture sites. This chapter quantifies mangrove forest cover and monitors their changes in the Persian Gulf and the Gulf of Oman from 1977 to 2017, using Landsat Thematic Mapper satellite imagery. The mangrove distribution maps are created by using remote sensing (RS) and Geographical Information Systems (GIS). An image processing technique entitled normalized differential vegetation index (NDVI) was used to detect mangrove forests in six countries located within the border study area. Even though the results show negative effects of human activity in some regions, overall, mangrove forests increased during 1977–2017. This increase in mangrove forests in the Persian Gulf and the Gulf of Oman represent sustainable development and management for mangrove ecosystems in this region.

Keywords

Mangrove forest · The Persian Gulf and The Gulf of Oman · Coastal Management · Valnerability

3.1 Introduction

Coastal environments and the diverse ecosystems they support have attracted considerable attention within the natural resource management community in recent years. This is largely attributed to the rapid increase in coastal activities such as maritime commerce, rapid expansion in the population of coastal communities, and the worldwide growth in the exploration of near/on-shore petroleum mineral resources in the last few decades. The impact of these activities on coastal wetlands in many parts of the world is poorly understood, yet wetlands are among the most highly altered coastal ecosystems worldwide. Coastal wetlands are renowned for their rich fauna and flora biodiversity. They support unique plant ecosystems, including Mangrove forest stands. Mangroves are woody plants that are widely distributed in estuaries and intertidal zones in the tropics and subtropics regions (Field et al. 1998; Sherrod and McMillan 1985).

There are many benefits to mangrove communities that include serving as breeding grounds to many fish, shellfish, birds, and other wildlife; their physical stability helps to prevent shoreline erosion, providing protection to inland areas during hurricanes and tidal waves (Badola and Hussain 2005; Ewel et al. 1998; Quarto 2005; Stutzenbaker 1999).

Mangrove are largely confined to regions between 30° north and south of the equator. Notable extension beyond this area includes the subtropics and, in some cases, even warm temperate zones such as Bermuda (32° 20′N) and Japan (31° 22′N) to the north, and New Zealand (38° 03′S) and Australia (38° 45′S) to the south (Spalding et al. 1997; Gao 1999).

A. S. Milani (✉)
Physical Geography Department, Earth Science Faculty, Shahid Beheshti University, Tehran, Islamic Republic of Iran
e-mail: Ar_salehipour@sbu.ac.ir

C. Makowski, C. W. Finkl (eds.), *Threats to Mangrove Forests*, Coastal Research Library 25,
https://doi.org/10.1007/978-3-319-73016-5_3

Coastal habitats across the world are under heavy population and development pressures, and are subjected to frequent storms. The continued decline of the forests is caused by conversion to agriculture, aquaculture, tourism, urban development and overexploitation (Alongi 2002; Giri et al. 2008). About 35% of mangroves were lost from 1980 to 2000 (MA 2005), and the forests have been declining at a faster rate than inland tropical forests and coral reefs (Duke et al. 2007). Relative sea-level rise could be the greatest threat to mangroves (Gilman et al. 2008). Predictions suggest that 30–40% of coastal wetlands (IPCC 2007) and 100% of mangrove forests (Duke et al. 2007) could be lost in the next 100 years if the present rate of loss continues.

As a consequence, important ecosystem goods and services (e.g. Natural barrier, carbon sequestration, biodiversity) provided by mangrove forests will be diminished or lost (Duke et al. 2007).

The world mangrove experts are of the opinion that the long-term survival of mangroves is at great risk due to fragmentation of the habitats and that the services offered by the mangroves may likely be totally lost within 100 years (Duke et al. 2007). Mangrove forests continue to disappear all over the world. They were estimated to cover 18.1 million km^2 worldwide (Spalding et al. 1997) but a more recent estimate indicates that the figure may now be below 15 million km^2 (www.fao.org/forestry/mangroves). With human populations increasing at an astonishing rate in coastal areas, it is crucial that additional and improved management practices and restoration programs are implemented throughout all mangrove-inhabited nations. According mangrove forest decline, Mapping and monitoring mangrove forests is an urgent need for many countries. Mangrove forests represent important ecological resources for the Persian Gulf and the Gulf of Oman countries, as well as worldwide. Large industrial and residential developments have happened in these regions during the last 40 years, affecting the mangrove communities. One of the earliest documented impacts on mangroves was the oil spill resulting from a pipeline breaking near the north-west shore of Tarut bay (Saudi Arabia) during a storm in April 1970 (Spooner 1970). Mangrove loss over time is mainly due to land conversion for shrimp cultivation and for infrastructure development, such as roads and harbors, to the excessive utilization for fuelwood and fodder, and over-grazing by camels.

This research investigates in more details the historical changes in the mangrove communities in these areas including Iran, Saudi Arabia, Bahrain, Qatar, United Arab Emirates, Oman Govatre bay in Pakistan. Using Remote Sensing and landsat image processing, providing an update on their current status by focusing on the 1987–2017 period and looking for potential causes and consequences of mangrove change since 1977.

3.1.1 Application of Remote Sensing for Mangrove Forests

The earth observation satellite data (such as Landsat) is useful for change detection applications. The distribution and abundance of mangrove in different regions of the world have been assessed with a variety of techniques. The local variability of studies spans all continents. Several studies have been carried out to investigate and compare the suitability of various classification algorithms for the spectral separation of mangrove. The main motivation for the characterization of these ecosystems are to monitor and manage them. Generally, all the remote sensing application in mangrove ecosystems can be categorized into three broad areas. These broad categories have been identified as resource inventory; change detection; and selection and inventory of aquaculture sites (Green et al. 2000; Vaiphasa et al. 2006). Change detection is a powerful tool to visualize, measure, and better to understand a trend in mangrove ecosystems. It enables the evaluation changes over a long period of time as well as the identification of sudden changes due to natural or dramatic anthropogenic impacts (for example: tsunami destruction or conversion to shrimp farms). Thus distribution, condition, and increase or decrease were the measured features used in the change-detection applications of mangrove.

Remote Sensing and GIS provide quantitative information on understanding the spatial distribution of mangrove forests (Sremongkontip et al. 2000). Remote sensing technology has been applied in various ways to characterize mangrove ecosystems. Some of the documented applications include mapping the area extent, detecting individual species, and providing estimates of structure and parameters such as leaf area, canopy height, and biomass (Heumann 2011). Remotely sensed data is one of the best source of information that can show the location of all areas that have been deforested or degraded or still healthy. Moreover, Landsat Thematic Mapper data can be used to define the degree of degradation of the mangrove forest (Giri 2008). In the field of vegetation mapping, especially the most commonly applied sensors include Landsat (mainly TM and ETM+), SPOT, MODIS, NOAA–AVHRR, IKONOS and QuickBird. But the most important satellite that used to vegetation indexes was Landsat sensor imagery.

Over the past decades, major steps have been taken toward estimating the amount of vegetation cover using Landsat sensor imagery (Hansen et al. 2002).

The Landsat might have the longest history and widest use for monitoring the earth from space. Since the first Landsat satellite was launched in 1972, a series of more sophisticated multispectral imaging sensors, named TM—Thematic Mapper, have been added ranging from Landsats 4 (1982), 5 (1984), 6 (1993, launch failed) to 7 (1999) (Enhanced Thematic Mapper Plus, ETM+). The Landsat TM and ETM + imaging sensors have archived millions of images with a nearly continuous record of global land surface data since its inception. Landsat 8 is an American Earth observation satellite launched on February 11, 2013. This sensor provides update satellite images and make possible detect and monitoring mangrove changes up to now. Because of this advantages Landsat sensor imagery used as one of the main resources in collecting spatial information for management and planning of land and natural ecosystems, digital satellite data have been greatly developed in the world over the past decade (Darvishsefat 1997). Several studies have been carried out worldwide to develop mangrove forest cover maps using various satellite images (Jensen et al. 1996; Kairo et al. 2002; Aschbacher et al. 1995; Vits and Tack 1995; Rasolofoharinoro et al. 1998; Ramsey and Jensen 1996; Green et al. 1998; Dahdouh-Guebas 2005; Tong et al. 2004; Kovacs et al. 2005).

3.1.1.1 Remote Sensing Methods in Mangrove Forest Detection

Spectral vegetation indices are frequently used to estimate vegetation biophysical/biochemical characteristics. In general, they have been proposed to reduce spectral effects caused by external factors such as the atmosphere and the soil background. A Vegetation Index (VI) is a spectral transformation of two or more bands designed to enhance the contribution of vegetation properties and allow reliable spatial and temporal inter-comparisons of terrestrial photosynthetic activity and canopy structural variations (Huete et al. 2000). There are many Vegetation Indices (VIs), with many being functionally equivalent. Many of the indices make use of the inverse relationship between red and near-infra red reflectance associated with healthy green vegetation. Since the 1960s scientists have used satellite remote sensing to monitor fluctuation in vegetation at the Earth's surface.

Vegetation Indices (VIs) have been historically classified based on a range of attributes, including the number of spectral bands (2 or greater than 2); the method of calculations (ratio or orthogonal), depending on the required objective; or by their historical development (classified as first-generation VIs or second-generation VIs) (Bannari et al. 1995).

For predicting leaf area index, five widely used vegetation indices were investigated. Narrow band vegetation indices involving all possible two band combinations of RVI, NDVI, PVI, TSAVI, and SAVI2 were computed.

The Perpendicular Vegetation Index (PVI)

The Perpendicular Vegetation Index (PVI) of Richardson and Wiegand (1977), used the red and near infrared bands to calculate the perpendicular distance between the vegetation spot on the NIR-Red scatterplot and the soil line. Since vegetation has higher near-infrared and lower red reflectance than the underlying soil, the vegetation spot will be on the top left corner of the scatterplot. As vegetation is increasing in density, the vegetation spot will be moving further towards the top left, away from the soil line. The PVI equation is expressed as the function of the slope and intercept of the vegetation images of the R and NIR band, and the soil images of the R and NIR band (Thiam and Eastmen 1999).

$$PVI = \sqrt{(\rho Gir, s - \rho Pir)^2 + (\rho Gr, s - \rho Pr)^2}$$

where ρr, s and ρir, s is reflectance of soil background in R and NIR bands, respectively; and ρr and ρir = reflectance of vegetation in R and NIR bands, respectively. PVI is determined using the distance between the intersection point (ρGir,s and ρGr,s) and the vegetation image pixel coordinate (ρPir and ρPr) by the Pythagorean Theorem.

Soil-Adjusted Vegetation Index (SAVI)

The weakness of the PVI is the assumption that there will be only one type of soil beneath the vegetation. However, this is not always the case, as there are environments where a mixture of soil types (a mixture of soil and rocks for example) can be found within a very small area. Huete (1988) proposed the Soil-Adjusted Vegetation Index (SAVI) to deal with this problem. SAVI is a hybrid between a ratio index (NDVI) and a perpendicular index (PVI). Its formula resembles the former:

$$SAVI = [(NIR - Red)/(NIR + Red + L))^* \ (1 + L)]$$

L is a correction factor and its value is dependent on the vegetation cover. For total vegetation cover it receives a value of zero, effectively turning SAVI into NDVI. For very low vegetation cover, it receives the value of 1. Huete (1988) suggested that the value of 0.5 is used when vegetation cover is unknown, as 0.5 represents intermediate vegetation cover. Other indices, such as the Transformed Soil-Adjusted Vegetation Index (TSAVI) (Baret et al. 1989) and Modified Soil-Adjusted Vegetation Index (MSAVI; Qi et al. 1994), followed similar rationale, but proposed different adjustment factors, that appeared to perform better than SAVI in certain cases. Additionally, specific vegetation indices have been proposed in the past, using high spectral resolution data (narrow bands) and looking specifically at chlorophyll and plant vigor. These include the Photochemical Reflectance Index (PRI) and the Normalized Pigment Chlorophyll Ratio Index (NPCI; Peñuelas et al. 1994), as well as the canopy chlorophyll content index (CCCI; Barnes et al. 2000).

The Normalized Difference Water Index

Following the same rationale as the NDVI, the Normalized Difference Water Index, employs the near-infrared band and a band in the short-wave infrared (SWIR) (Gao 1996). Instead of using the red band, the reflectance at which is affected by chlorophyll, a short-wave infrared band in the region between 1500 and 1750 nm is used instead, a region where water has high absorption. The near-infrared band is the same as with NDVI, as water does not absorb in this region of the electromagnetic spectrum.

The NDWI index is expressed with the following equation:

$$[\mathrm{NDWI} = (\mathrm{NIR} - \mathrm{SWIR})/(\mathrm{NIR} + \mathrm{SWIR})]$$

The black and white NDWI image on the right above represents the NDWI values for each pixel. As with the NDVI image, dark pixels have low values, white have high and gray have average. Notice that in the NDWI image the sea has high values, whereas for the NDVI image it appears dark. In general, it appears that all agricultural fields give high NDWI values, which suggests that the fields are adequality irrigated. Barren fields, appear dark in the NDWI image, which shows that those fields are drier, since they do not receive any irrigation.

Mangrove Plant Reflectance and Normalized Difference Vegetation Index (NDVI)

Normalized Difference Vegetation Index (NDVI) is the most widely used vegetation index to distinguish healthy vegetation from others or from non-vegetated areas (Manandhar et al. 2009). Reflectance is the ratio of energy that is reflected from an object to the energy incident on the object. Spectral reflectance of a crop differs considerably in the near infrared region (λ = 700–1300 nm) and in the visible red range (λ = 550–700 nm) of the electromagnetic spectrum (Kumar and Silva 1973). Healthy plants absorb most of the visible light and reflect the large amount of the far-red and near-infrared light (Fluor Cam 2014). Plants generally have low reflectance in the blue and red portion of the spectrum because of chlorophyll absorption, with a slightly higher reflectance in the green, so plants appear green to our eyes. Near infrared radiant energy is strongly reflected from the plant surface and the amount of this reflectance is determined by the properties of the leaf tissues: their cellular structure and the air cell wall-protoplasm-chloroplast interfaces (Kumar and Silva 1973). These anatomical characteristics are affected by environmental factors such as soil moisture, nutrient status, soil salinity, and leaf stage (Ma et al. 2001). The most commonly used is the Normalized Difference Vegetation Index (NDVI). It relies upon differential reflectance of the mangrove canopy (which is always green) in the red wavelength (here the response is mostly determined by the absorption band by the chlorophyll) and in the NIR wavelength, where the response is the result of scattering determined by the cuticles of leaves and the density of the cover.

$$NDVI = \frac{NIR - VIS(red)}{IR + VIS(red)}$$

Couple of studies have been used NDVI index in to mangrove forests in Bangladesh (Emch and Peterson 2006) Taiwan (Tsai-Ming and Hui 2009), Mexico (Arturo et al. 2010), and Alabama (Rodgers et al. 2009).

This study attempted to assess the distribution and changes in the area of forests in the Persian Gulf and the Gulf of Oman using an integrative method combining NDVI. The results could be used as a model for future planning in landscape management and planning of mangrove forests in the Persian Gulf and the Gulf of Oman.

3.2 Discussion

3.2.1 Study Area

The study areas are the Persian Gulf and the Gulf of Oman. Mangrove forests are found as isolated units of varying length and width along the coast lines of this regions. These forests scattered in intertidal and emergent plant communities that occur along the land-sea margins Iran, Saudi Arabia, Bahrain, Qatar, United Arab Emirates, Oman and small parts of Pakistan in Govatre Bay.

The Persian Gulf is inland sea of some 251,000 km^2 (96,912 mi^2) is connected to the Gulf of Oman in the east by the Strait of Hormuz; and its western end is marked by the major river delta of the Arvand River, which carries the waters of the Euphrates and the Tigris and karoon. Its length is 989 km (615 mi), with Iran covering most of the northern coast and Saudi Arabia most of the southern coast. The Persian Gulf is about 56 km (35 mi) wide at its narrowest, in the Strait of Hormuz. The waters are overall very shallow, with a maximum depth of 90 m (295 ft) and an average depth of 50 m (164 ft). Countries with a coastline on the Persian Gulf are (clockwise, from the north): Iran, Oman's exclave Musandam; the United Arab Emirates; Saudi Arabia, Qatar, on a peninsula off the Saudi coast; Bahrain, on an island; Kuwait; and Iraq in the northwest. Various small islands also lie within the Persian Gulf.

The Persian Gulf presents stressful environmental conditions: extreme heat (>30 °C) in summer and cold (<10 °C) in winter (Khosravi 1992; Behrouzirad 1998; Zahed et al. 2010). Annual rainfall is less than 200 mm in north parts of the Persian Gulf and about 78 mm in Qatar in South of The Persian Gulf (Al Mamoon et al. 2016). The salinity of Persian Gulf is also high: between 38 and 50 g/L in mangrove area. Petroleum hydrocarbons are detectable in nearly all parts of the region. It is believed that *Avicenna marina* tolerance for temperature and salinity changes contributes to its being the dominant specie in Iranian mangrove forests and Persian Gulf.

The Gulf of Oman, are located in northwest of the Arabian Sea, between the eastern portion (Oman) of the Arabian Peninsula to the southwest and Iran to the north. The gulf is 200 mi (320 km) wide between Cape al-Ḥadd in Oman and Govatre Bay on the Pakistan–Iran border. It is 350 mi (560 km) long and connects with the Persian Gulf to the northwest through the Strait of Hormuz. Of the 45-mangrove species in the world, *Avicennia marina* is the most commonly found species in Oman (Nizwanet 2014) (Fig. 3.1).

3.2.2 Mangrove Forest Distributions

3.2.2.1 Iran

Mangrove forests in Iran are located in the coastal areas along intertidal zone as discrete or connected small and large communities. These forests are conceived as the western most mangrove forests in Southwest Asia. the highest mangrove distribution and extend of mangrove forests are find along 1250 km in Coastal Zone of Iran. Mangrove forests in Iran begin from the easternmost part of the border in the Oman Sea (Govatre Bay) and continues to the western parts of the Persian Gulf ending in Poozeh-Mashe area (Bushehr Province).

Iranian mangrove forests occur between 25°, 11′ and 27°, 52′, in the north part of the Persian Gulf and Oman Sea. The areas of maximum area Iranian mangrove forests are located between Khamir Port and northwest of Qeshm Island. The entire area of the Khouran Straits (including Qeshm Island and KHouran Rivers mouth) is protected under several international convention and reserves, Ramsar Site in 1975, UNESCO (MAB) Biosphere Reserve in 1976 and "Important Bird Area" (identification by Bird Life International in 1994) (Mahdavi 2001; Scott 1995; Shariatnegad 2005).

- Bushehr province
  Zone 1: In the Poozeh-Mashe area, vicinity the Mel-e-Gonze village and near of the Omolgarm Island, occurring south of the Bordekhoon area. Zone.
  Zone 2: In the Bardestan estuary, adjacent the Dayer harbor.
  Zone 3: In the Nayband Bay, occurring in the Assaluyeh and Basatin estuaries.
- Hormuzgan Province
  Zone 4: Within the confines of the Khamir estuaries and mouth of the Mehran River to the Persian Gulf.

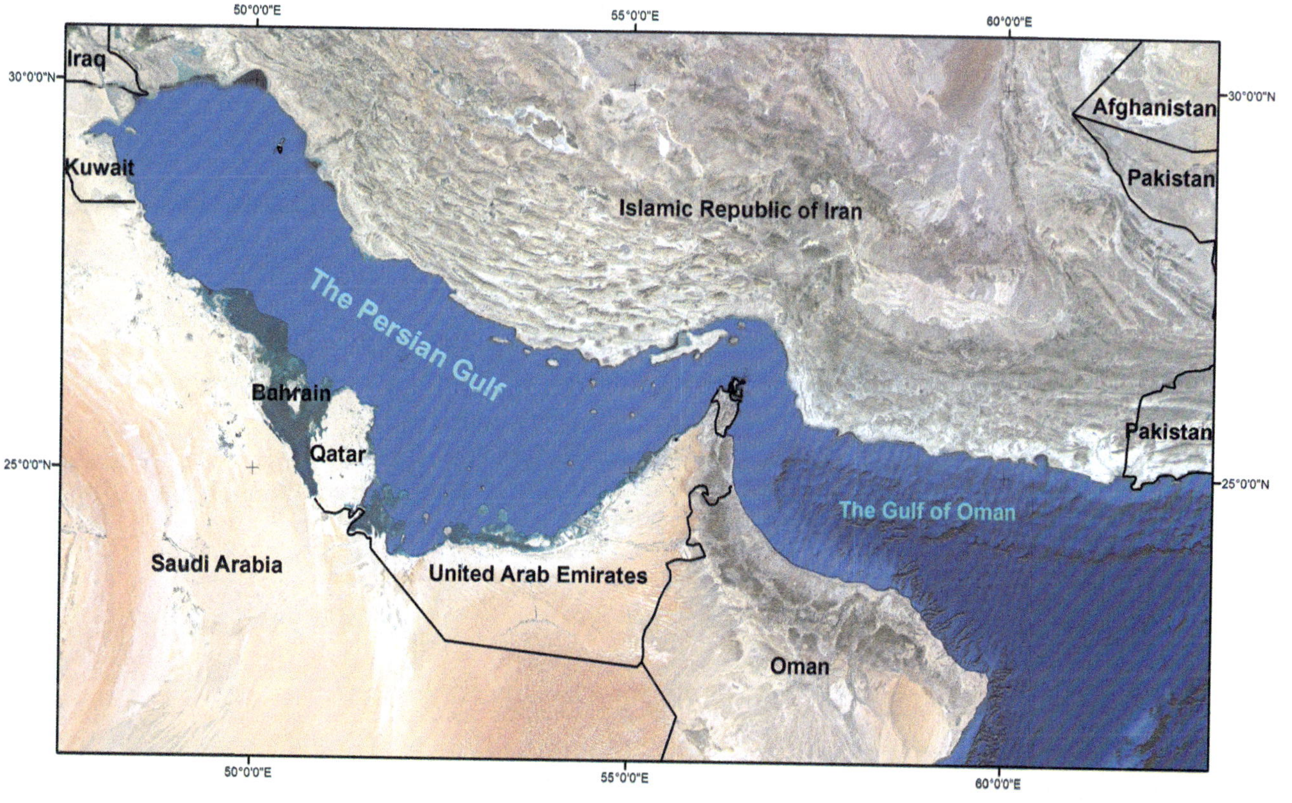

Fig. 3.1 The Persian Gulf and the Gulf of Oman study area (International Border in all maps do not have much accuracy)

Zone 5: Within the northwest estuaries of the Qeshm Island, in Khoor-e-Khooran and sandy Islands opposite villages of Tabl and Laft to Gooran (Harra Protected Area).

Zone 6: Close to the east of the Bandar Abbas, adjacent to the Kouleghan areas occurring within Jalabi and Hassan- Langi region and mouth of the Shoor River.

Zone 7: In the Tiyab and the Kolahi regions within estuaries of the Moshdar, Behine, Kargan, and the Minab.

Zone 8: In Sirik region, between the Gaz River and the Harra River (or Hiva River) occurring along 12 Km within estuaries of the Pachvar, Nakhl-e-Ziyarat, Garandhoo, Ziyarat, Ganari and the Kartan.

Zone 9: In Jask region, within the Gabrik and Jegin estuaries, the mouth of Shahr-e-Now River (in coast of the Lash village) and estuaries of the Yekdar, Surgolom and lagoons of the Kashi River and in part of the west of Jask, in area of ancient Jask city.

- Sistan and Balouchestan province

Zone 10: Chabahar Bay, East of Chabahar bay, Estuary of Shor River.

Zone 11: Within Govatr Bay, occurring in delta of the Bahookalat River – part of Gandoo Protected Area and Bahoo wetland (Behrouzi 1996, Danehkar 1998, 2001; Khosravi 1992) (Fig. 3.2).

Avicenna marina is the dominant specie within Iranian mangrove forests, as *Rhizophora mucronata* is found only in the Sirik region (Khosravi 1992; Danehkar 1996, 2001), *Rhizophora mucronata*. Macrunata trees occur in estuaries within the Sirik region (center part of the Tiyab Tiyab region) along the coasts of Garandhoo, Hookze, Badgard, Ganari, Ziyarat, Pachvar, Kartan, and Sorgolom villages (Danehkar 1994, 2001).

3.2.2.2 Pakistan

- Zone 12. During this research, mangrove forests in the Lura Region of Pakistan's Govatre Bay are also studied as mangroves of the Gulf of Oman Boundary.

3.2.2.3 Oman

Oman has a coastline stretch of 3165 km which had been perceived as entirely covered by mangroves long time ago. In Oman, Khawrs and mangroves are considered very special environments since they can be productive and valuable source for fish breeding and nursery areas in addition to their natural beauty for recreation. Mangrove vegetation is spread sporadically in the coastal areas of the country. Mangrove forests are scattered in Northern Batinah, Muscat, Eastern Sharqiyah, Mahawt Island and Salalah. It now covers

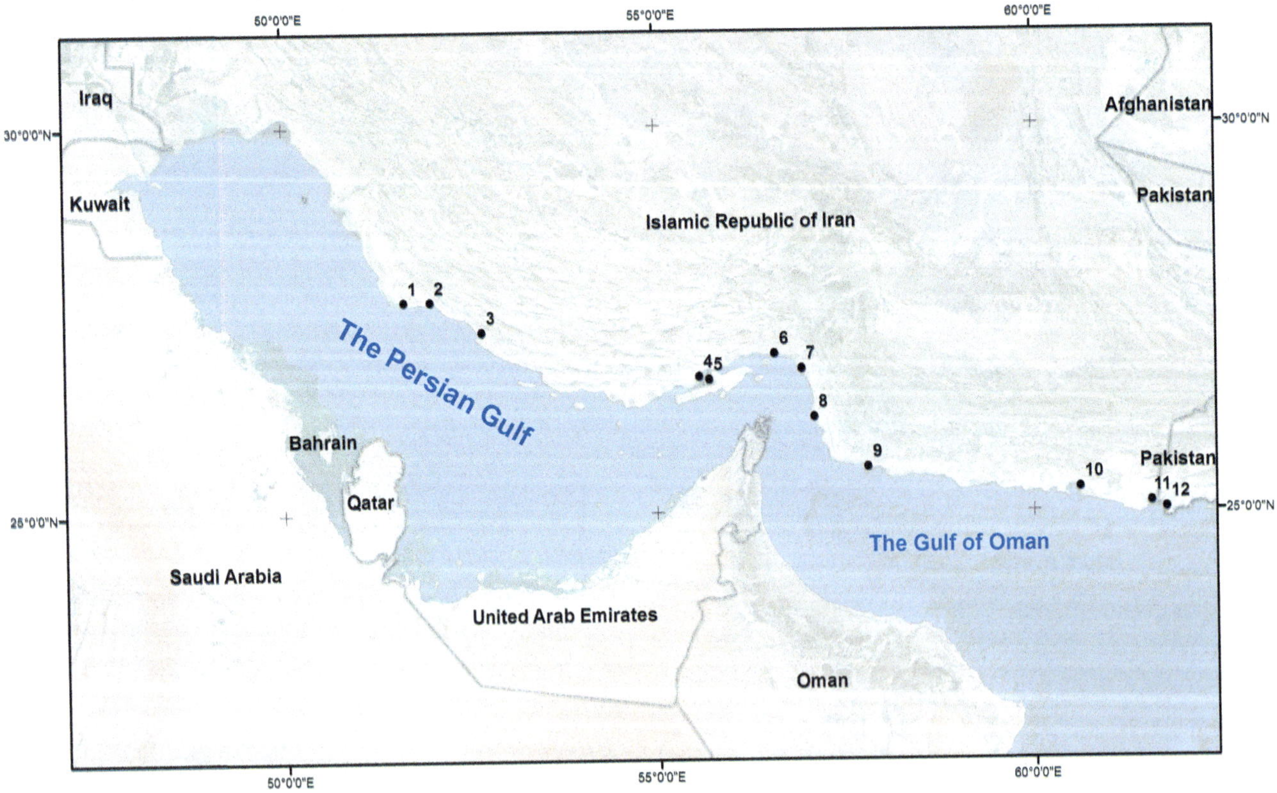

Fig. 3.2 Iran and Pakistan mangrove forest locations in the Persian Gulf and The Gulf of Oman

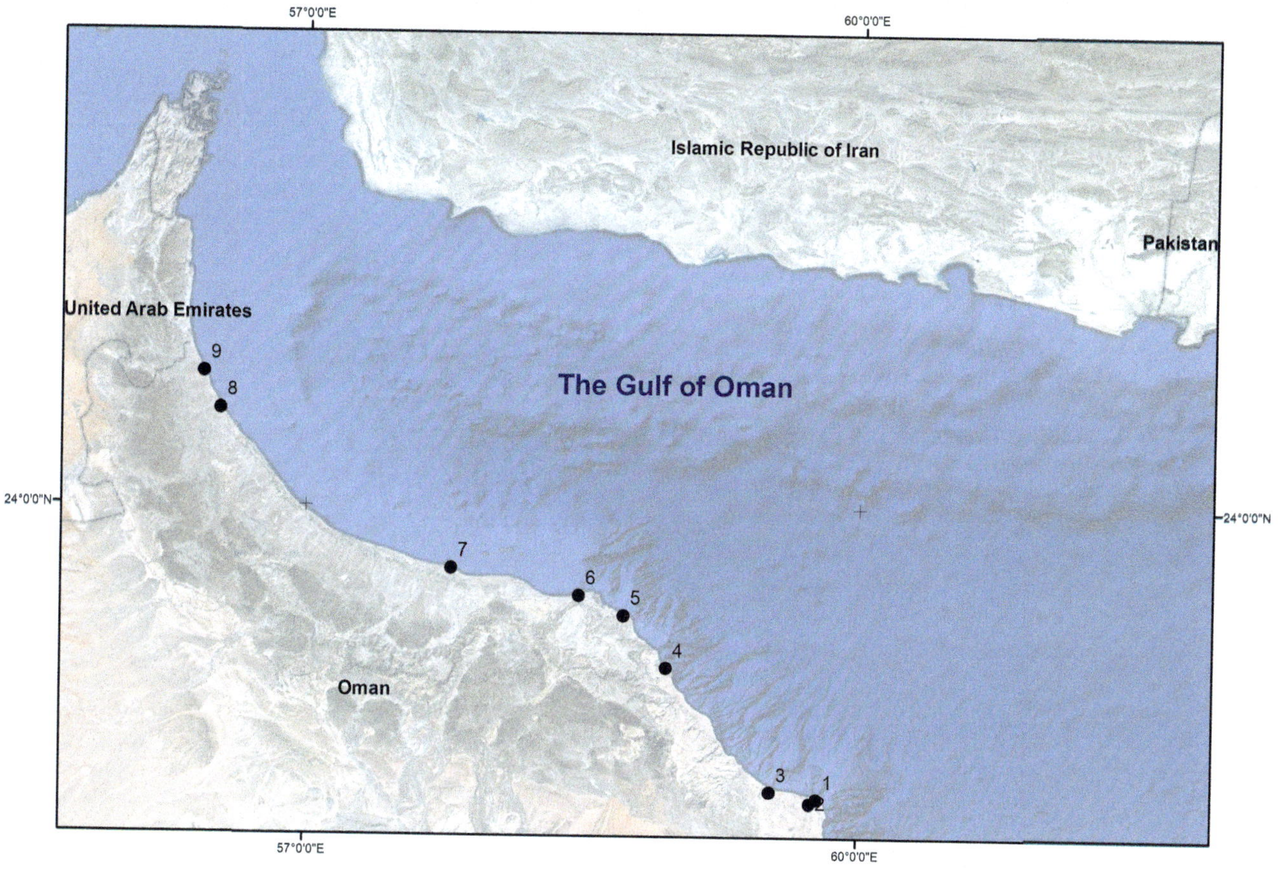

Fig. 3.3 Oman mangrove forest locations in The Gulf of Oman

a total of 1100 ha not only in The Gulf of Oman but also in Arabian Sea. There are few outstanding islands in Oman which include Dimaniyat, Masirah and KuriaMurai (Hallniyat). Except for Dimaniyat, all other islands are still in the proposal state to become protected areas (Directorate-General of Nature Conservation 2010). We found nine regions and enclose soundings that mangrove forest grew in it. All regions are located in intertidal channel from east to west including Zone 1 and 2 in Ras Al Hadd, 3; Sur, 4: Daghmar, 5: Khayran, 6: Dimaniyat, 7: Barka, 8: Liwa and 9: Shinas (Fig. 3.3).

3.2.2.4 United Arab Emirate

Mangrove forest area in the Emirates have natural and planted mangroves, which provide a rich natural habitat and safe breeding ground for several fish species, sea snakes, turtles and commercially important shrimp, snapper, grunt fish and sea bream. Sea birds also find safe nesting and egg laying areas in the vast mangrove forests. The average height of the trees found in Abu Dhabi is 3–5 m, and can reach up to a maximum of 8 m in some areas. Grey mangrove (*Avicennia marina*) is the only mangrove species that grows widely in the Emirates. However, the Environment Agency-Abu Dhabi (EAD) has succeeded in reintroducing the mangrove species (*Rhizophora mucronata*) which once grew in the Emirates and became extinct due to overexploitation. (EAD) has been able to cultivate a large number of these seedlings in the waters of Ras Ghanada Island after 100 years of its extinction. Besides, Mangrove National Park Located along the Eastern Ring Road, the Mangrove National Park, which opened its doors to the public on October 1, 2014, is the closest mangrove forest to Abu Dhabi City. there are vast areas of mangrove forests in Emirate, only kalba mangrove forests regions are located in the Gulf of Oman Coast (Zone 1), Other mangroves located in The Persian Gulf Coast including: Zone 2: Qurm, Zone 3 Mina Al Arab, Zone 4: Oum Al Quvain, Zone 5: Ajman,Zone 6: Ras Al Khor (Dubai), Zone 7: Ghanadah, Zone 8: Khorfarideh, Zone 9: Al Dhabeia, Zone 10: Al Aryam Island, 11: Al Dhabeia, Zone 12: Abu Al Abyadh Island,Zone13: Bu Tinah, Zone 14:L Al-Rawis and some little area enclosing these regions in The Emirate (Fig. 3.4).

3.2.2.5 Qatar

Qatar is home to the *Avicennia Marina* species; it is known as the grey or white mangrove trees, with the largest eight forests located in the east coast of the country. The oldest and largest mangroves can be found at Al Dhakira and Al

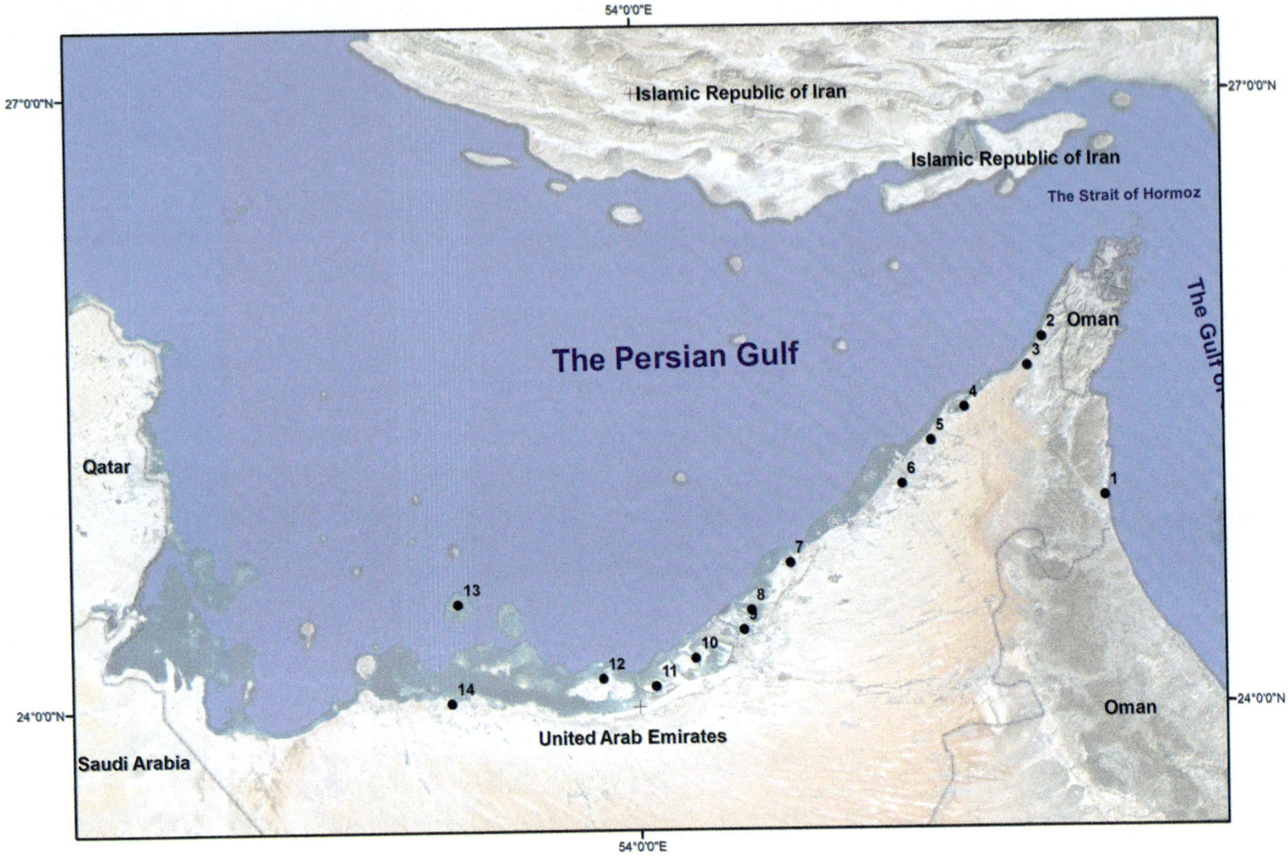

Fig. 3.4 United Arab Emirate mangrove forest locations in The Persian Gulf and The Gulf of Oman

Khor. A well-developed mangrove swamp stands around Al Dhakira and large population of resident and migrant birds reside in this area. *Avicennia marina*, the only mangrove species exists in Qatar. Along the northern and eastern coast of Qatar, mangrove vegetation was sparse and mixed with the salt marsh (Sabkha) frontier vegetation (Although the government have starting a replanting project around the country, the mangrove lake at Al Wakra was recently uprooted for development. In a country where the harsh desert conditions limit the vegetation growth, mangroves in Qatar provide a heaven for birds, fish and mammals. Recent studies have shown that *Avicennia Marina* populations have the ability to adapt to the varying weather along the Qatar coastline through the evolution of genetic variations in the different mangrove forests (Khader 2015) (Fig. 3.5).

3.2.2.6 Bahrain

There are three areas in Bahrain (so far) designated as Protected Areas; Mangroves at Ras Sanad (Tubli Bay), Hawar Islands and Al-Areen Wildlife Park. Tubli Bay is an inshore coastal area situated in the north-east of Bahrain. Manama, the capital city, lies north of the Bay. Tubli Bay is characterized by its unique ecology. It is one of the only sheltered, low-energy areas in Bahrain and rich ecologically. It provides an important coastal habitat. The main plant species that Tubli Bay is mostly recognized for are mangrove trees. The Bay (and closely surrounding areas) is the only place in Bahrain where these ecologically important species still survive. Only one type of mangrove species exists and that is known as the Black Mangrove; *Avicennia marina*. The protected mangrove reserve at Ras Sanad has a tree density of 100%. The highest mangrove tree was approximately 3.5 m tall. (Fig. 3.5).

3.2.2.7 Saudi Arabia

Saudi Arabia mangrove ecosystems along Persian Gulf are found in tiny patches in the North, on Jazeerat Qurma (which means Mangrove Island) and in the Damman area (Tarut Bay), where they form up well developed communities consisting of *Avicennia marina* (Fao 2005).

Tarut Bay are located in Northwestern of Bahrain (Fig. 3.5) is characterized by its heavy production and was reported (Czudek 2006) as the most important site on the Saudi area of The Persian Gulf Coast for wintering and migrating waders and other water birds, with a total of 58,000 water birds. The mangrove in Tarut Bay consists mainly *Avicennia marina*, which are capable of enduring extreme environmental conditions. These species are present

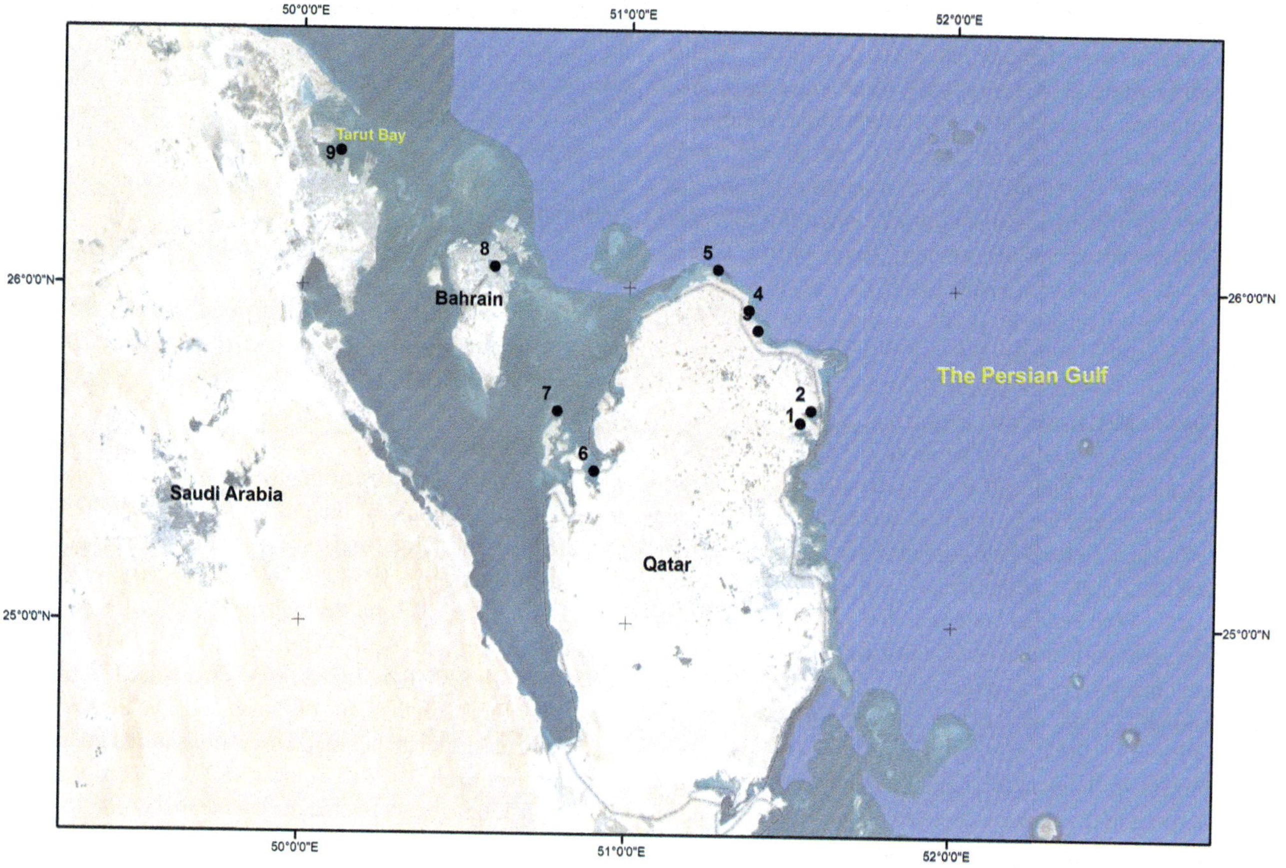

Fig. 3.5 Qatar mangrove forest locations 1: Al khor,2: Al Thakira,3: Sidriyat,4: Fuwayrit,5: Al Mafjar,6: Zekret and in Bahrain Country,6: Hawar Island and 7: Tubli Bay and 9: Tarut Bay in Saudi Arabia

in Tarut Bay in relatively higher density when compared to elsewhere in the world.

There are no native mangroves in Kuwait; several experimental plantations have been established since 1968 to study the adaptability and growth performance of *Avicennia marina*. However, the first relevant success was reached only in 2001 and mangroves are currently present in the localities of Shuwaikh, Doha, Sulaibhikhat and Sabiya (Samira 2004).

3.2.3 Mangrove Forest Threats in Study Area

The mangrove forests of study area in the Persian Gulf and the Gulf of Oman coast are remarkably tolerant to extreme environmental conditions and are highly productive (Field et al. 1998). Besides pollution through riverine inputs from adjacent countries (Iran, Iraq, Kuwait, Saudi Arabia, United Arab Emirate and the Emirates of Bahrain, Qatar and Oman), the Persian Gulf has been exposed to various additional contaminants as a consequence of marine transports, marine accidents in recent years. Industrial areas situated around the main ports on the coast are the main potential pollution sources. In the Persian Gulf mangroves have been principally affected by the large oil spill from the Persian Gulf War (1991) and even though they have mostly recovered from this event, oil pollution still represents one of the main threats. Camel grazing has been reported as another relevant threat, while shrimp aquaculture represents one of the emerging threats to this ecosystem.

In Iran, mangrove loss in some years was mainly due to land conversion for shrimp cultivation and for infrastructure development, such as roads and harbors, to the excessive utilization for fuelwood and fodder, and over-grazing by camels. On the other hand, some mangrove reforestation has been undertaken in the early 1990s (1991 and 1993) on around 400 ha in the coastal area of the Persian Gulf and of the Oman Sea for rehabilitation purposes.

Conversely many areas in emirates are increasingly threatened by oil and other kind of pollution. Landfill associated with urban and industrial development also threatened mangroves in many areas in United Arab Emirates.

In Saudi Arabia, mangroves have been principally affected by the large oil spill from the Persian Gulf War (1991), and even though they have mostly recovered from this event, oil pollution still represents one of the main threats in the country. Camel grazing has been reported as another relevant threat, especially at Khor Itwad in Saudi Arabia. Significant cutting has been reported in different sites, leading to reduction in tree density, abundance of multi-stemmed trees, and passages to the inner parts of the forests for the camels. Mangrove health is also compromised by the modifications of the coastal area topography, desertification and infilling caused by infrastructure development. Consequently, even if the area change over the last two decades may have been relatively small, several sites have been severely degraded and trees are suffering of a high mortality rate (Böer 2005; PERSGA/GEF 2004; Saenger 1993).

In Qatar, Mangrove use in this country is very small, although they are used in some areas as fodder for camels, goat and sheep, for fuel, and as stakes for fishing. These uses are probably decreasing in line with increasing wealth in the region. Conversely many areas are increasingly threatened by pollution, especially caused by oil, which is widespread in many areas of the Persian Gulf and Straits of Hormuz. Landfill associated with urban and industrial development also threatens mangroves in many areas (e.g. Al Khor and Al-Dhakirah) (Supreme Council for the Environment and Natural Reserves 2004).

Tubli Bay in Bahrain, location of the only remaining stand of mangroves in the country, is located in an urban and industrial area with high human pressure, such as land-fill and land reclamation for infrastructure development, uncontrolled fishing, sewage disposal and discharge. Oil spills, oil shipping and loading operations represent major threat to the wetlands. Ras Sanad mangroves were declared Wildlife Reserve in 1988 and reforestation activities were undertaken in order to restore former areas, but with limited success. Despite its protected status, the area is still under threat of damage.

In the Gulf of Oman there is very little human use of the mangroves, although they are used in areas as fodder for camels, for fuel, and as poles or stakes for fishing. These uses are probably decreasing in line with increasing wealth in the region. Conversely many areas are threatened by pollution, especially caused by oil and landfill associated with urban and industrial development. Natural regeneration and reforestation for environmental protection purposes seem to have balanced the loss of mangroves especially at Batinah, Sharquiya and Dhofar Regions. In addition some attempts of *Rhizophora stylosa* plantations have been made during the last 20 years on experimental plots in Qurm creek and near Salalah (Alkathiri 2004; Scott 1995; UNEP 1999).

Environmental conditions in Kuwait are dominated by harsh arid growing conditions, which do not advantage mangrove's growth. The soil and the sea salinity reach very high values and in addition to this the oil spill decrease mangrove's growth performance. During the Persian Gulf War in 1991, a large oil spill polluted over 700 km of coastline from southern Kuwait to Abu Ali Island, endangering most of local fauna and flora. Around 50% of the mangrove trees present at that time were affected by the oil and some 30% died off. However, restoration in these forests was quicker than expected and natural regeneration started just only 2 years after the impact. Nowadays only few dead trees remain to document that devastating oil spill (Barth 2001).

3.2.4 Mangrove Forest Change Detection in the Persian Gulf and the Gulf of Oman (1977–2017)

3.2.4.1 Methodology

In order to develop maps of mangrove extent over 40 years ago, all available Landsat images since 1977 were reviewed for quality, and cloud-free scenes selected for analysis. Landsat imageries in 1973, 1977, and Feb 2017 were collected from the USGS via the website. Most of Image acquisition date were in April. The Landsat Mss (1977), TM (1989), ETM+(2000) and OLI (2017) were used to monitor mangrove forests in the Gulf of Oman and The Persian Gulf. The Multispectral Scanner (MSS) sensor acquired imagery of the earth from July 1972 to January 1999 on board Landsats 1 through 5. Imagery collected during the first decade of these missions predates most other operational imaging efforts, providing a unique view of the earth not available elsewhere. These early missions also offered the first digital remote sensing products to the earth science community.

As the results of image acquisition, the date determined image quality. Complete list of the Landsat Multi-Spectral Scanner (MSS), Thematic Mapper (TM), and Enhanced Thematic Mapper Plus (ETM+) data used in this study is shown in Table 3.1. Thus, reflectance normalization was performed with a histogram matching model which was developed on 56 satellite images by using ERDAS IMAGINE.

Geometric correction of the imagery is required because raw satellite imagery usually contain geometric distortions. These distortions must be corrected before the satellite scenes can be used as a geographically accurate map. Geometric correction of the imagery is required because raw satellite imagery usually contain geometric distortions. These distortions must be corrected before the satellite scenes can be used as a geographically accurate map. Multiple images in to a single image. Color corrections were done via color balancing and histograms matching to make the images being mosaiced correspond. Spectral bands (Red and NIR) combination used in the calculation of NDVI has unique spectral characteristics for detection of stress vegetation. For example, red band is sensitive to changes in chlorophyll contents in the visible spectrum and NIR has a capacity to characterize vegetation varieties and conditions. The index has been successfully used for characterizing vegetation

Table 3.1 Satellite images were used in this study

S. No.	Date of images acquisition	Satellite	Sensors	Resolution (m)	Bands	Number of satellite images were process in study area for each country	Total satellite images
1	1977/4	Landsat4	Mss	80	4	Iran (7), Oman (2), Emirate (3) Qatar and Bahrain (2), Saudi Arabia (1)	15
2	1989/4	Landsat -5	TM	30	7	Iran (8), Oman (3), Emirate (3) Qatar and Bahrain (2), Saudi Arabia (1)	17
3	2000/4	Landsat 7	ETM	30	9	Iran (8), Oman (3), Emirate (3) Qatar and Bahrain (2), Saudi Arabia (1)	17
4	2017/4	Landsat -8	OLI/ TIRS	30	11	Iran (8), Oman (3), Emirate (3) Qatar and Bahrain (2), Saudi Arabia (1)	17

Table 3.2 Mangrove forest area and changes from 1977 to 2017 in the Persian Gulf and the Gulf of Oman

	Mangrove forest area (ha) in 1977	Mangrove forest area (ha) in 1989	Mangrove forest change (ha) 1977–1989	Mangrove forest area (ha) in 2000	Mangrove forest change (ha) 1989–2000	Mangrove forest area (ha) in 2017	Mangrove forest change (ha) 2000–2017	Mangrove forest change (ha) 1977–2017
Iran	4735	6052	1318	8015	1962	9403	1388	4667
Emirate	1327	1949	622	4777	2828	7926	3149	6599
Oman	206	153	−53	168	16	530	362	324
Saudi Arabia	97	107	10	146	39	837	691	740
Qatar	16	184	168	283	100	1002	719	986
Bahrain	31	25	−6	43	17	48	5	16
Total	6411	8470	2058	13,432	4962	19,746	6314	13,334

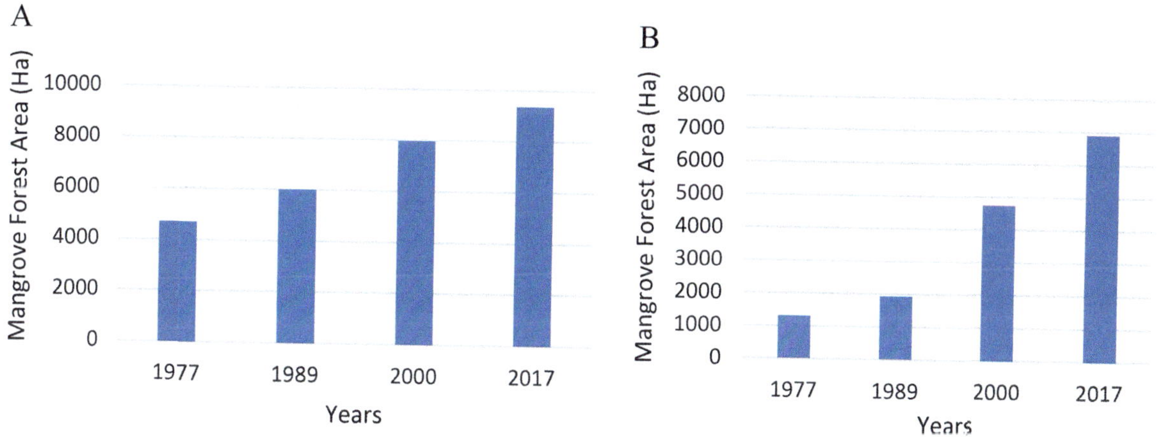

Fig. 3.6 Mangrove forest area change in Iran1977–2017. (**a**) Iran, (**b**) Emirates

cover and it is still considered a great potential for application in environmental monitoring because of it low cost compared to hyperspectral data.

3.2.4.2 Mangrove Forest Changes in Study Area (1977–2017)

Mangrove Forest in 1977

Multispectral scanners (MSS) onboard Landsat satellites 1–5 provide the earliest record of mangrove forests in the Persian Gulf and the Gulf of Oman. In this research, we use MSS image data in April and march 1977 in study area, totally 15 satellite Images used to detection, mangrove forest distribution. Area of mangrove forests in Iran were about 4735 ha (47.3 km (Table 3.2, Fig. 3.6a). maximum coverage of mangrove forest area in Iran are located in Zone 4: Within of the Khamir estuaries and 5: northwest estuaries of the Qeshm Island (Harra Protected Area). In this area of mangrove forests are about 3744.15 ha (37.4 km^2). This area (Qeshm Island) is the biggest community of the mangrove forests in the Persian Gulf and the Gulf of Oman.

In Oman Country in 1977, area of mangrove forests was about 206 ha (2.06 km^2), and maximum area belongs to Zone

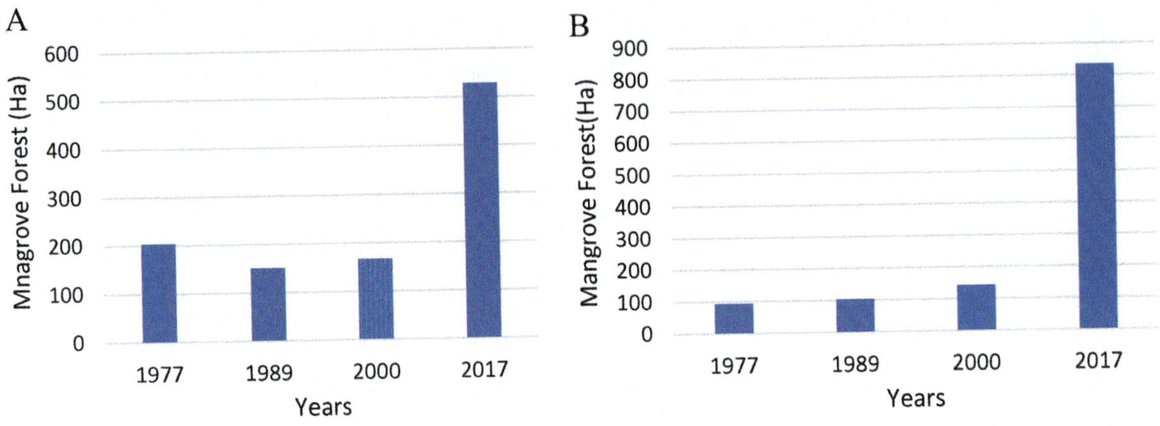

Fig. 3.7 Mangrove forest area changes 1977–2017, (**a**) Oman, (**b**) Saudi Arabia

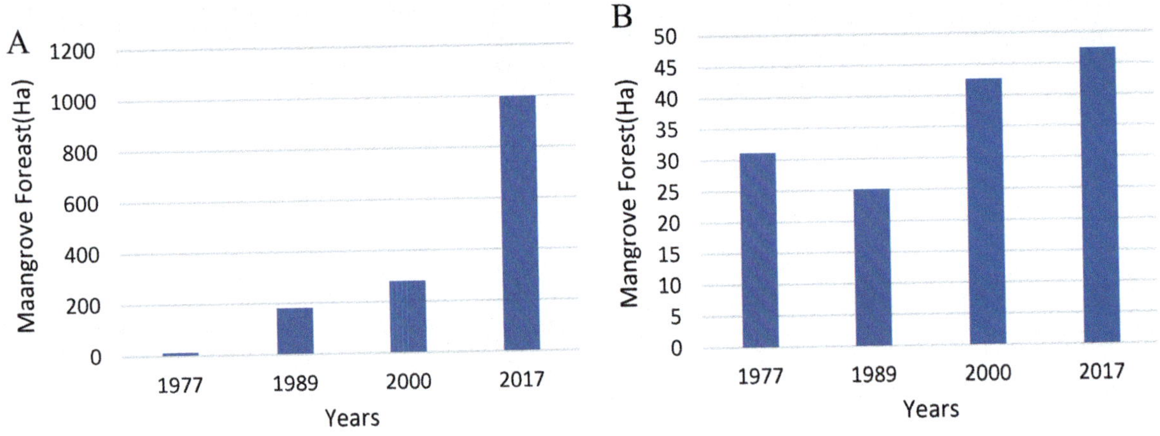

Fig. 3.8 Mangrove forest area changes 1977–2017, (**a**) Qatar, (**b**) Bahrain

5; Khayran Estuary with15 ha (0.15 km^2) and Qurm Bay (Table 3.2, Fig. 3.7a).

United Arab Emirate has about 1327 ha (13.14 km^2) in 1977. After Iran, this country has maximum distribution and area of mangrove forest in the study area. Al Dhabeia estuary near Abu Dhabi has maximum mangrove forests area 203 ha (2.03 km^2) in this country (Table 3.2, Fig. 3.6b).

In Qatar satellite can detect about 16-ha (0.16 km^2) mangrove forest especially in al Dhakira and al Khor in eats part of this peninsula (Table 3.2, Fig. 3.8a) In Bahrain, there were about 31 ha (0.03 km^2) in Howar Island and Tubil Bay (Table 3.2, Fig. 3.8b).

In this research, we can detect about 100 ha (1 km^2) in Tarut Bay in Saudi Arabia (Table 3.2, Fig. 3.7b). Total mangrove forest area within these years in the Persian Gulf and the Gulf of Oman was about 6411 ha (64.11 km^2) (Table 3.2, Fig. 3.9).

Mangrove Forest in 1989

NDVI processing on TM Satellite images on 1989 in study area indicated, the mangrove forests cover increased during 1977–1989, in study area and reached 8470 ha (84.70 km$^{2)}$. (Table 3.2, Fig. 3.9). The mangrove forests of Iran increased by 27.8% (1318 ha, 13.18 km^2) and reach to 6052 ha (60.52 km^2) from 1977 to 1989 (Table 3.2, Fig. 3.6a). In Qeshm Island (Zone 4 and 5) area of mangrove forest increased about 1120 ha (11.2 km^2) and developed more than 30% compare to 1977 (Fig. 3.10a). But the highest increase in percentage found in Govatre bay (Zone 11 and 12(in Pakistan Border). Images processing indicate mangrove forests in this bay increase from 77 ha in 1977 to 252 ha in 1989 and shows 326% increase in area of mangrove forest during these times. However, in some mangrove forest area including Zone 11 mangrove forests decrease.

In Oman mangrove forest area decrease about 53 ha and (26%) during 1977–1989. In 1990 mangrove forest reach to 153 ha (1.53 km^2) while in 1977 area of mangrove forests in

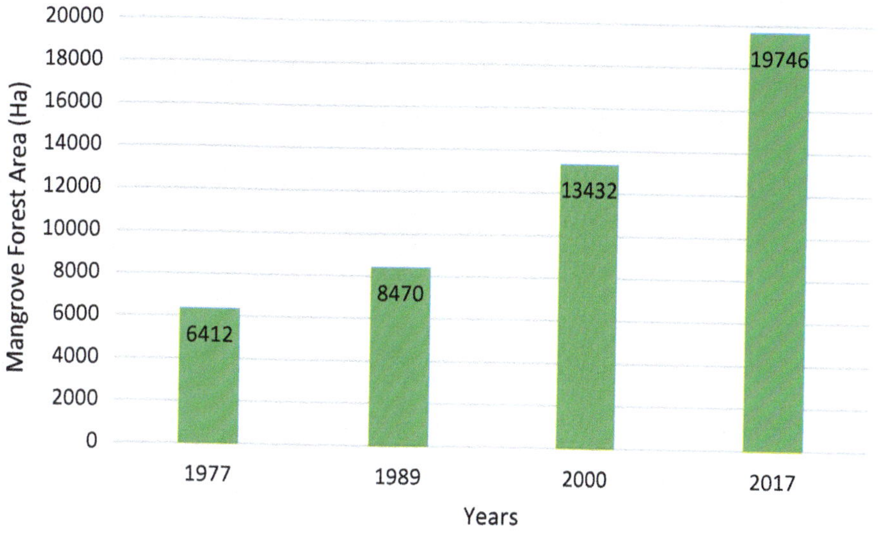

Fig. 3.9 Mangrove forest changes in The Persian Gulf and the Gulf of Oman 1977–2017

this country was 206.6 ha (2.06 km) (Table 3.2) but in some regions, like as Qurm Bay in mangrove forests increase during these times (Fig. 3.11a).

Satellite images present Mangrove forest area in United Arab Emirate increased from1977 to 1989 especially in Abu Dhabi region (Fig. 3.12a). During this time about 622 ha (46%) mangrove forest develop in this country and area of this forests increase to 1949 ha (1224 ha in 1977).

The maximum increase percentage of mangrove forests between countries of study area belongs to Qatar. Area of Qatar's mangrove forests in 1989 was about 183 ha (1.83 km^2), while in 1977 we detect just 16 ha and shows 168 ha (1052%) increase in mangrove forests in this country especially in Al Dhakira Bay (Fig. 3.13a).

Mangrove forests decrease from 31 ha in 1977 to 25 ha from 1977 to 1989 in this country especially in Tubli Bay. In Saudi Arabia (Tarut Bay) mangrove forest in 1989 reach to 107 ha (1.07 km^2) and shows increase about 10 ha during this time (Fig. 3.14a).

Mangrove Forest in 2000

At this time, same as early period (1977–1989), ETM satellite images processing indicate increasing mangrove forests in the Persian Gulf and the Gulf of Oman in all countries mentioned. Mangrove forests increased during this time about 4692 ha (46.92 km^2) and reached an overall area of 13,432 ha (134.3 km^2), which showed about 59% increase compared to before 1989 (Table 3.2, Fig. 3.9).

Area of Mangrove forests in Iran rose about 1962 ha (19.62km^2) between 1989 and 2000 and reached 8015 ha (80.15 km^2). All mangrove forests zones showed increases in Iran especially in Zone 6 (north of Striate of Hormoz and Govatre Bay) and Zones 11 and 12 (Pakistan Border) (Fig. 3.6a).

NDVI analysis in Oman showed about a 10% increase (16 ha) in mangrove forests compared to before 1989. The area of mangrove forests in 2000 was about 168 ha (1.68 km^2) and the maximum concentration of mangrove forests was located in the Quriyat area (Table 3.2, Fig. 3.7a).

Maximum increase in mangrove forests between all countries in the study area belongs to Emirates. Area of these forests increased about 2828 ha (28.28 km^2) between 1989 and 2000, about 145% increase when compared to before 1989. During this time, mangrove forests in the Emirates was about 4747 ha (47.47 km^2). Mangrove forests increases were about 39 ha in Saudi Arabia's Tarut and Aldaid regions (South of Qatar; Total Area 146 ha), 17 ha in Bahrain (Tubli bay and Howar Island; Total Area 43 ha), and 100 ha in Qatar, especially in Al Dhakira (Total Area 283 ha) (Table 3.2, Figs. 3.7b and 3.8a, b).

Mangrove Forest in 2017

According to NDVI analysis on OLI8 Landsat images for mangrove forest distribution and area in the Persian Gulf and the Gulf of Oman in April 2017, area of this forest increased from 2000 to 2017. Mangrove forest area in this year in study area was about 19,746 ha (197.46 km^2), which indicated a 6314 ha (64.14 km^2) increase from 17 years ago.

Mangrove forests in Iran at this time are about 9402 ha (94.02 km^2) and increased about 1388 ha (13.88 km^2) (Fig. 3.6a). Maximum increase in mangrove forest of Iran during this time belongs to Qeshm Island (Fig. 3.10b).

Mangrove forests in Oman also increased about 362 ha (3.62 km^2) and in 2017, the area of mangrove forests in this country was about 530 ha (5.30 km^2) (Fig. 3.7a). Qurm Bay is an example of this increase (Fig. 3.11b).

The maximum increase in mangrove forest area between countries of the Persian Gulf and the Gulf of Oman from 2000 to 2017 belongs to Emirates with 3149 ha (31.49 km^2). The area of mangrove forests in the Emirates in 2017 reached about 7926 ha (79.26 km^2) (Table 3.2, Figs. 3.6b and 3.12b).

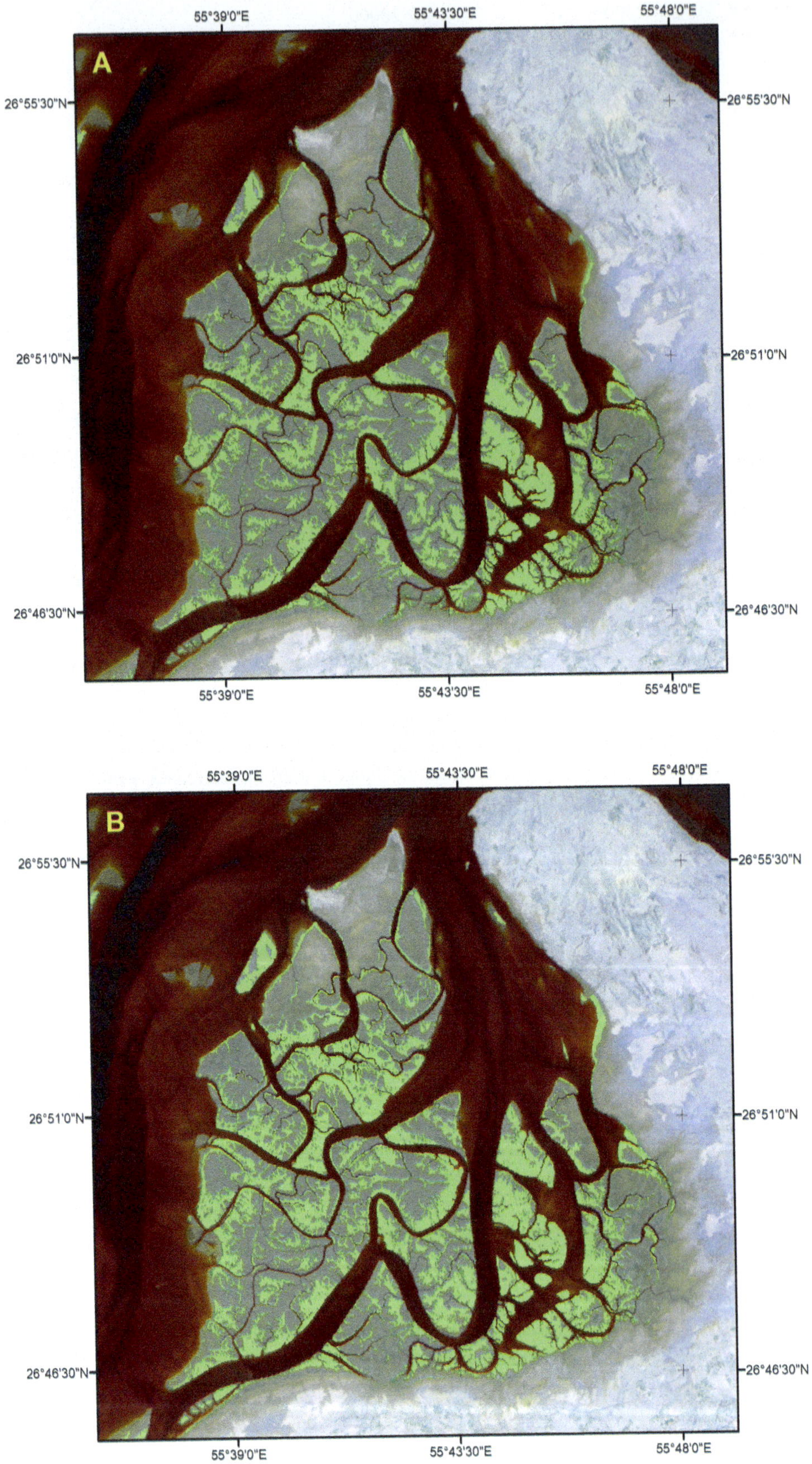

Fig. 3.10 Mangrove forests changes in Qesham Island in Iran, (**a**) 1989, (**b**) 2017

Fig. 3.11 Mangrove forests changes in Qurm region in Oman, (**a**) 1989, (**b**) 2017

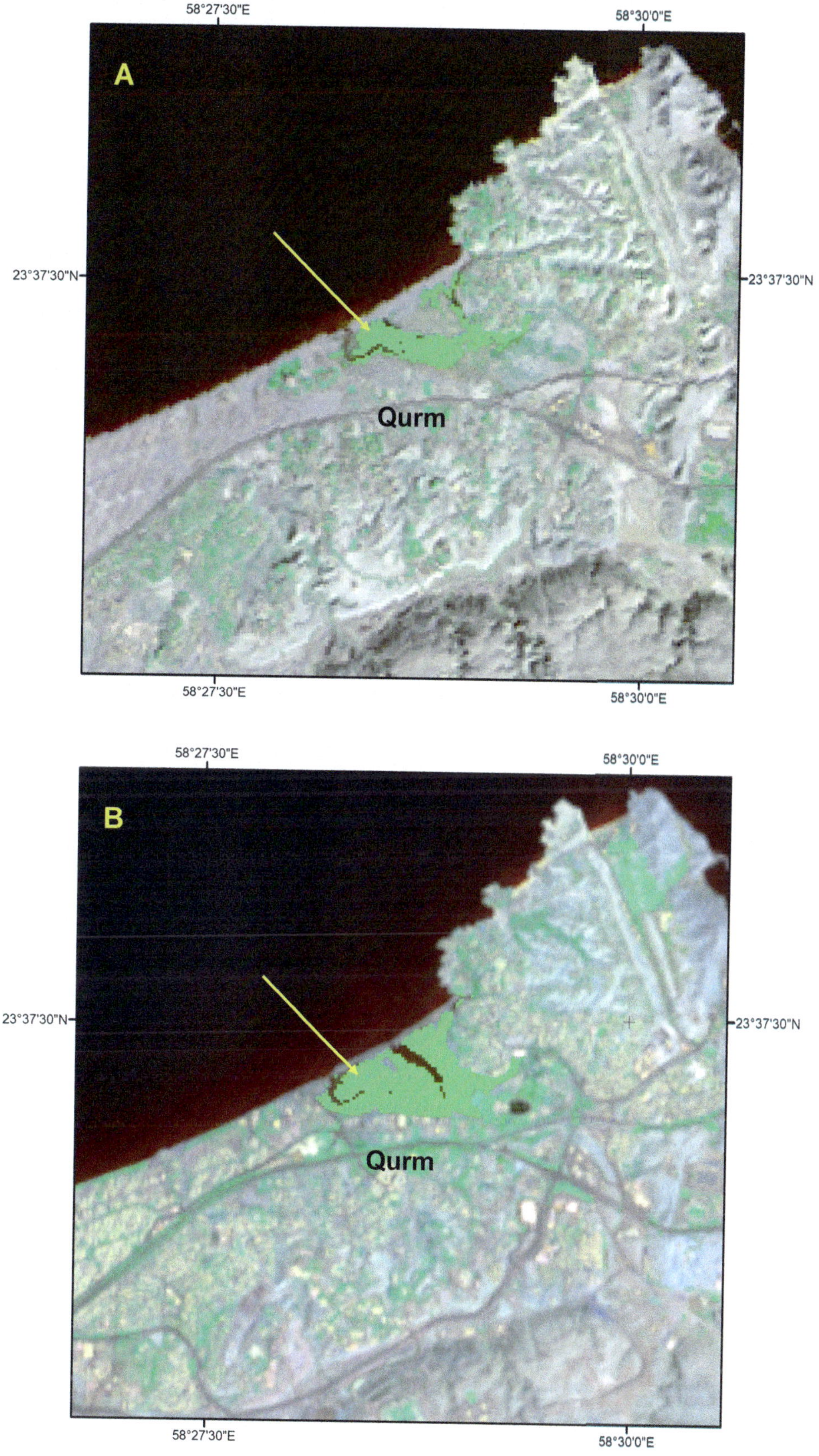

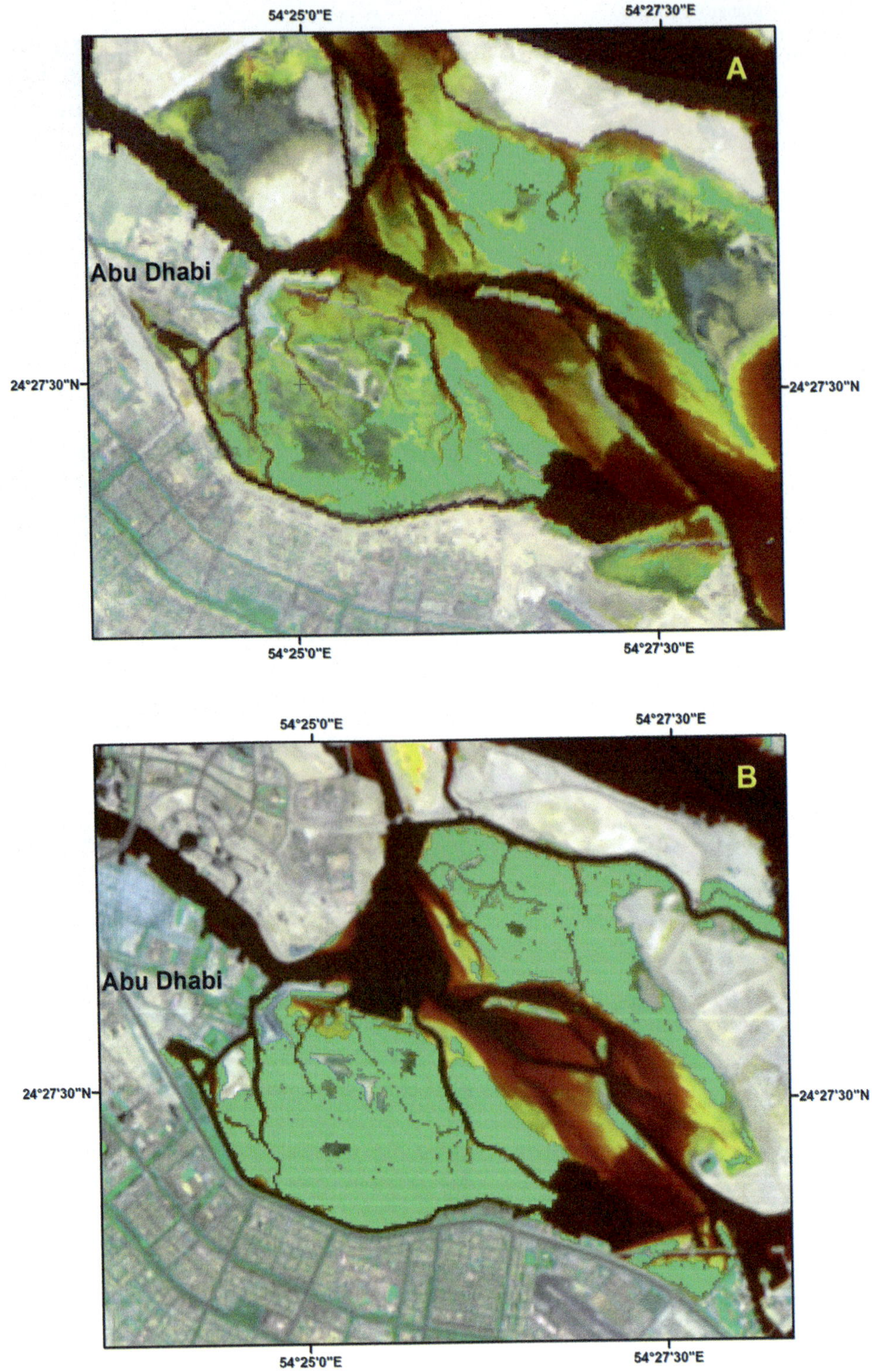

Fig. 3.12 Mangrove forests changes in Abu Dhabi, Emirates (**a**) 1989, (**b**) 2017

However, we can also see increasing mangrove forests in other countries. In Qatar, the area of mangrove forests increased about 719 ha (7.19 km^2) and reached about 1002 ha (10 km^2) (Table 3.2, Figs. 3.8a and 3.13b). Mangrove forests in Bahrain in 2017 was about 48 ha (0.48 km^2) and increased about 5 ha from 2000 to 2017 (Table 3.2, Fig. 3.8b). In Saudi Arabia, Tarut and Aldaid (South of Qatar) recorded total mangrove forests in 2017 of about 837 ha (8.37 km^2), an increase of about 691 ha (6.91 km^2) during this time (Table 3.2, Figs. 3.7b and 3.14b).

Table 3.2 and Fig. 3.9 indicate mangrove forest and their changes in the Persian Gulf and the Gulf of Oman between

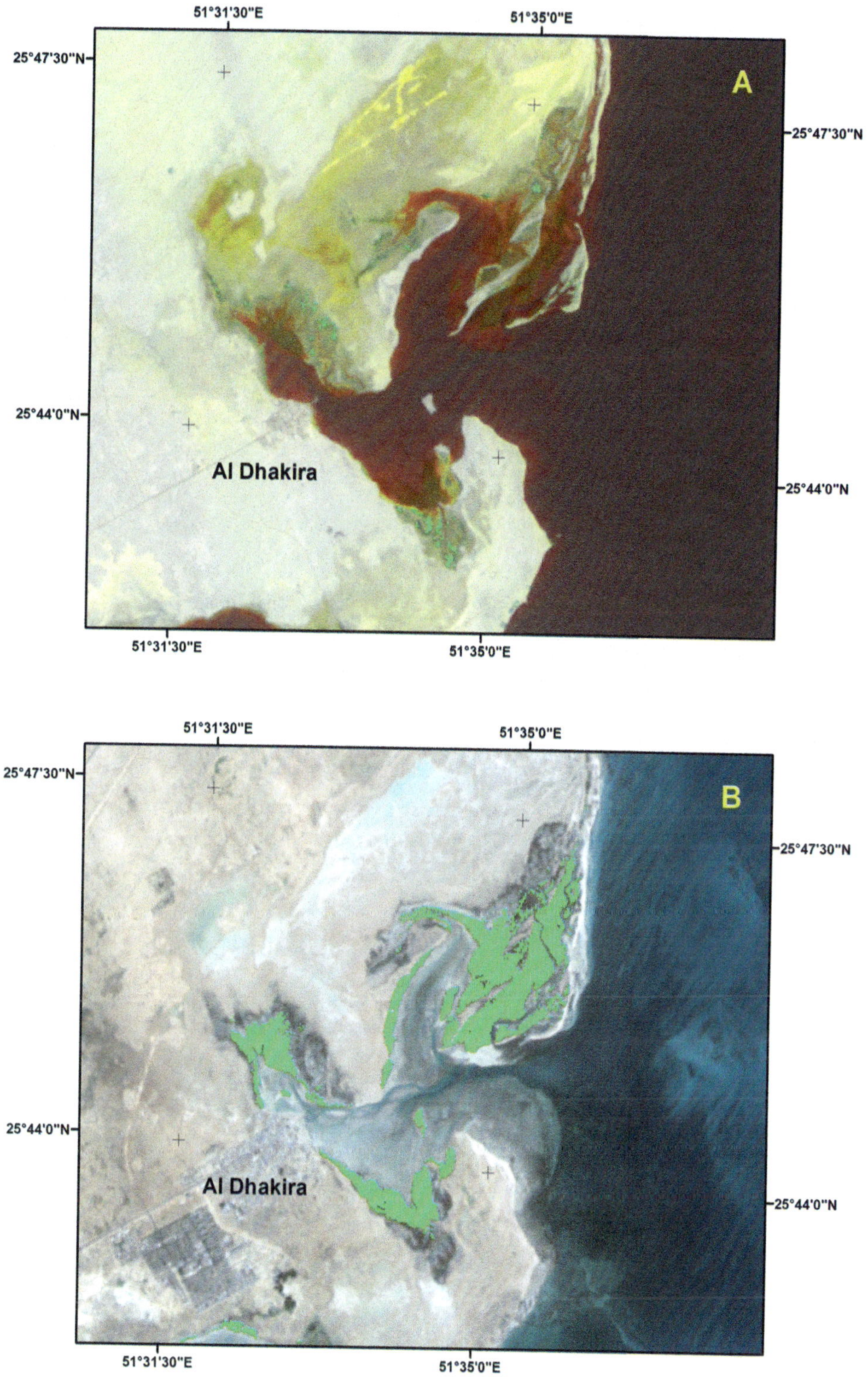

Fig. 3.13 Mangrove forests changes Al Dhakira, Qatar. (**a**) 1989, (**b**) 2017

1977 and 2017. Mangrove forests in developed and increased during the time. While in some region and some years we are faced to decrease mangrove forest, but dominate trend are increasing. In 1977, the area of mangrove forest within the study area was about 6411 ha (64.1 km^2) and about 40 years later this area increased about 13,334 ha (133.34 km^2) to approximately 19,746 ha (197.5 km^2). This study indicates mangrove forests in the study area increase about 203% from 1977 to 2017 (Table 3.2 and Fig. 3.9).

Maximum increase in mangrove forests area from 1977 to 2017 belongs to Emirates (6599 ha), followed by Iran (46.67 ha). However, we should also answer this question:

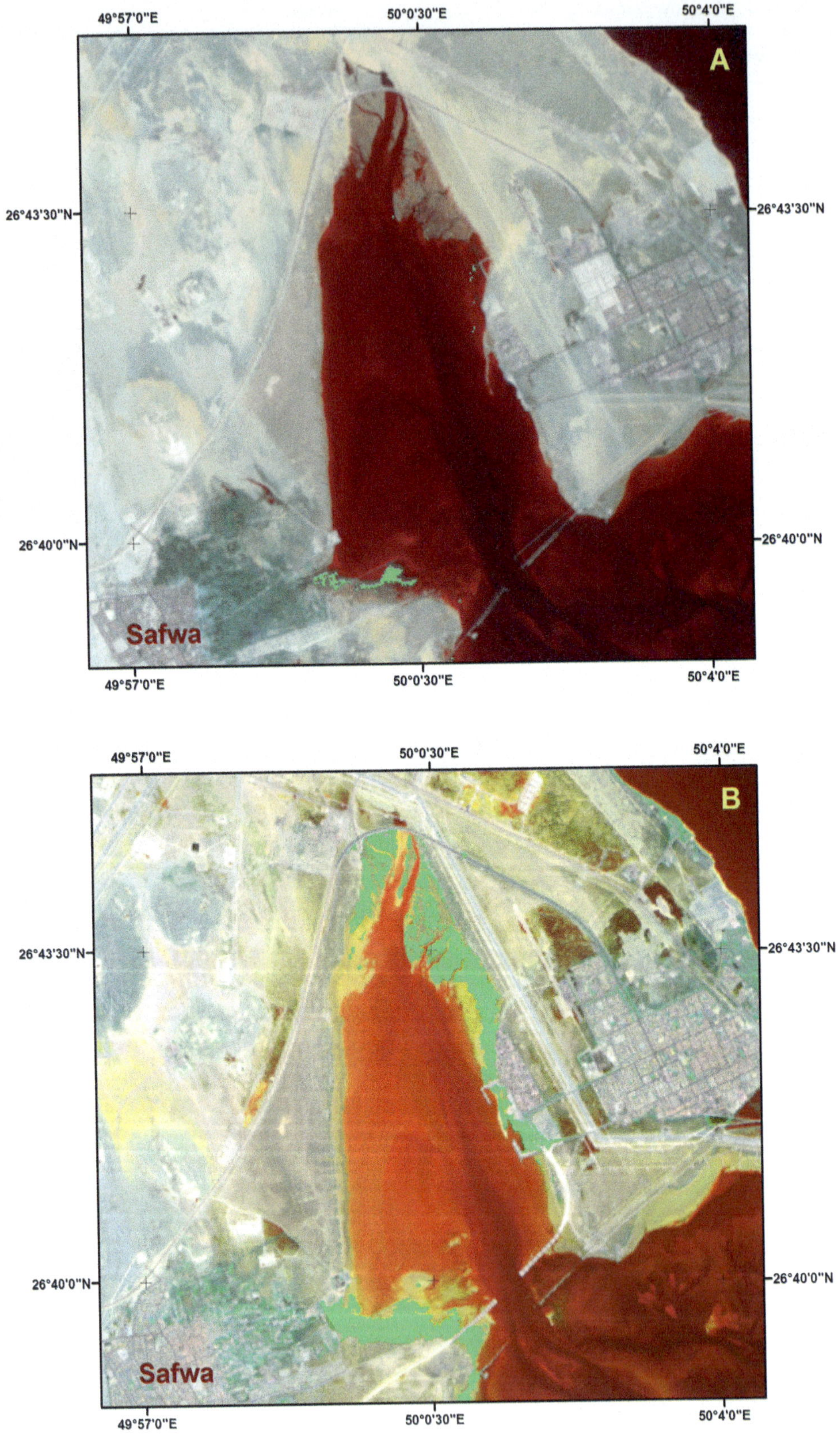

Fig. 3.14 Mangrove forests changes in North part of Tarut Bay, Saudi Arabia. (**a**) 1989, (**b**) 2017

What is the reason of this increase in mangrove forest area in the Persian Gulf and the Gulf of Oman? One of the main reasons for the positive development of mangrove forests in study area could be extensive forest management and conservation programs launched in these regions.

3.2.5 Mangrove Forest Management and Conservation Programs in the Persian Gulf and the Gulf of Oman Countries

3.2.5.1 Iran

Iran has areas of mangrove forest in the Persian Gulf and the Gulf of Oman. Mangrove forest plantation and developments in Iran started in 1914. Afforestation in Iran has its roots in traditional and national customs, and in the religious beliefs of Iran's people. With the growth of civilization in Iran, people began to plant trees for different purposes. Therefore, tree planting has a long history in Iran. The first modern afforestation was conducted with the planting of a mangrove species, *Rhizophora mucronata*, on the southern coasts in 1914–1918 during World War (Ajorlo 2003).

Developments of mangrove forests in southern Iran is one of the goals of the forests service. Recently, conservation of fragile ecosystems in dry, sub-dry and sub-humid areas is specifically taken into consideration and restoration of these ecosystems is given high priority in executive programs. For instance, some degraded mangrove forests are rehabilitated in recent years. Forests, rangelands and watershed organization of Iran have improved the mangrove plantation in south coasts of Iran in Bushehr, Hormozgan and Sistan and Baluchistan provinces to increase the socio-eco-environmental products of mangrove habitat. For example, the mangrove plantation around Qeshm Island has been established to develop those above-mentioned objectives. There is large area with potential for expansion of mangrove forests stands, for example total area of present mangrove in Hormozgan provinces is almost 12,500 ha, but swamps and tidal flat in this province amount to 28,000 ha. Within 3 years (2005–2007), more than three million seedlings of *Avicennia marina* trees have been produces and planted in Hormozgan Provinces only (Safiari and Nasouri 2009).

3.2.5.2 Oman

In Oman, mangrove forest area decreased from 206 to 153 ha (53 ha, 26%) from 1977 to 1989. The area of mangrove forest has been decreased due to the damages by the inhabitant to be used as fuel wood and overgrazing during long period of time. However, Mangrove was planted in Oman at various times from 1983 to 1985, however they did not produce concrete results. The reason they did not succeed is the lack of experience on technical knowledge and skill for mangrove afforestation (MRMEWR 2004). However extensive program called 'mangrove afforestation' have launched (MECA 2010). According to this program was undertaken on local use of mangroves and threat to the ecosystems. This program provided the basis for development of a conservation strategy and detailed information along with site-specific management measures, including protection, development of mangrove nursery and the provision of facilities for recreation as well as environmental education. Based on this program, considering the special value of mangroves in coastal environments in Oman, the Ministry of Regional Municipalities and Water Resources (MRMWR) and Ministry of Environment and Climate Affairs (MECA) have launched 'mangrove afforestation' program (MECA 2010). Nowadays there are 12 existing marine protected areas covering a range of environment, species which provide great opportunities for research as well as management. According to IUCN (1986, 1989) these marine protected areas are studied with interest on mangroves at Qurm Nature Reserve, Ras al Hadd Turtle Reserve for turtle conservation, Dimaniyat Islands Nature Reserves for the protection of coral reefs, hawksbill turtles and nesting birds along the Salalah coast, nine Khawrs (coastal lagoons) for mangroves and migratory birds. The mangrove transplantation project was initiated by MECA in collaboration with JICA in 2000. Four nursery establishments in Qurum Nature Reserve (QNR), Sur and Salalah were sources of seedlings which were distributed in various areas of the country. The project was envisioned to produce seedling in the nurseries and transplant them in khwars to enhance existing mangrove vegetation or reforest bare lagoons. QNR nursery in Muscat was the first pump irrigated nursery in the country and is currently producing 24,000 seedlings annually with about 85% survival. Participation in transplanting of seedlings to the site by local resident volunteers especially women and school children had been very encouraging. To date, less than half a million *Avicennia marina* seedlings had been successfully transplanted in various coastal locations of the country. Signs of mangrove regeneration are evident in many areas (Choudri et al. 2016). Results of this efforts caused increase mangrove forests morae than 2.5 times (250%) from 1977 to 2017).

3.2.5.3 United Arab Emirates

The maximum increase in mangrove forest area in the Persian Gulf and the Gulf of Oman belongs to the Emirates from 1977 to 2017. In the Emirates, despite numerous threats, mangrove extent has been increasing over the last 20 years due to natural growth and successful afforestation program activities, which were started by the Ministry of Agriculture and Fisheries in 1972 in Abu Dhabi (Anwahi 2005: Böer 2002; Khan 1982; Scott 1995; Spalding et al. 1997).

In United Arab Emirates, mangrove plantation NOC projects were launched. A No Objection Certificate (NOC) for planting mangroves is required for any mangrove planting project in Abu Dhabi Emirate. NOCs for mangrove planting are issued by the Environment Agency-Abu Dhabi (EAD). EAD's goal, in line with the targets outlined in the Emirate of Abu Dhabi Biodiversity Strategy (2015–2019), is to protect existing mangrove ecosystems, and ensure that environmental impacts due to development are avoided, minimized and adequately mitigate Excessive human activities have adversely affected the mangrove ecosystem, (EAD) urges developers to rehabilitate the affected areas through large-scale cultivation programs. A case in point is Saadiyat Island in Emirate where (EAD) partnered with the Tourism Development and Investment Company (TDIC) to plant 750,000 saplings of mangroves on 25% of the island, which is currently being developed as a cultural hub of Abu Dhabi. The move is aimed at mitigating the environmental damages caused by the massive development on the island. Property developers are also urged to protect this endangered natural treasure by giving the environment proper consideration from the early planning stages of their development. (EAD) has also urged the public not to litter around mangroves, especially as plastic bags can prevent the growth of mangrove trees and release pollutants that could be harmful for their health. (EAD) works to rehabilitate and protect mangrove forests in seven key sites in the Emirate: Saadiyat Island, Jubail Island, Marawah Marine Biosphere Reserve (which also comprises of Bu Tinah Island), Bu Syayeef Protected Area, Ras Gharab, the Mangrove National Park and Ras Ghanada (Environmental Agency Abu Dhabi 2015).

3.2.5.4 Qatar

Qatar is gradually increasing the level of protection of the country's mangroves, with 40% of the country's coastline now protected. Organizations such as Conservation International have begun mapping out the mangroves locations and data in Qatar and around the globe in order to assess the population distribution and threatened areas. With further enforcement and data tools, the mangrove forests of Qatar can be restored, and continue to provide immense benefits to this harsh desert environment. The management of Ras Laffan Industrial City initiated a mangrove conservation program which includes the protection of the existing plants and planting and sowing of additional seedlings and seeds respectively. Currently the mangrove population has increased 10,000 in numbers (Ras Luffan Industrial City 2005). As of November 2007, FEC, with the aid of representatives of Birdlife International, completed an extensive survey confirming three of the five suggested IBAs as indeed being prime bird conservation sites in Qatar. Those sites include Al Thakira mangroves, Khor Al Udaid, and Alshat Island. The two remaining sites have yet to be surveyed. In addition, five other sites were recommended for IBA consideration. These include Um Tais, Ras Rukon, Abu Nakhla pond, Ashat Island, and Shahanniya (Richer 2009). this activity lead to conservation of mangrove habitat and increasing area of mangrove forests in Qatar from 15 h in 1977 to 1002 ha in 2017. This increase in mangrove forest area indicates successful programs in this countries during the times.

3.2.5.5 Bahrain

Tubli Bay is an inshore coastal area situated in the northeast of Bahrain and mangrove forest in Bahrain concentrated in this region (Besides Howar Bay). For the past 40 or so years, Bahrain coasts especially the Tubli Bay in Bahrain (maximum concentration of mangrove forest in Bahrain) has been under threat from human activity. Infilling the coast for development has caused the size of the Bay to decrease from 23.5 km^2 in 1956 to 16.1 km^2 in 1996 (Environmental Affairs 1996). This on-going activity is causing even more degradation/destruction to the Bay's vulnerable ecology. Tubli Water Pollution Control Centre discharges 160,000 m^3 (Personal Communication 1996) into the Bay every day. Sand washing plants continue to discharge silt into the Bay daily, and the dumping of rubble and litter within its surroundings still remain a point of concern. In 1988 The Mangrove Stand, protected by the Environmental Affairs of Bahrain, is situated at Ras Sanad in the south-western part of the Tubli Bay. Fishing activities are carried out in the Bay, even though it represents a nursery ground for commercially important species. According to this programs mangrove forests area from 1989 to 2017 from 31 to 48 ha. The area is also RAMSAR listed internationally (Convention on the Conservation of Wetlands of International Importance especially as Waterfowl Habitat). It is characterized by its unique ecology as it provides a habitat for important coastal ecosystems, such as mangroves and seagrass. It is the only place in Bahrain with an ecological interaction between mangroves, seagrass and corals. These three systems are important for productivity of coastal fisheries that society depends on.

3.2.5.6 Saudi Arabia

Mangrove forests along Saudi Arabia coasts in the Persian Gulf increased from 97 ha in 1977 to 838 ha in 2017. In Saudi Arabia, shrimp aquaculture represents one of the emerging threats to this ecosystem. Farming first started in Al-Lith in 1983 on a small area of 20 ha, which has increased up to 1000 ha in 1998. Additionally, other shrimp farms are now being planned at Shuqaiq mangrove areas.

In Tarut Bay in Saudi Arabia Saudi Aramco, the ministry of agriculture and NCWCD have established several mangrove plantations since 1996 and this stand are thriving (Allen 1998). Some of this plantation are probably established too far landward too obtain maximal growth, but they are never less in good condition. The ecosystem of

Tarut Bay in Saudi Arabia is exposed to many environmental pressures and if the pollution and Urban utilization continues will lead to erosion of the Bay and near in the future will be burden on the environment and not the source of the wealth of fish and the most important site for the passage of winter birds in the Saudi sector on the coast of The Persian Gulf.

3.3 Conclusion

Mangroves were once recognized as worthless wastelands. This attitude made it acceptable for people to exploit mangroves as a source of land for constructing ports, condos, hotels, aquaculture ponds, and expansive infrastructure for the tourism and fisheries industries.

If improvements to mangrove management and restoration programs were not made soon, there would have been great losses in commercial fisheries and local economies, as well as increased erosion and shoreline instability in countless coastal communities.

The rate of loss of mangrove forests remains substantially higher than that of other terrestrial forests, even as management and conservation programs have become widespread (FAO 2007: Lal 2003). Humans are greedy for coastal land and the profits that coincide with its development. Increasing population pressure in the coastal zone and growing demand for development appear to be the key drivers of mangrove forest destruction. Where legislation and management are absent, the problem of mangrove overuse becomes more prominent (ISME 2004). Mangrove health, as well as that of all marine and terrestrial ecosystems, is directly influenced by the effectiveness of their management and conservation (FAO 2007).

Mangrove forests are unique wetland ecosystems that are established directly in the intertidal zone in the Persian Gulf and the Gulf of Oman. High population pressure in coastal areas could led to the conversion of many mangrove areas to other uses and numerous case studies describe mangrove losses over time. However, information on status and trends of area and distributions in the mangrove forests in the study area is necessary. While some studies had been done for mangrove forests in each country in the Persian Gulf and the Gulf of Oman, most obtained data are scattered and didn't normalize properly.

This research is first attempt at estimating the total mangrove area and their changes in the study area with the one of the powerful method for mangrove forests detection during same periods of tomes for 40 years from 1977 to April 2017. In this study emphasis is given over the NDVI technique, which has been applied to classify satellite images. Basically, through NDVI technique mangrove vegetation can easily be differentiated from other forest due to its different reflectance value. Landsat satellite images form 1977 and 2017 have been applied to assess the decadal land use/land cover changes in the study area. The results of this research show that Landsat data are very useful in monitoring changes in forest cover over moderate time intervals (40 years in this study) at a broad level of classification such as forest/non-forest for a regional area or state. However, the NDVI difference produced fewer visually identifiable errors of omission and commission than the maps of deforestation and reforestation, classification of the vegetation change map into forest change is required.

Results indicate that in 1977, the area of mangrove forest in the study area was about 6412 ha (6.91 km^2). About 40 years later, this same area increased to 13,334 ha (133.34 km^2) and reached a total of about 19,746 ha (197.5 km^2). This increase in mangrove forests in the Persian Gulf and the Gulf of Oman represent sustainable development and management in mangrove ecosystems for all countries in this region.

These successful programs will ultimately rely on the cooperation of numerous governmental bodies as well as local stakeholders striving for a common goal: the protection and restoration of all remaining mangrove habitats. Finally, mangrove forest management and development in the Persian Gulf and the Gulf of Oman could be a pattern for mangrove ecosystems in the world.

Acknowledgement This project was funded by Faculty of Earth Science in Shahid Beheshti University, Iran. We would like to express our appreciation to all colleagues who provided us with their valuable support and comments.

References

Ajorlo M (2003) The history of afforestation in Iran. XII forestry congress, Québec, Canada

Al Mamoon A, Noor S, Rahman A, Almasri S (2016) National sea outfall assessment in Qatar: opportunities and challenges. 37th Hydrology & Water Resources Symposium 2016: Water, Infrastructure and the Environment:295

Alkathiri AMM (2004) Information provided for the Global Forest Resources Assessment (FRA) 2005 thematic study on mangroves. Unpublished

Alongi DM (2002) Present state and future of the world's mangrove forests. Environ Conserv 29:331–349

Allen JA (1998) Mangroves as Alien species: the case of Hawaii. Glob Ecol Biogeogr Lett 7(1):61

Anwahi AA (2005). Information provided for the Global Forest Resources Assessment (GFRA) 2005 thematic study on mangroves. Unpublished

Arturo RL, Aimeé Cervantes E, César B (2010) Assessing distribution patterns, extent, and current condition of Northwest Mexico mangroves. Wetlands 12., 2010:717–723

Aschbacher J, Ofren R, Delsol JP, Suselo TB (1995) An integrated comparative approach to mangrove vegetation mapping using advanced remote sensing and GIS technologies: preliminary results. Hydrobiologia 295., 1995:285–294

Badola R, Hussain SA (2005) Valuing ecosystem functions; an empirical study on the storm protection function of Bhitarkanika mangrove ecosystem India. Environ Conserv 32:85–92
Bannari A, Morin D, Bonn F, Huete AR (1995) A review of vegetation indices. Remote Sens Rev 13(1–2):95–120
Baret F, Guyot G, Major DJ (1989) TSAVI: a vegetation index which minimizes soil brightness effects on LAI and APAR estimation. Geoscience and remote sensing symposium, 1989. IGARSS'89. 12th Canadian Symposium on Remote Sensing. 1989 International, pp 1355–1358
Barnes EM, Clarke T, Richards SE, Colaizzi PD, Haberland J, Kostrzewski M, Waller P, Choi C, Riley E, Thompson T, Lascano RJ, Li H, Moran MS (2000) Coincident detection of crop water stress, nitrogen status and canopy density using ground-based multispectral data. In: Robert PC et al (eds) Proceeding of the International conference on Precision Agriculture, 5th, Bloomington, MN. 16–19 July 2000. ASA, CSSA, and SSSA, Madison
Barth H (2001) Preliminary report on: the coastal ecosystems 10 years after the 1991 Gulf Waroilspill. http://www.uniregensburg.de/Fakultaeten/phil_Fak_III/Geographie/phygeo/downloads/barthcoast.pdf
Behrouzi RB (1996) Gandoo protected area. Environ Sci Q J 8:28–35
Böer B (2002) The coastal and sabkha flora of the United Arab Emirates. In: Sabkha ecosystems, pp 303–309
Böer B (2005) Information provided for the global forest resources assessment (FRA) 2005 thematic study on mangroves. Unpublished
Choudri BS, Said BM, Mushtaque A (2016) An overview of coastal and marine resources and their management in sultanate of Oman. J Environ Manage Tour 1:21
Czudek R (2006) Wildlife issues and development prospects in West and Central Asia Thematic study for the Forestry Outlook Study for Africa. The wildlife management working paper 9, of Food and Agriculture Organization of the United Nations, 71 pp
Dahdouh-Guebas FM (2005) How effective were mangroves as a defense against the recent tsunami? Curr Biol 15(12.), 2005):443–447
Danehkar A (1994) Study on Sirik region mangroves. MS Thesis, Faculty of Natural Resources-Tarbiat Modares University, Noor
Danehkar A (1996) Iranian mangroves forests. Environ Sci Q J:8–22
Danehkar A (1998) Marine sensitive areas of Iran. Environ Sci Q J 24:28–38
Danehkar A (2001) Mangroves forests zonation in Gaz and Harra international wetlands. Environ Sci Q J 34:43–49
Darvishsefat A, Zobeiri M (2007) Potentiality of Aster images for forest density mapping in Zagros forests of Iran. Geomatica Barselona, Barcelona, February pp 20–23
Directorate-General of Nature Conservation (2010) Fourth national report to the convention on biological diversity. Ministry of Environment and Climate Affairs Sultanate of Oman
Duke NC, Meynecke J-O, Dittmann S, Ellison AM, Anger K, Berger U, Cannicci S, Diele K, Ewel KC, Field CD, Koedam N, Lee SY, Marchand C, Nordhaus I, Dahdouh-Guebas F (2007) A world without mangroves? Science 317(5834):41–42
Emch ME, Peterson M (2006) Mangrove forest cover change in the Bangladesh sundarbans from 1989–2000. A remote sensing approach. Geocarto Int 21(1):5–12
Environmental Agency Abu Dhabi (2015) Technical guidance document for mangrove planting permitting and management plan
Ewel KC, Ong JE, Twilley RR (1998) Different kinds of mangrove swamps provide different goods and services. Glob Ecol Biogeogr Lett 7:83–94
FAO (2005) Global forest resources assessment 2005 thematic study on mangroves (Saudi Arabia)
Field CB, Osborn JG, Hoffman LL, Polseaberg JF, Ackerly DD, Berry JA, Biorkman O, Held A, Matson PA, Mooney HA (1998) Mangrove biodiversity and ecosystem function. Glob Ecol Biogeogr Lett 7:3–14
Fluor Cam PAR (2014) Absorptivity module & NDVI measurement instruction manual, product .PSI, spol. s r. o., Drásov 470, 664 24 Drásov, Czech Republic
Food and Agriculture Organization of the United Nations (2007) The world's mangroves 1980–2005, FAO Forestry Paper No. 153. FAO Forestry Department, Rome, p 77
Gao B-C (1996) NDWI – a normalized difference water index for remote sensing of vegetation liquid water from space. Remote Sens Environ 58:257–266
Gao J (1999) A comparatives study on spatial and spectral resolutions of satellite data in mapping mangrove forests. Intern J Remote Sens 20:2823–2833
Gilman E, Ellison J, Duke NC, Field C (2008) Threats to mangroves from climate change and adaptation options: a review. Aquat Bot 89 (2):237–250
Giri C, Zhu Z, Tieszen LL, Singh A, Gillette S, Kelmelis JA (2008) Mangrove forest distributions and dynamics (1975–2005) of the tsunami-affected region of Asia. J Biogeogr 35:519–528
Giri C, Ochieng E, Tieszen LL, Zhu Z, Singh A, Loveland T, Masek Jand Duke N (2011) Status and distribution of mangrove forests of the world using earth observation satellite data. Glob Ecol Biogeogr 20(1):154–159
Green ER, Mumb PJ, Edwards AJ, Clark CD, Ellis AC (1998) The assessment of mangrove areas using high resolution multispectral airborne imagery. J Coast Res 14:433–443
Green EP, Clark CD, Edwards AJ (2000) Image classification and habitat mapping. In: Edwards AJ (ed) Remote sensing handbook for tropical coastal management. UNESCO, Paris, pp 141–154
Hend K (2015) Mangroves in Qatar: perspectives. https://www.ecomena.org/mangroves-in-qatar/
Heumann BW (2011) An object based classification of mangroves using a hybrid decision tree – support vector machine approach. Remote Sens 3:2440–2460
Huete AR (1988) A soil-adjusted vegetation index (SAVI). Remote Sens Environ 25:295–309
Huete A, Didan K, Miura T, Rodriquez EP, Gao X, Ferreria LG (2000) Overview of the radiometric and biophysical performance of the MODIS vegetation indices. Remote Sens Environ 83:195–213
International Society for Mangrove Ecosystems and International Tropical Timber Organization. ISME (2004) Mangrove action plan for the sustainable management of mangroves 2004–2009. ISME, Okinawa
International Union for Conservation of Nature (IUCN) (1986) Sultanate of Oman. Proposals for a system of nature conservation areas. Report to Diwan of Royal Court, Sultanate of Oman. IUCN, Gland
International Union for Conservation of Nature (IUCN) (1986–89) Coastal Zone Management Plan. Reports to the Ministry of Commerce and Industry, Oman
IPCC (2007) Summary for policymakers. Climate change 2007: impacts, adaptation and vulnerability. Contribution of working group II to the fourth assessment report of the intergovernmental panel on climate change (ed by Parry ML, Can-ziani OF, Palutikof JP, van der Linden PJ, Hanson CE) Cambridge University Press, Cambridge, pp 7–22
Jensen JR, Lin H, Yang Y, Ramsey E, Davis BA, Thoemke CW (1991) The measurement of mangrove characteristics in Southwest Florida using SPOT multispectral data. Geocarto Int 2:13–21
Kairo JG, Kivyatu B, Koedam N (2002) Application of remote sensing and GIS in the management of mangroves forests within and adjacent to Kiunga Marine Protected Area. Lamu Kebya. Environ Dev Sustain 4:153–166
Khan MIR (1982) Mangrove forest of the United Arab Emirates. Pak J For 32(2):36–39
Khosravi M (1992) Ecological studies plan of Iranian mangrove forests. Report of mangrove forests identification phase. Department of the Environment, Tehran

Kovacs JM, Wang J, Verdugo F (2005) Mapping mangrove leaf area index at the species level using IKONOS and LAI-2000 sensor for the Agua Brava Lagoon, Mexican Pacific. Estuar Coast Shelf Sci 62:377–384

Kumar R, Silva L (1973) Light ray tracing through a leaf cross- section. Appl Opt 12:2950–2954

Lal P (2003) Economic valuation of mangroves and decision-making in the Pacific. Ocean Coast Manag 46(9-10):823–844

MA (Millennium Ecosystem Assessment) (2005) Mille ecosystem and human well-being: synthesis. Island Press, Washington, DC

Ma BL, Dwyer LM, Costa C, Cober ER, Morrison MJ (2001) Early prediction of soybean yield from canopy reflectance measurements. Argon J 93:1227–1234

Mahdavi A (2001) Investigation on quantitative and qualitative alterations of mangrove forests in Gheshm area by using of 1967 and 1994 aerial photos. University of Tehran

Manandhar R, Odeh IOA, Ancev T (2009) Improving the accuracy of land use and land cover classification of landsat data using post-classification enhancement. Remote Sens 1(3):330–344

Ministry of Regional Municipalities, Environment and Water Resources (MRMEWR)(2004a)

Ministry of Regional Municipalities, Environment and Water Resources (MRMEWR) (2004b) The master plan study on restoration, conservation and management of mangrove in the sultanate of Oman

Ministry of Environment and Climate Affairs (MECA) (2010) Fourth National report to the Convention on International Union for Conservation of Nature (IUCN) (1986) Sultanate of Oman. Proposals for a System of Nature Conservation Areas. Report to Diwan of Royal Court, Sultanate of Oman. IUCN, Gland, Switzerland

Nizwa Net (2014) Turtle species of Oman. http://www.nizwa.net/env/turtles/turtles.html. Accessed 3 Sept 2014

O'Grady AP, McGuinness KA, Eamus D (1996) The abundance and growth of Avicenia marina and Rhizophora stylosa in the low shore zone of Darwin Harbor, Northern Territory. Aust J Ecol 21:272–279

Peñuelas J, Gamon J, Fredeen A, Merino J, Field C (1994) Reflectance indices associated with physiological changes in nitrogen- and water-limited sunflower leaves. Remote Sens Environ 48:135–146

PERSGA/GEF (2004) Status of mangroves in the Red Sea and Gulf of Aden, PERSGA Technical Series No 11. PERSGA, Jeddah

Qi J, Chehbouni A, Huete AR, Kerr YH, Sorooshian S (1994) A modified soil adjusted vegetation index (MSAVI). Remote Sens Environ 48:119–126

Quarto A (2005) Mangrove restoration-natural protection from natural disasters. Biodiversity (Ottawa) 6:3–12

Ramsey E, Jensen JR (1996) Remote sensing of mangrove wetlands: relating canopy spectra to site-specific data. Photogramm Eng Remote Sens 62:939–948

Ras Luffan Industrial City (2005) Mangrove conservation program

Rasolofoharinoro R, Blaso F, Bellan MF, Aizpuru M, Gauquelin T, Denis J (1998) A remote sensing based methodology for mangrove in Madagascar. Int J Remote Sens 19:1873–1886

Richardson AJ, Wiegand CL (1977) Distinguishing vegetation from soil background information. Photogramm Eng Remote Sens 43:1541–1552

Richer R (2009) Conservation in Qatar: impacts of increasing industrialization. Center for International and Regional Studies, Doha

Rodgers JC, Adam Murrah W, William H, Giri C, Bruce C (2009) monitoring mangrove forest dynamics of the Sundarbans in Bangladesh and India using multitemporal satellite data from 1973 to 2000. Estuar Coast Shelf Sci 73(1–2):91–100

Saenger P (1993) Management of mangroves in the Kingdom of Saudi Arabia. Saline Water Conversion Corporation and Ministry for Agriculture and Water, Riyadh

Safiari Sh, Nasouri M (2009) Mangrove forests development. Ministry of Jihad-e Agriculture, Forest, Range and Watershed

Samira O (2004) Information provided for the Global Forest Resources Assessment (FRA) 2005 thematic study on mangroves. Unpublished

Scott DA (1995) A directory of Wetlands in the Middle East. IUCN, Gland. Switzerland and IWRB, Slimbridge, U.K. xvii + 560pp, 13 maps

Shariatnegad S (2005) Information provided for the Forest Resources Assessment (FRA) thematic study on mangroves. Unpublished

Sherrod CL, McMillan C (1985) The distributional history and ecology of mangrove vegetation along the northern Gulf of Mexico coastal region. Contrib Mar Sci 28:129–140

Spalding M, Blasco F, Field CD (1997) World mangrove atlas. The International Society for Mangrove Ecosystems, Okinawa

Spooner M (1970) Oil spill in Tarut Bay, Saudi Arabia. Mar Pollut Bull 1(11):166–167

Sremongkontip S, Hussin YA, Groenindijk L (2000) Detecting changes in the mangrove forests of southern Thailand using remotely sensed data and GIS. Int Arch Photogramm Remote Sens 33:567–573

Stutzenbaker CD (1999) Aquatic and wetland plants of the western gulf coast. Texas Parks and Wildlife Press, Austin. 465 p

Supreme Council for the Environment & Natural Reserves (2004) A national biodiversity strategy and action plan – State of Qatar. Final National Biodiversity Strategy and Action Plan. www.biodiv.org/doc/world/qa/qanbsap-01-en.doc

Thiam S, Eastmen RJ (1999) Chapter on vegetation indices. In: Guide to GIS and image processing. 2; Idrisi Production. Clarke University, Worcester, pp 107–122

Tong PHS, Auda Y, Populus J, Aizpura M, Habshi AAL, Blasco F (2004) Assessment from space of mangrove s evolution in the Mekong Delta, in relation to extensive shrimp farming. Int J Remote Sens 25:4795–4812

Tsai-Ming L, Hui CY (2009) Applying remote sensing techniques to monitor shifting wetland vegetation: a case study of Danshui River estuary mangrove communities. Taiwan Ecol Eng 35 (4):487–496

UNEP (1999) Overview on land-based sources and activities affecting the marine environment in the ROPME Sea Area. UNEP Regional Seas Reports and Studies No. 168. 127 pp.

Vaiphasa C, Skidmore AK, de Boer WF (2006) A post-classifier for mangrove mapping using ecological data. ISPRS J Photogramm Remote Sens 61:1–10

Vits C, Tack J (1995) The use of remote sensing as information source for environmental accounting of coastal areas in Kenja, Feasibility Study Reference. University of Ghent No. T3/02/603, 1e45

Zahed A, Rouhani M, Mohajeri S, Bateni F (2010) An overview of Iranian mangrove ecosystems, northern part of the Persian Gulf and Oman Sea. Acta Ecol Sin 30:240–244

Current Status of Mangrove Wetlands in Sinaloa: A Biological Corridor Along the Eastern Margin of the Gulf of California, México

4

Marlenne Manzano-Sarabia, Olivia Millán-Aguilar, Francisco Flores-Cárdenas, Lidia Rodríguez-Arredondo, Mayra I. Grano-Maldonado, and Mario Nieves-Soto

Abstract

The state of Sinaloa, located in northwestern Mexico, possesses significant mangrove coverage (ranked fourth in this country), which can be considered a biological corridor because of its ecological relevance for many resident and migratory species. Several ecosystem services are provided by mangrove wetlands; however, diverse drivers of change, mainly anthropogenic stressors, e.g. urban development, pollution, agricultural and aquaculture activities, have modified this large ecosystem during the last decades. It has been reported that Sinaloa has lost more than 5000 ha since 1985, yet the impacts on mangrove structure and functioning are still poorly understood. Furthermore, the frequency and intensity of climate phenomena like "El Niño" events might impact the phenology of mangroves in this region and deserve further studies. Bioaccumulation of contaminants, land cover change, aquaculture, hydrological changes, low and increasing temperature conditions and the impact of hurricanes on mangroves are discussed in this chapter in addition to recommendations for future studies, e.g. the impact of plagues and phytopathogens on mangroves in Sinaloa.

Keywords

Mangroves · NDVI · Remote sensing · Land cover change

4.1 Introduction

Mangrove studies have increased worldwide due to their biological, ecological, economic and social relevance. Although these ecosystems have characteristics of both marine and terrestrial environments, they also have their own characteristics and adaptation mechanisms that define them (e.g. viviparity, root type, salt gland secretion, pneumatophores; Tomlinson 1986). The distribution of mangroves is limited mainly by temperature, but also by geographic barriers. Globally, they are located in tropical and subtropical zones between 30°N and 30°S latitudes, although mangrove patches have been documented in higher latitudes and explained by the arrival of warm currents. These ecosystems are considered among the most productive on earth due the amount of organic matter they produce (Vega-Rodríguez 2008). For this reason, a large variety of animal species—including those with commercial value, e.g. penaeid shrimps—use mangroves as a refuge as nursery areas to develop in these ecosystems during early stages of their life cycle. Further, mangroves provide diverse ecosystem services, e.g. provisioning (food), regulating/supporting (storm protection) and cultural (providing aesthetic and recreational landscape) (Brander et al. 2012).

4.1.1 Mangroves in Mexico

Mexico is considered a megadiverse country. Its geographical position between the Nearctic and Neotropical regions places it among the top twelve sites with the highest biodiversity in the world because of its vast tropical forests and wetlands, the extension of its coastline and the wide variety of flora and fauna, reporting many endemic species. This country is considered among the six countries with the largest mangrove coverage in the world, ranking fourth with 764,486 ha (which represents 0.4% of the total area). Mangroves in Mexico are distributed from Chiapas to Baja California Sur in the coast of the Pacific Ocean, and also from southern Quintana Roo in the Caribbean to Laguna Madre in Tamaulipas in the Gulf of Mexico (Fig. 4.1). Eighty one priority sites have been identified because of their biological relevance, most of them located in the Yucatán Peninsula (Rodríguez-Zúñiga et al. 2013).

M. Manzano-Sarabia (✉) · O. Millán-Aguilar · F. Flores-Cárdenas
L. Rodríguez-Arredondo · M. I. Grano-Maldonado · M. Nieves-Soto
Facultad de Ciencias del Mar, Universidad Autónoma de Sinaloa, Mazatlán, Sinaloa, Mexico
e-mail: mmanzano@uas.edu.mx

C. Makowski, C. W. Finkl (eds.), *Threats to Mangrove Forests*, Coastal Research Library 25,
https://doi.org/10.1007/978-3-319-73016-5_4

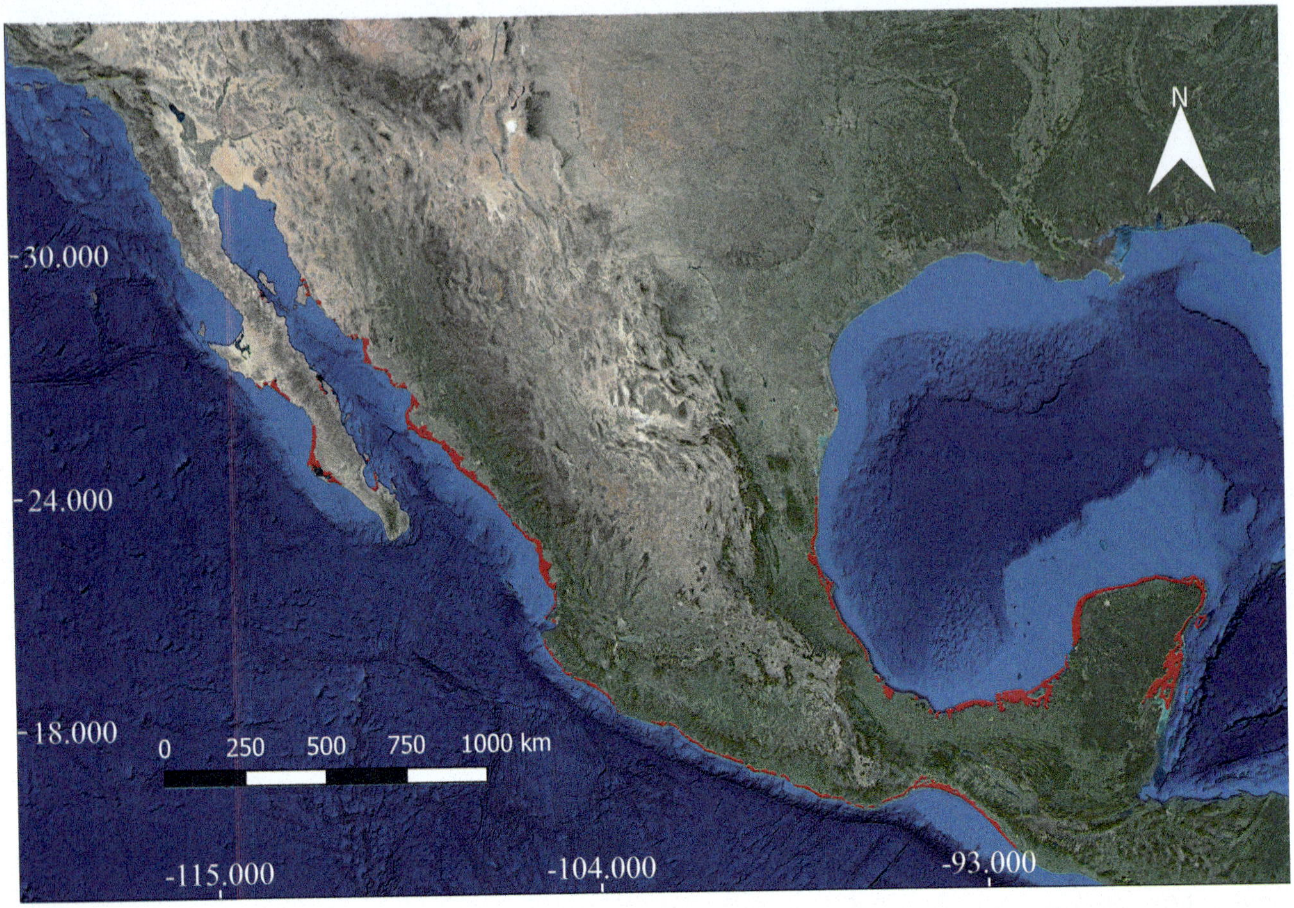

Fig. 4.1 Mangrove distribution in Mexico (red areas) (Data from: CONABIO (http://www.conabio.gob.mx/informacion/gis/), Google Earth and Quantum GIS)

Four mangrove species are mainly reported in Mexico: *Avicennia germinans* (black mangrove), *Rhizophora mangle* (red mangrove), *Laguncularia racemosa* (white mangrove) and *Conocarpus erectus* (buttonwood mangrove). All are considered at risk and under protected categories according to official laws (e.g. NOM-059-SEMARNAT-2010). The development of these four species of mangroves depends on different environmental conditions, e.g. microtopography, weather conditions, precipitation levels, and salinity, among others. There are anthropogenic activities which are forcing change to the landscape in Mexico, as reported in many regions globally, i.e. agriculture, livestock breeding, urban development and aquaculture infrastructure. Landscape changes within the mangrove areas were more evident in the decade of 1990s, when significant modifications in the Mexican Constitution laws related to land tenure occurred (Morzaria-Luna et al. 2014).

4.1.2 Current Situation of Mangroves Corridor in the State of Sinaloa

Biological corridors are areas or extensions along a territory that link ecosystems with populations that need to be protected, e.g. Natural Protected Areas (NPA) that need to be preserved. Biological corridors are established to protect ecosystems and to provide direct benefit to humans, being a platform for conservation throughout the sustainable management of natural resources. The integration of biological corridors in Mexico[1] began in 2002 as a response to the observed loss of biodiversity mainly in protected areas affected by fragmentation processes that cause isolation among populations. Wetlands in northwestern Mexico are also considered a biological corridor, and Sinaloa ranked fourth on mangrove coverage in Mexico (10.2% of the total coverage) according to an analysis performed in 2005 using multi-sensor data (SPOT-5 and Landsat-7 ETM) (Rodríguez-Zúñiga et al. 2013). A fragmented landscape represents a disturbed or stressed ecosystem which affects the associated biodiversity within. When an ecosystem—like the mangrove wetlands—is more fragmented or experiences high deforestation rates, the result may be population isolation, affecting the ecology of the species. According to Rodríguez-Zúñiga

[1] e.g. the Mesoamerican Biological Corridor which comprises Central America and some States of Mexico (Chiapas, Oaxaca, Tabasco, Yucatán, Campeche and Quintana Roo).

Table 4.1 Comparison of mangrove coverage (ha) over time in northwestern Mexico highlighting Sinaloa, the focus of the current study

AREA	1985	2005	2010
Baja California	26,760	26,555	26,730
Sonora	10,940	11,098	11,334
Sinaloa	**82,074**	**79,065**	**77,088**
Nayarit	78,024	69,784	66,920
TOTAL	197,798	186,502	182,072

Source: Rodríguez-Zúñiga et al. (2013)

et al. (2013), Sinaloa has lost ~5000 ha of mangroves since 1985 (Table 4.1), with a considerable increase in mangrove disturbance (2229 ha was reported for 2010) mainly explained by excessive anthropic development. Sinaloa mangroves showed signs of fragmentation from 1985 to 2005, meaning that large mangrove patches disintegrated causing more fragments and mangrove loss (3009 ha). In general, Sinaloa state has the ideal geographical conditions for aquaculture, agriculture and fishery activities, which has led to large affected mangrove areas. Natural phenomena (e. g. hurricanes) are another cause of transformation of these ecosystems in Sinaloa, as will be discussed in the following sections. In addition, climate phenomena are another source of disturbance in mangrove ecosystems, which has caused certain deterioration in these areas over the past decades. Because of these stressors, the mangrove wetland corridor of northwestern Mexico, including Sinaloa, must be preserved to protect the species that inhabit this area, many of them endemic and threatened.

4.2 Hazards and Vulnerability

4.2.1 Bioaccumulation of Contaminants

Mangroves may be used as bio-indicators of heavy metals (e. g. mercury, cadmium, zinc, chromium) as those are accumulated in sediments where this vegetation is established. Gold and silver mining activities are performed in the mountain ranges of Sinaloa where mercury is often used to optimize the separation of metals from the sediment (Castro-Díaz 2013). Despite the established protocols to control mining waste, there are records of contamination by spills in rivers (e.g. Concordia municipality); therefore, the transport of heavy metals from the mountain ranges to the mangrove wetlands may be considered a hazard for several local and resident species. Recent studies have analyzed the concentration of heavy metals in oysters *Crassostrea spp*. farmed in coastal aquatic ecosystems from Sinaloa and reported concentrations of 0.05–0.37 μg/g of total mercury (Delgado-Álvarez et al. 2015), not exceeding the Mexican legislation yet but still must be monitored because of the increasing human activities along the coastline. Similarly, other studies have been performed in the region with mine influence on the presence of trace metals (arsenic, cadmium, copper, lead and zinc) in plants (halophytes and mangrove leaves). By sampling the mangrove soil, they reported trace metals in roots and leaves of the analyzed plants, and while this represents no risk for them, this finding does not exclude bioaccumulation in organisms such as: mollusks and some fish species (Sánchez-Martínez et al. 2017).

4.2.2 Land Cover Change

Urban development, pollution, sedimentation, agricultural and aquaculture activities have contributed significantly to transform ecosystems into human landscapes (Drexler and Ewel 2001; Kovacs et al. 2001; Aburto-Oropeza et al. 2008) affecting mangroves and associated biota. According to official estimates, in the decade of the 1980s, mangroves covered 19.8 million hectares in the world's coastal zones, but ~25% has been lost during the last 20 years (CONABIO 2008). Mexico is not an exception, as human activities (*human dimensions*) have impacted coastal ecosystems significantly. Although mangroves are considered threatened or at risk under Mexican law (NOM-ECOL-059-2010), these regulations to protect them have been insufficient to ensure their conservation (e.g. mangrove logging is an illegal activity; however, it is still a common practice performed mainly by local residents to build "palapas", a traditional shelter roofed with branches on the beach; Fig. 4.2).

Since the 1970s, satellite-derived studies have been used to analyze mangrove coverage in Mexico (Rollet 1974). Using this data, inventories of mangrove wetlands have been conducted at the national level by the National Commission for the Knowledge and Use of Biodiversity (CONABIO) in order to identify their distribution and changes over time. In Mexico, mangroves covered an area of 774,090 ha in 2005 (Rodríguez-Zúñiga et al. 2012). Sinaloa recorded a considerable decrease of mangrove coverage during 1985–2005, decreasing from 82,073 to 79,064 ha, while disturbed areas increased 193 ha. Mangroves in other parts of northwestern Mexico had a similar trend. Coverage in Baja California remained unchanged from 1982 to 2005 (35.7 ha), however, anthropic development increased 1.67%. Mangroves in Baja California Sur decreased ~200 ha from 1975 to 2005, and anthropic development increased considerably by 3000 ha. The largest mangrove complex in the coast of the Mexican Pacific Ocean, and located between the states of Sinaloa and Nayarit, is "Marismas Nacionales". This zone registered a decrease of more than 8000 ha (1970–2005) due the opening of the Cuautla channel in 1972 (Rodríguez-Zúñiga et al. 2012). According to Rodríguez-Zúñiga et al. (2013), mangrove coverage comprised 764,486 ha in 2010 with a loss of 9604 ha. For

Fig. 4.2 Mangrove logging in Huizache-Caimanero lagoon, Sinaloa, Mexico (January 2013)

the northwestern region, mangroves covered 170,738 ha, ~5000 ha less than those reported for 2005.

4.2.3 Aquaculture

Remote sensing techniques are a common method to evaluate the extent of aquaculture in a specific area (Green et al. 2000). According to analysis using remote sensing techniques, 35% of the worldwide mangrove areas were lost during the last two decades derived from aquaculture (Valiela et al. 2001). One of the main threats to wetlands—particularly mangroves—in Mexico is the increase of aquaculture infrastructure due to modifications related to land tenure in the Mexican legislature during the decade of the 1990s. Alatorre et al. (2016) analyzed changes in mangrove coverage in the Gulf of California and found a direct relationship between increased aquaculture area and declining health status of mangrove ecosystems (negative trend in Normalized Difference Vegetation Index or NDVI). Aquaculture ponds have increased in coastal lagoons in Sinaloa (Fig. 4.3). According to official records of established shrimp farms in Sinaloa (CESASIN, unpublished data), Ahome, a municipality in northern Sinaloa has registered the greatest extension dedicated to this activity (10,658 ha). Additional impacts derived from this activity are related to water discharges into coastal lagoons which modify water quality affecting terrestrial and aquatic biota.

4.2.4 Hydrological Changes

Dams increase vulnerability of wetlands and mangrove ecosystems, as they disturb biogeochemical cycles, e.g. the flow of organic carbon (Cooke et al. 2005), nutrients balance, dissolved oxygen, salinity, and temperature (Friedl and Wüest 2002). For example, the balance in the estuaries depends of the interaction of freshwater inflow of the rivers with the influence of the ocean. Changes in the inflow of rivers will eventually increase salinity in coastal lagoons and patterns of sediment transport affecting the health of mangroves (Twilley and Rivera-Monroy 2009). From the Sierra Madre Occidental—the largest mountain range in Mexico—to the coast of Sinaloa, seven main rivers flow (Table 4.2), which are classified under hydrological regions 10 and 11 (CONAGUA 2015), named Sinaloa and Presidio-San Pedro basins, respectively. Most of these rivers were impounded, starting in 1948, with the Sanalona dam over the Culiacán river (Tables 4.2), in order to promote agricultural development in this region. Until 1995, nine more dams were built in the north-central part of the state. Recently, the Picachos dam (impounding the Presidio river in the southern region) was concluded in 2012; additionally, the Santa María dam (impounding the Baluarte river) is estimated to be concluded in 2018 (Table 4.3).

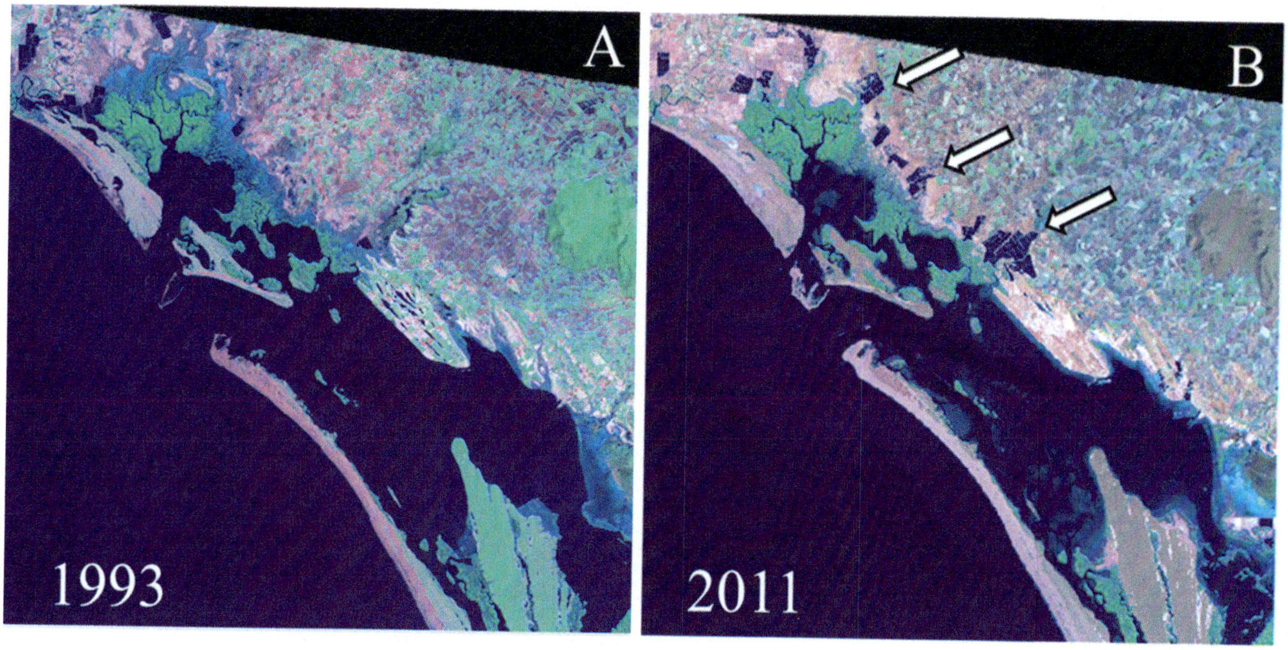

Fig. 4.3 Comparison of established shrimp ponds (white arrows) located in northern Santa María-La Reforma lagoon in Sinaloa, Mexico (Source: Landsat TM 5 images)

Table 4.2 Surface drainage of main rivers in Sinaloa during 2014

River	Surface drainage (Millions of m^3 per $year^{-1}$)	Basin area (km^2)	River length (km)
Sinaloa	2100	12260	400
Culiacán	3122	15731	875
San Lorenzo	1665	8919	315
Elota	463	2324	–
Piaxtla	1406	11473	220
Presidio	1084	6479	
Baluarte	1830	5094	142

CONAGUA (2015)

It is expected that the vulnerability of coastal lagoons to higher salinity conditions will increase over time. For example, while the Baluarte river in southern Sinaloa has not been impounded yet, currently, the Caimanero lagoon—influenced by this river—registers changes in salinity with near euhaline conditions most of the year, contrasting with old reports that described predominant mixohaline conditions (Flores-Cárdenas et al. 2016). It is expected that vulnerability to higher salinity conditions will increase once the Baluarte river is impounded.

4.2.5 Low Temperatures

Mangrove vegetation is susceptible to low winter air temperature (Lonard and Judd 1991). The effect is greater when low temperature periods are prolonged because this vegetation possesses limited capacity to withstand frozen temperatures for long time; therefore, latitudinal distribution is strongly linked to extreme low temperatures events (Yando et al. 2016). Although Sinaloa is located within tropical and subtropical climates, there are records of temperatures lower than 3 °C in winter, mainly caused by winter seasonal storms and cold fronts that influence northwestern Mexico. These conditions may cause severe damage in vegetation, including crops and mangroves. During the winter of 2014–2015, cold fronts were reported for northern Sinaloa causing extreme low temperatures in Agiabampo lagoon (northern lagoon). According to in situ prospection (Fig. 4.4) and satellite data (not shown), dead mangrove patches were identified after this event. It is not clear if the frequency and/or intensity of cold events during winter months in northwestern Mexico will increase in the near future, thereby threatening mangrove coverage in northern Sinaloa.

4.2.6 Increasing Temperature

Increasing temperature is a concern for the scientific community worldwide. Whether the origin is related to climate

Table 4.3 Main dams, volumetric capacity and start-up operations in the state of Sinaloa

Dam	River	Municipality	Year	Volume (hm^3)
Adolfo López Mateos	Badiraguato	Badiraguato	1969	3086
Miguel Hidalgo y Costilla	El Fuerte	El Fuerte	1956	2921
Luis Donaldo Colosio	El Fuerte	Choix	1995	2908
José López Portillo	San Lorenzo	Cosalá	1983	2580
Gustavo Díaz Ordaz	Sinaloa	Sinaloa de Leyva	1981	1737
Sanalona	Tamazula	Culiacán	1948	673
Josefa Ortiz de Dominguez	El Fuerte	El Fuerte	1967	514
Ing. Aurelio Benassini Vizcaíno		Elota	1988	403
Lic. Eustaquio Buelna	Mocorito	Guamuchil	1972	90
Ing. Juan Guerrero Alcocer	El Bledal stream		1994	55
Picachos	Presidio	Mazatlán	2012	322
Santa María (in progress)	Baluarte	El Rosario	2018	753

Data obtained from CONAGUA (2015)

Fig. 4.4 Patches with dead mangroves (*) after extreme cold conditions affecting the Agiabampo lagoon system in northern Sinaloa

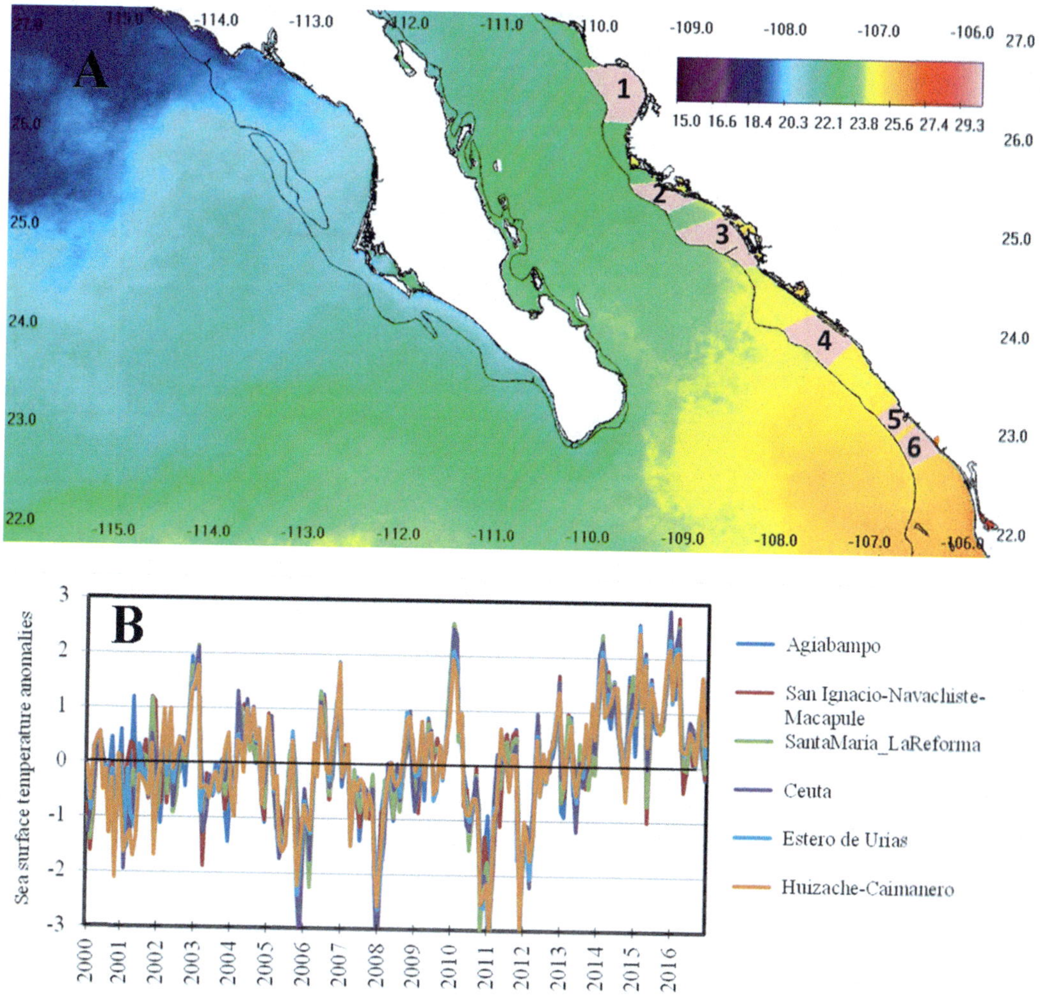

Fig. 4.5 (**a**) Selected areas adjacent to coastal lagoons* and within the continental shelf of Sinaloa and (**b**) calculated anomalies of sea surface temperature. *1: Agiabampo, 2: San Ignacio-Navachiste-Macapule, 3: Santa María-La Reforma, 4: Ceuta, 5: Estero de Urías, 6: Huizache-Caimanero

change and/or global warming, recent trends of sea surface temperature in the continental shelf of Sinaloa is the focus here. Temperature conditions may influence adjacent coastal lagoons and, therefore, mangroves forests.

Temperature influences mangroves in several ways, e.g. latitudinal distribution, species composition and phenology, in addition to affecting the assimilation of CO_2 due to limitations to biochemical reactions (Ward et al. 2016). Mangrove species are known to be resilient to anomalous warm periods; however, extended conditions with higher temperatures may increase their vulnerability, affecting survival. The temporal evolution of sea surface temperature has been similar for six studied areas, showing an increasing trend after 2010 (Fig. 4.5). The highest anomalies were observed by the end of 2015, matching with an "El Niño" event, one of the strongest on record.

Similarly, interstitial temperature recorded with a data logger in a mangroves area of the Caimanero lagoon in southern Sinaloa followed a comparable evolution (Fig. 4.6). The highest monthly mean values occurred during September–October 2015 and August–September 2016, matching with the expected warm conditions following "El Niño" in 2015. However, the impacts of this event on mangroves are not yet determined.

4.2.7 Hurricanes

Hurricanes are natural phenomena that may impact mangrove structure through wind damage, storm surges and sediment deposition (Smith et al. 2009), and its effects might be evident for several years afterward (Doyle et al.

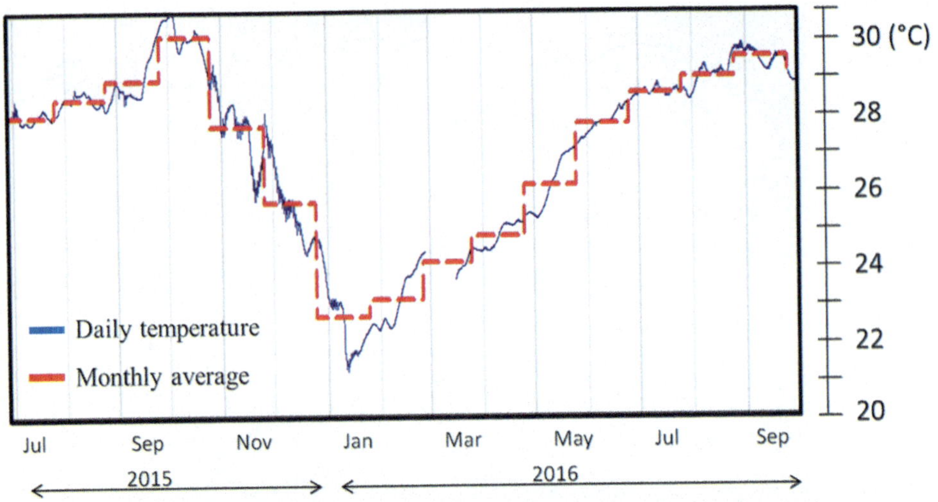

Fig. 4.6 Interstitial temperature in a mangrove zone in the Caimanero lagoon, Sinaloa, recorded with a data logger

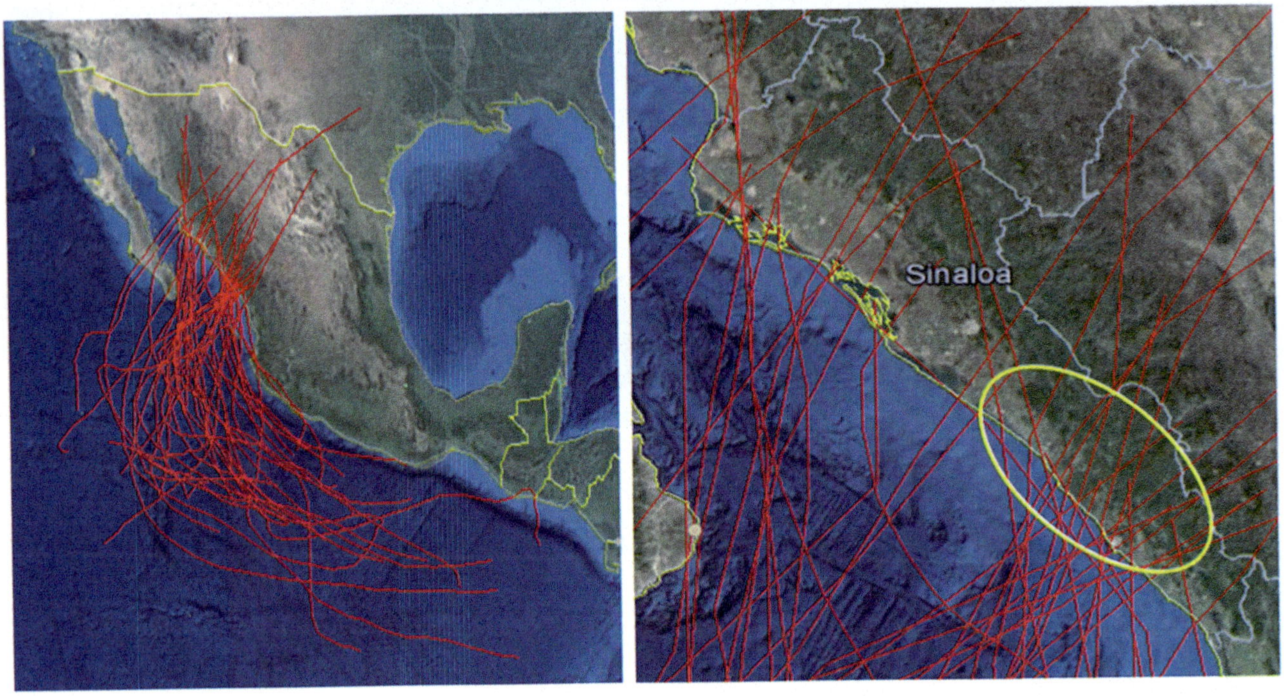

Fig. 4.7 Cumulative trajectories of hurricanes and tropical storms affecting Sinaloa state (1949–2013). Yellow circle shows the most vulnerable area (From Ceuta to Huizache-Caimanero lagoon system) to hurricanes trajectories

2009). Sinaloa is located in the eastern margin of the Gulf of California, under the hurricanes corridor of the Mexican Pacific. Seasonally, these events register trajectories that influence northwestern Mexico, mainly the Baja California peninsula. In central and southern Sinaloa, historical records (1949–2013) of hurricanes were analyzed, and the Ceuta and Huizache-Caimanero coastal lagoons were found to be the most vulnerable to these events (Fig. 4.7). However, the direct and indirect impact of hurricanes on mangrove forests in Sinaloa, unfortunately, remains unknown.

4.3 Discussion

The ecological relevance of the mangrove wetland corridor in Sinaloa, Mexico, has been highlighted through diverse designated categories, e.g. Ramsar sites[2] (Agiabampo,

[2] A Ramsar site refers to wetlands with ecological relevance in the world and named under the Ramsar Convention. An area of 8,643,579 ha (142 sites) have been designed as Ramsar sites in Mexico.

Santa María-Topolobampo-Ohuira, San Ignacio-Navachiste-Macapule, Santa María-La Reforma, Ceuta, Huizache-Caimanero, Marismas Nacionales). Many resident and migratory species use this biological corridor in at least one part of their life history. Diverse stressors have affected the corridor beginning decades ago, causing a reduction in coverage and, likely, lesser health status. Consistent with worldwide reports, the main identified drivers of change of mangroves in Mexico are: aquaculture, livestock, urban development, agriculture and climate change, with fewer reports for the last topic. Particularly, the impact of climate phenomena like the "El Niño-Southern Oscillation" must be investigated to identify the vulnerability of mangrove wetlands to these events.

The many coastal lagoons in Sinaloa are located in areas characterized by a myriad of differences among them, e.g. climate conditions, hydrology, precipitation patterns, and temperature, among others. Therefore, it is necessary to study the ecological history for each lagoon and mangrove wetland. For instance, Huizache-Caimanero in southern Sinaloa is one of the most studied lagoon systems in the northwest of Mexico during the last 50 years, due to its relevance as breeding and natural habitat of penaeid shrimp species of commercial importance. Although it was designed as a Ramsar site (http://www.ramsar.org/wetland/mexico), its level of protection is low compared to its ecological and economical relevance. Notorious signs of disturbance by human activities are evident in this wetland, not only at the landscape level, but also genetic disturbance at the molecular level has been identified (Millán-Aguilar et al. 2016). Ecological and anthropological studies have reported diverse ecosystem services that Huizache-Caimanero has offered to local residents, even since pre-Columbian periods. Consequently, this ecosystem has suffered from an important anthropic influence for centuries. For example, contemporary fishermen as well as ancient ones used seasonally fixed fishing structures named "tapos" to trap shrimp in early their phases and keep them enclosed in an area until reaching a larger length size. This physical barrier restricts propagation of mangrove's propagules, from the interior and exterior, causing both physical and genetic isolation (Millán-Aguilar et al. 2016). Of particular concern is the fragmentation of mangrove wetlands and the impact on pollinators as their movement patterns might be affected on fragmented landscapes, reducing pollen dispersal and, therefore, increasing vulnerability to biparental inbreeding (Cresswell 1997).

Regarding the social perception of the users of these ecosystems, there is a general lack of knowledge of the ecological relevance of mangrove wetlands, e.g. a 2013 study showed that 56.5% of surveyed fishermen did not know the designation of Huizache-Caimanero as a *Ramsar site* nor its meaning (not published). This lack of awareness by direct users further emphasizes the immediate need to promote the diversity of ecosystems services that are provided by mangrove wetlands in order to protect and preserve them. Education on a broad, societal level is clearly necessary. In addition, another primary concern is to obey and monitor the Mexican laws and international agreements concerning the conservation of these valuable ecosystems.

In recent years, the multiple options of media resources make dissemination easier regarding the ecology of mangrove wetlands, its relevance, and function. An important platform for such purposes is *Google* and its related applications, e.g. *Google Earth* and *Street View* (Fig. 4.6a). The latter allows virtual journeys to sites of ecological, cultural, historical significance and urban reference through a special collection using the *Street View Trekker* (Fig. 4.6b). Since December 2016, a virtual 360° tour into mangrove areas in Sinaloa is available through *Street View.*[3] This represents a unique opportunity for dissemination purposes to the general public, but it also offers scientific relevance. It provides a spatial and temporal record of the structure of mangrove wetlands using 360 degree panoramic photography as a reference for future studies, e.g. landscape ecology (Fig. 4.8).

Phytopathogen studies are of interest because mangrove forests are widespread in subtropical and tropical coastal ecosystems and are frequently surrounded by agricultural fields. However, the narrow focus of the study and occurrence of several phytopathogens on mangrove forests has not been explored deeply. The current knowledge regarding studies on diseases and plagues on mangroves is scarce. Few studies have focused on outbreaks of locust plague, fungus, among others (Todd et al. 2002; Reef et al. 2012; Wanderley-Costa et al. 2012). Research on mangroves plagues is important to understand the basic interactions between hosts, vectors and their parasites, i.e. migrant pest (locust), bacteria, consumption of leaves by insects or mollusks, plant fitness, prevention of insect herbivory on mangrove leaves, the diversity of endophytic fungi including its distribution, host and tissue preference or community composition. All this, unfortunately, remains practically unknown[4] in Mexico. Therefore, there is a serious need to increase the knowledge about mangrove-associated parasites, the diagnosis, management and control of phytopathogens and the identification of the factors that influence their prevalence.

[3] A project supported by *Google* and Universidad Autónoma de Sinaloa. Source: M. Manzano-Sarabia, Principal Investigator.

[4] A literature search performed in June, 2017, on *Web of Science* showed no records when combining the words "parasites, mangroves, Mexico" or "plagues, mangroves, Mexico" in a basic search.

Fig. 4.8 The Huizache Lagoon, Sinaloa. (**a**) 360 degree panoramic photography from *Street View, Google*. (**b**) Field work using the *Street View Trekker*. Technology featured by *Google* provides 360 degrees panoramic views. This equipment has 15 cameras which take photographs simultaneously every two seconds. This photographic device is placed on the back of the user

4.4 Conclusions

The mangrove wetland corridor along the state of Sinaloa possesses high ecological relevance for resident and migratory species. However, anthropogenic and climate related stressors are modifying its coverage and structure, affecting biodiversity and ecosystem services. While diverse efforts to protect and preserve the corridor are currently taking place, they are not yet sufficient to guarantee its conservation. In addition, diverse studies have been performed related to landscape ecology, mangrove structure, and the impact of human dimensions; however, there is a lack of knowledge about the impacts of other stressors, e.g. the effects of plagues, hurricanes and climate change on this vegetation, which might be a motivation for future studies.

Acknowledgments Authors acknowledge financial support provided by PROMEP UAS-NPTC UAS-PTC-039, PROFAPI 2011/100, 2012/098, 2013/106, Infraestructura-CONACyT 280994 and CONABIO LM004 grants.

References

Aburto-Oropeza O, Ezcurra E, Danemann G, Valdez V, Murray J, Sala E (2008) Mangroves in the Gulf of California increse fishery yields. Environ Sci 105(30):10456–10459

Alatorre LC, Sánchez-Carrillo S, Miramontes-Beltrán S, Medina RJ, Torres-Olave ME, Bravo LC, Wiebe LC, Granados A, Adams DK, Sánchez E, Mario U (2016) Temporal changes of NDVI for qualitative environmental assessment of mangroves: shrimp farming impact on the health decline of the arid mangroves in the Gulf of California (1990–2010). J Arid Environ 125:98–109

Brander L, Wagtendonkb AJ, Hussainc SS, McVittie A, Verburg PH, de Groot RS, van der Ploeg S (2012) Ecosystem service values for mangroves in Southeast Asia: a meta-analysis and value transfer application. Ecosyst Serv 1(1):62–69

Castro-Díaz J (2013) Commission for environmental cooperation an assessment of primary and secondary mercury supplies in Mexico. Commission for Environmental Cooperation, Quebec. 98p

CONABIO (2008) Manglares de México. México. 38p

CONAGUA (2015) Atlas del agua en México. Comisión Nacional del Agua, México. 137p

Cooke GD, Welch EB, Peterson SA, Nichols SA (2005) Restoration and management of lakes and reservoirs, Third edn. CRC Press, Boca Raton. 616p

Cresswell JE (1997) Spatial heterogeneity, pollinator behavior and pollinator-mediated gene flow: bumblebee movements in variously aggregated rows of oil-seed rape. Oikos 78:546–556

Delgado-Álvarez CG, Ruelas-Inzunza J, Osuna-López JI, Voltolina D, Frías-Espericueta MG (2015) Total mercury content in cultured oysters from NW Mexico: health risk assessment. Bull Environ Contam Toxicol 94(2):209–213

Doyle TW, Krauss KW, Wells CJ (2009) Landscape analysis and pattern of hurricane impact and circulation on mangrove forest of the everglades. Wetlands 29(1):44–53

Drexler JZ, Ewel KC (2001) Effect of the 1997–1998 ENSO-related drought on hydrology and salinity in a micronesian wetland complex. Estuaries 24(3):347–356

Flores-Cárdenas F, Hurtado-Oliva MA, Doyle TW, Nieves-Soto M, Díaz-Castro S, Manzano-Sarabia M (2016) Litterfall production of mangroves in Huizache-Caimanero lagoon system, Mexico. J Coast Res 33(1):118–124

Friedl G, Wüest A (2002) Disrupting biogeochemical cycles – consequences of damming. Aquat Sci 64(1):55–65

Green EP, Mumby PJ, Edwards AJ, Clark CD (2000) Remote sensing handbook for tropical coastal management. UNESCO Publishing, Paris

Kovacs JM, Wang J, Blanco-Correa M (2001) Mapping disturbances in a mangrove forest using ulti-date Landsat TM imagery. Environ Manag 27(5):763–766

Lonard RI, Judd FW (1991) Comparison of the effects of the severe freezes of 1983 and 1989 on native woody plants in the Lower Rio Grande Valley, Texas. Southwest Nat 36:213–217

Millán-Aguilar O, Manzano-Sarabia M, Nettel-Hernanz A, Dodd RS, Hurtado-Oliva MÁ, Velázquez-Velázquez E (2016) Genetic Diversity of the Black Mangrove Avicennia germinans (L.) Stearn in Northwestern Mexico. Forests 7:197

Morzaria-Luna HN, Castillo-López A, Danemann GD, Turk-Boyer P (2014) Conservation strategies for coastal wetlands in the Gulf of California, Mexico. Wetl Ecol Manag 22(3):267–288

Reef R, Ball M, Lovelock C (2012) The impact of a locust plague on mangroves of the arid Western Australia coast. J Trop Ecol 28 (3):307–311

Rodríguez-Zúñiga MT, Troche-Souza C, Vázquez-Lule AD, Márquez-Mendoza JD, Vázquez-Balderas B, Valderrama-Landeros L, Velázquez-Salazar S, Uribe-Martínez A, Acosta-Velázquez J, Díaz-Gallegos J, Cruz-López MI, Ressl R (2012) Los manglares de México: estado actual y establecimiento de un programa de monitoreo a largo plazo: 2ª y 3era etapas, Informe final SNIB-CONABIO proyecto No. GQ004. Comisión Nacional para el Conocimiento y Uso de la Biodiversidad, México

Rodríguez-Zúñiga MT, Troche-Souza C, Vázquez-Lule A, Márquez-Mendoza JD, Vázquez-Balderas B, Valderrama-Landeros L, Velázquez-Salazar S, Cruz-López MI, Ressl R, Uribe-Martínez A, Cerdeira-Estrada S, Acosta-Velázquez J, Díaz-Gallegos J, Jiménez-Rosenberg R, Fueyo-MacDonald L, Galindo-Leal C (2013) Manglares de México/Extensión, distribución y monitoreo. Comisión Nacional para el Conocimiento y Uso de la Biodiversidad, México. 128p

Rollet B (1974) Introduction a l'Etude des Mangroves du Mexique. Part 2. Rev Bois Forest Tropiques 157:53–74

Sánchez-Martínez MA, Riosmena-Rodríguez R, Marmolejo-Rodríguez AJ, Sánchez-González A (2017) Trace elements in two wetland plants (*Maytenus phyllanthoides* and *Salicornia subterminalis*) and sediment in a semiarid area influenced by gold mining. Regi Stud Mar Sci 10:65–74

Smith TJ, Anderson GH, Balentine K, Tiling G, Ward GA, Whelan KRT (2009) Cumulative impacts of hurricanes on Florida mangrove ecosystems: sediment deposition, storm surges and vegetation. Wetlands 29(1):24–34

Todd MC, Washington R, Cheke RA, Kniveton R (2002) Brown locust outbreaks and climate variability in southern Africa. J Appl Ecol 39:31–42

Tomlinson PB (1986) The botany of mangroves. Cambridge Univ. Press, Cambridge. 413p

Twilley RR, Rivera-Monroy V (2009) Sediment and nutrient tradeoffs in restoring Mississippi River Delta: restoration vs eutrophication. J Contemp Water Res Educ 141(1):39–44

Valiela I, Bowen JL, York JK (2001) Mangrove forests: one of the world's threatened major tropical environments. Bioscience 51:807–815

Vega-Rodríguez M (2008) Estimating primary productivity of red mangroves in southwestern Puerto Rico from remote sensing and field measurements. Ms. Sci.Thesis. Puerto Rico

Wanderley-Costa IPM, Costa-Maia L, Auxiliadora-Cavalcanti M (2012) Diversity of leaf endophytic fungi in mangrove plants of northeast Brazil. Braz J Microbiol 43:1165–1173

Ward RD, Friess DA, Day RH, MacKenzie RA (2016) Impacts of climate change on mangrove ecosystems: a region by region overview. Ecosyst Health Sustain 2(4):e01211

Yando ES, Osland MJ, Willis JM, Day RH, Krauss KW, Hester MW (2016) Salt marsh-mangrove ecotones: using structural gradients to investigate the effects of woody plant encroachment on plant–soil interactions and ecosystem carbon pools. Aust J Ecol 104:1020–1031

Geospatial Assessment of Spatio-Temporal Changes in Mangrove Vegetation of Pichavaram Region, Tamil Nadu, India

5

M. Vani and P. Rama Chandra Prasad

Abstract

The present work is a multi-temporal satellite based spatial dynamic study of an important coastal habitat, the Pichavaram mangrove ecosystem, over a period of 15 years. The study discusses the importance and the status of mangroves on both global and regional scales. Maximum likelihood estimate method of supervised classification technique has been used to classify the land use-land cover changes in the Pichavaram Reserve Forest, Killai Reserve Forest and Pichavaram Extension. The status of the classes has been monitored using Landsat ETM+ of 2000, 2006, 2011, and OLI of 2016 satellite imageries. The total area of the Pichavaram mangrove showed a net increase of approximately 11.41% of the total study area within a span of 15 years (2000–2016).

Keywords

Mangroves · Remote sensing · Supervised classification · Spatial variability analysis · Inter-class variability analysis · Coast

5.1 Introduction

India has a long coastline of about 7516 km including its island territories, which consists of a variety of coastal habitats such as estuaries, mangroves, coral reefs, mudflats, lagoons (Venkataramana 2007; Mandal and Naskar 2008). These coastal areas are considered as critical habitats as they are unique, fragile and exhibit high biodiversity by supporting several coastal and marine ecosystems (Burbridge and Koesoebiono 1981; Salm and Clark 1984; Sivakumar 2013). However, the combined pressure of natural processes and human activities lead to deterioration and loss of these critical habitats over years (Valiela et al. 2001; Polidoro et al. 2010; Prasad et al. 2010, 2017). The United Nations Conference on Environment and Development (UNCED 1992) emphasises the need for protection of coastal and marine environment through Agenda 21. There is a need to protect these productive areas from natural and human interventions while ensuring sustainable, rational utilisation of their resources and services (Giuseppe Barbaro 2013).

5.2 The Mangroves

Mangroves, the inter-tidal halophytic vegetation are one among the most productive forest ecosystems providing wide variety of ecological goods and services (Barbier et al. 2004; Kumar et al. 2012; Singh et al. 2012; Prasad et al. 2017). Apart, they also form excellent source of livelihood for coastal population (Dahdouh-Guebas et al. 2000; Hussain and Badola 2010; Orchard et al. 2016). They are found along sheltered coastlines, shallow-water lagoons, estuaries, rivers or deltas of tropical and sub-tropical countries (FAO 2007) covering 75% of global area (Spalding et al. 1997).

The word mangrove may be derived from Portuguese and English. 'Mangue' means tree, and 'grove' means group of trees (Dawes 1998; Dholakia 2004). Mangroves are of two dominant types, the riverine-type that fringes rivers and tidal creeks and the open water type that are directly exposed to waves (Lugo and Snedaker 1974). The former type is the most common. The vegetation composition of mangroves consists of many species with varied ability to live in brackish to seawater conditions (Peter 2015).

M. Vani (✉) · P. Rama Chandra Prasad
Lab for Spatial Informatics, International Institute of Information Technology, Hyderabad, India
e-mail: m.vani@research.iiit.ac.in; rcprasad@iiit.ac.in

C. Makowski, C. W. Finkl (eds.), *Threats to Mangrove Forests*, Coastal Research Library 25,
https://doi.org/10.1007/978-3-319-73016-5_5

5.2.1 Importance

Mangroves protect the land behind as a barrier from cyclones and tsunami (Ghazali et al. 2016). They also act as a buffer against floods and prevent soil erosion by trapping fine sediments that are carried into coastal zone by floodwaters (Van Santen et al. 2007; Barbier 2011). Thus these inorganic nutrients from the terrestrial runoff are circulated within the mangrove environment (Salim et al. 2015). Leaves littered by mangroves contribute largely to the organic matter available to the ecosystem (Robertson 1988; Numbere and Camilo 2017). Due to their productive nature, they serve as nurseries for prawns, crabs, lobsters and various fishes (Chong et al. 1990; Jeyaseelan and Krishna Murthy 1980).

They also serve as a shelter for a number of endangered species such as crocodile, turtle and pelican (Wilkinson et al. 1994). They offer a variety of commercial utilities in the form of wood for timber and fuel, fodder for cattle and with substances of commercial value such as lignin, tannin, etc. (Cherian et al. 2014; Shyam et al. 2015; Razzaque 2017). As per FSI (2001) report on mangrove cover, mangroves fix more carbon per unit area than other phytoplankton in tropical oceans.

Owing to their high productivity and services they offer, they are threatened by anthropogenic interventions in addition to natural processes and disasters (Chaudhuri et al. 2015; Masood et al. 2015). As a result, high depletion in mangrove areas is noted across the world (Mesta et al. 2014). Hence there is a need to protect and restore these resourceful critical habitats (Prasad et al. 2017).

5.2.2 Status of Global Mangroves

Mangrove forests are distributed throughout the tropical and subtropical coasts of the world and are particularly well developed in estuarine areas of the tropics (Satyanarayana et al. 2011; Kannaiah et al. 2015). The most extensive mangrove areas are found in Asia, followed by Africa and America. World-wide mangroves are disappearing at an alarming rate (Giri et al. 2007). In some developing countries about 80% of mangroves were lost in the last three decades (MoEF 1989, 2008). Mangrove forests, that once covered 75% of the coastlines of tropical and sub-tropical countries, dropped to less than 50% of their spatial extent and mostly in degraded condition (Spalding et al. 1997). This leads to decline of fisheries, salinization of coastal soils, erosion, and land subsidence, as well as reduction in trapping carbon dioxide.

5.2.3 Status of Indian Mangroves

Mangroves occur in the coastal states and islands of India (Fig. 5.1). As per FSI (2015) report, mangroves in India cover an area of 4740 km^2 making up 3% of world mangrove areas. Majority of mangroves are found along east and west coast, followed by Andaman and Nicobar islands. Out of the total area, 57% of the mangroves are found on the east coast, 23% on the west coast and the remaining 20% on the Bay Islands (Andaman and Nicobar). Of the total 2764.55 km^2 of mangroves along the east coast of India, West Bengal has 2125 km^2, Orissa 215 km^2, Andhra Pradesh 397 km^2 and Tamil Nadu 27.55 km^{2}of total mangrove cover along the coast (Selvam et al. 2003).

The largest extent of mangroves occurs in West Bengal (Sundarbans) spread over 4200 km^2, followed by Andaman and Nicobar islands- 1190 km^2 (MoEF 1989, 2008). Small patches of mangroves are reported to be found in Gujarat, Maharashtra, Andhra Pradesh, Goa, Odisha, Tamil Nadu, Karnataka and Kerala. About 45 species of mangroves have been reported along the Indian coast line and the most dominant genera to which they belong are *Rhizophora, Avicennia, Bruguiera, Sonneratia, Conocarpus, Heretiera, Xylocarpus, Ceriops* and *Excoecaria* (MoEF 2008). Both natural forces and human interventions have caused the degradation of mangroves in India (Chaudhuri et al. 2015). Realising the importance of mangroves, the Government of India and many Non-Governmental Organizations have taken up measures for conservation of mangrove in the country (ICMAM 2000). In Tamil Nadu, mangroves are well developed in Pichavaram and Muthupet that extend between Vellar and Coleroon estuaries.

5.3 Remote Sensing Study of Mangroves

Mapping and monitoring of mangroves at regular intervals helps in analysing their status (decrease or increase) for better protection and implementing conservation strategies (Prasad et al. 2010). Replacing traditional methods, geospatial technology with use of remote sensing satellite data makes it easy to map, monitor and assess the status of mangrove forest ecosystems (Mesta et al. 2014). With the availability of up-to-date information on the status and conditions of this important ecosystem, remote sensing plays an important and effective role to assess, manage, and monitor the mangrove forest cover changes. Landsat MSS, Landsat Thematic Mapper (TM), and System Pour l'Observation de la Terre (SPOT) multi-spectral data are used for management practices.Several researchers used varied spatial and spectral satellite data in mapping and monitoring mangrove areas adopting different classification approaches to suggest possible mangrove restoration areas for future management and conservation (Baltzera et al. 2004; Bird et al. 2004; Fromard et al. 2004; Kovacs et al. 2004; Kovacs and Flores-Verdugo 2005; Nayak 2002; Nisansala et al. 2015; Mesta et al. 2014).

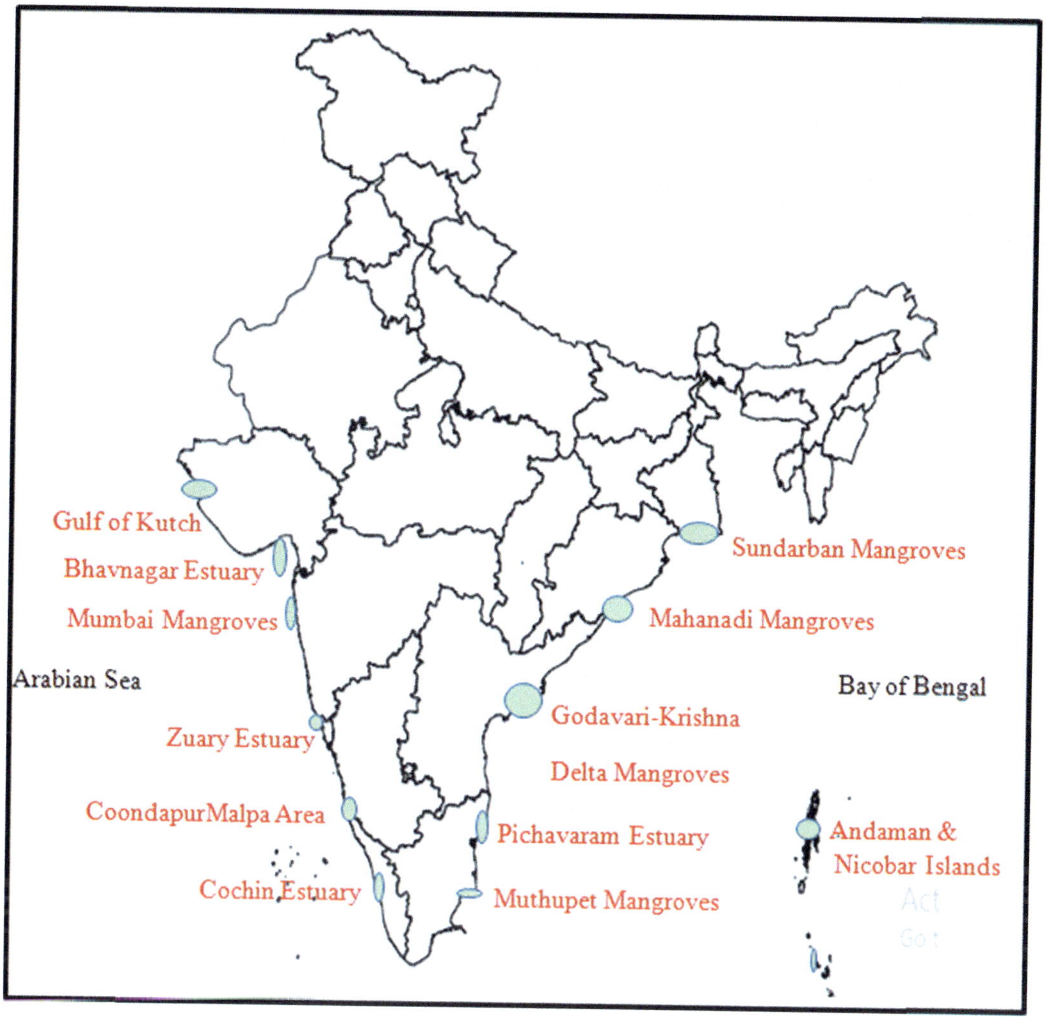

Fig. 5.1 Distribution of mangroves in India

High resolution images are preferable in change detection studies (Dixon et al. 2014).

While remote-sensing data analysis does not replace field inventory, it provides efficient and quick supplementary information. Large and inaccessible areas can be easily mapped and monitored. The use of remotely sensed data offers many advantages including across-the-board coverage, less expensive or free and historical satellite data for time-series analysis. In addition, recent advances in hardware and software used to process large volume of satellite data have enhanced the effectiveness of remotely sensed data (Prasad et al. 2010).

Remote sensing images of mangroves are for the most part reliable, accurate and collected consistently in space and time furthermore, the information can be obtained very quickly in near real time (Mesta et al. 2014). Other benefits of remote sensing in mangrove mapping are that data can be collected in a non-destructive manner since it is not necessary to move around in the mangrove areas to obtain data. Large, remote and inaccessible areas can therefore be easily mapped and monitored. Without doubt, the remote sensing technique is a serious alternative to the traditional field monitoring for large-scale tropical mangrove management (Srinivasa et al. 2011).

5.4 Study Area

Pichavaram mangroves form the major part of the mangrove forest in Cuddalore district of Tamil Nadu, located between 11°20′ to 11°30′ north and longitudes 79°45′ to 79°55′ east; and is one of the unique eco-tourism spots in south India (Fig. 5.2). It supports the existence of many rare varieties of economically important shell and finfishes, and serves as a home to local and migratory birds (MoEF 2008). The area was declared as a reserve forest in 1987 and covers an area of about 11 km^2 of which 50% constitute forest, 40% water and the rest covered by sand-flats and mud-flats (Krishnamurthy and Prince 1983). The area is surrounded by fishing villages, croplands, and aquaculture ponds.

As per the report from MoEF 2008, the Pichavaram mangrove wetland has 51 islets and the total area of the Vellar-

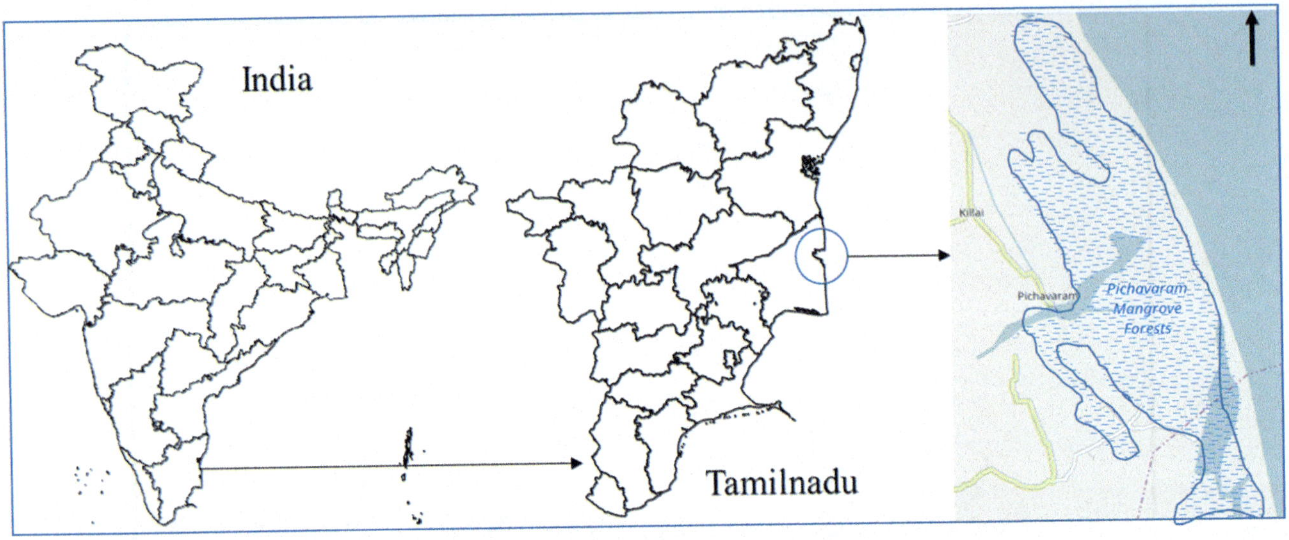

Fig. 5.2 Geographical location and extent of Pichavaram mangroves

Pichavaram-Coleroon estuarine complex is 23.355 km^2 of which only 2.41 km^2 is occupied by dense mangrove vegetation. Nearly 5.93 km^2 of this wetland are occupied by halophytic vegetation like *Suaeda*, 2.625 km^2 by barren mud flats and 12.385 km^2 by barren high saline soil. The report also says that the mangrove wetland of this total area occupies only 11 km^2, comprising the entire mangrove vegetation located in the middle portion of the Vellar-Pichavaram-Coleroon wetland.

A study conducted by Kathiresan et al. in 2005 reports that the aquaculture practise emerged in Pichavaram around 1980s and occupied about 3.99 km^2 in 1984; further increasing to 6.99 km^2 in 1996. It serves as a good example for degradation of mangroves due to aquaculture practices (Ranjan et al. 2008a). Furthermore, the anthropogenic (domestic and industrial) discharges from the nearby populated areas contribute to the ecological vulnerability of the ecosystem (Prasad 2005; Ranjan et al. 2008a, b).

While reports from MoEF (2008) shows accelerated degradation in the mangroves extent from 1970 through 1998, the study of Selvam et al. (2003) shows that there was an increase in the mangrove forests by 90% between 1986 and 2002 owing to restoration measures. A study conducted by Kathiresan and Rajendran (2005) aftermath the tsunami on 26 December 2004 indicates that the coastal mangrove forests can act as effective shield against tsunami. It is therefore an urgent call to initiate conservation and restoration measures detecting the hot spot areas in the mangrove forest.

5.4.1 Objective

In view of above context, in the current study we address spatio temporal changes in the mangrove vegetation of Pichavaram region of Tamil Nadu, India using different temporal satellite data sets.

Table 5.1 Description of satellite data used in the study

Satellite	Sensor	Spatial resolution	Date of acquisition
Landsat 7	ETM+	30 m	28th October 2000
Landsat 7	ETM+	30 m	30th January 2006
Landsat 7	ETM+	30 m	25th September 2011
Landsat 8	OLI	30 m	18th January 2016

5.4.2 Gaps Addressed in the Present Study

While some of the above works proved the effectiveness of remote sensing for studying the mangroves, there has been no significant study so far that addresses the change pattern in mangroves with a short period time interval of 5 years. The present study is an attempt in this direction; where in remote sensing techniques have been used to study the spatial dynamic of Pichavaram mangrove ecosystem over a period of 15 years. This methodology can be used for routine monitoring of the mangrove cover and plan restoration measures, wherever required.

5.4.3 Materials and Methods

Multi-temporal cloud-free satellite data of four periods viz., Landsat ETM+ of 2000, 2006, 2011 and Landsat OLI data of 2016 were downloaded from earthexplorer.usgs.gov (Table 5.1 and Fig. 5.3).

False Colour Composite (FCC) image of Pichavaram mangrove was generated with the band combinations of 2, 3, 4 and 3, 4, 5 for ETM+ and OLI data respectively

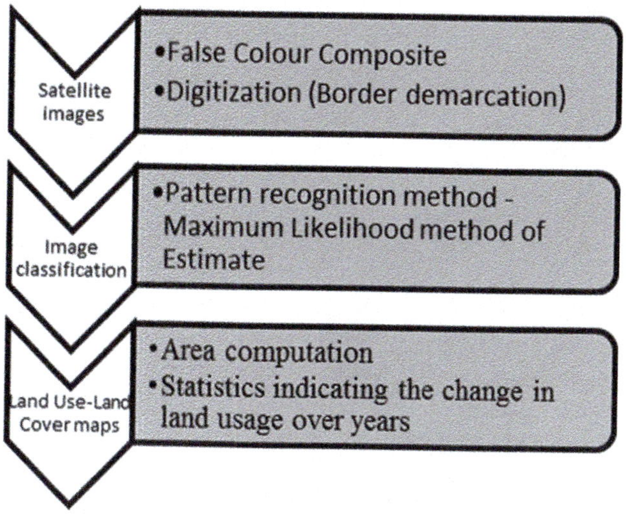

Fig. 5.3 Methodology for study on the spatio-temporal dynamic of the mangroves

(Fig. 5.4). Based on available literature data and maps, boundary of the study area was demarcated using QGIS. This vector data is used to clip the area of interest from the available satellite data of four periods. All the data sets were processed and subjected to image enhancement techniques to make the data more distinct spectrally for classification. Enhanced FCC aided in clearly distinguishing mangrove vegetation from other classes (water, mudflat, other vegetation, and sand). For classification of satellite data to create land use and land cover (LULC) maps, the clipped satellite images were imported into ILWIS software. Training sets were created for various LULC classes identified in satellite image, and was classified using maximum likelihood classifier. Finally area statistics were computed for four temporal LULC maps (Figs. 5.5, 5.6 and 5.7).

Fig. 5.4 False colour composite of remotely captured data in 2000

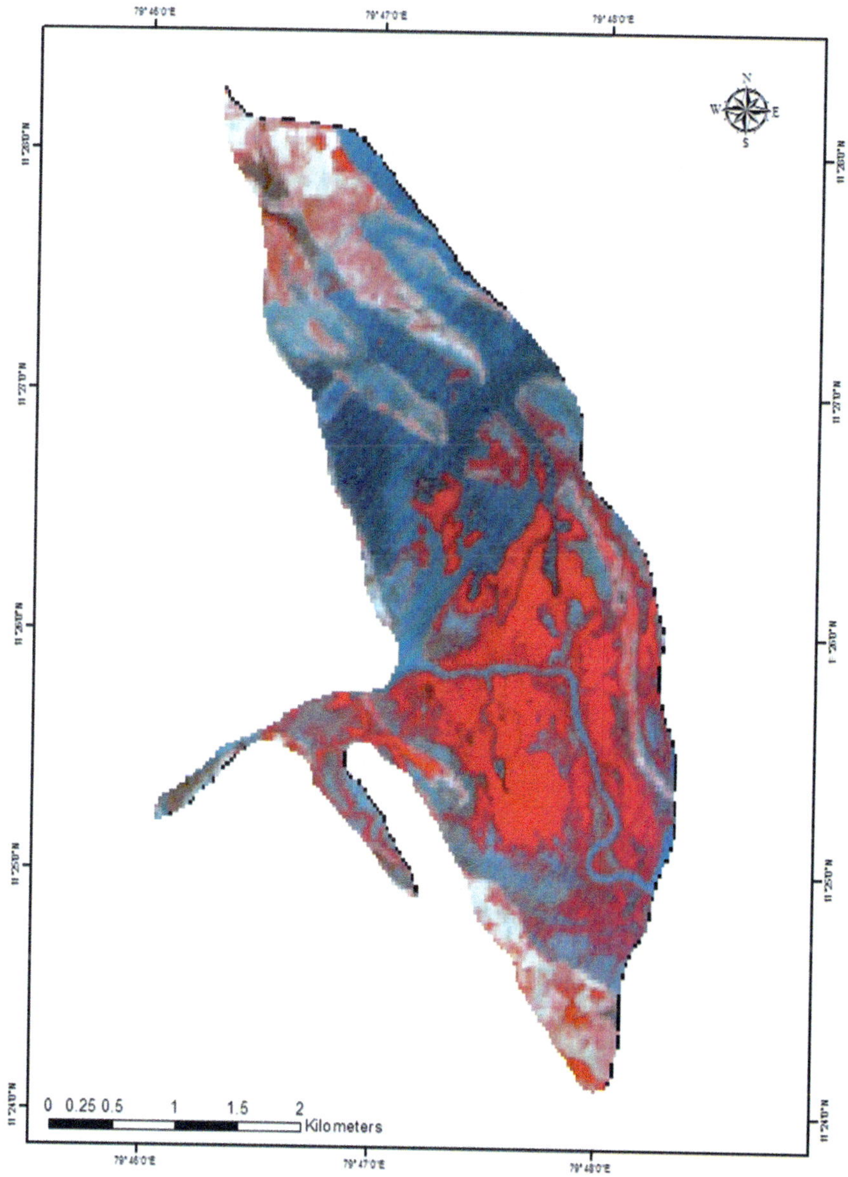

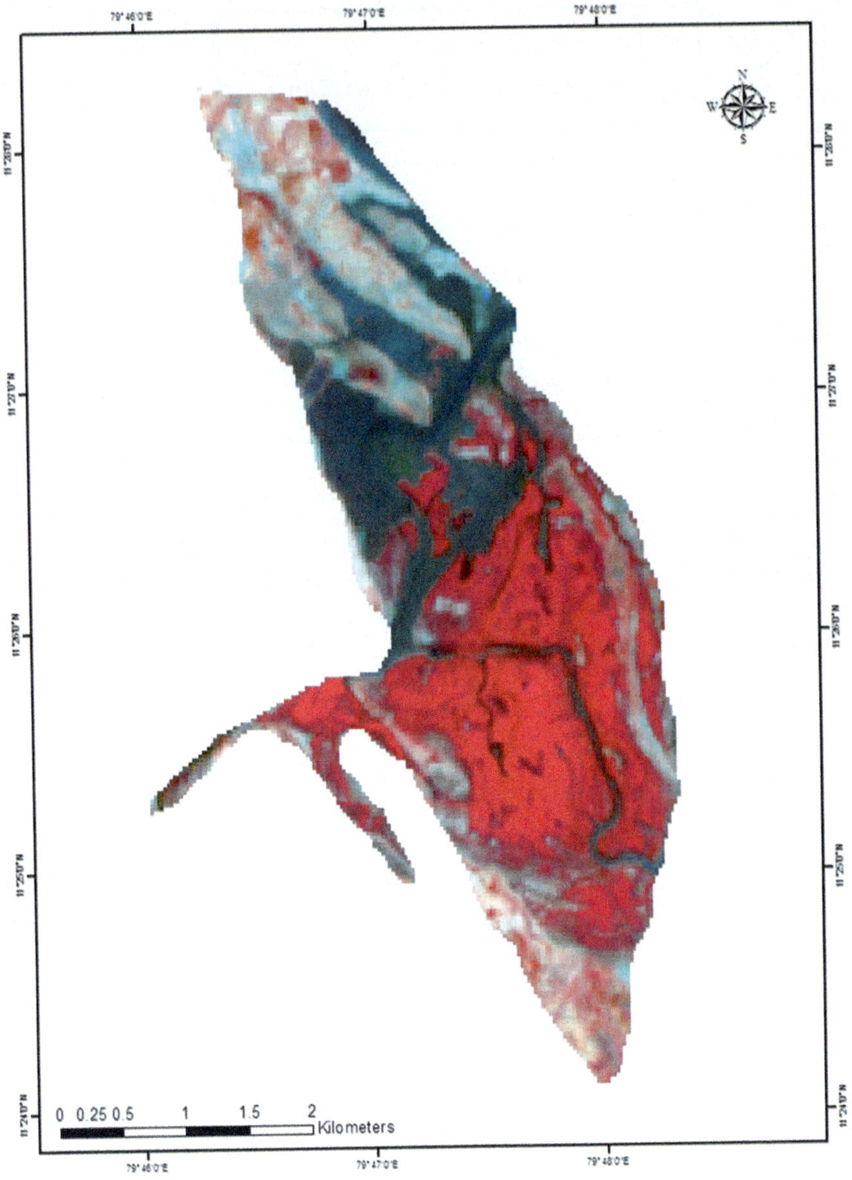

Fig. 5.5 False colour composite of remotely captured data in 2006

5.4.4 Results and Discussion

The spatio-temporal distribution of mangroves is shown in Figs. 5.8, 5.9, 5.10 and 5.11 respectively.

The statistics on spatial and temporal variation of the different classes for the study period is given in Figs. 5.12 and 5.13. The change analysis indicates that there is an increase in the mangrove cover from 2000 to 2016.

The total spatial extent of the mangroves is found to be increasing from 20.45% (332.43 km^2) in 2000, 22.37% (363.64 km^2) in 2006, and 27.19% (442 km^2) in 2011 to 31.86% (517.91 km^2) in 2016 of the total study area. However, the maximum rate of increase in mangrove cover is observed during 2006–2011 with 11.41% percentage increase in area. It is also noted that the class mudflat has also increased by about 14.74% indicating the spatial increase of the platform for mangrove regeneration.

Earlier studies indicated that mangroves have decreased from 6.4 to 3.72 km^2 between 1970 and 1987 (Selvam et al. 2002). This was mainly due to changes in the topography due to coupe felling system of management followed by various government agencies since 1911 (ICMAM 2000). In this system, healthy mangrove forests were clear felled for revenue generation in 15 and 20 years rotation. This caused various topographical changes in the biophysical condition of the mangrove wetlands leading to development of hyper saline condition that prevented natural regeneration of

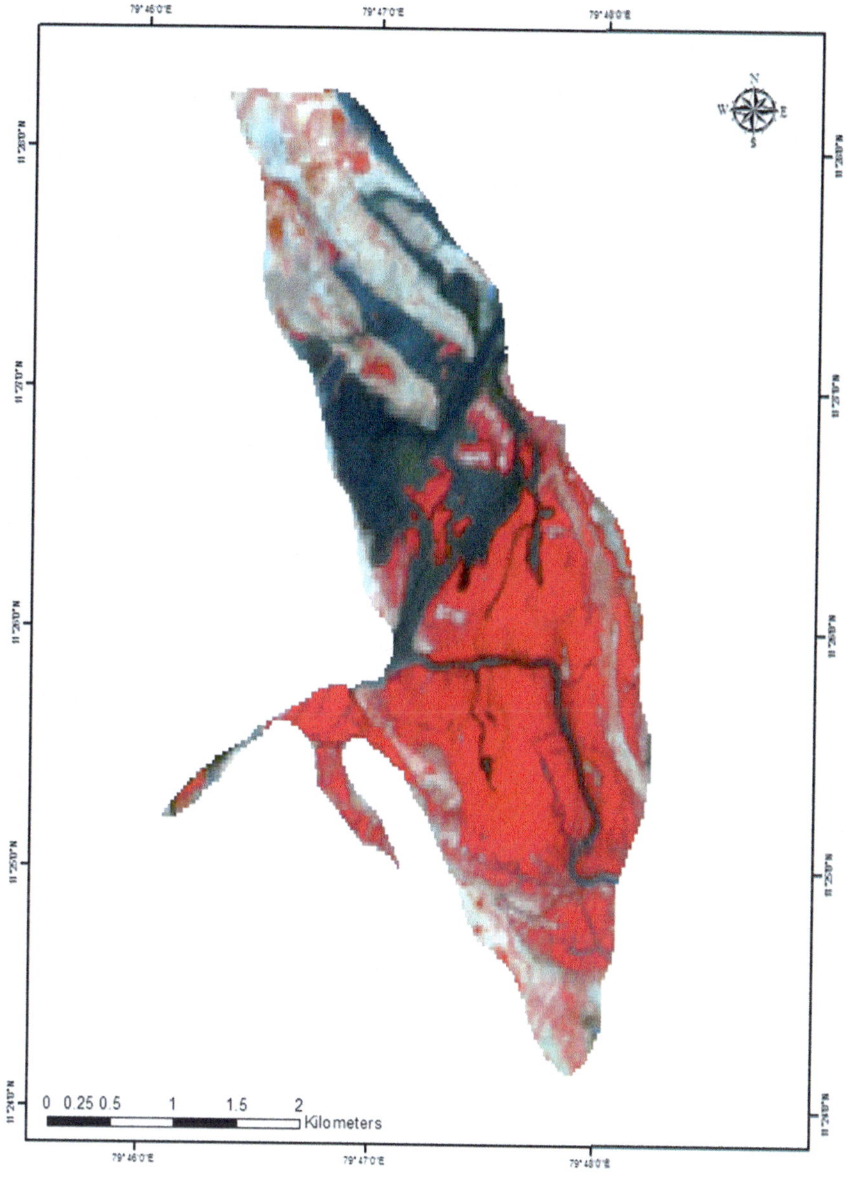

Fig. 5.6 False colour composite of remotely captured data in 2011

mangroves (Selvam et al. 2002; Prasad et al. 2017). In areas covered by coupe felling, the topography becomes trough shaped; tidal water enters and becomes stagnant leading to the development of hypersaline condition. This hyper saline condition was the major reason for the degradation of mangroves. As per the reports of MoEF (2008), cattle grazing may be a possible factor for mangrove degradation. It is also quoted that grazing continued on the peripheral areas of mangroves.

Kathiresan (2000) reported that within the last century, there has been a decline in 75% of the mangrove cover in Pichavaram and of the existing 25%, only 10% has dense vegetation, while the remaining 90% of the area has been degrading. The current study, however, indicates an increasing trend in the spatial extent of the mangrove cover of the region. During the entire period of study, it is observed that the total area of Pichavaram mangrove has gradually increased and a significant reduction in the degraded mangroves extent from 19.80% in 2000 to 7.93% of the total study area in 2016. This increase in the spatial extent of the healthy mangroves from 2000 to 2016 is attributed to the efforts of the State Forest Department along with M. S. Swaminathan Research Foundation (MSSRF), a Non-Governmental Organization who flushed the tidal waters into the areas where it was not reachable earlier (ICMAM 2000; MoEF 2008). This resulted in recruiting of young mangrove saplings into the area.

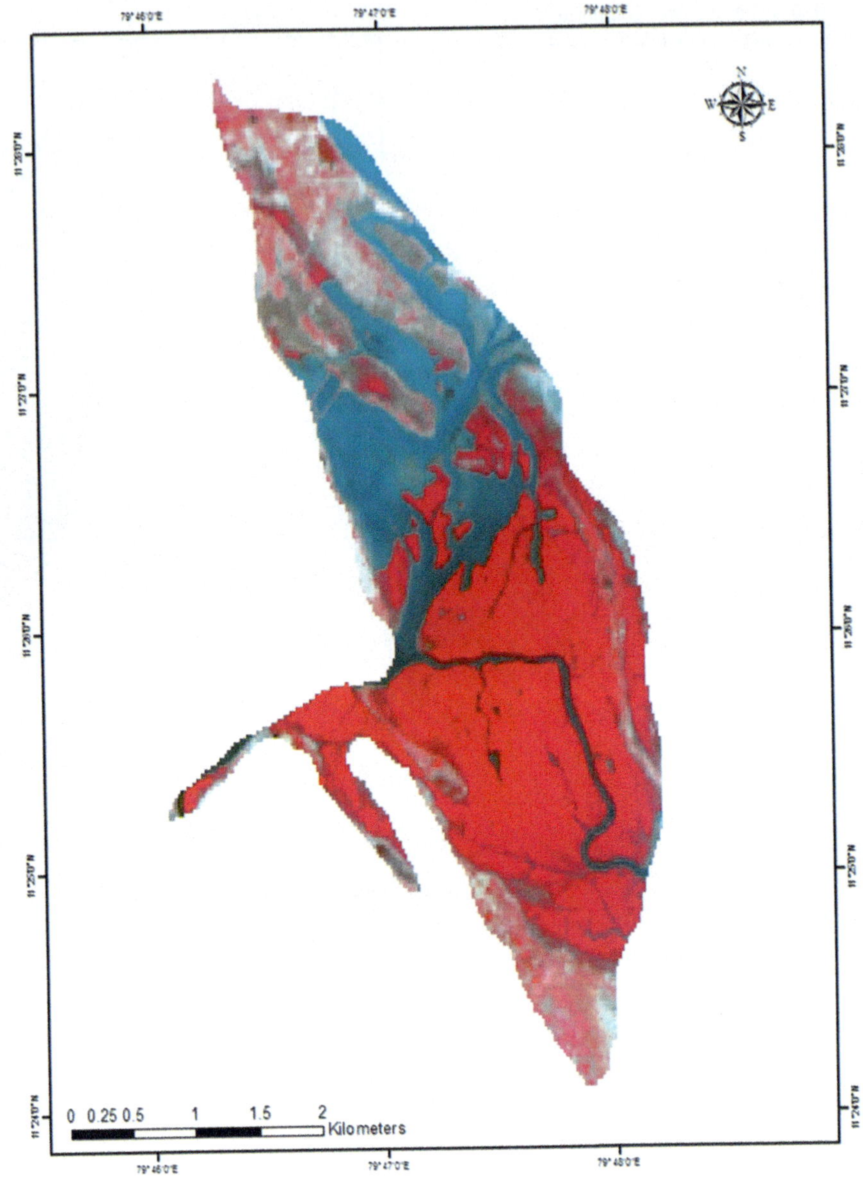

Fig. 5.7 False colour composite of remotely captured data in 2016

The MoEF (2008) report states that unsustainable exploitation of mangroves led to its degradation before the Indian Ocean tsunami in December 2004. It was that disastrous event that made the locals understand the real protecting capacity of the mangroves and the need to conserve them. Efforts to establish a nursery of endangered mangrove species has been taken up by the Tamil Nadu Forest Department. The report also states that mangroves were planted on the shore along the mangrove forest area; in over 0.5 km^2 of marshy areas and 0.02 km^2 of saltpan with dug out canals to ensure good tidal flushing.

5.5 Conclusions

Remote Sensing and GIS have been found to be extremely useful for understanding the coastal habitats, especially mangroves. The study demonstrated that by analysing the LU-LC of a location using remote sensing data and by studying their temporal variations, it is possible to get a reasonably accurate picture of the current status and change trend of the location over years. Arriving at community-level maps would require specific and more complicated techniques in conjunction with extensive ground-truthing which is time consuming. Contrary to some of the earlier

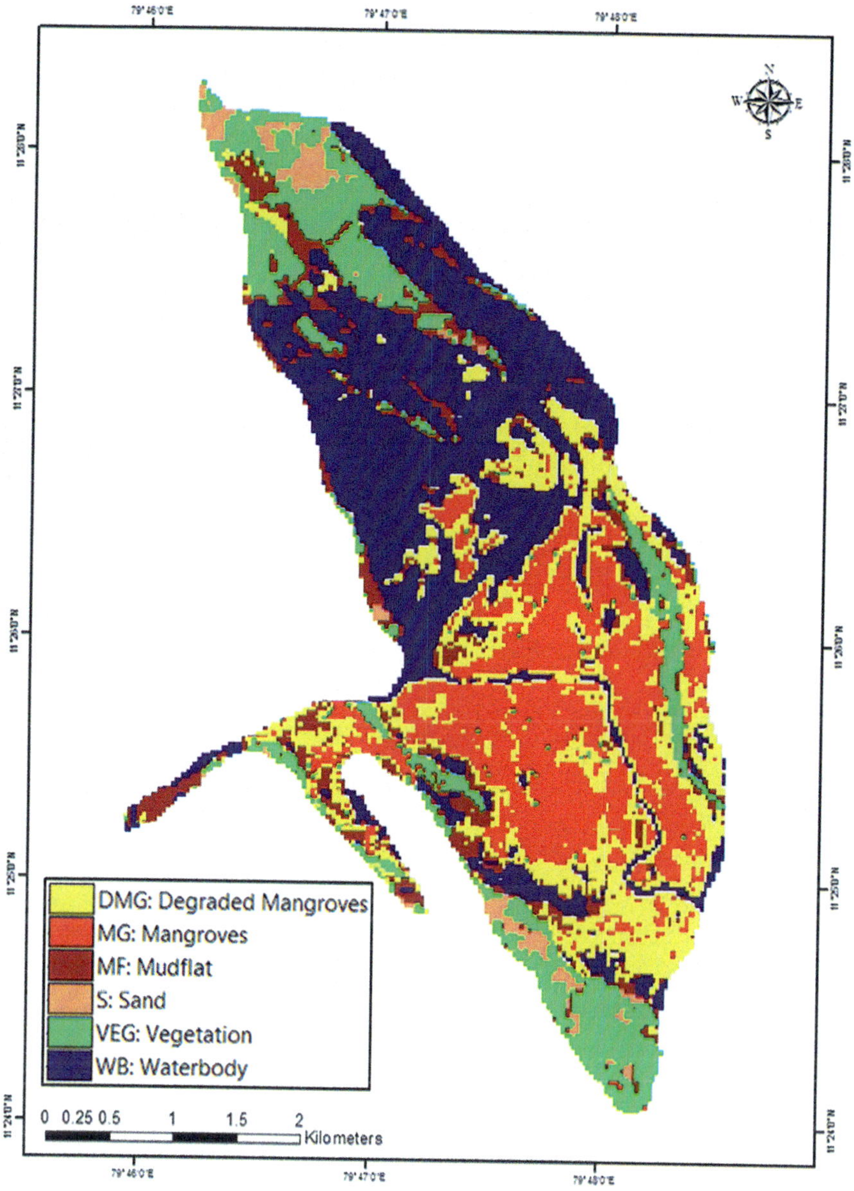

Fig. 5.8 Spatial distribution of mangroves in 2000

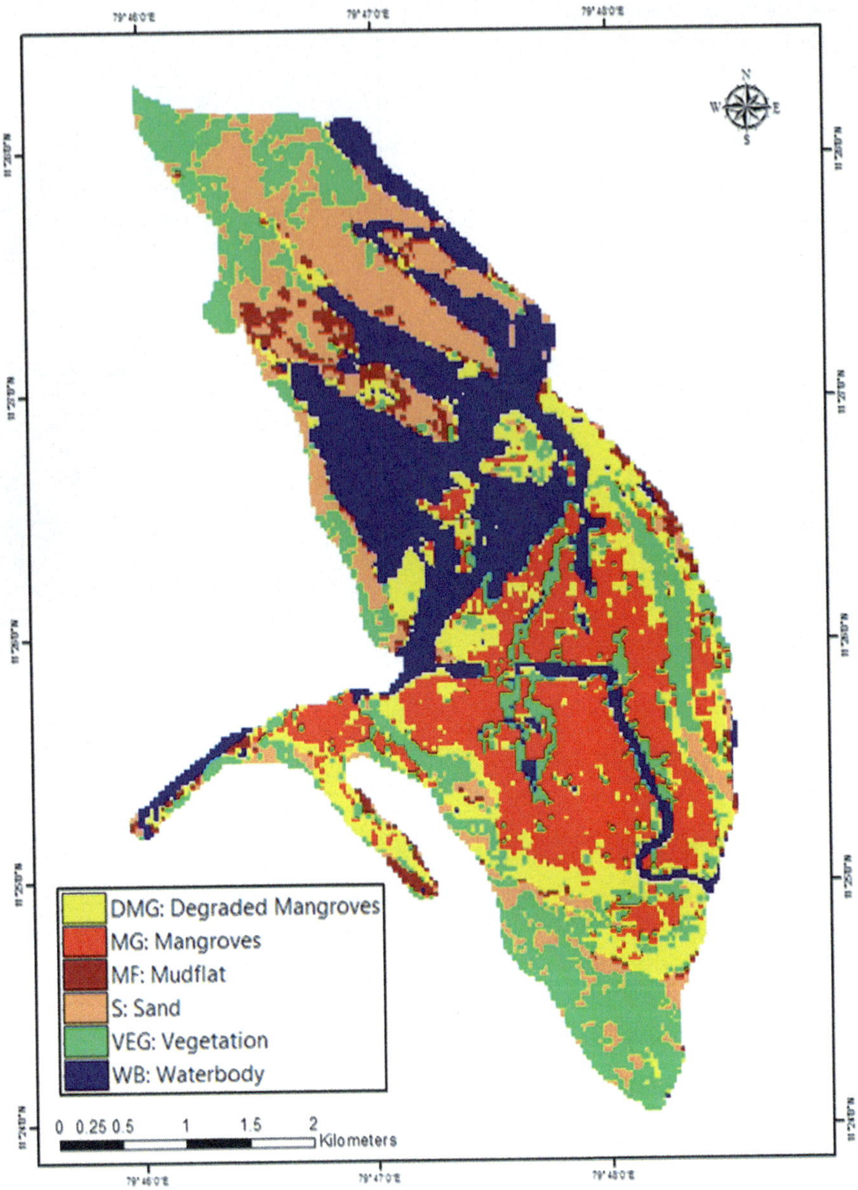

Fig. 5.9 Spatial distribution of mangroves in 2006

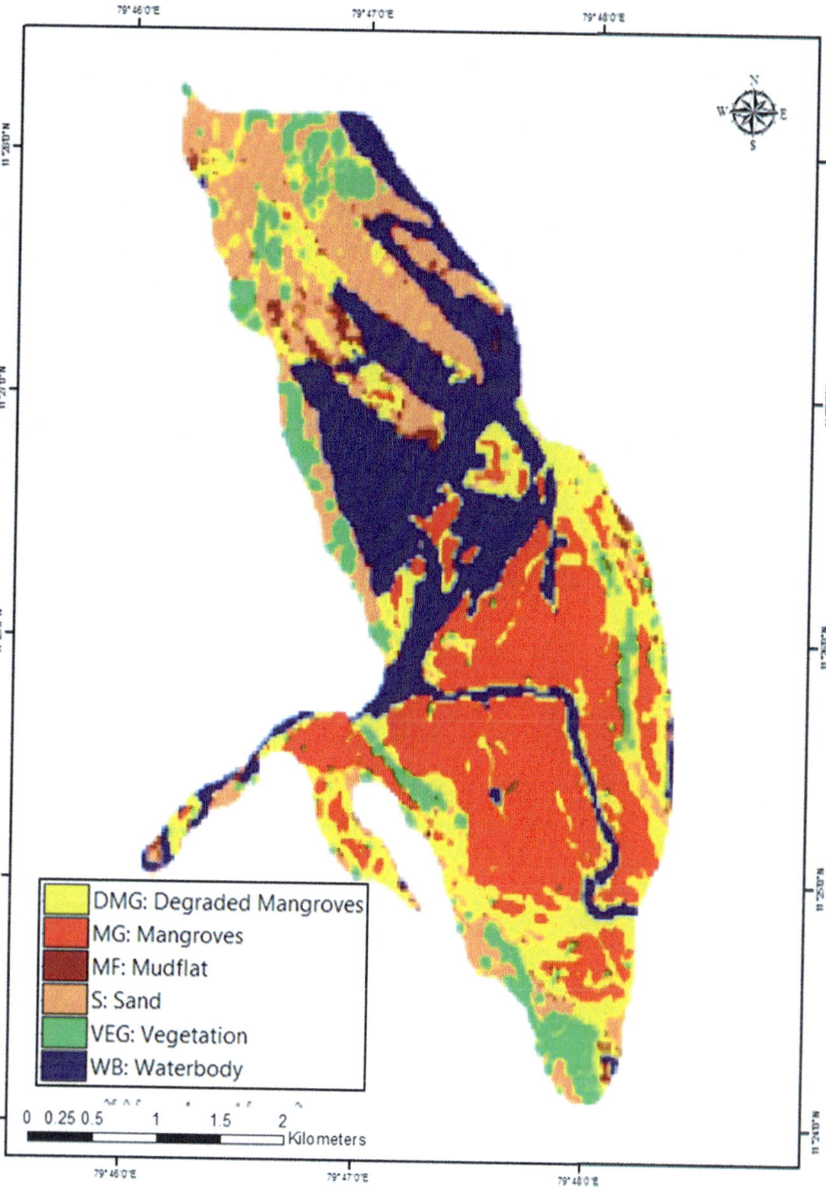

Fig. 5.10 Spatial distribution of mangroves in 2011

Fig. 5.11 Spatial distribution of mangroves in 2016

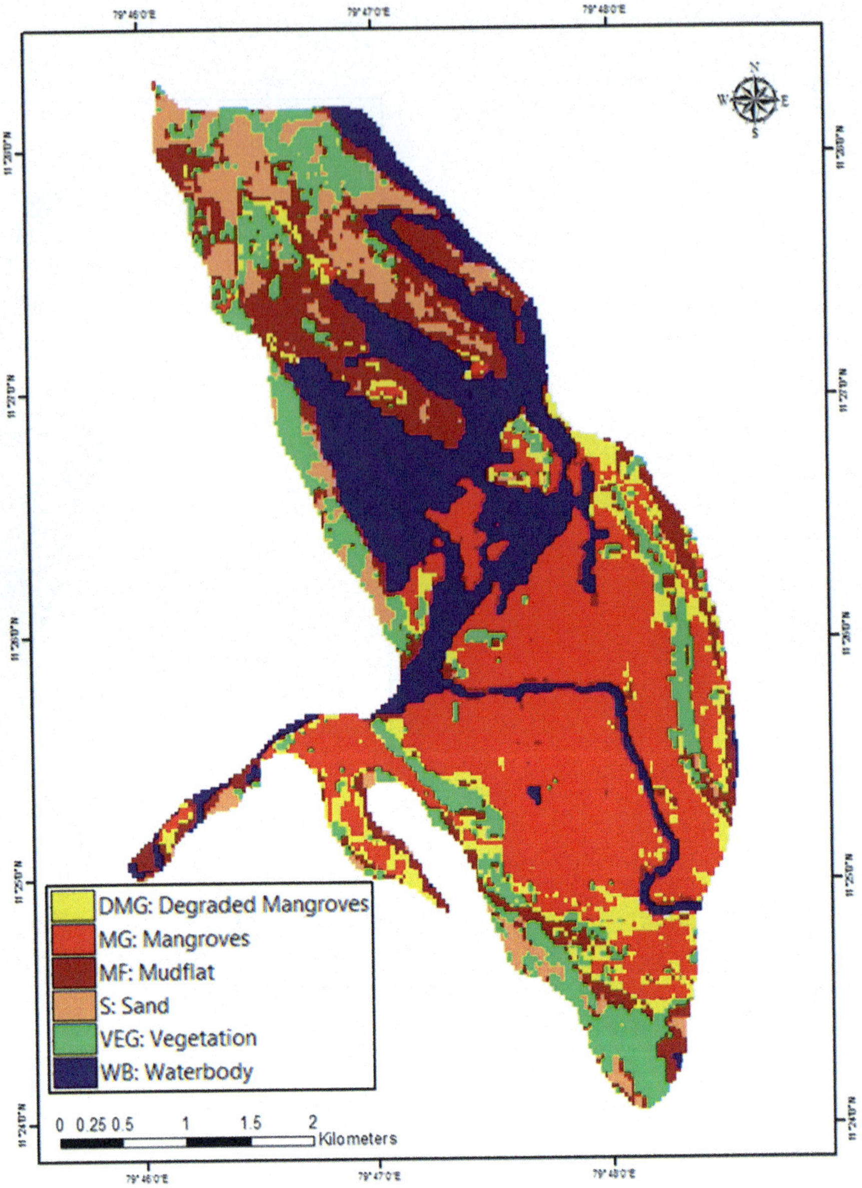

Fig. 5.12 Bar chart showing the percentage of spatial variation of different LU-LC classes from 2000 to 2016

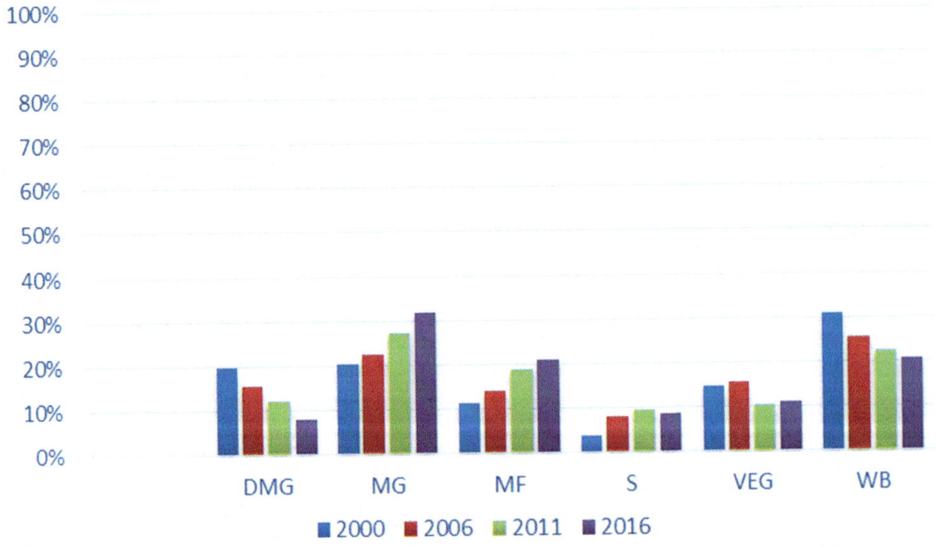

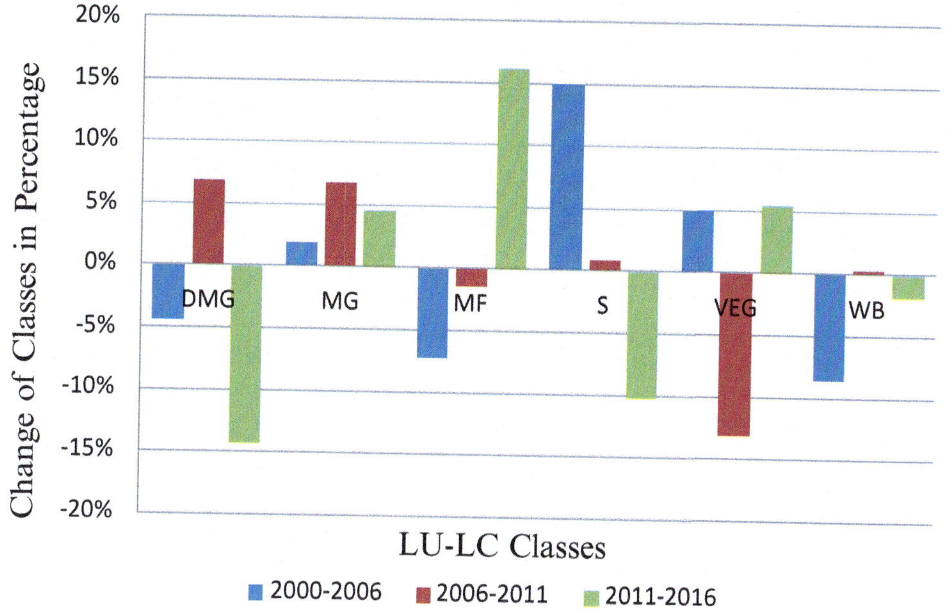

Fig. 5.13 Bar chart showing the percentage of spatio-temporal variation of different LU-LC classes from 2000 to 2016

studies, it is evident from the study that the total area of the Pichavaram mangrove showed a net increase of 11.41% within a span of 15 years (2000–2016). This is due to the identification of causative factors of mangrove degradation and subsequent conservational measures taken up by the State Forest Department.

References

Baltzera F, Allison M, Fromard F (2004) Material exchange between the continental shelf and mangrove-fringed coasts with special reference to the Amazon- Guianas coast. Mar Geol 208:115–126

Barbaro G (2013) Management and protection of coastal area, the importance of coastal processes during the planning phase. Air Soil Water Res 6:103–106

Barbier EB, Cox M, Sarntisart I (2004) Household use of mangrove and mangrove conservation decisions. In: Barbier EB, Suthawan S (eds) Shrimp farming and mangrove loss in Thailand. Edward Elgar, London, pp 115–130

Barbier EB, Hacker SD, Kennedy C, Koch EW, Stier AC, Silliman BR (2011) The value of estuarine and coastal ecosystem services. Ecol Monogr 81(2):169–193

Bird M, Chua S, Fifield LK, The TS, Lai J (2004) Evolution of the SungeiBuloh–Kranji mangrove coast, Singapore. Appl Geogr 24:181–198

Burbridge PR., Koesoebiono (1981) Coastal zone management in south-east Asia. In: Chia LS, MacAndrews C, (eds) Southeast Asian seas: frontiers for development. McGraw Hill, pp 110–13e5

Chaudhuri P, Ghosh S, Bakshi M, Bhattacharyya S, Nath B (2015) A review of threats and vulnerabilities to mangrove habitats: with special emphasis on East Coast of India. J Earth Sci Clim Change 6:270

Chong VC, Sasekumar A, Leh MUC, D'Cruz R (1990) The fish and prawn communities of a Malaysian coastal mangrove system, with comparison to adjacent mud flats and inshore waters Estuarine. Coast Shelf Sci 31:703–722

Dahdouh-Guebas F, Mathenge C, Kairo JG, Koedam N (2000) Utilization of mangrove wood product around Mida Creek (Kenya) amongst subsistence and commercial users. Econ Bot 54(4):513–527

Dawes CJ (1998) Marine botany. Wiley, New York. 27-Feb-1998 – Science – 480 pgs

Dholakia AD (2004) Fisheries and aquatic resources of India. Daya Publication House, Delhi, p 413. isbn:ISBN 8170353394

FAO (2007) The world's mangrove 1980–2005: a thematic study prepared in the frame work of the Global Forest Resources Assessment, 2005, FAO Forestry Paper 153. FAO, Rome

Fromard F, Vega C, Proisy C (2004) Half a century of dynamic coastal change affecting mangrove shorelines of French Guiana. A case study based on remote sensing data analyses and field surveys. Mar Geol 208:265–280

FSI (2001) State of the forest report: 2001. Forest Survey of India, Dehradun, 130 p

Gevana D, Camacho L, Carandang A, Camacho S, Im S (2014) Land use characterization and change detection of a small mangrove area in Banacon Island, Bohol, Phillipines using a maximum likelihood classification method. For Sci Technol 11:197–205

Ghazali N, Zainuddin K, Zainal MZ, Dali HM, Samad AM, Mahmud MR (2016) The potential of mangrove forest as a bioshield in Malaysia. Signal Processing & Its Applications (CSPA), IEEE 12th International Colloquium, pp 322–327

Giri C, Pengra B, Zhu Z, Singh A, Tieszen LL (2007) Monitoring mangrove forest dynamics of the Sundarbans in Bangladesh and India using multi-temporal satellite data from 1973 to 2000. Estuar Coast Shelf Sci 73:91–100

Hussain SA, Valuing BR (2010) Mangrove benefits: contribution of mangrove forests to local livelihoods in Bhitarkanika Conservation Area, East Coast of India. Wetl Ecol Manag 18:321–331

Integrated Coastal and Marine Area Management (ICMAM) (2000) Report on critical habitat information system using GIS for Pichavaram Mangrove South India

Jeyaseelan MJP, Krishnamurthy K (1980) Role of mangrove forests of Pichavaram as fish nurseries. Proc Indian Natn Sei Acad B 46:48–53

Kannaiah KD, Sheikhi A, Cracknell AP, Goh HC, Tan KP, Ho CS, Rasli FN (2015) Satellite images for monitoring mangrove cover changes in a fast growing economic region in Southern Peninsular Malaysia. Remote Sens ISSN 2072–4292(7):14360–14385

Kathiresan K (2000) A review of studies on Pichavaram mangrove, southeast India. Hydrobiologia 430:185–205

Kathiresan K, Rajendran N (2005) Coastal mangrove forests mitigated tsunami. Estuar Coast Shelf Sci 65:601–606

Kovacs JM, Flores-Verdugo F (2005) Mapping mangrove leaf area index at the species level using IKONOS and LAI-2000 sensors for the Agua Brava Lagoon, Mexican Pacific. Estuar Coast Shelf Sci 62:377–384

Kovacs JM, Flores-Verdugo F, Wang J, Aspden LP (2004) Estimating leaf area index of a degraded mangrove forest using high spatial resolution satellite data. Aquat Bot 80:13–22

Krishnamurthy K, Prince Jeyaseelan MJ (1983) The Pichavaram (India) mangrove ecosystem. Int J Ecol Environ Sci 9:79–85

Kumar Srinivasa T, Mahendra RS, Shailesh N, Radhakrishnan K, Sahu Kali C (2012) Identification of hot spots and well managed areas of Pichavaram mangrove using Landsat TM and Resourcesat—1 LISS IV: an example of coastal resource conservation along Tamil Nadu Coast, India. J Coast Conserv 16(1):1–12

Lugo AE, Snedaker SC (1974) The ecology of mangroves. Annu Rev Ecol Syst 5:39–64

Mandal RN, Naskar KR (2008) Diversity and classification of Indian mangroves: a review. J Trop Ecol 49(2):131–146

Masood H, Afsar S, Zamir UB, Kazmi JH (2015) Application of comparative remote sensing techniques for monitoring mangroves in Indus Delta, Sindh, Pakistan. Biol Forum 7(1):783–792

Mesta PN, Setturu B, Subash Chandran MD, Rajan KS, Ramachandra TV (2014) Inventorying, mapping and monitoring of mangroves towards sustainable management of west coast, India. J Geophys Remote Sens 3:130

MoEF, Sunderban Biosphere Reserve (1989) Project Document No. 10. Ministry of Environment and Forests. Government of India, New Delhi

Ministry of Environment and Forests (2008) Report on visit to Pichavaram in Tamil Nadu

Nayak S (2002) Use of satellite data in coastal mapping. Indian Cartographer CMMC-01:147–156

Nisansala RMNHK, Weragodatenna DDD, Gunaratne ABAK, Rathnayake RM, Jayamanne SC (2015) Change detection of mangrove cover in Puttalam lagoon using remote sensing and GIS technology. Proceedings of the National Aquatic Research and Development Agency (NARA), Scientific Session

Numbere AO, Camilo GR (2017) Mangrove leaf litter decomposition under mangrove forest stands with different levels of pollution in the Niger River Delta. Niger Afr J Ecol 55:162–167

Orchard SE, Stringer LC, Quinn CH (2016) Mangrove system dynamics in Southeast Asia: linking livelihoods and ecosystem services in Vietnam. Reg Environ Chang 16:865–879

Peter JH (2015) The biology of mangroves and seagrasses: biology of habitats. ISBN 979-0-19-102591-4

Polidoro BA, Carpenter KE, Collins L, Duke NC, Ellison AM, Ellison JC, Farnsworth EJ, Fernando ES, Kathiresan K, Koedam NE et al (2010) The loss of species: mangrove extinction risk and geographic areas of global concern. PLoS One 5:e10095

Prasad MBK (2005) Nutrient dynamics in Pichavaram mangroves, southeast coast of India. Ph.D. thesis. Jawaharlal Nehru University, New Delhi

Prasad RCP, Pattanaik C, Prasad SN, Dutt CBS (2010) Analysis of spatial and temporal changes in mangroves along Mumbai coast using remote sensing and GIS. IUP J Environ Sci 4:52–59

Prasad RCP, Karuna CVL, Asha Kumari J (2017) Evaluating mangroves of Krishna Wildlife Sanctuary in relation to the general status of mangroves in Andhra Pradesh, India. Int J Environ Stud 74 (2):214–229

Ranjan RK, Al R, Singh G, Chidambaram S (2008) Assessment of metal enrichments in tsunami genic sediments of Pichavaram mangroves, southeast coast of India. Environ Monit Assess 147:389–411

Ranjan RK, Ramanathan AL, Singh G (2008b) Evaluation of geochemical impact of tsunami on Pichavaram mangrove ecosystem, southeast coast of India. Environ Geol 55:687–697

Razzaque J (2017) Payments for ecosystem services in sustainable mangrove forest management in Bangladesh. Transl Environ Law 6(2):309–333

Robertson AI (1988) Decomposition of mangrove leaf litter in tropical Australia. J Exp Mar Biol Ecol 116(3):235–247

Salm RV, Clark JR (1984) Marine and Coastal Protected Areas: a guide for planners and managers, IUCN, Gland, Switzerland, xiii + 302 p

Shyam S Salim, Shridhar, Nivedita, Rahman M Ramees (2015) Economic valuation of marine ecosystem services: methodological issues and challenges. Summer School on Recent Advances in Marine Biodiversity Conservation and Management, 16 February – 8 March 2015. Central Marine Fisheries Research Institute

Satyanarayana B, Mohamad KA, Idris IF, Husain ML, Guebas FD (2011) Assessment of mangrove vegetation based on remote sensing and ground-truth measurements at Tumpat. Int J Remote Sens 32:1635–1650

Selvam V, Gnanappazham L, Navamuniyammal M, Ravichandran KK, Karunagaran VM (2002) Atlas of mangrove wetlands of India. MSSRF, Chennai, p 102

Selvam V, Ravichandran KK, Gnanappazham L, Navamuniyammal Taramani M (2003) Assessment of community-based restoration of Pichavaram mangrove wetland using remote sensing data. CurrSci 85(6):795–797

Singh AK, Abubakar Ansari, Dinesh Kumar, Sarkar UK (2012) Status biodiversity and distribution of mangroves in India: an overview on International day for biological diversity. Uttar Pradesh State Biodiversity Board

Sivakumar K (2013) Coastal and Marine Biodiversity Protected Areas in India: Challenges and way forward. In: Venkataraman K, Sivaperuman C, Raghunathan C (eds) Ecology and conservation of Tropical Marine Faunal Communities. Springer, Berlin/Heidelberg, pp 463–476

Spalding MD, Blasco F, Field CD (1997) World mangrove atlas. IntSoc Mangrove Ecosys, Okinawa

Srinivasa Kumar T, Mahendra S, Nayak RS, Radhakrishnan K, Sahu KC (2011) Identification of hot spots and well managed areas of Pichavaram mangrove using Landsat TM and Resourcesat—1 LISS IV: an example of coastal resource conservation along Tamil Nadu Coast, India. Coast Conserv 16:1–12

Susan C, Anu J, Nisha V (2014) Sustainable resource management – tools and techniques. ISBN: 978-81-7255-066-9

The United Nations Conference on Environment and Development (UNCED) (1992) The Rio declaration on environment and development. 3-14 June, Rio de Janeiro.

Valiela I, Bowen JL, York JK (2001) Mangrove forests: one of the world's threatened major tropical environments. Bioscience 51:807–815

Van Santen P, Augustinus PGEF, Janssen-Stelder BM, Quartel S, Tri NH (2007) Sedimentation in an estuarine mangrove system. J Asia Earth Sci 29:566–575

Venkataraman K (2007) Coastal and marine wetlands in India. Proceedings of Taal 2007: 12th World Lake Conference, pp 392–400

Wilkinson C, Sudara S, Chou LM (1994) Proceedings, Third ASEAN-Australia Symposium on Living Coastal Resources. Vol. 1: status, Reviews. Chulalongkorn University, Bangkok

Part II

Threats, Vulnerability, and Impacts

6 Natural Threats and Impacts to Mangroves Within the Coastal Fringing Forests of India

Ashis Kr. Paul, Amrit Kamila, and Ratnadip Ray

Abstract

Mangrove forest acts as a buffer against the storm waves, tidal waves, tsunami waves, littoral currents and shoreline erosion of the coastal fringes in intertidal region. However, the natural impacts are strongly felt on the fringing forests of different physiographic settings on India's East Coast and West Coast at present. The different physiographic settings for mangrove ecosystems in the eastern and western coasts of India include carbonate platforms of the coral fringed coasts, estuarine deltaic islands with swampy clay surface, clay dominated swale topography of inter-dune sand ridges of deltaic chenier coasts, various embayments and estuaries, and back waters.

Some events of cyclone landfalls, occurrences of seismic activities with tsunami event, increased hyper salinity, and impacts of current sea-level rise have produced a great damage to the mangrove forests of the intertidal region of the coast of India in the previous decades.

The entire work is carried out on the basis of the field observations, analysis of the physicochemical characters of mangrove substrates of different physiographic settings, time series analysis of wave and tide data and identification of mangrove zonations and areas of wetland changes using geospatial technology.

The present study reveals that some natural impacts are responsible for the loss of mangrove forests from different physiographic settings of India's East Coast. Hypersalinity in the inner parts of the islands due to drainage loss and exposure to evaporation, and storms driven inundation and erosion by frequent landfall of cyclones are producing damages to the fringing forests of Sundarban deltaic islands. Storm drifted over wash sand fan lobes into the swales across beach ridge cheniers, sheet erosion of uncovered compact mangrove mud banks by tidal waves associated with lateral shifting of sediment filled tidal channels, and hyper salinity of higher marshes have produced significant damages to the mangrove forests of the barrier back tidal basin of beach ridge chenier coasts and estuary fringes of Subarnarekha delta. Finally, the upliftment and subsidence of carbonate platforms by seismicity, and tsunami incidences with storm induced log drift have generated natural impacts of the loss of mangrove forests in carbonate settings of Andaman group of islands. Mangroves of Gujarat and Maharashtra coasts are threatened due to shortages of river flows into the coastal zones, aridity of the inner parts and hyper salinity.

Conservation through environmental zoning and construction of artificial drainage ditches to restore the frequency of inundation, and afforestation in the existing tidal mud flats are immediately needed to restore the threatened mangrove wetlands from the natural impacts along the shoreline of the Bay of Bengal and some parts of the Arabian Sea.

Keywords

Mangrove wetland · Intertidal region · Swampy clay surface · Hyper salinity · Estuarine delta · Subsidence of carbonate platforms · Artificial drainage ditches

6.1 Introduction

Mangroves act as natural settings of bio-engineered wall against the marine forcing factors (e.g. swell waves, tidal currents, storm breaks, tidal waves and tsunami process) in the coastal zones dominated by estuaries, embayments, delta plains, inter distributary flood basins, and carbonates

A. K. Paul (✉)
Department of Geography and Environment Management, Vidyasagar University, Midnapore, West Bengal, India

A. Kamila · R. Ray
Department of Remote Sensing and GIS, Vidyasagar University, Midnapore, West Bengal, India

C. Makowski, C. W. Finkl (eds.), *Threats to Mangrove Forests*, Coastal Research Library 25,
https://doi.org/10.1007/978-3-319-73016-5_6

platforms. They are distributed within three climatic zones of the global coasts, as in the equatorial zone (10° N and 5–10 ° S), in the tropical summer rainfall zone (25–30° N and S), with subtropical dry zone of the desert coast, and in warm temperate climates with absence of cold winters (Walter 1977). In India, mangroves are well distributed within the latitudinal extension of 6° 45′ N. and 23°30′ N along the sea shores of the Bay of Bengal, Arabian Sea and Andaman Sea under the influences of equatorial environment and tropical hot and humid environment with summer rainfall zones (Fig. 6.1).

Mangroves are particularly lying within the inter-tidal environment fringed with specific physiographic settings and processes, coastal configurations (Thom 1979) and areas of sediment supply into the coastal zones under definite geological settings (Paul 1999, 2002a). However, three physical mangrove settings are identified by Woodroffe et al. (1993) in Indo-Pacific and Atlantic systems which are restricted to lower latitudes particularly between 32° N and 38° S, as river dominated setting, tide dominated setting, and carbonate setting. Mangroves also can show a range of adap-

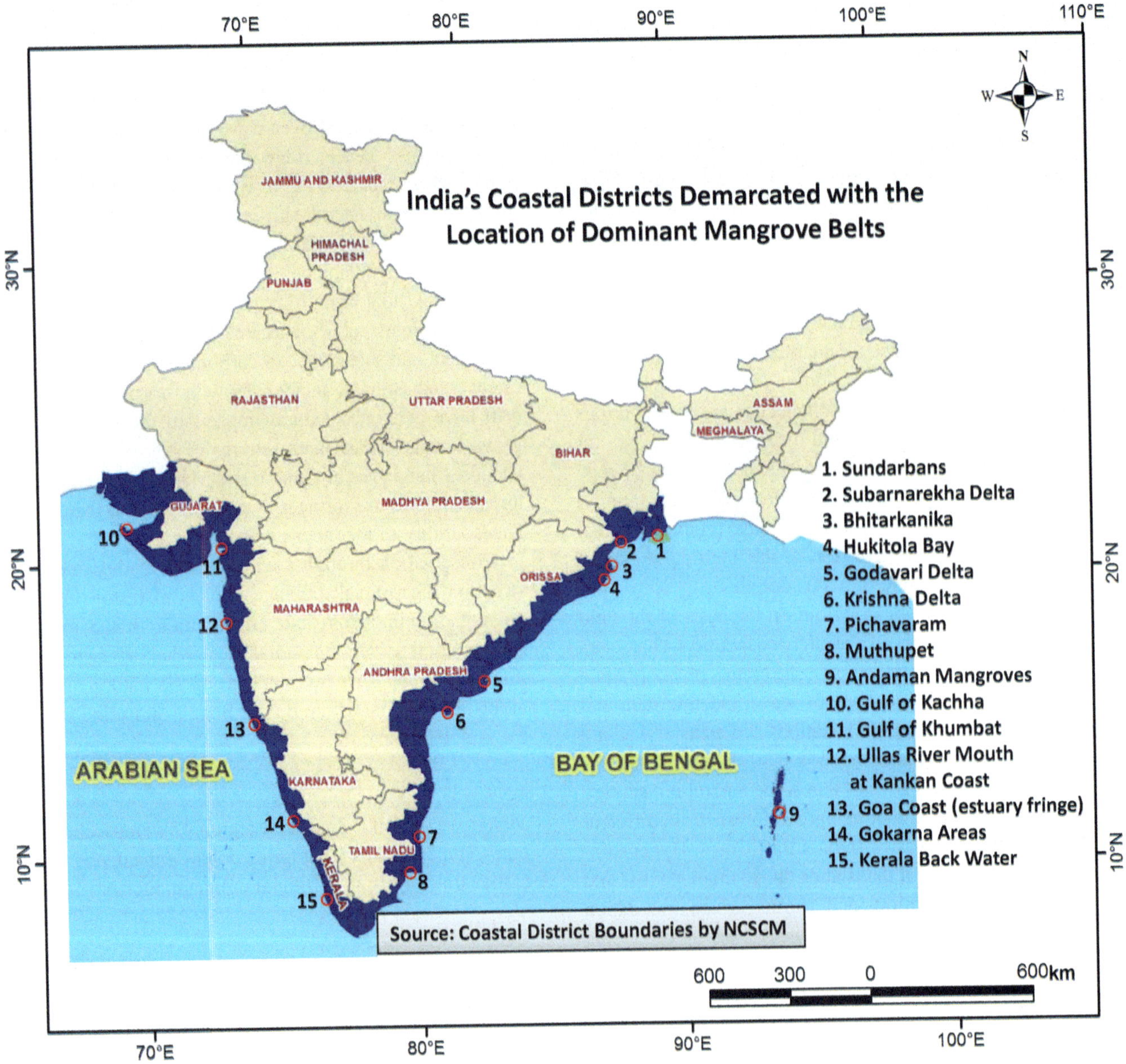

Fig. 6.1 The location map of mangrove belts with the coastal Districts of India (The Coastal District boundaries are drawn by NCSCM)

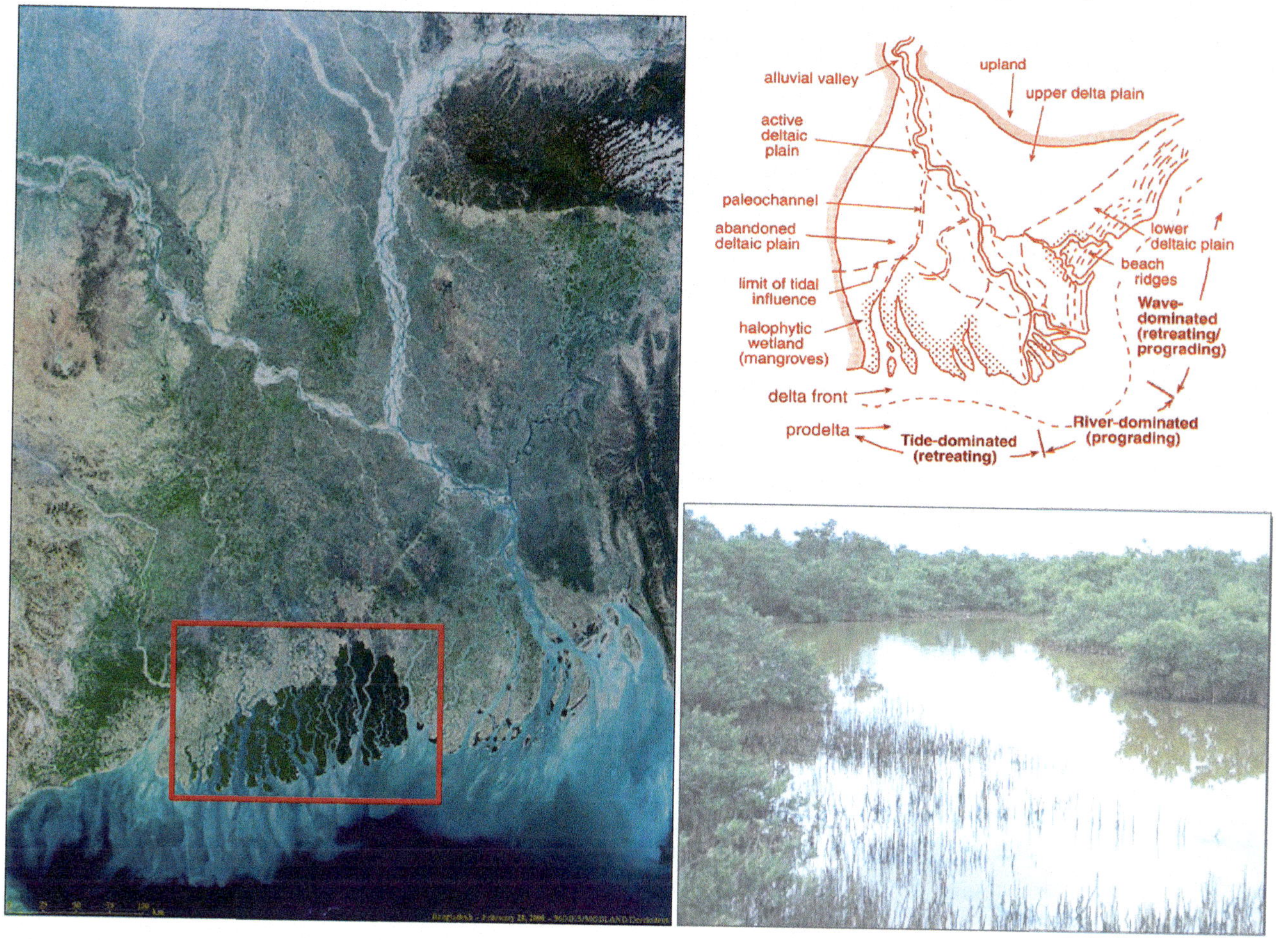

Fig. 6.2 The single largest mangrove forest belt of Sundarbans

tation like tidal root systems (Pneumatophores, Stilt roots and Knee bend roots), vivipary germination systems, and stunted growth under stress with saltwater spray at the sea front position and shortages of nutrients.

There is an emerged view that the occurrences of less intense storms are needed to sustain the productivity of mangrove forests, otherwise the forest habitats become affected by high salinity, low availability of nutrients and inundation by sea-level rise. The global warming related Sea Surface Temperature (SST) and increased frequency of extremely intense storms however, may disturb or produce more drastic changes over the mangrove forest habitats of the shore fringed areas with climate variabilities in the coming decades.

The productivity of deltaic mangrove forests always becomes renewed with supply of Annual River born sediment inputs and fresh water flows into the coastal zones by river floods, and reduction of salinity by fresh water inundations and their mixing with salt waters in the coastal habitats dominated by mangroves. The Sundarbans of the river deltas of Ganga and Brahmaputra provide most favourable environment with multiple loads of sediment and fresh water supply for growth and diversity of mangroves in South Asia (Fig. 6.2).

6.2 Materials and Methods of the Study

The present work is based on the consideration of monitoring changes of mangrove shores by Direct Field Observation Methods (DFOM), Remote Sensing techniques, various sources of secondary data on mangroves (Hydrological Hand Book, CWC, New Delhi; Shoreline Change Atlas of the Indian States, NCSCM, Chennai; Ocean Information, INCOIS, Chennai; Cyclone Hazard Map of India, NCRMP, New Delhi; Cyclone Data of India, IMD, Pune; and Exiting literatures) and related environmental information, measurement of a few water quality samples in the field with equipment, displayed Google Earth Images, and field photographic documents of the Coastal Zones of India as well as the use of

Table 6.1 Details of satellite data

Landsat 8 Operational Land Imager (OLI) and Thermal Infrared Sensor (TIRS)	Bands	Wavelength (μ)	Spatial Resolution (m)	Path/Row of Imagery Tiles	Radiometric Resolution
Imagery date: 2013 abd 2017	Band 1 – Coastal aerosol	0.43–0.45	30	(1) Path-138 Row-45	16-Bits
	Band 2 – Blue	0.45–0.51	30		
	Band 3 – Green	0.53–0.59	30		
	Band 4 – Red	0.64–0.67	30	(2) Path-148 Row-45	
	Band 5 – Near Infrared (NIR)	0.85–0.88	30		
	Band 6 – SWIR 1	1.57–1.65	30	(3) Path-150 Row-44	
	Band 7 – SWIR 2	2.11–2.29	30		
	Band 8 – Panchromatic	0.50–0.68	15		
	Band 9 – Cirrus	1.36–1.38	30	(4) Path-151 Row-44	
	Band 10 – Thermal Infrared (TIRS) 1	10.60–11.19	100 * (30)		
	Band 11 – Thermal Infrared (TIRS) 2	11.50–12.51	100 * (30)		

TIRS bands are acquired at 100 m resolution, but these are re-sampled to 30 m in delivering data product

40 years of professional experiences of Coastal Research in the Country by the present Author (Ashis Kumar Paul).

Two cloud-free Landsat digital data of Landsat 8, OLI/TIRS sensors for the years of 2013 and 2017 covering the study area were downloaded freely from the U.S. Geological Survey's (USGS) Earth explorer website (http://earthexplorer.usgs.gov/). The Landsat scenes were chosen for this study due to their affordability, availability, and medium to high spatial resolution. Details about the data are given in Table 6.1. These four Landsat digital data are level-one terrain-corrected (L1T) in WGS 84 geodetic datum, Universal Transverse Mercator map projection (UTM, Zone 45N), and north-up image orientation.

6.2.1 Radiometric Correction

Radiometric correction of satellite imagery falls into two broad categories like absolute and relative. Absolute correction converts DN of pixel to percentage reflectance value using several transformation equations (Richter 1990) whereas relative correction normalizes multiple satellite scenes to each other.

In this present study absolute correction process has been applied. The parameters used in different environmental condition detection algorithms, it require physical units, such as at-sensor radiance or top-of-atmosphere (TOA) reflectance, rather than the raw quantized calibrated pixel value (DN). TOA reflectance can be obtained from the quantized calibrated pixel value, as given by (Chander et al. 2009):

$$\rho_\lambda = (\pi * L_\lambda * d^2)/ESUN_\lambda * \cos\theta \quad (6.1)$$

where ρ_λ is the TOA reflectance of wavelength λ (unit less), d is the earth-sun distance (astronomical units), $ESUN_\lambda$ is mean exo-atmospheric solar irradiance (W/ ($m^2 \cdot \mu m$)), θ is the solar zenith angle (degrees), and L_λ is the spectral radiance at wavelength λ at the sensor's aperture [W/($m^2 \cdot sr \cdot \mu m$)]. L_λ can be obtained from the quantized calibrated pixel value also as given by (Chander et al. 2009):

$$L_\lambda = (L_{MAX} - L_{MIN}/QCAL_{MAX} - QCAL_{MIN}) * (DN - QCAL_{MIN}) + L_{MIN} \quad (6.2)$$

where L_{MAX} is the spectral at-sensor radiance that is scaled to $QCAL_{MAX}$ [W/(m^2srμm)], L_{MIN} is spectral at-sensor radiance that is scaled to $QCAL_{MIN}$ [W/($m^2 \cdot sr \cdot \mu m$)], $QCAL_{MAX}$ is the maximum quantized calibrated pixel value corresponding to L_{MAX} (DN), $QCAL_{MIN}$ is the minimum quantized calibrated pixel value corresponding to L_{MIN} (DN), and QCAL is the quantized calibrated pixel value (DN). The parameters in Eqs. (6.1) and (6.2) can be read from the header files of the ALI, TM, and ETM+ datasets or be retrieved from the USGS website.

6.2.2 Identification of Mangrove Health

In this present study, after the radiometric normalization techniques, to assess the mangrove forest cover dynamics in terms of mangrove health status, an object oriented enhancement algorithm has been designed using mathematical operators, which is supervised in nature and expressing the characteristics of plant chlorophyll 'a'.

Two considerations have been taken in this respect like, pixel value having chlorophyll influence will be greater in near-infrared band than red band and pixel value having no chlorophyll influence will be greater in red band than near-infrared band. On the basis of these considerations following supervised enhancement algorithm has been developed and is named as Modified Advance Vegetation Index (MAVI) (Ray et al. 2013). That can be calculated as,

$$[\{(\rho NIR + L)*(256 - \rho RED)\}*(\rho NIR - \rho RED)]^{(1/3)} \quad (6.3)$$

where, ρRED = image of Red band of TM5 Sensor, ρNIR = image of Near-infrared band of OLI Sensor, L= Coefficient, varies with the vegetation cover, Here L= correlation slope between NIR and Red band.

It differs from NDVI in the perspective of physiognomic vegetation classes though there is almost positive relationship between these two. It is more sensitive than NDVI as it is using the power degree of Infrared response.

6.2.3 Identification of Mangrove Vegetation Density

In the second phase the vegetation density (VD) has been calculated for both of images of year 2000 and 2010 by synthesizing the Eq. 6.3 and bareness index (BI) on Principle Component (PCA) basis and it is level sliced into 5 density zones having distinct areal extent. The bare soils index (BSI) is calculated as,

$$\begin{aligned}&[(\text{Band5} + \text{Band3})\\&\quad - (\text{Band 4} + \text{Band1})/(\text{Band 5} + \text{Band 3})\\&\quad + (\text{Band 4} + \text{Band 1})]\end{aligned} \quad (6.4)$$

6.3 Physiographic Settings of Indian Mangroves

The growth characters of mangroves are driven by physiographic components (geophysical and geomorphic) of the coast. Among them geophysical components include relative sea level variations of the coast, climatic characteristics, and tidal properties with their behaviors of the coastal landscapes (Thom 1984). However, the geomorphic components can be categorized into three sets of multiple factors; e.g. sedimentation characters (carbonates, river alluviums, terrigenous sediments etc.) of the coast, processes involved within the coast (e.g. waves, tides, river flows, storms etc.), and topographic expressions of the coast or the composition of landform assemblages along the coastal belt. Finally, the biologic components of the coastal environmental settings denote the role of ecologic factors in determining the occurrences of mangrove species of a particular region.

All the above components jointly produce some physiographic settings in which mangroves can grow in response to their activities reflected on the coast. In Indian coasts, the physiographic settings consist of (i) the low lying deltaic depositional surface with intervened distributaries and tidal creeks, (ii) tide dominated estuaries with islands and sub tidal shoals, (iii) barrier lagoons and coastal embayments (iv) carbonate banks, and (v) drowned bed rock valleys or 'ria' type shores.

Mangroves of the river deltas were adjusted with extensive sediment supply and annual freshwater flows from the catchment of river basins into the coastal zone in the near past. Monsoonal rainwater was also partially accumulated in the micro topographic surface depressions of the delta plain to spread the wetlands towards the wide hinter land. Differential rate of dilution of encroached tide water under such conditions supported favourable environmental gradients for growth and diversity of mangroves with evergreen and semi evergreen characters. Nutrient loads were sufficient to maintain the balance of mangrove growths in the depositional alluviums under the impacts of marine forcing factors and land derived flood waters and sediments reaching of the shore fringe areas of the river deltas.

Soil and pore water salinity, climate of tropical hot and humid conditions, tidal elevations and mineral component of the soil played a dominant role in determining the spread of mangroves in the low lying deltas (Ganga delta, Bhitarkanina delta, Mahanadi delta, Godavari delta and Krishna delta) of the coast of India (Selvam 2003) around Bay of Bengal (Table 6.2).

Estuaries are traditionally considered as geological sinks for downstream supply of sediments and encroachment of tidal sediments into the embayments from the seaward side. As a result of bidirectional flows of waters and sediments wide shore fringed tidal flats and islands become the favoroubIe sites for mangroves and salt marshes. Hugli river mouth, Subarnarckha river estuary, Brahmani-Baitarani river mouths, Mahanadi river mouth, Goutami Godavari estuary and Krishna river estuary belong to such category of physiographic setting in India.

Physiographic setting dominated by barrier bar lagoons and coastal embayments support the mangrove habitats up to a certain extent as long as the riverine supply of the fresh waters and finer sediments continue into the arms of the sea surrounded by lands or protected by the barrier bars. Mixing diminishes, and salinity increases into the lagoons or coastal embeyments if the landward connected distributary channels

Table 6.2 The status of mangrove forests and their characterization diversity in India

Mangrove belt	Area occupied (FSI 1999)	Species richness	Tidal range/ wave heights	Endangered species
Sundarbans	2125 km^2	65 (Naskar and Guha)	5.50–6.00/ 0.9–1.2 m	Heriteria fomes, Nypa fruticans
Mahanadi (Hukitola Bay, Bhitarkanika with Dhamara R. Mouth)	205 km^2	35–55 (Upadhyay et al. 2015)	2.80–4.50/ 1.2–1.5 m	NA
Devi R. Mouth	10 km^2	11	3.50–6.00/ 0.9–1.2 m	NA
Godavari Delta	241 km^2	16 (Banerjee et al. 1989)	1.34/ 1.0–1.5 m	Avicennia officinalis, Excoecaria agallocha, Luminitzera reacemosa
Krishna Delta	156 km^2	19 (Banerjee et al. 1989)	0.90/ 0.5–1.0 m	NA
Pichavaram	9.0 km^2	20 (Banerjee et al. 1989)	0.64/ 0.6–0.9 m	Xylocarpus granatum, Sonneratia apetala, Kandelia candel, Bruguiera gymnorhiza
Muthupet	12.0 km^2	20 (Banerjee et al. 1989)	0.70/ 0.6–0.9 m	Rhizophoraceae
Andaman group of Islands	929.0 km^2	40 (Singh et al. 1986)	1.90/ 1.0–1.5 m	Avicennia marina
Nicobar group of Islands	37.0 km^2	20–26 (Banerjee et al. 1989)	1.70/ 1.5–2.0 m	NA

become abandoned in the coastal hinterlands. Subarnarekha barrier back tidal lagoon, Hukitola bay, Chilika spit back islands, Corringa bay, Pichabharam backwater system, Muthupet of Cauvery delta are most important sites for such mangrove vegetation. Similar physiographic setting is also found in and around the coastal parts of Kerala and Karnataka on the side of Arabian Sea. The stability of sand spits and barrier bars control the environmental gradient available for mangrove growth and distribution over the tidal shoals or bars behind the protection provided by such barriers.

The carbonate banks of shore fringes and inner creeks of Andaman and Nicobar group of islands support a wide range of environmental gradients for mangroves of evergreen characters. Freshwater breaches or the seepage flow of water into the creek margins and shore fringes produce the environment of additional fresh water supply which is mixing with salt waters to develop the mangrove ecosystem under fragile conditions. Sediments of the carbonate banks are supplied from coralline debris, chemogenic cements or micrites, low amount of siliciclastic materials and other bioclasts. The calcium carbonate rich sediments with bio-encrusted surface and relatively sheltered shores of Andaman and Nicobar islands have provided some favourable areas for mangrove forests.

Drowned bedrock valleys or 'ria' type shores of Karnatak Coast, Goa Coast and Andaman and Nicobar Islands also support a type of physiographic setting for mangrove vegetation growth and distribution. Tectonically, such physiographic setting is unstable in nature due to occurrences of earthquakes and tsunamigenic impacts. Uplift and subsidence events produce the shifting of inter tidal, sub tidal and supra tidal zones of the unstable coasts. Both the submerged and emerged forests are found in such environments. During high tides, the hydraulic head of salt water pushes landward with transmission of pore water salinity during tidal pumping to support mangroves in the emerged platforms of carbonate sediments. During low tide, the fresh water breaches into the modern platform from the landward sides to reduce the pore water salinity of the shore fringes along the islands, and for which mangroves of less saline tolerant species can exist in the upper middle parts of the carbonate shore platform in Andaman and Nicobar group of islands.

Drowned bedrock valleys or 'ria' type shores are controlled by encroached sea water with tides and swell waves that transport marine sediments into the water accumulated valleys to fill the shallow basins generated by subsidence in the coastal belts. The submerged forests are trapped under the blanket of sediments into such physical settings of Andaman and Nicobar islands and in some parts of West Coast of India.

6.4 Mangrove Belts of the East Coast of India

Mangroves are distributed in the river deltas dominated by tidal ranges, river deltas with beach ridge cheniers and barrier lagoons dominated by mixed wave-tide environment, carbonate platforms and 'ria' type shores dominated by micro tidal and mixed wave-tide environment along the East Coast of India fringed with Bay of Bengal and Andaman Sea. Indian Sundarban occupies the southern part of the river delta of Ganga in which the coastal fringe forest or mangrove forest is distributed over 4264 km^2 area of tide dominated deltaic

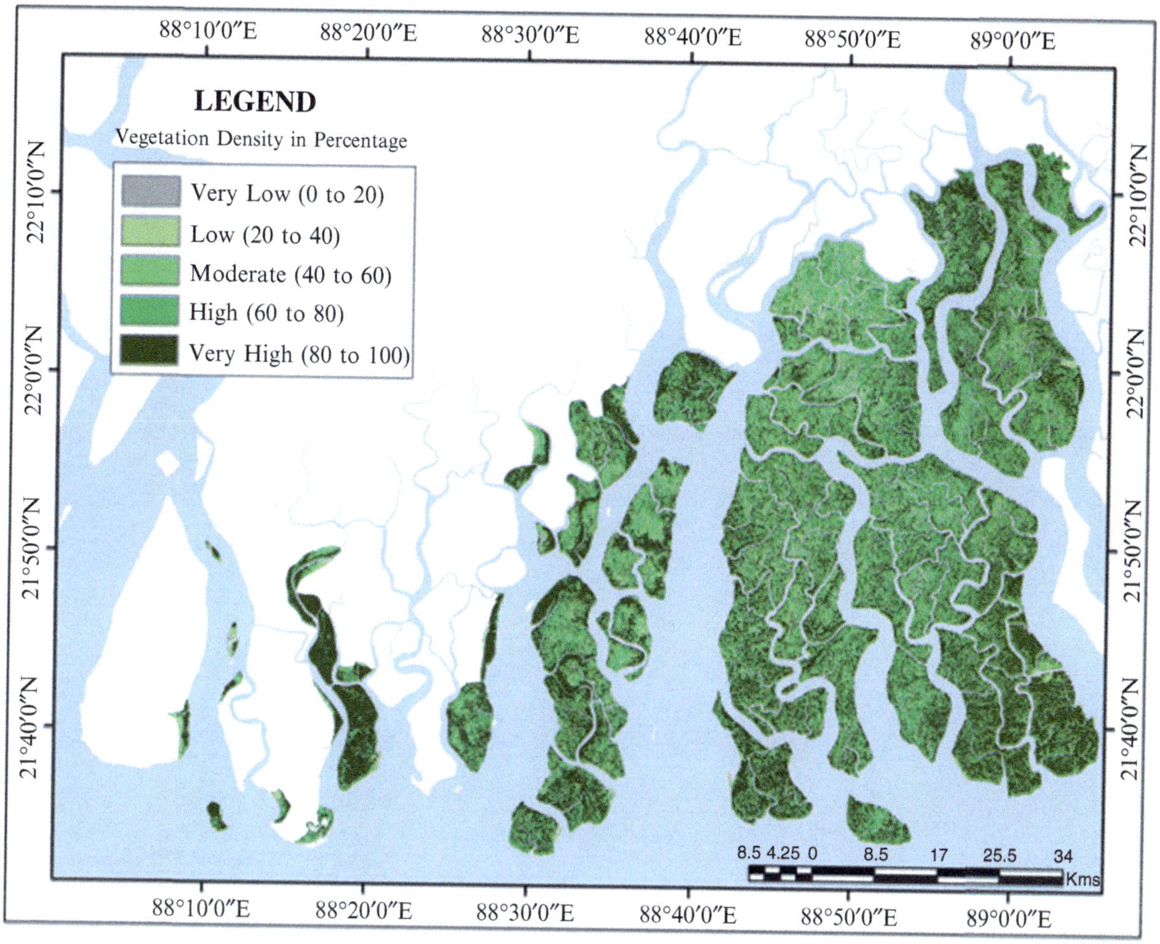

Fig. 6.3 Diversity of mangrove density in the parts of Sundarbans (2017) in West Bengal, India

depositional surface with intervened tidal creeks or channels. The entire deltaic plain covered by mangroves is separated into 54 islands of forested parts and broadly again separated into inter distributary flood basins by number of distributary channels extended from north to south (Figs. 6.3 and 6.4). Tidal range varies from 5.00 to 5.50 m in the Sundarban and the high load of sediment transport produces dynamic depositional surface in which the mangroves adjust their colony with variable environmental gradients (Paul 2002a).

The barrier back tidal lagoons and estuary fringe tidal flats of Subarnarekha beach ridge chenier delta are colonized by dense mangroves and saltmarsh vegetation. The mixed wave tide environment (tidal range of 3.50 m and wave height of 0.90–1.20 m) with discharges of sediment and fresh water flows into the coastal zones by the high magnitude floods of River Subarnarekha, and landfall of tropical cyc lones modify the topography of beach ridge chenier delta in which the sheltered flats provide geographical habitat for mangroves. Numerous tidal creeks are extended over the tidal flat of back barrier lagoon in the delta. The low height tidal flats around the major tidal rivers and islands or bars of Subarnarekha estuary and the creek banks of lagoonal flats are densely covered by mangrove. The higher flats are mostly colonized by saltmarsh vegetation. The storm effects and sand carrying winds of pre-monsoon months pushed the sand deposition lobes towards the lagoonal marshes behind barrier beach sand ridges in many parts of the delta.

The portion of shore fringe sheltered tidal flat of Bhadrak District, Islands of Brahmani River mouth (Kalidiania etc.), and majority of Bhitarkanika tidal flats separated by river systems of Baitarani- Bhrahmani Delta (Kendrapara District) offer a favourable habitat for mangrove and salt marsh vegetations (Fig. 6.5). The mangroves of the region are sheltered by shore fringed protective barrier spit with sand dunes all along the low lying deltaic flat. The area receives 1400–1500 mm rainfall annually with arrival of south-west monsoon winds in the northern Bay of Bengal like other areas of West Bengal (Indian Sundarban) and Odisha coast (Subarnarekha beach ridge chenier delta). The mangrove forests of the region are separated into different island units by extensive tidal rivers, creeks and river estuaries (Table 6.3).

The open shore tidal flats and swale topography in between shore parallel beach ridges of Bhadrak District of

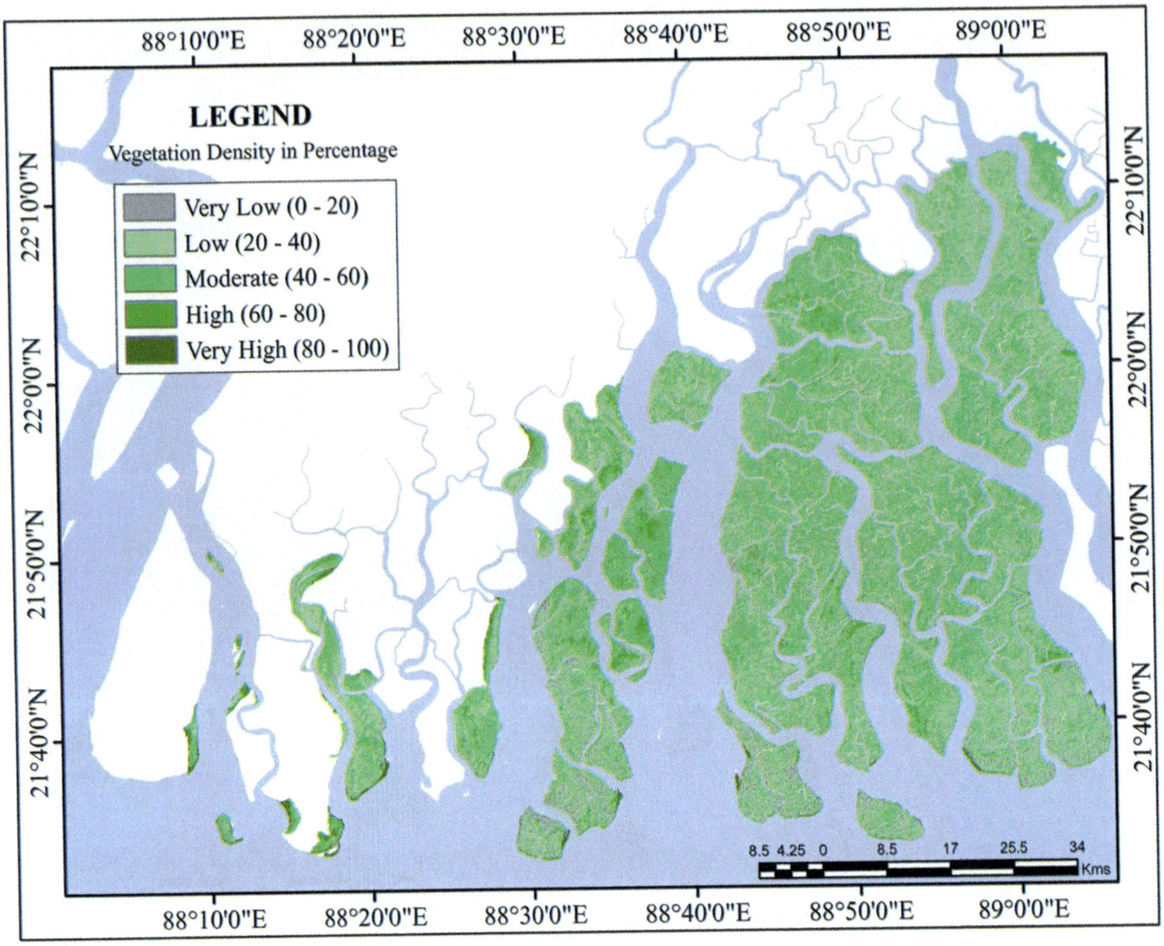

Fig. 6.4 Diversity of mangrove density in the parts of Sundarbans (2013) in West Bengal, India

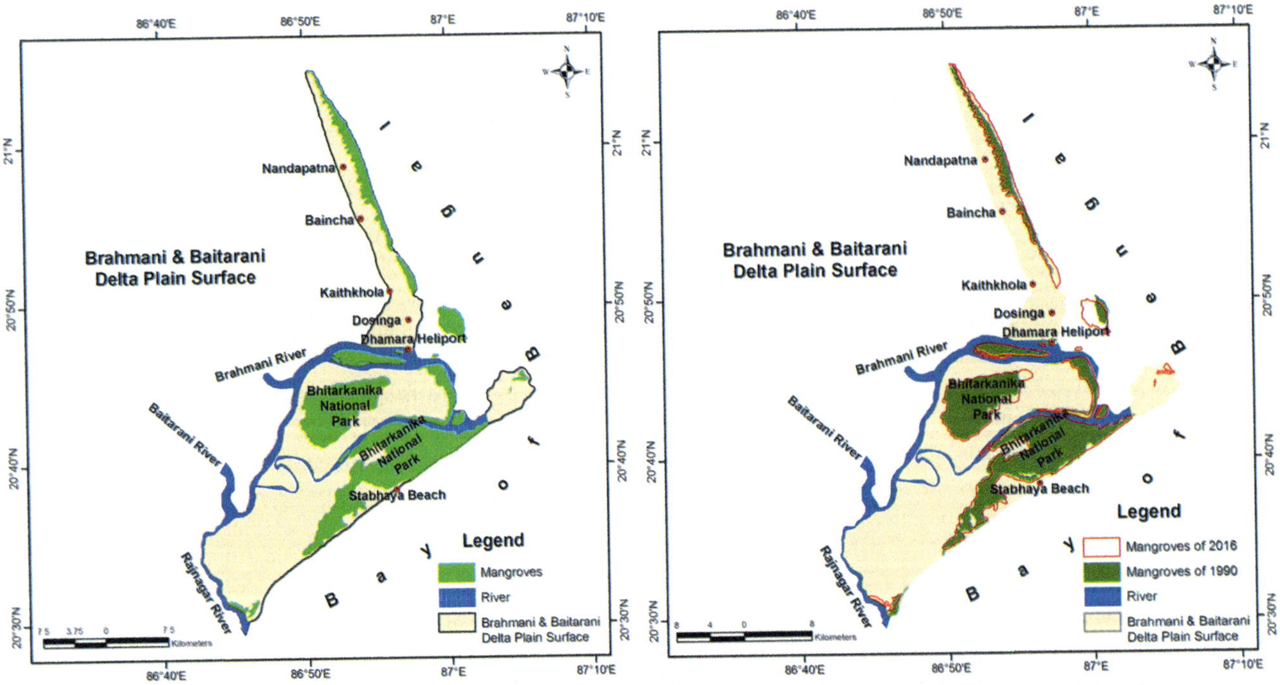

Fig. 6.5 Mangrove area of Bhitarkanika (Left Side) and areal change of Mangrove belt in Bhitarkanika (Right Side)

Table 6.3 The conservation of mangroves by afforesting newer areas of the coastal habitats during the period between 1990 and 2016 through the exploration of temporal remote sensing data on the aerial coverage of mangrove forests

Name of the Mangrove Belt	Areal coverage		Growth (Area in km^2)
	1990 (Area in km^2)	2016 (Area in km^2)	
Bhitarkanika	159.13	187.73	28.6
Corringa	184.42	193.13	8.71
Hukitola Bay	31.87	51.63	19.76
Krishna Delta	83.14	147.17	64.03
Muthupet	22.00	32.55	10.55
Pichavaram	3.26	8.27	5.01

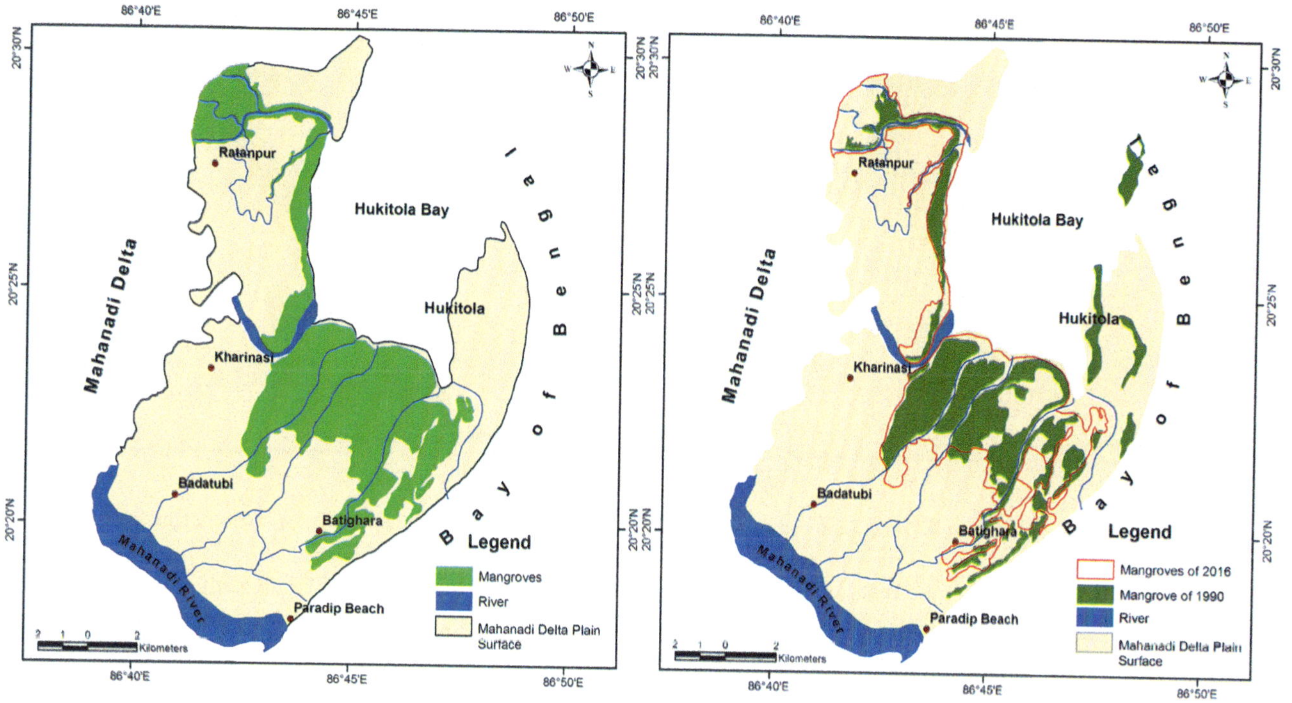

Fig. 6.6 Mangrove area of Hukitola Bay (Left Side) and areal change of Mangrove belt in Hukitola Bay, Mahanadi Delta (Right Side)

Odisha state occupy an extensive mangrove forest and mangrove associates particularly at the sheltered flats towards the north of Dhamara River. The curved shoreline of the Bay of Bengal provides a sheltered environment for deposition of finer alluviums to support such tidal forests with meso-tidal range (2–2.5 m).

Hukitola Bay mangroves are located within the coastal embayment of Mahanadi delta. Number of sand spits and barrier bars protect the waves and swell waves of the open marine environment by acting as sand ridge barriers along the sea front position to provide a sheltered water bodies behind them keeping with wide opening mouth towards the sea at and around the low energy areas of the shore fringe water space. Such embayment of Hukitola bay is also characterized by development of seaward advancement of prograded tidal flat behind the back barrier water space with fluvio-tidal deposition of distributary channels (Fig. 6.6). This wide and extensive tidal flats crisscrossed by tidal channels with meso-tidal range (>2 m) has occupied a large area of mangrove forest in northeastern shores of Mahanadi delta. Mangroves and salt marshes are also distributed along the estuary fringed tidal flats of Devi River mouth and Mahanadi River mouth of this area.

In Godavari delta of Telengana Province (coastal state) mangroves or the shore fringe forests are distributed in Corringa bay behind the Godavari spit, and along the low lying delta plain tidal flats fed by distributary channel systems. The Corringa Reserve Forest of mangroves and mangrove associates is located over the deltaic depositional tidal flats within the embayment settings and protected by Godavari spit from open marine environment (Fig. 6.7). The number of species diversity is relatively poor in compared to the tidal flats of Mahanadi delta.

As per estimation of forest survey of India (FSI), 1999, the Godavari delta covers an area of 241 km^2 of mangroves. The period between 1975 and 1986 was noted for degradations of

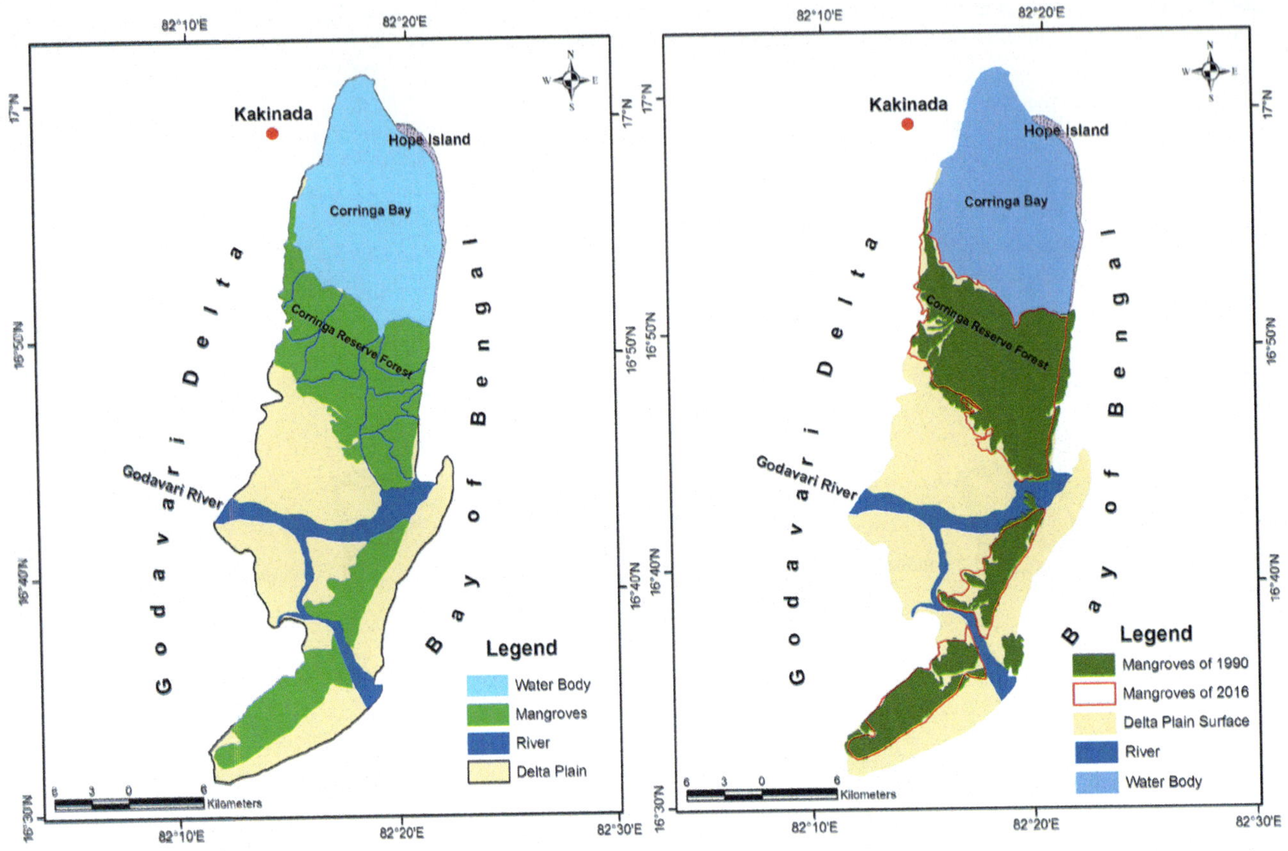

Fig. 6.7 Mangrove area of Corringa RF (Left Side) and areal change of Mangrove belt in Corringa RF, Godavari Delta (Right Side)

mangroves which were greater than the newly form mangroves of that time. However, the net increase of 367 ha of mangrove area was recorded within the period of 1986 and 2001 in the deltaic tract due to implementation of afforestation programs.

The low-lying deltaic tracts, estuary fringes, islands and back barriers are important settings of mangrove forests and related wetlands in Krishna Delta. Mangrove areas were distributed over 156 km^2 wetland surface of the delta (FSI 1999); but a large area of island mangroves is degraded due to erosion of island platform in the Maslipatnam Bay, and also by shoreline changes towards the delta front areas of Krishna River. In overall perspective, the mangroves are degraded in 59.91 km^2 areas of Godavari Krishna Deltas (MOEF 2012, in: Coastal Zones of India) of the coastal state. The deltaic shores are liable to wave influenced erosion and cyclone prone dynamic environment (NCRMP 2016) with significant height of storm surges. The mangroves are also vulnerable to sea level change at this part of Bay of Bengal (NIO, Goa: 2014). The number of mangrove species ranges from 32 to 35 in the delta. Saltpans are extensively distributed towards the land margins of mangrove forests at present (Fig. 6.8).

Pichavaram mangroves of Tamil Nadu coast are located in a barrier back tidal basin fed by Coleroon River (Fig. 6.9). The enormous reduction in discharge of fresh water into the coastal zones by Coleroon River created the favourable environmental gradient for survival of salt tolerant (Selvam 2003) mangroves in the settings. The areas of mangrove coverage are compared with Remote Sensing studies of three different years (1990, 1999 and 2016) to estimate the change detection of fringe forests (Table 6.4).

The Muthupet area of Cauvery delta is also located behind the barrier that is fringed with a tidal basin. About 12.00 km^2 areas of such tidal basin were occupied by mangrove forests in 1999 and though the coverage area is increased to 32.55 km^2 in 2016 by measures of restoration programmes but it is liable to cyclone vulnerabilities.

The event of 1952 cyclone was associated with a tidal wave or a storm surge wave of over 2 m height in the areas of Muthupet and Chattram forests in Cauvery delta. As a result of such submergence by salt waters for a few days, the frontal mangroves dominated by *Avicennia* were killed over extensive areas of the settings except the survival of younger growth of mangroves along the channel fringes on narrow belts (Venkatesan 1966; p. 30) of the coastal region. About

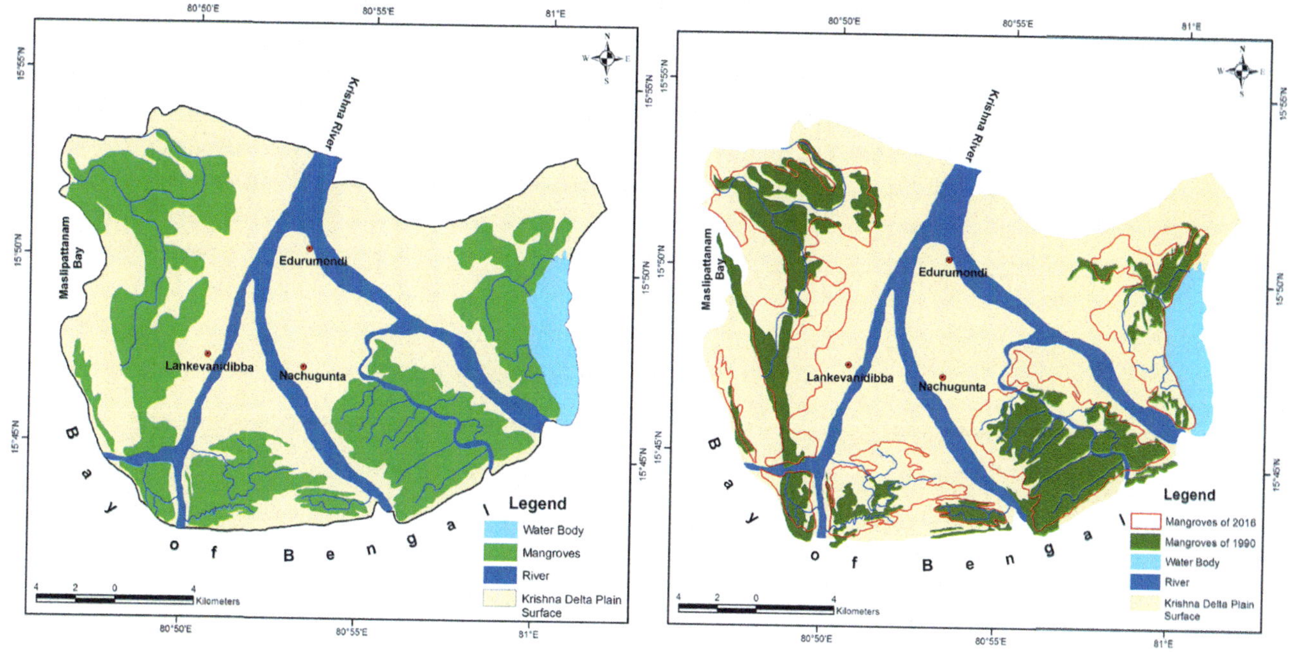

Fig. 6.8 Mangrove area of Krishna Delta (Left Side) and areal change of Mangrove belt in Krishna Delta (Right Side)

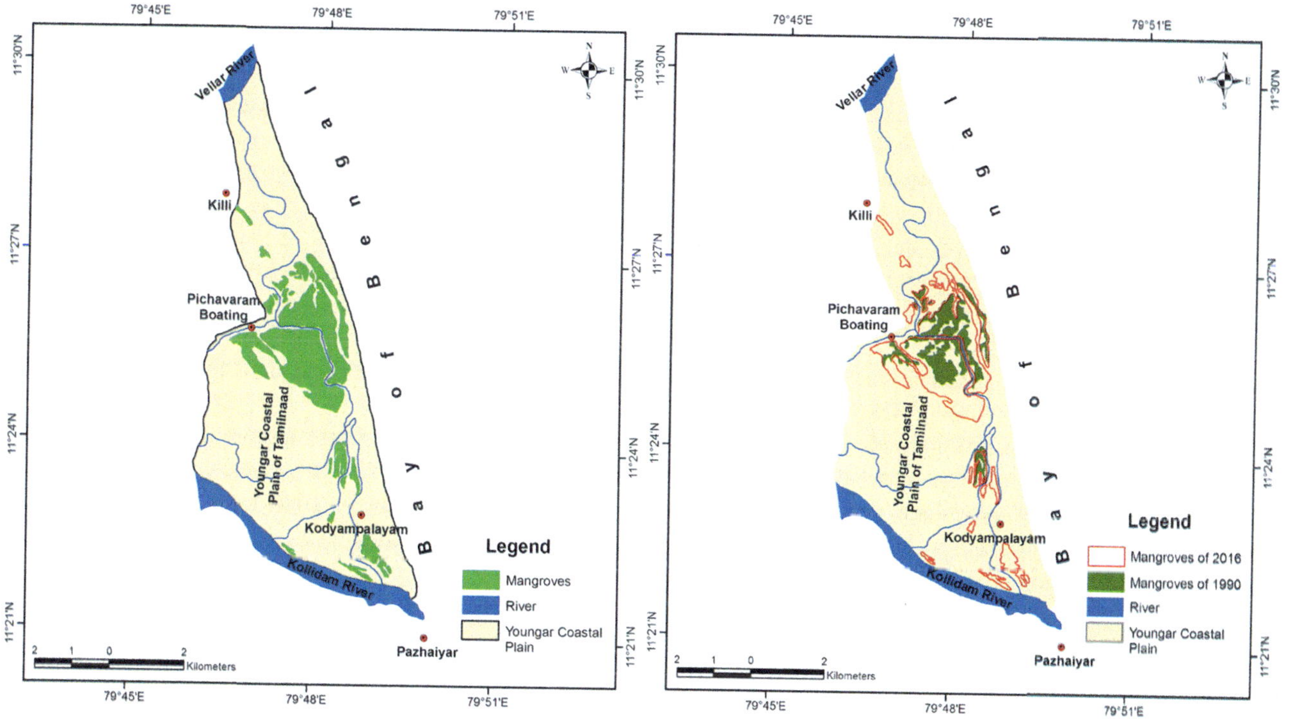

Fig. 6.9 Mangrove area of Pichavaram (Left Side) and areal change of Mangrove belt in Pichavaram (Right Side)

20 different mangrove species are now available in Muthupet area of Cauvery delta (Fig. 6.10).

Various embayments, channel straits, estuaries and shore fringed platforms of Andaman and Nicobar group of Island are also noted for mangrove coverage of 966 km^2 (FSI 1999) in the shore fringes of Bay of Bengal and Andaman Sea. About 23.77 km^2 areas of mangroves were lost in these island fringes due to land subsidence, emergence and tsunami event of the year 2004, December (MOEF 2012). The low tidal range (0.50–1.00 m), significant wave height (over 1.5 m), salt water influences with average annual rainfall of 3000 mm amount provide the favourable environmental gradient for

Table 6.4 Mangrove coverage changes and reduction of fresh water discharges by Coleroon River into the physiographic settings of Pichavaram systems (Tamil Nadu Coast)

Year/decades	Discharges of fresh water by Coleroon River	Year	Mangrove coverage area in Pichavaram
1920s	73,000 million ft^3	1990	3.30 km^2
1980s	31,000 million ft^3	1999	9.00 km^2
1990s	5000 million ft^3	2016	8.28 km^2

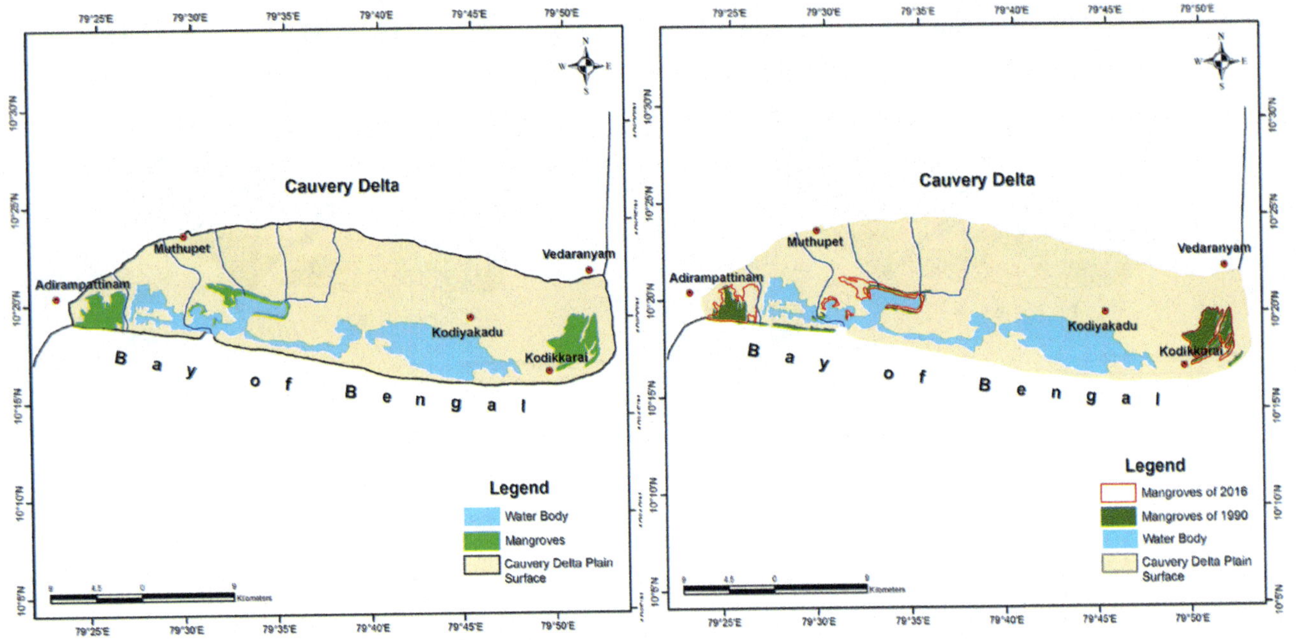

Fig. 6.10 Mangrove area of Muthupet (Left Side) and areal change of Mangrove belt in Muthupet, Cauvery Delta (Right Side)

colonization of mangroves over the carbonate platforms. Mangroves are wide and extensively distributed along the sheltered areas of channel strait fringes, shallow bays, tidal creeks, and embayed shores of the island groups. The open shore platforms are under direct wave attacks and do not support the favourable environmental gradients for dense colonization of mangroves. Mangroves were killed mostly in such physiographic settings after the incidences of earthquakes, tsunami waves, cyclone breaks and by sediment drifts in to the shore platforms at high wave energy levels. Mangrove regeneration process is affected on the shore platforms due to abrupt cementation of carbonate sediments in many places. About 20 species predominate in the mangrove forests of Andaman and Nicobar group of islands (Banerjee et al. 1989).

6.5 Mangrove Belts of the West Coast of India

Mangroves are mostly distributed in the back waters fed by river flows sheltered flats of the inner parts of estuaries, larger embayments and sheltered bays along the shore fringes of the Arabian Sea on the West coast of India. Very dense coverage of mangroves is found only in the shore fringes of Gujarat coast, Maharashtra coast and Goa coast, but the coverage becomes open and fragmented in Karnataka coast and Kerala coast. The composition of species compared with river deltas of East coast is not diverse in the coast fringed by Arabian Sea.

The mangroves of Gujarat coast are mostly distributed in the tidal flats of Gulf of Cambey, Gulf of Katchh and Gulf of Khambat. There are about 1026 km^2 areas of mangrove coverage (FSI 1999) in Gujarat coast. They are categorized as dense mangroves (732.20 km^2), very dense mangroves (248.41 km^2), degraded mangroves (29.789 km^2) and dense salt marsh vegetations (18.25 km^2) in the coast (Ministry of Environment and Forests, Gov. of India, MOEF 2012). However, the landward margins of such fringe forests are also characterized by location of saline flats and degraded scrubs. The fresh water flows of Mahi River, Luni River and Sabarmati River are not reaching to the coastal zones due to aridity of the inland areas except a little discharge by Narmada River and Tapi River at the coast. There are about eight to ten mangrove communities and seven to eight non-mangrove halophytes (Banerjee et al. 1989) in the coastal fringe forests of Gujarat coast (Fig. 6.11).

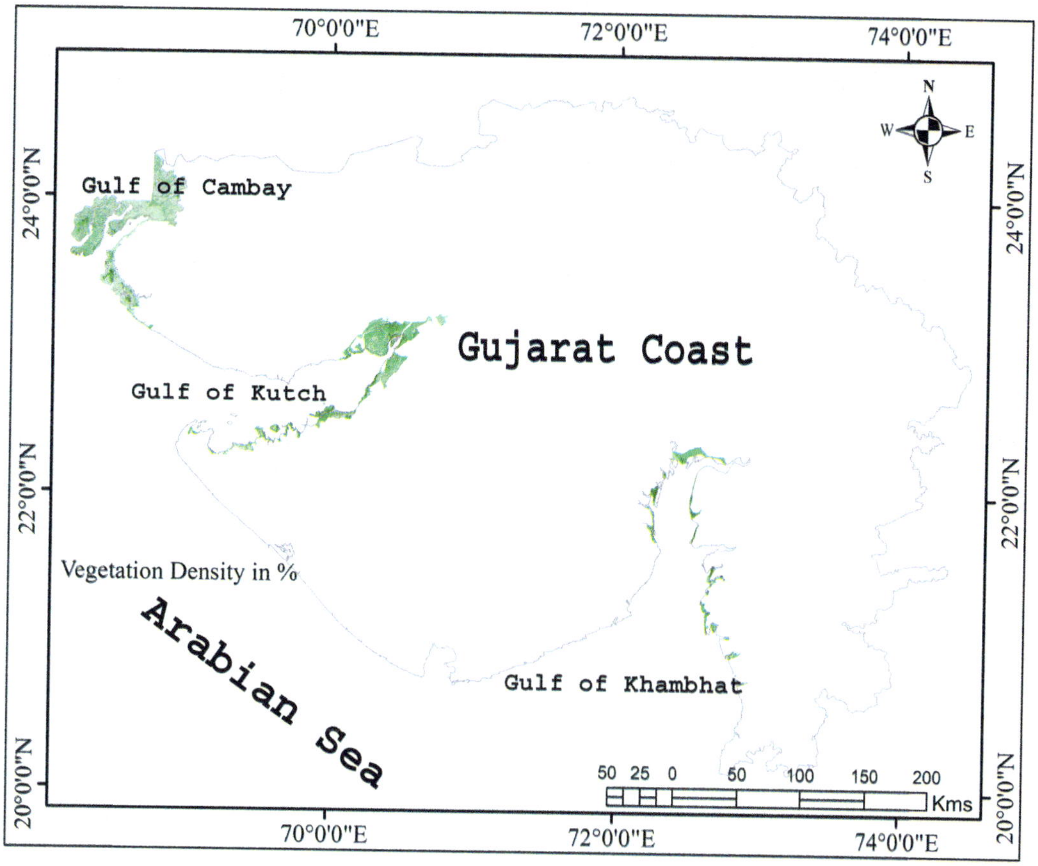

Fig. 6.11 The major mangrove forest sites in the Gulf of Khambhat, Kutch & Cambay of Gujarat coast

The estimated mangrove coverage of Maharashtra coast was about 300 km^2 but it reduces to 212.60 km^2 (MOEF 2012) areas of mangrove forests with additional location of extensive saltpans (67.05 km^2) at present. The estuary fringes of Ullhas River and Vaitarni River towards north and Mashala creek and related shore fringes of Raigad District to the south of Konkon coast represent the constituent of mangrove and salt marsh communities. Mangroves are also degraded over 31.39 km^2 areas of the coast at present. About 11 mangrove species are reported from the areas of Maharashtra coast.

Mangroves of Goa coast constitute about 339.04 km^2 areas of scrub and 27.78 km^2 areas of dense forest at present though the areas of 5.48 km^2 are degraded in the mangrove zones (MOEF 2012). There are seven estuaries in Goa coast and among them Mandovi, Zuari and inter connecting Camberjua constitute about 75% of mangroves in the coast. The total area of healthy mangroves was distributed over 200 km^2 areas in Goa coast in the past (Anon 1987). About 20 species of mangroves and non-mangrove halophytes are reported from coastal settings of Goa.

The rivers of western Karnataka (northern part of Malabar Coast) have descended down to the coast and ponded over wide low lying areas behind the shore fringed sand ridge barriers along the shoreline of Arabian Sea. The tidewater's encroach into the back barrier wetlands through the river mouth estuaries and spread over the basin with supply of fluvio-marine sediments (fluvial and tidal supply) and available nutrient loads support the mangrove habitats for the coastal zones of Karnataka State. The number of species recorded from this zone is about 11 (eleven), they are dominated by salt tolerant mangroves, such as *Rhizophora* sp., *Ceriops* sp., *Kandelia* sp. and *Avicennia alba* species.

In Kerala back water fringes, area of mangrove coverage is extended over 7.00 km^2 of the tidal flats along the shoreline of Arabian Sea. The reason constitutes the nine species of true mangroves and a few non-mangrove halophytes and other mangrove associates (Table 6.5).

6.6 Types of Natural Threats on the Coastal Fringing Forest Dominated by Mangroves in India

Mangrove ecosystem can be restored partially by afforestation and by improving the tidal drainage through the extension of artificial channels or ditches fed by high tide water into the inner parts of mangrove swamps. In the Tropical

Table 6.5 The location of degraded mangroves and saltpans on the West Coast and East Coast of India as per the records of MOEF database (2012)

Sl. No.	Mangrove areas of the coastal belt	Area in sq km (Degraded)	Area in sq km (Saltpans)
1	Andhra Pradesh (Telengana State)	59.91	222.94
2	Odisha Coast	30.35	19.11
3	West Bengal Coast	292.63	29.01
4	Andaman and Nicobar Islands	23.77	NA
5	Gujarat Coast	29.78	1115.60
6	Maharashtra Coast	31.39	67.05
7	Goa Coast	5.48	NA
8	Karnataka Coast	0.20	1.99
9	Kerala Coast	1.45	0.09
10	Tamil Nadu Coast	7.83	293.35
Total		482.79 (km^2)	1749.14 (km^2)

coastal environment of India, mangroves are mostly lost due to human uses of natural resources and land use conversion for rice paddy farming in the deltaic alluviums (Richards 1993; Blasco 1984; Banerjee et al. 1989; Naskar and Guha Bakshi 1987; Ashis Kr. Paul 2002a, b). However, it is very difficult to restore the fringe forests of the coastal belt if they are naturally degraded by marine forcing factors, tectonics related land subsidence and uplift, other hydrological stress from the drainage catchment areas of river deltas, coastal morphodynamics under the impacts of hydrological stress and sediment starvation, hyper salinity, eutrophication, shifting river banks in the deltaic areas, storm catastrophe, over wash sand fan lobes encroachment, erosion at the sea face with shoreline retreat, active barrier rollover process, and shortages of rainfall as well as fresh water discharges into the coastal zones (Table 6.6).

The natural death or threats of mangrove forests or the coastal fringe forests are studied by direct field observation (DFO) method and using Remote Sensing techniques in the present work. The reasons of natural death of Indian coastal fringe forest are categorized in the following ways.

6.6.1 Marine Forcing Factors (Sea Level Rise Related Effects)

The exposure of erosional shore platforms of the deltaic tidal flat at their sea face reveals some clear development of tidal notches of fingering gully heads, significant amount of sheet erosion over the near horizontal tidal platforms, and multiple stepped banks on the sea ward edge of the tidal flat in the form of erosional features along the sea face of the Sundarbans, Subarnarekha delta, Bhitarkanika, Godavari and Krishna delta of the East Coast of India. The peat layers or the paleo marsh deposits at the lower mid-littoral part, and the clear location of vertical stands of degraded older tree stumps with pneumatophores on the upper part of mid-littoral zone along the stiff (compact) clay banks of sea facing islands are currently visible in the extreme low tides (Paul and Kamila 2016).

Thus, the above features indicate a significant loss of wetlands with halophytic vegetations at or along the coastal fringe by direct impact of sea level rise related change in tidal current energy and wave energy levels. As some areas of the coastal belt of the Sundarbans and Bhitarkanika are developed and protected from the rising sea by earthen embankments, the mangrove wetlands of frontal shore platforms are not able to advance inland, and in many cases, they are squeezed between a rising sea and the protective embankment structures (e.g. Sagar Island, Tat Island, Henry's Islands in the Sundarbans; and Talchua Island, Kalidiania Island etc. in Bhitarkanika). Other physical effects of sea level rise are identified as sediment (sand size) infilling into the tidal creeks and salt water inundations into the low lying areas of supra tidal or supra littoral zones of the coastal belt. The mangrove stands are degraded along the banks of such tidal creeks impacting over the newly inundated tidal wetlands dominated by mangroves and other halophytes.

6.6.2 Neotectonics at the Coast (Subsidence and Emergence)

In 2004, 26th December, the entire Andaman and Nicobar groups of islands were affected by a severe earthquake (9.1 Rhicter Scale) related submergence and emergence of the coastal fringes and also by Tsunami deposited sediments into the bays, coves, and embayments along the affected shorelines (Paul 2005; Das and Pranamik 2011). By this tectonic activity, the shore fringe areas of North Andaman, Middle Andaman and parts of South Andaman were uplifted up to over 1.0 m, and Great Nicobar, Little Nicobar, Car Nicobar with some parts of south Andaman were submerged at a rate between 0.50 and 2.00 m in different sites. The present author visited the Andaman group of islands since 2003–2017 (once in a year) to watch the shore fringe forest

Table 6.6 Natural threats and impacts to mangroves of the low lying coastal belts of India

Habitat dynamics	Process involved	Abrupt adjustment with irregularities in the system
1. Shorefront instability	Exposure of the mud bank to the wave attack	Change in frequency and intensity of storms
2. Estuary fringe cliffing	Erosion in the horizontal plane	Change in mean tide elevation in inner part of estuary
3. Younger depositional islands	Liable to attack of tidal waves and storm surge heights	Change in areas of vertical aggradations and retro gradations
4. Formation of saline blanks	Inner part of swamp under the impacts of astronomic tides, moderate storm regimes and evaporation rate	Change in micro topography of the swamp with uneven rate of siltation and auto compaction
5. Restoration of former swamp and regeneration of mangroves or marsh vegetation by improving tidal drainage and mangrove plantation	Under high energy storm event with surge elevation	Change in growth form of mangroves
6. Vulnerability of reclaimed swamp by salt water flooding	Attempt to restore the former wetlands	Change in elevation of the reclaimed swamp in compare to open swamp
7. Wash over fan lobes with supply of ex-situ sediments in back barrier marsh and deltaic marsh	Swampy surface under the blanket of sand sized sediments	Change in beach height, beach width and beach slope in different shore line sections due to shortages of sediment supply
8. Embayment marsh in estuary margins with restricted drainage entrance	Abundant supply of sediments into the lagoonal setting	Change in bay tide elevation and open sea tide elevation
9. Emergent swamp of the ria head	Uplift with seismic energy	Change in local geological factors under tectonic setting
10. Submergent swamp of the coastal embayment	Subsidence and tsunami run up	Change in local geological factors and hydrological factors at the land sea interface

loss and related coastal dynamics over the island fringed shore platforms (Paul 2003).

The upper part of inter tidal zone is converted into supra tidal zone in the areas of sudden uplift, and the resultant fringe forest were relatively shifted towards the inland without further inundations by regular high tides. As a result of such shift in environmental gradient of uplifted inter tidal shore platforms, mangrove stands were lost and only a few species (mostly evergreen character) were adapted with seepage flow of saltwater during high tides and fresh water supply during rainy months at the landward parts.

However, the loss of mangrove forests and other littoral forests is very significant in the areas of subsidence (Katchal Island, Trinkart Island, Campbell Bay, Garashara, etc.) particularly by permanent inundations and tsunamigenic sediment filling into the subsided estuaries, and into other embayments of the affected shores. Vertical standing of tree stumps and trunks in degraded conditions, and large amount of log drifts over the shore platforms are still visible in the areas of partial subsidence.

6.6.3 Hydrological Stress from the Drainage Catchment Areas

The occurrences of catastrophic floods in the river deltas (Subarnarekha River 1950, 1978, 2008; Ganga River 1978, 2000; Mahanadi River 2003, 2008, 2011, 2013) have produced several impacts over the coastal fringe forests. They include as bank erosion, log drift, active sand splays on the bank margins, reduction of soil pore water salinity, and supply of pollutants and large volume of sediment loads (submerge the estuary fringe tidal flats). Such effects of hydrological stress are visible on the channel fringe mangroves and on island fringe mangroves of the river deltas in the events of major floods.

6.6.4 Coastal Morphodynamics Under Sediment Starvation

The morphological response to the change in environmental variables at the sea face due to shortages of sediment supply along the stretch of abandoned portion of the delta has exposed the area directly to wave attacks to generate the local source of sediment supply in the system. The loss of wetlands by wave erosion and high water submergence are very common at the sea face between Saptamukhi River mouths to Gosaba River mouth in West Bengal portion (India) of the Sundarbans. Long shore sediment transport supply from the western shores cannot reach the area for local adjustment of morphology, and which reason, the parts of Sundarban sea face are rapidly getting wave abraded. The vegetated tidal flat platform bears the signatures of such wave attacks by pushing the sand size sediment into the inner parts of the mangrove swamps and by reducing the buffer areas of low tide shore platform along the base of cliffed mud banks. In many cases, the vegetation at the shore face are stunted and partially degraded because of active saltwater sprays at the time of landward advancement of wave breakers particularly during high spring tides. Ample of evidences are located in the sea face of Lothian Island, Tat Island, Dhanchi Island, Bhangaduni Island, etc. in Indian Sundarbans (Figs. 6.12 and 6.13).

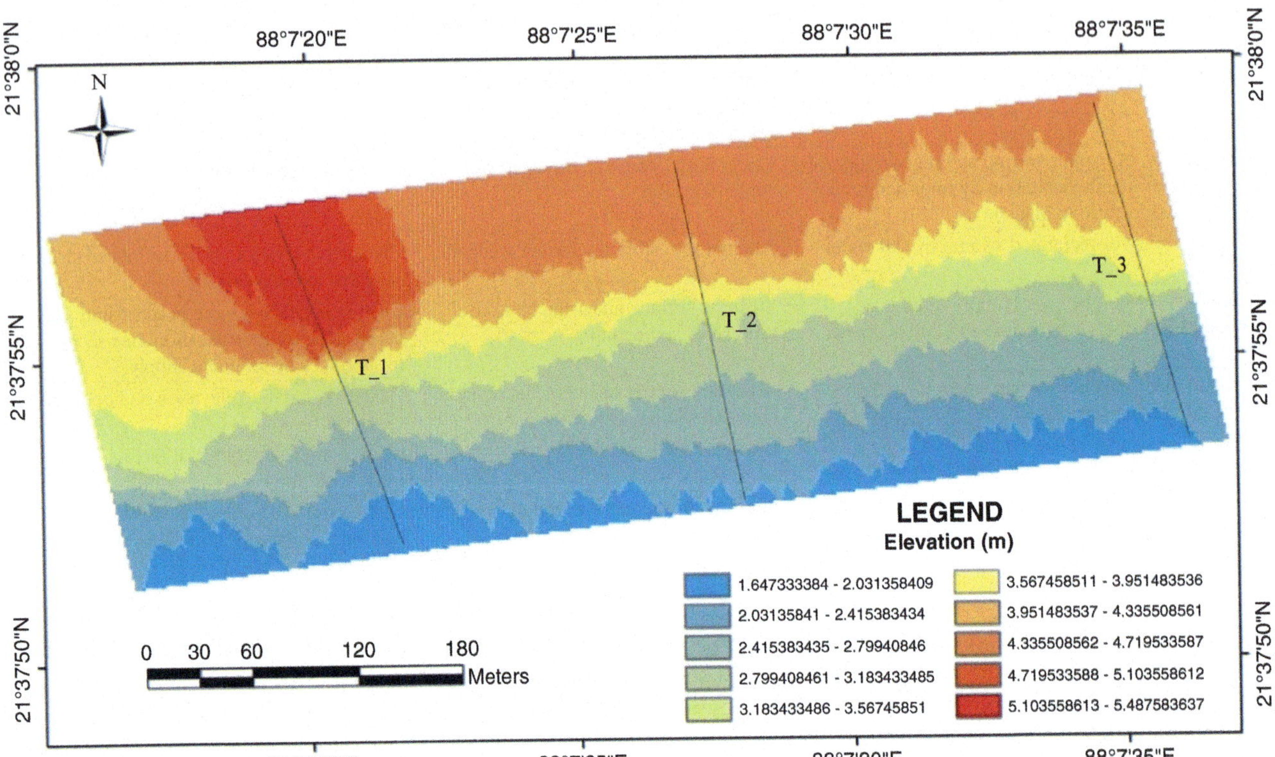

Fig. 6.12 Contour plan of Boatkhali station along the shoreline of Bay of Bengal (On the basis of Total Station survey conducted in the area on 15th April, 2015)

Fig. 6.13 Gully erosion of the exposed mud banks once covered by mangrove forests in the parts of Indian Sundarbans

6.6.5 Hypersalinity in the Inner Parts of the Mangrove Swamps (a Greatest Threat to the Fringe Forests of the Coastal Belt)

The inner parts of islands lie far away from the major tidal rivers, though they provide a platform for cover of mangroves in the river deltas. Sometimes, the mature platforms may be affected by tidal drainage loss due to their higher elevation, and distance from nearest tidal rivers. As tidal drainage cannot reach the areas during daily high tides, they become uncovered by tides frequently. The sub aerial exposures of such tidal flats are sensitive to evaporation under tropical hot and humid coast with alternate dry and wet periods. As the area is devoid of further siltation process due tom the absence of tidal drainage it remains as low or as a wide flat with concave upward basinal depression in the swampy forest. The areas are now only flooded by astronomic tides and particularly by storm generated saltwater floods across the island platforms. Gradually, the entire area of the island platform becomes uncovered by saltwater at low tide when the phases of astronomic tides and storm events will over in and around the coastal belt. However, the topographic lows developed by above reasons surrounded by mangrove forests will hold the salt water pockets or pools of tide water within the island platforms. In the next stage, when they will be evaporated by scorching sunlight with 30° centigrade plus temperature, the silty-clay surface with high soil moisture will be desiccated and encrusted by salts. Under such conditions, the mature halophytic vegetation is also killed by salt affected soils in the inner parts of island platforms. They are termed as saline blanks, natural saltpans or hypersaline patches within the mangrove forest of the river deltas in Indian East Coast. (Figs. 6.14 and 6.15).

Such saline blanks are very well developed in Sundarbans, Subarnarekha delta and in parts of Bhitarkanika. Further study shows that the blanks are categorized on the basis of their status.

From the Remote Sensing studies of coastal wetlands of India (National Wetland Atlas, 2012 and MOEF) they are categorized or identified as salt pans, degraded mangroves and partially degraded mangroves. In reality, such development of saline blanks may be expanded if the amount of rainfall decreases, extent of dry phase increases and the recurrence interval of major floods becomes longer in the coastal belt of river deltas (Table 6.7).

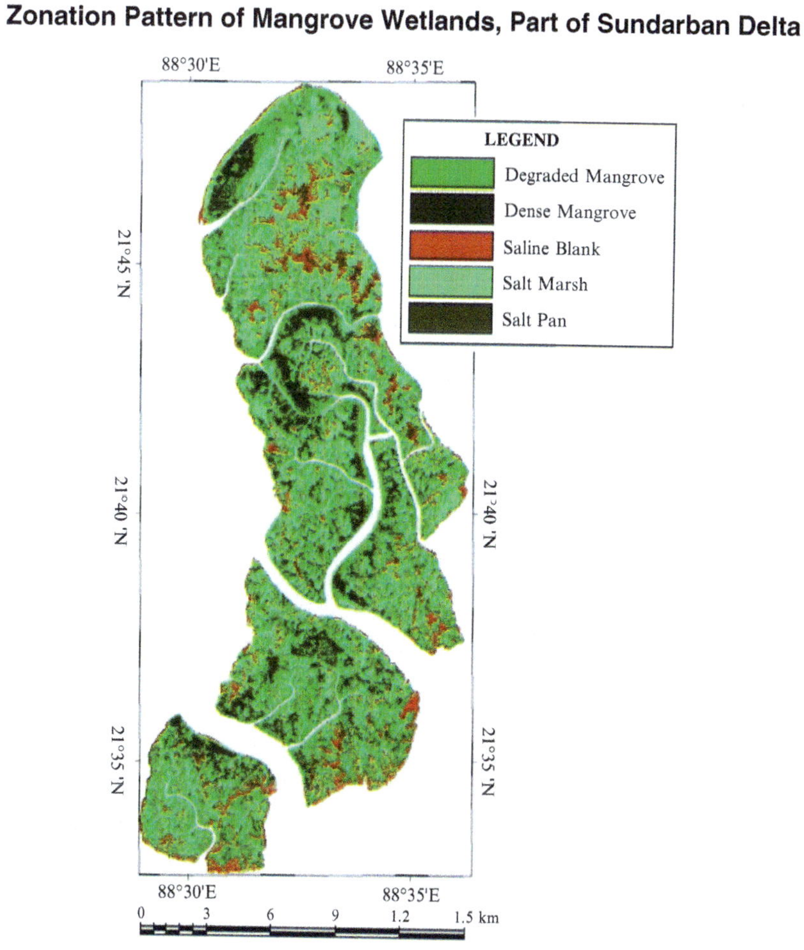

Fig. 6.14 Category of mangrove wetlands in the parts of Sundarbans in West Bengal, India

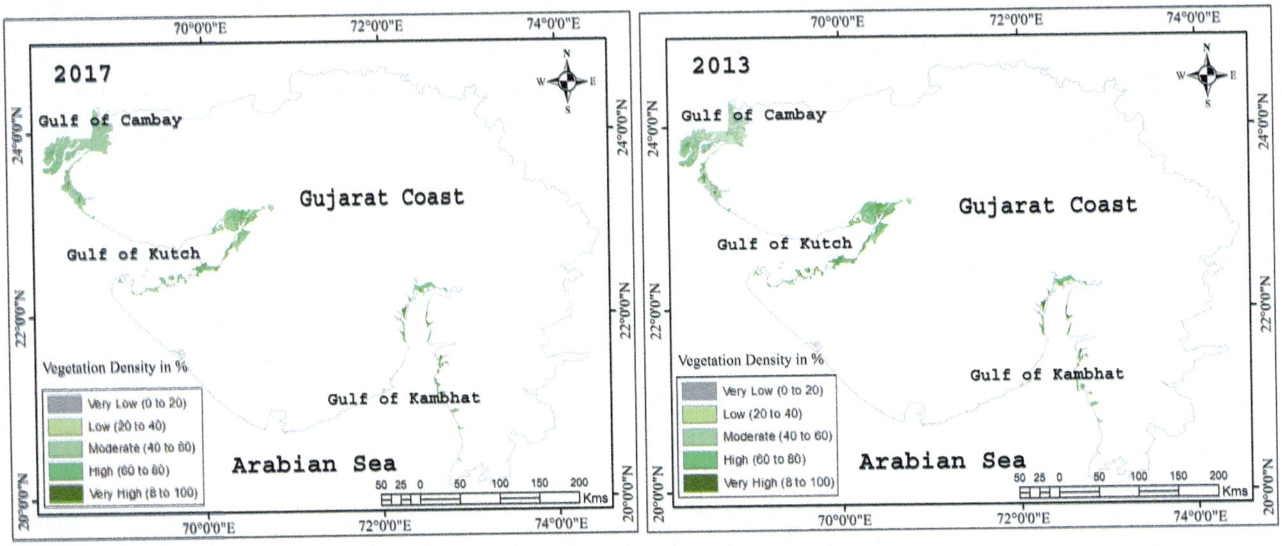

Fig. 6.15 Category of mangrove wetlands in the parts of Gujarat coast (2017 and 2013), India

Table 6.7 Categories of hypersalinity in the inner parts of mangrove swamps (Sundarbans)

Stage	Types	Status
Stage- A	Primary pans	Pan shaped depression with ponded salt water bodies
Stage- B	Hypersaline patch with Salicornia marsh	Halophytic heath lands over the sodium clayey/rich surface
Stage- C	Hypersaline patch with stunted or dwarfed mangroves	Salt affected surface with infrequent location of salt tolerant mangroves
Stage- D	Saline Blanks	Dry and desiccated surface with shrinkage cracks and salt encrustation
Stage- E	Salt Flats	Higher flats with salt dust and desiccated surface

The saline blanks developed over Lothian Island, Dhanchi Island, Kalas Island, Henrey's Island and inner parts of Sundarban islands (Indian part), reveal that in many cases almost all of the vegetation are denuded from such barren soils within the existing mangrove stands. In rainy months, their surface becomes encrusted by algae mats.

6.6.6 Eutrophication as Driver of Coastal Wetland Loss

It is the result of stagnation of inter tidal mangrove wetland ecosystem by oxygen depletion brought about by or supplied by untreated nutrients from the mangrove fringe fishery activities and encroached rice paddy farm plots in such a level that has over stimulated the biological production of the aquatic environment. The spread of algal blooms resulted by eutrophication has disturbed the mangrove regeneration through protecting the drifted seeds from soil touch and anchorage into swampy surface. It is clearly observed in the lagoon fringes of Chilika (Odisha State) in which the initial growth of mangroves is also severely affected by eutrophicated oxygen depleted soil surface. It is also reported from Godavari deltaic mangroves, Sundarbans, and Mahanadi Mangroves in eastern coastal belt of India.

6.6.7 Shifting Riverbanks in the Deltaic Areas

The intervened tidal channels within the mangrove dominated delta plains migrate as the channel beds are partially filled by sediments brought about by compensatory supply of silts into relatively sheltered areas of the dynamic environment. If the channel beds become flatter within channel siltation in expanse of deepening process the steeper banks will be undermined by flood flows of tidal current and above all the lateral shifting of channels will be the result.

Currently, the sediments supplied from the erosion of the sea face of abandoned delta (Indian Sundarbans) are brought into the inner parts of estuaries and major tidal channels of the Sundarbans to settle down.

The channel banks may oscillate with the occurrences of the episode of tidal scour and episode of tidal accretion in response to the variations in tidal volumes during extreme high tides and extreme low tides within the delta plain areas. As a result of accretion in one bank will develop point bars in which mangrove seedlings may colonize the accretionary sediment surface, but the existing mangrove trees on the opposite bank will be undermined by advancing channel induced bank collapse process. The local slumping blocks of compact muds with clumps of mangrove vegetation are often visible into the tidal creeks of river deltas.

Many fallen trees and trunks of mangroves and their branches lie decaying on the floor of steeper banks with remaining thickets. Such destruction of mature and immatured mangroves along the course of Bidya River, Durgadoani River of Sundarbans; Talsari Subarnapur River of Subarnarekha delta, Bramhani-Baitarani River banks and other tidal rivers of Bhitarkanika; and also around the tidal river of Corringa Reserve Forest are visible at present.

6.6.8 Storm Catastrophe

Mangrove forests are very sensitive to high magnitude storm attacks. Cyclone 'Ceder' (2007) in Sundarbans, cyclone 'Aila' (2009) and 1988 cyclone in Sundarbans, 1989 cyclone in Bhitarkanika, 1999 Paradip cyclone in Mahanadi delta, and 1978 cyclone in East and West Godavari districts were notorious due to mangrove damages by strong wind speed (>200 km per hour) and prolonged rate of salt water inundations (>3 days) in the coastal habitats of river deltas (Table 6.8).

Currently, Indian Meteorological Department (IMD) has produced the cyclone vulnerability atlas for Indian Coastal Districts (96 coastal districts). The study shows that around 28 coastal districts on the both sides of Bay of Bengal and Arabian Sea are estimated as highly vulnerable due to high frequency and magnitude of cyclone landfalls and storm surge elevations (Fig. 6.16).

Mangroves are damaged by top breakage, leaf falls, vertical uprooting from the collapsed banks, and also by rear occurrences of burning by fire ball activities resulted from cyclone landfalls in the low lying coastal belt with high wind speed (Table 6.9).

As time will goes, the frequency of high magnitude cyclones will be increased in near future if the global warming induced Sea Surface Temperature (SST) increases rapidly in the tropical seas.

6.6.9 Encroachment of Overwashed Sand Fan Lobes into the Mangrove Swamp

This is a significant threat to the fringe forests of the coastal belt recorded recently in and around the low lying beach dune fringed shores with vulnerable backshore mangroves, and also in the wind ward sides of the shore platforms of Andaman group of islands.

The storm surge elevations are predicted for the coastal zones of Bay of Bengal and Arabian Sea by IMD in its cyclone vulnerability atlas (2016). As the low lying delta fringe sea beaches and sand dunes are unable to hold the advanced surge waves during cyclone landfalls, the over washed dune sands and beach sands are encroached into the mangrove wetlands in the form of fan lobe terrace by killing frontal mangroves at the shore platforms at present.

The *Avicennia* species with their extensive breathing roots or vertical standing pneumatophores maintain their physiological adaptation in response to high rate of soil pore water salinity, active bio tidal accretion and mesotidal ranges at the sea front positions under diverse physiographic settings of the coastal zones. However, the most cases, the breathing roots are buried or entirely trapped by deposition of thicker sand sheets (0.5–1.50 m) of over washed fan lobes along the seaward sides, island platforms and estuary fringes of river deltas (Sundarbans, Subarnarekha delta, Bhitarkanika, Mahanadi delta, etc.) at present probably due to the impact of transgressive seas.

A recent cyclone disturbance took place (December, 2016) in South Andaman group of islands. The direct field observation on the shore platforms of a few islands (Neil island, Havelock island, Richis's Archipelago) in the post cyclone visit reveals that not only the vertical breathing roots of *Avicennia* sp. are trapped by sand drifts if thick deposits, but also in many cases the stilt roots of *Rhyzophora* sp. are mostly trapped by sand deposits over the shore platforms. Under such adverse conditions of substrate

Table 6.8 High magnitude cyclone landfalls in the Bay of Bengal coast since 1885

SL. No	Year	Month	Place of landfall	Intensity	Lowest pressure (hpa)	Max wind speed (kmp)	Storm surge (m)	Loss of human lives
1	1885	September	Paradip	Super	918.9	252	7	5000
2	1897	October	Bangladesh	Super	922.0	250	<7	175,000
3	1927	November	Nellore	Super	919.3	250	6	6000
4	1942	October	Midnapore	Super	935.0	225	5	15,000
5	1970	November	Sunderban	T7.0	920.0	260	<7	500,000
6	1977	November	Chirala	T7.0	911.0	259	5	10,000
7	1989	November	Kavali	T6.5	930.0	235	7	56
8	1990	May	Muslipattanum	T6.5	930.0	235	7	1000
9	1991	April	Bangladesh	Super	932.0	220	<6	138,000
10	1999	October	Paradip	T7.0	926.0	260	7.8	>9500
11	2007	November	Bangladesh	Super	930.0	220	7	10,000
12	2008	May	Yangon	Super	930.0	240	6	22,000

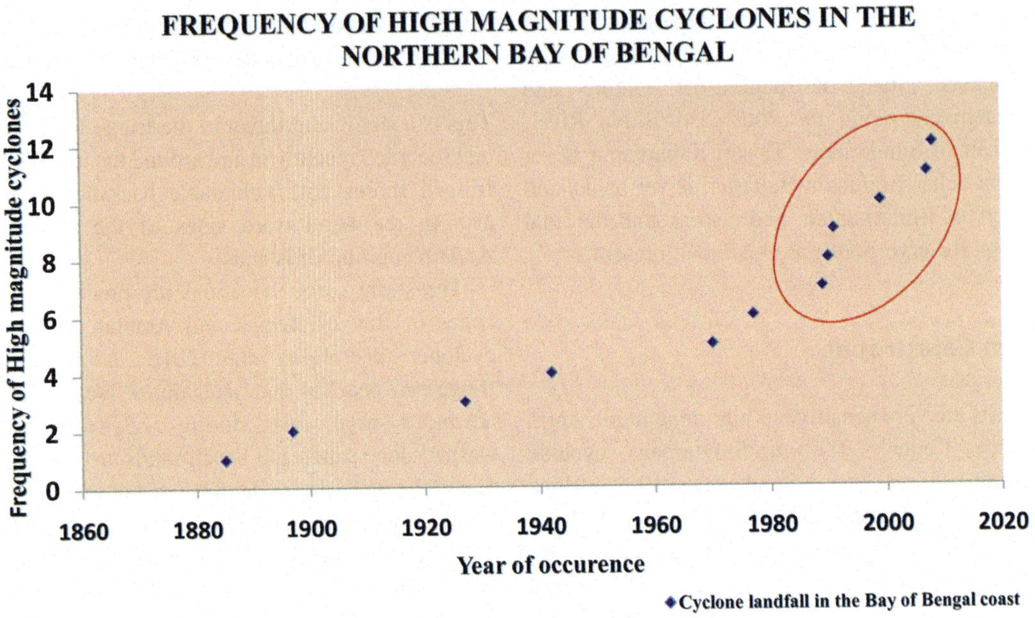

Fig. 6.16 Frequency of high magnitude cyclones in the northern Bay of Bengal since 1885

Table 6.9 Mangroves damaged by the natural threats of super cyclones in the coastal parts of India and adjoining Bangladesh

Name and year of cyclone landfall	Amount of vegetation loss along the coastal fringe forests
1. 1988 Sundarban Cyclone	76,560 ha (Md. Tariqul Islam 2014)
2. 1999 Paradip Cyclone	12,866 ha (Soudamini Das and Vincent 2009)
3. 2007 Cyclone 'Ceder'	30,000 ha (heavily damaged); 80,000 ha (partially damaged); (Mizanur Rahman Bijoy 2007)

modification of the near horizontal shore platforms, the vertical mangrove stands are mostly killed in the frontal positions (Fig. 6.17).

6.6.10 Erosion at the Sea Face and Active Shoreline Retreat

The shoreline change atlas of Indian coastal states produced by NCSCM reveals that almost all the states are affected by moderate to high rate of coastal erosion. The episodic erosion is also observed in many coasts of India by the present worker through temporal analysis of coastal change in between pre events and post events of tropical cyclone landfalls with the remote sensing techniques (Figs. 6.18, 6.19, 6.20, 6.21, 6.22 and 6.23).

The mangrove fringed shoreline of Krishna delta, Godavari delta, Hukitola Bay, Bhitarkanika, Subatnarekha estuary fringes, Sundarbans are largely affected by shoreline retreat and damage of the coastal fringe forest in near past. The loss of deltaic platforms with vertical stands of mangroves is studied with comparison of SOI Toposheet (1973) and Landsat ETM+ Image of 2008 for the coastal belt of Sundarbans. The erosion of deltaic island shores has reduced a significant amount of mangrove forests in the Sundarbans. The eroded shore fringe mud banks with exposures of stiff or compact clay beds and vertical stands of mangrove tree stumps reveal the effect of shoreline retreat over the fringe forests. Such features are observed in the islands of Sundarbans (Lothian Island, Henry's Island, Dhanchi Island, Kalas Island, Bhanaduni Island and Tat Island) which has left the impact of episodic signatures of land erosion along the deltaic shorelines (Fig. 6.24).

6.6.11 Barrier Rollover Process

The barrier sand spits of Talsari (Subarnarekha Ribver mouth) Gahirmata (Bhitarkanika), Hukitola (Mahanadi river delta), Godavari, Krishna and Pichabharam coast are moving

Fig. 6.17 Natural threat to shore fringed mangrove forests of deltaic habitat of the Bay of Bengal Coast. (1) Over wash fan encroachment into the mangrove forests of the Sundarbans, (2) Sand filled mangrove forests of Rasulpur River mouth habitat (*Avicennia* sp.), (3) Advancing sand fan lobes towards the barrier back mangrove swamps of the Subarnarekha Delta destroying the matured trees, (4) Open salt marsh tract submerged under the advancing sand sheet in estuarine island of Subarnarekha River, (5) Older tree stumps with residues of pneumatophores of *Heritiera fomes* on the eroded shore platform of Henry's Island, (6) Degraded shore fringed mangrove trees of Subarnarekha barrier back shore face habitat during the occurrences of fireball activities in 1989 cyclone, (7) Standing salt water pools during the storm surge of 1988 cyclone landfall at the shore face of Sundarbans and (8) A developed hypersaline patch within the mangrove forests of Henry's Island

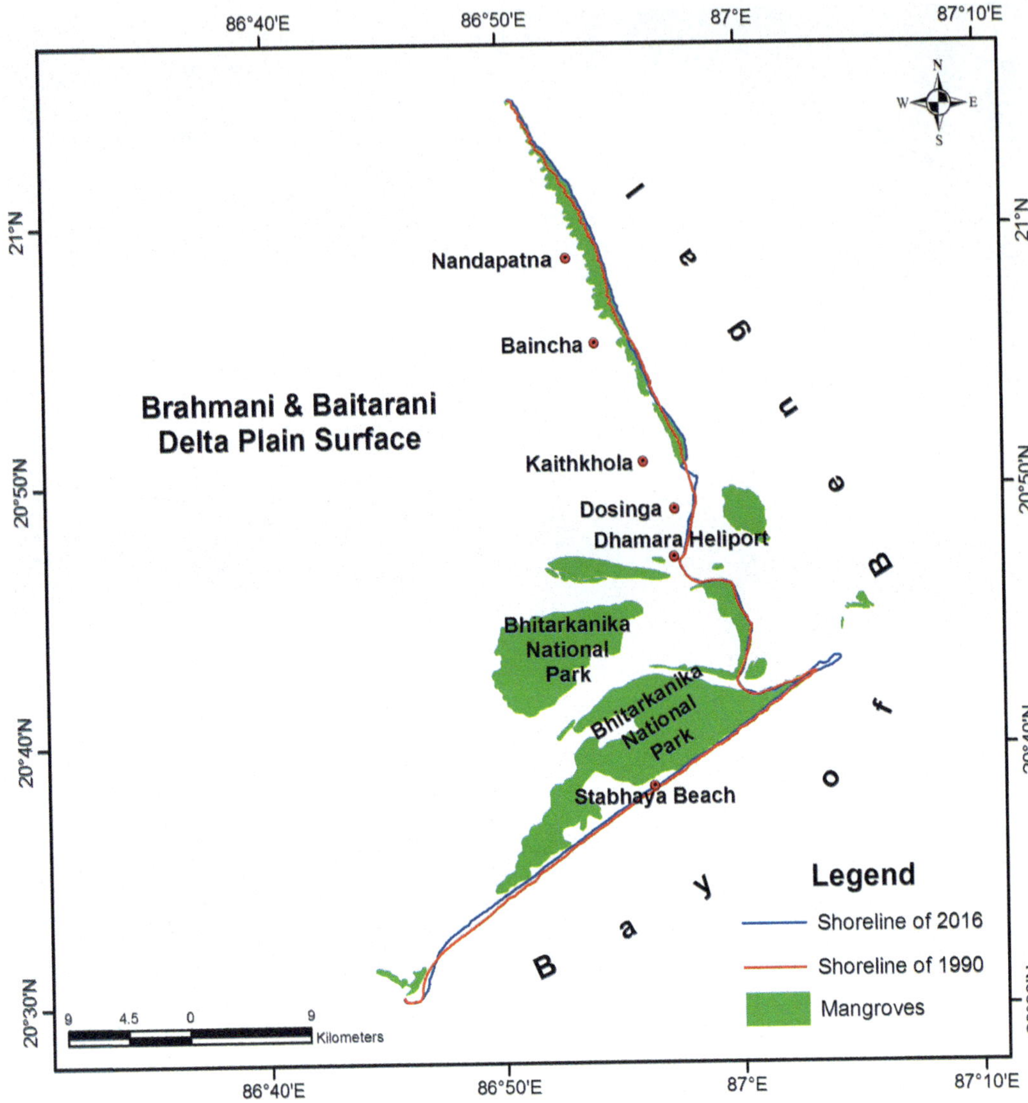

Fig. 6.18 Shoreline change detection of Bhitarkanika fringed with mangrove forests

landward by rollover process by submerging the back barrier mangroves at significant rate in response to sea level rise, and frequent attacks of swell waves. Barrier islands and barrier sand spits are sensitive to the strong winds of cyclonic storms, associated significant waves and elevated sea levels with increased frequency of high magnitude storms in the costal belt of India. Remote Sensing studies reveal that as the most of barriers migrate landward under the influence of rising sea level they are termed as trangressive barriers at the coastal belt. Barriers of Godavari spit, Chilika spit and Talsari spit are stratigraphically examined. They are affected by over wash process in the near past to the land ward sides as paleo overwash deposits in the coast. The mangroves of such back barrier sheltered environment are trapped by advancing sand ridge barriers of the coast. Lots of mangroves are killed by such barrier advancement.

6.6.12 Shortages of Rainfall in the Tropical Hot and Humid Coast

The tropical evergreen species *Heriteria fomes* were widely distributed to the inland portion of Sundrabans due to the occurrences of monsoonal rainfall with collection of rainwater in the smaller basinal depressions through much of the year in the delta plain surface. However, the coastal parts of Sundarbans displayed the mangrove forests dominated by salt tolerant species, and a large area (as much as quarter of

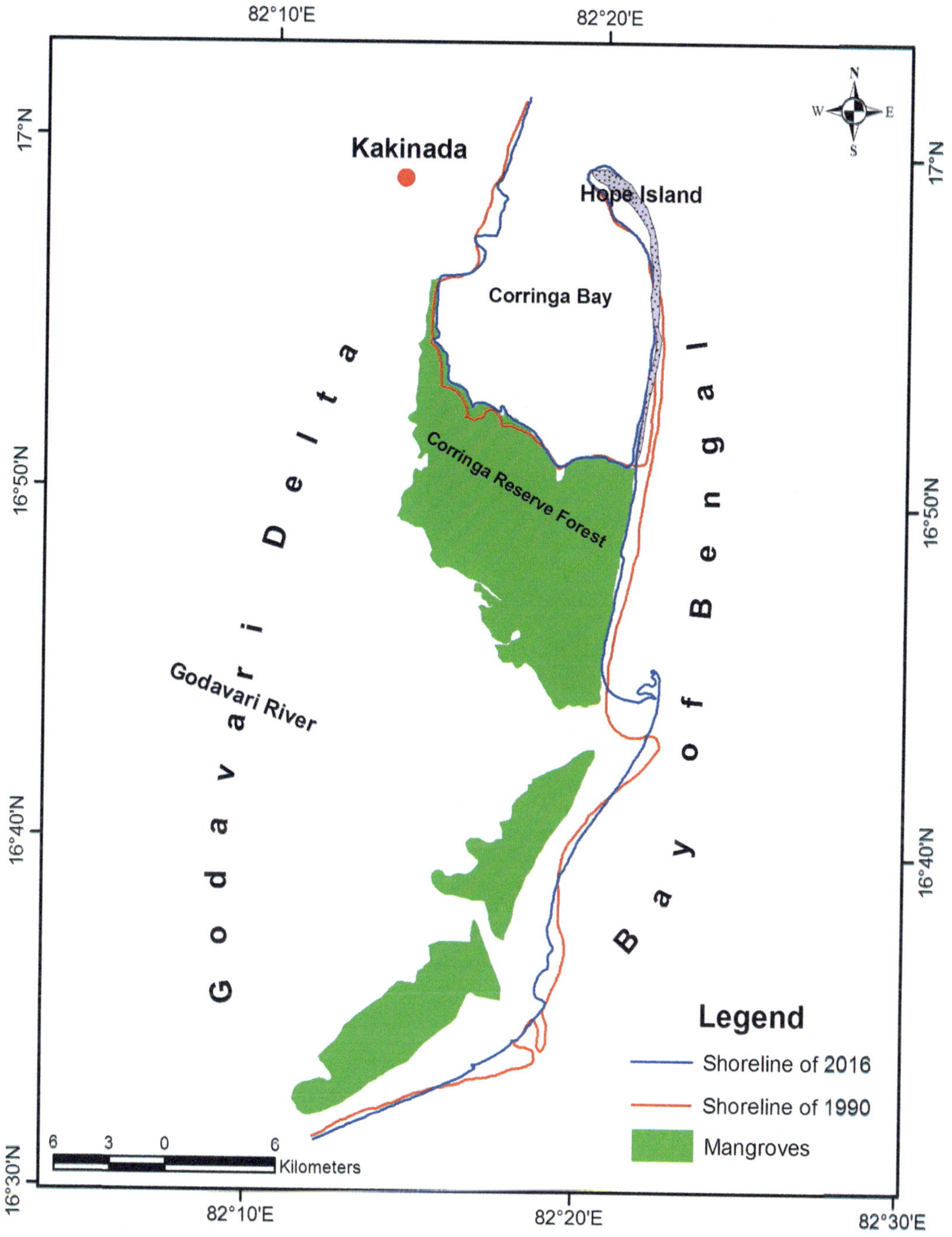

Fig. 6.19 Shoreline change detection of Corringa fringed with mangrove forests

the land area) was under fresh water and brackish water patches during rainy season (Richards 1993). It is also evidenced from the available rainfall data of a few stations of South Bengal in period of 40 years between 1976 and 2015 (Sagar Island, Kolkata, Midnapore) that rainfall has sharply decreased in Sagar Island, and Kolkata displays a little bit bellow mean, however, Midnapore station has slightly increased trend of rainfall.

The significant amounts of rainfall reduces the soil pore water salinity and washes out the surface encrusted salts by runoff process during rainy months and maintain the balance of productivity of mangrove wetlands. The current increased

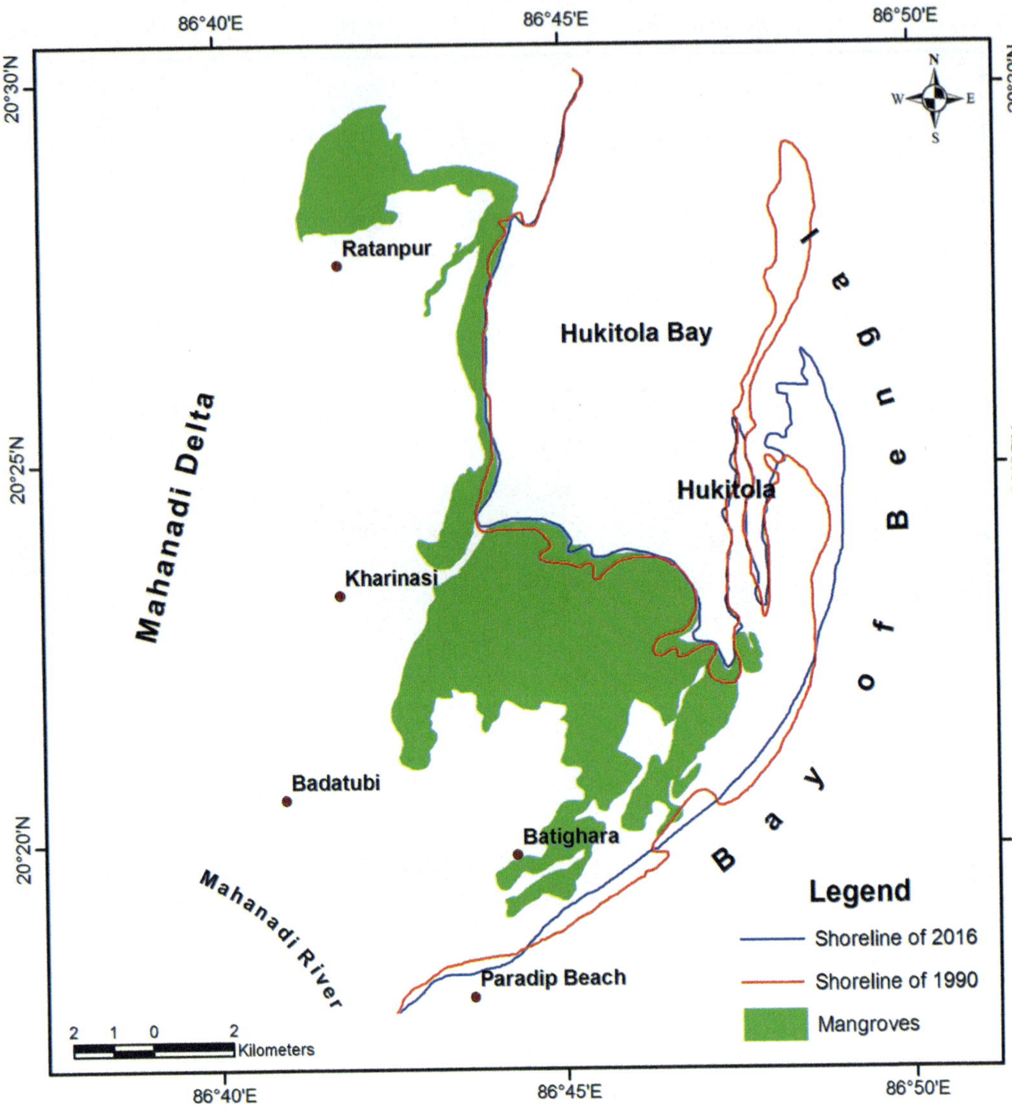

Fig. 6.20 Shoreline change detection of Hukitola Bay fringed with mangrove forests

rate of hyper salinity and growth of saline blanks within the inner parts of mangrove forests are mostly responsible for shortages of local rainfall (Ray et al. 2013) in the coastal belt of South Bengal.

The variability of climatic parameters may affect mangrove forests of the coastal belt in a number of ways as stated by (Blasco 1984) in one of his paper such as Physiological impacts of persistent dry winds, increase in water and soil salinity and also the chemical changes of ground water due to shortages of rainfall.

6.6.13 Shortages of Freshwater Discharge into the Coastal Zones by River System

The fresh water shortages are felt at the coastal zones due to over use of river waters in the drainage catchment areas of the East coast of India. Under arid conditions of Gujarat, where the influence of the freshwater is almost nonexistent, it increases the rapidity of evaporation of water between two high tides and which leads to the concentration of salts into

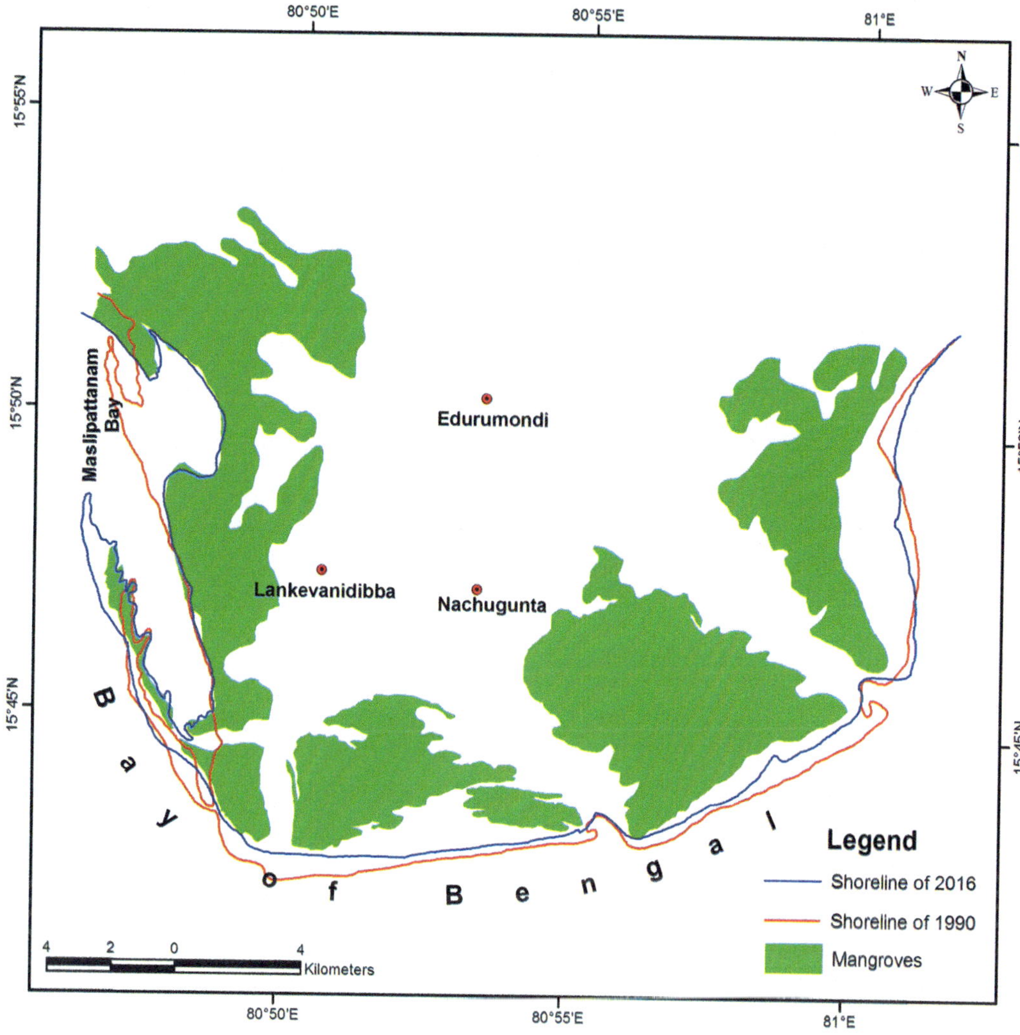

Fig. 6.21 Shoreline change detection of Krishna Delta fringed with mangrove forests

the inner parts of mangrove forests (Blasco 1984) as well as the degradation of mangroves.

The amount of freshwater reaching into the Sundarbans through Padma (Ganga) and Bhagirathi-Hugli distributary systems of the Ganga River are sharply declined after construction of Farakka Barrage in 1975. The freshwater supplies are also reduced into the Corringa Reserve Forest through Goutami-Godavari river system at present due to the construction of a series of cross dams along the drainage courses of Godavari river basin in Andhra Pradesh and Telengana provinces. The other river systems of the East Coast of India are also discharging very little amount of freshwater into their coastal areas except the rainy months. Such reduction of freshwater has led to develop the conditions of saline water and soil characters of river deltas which are not favourable for the growth and existence of less salt tolerant mangrove species.

In the near future, the upstream movements of saline seawater during high water stage will occupy the low lying delta plains as well as the estuary fringe areas dominated by mangrove stands, if the predicted sea-level rise continues to increase as per the IPCC estimation for the East Coast and West Coast of India (Current Working Report, 2016). The water salinity ranges from 20.0 to 21.1 gm/l during monsoon months and it rises up to 33.7 gm/l during summer months at Gahirmata section of Bhitarkanika Reserve Forest in the Bramhani- Baitarani river delta (Table 6.10).

Currently (March, 2016; April, 2017), the water salinity is measured with the Refactrometer in a few sites of back water settings of Haripur wetlands behind the Junpute spit (West

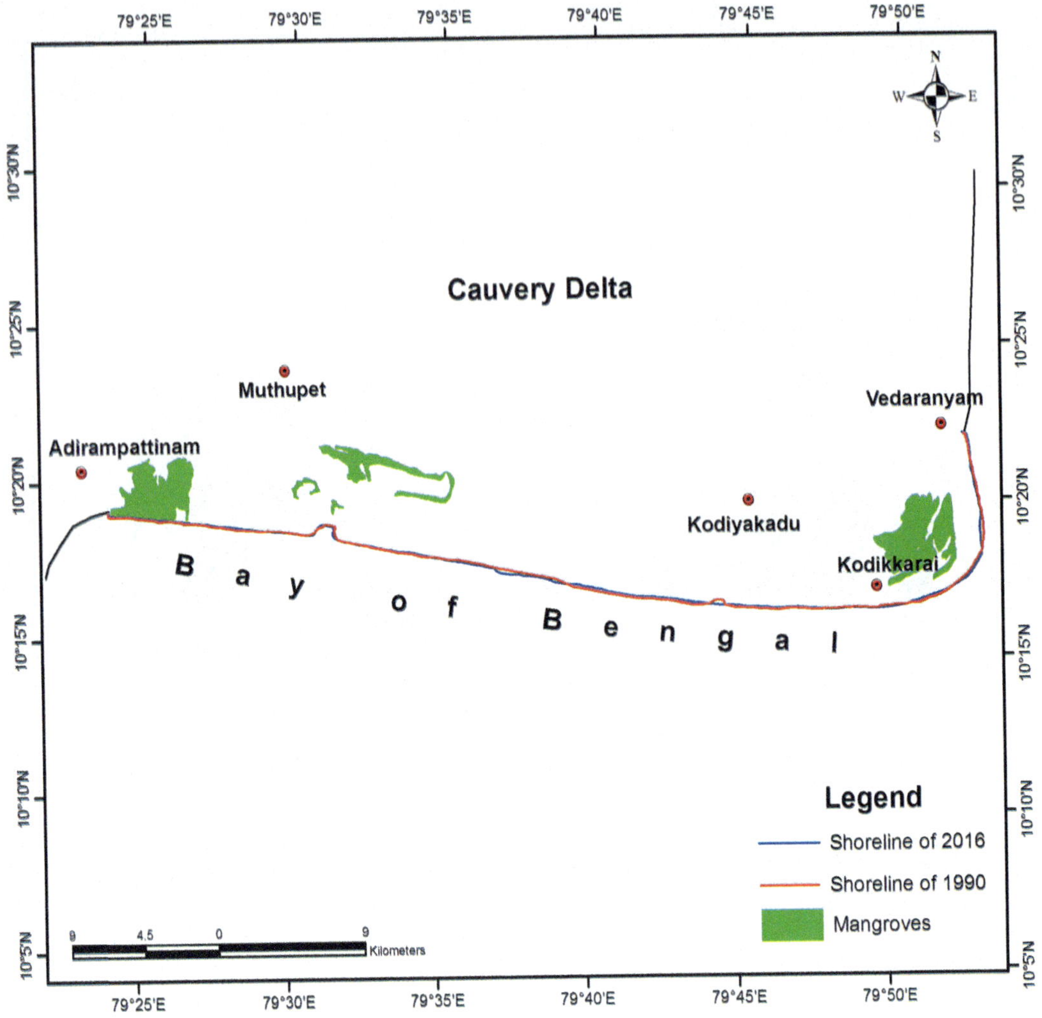

Fig. 6.22 Shoreline change detection of Muthupet fringed with mangrove forests

Bengal Coast), eastern Talsari wetlands behind the Talsari spit (Odisha State), and Maipura channel site behind the Gahirmata spit (Odisha State) in which mangroves are almost are degraded and eliminated. The result shows the presence of a significant amount of water salinity in the back water sites (ranges from 34 to 35 ppt) in compared with open marine near shore waters in front of the above mentioned sand spits (29–30 ppt). Therefore, it proves the absence of fresh water supply through overspill waters of the nearby river systems into the backwater sites particularly during pre-monsoon summer months (February to May) in the northern Bay of Bengal coasts.

6.6.14 Degradation of Mangroves Due to the Effects of Pests and Insects Infestation

Mangroves at their inland fringes are affected by pest infestation reaching nearby the rice paddy fields of the deltaic platforms. About 7.9% death of mangroves is recorded in Bhitarkanika by pest infestation, wind injury and riverbank erosion (Upadhyay et al. 2015). Areas largely uncovered by high tides in the winter months are affected by pests and insects infestation in Dungmole section of Bhitarkanika Reserve Forest.

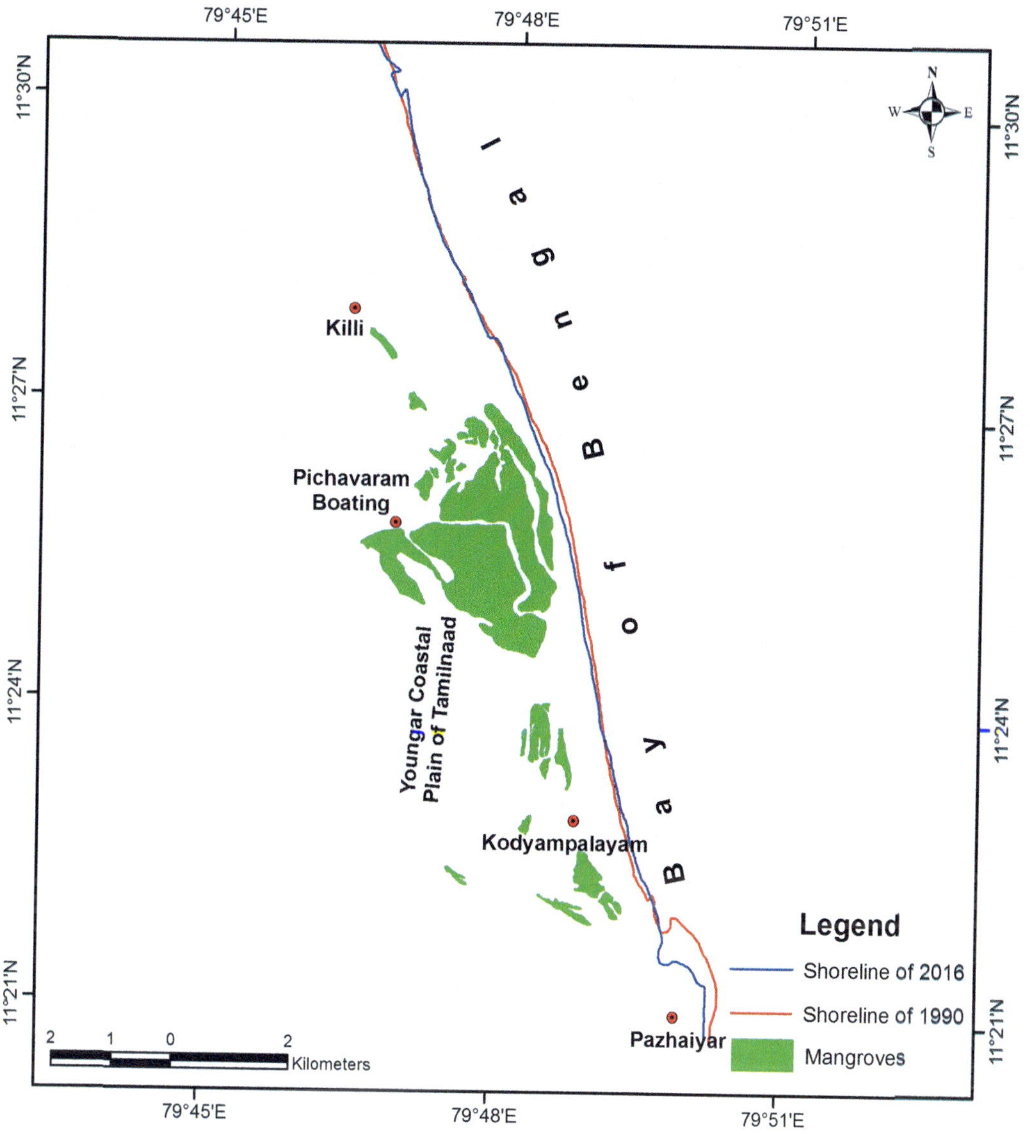

Fig. 6.23 Shoreline change detection of Pichavaram fringed with mangrove forests

It is reported from several studies (Santhakumaran 1985b; Roonwal 1954; Ganapati and Rao 1959) that activity of marine woodborers also can be the reasons in ultimate elimination of mangrove woody plants in the coastal belts of India. They have identified three major species of woodborers for such activities played by them in Indian mangroves. The intensity of woodborers on the living mangroves and dead mangrove stumps is reported from Goa coast to identify the severity of the problem of mangrove degradation (Santhakumaran 1985a, b). The severity of such problem is also reported from the Sundarbans, Godavari estuaries, Bhitarkanika and Andaman Group of islands (Table 6.11).

Though the above reasons are producing the major threats to the mangrove forests of the Indian coasts, but the degraded water quality and soil quality of the coastal zones also reduces the chlorophyll content within the mangrove vegetation in different coastal habitats as a reflectance of degradation of environmental quality. The health status of mangroves is estimated from the Remote Sensing techniques to identify such areas of degraded environmental quality in the coastal

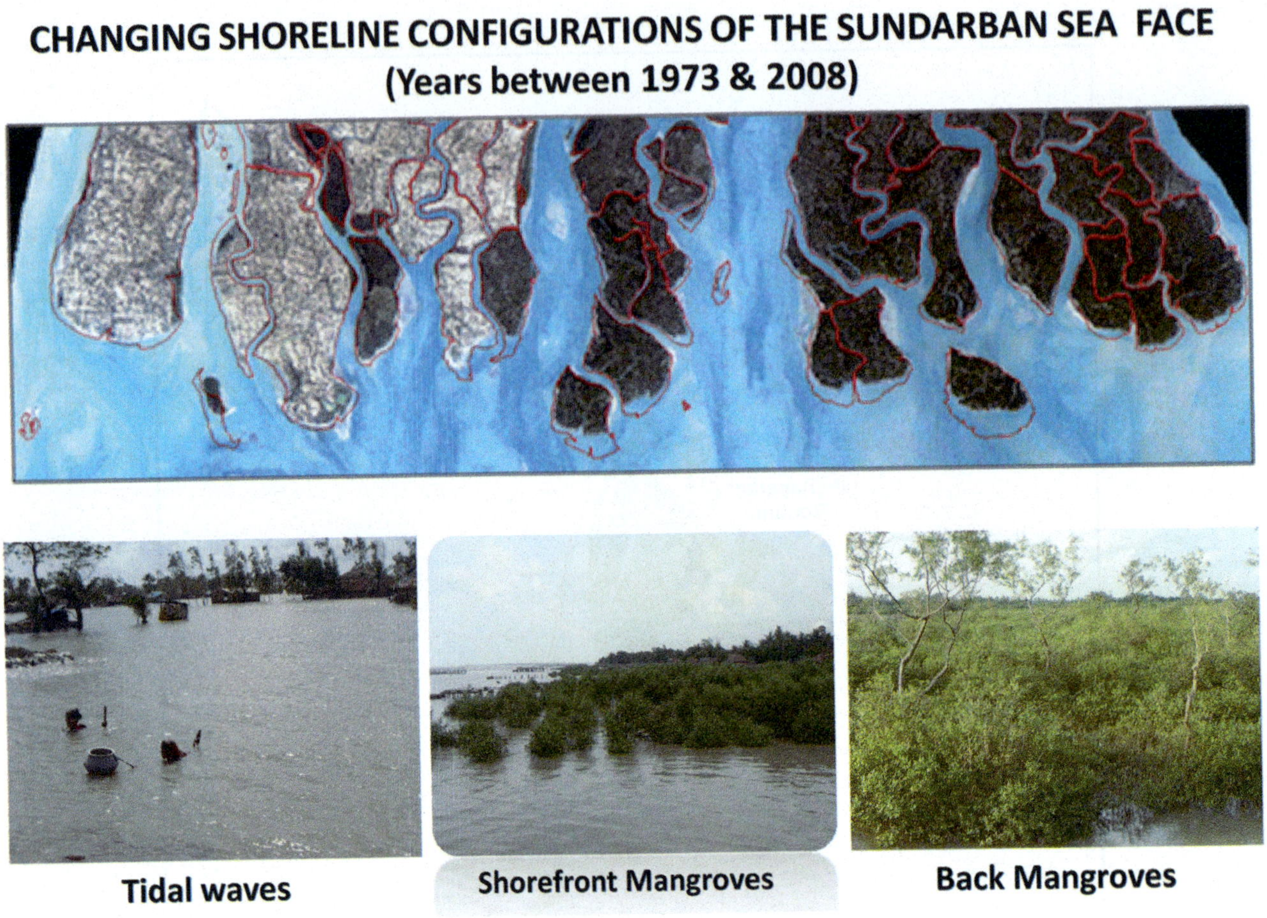

Fig. 6.24 Shoreline change detection of the Sundarbans fringed with mangrove forests

zones of Sundarbans, Subarnarekha Delta and Gujrat coast (Figs. 6.25, 6.26, 6.27, 6.28 and 6.29).

6.7 Mangrove Vulnerabilities in India

The coastal belts of India have been considered by Indian Meteorological Department (IMD) to identify the cyclone vulnerabilities on the basis of the assessment of frequency of high magnitude cyclone attacks and related storm surge heights on the Bay of Bengal and Arabian Sea (National Cyclone Risk Mitigation Project [NCRMP] 2016). The ocean information systems of the northern Indian Ocean (Bay of Bengal and Arabian Sea) by Indian National Centre for Ocean Information Services (INCOIS) provide wave and wind energy related processes of the coastal near shore environment of the country as well as the temporal Remote Sensing data of regional mangrove belts of the East Coast and West Coast is available from USGS source for such study. The data on the rate of sea-level rise scenario for the country is available from Permanent Service for Mean Sea Level (PSMSL) sources for the final assessment of vulnerabilities of Indian mangroves under natural threats. The field signatures of mangrove elimination are also providing important evidences of mangrove vulnerabilities in the East Coast and West Coast of India (Figs. 6.30 and 6.31).

6.8 Conclusions

Mangroves are distributed in the open coastal, relatively sheltered coastal, and protected coastal tidal flats of river deltas of the East Coast of India. Mangroves of carbonate settings with shore fringed platforms, channel strait margins and 'ria' type embayed shores are found in Andaman-

Table 6.10 Annual average discharges of the river systems of the East Coast and the West Coast of India (water reaching into the coastal zones after multiple uses)

Sl. No.	River system	Annual Average Discharges 1975 (K. L. Rao)	2005 (CWC)
1	Ganges-Brahmaputra River	493,400 million m^3	NA
2	Subarnarekha River	7940 million m^3	8411 million m^3
3	Baitarani River	5755 million m^3	14240 million m^3 (jointly)
4	Brahmani River	18,310 million m^3	
5	Mahanadi River	66,640 million m^3	42803 million m^3
6	Godavari River	105,000 million m^3	72957 million m^3
7	Krishna River	67,675 million m^3	27342 million m^3
8	Pennar River	3,238 million m^3	1516 million m^3
9	Cauvery River	20,950 million m^3	12175 million m^3
10	Vellar River	850 million m^3	NA
11	Kunleru River	420 million m^3	NA
12	Vaitarna River	4350 million m^3	NA
13	Ulhas River	3014 million m^3	NA
14	Savitri River	1467 million m^3	NA
15	Mandavi River	1320 million m^3	NA
16	Kalinadi River	6537 million m^3	NA
17	Gangavati or Bedti River	4925 million m^3	NA
18	Sharavati River	4545 million m^3	NA
19	Netravati River	4615 million m^3	NA
20	Ghaliar or Beypore River	5200 million m^3	NA
21	Periyar River	12,300 million m^3	NA
22	Pamba River	6300 million m^3	NA

Source: Estimated after the consideration of hand book on water and related information, 2005 (CWC, New Delhi), and India's Water Resources by K. L. Rao (1975)

Table 6.11 Observation of the types of natural threats towards the coastal fringe forests dominated by mangroves in Indian coastal belt

Sl. No.	Reasons of natural threats	Name of the coastal fringe forests affected	Severity of the problem: Very High (VH), High (H), Medium (M), and Low (L)
1	Marine forcing factors (sea level rise related effects)	Sundarbans and Bhitarkanika	H
2	Neo tectonics related subsidence and up-liftment	Andaman and Nicobar Group of Islands	H
3	Hydrological stress from the drainage catchment areas	Subarnarekha Delta and Mahanadi Delta	L
4	Coastal morphodynamics under sediment starvation	Sundarbans (Abandoned Delta); Subarnarekha Delta and Godavari Delta	M
5	Hyper salinity in the inner parts of the mangrove swamps	Sundarbans, Gujarat Coast and Godavari Delta	VH
6	Eutrophication as driver of coastal wetland loss	Sundarbans	L
7	Shifting river banks in the deltaic areas	Sundarbans, Bhitarkanika ans Subarnarekha River Estuary	M
8	Storm catastrophe	Sundarbans, Bhitarkanika, Subarnarekha Delta, Muthupet and Godavari Delta	VH
9	Over washed sand fan lobe encroachment into the swamp	Sundarbans, Subarnarekha Deltaic Shores, Shore Platforms of Andaman and Nicobar Group of Islands	H
10	Erosion at the sea face and shoreline retreat	Sundarbans, Krishna Delta, Godavari Delta and Hukitola Bay	VH
11	Active barrier roll over process	Talsari Spit, Gahirmata Spit, Hukitola Spit, Chilika Spit and Godavari Spit	H
12	Shortages of annual rainfall in the tropical hot and humid coast of India	Sundarban Fringes and Gujarat Coast	M
13	Shortages of fresh water discharge into the coastal zones by river systems	Godavari R., Krishna R., Cauvery R., Brahmani-Baitarani R., Mahi R., Luni R. and Sabarmati R.	H
14	Impact of insects	Bhitarkanika, Godavari Delta, Sundarbans and Goa Coast	H

Severity of the problem (Rated as): VH=3, H=6, M=3, L=2

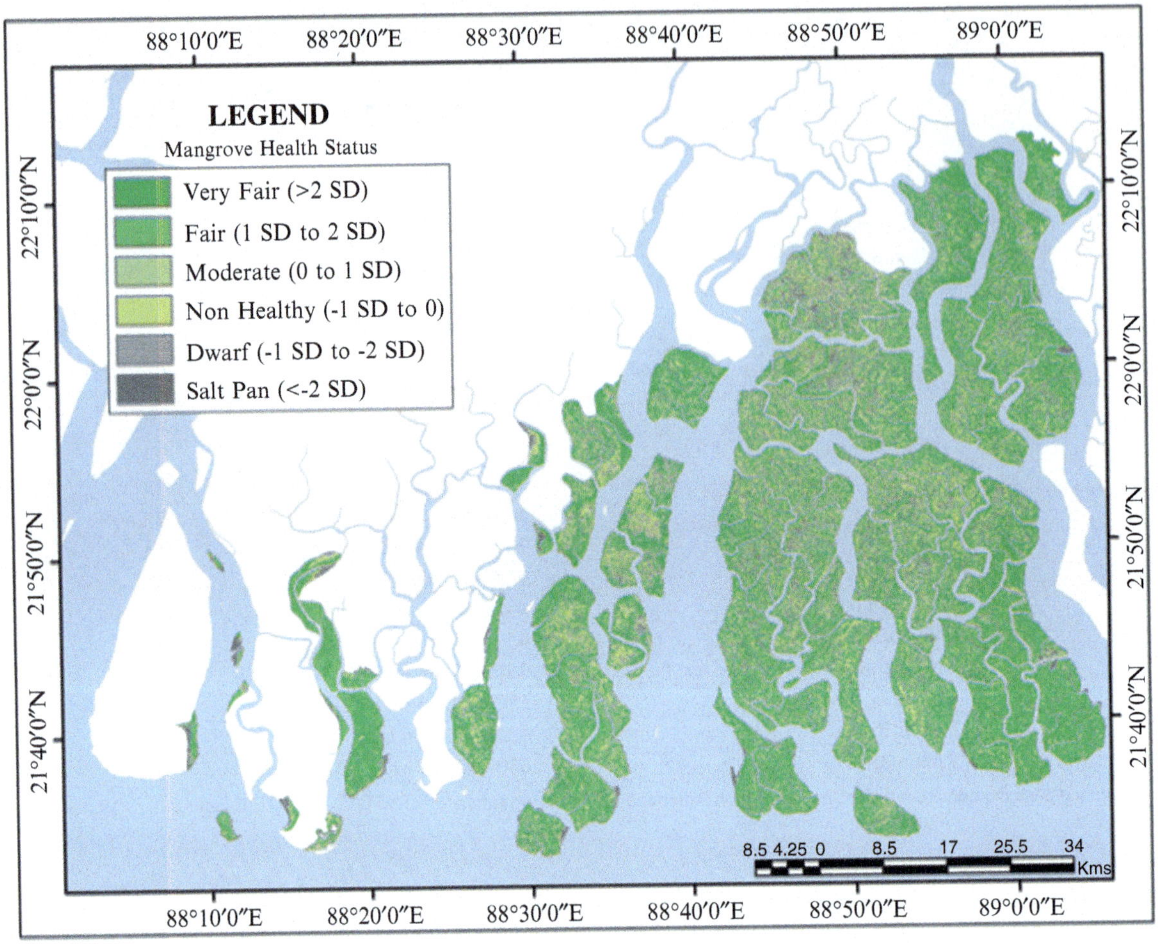

Fig. 6.25 Status of mangrove health in the parts of Sundarbans (2017) in West Bengal, India

Nicobar group of islands and in tiny islands of the Gulf of Mannar (Tamil Nadu Coast) over the Bay of Bengal. However, the mangroves of the west coast of the India are concentrated in the barrier back settings, in areas of estuary fringes, and in the sheltered tidal flats of Gulf areas in smaller pockets.

Several marine forcing factors related to sea level rise, cyclone vulnerabilities, changing nature of climate variability, hypersalinity and reduction of freshwater discharges by river systems are impacting over the mangrove forests and other fringe forest on the eastern coast of India fringed with the Bay of Bengal in the previous decades. The above factors also estimated as drivers of many other threats of secondary origins towards the conditions liable to mangrove degradations in the coastal belts. Mangroves and littoral forests are mostly affected by dynamic emergence and submergence of shore platforms induced by tectonics particularly in Andaman and Nicobar group of islands in 2004.

There are 14 major threats to the mangroves and other littoral forests in Indian coasts so far have been identified by the above study. The severity of such problems is rated as Very High, High Medium and Low on the basis of spatial impacts. Among them about 9 problems out of 14 are under High and Very High categories. The assessment represents that the mangroves of East Coast and West Coast of India are vulnerable to multiple natural factors and their variability on the basis of diverse physiographic settings. If the natural elimination of mangroves from the water fronts and inner parts of the swampy tracts continue at current rates parallel

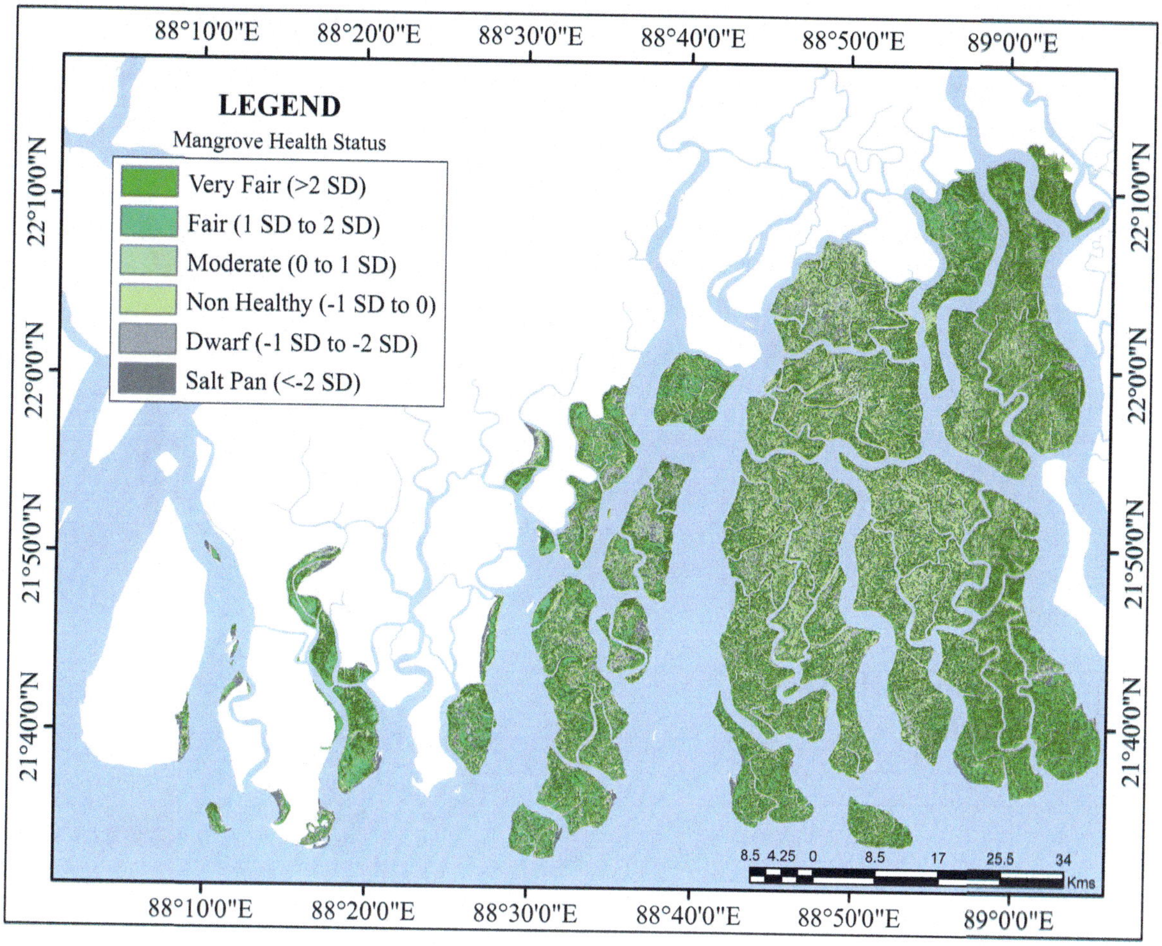

Fig. 6.26 Status of mangrove health in the parts of Sundarbans (2013) in West Bengal, India

with human modifications through reclamations from the landward sides of similar forests dominated by mangroves and non-mangrove halophytes, the coastal belts of India will be directly exposed to all type of marine hazards.

Various innovative measures are immediately needed to restore the mangrove forests at present in the country. Promotion of the mangrove forestry through direct plantation and also through agro forestry in low lying coasts may keep the balance between elimination and newer emerging areas of mangrove forests in the dynamic coastal environment of Tropical India. Further research on the methods of restoration of mangrove forests is needed for the East Coast and West Coast of India.

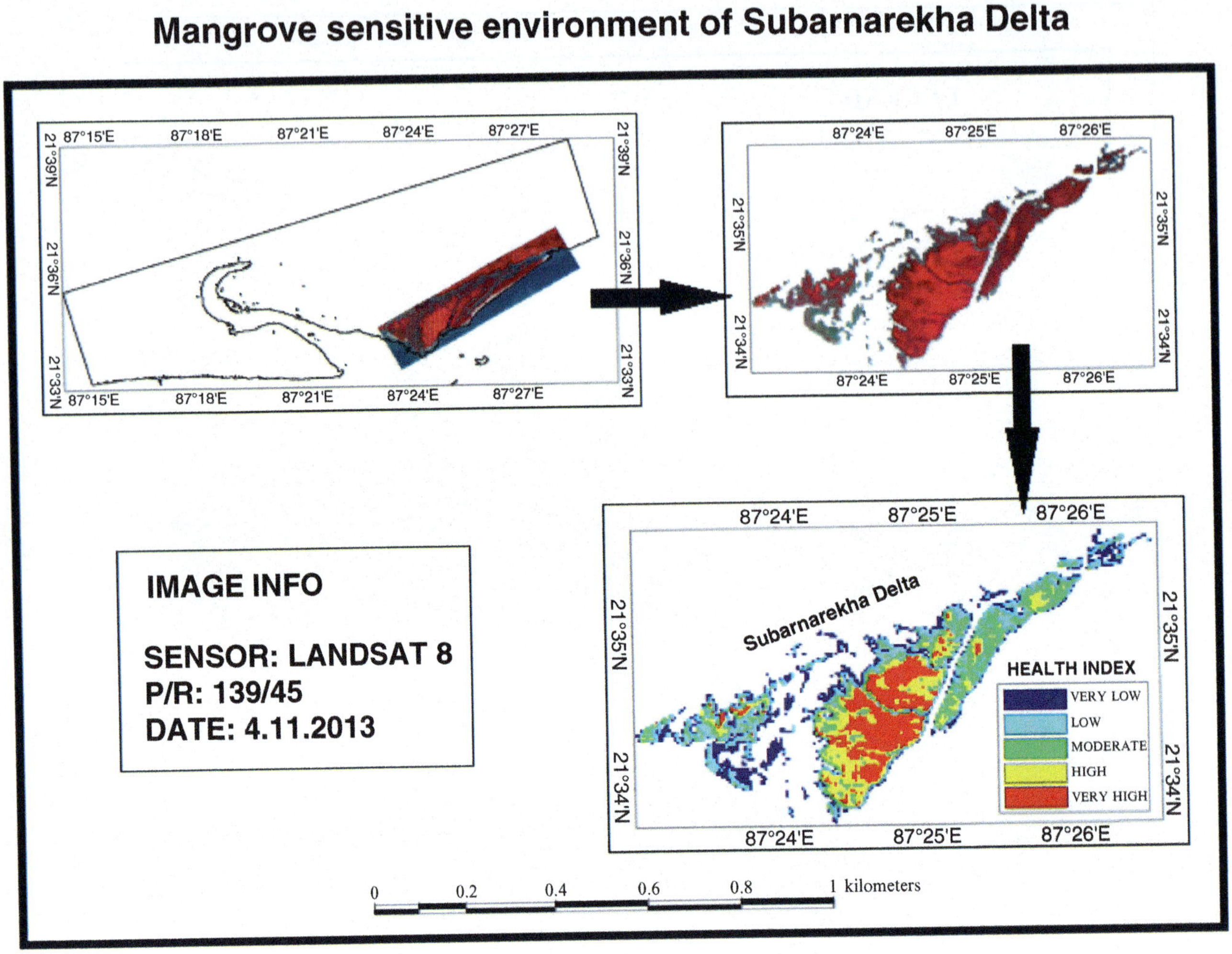

Fig. 6.27 Mangrove areas and their status of health in Subarnarekha Delta at barrier back environment, Odisha Coast, India

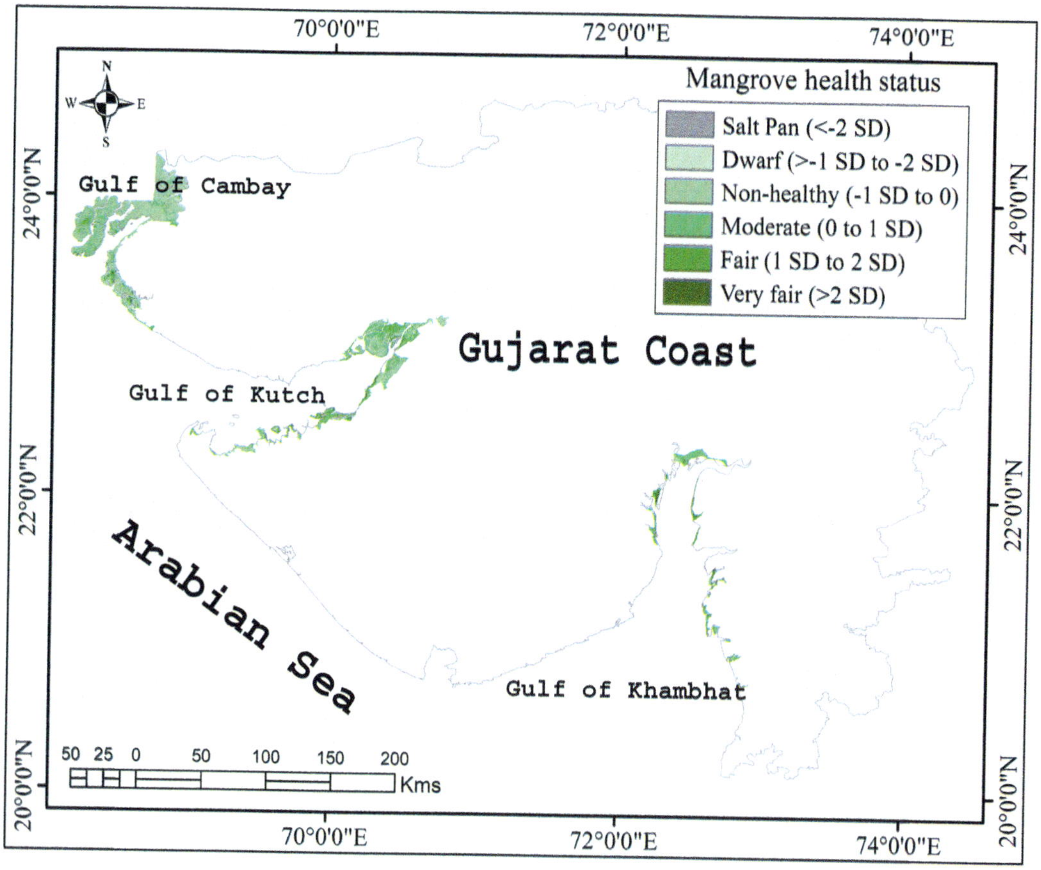

Fig. 6.28 Status of mangrove health in the parts of Gujarat coast (2017) in India

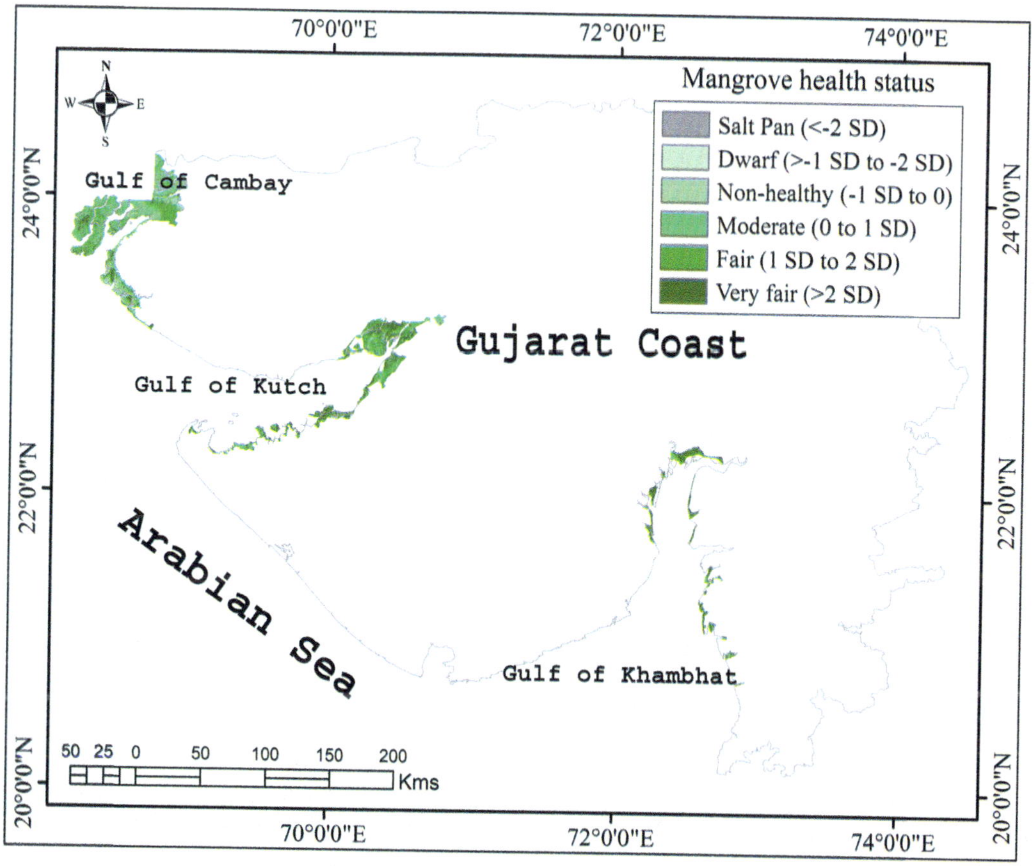

Fig. 6.29 Status of mangrove health in the parts of Gujarat coast (2013) in India

Fig. 6.30 Degradation of mangrove habitat of the carbonate platforms in Andaman and Nicobar group of Islands. Banerjee et al. (1989) Cemented beach rocks of Neil Island as unfavourable platform for mangrove growth, (Bird 2005) Degradation of frontal mangroves in the shore platforms (*Avicennia marina*), (Blasco 1984) Storm over wash sand deposit and natural death of shore fringed forest, (Chander et al. 2009) Storm damaged mangroves over the wave truncated shore platform (*Rhizophora* sp.), (Coastal Zones of India 2012) Prolonged salt water inundations in the shore platforms during the storm breaks in the Andaman Island and resultant uprooted mangrove trees in the shore fringed areas, (Saudamini et al. 2009) Natural damage of mangroves due to subsidence during earthquake of 2004 in and around Port Blair, (Das and Pramanik 2011) Subsidence related submerged forest in Katchal Island of Little Nicobar after the earthquake of 2004 and (Ellison and Stoddart 1991) Photographic document of submerged forest in South Andaman

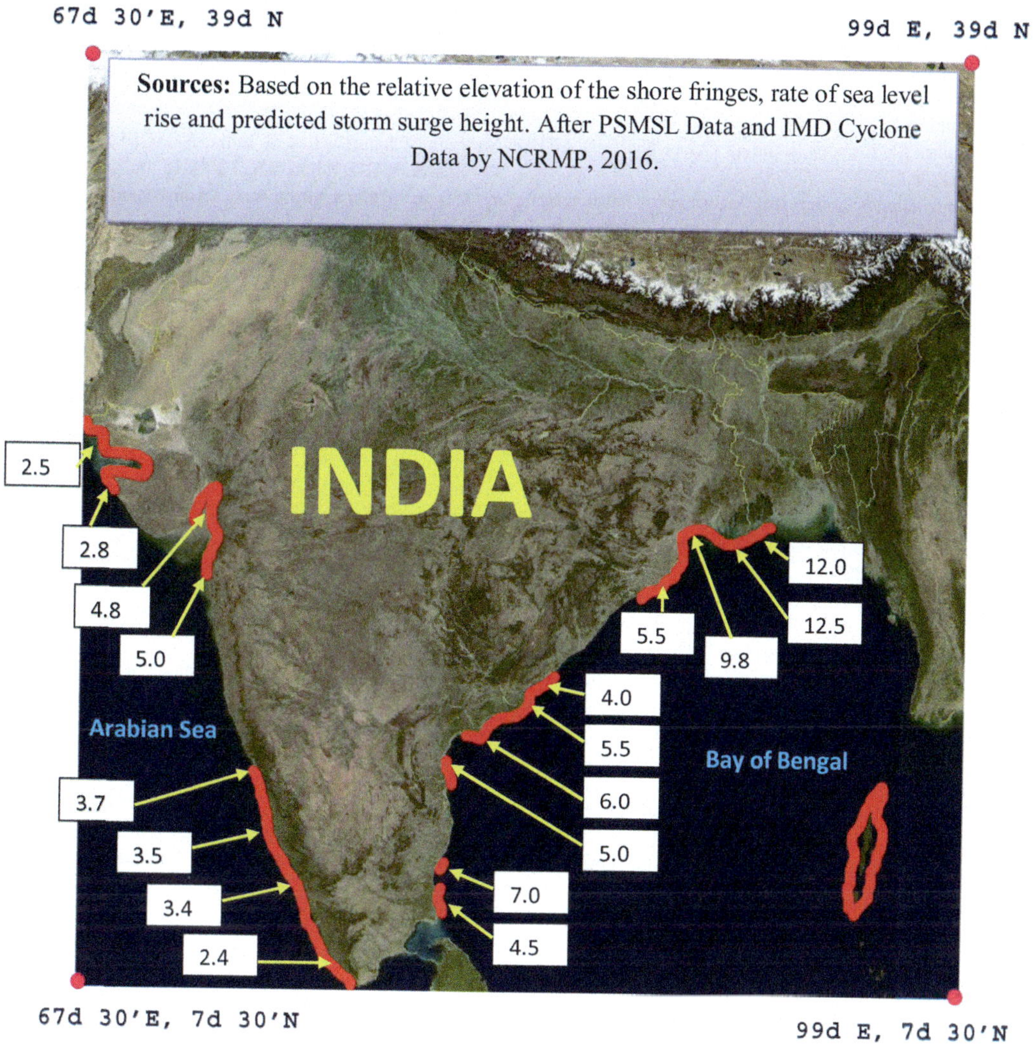

Fig. 6.31 Mangroves vulnerable to natural threats of cyclones, storm surge height (in m), sea level rise and relative location of shore fringes of India

References

Banerjee LK, Sastry ARK, Nayar MP (1989) Mangroves in India: identification manual. Botanical Survey of India, pp 113

Bird ECF (2005) Mangroves, geomorphology. In: Schwartz ML (ed) Encyclopedia of coastal science. Springer, Heidelberg, pp 611–613

Blasco F (1984) Climate factors and the biology of mangrove plants. In: Snedaker SC, Snedaker JG (eds) The mangrove ecosystem: research methods. UNESCO/SCOR, Paris, pp 18–35.

Chander G, Markham BL, Helder DL (2009) Summary of current radiometriccalibration coefficients for Landsat MSS, TM, ETM+, and EO-1 ALI sensors. Remote Sens Environ 113:893–903

Coastal Zones of India (2012) Project sponsored by MOEF, Government of India. Space Application Centre, ISRO, Ahmedabad, Gujarat

Das S, Vincent JR (2009) Mangroves protected villages and reduced death toll during Indian super cyclone. Proc Natl Acad Sci U S A 106 (18):7357–7360. Published online 2009 Apr 20. doi: 10.1073/pnas.0810440106, PMCID: PMC2678660

Das DP, Pramanik K (2011) Evaluation of impact of earthquake and tsunami on the coastal morphology of Andaman-Nicobar Island using multisensory temporal satellite data. Indian J Geosci 65 (Bo. 1):9–22

Ellison JC, Stoddart DR (1991) Mangrove ecosystem collapse during predicted sea level rise: Holocene analogues and implications. J Coast Res 7:151–156

FSI (1999). Final SFR, 2003. Pp. 39–43. http://fsi.nic.in/documents/sfr_2003_hindi.pdf)

Ganapati PN, Rao ML (1959) Incidence of marine borers in the mangroves of the Godavari estuary. Curr Sci 28(8):332–333

Integrated hydrological data book (2006) Hand book on water and related information, 2005. CWC, New Delhi, pp 85–87

Islam MT (2014) Vegetation changes of Sundarbans based on Landsat Imagery analysis between 1975 and 2006. Acta Geograph Debrecina. Landscape & Environment Series 8(1):1

Mizanur Rahaman Bijoy (2007) Cyclone effects on Sundarbans Presented by Mizanur Rahman Bijoy Network on Climate Change, Bangladesh (NCC,B)

Naskar K, Guha Bakshi DN (1987) Mangrove swamps of the Sundarbans. Naya Prakash, Calcutta, p 282

National Cyclone Risk Mitigation Project (NCRMP) 2016 Implemented through World Bank Assistance (Map showing coastal areas affected by cyclones). Based on IMD data on cyclone hazards

Paul AK (1999) Geomorphic significance of mangrove Swamps- a study in the Sundarban. In: Naskar KR, Sanyal P (eds) Mangal of the Sundarban. Naya Prokash, Calcutta, pp 39–52

Paul AK (2002a) Coastal geomorphology and environment. ACB Publications, Calcutta, p 582

Paul AK (2002b) Tsunami the disaster in Indian Ocean Nations. ACB Publications, Kolkata, p 125

Paul AK (2005) Oceanic Islands under acute shortage of fresh water resource, a case study in Andaman and Nicobar Archipelago. DELF/FAEA II:41–49. Article 7

Paul Ashis Kr, Kamila A (2016) Coastal Mud Banks of Subarnarekha Delta with Special Reference to Degradation and Accretion under Physical Processes. Procedings volume of 36th INCA International Conference, Santiniketan Visva- Bharati University, West Bengal, India. (in press)

Ray R, Mondal S (2014) Multi criteria based forest cover type mapping using Landsat ETM+ digital data – a case study on Eastern part of Jungle Mahal Rgion, West Bengal. Indian J Reg Sci XLVI:77–84

Ray R, Paul AK, Basu B (2013) Application of supervised enhancement technique in monitoring the mangrove forest cover dynamics – a case study on Ajmalmari Reserve Forest, Sundarban, West Bengal. Int J Remote Sens Geosci (IJRSG) 3(1):16–21

Richards JF (1993) Agricultural impacts in tropical wetlands: rice paddies for mangroves in South and Southeast Asia. In: Williams M (ed) Wetlands: a threatened landscape. Blackwell, Oxford, pp 217–233

Richter R (1990) A fast atmospheric correction algorithm applied to Landsat TM images. Int J Remote Sens 11:159–166

Roonwal ML (1954) Proc Vol Zool Soc (Calcutta) 7:91–105

Santhakumaran LN (1985a) Destruction of mangrove vegetation by marine Wood Borers aling Goa coast, with notes on their distribution in mangroves along the Indian coasts. In: Vhosale LJ (ed) The mangroves. Proceedings Vol. National Symposium on Biology, Utilization and conservation of mangroves. Shivaji University, Kolhapur, 1986, pp 492–498

Santhakumaran LN (1985b) Marine Wood Borers of India, an annotated bibliography, 147p. Wood Preservation Centre (Marine), National Institute of Oceanography, Dona Paula, Goa, India

Selvam V (2003) Environmental classification of mangrove wetlands of India. Curr Sci 84(6):757–765

Shoreline Change Atlas of India, NCSCM (2014) Space Applications Centre (ISRO) Ahmedabad 380015 and Coastal Erosion Directorate, Central Water Commission, Ministry of Water Resources. Government of India, New Delhi

Singh U, Wadhwani AM, Johri BM (1986) Dictionary of economic plants in India. Indian Council of Agricultural Research (ICAR), New Delhi, p 288

Thom BG (1979) Mangrove ecology: a geomorphological perspective. Mangrove Ecosystems in Australia: structure, function and management. Ed. by BFClough cape Ferguson 18–20 Apr, 1979.

Thom BG (1984) Coastal landforms and geomorphic processes. In: Snedaker SC, Snedaker JG (eds) The mangrove ecosystem: research methods. UNESCO/SCOR, Paris, pp 3–17

Thom BG, Wright LD, Coleman JM (1975) Mangrove ecology and deltaic estuarine geomorphology: Cambridge gulf ord river, Western Australia. J Ecol 63:203–232

Upadhyay VP, Mishra PK, Sahu JR (2015) Distribution of mangrove species within Bhitarkanika National Park in Orissa, India

Venkatesan KR (1966) The mangroves of Madras state. Indian Forester 92(1):27–34

Walter H (1977) Climate. In: Chapman VJ (ed) Ecosystems of the world. Wet coastal ecosystem. Elsevier, New York, pp 61–71

Woodroffe CD, Mulrennan ME, Chappel J (1993) Estuarine infill and coastal progradation, Southern van Diemen Gulf, Northern Australia. Sediment Geol 83:257–275

Woodroffe CD, Lovelock CE, Rogers K (2014) Chapter 11: Geomorphology and ecology of mangrove shorelines. In: Masselink G, Gehrels R (eds) Coastal environments and Global Change. Wiley-Blackwell, Chichester, pp 251–267

An Assessment of Vulnerability and Adaptation of Coastal Mangroves of West Africa in the Face of Climate Change

7

Isaac Boateng

Abstract

This chapter seeks to evaluate the current status of West Africa's mangroves. It assesses Climate Change vulnerability and adaptation options for mangroves in West Africa. West African mangroves contribute a wide range of environmental services, economic goods and social services. In spite of the important contributions of mangroves in the region, they experiencing high rate of degradation. It is estimated that the degradation and the deforestation of mangroves in the region have resulted from their uncontrolled anthropogenic exploitation due to urbanisation, population growth, salt production, industrial pollution and the cutting of mangroves for firewood. Besides the afore-mentioned anthropogenic impacts on the mangroves, the anticipated effects of climate change such as increased temperatures, sea level rise, increased intensity of storm and precipitation are likely to have the most severe impacts on mangrove ecosystems. Climate change and the anthropogenic driven variations of these environmental forces will inevitably have a profound effect on coastal zones and mangroves. The challenge of reversing the degradation of mangrove ecosystems in the face of uncontrolled exploitation and impacts of climate change seems to be a very complex problem. This assessment has identified that both the past and the present vulnerability were more controlled by anthropogenic activities than the effects of climate change, though it is expected that climate change may be the major driving force in the long-term. However, many adaptation options exist to enhance specific ecosystem services in ways that reduce negative trade-offs, but these involve changes in policies, institutional framework and better practices for exploitation, and good management strategies. The chapter concludes that West Africa should implement adaptation policy options including reducing anthropogenic impacts, maintaining coastal buffer zones, restoration of mangroves, catchment management, establishing regional monitoring and regulations and education and local participation to enhance sustainability.

Keywords

Climate change · Mangrove ecosystem services · Vulnerability · West Africa · Mangroves adaptation

7.1 Introduction

Mangroves have been described by different names including coastal woodland, tidal forest and mangrove forest. However, mangroves are complex and unique forests that thrive at the interface between the land and the sea, mostly around where rivers and lagoons enter the sea in the tropical and subtropical regions of the world. They have adaptation mechanisms that enable them to endure high concentrations of salt and consistent inundation of their root systems by a mix of tidal and fresh water. Mangroves require freshwater inflow, which brings silt with it as substrate for support and nutrients from upstream (Kathiresan and Bingham 2001; AFROL 2002).

According to the IUCN WCPA (2005), mangroves and associated lagoons provide ecosystem services, which include provisioning services (food and water); regulating services (flood and pollution control); cultural services (spiritual, recreational, and cultural benefits); and supporting services, (nutrient cycling) that maintain the conditions for life on Earth. These ecosystem services have economic, protection, environmental and social values that support local and national economies of many countries that have mangroves.

In West Africa, mangroves play a key role in coastal fisheries, which is estimated to contribute over $400 million

I. Boateng (✉)
School of Civil Engineering and Surveying, University of Portsmouth, Portsmouth, UK
e-mail: isaac.boateng@port.ac.uk

C. Makowski, C. W. Finkl (eds.), *Threats to Mangrove Forests*, Coastal Research Library 25,
https://doi.org/10.1007/978-3-319-73016-5_7

annually to the regional economy (USAID 2014). Besides the fisheries contribution, West African mangroves contribute to a wide range of environmental, economic goods and social services. In spite of these important contributions of mangroves in the region, they are experiencing deforestation at a rate of 1.7% per year (USAID 2014; Feka and Ajonina 2011). It is estimated that the degradation and the deforestation of mangroves in the region have resulted from their uncontrolled anthropogenic exploitation due to urbanisation, population growth, salt production, industrial pollution and the cutting of mangrove for firewood. These have led to the loss of 20–30% of mangrove over the past 25 years (Ajonina and Usongo 2001; Ajonina et al. 2005).

Previous assessments have identified that humans have changed ecosystems more rapidly and extensively over the past 50 years than in any comparable period of time in human history, largely to meet the increasing demands for food, fresh water, timber, fibre, and fuel (IUCN WCPA 2005; Ellison 2015). The overexploitation has resulted in a substantial and largely irreversible loss in the diversity of life on Earth (IUCN WCPA 2005). The challenge of reversing the degradation of mangrove ecosystems in the face of uncontrolled exploitation and increasing demands for their services seems to be a complex and an unsurmountable problem. However, many options exist to enhance specific ecosystem services in ways that reduce negative trade-offs, but these involve changes in policies, institutional frameworks and better practices for exploitation, and good management strategies.

Further, apart from the afore-mentioned anthropogenic impacts on mangrove, the anticipated effects of climate change such as increased temperatures, sea-level rise (SLR), increased intensity of storms and precipitation are likely to have the most severe impacts on mangrove ecosystems (Intergovernmental Panel on Climate Change [IPCC] 2014; Ranasinghe 2016). The world's coastlines are shaped by mean sea level, wave conditions, storm surge, and river flows. Climate change and the anthropogenic driven variations of these environmental forces will inevitably have a profound effect on coastal zone and mangroves (Ellison 2015; Ranasinghe 2016; Nicholls et al. 2014). These impacts are exacerbating the ecological damage of mangroves and threaten not only the goods and services obtained from the coastal mangroves, but the coastal communities that subsist on mangrove resources. Feka and Ajonina (2011) have identified that as these mangroves dwindle, the livelihoods and well-being of millions of vulnerable coastal communities and local governments that directly or indirectly depend on their resources are at risk.

Nonetheless, local knowledge on the vulnerability and adaptation of mangroves to Climate Change and its implications for environmental sustainability of ecosystem services provided by mangroves are limited in West Africa (USAID 2014). Is the present exploitation of mangrove in West Africa sustainable? What policies and institutions are available to ensure sustainable mangroves exploitation and management? How could we reverse the past degradation of mangrove ecosystems, while meeting the current increasing demands for their services? What are the present and future impacts of climate change on mangroves in West Africa? How vulnerable are West African mangroves to climate change? And finally what are the adaption options available for implementation to facilitate sustainability? There is the need to conduct detailed assessment of the vulnerability and impacts of climate change on West African's mangroves in order to determine how to facilitate the development of adaptation policies to reduce the effects. This chapter seeks to evaluate the current status of West Africa's mangroves. It assesses vulnerability and adaptation options for mangroves in West Africa to climate change.

7.2 West African Mangroves

West Africa is mainly a political and economic region, which is located west of the Africa continent. It includes all the countries in the region except Cameroon, Chad, Equatorial Guinea, and the Northern Saharan part – Mauritania and Niger. West Africa covers an area of 5,112,903 km^2 and has an estimated population of over 340 million. The region is governed by a regional body called the Economic Community of West African States (ECOWAS). ECOWAS was established in 1975 and has 15 member countries of which 11 are maritime countries with variable areas of mangrove forests along their coast and one is an archipelago (Cape Verde islands). West Africa mangroves represent 13% of mangrove forests in the world and cover more than 2.4 million hectares (USAID 2014). Nigeria and Guinea Bissau – two of the world's most mangrove-rich countries are located in West Africa (Fig. 7.1).

There is a significant amount of recent literature on mangroves in the sub-region (Table 7.1). At least 34 titles of papers have been published on Mangroves in West African as a whole or on countries within the regions, since 2000 (Table 7.1). Of the 34 publications only two of them focus on the impact of climate change on Mangroves – one covers the whole West Africa and the other focus on Ghana mangroves only (Table 7.1). This highlights the importance of this assessment.

Accrording to Feka and Ajonina (2011), the indigenous mangrove species in West Africa belongs to three families, including *Rhizophora racemosa*, *Rhizophora mangle*, and *Rhizophora harrisonii*, the white *Avicennia germinans* and *Laguncularia racemosa*. *Nypa fruticans* is an exotic species introduced in Nigeria from Asia, while *Astrotichum aureum* and *Cornocarpus erectus* are associates. Table 7.2 shows the

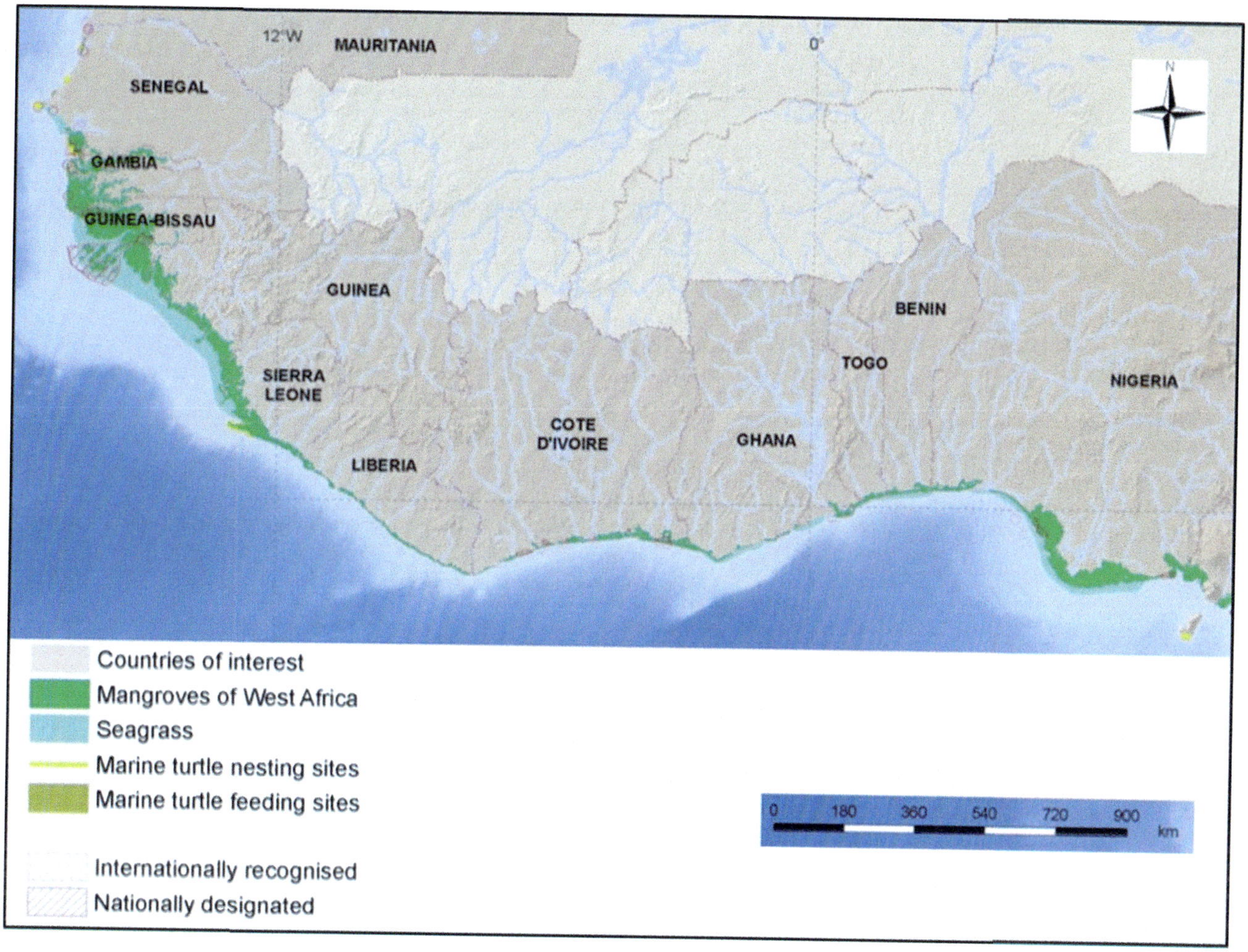

Fig. 7.1 Mangroves cover in West Africa (After UNEP-WCMC 2007)

distribution of mangrove species across the maritime countries in the region. There is little variation of mangrove species across the sub-region. In all, there are seven main species of mangroves (Table 7.2). These mangroves grow in the intertidal zone of West Africa and are subject to the influence of various factors: oceanographic, sedimentary, geomorphologic and anthropic (UNEP-WCMC 2007; Goussard and Ducrocq 2014).

Previous studies have indicated that West African mangroves have diminished from 20,500 km^2 in 1980 to their current 9000 km^2, dues to the increasingly effects of deforestation and overexploitation (Bojang and Ndeso-Atanga 2009; Feka and Ajonina 2011). The degradation of mangroves species and associated ecosystem services has direct economic consequences for human livelihoods, especially in areas with low ecosystem resilience to species loss (Polidoro et al. 2010).

Historically, mangrove resources were exploited by local coastal communities on a subsistence scale. However, recently both mangrove resources and the land covered by mangroves are exploited for economic purposes. The large-scale commercial exploitation of mangrove resources and mangrove lands threatens the livelihood of the local subsistence coastal communities, but contribute significantly to the Gross Domestic Products (GDP) of the local economies.

7.3 Direct and Indirect Contribution of Mangroves to West African Countries

West Africa has a rich mix of natural resources both on land and at sea, that provide many ecosystem services. These resources support local economy, provide resilience in the face of climate change, and support the livelihoods of many coastal communities. In fact, the coastal area between Nigeria and Senegal, offers the richest variety of marine biodiversity. The fishing industry earns the countries along this section of Atlantic coast an estimated $4.9 billion per year, increasing GDP at the national level and sustaining incomes, and providing food security for fishing communities at the local level (Worldbank 2013). These fisheries are very well supported by the coastal mangroves

Table 7.1 Recent literature related to understanding and management of mangrove forest in West Africa

Authors	Year of publication	Location or study area	Main topic area	Total number of publication
USAID	2014	West Africa	A policy brief on mangroves and climate change	9
Tang et al.	2014		Biomass and carbon of mangroves	
Tang et al.	2016		Mangrove biomass and carbon	
Corcoran et al.	2007		Distribution and biodiversity of mangrove	
UNEP	2007		Biodiversity and importance of mangrove	
Feka	2015		Sustainable management of mangrove biodiversity in rice fields and mangroves	
Bos et al.	2006		West African mangroves as habitat for wintering European warblers	
Zwarts et al.	2014		Mangrove ecosystem	
Carney et al.	2014			
Luiselli and Akani,	2002	Nigeria	Snake communities in the mangroves	4
James et al.	2013		Social valuation of mangroves	
Ainodion et al.	2002		Mangrove restoration in the Niger Delta	
Adekanmbi and Ogundipe	2009		Mangrove biodiversity, restoration and sustainability	
Laleye	2000	Benin	Acadja fisheries enhancement systems	1
N/A	N/A	Togo	N/A	0
Mensah	2013	Ghana	Mapping mangroves depletion	7
Ntyam et al.	2014		Importance of mangrove litter production	
Boatemaa et al.	2013		Impacts of sea level rise on mangroves	
Agyeman et al.	2007		Inventory and classification of mangrove	
Adotey	2015		Carbon stock assessment of Mangrove	
Ntyam	2014		Mangrove ecosystems and ecological value	
Nortey et al.	2016		Assessment of mangrove biomass and fish assemblages	
Koné	2008	Côte d'Ivoire	Methane in the mangroves of Côte d'Ivoire	3
Egnankou	2009		Rehabilitation of mangroves	
Kwassi and Blivi,	2004		Mapping mangrove forest	
N/A	N/A	Liberia	**Not available:** reason may be due to civil war	0
N/A	N/A	Sierra Leone	**Not available:** reason may be due to civil war	0
Kovacs et al.	2010	Guinea	Mangroves assessment classification and mapping mangroves	3
Flores De Santiago	2013		Rice farming in the mangrove-fringed and salinity intrusion	
Wolanski and Cassagne	2000			
N/A	N/A	Guinea Bissau	N/A	0
Satyanarayana et al.	2013	Gambia	Socio-ecological assessment of mangrove	2
Crow, B., and Carney,			Mangrove conservation and female oyster collectors	
Coberly	2014	Senegal	Mangrove dependencies and participation in rehabilitation	5
Hiraldo	2013		Mangrove conservation	
Vidy	2000		Estuarine and mangrove systems and the nursery concept	
Conchedda et al.	2007		Monitoring changes in mangrove ecosystems	
Sakho et al.	2011		Natural and anthropogenic factors on mangrove dynamics	

(Fig. 7.1), which serve as spawning grounds and habitats for several species of fish.

In coastal communities across West Africa, the ocean is a way of life. It is key to providing incomes and a critical source of nutrition, especially for the poor. Yet overfishing, poor fishing practices, and pollution are depleting fish stocks in some of the world's most important ocean habitats. Nevertheless, we hardly estimate the contribution of the mangroves (Table 7.1). The contributions of mangrove to the GDP of these countries are often captured either in the fisheries value or in the value of forestry resources (Feka and Ajonina 2011; Salem and Mercer 2012). Hence, the exact contribution of mangroves are difficult to obtain. This is based on the reason that a number of services offered by mangrove have no market value. Mangrove forests are key to protection of coastal areas from storm surges and coastal

Table 7.2 Species of mangrove in West Africa

Countries	*Acrostichum aureum,*	*Avicennia germinans*	*Conocarpus erectus*	*Laguncularia racemose*	*Nypa fruticans*	*Rhizophora harrisonii*	*Rhizophora mangle*	*Rhizophora racemosa*	Total species
Nigeria	x	x	x	x	x	x	x	x	8
Benin	x	x	x	x	–	–	x	x	6
Togo		x	x	–	–	–	–	x	3
Ghana	x	x	x	x	x	x	–	x	7
Côte d'Ivoire	x	x	x	x	–	–	–	x	5
Liberia	x	x	x	–	–	x	x	x	6
Sierra Leone	–	X	X	x	–	x	x	x	6
Guinea	x	x	x	x	–	x	x	x	7
Guinea Bissau	–	x	x	x	–	x	x	x	6
Gambia	x	x	x	x	–	x	x	x	7
Senegal	x	x	x	x	–	x	x	x	7

After UNEP (2007), Polidoro et al. (2010), and Goussard and Ducrocq (2014)

Table 7.3 The estimated value of mangroves to West African countries

Countries	The values of mangroves per Km^2	Current mangrove area in km^2	Total annual contribution of mangroves in $
Nigeria	$ 200,000	7386	1,477,200,000
Benin	$ 200,000	66	13,200,000
Togo	$ 200,000	11	2,200,000
Ghana	$ 200,000	137	27,400,000
Côte d'Ivoire	$ 200,000	99	19,800,000
Liberia	$ 200,000	110	22,000,000
Sierra Leone	$ 200,000	1050	210,000,000
Guinea	$ 200,000	2039	407,800,000
Guinea Bissau	$ 200,000	2999	599,800,000
Gambia	$ 200,000	581	116,200,000
Senegal	$ 200,000	1287	257,400,000
West Africa	$ 200,000	15,767	3,153,400,000

flooding including saline intrusion of coastal aquifers; densely populated coastal areas are particularly vulnerable and likely to become more so as climate changes over time. In addition, mangroves constitute key habitats of juvenile fish for the coastal and marine fisheries, which are of great economic importance in West African countries (Bromhead 2012).

There are many direct and indirect economic contributions of mangroves and therefore it is necessary to estimate the real contribution of mangroves to the national economy. This will strengthen the case for protection and conservation of mangrove forest. This chapter attempts to estimate the monetary value of mangrove services to the countries in the sub-region. UNEP-WCMC (2006) estimate suggests that the annual values of the benefits and services provided by one kilometre square of mangrove range from US$ 200,000 to 900,000. Based upon this value, we can estimate the value of mangroves services to a country if we know the area of mangroves coverage in square kilometres. Table 7.3 presents estimated monetary values of mangroves forests to each of the countries. Based on the current area of mangrove coverage in the each country (Feka and Ajonina 2011), the lowest range (US$ 200,000) of the 1 km^2 values of mangroves (UNEP-WCMC 2006; Corcoran et al. 2007).

The estimates in Table 7.3 show that mangroves contribute over $3 billion to the economies of West Africa annually, with Nigeria leading with over a billion dollars per annum. This estimate is based on the lowest annual yield from per square kilometre of mangrove. Thus, some countries may be receiving more than the estimates on Table 7.3. The various services and benefits derived from mangroves, upon which these monetary valuation have been estimated are outlined in Table 7.4.

Table 7.4 clearly shows that the exploitation of services and benefits from mangroves are not without a threat to the resource. Hence the need to assess the impacts various usage of the mangrove ecosystems and develop the appropriate and sustainable way of getting the benefits at present without destroying a similar benefit to the future generations.

Table 7.4 Services and benefits derived from mangroves in West Africa

Economic/non-economic uses	Details benefit of mangroves	Threat associated with the usages: direct, neutral or Indirect threats
Aquaculture	Coastal mangroves in the region are cleared to create fish ponds for commercial aquaculture activities	**Direct threat:** aquaculture pond destroy mangroves
Fisheries	More than 3 million people in the coastal areas of WA are dependent on fisheries as their source of food (protein) and their livelihoods	**Indirect threat:** mangroves are used as fuel (i.e. wood) for fish smoking
Hunting and harvesting	Mangrove forests serve as a habitat for non-aquatic animals and also provide edible fruits and plants that are hunted and harvested by the local for food	**Neutral threat:** hunting and harvesting do not cause much damage
Fisheries spawning grounds and habitat	Mangroves and associated wetland in WA serves as a habitat and a spawning grounds for several species of fish, shellfish and wintering migratory birds from Europe	**Neutral threat:** no threat associated if the carrying capacity is not exceeded
Cultural use	Mangroves have cultural and spiritual value/usage. They provide special materials for cultural and festival rituals	**Neutral threat:** it does not cause degradation
Oil exploration and production	Oil exploration, production and associated development in Niger Delta Nigeria and at the western coast of Ghana are occurring in areas covered by mangrove	**Direct threat:** it causes clearing and deforestation of mangroves
Timber and firewood usage	Mangroves are harvested by locals and use as fuel wood for fish smoking, salt production, and construction of boats, houses and fences as well as production of tools	**Direct threat:** it causes cutting and degradation of mangroves
Tourism	The diversity of life inhabiting mangrove systems, and their proximity in many cases to other attractions such as estuaries and sandy beaches, offers unique attractions	**Neutral threat:** facilitates mangrove conservation for tourism purposes
Salt production	Coastal mangroves in the region are cleared to create salt pans for commercial salt production	**Direct threat:** through the clearing and cutting of mangroves for firewood
Handy Craft production	Mangroves and other affiliated vegetation's are harvested by local's core domestic and commercial handcraft e.g. Cane chairs, bags, bird cage etc	**Direct threat:** it causes cutting and degradation of mangroves
Medicine	Mangroves have a number of therapeutic uses. For instance the leaves, stem and roots are used for the treatment of malaria, diarrhoea, ulcer, skin infections, diabetes and snake bite	**Neutral threat:** the amount of mangrove use for the this purpose is very small
Pollution filter	Mangroves do not only trap nutrients, but also polluted runoff, treated and untreated sewage and heavy metals and facilitate improved water quality	**Indirect threat:** provide filter, but excess pollution affect water quality and habitat
Carbon sequestration	**Mangroves** absorb CO_2 out of the atmosphere through photosynthesis and store it away in their rich soils. Mangroves **are natural carbon cleanser**	**Neutral threat:** help scrub excess carbon and reduce climate change
Coastal protection: Storm break and shoreline stability	Mangroves act as a barrier against storm in coastal areas, also their root systems and peat facilitate the interception and storage coastal sediment and thus improve coastal stability	**Neutral threat:** help to prevent coastal erosion and provide coastal protection

7.4 Threats to West African Mangroves

Many anthropogenic activities in the coastal areas such as exploitation of mangroves and some natural hazards have threatened the sustainability of mangroves forest of the world in the twenty-first century (Gilman et al. 2008; Polidoro et al. 2010; Ellison and Zouh 2012; Chaudhuri et al. 2015) the following are the specific threats that effects mangroves in west African:

- **Clearing of mangrove for development:** Population growth, migration to coastal zone, economic growth and beach tourism development have led to the clearing of mangrove forests to make room for human settlements, agricultural and infrastructure (such as harbours, hotels, aquaculture, and industrial development). In Ghana and Sierra Leone mangroves are cleared to create salt pans and shed to support commercial salt production (Thompson 2008; Affam and Asamoah 2011). Figures 7.2 and 7.3 present evidence of the conflict between mangrove conservation and development. While Fig. 7.2 shows the competition between mangrove and property development around Kpeshie lagoon in Accra, Ghana, Fig. 7.3 shows the clearing of mangrove for artisanal salt production at Nyanyano west of Accra, Ghana.

- **Pollution:** In West Africa many types of sewage flows untreated into the sea. The coastal lagoons, rivers and associated mangroves suffer from land based pollution (domestic, industrial and agriculture) (Ukwe et al. 2006), which weakens their ability to produce the ecosystem services, For instances, fertilizers, pesticides, and other toxic man-made chemicals applied in farms and the solid

Fig. 7.2 Mangrove squeeze in Accra, Ghana

Fig. 7.3 Clearing of mangrove for salt production at Nyanyano, West of Accra, Ghana

Fig. 7.4 Pollution of Mokwe Lagoon and associated mangroves in Ghana

waste like plastics (Fig. 7.4) are washed and carried by river systems from sources upstream to the mangrove wetlands downstream. The pollution can kill many species that use the mangrove forests as their habitat. For instance, oil pollution can smother mangrove roots and suffocate the trees. This severe pollution threat not only affect mangroves, but they also contaminate various species of fish and could enter the food chain.

- **Overharvesting**: the use of mangrove trees as both domestic and commercial fuel in the form of firewood, construction prop, wood chip and charcoal is a common practice in West Africa. The cutting of mangrove for this purposes have occurred for centuries, in this region, but it is no longer sustainable, as harvesting threatens the future of the Mangrove forests.

- **Dams and irrigation**: mangroves require a certain amount of freshwater to flourish. However, dams and irrigations upstream of coastal rivers reduce the amount of freshwater reaching mangrove wetland, which in return, increases the salinity levels and negatively affects the growth of the mangrove forest. Conversely, freshwater diversions for irrigation can also lead to mangrove wetland drying out. Dam and irrigation reduce the fertility and the filtering ability of mangrove forest and that affect the growth and ecosystem services of mangroves.

- **Overfishing**: Mangrove forests provide a large variety of fish and shellfish. These fisheries form an essential source of food for thousands of coastal communities in West Africa. The mangroves forests also serve as nurseries for many fish species. The increased global population and the corresponding increased demand for food has led to overfishing globally and for that West Africa the traditional artisanal fishing (Fig. 7.5), have been displaced by international commercial fishing that applies larger fishing vessels and bad fishing practices like par-trawling, and pirate fishing due to poor surveillance and enforcement of local and international fishing regulation in the region (Daniel 2016). These do not cause only overfishing, but also affect the ecological balance of food chains and mangrove fish communities (Fig. 7.6).

- **Climate change**: The effects of climate change such as increased temperatures, sea-level rise (SLR), increased intensity of storms and precipitation are likely to have the most severe impacts on mangrove ecosystems (Alongi 2008; Ellison 2015; Ranasinghe 2016). Mangrove forests require stable sea levels, a certain amount of freshwater and temperature for long-term survival. They are therefore extremely sensitive to current rising sea levels, rainfall induced flooding and higher temperature caused by global warming and climate change.

Fig. 7.5 Harvested mangroves for sale in a Market in Ghana

Fig. 7.6 Artisanal fishing landing sites along a mangrove shoreline of Ghana

- **Bush Fires**: Slash and burn is one of the common farming practices in West African. With this farming technique, the farmers clear the land through slashing the vegetation with a cutlass, wait for the weeds to dry, and then burn the dry slashed vegetation before planting. In recent times, due to the rise in temperatures, most peripheral part of the mangrove wetland dry up during the dry season. In most cases, the fire trespass from the burning of the slashed farmlands to the dry mangrove forest areas and destroyed them. This problem is becoming rampant. Thousands of hectors of mangroves in West Africa are burnt during the dry season through the activities of farmers.

7.5 Vulnerability and Impact of Climate Change on West African Mangroves

The vulnerability of both mangrove ecosystems and societies is interlinked (Pramova et al. 2015). This is because coastal communities depend on ecosystem services of mangroves for subsistence, livelihoods and coastal protection. Therefore, factors that negatively impact on mangrove ecosystem services and make them vulnerable, will inevitably make coastal communities vulnerable as a result of the loss of ecosystem services from the mangroves (Gilman et al. 2008; Ellison 2015). In addition, the impacts of the loss of ecosystem services on socio-economic systems could lead to unsustainable mangroves exploitation and management due to possible displacement and loss of livelihood associated with depletion of resources and livelihood. The unsustainable management of mangroves would increases their vulnerability. Thus there will be no win situation in those situations.

This implies that climate change and associated forces affecting communities living in coastal areas in West Africa are comparable to the ones affecting mangrove ecosystems. Sea-level rise, coastal storms and changes in precipitation will cause severe impacts on settlements, economic sectors, livelihoods, health, and well-being (IPCC 2014). However, local communities and economic sectors will be negatively impacted by changes in the ecosystems, which in turn affect the flow of ecosystem services that societies depend on. This will renders coastal societies even more vulnerable to climate hazards and change.

Sea-level rise will impact on people and economic sectors through land loss and inundation resulting from rising seawaters. This will force the shoreline to retreat and will decrease the land available for settlements, economic activities such as tourism, agriculture and livelihoods. Sea-level rise will also intensify coastal erosion, which is already a major problem in many low-lying coastal areas of West Africa (Ellison 2015). Salinity intrusion will decrease agricultural production by decreasing fresh water content in the soils. Sea-level rise will also increase the risk of coastal flooding, especially in combination with stronger tides, storms and precipitation. All these factors will force people to migrate landwards or elsewhere, which could lead to conflicts and further resource degradation.

According to Gilman et al. (2008), Pramova et al. (2015), and Ellison (2015), the main climate change related stressors that impact on mangrove ecosystems are: sea level rise, increase temperature, storms surge, and changes in precipitation, especially in relation to decreases in rainfall.

7.5.1 Sea-Level Rise

Sea-level rise is considered to be the biggest threat to mangrove ecosystems (Pramova et al. 2015; Gilman et al. 2008). According to Boateng et al. (2017), relative sea-level rise per year over the next 100 years would be about 3 mm/year in the region. Mangrove systems do not keep pace with changing sea level. Already, several of the anticipated impacts of accelerated sea level rise identified by the IPCC (2014) have been experienced in Ghana and other countries in the ECOWAS sub-region (Boateng et al. 2017). In fact, climate change and associated impacts of rising sea level, increased storm and torrential rainfall and flooding are already having impacts on coastal mangroves. This is because the changes have led to increased erosion and weakening of root structures of mangroves. There are also increases in salinity and frequency/depth of mangrove inundation beyond mangrove tolerance levels. In order to maintain the preferred hydro periods and salinity, the mangroves will need to migrate landward so that they can maintain their preferred condition. However, the success of the mangroves landward migration will depend on factors such as the availability of land/space and the slope at the back shore to allow the migration, ability of individual species to colonize new habitats at required rates (relative to sea-level rise) (Alongi 2008). The increase coastal population in West Africa, which began from the trading with Europeans from the fourteenth century and the associated development in the coastal zone has led to urbanisation in the coastal zone, coastal protection structures and development in the coastal areas that restrict the landward migration of mangroves in most places.

7.5.2 Increased Temperatures

The African Centre of Meteorological Application for Development (ACMAD) (2016) has projected that warming of the African continent will be 1.5 °C by 2050 and could rise to about 3 °C by 2100. This will aggravate the effects and intensify the impact of climate change on the continent. These could increase the intensity and frequency of tidal wave and storm and potentially cause inundation of the low-lying coastal areas through tidal waves induced by temperature and pressure imbalances between the land and the sea. The impacts of increased temperature is expected to cause changes in species composition, changes in the timing of flowering and fruiting, change in the mangrove productivity and expanding mangrove ranges to higher latitudes (Ellison 2000).

7.5.3 Storm

The southwest Monsoon winds and tidal waves develop tidal waves, as a result of atmospheric pressure imbalance between the land and the sea. These waves can have an impact on mangroves as they induce larger and more intense waves, increase wind speed, and cause changes in water levels. Climate change is increasing the intensity of these storm systems and their impact on mangroves. The negative impacts of storms on mangroves include defoliation, uprooting and mortality (Gilman et al. 2008; Pramova et al. 2015). Tidal surges, coastal flooding and strong wave currents induced by storms also impact sediment elevation through soil erosion or compression, and soil deposition (Alongi 2008). The intense rains events caused by the southwest monsoon winds will lead to flood causing erosion and debris flow to accumulate in the habitat of mangroves. These effects will increase vulnerability of mangroves due to the changes in hydrology and sediment elevation.

7.5.4 Precipitation

Globally, rainfall is predicted to increase by about 25% by 2050 in response to climate change (Gilman et al. 2008). In West African sub-region, it is anticipated that climate change will cause extreme dry seasons (haematin) and wet seasons (monsoon rainfall) in West Africa (IPCC 2014; ACMAD 2016). Changes in precipitation are expected to affect mangrove growth and distribution (Ellison 2000). Increases in rainfall will lead to mangrove area expansion, as mangroves will be able to colonise previously unvegetated areas at landward fringes in tidal wetlands. Areas with higher rainfall have also been shown to have higher mangrove diversity and productivity in comparison to areas with less rainfall, due to the higher supply of fluvial sediment and nutrients and reduced exposure to salinity and sulphate (Pramova et al. 2015). However, decreases in rainfall and increases in evaporation associated with hammattan will lead to increases in salinity due to lower groundwater tables and less surface freshwater input in to the mangrove ecosystem. This will lead to net losses of peat, as increases in seawater sulphate augment the anaerobic decomposition of peat. This effect will increase mangrove vulnerability to relative sea-level rise even further. Increases in soil salinity will also decrease net primary productivity and growth, as mangroves will experience increased tissue salt levels. Seedling survival will also be compromised, resulting in changes in competition between mangrove species.

7.6 Management and Adaptation Policies for West Africa

At a regional level, there has been some effort to conserve and protect mangrove forests from further destruction. Many West African countries have signed international conventions, including the Ramsar Convention on the Conservation of Wetlands, the Convention on Climate Change, the Convention on Biodiversity, and the Convention on Ozone Layer. Mangrove restoration efforts have been conducted in almost all the coastal nations along the West African coast to help communities restore Mangroves and sustainably harness their ecosystem services.

Despite these efforts there is still inadequate policy, law and institutional provision for mangrove forests management and adaptation. Many countries in West Africa have no clearly defined mangroves management structures or plans and tend to place mangrove forests under many institutions with conflicting roles. For instance, in Ghana it is under the responsibility of many ministries including Ministry of Forestry and Wildlife, Ministry of Environment and Ministry of Fisheries. National and international Non-Governmental Organisations also contribute through different community projects on the restoration, conservation and sustainable management of Mangroves. Besides the conflicting institutional role, the institutional framework for managing Mangroves in the sub-region tends to be top-down instead of bottom-up or lateral/collaborative (Boateng 2006). The top-down institutional framework applied in this region tends to make policy implementation ineffective at the operational/local level. This happens because the resources users at the local level are not invited to participate in the policy making. This normally either creates inefficient policy or poorly accepted policy. In most cases it leads to misunderstanding and conflict between the policy makers at the top and the resource users at the bottom/local level.

The key adaptation option for mangroves are:

(i) **Reduce anthropogenic impacts**: This policy focuses on reducing current anthropogenic stressors that are depleting mangroves ecosystem services and increasing vulnerability. Anthropogenic responses to climate change have the potential to exacerbate the adverse effects of climate change on mangrove ecosystems (Gilman et al. 2008; Ellison 2015). For example, human response to sea level rise could lead to increase in the construction of seawalls and other coastal erosion control structures adjacent to mangrove landward margins. The hard defence structures may cause erosion and scouring of the mangrove immediately fronting and down-current from the structure. The reduction of the human impacts on mangroves such as hard coastal defence structures, deforestation, burning during dry season, conversion for aquaculture and pollution will increase the overall ecosystem health and resilience and thus reducing the vulnerability on mangroves to climate change.

(ii) **Catchment management**: The impacts of sea-level rise on mangroves can be adapted by managing mangrove catchment area to minimize reductions in sediment supply. These can include, for example, limiting the development of impervious surfaces within the mangrove catchment and managing rates and locations of groundwater extraction (Pramova et al. 2015). Such activities can reduce alterations to natural groundwater recharge, which influences mangrove elevation. Limiting human activities that reduce mangrove soil organic matter accumulation, such as deforestation and pollution inputs, can contribute to maintaining relatively natural controls on trends in sediment elevation. Depending on the tree species and nutrient added, nutrient enrichment can also have a positive effect. Enhancement of mangrove sediment accretion rates, such as through the beneficial use of dredge spoils, could also augment mangrove sediment elevation to enhance resilience.

(iii) **Maintain coastal buffer zone (manage retreat)**: Unlike in the developed countries where coastal towns are built very close to the sea and buildings are protected by concrete structures, many coastal areas and towns in West African have buffer lands between the sea and the developed area. Though population growth and urbanisation have led to severe pressure and squeeze on the coastal buffer lands (setback zones), there is still opportunity for West African countries to protect coastal buffer zones from development pressures (Boateng 2009). This can be achieved through coastal landscape planning. The coastal buffer zones will allow the long-term landward migration of both the mangroves ecosystems and the coastal development to accommodate future impact of climate change.

(iv) **Artificial landward migration and refugia**: Protected areas can be established and managed to maintain mangrove representation, replication and refugia. Ensuring representation of all mangrove community types in a protected area network and replication of identical communities to spread risk can increase the chances of mangrove ecosystems surviving climate change and other stressors (Gilman et al. 2008; Pramova et al. 2015). Protected area selection can include mangrove areas that act as climate change refugia, which means selecting communities that are likely to be more resilient to climate hazards, such as mature mangrove communities. Protecting refugia areas that resist and recover quickly from disturbance can serve as a source of recruits to recolonize areas that are lost or damaged. Protected area networks should also account for likely movements of habitat boundaries and species ranges over time under different climate change scenarios, and also consider the role of ecosystems in the adaptation of society. Facilitated landward migration of mangrove may be necessary, as discussed in the previous point, and the connectivity between coastal ecosystems, including mangroves, should be protected to enhance overall resilience and maintain functional links.

(v) **Restoration**: Finally, mangrove restoration in areas where mangrove habitats previously existed can offset losses by climate change and consistent monitoring activities could help assess gradual changes (Coberly 2014). Regional networks, sharing of best practices using standardized techniques will enable the separation of site-based influences from global changes for a better understanding of mangrove responses to different stressors and enhance overall adaptive capacity in coastal socio-ecological systems.

(vi) **Regional monitoring and regulation**: There are some uncertainties in climate change projections and anticipated impacts (IPCC 2014). Therefore some of the predicted response of mangroves ecosystems may be misjudged. There is a need for continuous monitor changes on both local and regional scale. To achieve this, there the need to develop a regional regulation and standard techniques for assessing and establishing mangrove baselines, and monitoring changes using regional standardised tools and regulation to develop a better understanding of mangrove responses to sea-level and global climate change. This will facilitate the development of progressive adaptation and mitigation measures based upon the monitored changes.

(vii) **Education and local participation**: to achieve a successful implementation of any adaptation policy, there is the need to educate the local coastal communities, the impacts of climate change on mangroves the potential consequences if they do not apply or abide by the adaptation policies. There is the need to develop in the locals' knowledge and understanding of the present changes and their personal observations about the dwindling resources from the mangrove through community participation in the monitoring and the decision-making processes. Community participation can increase the support for adaptation actions. According to Gilman et al. (2008) education and outreach programmes are an investment to bring about behaviour change and attitudes by having a better-informed community of the value of mangroves and other ecosystems.

7.7 Conclusion

This chapter has assessed the status of West African mangroves and identified that over 20% of the mangroves have been lost over the last century through exploitation and other anthropogenic activities. The existing mangroves are still contributing to the economic development and the sustenance of the local coastal communities whose livelihood still depends on the mangroves. It has been identified that harnessing various services and benefits from the mangroves poses a varying threats to the mangrove ecosystem (Table 7.4). However, the combined effects of the anthropogenic threat and climate change impacts may have deleterious effects on coastal mangroves if adaptation measures are not developed and implemented well in advance to reduce, if not to prevent some of the anticipated catastrophic impacts of the combined effects of both anthropogenic and climate change impacts on coastal mangroves. Some of the adaptation measures suggested include reducing anthropogenic impacts, maintaining coastal buffer zones, restoration of mangroves, catchment management, establishing regional monitoring and regulations. There is also the need for education and local participation in the formulation and the implementation of the mangroves adaptation policies.

References

Abuodha PA, Kairo JG (2001) Human-induced stresses on mangrove swamps along the Kenyan coast. Hydrobiologia 458:255–265

Adekanmbi OH, Ogundipe O (2009) Mangrove biodiversity in the restoration and sustainability of the Nigerian natural environment. J Ecol Nat Environ 1(3):064–072

Adotey J (2015) Carbon stock assessment in the Kakum and Amanzule estuary mangrove forests. Ghana (doctoral dissertation, department of fisheries and aquatic sciences, school of biological sciences, university of cape coast)

Affam M, Asamoah DN (2011) Economic potential of salt mining in Ghana towards the oil find. Res J Environ Earth Sci 3(5):448–456

African Centre of Meteorological Application for Development (ACMAD) (2016) Status report on ACMAD-MESA on result 4 policy support. Presented by Dr Benjamin Lamptey (Deputy Director, ACMAD) at MESA Policy Dialogue Workshop. Kigali, Rewanda, 19–20 September

AFROL (2002) Mangroves of Western Africa threatened by global warming. In: Afrol news. Available at http://www.afrol.com/Categories/Environment/env019_mangroves_threatened.htm

Agyeman YB, Akpalu L, Kyereh B (2007) Preliminary inventory of selected mangrove sites in GHANA

Ainodion MJ, Robnett CR, Ajose TI (2002) Mangrove restoration by an operating company in the Niger Delta. In: SPE international conference on health, safety and environment in oil and gas exploration and production. Society of Petroleum Engineers, Kuala Lumpur. https://doi.org/10.2118/74033-MS

Ajonina GN, Usongo L (2001) Preliminary quantitative impact assessment of wood extraction on the mangroves of Douala-Edea Forest Reserve, Cameroon. Trop Biodivers 7(2 & 3):137–149

Ajonina PU, Ajonina GN, Jin E, Mekongo F, Ayissi I, Usongo L (2005) Gender roles and economics of exploitation, processing and marketing of bivalves and impacts on forest resources in the Sanaga Delta region of Douala-Edea Wildlife Reserve, Cameroon. Int J Sustain Dev World Ecol 12(2):161–172

Alongi DM (2008) Mangrove forests: resilience, protection from tsunamis, and responses to global climate change. Estuar Coast Shelf Sci 76:1–13

Boatemaa MAA, Kwasi AA, Mensah A (2013) Impacts of shoreline morphological change and sea level rise on mangroves: the case of the keta coastal zone. J Environ Res Manag 4(11):359–367

Boateng I (2006) Institutional frameworks in the administration of coastal and marine space in Africa. In: Sutherland M (ed) Administering marine spaces: international issues. The International Federation of Surveyors, Frederiksberg, Publication no. 36

Boateng I (2009) Development of integrated shoreline management planning: a case study of Keta, Ghana. Proceedings of the International Federation of Surveyors Working Week 2009-Surveyors Key Role in Accelerated Development, TS 4E, Eilat, Israel, 3–8 May

Boateng I, Wiafe G, Jayson-Quashigah PN (2017) Mapping vulnerability and risk of Ghana's coastline to Sea Level Rise. Mar Geod 40 (1):23–39. https://doi.org/10.1080/01490419.2016.1261745

Bojang F, Ndeso-Atanga A (2009) The relevance of mangrove forests to African fisheries, wildlife and water resources. FAO, Acca

Bos D, Grigoras I, Ndiaye A (2006) Land cover and avian biodiversity in rice fields and mangroves of West Africa. Wetlands International, Wageningen

Bromhead MA (2012) Forest, trees, and woodlands in Africa: an action plan for World Bank engagement. Forest, trees, and woodlands in Africa: an action plan for World Bank engagement. UNEP-WCMC, Cambridge. Arlington, USA: TNC, Tokyo, Japan: UNU, New York, USA

Carney J, Gillespie TW, Rosomoff R (2014) Assessing forest change in a priority West African mangrove ecosystem: 1986–2010. Geoforum 53:126–135

Chaudhuri P, Ghosh S, Bakshi M, Bhattacharyya S, Nath B (2015) A review of threats and vulnerabilities to mangrove habitats: with special emphasis on East Coast of India. J Earth Sci Clim Chang 6:270. https://doi.org/10.4172/2157-7617.1000270

Coberly, L.C. (2014) Mangrove dependencies and participation in rehabilitation. A case study of the Sine-Saloum Delta, Senegal. State University of New York College of Environmental Science and Forestry

Conchedda G, Durieux L, Mayaux P (2007) Object-based monitoring of land cover changes in mangrove ecosystems of Senegal. In: Analysis of multi-temporal remote sensing images, 2007. MultiTemp 2007. International Workshop on the. Leuven: IEEE, pp 1–6, DOI 10.1109/MULTITEMP.2007.4293039

Corcoran E, Ravilious C, Skuja M (2007) Mangroves of Western and Central Africa (No. 26). UNEP/Earthprint, Cambridge

Crow B, Carney J (2013) Commercializing nature: mangrove conservation and female oyster collectors in The Gambia. Antipode 45 (2):275–293

Daniel A (2016) Western Africa's missing fish. Overseas Development Institute, London

Egnankou, W. M. (2009). Réhabilitation des mangroves comprises entre Fresco et Grand-Lahou en Côte d'Ivoire: zones importantes pour la pêche in Nature et faune, volume 24, n 1. FAO, Rome, pp 85–93

Ellison J (2000) Chapter 15: How South Pacific mangroves may respond to predicted climate change and sea level rise. In: Gillespie A, Burns W (eds) Climate change in the South Pacific: impacts and responses in Australia, New Zealand, and Small Islands States. Kluwer Academic Publishers, Dordrecht, pp 289–301

Ellison JC (2015) Vulnerability assessment of mangroves to climate change and sea-level rise impacts. Wetl Ecol Manag 23:115–137. https://doi.org/10.1007/s11273-014-9397-8

Ellison JC, Zouh I (2012) Vulnerability to climate change of mangroves: assessment from Cameroon, Central Africa. Biology 1(3):617–638

Feka ZN (2015) Sustainable management of mangrove forests in West Africa: a new policy perspective? Ocean Coast Manag 116:341–352

Feka NZ, Ajonina GN (2011) Drivers causing decline of mangrove in West-Central Africa: a review. Int J Biodivers Sci Ecosyst Serv Manag 7(3):217–230

Flores De Santiago F, Kovacs JM, Lafrance P (2013) An object-oriented classification method for mapping mangroves in Guinea, West Africa, using multipolarized ALOS PALSAR L-band data. Int J Remote Sens 34(2):563–586

Gilman EL, Ellison J, Duke NC, Field C (2008) Threats to mangroves from climate change and adaptation options: a review. Aquat Bot 89 (2):237–250

Goussard JJ, Ducrocq M (2014) West African coastal area: challenges and outlook. In: Diop S et al (eds) The land/ocean interactions in the coastal zone of West and Central Africa. Springer International Publishing Island Press, Washington, DC, pp 9–21

Hiraldo R (2013) Democracy and the social organisation of capitalist production: mangrove conservation in Senegal

IPCC (2014) Climate change: impacts, adaptation, and vulnerability. Part A: global and sectoral aspects, Contribution of Working Group II to the Fifth Assessment Report of the Intergovernmental Panel on Climate Change. Cambridge University Press/Island Press, Cambridge/Washington, DC

IUCN WCPA (2005) Millennium ecosystem assessment; ecosystems and human well-being: synthesis. Ocean protection: present status and future possibilities. Agence des aires, Brest

James GK, Adegoke JO, Osagie S, Ekechukwu S, Nwilo P, Akinyede J (2013) Social valuation of mangroves in the Niger Delta region of Nigeria. Int J Biodivers Sci Ecosyst Serv Manag 9(4):311–323

Kathiresan K, Bingham BL (2001) Biology of mangroves and mangrove ecosystems. Adv Mar Biol 40:81–251

Koné YJM (2008) Dynamics of carbon dioxide and methane in the mangroves of Vietnam, and the rivers and the lagoons of Ivory Coast Dynamique du dioxyde de carbone et du méthane dans les mangroves du Vietnam, les rivières et les lagunes de la Côte d'Ivoire

Kovacs JM, de Santiago FF, Bastien J, Lafrance P (2010) An assessment of mangroves in Guinea, West Africa, using a field and remote sensing based approach. Wetlands 30(4):773–782

Kwassi A, Blivi A (2004) Use of landsat TM imagery as a tool for mangrove forest mapping; a case study of Ehotile Island in Côte d'Ivoire. In: 35th COSPAR scientific assembly, vol 35. p 3524

Luiselli L, Akani GC (2002) An investigation into the composition, complexity and functioning of snake communities in the mangroves of south-eastern Nigeria. Afr J Ecol 40(3):220–227

Mensah JC (2013) Remote sensing application for mangrove mapping in the Ellembelle district in Ghana. Doctoral dissertation, University Of Rhode Island

Nicholls RJ, Hanson S, Lowe JA, Warrick RA, Lu X, Long AJ (2014) Sea-level scenarios for evaluating coastal impacts. Wiley Interdiscip Rev Clim Chang 5(1):29–150

Nortey DD, Aheto DW, Blay J, Jonah FE, Asare NK (2016) Comparative assessment of mangrove biomass and fish assemblages in an urban and rural mangrove Wetlands in Ghana. Wetlands 36(4):717–730

Ntyam SCO (2014) Comparative study of the Douala-Edea Reserve (Cameroon) and Songor Ramsar Site (Ghana) using parameters of ecological value. Doctoral dissertation, University of Ghana

Ntyam SCO, Armah AK, Ajonina GN, George W, Adomako JK, Elvis N, Obiang BO (2014) Importance of mangrove litter production in the protection of Atlantic coastal forest of Cameroon and Ghana. In: The land/ocean interactions in the coastal zone of West and Central Africa. Springer International Publishing, Dordrecht, pp 123–137

Polidoro BA, Carpenter KE, Collins L, Duke NC, Ellison AM et al (2010) The loss of species: mangrove extinction risk and geographic areas of global concern. PLoS One 5(4):e10095. https://doi.org/10.1371/journal.pone.0010095

Pramova E, Chazarin F, Locatelli B (2015) Mangrove forests for adaptation: potential and vulnerability. Cirad Fr, Bogor

Ranasinghe R (2016) Assessing climate change impacts on open sandy coasts: a review. Earth Sci Rev 160. (2016):320–332

Sakho I, Mesnage V, Deloffre J, Lafite R, Niang I, Faye G (2011) The influence of natural and anthropogenic factors on mangrove dynamics over 60 years: the Somone Estuary, Senegal. Estuar Coast Shelf Sci 94(1):93–101

Salem ME, Mercer DE (2012) The economic value of mangroves: a meta-analysis. Sustainability 4(3):359–383

Satyanarayana B, Bhanderi P, Debry M, Maniatis D, Foré F, Badgie D, Dahdouh-Guebas F (2012) A socio-ecological assessment aiming at improved forest resource management and sustainable ecotourism development in the mangroves of Tanbi Wetland National Park, The Gambia, West Africa. Ambio 41(5):513–526

Tang W, Feng W, Jia M, Zuo H (2014) Assessment of biomass and carbon of mangroves in West Africa: USAID final report

Tang W, Feng W, Jia M, Shi J, Zuo H, Trettin CC (2016) The assessment of mangrove biomass and carbon in West Africa: a spatially explicit analytical framework. Wetl Ecol Manag 24(2):153–171

Thompson F (2008) Current status and conservation of mangroves in Africa: an overview. Retrieved from https://wrmbulletin.wordpress.com/2008/08/25/current-status-and-conservation-of-mangroves-in-africa-an-overview/

Toropova C, Meliane I, Laffoley D, Matthews E, Spalding M (2010) Global UNEP-WCMC. 2006. In the front line: shoreline protection and other

Ukwe C, Ibe C, Sherman K (2006) A sixteen-country mobilization for sustainable fisheries in the Guinea current large marine ecosystem. Ocean Coast Manag 49:385–412

UNEP-WCMC (2006) In the front line: shoreline protection and other ecosystem services from mangroves and coral reefs. UNEP-WCMC, Cambridge. 33 pp

UNEP-WCMC (2007) Mangroves of Western and Central Africa. United Nations Environment Programme (UNEP) Regional Seas Programme/UNEP –World Conservation Monitoring Centre. WCS, Cambridge, UK. 96pp

USAID (2014) Workshop report: West Africa regional mangroves and climate change. Retrieved from https://sites.google.com/site/mangroveworkshop

Vidy G (2000) Estuarine and mangrove systems and the nursery concept: which is which? The case of the Sine Saloum system (Senegal). Wetl Ecol Manag 8(1):37–51

Wolanski E, Cassagne B (2000) Salinity intrusion and rice farming in the mangrove-fringed Konkoure River delta, Guinea. Wetl Ecol Manag 8(1):29–36

Worldbank (2013) West Africa: fishing communities restore health to ocean habitats. June 5, 2013 http://www.worldbank.org/en/news/feature/2013/06/05/west-africa-fishing-communities-restore-health-to-ocean-habitats

Zwarts L, van der Kamp J, Klop E, Sikkema M, Wymenga E (2014) West African mangroves harbour millions of wintering European warblers. Ardea 102(2):121–130

Historical Losses of Mangrove Systems in South America from Human-Induced and Natural Impacts

8

Daniel Gorman

Abstract

Mangrove forests are an intrinsic part of the coastal landscapes of tropical and subtropical South America. Although less well studied than their North American, Southeast Asian and Australian counterparts, they cover large expanses (approximately 11% of global mangrove cover) and perhaps represent a greater proportion of their respective coastline than other ecosystem types. The last century has been witness to a large but ultimately unknowable loss of mangrove forests across the continent through intensified use of coastal zones by humans. Indeed it has been estimated that more than 11% of the mangroves existing in the 1980's have been lost or severely degraded. Additionally, while protection for mangroves in many parts of the world has increased considerably, management in South America remains complex and the lack of regular change indicators complicates status assessments at both national and international levels. This chapter presents a historical and contemporary background to losses of mangrove systems in South America with a focus on those countries that have the most up-to-date and accessible data (i.e., Brazil, Ecuador and the Guineas). The goal was to present what is known about the drivers and degree of historical loss, the current and future threats, and to document efforts at restoring degraded forests. The chapter concludes with advice on how to address important knowledge gaps and facilitate effective management to improve the conservation outcomes for South American mangrove systems.

Keywords

Wetlands · Urbanization · Aquaculture · Modification · Sea-level rise · Climate change

D. Gorman (✉)
Oceanographic Institute, University of São Paulo, São Paulo, Brazil
e-mail: dgorman@usp.br

8.1 Introduction

8.1.1 Ecological Value of Mangroves

Mangrove forests are an important component of nearly all earth's tropical and subtropical coastlines and signal the frontier between terrestrial and marine environments (Alongi 2008). Although historically viewed as unproductive and even unhealthy wastelands, greater research and experience has shown mangroves to be amongst the most important and productive ecosystems on our planet (Fig. 8.1). The three-dimensional structure of these coastal forests provide numerous ecosystem services including; renewable resources for human populations (Alongi 2002), nursery grounds for a multitude of ecologically and commercially important fish and crustacean species (Nagelkerken et al. 2008), a vital role in carbon export and sequestration (Lee et al. 2014; Alongi 2012), offsets to atmospheric CO_2 emissions (Alongi 2002) and an effective means of shoreline protection (Everard et al. 2014). Despite these numerous services, mangrove forests remain at risk due to continued deforestation, land reclamation and pollution. This is especially concerning given that the structure, ecology and genetic diversity of these systems is still not fully understood. Recognizing the importance of conserving mangrove forests across South America is vital given that they are believed to harbor ca. 11% of the planet's remaining cover and because the continent is experiencing continued economic development that can overwhelm the capacity to cope with the challenges of local and global change.

8.1.2 South American Mangrove Systems

The continent of South America comprises twelve independent countries and three dependent territories, with only Bolivia and Paraguay lacking maritime borders. Together, the combined coastline extends for more than 31,000 km

C. Makowski, C. W. Finkl (eds.), *Threats to Mangrove Forests*, Coastal Research Library 25,
https://doi.org/10.1007/978-3-319-73016-5_8

Fig. 8.1 Mangrove forests provide numerous ecosystem services to coastal environments as well as local human populations. The area of mangroves in South America (ca. 11% of global total; Ward et al. 2016) is not well defined and much of it lacks adequate legal protection (Photo: mangrove forest in Lagoa de Guaraíras, RN, Brazil, Daniel Gorman, 2016)

and includes tropical, subtropical and temperate biomes that fringe the Pacific, Caribbean and Atlantic Oceans (FAO 2007). Across this range, there exists a great diversity of coastal environments including; coral reefs, sandy beaches, mudflats, rocky reef, saltmarshes and mangrove forests (Miloslavich et al. 2010). The total extent of South American coastline covered by mangroves has been estimated at just under 2 million hectares, down from 2.2 million hectares in 1980 (FAO 2007). Because of the vastness and relative isolation of much of the coastline (particularly in the tropical north) large expanses of mangrove forest remain relatively unexplored and unstudied. Indeed, a basic *Web of Science* (database) search spanning the years 1900–2017, showed that the number of studies done on mangrove systems in South-east Asia and Australia was 1.8 and 7.7 times greater respectively, than the number done in South America.

Mangrove forests occur in all but the three coldest countries of Argentina, Chile and Uruguay (Lacerda and Schaeffer-Novelli 1999). This tropical-subtropical distribution is governed by a combination of climatic (i.e., mainly temperature and rainfall) and oceanographic factors (e.g., currents and sea surface temperature) and leads to a noticeably uneven distribution across the continent, with the Atlantic and Caribbean coasts harboring nearly 70% of the total. On the Pacific west coast, upwelling of cold water from the Humboldt current combine with a semi-arid climate and limited freshwater inputs to restrict the southern extension of forests to the Piura River in northern Peru (approximately 5°32′ S; Fig. 8.2). Along the Atlantic east coast, mangroves extend much further south to the Santo Antônio Lagoon, Santa Catarina, Brazil (approximately 28°28′S). This present limit is thought to be defined by the coldest mean monthly temperature of 20 °C and a thermal amplitude of less than 5 °C (Soares et al. 2012).

Across their South American distribution, mangroves occur within relatively sheltered muddy environments that span brackish, estuarine and intertidal zones. The group comprises a range of salt-tolerant trees and other plant species from an array of families. Members are highly adapted to periodic immersion and exposure to high tides, fluctuating salinities, low oxygen concentrations and exhibit a number of features including aerial roots, viviparous seeds and salt exclusion/excretion mechanisms (Tomlinson 1986). Compared to tropical south-east Asia where mangroves reach their greatest diversity (Polidoro et al. 2010), South American mangrove forests comprise only ten species that have different representation across the continent (Table 8.1). Species

Fig. 8.2 South Americas mangrove forests extend from the Pacific north coast of Peru, through the Caribbean and down to the southern Atlantic coastline of Brazil. Data derived from recent estimates for each country based on the model of Hutchison et al. (2014), available through the United Nations Environment Programme World Conservation Monitoring Centre (UNEP-WCMC) website (www.unep-wcmc.org)

diversity is highest in the tropics and typically declines with increasing latitude (Lacerda and Schaeffer-Novelli 1999).

Forests reach their greatest levels of development in the stable and humid tropical regions of northern Brazil, Columbia, Venezuela and the Guianas (i.e., Guyana, French Guiana and Suriname) (Fig. 8.2). Within these areas, forests can be many kilometers deep and exhibit strong horizontal zonation from fringing forests to inland river systems and lagoons. Within these systems, forests comprising the red mangrove (*Rhizophora mangle*) can reach heights of 40–50 m, with trunk girths exceeding 1 m in diameter (Lacerda and Schaeffer-Novelli 1999). Well-developed black mangrove forests (*Avicennia germinans*) regularly attain heights of up to 30 m on the coasts of the Guianas and Northern Brazil (Lacerda and Schaeffer-Novelli 1999). About half the mangrove area of the region is found in Brazil, which also has the third largest mangrove area in the world (FAO 2007).

Despite the important role of mangroves in the structure and function of coastal ecosystems, data on their distribution and status in South America is not well described. This has made understanding historical losses and rates of deforestation notoriously difficult (Blanco et al. 2012; Ellison and Farnsworth 1996). While an attempt to compile comprehensive long-term data by country was undertaken by the Food and Agriculture Organization of the United Nations (FAO 2007), this assessment was based on a relatively small number of sites stretching back to the 1980's; and its current relevance is uncertain. More recent assessments of the global status of mangroves, including those in South America, have been derived from LANDSAT imagery (Giri et al. 2011) and through the use of modeling approaches (Fig. 8.3).

Another challenge to assessing the status of South American mangrove systems as a whole (i.e., at the continental scale), has been the realization that much of the scientific research has remained stranded in the grey literature (i.e.,

Table 8.1 Mangrove species represented in South American countries (Table modified from FAO (2007))

Species	Brazil	Colombia	Venezuela	Ecuador	Suriname	Guyana	French Guiana	Peru
Acrostichum aureum		☆		☆	☆			
Avicennia bicolor		☆ (uncertain)						
Avicennia germinans	☆	☆	☆	☆	☆	☆	☆	☆
Avicennia schaueriana	☆		☆			☆		
Conocarpus erectus	☆	☆	☆	☆		☆		☆
Laguncularia racemosa	☆	☆	☆	☆	☆	☆	☆	☆
Pelliciera rhizophorae		☆		☆				
Rhizophora harrisonii	☆	☆	☆	☆				☆
Rhizophora mangle	☆	☆	☆	☆	☆	☆	☆	☆
Rhizophora racemosa	☆	☆ (uncertain)	☆				☆	
Total no. of species	7	9	7	7	4	5	4	5

☆ = uncertain

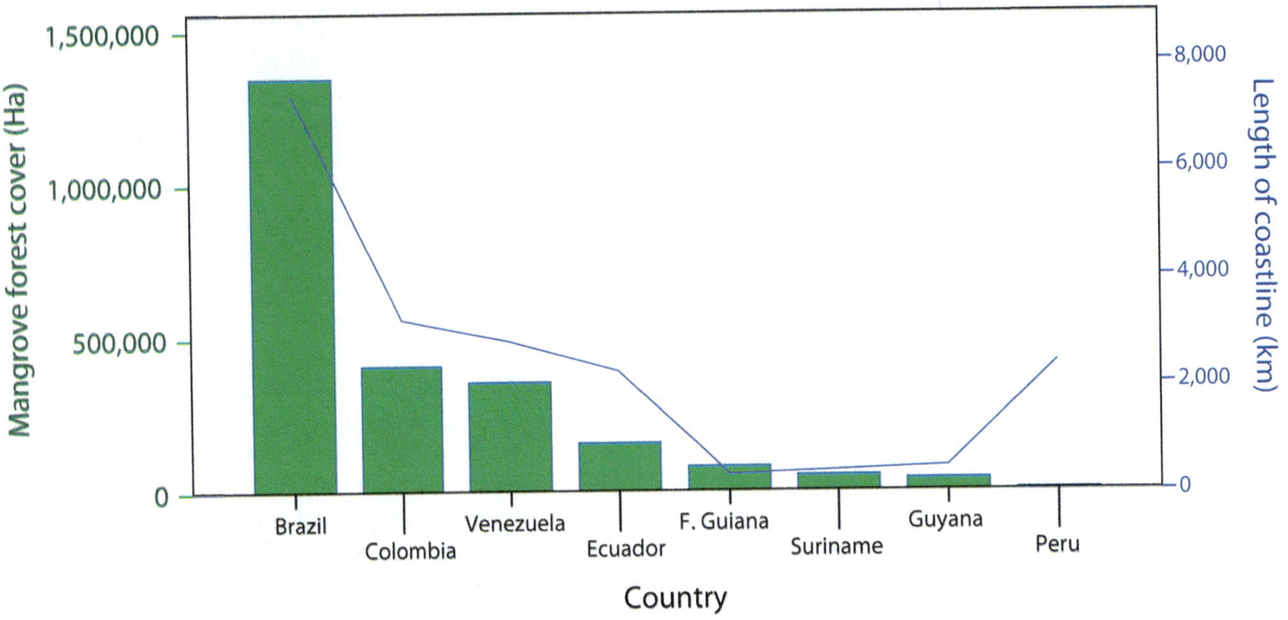

Fig. 8.3 Estimated mangrove forest cover and total length of coastline for South American countries. Bars are the most recent estimates for each country based on the model of Hutchison et al. (2014) available through the United Nations Environment Programme World Conservation Monitoring Centre (UNEP-WCMC) website (www.unep-wcmc.org); line represents respective coastline length (CIA 2013)

forestry documents, national or state government reports, theses, etc.) and typically has not been translated into the broader canon of scientific literature (primarily in English). Difficulties in accessing data complicates the derivation of historical trends for mangrove habitats (as it has for other coastal marine systems; Connell et al. 2008). Although the problem is being addressed through efforts at international collaboration, there is still a wealth of data that will become increasingly important to understand current and future trends.

8.2 Historical Drivers of Loss

8.2.1 Natural Factors

The coastal wetlands that support mangroves can change in response to a variety of geomorphic, hydrologic and climatic factors (Phillips 1999). Understanding the long-term stability of mangroves, therefore requires an understanding of the background prevalence of natural phenomena such as floods, tsunamis, storms and hurricanes (Alongi 2002). Because these disturbances vary considerably across the South American continent, the risks they pose need to be considered on a regional basis. Hurricanes for example, have been shown to cause significant damage to mangrove forests along the Caribbean coastline of Columbia (Bernal et al. 2016) and Venezuela (Miloslavich et al. 2003) but generally do not occur along the Atlantic or Pacific coastlines. Contrastingly, tsunamis might pose a major, albeit rare, risk to forests occurring along the coastlines of Peru and Ecuador. Studies done on the impacts of the 2004 tsunami in Banda Aceh, suggest ca. 50% losses of trees directly facing the tsunami flow, with the greatest mortality seen in those species lacking prop roots (Yanagisawa et al. 2010). Given that three out of the five mangrove species in Peru fall within this category, there might be cause for concern about the scale of damage from such an event.

The absence of cyclones and tsunamis along the tropical coastlines of the Guianas and northern Brazil, means that storms and especially flooding (from the Orinoco and Amazon river systems) are the principal drivers of natural loss. Flooding for extended periods, can cover aerial roots with water and/or sediments that lead to oxygen deprivation of plant tissues and subsequent changes to leaf function (Naidoo 1983). Further south, along the subtropical coastline of Brazil, low pressure Atlantic storms are generally responsible for the greatest natural losses of mangroves, especially forests that directly fringe the coastline. The actions of wind and waves can break branches, damage foliage and in some cases undermine and uproot entire trees especially at the seaward margin (Smith et al. 1994).

Another risk for mangrove systems in South America is that of inter-annual climate variability driven by El Niño Southern Oscillation (ENSO). El Niño events typically alter rainfall patterns in ways that lead to extreme flooding in some regions but droughts in others. During ENSO years, most of tropical South America receives lower than average rainfall and a decrease in extreme precipitation events (Grimm and Tedeschi 2009). Along the semi-arid coastline of Peru however, strong ENSO events like the one that occurred between 1982 and 1983 have been shown to cause mean sea-level rise of up to 40 cm and increase rainfall by up to ten time the usual. In this way, ENSO events are likely to have considerable, though probably quite regionally specific consequences for mangrove health through changes in salinity, inundation cycles and sediment budgets (Ellison 2000).

8.2.2 Human-Induced Factors

The impacts of human activity on South American mangrove ecosystems have been long and varied. The earliest evidence for exploitation dates back ca. 9000 years to pre-Columbian human groups (Lacerda et al. 1993). These were simple activities however, that involved the traditional use of timber and wood for energy production and in this way the degree of impact was governed by the specific characteristics and size of populations. For coastal communities, there was also a sustained reliance on the food resources generated by mangroves, as evidenced by the large mounds of shells and other organic debris, called "conchales" or "concheros" (Colombia) or "sambaquis" (Brazil). Such archaeological remnants provide important information on the characteristics of these populations including food habits and resource utilization (Prahl et al. 1990).

The earliest contemporary writings of mangroves from South America are attributed to the Portuguese noble Gabriel Soares de Souza in his 1587 work titled *Tratado Descritivo do Brasil* (Treaty of Brazil), in which he describes the physical structure of trees (remarking on their curious roots) and the seemingly endless 'riches of the sea' by which he refers to the productivity of oysters associated with these same roots. European colonization intensified the pressure on mangrove ecosystems principally as a result of the development of ports and settlements to facilitate the exchange of people and goods with European centres and the African continent (i.e., the slave trade). The earliest recorded regulation relating to the use of mangroves in South America, was a law imposed on Brazil in 1706 by King Jose of Portugal who made it illegal to fell mangrove trees under financial penalty (or even a 3 month jail sentence), without first harvesting the bark (Hamilton and Snedaker 1984). Although it would be nice to imagine this as an early example of environmentally sound management, it was instead driven by the concerns of leather merchants eager to guarantee the supply of high quality material for their tanneries. The supply of tannins was important for the production of parchment for manuscripts, books and diplomas, with the South American supply becoming increasingly important as Portuguese colonial power declined in Asia.

Fast-forward to the early twentieth Century, and the general attitude towards mangrove ecosystems had changed little. Mangroves and wetlands were still widely viewed as wastelands or useless swamps and were neglected and abused well up until the late 1960s (Lugo and Snedaker 1974). Even later when their importance began to be realized through studies done in the Caribbean and North America (Golley et al. 1962; Odum 1971), attitudes toward mangrove forests in South America changed only slowly and mainly through the activities of NGO's such as UNESCO and FAO who implemented programs of research, conservation and management (Lacerda and Schaeffer-Novelli 1999). Unfortunately, this period coincided with a period of socio-economic crisis across Latin America that drove a large segment of the population into poverty and placed increasing pressure on the environment. As a result, large areas of mangrove forest were lost during the latter part of the twentieth century through a combination of demographic and economic factors linked to coastal population growth and the urgent need to generate economic opportunities (Duke et al. 2007). Table 8.2 provides a summary of the main drivers of exploitation and loss of South American mangrove forests.

It has been estimated that more than 11% of the total expanse of mangrove forests in South America were cleared between 1980 and 2005 (data derived from; FAO 2007), however when added to historical losses, this figure is likely to be much greater. It is also important to note that losses were not uniform across countries (Fig. 8.4), with estimates showing that Peru lost by far the greatest proportion of its

Table 8.2 Summary of the historical drivers of mangrove loss across South America with stars indicating the degree of complicity

Category	Agent of change	Loss
Resource exploitation	Bark tannin	☆☆
	Wood for construction	☆☆
	Charcoal	☆☆
	Traditional use/harvesting	☆
	Animal fodder	☆
	Medicines and food (e.g., fish, shellfish)	☆
Competition for space	Conversion for food production	☆☆☆
	Industrial development	☆☆☆
	Conversion to salt production	☆☆
	Urban development (e.g., condominiums)	☆☆
	Infrastructure (e.g., ports, harbors, airports)	☆☆
Human modification	Diversion of freshwater resources	☆☆☆
	Land fill, solid waste disposal	☆☆
	Eutrophication (sewage and agriculture)	☆☆
	Pollution (chemical, organic and oil spills)	☆
Management failure	Inappropriate planning policies	☆☆☆
	Failure to recognize ecosystem value	☆☆☆
	Illegal land encroachment	☆☆
	Failure of community involvement	☆☆
	Land ownership and access	☆
	Inadequate enforcement	☆

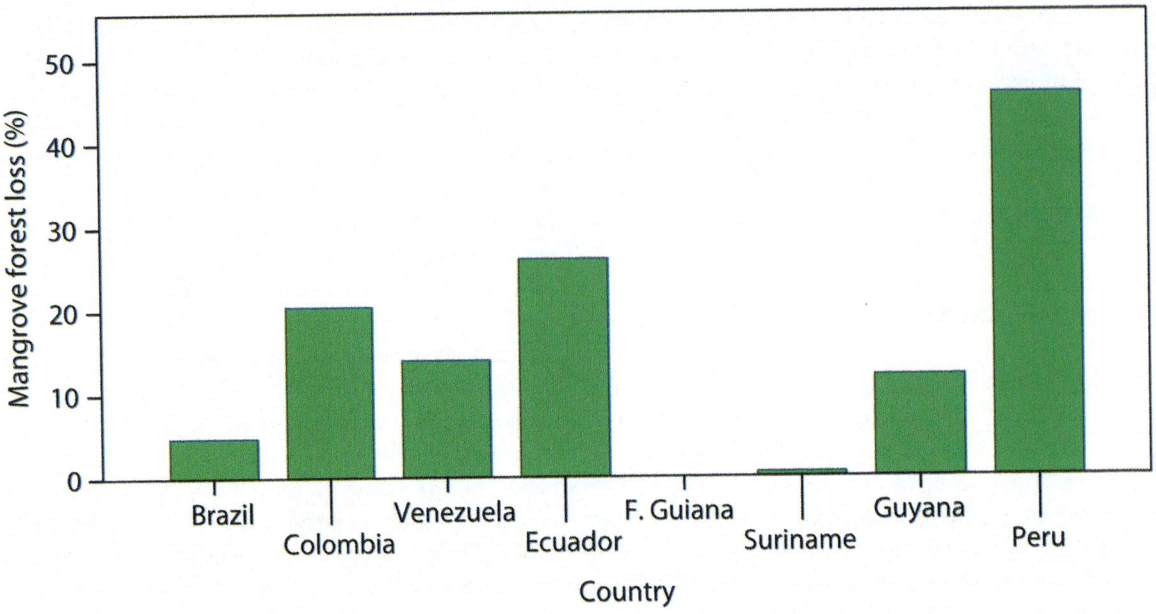

Fig. 8.4 Estimated loss of mangrove forest by South American country. Data from the Food and Agriculture Organization of the United Nations (FAO 2007)

original forests, followed by Ecuador and Columbia, Venezuela, Guyana and Brazil, and finally the lowest rates seen for Suriname and French Guiana.

8.2.3 Resource Exploitation

Although the traditional use of mangrove resources in South America is thought to be less than that which occurred in southeast Asia (Hong and San 1993), expanding coastal populations, particularly following European colonization, led to over-harvesting and a decline in natural resources in certain areas. Exploitation of mangroves for timber, charcoal, bark tannin, animal fodder and traditional medicines have all negatively impacted South American mangrove forests in a similar way to many other regions (Alongi 2002). Beginning in the sixteenth century, mangrove timber was exported from Colombia to Peru and reached in excess of 6000 poles per year by the seventeenth century (Lacerda et al. 1993). Likewise, in 1677 thousands of mangrove poles were exported to Cuba from the Pacific coast of Colombia (Prahl et al. 1990). Commerce in bark was also intense, with Ecuador for example, exporting more than 600 tons per year to neighboring countries between 1879 and 1906 (Lacerda et al. 1993). In Venezuela, extractive logging of *Rhizophora mangle* in the western part of the country for power utility poles, similarly led to marked declines in mangrove population numbers (Hamilton and Snedaker 1984). In Colombia mangrove resources have traditionally been used for construction, firewood, poles, coal, piles, pulp, to obtain tannin, and for fishing gear, cattle pens, etc. (Sánchez-Páez et al. 1997).

8.2.4 Conversion and Development

Conversion of coastal wetlands as a result of economic development and urbanization, has undoubtedly been the greatest driver of mangrove loss in South America over the past 30 years (FAO 2007; Hamilton 2013). This was driven by the imperative for economic opportunities and a need to increase food production for rapidly growing populations. Because they occur in coastal areas that provide direct access to saltwater, forests were regularly converted to provide habitat for shrimp, clams and other seafood products (Fig. 8.5a; Valiela et al. 2001). Nearly 50% of Ecuador's mangroves were converted between 1980 and 2000 as a result of shrimp farm development (Lacerda et al. 2002). In fact, the industry grew to become one of the countries three largest exports (second only to oil and just before bananas). Construction of ponds at first involved reclaiming coastal saltmarsh, but it soon started encroaching on adjacent mangrove forests. The outcome of this devastation soon became manifest in marked declines in the catches of commercially important prawns that directly relied on these mangroves as a nursery (Turner 1977). This matches the ample evidence for a direct relationship between mangrove area and offshore shrimp and fish catches in many regions (e.g., Philippines, Malaysia, Indonesia, and Australia; Primavera 1997). A similar proportion of the Tumbes mangrove forest in Peru was cleared for mariculture production between 1945 and 1985.

Fig. 8.5 Large areas of mangrove forest have been lost across South America because of a multitude of factors including; (**a**) conversion of wetlands for mariculture (photo: João Lara Mesquita, Mar Sem Fim, 2015), (**b**) urban development (photo: Portal da Copa ME, 2013), (**c**) altered hydrography (photo: Aaron Machado, Australian Marine Wildlife Research & Rescue Organisation, 2007), (**d**) anthropogenic inputs (photo: US Department of Agriculture, 2006), (**e**) widespread dumping of industrial and residential waste, and (**f**) entanglement by marine debris (photo: Daniel Gorman, 2017)

More recently in northeastern Brazil, a similar cycle occurred with a boom in the Pacific shrimp *Litopenaeus vannamei* mariculture (Lacerda et al. 2002; Maia et al. 2006) of which ca. 15% of the area is thought to have been converted from mangrove forests (Ferreira and Lacerda 2016). In addition to the direct devastation caused by these operations, there can also be long-term chronic effects for remaining forest caused by the nutrient-rich effluents produced by these activities (Suarez-Abelenda et al. 2014).

In addition to mariculture, there are a range of other activities that have caused substantial losses or degradation of mangrove forests in South America. One of these is the harvesting of salt from coastal impoundments or ponds, which, in addition to direct clearing often increases the salinity of adjacent water bodies (Silva et al. 2003). In Colombia significant growth in cattle ranching, banana farming and an expansion of the logging industry have likewise impacted mangroves, particularly those in the important area of Urabá. There have also been losses through the exploitation of fishing resources and the establishment of coconut crops.

Coastal population growth and expansion of the tourism industry during the 1980's were other significant drivers of forest loss in many parts of South America (Fig. 8.5b; Barbier and Cox 2003). These activities can have both direct effects (i.e., clearing) as well indirect consequences for forest health, brought about by the novel conditions created by humans (e.g., watershed contaminants; Peters et al. 1997). The desire for residential developments (e.g., condominiums), tourism infrastructure (e.g., hotels and marinas) and the development of ports, harbors, airports and coastal mining activities led to significant and ongoing losses of forests across the continent. In Brazil for example, mangrove losses associated with the larger population centers of Rio de Janeiro, Santos and Florianopolis have been considerable. It has been suggested, that of the forests existing around the Ilha Grande Bay (State of Rio de Janeiro, Brazil) in the 1980's, almost 80% were reclaimed for industry, residential developments and marinas (Lacerda and Schaeffer-Novelli 1999).

8.2.5 Human Modification of the Abiotic Environment

Although more subtle and often unambiguous, mangrove ecosystems have regularly suffered the effects of human modification of the abiotic environment (Fig. 8.5c). Changes to the structure and composition of forests have been associated with modifications to tidal and salinity regime, as well as changes to hydrology (i.e., freshwater inputs) and geomorphic processes (Cunha-Lignon et al. 2009; Marchand et al. 2004). Diversion of freshwater for irrigation has been recognised as a major threat to mangroves in a number of countries (e.g., Colombia, Restrepo and Kettner 2012; Peru, Burdette 2017; and Brazil, Schaeffer-Novelli et al. 2016). Mangrove communities are more susceptible to disease and pests when they are stressed by changes to salinity, inundation cycles, sedimentation and soil chemistry. Another related issue has to do with the widespread alteration of river flow through the construction of dams. In South America, 22 of the 38 large river systems currently dammed are considered to be either strongly or moderately effected (Nilsson et al. 2005), with the resulting changes to the downstream transport of sediments and freshwater likely to affect mangrove communities.

Other drivers can be more subtle, indirect and sub-lethal. For instance, the encroachment of human populations into coastal areas typically increases the wastes dumped into mangroves and adjacent coastal waterways (Figs. 8.5d and e; Mani-Peres et al. 2016). Although mangroves can assimilate some excess nutrients, this capacity is unknown for most systems and is likely to depend on the nature of discharge and the characteristics of the receiving system (Trott and Alongi 2000). Mangrove plants exhibit reduced growth when exposed to dissolved heavy metals, particularly when concentrations greatly exceed those of pristine mangrove soils (Yim and Tam 1999). In Colombia, large mangrove losses have been directly attributed to contamination of water with domestic and industrial wastes and in particular accidental spills of hydrocarbons (Guevara-Mancera et al. 1998). Studies of oil spills in the Caribbean have shown that mangroves exhibit increased mutation rates and long (approximately 20 years) recovery times after repeated exposure (Burns et al. 1993; Klekowski et al. 1994). Along the Colombian-Ecuadorian border, more than 10,000 people are engaged in gold prospecting, an activity that threatens downstream mangrove systems with mercury pollution, changes to sediment transport and freshwater flow.

Another emerging threat for the survival of mangrove forests, is the increasing volume of marine debris that enters coastal wetlands (Debrot et al. 2013). The risks posed by these materials are twofold, firstly entanglement by items such as plastic bags can lead to the mortality of juvenile plants (Fig. 8.5f; Gorman and Turra 2016) and secondly, accumulation on and around pneumatophores can suffocate adult trees. In this way, marine debris can act to reduce the health of existing forests while at the same time reducing the degree of successful recruitment. Given the enormous scale of this emerging problem, it is likely that the full implications of marine debris for mangrove forests are yet to be realized (Derraik 2002).

The cumulative effects of urbanization and changes to the abiotic environment can have major consequences for coastal wetlands. The mangroves of Guanabara Bay (Rio de Janeiro) for example, which are thought to have exceeded 50 km^2 at the beginning of the century, have been largely degraded through the modification of creeks and river banks, oil spills, solid waste dumping and decreased freshwater inputs, so that now less than 15 km^2 can be considered as relatively pristine. Along the coastline of São Paulo (Brazil) remnant mangrove stands have been heavily impacted through land reclamation, solid waste disposal, industrial and domestic effluents, chemical, organic contamination and oil spills from nearby ports and oil terminals (see the following case study of Araçá Bay).

8.2.6 Management Failure

The degradation of many coastal ecosystems has been exacerbated by the slow recognition that they need and deserve to be actively conserved (Lotze et al. 2006). Economic imperatives have historically outweighed consideration of the full suite of services provided by healthy wetlands and mangroves (IUCN 2016). Because of this, large and often inappropriate developments have regularly been approved across much of South America, resulting in substantial forest areas being destroyed, degraded or fragmented. Indeed, there is a direct negative relationship between areas of remaining forests and the level of primary sector (i.e., agricultural) activity (Barbier and Cox 2003). Most of the estimated 50,000 ha of forest cleared in Brazil over the past 25 years occurred along the more populous southern coastline, despite the greater areas of mangrove occurring in the north (Rovai et al. 2012). The severity of loss is also a function of the level of environmental protection, the accessibility of mangrove areas, and other institutional considerations (such as political stability), that vary widely both within as well as across countries.

An example of inept management in Brazil is given by the case study of Araçá Bay; a small but diverse coastal inlet located on the developed coastline of São Paulo (for a complete background see: Amaral et al. 2010). The changes to this bay clearly show how the cumulative effects of inappropriate development, pollution and socio-economic factors can lead to the degradation and possible extinction of mangrove ecosystems in certain areas. The bay, which lies in the shadow of Latin America's largest petroleum shipping terminal, contains some of the last stands of mangroves along this section of subtropical coastline (Amaral et al. 2016). Economic imperatives led to the reclamation of ca. 50% of the original bay for the construction of a port facility (beginning circa 1936; Francisco and Carvalho 2003). In addition to the direct effects on habitats, there were also major changes to the hydrodynamics of the area, which altered patterns of sediment accretion and erosion. Population growth necessitated the installation of a sewage outfall in 1990 (Mani-Peres et al. 2016), which rather than being sited offshore to avoid plumes entering the bay (Gorman et al. 2017), was inappropriately placed at its southern entrance. This led to increased inputs of organic contaminants and other pollutants and created persistent problems of eutrophication and anoxia (Gubitoso et al. 2008). These impacts combined with the widespread dumping of rubbish, residential and industrial wastes have resulted in losses of up to 65% of the historical mangrove area present in the 1960's (Fig. 8.6). What is particularly troubling about this case, is that certain sectors of industry and government are still trying to push ahead with further port amplification in the name of 'public good', which given the age of the remaining trees and critically low levels of recruitment, would likely lead to the extinction of mangroves from the area.

8.3 Current and Future Threats

Mangrove ecosystems are expected to face unprecedented challenges over the coming decades (Duke et al. 2007). While perhaps more subtle than some of the historical drivers of change, future threats are likely to be pervasive and highly consequential for species that are tightly coupled to both atmospheric and oceanographic processes (Schaeffer-Novelli et al. 2016). There is broad scientific consensus that continued human development and population growth will alter the global climate over the next hundred years, with major consequences for mangrove ecosystems, including those of South America (Ward et al. 2016). Large-scale phenomena such as sea-level rise, global warming, increased frequency of hurricanes and storms and changes to precipitation patterns, will act in concert with a range of more localized human impacts to further test the ability of mangroves to persist (Table 8.3).

8.3.1 Sea-Level Rise

Sea-level rise is widely considered to be the most serious threat to mangrove ecosystems globally (Field 1995; Gilman et al. 2008). Forests occurring in low lying tropical and sub-tropical regions might be expected to show some of the earliest indications of change. There is concern that areas in the northeast of South America may be susceptible to the anticipated long-term trends in sea-level rise (e.g., notably the Guianas; Anthony and Gratiot 2012). The coastline of Guyana in particular, lies between 0.5 and 1.0 m below high spring tide level and because of historical destruction of the mangrove fringe, now needs to be protected by a continuous 238 km long sea defense system consisting of walls and embankments (Evans 1998). Although the effects of sea-level rise have so far been negligible when compared to some of the historical drivers of loss (outlined previously), it is likely that any effects will intensify over the coming decades (Church and White 2011, but compare Houston and Dean 2011). The risk to mangrove communities is that

Fig. 8.6 Photos showing the extent of mangroves in Araçá Bay during the 1960's (top panel) with polygons highlighting subsequent areas of loss (Cultural Heritage Department Collection São Sebastião), and the present degraded condition of remnant stands (bottom panel) that has resulted from the cumulative effects of multiple human-induced stressors (Daniel Gorman 2017)

Table 8.3 Summary of the current and future threats facing mangrove forests across South America and the world

Category	Agent of change	Threat level
Sea-level rise	Inundation and drowning	☆☆☆
Global warming	Range shifts, altered productivity	☆☆☆
Freshwater diversion	Salinity fluctuations	☆☆
	Reduced sediment replenishment	
Changing climate	Increased frequency of hurricanes and severe storms	☆☆☆
	Changes to precipitation (both increases and decreases)	☆☆
	Flooding and erosion	☆☆
Coastal urbanization	Direct clearing for urban development	☆
	Land conversion for condominiums ('sea change')	☆☆
	Construction of marinas and tourism facilities	☆☆
Economic development	Development of ports, roads and large-scale infrastructure	☆☆
	Mariculture and resource harvesting	☆☆
	Pollution and marine debris	☆☆
Genetic isolation	Loss of genetic diversity	☆
Atmospheric CO_2	Changes in physiology and productivity	☆
Biological agents	Diseases and invasion	☆

because they are typically slow growing, some populations will be unable to adapt to changes in sediment elevation, accretion, erosion, soil composition and compaction (review Gilman et al. 2008). Indeed, the consequences could be considerable, particularly in areas where rising sea level combines with reduced sediment delivery owing to anthropogenic activities such as the damming of rivers (Lovelock et al. 2015). There is recent evidence however, that well-planned restoration efforts can ensure that forests will not only keep pace and persist under medium sea-level rise, but may actually expand to increase the carbon storage capacity of these systems (Krauss et al. 2017).

8.3.2 Global Warming

Because the distribution of mangroves along both sides of the South American continent is strongly influenced by climate and temperature (e.g., Soares et al. 2012), the consequences of global warming may have unpredictable effects for these systems. While mangroves are not expected to be adversely affected by changes to sea surface temperature (Field 1995), this may be location and species specific (depending on local currents and upwelling). In terms of air temperature, most species demonstrate increased rates of productivity up to temperatures of 25 °C, so it is possible that forests in some regions may actually benefit from warming. Prolonged exposure to higher temperatures however, can lead to lower rates of leaf formation and the cessation of photosynthetic activity. An interesting aspect of global warming with respect mangroves in South America, is the likelihood of a southern extension of the current distribution limit. This links with an increasing body of work that implicates warming ocean currents, and the consequent tropicalization of temperate seas, as a driver of species range-shifts in the marine realm (Molinos et al. 2017). Warming of temperate parts of the Brazilian and Peruvian coastlines could facilitate the southward migration of mangroves, though this will depend on a number of additional factors.

8.3.3 Extreme Weather Phenomenon

Storms and extreme weather events can significantly influence the productivity and health of mangrove ecosystems. Indeed, large storms have resulted in mass mortality in ten Caribbean mangrove forests over the last 50 years (Jimenez et al. 1985). While mangroves are typically resilient to regular storm events (Alongi 2002), any increase in the frequency and/or magnitude of extreme events (as predicted by the IPCC 2014) would be likely to magnify the scale of direct storm damage (e.g., defoliation and tree mortality; Gilman et al. 2008) and exacerbate flooding, coastal erosion, saline intrusion and changes to sediment dynamics. Similarly, changes to precipitation patterns may have additional and regionally variable consequences. Decreased precipitation can restrict productivity, decrease growth and seedling survival and favor regime-shifts to more salt tolerant species (Ellison 2000). In contrast, increased precipitation, as has been predicted for large parts of the tropical north of South America (Barreiro 2010), may increase mangrove area, diversity and growth rates (Field 1995).

8.3.4 Continued Exploitation

Mangrove resources in South America will continue to be exploited both directly and indirectly. In addition to the political and scientific dimensions, there are the social issues linked to good or bad catchment management that will influence conservation outcomes (Barletta et al. 2010). The levels of social and economic development of countries and their different regions are a key determinant of wetland health and conservation options. Many coastal communities still rely on mangrove resources for their livelihood, collecting timber and wood for fuel in many parts of Brazil, Colombia, Ecuador, Guyana, Peru and Venezuela. Mangrove forests will also need to incorporate new and changing uses. A recent example, is the expansion of a new method of harvesting crabs in Brazil called *redinha*, which is causing renewed damage to mangrove forests (Nascimento et al. 2011). The approach involves cutting a section of mangrove root, which is used one time to set a snare. The increasing economic return on crabs means that this incremental damage to mangroves is certain to continue. On a larger-scale, are the challenges of providing clean power generation and sufficient food for the populations of South American countries. Continued clearance of coastal wetland for agricultural production and changes to hydrological cycles for power generation are likely to have major effects. The proposed construction of 30 major dams in the Brazilian Amazon alone (Fearnside 2014) raises the likelihood of major impacts on the extensive areas of downstream mangrove forest in the Amazon delta (Latrubesse et al. 2017).

8.3.5 Other Factors

There are a range of other factors likely to influence the sustainability of mangrove forests in unpredictable ways. The combined effects of enhanced atmospheric CO_2 and increased nutrient availability on the growth of mangroves is still largely unknown (but see, Reef et al. 2016), although there is some evidence that not all species will respond in a similar way, which could lead to changes in species composition. Another often overlooked aspect of mangrove health relates to genetic diversity and its implications for the ability of forests to cope with future environmental change. Understanding how genetic diversity varies for populations at their current distributional range limit can provide valuable insights into how species may contend with changing environmental and demographic factors (Arnaud-Haond et al. 2006). Genetic connectivity can also be a means of assessing re-colonization pathways from refuge populations (Kennedy et al. 2016) and can therefore help to inform management about the likelihood of dispersal. Although disease and other biotic factors have not historically been major causes of mangrove loss (Jimenez et al. 1985) they may become factors that can effect forests that have been weakened by changes in the physical environment.

8.4 Managing Forests for the Future

Given the diversity of threats facing mangrove ecosystems across South America, conservation will become an increasingly important and pressing issue. Economic and social pressures are likely to intensify as the continent becomes more integrated with the global economy. Safeguarding remaining forests will require a range of responses from governments and local communities (Ward et al. 2016). Priority should be given to improving scientific understanding to better anticipate current and future trends and to undertake programs to reverse past losses and build resilience. There is also a growing impetus for 'blue carbon' initiatives which have the goal of preventing further CO_2 emissions (e.g., Brazilian mangroves have the fourth largest potential for emissions under senarios of continued deforestation; Atwood et al. 2017), while at the same time protecting and restoring these habitats to take advantage of their huge potential to sequester carbon in the context of combating climate change (Mcleod et al. 2011).

There remains a clear lack of up-to-date information, in some cases at the level of entire countries (e.g., Venezuela), describing the distribution and status of forests. Typically change indicators are presented at a continental scale (e.g., Spalding et al. 1997) or focused on specific countries (e.g., Brazil as reported in; Giri et al. 2011) with fewer assessments from some of the other individual countries (other than that of; FAO 2007). Indeed, comparing past estimates of mangrove covers across South America with those derived using recent modeling approaches (Hutchison et al. 2014) reveal major inconsistencies for certain countries (e.g., both significant overestimates as well as underestimates). Concern over such uncertainty is warranted if the estimates are correct that South America harbors ca.11% of the globes remaining forests. In this context, remote sensing may provide the spatio-temporal information necessary for regular assessments of the distribution, status and any changes to mangrove populations (Kuenzer et al. 2011; Nascimento et al. 2013).

Another requirement is the improved integration of data across institutional, national and international boundaries. Much of what is known remains held by governments and universities and dissemination has not kept pace with the needs of conservation and management (Macintosh and Ashton 2002). In particular, there is a lack of understanding about the ecology, biology and taxonomy of wetland ecosystems in the more remote Amazon basins despite these areas being acknowledged as strongholds for mangroves both at a continental and global scale. There is a need for immediate efforts to continue fostering international collaborations between scientists, engineers, economists and social scientists to identify priority issues and strategies.

8.4.1 Building Resilience and Resistance

One effective means of addressing past losses of mangrove ecosystems is through the active restoration of areas that were previously degraded (Kaly and Jones 1998). This is particularly relevant for the large expanses of coastal wetland that were degraded through mariculture activities and which now lay abandoned. While self-recuperation of mangrove habitats can occur simply by preventing new impacts to allow natural recovery (Field 1996), outcomes can be improved substantially when supported by active measures such as hydrological restoration (Lewis and Gilmore 2007). Guyana represents a good example of successful restoration initiatives in South America (Anthony and Gratiot 2012). The successful collaboration between the Guyana Mangrove Restoration Project

Fig. 8.7 Restoration and augmentation of mangrove habitats can help to build the resilience of wetlands threatened by local and global change. Upper plate shows the successful outcome of a large-scale *Avicennia germinans* revegetation program along the coastline of Guyana (photo: Ian Brierley, CATS, 2012), lower plate shows small-scale revegetation plots that can restore the species richness and abundance of mangrove associated faunal communities over relatively short periods (Photo: *Avicennia schaueriana* revegetation plots in Araçá Bay, Daniel Gorman, 2015)

and the European Union led to the planting between 2010 and 2012 of more than 330,000 *Avicennia germinans* seedlings across nearly 6 km of previously exposed coastline (Fig. 8.7 upper plate; Machin, 2013). In Brazil, similar mangrove restoration efforts have mitigated some of the historical losses that occured over the last three decades (ca. 5% Ferreira and Lacerda 2016). Although revegetation initiatives are often done primarily for the purposes of shoreline fortification (Gedan et al. 2011), they have also been shown as an effective means of restoring the diversity and abundance of associated faunal communities over relatively short periods (Fig. 8.7 lower plate; Gorman and Turra 2016).

There is a clear imperative to build the resilience of mangrove ecosystems to future environmental change (Mcleod and Salm 2006). Adaption measures can help to offset anticipated losses of forests and improve their resistance and resilience to both local and global change (Gilman et al. 2008). For example, mangroves are likely to be able to adapt to sea-level rise if it occurs slowly enough (Ellison and Stoddart 1991) and if there is available space for inland expansion. This will require the setting aside of land-ward buffer zones free of other activities to permit the unhindered migration of mangroves under such scenarios. Efforts will also be required to ensure sediment delivery that is sufficient to maintain rates of vertical accretion and permit the accumulation of organic matter which plays a crucial role in the surface elevation gains required for successful adaptation (Lovelock et al. 2015).

Another hope for the continued persistence of mangroves in South America is that some populations will expand southward into higher latitudes. There is evidence that certain species are already proliferating and extended the their pole-ward range limit on at least five continents (Saintilan et al. 2014). In Brazil, studies on the thermal tolerance of *Avicennia germinans* suggest that this species could expand southwards under established global warming scenarios (Soares et al. 2012). Given the long and continuous coastline on both the western (Pacific) and eastern (Atlantic) sides of South America, mangrove forests could expand considerably given suitable estuarine conditions. The introduction of mangrove forests into areas where they did not previously occur may be an effective, or even necessary way of compensating for ongoing losses within other parts of their current distribution.

Fig. 8.8 Given the multitude of threats facing South Americas mangrove ecosystems; effective governance and community engagement will be vital to ensure the continued provision of ecosystem services by these valuable natural ecosystems (Photo: Daniel Gorman)

8.5 Conclusions

The last century has been witness to a large but ultimately unknowable loss of mangrove forests across the continent of South America. The losses that can be reliably estimated (i.e., since the 1980's) suggest that much of this was driven by the intensified use of coastal zones by humans in terms of resource extraction (fuel and wood), competition for space (aquaculture and coastal development) and changes to the abiotic characteristics of coastal wetlands (e.g., sedimentation, eutrophication, and pollution). The failure of management to acknowledge these losses along with a preoccupation with economic development has led to major loss, degradation and fragmentation of these importance ecosystems to the detriment of the many ecosystem services they provide (e.g., shoreline protection, nurseries for commercial species, etc.). To conclude on a positive note, the large areas of mangrove forest (particularly in the North of Brazil and the Guianas) and their remoteness offer substantial conservation opportunities. There is a clear imperative for immediate human action to safeguard South American mangrove systems so that future generations are not deprived of the services offered by these unique and important environments (Duke et al. 2007) (Fig. 8.8).

References

Alongi DM (2002) Present state and future of the world's mangrove forests. Environ Conserv 29:331–349

Alongi DM (2008) Mangrove forests: resilience, protection from tsunamis, and responses to global climate change. Estuar Coast Shelf Sci 76:1–13

Alongi DM (2012) Carbon sequestration in mangrove forests. Carbon Manag 3:313–322

Amaral ACZ, Migotto AE, Turra A, Schaeffer-Novelli Y (2010) Araçá: biodiversity, impacts and threats. Biota Neotrop 10:219–264

Amaral ACZ, Turra A, Ciotti AM, Wongtschowski CLDBR, Schaeffer-Novelli Y (2016) Life in Araçá Bay: diversity and importance. Lume, São Paulo

Anthony EJ, Gratiot N (2012) Coastal engineering and large-scale mangrove destruction in Guyana, South America: averting an environmental catastrophe in the making. Ecol Eng 47:268–273

Arnaud-Haond S, Teixeira S, Massa SI, Billot C, Saenger P, Coupland G, Duarte CM, Serrao EA (2006) Genetic structure at range edge: low diversity and high inbreeding in Southeast Asian mangrove (*Avicennia marina*) populations. Mol Ecol 15:3515–3525

Atwood TB, Connolly RM, Almahasheer H, Carnell PE, Duarte CM, Lewis CJE, Irigoien X, Kelleway JJ, Lavery PS, Macreadie PI, Serrano O, Sanders CJ, Santos I, Steven ADL, Lovelock CE (2017) Global patterns in mangrove soil carbon stocks and losses. Nat Clim Chang 7:523–528

Barbier EB, Cox M (2003) Does economic development lead to mangrove loss? A cross-country analysis. Contemp Econ Policy 21:418–432

Barletta M, Jaureguizar AJ, Baigun C, Fontoura NF, Agostinho AA, Almeida-Val VMF, Val AL, Torres RA, Jimenes-Segura LF, Giarrizzo T, Fabre NN, Batista VS, Lasso C, Taphorn DC, Costa MF, Chaves PT, Vieira JP, Correa MFM (2010) Fish and aquatic habitat conservation in South America: a continental overview with emphasis on neotropical systems. J Fish Biol 76:2118–2176

Barreiro M (2010) Influence of ENSO and the South Atlantic Ocean on climate predictability over Southeastern South America. Clim Dyn 35:1493–1508

Bernal G, Osorio AF, Urrego L, Pelaez D, Molina E, Zea S, Montoya RD, Villegas N (2016) Occurrence of energetic extreme oceanic events in the Colombian Caribbean coasts and some approaches to assess their impact on ecosystems. J Mar Syst 164:85–100

Blanco JF, Estrada EA, Ortiz LF, Urrego LE (2012) Ecosystem-wide impacts of deforestation in mangroves: the Urabá Gulf (Colombian Caribbean) case study. Int Sch Res Netw ISRN Ecol 2012:1–14

Burdette C (2017) Western South America: Northwestern coast of Peru [Online]. WWF. Available: https://www.worldwildlife.org/ecoregions/nt1429

Burns KA, Garrity SD, Levings SC (1993) How many years until mangrove ecosystems recover from catastrophic oil spills. Mar Pollut Bull 26:239–248

Church JA, White NJ (2011) Sea-level rise from the late 19th to the early 21st century. Surv Geophys 32:585–602

CIA (2013) The world factbook 2013–14 [Online]. Central Intelligence Agency, Washington, DC. Available: https://www.cia.gov/library/publications/the-worldfactbook/index.html

Connell SD, Russell BD, Turner DJ, Shepherd SA, Kildea T, Miller DJ, Airoldi L, Cheshire A (2008) Recovering a lost baseline: missing kelp forests on a metropolitan coast. Mar Ecol Prog Ser 360:63–72

Cunha-Lignon M, Coelho C, Almeida R, Menghini R, Correa F, Schaeffer-Novelli Y, Cintron-Molero G, Dahdouh-Guebas F (2009) Mangrove forests and sedimentary processes on the South of Coast of São Paulo State (Brazil). J Coast Res 56:405–409

Debrot AO, Meesters HWG, Bron PS, De Leon R (2013) Marine debris in mangroves and on the seabed: largely-neglected litter problems. Mar Pollut Bull 72:1–1

Derraik JGB (2002) The pollution of the marine environment by plastic debris: a review. Mar Pollut Bull 44:842–852

Duke NC, Meynecke JO, Dittmann S, Ellison AM, Anger K, Berger U, Cannicci S, Diele K, Ewel KC, Field CD, Koedam N, Lee SY, Marchand C, Nordhaus I, Dahdouh-Guebas F (2007) A world without mangroves? Science 317:41–42

Ellison AM (2000) Mangrove restoration: do we know enough? Restor Ecol 8:219–229

Ellison AM, Farnsworth EJ (1996) Anthropogenic disturbance of Caribbean mangrove ecosystems: past impacts, present trends, and future predictions. Biotropica 28:549–565

Ellison JC, Stoddart DR (1991) Mangrove ecosystem collapse during predicted sea-level rise – Holocene analogs and implications. J Coast Res 7:151–165

Evans IJ (1998) The restoration of mangrove vegetation along the coastal belt of Guyana. Master of science. University of Aberdeen, Aberdeen

Everard M, Jha RRS, Russell S (2014) The benefits of fringing mangrove systems to Mumbai. Aquat Conserv Mar Freshwat Ecosyst 24:256–274

FAO (2007). Food and agriculture organization of the United Nations. The World's Mangroves 1980–2005. FAO Forestry Paper, Rome

Fearnside PM (2014) Impacts of Brazil's Madeira River Dams: unlearned lessons for hydroelectric development in Amazonia. Environ Sci Pol 38:164–172

Ferreira AC, Lacerda LD (2016) Degradation and conservation of Brazilian mangroves, status and perspectives. Ocean Coast Manag 125:38–46

Field CD (1995) Impact of expected climate change on mangroves. Hydrobiologia 295:75–81

Field C (1996) La restauracion de ecosistemas de manglar. International Tropical Timber Organization and International Society for Mangrove Ecosystems, Okinawa

Francisco J, Carvalho PF (2003) Desconstrução do lugar: o aterro da praia da frente do centro historico de São Sebastião (SP). UNESP, Rio Claro

Gedan KB, Kirwan ML, Wolanski E, Barbier EB, Silliman BR (2011) The present and future role of coastal wetland vegetation in protecting shorelines: answering recent challenges to the paradigm. Clim Chang 106:7–29

Gilman EL, Ellison J, Duke NC, Field C (2008) Threats to mangroves from climate change and adaptation options: a review. Aquat Bot 89:237–250

Giri C, Ochieng E, Tieszen LL, Zhu Z, Singh A, Loveland T, Masek J, Duke N (2011) Status and distribution of mangrove forests of the world using earth observation satellite data. Glob Ecol Biogeogr 20:154–159

Golley FB, Odum HT, Wilson RF (1962) The structure and metabolism of a Puerto Rican red mangrove forest in May. Ecology 43:9–19

Gorman D, Turra A (2016) The role of mangrove revegetation as a means of restoring macrofaunal community structure along subtropical coasts. Sci Total Environ 566:223–229

Gorman D, Turra A, Connolly RM, Olds AD, Schlacher TA (2017) Monitoring nitrogen pollution in seasonally-pulsed coastal waters requires judicious choice of indicator species. Mar Pollut Bull 122:149–155

Grimm AM, Tedeschi RG (2009) ENSO and extreme rainfall events in South America. J Climate 22:1589–1609

Gubitoso S, Duleba W, Teodoro AC, Prada SM, Rocha MM, Lamparelli CC, Bevilacqua JE, Moura DO (2008) Estudo geoambiental da região circunjacente ao emissário submarino de esgoto do Araçá, São Sebastião, SP. Rev Bras Geocienc 38:467–475

Guevara-Mancera OA, Sánchez-Páez H, Murcia-Orjuela GO, Bravo-Pazmiño HE, Pinto-Nolla F, Alvarez-León R (1998) Conservación y Uso Sostenible de los Manglares del Pacífico colombiano. Proyecto 171/91 Rev. 2 (F) Fase II (Etapa I) "Conservación y Manejo para el Uso Múltiple y Desarrollo de los Manglares en Colombia". MMA/OIMT/ACOFORE, Santa Fe de Bogotá D.C, Colombia

Hamilton S (2013) Assessing the role of commercial aquaculture in displacing mangrove forest. Bull Mar Sci 89:585–601

Hamilton LS, Snedaker SC (1984) Chapter one - overview. In: Hamilton LS, Snedaker SC (eds) Handbook for mangrove area management. United Nations Environment Programme and East-West Center, Honolulu

Hong PN, San TH (1993) Mangroves of Vietnam. International Union for the Conservation of Nature (IUCN), Regional Wetlands Office, Gland, p 173

Houston JR, Dean RG (2011) Sea-level acceleration based on US tide gauges and extensions of previous global-gauge analyses. J Coast Res 27:409–417

Hutchison J, Manica A, Swetnam R, Balmford A, Spalding M (2014) Predicting global patterns in mangrove forest biomass. Conserv Lett 7:233–240

IPCC (2014) Climate change 2014: impacts, adaptation, and vulnerability. Part A: global and sectoral aspects. In: CB F, Barros VR, Dokken DJ, Mach KJ, Mastrandrea MD, Bilir TE, Chatterjee M, Ebi KL, Estrada YO, Genova RC, Girma B, Kissel ES, Levy AN, Maccracken S, Mastrandrea PR, White LL (eds) Contribution of working group II to the fifth assessment report of the intergovernmental panel on climate change. Cambridge University Press, Cambridge/New York

IUCN (2016) The first global integrated marine assessment. World ocean assessment I. In: Webber M, Calumpong H, Ferreira B, Granek E, Green S, Ruwa R, Soares M (eds) Chapter 48 Mangroves. Cambridge University Press, Cambridge

Jimenez JA, Lugo AE, Cintron G (1985) Tree mortality in mangrove forests. Biotropica 17:177–185

Kaly UL, Jones GP (1998) Mangrove restoration: a potential tool for coastal management in tropical developing countries. Ambio 27:656–661

Kennedy JP, Pil MW, Proffitt CE, Boeger WA, Stanford AM, Devlin DJ (2016) Postglacial expansion pathways of red mangrove, *Rhizophora mangle*, in the Caribbean Basin and Florida. Am J Bot 103:260–276

Klekowski EJ, Corredor JE, Morrell JM, Delcastillo CA (1994) Petroleum pollution and mutation in mangroves. Mar Pollut Bull 28:166–169

Krauss KW, Cormier N, Osland MJ, Kirwan ML, Stagg CL, Nestlerode JA, Russell MJ, From AS, Spivak AC, Dantin DD, Harvey JE, Almario AE (2017) Created mangrove wetlands store belowground carbon and surface elevation change enables them to adjust to sea-level rise. Sci Rep 7:11

Kuenzer C, Bluemel A, Gebhardt S, Quoc TV, Dech S (2011) Remote sensing of mangrove ecosystems: a review. Remote Sens 3:878–928

Lacerda LD, Schaeffer-Novelli Y (1999) Mangroves of Latin America: the need for conservation and sustainable utilization. In: Lara-Domínquez AL and Yáñez-Arancibia A (ed) Ecosistemas de Manglar en América Tropical 1999. Instituto de Ecología A.C/ UICN/ORMA/NOAA/NMFS Silver Spring, México/Costa Rica/ MD USA, p 380

Lacerda L, Conde J, Alarcon C, Alvarez-León R, Bacon P, D'croz L, Kjerfve B, Polaina J, Vannucci M (1993) Mangrove ecosystems of Latin America and the Caribbean: a summary. The United Nations Environment Programme, Nairobi

Lacerda LD, Kremer HH, Kjerfve B, Salomons W, Marshall-Crossland JI, Crossland JC (2002) South American Basins: LOICZ Global Change Assessment and Synthesis of River Catchment, Coastal Sea Interaction and Human Dimensions. LOICZ Reports & Studies No. 21. LOICZ International Project Office

Latrubesse EM, Arima EY, Dunne T, Park E, Baker VR, D'horta FM, Wight C, Wittmann F, Zuanon J, Baker PA, Ribas CC, Norgaard RB, Filizola N, Ansar A, Flyvbjerg B, Stevaux JC (2017) Damming the rivers of the Amazon basin. Nature 546:363–369

Lee SY, Primavera JH, Dahdouh-Guebas F, Mckee K, Bosire JO, Cannicci S, Diele K, Fromard F, Koedam N, Marchand C, Mendelssohn I, Mukherjee N, Record S (2014) Ecological role and services of tropical mangrove ecosystems: a reassessment. Glob Ecol Biogeogr 23:726–743

Lewis RR, Gilmore RG (2007) Important considerations to achieve successful mangrove forest restoration with optimum fish habitat. Bull Mar Sci 80:823–837

Lotze HK, Lenihan HS, Bourque BJ, Bradbury RH, Cooke RG, Kay MC, Kidwell SM, Kirby MX, Peterson CH, Jackson JBC (2006) Depletion, degradation, and recovery potential of estuaries and coastal seas. Science 312:1806–1809

Lovelock CE, Cahoon DR, Friess DA, Guntenspergen GR, Krauss KW, Reef R, Rogers K, Saunders ML, Sidik F, Swales A, Saintilan N, Thuyen LX, Triet T (2015) The vulnerability of Indo-Pacific mangrove forests to sea-level rise. Nature 526:559–U217

Lugo AE, Snedaker SC (1974) The ecology of mangroves. Annu Rev Ecol Syst 5:39–64

Machin J (2013) Restoring mangroves in a challenging environment, Guyana. 1st Guyana mangrove forum: restoring and managing mangrove ecosystems in a changing world, 2013 Georgetown, Guyana

Macintosh DJ, Ashton EC (2002) Review of mangrove biodiversity conservation and management. Centre for tropical ecosystems research. University of Aarhus, Aarhus

Maia LP, Lacerda LD, Monteiro LHU, Souza GM (2006) Atlas dos Manguezais do Nordeste do Brasil. SEMACE, Fortaleza

Mani-Peres C, Xavier LY, Santos CR, Turra A (2016) Stakeholders perceptions of local environmental changes as a tool for impact assessment in coastal zones. Ocean Coast Manag 119:135–145

Marchand C, Baltzer F, Lallier-Verges E, Alberic P (2004) Pore-water chemistry in mangrove sediments: relationship with species composition and developmental stages (French Guiana). Mar Geol 208:361–381

Mcleod E, Salm RV (2006) Managing mangroves for resilience to climate change. IUCN, Gland

Mcleod E, Chmura GL, Bouillon S, Salm R, Bjork M, Duarte CM, Lovelock CE, Schlesinger WH, Silliman BR (2011) A blueprint for blue carbon: toward an improved understanding of the role of vegetated coastal habitats in sequestering CO. Front Ecol Environ 9:552–560

Miloslavich P, Klein E, Yerena E, Martin A (2003) Marine biodiversity in Venezuela: Status and perspectives. Gayana (Concepc) 67:275–301

Miloslavich P, Diaz JM, Klein E, Jose Alvarado J, Diaz C, Gobin J, Escobar-Briones E, Cruz-Motta JJ, Weil E, Cortes J, Bastidas AC, Robertson R, Zapata F, Martin A, Castillo J, Kazandjian A, Ortiz M (2010) Marine biodiversity in the Caribbean: regional estimates and distribution patterns. PLoS One 5:25

Molinos JG, Burrows MT, Poloczanska ES (2017) Ocean currents modify the coupling between climate change and biogeographical shifts. Sci Rep 7:9

Nagelkerken I, Blaber SJM, Bouillon S, Green P, Haywood M, Kirton LG, Meynecke JO, Pawlik J, Penrose HM, Sasekumar A, Somerfield PJ (2008) The habitat function of mangroves for terrestrial and marine fauna: a review. Aquat Bot 89:155–185

Naidoo G (1983) Effects of flooding on leaf water potential and stomatal-resistance in *Bruguiera gymnorrhiza* (l) lam. New Phytol 93:369–376

Nascimento DM, Mourão JDS, Alves RRN (2011) A substituição das técnicas tradicionais de captura do caranguejo-uçá (*Ucides cordatus*) pela técnica "redinha" no estuário do rio Mamanguape, Paraíba. Sitientibus série Ciências Biológicas 11:113–119

Nascimento WR, Souza PWM, Proisy C, Lucas RM, Rosenqvist A (2013) Mapping changes in the largest continuous Amazonian mangrove belt using object-based classification of multisensor satellite imagery. Estuar Coast Shelf Sci 117:83–93

Nilsson C, Reidy CA, Dynesius M, Revenga C (2005) Fragmentation and flow regulation of the world's large river systems. Science 308:405–408

Odum WE (1971) Pathways of energy flow in a south Florida estuary. University of Miami Sea Grant Program, Miami

Peters EC, Gassman NJ, Firman JC, Richmond RH, Power EA (1997) Ecotoxicology of tropical marine ecosystems. Environ Toxicol Chem 16:12–40

Phillips JD (1999) Earth surface systems: complexity, order and scale. Blackwell Publishers, New York

Polidoro BA, Carpenter KE, Collins L, Duke NC, Ellison AM, Ellison JC, Farnsworth EJ, Fernando ES, Kathiresan K, Koedam NE, Livingstone SR, Miyagi T, Moore GE, Vien Ngoc N, Ong JE, Primavera JH, Salmo SG III, Sanciangco JC, Sukardjo S, Wang Y, Yong JWH (2010) The loss of species: mangrove extinction risk and geographic areas of global concern. PLoS One 5:1–10

Prahl HV, Cantera JR, Contreras R (1990) Manglares y hombres del Pacífico colombiano. Fondo FEN Colombia, Bogotá

Primavera JH (1997) Socio-economic impacts of shrimp culture. Aquac Res 28:815–827

Reef R, Slot M, Motro U, Motro M, Motro Y, Adame MF, Garcia M, Aranda J, Lovelock CE, Winter K (2016) The effects of CO and nutrient fertilisation on the growth and temperature response of the mangrove *Avicennia germinans*. Photosynth Res 129:159–170

Restrepo JD, Kettner A (2012) Human induced discharge diversion in a tropical delta and its environmental implications: the Patia River, Colombia. J Hydrol 424:124–142

Rovai AS, Menghini RP, Schaeffer-Novelli Y, Molero GC, Coelho C Jr (2012) Protecting Brazil's coastal wetlands. Science 335:1570–1572

Saintilan N, Wilson NC, Rogers K, Rajkaran A, Krauss KW (2014) Mangrove expansion and salt marsh decline at mangrove poleward limits. Glob Chang Biol 20:147–157

Sánchez-Páez H, Alvarez-León R, Guevara-Mancera OA, Zamora-Guzmán A, Rodríguez-Cruz H, Bravo-Pazmiño HE (1997) Diagnóstico y zonificación preliminar de los manglares del Pacífico de Colombia. Proyecto PD 171/91. Rev 2 (F). Fase I "Conservación y Manejo para el Uso Múltiple y el Desarrollo de los Manglares en Colombia". MMA/OIMT, Santa Fe de Bogotá

Schaeffer-Novelli Y, Soriano-Sierra EJ, Do Vale CC, Bernini E, Rovai AS, Pinheiro M a A, Schmidt AJ, De Almeida R, Coelho C, Menghini RP, Martinez DI, Abuchahla GMD, Cunha-Lignon M, Charlier-Sarubo S, Shirazawa-Freitas J, Cintron-Molero G (2016) Climate changes in mangrove forests and salt marshes. Braz J Oceanogr 64:16

Silva CAR, Rainbow PS, Smith BD (2003) Biomonitoring of trace metal contamination in mangrove-lined Brazilian coastal systems using the oyster Crassostrea rhizophorae: comparative study of regions affected by oil, salt pond and shrimp farming activities. Hydrobiologia 501:199–206

Smith TJ, Robblee MB, Wanless HR, Doyle TW (1994) Mangroves, hurricanes, and lightning strikes. Bioscience 44:256–262

Soares MLG, Estrada GCD, Fernandez V, Tognella MMP (2012) Southern limit of the Western South Atlantic mangroves: assessment of the potential effects of global warming from a biogeographical perspective. Estuar Coast Shelf Sci 101:44–53

Spalding M, Blasco F, Field C (1997) World mangrove atlas. The International Society for Mangrove Ecosystems, Okinawa

Suarez-Abelenda M, Ferreira TO, Camps-Arbestain M, Rivera-Monroy VH, Macias F, Nobrega GN, Otero XL (2014) The effect of nutrient-rich effluents from shrimp farming on mangrove soil carbon storage and geochemistry under semi-arid climate conditions in northern Brazil. Geoderma 213:551–559

Tomlinson PB (1986) The botany of mangroves. Cambridge University Press, New York

Trott LA, Alongi DM (2000) The impact of shrimp pond effluent on water quality and phytoplankton biomass in a tropical mangrove estuary. Mar Pollut Bull 40:947–951

Turner RE (1977) Intertidal vegetation and commercial yields of penaeid shrimp. Trans Am Fish Soc 106:411–416

Valiela I, Bowen JL, York JK (2001) Mangrove forests: one of the world's threatened major tropical environments. Bioscience 51:807–815

Ward RD, Friess DA, Day RH, Mackenzie RA (2016) Impacts of climate change on mangrove ecosystems: a region by region overview. Ecosyst Health Sustain 2(4):e01211

Yanagisawa H, Koshimura S, Miyagi T, Imamura F (2010) Tsunami damage reduction performance of a mangrove forest in Banda Aceh, Indonesia inferred from field data and a numerical model. J Geophys Res Oceans 115:11

Yim MW, Tam NFY (1999) Effects of wastewater-borne heavy metals on mangrove plants and soil microbial activities. Mar Pollut Bull 39:179–186

9 Australian Mangroves: Anthropogenic Impacts by Industry, Agriculture, Ports, and Urbanisation

V. Semeniuk and I. D. Cresswell

Abstract

Australia has an extensively mangrove-lined coast and even though it has a relatively low human population compared to the length of its coastline, there have been significant impacts on its mangroves. Belonging to the Indo-Malesian Group of the Old World Mangroves, the most species-rich region of the World, Australia uniquely carries the Old World mangroves into a Tropical arid climate along its west coast and to a humid Tropical to Temperate climate along its east coast. In addition to being species-rich, Australia manifests a large variability of coastal types that result in a multitude and heterogeneity of suitable habitats and a richness in mangrove assemblages. As such, industrial, urban, port, and agricultural impacts are occurring on mangrove systems that globally have comparatively high environmental, ecological, and bioheritage value. Mangrove ecosystems are complex involving interactions between stratigraphy, hydrology, hydrochemistry, biological components, and anthropogenic activities which interface with these as overt and covert impacts. Overt impacts involve obvious destruction or alteration of mangroves such as clearing, burial by landfill for industrial, urban, recreational, and agricultural estates, cutting of channels, construction and expansion of harbours, bunding of natural channels and oil spills. Covert impacts are not as visible, and usually involve the alteration of hydrology and hydrochemistry of mangrove system by drainage basin contamination, and the alteration of salinity regimes, amongst others.

Twenty areas of mangrove coasts, selected from a variety of coastal types, climate and oceanographic settings, and various mangrove biodiversity settings, are provided to illustrate the range of impacts that have occurred. The impacts are graded as to being major, medium, minor, or innocuous, based on the extent and intensity of the impact(s), and the regional, national or global significance of the mangroves at the site. The extent and intensity of the impact on mangrove vegetation are assessed taking into account the spatial extent of mangrove vegetation, the changes sedimentologically, geochemically, and hydrologically/hydrochemically, and any changes in the benthic biota in terms of their composition and abundance. Of the 20 sites, nine areas are assessed as having had major impacts on their local mangroves.

Keywords

Mangroves · Australia · Anthropogenic Impacts · Industry · Agriculture · Ports · Urbanisation

9.1 Introduction

Australia is unique globally with respect to its mangrove-lined coast. As a single federated nation, with 36,735 km of coast, it is a country with one of the longest coastlines in the world, traversing tropical humid climates in the north to temperate humid in the south, but with a tropical arid climate along its mid west that comprises one of nine arid coasts globally (Semeniuk 1996). Located in the region of the Old World Mangroves (Semeniuk 1993), the mangroves of Australia are considered by Chapman (1977) and Tomlinson (1986) as belonging to the Indo-Malesian Group of the Old World Mangroves centred in the Indian-Pacific area, which is the most species-rich region of the World. In addition to it being species-rich, the mangrove coasts of Australia manifest a large variability of coastal types that result in a plethora of mangrove habitats and a richness in mangrove assemblages.

V. Semeniuk (✉)
V & C Semeniuk Research Group, Warwick, WA, Australia

I. D. Cresswell
CSIRO Land and Water, Hobart, TAS, Australia
e-mail: ian.cresswell@csiro.au

C. Makowski, C. W. Finkl (eds.), *Threats to Mangrove Forests*, Coastal Research Library 25,
https://doi.org/10.1007/978-3-319-73016-5_9

Further, for arid coasts worldwide, the west coast of Australia is the most mangrove-species-rich arid coast globally. In this context, some of the urban, industrial, and agricultural impacts have greater significance because they are occurring within globally-significant mangrove systems.

Australia is also uniquely placed globally in that it has a relatively low population compared to its extensive mangrove-lined coast. On the other hand, with its focus on exploiting minerals, fossil fuels, and primary industry to keep the nation wealthy (Horne 1964; Brocx 2008), as will be developed in this Chapter, Australia for a relatively small population has had significant impact on mangrove systems that globally have comparatively high environmental, ecological, and bioheritage value.

Unlike other countries where major impacts on mangrove-lined coasts are predominantly from silviculture, harvesting for fuel, and clearing for aquaculture (Giri et al. 2015; Saenger 2002), in Australia major impacts occur from expanding human population growth centres and, as a mineral-and-energy-rich nation, impacts of a variety of industries and export ports on its mangrove-lined coast.

This Chapter focuses on direct anthropogenic impacts on significant mangrove-lined coasts in Australia (Fig. 9.1), outlining the natural history resource that constitutes the mangroves, the types of impacts that have occurred, some of the reasons for these impacts, an assessment of the scale and severity of the impacts, and the significance of the impacts both (Australian) nationally and globally. The Chapter outlines the anthropogenic impacts on mangroves in some 20 localities in a context of the national diversity of mangroves and their habitats systems, and then specifically focuses on Port Hedland as a case study.

In order to understand the anthropogenic impacts that are manifold and complex in their interaction with the mangrove environment, some description of the mangrove ecosystems is provided from a coastal, stratigraphic, hydrological, hydrochemical, and biological perspective to highlight the processes of these ecosystems that can be affected by a range of anthropogenic impacts. This chapter deals with direct anthropogenic impacts that may be overtly or covertly manifested, but does not deal with impacts on mangroves that may be due to stress from variation in Earth climate systems as it is not possible to clearly differentiate natural climate variability effects from any broader longer term effects.

9.2 Distribution and Diversity of Australian Mangroves and Their Habitats

The Australian continent resides at the south-eastern end of the globally species-rich Old World mangroves that are centred on the Malaysian Archipelago (Fig. 9.2a). In this situation, the Australian coastline uniquely carries the cline of species richness from its northern extremity along the coast of the Northern Territory and the Kimberley (with the most humid areas located in the north-eastern part of the Northern Territory and north-western Kimberley) through to a temperate climate along its humid eastern seaboard and

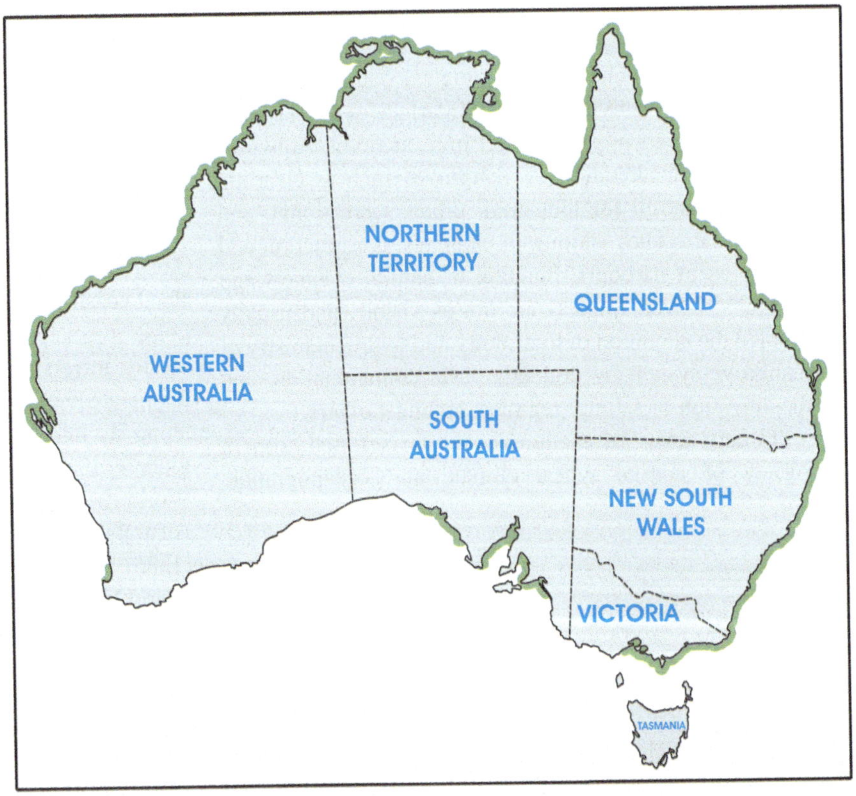

Fig. 9.1 The distribution of mangroves around the Australian coast

through to an arid climate along its western seaboard (Fig. 9.2b).

Various authors record different numbers of mangrove species in Australia depending on the definition of mangrove which varies from just woody trees and shrubs to a broader definition of woody trees and shrubs, creepers, and virtually any plant that grows in the tidal zone (e.g., ferns, creepers, and herbaceous plants). As such, Bunt et al. (1982) recognise 45 mangrove taxa, Busby and Bridgewater (1986) recognise 38 species, Smith and Duke (1987) recognise 40 species, Duke et al. (1998) recognise 39 species, Hutchings and Saenger (1987) recognise some 30 species of trees and shrubs, and Wightman and Andrews (1989) using a very broad definition that includes woody plants, climbers, grasses, ferns and palms recognise 51 species for the Northern Territory alone. Cresswell and Semeniuk (2011) focused only on woody trees and shrubs recognise 17 mangrove species in Western Australia, and 31 species in Australia.

While there is a cline from north to south of decreasing species richness asymmetrically distributed in Australia (i.e., west coast versus east coast), overarching there is a marked heterogeneity of coastal types determined by Australian mainland geology and geomorphology, coastal sedimentology/geomorphology, by oceanographic setting (wave-dominated, tide-dominated, mixed wave- and tide-influenced, wind-dominated), by tidal setting (macrotidal, mesotidal, and microtidal), and by climate (rainfall, evaporation, and wind) (Cresswell and Semeniuk 2017, 2018). In this context, the Australian coast presents rocky and bouldery shores, beach/dune shores, barrier coasts, sandy and/or muddy embayments, deltas, ria coasts, and estuaries located in tropical to temperate settings, in oceanographically diverse settings, and in humid to arid climates. Within the north-south gradient of decreasing species richness there is a plethora of coastal types at megascale and a multitude of progressively smaller-scale habitats within which mangroves species are selected and/or environmentally determined to develop diverse assemblages, variable structural formations, and physiognomic expressions.

For Australia the best studied coastal system in terms of types of diversity of mangrove coasts and their mangrove habitats is the Western Australian coast, as shown in Fig. 9.3. This illustrates megascale coastal types which form the template for the smaller scale habitats that occur within each of these coastal forms. For instance, in the ria dominated Kimberley Coast (Brocx and Semeniuk 2011), there are various sizes of v-shaped embayments that host finer-scale mangrove habitats (Fig. 9.2c, d) that include tidal flats, tidal creeks, spits and cheniers, hinterland-edge colluvial sheets, high-tidal alluvial fans, and rocky/bouldery shores. Each embayment has its own range of environmental factors that select mangrove species from the regional species pool and, through gradients of inundation, sediment types, and salinity,

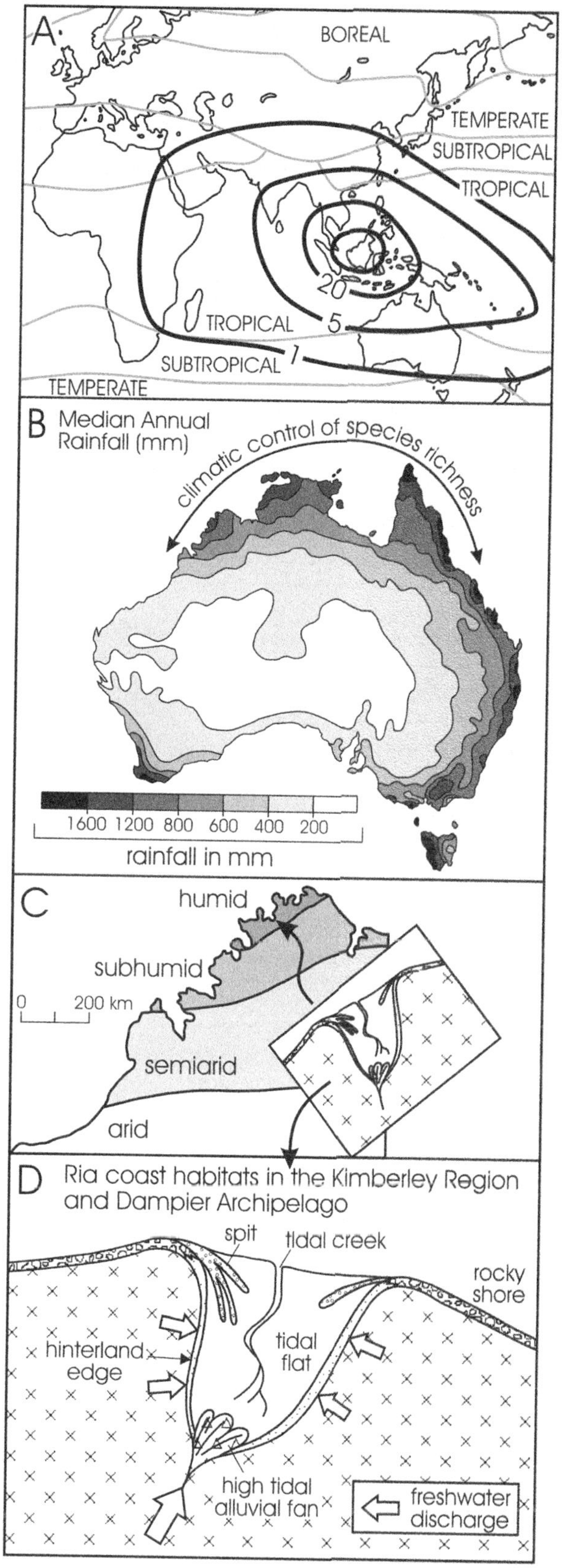

Fig. 9.2 The setting of Australian mangroves in their global context (**a**), in their climate setting (**b** and **c**), and example of the smaller scale habitats in a ria coastal setting that creates heterogeneity of floristic assemblages (**d**)

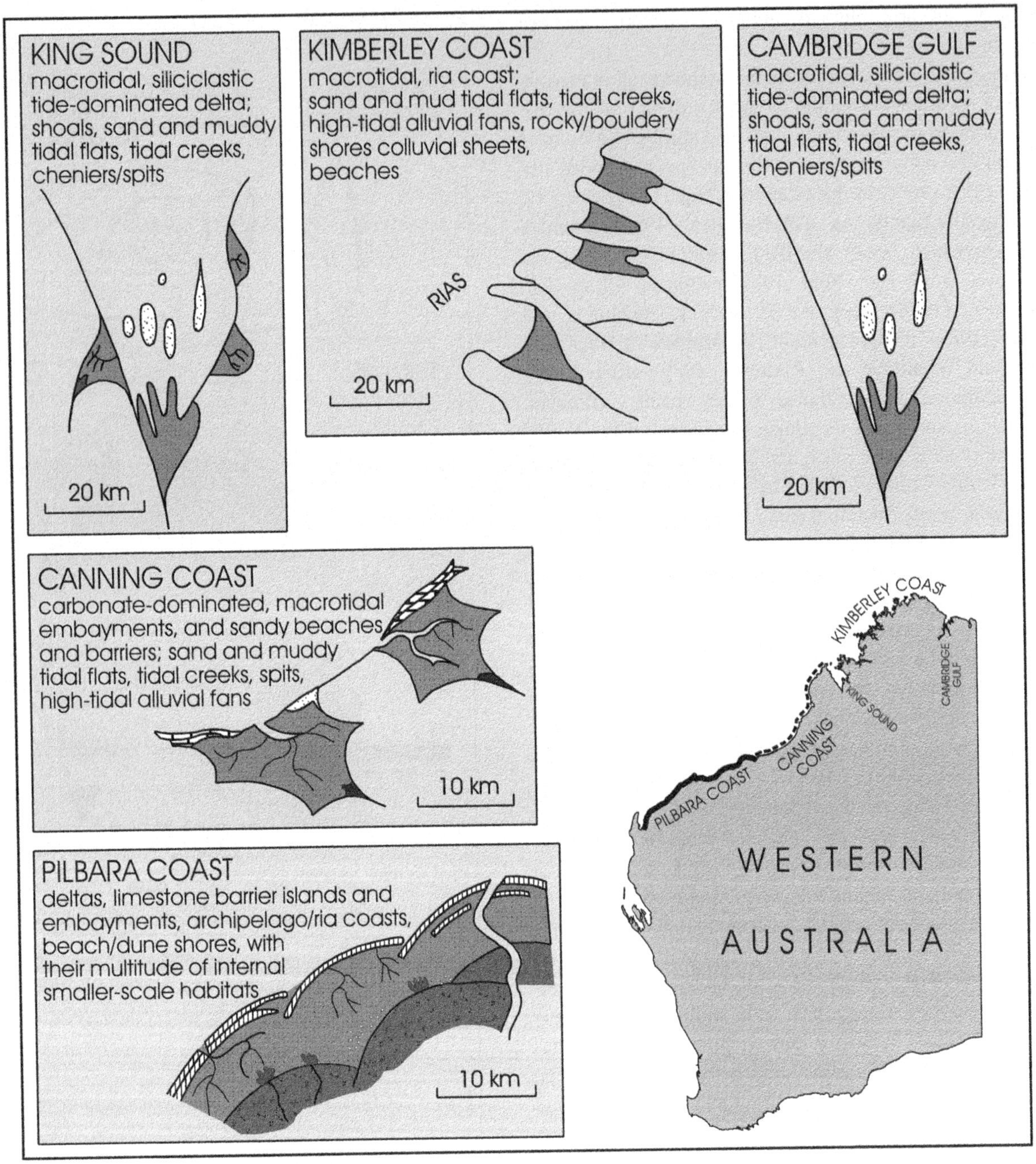

Fig. 9.3 Five major coastal types around north-western Australia. These form the template for the myriad of habitats for mangroves as listed in the diagram boxes

generate species assemblages (communities) (Semeniuk 1985a). Given the marked gradient of physicochemical factors within habitats, the mangrove assemblages colonising them are zoned (Semeniuk 1985a).

Superimposed on the multi-scale, geomorphic/stratigraphic control of mangrove distribution, assemblage composition and zonation, the climate gradient determines the local species pool, the width of the mangrove coast, the luxuriance of the mangroves, and types of mangrove population maintenance/regulation (Semeniuk 1983). For instance, in the more-humid northern regions of Western Australia there are up to 17 mangrove species which form luxuriant, tall, wide formations. In contrast towards the south, mangrove species richness falls to 3 or 4 species, and mangroves form narrow, shrubby formations backed by salt flats (Semeniuk et al. 1978; Semeniuk 1983). As a result of the climate gradient, similar habitats in similar geomorphic settings can have mangrove assemblages quite different in terms of composition, structure, width and maintenance (Semeniuk 1985a, b). This has important implications when assessing the conservation significance of individual mangrove units at subcontinental to regional scale.

Details on coastal types, descriptions of mangrove habitats, and aerial views of mangroves in the coastal regions of Western Australia are presented in Semeniuk et al. (1978, 1982, 2000a, b), Semeniuk (1980, 1983, 1985b, c, 1986a, 1993, 1994, 1996, 2008, 2011), Brocx and Semeniuk (2011), Semeniuk and Brocx (2011), Cresswell and Semeniuk (2011, and Cresswell et al. (2011)). The heterogeneity of coastal types and patchy distribution of habitats explain why mangrove species along the Australian coast commonly show an uneven, patchy occurrence and abundance. Figure 9.4 illustrates the variety of coastal forms and mangrove occurrences with a heterogeneous distribution of species along the coast. While focused on the patterns of the Western Australian coast, the same principles of coastal heterogeneity and habitat variability apply to the northern coast and eastern to south-eastern coast of Australia (Cresswell and Semeniuk 2017, 2018).

The cline in both climate and biodiversity, and the variability in coastal forms that generates habitat variability and along-shore heterogeneity in combination have created a globally distinct and significant continental-scale system of mangroves.

9.3 Sedimentological, Stratigraphic, Geochemical, Hydrological, Hydrochemical, and Biological Functioning of Mangrove Ecosystems

An understanding of the sedimentological, geochemical, hydrological, hydrochemical, and biological processes and functioning of mangrove ecosystems is required to describe key processes that maintain these ecosystems. Knowledge of these ecosystem processes is required to predict how anthropogenic impacts will interact and/or alter one or more of these aspects of the mangrove ecosystem in direct or indirect ways.

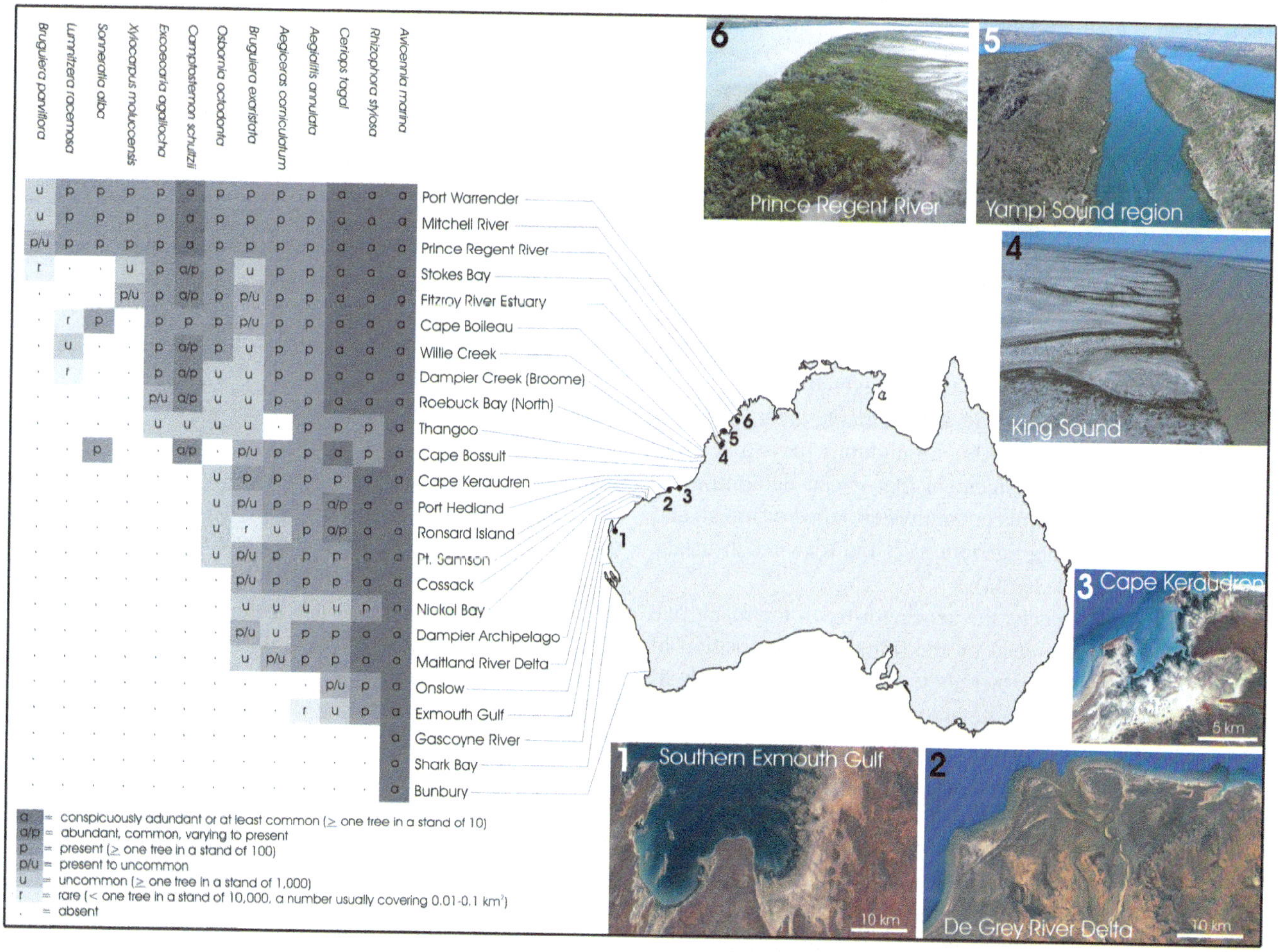

Fig. 9.4 Six examples of major coastal types around north-western Australia. The diagram illustrating the distribution and abundance of species recorded at 24 sites (Modified after Semeniuk 1983) shows patchy occurrences and variable abundances. This reflects the coastal control and heterogeneity of habitats that result in complex distribution of mangrove species

Numerous interrelated processes interact to maintain and regulate mangrove environments, including geomorphic/sedimentologic processes such as spit accretion, to hydrologic/hydrochemical processes, and to biologic processes. Understanding the main processes of maintenance and regulation is required for assessment of any given impact as some impacts may directly or subtly interfere with the main processes or pathways. Two examples illustrate this principle: (1) alteration of upland drainage may have impacts on those mangroves requiring seasonal recharge of freshwater; and (2) depletion of benthic fauna may have an effect on mangroves via the reduced rate of soil turnover. Misunderstanding or ignorance of the processes of maintenance and regulation leads to underestimating the effect of an impact.

Sedimentological aspects: the sedimentology of the mangrove environment is a primary determinant of the transport and accumulation of sediment usually as sand or mud which as substrates directly affect mangrove vegetation and the benthic biota in a sedimentological/ecological interaction in terms of erosion and sedimentation. Furthermore, it also determines the mineral content of the stabilised sediments, (i.e., whether they are siliciclastic, or carbonate-dominated, or organic-matter-dominated, whether they are gravel, sand or mud in texture), as well as the hydrological and hydrochemical attributes of the aquifers (Semeniuk 1981, 1985a, b, 1993; Semeniuk and Wurm 1987), all of which influence the floristic composition of the mangrove community, their structure, and the physiognomic expression of species.

Stratigraphic aspects: the stratigraphy of the mangrove environment determines its hydrogeology, i.e., whether it is homogeneous mud, or homogeneous sand, or whether it is stratigraphically multi-lithologically layered so that there will/can be preferred recharge and discharge layers/horizons (cf. Unno and Semeniuk 2009). Stratigraphy plays a key role in the mangrove environment in that it can determine the extent that sediments/soils remain waterlogged or moist, and how rapidly the ebbing tide returns to the sea via subsurface processes (Semeniuk 1985a).

Geochemical aspects: the geochemistry of the mangrove environment is determined by the (mineral) composition of the sediments, i.e., whether they are siliciclastic (such as quartz sand, or phyllosilicate mud), or carbonate-dominated, or organic-matter-dominated. This influences species occurrence of mangroves, their physiognomic expression, and the extent that they can take up nutrients. Geochemical aspects also involve the diagenetic processes operating in the mangrove environment, e.g., biomediated generation of iron sulphides, the precipitation of minerals such as carbonates or gypsum, and the dissolution of minerals (Semeniuk 2015a); the geochemical nature of the sediment as quartz sand, or carbonate mud/sand, or phyllosilicate mud has effect on the nutrient retention and capacity of the sediments/soils.

Hydrological aspects: the hydrology of the mangrove environment involves several interacting hydrological processes (Semeniuk 2015b, c): (1) the dynamics of recharge by seawater on neap tides versus spring tides, (2) the through-flow of tidal groundwater towards the low-tide zone downwards and seawards through the stratigraphic pile that underlies the vegetation, (3) the recharge and through-flow of freshwater from the hinterland via surface drainage channels and subterranean seepage (Semeniuk 1983, 1985a, b), (4) rainfall recharge, and (5) discharge of water effected by evaporation either by solar radiation or by wind; it also involves the complex interchange between flooding seawater and ebb-tide discharging seawater and/or freshwater through the myriad of conduits such as burrow networks and through preferred stratigraphic pathways (cf. Unno and Semeniuk 2009).

Hydrochemical aspects: the hydrochemistry of the mangrove environment involves several interacting processes (Semeniuk 2015a, b): (1) the sloping gradient of the tidal flat surface with its spring tide flooding and neap tide flooding coupled with solar-induced evaporation and wind-induced evaporation results in a gradient of increasing tidal flat salinity trending towards hypersalinity to landward (Semeniuk 1983, 2015b); (2) freshwater discharge from the hinterland whether as surface flow, or subterranean seepage, or via specific discrete aquifers results in a dilution of the high-tidal hypersaline groundwater (Semeniuk 1983); (3) at the site specific level, the hydrochemistry of the tidal flat groundwater at low tide or trapped as resident sediment-interstitial phreatic water at high tide undergoes a series of transformations in response to biological processes (e.g., anoxification, release of nutrients by leaching, amongst others).

Biological aspects: the biological aspects of the mangrove ecosystems present a complex suite of processes and ecosystem functions – there are a large range of biological activities occurring within the mangrove environment that occur at various scales from the macroscale to the microscopic scale and they are also interactive in the functioning of the mangrove ecosystem. Major components include: (1) at the macroscale level the primary production of organic matter by the vegetation through photosynthesis and uptake of nutrients and water from the sediments and soils (Pace and Lovett 2013); (2) leaf litter and other plant material at the sediment/soil surface is biologically degraded/decayed or comminuted to provide detritus, and recycled nutrients to the mangrove environment, or exported to seawards (Robertson and Daniel 1989; Robertson and Alongi 2016); (3) organic matter is biomediated or processed to produce available S, N, and P (as reviewed by Semeniuk 2015a); (4) the decaying organic matter is utilised by foraging benthos as a food source (Blaber 1997; Nagelkerken et al. 2008); (5) leaf litter is directly foraged by benthos such as crabs and

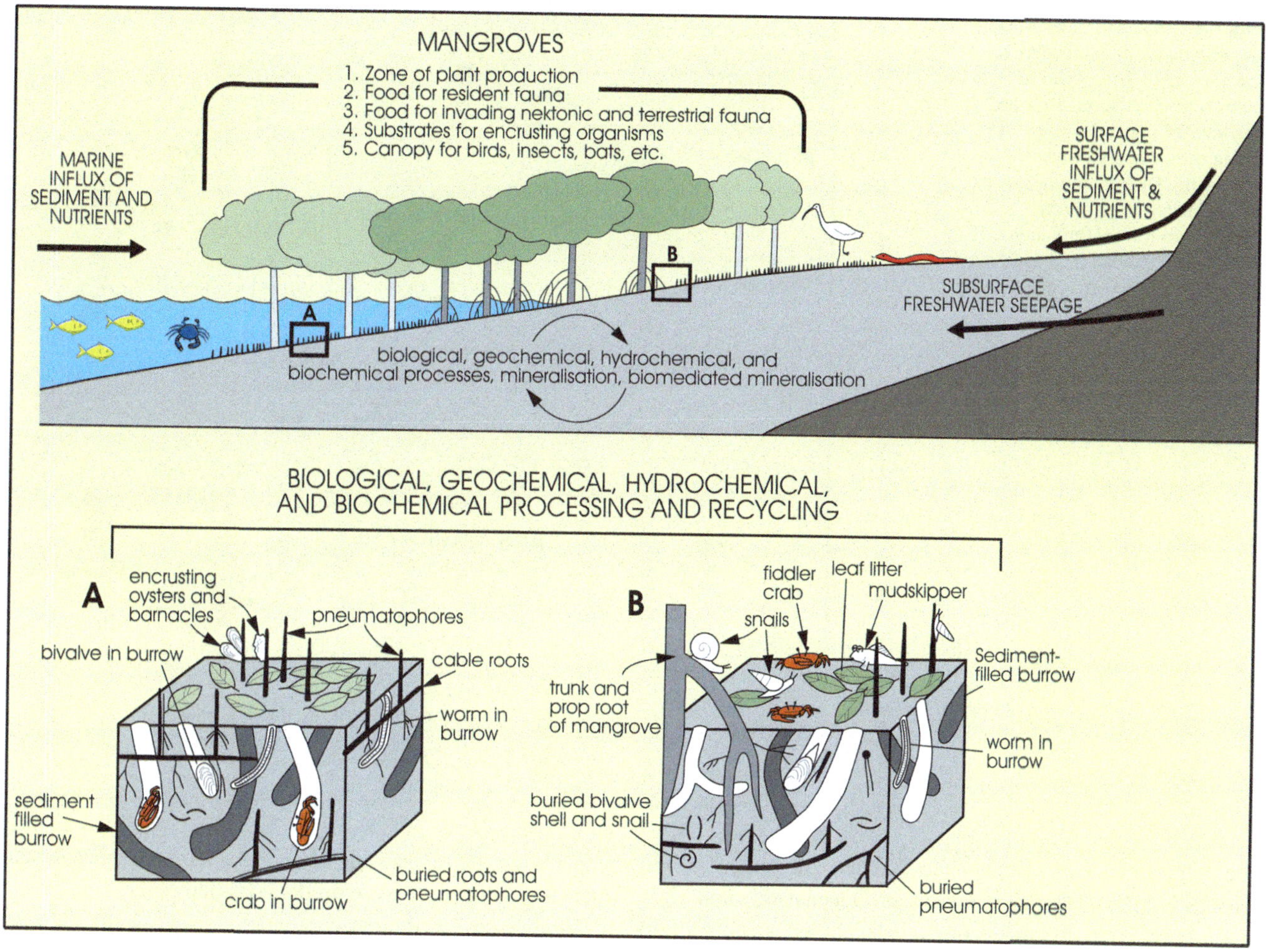

Fig. 9.5 A summary of key natural processes and functions of a mangrove ecosystem – there are biological, biochemical, geochemical, hydrochemical, hydrological processing and recycling; there is influx of nutrients and sediments from the sea, and carried by freshwater from the land; the mangrove zone as a site of primary production and its associated benthos, provides food for resident fauna, invading nekton and terrestrial animals, canopy for birds, insects, bats, amongst others. These natural processes are perturbated or destroyed by anthropogenic impacts and this diagram sets the template for understanding the types and levels of impacts that can occur overtly and covertly

gastropods (Odum and Heald 1972; Nordhaus et al. 2011); (6) excreta from the benthos is dissolved, disaggregated and dispersed to return nutrients to the ecosystem or the immediate nearshore waters (Semeniuk 2012); (7) organic matter is biomediated by bacteria to produce iron sulphide aggregates (often as framboids; see review in Semeniuk 2015a); (8) burrowing benthos recycle and bioturbate the sediments and soils to aerate the substrate and recycle nutrients (Stieglitz et al. 2000; Andreetta et al. 2014); (9) fauna such as avifauna, fish, snakes, and crocodiles invade the mangrove environment feeding on the benthos and plant life either during the high tide or at low tide, and contribute to the nutrient recycling or to biomass export (Blaber 1997; Nagelkerken et al. 2008).

An idealised summary of the essential components of the processes, functions, and products within the mangrove environment described above is illustrated in Fig. 9.5 focusing on the functions of the mangrove zone, the main forms of influx into the mangrove system, the main biological components of the mangrove habitat, and the biological, geochemical, hydrochemical, biochemical processes, and the biomediated mineralisation that operate in the sediments/soils of the mangrove environment. Anthropogenic impacts on mangrove environments will effectively interface and alter many of these features and will consequently have immediate effects, or in many cases longer term effects. Anthropogenic impacts on mangrove environments are dealt with in this chapter using the components of Fig. 9.5 as a template.

9.4 Human Settlements: Urbanisation-Driven Versus Industry-Driven

Anthropogenic impacts on mangroves axiomatically derive from two overlapping types of human activities: (1) growth of coastal settlements that start as small towns and expand to

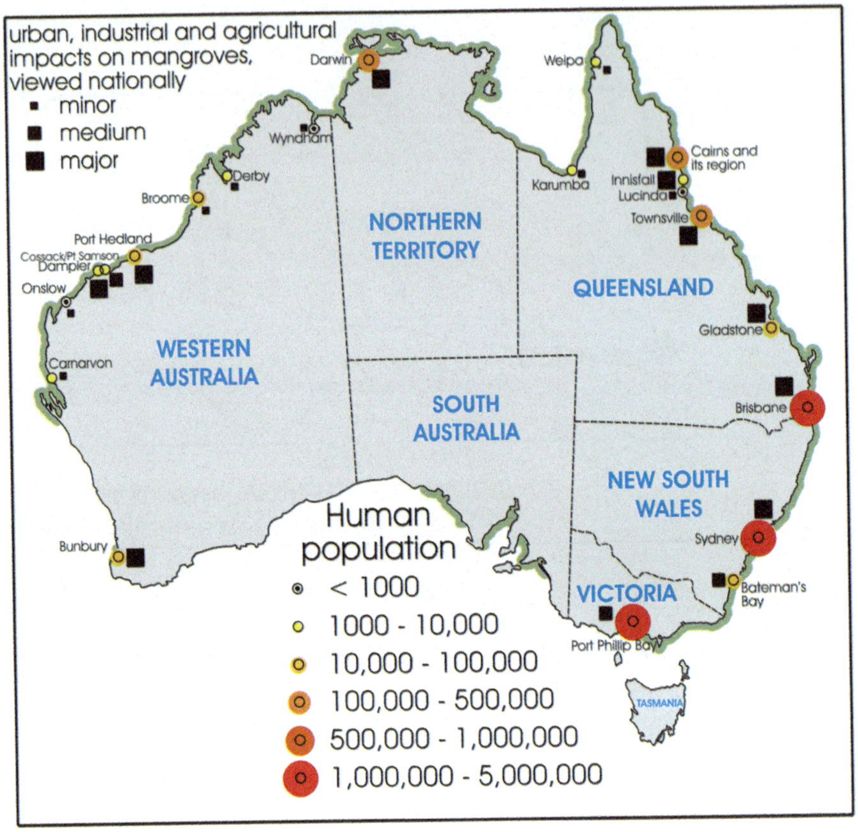

Fig. 9.6 The major sites where impacts on Australia mangroves have occurred – these are set in a context of population sizes, and the level of impact on the mangroves when viewed nationally. Distribution of sugar cane plantations has been generalised from information from the Australian Sugar Milling Council (http://asmc.com.au/industry-overview/)

be major cities (e.g., Bunbury in Western Australia, Darwin in the Northern Territory, Cairns in Queensland) and (2) - industry-driven settlements that are developed specifically to support mining, mineral or oil/gas extraction and export, agricultural products, including port facilities for export for any number of commodities (e.g., Dampier in Western Australia). The overlap occurs because, in time, if there is later development in the hinterland in relationship to ore extraction, oil/gas development, or other commodities (such as wheat grain or cattle), the original coastal settlement with its population, hospitals, schools, and other infrastructures may then be intentionally targeted as an export site, and consequently ports/harbours will develop (e.g., Onslow in Western Australia initially a coastal town became a port for export of solar-generated salt when these resources were later exploited in the region).

Industries develop population growth centres which develop further industries such that in a given period of time, an isolated centre may contain a varied and complex industrial system. For instance, the broader Dampier area in Western Australia within the past 50 years has developed an iron ore export facility, a solar salt production industry, and a natural gas extraction/export industry. Furthermore for the future there are proposals for a steel mill and perhaps a petrochemical plant, both of which will exploit the iron ore, natural gas energy and salt of this region. The result is a concentrated increase in industrial impacts and urbanisation impacts such as sewage disposal, refuse disposal and increased people pressure on the coastline.

In all cases, as human populations grow, or the industry intensifies, there will be growth of settlements and land required for urbanisation, roads, industrial estates, hospitals, commercial centres, harbours, and airports. Herein lie the reasons for the overlap of urbanisation-driven versus industry-driven human settlements and the expansion of developments along the coast, with the consequent pressures on coastal mangrove-inhabited land.

Figure 9.6 shows the main sites of impacts on mangroves along the Australian coast as described in this Chapter, the size of human settlements at these sites, and the relative size of impact on the mangroves based on area of the industrial, urban, and agricultural estates, the size of the ports, and extent of shipping at these sites. While it is clear that human population size (with its need for land and infrastructures) is an underlying factor to impacts on the coast and on mangroves, it is also the case that large industries with a relatively small human population will have large impacts on coasts and their resident mangroves.

Where human settlements are located on terrain that has some relative relief above MSL (e.g., the uplands surrounding Sydney Harbour (Port Jackson), which is a ria coast; Brocx and Semeniuk 2011), there is often little impact on the coastal zone and mangroves. That is, urbanisation, hospitals, schools, commercial centres, industrial estates, airports, and other infrastructures can be built on high ground away from the mangroves. On the other hand, where human settlements are sited on terrain that has low relative relief (e.g., stranded tidal flats, stranded flood plains, alluvial plains), as is the case for Brisbane that is built on alluvial plains and former floodplains), the need for land for urbanisation, industry, airports, and infrastructures, will commence on the lowlands and often encroach into the coastal zone, reclaiming and infilling tidal flats and encroaching on mangrove zones because of the developing need for ports, industrial estates, airports, and all their accompanying infrastructures. Moreover, relatively flat terrain such as alluvial plains and tidal flats (with mangroves), from an engineering perspective, are attractive lands to build airports, industrial estates, and urban areas because, apart from some land-filling to elevate the terrain above a high-water level, or flood level, such areas do not require major land modification.

9.5 Intensity/Severity of Impacts on Mangroves, and Spatial Scales of Impacts on Mangroves

Four general measures can be used to assess adverse anthropogenic impacts on mangrove ecosystems and the mangrove environment:

1. The intensity of impact(s) on mangrove vegetation;
2. The spatial extent of impact(s) on mangrove vegetation;
3. Changes sedimentologically, geochemically, and hydrologically/hydrochemically;
4. Changes in the benthic biota in terms of their composition and abundance.

These assessments of impacts are determined wholly on the effect on the mangrove vegetation and their resident benthic fauna, and not on the (largely unknown) effects on nearshore biota that depends on supply of detritus and nutrients from the mangroves.

The intensity of impact(s) on mangrove vegetation can be measured as change in community structure and composition, or debility of the plants (e.g., loss of leaves, yellowing of leaves, necrosis of leaves), or death. To assess the severity of this measure, there needs to be measurement of the grades of the effects, mapping of these grades spatially, and assessment/comparison against mangroves in the region. Total death of mangroves can be locally catastrophic (if the mangroves are regionally or nationally important), but it also has to be assessed in terms of overall areal extent. The impact in some cases may not result in widespread death but rather in tree debility – the plants loose leaves, or leaves become partially necrotic, but the plant does not suffer mortality; alternatively the plants may not grow as vigorously and luxuriously as they would in a natural environment. This may impact on long term viability of the ecosystem.

In terms of the spatial extent of impact(s) on mangrove vegetation, total death of mangrove plants over extensive areas is an intense and major impact. For instance, the permanent inundation of 2500 ha of mangroves for the purposes of creating a solar salt industry south-west of Dampier, Western Australia in the 1960s was an intense, major and permanent impact. Similarly, the infilling of a mangrove-vegetated linear embayment some 300 ha in area for the purposes of constructing port facilities at Port Hedland in the 1960s was another intense, major and permanent impact. In addition, an area of some 20 ha of a mangrove embayment north of Carnarvon, Western Australia was accidently permanently inundated by coastal mismanagement in the early 1990s – this also was an intense and major impact. All these impacts resulted in the wholesale death of mangroves and can be viewed as major impacts. The overall areal extent impacted needs to be understood in relation to the availability of habitat in that area, i.e., the overall impact of 20 ha impacted near Carnarvon destroyed a major component of the overall habitat available in that region, with potentially major flow-on effects to avifauna, and other animals such as fruit bats.

In the Dampier area, in Western Australia, dredge spoil was accidentally disposed of in two separate small embayments; the dredge spoil smothered and killed the mangroves covering some 2.2 ha and 1 ha of vegetation, respectively. While there was total death of the vegetation, relative to what mangroves remained in the embayments themselves (some 30–50%), and within the region (Semeniuk and Wurm 1987), the death of the mangroves in these embayments can be viewed as a medium-scale impact.

A measure of anthropogenic impacts on the mangrove environment can also be assessed by changes sedimentologically, geochemically, and hydrologically/hydrochemically. These factors can directly impact on mangroves, and can be readily measured. Alteration of hydrodynamics and groundwater salinity of high-tidal mangrove scrub in the Pilbara region by causeway/road construction caused the slow debility and then death of hundreds of hectares of mangroves (Gordon 1988).

Changes in hydrology, hydrochemistry, or sedimentology in the mangrove environment can alter the ecological structure of the benthic biota in terms of their composition and abundance.

All these changes noted above can occur at different spatial scales. Widespread anthropogenically-induced mortality covering hundreds to thousands of hectares of mangrove vegetation spatially should be viewed as a major impact, whereas at the other extreme, mortality of several square metres of mangroves in a stand that itself covers hundreds to thousands of hectares of vegetation can be viewed as a minor or innocuous impact. The measure of the severity of the impact in spatial terms, therefore, has to be gauged as to what is lost against what has remained. It also has to be viewed in the context of the biogeographic limit of a given species, and the relative rarity of the species at a given locality. For instance, the total loss of the 1 ha of *Avicennia marina* in Bunbury in Western Australia (Semeniuk et al. 2000a, b), a species that is widespread and common throughout Australia, but occurs as an isolated remnant 500 km from its nearest northern neighbour would represent a major loss. It presents a small but most southern isolated occurrence of the species in the State. So too, the total loss of the <1 ha of *Rhizophora stylosa* at Yardie Creek, in Western Australia (Semeniuk et al. 1978), a species that also is widespread and common throughout Australia, but occurs as a southernmost isolated remnant would represent a major loss.

It is also useful to differentiate between overt and covert impacts (Figs. 9.7 and 9.8). Overt impacts are those that are clearly visible and are easily discernible as a direct impact of human activity – they include dredge spoil disposal, clearing of mangrove vegetation, cutting of port channels, refuse disposal, and other activities that clearly are manifest on the surface and clearly are having an impact of removal, or death, or debility of the mangroves (Fig. 9.7). Covert impacts are 'hidden impacts' and may include indirect impacts which are more difficult to see. These include impacts on groundwater hydrodynamics, groundwater salinity, geochemistry, and microbiota, contamination by toxic effluents, or toxic gases that affect mangroves, or affect their resident biota.

The smothering or blanketing of mangroves and mangrove soils by industrial dust (such as mineral ore dust or coal dust) can be assigned to an overt or to a covert impact.

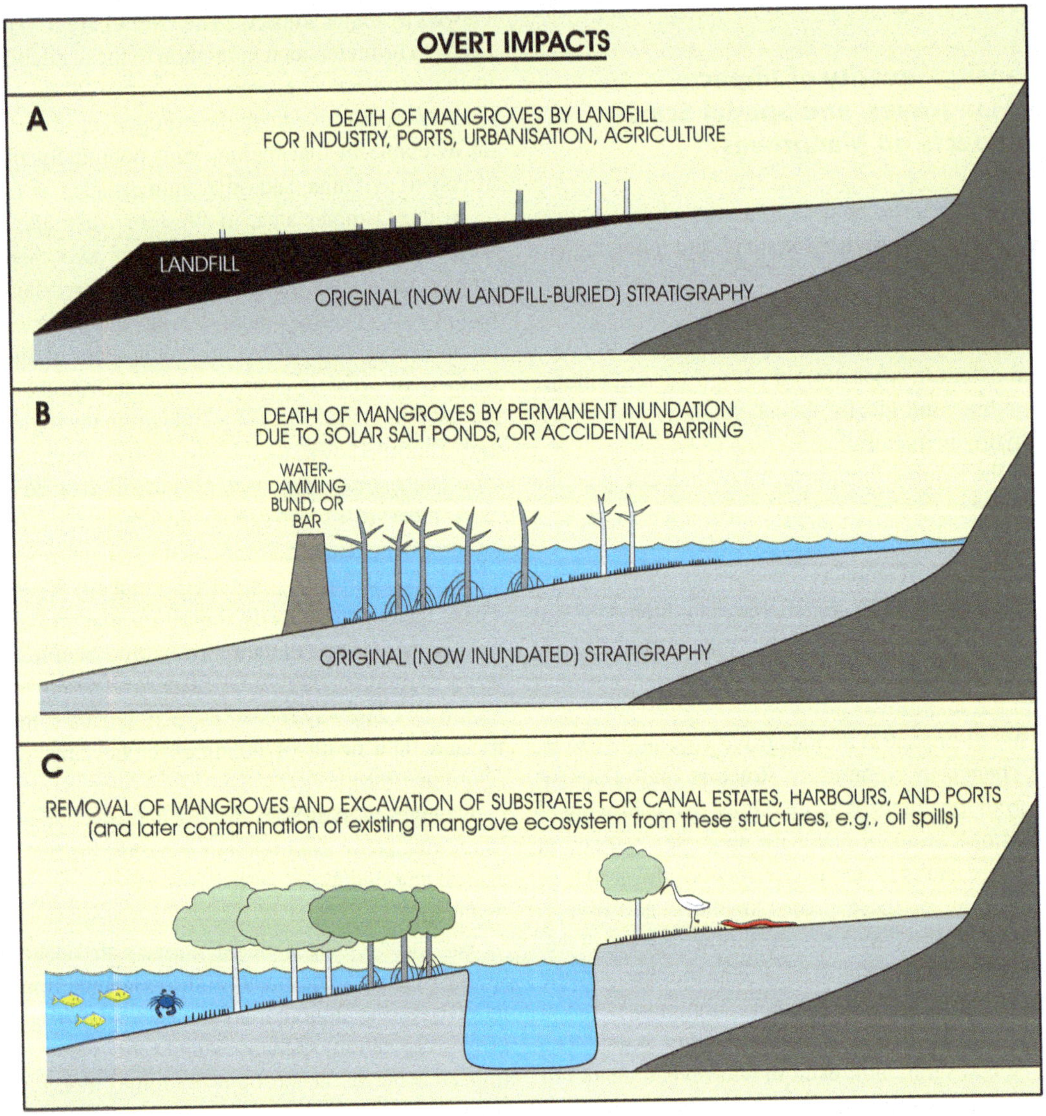

Fig. 9.7 Overt impacts on mangroves – view this diagram in conjunction with Fig. 9.5 which illustrates natural processes and functions

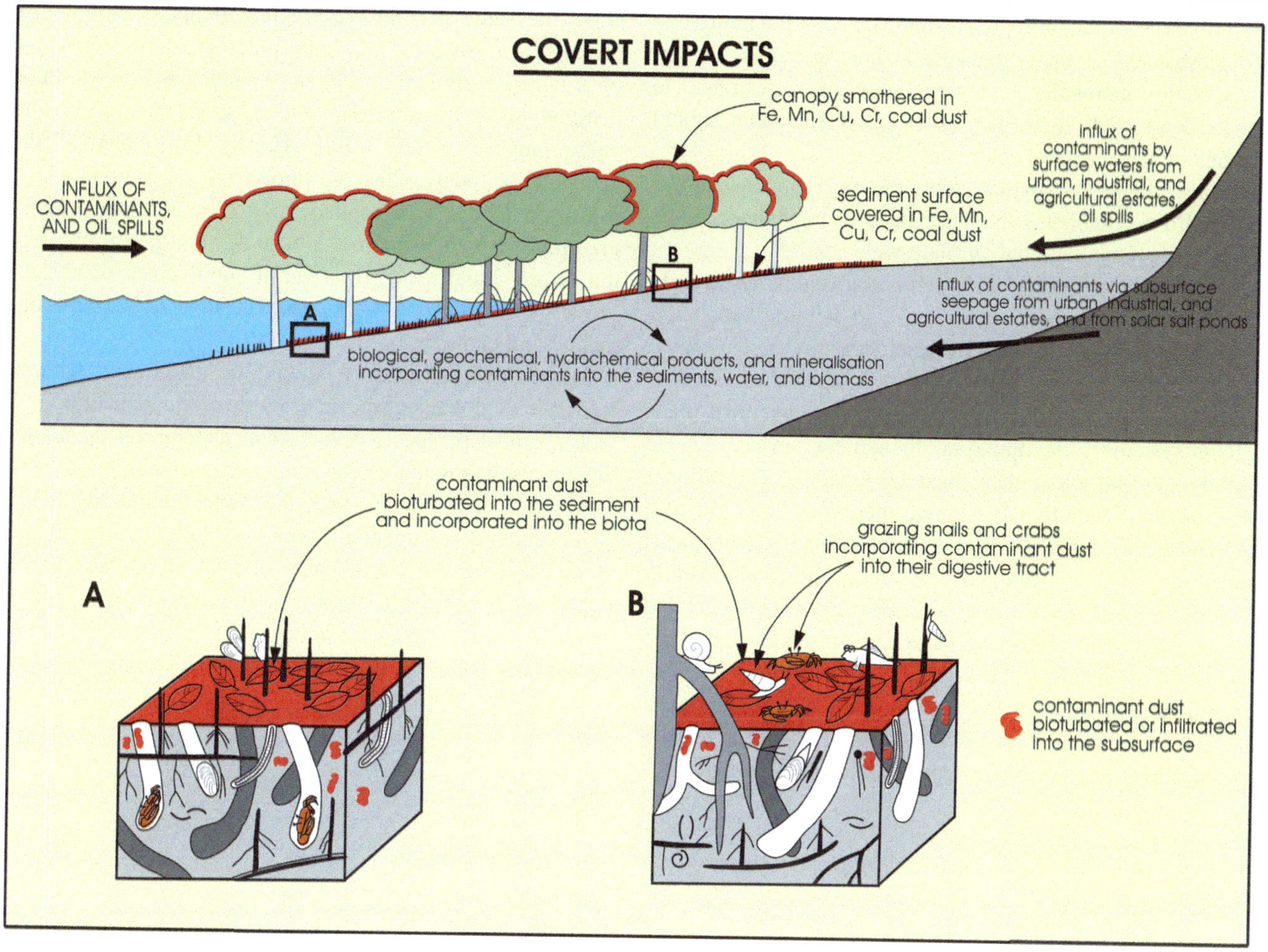

Fig. 9.8 Covert impacts on mangroves – view this diagram in conjunction with Fig. 9.5 which illustrates natural processes and functions

Such blanketing can be sparse and light, or heavy and thick. When dust heavily coats mangrove leaves and mangrove soils, it can appear as an obvious phenomenon and an overt impact, but it is not as clearly overt as landfill or clearing. However, regardless of whether the environment is blanketed by a light or heavy coating of dust, there are a multitude of subtle effects that are operating as covert processes (e.g., the infiltration and biochemical changes within the leaf; changes in hydrochemistry in groundwater and uptake of this contaminated water by mangroves and incorporation of exotic cations into plant tissue; grazing of dust-contaminated algal/diatom/fungal film by snails and crabs and the incorporation of exotic cations into animal tissue; Semeniuk 2012). We have assigned blanketing of mangroves by industrial dust to covert impact.

The elements and processes of the covert impacts are summarised in Fig. 9.8.

9.6 Assessment of Significance

The assessment of an impact balanced against the conservation significance of a mangrove system is complex. Some parts of society would argue all coastal wetlands are important, whereas others pointing to the economic benefits to the broader community would argue that a proposed impact on mangroves is acceptable. To rationally assess the significance of a mangrove system and then to assess the effect of an impact on that system, a standard set of assessment guidelines that are acceptable to the broader community is required.

Semeniuk (1986b) and Semeniuk and Semeniuk (1998) propose guidelines that assess the ecological significance of a given system (including mangroves) to address the conservation, social, or scientific value of a natural system, and are based on criteria of:

- Global, national, or regional significance of the mangrove ecosystem – is it widespread and common, is it significant globally, nationally, or regionally, and is it restricted to local areas? Is it, in fact, a heritage feature that should belong to the global community?
- Does the mangrove coastal system contain unique landforms (habitats), or other natural features that provide it with global, national, or State-wide status?
- Does the system offer the potential for scientific/research, educational, social, recreational, and wildlife sanctuary/habitat usage not offered by other areas?

In order to implement conservation assessment of mangrove systems it is important to address several issues. Firstly (1), sufficient data must be available for rational assessment. Secondly (2), whereas there may be conflicting demands for use of mangroves by social, government, developmental, educational, research, and conservation groups, any perceived value judgement needs to be viewed in perspective of some overall assessment of the value of the system. For instance, the scientific community may have significantly more information and an understanding about a natural system than the public and developers at large who are not aware of these environmental values and do not share the same perspectives. Alternatively, a community group in a residential area may place greater social value on a natural system in their area of residence than would be justified under an environmental assessment alone. Both value judgements are valid. Thirdly (3), many of the decisions of today will have impact on generations of the future and, decisions taken today should not unduly pre-empt or predetermine the attitudes and possibilities of the future. Finally (4), the overall significance of a mangrove system needs to be assessed against social, environmental and economic criteria.

No national assessment should be undertaken without reference to global significance. World Heritage listings provide a global perspective that some features on the Earth, though residing in a given nation, have outstanding universal values that, in effect, belong to the global community.

Taking into account all of the above addresses the problem of determining the significance of a given impact. Some impacts are, of course, totally destructive involving many hundreds or even thousands of hectares of a mangrove system as mentioned above. These may be viewed as major types of impacts. Thereafter, impacts may be ranked on a gradational scale, e.g.:

Table 9.1 Examples of types and level of environmental impacts on mangroves to date in Australia

Major impacts	Solar salt production
	Landfill by dredge spoil
	Landfill for ports, urbanisation, airports, industrial estates, recreation
	Harbour development
	Alteration of hydrology and hydrochemistry by various mechanisms
	Permanent inundation by barring of mangrove embayment
	Sugar cane plantation affecting downstream hydrology/hydrochemistry
Medium impacts	Alteration of hydrology and hydrochemistry
	Alteration of sedimentation patterns
	Altering harbour facilities (e.g., channel widening)
	Refuse disposal (sanitary landfill)
	Major sewage effluent discharge
	Effluent discharge from industries or from storm water
	Oil spills
	Harbour and jetty facilities
	Causeways and roads
	Voluminous Fe, Mn, Cu, Cr, and coal ore dust fall-out
Minor impacts	Hinterland edge landfill
	Small harbour/marina/jetty facilities
	Small causeways/roads
	Aquaculture
	Minor sewage effluent discharge
	Indiscriminate boat landings
Innocuous impacts	Food gathering by Indigenous communities and isolated residents
	Minor wood gathering and harvesting
	Subsistence fishing
	Cattle grazing

- Major
- Medium
- Minor
- Innocuous

The ranking of an impact would be based on its area extent and its permanence, and the likely interference of a given impact can only be rationally assessed when the impact is ranked and the conservation significance of a mangrove system is determined.

The ranking is a value judgement balancing size of impact, with severity/intensity of impact(s), with significance of the species involved, with significance of the coastal setting (globally, nationally, or regionally).

A ranking of common impacts on mangroves is provided in Table 9.1.

In addition to scale of impact, the assessment of any development or exploitation of mangrove systems also has to be gauged against a framework of intended use of that environment. If the mangrove is intended to remain as a wilderness because, for instance, it has global significance, then most forms of development would be unacceptable. If the mangrove system is to be used as a research/education unit, or as a nature reserve, then some forms of impact may be tolerated (e.g., access tracks, small boat harbours) as long as the development/exploitation does not interfere with the natural processes needed to establish research/education principles or to maintain it as a nature reserve. If the system is to be maintained as a habitat, nursery and zone of primary production for fisheries, then other forms of development may be acceptable up to the point where development impinges on the ability of the system to sustain productive yields, or the development impinges on the system such that it endangers public health.

9.7 The Types of Impacts on Mangroves Worldwide and in Australia

There is a large variety of types of impacts on mangroves recorded by other workers around the world and in Australia (Morton 1965; Ferguson-Wood and Johannes 1975; Odum and Johannes 1975; Percival and Womersley 1975; Clark 1977; Hegerl 1982; Jernelov and Linden 1983; Lewis 1983; Saenger et al. 1983; Tomlinson 1986; Semeniuk 1986a, b, 2012; Hutchings and Saenger 1987; Gordon 1988; Burns et al. 1993; Ellison et al. 1999; Saenger 2002; Naidoo and Chirkoot 2004; Lewis et al. 2011; Brocx and Semeniuk 2015). These include timber production, firewood and charcoal production, extraction of tannin, rayon, glue and dye, food production, clearing of forests, dredging and filling, impoundments, damming of rivers, aquaculture, sewage lagoons, solar salt ponds, industrial waste disposal, shipping wharves, canal developments, seepage of herbicides, chemical toxins and heavy metals, dust pollution from a variety of mineral ores (Fe, Mn, Cu, Cr) and from coal, oil spills, heated water discharges, drainage modification, and road works. Not all of these have occurred in Australia, but many of the impacts that have occurred have involved the permanent loss of substantial areas of mangroves and irrevocable alteration of their habitat.

The impacts on the mangrove environment are manifold if the focus is on the entire ecosystem i.e., involving the vegetation, the sediments/soils, the groundwater hydrology and hydrochemistry, the benthic fauna, and the microbiota. Indeed, since the mangrove ecosystem is an integrated holistic functioning system, impacts on any of these components can have an effect on the ecosystem. For instance, altering the hydrochemistry of the groundwater through salinity changes can affect benthic biota and affect mangrove species that are specifically linked to salinity field (cf. Semeniuk 1983, 1985a, b; Semeniuk and Wurm 1987). Altering the hydrochemistry of the groundwater can also affect the abundance and composition of the burrowing benthic fauna which in turn can affect changes in soil/sediment turnover, the extent of subsurface oxygen infusion, nutrient recycling, and microbial effects, amongst others. These effects may take months or years to manifest as debility, death, or structural and floristic changes in the vegetation.

Impacts on mangrove ecosystems, and on the mangrove vegetation can be dramatic and instant, e.g., bulldozer clearing of mangrove vegetation or the smothering and burial of mangrove ecosystems by landfill resulting in instant deaths, while permanent inundation of mangrove stands can result in near-immediate death and the smothering/coating and toxic effects of oil spills may also result in debility and/or death. Other impacts on mangrove ecosystems can be subtle and long term either resulting in slow decline and debility of vegetation and finally death, or resulting in structural and floristic changes. Subtle impacts on mangrove ecosystems through subsurface alteration of hydrology, the alteration of freshwater through-flow, changes hydrochemistry, or the infusion of excess nutrients, or toxic chemicals can affect the benthic biota and microbiota that can, in turn, alter the health and survivorship of the vegetation. This Chapter deals mainly on the direct impacts on mangrove vegetation but due to the close link between benthic fauna, microbiota, and vegetation, there is a need to describe impacts on these other associated biota – that is, where possible, the mangrove ecosystem is treated as a holistic, integrated, interacting unit but clearly not all subsurface and subtle effects that would have long-term effects on mangrove ecosystems have been exhaustively investigated or documented.

The clearing of vegetation, smothering and burial by landfill, permanent inundation, and the effects of oil spills clearly are overt impacts and readily and immediately evident. Alteration of soils, groundwater, benthic fauna, and microbial communities are covert impacts – often their processes are subsurface and commonly may take months or years to have an effect that is evident in the vegetation. While overt impacts are obvious, clearly covert impacts usually are not as evident.

The types of impacts on mangrove environments in Australia are listed below. They are graded generally from the most major and most common impacts in terms of size and severity, to minor impacts, to innocuous impacts:

1. Destruction of mangrove vegetation by landfill to create above high-tide surfaces for ports, urbanisation, recreational facilities, airports, industrial estates;
2. Destruction of mangrove vegetation by inundation to create solar salt evaporating basins;
3. Construction of causeways and roads that destroy mangroves, or impound tidal waters altering hydrodynamics and creating increased inundation, or altering salinity regimes;
4. Construction of harbours, ports, marinas, jetties, indiscriminate boat launching sites, and canal estates resulting in the clearing and destruction of mangroves;
5. Oil spills that have immediate effects on the vegetation and benthic biota, or that may have delayed effects;
6. Accidental or intentional discharge of toxic effluent resulting in mangrove debility, mangrove death, altered mangrove population dynamics, or altered mangrove community composition;
7. Sewage discharge in mangrove environments resulting in altered mangrove population dynamics, altered mangrove community composition, and changes in benthic biota;
8. Alteration of groundwater hydrodynamics and hydrochemistry as saline discharges from solar salt ponds resulting in mangrove debility, mangrove death, altered mangrove population dynamics, or altered mangrove community composition;
9. Dredge spoil disposal into extensive mangrove formations causing mangrove death;
10. Cutting or widening of shipping channels causing direct mangrove clearing;
11. Cutting or widening of shipping channels causing alteration of wave and tidal hydrodynamics and changes in mangrove/geomorphology/sedimentation patterns and mangrove/sediment interaction resulting in mangrove debility or mangrove death;
12. Remobilisation of muddy dredge spoil material into mangrove environments causing changes in mangrove/sediment interactions resulting in mangrove debility or mangrove death;
13. Coastal engineering works causing inadvertent changes in sand transport patterns and consequent sand-barring of mangrove embayments and causing permanent inundation and mangrove death;
14. Disposal of refuse into mangrove environments causing death;
15. Disposal of sewage, and effluent into mangrove environments causing debility, death, or changes in population dynamics;
16. Agriculture on alluvial plains (e.g. sugar cane plantations) landward of mangroves changing the hydrology/hydrochemistry of groundwater that discharges into the mangrove ecosystem;
17. Urban systems landward of mangroves changing the hydrology/hydrochemistry of groundwater that discharges into the mangrove ecosystem;
18. Coating of mangrove leaves, or in the extreme smothering of mangrove canopy by fine-grained dust of Fe, Mn, Cr, and Cu ores, and coal dust wind-mobilised from stockpiles resulting in debility and death of the down-wind vegetation, and in alteration of soil/sediment geochemistry by a veneer of the dust on the sediment/soil surface;
19. Innocuous impacts on mangroves include food gathering (usually of molluscs) by Indigenous communities of the regions, wood gathering and harvesting for fire-making by Indigenous communities and by travellers/campers, and subsistence fishing by various peoples;
20. Locally, wandering cattle and other livestock have ventured into mangroves on neap tides in macrotidal coasts when the terrain is traversable, and have foraged on leaves leaving the mangrove canopy neatly cropped at low levels; cattle also may litter the mangrove soils with manure which is dissolved and recycled into the soils.

Cattle and other livestock grazing on spring-tidal-level mangrove leaves during neap tide generally have innocuous impact on the mangroves, and the study by C A Semeniuk and Semeniuk (2014) shows that cattle manure within the mangroves is leached and/or washed away and has little impact on the vegetation.

The variety of impacts can be listed simply under four categories of decreasing importance (Table 9.2). The greatest number and the most severe impacts on mangroves, of course, often occur in the vicinity of the major population/industrialised centres but not always – for instance, Port Hedland is a major industrial port with major impacts on mangrove environments but compared to Sydney and Brisbane its has a relatively small (mining-town-dedicated) population. There is little commercial use made of the mangrove environment in Australia with the exception of

Table 9.2 Types of Impacts on mangroves geographically, and their level of impact (this list is not exhaustive)

Site	Major impacts	Medium impacts	Minor impacts	Innocuous impacts
Bunbury	Dredge spoil fill			Food gathering
Carnarvon			Boat marina Sanitary landfill Boat harbour Jetty Causeway	Food gathering Wood gathering and harvesting Subsistence fishing
Onslow			Partial destruction by solar salt ponds Boat marina/jetty	Food gathering Wood gathering and harvesting Subsistence fishing
Dampier	Solar salt ponds Dredge spoil fill	Causeway Boat harbour Iron ore dust	Indiscriminate boat landings	Food gathering Wood gathering and harvesting Subsistence fishing
Cossack/Point Samson		Causeway Indiscriminate boat landings Harbour facility		Food gathering Wood gathering and harvesting Subsistence fishing
Port Hedland	Solar salt ponds Dredge spoil fill	Harbour facility Sanitary landfill Causeway Mineral ore dust Induced erosion Oil spill	Indiscriminate boat landings	Subsistence fishing
Broome			Land filling Jetty facility Indiscriminate boat landings	Food gathering Wood gathering and harvesting Cattle grazing Subsistence fishing
Derby			Jetty facility Causeway Sewage disposal	Food gathering Wood gathering and harvesting Cattle grazing Subsistence fishing
Wyndham			Jetty facility Sewage effluent discharge	Food gathering Wood gathering and harvesting Cattle grazing Subsistence fishing
Darwin	Landfill for urbanisation, and industry	Ports Boat launching facility	Jetty facility Sewage effluent discharge Refuse disposal	Food gathering Wood gathering and harvesting Subsistence fishing

(continued)

Table 9.2 (continued)

Site	Major impacts	Medium impacts	Minor impacts	Innocuous impacts
Karumba			Land reclamation for roads Port facilities	Subsistence fishing
Weipa			Land reclamation for roads Port facilities	Subsistence fishing
Cairns and its region south to Innisfail	Landfill for urbanisation, airport, industry, roads, recreation Sugar plantations altering hydrology/hydrochemistry			Subsistence fishing
Innisfail	Landfill for urbanisation, airport, industry, roads, recreation Agriculture (sugar plantations) altering hydrology/hydrochemistry			Subsistence fishing
Townsville	Landfill for urbanisation, airport, industry, roads, recreation Urbanisation and agriculture alters groundwater hydrology and hydrochemistry Clearing for port facilities			Subsistence fishing
Gladstone	Landfill for urbanisation, industry, roads, recreation, ore stockpiles Urbanisation, industrial estates, stockpiles, and caustic soda ponds alter groundwater hydrology and hydrochemistry Clearing for port facilities Coal and alumina dust			Subsistence fishing
Brisbane	Landfill for urbanisation, airport, industry, roads, recreation			Subsistence fishing
Sydney (Botany Bay, Kurnell)	Landfill for industrial purposes Destruction of mangroves for port facilities Refuse disposal			Subsistence fishing
Batemans Bay		Landfill for urbanisation, industry, roads		Subsistence fishing
Port Phillip Bay		Landfill for urbanisation, industry, roads		Subsistence fishing

tourism, fishing and crabbing which, with proper management, is one of the least destructive uses of the mangrove environment.

9.8 A Tour of Anthropogenic Impacts on Mangroves Around Australia

The Australian coast is almost ringed by mangroves (Fig. 9.1), and generally the human population density is low compared to European, North American, and Asian countries, but all major population centres are coastally-located with over 85% of the Australian population living within 50 km of the coast. This has resulted in human population impacts on mangroves to be severe in some parts but overall of low intensity when viewed nationally. Australia, however, is mineral-rich and has large energy resources and, as a result, it has developed major port facilities and processing facilities to handle these commodities. Though often in remote areas, the construction of port facilities, processing facilities, and ancillary activities have resulted in population growth with the concomitant need for urbanisation, hospitals, schools, commercial centres, industrial areas, roads and causeways, and airports. Such developments have led to impacts on mangroves where there has been need for land and for port facilities along and amongst a mangrove-inhabited coastline. Where location of port facilities, processing facilities, and ancillary activities have coincided with major cities located on mangrove-inhabited coasts, the impacts on mangroves have intensified, firstly because of the location of a major city, and secondly, because of the industrial development.

Of course, not all of the mangrove-lined coast of Australia has been impacted by human activities, and considering the size of the continent, the length of the coastline, and the relatively low human population, most mangrove areas are remote and have remained relatively pristine. The Kimberley Coast stands as an example – it is one of the most remote regions on Earth, and much of its coastline (with its mangroves) is in a relatively pristine state (Brocx and Semeniuk 2011; Cresswell and Semeniuk 2011).

Where coastal developments centred on industry have taken place, the impacts on mangroves have been major.

It is useful to understand the anthropogenic impacts on mangroves around Australia, and in this context some 20 case studies have been selected to illustrate these impacts. They have been selected on the bases that they show some major impacts, that they range peripherally along the whole climatic, oceanographic, and species diversity of the Australian mangrove coast, that they illustrate a range of differing impacts, and that they illustrate that small areas of mangroves may have significance. The case study sites are (from Bunbury in south-western Australia, and proceeding in a clockwise manner): (1) Bunbury, (2) Carnarvon, (3) Onslow, (4) Dampier, (5) Cossack/Point Samson, (6) Port Hedland, (7) Broome, (8) Derby, (9) Wyndham, (10) Darwin, (11) Karumba, (12) Weipa, (13) Cairns, and its region south to Innisfail, (14) Innisfail, (15) Townsville, (16) Gladstone, (17) Brisbane, (18) Sydney (Botany Bay generally, and Kurnell within Botany Bay), (19) Batemans Bay, and (20) Port Phillip Bay.

The impacts at these sites, viewed nationally, range from minor to major. The sites of Carnarvon, Broome, Derby, Wyndham, and Weipa have minor impacts and involve mainly boat harbour construction and channels, some sewage outfall, and land reclamation for roads. The sites of Batemans Bay and Port Phillip Bay, viewed nationally, show medium impacts and involve landfill for urban and industrial estates and roads, use of mangrove environments as sewage outfall and refuse disposal, and constructing ports and harbour facilities. Major impacts have occurred at Bunbury, Dampier, Port Hedland, Darwin, Townsville, Cairns, Innisfail, Gladstone, Brisbane, and Sydney (Botany Bay generally, and Kurnell within Botany Bay) – here, there has been development of ports, reclamation of land for urban, industrial, and recreational estates, roads, airports, and refuse disposal. For the Queensland sites, there has been development of the hinterland for sugar cane plantations. Sugar cane plantations are particularly interesting as an ecological impact because generally they are located on the lowland or stranded floodplains that form the hinterland to mangrove embayments or coastal lowlands – consequently, they usually directly form the upslope drainage basin to the coastal mangrove system and hydrological and hydrochemical changes that they effect are transmitted to the mangrove ecosystems – as such they are "distal impacts".

In terms of areal extent, while there was not widespread clearing of mangroves at Bunbury in Western Australia, approximately 50% of the original stand was destroyed but this location is the most southern occurrence of *Avicennia marina* in the State (Western Australia) and represents a scientifically significant outlier (Semeniuk et al. 2000a, b). Further, the Bunbury mangroves inhabited a tide-dominated delta, which was an unusual (species-cum-habitat) situation for estuarine deltas in south-western Australia. Viewed from these perspectives, the removal of part of the Bunbury mangroves was a State-wide significant impact. In regards to the mangroves of Dampier, this area represents an occurrence of an archipelago/ria shore with its array of ria-style habitats in an arid zone (Semeniuk 1996) – it is the only occurrence globally of an arid-climate ria shore with species representing the arid-zone cline of mangroves of the Old World Mangrove setting – as such, though overall a small percentage of mangroves were destroyed by dredge spoil fill and by roads, the destruction was of a globally unique and significant stand. The destruction of mangroves in Darwin, though small in comparison to the total occurrence of mangroves in Port Darwin, was of a system that belongs to the humid-climate end of the spectrum of mangroves in ria coastal settings in Australia (Semeniuk 1985a, b). Similar to the argument presented for the Dampier mangroves, Darwin represents the humid-zone cline of mangroves of the Old World Mangrove setting in Australia – the destruction of these mangroves was of a nationally unique and significant system.

Figure 9.9 illustrates the range of impacts for the 20 case studies around Australia. Figure 9.10 illustrates an aerial view of the range of impacts on the mangroves of Bunbury and at Carnarvon. Figures 9.11 and 9.12 illustrate aerial views of the range of impacts on the mangroves of Dampier and Port Hedland. Figure 9.13 shows aerial views of the range of impacts on mangrove coasts at Darwin and at Brisbane.

9.9 A Specific Case Study: The Port Hedland Model

There is a focus on Port Hedland, Western Australia in this Chapter because it provides some details of the variety of activities that occur in industrial areas with time, the multiplicity of functions that ports and industrial areas evolve in time, and the overt and covert impacts that have affected mangroves in this area. Located in a globally significant coast (Semeniuk 1996; Brocx and Semeniuk 2015), the documentation of the industrial impacts become relevant in that they illustrate the often inexorable drive toward

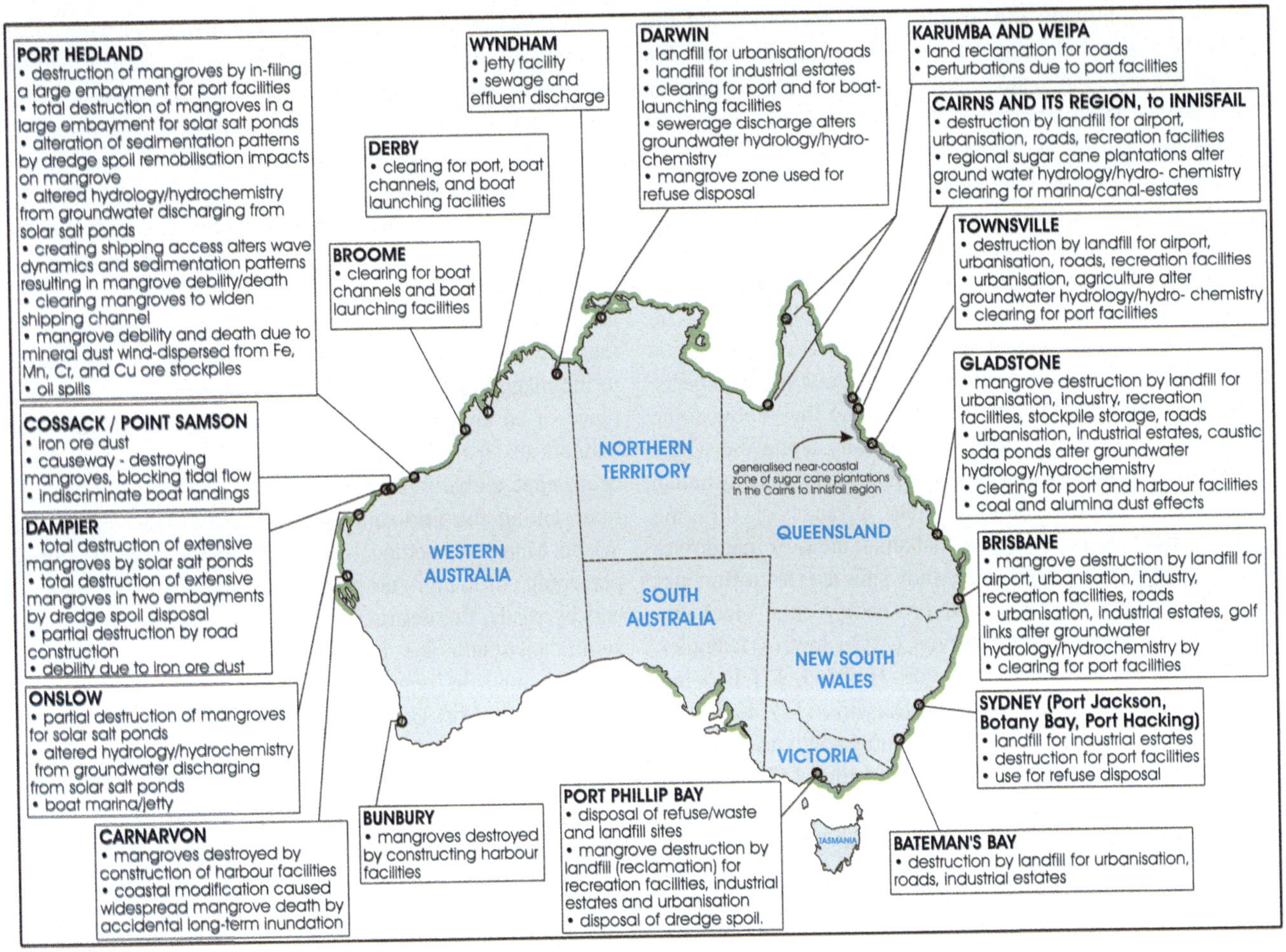

Fig. 9.9 The range of anthropogenic impacts on mangroves at the 20 sites around Australia

industrial development, mineral and energy exploitation, and national wealth creation to the detriment of the environment.

Port Hedland is located along the Pilbara Coast and is part of a shore-parallel limestone barrier complex in a macrotidal environment (Semeniuk 1996). The limestone barriers protect extensive mangrove-lined tidal creek complexes and salt flats from the open wave-dominated Indian Ocean, and between adjoining barriers there is a natural large tidal-exchange channel (an inlet). Semeniuk (1996) considered the Pilbara Coast as one of a limited number of arid coasts worldwide to be a globally significant coast and, in this context, considered the mangrove assemblages at Port Hedland that were part of a limestone barrier system also to be globally significant. The Port Hedland industrial complex resides wholly within this type of coast.

The climate of Port Hedland is arid, "pseudo-monsoonal" (Gentilli 1971). The summers are hot (daily maximum 37–40 °C (Bureau of Meteorology 1988) and rainfall is variable, falling mainly in the summer season. The critical factor in the climate in relation to industry is the strong wind: there are two wind seasons, a strong south-westerly through to north-westerly winds in summer, and in winter easterly through to southerly winds (Gentilli 1971; Bureau of Meteorology 1988; Steedman 1985; Steedman and Coleman 1985); the land-breeze/sea-breeze system operates throughout the year. The dryness of the climate and the strength of the winds have implications for ore stockpiles that are usually comprised of fine-grained crushed ore, or at least chunks of ore that have a significant fine-grained component.

Port Hedland has a long history as a port, as the potential of turning the tidal-exchange channel and tidal creek complex into a port was recognised early (in the late 1800s) – Peter Hedland arrived in the area and named the harbour 'Mangrove Harbour' and considered it would make a well protected harbour. It was first developed in order to service the needs of the local pastoral industry in the Pilbara, and the first jetty was built in 1896, which was extended in 1908 after the discovery of gold inland. The port was predominantly used to import goods and stores for the local industries and to export pearl, shell, wool, and livestock, and gold, tin and copper, and later manganese ore.

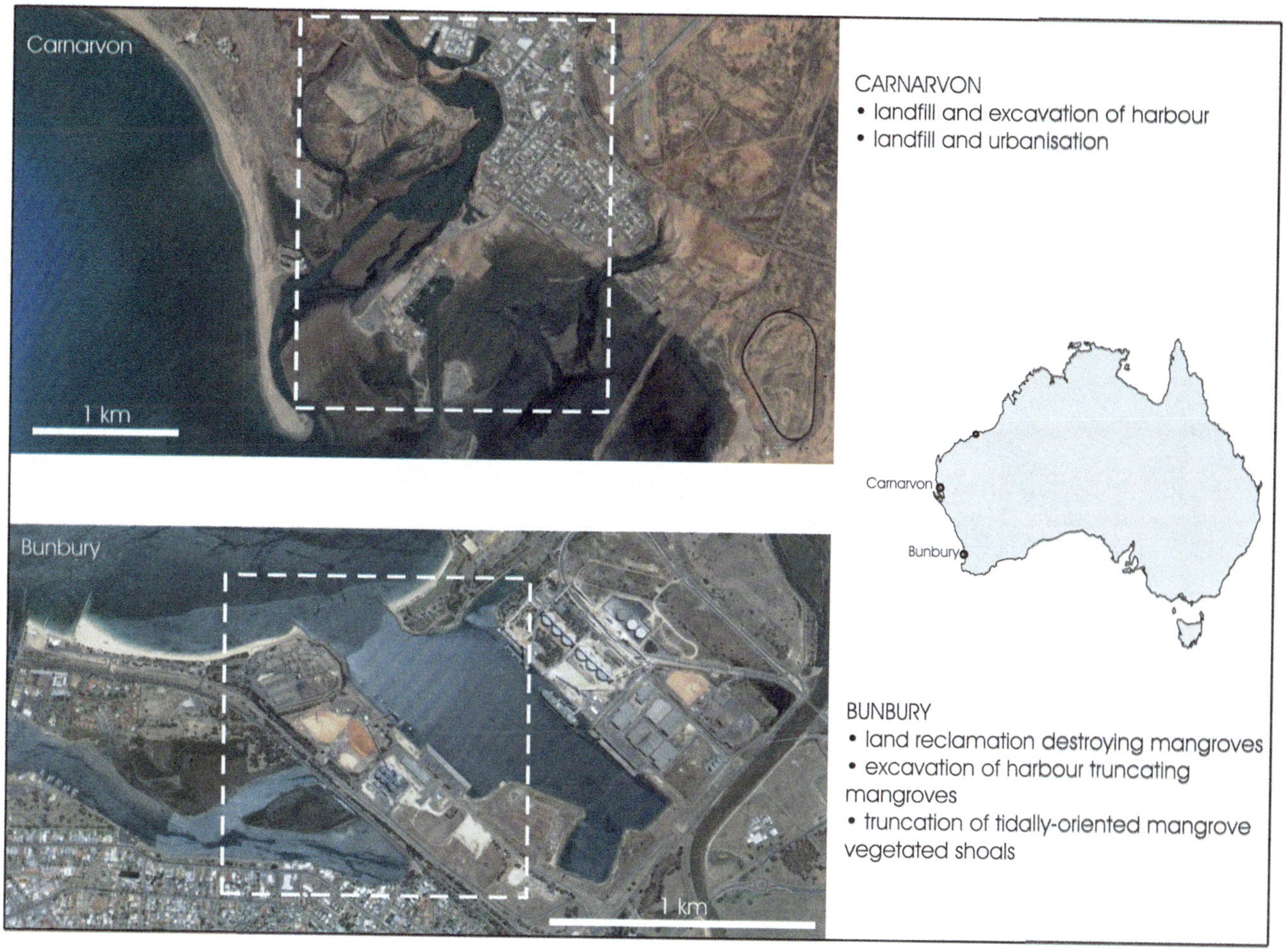

Fig. 9.10 Annotated aerial views of anthropogenic impacts on mangrove coasts at Bunbury and at Carnarvon

After iron ore became an important export commodity, Port Hedland markedly expanded and has been in existence as an enlarged port since 1960 (Hardie 1981; Shaw 2006), and shipment of ore began in 1966. Currently, Port Hedland is the largest tonnage port in Australia (Shaw 2006) and it has expanded its operations from its original inception to export iron ore to now export various mineral ores from the Pilbara hinterland (Fe, Mn, Cu, Cr), manufactured iron ore briquettes, and salt, and to import nitrate, cement, and oil. To date, there have been numerous mining companies that have used or that are using Port Hedland as an exporting facility. Over the years, many have been subsumed into larger companies, or have changed names, and the history of usage and company changes has been complex.

The effects of port development and other industrial developments on this mangrove coast have been: infilling of mangrove-vegetated tidal lagoons to create above-tide surface for port infrastructures, the dredging of shipping channels to accommodate large ships, the expansion and deepening of shipping channels and ship turning circles, the use of reclaimed terrain as ore stockpile storage sites, the use of tidal flat terrain as evaporation basins for producing solar salt (causing permanent inundation), the storage of salt in stockpiles, construction of roads and causeways, and the disposal of dredge spoil in locations where it has been or can be remobilised, and the storage of fuel facilities along the shore or near the port. Most of these developments have overt impacts on the mangroves of the region. The entire Port Hedland region of barrier islands, tidal creeks, and salt flats is now an extensive multiple-use industrial complex and in actuality it was/is located in the midst of a once significant mangrove complex.

Apart from the overt destruction and impacts of the industries on mangroves listed above, some of the other overt consequences of industrial developments have been: mobilisation of ore dust in this arid wind-dominated climate

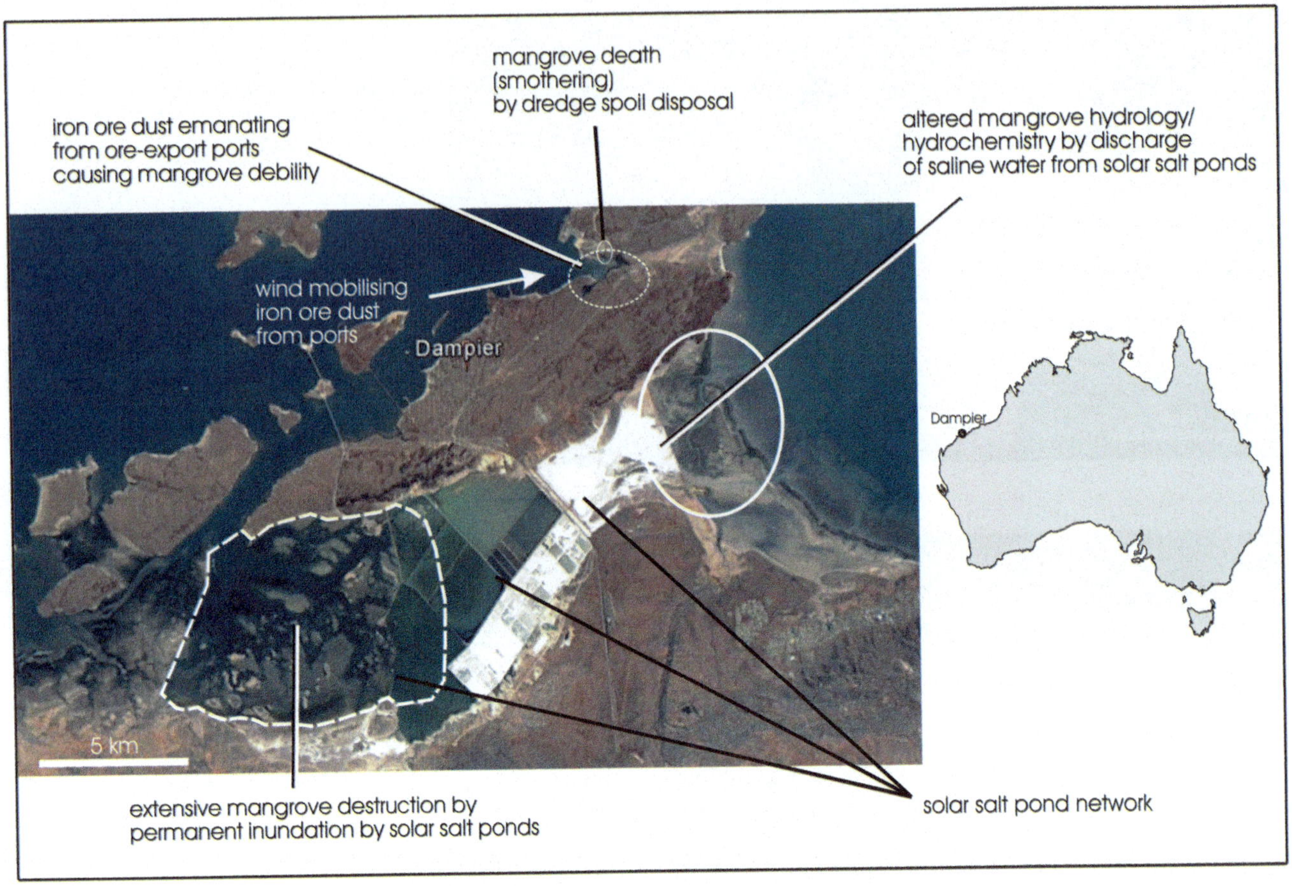

Fig. 9.11 Annotated aerial view of anthropogenic impacts on mangrove coasts at Dampier

which has coated the canopies of the mangroves causing debility and death (Semeniuk 2007, 2012) (Fig. 9.14), dredge and oil spills, and remobilisation of dredge soil that was placed in hydrodynamically inappropriate locations, or dredged spoil that has been trapped by construction of causeways that have arrested tidal-exchange.

The effects of industrialisation that are covert are as follows: ore dust settling on sediment/soil has been burrowed/bioturbated into the subsurface where it is now involved in geochemical, biochemical and hydrochemical exchanges; ore dust settling on sediment/soil dust has been inadvertently caught up in the feeding by snails and crabs and incorporated as exotic metals into their flesh; where chemical recycling has taken place exotic cations (Mn, Fe, Cr) in the groundwater environment under the mangroves have been taken up by the plants and incorporated into their leaf tissue; rainwater recharge through salt storage stockpiles, and ore storage stockpiles have picked up excess salt and the exotic cations and have delivered them to the underlying ground water; and finally, the various solar salt evaporating basins with their variable salinities and potential to crystallise carbonate, gypsum, and halite have, through subsurface discharge and seepage, contaminated and altered the natural hydrology and hydrochemistry of the tidal flats (Brocx and Semeniuk 2015). Figure 9.14 illustrates a range of impacts by industry in the mangrove environment.

9.10 Discussion and Conclusions

Australia has a rich biodiversity of mangroves grading from the cline located in the species-rich Old World mangroves. As described in this Chapter, Australia uniquely carried this species-rich ensemble of mangroves asymmetrically southwards through one of the most arid regions on Earth along its west coast and to a temperate climate along its humid east coast. This is a globally unique situation. Complementing this biodiversity and its asymmetric expression in west-coast versus east-coast, Australia presents by nature of its geology, oceanography, and climate a variable heterogeneous coast that has resulted in diverse mangrove habitats and diverse mangrove assemblages. Many of the locations of the

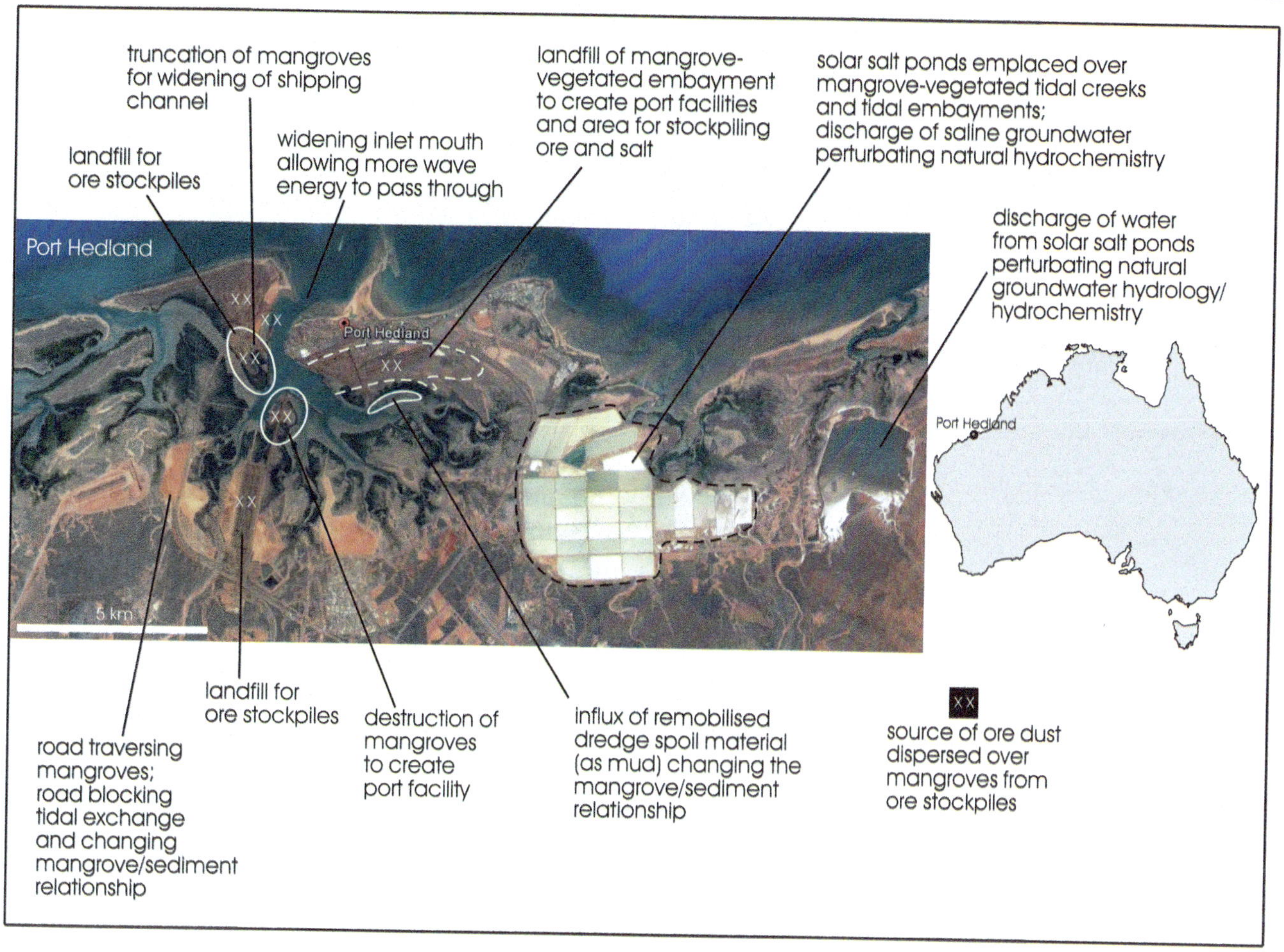

Fig. 9.12 Annotated aerial view of anthropogenic impacts on mangrove coasts at Port Hedland

Australian coast have high heritage value as unique and significant ensembles of mangrove. In this perspective, some of Australia's mangrove coasts have achieved World Heritage recognition areas as part of broader listings, but many other areas are of similar value for the global community. Australia, with its relatively low human population, has seriously impacted or destroyed valuable mangrove ecosystems that stood as World-class models of mangrove ecology akin, but vastly different to, the mangroves of the World Heritage area of the Sunderbans.

A tour of the Australian coast where there are major human populations and large industries paints a picture of exploitation and unsympathetic development, mangrove destruction, and/or major impacts on the mangrove systems. Largely, to date, there has been no recognition of the global heritage asset of Australia's mangrove coasts in terms of its coastal types and the expression of complex communities linked to the various coastal types, and in terms of the significance of assemblages (or communities) of mangroves linked to inter-related suites of habitats (such as the inter-related suite of habitats within ria coast embayments and their resulting mangrove communities; cf., Semeniuk 1985a; Cresswell and Semeniuk 2011). In order to achieve the appropriate recognition of the global, national and regional value of particular mangrove ecosystems it is necessary to fully document the suites of coastal types nationally that host mangroves and to assess their component mangrove systems nationally using a consistent approach. This, in turn, needs to lead to the development of a national approach for implementing consistent management. Cresswell and Semeniuk (2017) propose a national mangrove conservation strategy is required to hierarchically

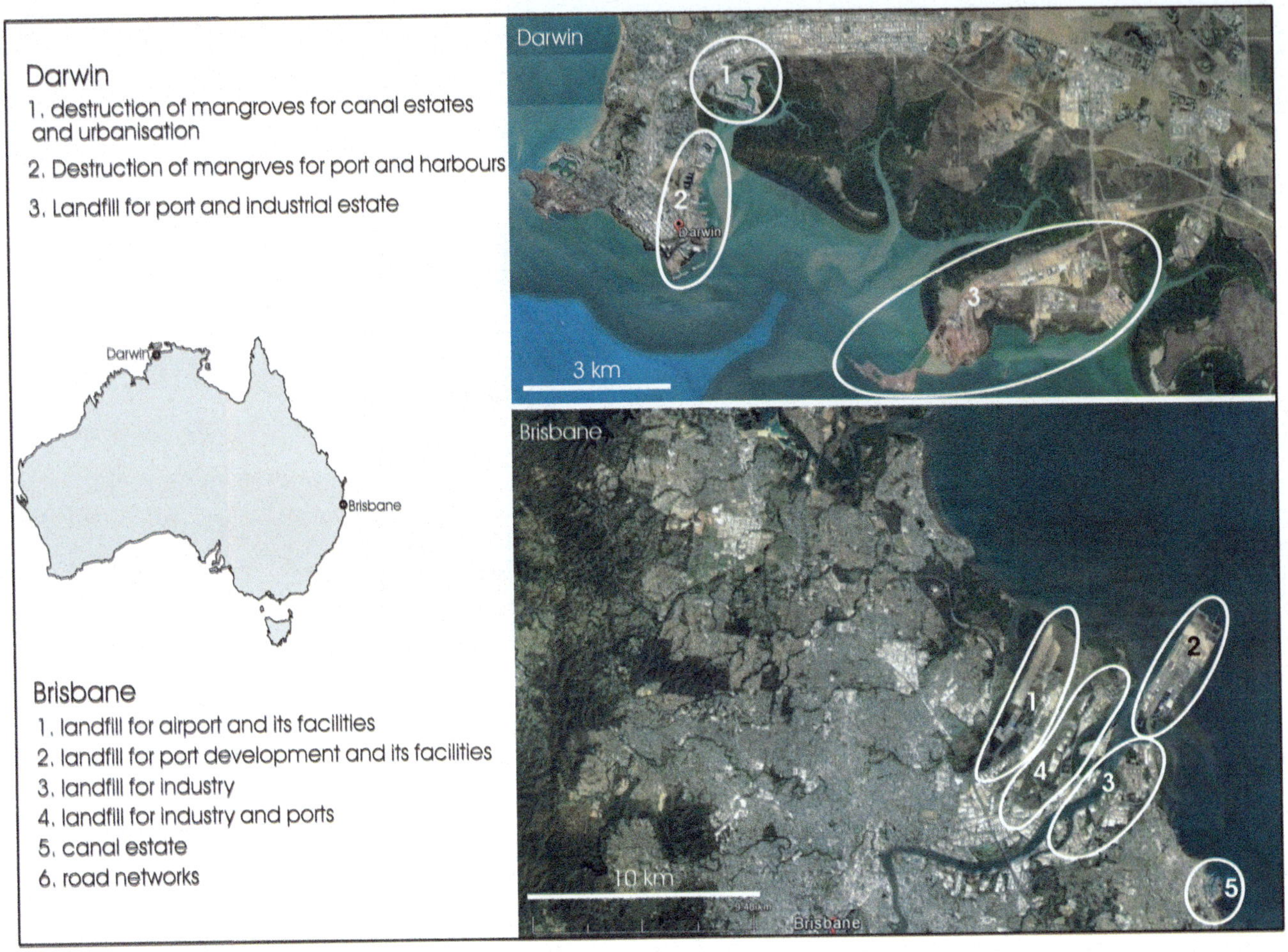

Fig. 9.13 Annotated aerial views of anthropogenic impacts on mangrove coasts at Darwin and at Brisbane

combine existing government biodiversity and coastal management strategies, policies and legislation to achieve both adequate mapping and conservation of Australia's mangrove systems.

While all the impacts documented in this Chapter have occurred in the past, mangroves are still being cleared, destroyed, or impacted by a range of anthropogenic activities. An important step to ensure appropriate conservation of mangrove ecosystems is to recognise the significance of the Australian mangroves as a global asset. Through a systematic mapping of coastal types and mangrove ecosystems, our coastal natural history resource can be better known and available for study and development assessment. With adequate knowledge it is then possible to plan for use of the coastline identifying globally significant areas for protection, and planning effectively the use of the land that already has been impacted. An important first step will be to document the current and planned destruction/alteration of mangroves including planned coastal infrastructure placement that will impact mangroves. These areas are the most urgent to be mapped and assessed such that management protocols can be put in place to best manage these mangrove coasts with a full understanding of their significance.

Finally, the effect of covert impacts needs to be fully understood and addressed in all coastal development. For instance, while solar salt production created obvious overt impacts, Brocx and Semeniuk (2015) highlighted the covert impacts that had not been recognised. It is only by fully identifying the values and the threats that it will be possible to maintain one of the most complex biological, geochemical, biochemical, hydrological, and hydrochemical environments on Earth, Australia's mangrove ecosystems.

Fig. 9.14 A range of impacts in the Port Hedland area. (**a**) Ship visible in the background berthed in a channel that has been cut through the mangroves; in the *middle* and background is *Avicennia marina* scrub that has been blanketed by Mn ore dust which appear *grey-green*; compare the colour with the *yellow-green* of the unaffected leaves in inset (**c**); to the *right* is the landfill that forms the base for ore stockpiles. (**b**) Plume of mineral Mn ore dust coating the mangroves; to the *right* the bright *yellow green* intact mangroves can be seen as a colour contrast; ship visible in the background berthed in a channel that has been cut through the mangroves. (**c**) Mangrove leaves not affected by Mn dust fall-out – leaves are bright *green* and healthy. (**d**) Mangrove leaves blanketed by Mn dust fall-out – leaves are debilitated, curling, and necrotic; compare this colour with the *yellow-green* of the leaves in inset (**c**). **d** Close-up of inset (**d**) – leaves are debilitated, curling, and necrotic; there is the beginning of scattered new growth of some *yellow-green* leaf shoots in response to this debility. (**e**) Surface of mangrove mud sediment floor covered in veneer of Mn ore dust (1); crabs have excavated fresh uncontaminated mud (2) from depth to the surface; note the feeding pellets of the crabs that are contaminated by Mn ore dust (3)

References

Andreetta A, Fusi M, Cameldi I, Cimo F, Carnicelli S, Cannicci S (2014) Mangrove carbon sink. Do burrowing crabs contribute to sediment carbon storage? Evidence from a Kenyan mangrove system. J Sea Res 85:524–533

Blaber SJM (1997) Fish and fisheries of tropical estuaries. Chapman & Hall, London

Brocx M (2008) Geoheritage: from global perspectives to local principles for conservation and planning. Western Australian Museum (Monograph), Welshpool. 175p. ISBN 978-1-920843-35-9

Brocx M, Semeniuk V (2011) The global geoheritage significance of the Kimberley Coast, Western Australia. J R Soc West Aust 94:57–88

Brocx M, Semeniuk V (2015) The solar salt ponds along the Pilbara Coast, Western Australia – a case study of a coastline of global geoheritage significance used for industrial purposes. In: Peppoloni S, Wyss M (eds) Geoethics: the role and responsibility of Geoscientists. The Lyell Collection, The Geological Society Special Issue 419, https://doi.org/10.1144/SP419.18

Bunt JS, Williams WT, Duke NC (1982) Mangrove distributions in North-East Australia. J Biogeogr 9:111–120

Bureau of Meteorology (1988) Climatic atlas of Australia. Department of Administrative Services, Australian Government Publishing Service, Canberra

Burns KA, Garrity SD, Levings SC (1993) How many years until mangrove ecosystems recover from catastrophic oil spills? Mar Pollut Bull 26:239–248

Busby JR, Bridgewater PB (1986) A preliminary atlas of Australian mangroves. AGPS, Canberra. 38pp

Chapman VJ (1977) Wet coastal ecosystems. Elsevier Scientific Pub. Co. 428p

Clark JR (1977) Coastal ecosystem management. Wiley, New York. 928 pp

Cresswell ID, Semeniuk V (2011) Mangroves of the Kimberley Coast, Western Australia: ecological patterns in a tropical ria coast setting. J R Soc West Aust 94:213–237

Cresswell ID, Semeniuk V (2017) Australian mangroves – their distribution and protection. In: Makowski C, Finkl CW (eds) Threats to mangrove forests: hazards, vulnerability, and management, Springer this volume, pp xx-yy

Cresswell ID, Semeniuk V (2018) Mangroves of Australia – their biodiversity in relation to climate, coastal setting, and heterogeneity of habitats. Book in preparation for Springer

Cresswell ID, Bridgewater PB, Semeniuk V (2011) A synopsis of the coastal vegetation of the Kimberley region. J R Soc West Aust 94:197–206

Duke NC, Ball MC, Ellison JC (1998) Factors influencing biodiversity and distributional gradients in mangroves. Glob Ecol Biogeogr Lett 7:27–47

Ellison AM, Proffitt CE, Devlin DJ, Bros SM (1999) Cumulative effects of oil spills on mangroves. Ecol Appl 9:1490–1496

Ferguson-Wood EJ, Johannes RE (eds) (1975) Tropical marine pollution. Elsevier Scientific Publishing Company, Amsterdam. 192 pp

Gentilli J (ed) (1971) Climates of Australia and New Zealand, World Survey Climatology, vol 13. Elsevier, Amsterdam. 405p

Giri C, Long J, Abbas S, Murali RM, Qamer FM, Pengra B, Thau D (2015) Distribution and dynamics of mangrove forests of South Asia. J Environ Manag 148:101–111

Gordon DM (1988) Disturbance to mangroves in tropical-arid Australia: hypersalinity and restricted tidal exchange as factors leading to mortality. J Arid Environ 15:117–145

Hardie J (1981) Nor'westers of the Pilbara breed. Shire of Port Hedland, Port Hedland. ISBN 0-9594155-0-5

Hegerl EJ (1982) Mangrove management in Australia. In: Clough BF (ed) Mangrove ecosystems in Australia: structure, function and management. A.N.U. Press, Canberra, pp 275–288

Horne D (1964) The lucky country. Penguin, Sydney

Hutchings P, Saenger P (1987) Ecology of mangroves. University of Queensland Press, St Lucia, Queensland

Jernelov A, Linden O (1983) The effects of oil pollution on mangroves and fisheries in Equador and Columbia. Biology and ecology of mangroves, Tasks for vegetation science, vol 8. Dr W Junk Publishers, The Hague, pp 185–188

Lewis RR III (1983) Impact of oil spills on mangrove forests. Biology and ecology of mangroves, Tasks for vegetation science, vol 8. Dr W Junk Publishers, The Hague, pp 171–183

Lewis M, Pryor R, Wilking L (2011) Fate and effects of anthropogenic chemicals in mangrove ecosystems: a review. Environ Pollut 159:2328–2346

Morton JF (1965) Can the red mangrove provide food, feed and fertiliser? Econ Bot 19:113–123

Nagelkerken I, Blaber SJM, Bouillon S, Green P, Haywood M, Kirton LG, Meynecke J-O, Pawlik J, Penrose HM, Sasekumar A, Somerfield PJ (2008) The habitat function of mangroves for terrestrial and marine fauna: a review. Aquat Bot 89:155–185

Naidoo G, Chirkoot D (2004) The effects of coal dust on photosynthetic performance of the mangrove, *Avicennia marina* in Richards Bay, South Africa. Environ Pollut 127(3):359–366

Nordhaus I, Salewski T, Jennerjahn TC (2011) Food preferences of mangrove crabs related to leaf nitrogen compounds in the Segara Anakan Lagoon, Java, Indonesia. J Sea Res 65(4):414–426

Odum WE, Heald EJ (1972) Trophic analysis of an estuarine mangrove community. Bull Mar Sci 22(3):671–738

Odum WE, Johannes RE (1975) The response of mangroves in man-induced environmental stress. In: Ferguson-Wood EJ, Johannes RE (eds) Tropical marine pollution. Elsevier Scientific Publishing Company, Amsterdam, pp 52–62

Pace ML, Lovett GM (2013) Primary production: the foundation of ecosystems. In: Weathers KC, Strayer DL, Likens GE (eds) Fundamentals of ecosystem science. Elsevier, Amsterdam, pp 27–51

Percival M, Womersley JS (1975) Floristics and ecology of the mangrove vegetation of Papua New Guinea, Botany Bulletin, vol 8. Department of Forests, Lae. 96 pp

Robertson AI, Alongi DM (2016) Massive turnover rates of fine root detrital carbon in tropical Australian mangroves. Oecologia 180 (3):841–851

Robertson AI, Daniel PA (1989) Decomposition and the annual flux of detritus from fallen timber in tropical mangrove forests. Limnol Oceanogr 34:640

Saenger P (2002) Mangrove ecology, silviculture and conservation. Kluwer Academic Publisher, Dordrecht

Saenger P, Hegerl EJ, Davies JDS (1983) Global status of mangrove ecosystems. Commission on Ecology Papers Number 3. International Union for Conservation of Nature and National Resources, Gland, Switzerland, 88 pp

Semeniuk V (1980) Mangrove zonation along eroding coastline, King Sound, NW Australia. J Ecol 68:789–812

Semeniuk V (1981) Sedimentology and the stratigraphic sequence of a tropical tidal flat, North-Western Australia. Sediment Geol 29:195–221

Semeniuk V, Chalmer PN, LeProvost IP (1982) The marine environments of the Dampier Archipelago. J R Soc West Aust 65:97–114

Semeniuk V (1983) Mangrove distribution in Northwestern Australia in relationship to freshwater seepage. Vegetatio 53:11–31

Semeniuk V (1985a) Development of mangrove habitats along ria shorelines in north and northwestern Australia. Vegetatio 60:3–23

Semeniuk V (1985b) Mangrove environments of Port Darwin, Northern Territory: the physical framework and habitats. J R Soc West Aust 67:81–97

Semeniuk V (1986a) Terminology for geomorphic units and habitats along the tropical coast of Western Australia. J R Soc West Aust 68:53–79

Semeniuk V (1986b) Threats to, and exploitation and destruction of, mangrove systems in Western Australia. In: Field CD, Dartnall AJ (eds) Mangrove ecosystems of Asia and the Pacific; status, exploitation and management. Proceedings of the Research for Development Seminar. Australian Institute of Marine Science, Townsville, pp 228–240

Semeniuk V (1993) The mangrove systems of Western Australia – 1993 Presidential Address. J R Soc West Aust 76:99–122

Semeniuk V (1994) Predicting the effect of sealevel rise on mangroves in northwestern Australia. J Coast Res 10:1050–1076

Semeniuk V (1996) Coastal forms and quaternary processes along the arid Pilbara coast of northwestern Australia. Palaeogeogr Palaeoclimatol Palaeoecol 123:49–84

Semeniuk V (2007) Mangrove habitats, Port Hedland Harbour 2007. Report by V & C Semeniuk Research Group to Port Hedland Port Authority. Copy of report residing in the Battye Library, Perth, Western Australia

Semeniuk V (2008) Sedimentation, stratigraphy, biostratigraphy, and Holocene history of the Canning Coast, north-western Australia. J R Soc West Aust 91:53–148

Semeniuk V (2011) Stratigraphic patterns in coastal sediment sequences in the Kimberley region, Western Australia: products of coastal form, oceanographic setting, sedimentary suites, sediment supply, and biogenesis. J R Soc West Aust 94:133–150

Semeniuk V (2012) Manganese ore dust contamination, Port Hedland, 2010–2012. Report by V & C Semeniuk Research Group to Port Hedland Port Authority. Copy of report residing in the Battye Library, Perth, Western Australia

Semeniuk V (2015a) Mineralization. In: Kennish J (ed) Encyclopaedia of estuaries. Springer, Dordrecht, p 438–447. ISBN: 9789401788021

Semeniuk V (2015b) Marine/freshwater mixing. In: Kennish J (ed) Encyclopaedia of estuaries. Springer, Dordrecht, p 404–417. ISBN: 9789401788021

Semeniuk V (2015c) Tidal flat salinity gradient. In: Kennish J (ed) Encyclopaedia of estuaries. Springer, Dordrecht, p 678–683. ISBN: 9789401788021

Semeniuk V, Brocx M (2011) King Sound and the tide-dominated delta of the Fitzroy River: their geoheritage values. J R Soc West Aust 94:151–160

Semeniuk CA, Semeniuk V (1998) Wetland regions of the southern Swan Coastal Plain 1998. Report to Water & Rivers Commission, Perth, Western Australia

Semeniuk CA, Semeniuk V (2014) Wetlands of the Roebuck Plains 2012–2013 – distribution, plant composition, ecological values, and status. Report by V & C Semeniuk Research Group to Nyamba Buru Yawuru Ltd. Copy of report residing in the Battye Library, Perth, Western Australia

Semeniuk V, Wurm P (1987) Mangroves of Dampier Archipelago, Western Australia. J R Soc West Aust 69:29–87

Semeniuk V, Kenneally KF, Wilson PG (1978) Mangroves of western Australia. Western Australian Naturalists Club, Perth. Handbook 12, 90 pp.

Semeniuk V, Semeniuk TA, Unno J (2000a) The Leschenault Inlet estuary – an overview. J R Soc West Aust. Special Issue on the Leschenault Inlet Estuary 83:207–228

Semeniuk V, Tauss C, Unno J (2000b) *Avicennia marina* in the Leschenault Inlet area. J R Soc West Aust. Special Issue on the Leschenault Inlet Estuary 83:317–334

Shaw M (2006) Moving mountains: the evolution of Port Hedland Harbour. Hesperian Press, Carlisle. ISBN 085905389X

Smith TJ III, Duke NC (1987) Physical determinants of inter-estuary variation in mangrove species richness around the tropical coastline of Australia. J Biogeogr 14:9–19

Steedman RK (1985) Air quality meteorology of the Burrup Peninsula, Western Australia, June 1982 to December 1984. Vol. 1 text and appendices A to G. R.K. Steedman Ltd. Report no. R271 August 1985, Unpublished Report to Woodside Offshore Petroleum Ltd, Perth.

Steedman RK, Coleman EJ (1985) On the joint probability of extreme winds, waves and currents at North Rankin 'A', north western Australia. In: E + P Forum (Exploration & Production Forum), The application of joint probability of Metocean phenomena in the oil industry's structural design work. The Oil Industry International Exploration & Production Forum, London

Stieglitz T, Ridd PV, Müller P (2000) Passive irrigation and functional morphology of crustacean burrows in a tropical mangrove swamp. Hydrobiologia 421:69–76

Tomlinson PB (1986) The botany of mangroves. Cambridge University Press, Cambridge. 413 p

Unno J, Semeniuk V (2009) The habitats of the Western Australian soldier crab Mictyris occidentalis Unno 2008 (Brachyura: Mictyridae) across its biogeographical range. J R Soc West Aust 92:289–363

Wightman GM, Andrews M (1989) Mangroves of the Northern Territory. Conservation Commission of the Northern Territory, Palmerston

10 Determining the Influence of Urbanization on Mangrove Zones of Northeastern Brazil: Characterization of Ceará State Coastal Zone Organic Matter Inputs

Stéphane Jean Louis Mounier, Rozane Valente Marins, and Luiz Drude de Lacerda

Abstract

Mangrove zones in Brazil are extremely sensitive to impacts from global climate change and urbanization. To understand the effects of urbanization on organic matter inputs in the coastal zone of Ceará State (northeastern Brazil), seasonal campaigns were carried out in two of the most environmentally significant river/mangrove systems within the region: the Cocó river, in the metropolitan region of the huge city of Fortaleza, and the Pacoti River, an environmental protection area, without urban influence, on the east coast of the state. Additionally, a spatial study was conducted along the Jaguaribe River, the largest river in the state connected a mangrove area. Organic matter characterization by dissolved organic carbon, spectrofluorescence and fluorescence quenching, and metal complexation capacity shows clearly that under urban pressure, the produced organic matter in mangrove areas is different, proving that carbon budget and pollutant fate have to be revised to take in account this difference.

Keywords

Mangrove · Organic matter · Fluorescence · Metal complexation · Urban pressure

10.1 Introduction

The global dispersion of man-made pollutants in pristine systems and the high levels of metals raise serious questions about their effects on ecosystems. These contaminants accumulate in organisms and undergo biomagnification in the food web, reaching harmful concentrations, including to human health by the consumption of animals of high trophic level species (Doney 2010; Bisi et al. 2012). The continent-ocean interface encompasses processes in the drainage basins of affluent rivers, ocean dynamics, both on the continental shelf and in the deep ocean and atmospheric processes. These define the magnitude and variability of flows and material dynamics through the land-sea gradient. These processes are affected by anthropogenic actions both locally and globally, and are also influenced by global climate change, which affects both continental and oceanic forcing (Jennerjahn et al. 2010). It is in the estuary, where all these processes interact, that environmental changes are more significant.

Estuaries are natural traps that "capture" continental materials transported by rivers. The gentle slope of the coastal strip creates extensive estuarine areas where the residence time of the waters is greater. During transport at the continent-ocean interface significant biogeochemical transformations of the transported materials occur and the intensity of these transformations is mostly related to the residence time of the water mass in the estuary, where the hydrochemical and hydrological dynamics is more complex. The residence time of the water body in estuaries is a function of the magnitude of the fluvial discharge and the intensity of the oceanic forcing on the continental shelf and on the estuarine region varying according to the interaction between these forces.

In mangroves, the residence time is greater than in estuaries and many pollutants are retained permanently. This accumulation leads to a specific biochemical environment that can induce changes in ecosystem's functioning. The objectives of this work are to investigate at the organic matter level the impact of urbanization and the tidal effect on the organic matter from local mangroves in river/mangrove systems along the coast of Ceará State in North-eastern Brazil.

S. J. L. Mounier (✉)
Laboratoire PROTEE, Université de Toulon, Toulon Cedex 09, France
e-mail: stephane.mounier@univ-tln.fr

R. V. Marins · L. D. de Lacerda
Universidade Federal do Ceara, Labomar, Fortaleza, Brazil

C. Makowski, C. W. Finkl (eds.), *Threats to Mangrove Forests*, Coastal Research Library 25,
https://doi.org/10.1007/978-3-319-73016-5_10

10.2 Study Sites

The metropolitan region of Fortaleza is located in the north-eastern state of Cearà, Brazil. It is made up of nine cities (Aquiraz, Caucaia, Eusébio, Fortaleza, Guaiúba, Itaitinga, Maracanaú, Maranguape and Pacatuba) and witnessed a great expansion in the recent years (Brandão 1995). This region has several river mouths summing up 1,522 ha of mangrove (Godoy 2015). These areas host several species that sustain biodiversity and also the local economy (SDU 1994).

The three most important mangrove areas of the metropolitan region are fed by the Cocó, Pacoti and Ceara rivers (Fig. 10.1), these areas has remained relatively stable from 2000 to 2004, as well as the total mangrove area in the State, totalling about 14,000 ha. Annual precipitation is 1,000 mm and occurs principally in March and June. In 2004, however, at the east coast of Ceará state, from Fortaleza to the Jaguaribe basin area, rainfall was abnormally high. In January, for example, month rainfall reached 404 mm and 378 mm, respectively, against a historical average of 105 mm and 82 mm, respectively (Xavier 2004). The average temperature ranges from 24 to 26 °C with a variability of 6 °C at maximum.

The fourth study area is the Jaguaribe river, the largest basin area to the state and the largest mangrove area in the eastern coast. It is also the major freshwater supply to the metropolitan region through a series of diversion canals. Table 10.1 summarizes the geographical information for the four river basins studied.

10.3 Description of the Studied Basins

10.3.1 The Cocó River

This river initiated in the Serra da Aratanha (Pacatuba) and has a length of 50 km with a basin area of 485 km^2. The river crosses the cities of Aquiraz, Pacatuba and Fortaleza. At the estuary, it meets with the Coaçu river, basin area of 375 ha, most of them covered by mangroves that expands up to 11 km inland (Miranda et al. 1988; SDU 1994) and covers about 452 ha (Godoy 2015). Human impact on the river is important because of garbage, cattle ranching, industrial activity and water use (Mavignier 1992), particularly over the last 24 km. Sixty percent of the metropolitan region is covered by the Cocó River basin. Detailed information is available in (Stanford 1986) and for climatology, geology and hydrology in the report of the "Study of the life of the Cocó River" AUMEF/Government of Ceará 1988. Mainly four types of geological formation are found in the basin: sand dunes covered by vegetation or already urbanized, clayey sediments rich in organic clay, sandy-clayey districts and in the interior with crystalline formation.

The mangrove vegetation at the estuary is composed by *Avecinia germinans* (black mangrove), *Laguncularia racemosa* (white magrove) and *Rhizophora mangle* (red mangrove). The land-mangrove interface harbours many species of grasses and shrubs, coconut, hose, cashew. Mangrove vegetation harbours a great variety of fish, 23 out of 25 are marine. Upstream the river is hypereutrophic, with algae blooms and abundant aquatic macrophytes. In the rainy season, the materials transported are deposited on beaches and are responsible for black spots generally confused with pollution by the population.

The industrial district of Maracanaú is situated 15 km from the centre of Fortaleza. It counts 41 companies (chemical, fabric, metallurgy, food, mineral and non-metallic exploitation, ...) (Mavignier 1992). The wastewater treatment station (WWTS) built since the beginning of the area development is already oversized. Since (1992) retention basins have been implanted. The last one of them leaks in the Maranguapinho River. An anaerobic lagoon followed by ripening lagoons (792,000 m^3) allows a residence time of 28 days with a flow rate of 840 L.s^{-1}. But it is not the totality of the urban effluents that are collected.

The contaminant content in the river seems low for fecal coliform, heavy metals and Biological Oxygen Demand (BOD). The general parameters observed in the Cocó River water are summarized in the Table 10.2.

There is an amplification of the concentrations of copper (Cu) and lead (Pb) in the animals across the food chain (Mavignier 1992). Leaching appears to be the main source of pollutants; however, metal concentrations in waters are lower than Brazilian legal standards (CONAMA standards). The leachate composition shows a very high organic matter concentration (Analyzed from 21/02/90 to 21/05/91).

10.3.2 The Pacoti River

The Pacoti river has its source in Baturite Mountain chain and has a length of 150 km. It 1,800 km^2 basin includes the contributory rivers Agua Verde, Jua and Guaiuba. The Pacoti river flows through some small towns and possesses many small islands. The mangroves in its estuary cover 158 ha (Miranda et al. 1988; SDU 1994). The influence of the sea is measured up to 15 km from the mouth in the dry season (Freire et al. 1991a, b, c). In this river human impact is weak because the activities are dominated by cattle and forest exploitation. This mangrove is chemically reductive because of the presence of large amount of organic matter (Freire et al. 1991a, b, c; Freire and Maia 1991).

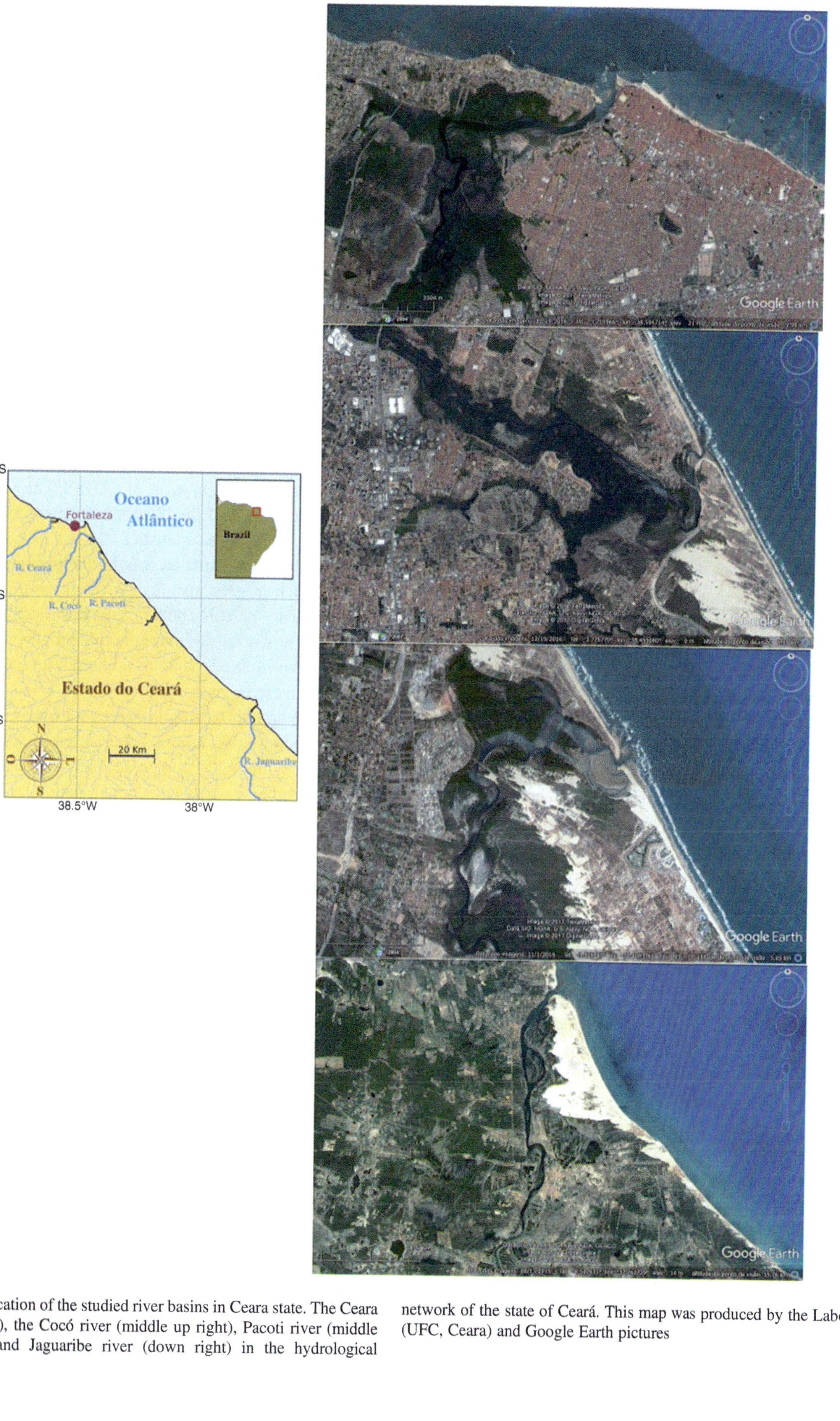

Fig. 10.1 Location of the studied river basins in Ceara state. The Ceara river (up right), the Cocó river (middle up right), Pacoti river (middle down right) and Jaguaribe river (down right) in the hydrological network of the state of Ceará. This map was produced by the Labomar (UFC, Ceara) and Google Earth pictures

Table 10.1 Extension and basin areas of the studied rivers and their respective mangrove areas

River	Extention (km)	Basin area (km^2)	Mangrove area (km^2)
Cocó	50	485	4.52
Ceará	60	900	7.26
Pacoti	150	1,800	3.40
Jaguaribe	610	75,669	7.29

Table 10.2 Annual anthropogenic emissions (ton.year-1) of N, P, Cu, Zn, Pb, Cd and Hg to the lower basins of the four estudied rivers (Lacerda et al. 2008; SEMACE 2005)

Fluxes	Cu	Zn	Pb	Cd	Hg	P	N
Ceará	7.09	102.2	7.37	0.77	0.072	393	815
Cocó	35.1	84.0	103.0	4.80	0.464	738	2,588
Pacoti	4.04	19.4	2.91	0.31	0.026	697	1,097
Jaguaribe	9.2	4.79	0.45	0.15	0.011	330	798

10.3.3 The Ceara River

The Ceara River is born in the Maranguape hill. Its length is 60 km with a basin of 900 km^2. The river crosses the cities of Maranguape, Caucaia and Fortaleza. The main tributary is the Maranguapinho with a confluence located about 7 km from the mouth. Total mangrove area reaches 726 ha (Godoy 2015) with an expanse up to 14 km inland (SDU 1994). The industrial district of Maracanaú is situated 15 km from the centre of Fortaleza. It harbours 41 companies (chemical, fabric, metallurgy, food, mineral and non-metallic exploitation) (Mavignier 1992). The waste water treatment station (WWTS) built since beginning of the area development is already oversized. Since 1992 retention basins have been implanted.

10.3.4 The Jaguaribe River

The basin of the Jaguaribe River is 75,669 km^2 for a length of 610 km. The river is much greater than the Pacoti, Ceara and Cocó rivers and does not allow a valid comparison among them. The population density was less than 30 hab.km^{-2} with a total population of 2,064,535 (data of 1996) (Gatto 1999). Main land uses are traditional agriculture together with livestock. Sixty percent of the basin has a semi-arid climate with locally with spots under moist, sub-humid and arid climate. Approximately 50% of the basin is in the state of Ceara, the other part is shared with the states of Piauí, Pernambuco, Paraíba and Rio Grande do Norte. The dominant vegetation is the Caatinga (Savanna). The most abundant species are *Anadenathera colubrina*, *Tabebuia chrysotricha*, *Auxemma oncocalyx*, *Astronium urundeuva*, *Caesalpinia pyramidalis*, *Aspidosperma pyrifolium*, *Bumelia sartorum*, *Erythrina velutina* *Croton* spp., *Mimosa tenuiflora*, *Combretum leprosum*, and cacti *Pilosocereus piauhiensis*, *Cereus jamacaru*. Most rains (80%) occur between January and May (600–1,200 mm) and varies in temperature between 24 and 27 °C.

In the rivers's lower basin and inshore waters; fishing, shrimp farming, lobster and crab are the major extractive activities. The Jaguaribe River is a well-regulated river that has many dams, both in the river proper or in its tributaries, of notice are the largest Castanhão reservoir with 6.7 million m^3, the Orós reservoir with 2.1 million m^3 of capacity and the Banabuiu, with 1.7 million m^3 capacity, these three reservoirs are responsible for over 95% of the total water storage capacity of the entire basin. Climatological monitoring is done by 17 rain stations, 4 on the main course and 13 on the tributaries. During low waters, the flow of the Jaguaribe (about 20 m^3 s^{-1}) is controlled downstream by the large Castanhão dam. As a result of extensive damming of the river, semi-arid climate and meso-tides make that the Jaguaribe estuary is frequently an importer of seawater and marine materials (Lacerda et al. 2013). The waters of the Jaguaribe River are of the bicarbonate-mixed type (Instituto Brasileiro Geografia e Estatistica, IBGE, 1996, http://www.ibge.gov.br/home/estatistica/populacao/contagem/). The Banabuiú River presents mixed waters with chloride and sodium. Eighty percent (80%) of the basin drains crystalline rock geology with weak hydraulic potentiality. The groundwater is chloro-mixed to chloro-sodium. The Jaguaribe River Basin receives untreated effluents from urban, agriculture, aquaculture and, to a certain degree, industries.

The general environmental characterized situation of the four rivers, relative to pathogenic micro-organisms; toxicology and sediment quality and contamination by nutrients and metals have been summarized by Nilin et al. (2013). A summary of anthropogenic emissions of nutrients and selected trace metals is presented in Table 10.2.

All the studied rivers show a varied state of human impact, particularly from urban areas and agriculture (Fig. 10.1). Major anthropogenic contaminants to the four rivers studied are trace metals of environmental significance and nutrients, mostly from urban sources, in the metropolitan area (Ceará and Cocó rivers) and agriculture and husbandry in the more rural basins (Pacoti and Jaguaribe rivers) (SEMACE 2005; Lacerda et al. 2008). Faecal coliforms and high biological oxygen demand (BOD) frequently occurs in the metropolitan area rivers. Annual loads of the most significant environmental contaminates to the four studied rivers are shown in Table 2.

10.3.5 Particularity of Mangroves

Mangroves are particularly sensitive ecosystems both ecologically (Blasco et al. 1996) and chemically (Otero and

Macías 2002). Most of the sediments carried by rivers reach the coastal zone with reduced speed, tides also contribute to choke the rivers, whereas mangrove trees trap the sediments as well as most pollutants associated with them, either metallic or organic (Da Silva et al. 1998). Mangrove vegetation is one of the most important driver of these process, also because of the large production of organic matter (Alberts et al. 1988; Lacerda et al. 1995). This sequestration is permanent due to the low remobilization of pollutants, as far as the natural mangrove biogeochemical conditions are preserved (Machado et al. 2002). This permanent sequestration has been reported even in highly polluted urban areas (Burke et al. 2000). However, changes in biogeochemical conditions of the sediments may result in remobilization of pollutants (Marins et al. 1997). The organic carbon concentration in mangroves ranges from 318 to 2,043 μM, depending on season and location and has no visible transition between the terrestrial and the marine end-members when observing the natural organic matter by fluorescence, for example (Jaffé et al. 2004). The modification of the uses in the basin can modify the flows of pollutants that enter the mangroves. Dam construction modifies physical-chemical conditions that alter river fluxes, producing an erosion of the coast and a migration or destruction of the mangroves (Lacerda et al. 2013; Lacerda and Marins 2002). For example the transformation of forest into agriculture causes the mercury trapped in the soils to be released (Lacerda et al. 2004). As mangroves only exert themselves in specific conditions, they are influenced by all modification of local and global environment conditions. A subtle change in the conditions can lead to an important response of the mangrove ecosystem that can die, move, and transform (Blasco et al. 1996).

It is therefore important to have geochemical indicators of the natural state. As organic material is an important mode of transmission of trace metals (Lu and Jaffe 2001; Mounier et al. 2001) and how little data exists on the fluorescence spectroscopy as a monitoring tool for mangroves (Alberts et al. 1988), this work proposed to use fluorescence spectroscopy on the Fortaleze city rivers to monitor the quality and quantity of the organic matter as an indicator of the impact of the soil use onto the mangroves geochemistry. Particular attention will be paid to the Pacoti, Cocó and Ceara rivers because their basins are of similar size but drastically different in soil uses.

10.4 Strategy and Comparison of the Cocó and Pacoti Rivers

Cocó and Pacoti rivers were sampled around the region of Fortaleze. Each river is divided into three zones: one near the source called "river zone", one inside the mangrove zone called "mangrove zone" and one at the mouth of the river to have the marine part called "marine zone".

Every month between May 2004 and September 2004 a sample of each zone was made. Filtration was done in the field with a 13 mm diameter burnt glass filter using a syringe and a HDPE filter holder. Six vacuum tubes, two by zone, were filled with the sample, two of them stabilized with 50 μL of 1 M NaN_3 for carbon, fluorescence quenching and anion concentration, and two treated with hydrochloric acid for cation analysis. The tubes were kept in the cold and dark until measurement. The filters were dried and used to measure the particulate organic carbon (POC) with a SSM SHIMADZU TOC instrument.

The physico-chemical parameters were measured at the same time of the water sampling, using a field multi-parameter instrument for pH, dissolved oxygen, conductivity. The probes were calibrated before each measurement. The results are presented in Table 10.3.

10.4.1 Analyzes

Measurements of the major elements, cations and anions, were done on a Dionex DX 120 ion chromatography with a ISOCRAT AS9-HC column and CS12A ISOCRAT column. The eluents are respectively Na_2CO_3 9 Mm, 950 $mg.L^{-1}$ and methane sulfonic acid (M.S.A) 20 mM, 1.3 $mL.L^{-1}$. The flow rate was set at 1 $mL.min^{-1}$ and the injection volume set at 100 μL.

Measurement of total organic carbon was done with a SHIMADZU TOC5000. The standard used was a solution of potassium phthalate. Each sample was acidified (100 μL of 5 N HCl in the 5 mL sample) and bubbled with nitrogen gas before measurement for eliminating the carbonates.

Fluorescence was measured with an HITACHI4500. Excitation Emission Matrix of Fluorescence (EEM) were made in a quartz cell at natural pH with the following parameters: 250–450 nm for excitation, 300–600 for emission, a scan speed of 30,000 $nm.min^{-1}$ and excitation and emission slit fixed at 5 nm.

To study the properties of complexation the fluorescence quenching was used. By placing the sample in a glass vessel, the pH was fixed to 5 with a buffer solution of $Na_2PO_4{\cdot}H_2O$, and addition of copper $Cu(NO_3)_2$ with 0.1, 1 and 10 mg L^{-1} of copper solution till a stabilization of the fluorescence signal extinction. The fluorescence measurement was done with an excitation spectrum from 200 to 400 nm and an emission wavelength fixed at 420 nm, all other parameters were identical to the EEM measurement.

Table 10.3 Values of the physico-chemical parameters of river water samples in the river, mangrove and marine zones of the Pacoti and Cocó rivers, Ceará State, north eastern Brazil

River	Field number	Date	Reference	T (°C)	Cond. (μS.cm^{-1})	pH	Sal.	O_2 (mg.L^{-1})	Zone
Cocó	1	07/05/04	Co101	29.5	764	6.14	0.4	6.1	River
Cocó	1	07/05/04	Co102	28	949	6.38	0.5	0.5	Mangrove
Cocó	1	07/05/04	Co103	29.1	46,600	7.8	30.0	2.5	Marine
Pacoti	1	25/05/04	Pa101	28.7	2,008	6.38	0.9	3.8	River
Pacoti	1	25/05/04	Pa102	29.5	54,300	8.2	35.4	4.6	Mangrove
Pacoti	1	25/05/04	Pa103	29.1	54,500	8.2	35.1	5.2	Marine
Cocó	2	09/06/04	Co201	26.4	770	6.54	0.4	6.3	River
Cocó	2	09/06/04	Co202	26.9	982	6.65	0.5	1.8	Mangrove
Cocó	2	09/06/04	Co203	27.5	53,100	8.12	35.0	3.2	Marine
Pacoti	2	09/06/04	Pa201	27	3,332	6.75	1.7	4.5	River
Pacoti	2	09/06/04	Pa202	27.4	56,600	8.23	35.7	4.6	Mangrove
Pacoti	2	09/06/04	Pa203	27.2	56,400	8.23	35.7	4.5	Marine
Cocó	3	30/06/04	Co301	28.3	756	6.45	0.4	9.8	River
Cocó	3	30/06/04	Co302	27.4	378	7.1	0.2	5.8	Mangrove
Cocó	3	30/06/04	Co303	28.3	52,700	8.1	34.6	4.7	Marine
Pacoti	3	30/06/04	Pa301	27.0	916	5.68	0.4	3.8	River
Pacoti	3	30/06/04	Pa302	28.9	12,220	6.25	6.4	3.6	Mangrove
Pacoti	3	30/06/04	Pa303	28.3	29,770	6.93	17.2	3.9	Marine
Cocó	4	04/08/04	Co401	25.8	753	7.07	0.3	5.0	River
Cocó	4	04/08/04	Co402	25.8	2,566	7.67	1.3	3.6	Mangrove
Cocó	4	04/08/04	Co403	25.8	52,100	8.21	32.9	3.3	Marine
Pacoti	4	04/08/04	Pa401	25.7	3,772	8.04	1.9	4.9	River
Pacoti	4	04/08/04	Pa402	25.5	54,000	8.2	34.3	4.0	Mangrove
Pacoti	4	04/08/04	Pa403	25.5	54,100	8.29	34.4	4.9	Marine
Cocó	5	27/08/04	Co501	–	878	6.62	0.4	5.5	River
Cocó	5	27/08/04	Co502	–	839	6.9	0.5	5.0	Mangrove
Cocó	5	27/08/04	Co503	–	30,670	7.01	20.5	5.0	Marine
Pacoti	5	27/08/04	Pa501	–	8,190	8.33	4.5	6.0	River
Pacoti	5	27/08/04	Pa502	–	41,420	8.0	26.6	4.5	Mangrove
Pacoti	5	27/08/04	Pa503	–	49,250	7.31	32.3	4.4	Marine
Cocó	6	17/09/04	Co601	27.6	752	6.99	0.4	4.9	River
Cocó	6	17/09/04	Co602	26.0	7,310	7.01	4.0	3.3	Mangrove
Cocó	6	17/09/04	Co603	27.2	51,400	8.09	33.7	3.5	Marine
Pacoti	6	17/09/04	Co601	27.5	8,550	8.12	4.7	4.5	River
Pacoti	6	17/09/04	Co602	26.9	53,500	8.13	35.3	5.2	Mangrove
Pacoti	6	17/09/04	Co603	26.7	53,300	8.18	35.2	5.4	Marine

10.4.2 Results

10.4.2.1 Dissolved and Particulate Organic Carbon

The results presented in Fig. 10.2 show that for the Cocó River the concentration of dissolved carbon is always higher in the mangrove area. For Pacoti, on the contrary, there is a systematic increase from the river zone to the marine zone. The salinity suddenly increases at the marine zone in the Cocó river and in the mangrove zone in the Pacoti rivers, except for the month 3 (June 2004) where a low salinity was observed probably due to a low tide conjugated to a high discharge. The pH always increases from the Cocó river zone (~7) to the marine zone (~8), while it change more stochastically in the Pacoti River. Concerning the particulate organic carbon (POC), there is no trend along the period and among the zones. The average POC concentration is about 0.2 mg of C L^{-1} (±0.1) except for the Cocó River at the river zone where a greater variability is observed leading to an average value of 0.5 ± 0.6 mg of C L^{-1}.

With these results it is already possible to say that these two ecosystem do not have the same dynamic. Apparently it seems to have a carbon production in the mangroves of the Cocó River not present in the Pacoti River.

10.4.2.2 Fluorescence Spectroscopy

The 2D Fluorescence

The measurement of the fluorescence intensity at peak A ($\lambda_{ex} = 250/\lambda_{em} = 450$) and peak C ($\lambda_{ex} = 350/\lambda_{em} = 450$) shows that for the Cocó River intensity is higher for the

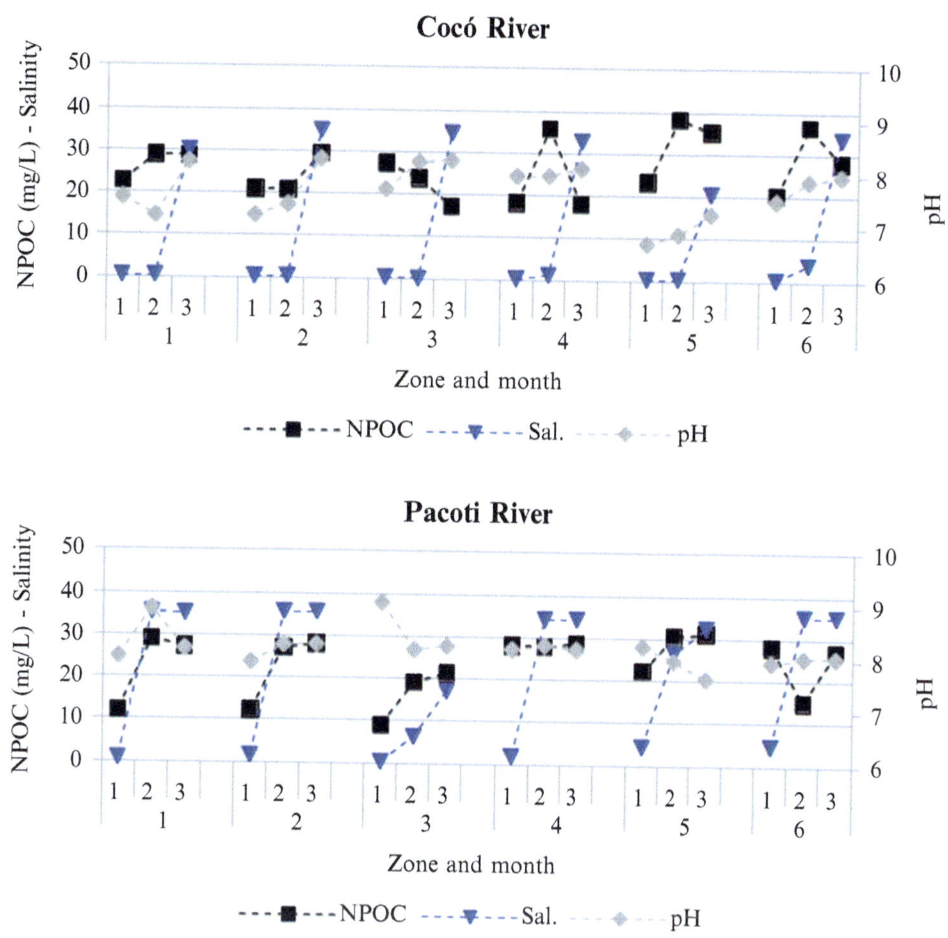

Fig. 10.2 Results of dissolved organic carbon concentration in ppm (black square), salinity (blue triangle) and pH (grey losange) for the Cocó and Pacoti rivers. Points 1, 2, 3 representing the river, mangrove and marine zones respectively

mangrove area (Fig. 10.3, solid lines), as for its organic carbon concentration. In the Pacoti River the only fields that behave in the same way are field 3 and 5 where salinity and pH behave differently.

Observing the ratio between the two intensities allows one to see the quality of the organic material (Mounier et al. 1999). When $r_{a,c}$ decreases, maturity increases as intensity from the type C increases more than the type A (Fig. 10.4). This ratio does not depend on the carbon concentration (Patel-Sorrentino et al. 2002). All the values are similar and around value of 1, except for the Cocó river in field 4 where the ratio is about 1.5. In the Cocó River there is a decrease from the river zone until the marine zone. This could be due to a maturation of the organic material during the transport. In contrast in the Pacoti River, the variation is not recorded with maximum in mangrove area except in field 2. For the Pacoti River, the maximum often appears in the marine zone, indicating a recent organic material (fields 1, 3, 4, 5 and 6).

3D Fluorescence

3D fluorescence is a map of all the photons emitted by the organic matter that fluoresce (Fig. 10.5). This measurement gives a quantification of the fluorescent organic material as well as qualitative information using the excitation and emission wavelength position. Generally the most popular peaks are the peak A ($\lambda_{ex} = 250/\lambda_{em} = 450$) and the peak C ($\lambda_{ex} = 350/\lambda_{em} = 450$). We can observe that the fluorescence is more intense in the Cocó River.

The peak T (tyrosyn-like) corresponding to the signal of the proteins is not observed. The waters are mainly composed of humic type (peak C) as presented in Figs. 10.6 and 10.7.

From the 3D fluorescence of all the excitation emission matrices of fluorescence presented above, the CP/PARAFAC separation algorithm was used. Two components were extracted (CORCONDIA 96%) and are represented by the two components in Fig. 10.8.

The first component shows a maximum at 300 nm of excitation and 400 nm of emission. Its shape is circular, and related to the fluorescence domain. It represents the humic-like type components linked to fulvic acids. The component C2 exhibit two excitation maxima one secondary at 280 nm and one principal at 350 nm. The emission wavelength is around 480 nm. This later component represents the humic-like substances coming from terrestrial origin. No more than two components where found meaning that

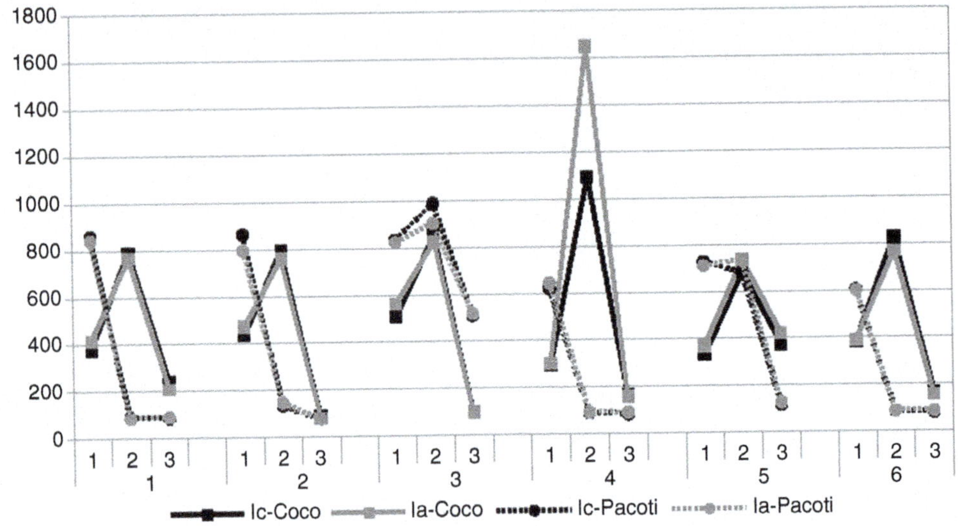

Fig. 10.3 Fluorescence of Ic and Ia (dark and grey solid, respectively) of the Cocó river and fluorescence Ic and Ia of the Pacoti river samples (dark and grey dashed, respectively)

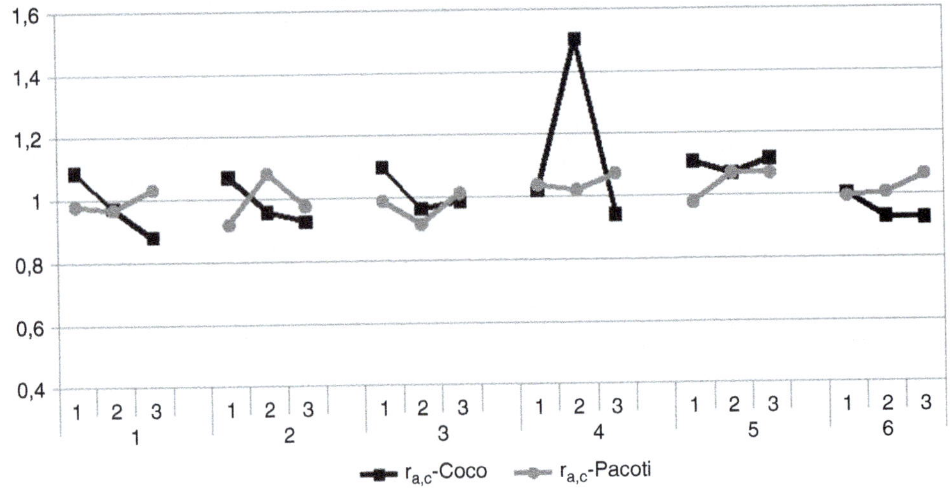

Fig. 10.4 Ratio for the fluorescence peaks of the Cocó and Pacoti rivers

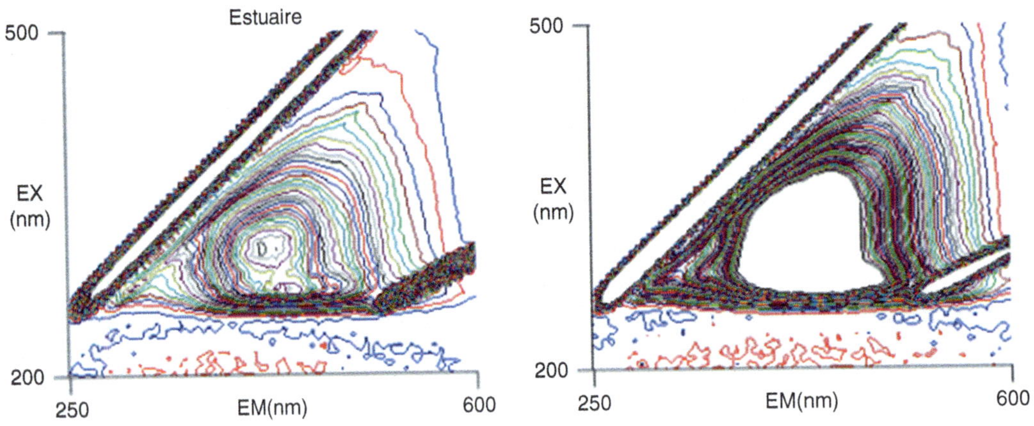

Fig. 10.5 Excitation emission fluorescence matrices example of the Cocó water river (left) and the Pacoti river (right)

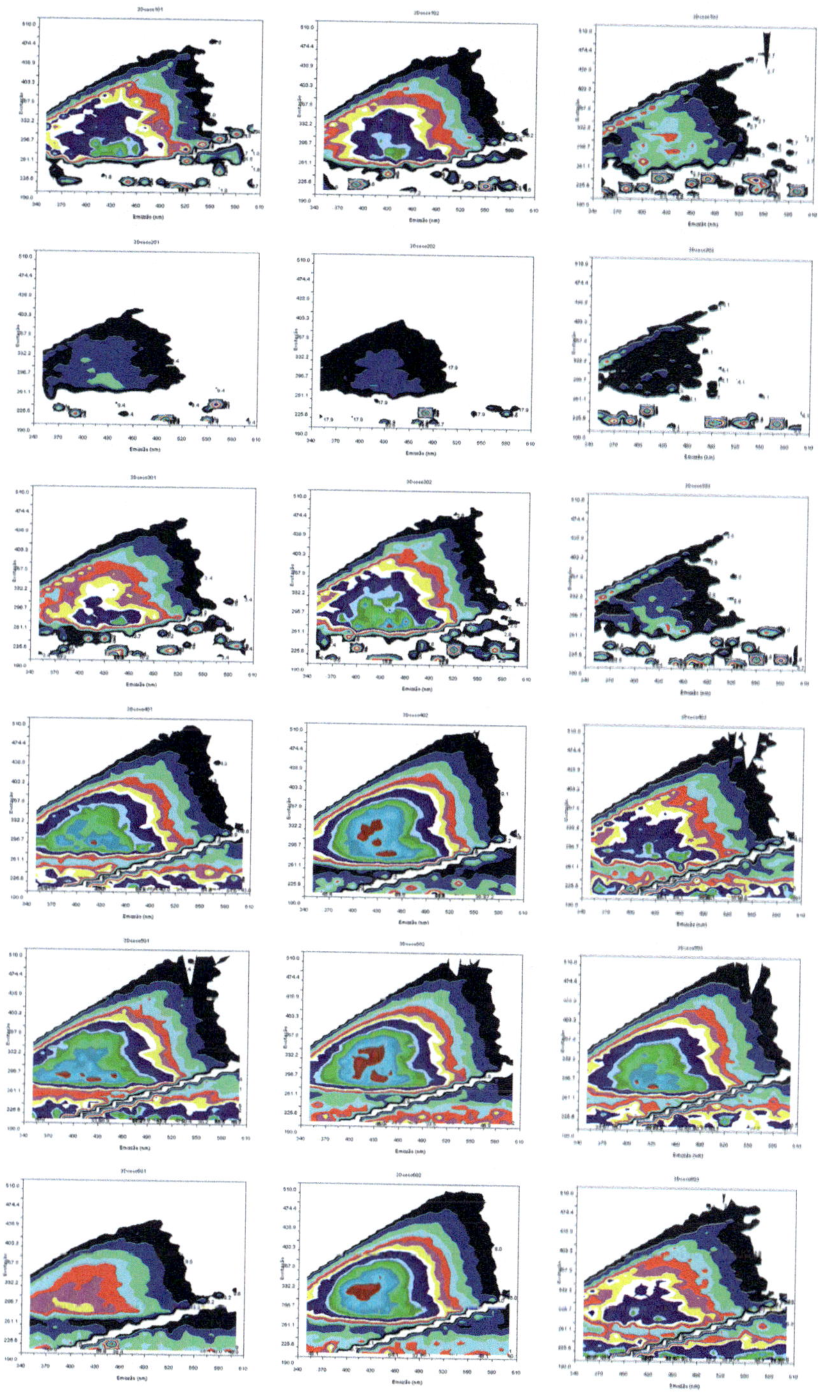

Fig. 10.6 Excitation Emission Matrices of the Cocó River false colour representation. Column 1, 2 and 3 represent the river, mangrove and marine zones and line represent the line correspond to field campaign 1, 2, 3, 4, 5 and 6

possible fluorescence biological activity was not statistically extracted.

Their evolution along the rivers and through time is shown in Fig. 10.9. The Cocó River presents a higher C1 contribution in the mangrove zone. Each sampling has the variation of the C1 contribution in mangrove zone higher than the C1 in river and marine zone (Fig. 10.9). The C2 component on the contrary does not vary. The urban impacted mangrove zone produced more fulvic components. This is confirmed by the $r_{a,c}$ ratio that is systematically higher in mangrove zone except for the campaign 1 and 2 (Fig. 10.9, grey line).

The Pacoti River shows a different trend from the Cocó River. Indeed, the fulvic contribution decreases regularly from river zone to the marine zone. Moreover the $r_{a,c}$ is also decreasing meaning that the organic matter become more and more terrestrial like during the transport to the mangrove

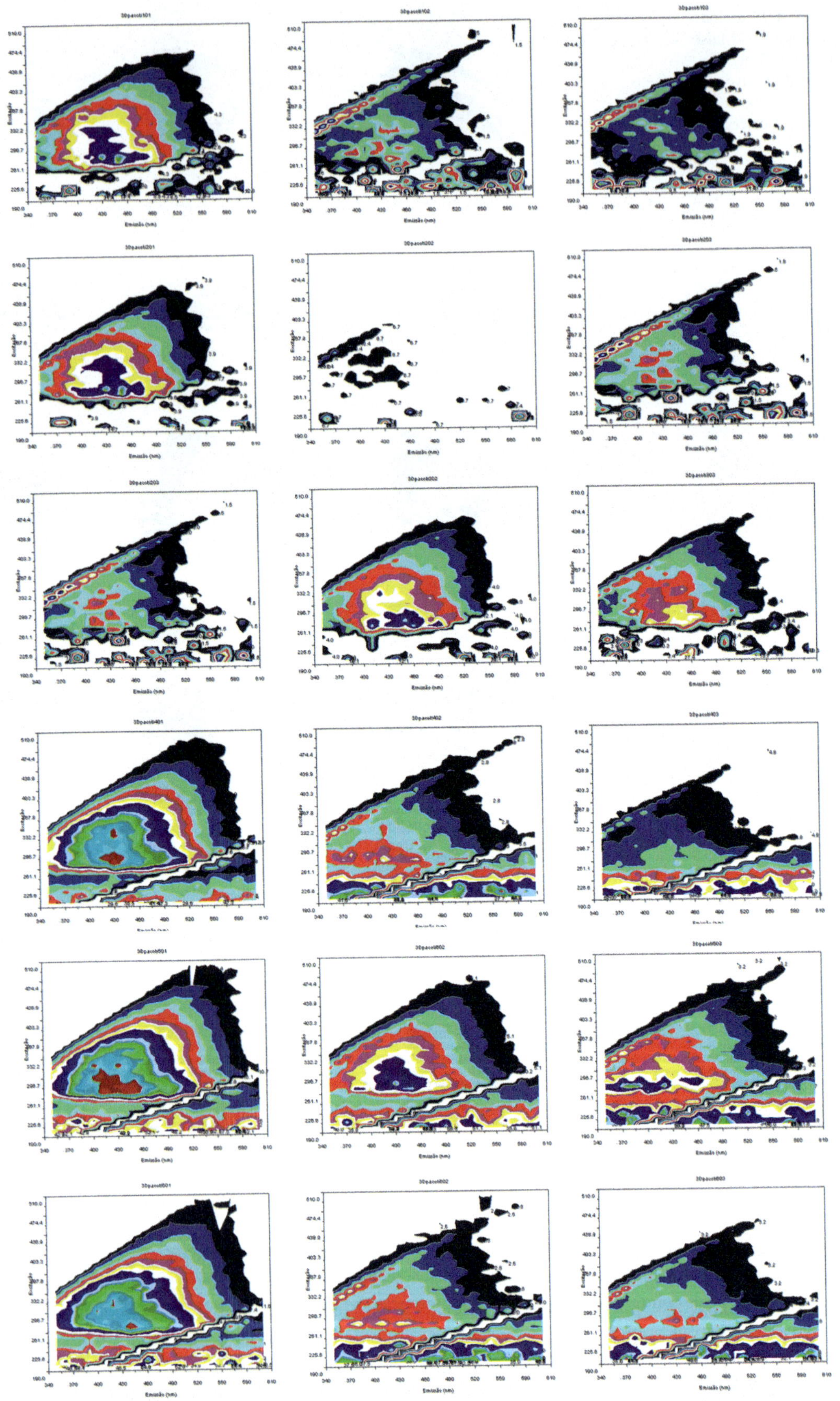

Fig. 10.7 Excitation Emission Matrices of the Pacoti River false colour representation. Column 1, 2 and 3 represent the river, mangrove and marine zones and line represent the line correspond to field campaign 1, 2, 3, 4, 5 and 6

zone. In field 1–3, the fulvic contribution was higher than the terrestrial one (C2), but in fields 4, 5 and 6 it is the contrary for mangrove and marine zones.

Hence by fluorescence measurement, it is observable that the organic matter produced by the mangrove differs a lot between urban impacted and non-impacted mangroves, except on month 3 when the fluorescence organic matter is similar to the Cocó River pattern. More fulvic like components are produced in the Cocó River, maybe as a defence mechanism, since herbivory tends to be more intense under more stressed conditions and fulvic components are frequently associated with plant defences against herbivores.

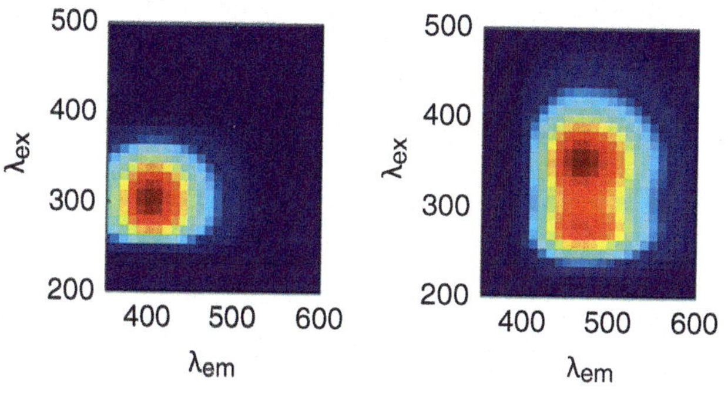

Fig. 10.8 PARAFAC component extracted from the Cocó and Pacoti rivers excitation emission matrices of fluorescence

10.4.3 Major Elements

10.4.3.1 The Cations

The concentrations of the cations (Fig. 10.10) present high values for the Na in the marine zone, as expected due to stronger sea influence. Indeed it is connected with the proportion of sea water in the system. In the other zones the dominant cation is Na^+ but also K^+ and Mg^{2+}. It only has anomaly during field campaign 4, where K^+ and NH_4^+ present a higher concentration than Ca^{2+}.

The patterns for the Pacoti River are different but have the same relation and influence as related to the sea, with an increase of the Mg^{2+} and Na^+ concentration. One can see that the Pacoti River is more invaded by the sea intrusion in the mangrove zone (Fig. 10.11) than the Cocó River.

10.4.3.2 The Anions

The anion concentrations in the Cocó River are shown in (Figs. 10.12 and 10.13). The increase of concentrations results from the influence of seawater. That is true except for nitrate and sulphate, which disappear when salinity increases.

The Pacoti River again does not presents the same variation as the Cocó River (Fig. 10.14). From these analyses it is difficult to interpret the results. But one can say that during the study, nitrate concentration is high at the river zone.

Fig. 10.9 Evolution of the contribution of the CP/PARAFAC components along the rivers. The component C1 and C2 are represented by the blue and red solid line respectively. The ratio of their contribution C1/C2 is represented by the grey dashed line.

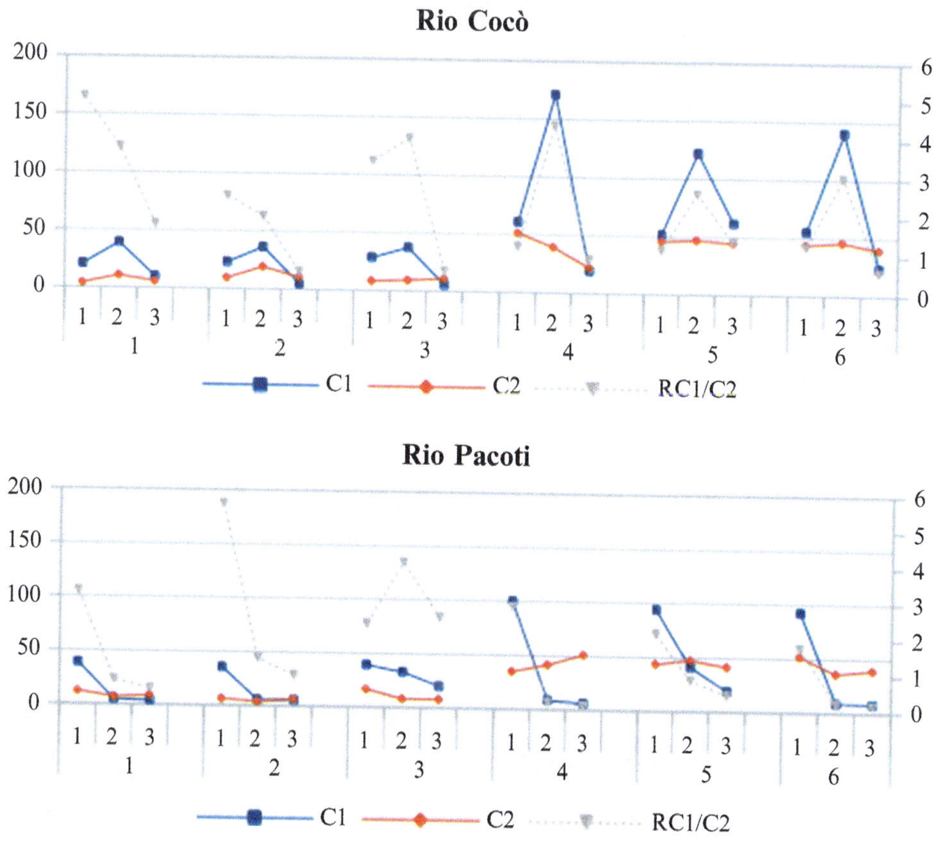

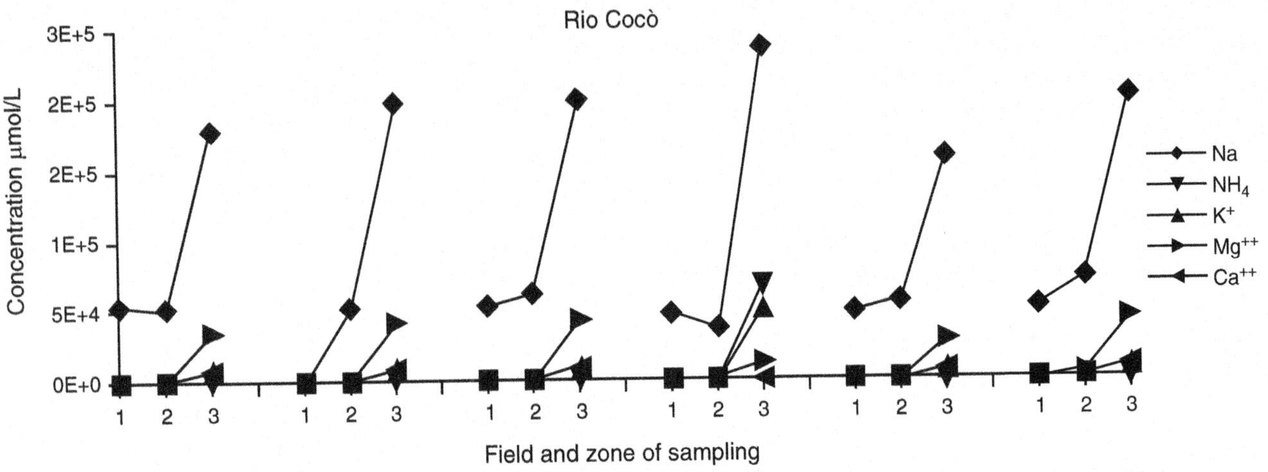

Fig. 10.10 Variation of cation concentration in the Cocó River

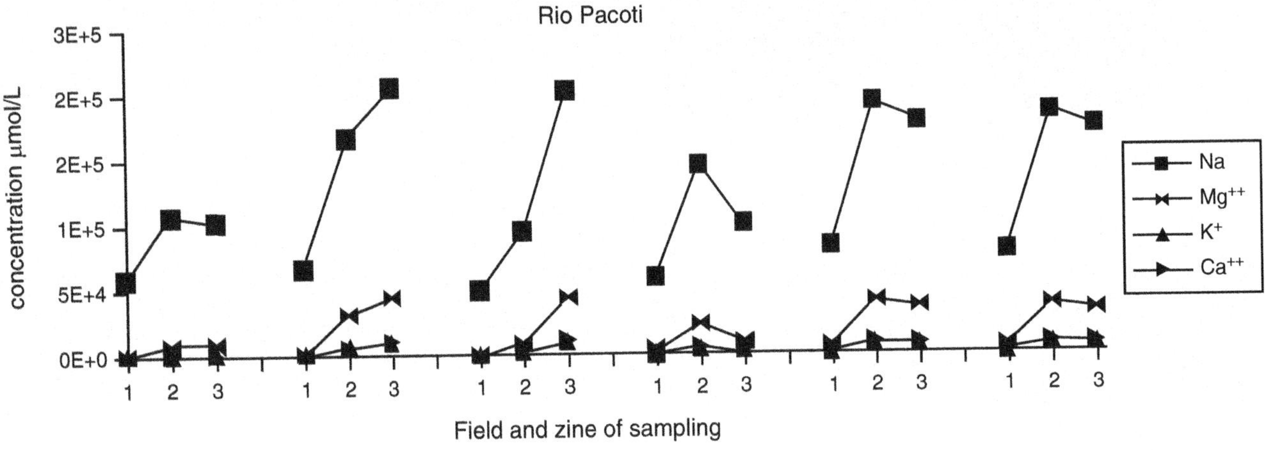

Fig. 10.11 Variations of cation concentrations in the Pacoti River

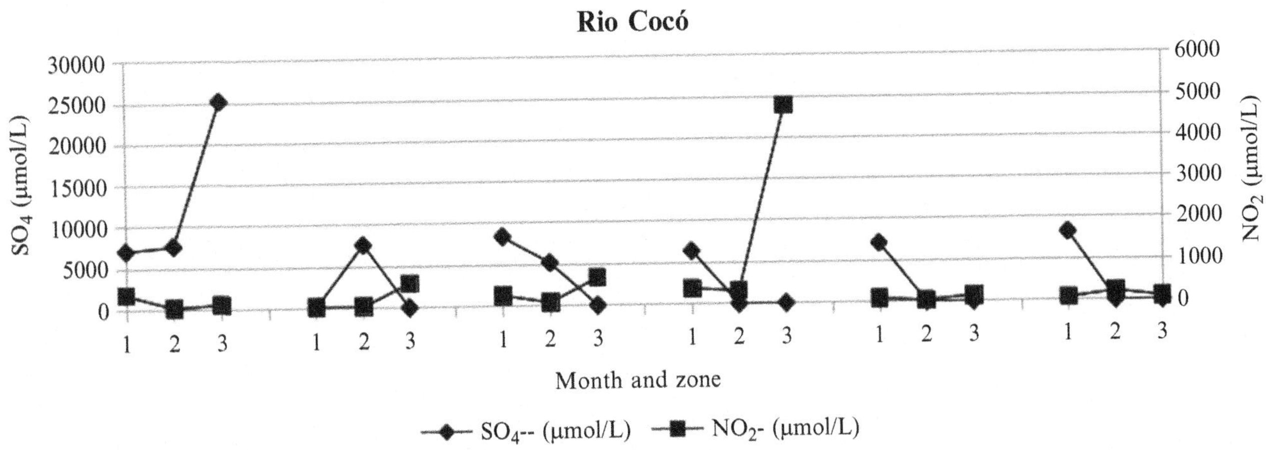

Fig. 10.12 Variation of sulphate and nitrate concentration in the Cocó River

10.4.3.3 Fluorescence Quenching

Complexing capacity is the ability of the natural organic matter (NOM) to complex the metals forming a metal-NOM complex. For a sample, the site concentration of complexation is expressed in µmol/L of metal, and the

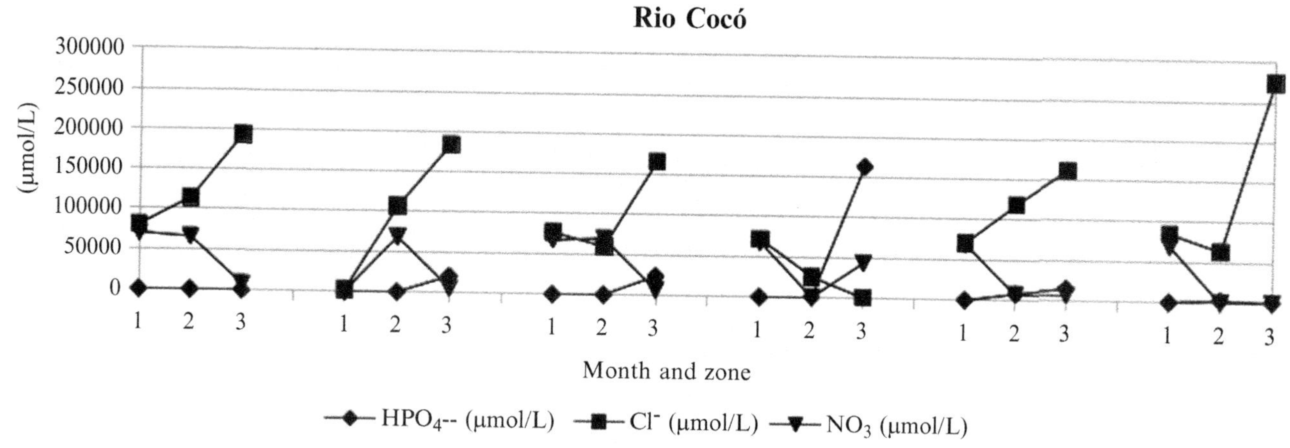

Fig. 10.13 Variation of phosphate, chloride and nitrate concentration in the Cocó River

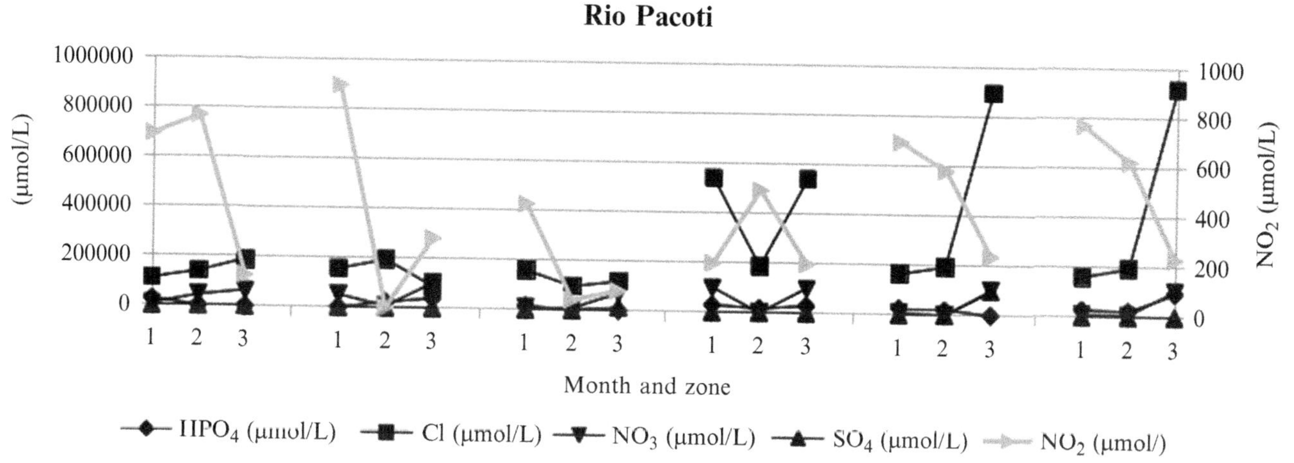

Fig. 10.14 Variation of anion concentrations in the Pacoti River

complexion capacity is the value in µmol of site/mg of carbon (µmol of Cu^{2+}/mg of C).

Using the Ryan and Weber (1982) model, it is possible to calculate the total site concentration L_T and the equilibrium complexation constant K_L, which represent the strength of the complex bound (Ryan and Weber 1982).

The obtained results are represented in Figs. 10.15 and 10.16. The Cocó River shows a greater capacity for the river zone and the marine zone. Systematically the complexation capacity is minor for the mangrove zone. This can be explained by a specific production of carbon with weak complexation properties. In the mangrove zone metals are less complexed than in the Cocó River mouth. At the contrary in the Pacoti River the complexation capacity always increases from the river zone until the marine zone. In Pacoti, organic matter site density increases continuously and metals have more and more probability to be complexed. There is no production of specific organic material in the mangrove zone of the Pacoti River. Based on the results of quenching of fluorescence it appears that the dynamic of the two rivers is not similar in term of NOM production. The Cocó River, more affected by anthropogenic activities, produces non complexant organic matter in the mangrove zone, while in the Pacoti basin the NOM complexant capacity increases. The results do not show trend on complexant capacity during the 6 months sampling period in both rivers.

10.5 Sampling Strategy on the Jaguaribe River

10.5.1 Spatial Sampling

Two fields campaigns were carried out (Table 10.4) in April and September 2004 to have variation of physical-chemical parameters of the waters of the lower Jaguaribe basin. The

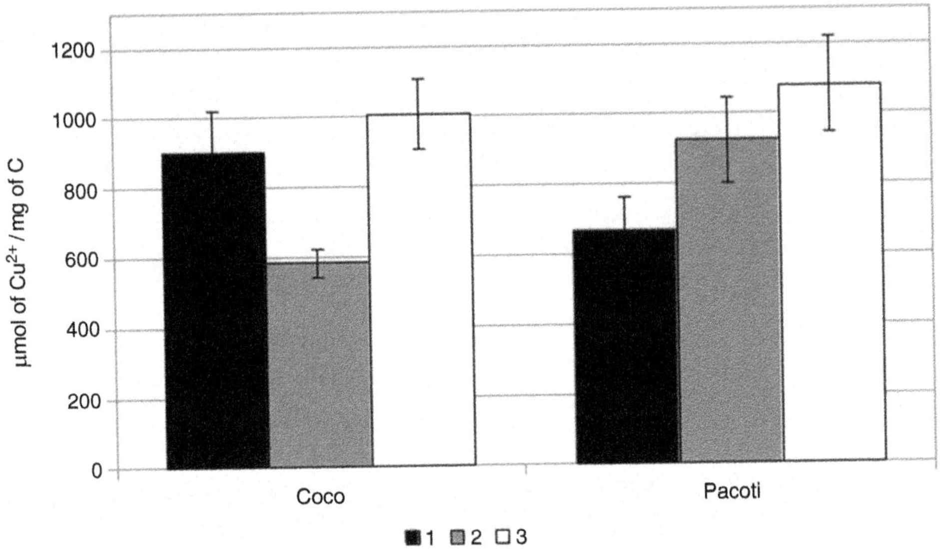

Fig. 10.15 Distribution of complexation capacity for copper in the Pacoti River and Cocó River along the river. *1* River; *2* mangrove; *3* marine zones

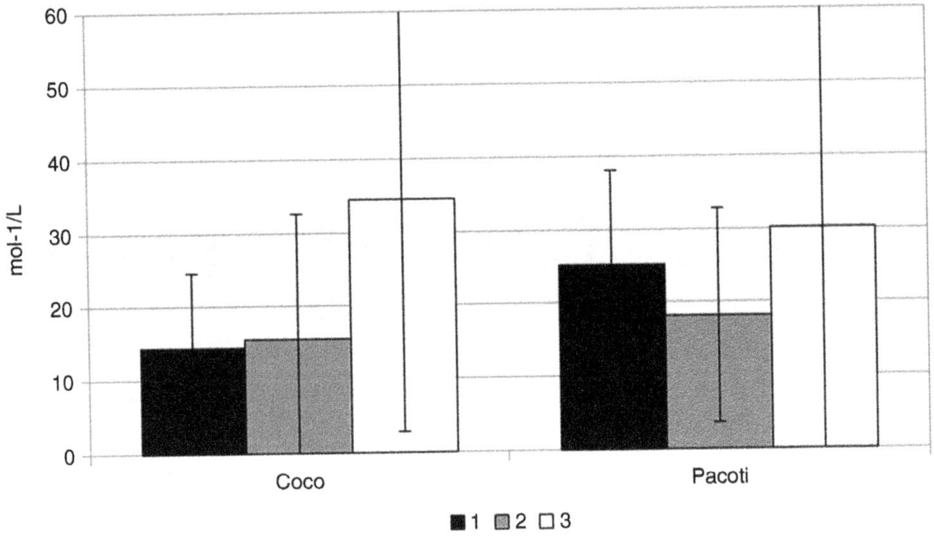

Fig. 10.16 Distribution of complexation constant for copper in the Pacoti River and Cocó River along the river. *1* River; *2* mangrove; *3* marine zones

mangrove zone of the Jaguaribe occurs between the point 2 (Barra) and 4 (Raimundinho) (Fig. 10.17).

10.5.2 Sampling During a Tide

The waters of the Jaguaribe River were sampled along time during a tide period. These 15 samples were sampled in front of the Ilha Grande Island (Fig. 10.17, Table 10.5).

10.5.3 Results of the Spacial Distribution Study

10.5.3.1 Dissolved Organic Carbon

It can be observed that the concentration of organic carbon in the Jaguaribe river basin increases in April and September. Figure 10.18 shows the concentrations observed from the site 1 (Marine) to Itaiçaba (River) throughout the continent-sea gradient. There is a decrease in carbon concentration showing dilutive effect from land to sea both in April and September. In April the carbon concentration is very low, particularly downstream the dam of Itaiçaba and in Aracati (downstream the bridge) while the values upstream the bridge are very high. If the value after the dam can be a false response, we can consider a large dilution after the Aracati point (after the bridge). The value of carbon concentration is higher in September than in April. The dilution effect is confirmed by the linear correlation between DOC and Cl^- concentration ($r^2 = 0{,}89$) (Fig. 10.18).

10.5.3.2 Fluorescence Spectroscopy

The fluorescence results (Figs. 10.19 and 10.20) show a very different fluorescence response in April and September. Clearly fluorescence is highest in April, the period with lower concentrations of dissolved organic carbon. Once reaching the sea (site 2) in April the fluorescence decreases

Table 10.4 Location of sampling points on the Jaguaribe River, and values of copper (*Cu*$^{2+}$), dissolved organic carbon (*DOC*) and inorganic carbon (*IC*)

River	Campaign	Date	Reference	Cu^{2+} µg L^{-1}	DOC mg of C L^{-1}	IC mg of C L^{-1}	Comment
Jaguaribe	1	30/04/04	Ja101	nd	7.1	21.1	Mar
Jaguaribe	1	30/04/04	Ja102	nd	7.0	26.2	Barra
Jaguaribe	1	30/04/04	Ja103	2.1	10.4	25.2	Ilha Grande
Jaguaribe	1	30/04/04	Ja104	3.52	10.8	27.1	Raimundinho
Jaguaribe	1	30/04/04	Ja105	2.6	11.7	25.8	Aracati (quando ver ponte)
Jaguaribe	1	30/04/04	Ja106	3.26	7.5	25.9	Aracati (depois da ponte)
Jaguaribe	1	30/04/04	Ja107	1.92	19.0	24.4	Cabreiro
Jaguaribe	1	30/04/04	Ja108	1.64	0.26	23.2	Itaiçaba (antes da Barragem)
Jaguaribe	1	30/04/04	Ja109	1.63	20.6	22.4	Itaiçaba (depois da Barragem)
Jaguaribe	2	19/09/04	Ja201	nd	12.4	27.4	Mar
Jaguaribe	2	19/09/04	Ja202	nd	12.3	25.6	Barra
Jaguaribe	2	19/09/04	Ja203	nd	13.9	24.0	Ilha Grande
Jaguaribe	2	19/09/04	Ja204	nd	16.0	26.1	Raimundinho
Jaguaribe	2	19/09/04	Ja205	nd	17.0	24.4	Aracati (quando ver ponte)
Jaguaribe	2	19/09/04	Ja206	2.11	17.1	22.4	Aracati (depois da ponte)
Jaguaribe	2	19/09/04	Ja207	2.27	17.6	21.6	Cabreiro
Jaguaribe	2	19/09/04	Ja208	1.21	19.3	19.5	Itaiçaba (antes da Barragem)
Jaguaribe	2	19/09/04	Ja209	0.99	19.2	18.7	Itaiçaba (depois da Barragem)

Fig. 10.17 The mangrove zone of the Jaguaribe River (Ilha Grande sampling stations)

drastically. From site 3–9 the fluorescence reaches the values of 500 u.a (arbitrary unit) and becomes quasi constant. In contrast, in September fluorescence values are weak (less than 400 u.a., Fig. 10.19) and one can observe from point 3 an increase in the fluorescence of peak A which represents the young organic matter. This is confirmed by graph (Fig. 10.20) where the ratio increases from point 3. One can also observe that the $r_{a,c}$ does not increase in April. This confirms a production of young organic matter at the end of the river path and at the end of the basin, where the influence of the sea is the greatest in the mangrove zone.

As the concentration of organic carbon increases in September, the $r_{a,c}$ increases only at three points in the basin near the sea. In other words, the carbon inputs are variable up and down. The production of young organic material only takes place during the September sampling campaign in the basin of the Jaguaribe.

10.5.4 Results of the Temporal Study

10.5.4.1 Organic Carbon, pH and Salinity

The salinity varies from 8 to 20 with a sinuous curve. The minimum salinity occurs at 11:35 p.m. and the maximums happen at 6:00 p.m. and 5:00 p.m. Maximun pH values were observed during highest salinity. The total organic carbon shows anomaly during the same moments (Fig. 10.21).

Table 10.5 Samples from the 20/06/04 field campaign for temporal studies durng a tidal cycle

River	Number	Hour	Reference	Cu^{2+} (µg L^{-1})	POC (mg of C L^{-1})	DOC (mg of C L^{-1})	IC (mg of C L^{-1})	T°C	Cond.mS cm^{-1}	Sal.	pH
Jaguaribe	1	05h40	TJa01	2.87	1.00	17.9	37.0	26.0	30.65	18.2	7.91
Jaguaribe	2	06h35	TJa02	2.86	1.65	19.2	34.5	27.0	14.8	8.1	7.88
Jaguaribe	3	07h30	TJa03	2.84	3.15	19.9	29.1	28.0	1.19	0.6	8.01
Jaguaribe	4	08h35	TJa04	–	1.89	4.3	31.1	–	35.16	20.4	8.05
Jaguaribe	5	09h35	TJa05	2.48	2.09	0.2	30.2	–	33.39	19.1	8.03
Jaguaribe	6	10h40	TJa06	2.65	2.59	0.3	–	–	29.29	16.7	7.93
Jaguaribe	7	11h35	TJa07	2.50	2.33	20.6	31.5	–	24.33	13.8	7.95
Jaguaribe	8	12h50	TJa08	2.77	1.75	–	27.6	29.0	18.00	10.4	8.02
Jaguaribe	9	13h35	TJa09	3.18	2.27	20.7	26.1	29.2	15.68	8.4	8.17
Jaguaribe	10	14h30	TJa10	3.10	2.24	22.7	26.7	–	14.29	7.6	8.38
Jaguaribe	11	15h35	TJa11	3.13	2.61	–	30.3	28.1	7.76	4.0	8.22
Jaguaribe	12	16h35	TJa12	3.00	2.16	0.23	28.9	–	19.70	10.8	8.14
Jaguaribe	13	17h35	TJa13	7.38	2.02	18.9	26.1	–	25.17	14.2	–
Jaguaribe	14	18h35	TJa14	–	3.28	20.4	25.6	–	29.31	16.6	8.09
Jaguaribe	15	19h25	TJa15	2.34	1.75	21.8	23.7	–	30.89	17.8	8.13

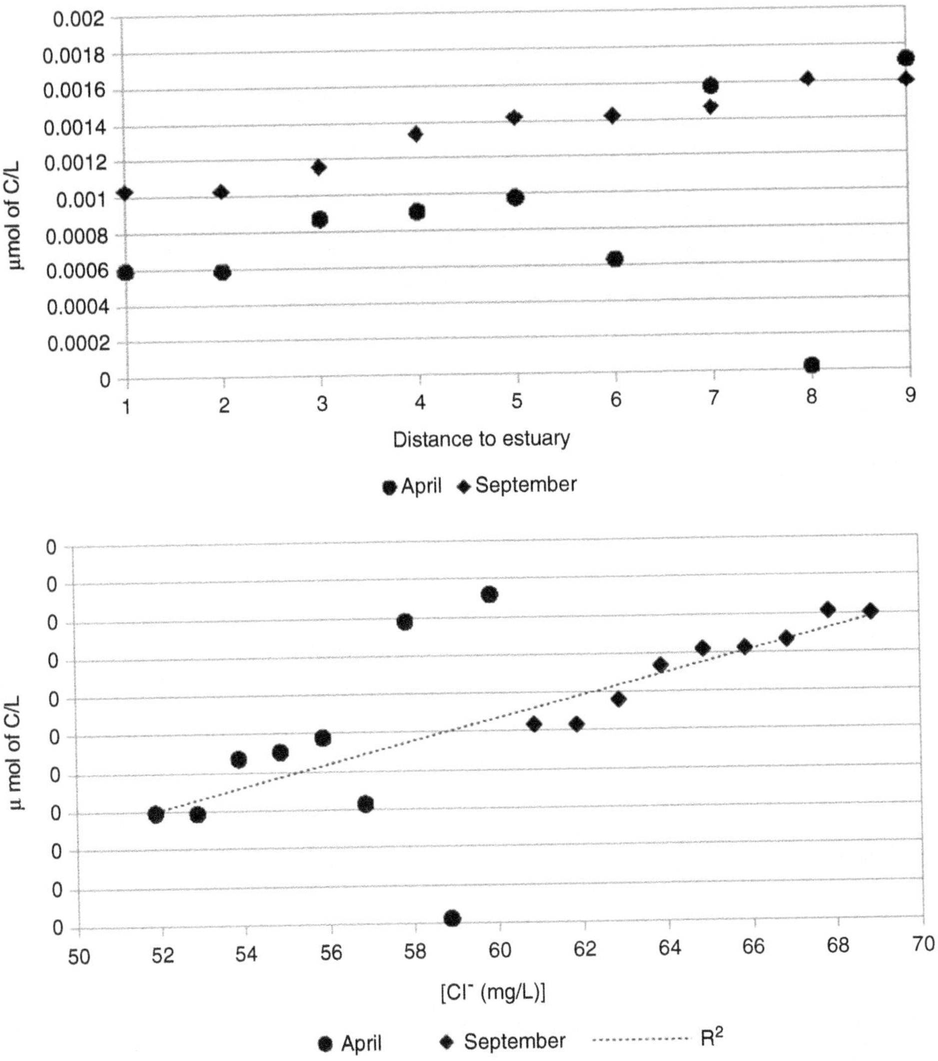

Fig. 10.18 Variation of the dissolved organic carbon during the two sampling campaigns and the relation to salinity

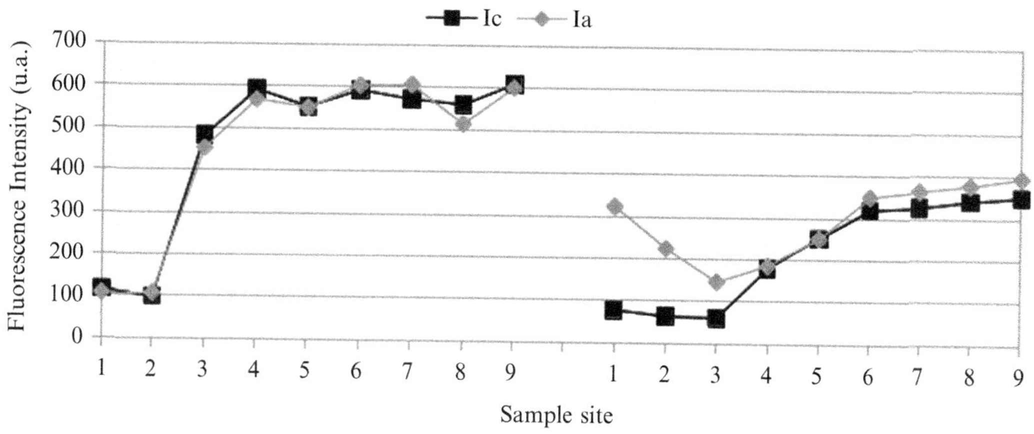

Fig. 10.19 Fluorescence of peaks A (dark square) and C (gray losange) (Coble 1996) in the Jaguaribe river basin, in April (left) and September (right)

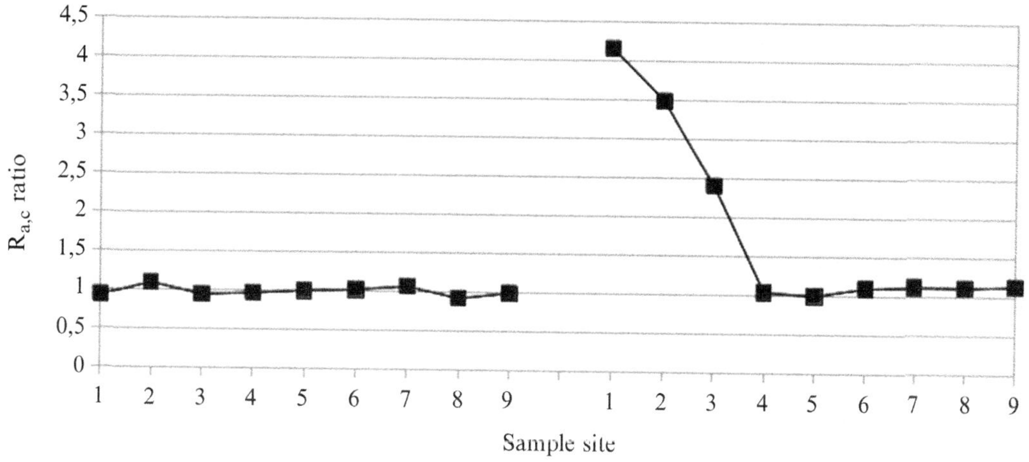

Fig. 10.20 Reason for fluorescence intensity of the Jaguaribe river basin in April (left) and September (right)

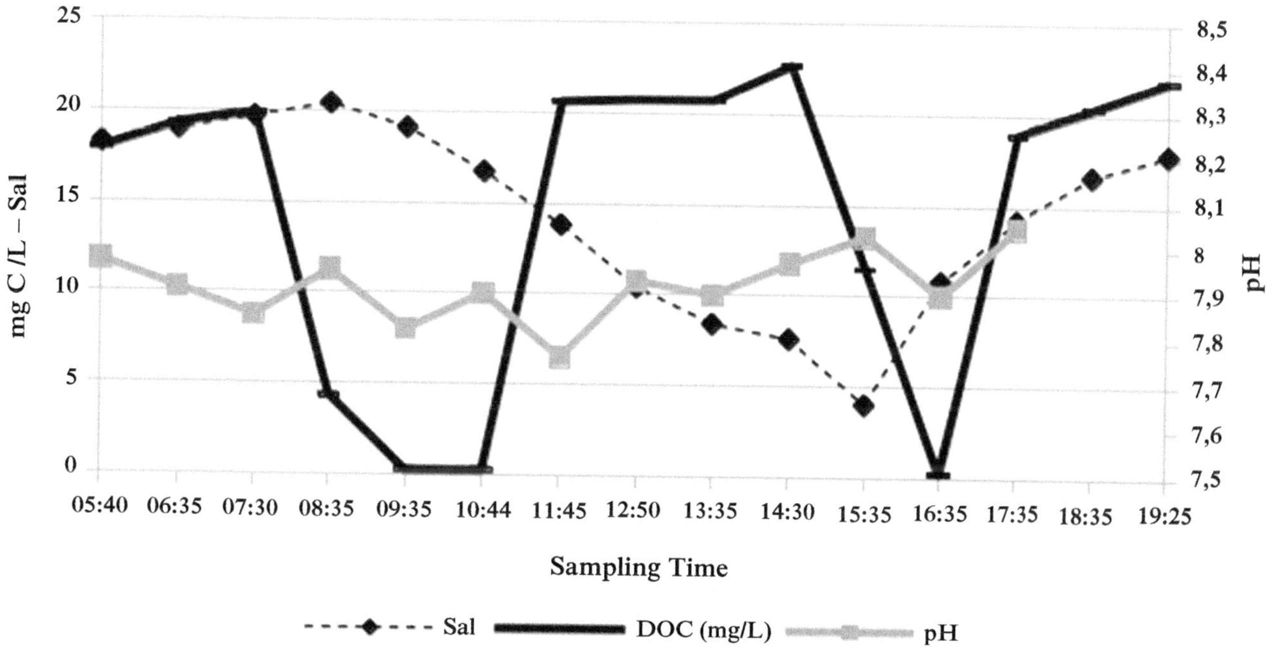

Fig. 10.21 Variation of carbon, pH and salinity during a tidal episode

10.5.4.2 Fluorescence Spectroscopy

The fluorescence of peaks A and C in the Jaguaribe River shows a maximum during low tide when the influence of the sea minor the salinity is lowest. The fluorescence increases to 500 versus 300 when the salinity is to the maximum (Fig. 10.22).

From the graph in Fig. 10.23, we can see that the ratio is quasi constant throughout the tidal period after 7H30 a.m. This means that there is no variation in the quality of the organic material, either in low or high tide. The variation of concentration comes solely from the dilution of river waters by sea waters and does not change the carbon quality. The rapid increase from 5H40 to 7H40 suggests a different biogeodynamic, less juvenil fluorescent dissolved organic matter during the nigth that should be investigated.

At least we can note that this variation does not match the variations of dissolved organic carbon. One can consider that during the tide the quality of carbon is constant during the event, even if its quantity changes.

The 3D fluorescence of Jaguaribe samples during the tide shows that the quality of the fluorescent organic matter is constant. There is a peak excited at 380 nm with an emission around 430 nm (Fig. 10.24). Only the intensity seems to change during the sampled tidal cycle. These data were treated by CP/PARAFAC to extract subtle variations that occurred during the event.

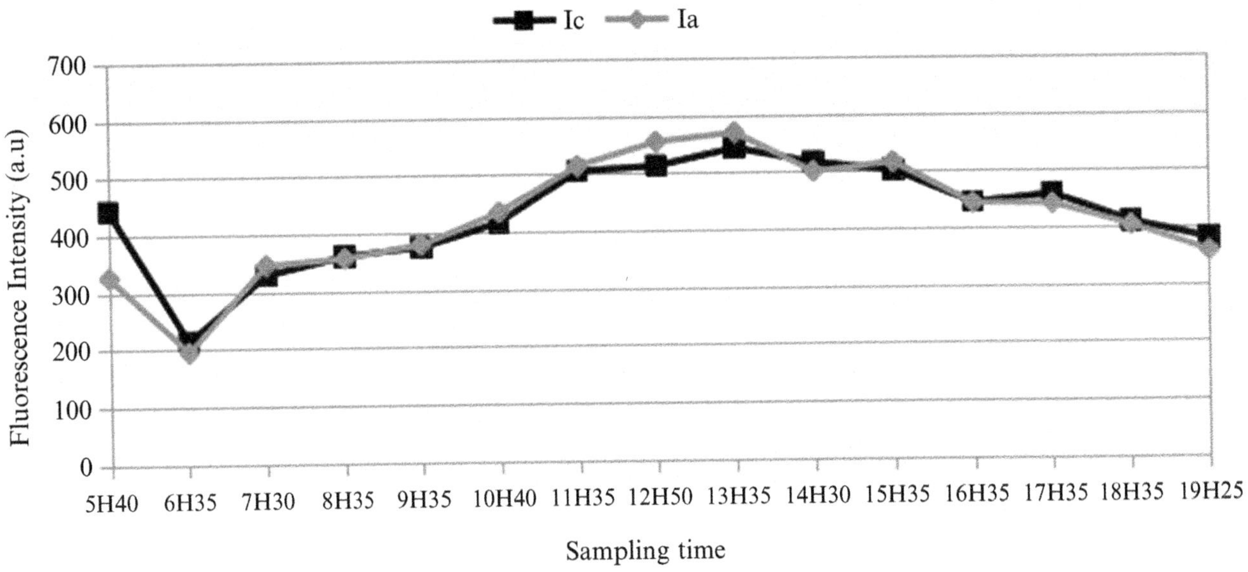

Fig. 10.22 Fluorescence of peaks A and C for the Jaguaribe River in April during the temporal sampling

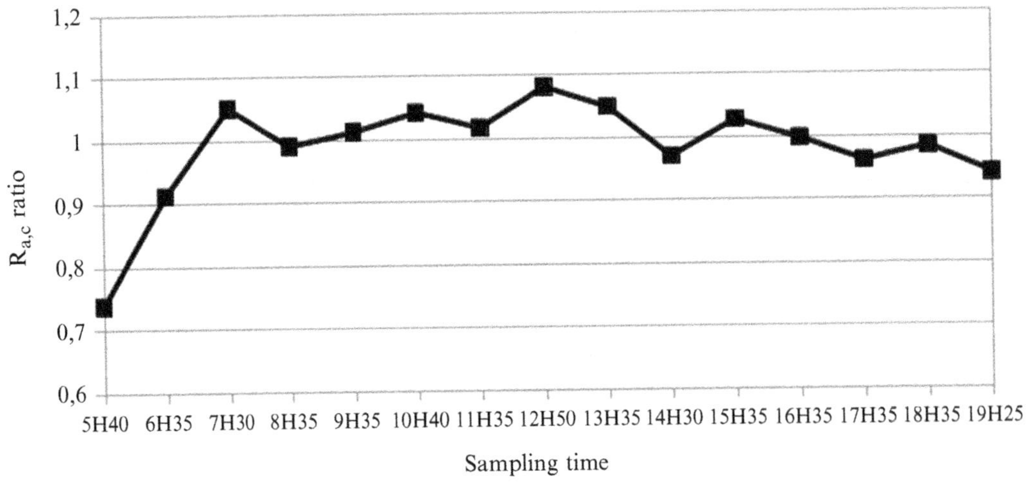

Fig. 10.23 Values the ratio $R_{a,c}$ for the Jaguaribe river in April during the temporal sampling

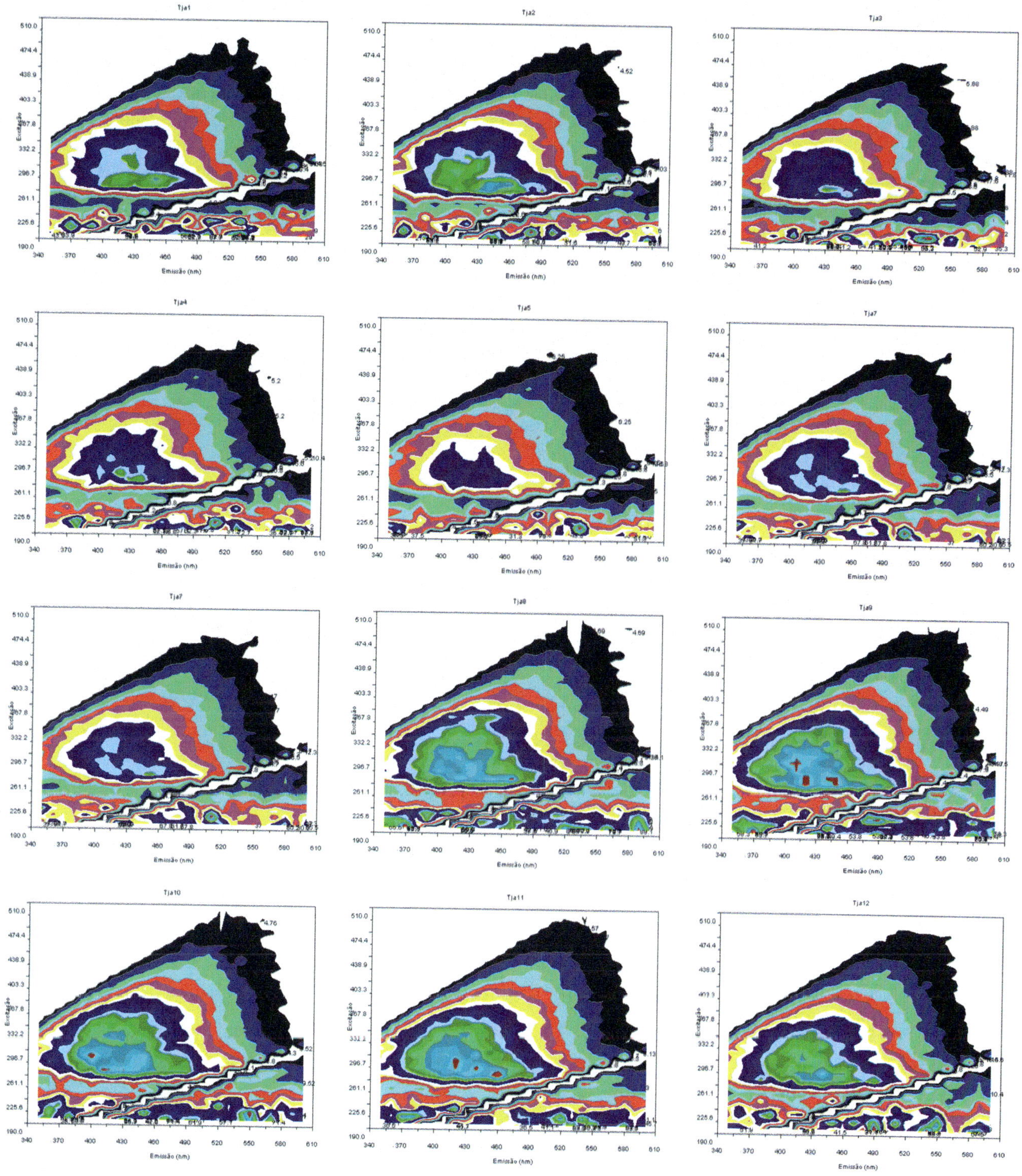

Fig. 10.24 Excitation Emission Matrices of fluorescence of the Jaguaribe representing fluorescent organic matter during tide event (in false colour).

10.5.5 Variation of the Major Elements

10.5.5.1 Space Strategy

Changes in cation and anion concentrations in the Jaguaribe River Basin in April and September 2004 are represented respectively in Figs. 10.25 and 10.26. We observed a higher mean concentration for all elements again in September. Particularly Na^+ with highest concentrations in point 6 (Aracati downstream the bridge). It also has an inland sea influence in September with a strong concentration of K^+, Na^+ and Ca^{2+} up to point 6 in September versus point 3 in April. Except for the anomaly of K^+ in point 4 in April (Raimundinho), the waters are therefore sodic.

Representing the K^+, Mg^{2+} and the Ca^{2+} concentration versus the Na^+ one (Fig. 10.26), it is observed that except for the river before the dam, sample site 9, the cations concentration are highly correlated to the Na^+, which represent the sea influence. The point 6 differs from its K^+ and Ca^{2+} concentration that confirms it out of sea influence behaviour.

The anions concentrations are also controlled by the sea influence for Cl^- and (HCO_3^-,CO_3^{2-}), except for site 2 in September (Fig. 10.27). For NO_3^- and SO_4^{2-} anions, it is

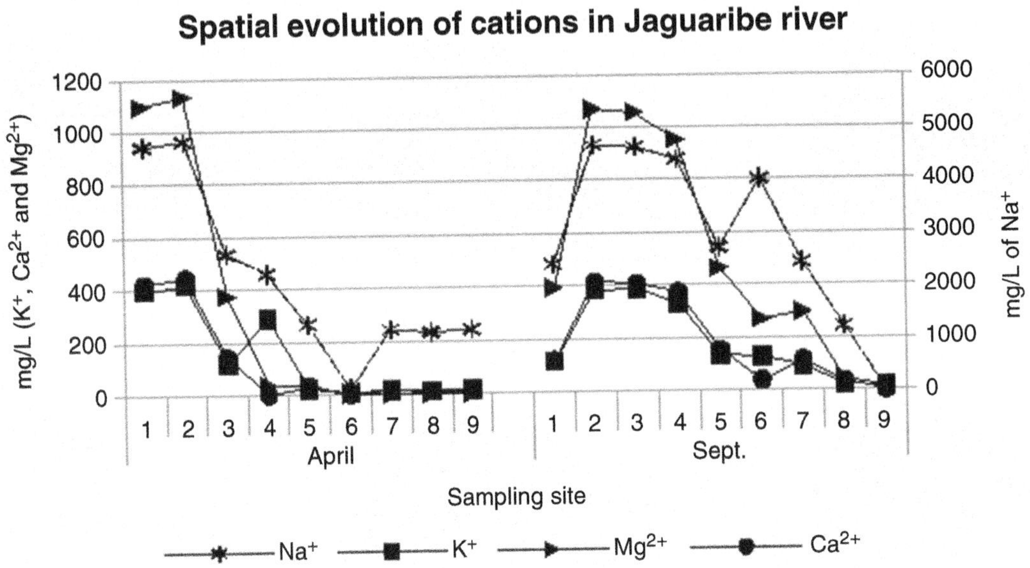

Fig. 10.25 Variation of cations in the Jaguaribe River basin in April (left) and in September (right)

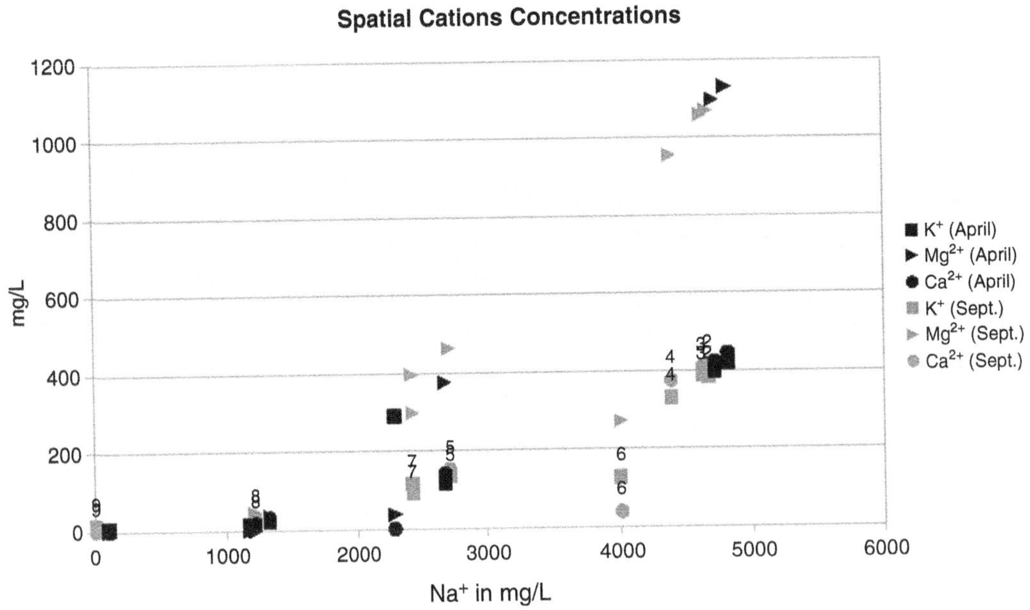

Fig. 10.26 Sodium to cations concentration relation for the Jagaribe River along downstream

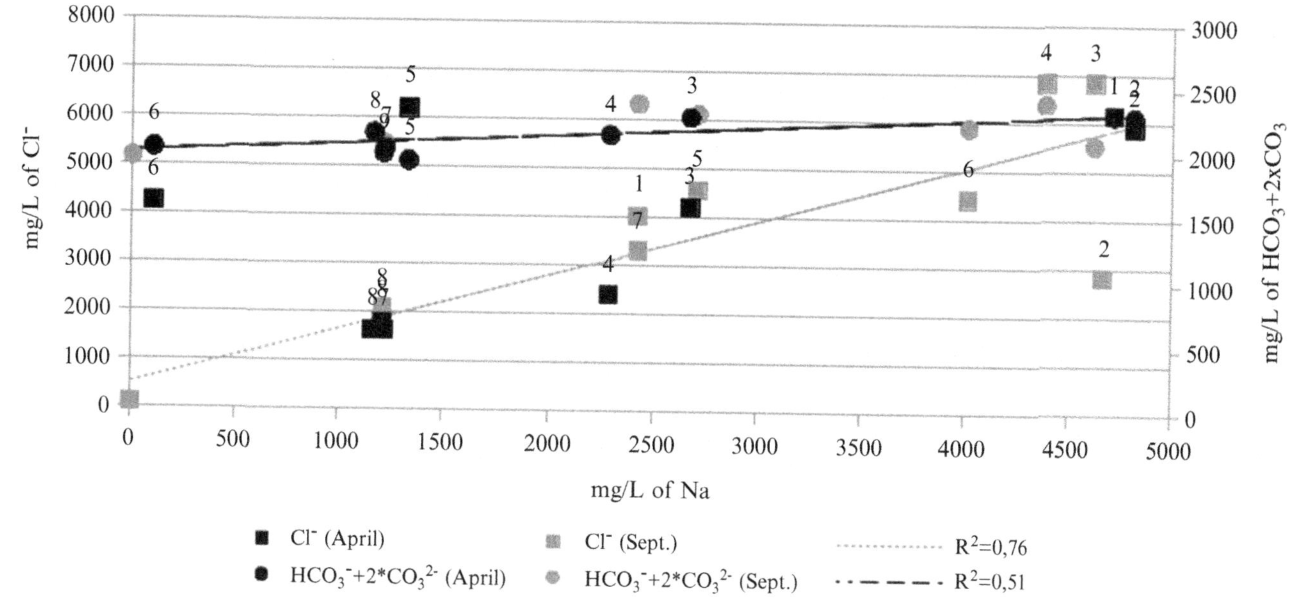

Fig. 10.27 Variation of the anions in the Jaguaribe basin in April (left) and in September (right)

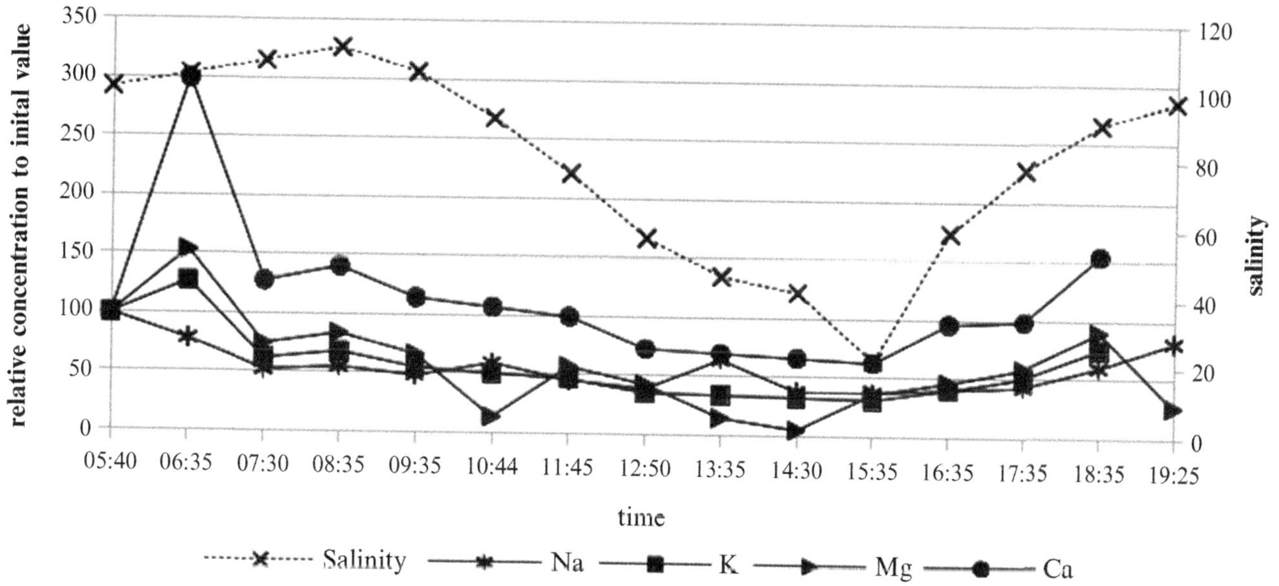

Fig. 10.28 Variation of the cations in the Jaguaribe basin during the temporal sampling

clear that the concentrations decrease after the sea influence. In the Jaguaribe, the carbonates are highly stable and represent with the Cl^- the major anions.

10.5.5.2 Time Strategy

The temporal sampling in April is presented in Fig. 10.28 for the cations and in Fig. 10.29 for the anions. The graph represents the percentage variation of the initial concentration to accentuate the relationship with the salinity. The salinity starts at 18.2 with a maximum of 20.4–08:30. The minimum of salinity occurs in the afternoon around 3:30 p.m. and has a salinity of 4. The results in concentrations are summarized in Table 10.6. Clearly the concentrations in cations and anions are normally associated with salinity. The most concentrated element is Ca^{2+} (Fig. 10.28). For the other elements concentration, it decreases until the minimum salinity is reached.

These results do not show an influence of the activities above the sampling point. Except the first point in the early

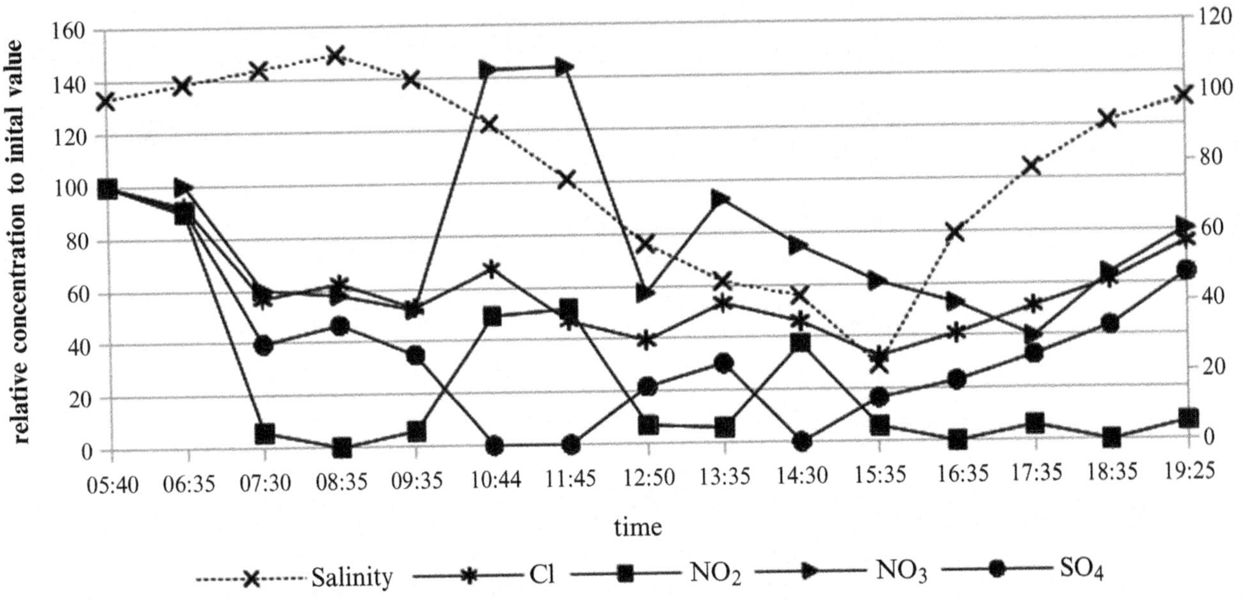

Fig. 10.29 Variation of the anions in the Jaguaribe basin during the temporal sampling

Table 10.6 Results on larger elements of temporal sampling

			Anion in mol L^{-1} (*1000)					Cation in mol L^{-1} (*1000)				
Ref	Hour	Sal.	HPO_4^{2-}	Cl^-	NO_2^-	NO_3^-	SO_4^{2-}	Na^+	NH_4^+	K^+	Mg^{2+}	Ca^{2+}
Tja01	05:40	18.2	ND	183	1	ND	18	193	ND	4	18	2
Tja02	06:35	–	ND	169	1	8	16	151	ND	5	28	6
Tja03	07:30	–	ND	104	ND	5	7	101	ND	3	14	3
Tja04	08:35	20.4	ND	114	ND	5	8	106	ND	3	16	3
Tja05	09:35	19.1	ND	97	ND	4	6	93	ND	2	12	2
Tja06	10:44	16.7	ND	123	ND	12	0	115	6	14	3	ND
Tja07	11:45	13.8	ND	86	ND	12	0	88	ND	2	11	2
Tja08	12:50	10.4	ND	72	ND	5	4	75	ND	1	8	2
Tja09	13:35	8.4	ND	98	ND	8	5	124	5	12	3	ND
Tja10	14:30	7.6	ND	84	ND	6	ND	74	ND	6	1	ND
Tja11	15:35	4.0	ND	60	ND	5	3	72	ND	1	7	1
Tja12	16:35	10.8	ND	75	ND	4	4	77	ND	2	9	2
Tja13	17:35	14.2	ND	94	ND	3	6	83	ND	2	11	2
Tja14	18:35	16.6	ND	112	ND	5	8	116	ND	3	17	3
Tja15	19,25	17.8	ND	138	ND	7	11	157	21	23	5	ND

morning that could by an outlier, there is a perfect correlation between the Na^+ concentration and the Ca^{2+}, Mg^{2+} concentration. Considering that Na^+ is principally coming from sea, this means that during the Jaguaribe tide event, there is only mixing of water. The two nitrate concentrations (10:00–11:00) should pointed out that there is a pulse during this period coming from the mangroves but this must be confirmed. The same trend is observed for the anions where correlation of Cl^- and SO_4^{2-} is about $r^2 = 0.88$ and $r^2 = 0.85$ respectively, while the carbonate is quite constant during the tide as it does not represent a good sea end member. In conclusion, there is no entry of major elements during the tidal episode sampled in this experiment.

10.6 Conclusion

This work allowed for the first characterization of natural organic matter within two zones of mangrove, one urban (Cocó River) and another less affected by urban activities (Pacoti River). A characterization of the Jaguaribe River basin was also made temporally and spatially. The results showed that the two mangrove zones in the Fortaleza metropolitan area are very similar from the point of view of larger elements and the 2D or 3D fluorescence. However, they showed slight differences in the quality of the organic matter in relation to the metals complexing and the intensity of fluorescence. The Cocó River basin seems to have a production of specific organic matter. The properties of complexation then can appear weak for the Cocó river with a higher production of young (fulvic-like) less complexing organic matter. Also, the ability to fluoresce is greater for the organic matter of this zone. These differences appear throughout the sampling period showing that it is not in relation to time, but to human activities, which mainly influence the quality and behaviour of the two mangrove areas.

For the lower part of the Jaguaribe River, measurements of dissolved organic carbon and major elements were made. A comparison of the results showed a strong seasonal variability, with concentrations generally higher in September (dry season) than in April (rainy season). There is a steady decrease in concentrations from more inland stations to sea-influenced ones. No external or unusual inputs were detected. The influence of the sea is visible up to point 5 (Aracati) in September but only to point 3 (Ilha Grande) in April. During the tide, there is only variation of the elements with the salinity. The fluorescence failed to detect any specific processes during the tidal cycle.

During this study many measures were performed. It has shown a strong influence of the urbanization on the geochemistry of the organic material produced in the mangrove zones. This modification may play an important role in the transport of pollutants such as metals.

In the future it would be necessary to do a more complete study on the flows coming in and out of the mangrove zone. Also a study on metals at the same time could be important to really understand what happens in that zone and the consequences on coastal geochemistry.

Acknowledgement This work was supported by the Brasilian Research Center (CNPq, Processo n° 309342/3003-4), the Toulon University (CRCT Campaign 2004) and PQ 309718/2016-3 to RVM. The FUNCAPCNPq PRONEX Project PR2-0101-00052.01.00/15, also supported part of this study.

References

Alberts JJ, Filip Z, Price MT, Williams DJ, Williams MC (1988) Elemental composition, stable carbon isotope ratios and spectrophotometric properties of humic substances occurring in a salt marsh estuary. Org Geochem 12:455–467. https://doi.org/10.1016/0146-6380(88)90155-6

Bisi TL, Lepoint G, Azevedo ADF, Dorneles PR, Flach L, Das K, Malm O, Lailson-Brito J (2012) Trophic relationships and mercury biomagnification in Brazilian tropical coastal food webs. Ecol Indic 18:291–302. https://doi.org/10.1016/j.ecolind.2011.11.015

Blasco F, Saenger P, Janodet E (1996) Mangroves as indicator of coastal change. Catena 27:167–178. https://doi.org/10.1016/0341-8162(96)00013-6

Brandão R de L (1995) Mapa Geológico da Região Metropolitana de Fortaleza

Burke DJ, Weis JS, Weis P (2000) Release of metals by the leaves of the salt marsh grasses Spartina alterniflora and Phragmites australis. Estuar Coast Shelf Sci 51:153–159. https://doi.org/10.1006/ecss.2000.0673

Coble PG (1996) Characterization of marine and terrestrial DOM in seawater using excitation-emission matrix spectroscopy. Mar Chem 51(4):325–346

Da Silva JCGE, MacHado AASC, Oliveira CJS, Pinto MSSDS (1998) Fluorescence quenching of anthropogenic fulvic acids by Cu(II), Fe (III) and UO2/2+. Talanta 45:1155–1165. https://doi.org/10.1016/S0039-9140(97)00224-5

Doney SC (2010) The growing human footprint on coastal and open-ocean biogeochemistry. Science 80:328

Freire GSS, Maia PL (1991) Regime Hidrodinâmico do Estuário do Rio Pacoti. Rev Geol 4:29–32

Freire GSS, Maia LP, Castelo Branco MPN (1991a) Frente salina e balanço do sal no estuário do Rio pacoti –Ceará. Rev Geol 4:33–41

Freire GSS, Maia PL, Andrade de Meireles AJ, Mota RF (1991b) Natureza do Material em Suspensão do Estuário do Rio Pacoti. Rev Geol 4:13–20

Freire GSS, Ottmann F, Melo O (1991c) Ocorrência de Pirita e Gipso no Estuário do Rio Pacoti. Euzébio/Ceará Rev Geol 4:5–12

Gatto LCS (1999) Diagnóstico Ambiantal da Bacia do Rio Jaguaribe Diretrizes Gerais para a Ordenação Territorial. IBGE

Godoy MDP (2015) Alterações nas áreas de mangue em estuários no Estuário do Ceará devido à mudanças nos usos do solo e mudanças climáticas. Universidade Federal do Ceará

Jaffé R, Boyer JN, Lu X, Maie N, Yang C, Scully NM, Mock S (2004) Source characterization of dissolved organic matter in a subtropical mangrove-dominated estuary by fluorescence analysis. Mar Chem 84:195–210. https://doi.org/10.1016/j.marchem.2003.08.001

Jennerjahn TC, Knoppers B, Souza W, Carvalho EV, Mollenhauer G, Hübner M, Ittekkot V (2010) The tropical Brazilian continental margin. In: Liu K-K, Atkinson L, Quiñones R, Talaue-McManus L (eds) Carbon and nutrient fluxes in continental margins a global synthesis. Springer, Berlin, pp 427–442

Lacerda LD, Marins RV (2002) River damming and changes in mangrove distribution. ISME/Glomis Electron J 2:1–4

Lacerda LD, Ittekkot V, Patchineelam SR (1995) Biogeo-chemistry of mangrove soil organic matter: a comparison between *Rhizophora* and *Avicennia* soils in South-eastern Brazil. Estuar Coast Shelf Sci 40:713–720. https://doi.org/10.1006/ecss.1995.0048

Lacerda LD, De Souza M, Ribeiro MG (2004) The effects of land use change on mercury distribution in soils of Alta Floresta, Southern Amazon. Environ Pollut 129:247–255. https://doi.org/10.1016/j.envpol.2003.10.013

Lacerda LD, Molisani MM, Sena D, Maia LP (2008) Estimating the importance of natural and anthropogenic sources on N and P emission to estuaries along the Ceará State Coast NE Brazil. Environ Monit Assess 141:149–164. https://doi.org/10.1007/s10661-007-9884-y

Lacerda LD, Dias FJS, Marins RV, Soares TM, Godoy JMO, Godoy MLDP (2013) Pluriannual watershed discharges of Hg into a tropical semi-arid estuary of the Jaguaribe River, NE Brazil. J Braz Chem Soc 24:1719–1731. https://doi.org/10.5935/0103-5053.20130216

Lu X, Jaffe R (2001) Interaction between Hg(II) and natural dissolved organic matter: a fluorescence spectroscopy based study. Water Res 35:1793–1803. https://doi.org/10.1016/S0043-1354(00)00423-1

Machado W, Silva-Filho EV, Oliveira RR, Lacerda LD (2002) Trace metal retention in mangrove ecosystems in Guanabara Bay, SE Brazil. Mar Pollut Bull 44:1277–1280

Marins RV, Lacerda LD, Gonçalves R, Ćosović B (1997) Effect of root metabolism on the post-depositional mobilization of mercury in salt marsh soils. Bull Environ Contam Toxicol 58:733–738. https://doi.org/10.1007/s001289900394

Mavignier AL (1992) Estudo Fisico, Quimico e Bacteriologico do Rio Coco

Miranda PTC, Martins MLR, Soares ZML (1988) Levantamento e quantificação das áreas de manguezais no Estado do Ceará (Brasil)

através de sensoriamento remoto. In: Simp V (ed) Brasileiro Sensoriamento Remoto, Brazil, Natal. Anais... São José dos Campos: INPE, 1988,. pp 90–94. ISBN 978-85-17-00050-8

Mounier S, Patel N, Quilici L, Benaim JY, Benamou C (1999) Fluorescence 3D de la matiere organique dissoute du Fleuve Amazone. Water Res 33:1523–1533. https://doi.org/10.1016/S0043-1354(98)00347-9

Mounier S, Lacerda LD, Marins RV, Bemaim J (2001) Copper and mercury complexing capacity of organic matter from a mangrove mud flat environment, Sepetiba Bay. Brazil Bull Environ Contam Toxicol 67:519–525. https://doi.org/10.1007/s001280154

Nilin J, Moreira LB, Aguiar JE, Marins R, Abessa DMS, Lotufo TMC, Costa-Lotufo LV (2013) Sediment quality assessment in a tropical estuary: the case of Ceará River, Northeastern Brazil. Mar Environ Res 91:89–96

Otero XL, Macías F (2002) Variation with depth and season in metal sulfides in salt marsh soils. Biogeochemisrty 61:247–268

Patel-Sorrentino N, Mounier S, Benaim JY (2002) Excitation-emission fluorescence matrix to study pH influence on organic matter fluorescence in the Amazon basin rivers. Water Res 36:2571–2581. https://doi.org/10.1016/S0043-1354(01)00469-9

Ryan DK, Weber JH (1982) Fluorescence quenching titration for determination of complexing capacities and stability constants of fulvic acid. Anal Chem 54:986–990. https://doi.org/10.1021/ac00243a033

SDU (1994) Composição Floristica e Estrutura dos Bosques de Mangue dos Rios Ceará, Cocó e Pacoti

SEMACE (2005) Secretaria de Meio Ambiente do Estado do Ceará, programa Zoneamento Ecológico e Econômico do Litoral do Ceará, Fortaleza (www.semace.gov.ce)

Stanford WP (1986) Hidrogeologia da região metropolitana de Fortaleza. n°1, Águas Subterrâneas, Associação Brasileira de Águas Subterrâneas, São Paulo, Brasil, ISSN 0101-7004

Vulnerability of Mangrove Forests and Wetland Ecosystems in the Sundarbans Natural World Heritage Site (Bangladesh)

11

Shafi Noor Islam, Sandra Reinstädtler, and Albrecht Gnauck

Abstract

The Sundarbans Natural World Heritage Site is lying within the Bangladesh coastal region, which is gifted with vast natural resources, a delta, tidal flat, mangrove forests, marches, lagoons, bars, spilt, estuaries and coastal ecological environment. These habitats, biotopes and ecosystems also serve as potential resources for anthropogenic communities: 36.8 million people are living within the coastal region of Bangladesh and being dependent on coastal water resources, for which the Sundarbans Natural World Heritage Site is giving some protective management support. Nevertheless the natural coastal resources are drastically reducing due to unplanned use by the community and the stakeholders, although the Ganges-Brahmaputra-Meghna Rivers are carrying 6 million m^3/s water. As a result, the Sundarbans mangrove forests and wetlands are vastly affected through these developments. The present situation demands that an integrated natural resource management plan is necessary for the protection of the mangrove coastal ecosystem. This chapter was prepared based on primary and secondary data sources, as the objectives were to analyze the present coastal mangrove natural resources management status. The study investigates the deltaic Sundarbans natural world heritage site with its mangrove forests and wetlands ecosystem development and management strategies to ensure less vulnerability and a sustainable development of coastal mangrove resources in the Ganges-Brahmaputra Rivers deltaic coastal floodplain region of Bangladesh.

Keywords

Coastal wetland · Ecosystem · Ganges-Brahmaputra delta · Degradation and management

S. N. Islam (✉)
Department of Geography, Development and Environmental Studies, University of Brunei Darussalam (UBD), Bandar Seri Begawan, Brunei Darussalam
e-mail: shafi.islam@ubd.edu.bn

S. Reinstädtler
Department of Environmental Planning, Brandenburg University of Technology Cottbus-Senftenberg, Cottbus, Germany

A. Gnauck
Department of Ecosystems and Environmental Informatics, Brandenburg University of Technology Cottbus-Senftenberg, Cottbus, Germany

11.1 Introduction

Bangladesh is a land of water, but water is also one the most critical and major problem in the country. The deltaic coastal areas of Bangladesh are especially impacted and more vulnerable. The deltaic coastal areas are about 710 km in length with extending along the Bay of Bengal from the mouth of Teknaf River in the south East to the Mouth of Raimangal River in the south west and including the greater districts of Chittagong, Noakhali, Barisal, Patuakhali and Khulna (Nishat 1988). The coastal region is composed of the land and the sea including estuaries and islands adjacent to the land water interface of south east Bangladesh and the coast can be divided into three distinct regions (Pramanik 1983). Most parts of the Bengal deltaic coastal areas of Bangladesh are an active or nascent delta with vigorous dynamism and the islands or chars are formed, eroded and reformed (Islam 2016). The on the outskirts of the coastal area located Swatch-of-No-Ground submarine canyon (SoNG)[1] being situated on the south of Ganges-Brahmaputra delta near to Dubla Char, is also a major feature only 24 km far from the coast water and from the Sundarbans mangrove forest (Rahman 1988, 2003) being interesting for understanding the geomorphological facts within the entire region.

[1] The first marine protected area is placed over here at SoNG in Bangladesh, which is also called Ganga Trough.

C. Makowski, C. W. Finkl (eds.), *Threats to Mangrove Forests*, Coastal Research Library 25,
https://doi.org/10.1007/978-3-319-73016-5_11

Focusing on the Sundarbans Natural World Heritage Site and its mangrove forests and wetlands, the major water challenges will be described in its vulnerabilities in exemplary form. The connection to existing ecosystem services will help in understanding the topical specifications of vulnerability and its focal areas for Sundarbans mangroves. The Sundarbans Natural World Heritage Site is lying in the coastal region of the Ganges-Brahmaputra Rivers Delta, which is the largest floodplain wetland region worldwide and is located in the South Asian Region (comp. Morgen and Mclntire 1959; Khandoker 1987; Sarker et al. 2003; Goodbred and Nicholls 2004; Islam 2016).

With naming some of the water challenges, in the first row the scarcity of upstream fresh water has created serious environmental problems in the coastal region of Bangladesh. This strongly needed upstream fresh water supply to the coastal region is playing a potential role to make a balance within the complex interplay of varied existing ecosystems. Also the oceanic saltwater and tidal pressures or storm winds are giving challenges to the coastal area. In this constellation the coastal mangrove forests and wetlands provide numerous important ecological services to humanity: the tropical and sub-tropical coastal mangrove wetlands provide important ecosystem services to the population living in the hinterland such as by sheltering it from storm winds, capturing salt spray and improving crop production in arid coastal areas (Wolanski 2007). Also the fact of proceeding higher flora and fauna species appearance in variety and population density is giving another multiplicating factor and ecosystem service for mangrove forests and wetlands: it is at the coast where the heavily sediment-laden river water with very little salinity meets the saline sea water. This mixing creates unique ecosystems with unique flora and fauna species appearance that exhibit their presence here in unique habitat forms such as these mangroves and enhance biodiversity. Amongst these mangroves the most important once are the here to be described Sundarbans with its unique tropical mangrove forest and the wild life, the fisheries dominated by economically important species such as Hilsa and Prawns (Rahman 1988, 2003). Mangrove forests and wetlands also safeguard the coast against erosion and guard against loss of capital infrastructure and human lives (Wolanski et al. 2009).

Coastal wetland environment and ecosystem services play also a very important role in the socio-economic development of the Bengal Deltaic coastal region of Bangladesh (Islam 2016). About 25% (35 million people) of population live in the deltaic coastal areas and most families are dependent on coastal natural resources for their livelihood (Hidayati 2000). So a coastal ecosystem such as a mangrove forest or wetland also supply these socioeconomic benefitting ecosystem services for the human population with including building material, fisheries, forage, fuel, timber, and the protection of commercial, recreational, and naval vessels (Williams 1919; Wolanski et al. 2009). Coastal wetland environment and ecosystem services play a very important role in socio-economic development of the Bengal Deltaic coastal region of Bangladesh (Islam 2016) and emphasize the tremendous importance of the established Sundarbans Natural World Heritage Site in a part of this coastal region as balancing spatial area. About 25% (35 million people) of population live in the deltaic coastal areas and most families are dependent on coastal natural resources for their livelihood (Hidayati 2000).

For creating a balance in the variety of existing ecosystems and creating a less vulnerable basis therefore sustainable deltaic coastal wetland ecosystem management is essential for the coastal community's livelihoods as well as other stakeholder's benefits in Bangladesh (Pramanik 1983; Rahman 1988; Costanza 1992; Islam 2007, Islam 2016, Islam and Gnauck 2009a) and especially within the multivariate existing ecosystem services within Sundarbans mangrove forests and wetlands and the protected area of Sundarbans Natural World Heritage Site.

With summarizing the overall research emphasize, the question for sustainable development is a key concept in strongest driving force of the water sector (comp. Grigg 1996; Gleick 1998; Costanza 1992; Islam 2007) and its specific problems and vulnerabilities in Bangladesh and especially in the Sundarbans Natural World Heritage Site. The viewpoint of peaks of vulnerability causes, pressures, states, impacts for mangrove forests and wetlands should guide into possible to be found solutions in decision-making for the specific and unique Sundarbans Natural World Heritage Site.

11.2 The Aims and Objectives of the Study

The aims of this study are to understand about unique characteristics and its vulnerabilities of the coastal deltaic mangrove forest and wetland region of the Sundarbans Natural World Heritage Site in Bangladesh. Special emphasize was placed on the interrelationship of the – through mangrove forests and wetlands origins – ecosystem services, its potentiality of coastal socio-economy and community development combined to vulnerability factors and the role for protecting Sundarbans coastal mangrove forest and wetland ecosystem.

There are some specific objectives of this study, which have been recognized such as following:

(i) To understand the characteristics and its vulnerability of the Sundarbans Natural World Heritage Site combined to its catchment area of Ganges-Brahmaputra Rivers and their deltaic coastal mangrove forest and wetland ecosystems as well as its role for the coastal region.

(ii) To investigate in the analysis of the vulnerable (with especially the example of water salinity intrusion) and degraded mangrove forest and wetland ecosystem in the deltaic floodplain regions in Bangladesh.
(iii) The study seeks the adequate coastal mangrove forest and wetlands ecosystem management strategies and
(iv) Making recommendations for future development policies for the coastal natural resources, protection and management of coastal deltaic mangrove degraded ecosystem in the Bangladesh coastal region.

11.3 Data and Methodology

The present study was carried out based on primary and secondary data sources. The primary data collected from field investigation in 2003 and 2008. PRA practices were arranged with the local people at Munchiganj near Sundarbans, Koira Upazila, and Mongla port area. The information were collected from Galachipa and Kalapara upazilas of Patuakhali district. Besides these, various reports and published articles in journals and conference proceedings have been used for this study. Published materials, reports and journals were collected from different governmental and non-governmental organizations in Bangladesh. For secondary data collection, CARDMA reports on coastal resource development and management were used very openly. The especial research reports of CARDMA, FAP 24 report, BWDB, CEGIS reports and BCAS reports were also used for this study. Beside these some interviews were arranged with water engineers, environmentalists, geographers, geologists and sociologists and expert people on river systems and its ecology. Besides, the secondary data were collected from the relevant published research works[2] in the country of Bangladesh as well as outside of the country. The collected data were processed, analyzed, and visualized through MS Excel interpolation, VISIO 32, ArcGIS 9.3 and 10.1 software. They were used for vulnerability approximation to investigate in the degraded deltaic coastal floodplain mangrove forest and wetland ecosystems in the Ganges-Brahmaputra Rivers Deltaic region in Bangladesh.

11.4 Geographical Location and Physical Characteristics of the Coastal Mangroves in Sundarbans Region

The Sundarbans Natural World Heritage Site and the nearly entire spatial extent of the country of Bangladesh is situated in the Ganges-Brahmaputra-Meghna Rivers catchments and the Bengal coastal zone, for which in the following the geomorphological development as well as its present catchment characteristics should be described in more detail in order to estimate the vulnerability factors in a more precise way (Goodbred and Kuehl 2000, Islam and Gnauck 2009a).

The Sundarbans and the deltaic coastal region of Bangladesh are located in the south of the country and its geology with lying within the Bengal basin began 130 million years ago. The initial breakup of Gondowana land possibly occurred in early Cretaceous (130 million years ago) with a separation of India from combined Australian and Antarctica, the sea floor was created in what is now the Bay of Bengal (Anwar 1988).

Also the Bangladesh coast with including Sundarbans constitutes the low-lying landmass of the Himalayan river basin ecosystems (Iftekhar 2006). The formation of the delta plain commenced some 12,000 years ago when the sea level started to rise rapidly due to the melting of ice sheets to the north. About 6000 years ago, sea level rise slowed down and coastal plains started to form (Islam 2001; Iftekhar 2006). The large rivers especially the Brahmaputra, Meghna and the Ganges are playing an important role in the coastal region. Biogeographically, the coast is part of the Bay of Bengal large marine ecosystem (Nishat et al. 2002). The Ganges fresh water plays a vital role in coastal economy, floodplain fertility and keeps the balance of mangrove forest and wetland ecosystem services in south western coastal part of Bangladesh (Islam and Gnauck 2009b).

The coastal morphology is further characterized by different features like vast networks of rivers and channels, heavy water discharge carrying sediments, many islands in the coastal channels, strong tidal waves, cyclone and surges, physiographically funnel shape (Figs. 11.1 and 11.2). Also the Swatch-of-No-Ground canyon is located in the continental shelf about 24 km south of the Bangladesh coast (Figs. 11.1 and 11.2). Coastal mangrove landscapes and forest as well as wetland ecosystems are characterizing the south western region of Bangladesh coast (Islam et al. 2017).

The deltaic coastal region of Bangladesh within this Bengal basin is vastly influenced by the GBM-catchment as well as by the oceanic influences. The coastal length is 710 km and the coastal area comprises 47,210 km^2 with a population of 36.8 million (BBS 1979, 1981, 1985, Islam 2016). There are existing 19 coastal districts, which are including the coastal region such as Jessore, Narail, Satkhira, Khulna, Bagerhat, Gopalganj, Shariatpur, Chandpur, Pirojpur, Jhalokati, Barguna, Patuakhali, Barisal, Bhola, Lakshmipur, Noakhali, Feni, Chittagong and Cox's Bazar (Figs. 11.2 and 11.1) (Islam et al. 2017). The coastline of the Ganges delta extends from the mouth of Hoogly-Bhagirathi river in the west to the mouth of the Padma-Meghna river in the east (Figs. 11.1 and 11.2) (Rob 1998). The coastal area of the delta is confined roughly in an area between longitudes 80° 0′E to 91° 0′E and 21° 30′N to 22°

[2] For instance published papers in journals and books.

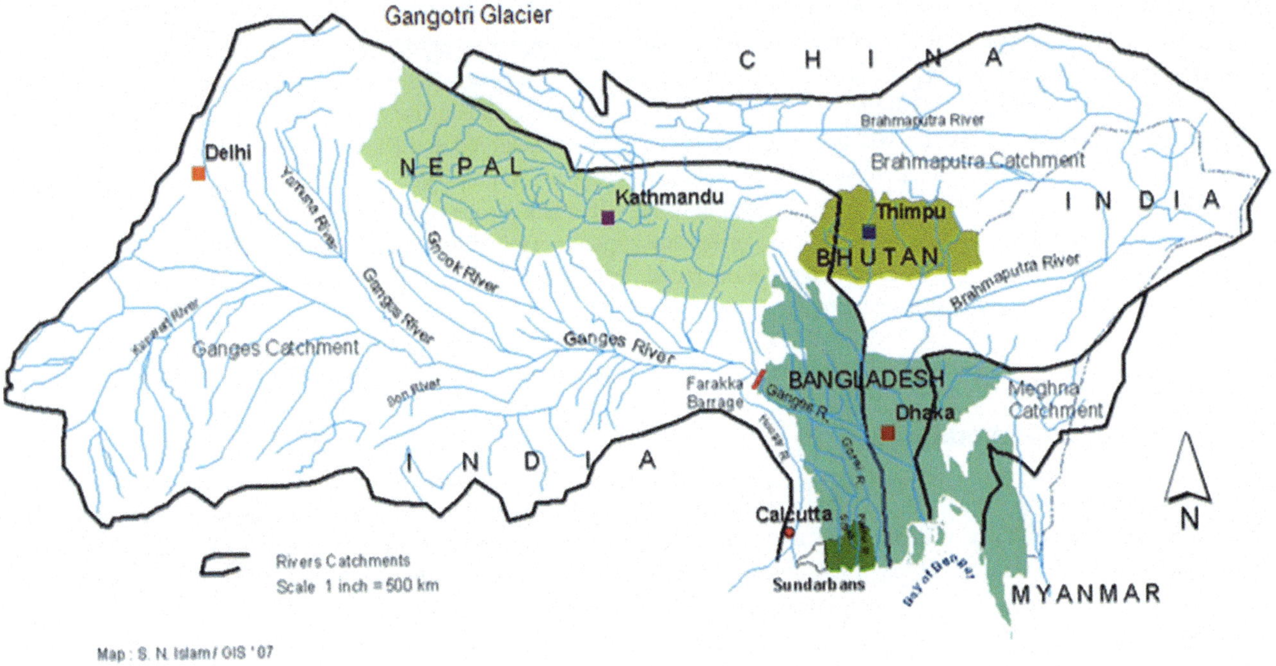

Fig. 11.1 Geographical Location of the Sundarbans Mangrove Wetlands in Ganges-Brahmaputra-Meghna-Rivers Catchment Area, Bangladesh (Map Source: Islam and Gnauck 2008)

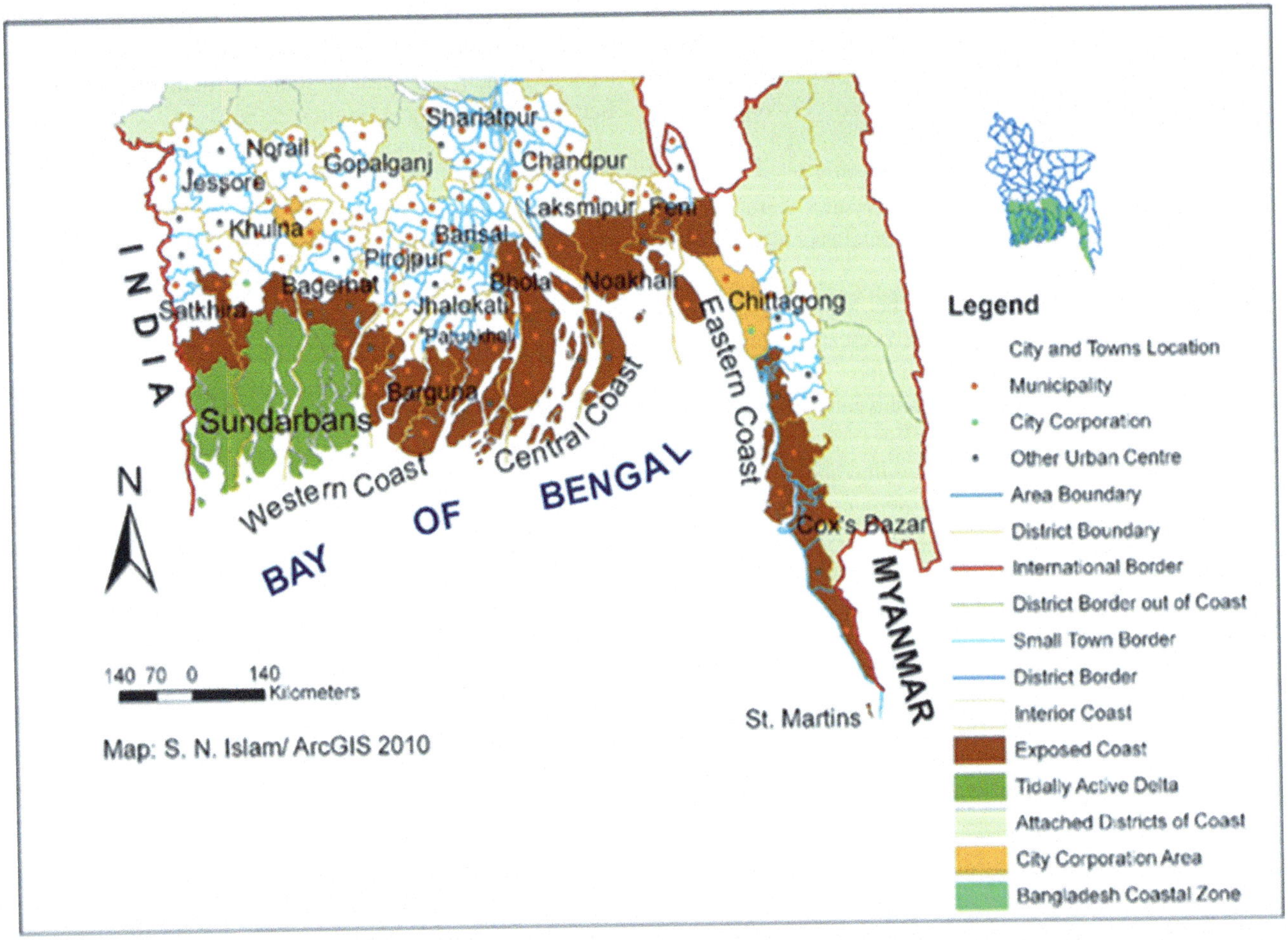

Fig. 11.2 Coastal Morphology of Sundarbans Mangroves and Surrounding other Ecosystems in Bangladesh (Source: Islam 2010)

30′N latitude. The Sundarbans implementing coastal area of Bangladesh is deltaic active and hydrologically dynamic. The coastal areas of Bangladesh have been defined and categorized by the Coastal Area Resource Development and Management Association (CARDMA) in three different types of coastal regions on the basis of physiographic characteristics. These three divisions are: (1) the Eastern Region, (2) the Central Region and (3) the Western Region (Figs. 11.1 and 11.2). The eastern coastline is extending from the Big Feni River to Badar Makam along Chittagong, which is regular, unbroken and protected along the sea by mud flats and submerged sands. For instance the Cox's Bazar sand beach, which is about 145 km long, is part of this coastline (Figs. 11.2 and 11.1).

The central region runs east from the Tetulia River to the big Feni River estuary and including the mouth of the combined flows of the Ganges-Brahmaputra-Meghna (GBM) Rivers (Akter et al. 2010; Islam et al. 2017). This is the reason for this regions characteristic, heavy sediment input, formation of new *chars* and river bank erosion and accretion. The western region covers the coastline westward from the Tetulia River to the international boundary located at the Hariabhanga River (Figs. 11.1 and 11.2). This region is mostly covered with dense mangrove forests with having a reduced river bank erosion. The rivers of the region are mostly stable; land accretion does not occur massively (Jalal 1988).

The coastal regions of Bangladesh are categorized by three types of divisions based on the physiographic and ecological characteristics, from which the first one of the Atlantic Type inhabits the Sundarbans Natural World Heritage Sites perimeter: this western region is part of the Ganges deltaic floodplain, which includes the Sundarbans as being largely covered by mangrove forests. This western part of the coast is more stable than other parts of the coastal region. This category is comparable to the Atlantic coastal characteristics (Figs. 11.1 and 11.2).

Next to this Atlantic Type or Western (coastal) Region the other two categories of the coastal region are named Central Region[3] and Pacific Type[4] or Eastern Region (Figs. 11.1 and 11.2).

The GBM systems further attributes, which influence the more or less vulnerable characteristics of the Sundarbans mangroves, are such as being discharged through the low-lying area of the central coastline, heavy sediment at inputs from the rivers result in a morphologically dynamic coastal zone. The central region has deltaic characteristics. So the GBM Rivers discharge is about 1.5 million m^3/s during the peak period (Hasan and Mulamoottil 1994).

11.5 Sundarbans Natural World Heritage Site

The Sundarbans Natural World Heritage Site is the one natural World Heritage site in Bangladesh next to two cultural World Heritage sites[5] with having a perimeter extent of 139,700 hectares (Islam 2003). The Bengali meaning of the Sundarbans is 'Beautiful forest'[6] and its principal tree species 'Sundari' or *Horitiera fomes* might be one of the reasons for the in this way named mangrove forest area (for further theories comp. Islam 2003).

Being located in the coastal zone, it can be stated that the coastal zone is a large and diverse area comprising a rich array of social, economic and environmental resources (Iftekhar 2006), which is applicable and influencing the Sundarbans in the same way. The influencing grade, which is also an important factor for vulnerability and degradation state evaluations (comp. Chap. 6.1) in the Sundarbans, can be estimated within the following facts of the Bangladesh coastal zone: (1) the coastal zone provides a natural and cultural heritage, the Sundarbans natural world heritage site, indigenous life style, and (2) tangible cultural heritage as well as (3) intangible cultural heritage and historic events. It further contains (4) natural resources that are a natural inheritance of immense economic, cultural and intrinsic value. (5) Coastal communities enjoy the coastal resources such as mangrove forests, indigenous cultural events, sand shore beaches, sun and the eco-tourism in the Sundarbans natural heritage site. (6) There are lots of cultural activities in the coastal region, which are dependent on coastal environment and its elements. (7) A total number of 600,000 people are employed to be about for 6 months. Besides this fact 120,000 tourists visit the Sundarbans heritage site yearly and 50,000 people work daily (Islam 2003).

Also being part of the estuary of the GBM Rivers catchment, the Sundarbans Natural World Heritage Site is the largest area of still functioning mangrove forest with spanning a spatial extent of 1 million hectares in South-West

[3]***Central Region:*** The Ganges-Brahmaputra-Meghna (GBM) River systems fall into the Bay of Bengal through the Meghna estuary. The Meghna estuary is located in the central region of the coastal area of Bangladesh. The central region is more unstable and most vulnerable. Coastal char formation, river erosion and accretions are the regular activities and building up the character of region (Figs. 11.1 and 11.2).

[4]***Pacific Type:*** The eastern region of Chittagong specially is a narrow strip with a long sandy beach in Cox's Bazar, known as Eastern Region. It is a long strip until Teknaf and St. Martins Coral Reef Island. The eastern part of Bengal coastal region is a hotspot of biodiversity and displaying coastal narrow floodplain and mountainous characters. This eastern region has Pacific characteristics (Figs. 11.1 and 11.2).

[5]The cultural World heritage sites are the Historic Mosque City of Bagherat (1985) and the Ruins of the Bhuddist Vihara at Paharpur (1985) (Islam 2003).

[6]'Sunder' means 'beautiful' and 'ban' 'forest' (Islam 2003).

Bangladesh and South Eastern portion of the state of West Bengal in India (Islam 2003). This massive area of mangrove forest inhabits 75% of its protected site in Bangladesh and 25% in India (Islam 2003).

The most fascinating and picturesque mangrove forests of Sundarbans can be observed at the southern extremity of the Gangetic delta and to the northern apex of the Bay of Bengal within the districts of Shatkhira, Khulna and Bagerhat (Islam 2003).

The mangrove forests of the Sundarbans Natural World Heritage Site cover an area of 6017 km^2, of which 4143 km^2 is located on land and the remaining 1874 km^2 are within water such as in form of rivers, canals and creeks of widths varying from a few meters to several miles (Islam 2003). The Sundarbans mangrove forests constitute about 51% of the forest area of the country of Bangladesh (Islam 2003).

11.6 Vulnerability and Degradation of Coastal Mangrove Ecosystems in the Sundarbans Region

The Sundarbans Natural World Heritage Site is part of the continental shelf of Bangladesh coast, which has to be observed for estimating a vulnerability and degradation state in an entire and non-separated way. A secluded assessment of the protected site would not bring forward integrative solutions: so this continental shelf of Bangladesh coast is about 66,000 km^2 with having coastal water being very shallow with less than 10 m depth and covering about 24,000 km^2 as well as up to 200 m depth and covering about 70,000 km^2. The area of an Exclusive Economic Zone (EEZ) is about 1, 64,000 km^2 which is larger than the land area of Bangladesh with about 1,47,570 km^2 extent (Islam 2016; Islam et al. 2017). The sector contributes 22% of the total animal protein intake of the country. Especially this sector is most prestigious in order to find optimized and less conflicting (land use) planning solutions in order to strengthen the protection against proceeding vulnerability and degradation as well as to give benefits to this sector (Islam 2016; Islam et al. 2017).

The rise of a vulnerability or degradation grade as well as general damages of coastal ecosystems such as the unique Sundarbans Natural World Heritage Site poses a direct threat to human, flora and fauna survival in the coastal region in Bangladesh. Therefore such negative impacting processes disturb in sorts of a balanced sustainable development: primarily coastal resources are an important economic and socio-cultural source of both food and income. In correlation to a further rise of the vulnerability or degradation grade in coastal zones and within or at its boundaries of the Sundarbans, a critical situation arises for business as well as for ecology and biodiversity through ports, industrial development and – from that businesses partially influenced – lower developed or disturbed coastal-marine biodiversity resources (Fig. 11.2). In recent years coastal areas such as Sundarbans are also becoming more attractive for tourism and recreation purposes, which are giving a further driving force for a higher vulnerability or even degradation process. Fish and sea food are primary sources of animal protection in the coastal region (Tan et al. 1997. In the coastal region of Bangladesh and especially within already degraded parts of the even more extended Sundarbans mangroves (comp. Islam 2003), shrimp farms are densely concentrated and created natural disasters, which have cost thousands of lives. The rise of the vulnerability grade, degradation or even the complete destruction of mangroves also alters the regional water system and coastal habitats, affecting fisheries coastal ecosystem balances and climate stability (Islam 2003, 2016; Islam et al. 2017), which is bringing up a doubled negative loop in direct, cumulative or synergistically impacting coastal areas and the Sundarbans mangroves.

All of these effects further on reduce the fertility of land located in nearby areas (Biksham and Andrea 1996) and so also the Sundarbans Natural World Heritage Site. It is estimated that 90% of all marine organisms spend some portion of their life cycle within mangrove system (Adeel and Pomeray 2002). The Sundarbans mangrove forests furnish ecosystem services, which are made all the more valuable within times of climate change. They protect coastal communities from cyclone and storm damage, and this function may become even more important as climate change intensifies. Globally, mangroves are being cleaned or damaged at the alarming rate of 1–2% annually and in Bangladesh case it is the same. The Chakoria Sundarbans mangrove forest of Cox's Bazar in Bangladesh was already cleaned up one decade ago: the area was occupied by shrimp culture and the vulnerability and degradation grade raised with visually looking like a saline desert. The Challenge of coastal environment and water resources management in the Sundarbans Natural World Heritage Site as well as in the entire country of Bangladesh is to identify the trends impacting on coastal and mangrove resources, balance the multiple and competing demands and respond with strategies to achieve sustainable management in the coastal areas (Islam et al. 2017).

11.6.1 Vulnerability of Mangrove Ecosystem and Biodiversity

The vulnerability of mangrove ecosystems has to be understood in context of the definition of a coastal zone, in which mangrove ecosystems are placed: a coastal zone is the area on both sides of the actual land-sea interface, where the influences of land and water on each side are still in determining factors-climatically, physiographical and ecologically (Ferdra and Feoli 1998). Also the location of the landward boundary of the coastal zone, which implements functions of three basic geophysical processes, is important for estimating vulnerability factors of mangrove forests: tidal fluctuations; salinity; and risk for cyclone and storm surges. The coastal zone of Bangladesh, which is being affected by these processes, cover an area of 47,210 km^2, or 32% of the country with serving as landmass of 19 districts (Fig. 11.2) and abound 36.8 million people, of whom 29% represent the population living in the coastal zone (MoWR 2005; Goodbred and Nicholls 2004; Islam 2016).

A higher vulnerability amplitude for Sundarbans mangrove forests and wetlands is coming from the fact that the bay of Bengal is affected by Tsunami incidences, for which the Sundarbans region is placed within the moderate Tsunami vulnerability belt (comp. Islam 2007). Further influencing processes for a higher predestination of vulnerability is the Sea level rise in general or the human influence through land use such as others. Up to that the coastal zone is highly affected by high spring tides when water is saline during the dry season (Islam and Gnauck 2007a, b). Also the coastal areas are highly prone to cyclone induced storm surges that bring about the most catastrophic damage here (Chowdhury 2009).

Next to the geomorphological, abiotical influences also other biotical influences next to the already named human land use habits are giving impacts to the vulnerability stage of the Sundarbans Natural World Heritage Site. Some abiotical facts are combined to the characteristics of the Ganges-Brahmaputra-Meghna (GBM) rivers system, which is the largest river system in the world and is passing through Bangladesh on its way to the Bay of Bengal and carries sediments 1.8–2.4 billion tons (Nishat 1988). These facts are subject to ongoing coastal dynamic processes being generated by the river flow, tide and wide actions and leading to vulnerability enhancing processes such as accretion and erosion in the coastal area of Bangladesh (Nishat 1988; Islam et al. 2017).

Besides there are some other delineates of the coastal region, which are high salinity, tidal waves, especially the influence of shrimp cultivation, and administrative management of the coastal region.

11.6.2 Degradation of Mangrove Ecosystem and Biodiversity

The salinity investigation results show that the south west Bengal coastal regions and in these the Sundarbans Natural World Heritage Site are carrying the highest rate of water salinity which is unbalancing the coastal ecosystem and general ecology. According to salinity approximation this high rate is harmful to rural and urban biodiversity as well as for the species of humans and their urban drinking water (Fig. 11.3) (Islam et al. 2017).

The Fig. 11.4 demonstrates the water salinity intrusion trends in the south and southwest region of the Ganges deltaic region, which includes the entire Sundarbans mangroves (Islam et al. 2017).

Fig. 11.3 The anthropogenic influences on coastal forests and wetlands in the Sundarbans mangrove in Bangladesh (Source: Islam and Gnauck 2007a)

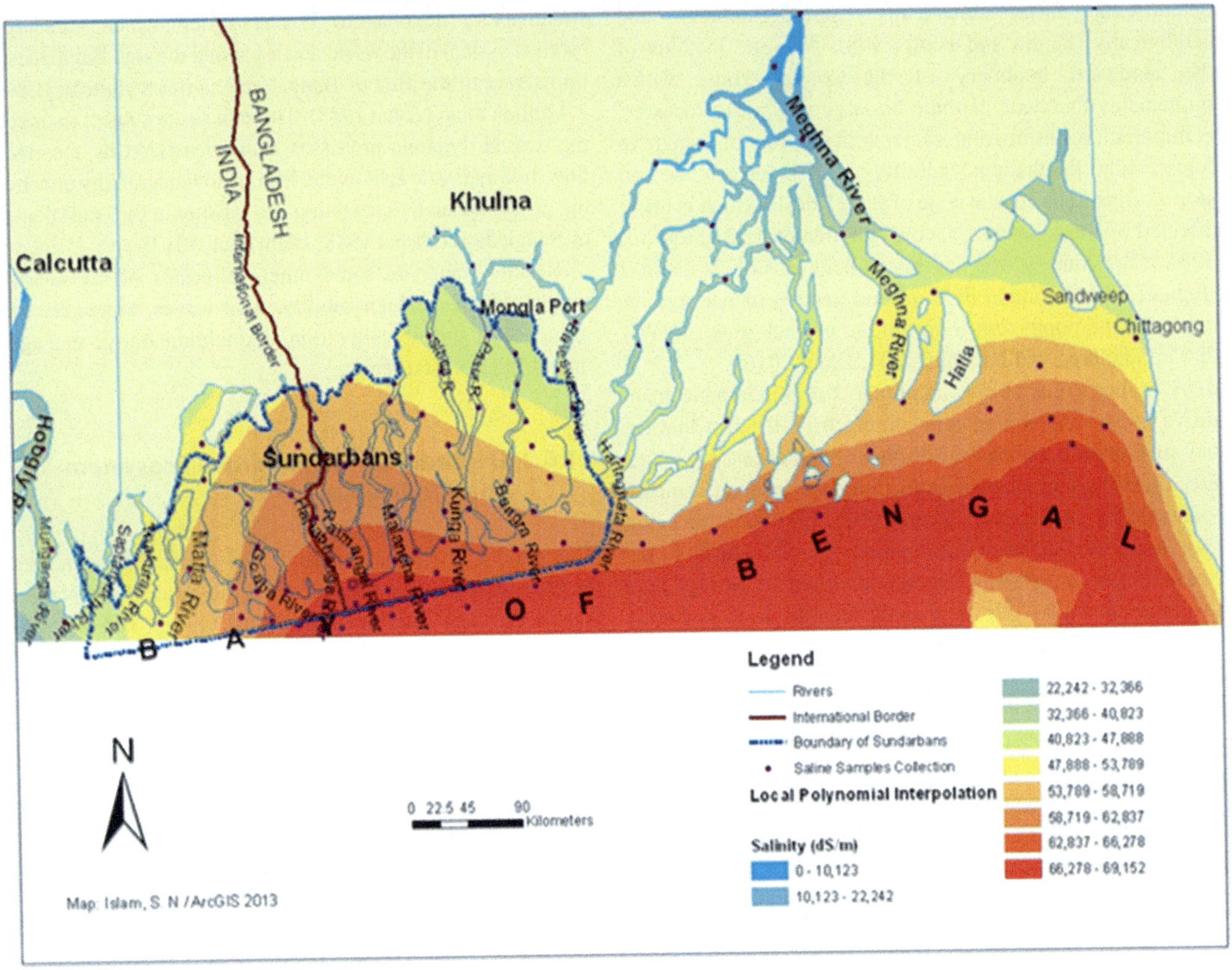

Fig. 11.4 The Salinity intrusion pattern in the Bengal coastal region (Source: After Islam 2014)

Four major cities and 136 small towns are located in the coastal region and a major portion of the inhabitants are dependent on mangrove resources in the coastal region. Most of the towns are effected through salinity intrusion and sea level rise impacts in the region as Sundarbans mangroves are also affected due to high salinity intrusion. Therefore, the investigation results of salinity modelling in the South and Southwest coastal deltaic regions are under threat for ecosystems and coastal rural and urban ecosystem goods and services. Also several observed districts face the greater challenge of higher salinization intrusion by comparing the data of an intense of affected land in hectare from 1973 and 2000 (comp. Sect. 7.1, Table 11.1) (Islam and Gnauck 2007a, b).

Especially the coastal mangrove and agro-biodiversity loss is a common scenario in the Ganges-Brahmaputra-Meghna Rivers deltaic region between Bangladesh and India. The Fig. 11.3 demonstrates the scenarios of the coastal mangrove forest and wetland region in the Sundarbans and its Sundarbans Natural World Heritage Site in Bangladesh. The quality of mangrove forest and wetland water and soil are rapidly degrading due to high saline water intrusion and anthropogenic influences (Figs. 11.3 and 11.4).

Table 11.1 The coastal land area affected by high salinity intrusion (in1973 and 2000)

Districts	1973 (ha)	2000 (ha)
Patuakhali	115,000	139,000
Barguna	103,000	104,000
Bhola	40,000	93,000
Pirozpur	20,000	28,000
Barisal	8,000	10,000
Jhalokhati	1,500	3,000

The study also found that the mangrove reduction rate is about 45% in both countries (Bangladesh – India). Deforestation is raising and land-cover is changing due to shrimp farming, salt farming, agricultural land extension, urbanization extension and settlement development (Adel 2001, ESCAP 1988, Milliman et al. 1989). These development processes are adversely, affect coastal fish production and lead to a loss of agrobiodiversity and coastal floodplain biodiversity and of livelihood, which means to negatively influence 3.5 million people, who are dependent on natural resources in the coastal region in Bangladesh (Miah 1989, Helcro 1993, Anon 1995). The mangrove wetland ecosystems are dependent on the water and soil salinity. Almost all mangrove forests need a freshwater supply from an upstream source. In the Sundarbans Ganges deltaic coastal region the two potential rivers such as the Passur-Mongla and Chunar-Munchigannj are carrying the high rate of salinity intrusion. The Fig. 11.4 shows the high salinity intrusion trends in the coastal mangrove forest and wetland region. The salinity model also demonstrates that the salinity trends are much higher in the southwestern region of the Sundarbans mangrove wetlands regions. The salinity rate was 42,000 dS/m in 2003, whereas in 2010 the salinity rate is 53,000 dS/m in the Passur-Mongla river point. On the other hand, the southwestern corner is showing the highest rate of water salinity in 2010, which is over 53,000 dS/m (Islam et al. 2017).

The salinity penetration in the upstream areas of the coastal zone is one of the main obstacles to maintenance of water quality for drinking, irrigation and fisheries purposes (Islam and Gnauck 2009a, b) as well as for the mangrove ecosystem and biodiversity in general. And already the coastal mangrove and wetland ecosystems have been recognized as a driving force for biodiversity conservation and coastal urban socio-economic improvement (Nishat 1993; Ahmed and Falk 2008; Islam et al. 2017).

Sixty six different species of mangroves (comp. Fig. 11.5) are growing in Sundarbans where 70 species have been recorded in the world, 12 species of plants and animals already vanished and Javan rhinoceros, Single horn rhinoceros, Water buffalo, Swamp deer, Mugger crocodile, Gaur and Hog deer are extinct animals (FAP4- FPCO 1993, FAP4-FPCO 1999).

Also in the Ganges-Brahmaputra rivers deltaic floodplain alone approximately 2.1 million ha of wetlands have been lost due to flood control, drainage, and irrigation development (Khan et al. 1994). Therefore, coastal urban wetlands biodiversity is facing serious challenges from salinity intrusion, environmental changes and anthropogenic impacts (Sarkar 1993; Sarker et al. 2003; Nair 2004; Ahmed and Falk 2008).

The mangrove forest and wetlands in the coastal region include rivers, estuaries, mangrove swamps, marsh (*haor*), oxbow lakes (*baor*) and *beels*, water storage reservoirs, fish ponds, and some other lands are also facing the similar environmental problems (Khan 1993; Hughes et al. 1994; Gopal and Wetzel 1995; Islam and Gnauck 2008; Islam et al. 2017). In such problematic complicated situation the smart-use of natural resources and salinity desalination processes can solve the coastal floodplains biodiversity and ecosystem problems in the Ganges-Brahmaputra coastal surface areas in Bangladesh.

Fig. 11.5 The mangrove vegetation and biodiversity in the coastal wetland areas in Sundarbans region in Bangladesh (Source: Islam et al. 2017)

11.7 Sundarbans Mangrove Forests and Wetland Management Strategies

The coastal wetland resources are totally dependent on the upstream fresh water supply and availability in the coast, therefore some potential issues should be incorporated in national water policy in Bangladesh: if the present coastal wetland resource management strategies are analysed and compared with the national policies, it will show the real coastal wetland resource scenarios. There are 257 rivers in Bangladesh and they have an important role for supplying fresh water to the coastal areas (Islam 2007, Islam 2006), which would be most important for an optimized protection of the Sundarbans mangroves and the Sundarbans Natural World Heritage Site. So the Ganges-Brahmaputra-Meghna Rivers carry almost 6 million m^3/s of water per day during flood season to the Bay of Bengal (Helcrow 1993, Goodbred and Kuehl 2000).

So the National water policy in Bangladesh (1999) is most important for optimized Sundarbans mangrove forests and wetland management strategies and also gives due importance on research and development. In article 3 the objectives of the national water policy (1999) has to be developed a state of knowledge and capability that will enable the country to design coastal water resources and wetland management plans by itself with economic efficiency, gender equity, social justice and environmental awareness. Article 4.15 of 'Research and Information Management' of this national water policy (1999) the involvement of public and private research organizations and universities should be strengthened and promoted and some specific objectives should be ensured such as (1) appropriate technologies innovation and implementation, (2) develop and promote coastal wetland and water resources management techniques and (3) produce skilled professionals for water resource management, which might also positively influence a better mangrove forest and wetland management.

The Ganges-Brahmaputra-Meghna Rivers carry also 13 million tons of suspended sediment per day during flood season to the Bay of Bengal[7]; it is three times the amount of borne by the Mississippi River (Coleman 1969; Anwar 1988). So it can be recorded that erosion and accretion are common phenomena in the coastal zone, where nearly a billion tons of sediment are brought by the rivers. About two thirds of these sediments are discharged into the Bay of Bengal, while the rest contributes to the formation of new land and islands (Iftekhar 2006). However river bank erosion is a severe problem also in sorts of mangrove forest and wetland management. The annual rate of erosion in the Meghna river estuary alone is around 3199 ha/year (Islam 2006). Aspects of the sedimentary history of the Bengal Basin since ca. 9000 years. BP can be interpreted from data provided in this study.

It is also undoubtedly true that the Ganges-Brahmaputra-Meghna River system has been carrying sediments from upstream and depositing them in the basin, but their depositional environment needs to be considered (Islam 2001). The Fig. 11.6 shows a model of integration of coastal mangrove forests and wetlands in the Bengal coastal region in Bangladesh.

Water related sectors such as agriculture, land use, water navigation sector, industrial sector, mangrove wetland ecosystems, forest sector, fisheries and livestock and coastal community livelihood are directly dependent on coastal mangrove forest and wetland resources (Fig. 11.6) (Islam et al. 2017).

Bangladesh produce 1 million tons of fishery products, which is more than twice of Indian aquaculture production per capita and Nepal, Pakistan and Sri Lanka lag far behind. The coastal Bangladesh aquaculture is emerging as a prime rural industry, contributing to employment, food security, poverty reduction and export earnings. The export of aquaculture products means for Bangladesh, to earn half a billion dollars per year, which is making them the country's second-biggest earner of foreign exchange after textile. All these water resources sectors and sub sectors are inter-connected for their own sectoral development and management. Therefore the coastal water resources sector is a potential sector in the coastal region and is serving as a driving force for sustainable development of the coastal region of Bangladesh.

In most developments, water resources management and sustainable livelihood discourses a need in form of integration at different levels are recognized and emphasized (Rahman 2003). The soil of the entire coastline is generally sediment deposit of the GBM and other rivers. Such soils are fertile, but recently it has been affected by high salinity intrusion. It has been estimated as being almost half of the coastal arable land and can be classified as saline zone. Saline affected areas such as the greater Khulna and Patuakhali districts are being the most affected areas in the coastal region of Bangladesh (Jalal 1988).

---

[7] It is estimated that some 1.5–1.8 billion tons of sediment is denounced in the Bay of Bengal per year (Nishat 1988; Rahman 1988) but Anwar (1988) and Jabbar (1979) mentioned that the mighty rivers Ganges, Brahmaputra and Meghna transports about 2.4 billion tons of sediments annually to the Bay of Bengal. Curray and Morre (1971, 1978) show an enormous sub aqueous delta beneath the Bay of Bengal formed by the sediment derived from the Ganges-Brahmaputra-Meghna Rivers, the Bengal Deep Sea Fan, and the largest deep sea fan in the world. Sediments deposited on top of the upper most peat are fluvial controlled. These are the sediments deposited by the Ganges-Brahmaputra-Meghna rivers system (Islam 2001).

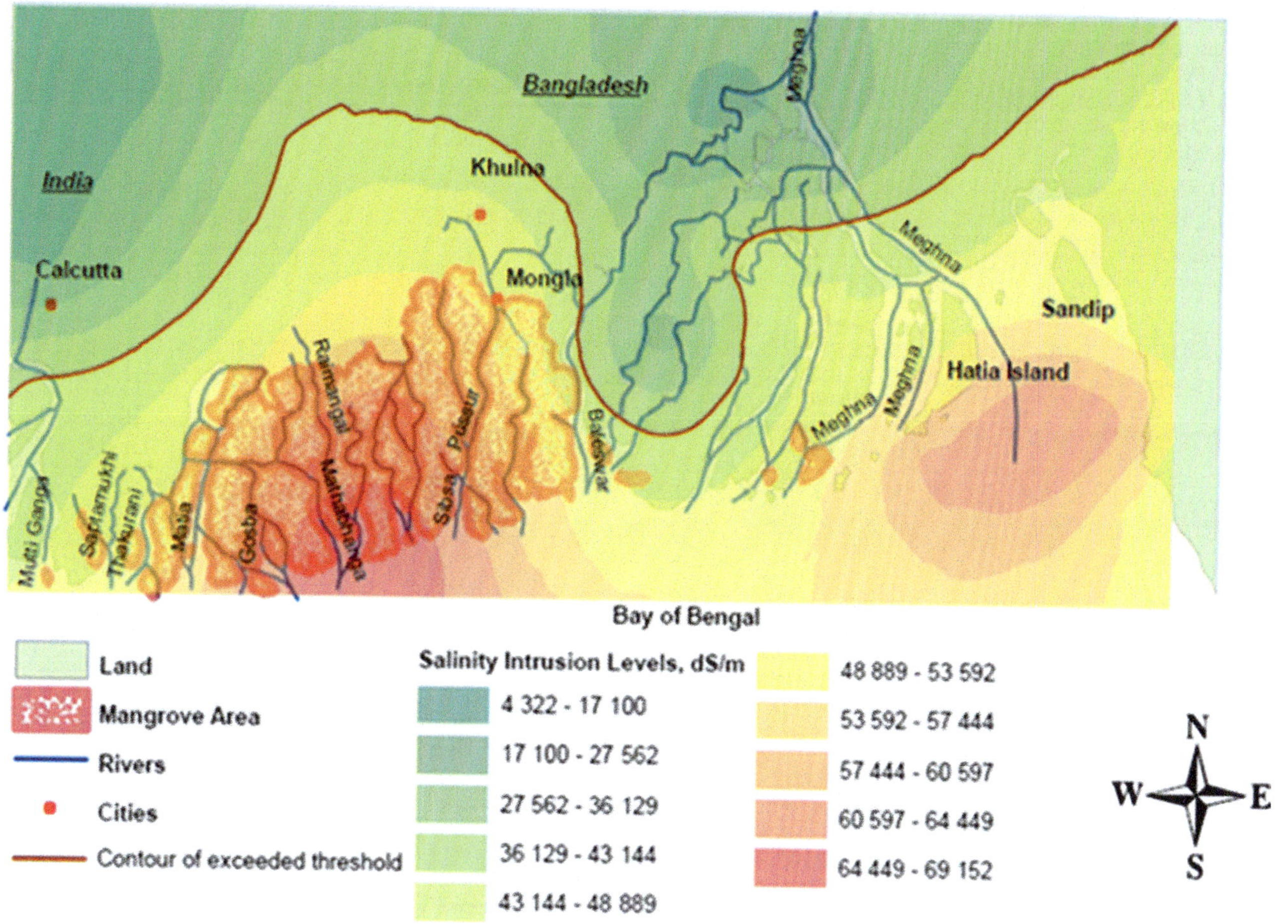

Fig. 11.6 Interlinkage of coastal wetlands, water resources, ecosystem services and livelihoods (Source: Islam 2016)

11.7.1 River Water and Sediment Flow Characteristics Influencing the Coastal Mangrove Forest and Wetland Ecosystems

As already stated, from most importance for the further protective and sustainable development of Sundarbans Natural World Heritage Site is the provision of freshwater resources. So the river water cycle and flow dynamics are from greatest interest: the Ganges river is one of the freshwater sources, which itself receives over 60% of its discharge from its tributaries. The Gomti, Damodar, Yamuna Rivers, Mohananda are the major tributaries of Ganges Rivers in the Indian portion (Helmer and Hespanhol 1997), which has to be acknowledged for a continuing, advantageous development of the Sundarbans mangroves. The main distributaries of the Ganges River in Bangladesh are the Baral, the Gorai, the Arial Khan, the Bhairab, the Mathabhanga, the Kumar, Chitra and the Ichamati River (Miah 2001). Their average annual combined discharge into the Bay of Bengal is 100,000–140,000 m^3/s carried by Ganges-Brahmaputra-Meghna (GBM) Rivers (EGIS 2000). As a whole the Brahmaputra (Jamuna) River carries water flow 60,000–100,000 m^3/s. The minimum water flow of the Meghna River is 3750 m^3/s (Fig. 11.7) (Fedra and Feoli 1999, Goodbred and Kuehl 2000). The Ganges River is the major source of silt deposition and delta formation in the Bay of Bengal (Joseph 2006; Islam and Gnauck 2008). The Brahmaputra carries every year, about 600 million tons of sediments, with alluvial sand constituting large proportion (FAP 24 1996). The sediment is deposited in form of dunes in the river bed, which is about 350, in some cases up to 600 meters long and they can move very fast as 17 m/s (FAP 24 1996; EGIS 1997).

The Meghna Delta is the main output of the Ganges, the Brahmaputra and the Meghna to the Bay of Bengal (Chowdhury and Haque 1990). The sediment load is extremely high, with suspended sediment load during flood stage reaching as high as 13 million tons per day (Coleman 1969). The strong south Asian monsoon and high Himalayan source area supports one of the world's largest riverine sediment loads (1.8–2.4 billion tons annually) for the Ganges-

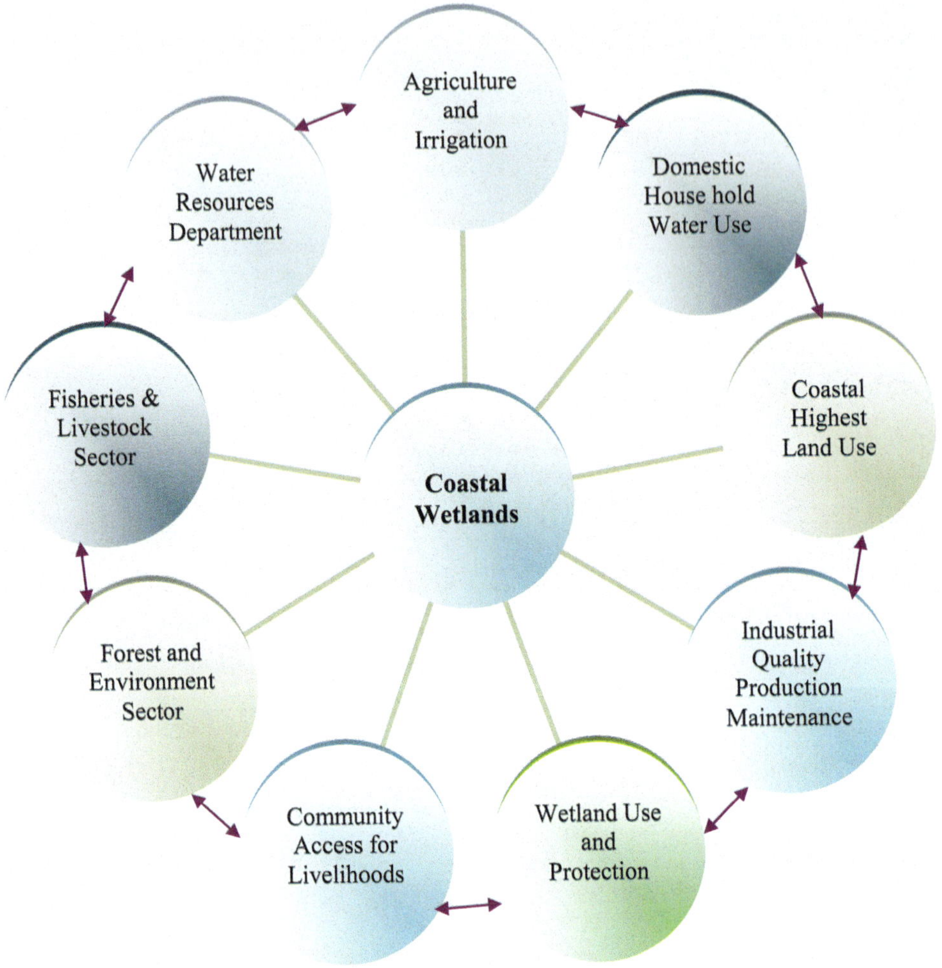

Fig. 11.7 Major rivers annual water discharge flow in the Ganges Deltaic region (Data Source: EGIS 2000; Miah 2001; Goodbred and Nicholls 2004; CEGIS 2010)

Brahmaputra-Meghna dispersal system (Anwar 1988; Umitsu 1993; Goodbred and Nicholls 2004).

11.7.2 Present Threats for the Coastal Mangrove Forest and Wetland Ecosystems

The study finding has asserted that India's diversion of fresh water had resulted in a loss of biodiversity. This loss of biodiversity became three times more in Bangladesh in the year of 2007 than the already observed loss from 1976 (Islam and Gnauck 2009a, b). The observed species decline was already in 1976 negatively coupled with the rice output of 236,000 metric tons, which is further correlating with the just in before closed up construction of Farakka Barrage in India. The previous study result of EGIS (Environmental Geographical Information Services) from the Centre for Environmental Geographical Information Services (CEGIS) showed, that the water quality of Sundarbans region had been degraded in the dry season (February–June): 60% of the water of Sundarbans rivers are poor quality with an EC (Electrical Conductivity) dS/m is 5532; the other 40% show up a good quality with an EC of dS/m 2766 (Islam and Gnauck 2009a, b). Coastal water pollution such as oil spills are a new threat and could cause immense damage, especially to aquatic fauna, seabirds, mangrove biodiversity and coastal agricultural cropping systems. The yearly natural calamities, global warming and its impacts are also new challenges and threats for coastal food security and agro-biodiversity (Islam and Gnauck 2009a, b; Islam et al. 2017).

As being declared through an earlier soil investigation conducted by SRDI in 1970, soil salinity was mainly found in the Ganges tidal floodplain of the coastal region. The Ganges River floodplain and the peat basins areas were classified as being non-saline. Soil salinity occurred south of Khulna and Bagerhat districts. The surface and subsurface salinization is very common with saline water intruding into freshwater aquifers. Each year, a new area of 146 km^2 is affected by salinity, whereby in the same process a significant reduction in biodiversity is observed (Dutta and Iftekhar 2004). Salinity range was between 8644 dS/m and 17,288 dS/m. A rise of soil and water salinity has been noticed in 1975 when Farakka Barrage was constructed on the Ganges

River and withdrawal fresh water from the basin (Islam and Gnauck 2007a, b). At present soil salinity level has been recognized south of Khulna and Bagerhat town, where it ranges from 17,288 dS/m to 32,415 dS/m during the dry season (November–May). Soil and water salinity is rapidly increasing in the coastal region and currently, river water salinity moves up as far as Kamarkhali River port in Magura district (SRDI 2000; EGIS 2000; Islam and Gnauck 2008). The salinity trends are higher in the Sathkhira, Khulna, Bagerhat, Borguna, Jhalokhati, Potuakhali, Bhola and the southern part of Noakhali districts (Table 11.1).

The salinization trends are higher in the south western region of Bangladesh. These trends urgently have to be implemented while developing and finding management strategies for the Sundarbans Natural World Heritage Site. Within the gradually eastern region comparatively less saline intrusion in the coastal region can be recognized.

At present, tidal waves, surges and coastal flooding are common annual natural phenomena in Bangladesh (Islam 2001). Most coastal regions show uneven sediment compaction which is controlled by differential sedimentation rates, sediment composition, water content, depth of overburden sediment layers and tectonic activities (Greensmith and Tucker 1986; Islam 2001). A national strategy for coastal resource management in Bangladesh is necessary to identify and analyse the problems and opportunities for future development. A study of ESCAP (Economic and Social Council for Asia and the Pacific) in 1988 and GOB (Government of Bangladesh) have referred five sets of constraints to the development of a strategy for coastal water and wetland resource management in Bangladesh such as:

- Policy making and strategies' implementation
- Planning for coastal water and wetland resources maintenance
- Integrated coastal wetland resources management
- Coastal wetlands and marine resources uses and sustainability
- Local environmental and coastal ecological perspective and
- Lack knowledge of coastal wetland environment and better understanding (Jalal 1988).

The water and soil salinity penetration in the upstream area of the whole coastal zone is one of the main obstacles to maintenance mangrove forests and wetlands such as the Sundarbans mangroves and in the same way the water quality for drinking, irrigation and fisheries purposes. The shortage of upstream freshwater supply to the coast and tidal waves are facilitating to penetrate salinity intrusion (SRDI 2000). In the coastal belt are 81,000 ponds existing with scattered locations, which are the main source of freshwater supply for domestic consumption in rural areas. The quality of wetland's water is the major constraint in water supply (Jalal 1988).

Another threat is correlated to the fact, that the coastal wetland lands are fertile for shrimp cultivation as the water is mostly saline. A total of nearly 150,000 ha were used by shrimp cultivation in the region, which is representing 11.8% of the land in the region respectively. Almost 60% of the biomass comes from agricultural residues in the coastal region. Around 200 industries in industrial zones of Chittagong discharge their untreated wastes to the Karnafuli River. Another 190 industries are located in three industrial zones of Khulna with discharging untreated toxic waste into the Bhairab River, which is carrying into the coastal region and therefore giving further threats to the Sundarbans mangroves. There are 50 coastal towns, which are located in the coastal region and are affected by high salinity intrusion in drinking water and arsenic contamination in ground water aquifers.

Beside these, there are two existing international ports situated in the coastal region. The ships at the ports have spillages of crude oil and discharge of ballast water, sewage from ships and other activities, which are causing severe coastal water pollution into the coastal marine ecosystems such as the Sundarbans mangrove forests and wetlands. The tourism activities in the coastal region of Bangladesh are also distracting water quality in the coastal region.

As a whole the Bangladesh coastal region is the most vulnerable coast in the world (Jalal 1988). There is none integrated management plan for the coastal natural resources or for the coastal communities existing. The coastal land cultivation, paddy and shrimp production policy is still dependent on the local initiatives. Freshwater flows, drinking water supply, water for irrigation and fishing are severely affected. Considering the present water resource management condition a newly policy framework guideline is essential to protect and maintenance the coastal water resources in Bangladesh and with it the Sundarbans mangrove forests and wetlands in the Sundarbans Natural World Heritage Site.

11.7.3 Sea Level Rise and Its Impacts on Coastal Region

Climate change and sea level rise, being induced by global warming, also compromise the ecological stability of the coastal zone; in sum, due to various natural and anthropogenic factors, the natural resource base of the zone is declining and the vulnerability as well as degradation grade is raising. Failing ecosystem productivity further degrades the quality of life for the local population (Dasgupta 2001) and all other floristic and faunistic species: relative Sea-Level Rise (SLR) movement has an immediate and direct effect on the coastal inter tidal ecosystems, particularly on

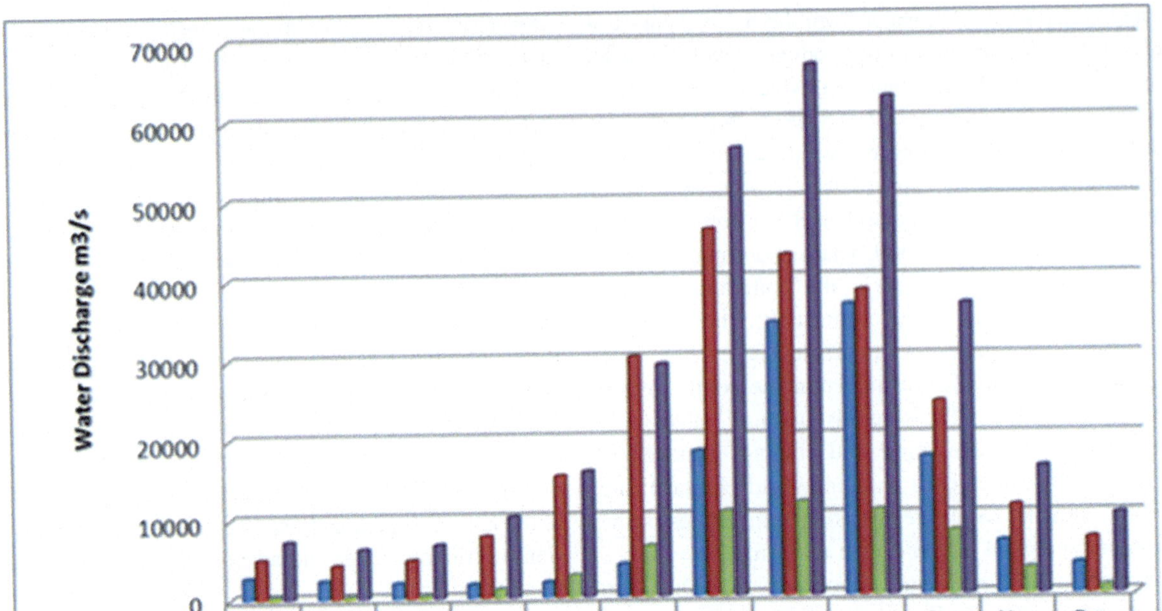

	Jan	Feb	Mar	Apr	May	Jun	Jul	Aug	Sep	Oct	Nov	Dec
Ganges	2744	2336	1984	1781	1971	4136	18324	34466	36699	17439	6682	3819
Brahmaputra	4944	4209	4786	7752	15280	30225	46242	42962	38418	24335	11137	6980
Meghna	370	264	362	1106	2778	6348	10691	11785	10722	8008	3240	919
Padma	7243	6242	6756	10220	15867	29337	56466	66980	62889	36716	16078	10104

Fig. 11.8 Climate change impacts in the coastal region of Bangladesh (Source: Akter et al. 2010)

vegetation. Arise of relative SLR decreases the influences of terrestrial processes and increases the influence of coastal marine processes (Islam 2001). The worlds' great deltas are among the most densely populated and most vulnerable of coastal areas, which are threatened by sea level rise (Broadus 1993). Global warming and sea level rise and vulnerability of coastal wetland ecosystems are factors that have to be considered the long term management strategy for dealing with the coastal mangrove forest and wetland issue also within the Sundarbans Natural World Heritage Site. The impacts of climate change in any given region depend on the specific climatic changes that occur in that region. Local changes can differ substantially from the globally averaged climate change (Harvey 2000). The global warming and climate change with its predicted sea level rise is further accelerating by global warming and therefore will cause a further 'Squeezing' of the natural tidal land. In Bangladesh case a sea level rise of 3 mm/year with being predicted as occuring before the year of 2030 and a resulting inundation of 2500 km^2 or 2% of the land have been projected by IPCC (2007) and MoEF. About 20% of the net cultivable area of Bangladesh is located in the coastal and offshore island (Fig. 11.8).

A very recent study on the coastal area in Bangladesh by IPCC 2007 shows that the mean tidal level at Hiron Point is showing an increase of 4.0 mm/year, which is higher than the global rate. Soils in this area are affected by different degrees of salinity (Rahman 1988). About 203,000 hectare very slightly, 492,000 hectare slightly, 461,000 hectare moderately and 490,200 hectare strongly salt affected soils are assessed in south western part of the coastal area (Fig. 11.8). The fact of climate change impacts are a new threat for the coastal area of Bangladesh (Fig. 11.8). In the case of Sundarbans mangrove and its World Heritage Site, sea level rise would result in saline water moving further into the delta, which would be the major threat for mangrove forest and coastal wetland ecosystems (IECO 1980).

The fate of the Sundarbans can be seen within different sea level rise and the potential impacts on environment: if 10 cm SLR will inundate, 15% of the Sundarbans will be affected; having a SLR of 1.5 m, about 17 million (15%) of the population will be concerned. The afflicted population will have to be displaced or getting homeless, whereas 22,000 km^2 (16%) of the land will permanently be inundated (Fig. 11.8).

The Fig. 11.8 shows a scenario with 3 m SLR, which would be even a worse scenario for Bangladesh with almost one third of the land –and for sure also the Sundarbans Natural World Heritage Site – could be inundated by saline water. The reduction rate of mangrove areas will be in between 50% to 75%, which would be even more harmful for coastal ecosystems in the estuaries (IPCC 2007). Besides, other environmental problems will arise in the coastal belt such as (1) water pollution and scarcity, (2) soil degradation, (3) deforestation, (4) solid and hazardous wastes, (5) loss of bio-diversity estuary landscape damage and (6) river bank erosion, which will create a lot of new challenging problems for human livelihood in the coastal region. In such situation it will further create an unstable agricultural crop production, damaging fisheries, livestock and food security in the coastal riverine islands in Bangladesh. Therefore climate change impacts on coastal region such as on the estuaries in the Sundarbans will create new threats for estuaries ecosystems and landscapes.

11.7.4 Changes of Mangrove Biodiversity in the Bengal Coastal Region

The salinity investigation results show on the one hand a greater vulnerability as well as potential for further degradation and on the other hand, that the south west Bengal coastal regions are carrying the highest rate of water salinity. This is misbalancing the existing ecosystems and their inhabited ecologies. According to salinity approximation this high rate is harmful to rural and urban biodiversity as well as for rural and urban drinking water (Fig. 11.8). The Fig. 11.8 demonstrates the water salinity intrusion pattern in the south west region of the Ganges deltaic region, where 4 major cities and 136 small towns are located in the coastal region in Bangladesh. Most of the towns are affected through salinity intrusion and sea level rise impacts in the region. Therefore, the investigation results of salinity modelling in the Southwest coastal deltaic regions show that ecosystems and coastal rural and urban ecosystem goods and services are under threat. As a result, coastal mangrove and agro-biodiversity loss is a common scenario in Bangladesh and India. The mangrove reduction rate for instance is about 45% in both countries. Deforestation and land cover is changing due to shrimp farming, salt farming, agricultural land extension, urbanization extension and settlement development arc adversely affecting coastal fish production and leading to a loss of agro-biodiversity and natural resources for coastal rural or urban biodiversity and 3.5 million people, who are dependent on a functioning coastal region in Bangladesh (Anon 1995).

Moreover a rich biodiversity of flora and flora support and contribute substantially to the socio-economic life of these millions of people in rural, remote areas of the world by providing opportunities of employment, food and nutrition, fuel, fodder, transportation and irrigation (Nishat 1993). The coastal forest and wetland resources in Bangladesh have suffered drastically from the impacts of growing human population, climate change and anthropogenic activities on natural resources. The water resources and mining resources are similarly used for an economic development. Furthermore natural resources are recognized as a driving force for biodiversity conservation and coastal socio-economic improvement (Nishat 1993; Ahmed and Falk 2008). So in the Ganges-Brahmaputra-Meghna floodplain alone approximately 2.1 million ha of wetlands have been getting lost due to flood control, drainage, and irrigation development (Khan et al. 1994). Therefore, coastal wetlands' biodiversity is facing serious challenges from salinity intrusion, environmental changes and anthropogenic impacts (Sarkar 1993; Sarker et al. 2003; Nair 2004; Ahmed and Falk 2008). The wetlands in the coastal region include rivers, estuaries, mangrove swamps, marsh (*haor*), oxbow lake (*baor*) and *beels*, water storage reservoirs, fish ponds, and some other lands with also facing the similar environmental problems named in before (Khan 1993; Hughes et al. 1994; Gopal and Wetzel 1995; Islam and Gnauck 2008). In such problematic, complicated situation the smart use of natural resources and salinity desalination processes can solve the coastal urban biodiversity and ecosystem problems in the coastal surface areas in Bangladesh.

11.7.5 Threatened Coastal Deltaic Forests, Forestry and Agricultural Crop Production

In the Ganges deltaic region of Bangladesh more than 60% people are dependent on agricultural crop production (Fig. 11.9). At present 30% of agricultural productions and 45% of mangrove forests are reduced in the coastal region due high to medium level of salinization.

The coastal region of Bangladesh is very important for the country because of natural resources, newly formed land area and mangrove forest, forestry, agricultural and fishery resources. A total of 6017 km^2 of mangrove forests is covering the perimeter of the Sundarbans Natural World Heritage Site (Islam 2003). These areas of the Sundarbans mangrove forests constitute about 51% of the forest area of the country of Bangladesh (Islam 2003). Comparing with the agricultural situation, a total of the Bengal coastal region with 47,000 km^2 of the land area is declared as coastal region where agriculture and aquaculture are the most popular and potential agro-crops and food production of the country. They are covering 34% of the country's food demand and a major portion of this 34% is placed within the coastal urban region.

The major role of mangrove forests and wetlands are their high nutrient factors and its retention/removal through a support for food chains, fisheries production. Also habitat for wildlife, recreation, natural heritage values, biomass production, water transport, biodiversity presentation and micro-

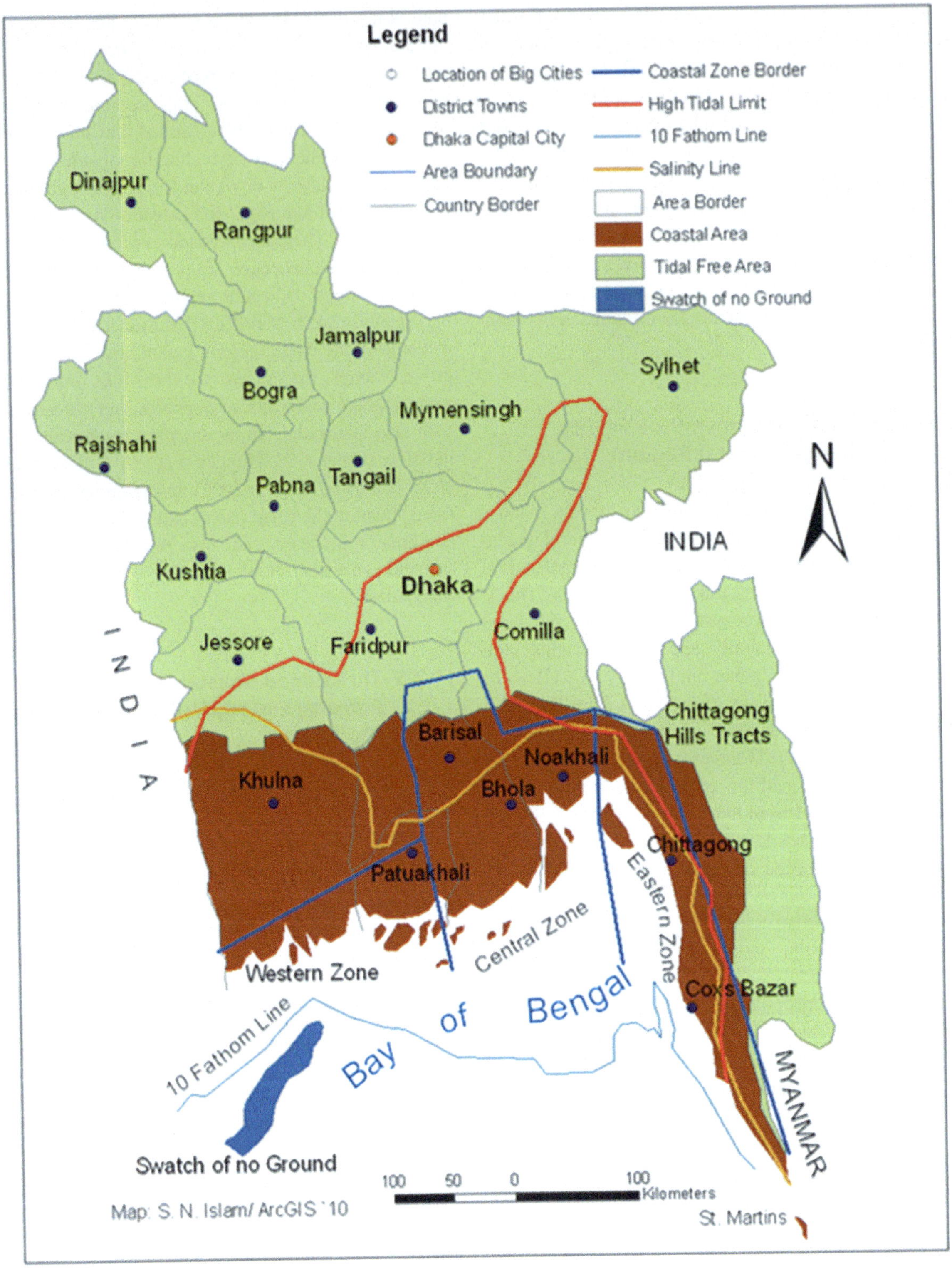

Fig. 11.9 The lower Ganges-Brahmaputra-Meghna Deltaic coastal mangrove forests and wetlands region is affected by salinity intrusion (Source: Islam 2014)

climate stabilization are some further important ecosystem services being placed within these for forests, forestry and agriculture most important areas.

The Fig. 11.9 shows the water and soil salinity intrusion pattern in the south western Ganges deltaic region of the Bengal coast. There the salinity trend is imposing in the upstream areas from south direction to north direction.

In the north direction most part of the mangrove forests converted already to agricultural lands, which are now again converting from agricultural crops to shrimp cultivation due to high salinity intrusion. As a result the soil fertility of former forest or wetland areas and agricultural lands are losing its fertility.

The major cities like Calcutta are still located in the upper north and Mongla, Bagerhap, Sathkhira, Khulna are located under this salinity threshold line (Fig. 11.9). On the other hand most of the coastal towns in the deltaic region are also affected due to high salinity intrusion: the traditional urban vegetation and crops are also changing due to salinity intrusion in the urban areas of the Bengal coastal region. In Fig. 11.9 the red line illustrates the salinity thresholds line for agricultural crops and vegetation productivity.

The threshold line is indicating the contour of an exceeded threshold value (43,220 dS/m), which is harmful and is giving threat to normal forests, forestry or agricultural crop production and coastal urban vegetable production. Further on this line divides up the areas with a lower or high threat for safe drinking water availability in the coastal towns in Bangladesh.

So the coastal region of the Bengal delta of Bangladesh with its Bengal coastal region of 47,000 km^2 of land is now under threat for (mangrove) forests, forestry, agricultural crop production and rural and urban biodiversity conservation (Islam et al. 2017). Also the present condition is penetrating the coastal rural and urban food security in case of a gradually reduction of agricultural productivity within the coastal region in Bangladesh and within the protected part of the Sundarbans mangroves.

11.8 Approach for Coastal Mangrove Ecosystems Management and Sustainability

The coastal water resource usages and maintenance is the fundamental management aspect for local communities also in the Sundarbans region. The fringing of this study identified a number of environmental obstructions for proper use, reuse and maintenance of their coastal natural resources. Most of the developing countries have not yet developed the coastal water and natural resources management policies and strategies. Therefore, shortage of upstream freshwater supply, saline water intrusion and climate change impacts on the coastal regions and here within the Sundarbans Natural World Heritage Site are the greater challenge for enduring mangrove forests, forestry and agricultural crop production with maintaining food security in the coastal region of Bangladesh. These preoccupations need to be recognized and each given due weight in order to integrate them into policy and decision making in coastal water resource management to fight against climate change impacts, loss of mangrove-biodiversity, food insecurity and poverty for ensuring sustainable development for the coastal communities as well as floristic and faunistic other species. This means to foster a development, which meets the needs of the present without compromising the ability of future generations to meet their own needs. For such development it is important to carry out a programme for making the people aware of the impending dangers and to develop along with them methods of coastal adaptation and migration to promote water resources management and proper use for food security in the coastal region of Bangladesh and especially within and at it outbounds of the Sundarbans mangroves while acknowledging and following the vulnerability adaptation framework.

Figure 11.10 is demonstrating a local level planning approach for sustainable local level natural resources management.

The model (Fig. 11.10) has been prepared based on the present situation of coastal mangrove forest and wetland resources scenarios in Bangladesh case. Considering the findings, this long-term management plan for a sustainable use of coastal water resources is necessary and is urgently needed (Islam et al. 2017).

This model could be implemented in the coastal region in Bangladesh or especially in the part of the Sundarbans Natural World Heritage Site as well as in other coastal regions, which have a similar character like Bangladesh. Besides such as following, some practical recommendations can be applied for a better management of coastal mangrove forest and wetland resources in Bangladesh:

- Therefore applied research, awareness raising education, monitoring and evaluation are the key potential issues of a successful coastal wetland resources management and conservation.
- An increase of salinity due to the lack of freshwater flows in the transboundary river basin has caused damage to the vast area of coastal region and its ecosystems. Adverse development initiatives should strengthen environmental stability and protect against further degradation.
- Upstream river water supply should be ensured for the coastal region in Bangladesh coast and the inhabited Sundarbans Natural World Heritage Site. Especially whenever Ganges water supply is a disputable problem between India and Bangladesh, therefore other rivers water and monsoon water should be harvested throughout for instance a reservoir in the upstream area and supply freshwater to the south western coastal region could solve high salinity problem (Begum 1987).
- A sharp rise in water levels generally begins in May and continues until July. Therefore, some potential steps

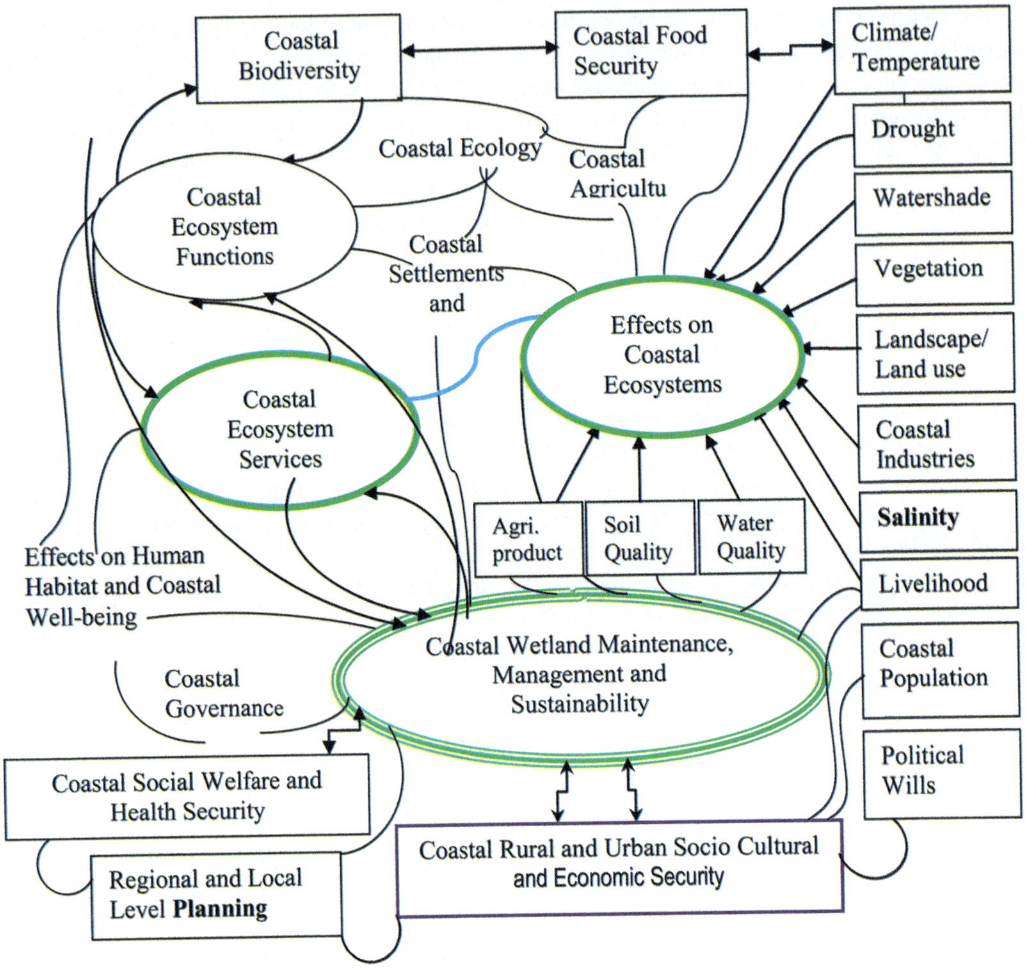

Fig. 11.10 Model of Coastal mangrove forest and wetland resources protection, management and sustainability (drawn by the authors after Islam 2015)

should be taken into consideration for sustainable water resources management in the coastal region such as:

- Monitoring of ground and surface water quality and
- Modelling for its sustainable management of coastal mangrove forest and wetland resources.

The coastal community plays a further major role and becomes the key to success in integrated coastal mangrove resources management. It has been realized that coastal natural resources rely on the active participation of the coastal communities in protecting, maintaining and conserving their surrounding resources. Coastal community base management could be the best strategy in integrated coastal management that could ensure the involvement of the coastal society, and stakeholder in implementation stage. The indigenous coastal community is ecologically friendly and is an important source of intellectual asset in management of the coastal ecology (Hidayati 2000). The community, as the prime user of the coastal mangrove forest and wetland resources, should be involved and local technology should be considered within protecting practices and maintaining mangrove forest and wetland resources in the coastal region of Bangladesh and especially in the Sundarbans Natural World Heritage Site.

11.9 Conclusions and Recommendations

Coastal mangrove forests and wetlands can be considered as driving force for community social and economic enhancement. They have the ability to focus tremendous energy and to generate significant creative and economic betterment. In general, the country's coastal mangrove forest and wetland natural resources as well as the ecosystems are degrading due to anthropogenic influences and natural calamities. Considering the scenarios of the Ganges-Brahmaputra deltaic coastal zones of Bangladesh have great importance for the country's economy, industrial, ecological, socio-economic and cultural context. Moreover the Sundarbans coastal mangrove forests and wetlands support rich biodiversity and are contributing substantially to the livelihoods of millions of people through creating opportunities of employments. The climate change impact has been notified as being an additional negative impact on coastal mangrove forests and wetlands as well as the local communities and stakeholders in the Ganges-Brahmaputra rivers deltaic region and its Sundarbans Natural World Heritage Site in Bangladesh.

The coastal region of Bangladesh encompasses about one fifth of the country's landmass and supports livelihood of 36.8 million people. The entire region with its Sundarbans Natural World Heritage Site is becoming more vulnerable due to natural and anthropogenic causes. Growing anthropogenic and climatic impacts have been found to put multifarious adverse impacts on coastal water resources, environmental and livelihood of the inhabitants of the coastal region. It is estimated that the requirement of the Ganges freshwater supply to the Sundarbans is 49 MAF and other rivers are also supplying the similar quantities of surface water to the Sundarbans. Any endeavour through to be suitable for the wellbeing of the people as well as for the coastal environment, has to be viewed with options in considering the socio-economic and cultural traditions of the society. Any particular adjustment to be included in the final plan ought to be within the technological capability of the people.

The study findings show that water and soil salinity intrusion is a severe threat for Sundarbans Natural World Heritage Site and its implemented mangrove ecosystems and its services such as agricultural crop production as well as coastal settlements and population. National and international political commitments and wills should be ensured properly. Therefore an integrated coastal natural resource management policy and a guideline framework for livelihood is necessary and urgently needed for Sundarbans Natural World Heritage Site and its surrounding Ganges – Brahmaputra rivers deltaic region in coastal Bangladesh.

Considering the influence, problems, prospect and their impacts on socio-economic condition of the coastal community and on the ecosystems of coastal natural resources, the strategies and intervention in planning is necessary. There is a potential realization that effective coastal water resources management at local level is an essential and appropriate management structure, which should be developed.

The study would suggest that most mangroves along the coast of Bangladesh would disappear under sea-level rise scenarios when rate of rise up to 1.5 cm year^{-1} have been predicted. So Sundarbans Natural World Heritage Site and its surrounding would be under greatest threat and danger (has been seen in Fig. 11.8). Therefore an integrated coastal water resources management policy and a guideline framework is necessary and has an emergency state. The finding of this study could help to the policy and decision makers to prepare a guideline framework for a heritage management plan as well as for a general integrated coastal mangrove forest and wetland resources protection, conservation and management in Bangladesh.

Besides the above elements and factors the following important issues should be included in the national coastal mangrove forest and wetland resources management and conservation plan, which should be developed based on the changing tendency of ecosystems in the deltaic coastal region of Bangladesh. So the following recommendations could be considered in making the policy framework for better management in order to achieve the goal for a sustainable coastal mangrove forest and wetland water resources management:

- Increase coastal community socio-economic welfare and alleviate poverty for sustainable livelihood in the vulnerable coastal region. Provide alternative income generation opportunities to the coastal communities that should be ecologically friendly and profitable.
- Increase public awareness concerning environmental training and education, particularly related to the importance of the sustainable use of coastal mangrove forest and wetland and other natural resources. Provide information to the coastal communities concerning natural resources collection, proper uses and long-term protection measures.
- Coastal mangrove forests and wetlands biodiversity maintenance in the region; mangrove ecosystem functions and services should be included in the maintenance of the coastal region for a better interest of coastal communities in the Sundarbans Natural World Heritage Site of the Ganges-Brahmaputra Rivers deltaic mangrove forest and wetland region.
- To develop and train up capacity building of the local community groups, local government, stakeholders, NGOs and national policy makers and planners those are involved in coastal resource management activities in the Sundarbans Natural World Heritage Site or its surrounding Ganges-Brahmaputra deltaic coastal mangrove forest and wetland region in Bangladesh.

References

Adel MM (2001) Effect of water resources from upstream water diversion in the Ganges Basin. J Environ Qual 30(March–April. 2001, (Environmental issue), USA):356–368

Adeel Z, Pomeray R (2002) Assessment and management of mangrove ecosystems in developing countries. Trees 16(2–3):235–238

Ahmed R, Falk GC (2008) 'Bangladesh': environment under pressure. Geographesche Rundschau Int Ed 4(1):13–19

Akter J, Islam SN, Gnauck A (2010) Water resources management in the coastal region of Bangladesh. In: Modelling and simulation of Ecosystems, Workshop Kolpinsee 2009. Shaker-Verlag, Aachen, pp 167–185

Anon D (1995) Integrated resource management plan of the Sundarbans reserve forest Vol. 1 draft final report of FAO/UNDP project

BGD/84/056 integrated resource development of the Sundarbans reserve forest, Khulna, Bangladesh, 585
Anwar J (1988) Geology of coastal area of Bangladesh and recommendation for resource development and management. In: National workshop on coastal area resource development and management, Part II. CARDMA, Dhaka, pp 36–56
BBS-Bangladesh Bureaue of Statistics (1979) Government of the People's Republic of Bangladesh, Dhaka, www.bbs.gov.bd
BBS-Bangladesh Bureaue of Statistics (1981) Government of the People's Republic of Bangladesh, Dhaka, www.bbs.gov.bd
BBS-Bangladesh Bureaue of Statistics (1985) Government of the People's Republic of Bangladesh, Dhaka, www.bbs.gov.bd
Begum K (1987) Tension over Farakka Barrage in South Asia – a techno political angel. University Press, Dhaka
Biksham G, Andrea FS (1996) What price Prawn? Shrimp aquaculture's impact in Asia. Environment 38(7):12–14.33
Broadus JM (1993) Possible impact of and adjustment to sea level rise: the cases of Bangladesh and Egypt. In: Warrick RA, Barrow EM, Wigley TML (eds) Climate and sea-level change; observations, projections and implication. Cambridge University Press, Cambridge, pp 263–275
CEGIS – Centre for Environmental and Geographical Information Services (2010) Identification mission K2K, water mondiaal delta alliance reconnaissence visit to Dhaka Bangladesh. Mission report (draft), 20–24 June 2010. Slingerland K, Bons K, Driel WV (eds) Wageningen University and Research Centre, The Netherlands, pp 1–22
Chowdhury NT (2009) Water management in Bangladesh: an analytical review. Water policy uncorrected proof, IWA publishing. pp 1–20
Chowdhury ZU Haque A (1990) Permissible water withdrawal based upon prediction of salt-water intrusion in the Meghna Delta. In: The hydrological basin for water resources management (Proceedings of the Beijing symposium, October 1990), IAITS Publ. no. 197, pp 111–117
Coleman JM (1969) Brahmaputra River: channel processes and sedimentation. Sediment Geol 3(2–3):129–238
Costanza R (1992) Toward an operational definition of health. In: Costanza R, Norton B, Haskell BD (eds) Ecosystem health -new goals for environmental management. Inland Press, Washington, DC, pp 239–256
Curray JR, Moore DG (1971) Growth of the Bengal deep sea and denudation in the Himalayas. Geol Soc Am Bull 82:563–572
Dasgupta P (2001) Human well-being and the natural environment. Oxford University press, Oxford
Dutta AK, Iftekhar MS (2004) Tree species survival in the homestead forests of salt affected areas: a perception analysis for Bangladesh. J Biol Sci 4(3):309–313
EGIS – (Environmental and Geographical Information Services) (1997) Morphological dynamics of the Brahmaputra-Jamuna River. Water Resources Planning Organization/ Delft Hydraulicc, Dhaka
EGIS – (Environmental and Geographical Information Services) (2000) Environmental baseline of Gorai River restoration project, EGIS-II. Bangladesh Water Development Board, Ministry of Water Resources, Government of Bangladesh, Delft, p 150
ESCAP – (Economic and Social Commission of Asia and the Pacific) (1988) Coastal environment management plan for Bangladesh, vol 2. Final: Report. ESCAP, Bangkok
FAP (Flood Action Plan) 24 (1996) River Servey Project- Morphology of Gorai off-take: Special Report no.10, Dhaka. WARPO.
FAP (Flood Action Plan) 4 (FPCO) (1999) Draft final report, vol 3: Morphological studies. Ministry of Irrigation, Bangladesh (unpublished)
FAP 4 (FPCO), 1993. Draft final report, vol 3: morphological studies. Ministry of Irrigation, Bangladesh (unpublished)
Fedra Kan Feoli E (1999) GIS technology and spatial analysis in coastal zone management; in Belfiore S (ed.,) Education and Training in integrated coastal area Management: The Mediterranean prospect, Ocean change pub. Italy
Gopal B, Wetzel G (1995) Limnology in developing countries, vol 1. International Association of Limnology, New Delhi, pp 1–230
Gleick P (1998) The world's water. Island Press, Washington DC
GOB – (Government of Bangladesh) (1999) National Water Policy, Ministry of Water Resources. Government of the People's Republic of Bangladesh, Dhaka
Goodbred SL, Kuehl SA (2000) Late quaternary evolution of the Ganges – Brahmaputra River delta: significance of high sediment discharge and tectonic process on margin sequence development. Sediment Geol 133:227–248
Goodbred SL Nicholls R (2004) Towards integrated assessment of the Ganges-Brahmaputra delta. Proceeding of the 5th International conference on Asian Marine Geology, and 1st Annual meeting of IGCP475 Delta and APN Mega-Delta, 13th February, 2004
Greensmith JT, Tucker EV (1986) Compaction and consolidation. In: van de Plassche O (ed) Sea-level research: a Manuel for the collection and evaluation of data, geo books, Norwich, pp 591–603
Grigg NS (1996) Water resources management principles, Regulation and cases. McGraw Hill, Washington, DC, pp 8–14
Halcrow (1993) Southwest area water resources management project (FAP 4) morphological studies. FPCO, Dhaka
Harvey LDD (2000) Climate and global environmental change. Pearson Education Ltd, London, pp 187–207
Hasan S, Mulamoottil G (1994) Natural resource management in Bangladesh. Ambio 23(2):141–145
Helmer R, Hespanhol I (eds) (1997) Water Pollution control. A guide to the use of water quality management principles. E & FN SPON, London, pp 304–316
Hidayati D (2000) Coastal management in ASEAN countries -the struggle to achieve sustainable coastal development. UN. University Press, Tokyo, pp 1–74
Hughes R, Ahmed S, Clayton BD (1994) Floodplains or flood plann ? A review of approaches to water management in Bangladesh. IIED, London, pp 1–12
IECO (1980) Southwest regional plan: supplement e: saline intrusion and tidal hydraulics; supplement f: environment and ecology (part I & II), supplement g: fishery (part I & II), supplement h: forestry (part I & II), supplement i: navigation; supplement j: augmentation of the dry-season flows of the Ganges. Bangladesh Water Development Board, Dhaka
Iftekhar MS (2006) Conservation and management of the Bangladesh coastal ecosystem: overview of an integrated approach. Nat Res Forum 30(2006):230–237
IPCC – (Intergovernmental Panel on Climate Change) (2007) Summary for policymakers. In: Climate change 2007: impact adaptation and vulnerability. Contribution of working group II to the fourth assessment report of the IPCC ML Parry OF Canziani, Palutikot PJ et al. Cambridge University Press, 1000 pp
Islam MS (2001) Sea-level changes in Bangladesh: last ten thousand year. Asiatic society of Bangladesh/Asiatic Civil Military Press, Dhaka, pp 1–185
Islam R (2007) Pre- and post-tsunami coastal planning and land-use policies and issues in Bangladesh. Book of proceedings of the workshop on coastal area planning and management in Asian tsunami-affected countries. Workshop from 27th–29th September 2006 in Bangkok, Thailand, session I, the regional perspective. Food and Agriculture Organization of the United Nations, Regional Office for Asia and The Pacific, Bangkok, Thailand
Islam S (ed) (2006) Encyclopedia of Bengal. Asiatic Society of Bangladesh/Asiatic Civil Military Press, Dhaka
Islam SN (2003) Sustainable eco-tourism as a practical site management policy. AH Development Publishing House, Dhaka, pp 1–175
Islam SN (2010) Threatened wetlands and ecologically sensitive ecosystems management in Bangladesh. In: Frontiers Earth Science in China, Springer- Verlag Berlin, Berlin 4(4): 438–448.

Islam SN (2014) Degradation of coastal wetlands ecosystems and mangrove biodiversity in Sundarbans in Bangladesh. J Bangladesh Natl Geogr Assoc (BNGA) 32(1&2):1–14

Islam SN (2015) Char-lands development policies for Settlement and livelihoods sustainability in the Ganges-Padma River Basin in Bangladesh: A case analysis. Int J policy stud (IJPS) 6(1):1–24

Islam SN (2016) Deltaic floodplains development and wetland ecosystems management in the Ganges-Brahmaputra-Meghna Rivers Delta in Bangladesh. Sustain Water Resour Manag 2 (3):237–256. https://doi.org/10.1007/s40899-016=0047-6)

Islam SN, Gnauck A (2007a) Effects of salinity intrusion in mangrove wetlands ecosystems in the Sundarbans: an alternative approach for sustainable management. In: Okruszko T, Jerecka M, Kosinski K (eds) Wetlands: monitoring modelling and management. Taylor & Francis/Balkema, Leiden, pp 315–322

Islam SN Gnauck A (2007b) Increased salinity in the Ganges delta and impacts on coastal environment in Bangladesh. In: Skvorsova LA (ed) Proceedings of the international scientific-practical conference logistics and economics of resource and energy- saving in industries (ISPC "LEREI -2-2007") 12th September to 15th September 2007 at Saratov State Technical University, Russia, pp 244–248

Islam SN, Gnauck A (2008) Mangrove wetland ecosystems in Ganges-Brahmaputra delta in Bangladesh. Front Earth Sci China 2 (4):439–448

Islam SN, Ganuck A (2009a) Threats to the sundarbans mangrove wetland ecosystems from transboundary water allocation in the Ganges basin: a preliminary problem analysis. In: International Journal of Ecological Eeonomics and Statistics (IJEES)13 (09):64–78.

Islam SN, Gnauck A (2009b) The coastal mangrove wetland ecosystems in the Ganges delta: a case study on the Sundarbans in Bangladesh. In: Proceeding of American Association of Petrolium Geologist-AAPG Hedberg conference on variations in fluvial-deltaic and coastal reservoirs deposited in tropical environments, 29th April to 2nd May, 2009, Jakarta, Indonesia, pp 26–29.

Islam SN, Reinstädtler S, Fredaush J (2017) Challenges of climate change impacts on urban water quality management and planning in coastal towns of Bangladesh. Int J Environ Sustain Dev 16 (3):228–256

Jabbar MA (1979) Land accretion in the coastal area of Bangladesh-report of Bangladesh landsat programme, Dhaka, 43 pp

Jalal KF (1988) Towards a national strategy on coastal resources management in Bangladesh. In: National workshop on coastal area resource development and management, Part II. CARDMA, Dhaka, pp 67–78

Joseph PS (2006) The environmental management the better supply of fresh water in Transboundary River: the Ganges could run dry. In: Perez JG (ed) Proceedings of III international symposium on transboundary waters management-overcoming water management boundaries. Universidad de Castilla-La Mancha. 30 May –2 June, Ciudad Real, Spain

Khan AA (1993) Freshwater wetlands in Bangladesh: opportunities and options. In: Nishat A, Hussain Z, Roy MK, Karim A (eds) Freshwater wetlands in Bangladesh issues and approaches for management. IUCN, Dhaka

Khan SM, Hoq E, Huq E et al (1994) Wetland of Bangladesh, Bangladesh Centre for Advanced Studies (BCAS). Holiday Printers Limited, Dhaka, pp 2–57

Khandoker RA (1987) Origin of Elevated Barind-Madhupur areas, Bengal basin, result of neotectonic activities, Bangladesh. J Geol 6 (2):1–9

Miah MM (1989) Siltation and coastal zone management; Bangladesh country study. Report BIDS, Dhaka

Miah MA (2001) Effect on water resources from upstream water diversion in the Ganges Basin. J Environ Qual 30(2):356–368

Milliman JD, Broadus JM, Gable F (1989) Environmental and economic implication of rising sea level and subsiding deltas: the Nile and Bengal examples. Ambio 18(6):340–345

Morgen JP, Mclntire WG (1959) Quarternary geology of the Bengal Basin, East Pakistan and India. Bull Geol Soc Am 70:319–342

MoWR – (Ministry Of Water Resource) (2005) The national coastal zone policy. Ministry of water Resources, Dhaka

Nair KS (2004) Wetlands management to meet the food and water cricis. In: Abstract book of 7th INTECOL International conference on Wetlands, 25–30 July 2004, Utrecht, The Netherlands, pp 1–217

Nishat A (1988) Review of the present activities and state of art of the coastal areas of Bangladesh. In: National workshop on coastal area resource development and management, Part II. CARDMA, Dhaka, pp 23–35

Nishat A, Huq SM, Barua I, Reza S, Ali AHM, Moniruzzaman KAS (2002) Bioecological zones of Bangladesh, IUCN – the World Conservation Union. Bangladesh Country Office, Dhaka

Pramanik MAH (1983) Remote sensing applications to coastal morphological investigations in Bangladesh, unpublished PhD thesis. Dept. of Geography and Environment, Jahangirnagar University, Savar, Dhaka, 227 pp

Rahman A (1988) Bangladesh coastal environment and management. In: National workshop on coastal area resource development and management, Part II. CARDMA, Dhaka, pp 1–22

Rahman A (ed) (2003) The degraded sundararbans. In: Paribesh patra (Bengali) development magazine, issue 3, 4 October–December 2002 and January–March 2003. Unnayan shamunnay, Dhaka, Ministry of Environment and Forest and UNDP, Dhaka, Bangladesh

Rob MA (1998) Changing morphology of the coastal region of Ganges Delta. Orient Geogr 4(2):49–64

Sarker MH, Huque I, Alam M, Koudstaal R (2003) Rivers, Chars and Char dwellers of Bangladesh. J River Basin Manag 1(1):61–80

Sarkar SU (1993) Faunal diversity and their conservation in freshwater wetlands. In: Freshwater wetlands in Bangladesh: issues and approaches for management. IUCN, Gland, pp 1–364

SRDI – soil Resource Development Institute (2000) Soil salinity in Bangladesh. Government of the People's Republic of Bangladesh/ Soil Resource Development Institute, Dhaka

Tan CY, Yahya J, Rahim KA, Ghee LT, Meyer AS, Othman MSH, Sivalingam G (1997) Environment and development in a resource-rich economy: Malaysia under the new economic policy. Harvard Institute for International Development, Cambridge

Umitsu M (1993) Late quaternary sedimentary environments and landforms in the Ganges delta. Sediment Geol 83(3–4):177–186

Williams CA (1919) History of the rivers in the Gangetic Delta 1750-1918. Bengal secretariat Press. Calcutta (Reprinted 1966, East Pakistan Inland Water Transport Authority, Dhaka), 99 p

Wolanski E (2007) Estuarine ecohydrology. Elsevier, Amsterdam, p 157

Wolanski E, Brinson MM, Cahon DR, Perillo ME (2009) Coastal wetlands: a synthesis. In: Coastal wetlands- an integrated ecosystem approach. Elsevier, New York, pp 1–62

Part III

Pollution and Contamination by Hydrocarbons and Heavy Metals

The Impact of Oil and Gas Exploration: Invasive Nypa Palm Species and Urbanization on Mangroves in the Niger River Delta, Nigeria

12

Aroloye O. Numbere

Abstract

The Niger River Delta is a world-acclaimed biodiversity hot spot according to the World Bank. Its mangrove forest is the largest in Africa and the Atlantic. This ecosystem provides firewood, building materials, medicinal herbs and food for the local population. But oil and gas exploration, deforestation, dredging, urbanization and invasive species over the years had converted it from a pristine to a disturbed state. The greatest damage to mangroves in the Niger Delta comes from oil and gas exploration activities, which began in 1956 in Oloibiri. Millions of crude oil spillages had occurred, since the striking of the first oil well from ruptured well heads, pipelines, and jetties, constructed on both onshore and offshore locations. This degraded condition had reduced the mangrove forest from highly dense to lowly dense and also to mixed forest. Exploratory activities had also resulted to additional problems, such as invasion by alien species and urbanization of mangrove areas. Urbanization is beneficial to man's development, but costly for the mangroves. The establishment of industrial and residential quarters to accommodate oil workers and their families had increased the urban sprawl around mangrove forest areas. These activities had reduced the resilience of mangroves against the invasion of nypa palm (*Nypa fruticans*). The palms were intentionally introduced into the Niger Delta in 1906, but for close to a century the mangroves had kept them in check. However, in the last 20 years, the palms had overwhelmed and completely colonized most mangrove forests.

A. O. Numbere (✉)
Department of Animal and Environmental Biology, University of Port Harcourt, Choba, Nigeria

Keywords

Niger Delta · Urbanization · Invasive species · Hydrocarbon pollution · Mangrove · Exploration · Seismic activities · Oil spillages · Pipelines

12.1 Introduction

Global loss of wetland is put at 30% with 10–15% decline of mangroves predicted by the year 2100 (Alongi 2008). The major causes of biodiversity loss include habitat loss, invasive species, pollution, over population and over exploitation (James et al. 2007). In the Niger Delta, mangroves are being lost as a result of the "evil quartet" of pollution, deforestation, invasion, and urbanization. Although, natural disturbances such as tsunami, hurricane, climate change and sea level rise significantly reduce global mangrove population (Ellison 2005; Webb et al. 2014), human-mediated effect far exceeds the natural effect in the Niger Delta. The delta serves as a habitat for numerous organisms ranging from microorganisms to invertebrates and vertebrates. The region has a rich supply of animals such as monkeys, hippopotami and crocodiles that have all existed in pristine form for centuries before the advent of man. The area is regarded as a global biodiversity hotspot because of its rich natural resources (NDES 1997). It has two national parks, namely, the Cross River National Park (Latitudes 8° 15′ and 9° 30′ E; and longitudes 5° 05′ and 6° 29′ N) and the Okomu National Park (Latitudes 6°.15′ and 6°.25′; and longitudes 5° 9′ and 5° 23′). The World Wide Fund for Nature classified it among 200 global eco-regions that are critically endangered (WWF 2001) while the World Bank regarded it as the second most sensitive environment in Africa (World Bank 1995). The resources of mangrove forest had been the source of livelihood for the local people who depend on them for wine, food, roofing sheets, baskets, hats, mats and fire wood, which are all locally produced from mangrove by-products (NDES 1997; Polidoro et al. 2010). In addition,

C. Makowski, C. W. Finkl (eds.), *Threats to Mangrove Forests*, Coastal Research Library 25,
https://doi.org/10.1007/978-3-319-73016-5_12

farmers derive manure from the swamps, which they use to fertilize their crops while fishermen get high catch from the numerous fisheries resources present within the surrounding waters. Out of all the causes of environmental degradation in the Niger Delta, oil and gas exploration has the most significant effect on the environment.

The first oil well in Nigeria was struck in a riverine community called Oloibiri in the Niger Delta in 1956 (UNEP 2011). The discovery of crude oil placed Nigeria on the global map of oil producing nations of the world. Nigeria is the largest producer of crude oil in Africa, and the sixth largest in the world (UNEP 2011). However, this enviable position has not come without a price. This is because the discovery had changed the undisturbed ecological state of the mangroves to a disturbed state. Many mangrove forests in the Niger Delta had been converted to waste land, which has defied all forms of restoration attempts because of the constant presence of oil spills. Oil and gas exploration activities had resulted to spillages, which severely damage mangrove ecosystem (Mastaller 1996; Ohimain et al. 2008), and contribute to their global decline (Duke et al. 1997). Similarly, the creation of a right of way (ROW) passage for crude oil pipelines that transport refined products to and from the refineries also leads to massive deforestation. Increase in industrial activities in rural areas leads to the migration of people from urban to rural areas, thus facilitating urbanization, which in turn decreases mangrove forest in the Niger Delta. For example projects such as construction of residential quarters, roads, hospitals, recreational sites and schools are done to cater for the expanding population of people migrating from the countryside to the rural areas in search of industry-based jobs. Urban settlements are often centers of occupational and production specializations, which lead to commercial activities. Influx of workers leads to increase in social vices (e.g. artisanal refineries, pipeline vandalism, militancy etc), reduction in traditional occupations (e.g. farming, fishing and hunting), and increase in third party agitations (e.g. loss of assets and properties, employments etc). These situations have negative effect on mangroves (Wang et al. 2016) by encouraging the entry of invasive nypa palms (*Nypa fruticans*) (CEDA 1997) and other alien species into mangrove forests.

Pollution is another major threat to mangroves, and it leads to the impairment of soil, water and air quality. It also leads to the reduction in biodiversity and increased morbidity for upper respiratory tract infection such as asthma and emphysema. These diseases are particularly prominent amongst children and adults as a result of constant inhalation of soot from gas flared from oil industries (Aquilera et al. 2010). Poor environmental conditions contribute to the decline in the average life span of the people to 50 years (UNEP 2011). Oil pollution on rivers had destroyed the fish industry that was once the hub of commercial activities in the area (Ohimain et al. 2008), and the few aquatic resources that survive pollution are contaminated with heavy metals.

The Niger Delta Environmental Survey (NDES 1997) revealed that within a period of 20 years over a million barrel of crude oil was discharged into the environment through numerous cases of oil spills. The demand for increased petroleum products locally and internationally is complicating the already existing environmental problems by making the oil industries to increase their production quota leading to more oil fields with more pipelines and more spillages. Clean-up operations are usually costly and difficult to embark on (IUCN 1993). The oil spills, which represent contaminants to the mangroves alter succession, productivity and nutrient cycling (Twilley 2008). Hydrocarbon pollution cause differential mortality of trees leading to long-term changes in mangrove community structure (Lugo 1980; Snedaker et al. 1996; Twilley 1988). There is a positive correlation between pollution and chlorophyll-deficient mutations in red mangroves (*Rhizophora mangle*) (Klekowski et al. 1994). Similarly, prolonged exposure to toxic substances in mangroves causes deformities that take a long time to recover (Levings and Garrity 1994; Odum and Johannes 1975). For instance, during a field work in Buguma a deformed red mangrove propagule (*Rhizophora mangle*) was found that has two rather than one growing tip, which is typical of this species. Red mangrove trees make up 65% of the mangrove forest in the Niger Delta (Numbere 2014). They are the species mostly subjected to long term contamination from hydrocarbon pollution. Hydrocarbon pollution could remain inert inside the swamps for decades, but when exposed becomes activated and poison newly growing plants. It takes a minimum of 20 years for the toxicity of crude oil to completely disappear (Burns et al. 1994), and another 30–40 years for mangroves to regenerate after a complete clearance (Adegbehin & Nwaigbo 1990). The loss of mangrove community leads to a decrease in area, which could result to extinction (Alongi 2008).

12.2 Description of the Niger Delta Geology

The Niger Delta is located in the southernmost part of Nigeria, and covers an area of 211,000 km^2 (UNEP 2011). It is made up of 853 km long coastlines surrounded by mangroves (UNEP 2011). The Niger Delta consist of fluvial and marine sediments built up during the cretaceous period over 60 million years ago. Its land structure is a coarsening upward regressive sequence of tertiary clastics that is developed around south of Anambra and Benue troughs (Ifeadi and Awa 1987), leading

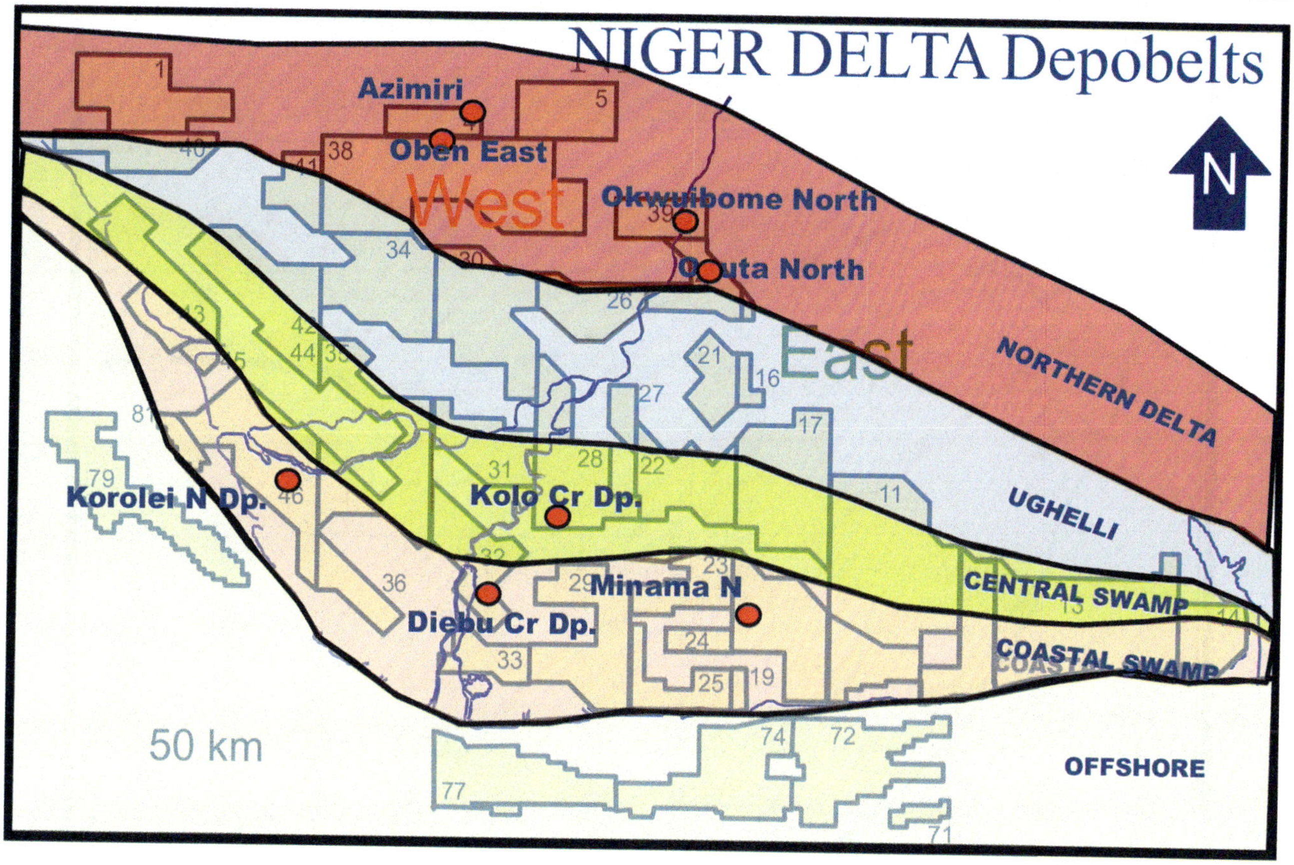

Fig. 12.1 Niger Delta depobelts (Dp) shows areas of crude oil deposits, where exploratory activities are carried out (Credit: A.O. Numbere)

to the formation of depobelts that represent the most active part of the delta at each stage of development (Fig. 12.1). Its age is estimated to be Eocene-Recent (i.e. 0–45 million years ago). The Niger Delta is divided into three lithostratigraphic units namely: Benin, Agbada and Akata formations. The primary source of rock is the geologically active Akata Formation, with contribution from overlying marine sand-shale sequence. It is a sedimentary rock formed through sedimentation process (SPDC 1999). Beneath the earth crust are large storages of crude oil resource, which is formed by decomposed plant and animal matter. The bulk of the plant material comes from dead mangrove vegetation trapped within sediments to form peat. The peat is weathered to form crude oil after being subjected to high temperature and pressure for millions of years.

The Niger Delta area is environmentally sensitive (Gundlach et al. 1981), and supports large amount of biodiversity, but exploratory activities had destabilized the growth and development of the mangrove forest. The delta region acts as a trap and a filter to pollutants that are carried by river into the estuarine mangrove forest. The continuous exploration, production and processing of crude oil and its transportation and marketing expose the mangrove environment to constant threat of oil pollution (Ohimain 2003). Oil and gas exploration activity is the major problem of mangrove in the Niger Delta (Fig. 12.2). Exploration activities cause domino effect by leading to hydrocarbon pollution, invasion and urbanization (James et al. 2007). These activities are mainly man-made hazards that cause accidents, risk or dangers in the environment, and could subject living things to a material risk of death, injury or impairment of health. These hazards can be physical, chemical, and toxic, and also corrosive, flammable and explosive (Withgott and Brennan 2011).

To fully account for the impact of exploratory activities on mangrove forest, the topics below are considered in this chapter.

12.3 Oil and Gas Exploration

Exploratory activity has adverse impact on plants and animals in the environment (Nellemann and Cameron 1998). It involves chain of events that adversely impact the mangrove ecosystem. These events are grouped into three

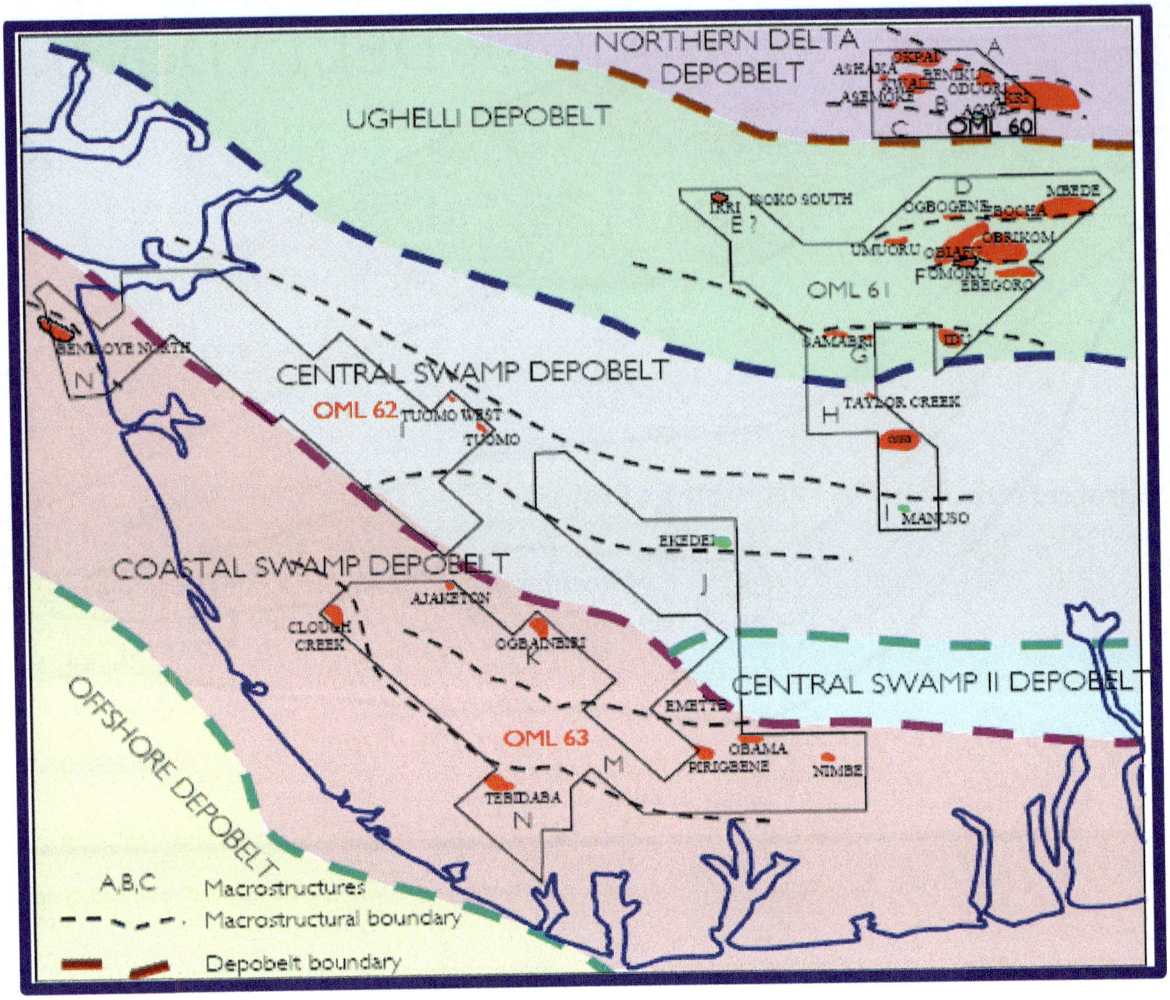

Fig. 12.2 Exploratory activities in the Niger Delta, Nigeria. A depobelt boundary shows areas of active oil and gas exploration. Different oil mining licenses (OML) are owned by both private and government firms. The OML give the companies the right to explore for crude oil at the depobelts (Credit: A.O. Numbere)

stages namely : (1) pre-exploratory; (2) exploratory; and (3) post-exploratory.

12.3.1 Pre-exploratory

Reconnaissance work is carried out within the mangrove forest before the commencement of a full blown exploration. This preliminary work is aimed at detecting the presence of crude oil within seismic layers of the earth crust. Mangrove trees are cut along the seismic route to create unrestricted access into the forest for the survey department of the exploratory team to go in and carry out their job (Ohimain et al. 2008). The survey department follows after the permit unit. The permit unit is involved in land acquisition, and initiates the exploratory process by securing the rights to explore from local and regional authorities. The function of the survey department is to transfer what is on the map to the ground, by determining the actual positions of the seismic lines. Once the seismic locations are identified, workers are assembled to clear the forest and lay the lines. The surveyors determine the actual width of seismic lines that will accommodate the workers and their materials. The environmental effect of this activity is that mangrove trees along seismic lines are cut down leading to massive deforestation (Fig. 12.3). The more the number of workers in the exploratory team, the wider the seismic lines, and the larger the amount of vegetation to be removed. This makes the interior of the mangrove forest to get exposed to further exploitation (Ohimain et al. 2008), and destruction of the mangrove community (Fagbami et al. 1988). This action is devastating as it takes the mangroves about 15 years or more to re-vegetate as compared to the rain forest that takes up to 5 years to grow to maturity. However, field observations had indicated that

Fig. 12.3 Massive deforestation to create right of way (ROW) passage for crude oil pipelines within the Niger Delta mangrove forest. This location is close to a jetty head, and the mangrove trees were cut to create room for the installation of oil and gas pipelines. The area is mangrove dominated but had been invaded by the palms as shown in the background (Credit: A.O. Numbere)

paths created within mangrove forest during exploratory work hardly or don't grow even after several years.

The Drill Department is the next, and comes in after the seismic lines had been successfully established by the surveyors. The duty of the drill department is to mark off and drill the holes where the geophones will be placed to take seismic responses or the dynamites will be placed and blasted. Two ways by which the drill department carry out their duty include: (i) drilling of a deep hole that measures 20–30 m in length with one single charge of 1 kg dynamite placed within the earth, and (ii) drilling of multiple (4–5) shallow holes that measures 3–5 m each in length. The dynamites are placed in the holes after the drilling, covered with soil and blasted. The hazardous effect of this action is that the blow-up destroys nearby mangrove trees and creates craters in the soil, leading to soil degradation. The holes created, if allowed to remain will be washed away by erosion. Iron cuttings from drilling activity also increase the heavy metal concentration of the soil (Lacerda et al. 1993; Mackey et al. 1992). Bentonite, a chemical, used for drilling, is often left on the soil surface to be washed away by rain deep down the soil profile or into nearby water body. Because of the small particle size of mangrove soil they accumulate heavy metals quickly (Clark et al. 1997), and transmit them to the root and stem of the mangroves (Yim and Tam 1999). The acoustic effect of the exploding dynamites scares away animals in the mangrove forests. The sound of the blasting dynamites triggers minor tremors and shock waves that could lead to artificial earthquakes, which destabilize the soil sub-structure. Furthermore, unexploded dynamites called "sleeping dynamites" are potent forces of destruction when activated by animals and humans that traverse the mangrove forest.

The laying of crude oil pipelines is an adventurous activity, but has hazardous effect on the mangrove forest. Seismic drilling and production operations had converted an estimated area of mangrove swamps totaling over 1000 km^2 into areas crisscrossed with oil and gas flow lines (Fig. 12.4).

A large tract of mangrove forest is often made bare of vegetation in other to create a right- of-way (ROW) passage

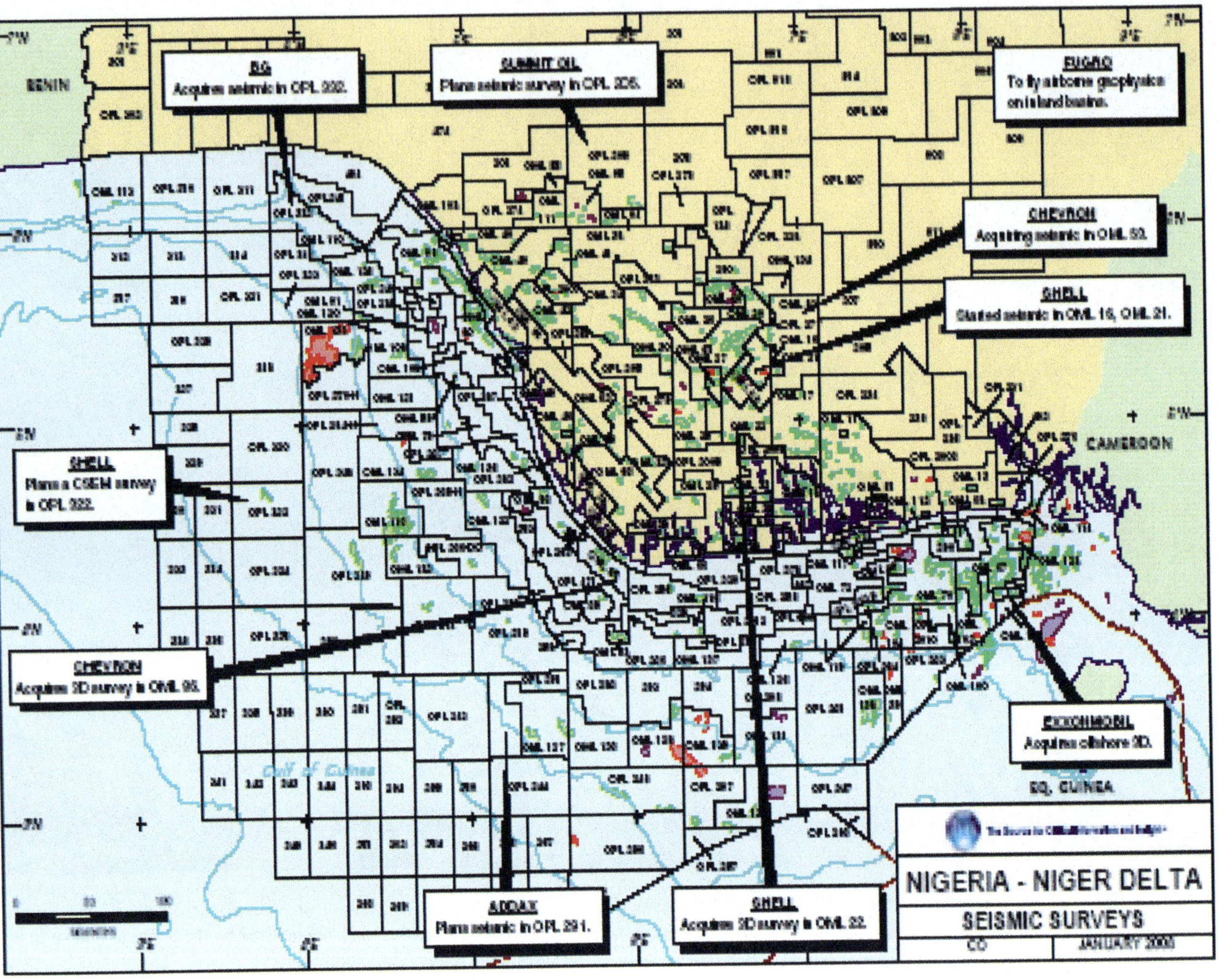

Fig. 12.4 On-shore and off-shore seismic survey points established by some local and international oil companies (e.g. Shell, Chevron, BG, Fugro etc.) running across or beneath land and water surfaces in the Niger Delta. This indicates the amount of perturbation experienced by the area since the first oil well was struck in 1956 (Credit: Niger Delta Seismic Surveys, 2008)

for pipelines (Fig. 12.5), well locations and contractor camps. The surface of the nickel/steel alloy pipes corrodes after several years' under the elements of the weather to contaminate the underlying soil. During welding of pipelines, metal scraps are released into the swamp or water body. The scraps corrode and change the chemistry of the mangrove soils. Destruction of the mangroves forest along the ROW creates opportunity for foreign species to invade and proliferate (UNEP 2011) and become host to diseases that cross over to the mangroves. Examples of other plant species that proliferate in the area include corn vine (*Dalbergia ecastophylum*), coco plum (*Chysobalamus icaco*) carpet grass (*Axonopus compressus*), elephant grass (*Pennisetum purpureum*), guinea grass (*Panicmum maximum*), goose grass (*Eleusine indica*) and goat weed (*Ageratum conyzoides*).

A typical pipeline ROW is about 20 m wide and runs for about 16 km and traverses flat land, stream and swamp locations. The numbers and dimensions of the pipes are two 14″, one 8″ and one 2″ pipes for natural gas liquid (NGL), vapor, and gas respectively. The pipes running through the swamp sit on concrete culvert in the land/residential area while pipe racks are used across the swamp area (EIA 2001). The concrete pipe racks are often constructed by first laying deep foundations, which contribute to the solidification and destabilization of the swamp.

Construction of jetty facilities also contributes to the removal of large sections of mangroves. The jetty contains the following parts: a concrete flare, a fiscal metering system, a vapor return compressor, 12″ NGL loading arm, an NGL analyser and 8″/10″ vapor unloading arm (EIA 2001).

Oil companies undertake dredging of canals at the jetty, which is a detrimental activity on the mangroves (Ohimain 2003). It is the process of soil excavation in swampy terrain for the purpose of widening existing water courses, creation

Fig. 12.5 Nickel/steel alloy pipelines (measuring between 8 and 10″) starting from the refinery to the loading jetty at Okrika in the Niger Delta area of Nigeria. It has an approximate length of 16 km. Mangrove forest is found on the left and right sides while the middle area, which formerly had mangrove trees is now made bare to create a right of way passage (ROW) for pipelines. The ROW contains about 10 pipelines in a row. These pipelines are over 30 years old and the process of corrosion has already begun as shown (Credit: A.O. Numbere)

of new canals, and a re-use of excavated materials for reclamation. The canal is dredged parallel to the new causeway to allow for the entry of barges for piling works and materials transportation. The dredging allows tankers to berth at the port without interference. The total dredging depth is ~12.10 m. The hazardous effect here is that the dredged materials called "dredged spoils" are dumped on existing fringing mangrove forest leading to stifling and poisoning, which cause the death of young mangroves trees (Ohimain 2004). The "dredged spoils" affect surface topography and hydrology, and cause acidification and accumulation of heavy metals in surface and ground water (UNEP 2011). Transportation of dredging equipment such as swamp boogies and barges disfigures the swamp terrain, destroys trees, and destabilizes surface and benthic organisms (Olgard and Gray 1995).

12.3.2 Exploration Stage

This is the stage where the actual drilling occurs, where raw crude oil and gas are pumped out from the earth crust and channeled through pipes via oil well heads (Fig. 12.6) to the refineries for further processing into petro-chemical products. Frequent crude oil spillages usually occur from failed pipes during this stage (Fig. 12.7). The crude oil contaminates and destroys the mangroves (Fagbami et al. 1988). The Niger Delta mangroves constantly sit on thick layers of crude oil as a result of several years of frequent oil spillages (UNEP 2011). Areas where crude oil spilled in the 1970s are still in a polluted state. For instance, a town called Ejama-Ebubu in the Niger Delta that experienced oil spillage 40 years ago is still in a contaminated state despite series of clean up attempts (UNEP 2011).

Fig. 12.6 Crude oil well-head located in a mangrove swamp in Buguma, Niger Delta Nigeria. It shows some sections of the mangrove forest already cleared to install rigs and pipes. Pipeline corrosion occurs several years after submergence under water (Credit: A.O. Numbere)

Crude oil is discharged into the river and mangrove forest during transfer into ocean liners. The discharged crude oil cuts off oxygen supply to the lenticels and pneumatophores of mangroves (Teas et al. 1993). Other sources of pollution include: noise from detonation of dynamites, movement of heavy and light vehicles in and out of the forests, rattling of drilling rigs 24 h daily, disposal of spent lubrications, diesel and domestic wastes, spent drilling mud and chemical and well blow-outs (Okoye et al. 1987). Part of the crude oil pipes that runs under the soil surface displaces soil organisms and truncates biotic processes. This also affects ground and soil water quality, which could be impaired by leachates from generated dredged spoil. The pipelines that pass under water also damage aquatic life such as zooplankton, phytoplankton, benthic organisms and fisheries. The disturbance of the river excites bottom sediments and increases the turbidity of the sea, which affects light penetration and photosynthesis. This could lead to the reduction in species composition and diversity of aquatic resources. The corrosion of pipes placed in aquatic environment triggers explosion leading to oil spillages, which contaminates the entire mangrove community (Snowden and Ekweozor 1987).

Crude oil well heads are pipeline connections that frequently experience engineering failures, sabotages by third party, accidents and natural causes leading to massive oil spillages (Fig. 12.7). These processes destroy large expanse of mangrove forest (Fig. 12.8).

Drilling activity has direct effect on mangrove through the generation of waste. For instance, some drill cuttings have no hazardous effect on the environment. But shale that

Fig. 12.7 The multi-colored oil film on the swamp floor indicates a fresh crude oil spillage that just occurred close to a ruptured oil pipeline at Okrika crude oil jetty location. Crude oil contaminates the swamps, asphyxiates mangrove roots and cause leaf yellowing and death. (Credit: Numbere A.O)

comes from beneath the earth crust most times contains heavy metals and radioactive elements that harm the mangrove forest (Goyer 2001). Similarly, radioactive fallouts occur during logging to assess for deep oil wells. Other waste generating activity in the field includes abandoned drilling pipes, stationeries and kitchen items etc. Splashed concrete used to reinforce enhanced recovery well hardens and prevents the growth of vegetation. The action of making bare the wetlands surface is what enables opportunistic species such as the nypa palms and other alien species to invade and colonize the area (UNEP 2011). Furthermore, construction and drilling activities lead to disturbance of soil dwelling organisms and affects the soil quality.

12.3.3 Post-Exploratory Stage

Oil wells after years of crude oil exploration get dried up, and are abandoned for other oil wells. These abandoned wells are later converted to enhanced recovery (ER) or fluid injection wells. The ER wells are used to recover some crude oil still remaining in the earth crust (Withgott and Brennan 2011). The recovery process involves the bombardment of the well with water to push out the remaining oil from the earth crust. This forceful ejection creates pressure and increases the stress regime of the rocks, which fractures the earth crust and creates seismic fault lines that affect the entire landscape. During this process artificial earthquakes, shocks and tremors do occur. An example is the earthquake in Denver, USA in 1967 caused by Chevron, which occurred when water was pumped into the ground to increase oil supply. In Nigeria no known earthquake had been witnessed, rather there had been some recorded cases of earth tremors. The first earth tremor was recorded in a town called Warri situated in the Niger Delta in 1933. The most recent earth tremor occurred in 2016 in Mbiama, a town in the Niger Delta. The News Agency of Nigeria (NAN) reported that the earth tremor occurred around some oilfields operated by Shell Petroleum Development Company (SPDC) located in two states in the Niger Delta (http://dailypost.ng/2016/09/25/earth-tremor-rivers-bayelsa-communities-cry-out/). These areas are oil producing communities and host several oil companies. There is a general belief that if artificial earthquake should ever occur in Nigeria it will be in the Niger Delta region

Fig. 12.8 Defoliated mangrove forest caused by an oil spill at an oil well head located in Buguma, Niger Delta, Nigeria. There was complete death of mangrove trees along the path of spillage while trees on the left side, outside the path of the spilled crude, were still alive (Credit: A.O. Numbere)

because of several years of oil and gas exploration activities that had created seismic faults within the earth crust in the area. Any earthquake incidence would be a disaster for both the mangroves and human communities along the coasts.

Other sources of pollution at the post-exploratory stage are refining and petrochemical processes. These activities lead to effluent discharges, gas flaring and emissions, tank leakages, valve malfunctioning and human errors, explosions and fire out-brakes. The pipeline and product marketing operation (PPMO) is also a major post exploratory activity embarked upon by the refinery to distribute refined products to marketers and end users (Nwankwo and Irrechukwu 1981). This activity impact the mangrove forest by acting as a source of pollution via pipeline leakages and ruptures, tank leakages and overflows, road tanker and sea tanker collisions, valve and pump malfunctions and hose ruptures at jetties or depots (Table 12.1).

12.3.4 Impact of Oil Pollution on Vegetation and Fisheries

Impact on vegetation (flora): Sheltered mangrove swamps and salt marshes at beach ridges are very sensitive shorelines. Thus floating oil transported by wave energy impairs the growth of mangrove vegetation by oiling them. This results to the yellowing and defoliation of the trees (Anderson and Lee 1995; Snowden and Ekweozor 1987). In some cases, the juvenile vegetation are killed outright as a result of the smothering action on the pneumatophores of the red (*Rhizophora racemosa*), black (*Laguncularia racemosa*) and white (*Avicennia germinans*) mangroves leading to oxygen starvation (Mastaller 1996).

Table 12.1 The potential impact of oil and gas exploration on the mangrove and human environment create numerous problems for the mangrove ecosystem

S/No.	Oil operations	Potential impact
A	Exploration	Destruction of vegetation, farm lands and human settlements, introduction of foreign and invasive species (E.g. *Nypa fruticans*)
1	Geo-physical investigation	Noise and seismic shooting, effects on animals and fisheries, pigging waste
2	Drilling	Noise and seismic shooting, drill cuttings, drill mud, craters and destruction of vegetation
B	Production/processing	Facility and pipeline installation, facility operation
1	Gas flaring	Air pollution from gas and oil processing, evaporation and flaring, heat production affects and destroys the mangroves
2	Tanker loading	Water pollution from ship ballast, destruction of seabed by dredging
3	Storage depots	Land clearance, Land pollution from effluents
4	Transportation	Destruction of seabed, sedimentation, air pollution, erosion and flooding
5	Refinery	Air, water and soil pollution
6	Health effects	Respiratory disease e.g. emphysema, asthma and bronchitis

Impact on fisheries (fauna): The categories of organisms mostly affected by oil pollution include different species of mollusks, crustaceans, echinoderms, polychaetes, cnidarians, oysters, scallops and periwinkles. The immotility of many benthic organisms as well as the free swimming larvae confined to the bottom of the river makes them especially vulnerable to oil pollution (Olsgard and Gray 1995). These organisms are very susceptible to oil pollution because many of them inhabit the intertidal zones where they easily become coated with oil, and smothered when spilled oil drifts ashore (Hayes and Gundlach 1979). Edible crabs that are found in these areas such as *Callinectes palidus*, *Uca tangeri*, *Ostrea tulipa* and the inedible crab, *Goniopsis pelii* are killed by oil pollution. These organisms will not exist without the mangrove serving as a sanctuary.

12.4 Invasive Nypa Palms (*Nypa fruticans*)

Four main drivers of global environmental change are nitrogen, climate change, land use change and invasive species (Aber et al. 2001). Invasive species compete for resources and modify the habitat in which they are found (Wootton 1994). They are ecosystem engineers as they change some fundamental aspect of the physical and chemical environment (Jones et al. 1997). They are the next major threat to mangroves after oil and gas exploration in the Niger Delta. The palms are monotypic genus that are widely distributed in Asia and the western world including India, Sri Lanka and northern Australia (Jones 1995). They are regarded as mangroves in some areas, but are not true mangroves (Kathiresan and Bingham 2001). They have inimical and domineering relationships to the true mangroves (e.g. *Rhizophora*). They were intentionally introduced into the Niger Delta in 1906 for the purpose of flood and erosion control (Keay et al. 1964). It was believed that they will act as a barrier against surging flood eating deep into the coastal regions. But this would not be achieved as they began expanding their range and started encroaching into the mangrove forest. They have presently occupied large parts of the mangrove dominated areas in the Niger Delta (Wang et al. 2016) as well as parts of West Africa (Sunderland and Morakinyo 2002) such as Cameroun. In some areas they had completely taken over the mangrove stands as the dominant species (Isebor 2003) (Figs. 12.9 and 12.10). For instance, within the Niger Delta region the following areas are highly infested with nypa palms: Bonny, Andoni, and Okrika estuaries, Marine base and Eagle Island creeks along Port Harcourt, Ogoni axis, New Calabar, Imo, Saint Barbara, Saint Bartholomew and Orashi Rivers along the Atlantic Coast. Large percentages of previously exclusive mangrove zones had been completely taken over by the palms in the last 20 years. In some areas there are mixed forest of mangroves and nypa palms (Wang et al. 2016), yet in some areas the mangroves had been encircled by the palms while in other areas no mangrove exist at both the upper and lower stream sections of the river. What is found in some areas are little "mangrove islands" that are left in the middle of the river waiting to go into complete extinction. The rapid growth of the palms is facilitated by anthropogenic perturbations of the mangrove soil whose organic content had risen drastically as a result of constant deposition of human waste and other

Fig. 12.9 A restoration site at Asarama in the Niger Delta, Nigeria where red mangrove species (*R. racemosa*) were existing before overtaken by the palms (*Nypa fruticans*). The red mangrove propagules are now planted within the nypa palm stands to recover the mangrove population. This area was mangrove dominated before the invasion of the palms. At the background are palms, which indicate that the mangroves still have a long way to go to recover their territory from the palms (Credit: A.O. Numbere)

pollutants (Okoye et al. 1991). In areas where both nypa palms and mangrove grow, the latter are out-competed (Numbere 2014; Numbere and Camilo 2016) and their positions taken over by the palms (CEDA 1997).

It was observed during field work that the nypa palms grow outwardly from the center of the mangrove forest in groups of multiple stands in a sort of center-spread fashion. The reason for this growth pattern is still being investigated, but it is believed that the growth of the nypa palms within the mangrove stands is for protection from surging ocean currents. This is because the nypa palms are poorly equipped to ward off tidal currents. In addition, the palms grow amongst the mangroves to absorb soil nutrients embedded within mangrove soils. The seeds of the nypa palms develop fibrous roots and absorb nutrients at a faster rate than the mangroves. This deprives the mangroves of nutrients leading to deficiency, outright migration or death. A recent study (unpublished) comparing the morphology of mangrove and nypa palm roots and their nutrient utilization capability shows that nypa palm roots are "straw-like" and more porous as compared to the mangrove roots that have thick and tightly packed cells (Fig. 12.11). The structure of the palms predisposes them to absorbing more nutrients when placed in the same environment as the mangroves.

In this study, it was found that there were more bacterial content (4.85 ± 0.6 cfu/g $\times$ 10^6) than fungi (0.35 ± 0.01 cfu/g $\times$ 10^6) in nypa palm and mangrove roots. Mangrove roots had more microbial content (2.85 cfu/g $\times$ 10^6) than nypa palm roots (2.35 cfu/g $\times$ 10^6). In the same vein, nypa palm roots (7.7 ± 3.2 mg/kg) had higher heavy metals than mangrove roots (2.7 ± 1.2 mg/kg) while average nutrient content (i.e. nitrogen, phosphorous and nitrates) was more in mangrove roots (123.9 ± 121.1 mg/kg) than in nypa palm roots (44.9 ± 40.7 mg/kg). This shows that mangrove roots had more un-transported nutrients than nypa palm roots, which implies that nutrient transportation was

Fig. 12.10 Nypa palm (*Nypa fruticans*) stands embedded within a previously exclusive mangrove forest. The soil chemistry and texture change in areas where the palms dominate, and this affects mangrove growth. There is also high subsidence in nypa palm dominated areas because of the inability of the nypa roots to compact the soil and prevent soil erosion as do the mangroves (Credit: A.O. Numbere)

slower in mangroves than in nypa palms. Therefore, mangrove trees will stand little or no chance of survival if placed in the same environment with the palms.

In another study, it was found that nypa palm seedlings had better growth and higher survival rate in mangrove soil than in nypa soil (Numbere 2014). In yet another experiment carried out to test the effect of propagule pressure in mangrove forest, it was found out that Nypa palm seedlings out-numbered mangrove propagules in a mangrove forest in a ratio of 27:1 (in press). These studies indicate that nypa palms are a major threat to mangroves in the Niger Delta. And if nothing is done to curb the aggressive colonization of mangrove forest by the palms in the next 50 years the mangrove population would have reduced to an endangered level, and might go into extinction in a century. The reason behind this prediction is that in 2002, I sampled a small mangrove forest measuring about 400 m^2 for my Masters work at the foot of a bridge near a small creek at Marine Base, after 15 years, in 2017, when I re-visited the area, I observed that no single mangrove was found in that location. This is because the area had been completely taken over by nypa palms as a result of deforestation and waste disposal in the forest. This same pattern of mangrove disappearances had been observed in other locations. Every year a mangrove territory falls to the rampaging palms. In other to stop this trend it will take a deliberate effort of the government or non-governmental agencies (NGO's) to come to the rescue of the mangroves or else we may be seeing the end of the mangroves in this region.

12.5 Urbanization

Urbanization is also a key product of deforestation that had decimated large amount of mangrove forest stands in the Niger Delta. One of the major causes of deforestation is

Fig. 12.11 (**a**) Mangrove and (**b**) Nypa palm roots. The roots of nypa palm are more fibrous, lighter and completely buried in the soil whereas the mangrove roots are adventitious, harder and protrude from the soil surface. The nypa roots are hollow and "straw-like" while the mangrove roots are non-hollow and have tightly packed cells. The roots of palms have the advantage of faster nutrient absorption than the mangrove roots (Credit: A.O. Numbere)

population explosion in Nigeria with about 182 million persons (NPC 2017). Nigeria is the most populous country in Africa, and the most populous black nation in the world. The Atlantic coastal areas are the highest centers of human habitation. For instance between Senegal and Nigeria an estimated 60 million people representing 25% of the population live within 60 km of the coastal areas (NOAA/NOS 2002). In Nigeria 20 million people live along the coastal zone (IPCC 2000) in areas locally called "water side". Increase in poverty and the need for shelter had pushed the people deeper into the mangrove forests to seek for shelter. These shelters are locally constructed with palm fronds and wood from the mangrove trees. But most times after clearing the mangrove forest on site, the swamp is solidified with thick mangrove soil locally called "chikoko" (Numbere 2014). This chikoko soil has the tendency of becoming hardened after reinforcement with wood and mud. It is on this hardened platform that the foundations for wooden or concrete buildings are erected. Human alterations of the ecosystem such as the clearing of large sections of the forest to build houses have rendered the mangrove susceptible to invasion by foreign species. Land reclamation in the form of sand dredging is often done around the mangrove swamp (Ohimain 2003). The area that is dredged and sand filled is left to lie fallow for several years. This activity alone had led to the loss of significant amount of mangrove forest communities in the Niger Delta (Fig. 12.12). In a questionnaire survey conducted in three local settlements built on cleared mangrove forest in the Niger Delta, it was found that 70% of the respondents said they like mangroves because of the benefits they derived from it. But when probed further on how much they will pay to preserve it, their response indicated that they cared less about its preservation (in press).

Increase in urban areas result to the conversion of high density to low density mangrove forest (Wang et al. 2016). Urban regions, which include construction site, road, urban vegetation, buildings and other infrastructural facilities, had taken over an area of mangrove swamp estimated at 185.7 km^2 out of 3459 km^2 of the total mangrove forest sampled. This makes 5.4% of the mangrove vegetation in the Niger Delta (Wang et al. 2016). This study indicates that urban areas increased rapidly from 1984 to 2007, and had negative relationship with mangrove forest and a positive relationship with mixed forest. Mixed forest includes a combination of mangrove and nypa palm trees. This study further indicates that there is a linkage between exploration and urbanization. Apart from the destruction of mangrove forest by pollution (Snowden and Ekweozor 1987), exploratory activities had resulted to an increase in urban areas and a decrease in mangrove swamps. Areas where mangroves are removed were quickly flooded with developmental projects

Fig. 12.12 Dredged and sand filled mangrove forest in Buguma, Niger Delta, Nigeria. There are still some mangroves that can be seen at the background. On the left of the picture is the fence of a secondary school. No mangrove tree has ever grown on this area since the sand filling in 1984. The area is now occupied by weeds and other alien plant species (Credit: A.O. Numbere)

such as industrial complexes, residential quarters and social amenities (Fig. 12.13). James et al. (2007) revealed that 210 km^2 of mangrove forest had been lost as a result of urbanization, dredging and oiling activities in the Niger Delta. Similarly, Wang et al. (2016) revealed that 5% of mangrove forests had been lost from 1984 to 2007 in the Niger Delta (Fig. 12.14). Exploration activities increased urban areas and reduced wet mud, leading to land fragmentation. Wet mud is the swampy areas, which serves as the natural habitat for mangroves and other wetland species. The disappearance of wet mud is an indication of mangrove loss and replacement with urban areas and nypa palms (Wang et al. 2016).

An example of dredging and sand filling activity occurred in a town called Buguma in the Niger Delta in 1984. This action removed a total of about 4.2 million m^2 of mangrove forests from the landscape (Fig. 12.15). The dredging was embarked upon as a program by the local authority to reclaim land for human habitation. The decision to reclaim more land was reached following increase in population in the small island town. Since this dredging activity took place over 30 years ago the area had remained bare of vegetation without any trace of mangrove re-growth.

12.6 Interaction Effect

At present, a far greater threat to mangrove forest in the Niger Delta that is not often mentioned in the literature is the interaction effect of hydrocarbon pollution, urbanization and invasion. The mangroves are naturally resilient and resistant to environmental perturbations (Boesch 1974; Gunderson et al. 2002; Horn 1974). But the interaction of these three threats have further denigrated their power of resilience, and had reduced their survival ability in the face of multiple threats. Continuous deforestation and replacement of the mangrove

Fig. 12.13 Urbanization in progress. The conversion of mangrove forest to urbanized area. At the far end are some mangrove trees still standing, while at the middle are residential buildings with one under construction in Buguma, Niger Delta, Nigeria (Credit: A.O. Numbere)

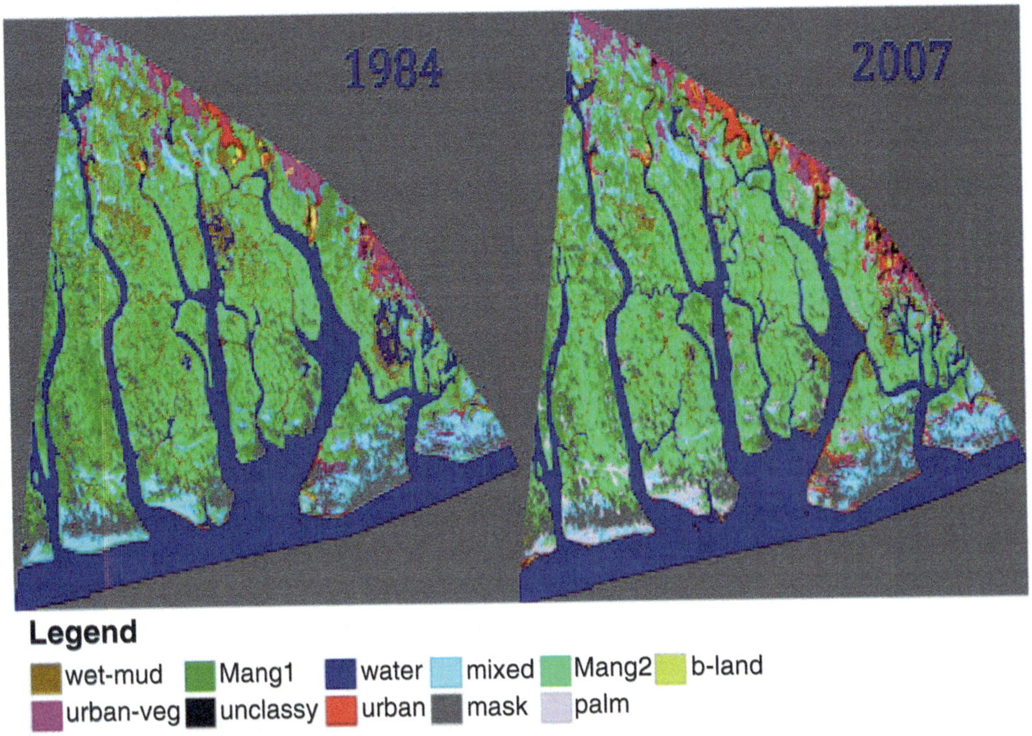

Fig. 12.14 Mangrove landscape change between 1984 and 2007 in the Niger Delta, Nigeria. Wet mud refers to swampy areas, Mang 1 refers to high density mangroves, Mang 2 refers to low density mangroves and urban refers to urbanized areas, and include industrial, residential complexes, and construction sites. (Source: Wang et al. 2016).

Fig. 12.15 Dredged and sand filled area in Buguma, Niger Delta, Nigeria. The white patches indicate the sand filled area while the green patches indicates mangrove forests that are still standing. The white patches sum up to give a total of 4.2 million m^2 of mangroves removed in 1984. This estimates was made using Arc GIS (ESRI 2006)

forest with urban infrastructure had given the mangroves no chance to exist in their natural habitat. And mangroves cannot co-exist with urban areas because they are habitat specialist that only thrives in wetland environment. Studies had shown that mangroves can survive limited hydrocarbon pollution (Numbere 2014), and recuperate if tidal flushing reduces the level of hydrocarbon pollution. Litter decomposition also take place on the forest floor in highly polluted mangrove forest (Numbere and Camilo 2017). Mangroves can also have limited growth in the presence of nypa palms, and can survive dredging to relaim land for urban development, as experienced in Buguma. But will have difficulty surviving when the three components are acting at the same time (Fig. 12.16).

12.7 Concluding Remarks and Conservation Measures

The key problem of conservation of mangroves in Africa is poverty. This is because indigenous people rely more on the numerous by-products of mangroves such as fire wood, medicinal herbs, and food, but have little or no benefit from the palms (UNEP 2007). The reason might be that they have not found any use of the palms yet. Therefore, for any mangrove conservation effort to succeed in the Niger Delta there need to be the inclusion of human management. The creation of zones of use, buffer and transition is advantageous to plants, animals and humans. At the core area no resource extraction is to be allowed, rather scientific activity can be permitted to monitor population dynamics of the mangrove ecosystem. If this design were adopted, it means at the buffer zone some level of resource use by people can be allowed. Tourist potentials can also be tapped to generate funds to run the reserve. The transition zone will have higher level of resource use such as commercial activities, building of residential quarters and ecotourism. Human habitation can be allowed to enable the people to protect and also act as custodians of the forest. The model recommended for the Niger Delta is similar to that of the Great Buffer Reef Marine Park Authority (GBRMP) in the east coast of Australia (Day 2002). Here use and preservation zones were established. The preservation zone is used for scientific study, but prevents tourism and commercial activities. The park allows the establishment of breeding sites for threatened and endangered species. In the GBRMP ten percent of the coral reef was designated as protection zone.

Another key restoration measure against hydrocarbon pollution that can be implemented in the Niger Delta is the planting of fresh mangrove propagules from nursery bed to the field. This is because studies had shown that high density mangrove can withstand hydrocarbon pollution better than low density mangrove (Wang et al. 2016). This same study indicates that oil and gas exploration significantly decreased the areas occupied by mangroves. In other to restore mangrove vegetation more mangrove propagules should be planted in dedicated areas that will be strictly protected to prevent human intrusion. The planting of more mangrove seedling is important to prevent propagule pressure from the palms, which had already overwhelmed the mangroves in many locations. In another context, a win-win conservation measure or limited protection can be adopted in areas that have large human population to ensure a balance between forest restoration and human utilization. In the same vein, the proliferation of urban areas and the reduction of swampy areas mean that mangroves are losing their natural habitat through land modification (Ellison 1994), which is not good for the survival of the mangroves. As a remedial measure against invasion, the palms should be destroyed with swamp boogies and the seeds removed to prevent re-growth and re-colonization (see Fig. 12.16). The disturbed soils should also be removed and replaced with mangrove soils to facilitate the growth of the mangroves after the removal of the palms. This is because mangrove seedlings have poor growth performance when planted in

Fig. 12.16 Eagle Island estuary, Niger Delta, Nigeria. A combination of pollution, urbanization and invasive nypa palms had led to a complete disappearance of a once thriving mangrove forest in this location. Twenty years ago this area was a mangrove dominated forest (*R. racemosa*). The nypa palms were uprooted by swamp boogies because they were blocking an adjoining canal. At the background are buildings (urban areas) constructed several years ago after the swamps were dredged and made bare (Credit: A.O. Numbere)

the same environment with the Nypa palms (Numbere and Camilo 2016).

Despite the threats of oil and gas exploration, urbanization and invasion all hope is not lost. This is because the situation can be reversed if the oil companies performed their social and environmental duties responsibly by ensuring that old oil facilities are maintained to avoid oil spillages. In addition, local land owners and communities from whom the exploratory lands were acquired should be adequately compensated to prevent acts of agitation and sabotage of the pipelines. Personnel from the oil companies should respond quickly to incidences of spillages and be pro active in detecting the cause of oil spillage. The oil companies can save the mangrove environment by embarking on less intrusive seismic refraction method rather than the use of a more intrusive method of using explosives like dynamite to explore for crude oil. The refraction method can be used for oil and gas exploration and exploitation, and applies the principle of refraction and reflection to locates oil deposits in the earth crust without adverse effect on the environment.

References

Aber J, Neilson RP, McNulty S, Lenihan JM, Bachelet D, Drapek RJ (2001) Forest processes and global environmental change: predicting the effects of individual and multiple stressors: we review the effects of several rapidly changing environmental drivers on ecosystem function, discuss interactions among them, and summarize predicted changes in productivity, carbon storage, and water balance. Bioscience 51(9):735–751

Adegbehin JO, Nwaigbo LC (1990) Mangrove resources in Nigeria: use and management perspectives. Nat Res Forum 26:13–21

Alongi DM (2008) Mangrove forests: resilience protection from tsunamis, and responses to global climate change. Estuar Coast Shelf Sci 76:1–13

Anderson C, Lee SY (1995) Defoliation of the mangrove *Avicennia marina* in Hong Kong: cause and consequences. Biotropica 27:218–226

Aquilera F, Mendez J, Pasaro E, Laffon B (2010) Review on the effects of exposure to spilled oils on human health. J Appl Toxicol 30 (4):291–301

Boesch DF (1974) Diversity, stability and response to human disturbance in estuarine systems. In Proceedings of the first international congress of ecology, Wageningen, 109–114

Burns KA, Garrity SD, Jorrisen D, MacPherson J, Stoelting M, Tierney J, Yelle-Simons (1994) The Galeta oil spill II, unexpected persistence of oil trapped in mangrove sediments. Estuar Coast Shelf Sci 38(4):349–364

CEDA (1997) Coastal profile of Nigeria. Federal Environmental Protection Agency, Abuja

Clark MW, Mc Conchie D, Lewis DW, Saenger P (1997) Redox stratification and heavy metal portioning in mangrove plant and sediment. Geochem Model Chem Geol 149:147–171

Day JC (2002) Zoning—lessons from the Great Barrier Reef marine park. Ocean Coast Manag 45(2):139–156

Duke NC, Pinzon M, Prada MC (1997) Large-scale damage to mangrove forests following two large oil spills in Panama. Biotropica 29(1):2–14

EIA (2001) Natural gas liquid export project by the Nigerian Agip Oil Company Limited. Report No. 0096. 13

Ellison JC (1994) Mangrove retreat with rising sea-level Bermuda. Estuar Coast Shelf Sci 37(1):75–87

Ellison JC (2005) Holocene palynology and sea level change in two estuaries in southern Irianjaya. Palaeogeogr Palaeoclimatol Palaeoecol 220:291–309

ESRI (2006) Arc GIS 9.1, Redlands, Ca.USA

Fagbami AA, Udo EJ, Odu CTI (1988) Vegetation damage in an oil field in the Niger Delta of Nigeria. J Trop Ecol 4:61–75

Goyer RA (2001) Toxic effect of metals In: Klassen C.D, Cassarett and Doull's Toxicology. McGraw-Hill, New York, pp 891–867

Gunderson LH, Holling CS, Pritchard L Jr, Peterson GD (2002) Resilience of large-scale resource systems. In: Gunderson LH, Pritchard L Jr (eds) Resilience and the behavior of large scale systems, SCOPE Series, vol 60. Island Press, Washington, pp 3–20

Gundlach ER, Hayes MO, Getter CD (1981) Sensitivity of coastal environment of oil spills. In: proceedings: "The Petroleum Industry and the Nigerian Environment" NNPC/FMOW & H, 0-12 Nov. 1981. Petroleum Training Institute Warri, Nigeria. 82–88

Hayes MO, Gundlach ER (1979) Coastal processes field manual for oil spill assessment, Prepared for NOAA, office of marine pollution assessment, Boulder, Colorado. Research Planning Institute, Inc, South Carolina

Horn T (1974) The ecology of secondary succession. Annu Rev Ecol Syst 5:25–37

Ifeadi CN, Awa AK (1987) Ground water contamination by hydrocarbon in Nigeria petroleum industry. Proceedings: seminar on the petroleum industry and the Nigeria environment. NNPC/FMOW & H, 9-12, Owerri, 22

IPCC (2000) Special report on the regional impacts of climate change: an assessment of vulnerability. African chapter. Available at www.grida.no/climate/ipcc/regional/006.htm. Accessed 29 July 2017

Isebor C (2003) The invasive and exotic species in the Nigerian coastal zone. Proceedings of the 13th Biennial Coastal Zone Conference, Baltimore, MD. 13-17th July

IUCN (1993) Oil gas exploration and production in mangrove areas. Guidelines for environmental protection. IUCN-The World Conservation Union, Gland, 47

James GK, Adegoke JO, Saba E, Nwilo P, Akinyede J (2007) Satellite-based assessment of the extent and changes in the mangrove ecosystem of the Niger Delta. Mar Geod 30:249–267

Jones DL (1995) Palms. Smithsonian Institution Press, Washington, DC

Jones CG, Lawton JH, Shachak M (1997) Positive and negative effects of organisms as physical ecosystem engineers. Ecology 78:1954–1957

Kathiresan K, Bingham BL (2001) Biology of mangrove ecosystems. Adv Mar Biol 40:81–251

Keay RWJ, Onochie CFA, Standfield DP (1964) Nigerian trees. Federal Department of Forestry Research, National Press Limited, Ibadan

Klekowski EJ, Corredor JD, Lowenfeld R, Klekowski EH, Morelli JM (1994) Using mangroves to screen for mutagens in tropical marine environments. Mar Pollut Bull 28(6):346–350

Lacerda LD, Carvalho CEV, Tanizaki KF, Ovalle ARC, Rezende CE (1993) The biogeochemistry and trace metals distribution of mangrove rhizospheres. Biotropica 25:251–256

Levings SC, Garrity SD (1994) Effects of oils spills on fringing red mangroves (*Rhizophora mangle*): losses of mobile species associated with submerged prop roots. Bull Mar Sci 54:782–794

Lugo AE (1980) Mangrove ecosystems: successional or steady state? Biotropica 12:65–72

Mackey AP, Hodgkinson M, Nardella R (1992) Nutrient levels and heavy metals in mangrove sediments from the Brisbane River, Australia. Mar Pollut Bull 24:418–420

Mastaller M (1996) Destruction of mangrove wetlands-cause and consequences. Nat Res and Dev 43–44:37–57

NDES (1997) The Niger Delta environmental survey (phase 1), environmental and socio-economic characteristics. Environmental Resources Managers Limited, Lagos

Nellemann C, Cameron RD (1998) Cumulative impacts of an evolving oil field complex on the distribution of claving caribou. Can J Zool 76:1425–1430

NOAA/NOS (2002) Filling critical gaps and promoting multi-site approaches to new nominations of tropical, coastal, marine and small island ecosystem West Africa. World Heritage Biodiversity Workshop 25 Feb–1 Mar 2002. Regional Papers. West Africa

NPC (2017) Current population of Nigeria. www.npc.org. Accessed 26 May 2017

Numbere AO (2014) Impact of hydrocarbon pollution on the mangrove ecosystem of the Niger River Delta, Nigeria. Dissertation, Saint Louis University

Numbere AO, Camilo GR (2016) Reciprocal transplant of mangrove (*Rhizophora racemosa*) and Nypa palm (*Nypa fruticans*) seedlings in soils with different levels of pollution in the Niger River Delta, Nigeria. Global J Environ Res 10(1):14–21

Numbere AO, Camilo GR (2017) Mangrove leaf litter decomposition under mangrove forest stands with different levels of pollution in the Niger River Delta, Nigeria. Afr J Ecol 55(2):162–167

Nwankwo JN, Irrechukwu DO (1981) Problem of environmental pollution and control in the Nigeria petroleum industry. Proceedings of the 1981 International Seminar on the Petroleum Industry and the Nigerian Environment, Nov. 9-12 (1981) Petroleum Training Institute,Warri, Delta State, Nigeria, pp 102–107

Odum WE, Johannes RE (1975) The response of mangroves to man-induced environmental stress. In: Ferguson-Wood EJ, Johannes RE (eds) Tropical marine pollution, Elsevier Oceanography series, vol 12. Elsevier Scientific Publ. Co., Amsterdam, pp 52–62

Ohimain EI (2003) Environmental impacts of oil mining activities in the Niger Delta mangrove ecosystem. Proceedings of the 8th International Mine Water Association (IMWA) Conference, Johannesburg. pp 503–517

Ohimain EI (2004) Environmental impacts of dredging in the Niger Delta. Options for sediment relocation that will mitigate acidification and enhance natural mangrove restoration. Terra et Aqua 97:9–19

Ohimain EI, Gbolagade J, Abah SO (2008) Variations in heavy metal concentrations following the dredging of an oil well access canal in the Niger Delta. Adv Biol Res 2(5–6):97–103

Okoye NV, O Sullivan AJ, Schouten JR (1987) Monitoring and evaluation of oil related pollution in NNPC operation. Proceeding of the petroleum industry and the Nigeria environment NNPC/FMOW & H, 9-12. Owerri, 17

Okoye BCO, Afolabi AO, Ajao AE (1991) Heavy metals in the Lagos lagoon sediments. Int J Environ Stud 37:35–41
Olgard E, Gray JS (1995) A comprehensive analysis of the effects of oil and gas exploration and production on the benthic communities of the Norwegian continental shelf. Mar Ecol Prog Ser 122:277–306
Polidoro BA, Carpenter KE, Collins L, Duke NC, Ellison EM et al (2010) The loss of species: mangrove extinction risk and geographic areas of global concern. PLoS One 5:e10095–e10095. https://doi.org/10.1371/journal.pone.0010095
Shell Petroleum Development Company (SPDC) (1999) EIA of Belema gas injection and oil field development project. 9–10
Snedaker SC, Biber PD, Aravjo RJ (1996) Oil spills and mangroves: an overview. Managing oil spills in mangrove ecosystems: effects, remediation, restoration and modeling 1
Snowden RJ, Ekweozor IKE (1987) The impact of a minor oil spillage in the estuarine Niger Delta. Mar Pollut Bull 18:595–599
Sunderland TCH, Morakinyo T (2002) Nypa fructicans, a weed in West Africa. Palms 46(3):154–155
Teas HJ, Lessard RR, Canevari GP, Brown CD, Glenn R (1993) Saving oiled mangroves using a new non-dispersing shoreline cleaner. In International Oil Spill Conference 1993 (1993, 1, 147–151). American Petroleum Institute, Washington, DC
Twilley RR (1988) Coupling of mangroves to the productivity of estuarine and coastal waters. In: Coastal-offshore ecosystems interactions: lecture notes on coastal and estuarine studies, vol 22. Springer-Verlag, Berlin, pp 155–180
Twilley RR (2008) Mangrove wetland. Ecosystems:2198–2208
UNEP (2007) Mangroves of Western and Central Africa. UNEP-Regional Seas Programme/ UNEP-WCMC. Available at www.unep-wcmc.org/resources/publications/UNEP_WCMC_ bio_series/26.htm. Accessed 20 May 2017
UNEP (2011) Environmental assessment of Ogoniland. United Nations Environmental Programme. Available online at www.unep.org/nigeria. Accessed 26 Sept 2016
Wang P, Numbere AO, Camilo GR (2016) Long term changes in mangrove landscape of the Niger River Delta, Nigeria. Amer J of Environ Sci 12:248–259
Webb EL, Jachowski NR, Phelps J, Fries DA, Than MM, Ziegler AD (2014) Deforestation in the Ayeyarwady Delta and the conservation implications of an internationally engaged Myanmar. Glob Environ Chang 24:321–333
Withgott JH, Brennan S (2011) Environment: the science behind the stories. Pearson Education, Ca. 616–640
Wootton JT (1994) The nature and consequences of indirect effects in ecological communities. Annu Rev Ecol Syst 25(1):443–466
World Bank (1995) Defining an environmental management strategy for the Niger Delta. Report prepared by Jasdip Smig, David Moffat and Olof Linden. Washington, DC: Industry and Energy Operations Division, West African Department, World Bank
WWF (2001) Niger River Delta. World Wild Foundation global 200 - eco-regions, 2001. Available online at http://www.worldwildlife.org/wildworld/profiles/g200/g155.html. Accessed 20 Apr 2004
Yim MW, Tam NFY (1999) Effects of waste water-borne heavy metals on mangrove plants and soil microbial activities. Mar Pollut Bull 39:179–186

Oil-Related Mangrove Loss East of Bonny River, Nigeria

13

Erich R. Gundlach

Abstract

This study documents the largest cumulative loss of mangroves due to oil-related activities, including 4415 ha due directly to oil spillage (operational and illegal activities) and 105 ha due to pipeline corridors. Additionally, 217 illegal refinery sites are found which destroyed 116 ha of high ground habitat adjacent to mangroves. Source information utilized includes satellite imagery (1999–2016), aerial videography and photographs (2000, 2010, 2015), and field surveys (1983, 2013, 2015). Mangrove losses began in 2008/2009 due to four spills (three caused by corrosion, one from illegal activity) that caused ~2000 ha of damage. Two of these spills are part of a legal settlement with the resident Bodo community involving a mangrove loss of ~1000 ha. Illegal tapping of the major north-to-south oil-transport pipelines along with a concurrently large increase in illegal refineries became evident in the eastern part of the study area in 2010 and 2011, causing an additional ~1000 ha loss. In 2010, illegal activities also increased in the north causing >300 ha of mangrove loss. After 2013, mangrove losses are small as military operations substantially decreased but did not stop illegal activities. Field studies in 2015, designed to provide guidance to cleanup operations and mangrove restoration, found large areas with very high concentrations of surface and subsurface oil that will inhibit mangrove recovery. Indications are that natural recovery will take much longer than 30 years and, in some areas, may never occur without intervention due to substrate changes that now inhibit seed settlement and growth.

Keywords

Mangrove · GIS mapping · Oil spill impacts · Remote sensing · Bonny River · Niger Delta · Nigeria

13.1 Introduction

This paper analyses the cause, extent, and time-sequence of mangrove loss along the eastern side of the Bonny River, Rivers State, Nigeria. This information is critical for proper undertaking of remedial and restoration methods as described by UNEP (2011) and in process of implementation by the Bodo Mediation Initiative (BMI 2016) and the Hydrocarbon Pollution Remediation Project (HYPREP) under the Nigerian Ministry of Environment (MOE 2017).

The area of interest lies along the eastern part of the Niger Delta (Fig. 13.1). The Niger Delta includes waterways having water flow from the Niger River as well as adjacent areas having similar conditions. The Niger River distributary system is defined as extending from the Benin River (5.7364 N, 5.0636 E) in the west to the Imo River (4.4544 N, 7.5908 E) in the east (NDES 1997). Mangrove forests are located along the entire Niger Delta in areas of saline to brackish waters extending up to 40 km toward the mainland from a series of barrier islands fringing the Atlantic Ocean in the south. The Niger Delta contains the largest extent of mangroves in Africa. NDES (1997) using 1981 FAO data estimates the total Niger Delta mangroves at 966,676 ha, but this is likely a high value as UNEP (2007) suggests that Nigerian mangroves have reduced by 26% since the 1980s due to development.

13.1.1 Regional Overview

The study area extends from the Bonny River eastward to the furthest extent of observable mangrove loss, and from Bonny

E. R. Gundlach (✉)
E-Tech International Inc, Boulder, CO, USA

C. Makowski, C. W. Finkl (eds.), *Threats to Mangrove Forests*, Coastal Research Library 25,
https://doi.org/10.1007/978-3-319-73016-5_13

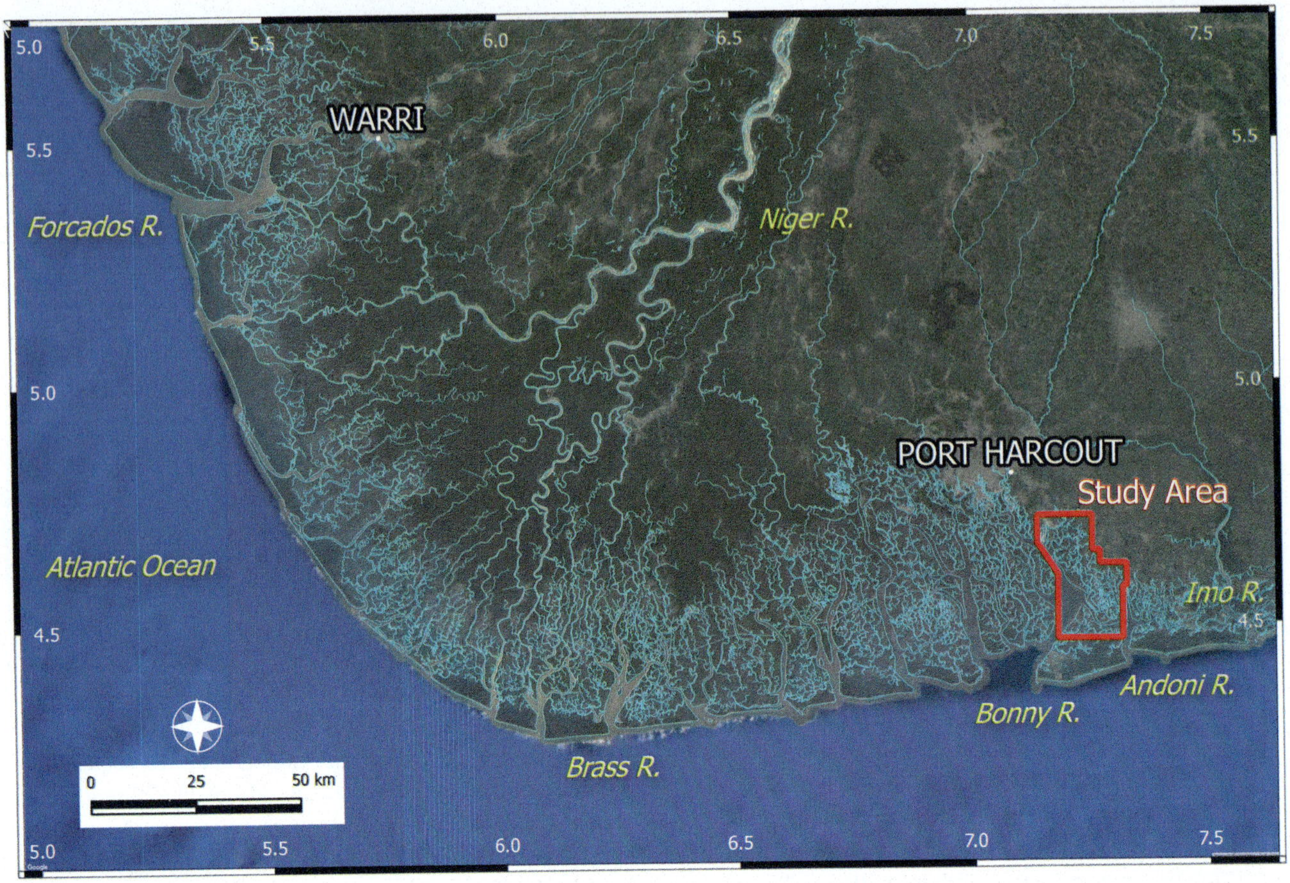

Fig. 13.1 Study area location in the eastern Niger River delta

Island in the south to the mainland in the north between Onne and Ogu (Fig. 13.2). Portions of five Local Government Areas of Rivers State are included: Andoni, Bonny, Gokana, Ogu Bolo, and Okrika. The northern part of the study area is part of Ogoniland.

Mangroves intersected by numerous intertidal channels and creeks are the dominant ecological habitat of the area. A few small villages are present on supratidal islands within mangrove areas. The mainland towns of Bodo, Ogu and Bolo are much larger. Ogu is similar in size to Bodo while Bolo is smaller. Bodo has a reported population of 69,000 (Pegg and Zabbey 2013).

Allen (1965) characterizes the eastern Niger Delta as having three geomorphological units: barrier islands formed by sand ridges, a broad tidal flat colonized by mangroves and marked by a reticulate drainage system, and an interior forested river floodplain. In the area of study, historic delta sediments dominated by silt and clay underlie most of the mangrove area, varying in stability from hard and compact to very soft and unstable where walking without sinking 40 cm or more is not possible. Within the mangrove habitat are supratidal islands composed of sand and silts and commonly forested by palm trees.

As measured at Bonny, rainfall in the area can be very heavy, sometimes over 1000 mm a month, with heaviest rains falling June to September (Worldweatheronline 2017). Average rainfall measured over several decades varies from 3500 to 4000 mm annually along the coastal belt, decreasing to 2000–2500 mm in the northern part of the study area (Adejuwon 2012). Temperatures range from 23 to 36 °C.

The tides of the study area are diurnal showing a range of 1.4–2.6 m. Upper areas are entirely drained during low tide, exposing large sand- or mud-dominated tidal flats. Tidal currents measured by the author show a range of 0.4–1.6 kn during flood conditions, and 0.6–1.2 kn during ebb, in channels between Bodo and south to Opobo channel and the Bonny River.

The inflow of freshwater to the area is relatively minor. The Bonny River extends 72 km from the Atlantic Ocean and has no further interior drainage. In contrast, the Imo River to the east extends far into the mainland and serves as a drainage basin for a large interior section, but this has no effect on the study area. Folorunsho and Awosika (2014) find that "The Bonny River system has the largest tidal volume of all river systems in the delta. There is generally a net flux of tidal

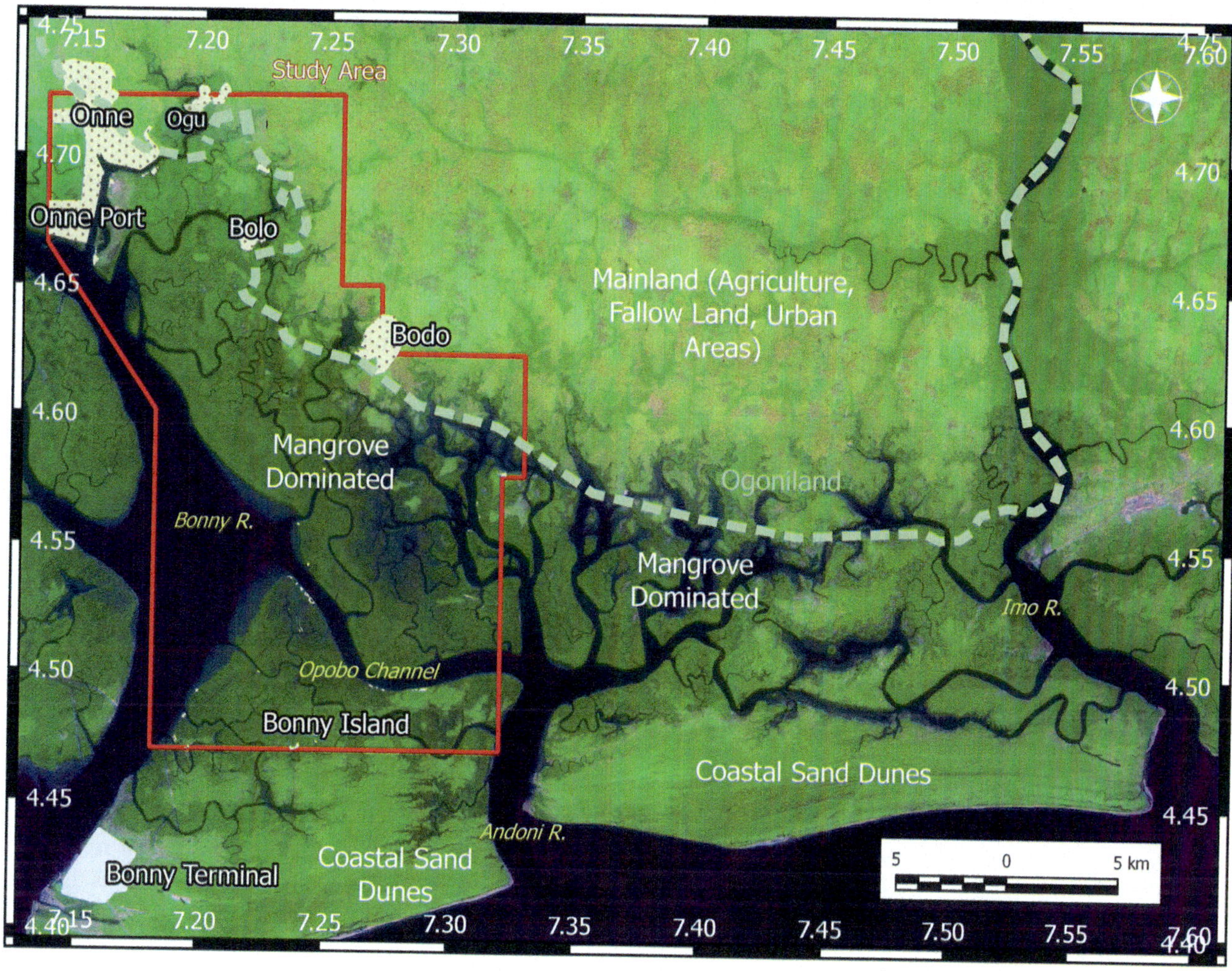

Fig. 13.2 Regional setting of the study area along the east side of the Bonny River. A SPOT image from 1999 is shown as background

water up the river, which disperses into various creeks and channels".

Salinities measured by Dublin-Green (1990) in the Bonny River of the area of study show a range of 18–20 during September (wet) and from 20 to 29 during February, May and December (dry). Field measurements by the author in the dry month of April from Bodo down to the Bonny River show only a minor difference in salinity (19.5–23.5).

The intertidal conditions, limited flushing by upland rainfall, and measured current velocities indicate that oil spilled into the creeks and lagoons potentially will cover large areas and remain within the intertidal mangrove habitats for a substantial period.

13.1.2 Mangroves and Associated Organisms

Six species of mangroves are present in the area: *Rhizophora mangle, R. harrisonii, R. racemosa, Avicennia germinas, Conocarpus erectus* and *Languncularia racemosa* (UNEP 2007, Zabbey and Tanee 2016). The red mangrove (*R. mangle*) is by far the most common.

Differences in salinity, duration of water coverage during the tidal cycle, and underlying soil conditions influence the presence and condition of resident mangroves and associated biota. High salinities and salt deposits on interior basins caused by limited water coverage during spring tides are likely to stress the plant and result in stunted mangroves and sparse surface coverage. Soil conditions are commonly dominated by mangrove root fibers forming a peaty clay called 'chikoko'.

Common aquatic/intertidal fauna include crabs, snails, polychaetes, barnacles and fish (NDES 1997; George et al. 2009; Zabbey et al. 2010). Resources used by the local population include mangroves as firewood, fish, periwinkles, cockles and crabs. Pre-impact analyses by Pegg and Zabbey (2013) at the site found 12 animal species on mangrove stems and stilt roots and 18 species on adjacent tidal flats, of which 34% were locally edible and the others occupied critical links in the food web. The loss of the mangrove forest therefore affects the local community relying on these resources (Fentiman and Zabbey 2015).

The nipa palm (*Nypa fruticans*) is an exotic species that is common in many parts of the study area and has become dominant in certain areas. It is highly opportunistic and found to out-compete mangroves in oiled and depleted soil areas (Sunderland and Morikinyo 2002). Productivity of the mangrove habitat is reduced with the introduction of nipa palm. Baker (1983) found nipa palm at only 1 of the 27 survey sites in the central west portion of the study area whereas now it is common throughout, particularly in the south and central parts of the study area.

Of all shoreline types, mangrove habitats have been rated as the most sensitive to spilled oil based on their high biodiversity, ability to trap and retain spilled oil, and the long time required for recovery (Gundlach and Hayes 1978; Gundlach et al. 1981; NOAA 2012; IPIECA 2012). Summaries of oil impact on mangroves include Duke (2016a), IPIECA (1993), Lewis et al. (2011), NOAA (2014), and USDOI (1997), among others.

13.1.3 Oil-Related Infrastructure

Development of the Bonny River area began by Shell Petroleum Development Company of Nigeria Limited (SPDC) in the 1950s. Bonny Terminal, the export hub for the region, opened in 1961. The majority share (55%) of SPDC is owned by the Nigerian government via the Nigerian National Petroleum Corporation. Shell (30%), Total (10%) and Eni (5%) have lesser shares.

SPDC eventually drilled at least 325 wells in Rivers State but suspended operations in the Ogoni area in 1993 due to civil unrest. All infrastructure was abandoned including oil and gas wells, gas plants, flow stations, manifolds and pipelines. Within the study area, there are 38 inactive oil wells and two manifolds (Fig. 13.3). The Trans Niger Pipeline (TNP) comprised of two (24-in. and 28-in.) oil pipelines remains active, crossing the area from the north to Bonny Terminal in the south. In addition to a gas line crossing the study area, there is a large-diameter pipeline operated by NNPC (Nigeria National Petroleum Corporation) which delivers crude oil for export via the Bonny Terminal. The specifics of the NNPC pipeline throughput and extent of operation are unknown.

13.1.4 Persistence of Bonny Light Crude Oil

Bonny Light crude oil is produced in the region. It is characterized as a light oil with low asphaltenes, low sulfur and relatively high concentrations of nickel and vanadium. Properties are shown in Table 13.1.

With processing, Bonny Light crude has a high yield for diesel and lighter products (Table 13.2), which makes it ideal for legal and illegal refining. Assuming that temperatures on the order of 350 °C are reached by the illegal refining process, then two thirds of the original crude oil input could be converted to saleable product by illegal refiners. As shown later, the heavier remaining oil and residuum is discarded into the surrounding environment.

When spilled, the light components of Bonny Light crude oil are exposed to a high degree of evaporation. Bonny Light is categorized as a Group 3 oil, which has a longevity of approximately a month to a few months when spilled on the water's surface (ITOPF 2011). However, once oil becomes incorporated into the surface or remains as a tarry residue, the persistence of the remaining oil is extended greatly. Gundlach (1987). estimated persistence of oil on different shoreline types using the following formula:

$$M_i = M_{io} + e^{-Kft}$$

Where:

M_i = mass of oil on beach segment i
M_{io} = mass of oil originally deposited on the beach
K_f = removal rate constant based on exponential decay
T = time in days since original deposition

As originally developed, information for oil removal in mangroves was not available, but using marshes as the closest habitat type indicates a removal rate factor of 0.001–0.1 (Etkin et al. 2007). Taking the slower value (0.001) as applicable to large expanses of oiled mangroves then the calculated oil persistence is on the order of 16 years, which is reasonable considering the extensive presence of both surface and subsurface oil in the area. This longevity is the case for oil alone. The persistence of damage to mangroves is discussed in detail later.

13.2 Methods

A detailed review of satellite imagery, photographs and video taken during low-level overflights, and ground surveys forms the basis of this study. Data were input and analyzed using an open-source geographic information system (QGIS version 2.18.4). Specific data sources and their utilization are described below.

13.2.1 Digital Imagery

Images available (Table 13.3) were reviewed in a re-iterative method, i.e. going back and forth in time using historic images to see when and where mangrove damage occurred.

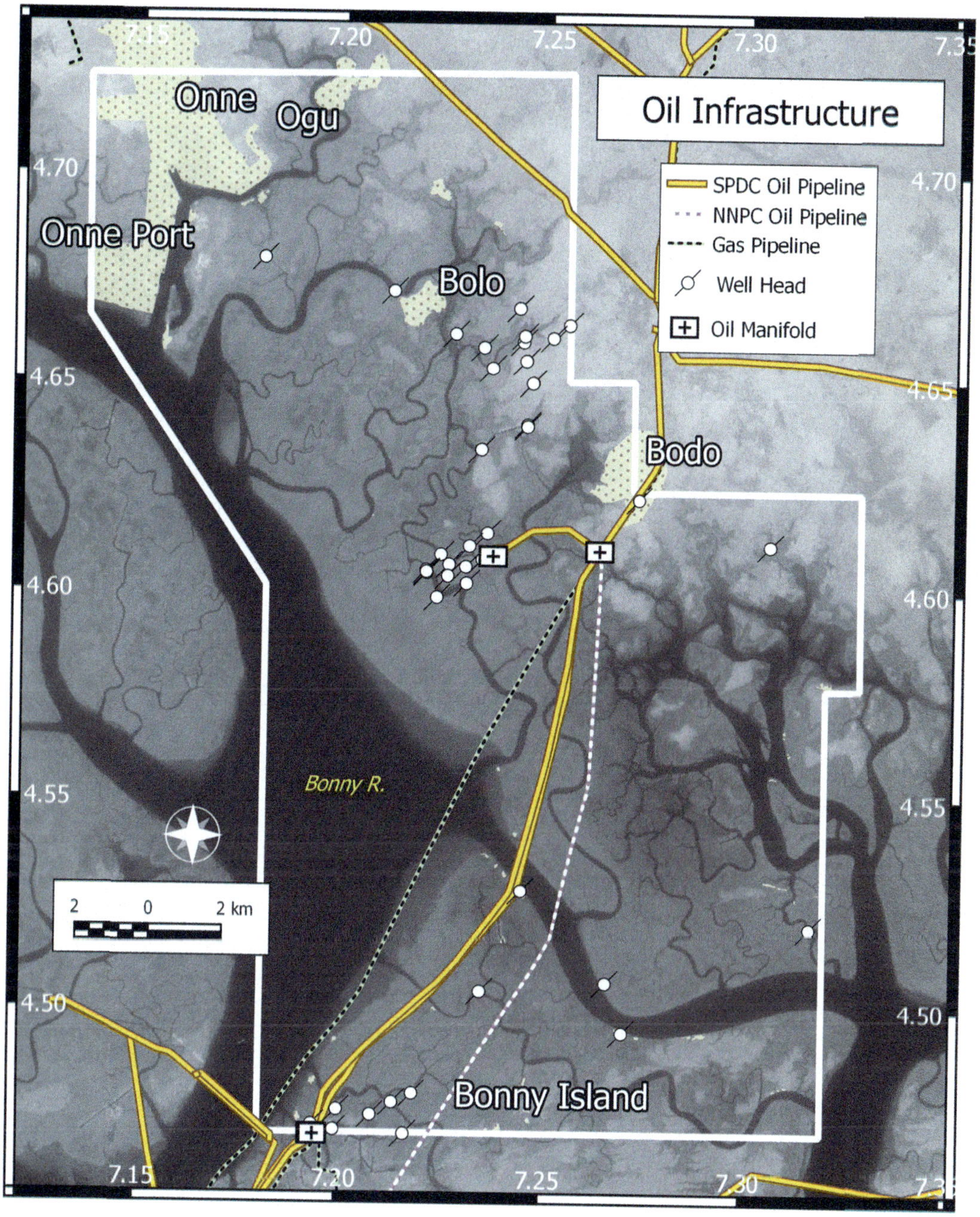

Fig. 13.3 Oil infrastructure within the study area

Table 13.1 Characteristics of Bonny Light crude oil (Brillpetco.com 2017)

Property	Measured
Gravity, API°	32.9
Gravity, SG	0.86
Total nitrogen, ppm	1170
Pour Point °C	−14.4
Viscosity at 40 °C	4.99
Vanadium, ppm	0.42
Nickel, ppm	4.16
Asphaltenes (H.C7) wt%	0.0032

Google Earth images, except for 2013, were reviewed outside of the GIS using Google Earth Pro. Each individual image listed in Table 13.3 covers a portion of the study area. Fortunately, in all cases pre- and post-impact images were available enabling assessment of mangrove loss minimally to 2013 and in most cases extended to 2015 or 2016. Imagery is available in true color and false-color infrared. As the condition of vegetation is clearer using the false-color infrared, these images are commonly selected for use in the text.

Table 13.2 Bonny Light crude true boiling point (TBP) yields (Modified from Toboc.com 2017)

TBP yields	Volume %	Total %
Butanes and lighter	0.92	0.92
Light gasoline (12.7–79.4 °C)	4.25	5.17
Light naphtha (79.4–148.9 °C)	13.73	18.90
Heavy naphtha (148.9–204.4 °C)	10.12	29.02
Kerosene (204.4–260.0 °C)	13.28	42.30
Atmospheric gas oil (260.0–343.3 °C)	22.69	64.99
Light vacuum gas oil (343.3–426.7 °C)	16.81	81.80
Heavy vacuum gas oil (426.7–565.6 °C)	13.26	95.06
Vacuum residuum (>565.6 °C)	4.94	100.00

Table 13.3 Digital imagery available for this analysis

Date	Source	Approximate resolution
30 Dec 1999	Google Earth	~20 m
Jan 2000	SPOT	20 m
12 Dec 2000	Ikonos-2	3.2 m
03 Dec 2006	Quickbird	1.0 m
03 Dec 2006	Google Earth	2.0 m
15 Jan 2007	SPOT	2.5 m
29 Jan 2008	SPOT	5.0 m
25 Jan 2009	Google Earth	2.0 m
26 Jan 2009	Quickbird	0.6 m
08 Jun 2009	Ikonos 2	0.8 m
24 Aug 2009	Ikonos 2	0.8 m
08 Nov 2009	Ikonos 2	0.8 m
19 Nov 2009	SPOT	2.5 m
27 Nov 2009	Google Earth	3.0 m
12 Dec 2009	Google Earth	2.0 m
06 Feb 2010	Google Earth	2.0 m
07 Feb 2010	Ikonos 2	0.8 m
11 Jun 2010	Google Earth	2.0 m
08 Jan 2011	Ikonos	0.8 m
23 Jan 2013	Google Earth	2.0 m
16 Dec 2013	Google Earth	2.0 m
18 Dec 2013	Google Earth	2.0 m
07 Jan 2014	Google Earth	2.5 m
12 Jan 2014	Google Earth	2.5 m
16 Jan 2014	Google Earth	2.0 m
07 Jan 2015	SPOT	1.5 m
19 Jan 2015	Google Earth	2.0 m
10 Feb 2016	Google Earth	2.5 m

13.2.2 Aerial Video and Photographic Surveys

Historic videography and photographs assisted understanding the time sequence of mangrove loss. Videotaping and still photographs from a low flying (~300 to 400 m) helicopter were taken by the author on 18 and 19 November 2000 and by Victor Imevbore of ERML (Lagos) on 11 September 2015. Aerial photographs from 29 November 2010 were provided by Thorsten Kallnischkies (contaminated land expert for UNEP's Ogoniland assessment) and from 15 April 2010 by Jonas Pålssen (World Maritime University). Figure 13.4 provides the flight tracks of the major aerial surveys.

13.2.3 Ground Surveys

Ground surveys confirmed observations made using imagery to determine mangrove and oiling conditions (Table 13.4). Surveys in 2013 and 2015 collected information on mangrove status (quantity dead, living or stressed), oiling level (surface: heavy, light or medium concentrations), subsurface oiling (silver, rainbow or black oil) as observed in a pit dug 20–30 cm, and the appearance of oil on adjacent surface waters. The 2010 UNEP survey provided oiling condition and chemical analyses (UNEP 2011; Linden and Pålssen 2013; Pålssen and Linden 2014). At 13 sites in 2015, samples were taken for chemical analysis. Results are described by Little et al. (2018) in this Coastal Research Library publication. Figure 13.5 provides field site locations showing that surveys concentrated in the center of the study area.

13.2.4 Oil Spills

Oil spill information from 1986 to June 2015 was derived from several sources, including the National Oil Spill Detection and Response Agency (NOSDRA https://oilspillmonitor.ng/), UNEP (2011) and SPDC (http://www.shell.com.ng/sustainability/environment/oil-spills.html).

Data are reported as incidents with date, location coordinates, text as to location (pipeline, wellhead, manifold, etc.) cause and amount discharged. This study uses three categories based on recorded cause: operational (corrosion, line failure, operation), illegal activities (sabotage, bunkering and theft) and unknown (unknown, not recorded and currently unavailable). Data related to the amount discharged are not considered reliable and not reported here. Among many reasons not to consider the recorded value as valid is that leak volume is difficult to estimate from a ruptured pipeline and spilled oil from a site is subject to tidal washing which can distribute spilled oil far from the source before an assessment team is able to perform a survey.

Each incident was verified as to location and duplicates in the database were eliminated. In a few cases, the location was corrected to place it on the pipeline or wellhead as confirmed by text data indicating site location.

Recent spill reports have significantly improved in quality and now contain an onsite Joint Investigation Report (prepared by government officials), response actions, photographs, and the response/cleanup closeout date. The spills reported herein show the timing, location and cause of the event, focusing on the potential impact of the event (s) on surrounding mangroves.

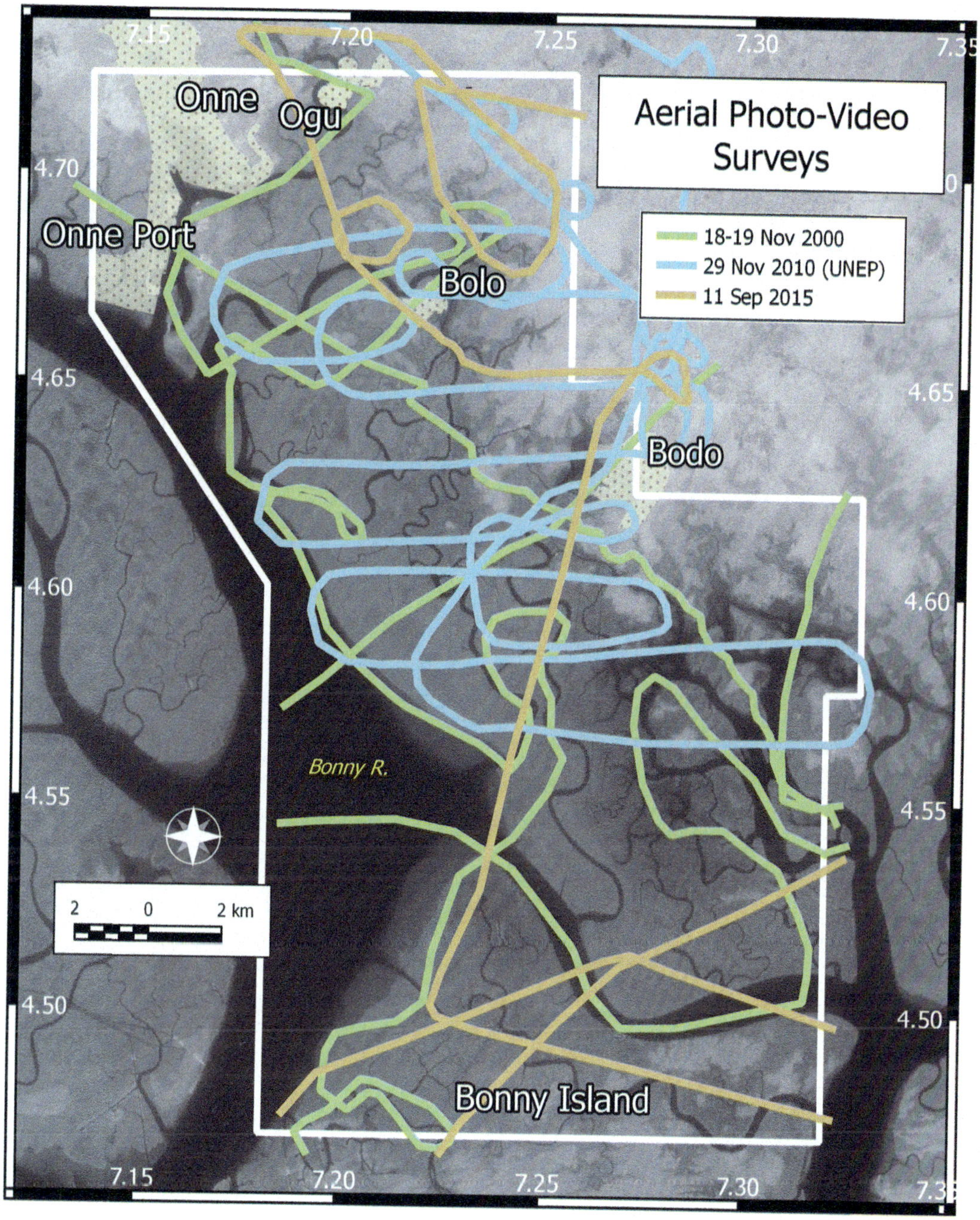

Fig. 13.4 Transit routes of primary low-altitude aerial video and photographic surveys that provided supporting information for this study

Table 13.4 Ground surveys used to assist assessment of mangrove loss. The author participated in the 2013 and 2015 surveys

Survey	Date	Stations	Data type
UNEP	20 Jun – 19 Nov 2010	16	Site oiling, chemistry
Rapid site assessment	9–13 Apr 2013	98	Surface/subsurface oiling, mangrove condition, site photographs
Pre-SCAT	1–8 May 2015	15	Surface/subsurface oiling, mangrove condition, samples taken, site photographs
SCAT	14–25 Aug 2015	35	Surface/subsurface oiling, mangrove condition, samples taken, site photographs

13.2.5 Habitat Classification

Habitats of the study area were manually digitized using QGIS software on a high-resolution monitor at an image scale of 1:500 or less. The following categories were classified.

- Mangrove Loss – dominated by dead mangrove remnants in former mangrove areas, confirmed by pre- and post-damage imagery. In some cases, the substrate has changed and no mangrove remnants remain, but former conditions were verified by imagery. Areas of infilled mangroves as

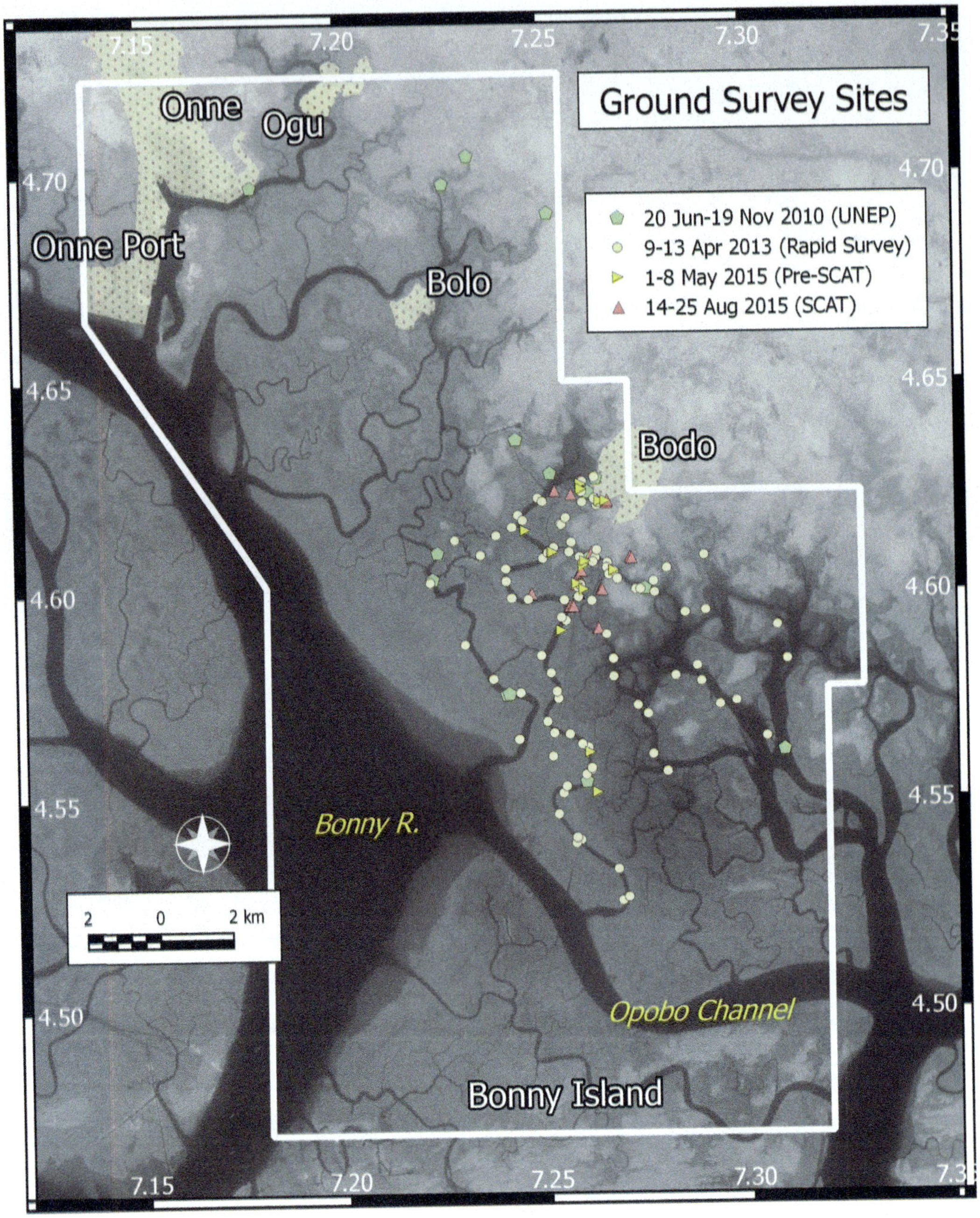

Fig. 13.5 Location of ground survey sites available to assist analysis of mangrove loss

for port or village development are purposely excluded. Within the study area, a single relatively small infilling in the easternmost part of the study area was found and excluded from the analysis.

- Mangrove No Loss – areas with little to no damage to mangrove plant coverage. Tidal flat areas within mangrove habitat having sparse, or in some cases, no plant coverage both during pre- and post-damage conditions, are considered part of this category. In other words, mangrove loss did not occur in areas where no mangroves previously existed.
- Pipeline Corridors – visibly cleared areas through mangrove habitat caused by placement of oil or gas pipelines.
- High Ground – supratidal islands within mangrove areas above the high-tide waterline, commonly dominated by palm trees.
- Villages and towns within and adjacent to mangrove habitat.
- Road – an unfinished sand road within the mangrove area,
- Illegal Refinery – area of illegal refining activity primarily on High Ground within the mangrove area.
- Water – channels and creeks in the mangrove areas. A mid-to-high tidal level is assumed such that intertidal flats

naturally devoid of mangroves are categorized as water and not counted as a former mangrove area. Images showing mid-to-high tidal levels are common and are used to confirm the high tide level,

Some small areas of mangrove loss in the study area are not measured in this study. These include pipeline corridors in former mangrove areas that are underwater during mid- to high tide, dredged slots made for bringing a drilling barge onsite and now contain a wellhead, and still visible seismic survey 'shot' lines, cut through the mangroves approximately 50 years ago for placing explosive charges and measuring subsurface geology for oil exploration. Oiled substrates without former mangroves, as along the shoreline fronting the town of Bodo, are also not included in this assessment.

The process used to define mangrove loss first starts with pre-loss conditions. Two primary pre-loss image sets were used: 3 December 2006 and 26 January 2009. Images prior to this were consulted to ensure that these were the latest before changes to mangrove condition occurred. Because each image does not cover the entire area or impacts occurred later, additional images from 2008 or later in 2009 and 2010 were used to define baseline conditions in certain areas. In the text, specific images are shown to support the analysis and indicate pre- and post-impact mangrove conditions.

From the base condition, mangrove loss was tracked using intermediary images (e.g. 2010, 2011) until 2013 at which time loss was digitized in QGIS using available high quality images. No images were available from 2012. The extent of loss after 2013 was visually determined by comparing 2013 to Google Earth images from 2015 or 2016 and oblique aerial photographs taken on 15 September 2015.

Determining mangrove loss when all mangroves are dead is straight forward to observe and digitize using the GIS. However, determining mangrove condition (alive versus dead) is complicated in cases where dead mangroves are interspersed among living mangroves and vice-versa. In most cases, areas of living mangroves were digitized such that the remaining areas (having mangrove stumps or dead mangroves) are then considered as mangrove loss. Because the dead areas are either specifically digitized, or the result of removing living mangroves, the mangrove loss category is considered to have a high degree of accuracy, estimated to be within 1% of the total area. Because of this methodology, it is more likely that areas characterized as 'Mangrove No Loss' have more dead areas than 'Mangrove Loss' areas have living mangroves. It is also recognized from ground surveys that even areas that appear totally dead from imagery may have some sparsely distributed surviving remnants.

In total, the resultant feature map of the study area was based on the digitizing of over 3800 polygons, each designated by category and having a GIS-calculated surface area. The 'Mangrove No Loss' category contains 1956 units and 1200 are defined as 'Mangrove Loss'. Figure 13.6 shows

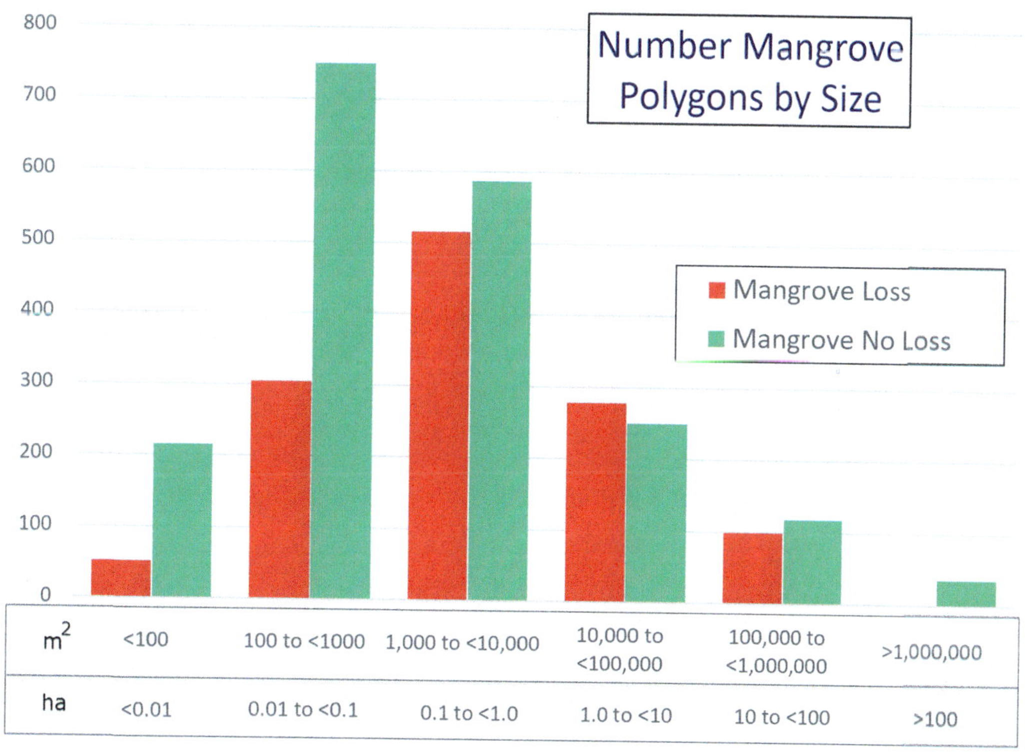

Fig. 13.6 Size of digitized polygons used to determine extent of living (mangrove no loss) and killed mangroves (mangrove loss)

the breakdown of these two categories by polygon size in hectares and square meters. Note there are several hundred areas less than 100 m^2 (e.g. 10 × 10 m) in size, an indication of the detail used during digitizing habitat features. As noted above, there are more polygons related to living mangroves ('Mangrove No Loss') because the focus was ensuring that living areas were excluded from being counted as having been killed.

13.3 Results

13.3.1 Spill Incidents

Records of 74 incidents were found for the study area extending from 1986 to June 2015. Incident locations are primarily along the SPDC pipelines but also include some wellheads (Fig. 13.7). There are likely gaps in the data as

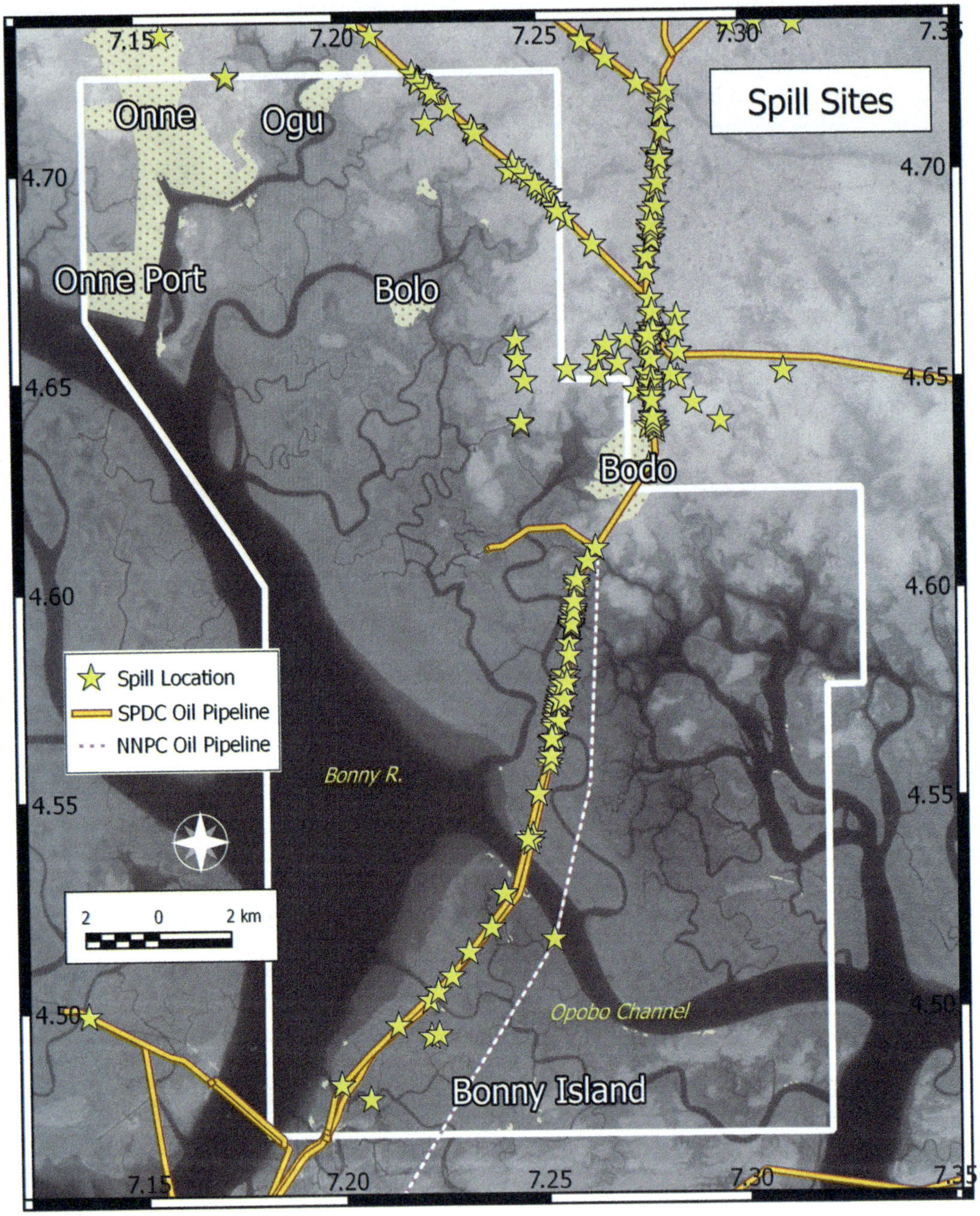

Fig. 13.7 Location of 74 spill incidents found to occur in the study area from 1986 to June 2015. Incidents primarily occur along the pipelines transporting crude oil from the north to Bonny Terminal in the south

Table 13.5 Number of spill records and cause 1986 to June 2015

Year	Incidents	Operational	Illegal	Other
1986–1997	13	1	10	2
1997–2004	0	0	0	0
2005–2007	4	4	0	0
2008	3	2	1	0
2009	2	2	0	0
2010	19	0	19	0
2011	12	0	12	0
2012	1	0	0	1
2013	9	0	9	0
2014	8	0	8	0
2015 (to June)	3	0	3	0

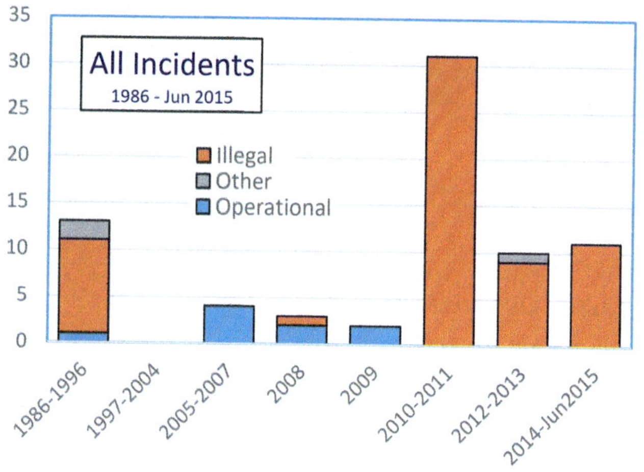

Fig. 13.8 Number of spill events and cause within the study area from 1986 to June 2015. Operational spills include those caused by pipeline or wellhead defects such as caused by corrosion. Illegal activities include those listed in the incident log as 'sabotage', 'illegal bunkering' and/or 'theft'. Other events are those listed as "other" or as 'currently unavailable'

there were no records of events from 1997 to 2004, and incidents in 2012 are abnormally low. Table 13.5 indicates that the vast majority (62) were from illegal activities.

The number of incidents within the study area jumped dramatically in 2010 reaching a high of 19 events. The years 2010–2011 showed 31 events after which the number of events dropped in half for the next 2-year periods, but remained much higher than pre-2010 conditions (Fig. 13.8).

13.3.2 Oil Theft and Illegal Refining

In the Niger Delta, there has been a systemic theft, processing and sale of stolen oil. A brief description of the process as it relates to environmental impacts follows. Images are available from The Atlantic (2013), among other sources.

Oil theft begins when crude oil is extracted from a pipeline actively transporting oil using a 'hot tap' during which the flow in the pipeline continues while a hole and valve are placed into the pipe (Fig. 13.9). The valve controls the pressure and outflow from the high-pressure/high-volume pipeline to a smaller hose for transport to an awaiting boat.

For efficiency, primary hot tapping sites are located close to a creek or channel for access by wooden fishing boats ~10 to 30m in length that will then transport the crude oil to an illegal refinery. The boats are unlined and commonly crude oil is uncontained within the vessel (i.e. open to the elements and not in storage containers). There is a high risk of oil spillage throughout the entire process.

Illegal refining sites are located almost exclusively on palm-dominated supratidal islands within mangrove areas. These sites have a channel for boat access but are above the high-tide water line so that the active refinery is not subject to intertidal flooding.

The operation at the site entails using a metal tank and high-temperature fire to boil off the oil's lighter components, which are then captured for transport and sale in the local community. Heavier oil components and residue are disposed in open pits and onto the surrounding land. Heavy smoke is visible during the distillation process due to the burning of crude oil as fuel to raise the temperature within the distillery tank.

On imagery, there are several identifying features of an illegal refinery site. The site's overall appearance is very black caused by asphaltic tar on the surface of the work site. Disposal pits give the site a mottled appearance. Adjacent mangroves are blackened by oil and likely to be dead, and lastly, an intertidal channel is visible leading to the site (Fig. 13.10). Other small sites In addition to the sites detected in this survey, there are likely to be others hidden under the tree canopy.

Between the time of the UNEP survey in 2010 and 2013, the Nigerian military formed a Joint Task Force (JTF) ostensibly to combat terrorism and oil theft. As observed during the 2013 and 2015 field surveys, the JTF keeps two large, multistory barges for personnel living quarters and several gunboats in the waters south of Bodo and in Opobo Channel. At the JTF barge site south of Bodo, the grounding and partial destruction of numerous fishing boats assumedly used for oil transport, and the lack of smoke from active sites at the time of imagery and surveys in 2013 and 2015, attests to their success in stopping many operations. It was reported that almost 2000 illegal refineries were destroyed in the Niger Delta in 2013, along with the capture of pumps,

Fig. 13.9 Example hot-tap valves from the 12″ Imo River 1 pipeline joint investigation report number 750226 (http://www.shell.com.ng/sustainability/environment/oil-spills.html)

drums of product and boats of various sizes (Information Nigeria 2013). In spite of this effort, there were still at least 11 illegal actions along SPDC pipelines between 2014 and June 2015, including one observed by the author during the August 2015 survey. As an example, the JTF broke up a very large processing center close to Port Harcourt (north of the study area) in September 2015 confiscating 5000 drums and four tanker trucks with illegally refined diesel and 21 large (>15 m) boats (Tori News 2015).

In total, this study found 217 illegal refinery sites having a total area of 115.2 ha. Locations are mostly in the center and northern sections of the study area (Fig. 13.11). Historically, illegal sites have expanded and combined. It is common for two or more small sites observed in 2010 to have merged into a single very large site in 2013.

13.3.3 Presence of Spilled Oil

Oiling of the channels and creeks of the area is widespread in spatial and temporal coverage, affecting both mangrove loss and their potential recovery. Surface water oiling in at least some portion of the study area is visible in all reviewed images and survey photographs from January 2009 to February 2016. Sheen and darker streaks are common. In most cases, surface oiling covers many kilometers of waterway. As an example, the extent of oil observed on surface waters on the 19 November 2009 image is shown in Fig. 13.12. Other examples are discussed later in the text.

13.3.4 Overview of Total Mangrove Loss

This study finds that 4414.6 ha of former mangrove areas were lost due to spilled oil and another 104.7 ha were lost due to corridors for pipeline placement. The measured value for mangrove loss is likely to be greater by a possible 1-to-2% (44–88 ha) due to digitizing limitations associated with dead areas interspersed within living mangroves. The locations of mangrove loss are shown on Fig. 13.13 and are discussed for each area (Bonny Island, Opobo Channel, Bonny River, Bodo West, Bodo East and Bolo) in the following sections.

13.3.5 Bonny Island Mangrove Loss

All of Bonny Island was reviewed for mangrove loss. Results show that losses are concentrated on the northeastern part of the island along the pipeline corridors (Fig. 13.14). The area of mangrove loss is shown in detail in Fig. 13.15, including the locations of 13 spill events that occurred in the area between 1997 and June 2015 (Table 13.6).

As seen on images, the first large-scale loss of mangroves occurred between 29 January 2008 and 19 November 2009 (Fig. 13.16). The spill event of 2 October 2008, recorded as caused by illegal activity, is located in the center of the mangrove loss area and is the most likely cause of damage. Comparing the loss from this spill to the total 2015 loss in the entire Bonny Island area indicates that approximately 80% is attributable to the October 2008 event. The remaining

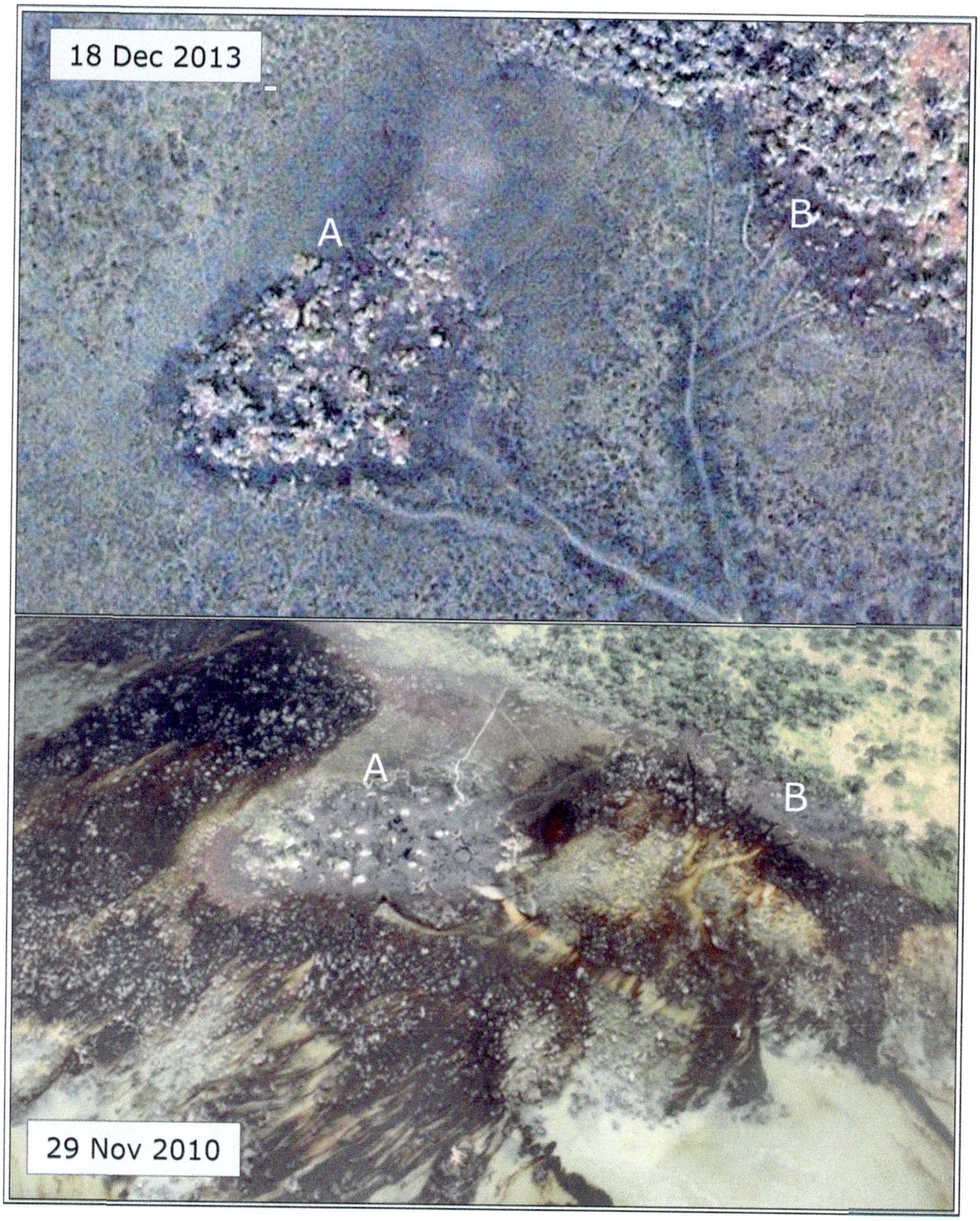

Fig. 13.10 Presence of illegal refineries ('A' and 'B', 4.593 N, 7.270 W) illustrated by blackened and mottled appearance on area above intertidal waters; image from 18 December 2013 and aerial photo by J. Kallnischkies on 29 November 2010

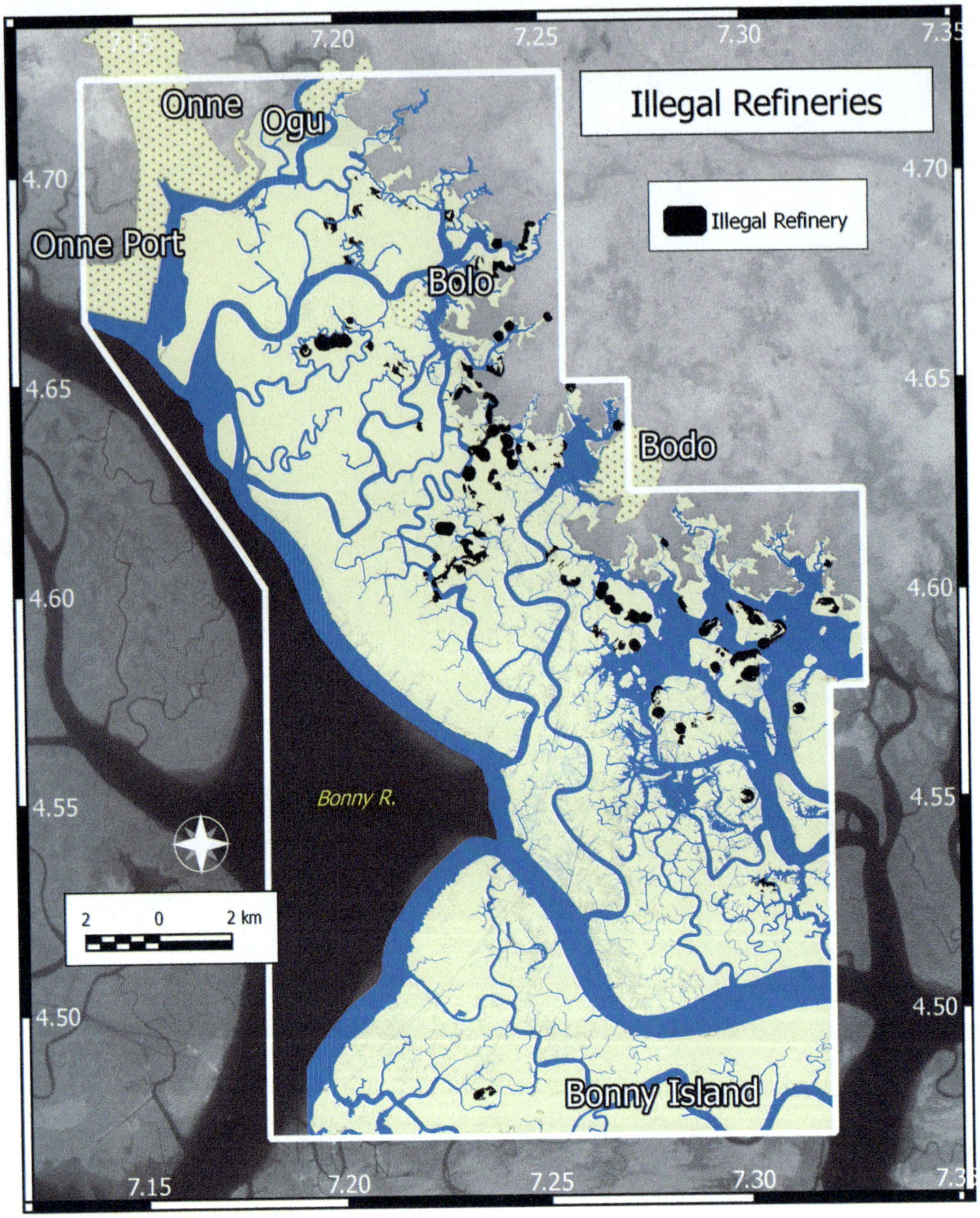

Fig. 13.11 Location of 217 refinery sites found in the study area

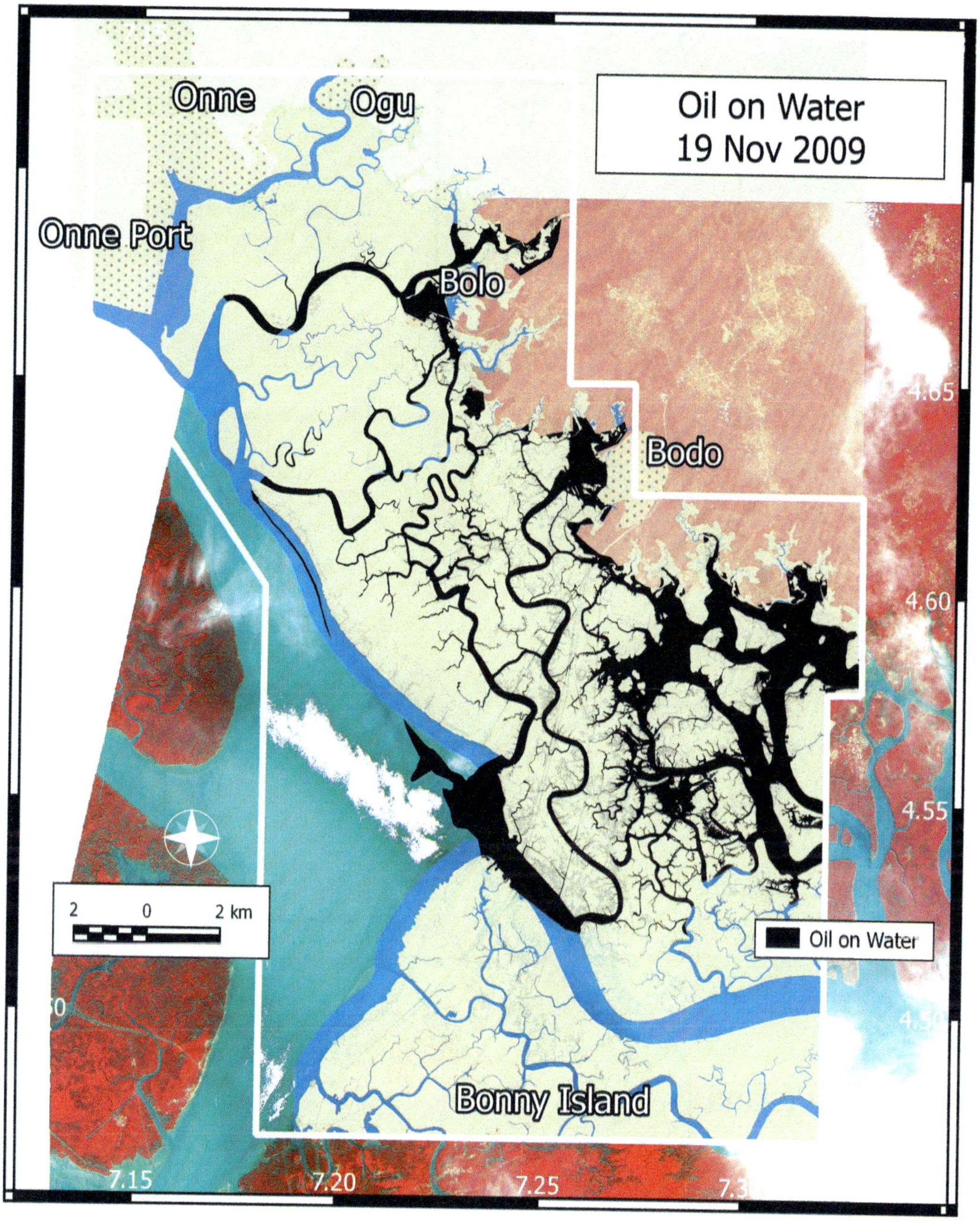

Fig. 13.12 Oiling visible on surface waters on 19 November 2009. The false color SPOT image used for the analysis is shown as background

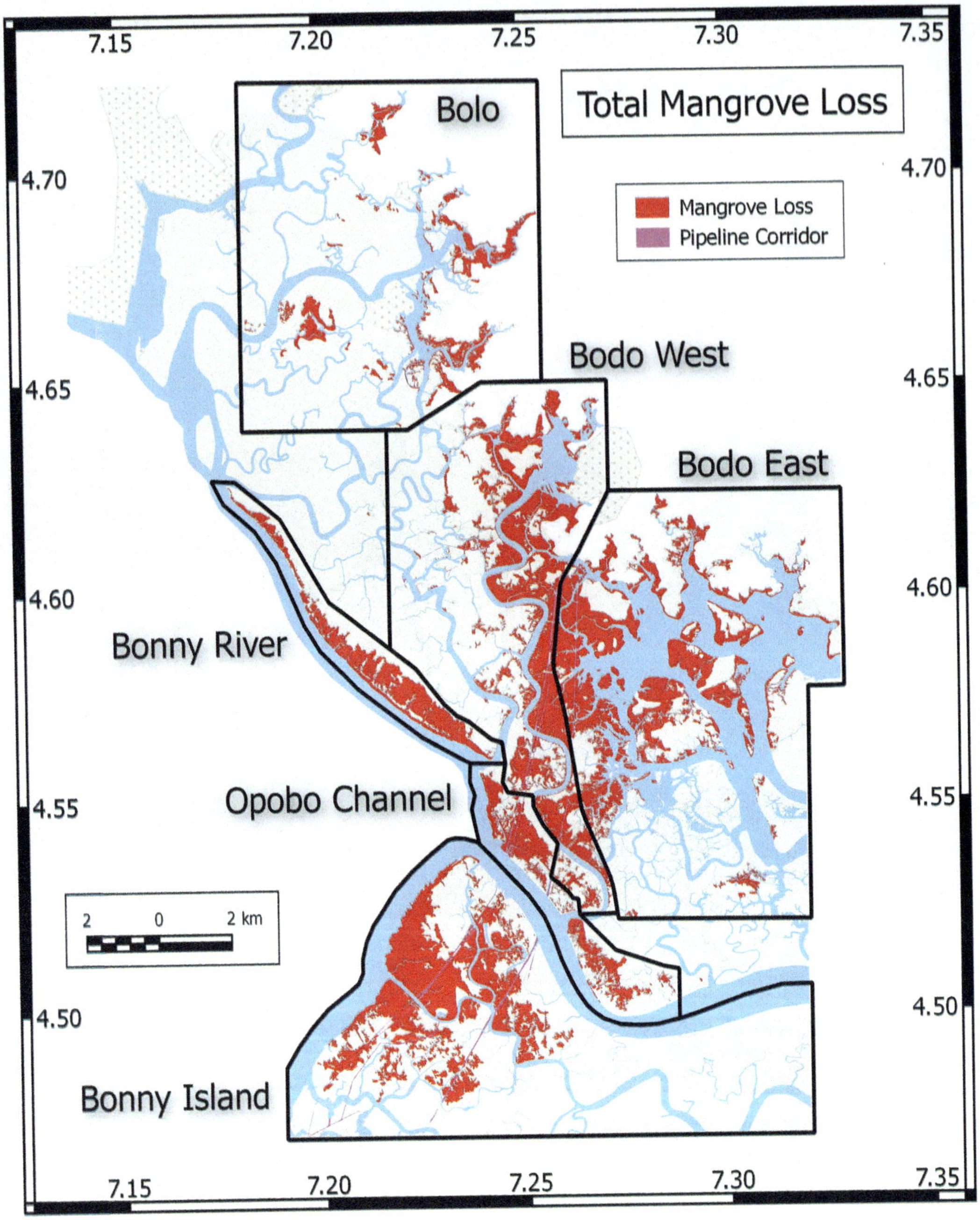

Fig. 13.13 Overview of measured categories of 'Mangrove Loss' and loss due to a 'Pipeline Corridor'. Each of the indicated six areas is shown in detail in the following text

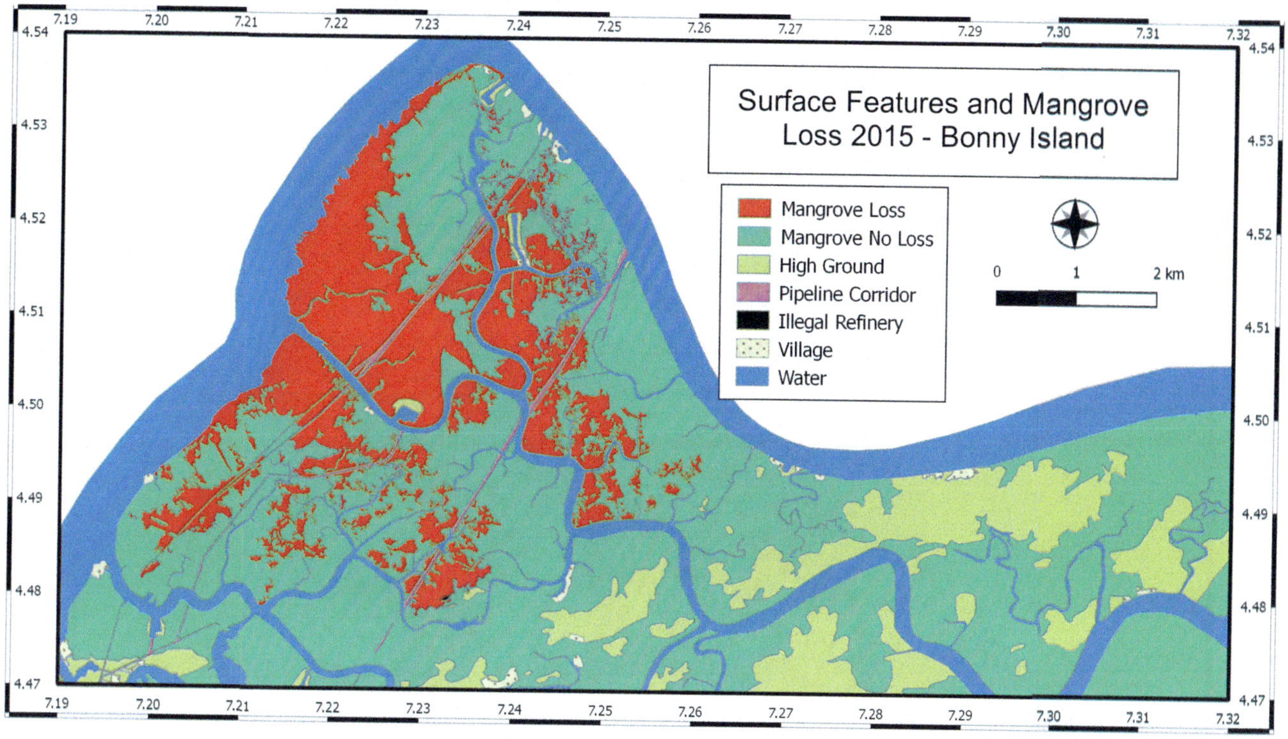

Fig. 13.14 Results of image comparisons to yield the 2015 condition of mangrove habitat and other surface features in the Bonny Island area

mangrove loss in the Bonny Island area is likely from the other spill events listed in Table 13.6.

Overview photographs of the area from 15 September 2015 provide an indication of mangrove loss (Fig. 13.17). Arrows on each photograph indicate mangroves that were alive in 2013 but are dead in 2015, showing that mangrove losses continued past 2013

Illegal refineries are not common in the Bonny Island area. Four were found in higher ground near the pipeline corridor in the south-central part of the study area (Fig. 13.18).

In summary, the majority of damage to the Bonny Island mangroves can be traced to losses following an oil spill on 2 October 2008. Mangrove losses have increased since then, including between 18 December 2013 and 15 September 2015. The total loss of mangroves, including that from pipeline corridors, is shown in Table 13.7.

13.3.6 Opobo Channel Mangrove Loss

Opobo Channel is an intertidal channel to the north of Bonny Island connecting the Bonny River to the Imo River. It contains three oil pipeline crossings and two wellheads in excavated slots along the north and south sides of the channel. The interior of several areas consists of open flats historically devoid of mangroves and are not included as measured mangrove loss. The review of imagery and aerial photos shows major mangrove losses occurred adjacent to the northern two oil pipelines and from the south bank of the Kpador Channel extending to interior mangrove areas to the south (Fig. 13.19).

In the northern area, marked as ‘A’ in Fig. 13.19, there are four recorded spills in the center of mangrove losses, two in 2009 and two in 2010 (Table 13.8).

The time sequence of mangrove loss is shown in Fig. 13.20. The December 1996 image shows no observable impact on mangroves. In August 2009, severe impacts and apparently recent tracks from vehicles are evident, likely caused by response to the spill incident in May 2009. There were no images available between 1996 and 2009, nor were there previous oil spills other than from May 2009. Images from November 2009 and February 2010 show unchanged conditions. However, additional mangrove loss occurred between February 2010 and December 2013, likely influenced by two spills recorded in May and August 2010 (Table 13.8).

The 2013 image also shows tracks from earth-moving equipment over much of the area as well as a conversion of

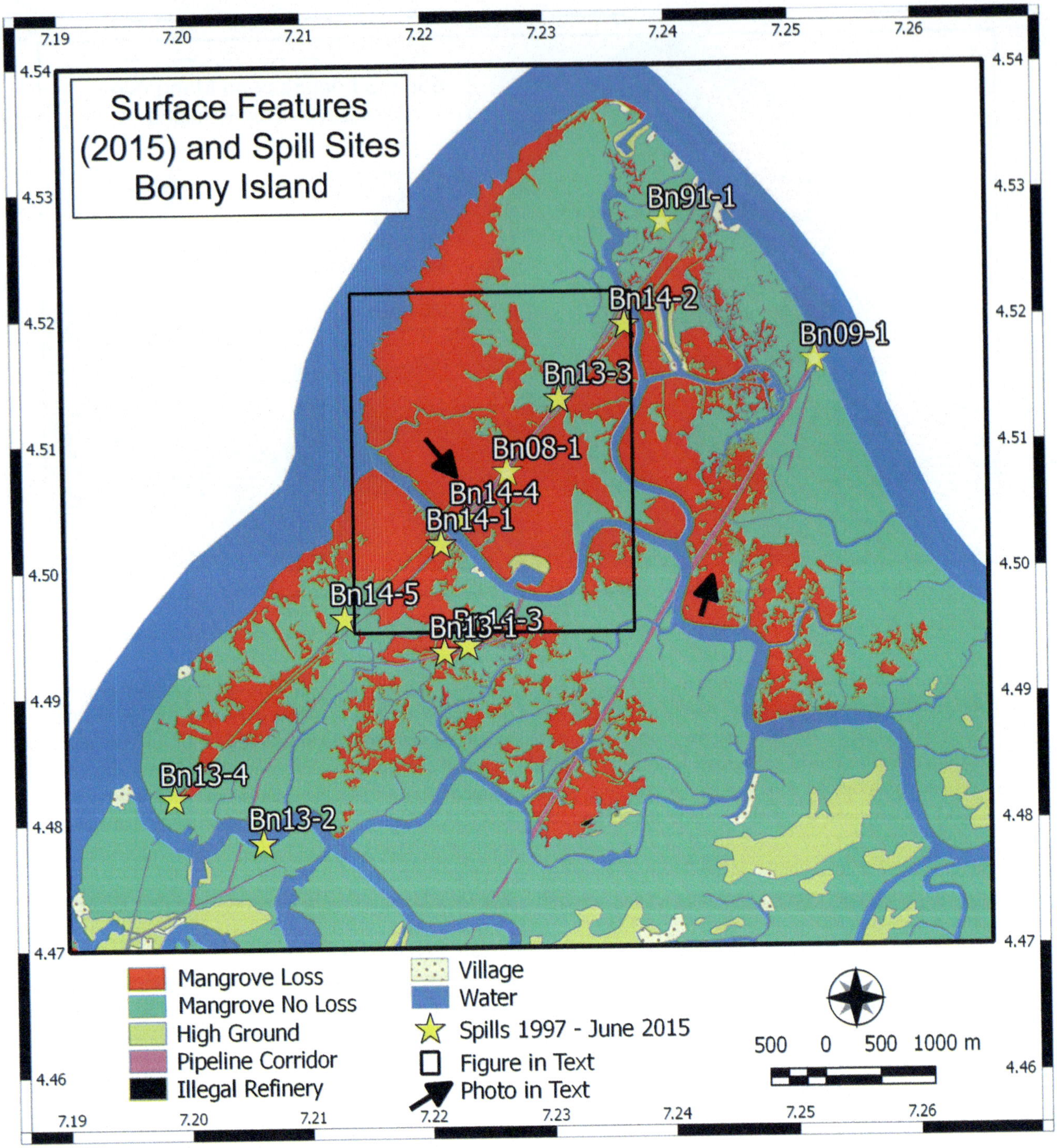

Fig. 13.15 Western portion of the Bonny Island area showing mangrove loss, oil spill sites and other surface features. The *boxed* area is further illustrated in Fig. 13.16. Spill sites (Bn13-1 etc.) are listed in Table 13.6

Table 13.6 Oil spill sites in the Bonny Island area, 1997 to June 2015. Locations are shown on Fig. 13.15

Site	Date	Cause	Latitude	Longitude
Bn97-1	10-Nov-97	Illegal activity	4.5275	7.2397
Bn05-1	23-Dec-05	Illegal activity	4.5275	7.2397
Bn08-1	2-Oct-08	Illegal activity	4.5077	7.2672
Bn09-1	16-Jul-09	Operational	4.5165	7.2520
Bn13-1	4-Apr-13	Operational	4.4934	7.2214
Bn13-2	17-Jul-13	Illegal activity	4.5275	7.2399
Bn13-3	27-Jul-13	Illegal activity	4.5134	7.2310
Bn13-4	19-Sep-13	Illegal activity	4.4820	7.1988
Bn13-5	23-Sep-13	Illegal activity	4.5020	7.2212
Bn14-1	16-Mar-14	Operational	4.5193	7.2365
Bn14-2	15-Aug-14	Operational	4.4939	7.2233
Bn14-3	22-Nov-14	Illegal activity	4.5077	7.2268
Bn14-4	24-Nov-14	Illegal activity	4.5041	7.2232

the former mangrove habitat to compact silt and clay devoid of mangrove remnants. After December 2013, four Google Earth images available to February 2016 essentially illustrate the same mangrove condition as in 2013. The influence of the 2009 and 2010 spills appears localized to the northern part of Opobo Channel, causing approximately 25% of the total damage to the area.

Opobo Channel area 'B' (in Fig. 13.19) is along the southern part of the channel and shows a later appearance of initial mangrove loss. Some minor impact occurred at the upper reaches of small tidal channels by February 2010, but the major change occurs between February 2010 and December 2013 (Fig. 13.21). Oiling seems likely to have originated from a combination of sources including the spill sites previously mentioned at area 'A', from Bonny Island, and from oil coming down the Kpador Channel from numerous spill sites in the north, discussed later as part of Bodo West. Floating oil then likely entered the small creeks of this area and impacted the interior mangroves.

The extent of mangrove loss is visually evident by comparing aerial photographs from 2000 and 2015 from the shoreline northwest of Opobo Channel area 'A' (Fig. 13.22).

Table 13.9 summarizes the extent of mangrove loss for the Opobo Channel area. Losses in the north are likely due to 2009 and 2010 spill incidents that resulted in large-scale losses observed in August 2009 and which increased further between 2010 and 2013. In the southern part of the area, losses cannot be tied to a specific spill event and occurred later (after February 2010). Floating oil from other areas, particularly Bodo West via the Kpador Channel seems plausible. No illegal refinery sites are found in the area due to its low-lying nature, i.e. flooded during high tides.

13.3.7 Bonny River Mangrove Losses

This area extends to the north of Bonny Island and Opobo channel along the eastern shore of the Bonny River (Fig. 13.23). The pattern of losses shows a strong decrease to the north, indicating that the origin of damages was to the south and is related to spill events discussed previously on Bonny Island and along the pipelines of Opobo Channel. The area consists of low-lying silt and clay sediments that rise gradually to a higher interior ridge.

The Bonny River area shoreline was historically covered entirely by mangroves. The sequence of mangrove loss is shown in Fig. 13.24. The first mangrove losses, varying between widths of 50 and 100 m, occurred along the outer fringe facing the Bonny River between January 2008 and June 2009. (The 8 June 2009 image is not presented but is essentially the same as November 2009). No additional losses were observed on a later image from 7 February 2010. Major additional losses then occurred by 2013 extending the zone of dead mangroves 800 m landward from the coast. Additional Google Earth images from January 2014 and February 2016 indicate no further obvious losses to mangroves after 2013.

In summary, the extent (greater in the south) and timing of the first losses is related to spills that occurred in areas to the south, including those in October 2008 and July 2009 on Bonny Island and along the north shore of Opobo Channel between May 2009 and August 2010. After 2010, losses cannot be tied to a specific event, but may have included oiling coming from nearby Opobo Channel or Bonny Island. Table 13.10 provides the summary of measured mangrove losses in the Bonny River area. Illegal refineries are not present as the area is low-lying and lacks access channels and proximity to an oil source.

13.3.8 Bodo West Mangrove Losses

The area around and to the south of Bodo is divided by an incomplete road into two areas, east and west. The road was partially constructed using dredged sand but never finished. The Bodo West area extends south to the Opobo Channel and Bonny River areas (Fig. 13.25). Transport between Bodo East and Bodo West is limited to a 26 m channel in the north and a 12 m channel in the south. The northern channel has a metal barge placed across it and preventing large watercraft from passing from one side to the other. Road construction destroyed mangroves within its path, but as there is water circulation from each side, appears not to have caused additional damage to mangrove habitat. Because

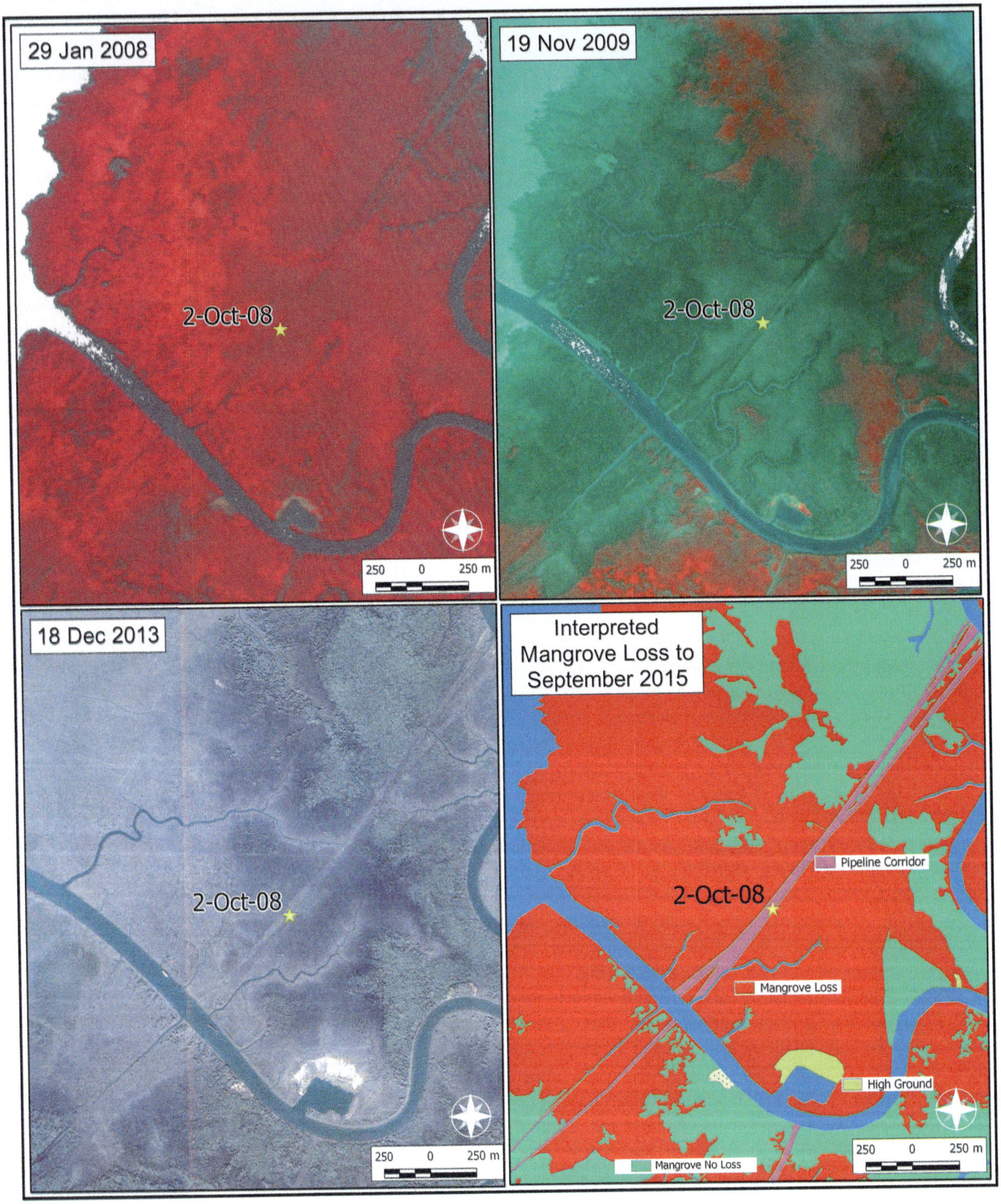

Fig. 13.16 Historic sequence at the 2 October 2008 spill site (located in the *boxed* area of Fig. 13.15) showing false color SPOT5 satellite imagery from 2008 to 2009 (before and after the spill), Google Earth imagery from 2013 and interpreted results of mangrove loss extended to September 2015 using results of the 2015 aerial survey

Fig. 13.17 Aerial photographs from 15 September 2015 of two sites in the Bonny study area showing the condition of former mangrove habitat. See Fig. 13.15 for location. *Arrows* indicate areas were alive in the image from 18 Dec 2013 but are dead in 2015

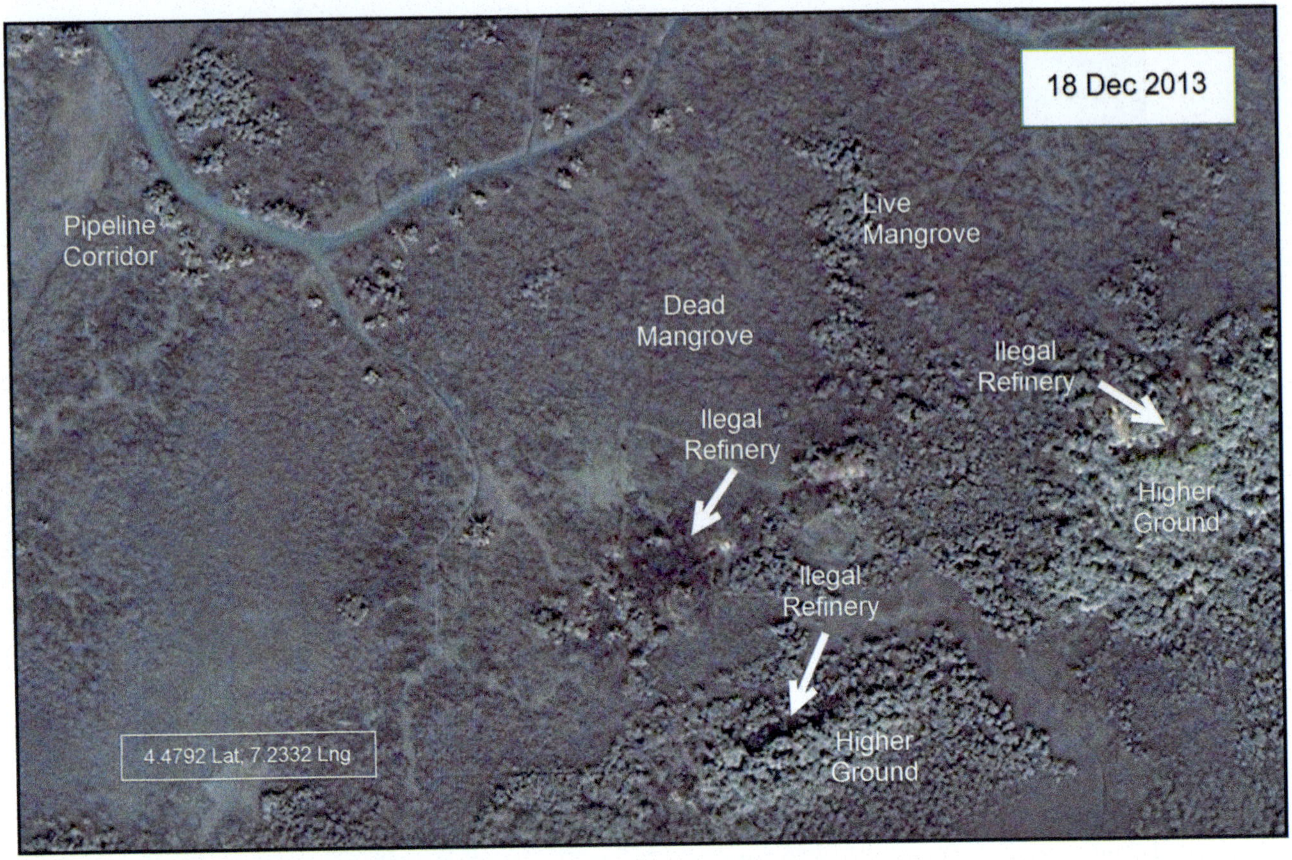

Fig. 13.18 Illegal refineries and associated habitat condition on Bonny Island, 18 Dec 2013

Table 13.7 Measured mangrove losses and illegal refineries on Bonny Island, 2008–2015

Component	Hectares	Number
Mangrove loss	1058.2	
Pipeline corridor	58.2	
Total mangrove loss	1116.4	
Illegal refineries	0.8	4

there is limited water circulation from one side to the other, there is a different oil-impact history on each side.

Bodo West has ten wellheads connected via inactive pipelines to a central gathering flow station and then to a main flow station/oil manifold in the north-central part of the area. This part of the system is out-of-service although UNEP (2011) reports that it has not been decommissioned. Two pipelines (24-in. and 28-in.) cross the area north-to-south and actively transport crude to Bonny Terminal.

The first mangrove losses occurred in 2009 and are attributable to 2008 operational leaks in areas ‘A’ and ‘B’ on Fig. 13.25. A summary of spill incidents from 1986 to June 2015 are shown for the Bodo West area in Table 13.11. Prior to 2010, recorded incidents were uncommon (ten during the period 1986–2010).

Figure 13.26 shows the historic sequence and location of spill incidents from 2008 to June 2015. From 2010 onwards, discharges from illegal activities became numerous and prevalent. Pipeline tapping to provide crude oil for the illegal refineries was common in areas of access to the Kpador Channel, the primary waterway leading to Bodo. A total of 87 illegal refinery sites were detected in the Bodo West area.

The operational spills of 2008, located within Bodo West areas ‘A’ and ‘B’, initiated mangrove losses in the area. Area ‘A’ is shown in Fig. 13.27. The loss is attributable to an operational leak that occurred in late 2008. There is debate on how long it took to stop the leak but it is likely that discharge from the leak continued for many weeks if not months (e.g. Guardian 2011; AI-CEHRD 2011).

The satellite image from 26 January 2009 shows visible oiling of sediments and waters in the area of the 2008 pipeline break (Fig. 13.28). Additional imagery is illustrated in AI-CEHRD (2011).

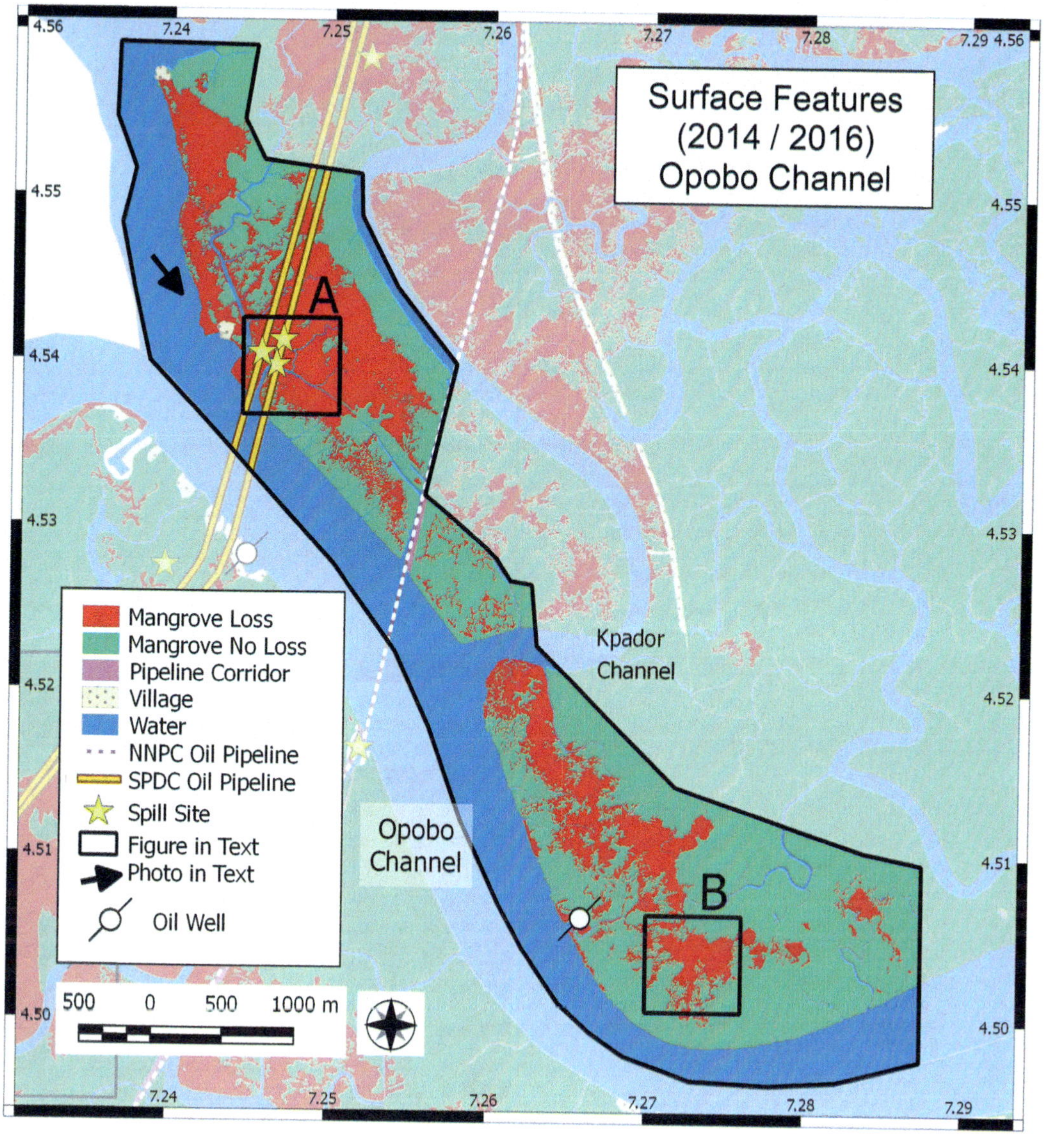

Fig. 13.19 Mangrove condition and surface features of the Opobo Channel area

Table 13.8 Oil spill sites in the Opobo Channel area, 1997 to June 2015

Date	Cause	Latitude	Longitude
8-May-09	Illegal activity	4.5412	7.2470
16-Oct-09	Operational	4.5397	7.2466
18-May-10	Illegal activity	4.5397	7.2466
5-Aug-10	Illegal activity	4.5404	7.2456

A comparison of the same area in 2000 (before spills) and 2013 (after multiple spills) illustrates the incurred extent of mangrove loss (Fig. 13.29).

Another operational pipeline discharge occurred in 2008 in Bodo West area 'B'. Like that illustrated previously for Bonny Island, the difference in mangrove coverage before and after is striking and centered at the spill site (Fig. 13.30). The spill occurred on 7 December 2008. The image from January 2008 shows no impact (before the spill) while that from August 2009 shows a large impact area centered close to the leak location. A post-impact ground photograph of the area is shown in Fig. 13.31.

In the Bodo West area, most mangrove loss is attributable to the 2008 spills. By 2009, by far the majority of mangrove loss seen in 2013 had already occurred. However, if mangroves did survive the first wave of impacts in 2008/2009, further losses occurred in response to illegal pipeline tapping and resulting oil spillage.

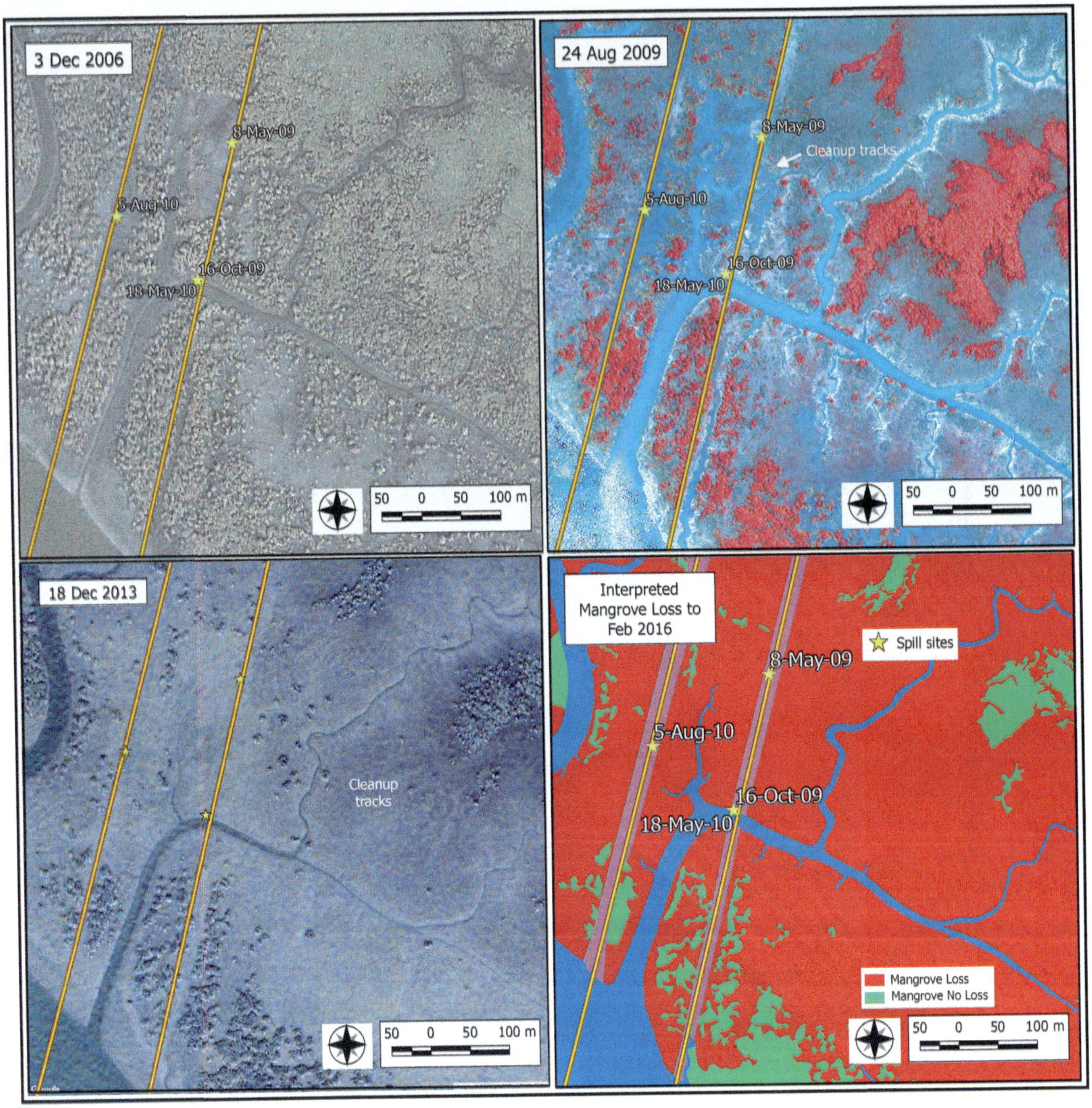

Fig. 13.20 Time sequence of imagery from Opobo Channel area 'A' (in Fig. 13.19) showing major impacts in August 2009 with additional losses noted in 2013. Google Earth imagery to February 2016 shows essentially the same conditions as in 2013, so the interpretation of mangrove loss is extended to that date

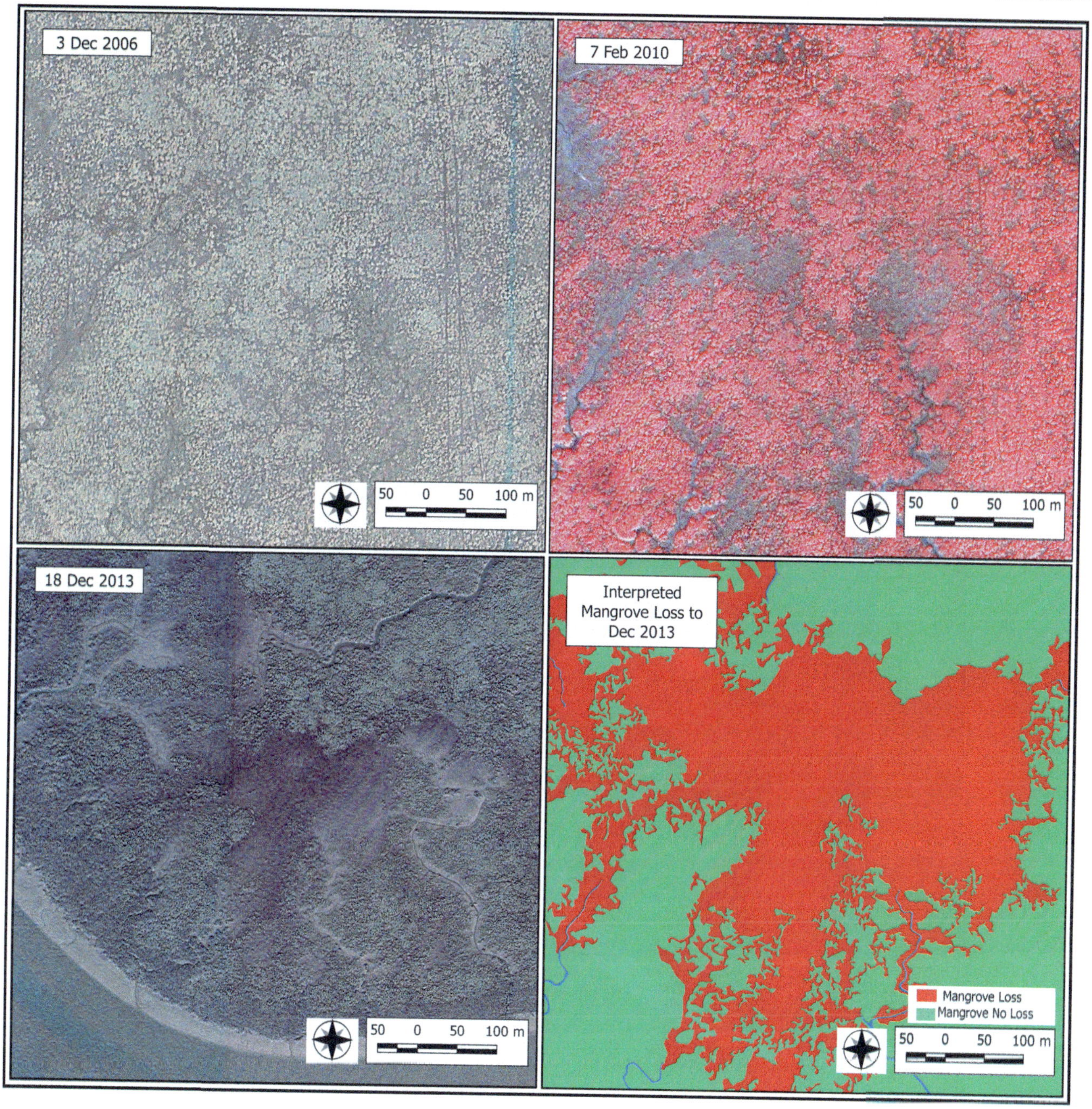

Fig. 13.21 Sequence of mangrove losses at Opobo Channel area 'B' in Fig. 13.19, showing some minor impact by February 2010 and major loss by December 2013

Fig. 13.22 Aerial photographs showing change in mangrove cover from 2000 to 2015. Site location is shown in Fig. 13.19. A coastal village and *arrows* indicate the same locations on each photograph

Table 13.9 Measured mangrove losses and illegal refineries in the Opobo Channel area, 2006 to 2013/2015 depending on image coverage

Component	Hectares	Number
Mangrove loss	280.2	
Pipeline corridor	9.3	
Total loss	289.5	
Illegal refineries	0	0

An example is provided by Bodo West area 'C' (Fig. 13.32). By 2009, most of the mangroves are killed, but some areas as along the western-most pipeline initially survived (top right image) but were subsequently lost by 2013. Imagery shows that the additional loss occurred between 7 Feb 2010 and 8 January 2011. A total of 15 spills (14 illegal, 1 unknown) are reported in this particular area between 2010 and June 2015. Two spills occurred in 2010,

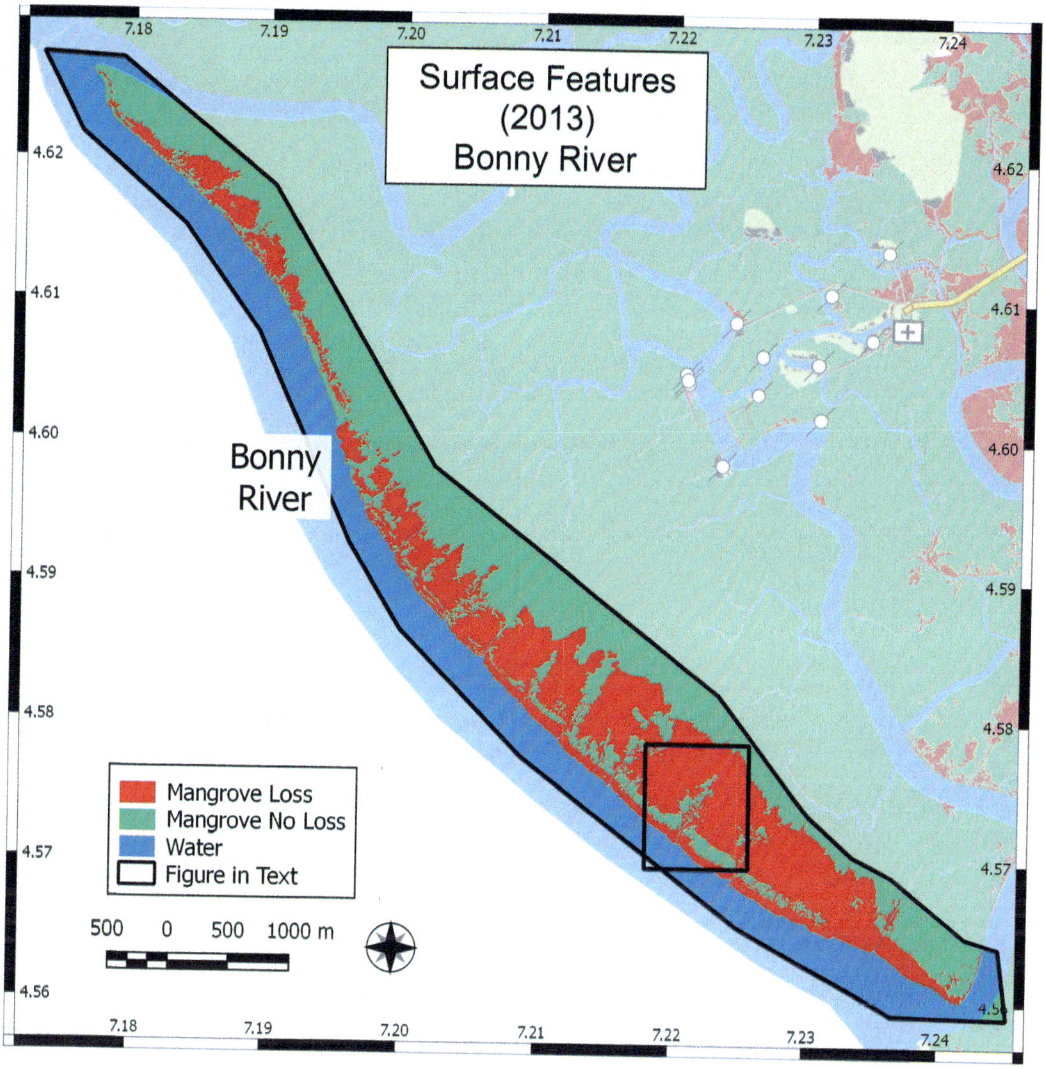

Fig. 13.23 Condition of mangroves in the Bonny River area of study in 2013

one of which is in the specific area where mangroves were lost between 2010 and 2011. No spills are reported in this area prior to 2010.

Bodo West area 'D' similarly shows major impacts originating in 2008 but worsening between 2010 and 2011, followed by additional losses by 2013 (Fig. 13.33). Oil is visible on the water in each image (2010, 2011 and 2013) indicating that the additional losses are from sources causing the chronic oiling of surface waters and not the two major spills of 2008.

In November 2010, the UNEP overflight captured extensive and heavy oil concentrations on the water in Bodo West area 'D' (Fig. 13.34).

In conclusion, major oiling and subsequent mangrove loss in the Bodo West area occurred because of two pipeline leaks caused by corrosion in late 2008. No leaks were reported for the area in 2009, after which numerous spill events almost exclusively caused by illegal activities began in 2010 and continuing at least until 2015. As noted in 2013 imagery, the events between 2010 and 2013 caused some additional mangrove losses over that observed from the 2008 spills. The oil observed in the water after 2009 was likely due to the combination of pipeline leaks at the source of illegal pipeline taps as well as from illegal refining activity and associated losses during the illegal transport of crude and refined product by

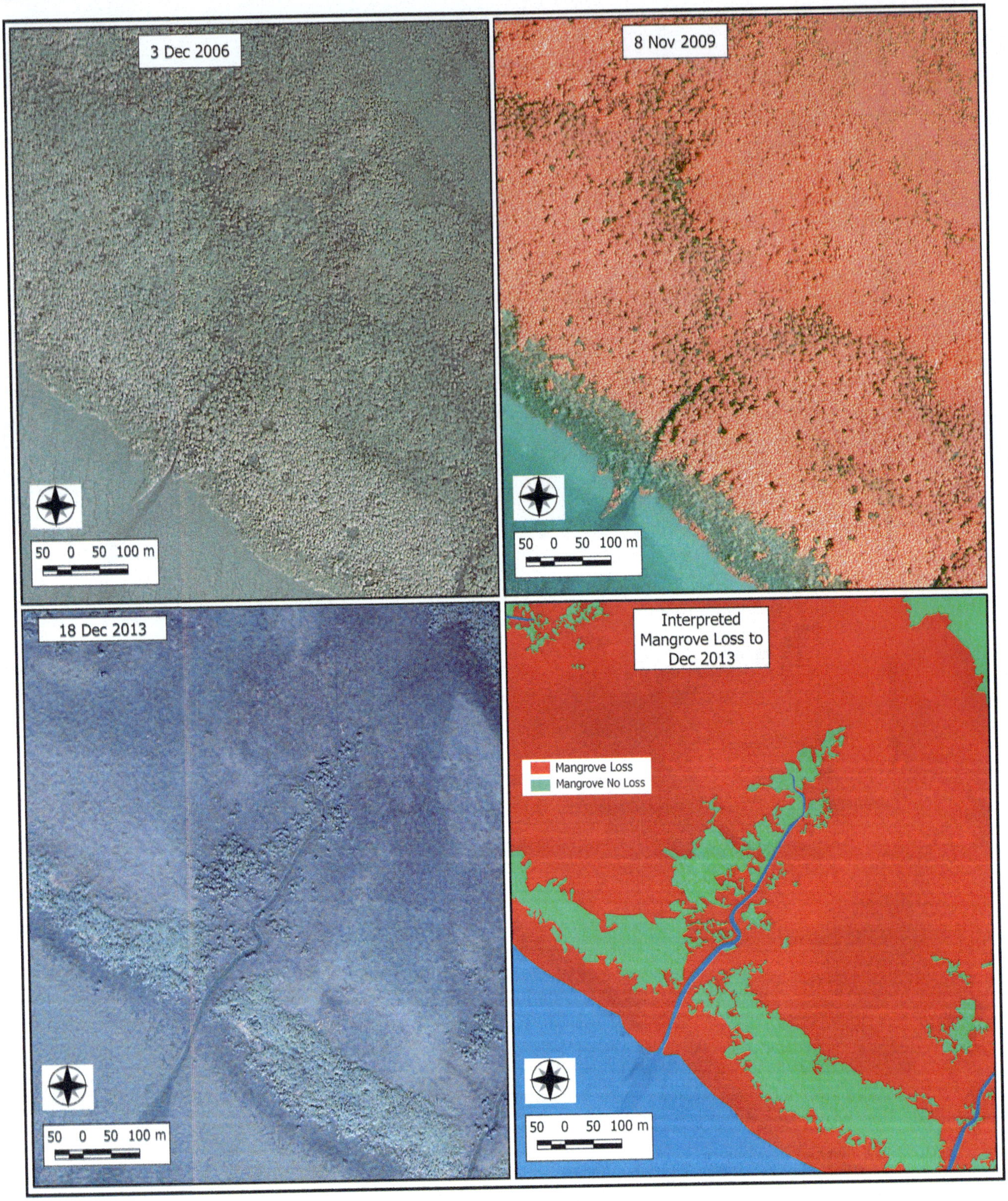

Fig. 13.24 Bonny River sequence of mangrove losses from 2006 to 2013. Other imagery shows that impacts had occurred along the outer fringe by June 2009 and was the same in February 2010. Therefore, the extensive losses shown in 2013 occurred after February 2010

Table 13.10 Measured mangrove losses and illegal refineries in the Bonny River area, 2008–2013. Losses from pipeline corridors are related to placement of a gas line

Component	Hectares	Number
Mangrove loss	464.9	
Pipeline corridor	1.7	
Total loss	466.6	
Illegal refineries	0	0

boats to and from the refining site. The summary of mangrove loss and illegal refineries is shown in Table 13.12.

13.3.9 Bodo East Area Mangrove Loss

Bodo East is separated from Bodo West by a narrow 26 m channel in the north and a smaller channel to the south. Oil

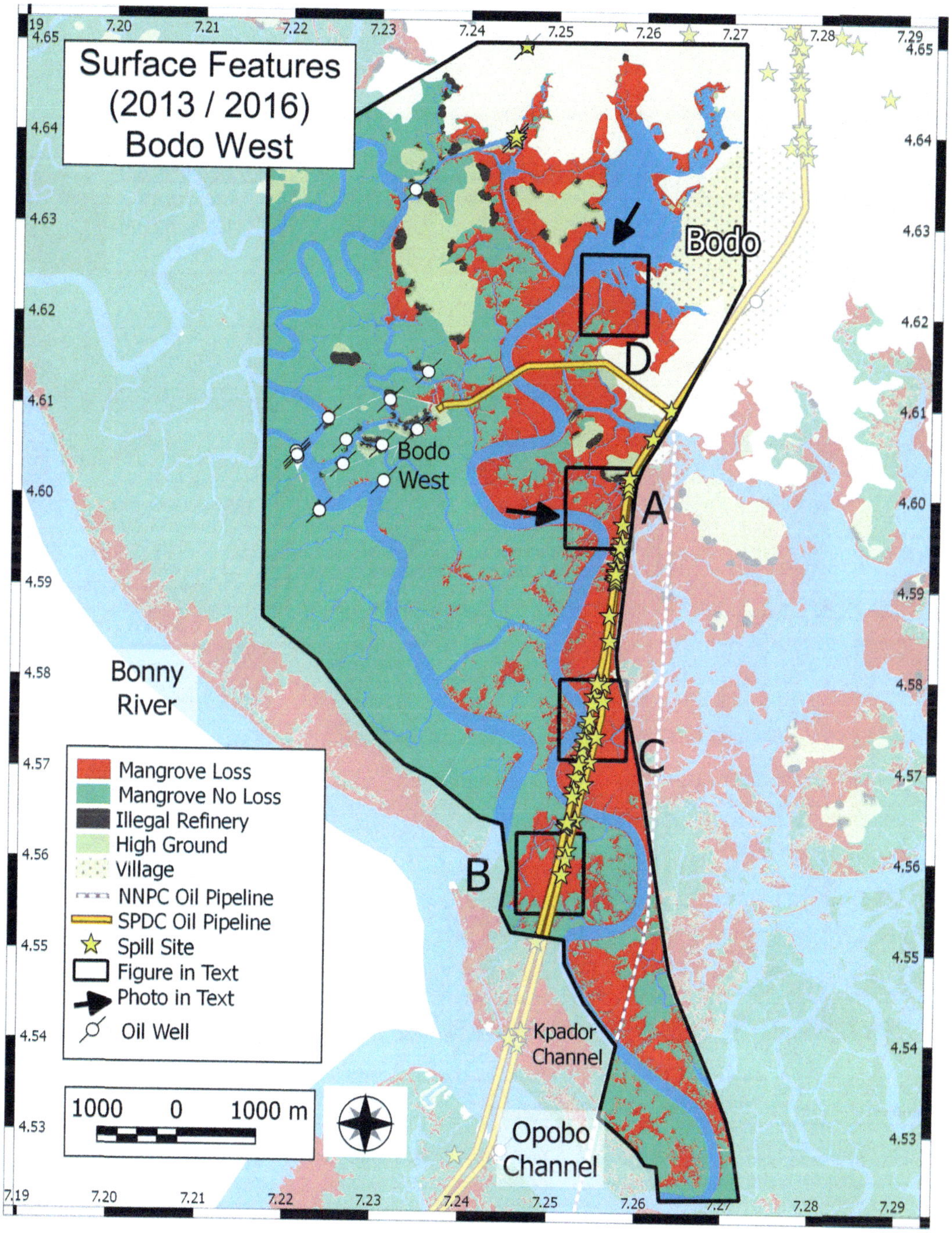

Fig. 13.25 Surface features and mangrove loss in the Bodo West area. The spill sites shown occurred between 2008 and June 2015. Illegal refineries are present in area 'A' and north, and around and to the north of oil wells in the Bodo West oil field

Table 13.11 Spill incidents and cause from 1986 to June 2015 in the Bodo West area. Two of the operational spills in 2008 continued into 2009

Year	Incidents	Operational	Illegal	Other	Location
1986–1997	6	1	3	2	Bomu oil wells
1997–2004	0	0	0	0	No incidents recorded
2005–2007	1	0	1	0	Oil pipelines to bonny
2008	3	2	1	0	Oil pipelines to bonny
2009	0	0	0	0	No incidents recorded
2010	16	0	16	0	Oil pipelines to Bonny
2011	12	0	12	0	Oil pipelines to Bonny
2012	1	0	0	1	Oil pipelines to Bonny
2013	9	0	9	0	Oil pipelines to Bonny
2014	8	0	8	0	Oil pipelines to Bonny
To June 2015	3	0	3	0	Oil pipelines to Bonny

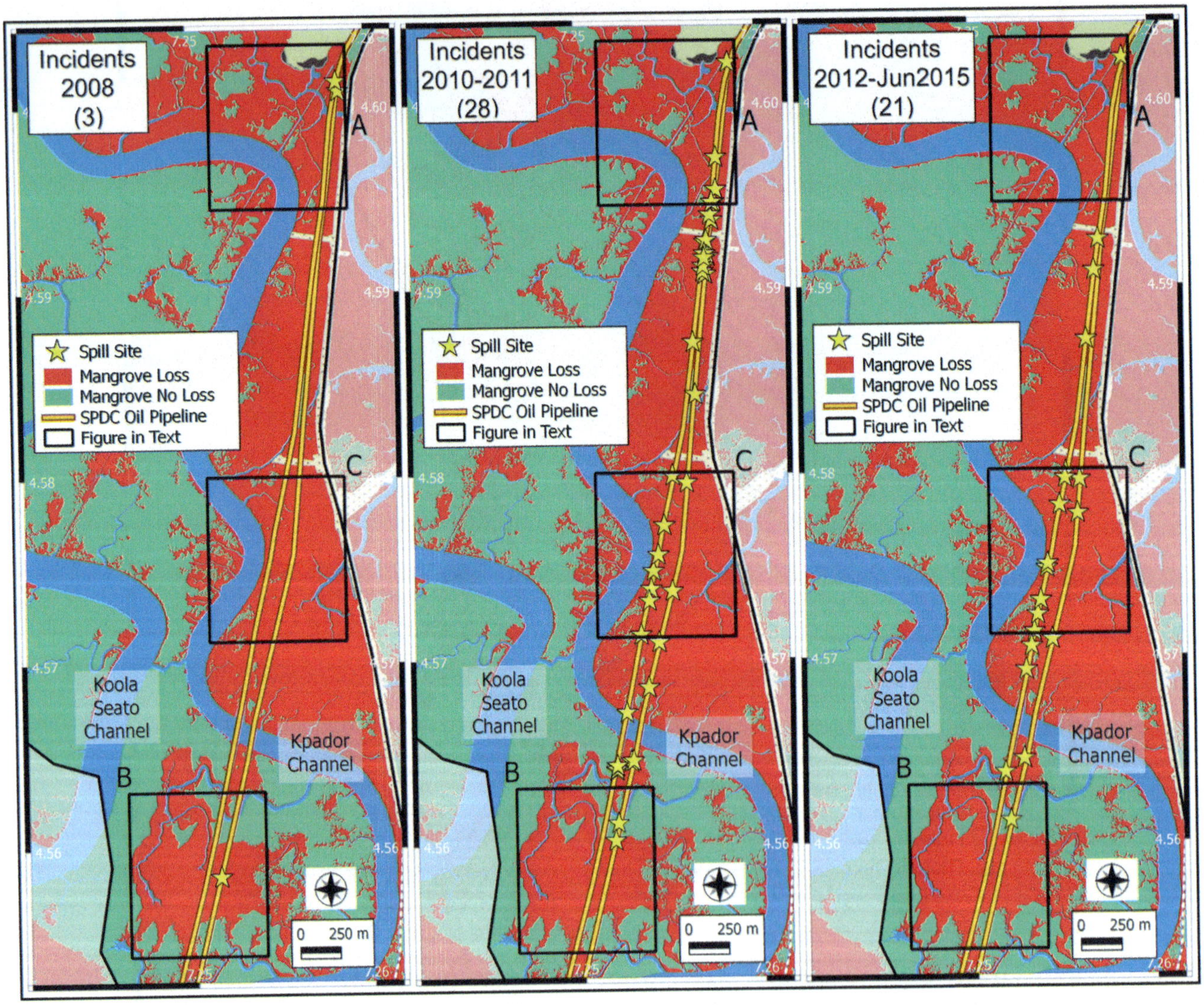

Fig. 13.26 Location of spill incidents in Bodo West 2008 to June 2015 shown for three time intervals

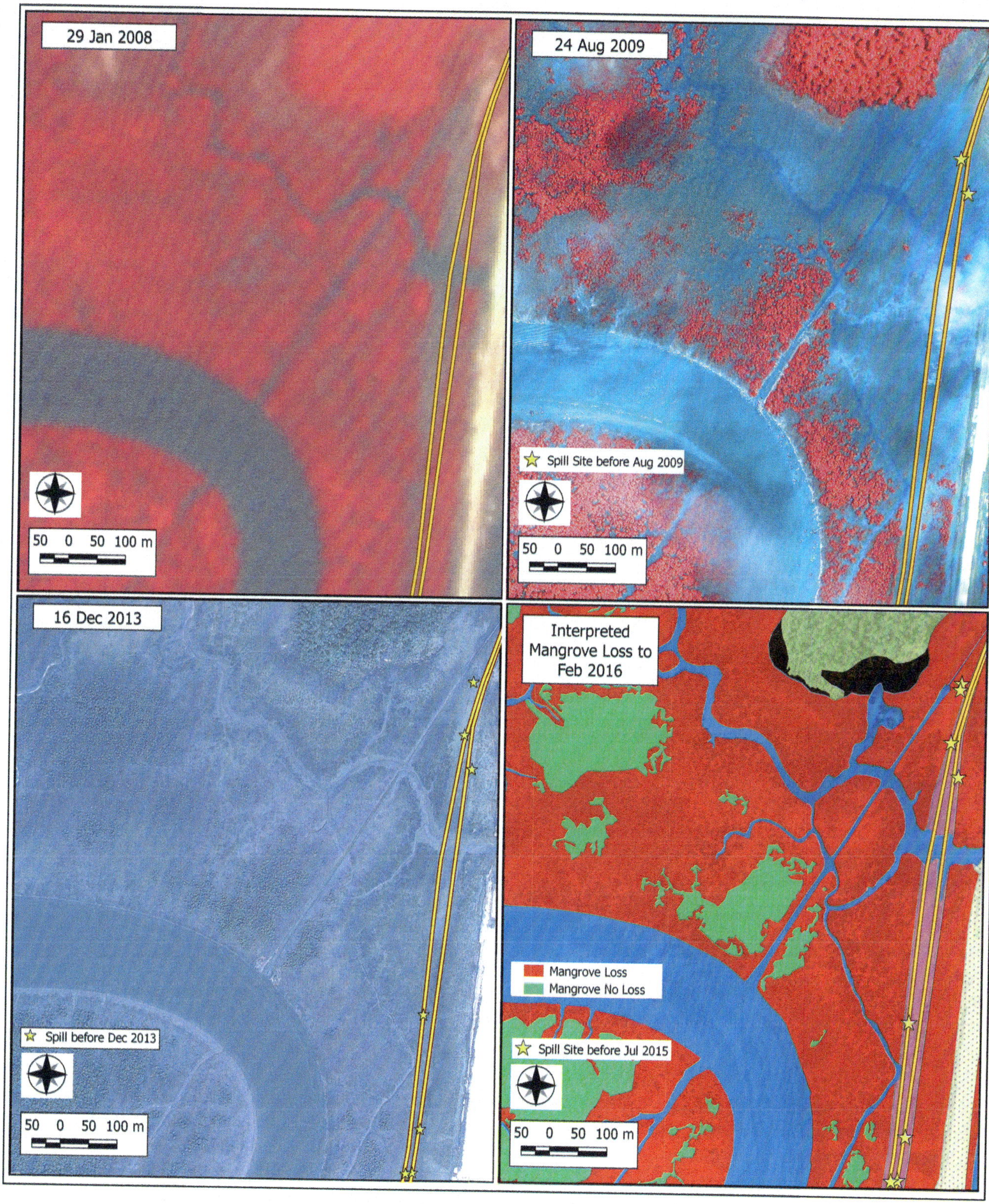

Fig. 13.27 Historic sequence of mangrove loss in Bodo West area 'A' beginning with an operational leak in the pipeline in late 2008 that resulted in large losses visible in 2009. Additional losses in 2013 are evident, after which losses stabilized to 10 February 2016. Site location is shown in Fig. 13.25

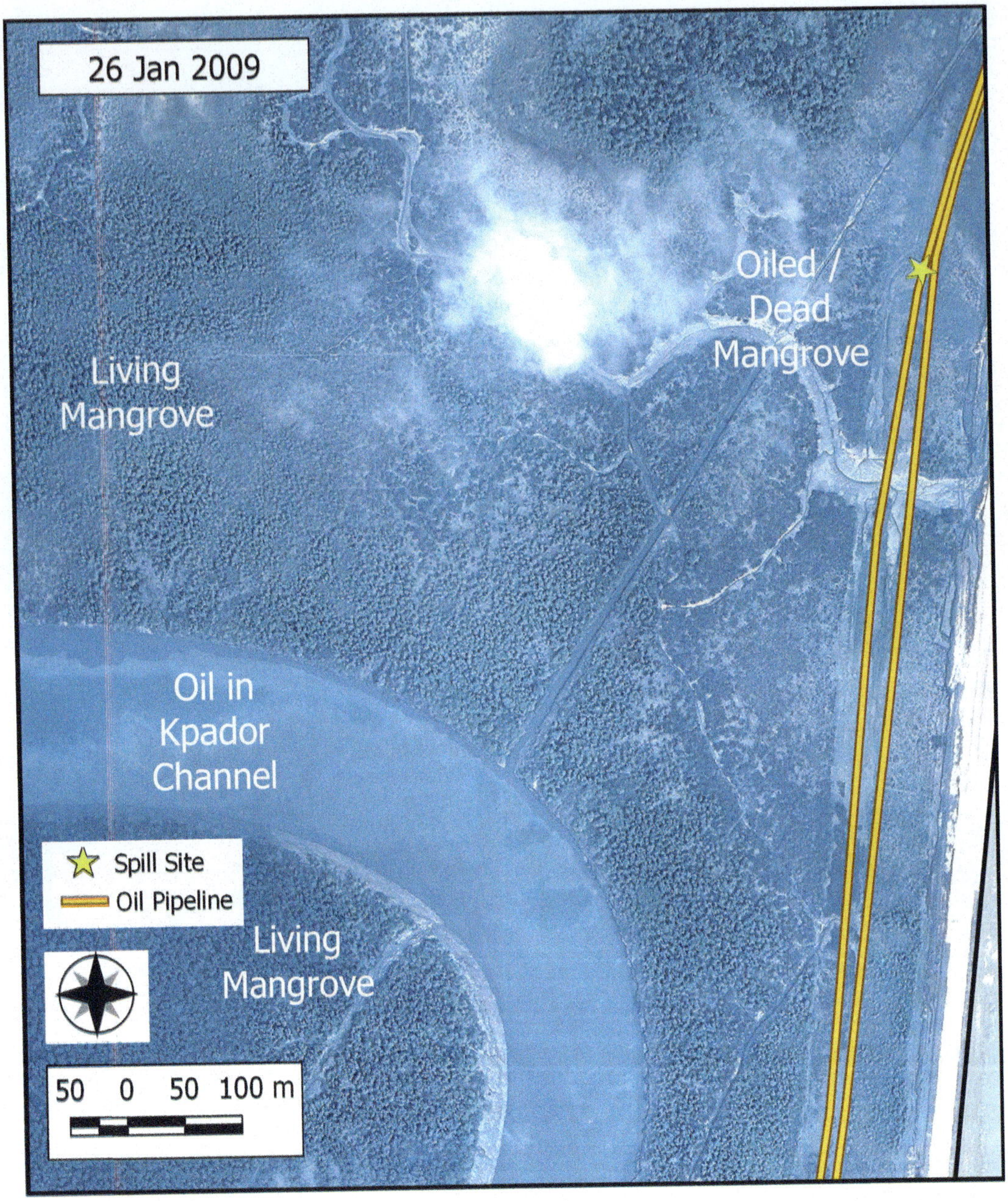

Fig. 13.28 Satellite image of area 'A' (Fig. 13.25) from 26 January 2009 showing mangrove condition and oiling after the operational spill from the indicated spill site

infrastructure is limited to an NNPC oil transport pipeline transiting from the north to the Bonny export terminal in the south. There are no wellheads or associated small pipelines in the area. The details of operation and leaks from the NNPC pipeline are unknown, although one spill (Bn09-1, Fig. 13.15) is reported at its entry point to Bonny Island far to the south. Surface features and mangrove loss in Bodo East are shown in Fig. 13.35.

Initial and major mangrove losses in Bodo East occurred between 24 August 2009 and 19 November 2009 (Fig. 13.36) and cannot be attributed to a specific event or events. Bodo West impacts occurred earlier and were already visible in January 2009. Imagery presented in the figure also shows that not all mangroves were killed by 10 February 2010, but all were lost within 11 months later on 8 January 2011. Images from 2013 to August 2015 show mangrove loss essentially the same as observed in 2013.

There are two possible sources for the damage to mangroves in Bodo East: (1) discharges from illegal refineries and (2) unreported pipeline leak(s). Review of the

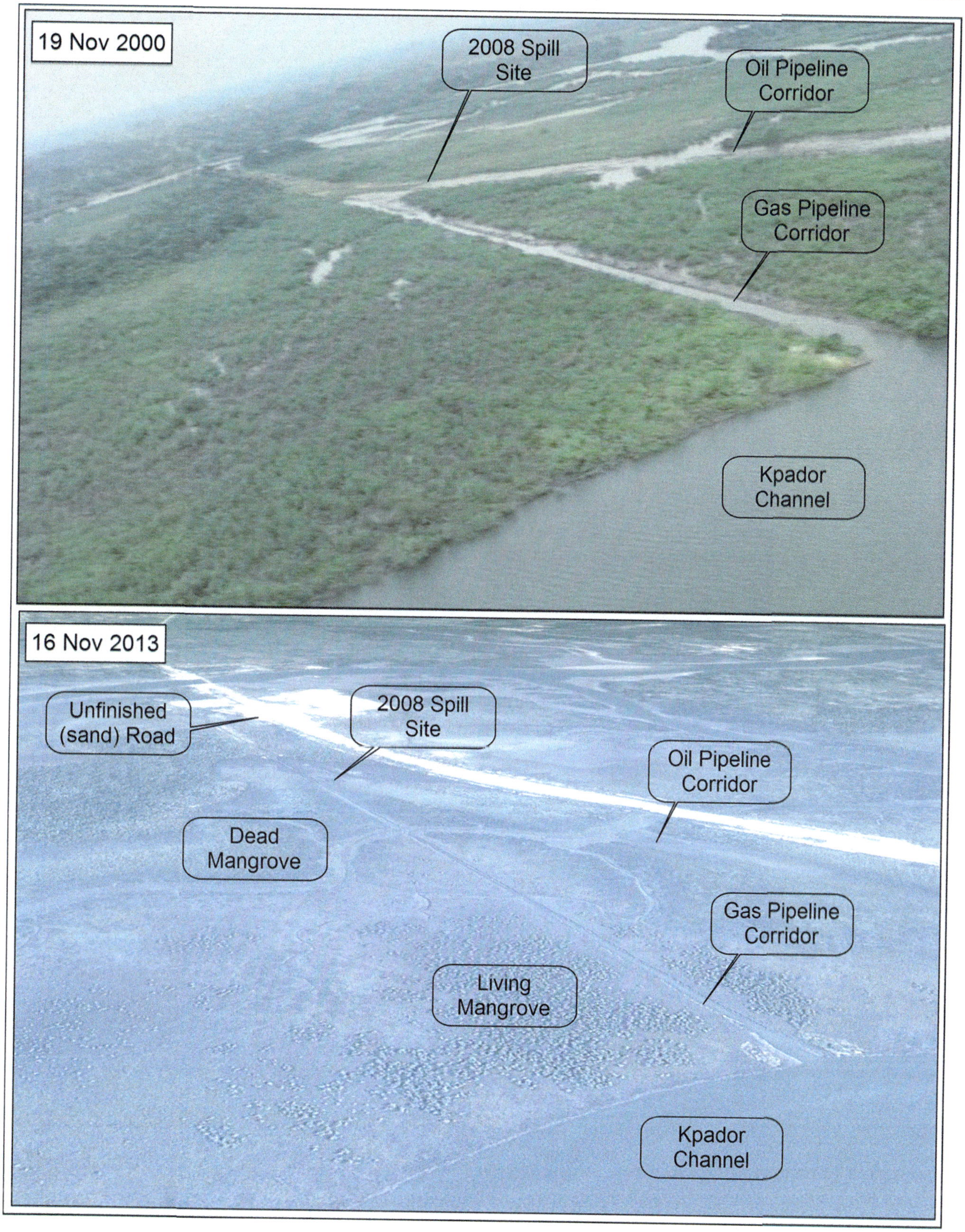

Fig. 13.29 Comparative oblique views of the 2008 spill site area in Bodo West area 'A' (Fig. 13.25) from 2000 to 2013 showing changes to mangrove coverage

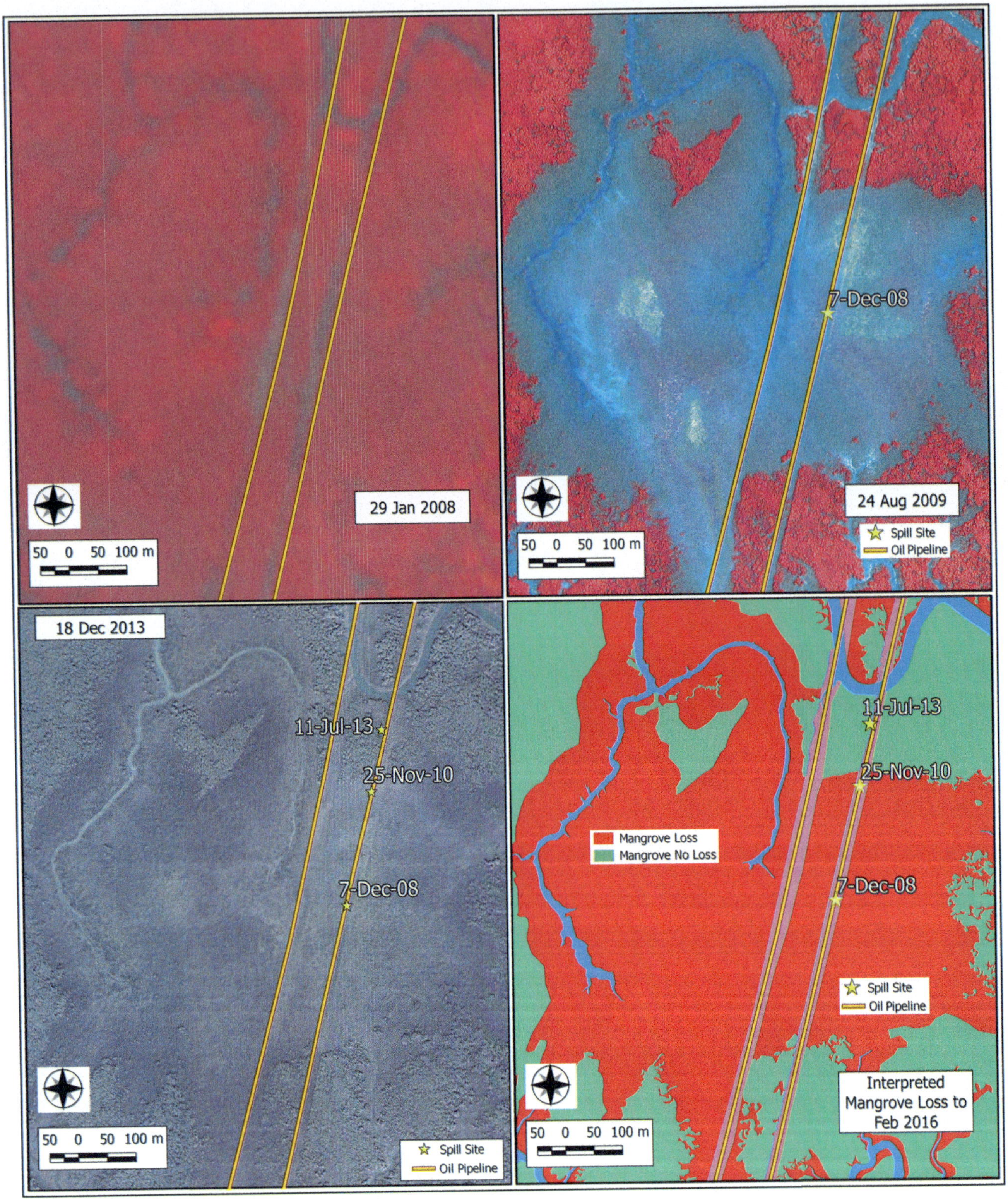

Fig. 13.30 Bodo West area 'B' sequence of mangrove loss 2008–2013, attributable to the 7 December 2008 oil spill. Additional images from Google Earth 2013 to Feb 2016 show mangrove losses remain similar to that in 2013. Figure 13.25 shows the site location

Fig. 13.31 View in 2013 across dead mangroves looking east at the site of the December 2008 operational spill site in Bodo West area 'B' from Fig. 13.25

proliferation of illegal refineries in the area show coincidence with the 2009–2011 mangrove loss. Table 13.13 shows the number of illegal operations increased from 0 to 38 by 2011, and increasing further to 60 in 2013. Because the image areal coverage varies, using the same size base (2011 image) shows that within the same area, the increase was to 52. The additional eight sites lie outside the 2011 coverage but are seen on the 2013 image.

Imagery with the location of illegal refineries from November 2009 and January 2011 area is shown in Fig. 13.37.

The UNEP survey completed an overflight of the area in November 2009 showing that oiling was very heavy in the area (Fig. 13.38). It is doubtful whether losses from illegal refineries and small boats transporting crude oil and refined products could cause such heavy oil concentrations in such a widespread area. The previous Fig. 13.34 from the same overflight in Bodo West also shows heavy concentrations.

As oil quantities are so great in the 2010 aerial photographs as well as in imagery from 2009 to 2011, it appears plausible that an unreported pipeline leak or multiple leaks were responsible for at least some of the oil observed. The most likely candidate spill site is within the slotted area made for the crossing of the NNPC pipeline, area 'B' in Fig. 13.35. Figure 13.39 compares imagery showing oil-on-water conditions in February 2010 and January 2011. Both show heavy oil concentrations and pooled oil in the dredged pipeline slot, particularly in 2011. The oil concentrations in

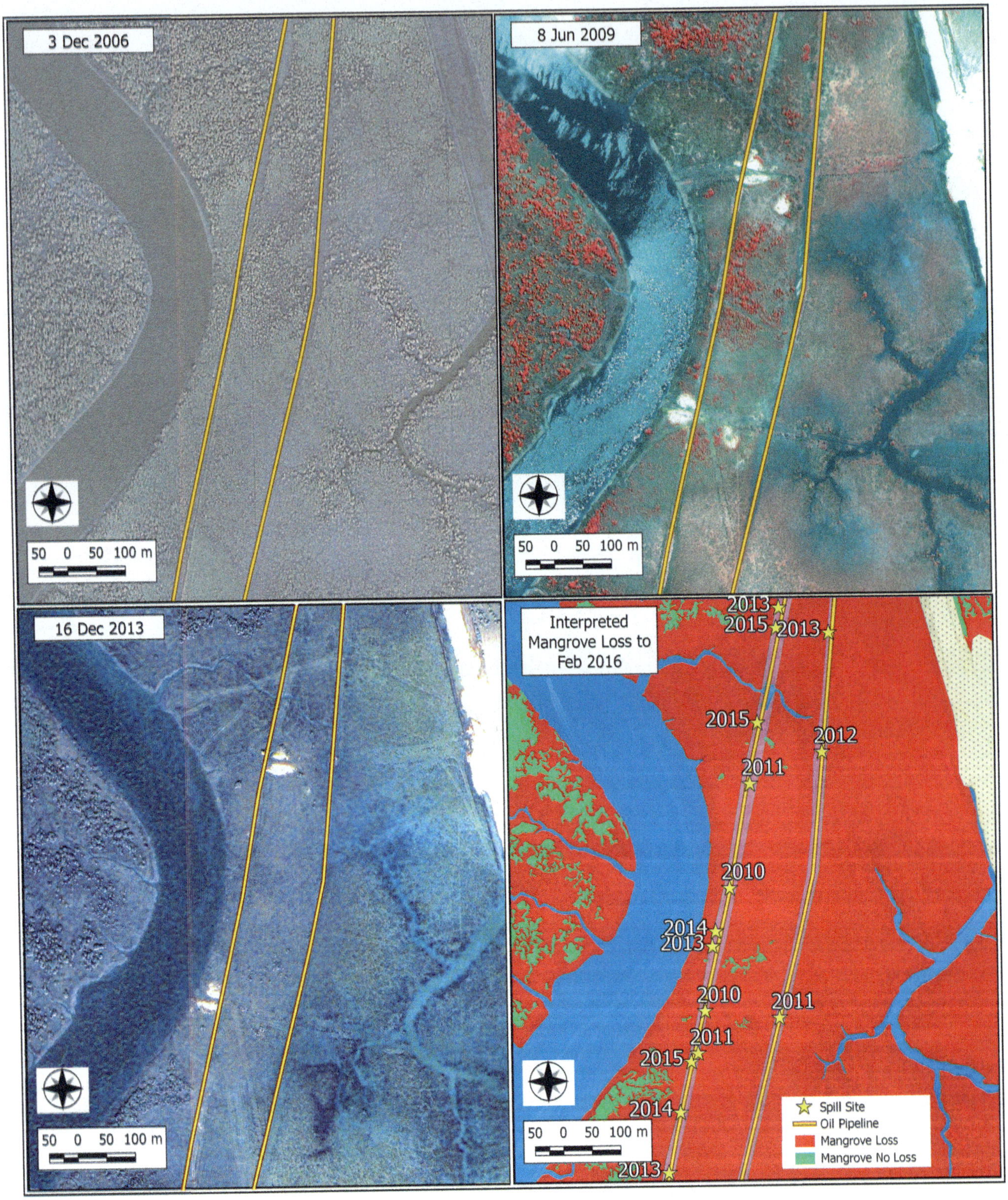

Fig. 13.32 Bodo West area 'C' sequence of mangrove loss from 2006 to 2013, updated to February 2016, with oil spill sites 2010 to June 2015. The site location is shown in Fig. 13.25. Surviving mangroves in 2009 (*bright red* coloration) were lost by 2013

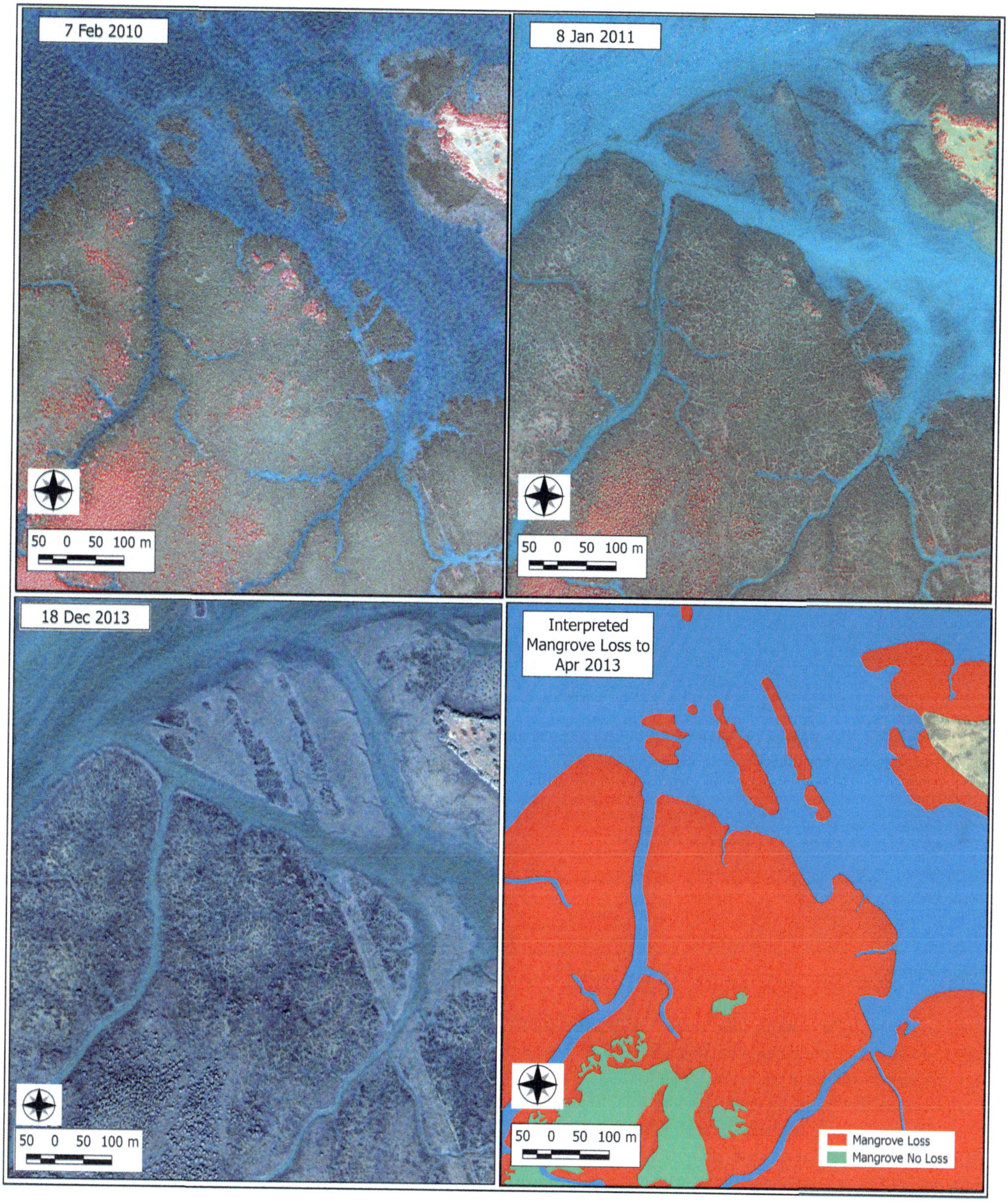

Fig. 13.33 Bodo West area 'D' sequence of mangrove loss, 2010–2013, showing an increase in loss from 2010 to 2011 and then from 2011 to 2013. The site location is shown in Fig. 13.25

Fig. 13.34 Aerial photograph on 29 November 2010 overlooking Bodo West area 'D' looking south. Large concentrations of black oil are common on the water's surface. The site location of area 'D' is shown in Fig. 13.25. T. Kallnischkies photographer

Table 13.12 Mangrove losses and high ground contaminated by illegal oil refining in the Bodo West area, 2008–2013/2016

Component	Hectares	Number
Mangrove loss	1096.7	
Pipeline corridor	28.1	
Total loss	1,1248	
Illegal refineries	28.9	87

February 2011 seem comparable to that in the November 2010 photographs indicating the potential that a pipeline leak may have continued for several months.

Further to the south in Bodo East area 'C', mangrove losses occurred later, between February 2010 and January 2011 (Fig. 13.40). Oil is present on the surrounding waters in both images but is particularly evident in on the 2011 image. During this time, the number of illegal refineries in the local area rose from 6 to 11 and substantially increased in size. The eastern shore area of Bodo East follows the same later oiling sequence, with massive changes occurring between 7 February 2010 and 8 January 2011.

In summary, the proliferation of illegal refineries from none in June 2009 to at least 60 in 2013 undoubtedly played a role in the loss of mangroves in the Bodo East area. No pipeline spills have been reported, but the quantity of oil on the water observed in photographs from November 2010 and imagery (particularly 7 February 2010 and 8 January 2011) indicates the possibility of major oil discharge(s) from the NNPC pipeline during this period. The summary of mangrove loss and illegal refining sites for Bodo East are presented in Table 13.14. This is the largest loss of the six areas reviewed in this study.

13.3.10 Bolo Area Mangrove Loss

The Bolo area is to the north of Bodo West. It is interconnected to the south by two waterways although not widely used. The economic and political focus of the area is toward the Local Government Area administration center of Ogu and north.

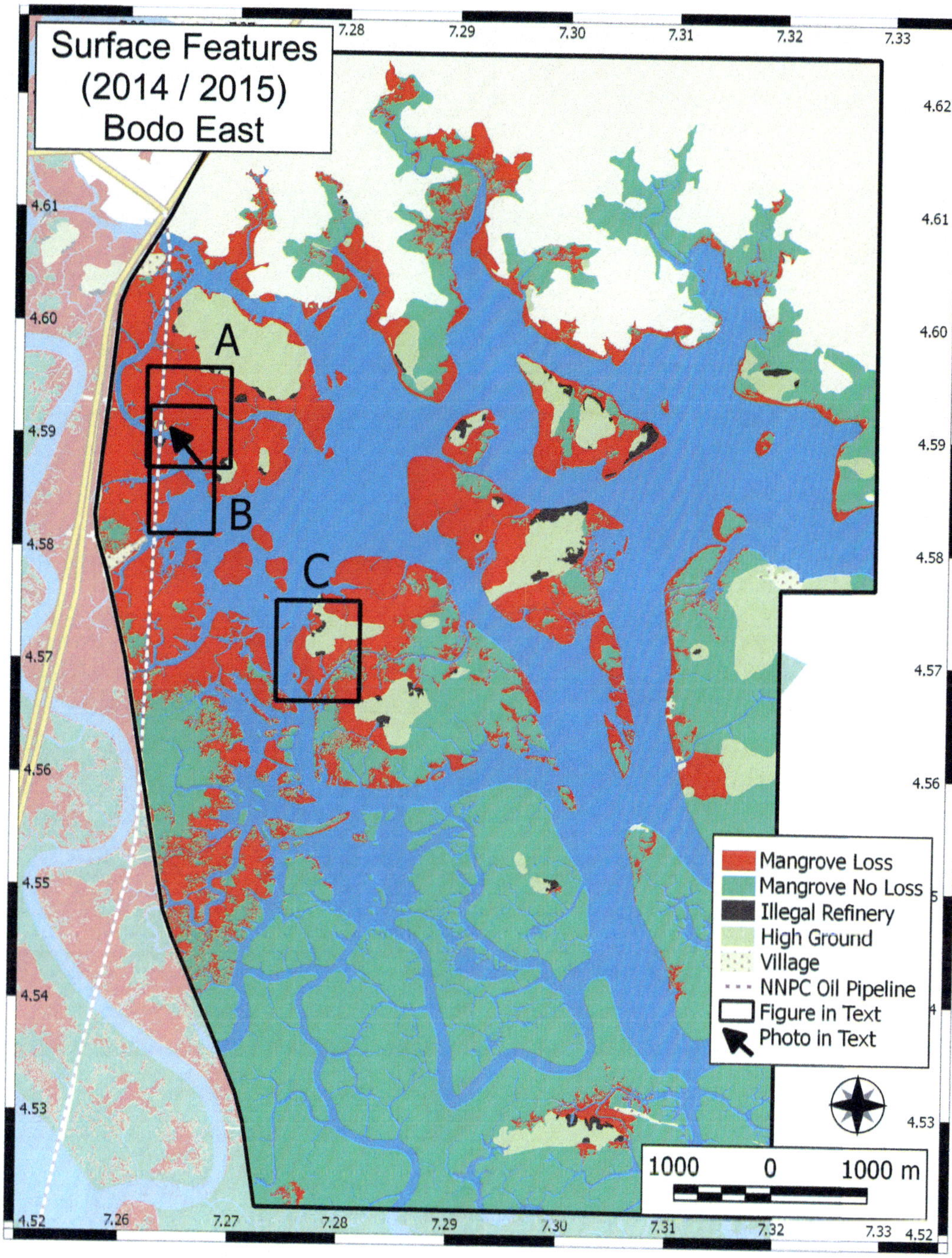

Fig. 13.35 Bodo East surface features and mangrove loss. Areas shown as 'A', 'B' and 'C' are discussed further in the text

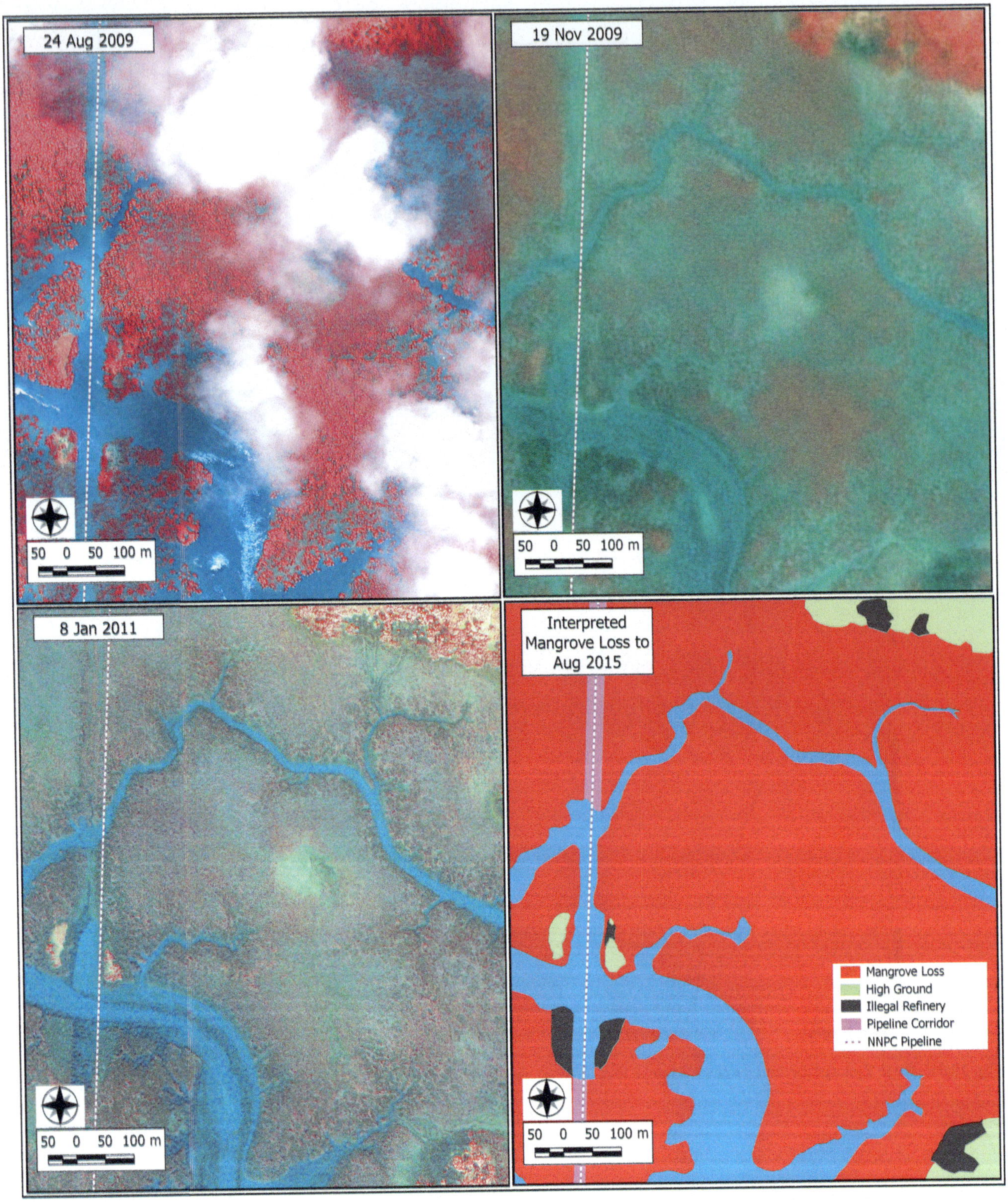

Fig. 13.36 Sequence of mangrove loss in Bodo East area 'A' (Fig. 13.35) indicating the first major loss occurred between 24 August and 19 November 2009, followed by complete loss by 8 January 2011. Other imagery not shown illustrates that complete loss occurred between the narrow window of 7 February 2010 and 8 January 2011

Table 13.13 Number of illegal refineries observed on images from 2009 to 2013 in the Bodo East area. As imagery areal extent varies among images, the 2011 image is selected to compare refinery increases within the same-sized area

Image date	Sites observed in multiple images	Sites observed in area of 2011 image
2009, 8 Jun	0	0
2009, 24 Aug	6	6
2009, 19 Nov	24	24
2010, 7 Feb	26	26
2011, 8 Jan	38	38
2013, 15 Apr	60	52

The overview of mapped surface features is provided in Fig. 13.41. There are 14 wellheads in the area, 7 of which are within or adjacent to mangrove areas. Small diameter pipelines connect to the wellheads, except for two isolated wellheads in the center-west of the area.

Comparing Bolo area 'A' imagery from 25 January 2009 to 29 November 2009 indicates that illegal refining activities started during this time (Fig. 13.42). The January 2009 image shows two refineries while five refineries are evident in the November image. Damage to mangroves was not evident in January 2009 and minimal in November 2009. However, major losses to mangroves are evident in 2013 as are six quite large illegal refineries. An image from January 2015 shows similar impacts as observed in 2013.

Low-altitude aerial photographs from Bolo area 'A' in 2010 are compared to 2015 in Fig. 13.43. In November 2010, mangroves were in good condition and refineries were small and few in number. The conditions in 2015 show a massive die off of mangroves. Oil is present in the channels in both photographs.

Area 'B' in Bolo shows a similar timeline, but more severe consequences (Fig. 13.44). In January 2009, mangroves are healthy and illegal refineries are not evident. Later in November 2009, two small illegal refineries have started. In 2013, most of the mangroves are dead and nine illegal refinery sites are evident along the north shore. A comparison of the 2013 and February 2016 images shows the same conditions, with few remaining remnants of the former mangrove habitat.

A 2015 aerial photograph of the northeast section of Bolo area 'B' is shown in Fig. 13.45. Oil is present on the surface of sediments as are scrape marks likely caused by cleanup operations. Heavy tar conditions indicative of an illegal refinery are shown by a white oval in the following figure. The damage to mangroves was likely caused by discharges from illegal refineries and possibly influenced from a pipeline leak upstream of the site where the Bomu-to-Bonny Terminal pipelines (24- and 28-in.) pipelines cross the estuary (upstream above area 'B' on Fig. 13.41). On 17 April 2010, there was reported an illegal activity that caused an oil discharge and fire at this location. Further to the north also on Fig. 13.41, mangroves to the east of Ogu are similarly damaged and likely to have been affected by reported spills along these pipelines.

In summary, mangroves were damaged in the Bolo area between late 2009 and 2013. Losses appear to have stabilized after 2013, as illustrated by 2015/2016 images that show similar losses to 2013. Most likely causes for all areas are illegal refining with the addition of potential discharges from illegal tapping of the SPDC pipelines in areas upstream of, or adjacent to, mangrove habitat. The summary of mangrove loss and identified illegal refineries is presented in Table 13.15.

13.3.11 Mangrove Loss Summary for All Areas

More than 4400 ha of former mangrove habitat were lost from spilled oil between 2008 and 2016 in the study area (Table 13.16). Another 104 ha was lost due to pipeline corridors cut through mangrove areas several decades ago. Additionally, 217 distinct illegal refineries sites were found, destroying 116 ha of supratidal (palm dominated) habitat adjacent to mangroves.

Four sites of pipeline leaks were responsible for causing initial, rapid and large-scale losses of mangroves in 2009. These include the 2008–2009 spills on Bonny Island (~940 ha), Opobo Channel (~170 ha) and the two spills in Bodo West (~1000 ha). The Bonny Island incident was attributed to illegal activity while the three others were operational (pipeline leaks) in nature.

The sequence of mangrove losses for the entire study area is shown in Fig. 13.46. Major losses occurred in 2009 from the spills described above, causing mangrove losses on Bonny Island, along the Opobo Channel and Bodo West. In 2010, there was an increase in the number of spills caused by illegal activities which, because the majority of mangroves were already killed in 2009, caused only a minor increase in mangroves lost, as along the Bonny River and into Bodo West. In 2011, due to illegal activities and a possible spill

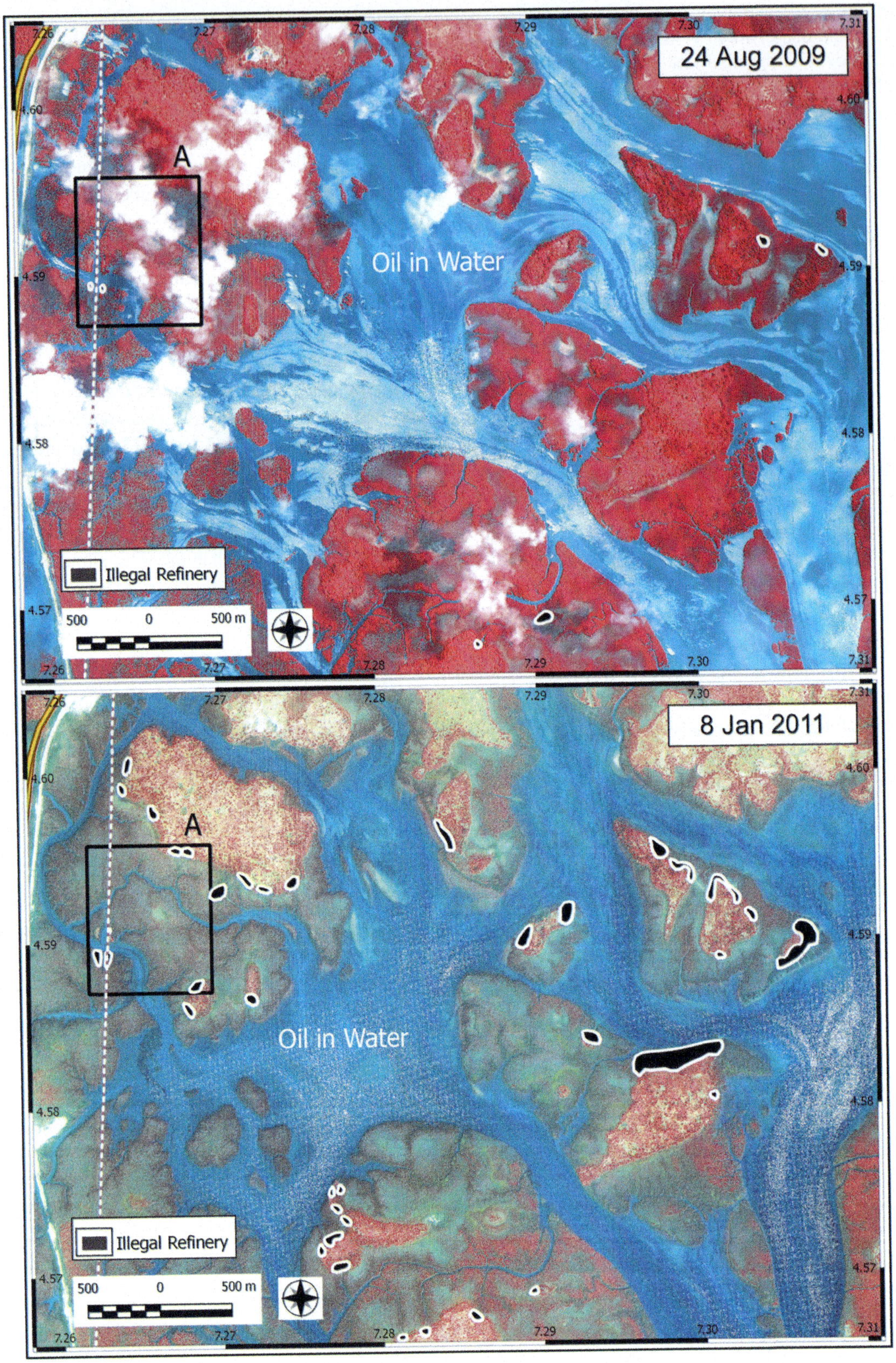

Fig. 13.37 False color infrared imagery of northern Bodo East illustrating mangrove condition and location of illegal refineries. In August 2009, mangroves were healthy and six refineries present. In February 2011, mangrove losses were extensive (*dark* coloration) and 38 refineries present. Area 'A' is shown in Figs. 13.36 and 13.38

Fig. 13.38 Aerial photograph from 20 November 2010 of the Bodo East area 'A' (Fig. 13.35) looking NNE showing heavy oil concentrations. T. Kallnischkies photographer

leak on the NNPC pipeline, oil-related damages spread into Bodo East, causing over 1000 ha of mangrove loss.

Additional damages observed in 2013 occurred in the Bolo area due to illegal refining and likely from pipeline leaks from the oil-transport pipelines passing at the head of mangrove embayments in the northeast of the Bolo area. Losses also increased along the Bonny River from <100 m in 2010 to over 800 m landward in 2013. Other smaller increases in Bodo East and West are associated with illegal refining activities. Losses observed after 2013 are small and primarily on Bonny Island.

13.4 Discussion

13.4.1 Mangrove Losses and Recovery at Other Sites

Worldwide, the historic losses of mangroves due to oil-related activities are commonly tied to a specific event, e.g. pipeline rupture, tank rupture or well blowout. In most cases, the extent of mangrove loss from these single events is relatively small. Sites having confirmed losses of greater than 5 ha are shown in Table 13.17. A few other sites are offered by Duke (2016a, b) including those in the area of this study. Only the Funiwa 5 well blowout off the coast of Nigeria in 1980 is even close to the scale of the losses reported herein.

The full recovery of large areas of mangrove loss has not been documented. In those areas where recovery is being monitored, studies indicate ranges of 20–30 years and continuing (Table 13.18).

A minimum estimate of 30 years for mangrove recovery is therefore expected, assuming that recovery is not affected by other factors such as additional oiling or toxic concentrations remaining in sediments. Duke (1999) projected a required recovery to 50 years based on the regrowth of trees, and "assumes recovery proceeds unimpeded, which it often does not." Applying the predictions of oil longevity and mangrove recovery supports the likelihood that mangroves in the study area will require several decades to recover, with a minimum of 30 extending to possibly 50 years. Confounding recovery potential is that oil concentrations within sediments in many areas are very high and at levels toxic to mangrove regrowth (Little et al. 2018, Chap. 14, this volume).

Additionally, many locations show that the substrate is substantially altered from previous conditions of high organic matter and prop roots to hard clay without plant debris or

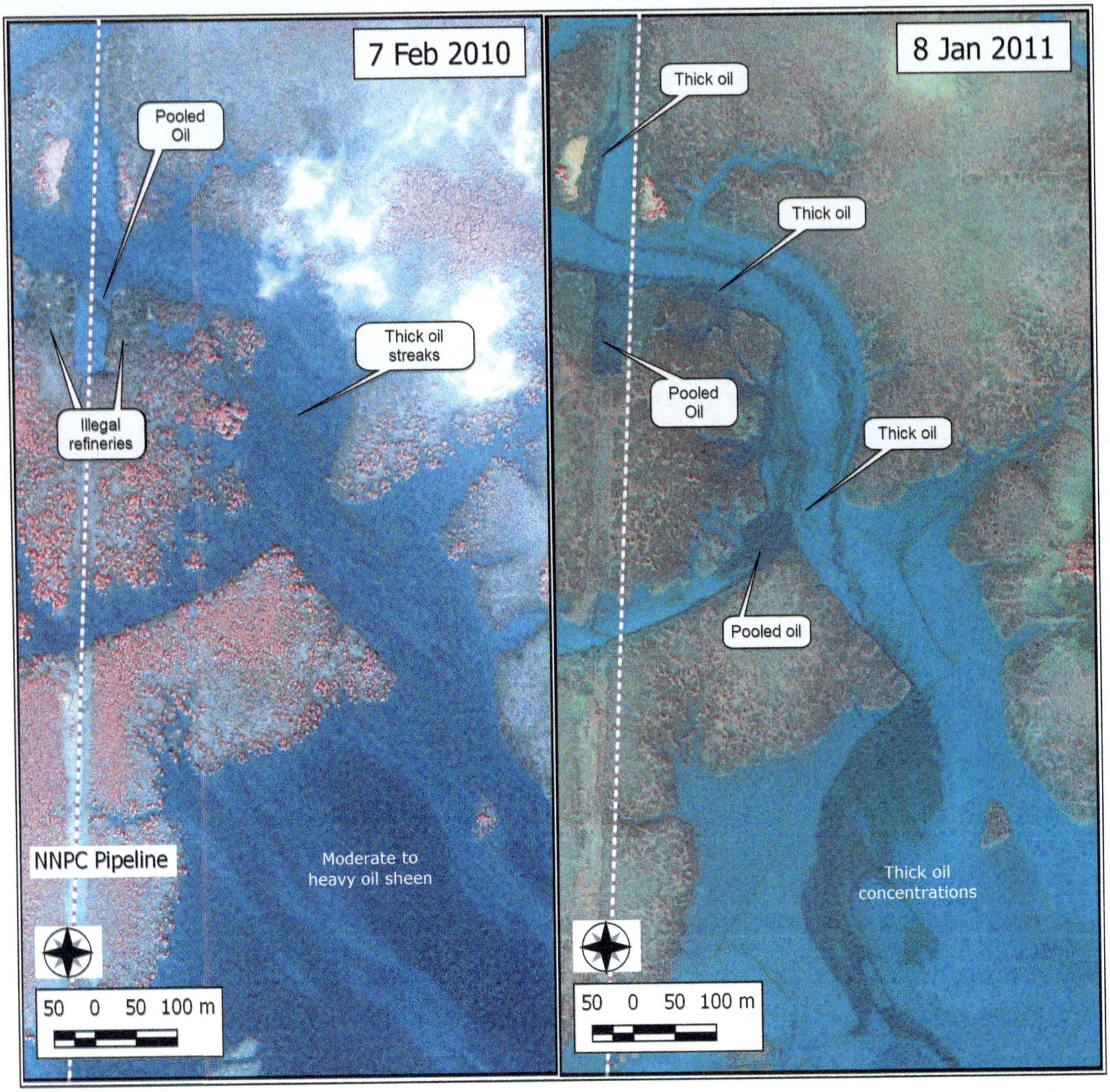

Fig. 13.39 Bodo East area 'B' images from February 2010 to January 2011 showing substantial oiling of the water's surface possibly originating from a pipeline leak in the slotted area where the NNPC pipeline crosses the channel (*upper left* of each image). Illegal refineries are located on the banks of the slots. Site location of Bodo East area 'B' is shown in Fig. 13.35

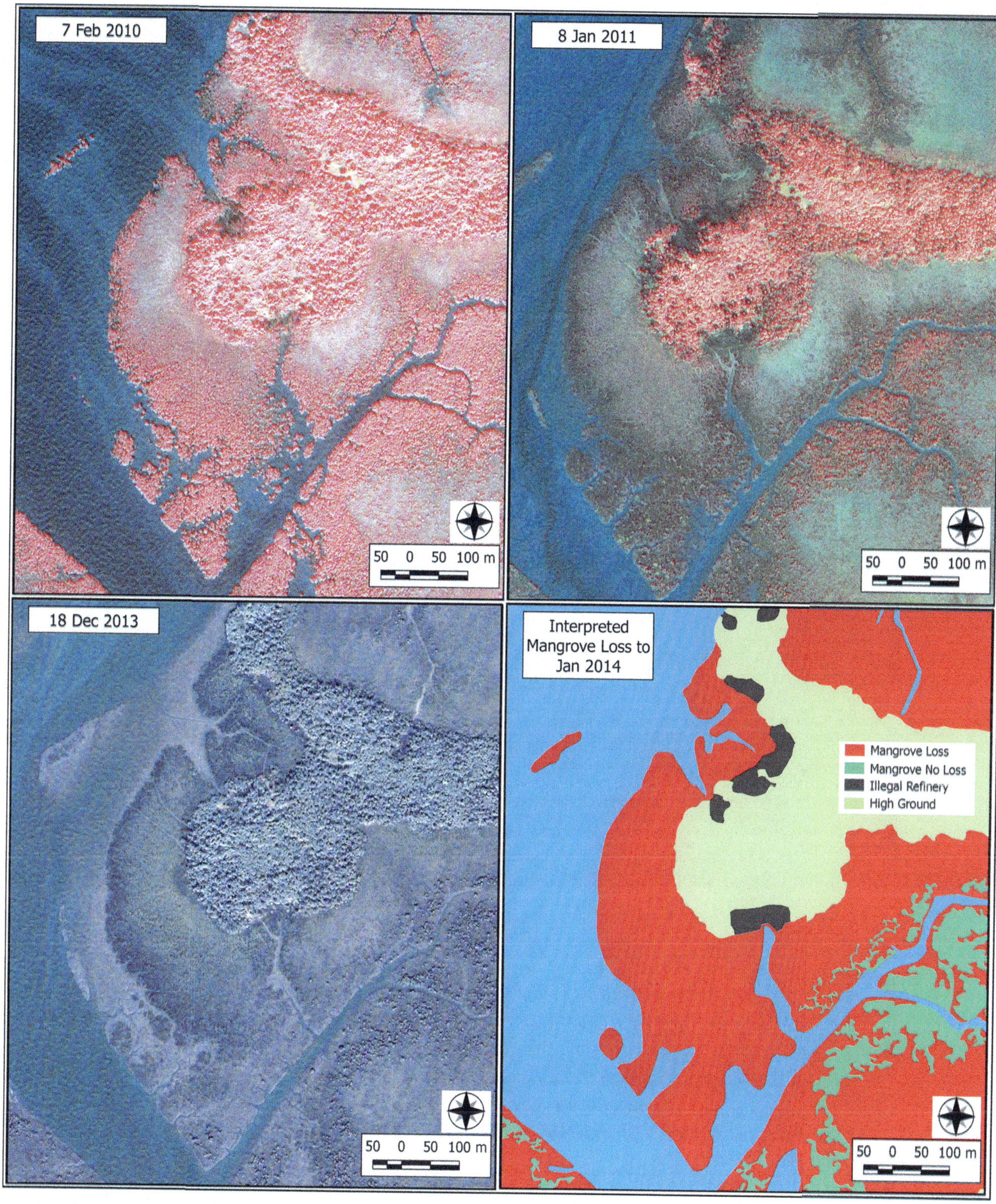

Fig. 13.40 Sequence of mangrove loss in Bodo East area 'C' from 2010 to January 2014. The location of area 'C' is shown in Fig. 13.35. A Google Earth image from 12 January 2014 shows similar conditions as observed from the 2013 image

Table 13.14 Mangrove losses and high ground contaminated by illegal oil refining in the Bodo East area, 2008–2015

Component	Hectares	Number
Mangrove loss	1175.9	
Pipeline corridor	7.4	
Total loss	1183.3	
Illegal refineries	35.4	60

organic matter. Mangrove resettlement will be inhibited and these areas may never return to their pre-oiling condition without additional effort.

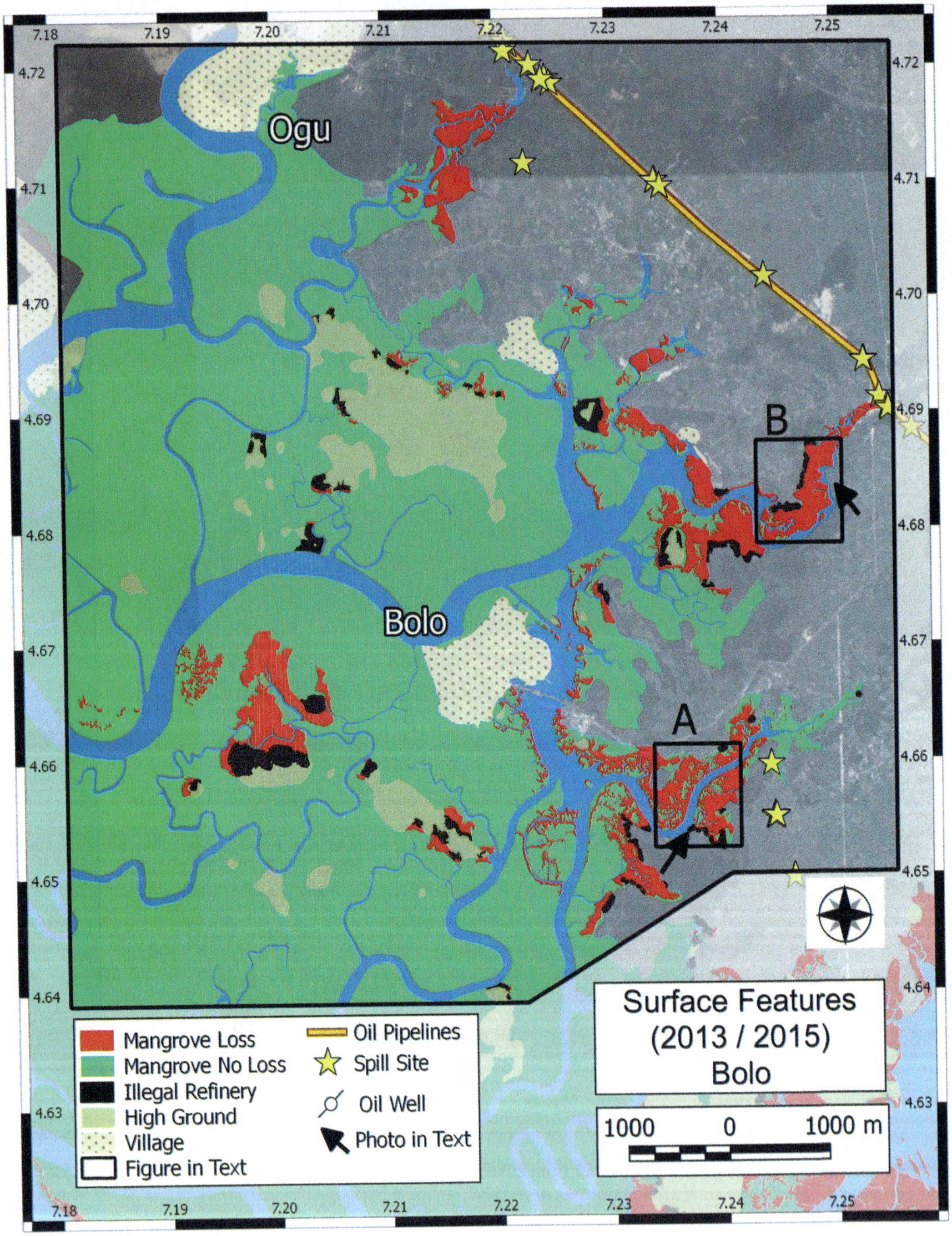

Fig. 13.41 Overview of surface features of the Bolo area based on imagery and photographs continuing to 2013 or 2015, depending on image coverage. Bolo areas 'A' and 'B' are discussed further in the text

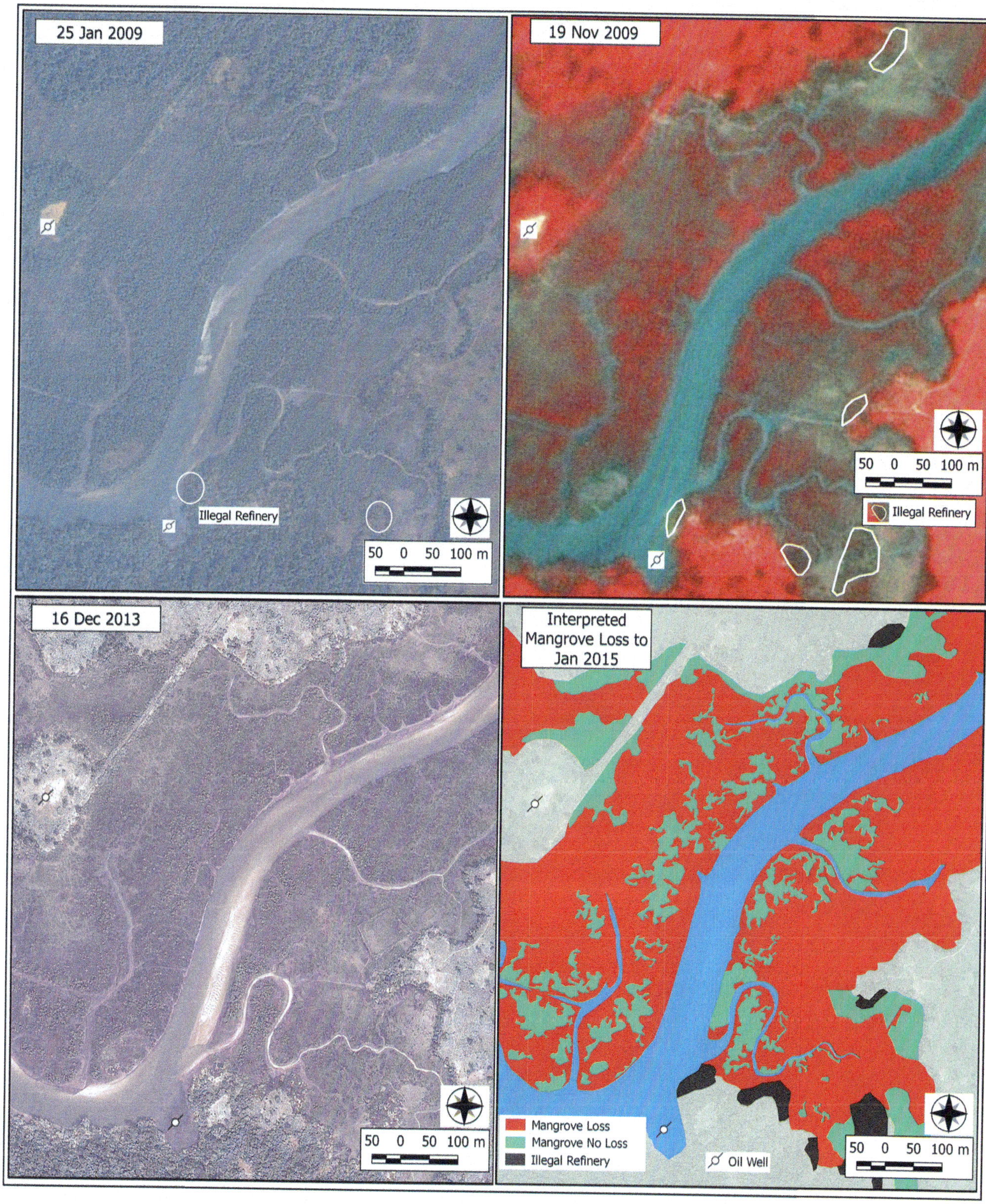

Fig. 13.42 Sequence of impacts in the Bolo area 'A' showing no mangrove loss in January 2009, to five illegal refineries and some mangrove loss in November 2009, and ending with extensive loss in 2013 and six large illegal refinery sites. Imagery from 19 January 2015 shows similar mangrove loss and refineries to 2013. The location of Bolo area 'A' is shown in Fig. 13.41

Fig. 13.43 Aerial photographs of Bolo area 'A' (Fig. 13.41) from 29 November 2010 to 15 September 2015 (Photos by T. Kallnischkies (*upper*) and V. Imevbore (*lower*)) Letters *A*–*C* show illegal refineries. The *white box* in the *upper photo* is the location of the *lower photo*. Smoke is seen at location *C* likely indicating active refining

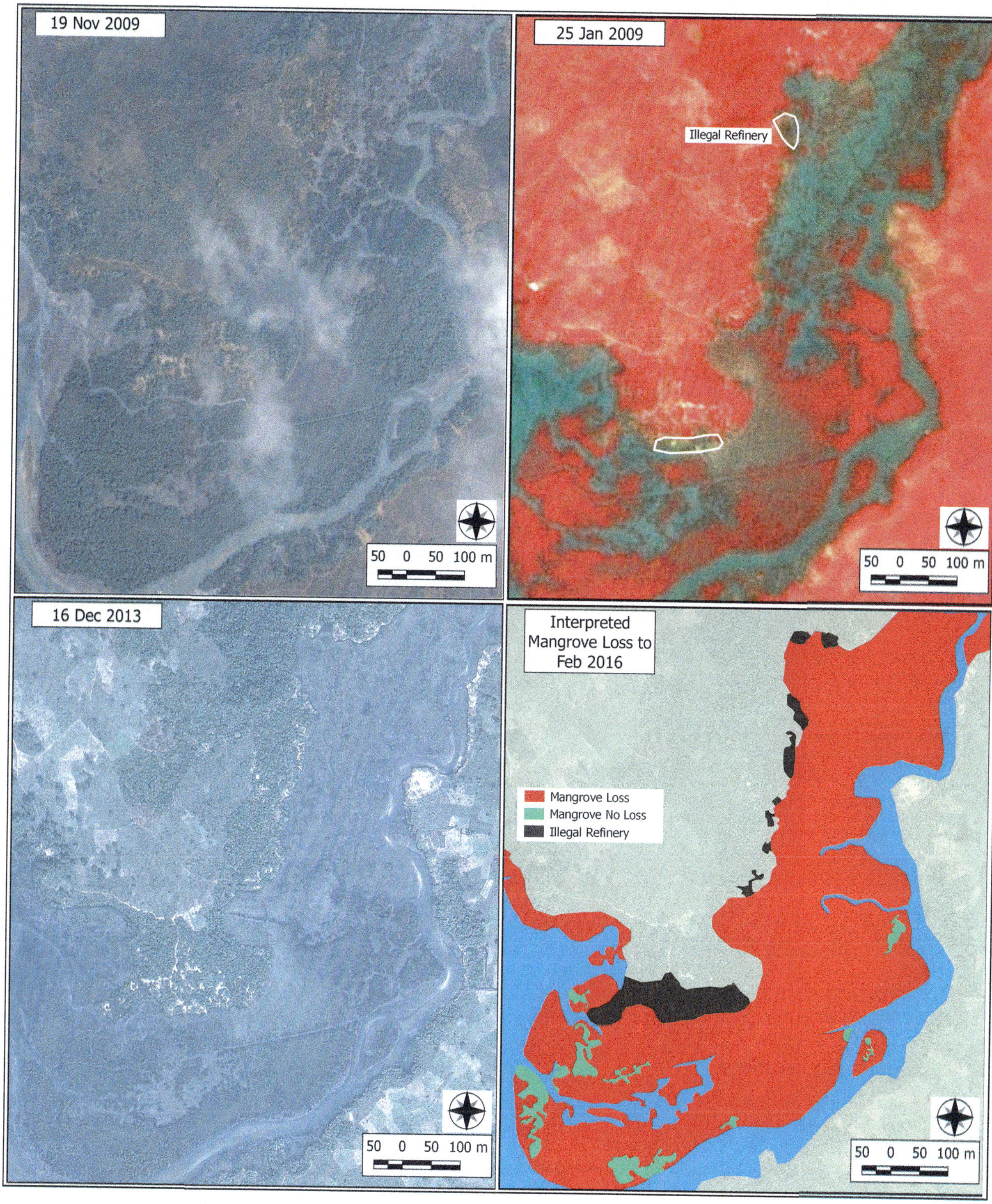

Fig. 13.44 Sequence of mangrove loss in Bolo area 'B' from 2009 (little to no loss with few refinery sites) to 2013 (major losses and nine refinery sites), extended to 10 February 2016 imagery which shows similar conditions to that of 2013. The site location of Bolo area 'B' is shown in Fig. 13.41

Fig. 13.45 Aerial photograph from 15 September 2015 of the northeastern section of Bolo area 'B' from Fig. 13.41. Scrape marks potentially from a cleanup operation are evident. Oil is in the water and along the shoreline. The *oval* indicates a likely illegal refinery. A pipeline discharge occurred upstream (to the *right*) in April 2010 (Photograph by V. Imevbore)

Table 13.15 Mangrove losses and high ground contaminated by illegal oil refining in the Bolo area, 2008–2015

Component	Hectares	Number
Mangrove loss	338.7	
Pipeline corridor	0	
Total loss	338.7	
Illegal refineries	51.2	66

13.4.2 Enhancing Recovery

The Shoreline Cleanup Assessment Technique (SCAT) teams fielded in May and August 2015 reviewed 32 sites in the study area and developed a series of recommendations designed to enhance recovery of the area. For reference, SCAT is a well-known and internationally accepted program to scientifically advise on the cleanup and recovery of oiled shorelines (Owens and Sergy 2003; IPIECA 2014; NOAA 2013).

Recommendations of the SCAT teams are focused on successfully restoring mangroves to enable full habitat functionality by successful application of a net environmental benefit analysis (IPIECA 2015). By bringing back the mangrove plants, the associated biota will follow. Recommendations include the following:

- *Do More Good than Harm:* Cleanup operations must be controlled so that damages are not increased, particularly where oil has deeply penetrated into soft muds. Operations must be altered or stopped until additional damage can be prevented. Natural recovery may be a preferred option in many areas.
- *Establishment of a Cleanup Endpoint:* The stopping point for active cleanup will be the reduction of oil in sediments to a level where mangrove seedlings will be able to survive and thrive, and for the removal of surface tar and asphalt. Even after extensive cleanup options, some level of oiling is likely to remain,
- *Initiation of Mangrove Planting:* Mangroves appear likely to survive in areas having some silver and rainbow sheen, and possibly even in areas where substantial brown or black oil is visible when sediments are agitated. In areas

Table 13.16 Summary of mangrove loss and illegal refineries for the study area east of the Bonny River

	Loss of mangroves habitat			Illegal refineries	
Area	Oil-related loss (ha)	Pipeline corridor (ha)	Total (ha)	Number	High ground (ha)
Bonny Island	1058.2	58.2	1116.4	4	0.8
Opobo Channel	280.2	9.3	289.5	0	0
Bonny River	464.9	1.7	466.6	0	0
Bodo West	1096.3	28.1	1124.8	0.87	28.9
Bodo East	1175.9	7.4	1183.3	60	35.4
Bolo	338.7	0	338.7	66	51.2
Total	4414.6	104.7	45,219.3	217	116.2

showing little-to-no oiling, mangrove planting can be initiated without physical cleanup.

- *Cleanup Prioritization:* Priority areas for treatment are contaminated sites within 5 km of human habitation and sites actively releasing moderate to heavy sheen or black oil to waterways.
- *Cleanup methods – Hard Substrates:* Rakes and shovels are preferred to remove oil and avoid sediment removal. Where liquid oil is found under surface oil, low-pressure flushing is advised.
- *Cleanup methods – Soft Substrates:* Includes soft muds where oil or heavy sheen becomes exposed upon agitation. Low-pressure flushing is likely to be effective at releasing oil.
- *Prevention of re-oiling:* Oleophilic snare, sorbents and boom can be used to prevent the spread of released oil.
- *Need for SCAT advice:* SCAT teams should be fielded to determine the level of oiling at each location, recommend and verify cleanup methods specific to each type of habitat, monitor cleanup effectiveness and recommend changes as needed, and to document cleanup activities and completion. SCAT can test and advise on alternative methods such as bioremediation or other environmentally friendly technology.

13.5 Conclusions

This report documents the analysis of multiple satellite images, aerial photographs and ground surveys to determine the probable cause and timing of over 4400 ha of mangrove loss along areas east of the Bonny River between Bonny Island in the south and the mainland areas of Ogu/Onne in the north. In addition, 217 illegal refinery sites damaging 115 ha of palm-dominated supratidal areas adjacent to mangroves are identified.

Outside of the operationally caused spills along Opobo Channel and in Bodo West in 2008, recorded spill incidents are reported as mostly due to illegal hot tapping of the SPDC pipelines transporting oil from the north to Bonny Terminal in the south. After 2009 and into 2010, most mangrove losses are related to spills resulting from hot tapping of pipelines and from the illegal refining and related transport of crude and refined product in open fishing boats.

Smaller areas of mangrove loss as from flooded pipeline corridors, wellhead slots and seismic shot lines (cleared during early exploration of the area) are not included in this analysis, nor are extensively oiled shorelines not having previous mangroves. While this report has located the extent of mangrove loss east of the Bonny River, areas to the west of Bonny River also show large areas of mangrove loss due to oil-related activities and will be a subject of a future study.

Imagery reveals that surface oiling of creeks and channels throughout the area was common from 2009 to 2016. Illegal activities apparently decreased in 2013, and less oil was evident on the water's surface in 2015 during aerial and ground surveys.

The sediments in many areas, particularly in Bodo West and East, show high levels of surface and/or subsurface oiling that will inhibit or prevent mangroves restoration (see Little et al. 2018, Chap. 14, this volume).

Restoration planning for the area is underway. The Bodo Mediation Initiative (BMI) sponsored Shoreline Cleanup Assessment Technique (SCAT) teams in 2015 (and continued in 2017/2018) that identified sites for potential cleanup and established a set of guidelines to ensure that cleanup activities will not cause more harm than good. Mangrove planting is an important part of these guidelines, and can be undertaken immediately in areas that have low or no remaining oil concentrations.

Other studies following the duration of time required for natural regeneration of oil-damaged mangrove areas indicate a time of 30 years and longer. Within the oil damaged study area, some regeneration is seen but is relatively sparse. Affecting factors include remaining oil on surface and subsurface sediments, new oiling spread by intertidal currents from land sites or heavily oiled areas, lack of adjacent seed stock, loss of original base sediments (conversion from organic rich to hard clay), and presence of algal mats. In many areas, the return to a healthy mangrove habitat will never occur without intervention.

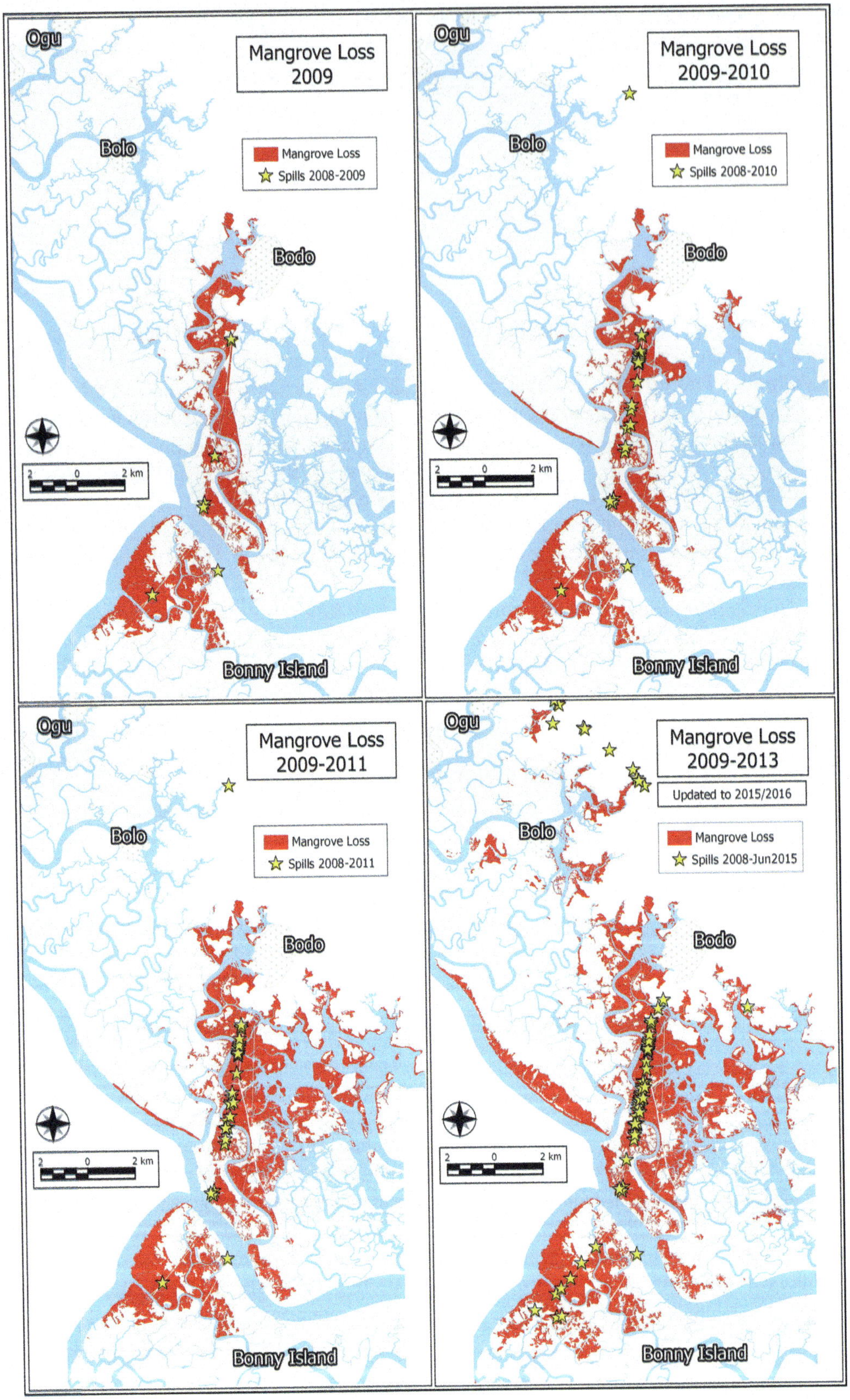

Fig. 13.46 Cumulative mangrove losses for the study area, 2009–2013, with updates in certain areas to 2015 or 2016. The sequence of spill events is from 2008 to June 2015

Table 13.17 Documented mangrove loss >5 ha from oil-related events

Mangrove loss	Event	Location	Citation
338 ha	Funiwa 5 well blowout, 57,150 t crude oil, 1980	Offshore Nigeria	Gundlach et al. (1993)
64 ha	Fuel tank rupture, 200–475 t, medium crude oil, 1986	Bahia Las Minas, Panama	Keller and Jackson (1993)
46 ha	*Witwater*, 2150–2850 t diesel and Bunker C, 1968	Bahia Las Minas, Panama	Duke and Pinzon (1991)
12 ha	Tank leak, JP-5 fuel 380 t, 1999	Roosevelt Roads, Puerto Rico	NOAA (2014)
10.5 ha	Pipeline break, 31,500 t crude oil, 1983	Southeast Brazil	Santos et al. (2012)
5.5 ha	Tank leak, JP-5 fuel 170 t, 1986	Roosevelt Roads, Puerto Rico	Ballou and Lewis (1989)
5 ha	Tanker spill at refinery, 11,250 crude oil, 1971	St. Croix, US Virgin Islands	Lewis (1983)

Table 13.18 Field studies indicating time to full recovery of oil-damaged mangroves

Duration	Characteristics	Location/loss	Citation
20 years and continuing	October 1983, 31,500 t crude oil within 300 ha. Mangroves show reduced structural development and different species composition	Southeast Brazil 10.5 ha	Santos et al. (2012)
23 years and continuing	*Witwater*, 1968. 2150–2850 t diesel and Bunker C	Panama, 46 ha	Keller and Jackson (1993)
31 years and continuing	TROPICS experimental sites treated with dispersed and undispersed Prudhoe Bay crude	Panama, two 900 m^2 sites	Renegar et al. (2017)
29 years and continuing	*Zoe Colocotronis*, 1973, basin mangroves still recovering in 2002, but possible damage from reoiling	Puerto Rico, 2.7 ha	Getter and Lewis (2003)
30 years	Regression curve based on several spills	Various	Duke (2016a)

Acknowledgements I would like to thank the BMI and SCAT team members for their efforts in overall direction and/or the collection of field data in 2015 including BMI Co-Chairs Bert Rohaar (retired, Dutch embassy) and Inemo Samiama (StakeholderDemocracy.org); Father Abel Agbulu (Bodo church leader); Ferdinand Giadom, John Aawa, Loveday Kpandei and Kabari Visigah (Bodo Community); Kay Hotzman (BMI Project Director, 2015); David Little (UNEP and independent technical advisor); Obiageri Nwokoro, Akpene Samiama (StakeholderDemocracy.org); Anthony Kinika (Rivers State Ministry of Environment); Philip Shekwolo, Augustine Igbuku, Igo Weli, Ogonnaya Iroakasi, Lede Kumannsee, Kingsley Osuh, Johnson Akinnalwonu, Franklin Igbodo and Jonathan Obasohan (SPDC); Raphael Saue (BMI); Sola Oladipo (NOSDRA); Nenibarini Zabbey (NACGOND, National Coalition on Gas Flaring and Oil Spills in the Niger Delta); Matthijs Bonte (Shell Global Solutions); Nebo Chika and Enuvie Akpokodje (UniPort); and Adebayo Oladeji, Prosper Ugbehe, Joel Olarenwaju Adesida, Adekunle Adesida (Geoterrain Nigeria Limited). David Little and Nenbarini Zabbey are especially thanked for their critical reading and comments on the manuscript. Nicolas Story assisted with identification of mangrove loss and illegal refinery areas. Thorsten Kallnischkies, Jonas Pålssen and Victor Imevbore are thanked for their site photographs. This analysis was self-supported.

References

Adejuwon JO (2012) Rainfall seasonality in the Niger delta belt, Nigeria. J Geogr Reg Plann 5(2):51–60. http://www.academicjournals.org/article/article1381913298_Adejuwon.pdf. https://doi.org/10.5897/JGRP11.096. Accessed 29 Mar 2017

AI-CEHRD (2011) The true 'Tragedy': delays and failures in tackling oil spills in the Niger Delta. Amnesty International/Center for the Environment, Human Rights and Development, Amnesty International, London. 54p. https://www.amnesty.org/en/documents/AFR44/018/2011/en/. Accessed 4 Apr 2017

Allen JRL (1965) Coastal geomorphology of eastern Nigeria: beach-ridge barrier islands and vegetated tidal flats. Geol Mijnb 44(1):1–21

Atlantic (2013) Nigeria's illegal oil refineries. The Atlantic. https://www.theatlantic.com/photo/2013/01/nigerias-illegal-oil-refineries/100439/. Accessed 30 Mar 2017

Baker JM (1983) Environmental study of shell-operated areas; mangrove ecosystems. Technical Summary Report FSC/OPRU/32/83 to SPDC, Nigeria. 62p

Ballou TG, Lewis RR III (1989) Environmental assessment and restoration recommendation for a mangrove forest affected by jet fuel. Proc int oil spill conf 1989(1):407–413. https://doi.org/10.7901/2169-3358-1989-1-407

BMI (2016) Final project report, Bodo Mediation Initiative. Report to the Embassy of the Kingdom of the Netherlands. 36p http://www.stakeholderdemocracy.org/wp-content/uploads/2016/06/Merged-Bodo-Mediation-End-Reportd.pdf. Accessed 30 Mar 2017

Brillco.com (2017) Bonny Light crude oil specifications (BLCO). 1p http://brillpetco.com/Bonny%20Light%20Crude%20Oil%20Specifications.pdf. Accessed 24 Mar 2017

Dublin-Green CO (1990) Seasonal variations in some physic-chemical parameters of the Bonny Estuary, Niger Delta, Technical Paper No. 59. Nigerian Institute for Oceanography and Marine Research, Lagos. 24p. http://www.oceandocs.org/bitstream/handle/1834/2408/NIOMR%20TP-59-.pdf?sequence=1&isAllowed=y Accessed 1 Apr 2017

Duke NC (1999) Introduction to the project. In: Duke NC, Burns KA (eds) Fate and effects of oil and dispersed oil on mangrove ecosystems in Australia. Final Report to the Australian Petroleum Production and Exploration Association. pp 236–239. https://www.researchgate.net/publication/37629636_Fate_and_Effects_of_oil_and_dispersed_oil_on_mangrove_ecosystems_in_Australia

Duke NC (2016a) Oil spill impacts on mangroves: recommendations for operational planning and action based on a global review. Mar Pollut Bull 109(2):700–715. https://doi.org/10.1016/j.marpolbul.2016.06.082

Duke NC (2016b) Oil spill impacts on mangroves – supplementary data. Mar Pollut Bull 109(2):700–715. https://doi.org/10.1016/j.marpolbul.2016.06.082

Duke NC, Pinzon ZS (1991) Chapter 6, mangrove forests. In: Keller, Brian D, Jackson JBC (eds) Long-term assessment of the oil spill at Bahia Las Minas, Panama, interim report, volume II. OCS Study MMS 90-0031 U.S. Department of the Interior. pp 153–173. http://www.data.boem.gov/PI/PDFImages/ESPIS/3/3664.pdf. Accessed 24 Mar 2017

Etkin DS, French-McCay D, Michel J (2007) Review of the state-of-the-art on modeling interactions between spilled oil and shorelines for the development of algorithms for oil spill risk analysis modeling. MMS OCS Report 2007-063. 157p. https://www.boem.gov/uploadedFiles/BOEM/Environmental_Stewardship/Environmental_Assessment/Oil_Spill_Modeling/2007-063-ModelingInteractions.pdf. Accessed 1 Apr 2017

Fentiman A, Zabbey N (2015) Environmental degradation and cultural erosion in Ogoniland: a case study of the oil spills in Bodo. Extr Ind Soc 2:615–624. https://doi.org/10.1016/j.exis.2015.05.008

Folorunsho R, Awosika L (2014) Morphological characteristics of the Bonny and Cross River (Calabar) Estuaries in Nigeria. In: The land/ocean interactions in the coastal zone of West and Central Africa. Part of the series: Estuaries of the World. Springer Publishing, pp 87–96. doi:https://doi.org/10.1007/978-3-319-06388-1_8

George ADI, Abowei JFN, Daka ER (2009) Benthic macro invertebrate fauna and physical-chemical parameters in Okpoka Creek sediments, Niger Delta, Nigeria. Int J Anim Vet Adv 1(2):59–65

Getter CD, Lewis III RR (2003) Spill response that benefits the long-term recovery of oiled mangroves. In: Proceedings international oil spill conf. 12p. http://ioscproceedings.org/doi/pdf/10.7901/2169-3358-2003-1-539. Accessed 1 Apr 2017

Guardian (2011) Shell oil spills in the Niger delta: 'Nowhere and no one has escaped'. https://www.theguardian.com/environment/2011/aug/03/shell-oil-spills-niger-delta-bodo. Accessed 12 Mar 2017

Gundlach ER, Hayes MO (1978) Classification of coastal environments in terms of potential vulnerability to oil spill damage. Mar Technol Soc J 12(4):18–27. https://www.researchgate.net/publication/245666054_Classification_of_coastal_environments_in_terms_of_potential_vulnerability_to_oil_spill_damage. Accessed 1 Apr 2017

Gundlach ER, Hayes MO, Getter CD (1981) Sensitivity of coastal environments to oil spills. In: Proceedings petroleum industry and the Nigerian Environment, Warri, Nigeria. pp 82–88. https://www.researchgate.net/publication/315736071_Sensitivity_of_Coastal_Environments_to_Oil_Spills_-_Nigeria_Conference_Proceedings. Accessed 1 Apr 2017

Gundlach ER (1987) Oil-holding capacities and removal coefficients for different shoreline types to computer simulate spills in coastal waters. Proc int oil spill conf 1987(1):451–457. https://doi.org/10.7901/2169-3358-1987-1-451

Gundlach ER, Neff JM, Little DI (1993) Evaluation of marine post-spill sites for long-term recovery studies. MRSC, Technical Report Series 93-001. 183p. https://www.researchgate.net/publication/303666736_Evaluation_of_Marine_Post-spill_Sites_for_Long-term_Recovery_Studies. Accessed 14 May 2017

Information Nigeria (2013) JTF reviews activities in N/Delta, says 1,857 suspected oil thieves arrested, 1,951 illegal refineries destroyed in 2013. https://www.informationng.com/2013/12/jtf-reviews-activities-in-ndelta-says-1857-suspected-oil-thieves-arrested-1951-illegal-refineries-destroyed-in-2013.html. Accessed 30 Mar 2013

IPIECA (1993) Biological impacts of oil pollution: mangroves. Report Series Volume Four. London. 24p. http://www.commissionoceanindien.org/fileadmin/resources/Autoroute%20maritime%20IPIECA/IPIECA%20Biological%20impacts%20of%20oil%20pollution%20mangroves%20Vol%204%20IPIECA.pdf. Accessed 19 Mar 2017

IPIECA (2012) Sensitivity mapping for oil spill response. IOGP Report Number 477. London. 40p. http://www.ipieca.org/umbraco/Surface/Media/Download?url=%2fmedia%2f2822%2fsensitivity_mapping_for_oil_spill_respons_2012_r2016.pdf. Accessed 24 Mar 2017

IPIECA (2014) A guide to oiled shoreline assessment (SCAT) surveys. OGP Report Number 504. London. 40p. http://www.oilspillresponseproject.org/wp-content/uploads/2016/02/GPG-SCAT.pdf. Accessed 25 Mar 2017

IPIECA (2015) Response strategy development using net environmental benefit analysis (NEBA), rev 2016. OGP Report Number 527. London. 44p. http://www.ipieca.org/resources/good-practice/response-strategy-development-using-net-environmental-benefit-analysis-neba-good-practice-guidelines-for-incident-management-and-emergency-response-personnel/

ITOPF (2011) Fate of marine oil spills, Technical Information Paper 2. International Tanker Owners Pollutions Federation Limited, London. 12p. http://www.itopf.com/knowledge-resources/documents-guides/document/tip-2-fate-of-marine-oil-spills/. Accessed 15 Mar 2017

Keller BD, Jackson JBC (1993) Long-term assessment of the oil spill at Bahia Las Minas, Panama, synthesis report, volume I: executive summary. OCS Study MMS 93-0047. U.S. Department of the Interior. 129p. https://invertebrates.si.edu/boem/reports/POSP_synthesis_vol1.pdf. Accessed 24 Mar 2017

Lewis RR III (1983) Chapter 19: impact of oil spills on mangrove forests. In: Teas H (ed) Biology and ecology of mangroves, tasks for vegetation science. Springer Publishing, pp 171–183. doi:https://doi.org/10.1007/978-94-017-0914-9_19

Lewis M, Pryor R, Wilking L (2011) Fate and effects of anthropogenic chemicals in mangrove ecosystems: a review. Environ Pollut 159 (10):2328–2346. https://doi.org/10.1016/j.envpol.2011.04.027

Linden O, Pålssen J (2013) Oil contamination in Ogoniland, Niger Delta. Ambio 42(6):685–701. https://doi.org/10.1007/s13280-013-0412-8. https://www.ncbi.nlm.nih.gov/pmc/articles/PMC3758819/ Accessed 4 May 2017

Little DI, Holtzman K, Gundlach ER, Galperin Y (2018) Sediment hydrocarbons in former mangrove areas, southern Ogoniland, eastern Niger Delta, Nigeria. In: Makowski C, Finkl CW (eds) Coastal Research Library (CRL): threats to mangrove forests: hazards, vulnerability, and management. Springer, Cham, pp 323–342

MOE (2017) Webpage 'Ogoniland' Ministry of Environment, Nigeria. http://www.environment.gov.ng/ogoni. Accessed 20 Mar 2017

NDES (1997) Niger Delta environmental survey. Final Report Phase 1. 300p. https://www.researchgate.net/publication/315800412_Niger_Delta_Environmental_Survey_1997_Vol_1_-_Environmental_and_Socio-Economic_Characteristics. Accessed 15 May 2017

NOAA (2012) Environmental sensitivity index guidelines, version 3. National Oceanic and Atmospheric Administration, Office of Response and Restoration, Technical Memorandum NOS OR&R 11. 192p. http://response.restoration.noaa.gov/sites/default/files/ESI_Guidelines.pdf. Accessed 25 Mar 2017

NOAA (2013) Shoreline assessment manual, 4th edn. National Oceanic and Atmospheric Administration, Office of Response and Restoration. 154p. http://response.restoration.noaa.gov/sites/default/files/manual_shore_assess_aug2013.pdf. Accessed 25 Mar 2017

NOAA (2014) Oil spills in mangroves, planning and response considerations. In: Hoff R, Michel J (eds) National Oceanic and Atmospheric Administration, National Ocean Service, Office of Response and Restoration. 96 p. http://response.restoration.noaa.gov/sites/default/files/Oil_Spill_Mangrove.pdf. Accessed 19 Mar 2017

Owens EH, Sergy GA (2003) The development of the SCAT process for the assessment of oiled shorelines. Mar Pollut Bull 47 (9–12):415–422. https://doi.org/10.1016/S0025-326X(03)00211-X

Pålssen J, Linden O (2014) Oil contamination in the Niger Delta, vol 2014. In: Proceedings international oil spill conf. pp 1706–1718. doi:https://doi.org/10.7901/2169-3358-2014.1.1706

Pegg S, Zabbey N (2013) Oil and water: the Bodo spills and the destruction of traditional livelihood structures in the Niger Delta. Community Dev J 48(3):391–401. https://doi.org/10.1093/cdj/bst021

Renegar A, Schuler P, Turner N, Dodge R, Riegl B, Knap A, Bera G, Jezequel R, Benggio B (2017) TROPICS field study (Panama), 32-year site visit: observations and conclusions for near shore dispersant use NEBA and tradeoffs. Proc int oil spill conf 2017(1):3030–3050. https://doi.org/10.7901/2169-3358-2017.1.3030

Santos LCM, Cunha-Lignon M, Shaeffer-Novelli Y, Cintron-Molero G (2012) Long-term effects of oil pollution in mangrove forests (Baizada Santista, southeast Brazil) detected using a GIS-based multitemporal analysis of aerial photographs. Braz J Oceanogr 60 (2):159–170. https://doi.org/10.1590/S1679-87592012000200006

Sunderland TCH, Morikinyo T (2002) *Nypa fruticans*, a weed in West Africa. Palms 46(3):154–155. http://www.palms.org/palmsjournal/2002/vol46n3p154-155.pdf Accessed 1 April 2017

Toboc.com (2017) Bonny Light crude oil assay. 2p. http://www.toboc.com/images/pdf/624149.pdf. Accessed 24 Mar 2017

Tori News (2015) Photos of massive oil bunkering site uncovered by Nigerian Army in Port Harcourt. http://www.tori.ng/news/8050/photos-of-massive-oil-bunkering-site-uncovered-by.html. Accessed 21 Mar 2017

UNEP (2007) Mangroves of western and central Africa. UNEP-Regional Seas Programme/UNEP-WCMC. 92p. http://www.oceandocs.org/bitstream/handle/1834/5474/Mangroves_of_Western_and_Central_Africa.pdf?sequence=1&isAllowed=y. Accessed 1 Apr 2017

UNEP (2011) Environmental assessment of Ogoniland. United Nations Environment Programme. 262p. http://postconflict.unep.ch/publications/OEA/UNEP_OEA.pdf. Accessed 1 Apr 2017

USDOI (1997) Managing oil spills in mangrove ecosystems: effects, remediation, restoration, and modeling; a review produced from a workshop convened August 1995 at McNeese State University. In: Profitt E (ed) US Department of the interior, OCS study MMS 97-003. 81p. https://www.data.boem.gov/pi/pdfimages/espis/3/3254.pdf. Accessed 19 Mar 2017

Worldwideweatheronline (2017) Bonny, rivers monthly climate average, Nigeria. https://www.worldweatheronline.com/bonny-weather-averages/rivers/ng.aspx. Accessed 1 Apr 2017

Zabbey N, Tanee FBG (2016) Assessment of asymmetric mangrove restoration trials in Ogoniland, Niger Delta, Nigeria: lessons for future intervention. Ecol Res 34(3):245–257. https://doi.org/10.3368/er.34.3.245

Zabbey N, Hart AI, Wolff WJ (2010) Population structure, biomass and production of the West African lucinid *Keletistes rhizoecus* (Bivalve, Mollusca) in Sivibilagbara swamp at Bodo Creek, Niger Delta, Nigeria. Hydrobiologica 654:193–203. https://link.springer.com/article/10.1007/s10750-010-0381-x Accessed 1 Apr 2017

Sediment Hydrocarbons in Former Mangrove Areas, Southern Ogoniland, Eastern Niger Delta, Nigeria

14

David I. Little, Kay Holtzmann, Erich R. Gundlach, and Yakov Galperin

Abstract

The sediments and former mangrove areas near the town of Bodo in the Niger Delta are highly contaminated in the upper 20 cm by oil residues. The oil pollution resulting from two spills of approximately 10,000 t of Bonny Light crude oil in 2008 from the Trans-Niger Pipeline (TNP) killed mangroves in more than 1000 ha of local creeks. The impact of pollution in 2008 was exacerbated from late 2009 by increased transport and artisanal refining of crude oil stolen from tapped pipelines. It has been controversial which source of impact is the more relevant. Areas still contaminated include traditional fishing resources used by local communities. After years of preliminary engagement to build trust among stakeholders, the Bodo Mediation Initiative (BMI) was constituted formally in October 2013 to conduct cleanup of sediments in the impacted mangrove areas in an act of community self-management. This was to be accomplished with assistance from Shell Petroleum Development Company (SPDC) and mediation by the Dutch mission, federal and local government agencies, with advice from the United Nations Environment Programme (UNEP). The BMI then planned and undertook a preliminary Shoreline Cleanup Assessment Technique (SCAT) survey of the residual oil. SCAT is an internationally accepted visual survey method using standardized field procedures, photography and qualitative record-keeping. SCAT was calibrated using quantitative sediment sampling. These parallel activities were to investigate the degree of contamination, to help plan the cleanup, and to compare the situation before and after cleanup.

Total, aliphatic and aromatic hydrocarbon concentrations were determined using Gas Chromatography with Flame Ionization Detector (GC-FID). The results show the highly contaminated status of the sediments near Bodo in August 2015. Only one (1) of thirty-two (32) surface sediment samples (0–5 cm) showed a concentration of Total Petroleum Hydrocarbons (TPH in the C_5–C_{44} hydrocarbon range) below the Nigerian regulatory limit of 5000 mg/kg. Six (6) samples showed concentrations between 5000 and 10,000 mg/kg, seven (7) samples ranged between 10,000 and 20,000 mg/kg, fifteen (15) between 20,000 and 100,000 mg/kg, and three (3) were contaminated at levels above 100,000 mg/kg. Fifty-five (55) samples were taken at two different depths (0–5 cm and 15–20 cm), including the surface dataset. Although concentrations of TPH, aliphatic and aromatic hydrocarbons in the equivalent C_{12}–C_{44} range (EC_{12}–EC_{44}) were elevated in surface and subsurface samples, thirteen (13) of the subsurface samples had TPH concentrations below the Nigerian regulatory limit of 5000 mg/kg. TPH concentrations in three (3) subsurface samples showed concentrations between 5000 and 10,000 mg/kg, ten (10) samples ranged between 10,000 and 20,000 mg/kg, and six (6) between 20,000 and 100,000 mg/kg. The differences between surface and subsurface mean concentrations of TPH, aliphatic and aromatics fraction were each statistically significant ($p < 0.01$). The decreased concentrations from the surface towards deeper strata within the fine-grained sediments were as expected in the absence of significant deposition of new sediment above the oil spill layer. For the lower molecular weight fraction of aromatics (EC_5–EC_{12}) the trends were different. This fraction was detected in fifty-

D. I. Little (✉)
Swavesey, Cambridge, Cambridgeshire, UK

K. Holtzmann
Berlin, Germany

E. R. Gundlach
E-Tech International Inc, Boulder, CO, USA

Y. Galperin
Environmental Geochemistry Consulting, Moorpark, CA, USA
e-mail: ygalperin@geochemconsulting.com

C. Makowski, C. W. Finkl (eds.), *Threats to Mangrove Forests*, Coastal Research Library 25,
https://doi.org/10.1007/978-3-319-73016-5_14

five (55) samples in much lower concentrations overall, and with only twenty (20) samples exceeding 1 mg/kg, and seven (7) exceeding 5 mg/kg of which six (6) were subsurface samples. Also, in the case of lighter aromatics, the mean concentrations increased significantly with depth ($p < 0.01$) due to their higher volatility and potential for penetrating fine-grained sediments compared to the heavier hydrocarbons.

The Polycyclic Aromatic Hydrocarbons (PAH) concentrations determined by Gas Chromatography-Mass Spectrometry (GC-MS) in 55 samples ranged widely with no significant difference with depth. Although none exceeded the Nigerian regulatory limit of 40 mg/kg, fifteen (15) samples of which seven (7) were in the subsurface, exceeded Environment Canada sediment quality guidelines for the sum of sixteen PAHs (ΣPAH_{16}). A forensics evaluation of the available PAH data suggests the following sources of PAHs: petroleum, combustion of petroleum, and combustion products from wood and grass. This mixture illustrates inputs from both oil industry and illegal refinery activities, overprinted on a baseline of traditional biomass fuels. The SCAT and hydrocarbons results can be used to design practical cleanup and monitoring actions and to assess health risks for people ingesting or exposed to oil above an acceptable daily intake. Recommendations include preparing for the mangrove restoration, addressing possible human health impacts, better law enforcement, and improving employment opportunities in Bodo.

Keywords

Bodo · Contaminated sediments · Intervention values · Oil spill cleanup · PAH isomer fingerprinting · Potential health risk · UNEP environmental assessment

14.1 Introduction

14.1.1 Background

Bodo is a town located in Rivers State, Nigeria with a population total of about 69,000 inhabitants including outlying settlements (Pegg and Zabbey 2013). The study area is in south Ogoniland, near Port Harcourt on the Bonny River in the eastern Niger Delta and bordering the Atlantic Ocean. Bodo is in Gokana Local Government Area (LGA) which has a population of 228,828 (2006 census). Figure 14.1 shows the location of Bodo.

Environmental reviews and surveys of the Niger Delta were carried out in 1983 for SPDC by Baker (1983), in 1983–1985 for the Nigerian National Petroleum Corporation (NNPC) by RPI (1985), and in 1996 by the Niger Delta Environmental Survey (1997), an industry-funded non-government organisation (NGO). In 2007, UNEP was asked by the federal government of Nigeria to conduct a detailed survey of oil industry environmental and potential health impacts in Ogoniland (LGAs Gokana, Eleme, Tai and Khana). The UNEP findings were published (UNEP 2011) and an overview of conditions on higher land and groundwater as well as intertidal mangrove areas is given below:

Box 14.1. UNEP Environmental Assessment of Ogoniland

- Sediment and groundwater were found to be contaminated by aliphatic and aromatic hydrocarbons mostly within and adjacent to the Right-of-Way (RoW) of onshore pipelines. During construction, the foundation bedding of the pipelines was composed of loosened and/or made ground (usually coarse sandy material). This bedding forms a hydraulic high compared to the ambient undisturbed sediment, which has a lower permeability. Due to this hydraulic gradient the RoW bed often functions as a trap for spilled oil.
- Regression modeling, correlation and non-hierarchical cluster analyses of the analytical data revealed that contamination with aromatic hydrocarbons was locally still very high in samples taken from both recent and years-old spill sites. Contamination was influenced by the intensity and age of the impact, and by the depth of the groundwater table (Thiergärtner and Holtzmann 2014).
- UNEP criticised the remediation procedures typically being used in Ogoniland. Remediation is carried out mostly by contractors which are not instructed to apply state-of-the-art techniques and have often not been supervised adequately. One of the most frequent remediation techniques applied by SPDC in Ogoniland is Remediation by Natural Attenuation (RENA). This is a form of landfarming, involving the excavation and/or aeration of polluted material and the degradation of the hydrocarbons within the sediment under atmospheric conditions, sometimes enhanced by nutrient fluids (bioremediation). In many cases, RENA has failed because insufficiently degraded hydrocarbons in backfilled material led to additional contamination hazards for the population (UNEP 2011; Thiergärtner and Holtzmann 2014).

(continued)

Box 14.1 (continued)

- Groundwater of the upper aquifer was found to be highly contaminated locally. However, a fully representative analytical screening of the water quality in hand dug wells, the main source for potable water in Ogoniland, was not performed.
- Although systematic health questionnaires were conducted and sources of health risk were correctly identified, no epidemiological effects of the contamination on the health of the population were proven. This was not achievable mainly because a comprehensive physical and medical examination of the local population was not included in the survey.
- Analytical investigation of the tissue of fish and shellfish from local markets showed that hydrocarbon concentrations were compliant with current environmental quality standards and generally lower than 1983 levels in samples collected in the field (Baker 1983). In contrast, the most contaminated fishing areas in 2010 were being avoided by fishermen.
- Plants and agricultural products were not analysed chemically. However, plants showed symptoms of stress, for instance degraded growth, which might be ascribed to contaminated sediment and soils close to the roots.
- UNEP recommended the establishment of an Environmental Restoration Fund for Ogoniland, with estimated costs for the first 5 years of 1 billion US $. This included implementation of: emergency measures (*e.g.*, alternative drinking water sources for those communities with a contaminated water supply); clean-up of contaminated land and intertidal sediments; mangrove transplants and restoration; and accompanying monitoring measures.

The studies described above were not baseline surveys designed to show before-after and control-impact (BACI) effects, because oil had already been discovered in 1956 at Oloibiri-1, and the TNP from Bomu to Bonny was constructed in 1965. Soon after, oil was a central cause of the disastrous Biafran war, and oil pollution and social problems were among its legacies. Eighty-nine (89) producing oil wells were completed in Ogoniland by 1993 when SPDC production was halted by civil unrest. There have been many false starts to the cleanup of oil spills in the Niger Delta, in particular in Ogoniland due to a history of protest against SPDC. Nevertheless, UNEP successfully prepared the way for the Bodo Mediation Initiative (BMI) during its intensive survey of Ogoniland, including the Bodo area (UNEP 2011). Implementation of the wider UNEP recommendations was announced on 2 June 2016, and on 4 August 2016 the President of Nigeria established a Governing Council and Trust Fund.

Major sources of oil contamination in the area include two operational spills in 2008 prior to the UNEP field work. These spills were caused by pipeline corrosion and resulted in mangrove losses along the TNP crossing Bodo (Gundlach 2013). The mangrove losses in the area of the present chemistry sampling are shown in Fig. 14.1 (Gundlach 2017 this volume). From 2012, the Embassy of The Netherlands in Abuja in conjunction with leading NGOs began the BMI to facilitate cleanup and restoration of sediments in former mangroves near Bodo instead of waiting for an unpredictable outcome from a legal process in a United Kingdom court. Supported by the Nigerian and Dutch governments, in October 2013 the BMI started the formal process for planning and implementation of cleanup. An out-of-court settlement was reached in December 2014 between the Bodo community and SPDC that led to payment of compensation for the entire affected area of the creek, including 1000 ha of mangroves. A remediation project team was established by the BMI with local, national and international experts, regulators, church leaders, NGOs and a UNEP observer, together with employees of SPDC. A primary objective of remediation is to reduce sediment contamination to a level that facilitates recovery of the mangrove habitat and local livelihoods, either naturally or by transplanting mangrove seedlings cultivated in nurseries.

The project area in the Niger Delta consists of intertidal sediments dominated by silts and clays, and bisected by tidal channels mostly accessible only by boat (Figs. 14.2, 14.3 and 14.4). To prepare for the cleanup, the BMI decided to investigate the difficult ground conditions using a visual survey methodology, consistent with the internationally-accepted Shoreline Cleanup Assessment Technique (SCAT; Owens and Sergy 1994; Owens and Sergy 2003). The BMI has successfully brought together the stakeholders and in 2015 conducted preliminary surveys using a multi-disciplinary SCAT approach. It also engaged international cleanup contractors supported by experienced national companies. Despite all field operations being stopped by violent agitation in October 2015, the BMI has maintained the technical and logistical capability to plan and implement the cleanup and restoration of the degraded Bodo mangrove ecosystem. SCAT fieldwork resumed in September 2017.

The BMI decided to verify and calibrate the SCAT findings by commissioning analyses of samples for hydrocarbons in sediments taken from the areas of affected mangroves near Bodo. The aim of this contribution is to

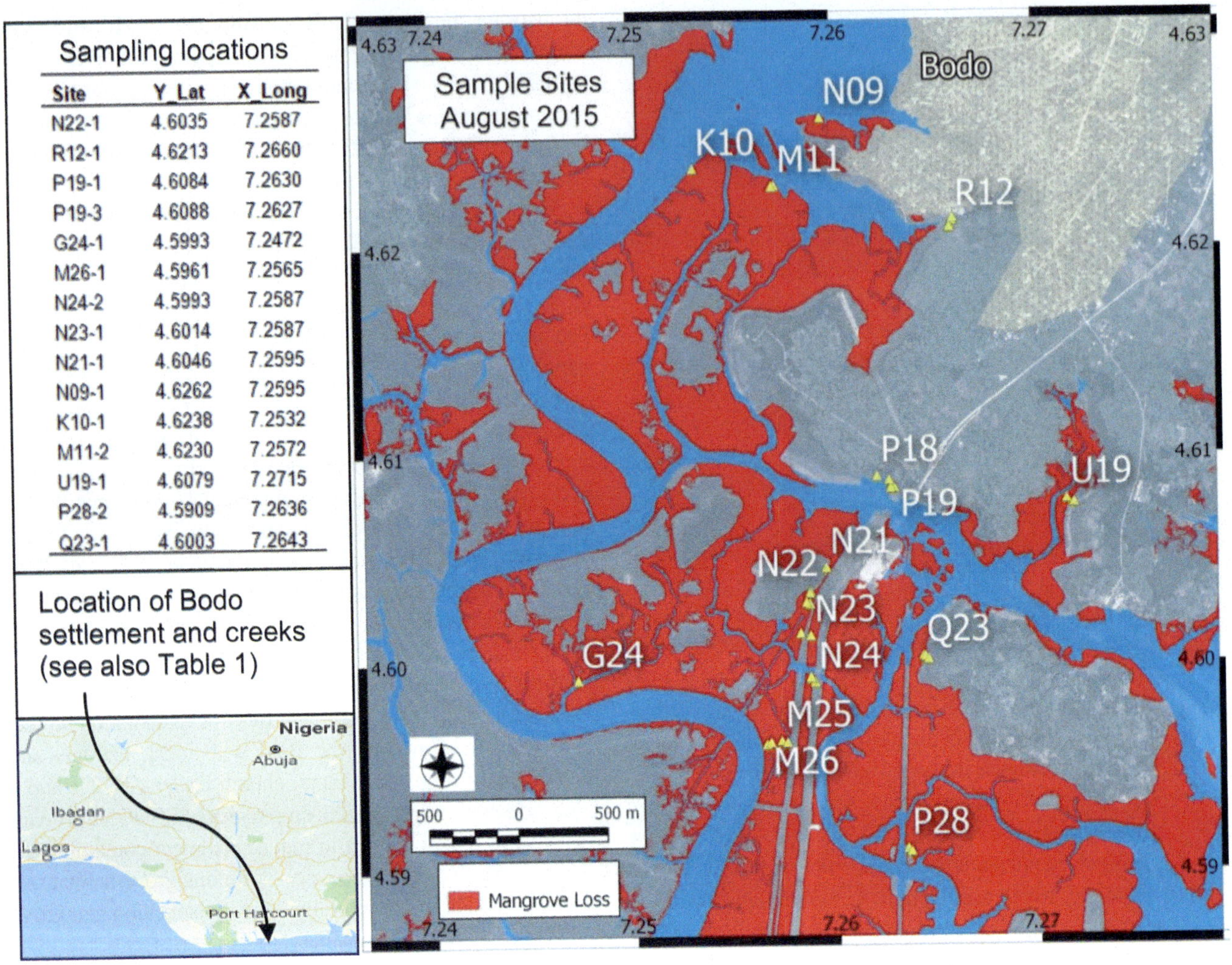

Fig. 14.1 August 2015 sampling sites near Bodo; mangrove loss defined by Gundlach (this volume)

Fig. 14.2 Typical views of oil-killed mangroves in the Bodo study area, 1 May 2015 (Credit: D. Little)

Fig. 14.3 Difficult working area on fine-grained and low shear-strength mudflats among dead mangroves, 7 May 2015 (Credit: M. Bonte)

Fig. 14.4 Oiled mangroves indicate tidal range; Nipa palms *Nypa fruticans* (Asian invasive species) and terrestrial vegetation visible (Credit: D. Little)

present and discuss the analytical results of the first sampling campaign in August 2015.

14.1.2 Natural Environment Baseline

In January–February 1983 an environmental study of five SPDC-operated areas in the Niger Delta was conducted by Baker (1983). This included the area of the present chemistry sampling on the intertidal mudflats near Bodo. The sediments were described as consisting of accretional poorly-sorted medium to very fine silts (59–99% <63 μm particle diameter). The Bodo area is part of the interconnected Bonny and Andoni rivers system, that are in fact tidal embayments unconnected to the Niger Delta distributary system (NDES 1997). Tidal range is 1.4 m at Bonny town and 1.8 m at Port Harcourt (RPI 1985). Water circulation is driven by flood-tide dominated tidal currents, and salinity is dominated by seasonal rainfall. Steep salinity gradients lead to clay flocculation in the channels, bringing nutrients and suspended muds of mainly marine origin to the intertidal flats and mangrove forest. Salinity is also a primary environmental gradient affecting the local biota, ranging during the early 1983 dry season from 13 to 30 parts per thousand (ppt) from the north of Bodo southwards to the confluence of the Kpador channel with Bonny River (Baker 1983). In the rainy season salinity decreases across all the estuary system (Dublin-Green 1990). For example in April 2013 salinity ranged between 19.5 and 23.5 ppt from Bodo to Bonny River (Gundlach this volume).

Before development of the oil industry in Bodo, the natural vegetation was dense mangrove forest. This formed a self-sustaining coastal defence against Atlantic storms, and supported diverse natural habitats supplying sustainable environmental services to the Ogoni people. In 1983, creeks and mangroves provided nutrient filtration, fish nursery and bird staging areas, and local transport corridors for human access to vital natural resources such as food and wood for building, fuel and making cultural artifacts (Fentiman and Zabbey 2015). For all these reasons mangrove habitats are ranked by the Environmental Sensitivity Index (ESI) as the type of marine shoreline most sensitive to persistent oil pollution world-wide (Gundlach and Hayes 1978), and in particular Nigeria (Gundlach et al. 1981; RPI 1985).

During the 1983 survey across five SPDC-operated areas, the red mangrove *Rhizophora racemosa* was the dominant baseline species along almost the entire forest shoreline of the Bodo West area, with the subordinate species *R. harrisonii* and black mangrove *Avicennia germinas* also found respectively at 14% and 27% of the twenty-two (22) rapid assessment sites. A further species of *Rhizophora* (presumably *R. mangle*) was found at 18% of sites. The intertidal mangrove community was complemented at 45% of sites by low shrubs of *Laguncularia racemosa,* occasionally dense near creek confluences towards the Bonny River in the south. In 1983, the invasive Nipa palm *Nypa fruticans* was observed mainly in the Forcados area, and only at one of a total of twenty-seven (27) sites in the Bodo area, where it is now more common. Since 1983 these and other ecologically important aspects of the Bodo area have been subject to habitat scale and cascading impacts intensifying after 2008 (Figs. 14.1, 14.2, 14.3 and 14.4). The mangroves formerly provided a firm substrate and shelter for a rich community of invertebrates such as the oyster *Crassostrea gasar*, the fiddler crab *Uca tangeri* and other crabs at 86–90% of sites. Additionally, the isopod *Sphaeroma terebrans* and snail *Littorina angulifera* were found at 33% of the sites, providing a traditional food resource for the community (Baker 1983).

In January–February 1983 surficial (0–5 cm) sediment samples (n = 36) were taken across all five SPDC-operated areas for determination of hydrocarbon concentrations. In the Bodo West area, the arithmetic mean TPH concentration in five (5) composite samples from six (6) intertidal sites in the locality was relatively low at 2690 mg/kg; well below the threshold (intervention) concentration of 5000 mg/kg set more recently by the Nigerian Department of Petroleum Resources (DPR 1991). The Bodo samples were found to range from 110 mg/kg of weathered and partly biogenic hydrocarbons, up to 6750 and 8100 mg/kg of degraded crude oil at two sites respectively near the Bodo wellhead and the confluence with the Bonny River. The first of these two samples that exceeded the current DPR limit had been affected by a spill from the Bodo West flow station in the year prior to the survey by Baker (1983).

Sediment samples (n = 37) in Ogoniland were taken from the 0 to 10 cm depth by UNEP teams in April–November 2010 (UNEP 2011; Lindén and Pålsson 2013; Pålsson and Lindén 2014). The range in TPH concentrations found at the nine (9) intertidal sites in Bodo and the surrounding Gokana LGA was from a minimum of 279 mg/kg (twice that found almost three decades earlier) up to 17,900 mg/kg. This maximum value was also twice that measured in Bodo in 1983, using comparable methods. Six (6) samples exceeded the DPR limit of 5000 mg/kg TPH. The mean value was 7500 mg/kg, almost three times the 1983 mean. Deeper sediment layers at five (5) sites were also sampled in 2010 (UNEP, unpublished). Four (4) of these sites were in the impacted mangrove areas near Bodo. Although the mean TPH concentration of 11,918 mg/kg (n = 4) in the surface 0-10 cm exceeded the DPR limit, the concentrations of TPH, aliphatic and aromatic hydrocarbons were much lower in the 20–30 cm and 40–50 cm layers than in the surface 0-10 cm. The same was true for PAH, except at one site (Bodo West) where the ΣPAH_{16} concentration was 8.09 mg/kg in the 40–50 cm layer. Unpublished data from two independent sampling events in 2014 show high TPH concentrations at 20–40 cm sediment depths. However, these 2014 data were obtained using different methods to the 1983 and 2010 data that better represent the baseline conditions for comparison with the August 2015 survey, albeit at a different set of sampling sites (see Discussion).

14.2 Methodology

14.2.1 Shoreline Cleanup Assessment Technique (SCAT)

The methodology for the May–August 2015 survey campaign was based on two multi-discipline SCAT surveys and an analytical chemistry investigation in parallel with SCAT. Initially used in 1988 in the 'Nestucca' spill and developed fully during the 1989 'Exxon Valdez' spill, SCAT is now an accepted, codified and internationally established survey methodology (Owens and Sergy 1994; Owens and Sergy 2003). Areas contaminated by oil spills are investigated by multi-disciplinary teams visually assessing factors such as colour, form and degree of oiling, depths of sediment penetration, and obvious biological impacts, but without necessarily taking numerous sediment samples for analyses. SCAT surveys provide a characterization of the severity of oil contamination and build consensus on a recommended cleanup methodology. In this case, the SCAT surveys utilized a 200 × 200 m grid pattern. The sediment sampling followed this plan in order to link the analytical chemistry results to the field oiling conditions, character and severity, and to justify the proposed cleanup strategy. Before and after cleanup SCAT surveys will be the primary line of evidence to document project success. It was proposed to select a number of sites representing the variability in severity of contamination (e.g., 'heavily oiled' sites to 'rainbow sheen only').

In May 2015, a preliminary ('pre-SCAT') campaign was conducted during 1 week in the Bodo area to plan the main campaign for August 2015 (Gundlach 2015a). SCAT was chosen to prepare the cleanup contractors for the field conditions in the project area and to enable them to apply the appropriate remediation technologies for different site conditions. It was thus the first attempt to develop a broad cleanup plan, including the segmentation and prioritisation of the affected area. SCAT also should enable the contractors to apply the optimal safety and precautionary measures for the work force, including personal protective equipment (PPE).

The aims of the survey undertaken in August 2015 "were to (a) define areas that can be cleaned during the first phase of cleanup, (b) provide an overview of different habitats and levels of oiling, and (c) continue SCAT field training of participants" (Gundlach 2015b). The data collected in both surveys were noted in the field on the 'Shoreline Oiling Summary Form', and transferred onto maps. In total, fifty-two (52) SCAT sites were surveyed in twenty-nine (29) grid cells during 2015.

14.2.2 Chemistry Sampling and Analyses

The BMI determined that sampling and chemical analysis should not be carried out for each SCAT location surveyed. Instead, samples were collected from a selection of sixteen (16) grid cells to establish representative pre- and post-remediation hydrocarbon concentrations. The BMI advised the project team to conduct the sampling of the designated 1000 ha remediation area in order to calibrate and verify the SCAT data, using multiple samples taken within the same grid pattern as used in the SCAT surveys. Another primary rationale for the chemistry survey was to assess the hydrocarbon concentration in the sediment below which mangroves can survive, regenerate or recolonize.

The 'Bodo Creek Chemical Sampling and Analysis Protocol' was prepared by a Shell employee with input and support from a technical sub-group drawn from the BMI (Bonte 2015). The aim of this exercise was to organize the operations according to international best practices and to be consistent with SPDC's 'Oil Spill Response and Remediation Management System' (SPDC 2013). The samples were collected by a Nigerian company (Geoterrain Nigeria Limited) according to a plan agreed by the BMI. Figure 14.1 and Table 14.1 show the sample site coordinates and dates, arranged primarily along the TNP. The total number of samples was 57, with less data being reported for some parameters due to 'non-detect' values in the laboratory. Samples were sent under chain-of-custody to ALcontrol an ISO 17025 accredited laboratory in the United Kingdom. Key analytical parameters included:

- Methyl tert-Butyl Ether (MTBE), Benzene, Toluene, Ethylbenzene, Xylenes (BTEX);
- Low molecular weight aromatics fraction (C_5–C_{12});
- Total Aliphatics (C_{12}–C_{44}) and aliphatics within specified carbon number ranges;
- Total Aromatics (EC_{12}–EC_{44}) and aromatics within equivalent carbon number ranges;
- Total Petroleum Hydrocarbons (TPH C_5–C_{44}); and
- Polycyclic Aromatic Hydrocarbons (PAH) as individual and total concentrations of the 16 parent PAHs from the US Environmental Protection Agency (USEPA) Priority Pollutant List (Naphthalene, Acenaphthylene, Acenaphthene, Fluorene, Phenanthrene, Anthracene, Fluoranthene, Pyrene, Benz(*a*)anthracene, Chrysene, Benzo(*b*)fluoranthene, Benzo(*k*)fluoranthene, Benzo(*a*)

Table 14.1 Surface (0–5 cm) sediment sampling locations, dates, contaminants (mg/kg) and SCAT data

Grid	Coordinates		Date	Laboratory data		SCAT fieldwork data		
Site	Lat	Long	2015	TPH	ΣPAH_{16}	Surface	Subsurface	Sediment type
N22-1	4.6035	7.2587	13/08	60,000	16.7	TB	BO, PFP	Chikoko mud
N22-2	4.6034	7.2587	13/08	99,500	16.7	TB	BO, PFP	Chikoko mud
N22-3	4.6029	7.2587	13/08	24,400	1.6	AP, FR	BO, PFP	Chikoko mud
N22-4	4.6029	7.2585	13/08	25,500	5.7	NO	BO, PFP	Chikoko mud
R12-1	4.6213	7.2660	14/08	7700	1.0	SS	SS, PFP	Mud + algal mat
R12-2	4.6210	7.2659	14/08	27,700	0.8	BO	BO, PFP	Soft mud over chikoko
R12-3	4.6214	7.2660	14/08	81,100	12.6	BO	BO, PFP	Soft mud + algal mat
P19-1	4.6084	7.2630	17/08	54,900	10.0	AP	BO, SS, PFP	Chikoko mud
P19-2	4.6084	7.2628	17/08	6410	1.0	TB	BO, PFP	Mud + dead organics
P19-3	4.6088	7.2627	17/08	75,000	9.0	BO	BRO, PFP	Mud mixed with leaves
P18-2	4.6089	7.2621	17/08	88,500	24.2	BO	BO, PFP	Chikoko mud + algal mat
G24-1	4.5993	7.2472	18/08	26,700	n/d	NO	BO, SS, PFP	Chikoko mud
M26-1	4.5961	7.2565	18/08	7110	n/d	NO	BO, SS, PFP	Chikoko mud
M26-2	4.5962	7.2567	18/08	17,700	0.6	NO	BO, SS, PFP	Chikoko mud
M25-3	4.5963	7.2572	18/08	16,400	4.6	NO	BO, SS, PFP	Chikoko mud
M26-4	4.5962	7.2575	18/08	15,000	5.7	NO	BO, OP	Chikoko mud
N24-2	4.5993	7.2587	20/08	111,000	19.9	TB	BO, OP	Mud + algal mat
N24-3	4.5993	7.2587	20/08	17,200	0.9	NO	BO, PFP	Chikoko mud + algal mat
N23-1	4.6014	7.2587	20/08	9570	0.2	SS	BRO, PFP	Silty sand, chikoko, algal mat
N23-2	4.6015	7.2582	20/08	66,600	n/d	TB, SS	BRO, OP	Chikoko mud + algal mat
N21-1	4.6046	7.2595	20/08	8720	7.0	FR	BO, OP	Sand + roots
N09-1	4.6262	7.2595	21/08	135,000	1.2	SS	BO, OP	Chikoko mud + algal mat
K10-1	4.6238	7.2532	21/08	22,500	0.6	SS	BO, PFP	Chikoko mud + algal mat
M11-1	4.6290	7.2571	21/08	117,000	0.6	TB, AP	BO, PFP	Clay
M11-2	4.6230	7.2572	21/08	96,400	19.8	NO	BO, BRO, PFP	Chikoko mud + algal mat
U19-1	4.6079	7.2715	25/08	12,800	0.6	SS	SS, PFP	Chikoko
U19-2	4.6077	7.2719	25/08	3650	n/d	NO	BO, PFP	Chikoko mud + algal mat
P28-1	4.5911	7.2634	25/08	80,700	1.2	AP	BO, OP	Clay + Chikoko + asphalt
P28-2	4.5909	7.2636	25/08	6130	0.6	NO	BO, SS, PFP	Chikoko mud + algal mat
Q23-1	4.6003	7.2643	25/08	16,300	1.3	SS	BRO, OP	Sand + chikoko
Q23-2	4.6003	7.2644	25/08	25,800	3.3	TB	BO, PFP	Sand, chikoko, asphalt, algal mat
Q23-3	4.6002	7.2646	25/08	12,600	3.8	FR	BO, OP	Brown silty sand

SCAT oiling codes: *TB* tar or tarballs, *AP* asphalt pavement, *FR* fresh oil, *OP* oil-filled pores, *PFP* partially-filled pores, *BO* black oil, *BRO* brown oil, *SS* silver/rainbow sheen, *NO* no oil, *Chikoko* mangrove peat, *n/d* not detected; sampling site grid locations corrected to those plotted by SPDC Geomatics Department

pyrene, Indeno(*1,2,3-cd*)pyrene, Dibenzo(*a,h*)anthracene, and Benzo(*g,h,i*)perylene).

The above scope was specified by the technical sub-group convened by BMI. The selected laboratory had been used by UNEP in 2010 for TPH determinations. Their methods were also comparable to those used by Baker (1983): pooling of five (5) composite surficial samples, storage at −20 °C, solvent extraction, separation, and analysis by GC-FID. Petroleum hydrocarbons were extracted using a microwave-assisted solvent extraction method (EPA 3546). The analysis of extractable total petroleum hydrocarbons (TPH, in the C_5–C_{44} range) was using GC-FID (EPA Method 8015 modified) and PAH was analyzed using GC-MS (EPA Method 8270 modified). The volatile petroleum hydrocarbons (C_5–C_{12} gasoline range) and BTEX were analyzed using static head space extraction (EPA Methods 8020 and 8260 modified). In addition, the laboratory reported TPH results in the risk-based specified carbon number ranges for aliphatic and aromatic hydrocarbons (TPHCWG 1997). Quality assurance/quality control (QA/QC) was in terms of percentage recovery of compounds from analytical QC samples run with the field samples: surrogate recoveries were between 81% and 111% and generally well inside the statistical lower and upper control limits.

The laboratory work also included some analytes typically used in petroleum fingerprinting studies. However, it was assumed that most of the pollution detected would be Bonny Light crude oil, and so standard USEPA-type procedures were followed, rather than specialised hydrocarbon fingerprinting methodology (Stout and Wang 2016). The data obtained did not allow for a detailed environmental forensics level of interpretation. For example, chromatographic resolution by carbon range numbers did not identify individual alkane peaks, preventing reliable evaluation of the petroleum type(s) in the samples analysed. In addition, since the gas chromatograms were presented in a format normalised to the highest peak, the chromatograms were often too small scale to be useful for fingerprinting by visual comparison. In some cases the only identifiable peaks were the added internal standards, and contaminant peaks characteristic of the sample were hard to discern.

The individual PAH concentrations in some of the samples were below the laboratory detection limits, and eleven (11) total PAH values were below the detection limits. In addition, the PAH analytes did not include alkylated homologues. This selection was based on the correct assumption that, apart from the alkylated naphthalene and dibenzothiophene, these compounds are present at very low levels in the non-weathered crude oil. However, illegal refineries exist in many parts of the Niger Delta, including 87 sites in Bodo West and 60 in Bodo East (Gundlach this volume). When the light ends of crude oil are distilled off then the residual materials contain elevated levels of alkylated homologues. Without the alkylated homologues, only parent high molecular weight PAHs are reported, which limits the fingerprinting value of the analytical data. In future, high resolution GC-FID chromatograms and GC-MS outputs with selected ion chromatograms of representative samples should be included, as has been done recently in Owaza, Niger Delta by Faboya et al. (2016). This would provide fingerprinting tools even for samples where most *n*-alkanes are biodegraded.

14.3 Results

14.3.1 Summary of SCAT Survey Results

During the pre-SCAT survey in May 2015 teams visited seventeen (17) impacted sites, documenting the variability of surface and subsurface oil occurrence and behaviour (Fig. 14.5). Observations included: "dead mangrove plants, hydrocarbon impacted sediments to varying degrees from oil sheen to weathered crude in form of asphaltic tar, burnt carbonized residue and free-phase oil, and evidence of regeneration of dead mangrove vegetation where re-oiling has been minimised" (Gundlach 2015a).

The August SCAT survey provided an inventory of phenotypes of contaminated mangroves in the area: "The very large extent of oiling and required cleanup was again noted by survey participants. All sites surveyed during August 2015 need some level of cleanup, particularly related to reducing the level of oiling in soft muddy sediment" (Gundlach 2015b). SCAT observations of oiling and sediment types in August 2015 are given in Table 14.1 for those sites sampled in parallel for chemistry (Fig. 14.1). Additionally, mangrove loss was observed in each of the thirteen (13) scat grid cells that were not yet sampled. All but two of these were within the designated 1000 ha remediation area.

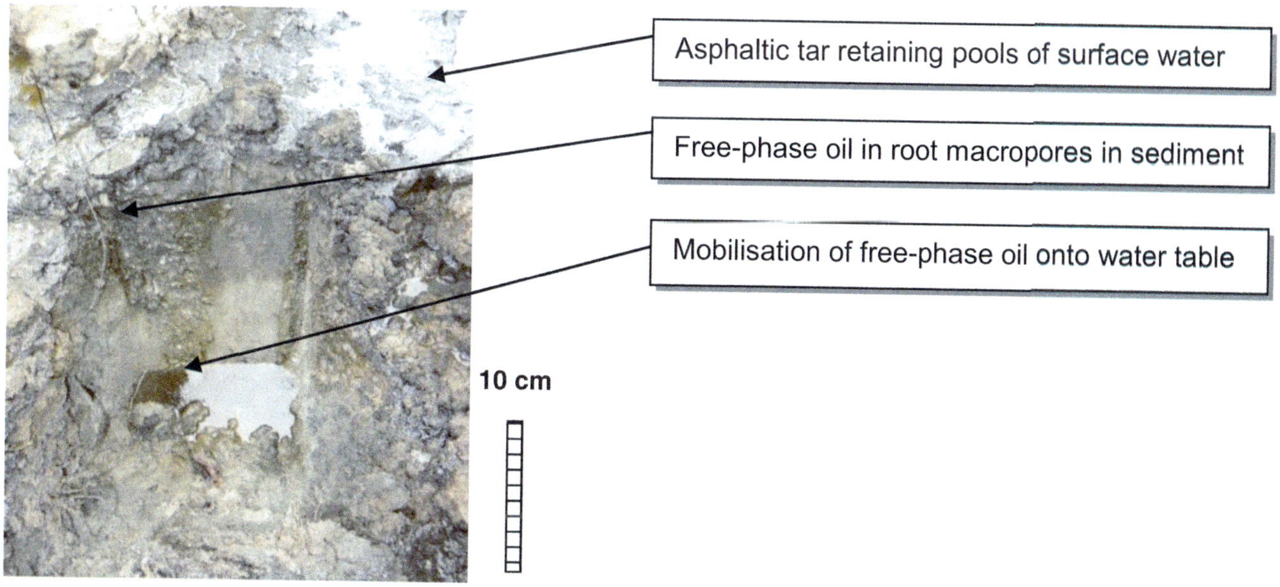

Fig. 14.5 SCAT pit showing surface asphaltic tar, mobilisation of free-phase oil into sediment macropores and onto the water table (Credit: M. Bonte)

Full results of the SCAT surveys carried out by the BMI in 2015 are given by Gundlach (2017).

14.3.2 Chemistry Results

14.3.2.1 Surface Samples

Site locations sampled from 13 to 25 August 2015 are shown in Fig. 14.1 and analytical results are given in Table 14.1 along with field information from SCAT.

Thirty two (32) sediment samples were taken from fifteen (15) grid locations at a depth interval of 0–5 cm. At these sites only one (1) of the 32 surface sediment samples showed a TPH (C_5–C_{44} range) concentration below the Nigerian regulatory limit value of 5000 mg/kg (see Discussion). Six (6) samples had concentrations between 5000 and 10,000 mg/kg; seven (7) samples ranged between 10,000 and 20,000 mg/kg; fifteen (15) between 20,000 and 100,000 mg/kg; and three (3) samples exceeded 100,000 mg/kg (Table 14.1).

There was general agreement between visual SCAT observations and chemistry data. However, due to oil weathering and the presence of algal mats, the sediment surface often appeared cleaner than when pits were dug to reveal fresh oil in the subsurface (e.g., especially all sites in grid cell M26, but also N22-4, G24 and P28-2). Table 14.1 shows that in the subsurface 15–20 cm layer, of thirty (30) observations that were made of 'black oil' or 'brown oil' (SCAT oiling codes BO and BRO), twenty-four (24) of them corresponded with TPH concentrations >10,000 mg/kg in the surface layer. This fully confirmed the SCAT finding of severe oiling.

14.3.2.2 Surface and Subsurface Samples Compared

Because TPH (as well as aliphatic, aromatic fractions and PAH) concentrations were also high in the deeper sampling interval, at thirteen (13) grid locations 55 samples were taken in two different depths (0–5 cm and 15–20 cm). These data from surface and subsurface are given in Tables 14.2 and 14.3 respectively. Data for TPH and the sum of PAH_{16} correspond with the above surface dataset (Table 14.1), but for N23 and N24 data were not available for aromatic and aliphatic fractions.

Of the subsurface total of 32 samples, thirteen (13) samples had TPH concentrations below the regulatory limit of 5000 mg/kg. TPH concentrations in three (3) subsurface samples were between 5000 and 10,000 mg/kg, ten (10) samples were between 10,000 and 20,000 mg/kg, and six (6) between 20,000 and 100,000 mg/kg (Table 14.3). In summary:

- ***In the surface 0–5 cm the mean concentrations were***: TPH 43,208 mg/kg; Aliphatics 25,045 mg/kg; Aromatics 17,015 mg/kg; EC_5–EC_{12} aromatics fraction 3.26 mg/kg; and PAH_{16} 6.11 mg/kg (n = 28, except TPH n = 32, and PAH n = 27).
- ***In the subsurface 15–20 cm the mean concentrations were***: TPH 14,601 mg/kg; Aliphatics 7702 mg/kg; Aromatics 4288 mg/kg; EC_5–EC_{12} aromatics fraction 9.28 mg/kg; and PAH_{16} 5.09 mg/kg (n = 27, except TPH n = 32, and PAH n = 23).
- ***In both depth strata combined the mean concentrations were***: TPH 28,904 mg/kg; Aliphatics 16,531 mg/kg; Aromatics 10,767 mg/kg; EC_5–EC_{12} aromatics fraction 6.16 mg/kg; and PAH_{16} 5.65 mg/kg (n = 55, except TPH n = 64, and PAH n = 50).

The TPH concentrations in the sediments of the Bodo area ranged over almost three (3) orders of magnitude; from 163 mg/kg at R12 (subsurface sample) to 135,000 mg/kg at N09 (surface sample). Although the low end of the present range in TPH concentrations is similar to the background concentrations found in 1983 and slightly less than in 2010, the high end of the range in oil content (13.5% of the sediment mass) is not an outlier, and indeed only one surface sample was below the regulatory limit of 5000 mg/kg. This emphasizes the severity and persistence of the 2008 Bodo pipeline oil spills, and the subsequent spills due to other illegal activity.

14.3.2.3 Further Data Evaluation

With such a wide range of data, it is difficult to express meaningful measures of central tendency, such as arithmetic means. Under these circumstances, to make valid comparisons between the surface and subsurface layers, samples need to be compared at the same locations and the variability of the TPH data needs further evaluation. The ratio of the standard deviation (SD) to the mean of samples in each depth layer at each site was calculated (where n = 2 or more, noting that sites P18, G24, K10, N09 and N21 had only one sample per depth).

When expressed as a percentage, each of these ratios represents a dimensionless coefficient of variation (CV) where CV = SD × 100/mean. For surface 0–5 cm samples the CV ranged from 13.7% (M11) to 122% (P28) and the mean was 67%. In the subsurface 15–20 cm samples the CV ranged from 20% (U19) to 132% (N23), with a similar mean of 58%. The overall similarity of the CVs in terms of depth in the sediment was encouraging, but the size and range of CVs underlined the intense patchiness of oil distribution in the mangrove sediments of the Bodo area. Although reduced by taking composite samples, this patchiness is normal in coastal sedimentary systems, due to the unpredictable oil distribution at the time of the spill and subsequently to variable oil penetration down macropores

Table 14.2 TPH, Aliphatics (C_{12}–C_{44}), Aromatics (EC_{12}–EC_{44}), Aromatics (EC_5–EC_{12}) fraction and ΣPAH_{16} data for surface (0–5 cm) sediment samples taken from 15 grids in August 2015

	mg/kg				µg/kg
Surface sample	TPH	Aliphatics	Aromatics	Aromatics EC_5–EC_{12}	ΣPAH_{16}
N22-1	**60,000**	35,400	24,600	0.14	**16,700**
N22-2	**99,500**	68,900	30,600	0.18	**16,700**
N22-3	**24,400**	12,500	11,900	0.59	1570
N22-4	**25,500**	17,000	8460	0.27	5660
R12-1	**7770**	4310	3460	0.62	1030
R12-2	**27,700**	16,000	11,700	1.68	794
R12-3	**88,100**	70,100	18,100	1.86	**12,600**
P19-1	**54,900**	36,400	18,400	0.19	**10,000**
P19-2	**6410**	4430	1990	0.32	962
P19-3	**75,000**	36,400	38,600	0.39	**8980**
P18-2	**88,500**	48,700	39,700	0.13	**24,200**
G24-1	**26,700**	14,900	11,800	0.06	n/d
M26-1	**7110**	4730	2380	0.12	n/d
M26-2	**17,700**	11,100	6590	0.07	590
M26-3	**16,400**	9310	7060	0.3	4600
M26-4	**15,000**	10,400	4590	0.87	5710
N24-2	**111,000**	n/a	n/a	n/a	**19,900**
N24-3	**17,200**	n/a	n/a	n/a	875
N23-1	**9570**	n/a	n/a	n/a	236
N23-2	**66,600**	n/a	n/a	n/a	n/d
N21-1	**8720**	5780	2660	73.54	6980
N09-1	**135,000**	73,200	61,900	0.82	1180
K10-1	**22,500**	13,400	9090	0.33	590
M11-1	**117,000**	61,200	55,500	0.12	590
M11-2	**96,400**	42,100	54,300	4.44	**19,800**
U19-1	**12,800**	8710	4080	0.03	590
U19-2	3650	2330	1320	0.02	n/d
P28-1	**80,700**	54,500	26,200	2.06	1180
P28-2	**6130**	3510	2620	0.95	590
Q23-1	**16,300**	10,300	6070	0.57	1340
Q23-2	**25,800**	17,900	7860	0.39	3330
Q23-3	**12,600**	7750	4890	0.11	3820
Means	**43,208**	25,045	17,015	3.26	6111

In bold: TPH concentration >5000 mg/kg limit value (EGASPIN) and PAH_{16} concentration >7070 µg/kg Probable Effects Level (Environment Canada; note that the EGASPIN limit value of 40,000 µg/kg was not exceeded)
n/d not detected, *n/a* data not available

such as plant root channels and infaunal burrows (Howard and Little 1987; Kingston et al. 2000).

To compare the widely-ranging contaminant concentrations in relation to the sample depth it was thus necessary to $\log_{10}$-transform the data. This was done to reduce the samples' variances in relation to their means, and helps to approach the normal distribution permitting use of a parametric t-test. Tables 14.2 and 14.3 respectively show the surface and subsurface data. Note that the individual data points were rounded by the laboratory to three significant figures, and so in Tables 14.2 and 14.3 the data for separated fractions of aliphatics and aromatics do not always sum exactly to the respective values for TPH (C_5–C_{44}).

14.3.2.4 TPH, Aliphatic and Aromatic Hydrocarbons Statistical Results

Although aromatic hydrocarbons concentrations were significantly higher in the surface than subsurface using the parametric t-test, the similar vertical differences in mean TPH and aliphatic hydrocarbons could not be tested because of the high variance ratios (shown by the F-test) of the samples even after $\log_{10}$-transformation. Nor could non-parametric alternatives to the t-test be used (e.g., Wilcoxon's test for matched pairs and Mann-Whitney U-test), because the frequency distributions differed in skewness between the surface and subsurface samples (see "Discussion"). However, when measurements constitute a matched pair, the matched pairs t-test can be used. Despite the necessary omission of the few

Table 14.3 TPH, aliphatics (C_{12}–C_{44}), aromatics (EC_{12}–EC_{44}), aromatics (EC_5–EC_{12}) fraction and ΣPAH_{16} data for subsurface (15–20 cm) sediment samples (n = 27) taken from 13 grids in August 2015

	mg/kg				μg/kg
Subsurface sample	TPH	Aliphatics	Aromatics	Aromatics EC_5-EC_{12}	ΣPAH_{16}
N22-2	2430	1610	815	0.48	539
N22-3	3070	1980	1090	3.31	976
N22-4	1610	1030	576	0.18	2890
R12-1	163	76.2	86.6	0.03	n/d
R12-2	4880	3150	1720	2.6	1020
R12-3	**19,600**	14,500	5140	2.64	2320
P19-1	**21,900**	16,500	5390	0.89	**7700**
P19-2	4710	2860	1840	0.74	2070
P19-3	**13,600**	8780	4820	0.51	1630
P18-2	2500	1590	910	0.2	320
G24-1	**17,100**	10,200	6890	2.89	2200
M26-1	**17,100**	10,600	6470	9.42	3880
M26-2	**23,300**	13,900	9310	9.99	2020
M26-3	**7820**	5530	2200	32.31	**8500**
M26-4	**6520**	3590	2910	8.02	521
N24-1	**17,000**	n/a	n/a	n/a	**10,300**
N24-2	**19,800**	n/a	n/a	n/a	**20,500**
N24-3	**89,500**	n/a	n/a	n/a	**22,800**
N23-1	531	n/a	n/a	n/a	590
N23-2	**16,000**	n/a	n/a	n/a	n/d
N21-1	**5070**	3010	1590	150.9	**7850**
N09-1	**71,700**	45,000	26,700	1.07	1180
K10-1	**14,200**	9320	4860	1.64	n/d
M11-1	**12,600**	8640	4000	1.16	1560
M11-2	**29,500**	17,900	11,600	3.96	n/d
U19-1	215	107	108	0.11	n/d
U19-2	287	174	113	n/d	n/d
P28-1	2050	1270	768	1.27	n/d
P28-2	1290	662	630	0.22	n/d
Q23-1	**25,300**	15,200	10,100	0.64	**14,200**
Q23-2	**11,700**	8380	3370	0.12	1540
Q23-3	4180	2390	1770	5.93	n/d
Means	**14,601**	7702	4288	9.28	5092

In bold: TPH concentration >5000 mg/kg limit value (EGASPIN) and PAH_{16} concentration >7070 μg/kg Probable Effects Level (Environment Canada; note that EGASPIN limit value of 40,000 μg/kg was not exceeded)
n/d not detected, *n/a* data not available

unpaired samples, this is a more sensitive test, and for consistency it was used on all the hydrocarbons data.

The t-test for matched pairs of $\log_{10}$-transformed data indicated that the difference between the mean TPH (C_5–C_{44}) concentrations of the surface 0–5 cm layer and the subsurface 15–20 cm layer was statistically significant at $p < 0.01$ (t = 5.73 at 30 degrees of freedom, d.f.). The difference between the two sediment layers in terms of mean C_{12}–C_{44} aliphatic hydrocarbons was statistically significant at $p < 0.01$ (t = 5.57, 26 d.f.). The vertical difference in mean EC_{12}–EC_{44} aromatic hydrocarbons was also statistically significant at $p < 0.01$ (t = 5.92, 26 d.f.).

14.3.2.5 Light Aromatics Fraction Statistical Result

Using the matched pairs t-test the difference between the two sediment layers in terms of mean aromatic hydrocarbons in the carbon number range between EC_5 and EC_{12} was found to be statistically significant at $p < 0.01$ (t = 3.01, 25 d.f.). However, in contrast to the TPH, aliphatic and higher molecular weight aromatic hydrocarbons, in the case of the EC_5–EC_{12} aromatics the subsurface layer was significantly more contaminated than the surface layer.

This is a function of the volatility of light aromatic compounds that would have evaporated quickly from the sediment surface after spillage, and partly due to the ability of lighter aromatics to penetrate fine-grained sediments more easily than heavier hydrocarbons. Recent rapid sediment deposition could also bury lighter hydrocarbons, but is not a likely factor because deposition rates since the spill would have had to be very high.

14.3.2.6 PAHs Results and Sources

In the surface layers of six sampling sites, the PAH_{16} concentrations were 24.2 mg/kg (P18), 19.9 mg/kg (N24), 19.8 mg/kg (M11), 16.7 mg/kg (two samples at N22), 12.6 mg/kg (R12), 10.0 and 8.98 mg/kg (P19). These PAH_{16} levels are very high and exceed international sediment quality guidelines (Environment Canada 7.07 mg/kg Probable Effects Level; PEL). Although sample variances of total PAH_{16} were sufficiently close to proceed with the t-test, the difference in mean concentrations between the surface and subsurface layers was found not to be statistically significant ($t = 0.98$). The matched pairs t-test also was not significant ($t = 0.64$). Indeed, subsurface PAHs concentrations were also found in excess of PEL as follows: 22.8, 20.5 and 10.3 mg/kg (N24), 14.2 mg/kg (Q23), 8.5 mg/kg (M26), 7.85 mg/kg (N21), and 7.7 mg/kg (P19). In contrast to these high values, there had been no PAH_{16} concentrations exceeding PEL measured in Bodo 0–10 cm sediments in 2010 (Lindén and Pålsson 2013).

The probable source types of the PAHs are shown in Fig. 14.6. The sectors of the graph pertaining to each potential source type such as petroleum, combustion products of petroleum and biomass burning are each marked according to the work of Yunker et al. (2002), as modified by Oros and Ross (2004), Costa and Sauer (2005), Stout and Wang (2016), and Little and Galperin (2017).

Despite many samples being at or below detection limits for individual PAHs, the double ratio cross-plots in Fig. 14.6 clearly show that in about half the samples the PAH was petroleum-derived, as would be expected near the SPDC pipeline spills of 2008 and oil theft and spillage of Bonny Light crude oil at Bodo. In addition, the geochemical influence of petroleum combustion at illegal refinery sites was seen at Q23 and at N22 (surface and subsurface at both sites). At Q23 the total PAHs concentration reached 14.2 mg/kg in one subsurface sample. In the surface sample at N22 (surface) and also at R12 (subsurface) near Bodo waterfont, isomer ratios transitional between petroleum and petroleum combustion were observed.

However, acute atmospheric fallout across the region is also produced by oil industry flaring and well tests, tyre burning (pers. com. Pålsson) and chronic emissions from vehicles and industries in Port Harcourt and Eleme such as oil refining and petrochemicals (UNEP 2011) and cement, paint and fertiliser (NDES 1997). Although all these are partly responsible for petroleum combustion as a secondary source of PAHs, their input is outweighed by oil spills from several pipeline tappings and at new illegal refineries causing further mangrove impacts documented in cumulative losses mapped using satellite imagery and aerial photography from late 2009 (Gundlach this volume).

A third PAHs source was biomass combustion from both natural and anthropogenic forest fires, and domestic fires. Although of minor interest to the problem of cleanup, biomass combustion was sub-dominant at N09, N21, Q23, M26, P18, P19 and U19, usually in surface samples, but at N09 and P18 in both surface and subsurface samples. This suggests that in addition to oil fires, the burning of wood and grass (and coal, although most unlikely in Bodo) has contributed to the PAHs geochemical signature for some considerable time. This is typical of rural areas undergoing forest clearance for agriculture, and with the use of wood fuel in domestic cooking. UNEP (2011) showed the health risk of indoor cooking fuels in Ogoniland.

14.3.2.7 Additional Contaminant Source Indicators

Many of the above site-by-site patterns agreed well with other source indicators such as MTBE and volatile aromatic hydrocarbons (BTEX). Although reported as being below the limit of detection at most sites, where they were detectable, these contaminants often occurred at much higher concentrations in the subsurface sediment layer.

For BTEX at N21 the subsurface concentration (0.052 mg/kg) was five times more than at the surface. Examples of MTBE enrichment include N21 (surface and subsurface) and M26 (subsurface, as high as 0.96 mg/kg). MTBE is a refined fuel blending component and therefore a precise indicator of contamination by spilled petrol. It was introduced to replace the lead additives in petrol that were banned in Nigeria in 2005. It is likely that the MTBE indicates the increasing use of petrol in the creeks since then by small vessels, including security craft, and perhaps those servicing illegal refineries.

For light aromatic EC_5–EC_{12} fractions at M26 the subsurface mean concentration (14.94 mg/kg, $n = 4$) was over forty (40) times more than in the surface. In both the surface and subsurface of M26 the gas chromatograms showed fresh petrogenic inputs in the range C_{12}–C_{16}. In addition, the range of relatively fresh oil input up to C_{21} was seen in chromatograms of samples from N21, G24, N22, M26, P19 and R12 (subsurface), and less obviously from N22, M26, P19 and R12 (surface).

Over time, all these lighter hydrocarbons not only evaporate, but they can also penetrate the sediment and could mobilise heavier molecules including PAHs into deeper sediment layers, further reducing their biodegradation rates.

Fig. 14.6 Ratios of Anthracene/(Anthracene + Phenanthrene) [AN/(AN+P)], Benz[*a*]anthracene/(Benz[*a*]anthracene + Chrysene) [BA/(BA+C)], Indeno(1,2,3-*cd*)pyrene/(Indeno[1,2,3-*cd*]pyrene + Benzo[*g,h,i*]perylene) [IN/(IN+BPE)] each plotted against Fluoranthene/(Fluoranthene + Pyrene) [FL/(FL+PY)] showing main PAH source types in the Bodo area sediments (after Yunker et al. 2002); for sample identities see Tables 14.2 and 14.3; data points in line are at limit of detection; all data are colour-coded: blue = petroleum from oil spills including those resulting from the 2008 pipeline failures and illegal refinery sites; black = petroleum combustion from regional oil industry flaring, vehicular and industrial emissions and local illegal refinery sites; and red = biomass combustion (wood, grass)

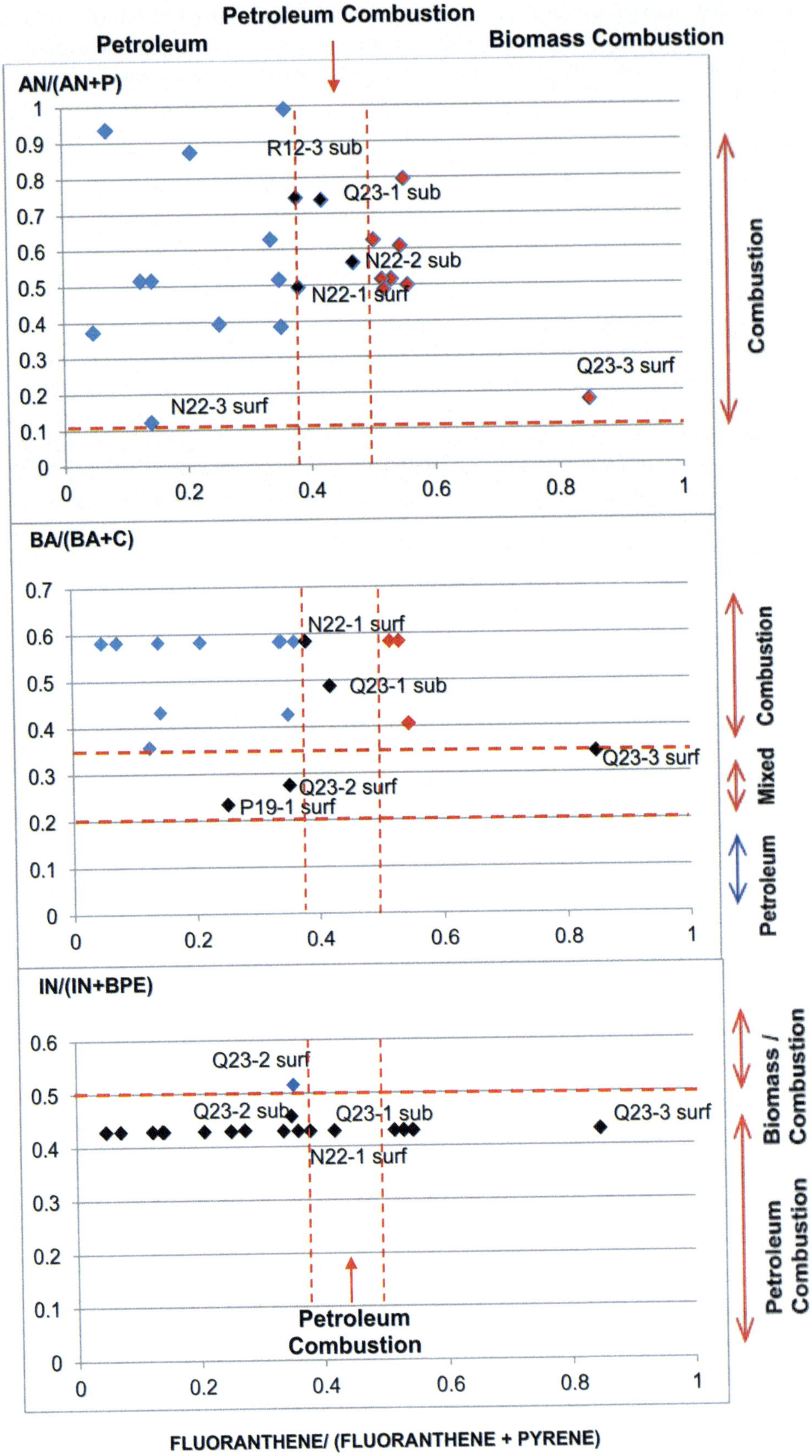

Fig. 14.7 Aerial view of Kpador Channel near Bodo showing continuously-oiled mangroves in the intertidal zone, with mobile free-phase oil in the creeks; also showing impacts from pipeline RoWs and failed road-building projects before the 2008 spills, and from illegal refineries since those spills (smoke and blackened trees left of center), 17 April 2010 (Credit: J. Pålsson)

14.4 Discussion

The surveyed area near Bodo which includes the typical substrates of the mangroves is extremely contaminated by residues of crude oil, either released into the environment after accidental oil spills and leaks, or due to illegal activities such as oil theft from pipelines and the artisanal refining of the stolen oil. In 2015 oil was observed across the intertidal mangrove habitat surface, and also draining into creeks, although not generally as seriously as was seen on 17 April 2010 (Fig. 14.7, showing the area of sampling stations N21 - N24 inclusive, see Fig. 14.1). However, it has not always been like this, and the contamination is patchy. This patchiness is important to understand if the future cleanup is to avoid adding to the impact of the original oil by physically removing too much of the relatively clean subsurface layers during cleanup, as found in other oil spill cleanups (Little et al. 2003). A range of cleanup methods will be required, often in combination, rather than total sediment removal (Gundlach 2017).

The mean TPH concentration appears to have increased by almost three times from 2690 mg/kg (1983) to 7500 mg/kg (2010), and another three-fold increase in mean TPH concentration to 28,904 mg/kg took place by 2015. To establish whether these apparent changes were due to the incomparability of the methods, the minimum TPH concentrations in Bodo sediments were compared between 1983, 2010 and 2015. The GC-FID traces give an indication of the background TPH status in the area, with a relatively low Unresolved Complex Mixture (UCM) of weathered petroleum and some biogenic resolved peaks. The background concentrations ranged from 110 mg/kg (1983), to 279 mg/kg (2010), and 163 mg/kg (2015). This background stability through time suggests reasonable comparability between the methods.

Whereas the sampling and analytical methods used were broadly comparable between all three of these surveys, the intertidal field site locations within Bodo were different, and thus not directly comparable, as follows:

- 1983: northern Bodo south to Kpador channel/Bonny River confluence (Baker 1983);
- 2010: across tidal regions of southern Gokana LGA (Lindén and Pålsson 2013); and

- 2015: the present survey of the Bodo TNP pipeline spills area targeted for cleanup (Fig. 14.1).

It was found in the analysis of over 660 samples from UNEP surveys across Ogoniland in 2010 that frequency distributions of contaminant concentrations were strongly left-skewed with 91% of all samples being less contaminated than the TPH limit value of 5000 mg/kg. This large number of samples included inland soils and deeper layers sampled using hand augers (Thiergärtner and Holtzmann 2014).

However, of those surface 0–10 cm depth samples taken in 2010 in the intertidal zone closest to Bodo, the equivalent percentage of left-skewness for TPH was 33% <5000 mg/kg (Lindén and Pålsson 2013). This was similar to the 44% <5000 mg/kg in the 15–20 cm samples in 2015. The left-skewed frequency distributions in 2015 show that the 15–20 cm layers included several samples tending to background TPH levels. In contrast, the surface 0–5 cm samples in 2015 were more generally contaminated, and the left-skewness was therefore almost absent (only 3.6% <5000 mg/kg). The skewness indicates a degree of comparability between the 2010 and 2015 frequency distributions of TPH in the subsurface layer, but not in the surface layer, and this shows the success of the SCAT surveys in visually targeting the most polluted areas for cleanup.

The chemical analyses in 2015 generally confirmed the visual results from the SCAT field surveys in 2015. The visual description of the contaminated sediment ranged from 'rainbow sheen only', through 'moderate sheen', 'substantial oiling', 'black oil present' and 'oil soaked mud 10–25 cm deep' to 'surface tar' mainly on illegal refinery sites (Gundlach 2015a, b). The advantage conferred by embedding the 2015 chemistry sampling within the SCAT framework is that the analytical chemistry data from selected sampling sites are better understood in their field context. Equally, the chemistry is necessary to confirm the degree and nature of the contamination, which is above regulatory intervention values in either surface and/or subsurface layers at every site sampled, even if not visible on the surface until the sediment is agitated or algal mats are removed.

Furthermore, from this dual approach and understanding, the 'hot-spots' may be more clearly demarcated for cleanup without recourse to taking many thousands of samples. In future, the selection of field locations to monitor will be partly determined by the progress and effectiveness of the cleanup, and as prioritised by the SCAT surveys. The monitoring will compare changes in oiling conditions due to cleanup (before-after); any incremental impacts arising from cleanup (control-impact) will also require selected oiled sites to be 'set-aside' from cleanup (i.e., controlled) for scientific study using a BACI design.

The high statistical significance of the differences in contaminant concentrations between 0–5 and 15–20 cm layers suggests in this case that cleanup should selectively remove the 0–5 cm layer of highly contaminated sediments. However, there are important caveats on the use of the preliminary chemistry data in calibration of SCAT when considering decisions on cleanup, such as: (a) if the most highly contaminated layer is between 5 and 15 cm; (b) if the most highly contaminated layer is deeper than 20 cm; and (c) if the contaminated site is in close proximity to a sensitive receptor (e.g., Bodo waterfront).

The above cleanup recommendation from the preliminary samples is not a prescriptive protocol and it should not be extrapolated to other areas. This is a key point because the cleanup planning effort was initially designed for free-phase oil removal; then, due to ongoing weathering processes, 'free-phase' oil cleanup was extended from flushing and skimming to include the removal of accessible surface tar. It is emphasised that in the full cleanup phase the purpose of the SCAT approach is to facilitate more precise cleanup recommendations tailored to the 'hot-spots' of each new cell in the grid (Gundlach 2017).

In the surface sediments the mean concentrations of 25,045 mg/kg (C_{12}–C_{44} aliphatics) and of 17,015 mg/kg (EC_{12}–EC_{44} aromatics) are generally extremely high. A very similar concentration (24,664 mg/kg) of aliphatics (C_{14}–C_{29}) was detected in Owaza by Faboya et al. (2016). In contrast, the light aromatic EC_5–EC_{12} fractions have been measured in Bodo at relatively low levels, with the only exceptions of both samples at N21 and the subsurface at M26. The light end aromatics were reduced relative to the total aromatics EC_5–EC_{44}. This is also seen in many of the gas chromatograms and appears to indicate weathering and therefore a lower acute eco-toxicity of the pollution.

However, the abundance of dead mangroves clearly shows the negative impact of the crude oil residues at least on the flora and associated fauna (e.g., bivalves on mangrove prop roots) under the daily tidal influence (Gundlach 2017); Gundlach this volume). This chronic toxicity is likely to be partly attributable to the persistence of high molecular weight hydrocarbons and in particular the PAHs. Although PAHs are low in concentration in fresh Bonny Light crude oil, they increase in relative concentration during weathering processes, and are highly enriched in those areas with illegal refineries, even in the subsurface sediments. In Bodo, between the years 2010 and 2015 the mean PAH_{16} concentration increased by more than three times from 1.3 to 5.65 mg/kg, similar to the increased TPH concentrations. It is probable that the PAHs inputs from petroleum combustion are mainly from the illegal refineries that proliferated after 2009. This may be stated because the oil industry production and any well-testing and flaring activities near Bodo have been shut down throughout this period.

Pyrogenic emissions from towns, industry and traffic in the wider Ogoniland area are also likely sources. The multiple sources of contaminants demonstrate the need for forensics quality hydrocarbon analyses in a subset of the sites sampled, not least because of the need to understand pathways and then prevent adverse impacts on human health. In the mangrove sediments of the Calabar River a similar mix of PAH sources was found using isomer ratios, but apparently without additional input from illegal refineries (Oyo-Ita et al. 2013).

In 1983, Bodo West was the most exploited of the five SPDC areas observed in terms of wood cutting (45% of sites), oyster gathering and fishing (50% of sites), activities that depend on sustainable reources. Apart from the physical presence of the oil production facilities and one spill site in Bodo West, no general biological impacts were attributed in 1983 to the oil industry (Dixon 1983). With evidence of oil industry presence at 32% of sites, Bodo West was less affected by oil industry disturbance than Forcados (37% of sites) and by urbanization and industry at Warri (27% of sites). The Bodo pipeline spills in 2008 and a subsequent increase in oil theft, transport and illegal refining have in combination caused catastrophic, habitat-scale and ongoing mangrove ecosystem impacts (Gundlach this volume).

The environmental situation in the Bodo area is clearly not beneficial for local people as well as the ecosystem supporting them. To put this subjective assessment in a comprehensive context, an environmental quality standard (EQS) could be derived scientifically and used for comparison with analytical results. There are intervention values used internationally for certain (parent) PAHs, and for single light aromatic compounds, such as benzene (EC_6). In The Netherlands an intervention value of 1 mg/kg benzene in soil is used (Into Dutch Soil 2014). However, there are no EQS for low molecular weight aromatics (EC_5–EC_{12}) fractions. The derivation of an EQS for chemical substances is a sophisticated process, influenced by chemistry, toxicology, epidemiology, phenomena such as metabolism, interaction of chemical compounds, and other factors that in combination affect the probability of adverse effects.

Official evaluation criteria for sediment contamination in Nigeria are set by DPR (1991): "Environmental Guidelines and Standards for the Petroleum Industry in Nigeria (EGASPIN)". Because of the relevance of the oil industry to environmental impacts and the dominance of the Dutch company Shell in the oil-producing areas, EQSs from The Netherlands have been generally adopted by EGASPIN (Ministerie van VROM 2000). EGASPIN defines intervention values as those that indicate "the quality for which the functionality of the sediment for human, animal and plant life is, or is threatened with being, seriously impaired. Concentrations in excess of the intervention values correspond to serious contamination". However, compared to the control of oil and chemical pollution, the conservation of biodiversity and management of adverse ecological impacts are not the focus of EGASPIN (IUCN 2013).

This was recognized by the BMI when they agreed that the aim was to reduce oiling levels to those that would support the establishment or transplantation and growth of mangrove seedlings. Ecological restoration will also depend on correct definition of the ecological goals, and this will require detailed understanding of the inter-dependent mangrove species assemblages, communities and habitats. A cause for optimism is that a group of whimbrel *Numenius phaeopus* was observed by the authors at Bodo on 1 May 2015, indicating the functional importance of the ecosystem as a stopover for long-distance migrant birds. Conversely, the increased role of alien invasive species such as Nipa palm in the ecosystem is a major concern. It is fortunate that much of the essential knowledge is to be found in Bodo (Zabbey et al. 2010, 2014, 2017, in press; Zabbey 2012; Zabbey and Malaquias 2013; Pegg and Zabbey 2013; Zabbey and Uyi 2014; Zabbey and Tanee 2016).

Further constraints on applying the above-mentioned intervention value include the fact that it was established for unweathered hydrocarbons. Additionally, it is based on the 'Dutch list' (underlying EGASPIN) and is defined for a standard terrestrial soil type with 10% organic matter and 25% clay. This is different from the deltaic sediments in the mangroves, which consist of marine and coastal fine-grained sediments. Although not so much the case in Bodo as in other areas of the delta (e.g., Warri), the sediments are often affected by dredge spoils or composed of made ground of widely differing organic content and granulometry (Fig. 14.7).

UNEP (2011) pointed out that not having separate values defined in EGASPIN for triggering and closing out remedial interventions was a loophole that allowed negligible cleanup to be accepted. Criticism of EGASPIN has also focused on the high TPH intervention value of 5000 mg/kg, and on the lack of limits on individual PAHs especially Benzo(*a*)pyrene (IUCN 2013). A valid starting point to protecting the local population is to focus on those areas, transfer pathways and activities presenting the highest potential human exposure to toxic and persistent hydrocarbons. The elevated hydrocarbon concentrations in sites R12 and N09 are within a few hundred meters of human receptors in Bodo.

The intervention value for TPH of 5000 mg/kg in sediment adopted by DPR for comparison with the measured concentrations gives a good sense of the magnitude of the pollution in the intertidal sediment. In contrast, the intervention value for PAH_{16} of 40 mg/kg is too high by international standards (Tables 14.2 and 14.3). By considering the environmental regulatory standards of Nigeria and in conjunction with SCAT, the analytical results will need to be interpreted ecologically to derive recommendations regarding the net environmental benefit of remediation of the oil-contaminated sediment. This focus will help to minimize any extra impacts from cleanup in soft sediments, and to avoid further unnecessary delays in the restoration of mangroves and livelihoods in the Bodo area.

14.5 Conclusions

Intertidal surface sediments near Bodo are widely and severely oiled since the 2008 pipeline spills, and in areas known to be subject to increases since 2009 in oil theft and illegal refinery activities. Of the total of sixty-four (64) samples taken at two depths (0–5 cm and 15–20 cm), in the surface sediment all but one sample had a TPH concentration well above 5000 mg/kg, the EGASPIN regulatory limit. In contrast, thirteen (13) samples in the deeper layer had TPH concentrations below the 5000 mg/kg limit. Although contamination by TPH, aliphatic and aromatic hydrocarbons was moderate to high in the deeper layer, it was shown statistically to be significantly lower than in the extremely contaminated surface layer.

Whereas TPH, aliphatic and aromatic hydrocarbon fractions decrease in concentrations from the surface towards deeper sediment as expected, the trends for the lower molecular weight aromatics differ. The latter are detected in much lower concentrations, with only twenty (20) out of 55 samples found at >1 mg/kg. In most samples, the concentrations of lighter aromatics increase with greater depth in contrast to the case of heavier hydrocarbons. This was as would be expected from surface evaporation and also due to chromatographic separation during oil infiltration, washing and draining through intertidal sediments.

Polycyclic aromatic hydrocarbons (PAH_{16}) concentrations range widely, with the higher values of PAH_{16} raising the possibility of potential health risk for any people exposed to the pollution. It should be recalled that seven (7) of the individual PAHs are classified by USEPA as probable human carcinogens, but this risk and the pathways of human exposure at Bodo are still not adequately quantified. Ecological risk is better understood in that the sediment PAH_{16} concentrations in 55 samples show that international sediment quality guidelines are exceeded in fifteen (15) samples, of which seven (7) are in the subsurface. This explains ongoing adverse ecological impacts.

A preliminary forensics evaluation of the available PAH data suggests the following likely sources of PAHs in the area: petroleum, combustion of petroleum, and combustion products from biomass (wood, grass). Such sources typify inputs from oil industry spills and illegal refinery activities, overprinted on a baseline of traditional biomass fuel usage. The petroleum and petroleum combustion inputs are consistent with the lingering effects of the 2008 SPDC spills and the increase in oil theft and illegal refining since then.

14.6 Recommendations

Using public health and demographic data (NPC 2004), recent work has compared siblings conceived before and after onshore Nigerian oil spills; the study showed that neonatal mortality rates doubled after spills nearby (Bruederle and Hodler 2017). Therefore, an answer is needed to this legitimate question: "Is the existing contamination of the Bodo area lethal only for mangroves and the associated fauna, or does it also endanger the health of the local people?" Independent health officials should evaluate the chemistry data as soon as possible to determine the best course of action to protect the local people from any ongoing exposure to toxins of unknown but potentially significant magnitude. This recommendation is made because of the very high sediment concentrations of aromatics (EC_5–EC_{44}) and other hydrocarbons, especially PAHs. The factors determining the hazard posed by toxic substances are a function not only of their concentrations and distribution, but also of pathways and duration of exposure. The pathways are demonstrable for mangroves, their epifauna, and sedentary infauna in contaminated sediments all under tidal submersion twice daily, but remain unclear for the Bodo human population (e.g., via skin contact, inhalation and ingestion). When the sediments begin to respond to remediation and the restored mangrove system undergoes recolonization by edible fauna, the present avoidance of contaminated areas by fishermen may cease, and exposure to food-borne contaminants could then resume. To the extent that anyone still relies on potentially-contaminated water for drinking and household use, they should be provided with clean water until remediation is complete, as recommended by UNEP (2011).

The first appropriate and apparently obvious response is to remediate the Bodo intertidal and shoreline sediments in accordance with applicable Nigerian laws and guidance (i.e. EGASPIN). From the work of the BMI, this chapter is a contribution to such a response. A properly designed confidential medical screening programme and epidemiological investigation might also be among the appropriate responses to contamination in Bodo, and similar recommendations were made for Ogoniland by UNEP (2011). The UNEP studies were not able to consider fully the exposure scenarios typical for the Bodo area such as the consumption of contaminated water and seafood, children fishing and bathing in the creeks, and local people working there now or in the past. The latter exposure scenarios should include those whose work on-site has been legal or illegal, with or without appropriate Personal Protective Equipment.

The PPE provided for the work force should be protective against a release of volatile hydrocarbons from subsurface layers once the surface layer (0–5 cm) has been removed during cleanup. The provision of hand-held gas detectors and face masks should follow international best practice and SPDC safety standards. This is needed because evaluation of the analytical data shows that volatile fractions penetrated further than the bulk of the oil, and are currently sequestered in the subsurface layer. Removal of the upper layer of contaminated sediment could lead to the renewed evaporation of volatiles into the breathable air.

The visual and photographic data from the SCAT surveys should continue to be calibrated by chemical analyses at selected sites. This is important because better risk assessment is both necessary and possible using high quality analytical methods in situations where remediation of the environment or the protection of people is required. The SCAT approach should continue to characterize the oiling conditions in the field, to develop cleanup strategy, and to design and monitor segment-by-segment tactical plans. Laboratory and field approaches should continue to be linked in the various stages of operations (planning, gridding, sampling, analytical, data processing, interpretation, public communications, and scientific publications).

In addition to SCAT activities, chemistry monitoring will require selected oiled sites to be set-aside as controls for scientific study of cleanup effectiveness and impact, to distinguish them from the impacts of the original oiling(s). Preparing for the mangrove restoration will require the establishment of reliable lines of supply of propagules from mangrove nurseries. Restoration can begin when sediment hydrocarbons and ground conditions are suitable, but crucially only when security is such that vandalism and pipeline tapping have ceased, and proper inspections can take place. Necessary actions also include improving youth employment and training opportunities to help alleviate poverty (e.g., Rivers State Sustainable Development Agency). If supported by better law enforcement, these actions would also directly reduce oil theft and illegal refining activities that otherwise might recur and continue to damage mangrove habitats and human health.

The forensic evaluation of the chemical 'fingerprints' of Bodo sediments should be enhanced by a fuller PAHs profile including the alkyl homologues of parent PAHs. Additionally, the hopane and sterane biomarkers characteristic of crude oil and weathered refined products should be analysed by GC-MS. The PAH forensics results from the present study indicate oil industry and illegal refinery sources consistent with the 2008 spills and the subsequent increase in oil theft and illegal refining. However, the methods used so far are not adequately discriminatory for fine-grained sediments with multiple chronic inputs, and should be refined to ensure understanding of human and biota exposure to individual hydrocarbons.

On the basis of the present data, only the more highly-contaminated surface sediment layer (0–5 cm) should be physically removed. It is emphasised that in the full cleanup phase the SCAT approach will facilitate more precise cleanup recommendations tailored to each cell in the grid. The preliminary data show that intrusive remediation measures should concentrate on the more heavily oiled surface 5 cm of sediment because limiting the physical removal to 5 cm would reduce oil content without undue risk of mudflat erosion. This is needed because oily waste volumes would be minimised and environmental outcomes would be improved by avoiding accidental inclusion of relatively uncontaminated subsurface material among the wastes. Considering the resulting lower costs/hectare, a larger area could be remediated within a given budget. Such a sustainable approach would reduce pollution in Bodo without increasing: (a) direct trampling and compaction of the sediments; (b) multiple handling of larger volumes of oiled wastes; and (c) carbon footprint impact of traffic emissions.

Acknowledgements We are very grateful to the people of Bodo who made this study possible. The preparation of this chapter was self-funded, and the views expressed are our own. UNEP supported the senior author's participation in the Bodo Mediation Initiative between 2013 and 2017. During 2015, SPDC supported the participation of the second and third authors in the BMI, and the field surveys and laboratory analyses were also organized by SPDC. With apologies to anyone omitted below, we thank all our colleagues for their collaboration: Bert Ronhaar, Inemo Samiama, Obiageri Nwokoro, Akpene Samiama, David Gorman, John Alawa, Prof. Ben Naanen, Livinus Barikor, Loveday Kpandei, Ferdinand Giadom, Nenibarini Zabbey, Kabari Visigah, Marvin Dekil, Noble Pepple, Biedima Oliver, Hope Nuka, Raphael Saue, Sola Oladipo, Opakirite Braide, N. Naazigha-Lue, Anthony Kinika, Messrs. Bouvay and Harrison, Sister Philomena Okwu, Fathers Abel Agbulu and Edward Oby (BMI); John Groffen, Maurice Paulussen, Angelique van der Made, and Fidelia Onoghaife (Dutch mission); ALcontrol Laboratories staff; Matthijs Bonte and Ilcana Rhodes (Shell); Igo Weli, Alice Ajey, Franklin Igbodo, Augustine Igbuku, Philip Shekwolo, Rik Prager, Andy Lee, Ian McBurnie, Johnson Akinnalwonu, Kingsley Osuh, Philip Mshebila, Ogonnaya Iroakasi, Lede Kumannsee, Jonathan Obasohan and Vincent Nwabueze (SPDC); Erik Solheim, Henrik Slotte, Muralee Thummarukudy, David Jensen and Chinyere Nweke (UNEP); Adekunle Adesida, Prosper Ugbehe, Adebayo Oladeji and Joel Olarenwaju Adesida (Geoterrain Nigeria Ltd); Geraint Williams and laboratory staff (ALcontrol Laboratories); and the following individuals: Jenifer Baker, Brian Dicks, Iain Dixon, Charo Gil, John Hartley, Sid Howells, Dave Levell, Jon Moore, Jonas Pålsson, Marjie Staggs, Pete Tibbetts, Donna Vorhees, and Mark Wilson.

References

Baker JM (1983) Environmental study of shell-operated areas: mangrove ecosystems. Technical Summary FSC/OPRU/32/83, unpublished report to SPDC, 62p

Bonte M (2015) Bodo creek chemical sampling and analysis protocol, Bodo Mediation Initiative unpublished memorandum, Port Harcourt, Nigeria, 9p

Bruederle A, Hodler R (2017) The effects of oil spills on infant mortality: evidence from Nigeria. CESifo Working Paper No. 6653, 40p

Costa HJ, Sauer TC (2005) Forensic approaches and considerations in identifying PAH background. Environ Forensic 6:9–16

Department of Petroleum Resources (1991) Environmental guidelines and standards for the petroleum industry in Nigeria (EGASPIN), Lagos. Revised edition 2002, 451p

Dixon IMTD (1983) Environmental study of shell-operated areas: mangrove ecosystems. Technical Report FSC/OPRU/23/83, unpublished report to SPDC, 219p

Dublin-Green CO (1990) Seasonal variations in some physico-chemical parameters of the Bonny Estuary, Niger Delta. Nigerian Institute for Oceanography and Marine Research, Lagos. Technical Paper No. 59, 24p

Faboya OL, Sojinu SO, Sonibare OO, Falodun OT, Liao Z (2016) Aliphatic biomarkers distribution in crude oil-impacted soils: an environmental pollution indicator. Environ Forensic 17(1):27–35. https://doi.org/10.1080/15275922.2015.1091400

Fentiman A, Zabbey N (2015) Environmental degradation and cultural erosion in Ogoniland: a case study of the oil spills in Bodo. Ext Ind Soc 2:615–624. https://doi.org/10.1016/j.exis.2015.05.008

Gundlach ER (2013) Impact analysis of two Bodo oil spills. E-Tech, Boulder. 69p

Gundlach ER (2015a) Pre-SCAT survey 01–08 May 2015. Report 4 June 2015 for BMI, E-Tech International, Boulder, 75p

Gundlach ER (2015b) Report on SCAT Findings 03–26 August 2015, Report 4 September 2015 for BMI, E-Tech International, Boulder, 13p

Gundlach ER (2017) 2015 Pre-SCAT and SCAT combined data report for Bodo area. Report 15 July 2017 for BMI, by E-Tech International, Boulder, 193p

Gundlach ER (this volume) Oil-related mangrove loss east of Bonny River, Nigeria. In: Makowski C and Finkl CW (eds) Coastal research library (CRL): threats to mangrove forests: hazards, vulnerability and management solutions. Springer Science, Dordrecht

Gundlach ER, Hayes MO (1978) Vulnerability of coastal environments to oil spill impacts. Mar Technol Soc J 12(4):18–27

Gundlach ER, Hayes MO, Getter CD (1981) Sensitivity of coastal environments to oil spills. In: Proceedings seminar on the petroleum industry and the Nigerian environment, 9–12 November 1981, Warri, Nigeria

Howard S, Little DI (1987) Effect of infaunal burrow structure on oil penetration into sediments. In: Proceedings international oil spill conference. American Petroleum Institute, Washington, DC

Into Dutch Soil (2014) Rijkswaterstaat Ministry of Infrastructure and the Environment, The Netherlands. 79p

IUCN (2013) Sustainable remediation and rehabilitation of biodiversity and habitats of oil spill sites in the Niger Delta: main report including recommendations for the future. Report to Shell Petroleum Development Company Ltd of Nigeria, by the Independent IUCN–Niger Delta Panel (IUCN–NDP), Gland, Switzerland, 71p

Kingston P, Little DI, Harkantra S (2000) Biological impacts of oil pollution: sedimentary shores, IPIECA Report Series, vol 9. IPIECA, London. 21p

Lindén O, Pålsson J (2013) Oil contamination in Ogoniland, Niger Delta. Ambio 42(4):685–701. https://doi.org/10.1007/s13280-013-0412-8

Little DI, Deakin T, Fichaut B, Meech RJ (2003) Some observations on heavy fuel oil spills: trends, impacts and comparisons with crude oil spills. In: Proceedings 26th Arctic and Marine Oilspill Program Technical Seminar, Ottawa, pp 971–998

Little DI, Galperin Y (2017) The assessment of hydrocarbon contamination in contrasting sedimentary environments. In: Heimann K, Karthikeyan OP, Muthu SS (eds) Biodegradation and bioconversion of hydrocarbons: research advances and recent developments. Springer Science+Business Media Singapore, Singapore. isbn:978-981-10-0199-4. http://link.springer.com/chapter/10.1007/978-981-10-0201-4_1

Ministerie van VROM (2000) Dutch target and intervention values 2000 (the New Dutch List). Staatssuitgeverij, Den Haag. 51p

NDES (1997) Niger delta environmental survey final report phase 1. Environmental Resource Managers Limited, Lagos. two volumes 301p and 128p

NPC (2004) Nigeria demographic and health survey 2003. National Population Commission Federal Republic of Nigeria and ORC Macro Calverton, Maryland, 348p. http://www.dhsprogram.com/pubs/pdf/FR148/FR148.pdf

Oros DR, Ross JRM (2004) Polycyclic aromatic hydrocarbons in San Francisco Estuary sediments. Mar Chem 86:169–184

Owens E, Sergy G (1994) Field guide to the description of oiled shorelines, 2nd edn 2000, Environment Canada, Ottawa, 66p

Owens EH, Sergy GA (2003) The development of the SCAT process for the assessment of oiled shorelines. Mar Pollut Bull 47(9–12):415–422

Oyo-Ita OE, Offem JO, Ekpo BO, Adie PA (2013) Anthropogenic PAHs in mangrove sediments of the Calabar River, SE Niger Delta, Nigeria. Appl Geochem 28:212–219

Pålsson J, Lindén O (2014) EPH and PAH levels in water and sediment in Ogoniland, Nigeria. In: Proceedings international oil spill conference. American Petroleum Institute, Washington DC

Pegg S, Zabbey N (2013) Oil and water: the Bodo spills and the destruction of traditional livelihood structures in the Niger Delta. Community Dev J 48(3):391–405

RPI (1985) Environmental baseline studies for the establishment of control criteria and standards against petroleum-related pollution in Nigeria. Research Planning Institute Inc., report to Nigerian National Petroleum Corporation, Lagos, Nigeria, 598p

SPDC (2013) Oil spill response and remediation management system (OSRRMS) v4. 2013

Stout S, Wang Z (2016) Standard handbook oil spill environmental forensics: fingerprinting and source identification. Academic, London. 1142p

Thiergärtner H, Holtzmann K (2014) Modeling and preliminary assessment of crude oil contaminated soil in Ogoni (Nigeria). Altlasten Spektrum, Berlin 2:61–71

TPHCWG (1997) Selection of representative TPH fractions based on fate and transport considerations, total petroleum hydrocarbons criteria working group series, 3. Amherst Scientific Publishers, Amherst. 109p. isbn: 1-884-940-12-9

UNEP (2011) Environmental assessment of Ogoniland, United Nations Environment Programme, Nairobi, 257. http://www.unep.org/nigeria

Yunker MB, Macdonald RW, Vingarzan R, Mitchell RH, Goyette D, Sylvestre S (2002) PAHs in the Fraser River basin: a critical appraisal of PAH ratios as indicators of PAH source and composition. Org Geochem 33:489–515

Zabbey N (2012) Spatial and temporal variability in interstitial water quality of soft-bottom flats at Bodo Creek, eastern lower Niger Delta, Nigeria. Trop Freshw Biol 21(1):83–103

Zabbey N, Malaquias MAE (2013) Epifauna diversity and ecology on intertidal flats in the tropical Niger Delta, with remarks on the gastropod species *Haminoea orbignyana*. J Mar Biol Assoc U K 93(1):249–257

Zabbey N, Tanee FBG (2016) Assessment of asymmetric mangrove restoration trials in Ogoniland, Niger Delta, Nigeria: lessons for future intervention. Ecol Restor 34(3):245–257

Zabbey N, Uyi H (2014) Community responses of intertidal soft-bottom macrozoobenthos to oil pollution in a tropical mangrove ecosystem, Niger Delta, Nigeria. Mar Pollut Bull 82:167–174

Zabbey N, Hart AI, Wolff WJ (2010) Population structure, biomass and production of the West African lucinid *Keletistes rhizoecus* (Bivalvia, Mollusca) in Sivibilagbara swamp at Bodo Creek, Niger Delta, Nigeria. Hydrobiologia 654:193–203

Zabbey N, Vincent-Akpu IF, Etela I (2014) Green economy: challenges and prospects for improved aquatic agricultural system (AAS) in Niger Delta communities. Agric For Fish 3(1):28–35

Zabbey N, Sam K, Onyebuchi AT (2017) Remediation of contaminated lands in the Niger Delta, Nigeria: prospects and challenges. Sci Total Environ 586:952–965. https://doi.org/10.1016/j.scitotenv.2017.02.075

Heavy Metal Distribution and Accumulation from Natural and Anthropogenic Sources in Tropical Mangroves of India and Bangladesh 15

Prabhat Ranjan, Karuna Rao, Alok Kumar, and A. L. Ramanathan

Abstract

The chapter presents a comparative account of trace metal distribution, their accumulation in sediments as well as in biota across the Indian and Bangladesh estuarine-mangrove complex through published literature. The study shows that trace metals like Fe and Cd show high contamination in most of Indian and Bangladesh mangrove ecosystem whereas other trace metals like Al, As, Cr, Co, Cu, Mn, Ni, Pb and Zn, show variable contamination in different mangrove settings with 'low' to 'moderate' value. Tsunamigenic sediment shows higher concentration of almost all trace metals due to waste and discharge brought by tsunami wave or sediment from deep shore of ocean. Speciation of trace metals show dominance of Fe, Cd, Cu, Ni, Pb and Zn mostly in residual fraction, thus making it unavailable in prevalent environments. Whereas, Mn can be found in exchangeable fractions that are readily available and potential risks. Trace metals accumulation in fishes shows the following order Cd>As>Ni>Cu>Pb>Cr, with highest accumulation of Cd; pelagic fishes exhibit lower values of heavy metals than the bottom dwelling fishes. Bioaccumulation of Cu and Zn in both gastropods and bivalves was higher than bioavailability. So, due to increased risk of trace metals pollution in the estuarine-mangrove complex and loss of mangrove biodiversity, there is dire need of in-depth study and better management practices in the tropical mangrove ecosystem of India and Bangladesh.

Keywords

Estuarine · Contamination · Bioaccumulation · Residual

15.1 Introduction

Mangroves are the plants that occur along the inter-tidal zone of tropical and subtropical areas between 30°N and 30°S, covering almost 75% of the coastal areas worldwide (Pernetta 1993). The common characteristic of all these plants is their tolerance to salt and brackish waters. They are known to perform many functions including protective, productive and economic. Mangrove wetlands act as habitat for various organisms which are ecologically, as well as economically, important, such as estuarine/marine fishes and shellfishes. Among coastal ecosystems, the mangrove ecosystem acts as a storehouse of biological diversity like the tropical rainforest (Swaminathan 1991).

Heavy metals have a natural occurrence, but the activities of humans have dramatically increased the background values of these heavy metals in the environment. The trace metal pollutants make a way into the mangroves and coastal environments through rivers. They are considered to be serious pollutants as they are persistent for years (non-biodegradable) and can accumulate in the aquatic organisms. The major sources of trace metals into the mangroves are tidal activities, rainfall and land runoff which carries a good amount of solid wastes and waste waters (Nriagu and Pacyna 1988).

To understand the bioavailability and extent of toxicity of metals, it is important to study the metal concentration in the mangrove sediments as well as on flora and fauna of mangrove ecosystem. While entering into the coastal environment, these metals get transferred to the sediments through many processes like adsorption onto the particles and co-precipitation with solid phases (Santschi et al. 1990; Sharma et al. 1999). The study of trace metals in mangrove

The original version of this chapter was revised. An erratum to this chapter can be found at https://doi.org/10.1007/978-3-319-73016-5_32

P. Ranjan · K. Rao · A. L. Ramanathan (✉)
School of Environmental Sciences, Jawaharlal Nehru University, New Delhi, India

A. Kumar
Department of Environmental Science, Central University of Rajasthan, Kishangarh, Ajmer, Rajasthan, India

C. Makowski, C. W. Finkl (eds.), *Threats to Mangrove Forests*, Coastal Research Library 25,
https://doi.org/10.1007/978-3-319-73016-5_15

ecosystem is important as they act as a potential sink for anthropogenic metal contaminants (Silva et al. 1990; Tam and Wong 1996; Clark et al. 1997). Metals are collected in these ecosystems only for the short term and will become remobilized and suspended back to the water column with the change in the environmental conditions or when the retention capacity of the surface soil mass is saturated (Tam and Wong 1993). The present study highlights the status of different estuarine mangrove complexes across the Indian and Bangladesh coast with reference to occurrence of trace metals in the sediments as well as biota of mangrove forest.

15.2 Different Mangrove Ecosystem of India and Bangladesh

Different mangrove ecosystem of India and Bangladesh (Fig. 15.1).

Sundarbans (West Bengal, India): Sundarban mangrove ecosystem lies between two countries India and Bangladesh. The Indian Sundarbans lies at the apex of the Bay of Bengal between *21°40′ N* and *22°40′ N* latitude and *88°03′ E* and *89°07′ E* longitude. It is situated on the lower deltaic plains of the Ganges–Brahmaputra Rivers. This wetland is tide-dominated, featuring a complex network of tidal creeks (Figs. 15.1 and 15.2) and bordering hundreds of tidal islands (Sarkar et al. 2008; Banerjee et al. 2012; Kumar and Ramanathan 2015).

Bhitarkanika (Odisha, India): Bhitarkanika mangroves (Fig. 15.3) is located in the Kendrapara district of Odisha lying between *20°04′–20°08′N* latitudes and *86°45′–87°50′ E* longitudes. It was announced as a wildlife sanctuary in 1975 with an area of 672 km^2. It is a tide-dominated mangrove with mean tide level of 1.5–3.4 m (Sarangi et al. 2002).

Coringa and Gaderu (Andhra Pradesh, India): Coringa and Gaderu mangrove ecosystem cover an area of 132 km^2 and is located to the north of Gautami-Godavari river lying between *82° 15′*and *82°22′ E* latitude; *16°43′* and *17°00′ N* longitude. The Gautami-Godavari River pours into Kakinada bay through the main distributaries Gaderu (11 km) and Coringa (26 km). The dense vegetation of mangrove (Fig. 15.4) is present between the area of Gautami-Godavari and Kakinada bay and belongs to Coringa wildlife sanctuary (Ray et al. 2006).

Pichavaram (Tamil Nadu, India): Pichavaram mangrove forest is situated between the Vellar and Coleroon estuaries (lying between *11° 02′ N* latitude and *79° 47′ E* longitude). The forest's northern part is located closer to the Vellar

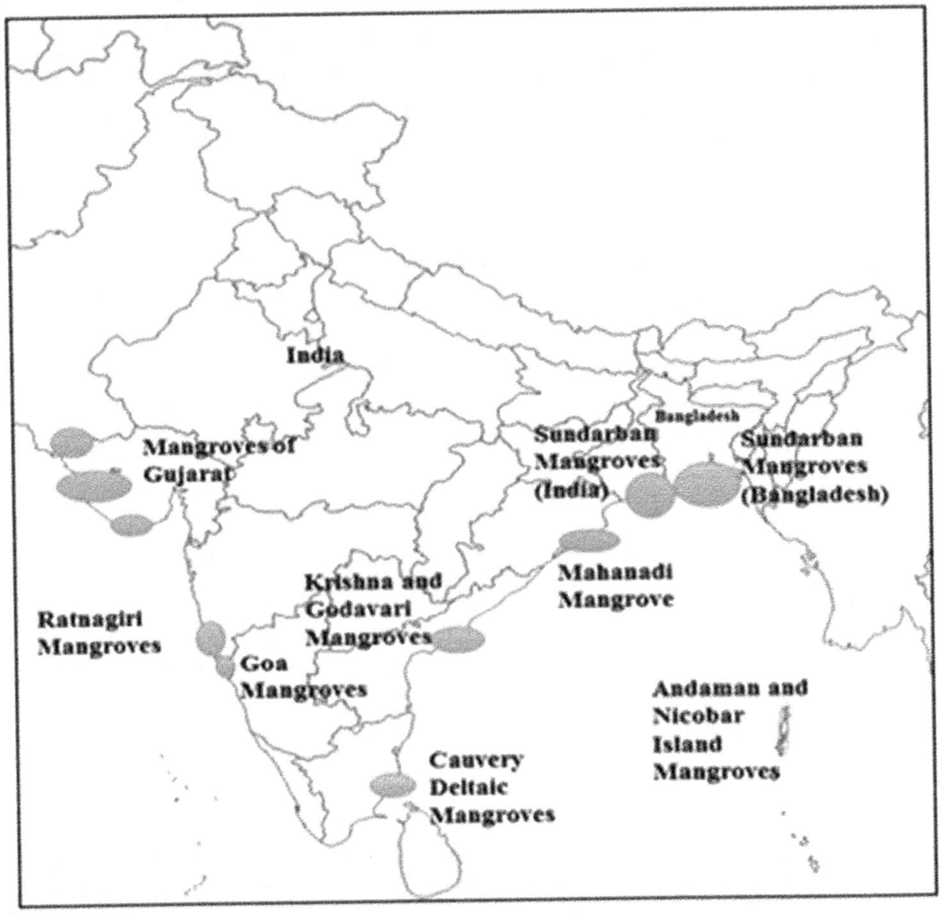

Fig. 15.1 Map locating mangrove forest in India and Bangladesh

Fig. 15.2 Sundarban mangroves

estuary and is dominated by mud-flats, while its southern part, which lies near the Coleroon estuary, is covered by dense mangrove vegetation (Fig. 15.5). The Pichavaram mangrove has three different water sources. (1) An irrigation channel (Khan Sahib Canal) and main channel of Coleroon River supplies fresh water to this mangrove ecosystem; (2) It also receives water from the Vellar and Coleroon estuaries; and (3) Neritic water from the adjacent Bay of Bengal. It occupies an area of 1100 ha, out of which 40% is waterways, 50% is forest, and the remaining mudflats and sand-flats (Ramanathan et al. 1999; Ranjan et al. 2013).

Kumarakom Mangrove (Kerala, India): Kumarakom mangrove (Fig. 15.6) covers an area of 5000 m^2 along the Vembanad estuary. It is located between *09°28′* and *10°10′ N* latitude and *76°13′ E* longitude. It has well-flourished mangrove ecosystem till the commissioning of Thannirmukham barrier constructed on Vembanad estuary; after that, it has severe regeneration problem caused drastic changes in the physicochemical conditions due to Thannirmukham barrier (Badarudeen et al. 1996).

Goa mangrove (Goa, India): Goa mangrove located between *14°53′–15°48′ N* latitudes and *73°40′–74°20′ E* longitudes on the central west coast of India. Goa mangrove (Fig. 15.7) along with Mandovi Zuari estuary complex covers an area of 12,000 ha (Attri and Kerkar 2011).

Mumbai mangrove (Maharashtra, India): The Mumbai city lies between *18°55′ N* and *19°20′ N* latitudes and *72°45′ E* and *73°00′ E* longitudes. It has the extensive network of creeks fringed with mangroves (Fig. 15.8) along both banks, having heavy industrialization and urbanization.

Gulf of Kachchh mangrove (Gujarat, India): The Gulf of Kachchh (GoK) mangrove (Fig. 15.9) is situated in Saurashtra in the western state of Gujarat occupying an area of 7300 km^2 is one of the most productive and diversified

Fig. 15.3 Bhitarkanika Mangroves

habitats along the west coast of India. Gujarat state has 991 km^2 of mangrove cover out of which 96% of total cover is contributed by Gulf.

*Andaman and Nicobar mangrove (Andaman and Nicobar Islands, India):*Andaman and Nicobar is an extensive group of islands comprises of around 572 islands, and islets lie in the lap of Bay of Bengal situated between *6°45′* and *13°41′ N* latitude and *92°12′* and *93°57′ E* longitude to the eastern side of India. Andaman and Nicobar is about 1200 km from the mainland of India and occupies a total land area of about 8249 km^2 which is surrounded by rich tropical rain forest and mangrove forest (Fig.15.10).

Sundarbans mangrove (Bangladesh): The mangroves of Sundarbans is the largest mangrove ecosystem in the world (Siddiqi 2001). The geographical location of Sundarban lies between the latitudes 21° 31′ N and 22° 30′ N and longitudes 89° 01′ E and 90° 18′ E. Among these, about 60% of these forests occur in Bangladesh, with the mangroves of Bangladesh located between 21°30′ and 22°30′ N and 89°00′ and 89°55′. It lies within the Khulna administrative division, spreadingover parts of Khulna, Bagerhat, and Satkhira districts. These areas are protected as Sundarban Reserve Forest, which comes under both Ramsar and World Heritage site of UNESCO since 1992 and 1997 respectively. The main mangrove plant species are *Heritiera fomes* which is locally called as Sundari or Sundri. Other plant species includes *Avicenniasps, Xylocarpus granatum, Xylocarpus mekongensis, Ceriopsdecandra, Sonneretiaapetala, Rhizophoramucronata, Bruguieragymnorrhiza, Aegiceras-corniculatum* and *Nypafruticans* (Fig. 15.11).

Trace metals in the coastal environments of Bangladesh mangroves may enter through diverse sources like natural weathering processes or by the anthropogenic activities (Kumar et al. 2016). The Sundarbans mangroves may receive trace metals through the various human activities like ferry services, fishing, sewage drainage, manufacturing of bricks and power plants, etc.

Fig. 15.4 Gautami-Godavari (Coringa) mangroves

Fig. 15.5 Pichavaram mangroves

Fig. 15.6 Kumarakom mangroves, Kerala

15.3 Trace Metal Distribution in Surface Sediment in Different Mangrove of India and Bangladesh-

Trace metal distribution in the different mangrove of India and Bangladesh is presented in Table 15.1. In all the studies done so far, Sundarban mangrove of Bangladesh, organic carbon has low value of 0.69% (Kumar et al. 2016) and 0.88% (Raknuzzaman et al. 2016) as compared to world average organic carbon value of 7.9% for mangrove (Donato et al. 2011; Rogers et al. 2013). It may be due to the export of locally produced organic matter to the Bay of Bengal as dissolved OM (Kumar et al. 2016; Rogers et al. 2013). As per Kumar et al. 2016, Sundarban mangrove ecosystem Cr, Ni, Fe, and As, shows high concentration whereas Cu, Mn, Cd, Pb, and Co is lower than other mangroves. Minimum organic carbon in Bangladesh Sundarban mangrove does not explain the fact that organic carbon governs TM distribution (Kumar et al. 2016; Kumar and Ramanathan 2015; Rogers et al. 2013). Sources of TM in Sundarban are aquaculture, agricultural runoff, industrial activity, like stainless steel, ship breaking, textiles, dyeing, leather, etc. (Kumar et al. 2016; Raknuzzaman et al. 2016; Rahman et al. 2011).

In Sundarban mangrove of India textural properties shows higher mud content which can be attributed to small fluvial discharge and mixing of fresh and saline water leading to flocculation and eventually settling of finer particles (Nair et al. 1982). Little organic carbon is present in all the study done so far suggesting OC does not play an important role in the accumulation of trace metal in Sundarban mangroves. The Indian part of Sundarban mangrove shows the lower concentration of As than reported in Bangladesh by Kumar et al. (2016). At most of the site of Sundarban Cu, Mn, Cd, and Co were found lower than other mangrove ecosystems of the world. Although its normalized value with upper continental crust shows an elevated amount of Cd, Cu, Pb, Mn, Fe and Zn which may be attributed to the mechanized boat, fishing, dredging activities, industrial effluent, and domestic sewage. Elevated value of Cr and As can be attributed to industrial activity and geogenic sources respectively.

Bhitarkanika shows the lowest concentration of Cu, Fe, Mn and Zn than other mangrove ecosystems of India (Saranagi et al. 2002). Coringa, Pichavarm, Goa, and GoK

Fig. 15.7 Goa mangroves

show higher contamination of Cd and Cu. Cd accumulation can be attributed to geogenic as well industrial and domestic sewage. A higher concentration of all trace metals in tsunamigenic sediment (Ranjan et al. 2008) might be due to waste brought by tsunami waves from nearby industries and domestic as well as sediment brought from deep ocean having trapped high Cd and other elements. Co and Fe are most elevated among Goa mangroves. Most of the trace metals are in low concentration at Muthupet, Vellar, Kumarakom, GoK and Sharavati mangroves.

15.3.1 Depth Profile of Trace Metal in Different Mangrove Ecosystem

The scenario of the average value of the vertical profile of trace metal is represented in Table 15.1. In the study done by Banerjee et al. (2012), on Sundarban, shows a sedimentation rate of 3.0–4.5 mm/year and mass accumulation rate of 0.41–0.78 gcm^{-2} year^{-1}, the high accumulation rate is due to frequent exposure to tidal inundation and high salinity condition. In a study done on Sundarban mangrove by Banerjee et al. (2012) and Chatterjee et al. (2009); organic carbon is lower than another mangrove, and it increases in the subsurface layer which can be attributed to the decomposition of leaf litter, high biological activity, and burial of microalgae (Volkman et al. 2000). In both the studies Fe and Mn shows a similar trend as of OC and other trace metals showing its vertical profile variation in relation with OC and Fe-Mn oxyhydroxide, which shows trace metal accumulation with Fe-Mn oxy-hydroxide (Waldichuk 1985). In the study of Banerjee et al. (2012) and Chatterjee et al. (2009), Zn and Cu; Cd and Cu show uniform concentration in vertical profile respectively which can be attributed to natural weathering as well as industrial effluent. Cr, Co, Cu, Mn, and Zn elevated contamination in the top layer of both studies shows the impact of industrial discharge and anthropogenic action.

Trace metal concentration in the depth profile of Pichavaram mangrove shows decreasing trend for Cr, Cu, Fe (except Ranjan et al. 2008, 2013 where Fe value increased with depth) and Pb with a peak value in sub-surface layer of depth profile and decreasing trend afterwards in most of the cores (Ramanathan et al. 1999; Ranjan et al. 2008, 2013); which can be attributed to leaching, uptake by vegetation and post-depositional remineralization. Fe and Mn in the study done by Ranjan et al. (2008), show increased value with depth up to middle profile then decrease afterward attributing to the post-diagenetic process including tsunami turbulence

Fig. 15.8 Mumbai mangroves

process. Al and Cd show increasing trend in the study done by Ramanathan et al. (1999), which can be attributed to anthropogenic inputs.

Contrary to Pichavaram, the study done on Muthupet by Janaki-Raman et al. (2007), Fe and Mn show decrease in concentration from surface till 10–20 cm after that it increases with vertical profile, similar to that Cr, Cu, Co, Ni, Zn and Cd varies vertically suggesting soluble complexes with reduced sulfur which have amplified the passage of these elements from sediments to the water column (Huerta-Díaz and Morse 1992). Diagenetic processes play a significant role in vertical distribution whereas similar to Pichavaram in the study done by Natesan et al. (2014), in Muthupet, Zn, Mn, Ni, Pb, Cr and Cd in almost all core shows a similar trend of increase in concentration up to subsurface layer then decrease with depth. Cu shows external input from sewage sludge and municipal waste discharge, Zn is from maritime activities, increase in Ni at a subsurface layer is due to manganese oxyhydroxide sorption of Ni, and Pb concentration is attributed to boat exhaust, oil spillage, mechanical boat for fishing and sewage effluent. The similar trend of high value at sub-surface layer and decreasing downward is shown for Fe, Mn, Al, Cr, Cu, Co, Pb and Zn in Sharavati (Fernandes and Nayak 2016) and Goa (Veerasingam et al. 2015). Other trace metals mostly follow the trend of Fe and Mn due to diagenetic processes.

15.3.2 Speciation of Trace Metal in Different Mangrove Ecosystem

Average values of trace metal speciation are given in Table 15.1. In Sundarbans of India and Bangladesh (Kumar et al. 2015), Fe and Zn was found to be dominant in the residual fraction followed by reducible, oxidizable and acid soluble fraction suggesting most of Fe and Zn are not bio-available in normal condition. Cu shows dominance in residual followed by reducible, also indicating unavailability under normal conditions whereas Mn shows dominance in acid soluble fraction followed by residual showing it

Fig. 15.9 Gulf of Kachchh mangroves

bioavailability as well as its potential risk for the aquatic environment. The high value of Mn in acid leachable fraction was also found by Jonathan et al. (2010), in Sundarban with Fe showing it diagenetic behaviour as well as elevated value of acid leachable Cr, Cu, Ni, Pb and As is also found which may be attributed to industrial effluent, domestic discharge as well as dredging.

In the speciation study on the vertical profile of sediments by Ranjan et al. (2013), it was found that Cd, Cu, Cr, Fe, Ni, Pb, and Zn were dominant in organic/sulfide bound fraction and Cd is little in a readily bound fraction. Total metal concentration Cd, Cr, Fe, Mn, Ni, Pb and Zn increases with depth showing dominance in residual fraction, with depth there is an increase in organic matter as well as a sulfur bound fraction.

Trace metal speciation by Dias and Nayak (2016) in Goa mangroves shows that almost all trace metals are mostly dominant in residual fraction and not bio-available under prevalent conditions which reduce the potential risk. A similar trend in Goa for Ni and Cu was found by Chakraborty et al. (2015) and for Fe and Cr in the study of Noronha-D'Mello and Nayak (2015). Mn show potential risk with next dominance in exchangeable fraction (Dias and Nayak 2016) and Fe-Mn oxide (Noronha-D'Mello and Nayak 2015) whereas other trace metals show next dominance in Fe-Mn oxide bound or organic bound (Dias and Nayak 2016). In all speciation studies of Goa mangrove sediments by Dias and Nayak (2016). Chakraborty et al. (2015) and Noronha-D'Mello and Nayak (2015) the depth profile of total trace metal is similar to another study with elevated value in subsurface layer and decreasing trend afterward for almost all trace metals. In the study of Noronha-D'Mello and Nayak (2015), trace metals also show elevated value in lower bottom indicating anthropogenic input.

15.3.3 Processes Governing Trace Metal Pollution

Trace metal distribution and accumulation in mangrove ecosystem is governed by many factors like sediment texture, organic carbon, hydrodynamic regimes, salinity, geomorphological settings, industrial discharge, sewage, terrestrial runoff etc. (Peters et al. 1997; Jonathan et al. 2010), controlled by many processes like bioturbation, flocculation-

Fig. 15.10 Andaman and Nicobar mangroves

deflocculation, Fe-Mn oxyhydroxide coating, erosion etc. (Bellucci et al. 2002).

Fine sediment particle has higher surface area, ionic attraction to form inorganic and organic complexation thus trace metals shows the close association with fine grain sediment thus grain size play a significant role in the distribution of trace metals (Zhang et al. 2009; Huang et al. 2014; Subramanian and Mohanachandran 1990). Most mangrove ecosystems of India and Bangladesh show the dominance of mud (silt+clay) which can be attributed to low fluvial discharge, estuarine mixing, suspension and resuspension processes and settling of suspended particles (Nair et al. 1982; Jonathan et al. 2010). Organic carbon plays a major role in trace metal distribution. It has a more affinity towards trace metals through adsorption and complex formation in aquatic environments (Badarudeen et al. 1996; Zhang et al. 2004; Mukherjee et al. 2009; Marchand et al. 2011) thus it acts as a metal carrier (Ray et al. 2006). In Sundarbans mangrove, organic carbon does not play the major role in trace metal distribution as it has low organic carbon than other mangrove ecosystems. Low organic carbon is attributed to marine sedimentation, mixing process at sediment water interface, rapid tidal export of locally produced organic matter (Rogers et al. 2013; Kumar and Ramanathan 2015), high rate of degradation by microbial mediated processes, poor adsorbability of organic matter on negatively charged grains. The high value of organic carbon is mainly due to decomposed detritus and anthropogenic inputs.

The anaerobic condition favors sulfides formation of Fe and Mn (Silva et al. 1990). The high value of Fe in most of the mangrove ecosystem can be attributed to Fe accumulation as sulfide, under anaerobic conditions of mangrove sediment (Anu et al. 2009; Nobi et al. 2010), other trace metals also precipitate with Fe forming polysulphide minerals (Sarkar et al. 2004). The Fe-Mn complex also referred to as sink of trace metals due to its affinity with other trace metals; even vertical profile of Fe and Mn in several mangroves suggest diagenetic enrichment (Froelich et al. 1979; Santschi et al. 1990; Chatterjee et al. 2009; Kumar and Ramanathan 2015). Fe-Mn oxyhydroxide in sediment layer form Fe^{2+} and Mn^{2+}

Fig. 15.11 Sundarban (Bangladesh) Mangroves

which move upward in sediment profile and get precipitated in the oxic-suboxic layer. Fe-Mn oxyhydroxide has the affinity to accumulate other trace metal as they pass through water to sediments (Waldichuk 1985).

Very higher enrichment of Cd in Pichavaram, Coringa, and Gaderu, GoK can be attributed to the geogenic origin. In Pichavaram mangrove, tsunamigenic sediment, the high concentration of Cd (Ranjan et al. 2008) is attributed to marine sediment trapped in the deep ocean rich in Cd, get into the sediments after the tsunami. The high Cd concentration in Pichavaram mangrove was attributed to the combined effect of both natural sources and anthropogenic inputs (like fertilizers brought by Vellar, Coleroon and Uppanar River).

Extremely high enrichment of Pb in Pichavaram and Bhitarkanika mangrove ecosystem can be attributed to Pb precipitation around redox boundaries (Lee and Cundy 2001) and its downward flux due to biogenic particles (Lambert et al. 1991). A higher concentration of Pb at Manakudy estuary mangrove (161.254 μg/g) was due to the local redox condition, allowing Pb precipitation with the Mn-oxide formation (Lee and Cundy 2001). In Tsunamigenic sediment, high enrichment of Pb can be attributed to discharges from nearby industries, aquaculture, domestic sewage, etc. brought back by tsunami waves (Purvaja and Ramesh 2000; Subramanian 2004). In Bangladesh Sundarbans, high Pb concentration was contributed due to Pb smelting factories, ship breaking industries and automobile industries (Rahman et al. 2011).

The contamination of As has been observed by many authors in Sundarbans (Dowling et al. 2002; Massolo et al. 2012) which could be geogenic in origin like coal seams and arsenic minerals in rock or it can be due to fertilizer industry (Mirlean et al. 2003), treating wood by using copper arsenate and other industrial activity occurring the in drainage basin (Wasserman et al. 2001). The contamination of Ni may be due to weathering of rocks (Achyuthan and Richardmohan 2002) or by stainless steel factory, plating, dye and textile industries (especially around Sundarbans of Bangladesh), Ni-Cd rechargeable batteries, ship breaking industries and other anthropogenic sources.

Thus, there are many points and non-point sources for trace metals distribution and accumulation in mangrove setting and only one factor may not play a role in trace metals distribution whereas it is a complex process where all factors play an important part.

Table 15.1 Concentration of trace metals in Surface and core samples of different mangrove ecosystems of India and Bangladesh

Location	Type of sample	Method	Al	As	Cd	Co	Cr	Cu	Fe	Mn	Ni	Pb	Zn	OC	Sand	Silt	Clay	Ref.
Sundarbans (Bangladesh)	Surface	Total TM	1.22	14.01	0.07	14.44	38.69	31.73	3.43	0.06	167.29	17.88	58.59	0.63	8.73	75.94	15.33	1
Sundarbans (Bangladesh)	Surface	Total TM		2.66	0.069		26.43	25.63			23.8	28.3	44.6	0.88				2
Sundarbans (Bangladesh)	Surface	Total TM			1.79	13.65		20.1416	2.66	0.05	34.383	11.6						3
Sundarbans (Value in Range)	Surface and core	Speciation						36–82	3.9–5.3	0.0424–0.0770			55–83	0.601–0.932	2.46–7.62	64.14–73.92	19.60–31.22	4
Sundarbans (Value in Range)	Surface	Total TM		4.12–11.46	0.02–4.4		21.2–60.9	11.6–102.47			19.10–52.6	7.09–183.88		0.47–0.91	0.51–37.40	18.20–66.64	23.18–77.80	5
Sundarbans	Surface	Total TM		3.82	0.21	7.67	28.3	38.29	0.29	0.06	34.5	15.8	34.42	0.60–0.66	1.80–15.45	32.58–38.93	51.98–59.28	6
Sundarbans	Core and Surface	Total TM			1.88	23.48	44.13	38.47	3.75	0.0574	50.35	30.28	75.87	0.786				7
Sundarbans	Surface and Core	ALTM		8.09	0.18		99.01	28.94			51.86	23.01		1.552	28.47			8
Sundarbans	Surface and Core	Speciation	6.33	8.3	0.15	10.41	55.98	25.74	3.12	0.58	30.17	65.59	22.8	0.625	15.34	48.24	36.41	9
Sundarbans	Surface	Total TM						90.765				38.175	303.75					10
Sundarbans	Surface	Total TM	2.46	3.5	0.1	12.46	36.44	35.47	3.08	0.14	33.46	17.2	74.18	0.457	21.684	56.54	21.76	11
Bhitarkanika	Surface	Total TM						3.78	0.003	0.0009			1.01					12
Coringa- Gaderu	Surface	Total TM			10.9	28.8	2.2	47.8	0.46	0.11	25.7	55.8						13
Pichavaram	Surface and Core	Total TM			23		152	34	3.80	0.033	51	21	16					14
Pichavaram	Surface and Core	Total TM			34.74		6200	132.3	2.50	0.08	252.1		106					15
Pichavaram	Surface and Core	Speciation			6.6	35.3	141.2	43.4	3.25	0.09	62	11.2	93					16
Muthupet	Surface and Core	Total TM			0.24	11	117.3	18.1	3.21	0.0225	53.2	21.6	115.2					17
Muthupet	Surface and Core	Total TM										43.4	261.88		11.9			18
Vellar estuary mangrove	Surface	Total TM	0.0542		9.15		9.44	16.28	1.00	0.0178	1.64	0.98	39.28					19
Kumarakom	Surface	Total TM					9	13	1.83	0.003			84	2.3	45.31	50.75	4.07	20
Sharavati	Surface and Core	Total TM	3.4			4	52	9	1.6	0.013	11		17	0.4	82	4	14	21
Goa	Surface and Core	Speciation				114.35		64.99	8.72	0.25	118.56		277.9	1.89	55.27	20.88	24.07	22

Goa	Surface and Core	Speciation						49.36	9.94	0.29	64.02	18.74		5.06	44.06	34.48	21.5	23
Goa	Surface and Core	Total TM				24.41	147.75	47.32	19.25	0.154		24.5	66.62		<20			24
Goa	Surface and Core	Speciation	8.38			22.85	271.18	36.26	14.53	0.512	47.15		104.34	2.13	5.82	48.55	45.64	25
Goa	Surface	Total TM				34.36	17.32	45.34	12.18	0.16		22.51	72.853	2.63	26.74	20.69	52.28	26
Gulf of Kachchh	Surface	Total TM			6.113		8.52	5.04	0.19	0.005	2.708	0.37	231					27
Gulf of Kachchh (Value in Range)	Surface	Total TM			0.173–23.08	18–26	120–255	44–138	0.11–8.23	0.085–0.11	51–78	14–34	0.29–160	1.47–3.25	77.2–97.93	0.60–21.45	0.1–2.5	28

1- Kumar et al. (2016), 2- Raknuzzaman et al. (2016), 3- Rahman et al. (2011), 4- Kumar and Ramanathan (2015), 5- Antizar-Ladislao et al. (2015), 6- Chowdhury et al. (2015), 7- Banerjee et al. (2012), 8- Jonathan et al. (2010), 9- Chatterjee et al. (2009), 10- Sarkar et al. (2008), 11- Sarkar et al. (2004), 12- Sarangi et al. (2002), 13- Ray et al. (2006), 14- Ranjan et al. (2013), 15- Ranjan et al. (2008), 16- Ramanathan et al. (1999), 17- Raman et al. (2007), 18- Natesan et al. (2014), 19- Palpandi and Kesavan (2012), 20- Badarudeen et al. (1996), 21- Fernandes and Nayak (2016), 22- Dias and Nayak (2016), 23- Chakraborty et al. (2015), 24- Veerasingam et al. (2015), 25- Noronha-D'Mello and Nayak (2015), 26- Attri and Kerkar (2011), 27- Kumar et al. (2015)

TM Trace Metal, *ALTM* Acid Leachable Trace Metals

*Al, Fe, Mn, Sand, Silt, Clay and Organic carbon are in %; other elements are in mg/kg

15.4 Trace Metals in Different Biota of Mangrove Forest

Since long back, the surface sediments has been acknowledged as a sink of many pollutants, due to which it generates a risk to the quality of aquatic ecosystem by many biogeochemical exchanges with the overlying surface water (Massoudieh et al. 2010). There are many heavy metals like Cr, Ni, Cu, As, Pb and Cd which are commonly found in the aquatic environment (Wong et al. 2001; Zhang et al. 2012; Ahmed et al. 2015) and causes contamination of water, sediments and surrounding biota (flora and fauna) due to their toxicity (Olias et al. 2006), non-degradable nature and ability to get bio accumulated in the biota (Soto et al. 2011; Tao et al. 2012). In the coastal ecosystem, aquatic products like fishes, crabs, molluscans etc. act as the main source of food for the local people. These aquatic biota living in the contaminated aquatic environment, when consumed by humans, causes the severe health hazard to human beings.

Thus to calculate the capacity of the heavy metals to get bioaccumulated in the tissues of aquatic organisms, the Biota Sediment Accumulation Factor (BSAF) is calculated (Table 15.2).

Trace metal accumulation in biota of different mangrove ecosystem from India and Bangladesh is presented in Table 15.3, with the help of different biological indexes. In the study done by Islam et al. (2017); Bangladesh Sundarban show, mean BSAF in fishes follow the order as Cd>As>Ni>Cu>Pb>Cr. Among these trace metals, the highest BSAF value is shown by Cd, which indicates that the rate of accumulation of Cd is higher in fish species. Few fishes like *A. testudineus* and *H. Fossili* shows the high value of BSAF due to ingestion of sediment and feeding behavior of these fishes. Thus these species of fishes can be used as a possible bio-indicator for the estimation of contamination of heavy metals in this aquatic ecosystem (Tao et al. 2012; Islam et al. 2014). Through the concentration of heavy metals on different fish species of mangrove and BSAF value, it was found that the pelagic fishes exhibit lower values of heavy metals than the bottom dwelling fishes (Gupta et al. 2009). Low values of Cr pollution are reported in the industrial area, so its accumulation rates in biota are also very low. It has been reported that higher uptake of Cr occurs in young fishes, but its values decline with the growing age of the fishes, which may be due to rapid removal of Cr (Rahman et al. 2012). The value of Ni contamination was greater than the maximum allowable concentration (MAC), showing that Ni contamination is a matter of risk to human health by intake of these contaminated fishes.

The As pollution is also very high in these aquatic ecosystems due to natural and anthropogenic sources. From the above table, it was found that the mean values of Cd were higher in most fish species than MAC (0.10 mg/kg) revealing that their consumption may lead to potential health hazards.

Table 15.2 Bio-contamination index

Bio-contamination index	Definition	Formula	Index		References
Biota Sediment Accumulation Factor (BSAF)	BSAF used to calculate the accumulation profile of heavy metal pollutant	$BSAF=C_{fish}/C_{sediment}$	BSAF<1	No accumulation	Abdallah and Abdallah (2008)
		C_{fish}=concentration of metal in fish (mg/kg dw) and $C_{sediment}$=metal concentration in sediment (mg/kg dw)			
Target Hazard Quotient (THQ)	THQ used to assess the non-carcinogenic risk for metals by the consumption of biota	THQ=[(EFr X ED X FIR X C)/RfD X BW X AT] X 10^{-3}	THQ <1	Safer consumption limit	USEPA (2010a, b)
	C=concentration of metal in samples [mg/kg, fresh weight (fw)]	EFr=frequency of exposure (365 days/year)	THQ≥1	Potential health risk	
	RfD=Oral Reference Dose (mg/kg/day)	ED=duration of exposure (70 years)			
	BW=Average body weight (adult, 60 kg)	FIR=Ingestion rate of fish (g/person/day)			
	AT=AT is the averaging time for non-carcinogens (365 days/year × number of exposure years, assuming 70 years)				
Bio-concentration Factor (BCF)	Accumulation of heavy metal in or on an organism when the source of chemical is solely water	BCF=*Corg*/*Csed*	BCF<1	No bioaccumulation (safe)	Vassiliki and Konstantina (1984)
	Corg=metal concentration in the organisms	Csed=same metal concentration in the sediment	BCF>1	Bioaccumulation (unsafe)	

Table 15.3 Trace metals accumulation in different biota of the mangrove ecosystem

	Species	Index	Cr	Ni	Cu	As	Cd	Pb	Fe	Mn	Zn	Co	MPI	Hg	Reference
Bangladesh Sundarban	Channapunctata	BSAF	0.007	0.038	0.037	0.106	0.073	0.029							Islam et al. (2017)
		THQ	0.00008	0.034	0.015	1.38	0.008	0.113							
	Awaousgrammepomus	BSAF	0.01	0.045	0.047	0.089	0.067	0.015							
		THQ	0.00016	0.041	0.019	1.47	0.008	0.059							
	Anabas testudineus	BSAF	0.014	0.094	0.092	0.254	0.459	0.061							
		THQ	0.00022	0.093	0.039	2.97	0.054	0.218							
	Heteropneustesfossilis	BSAF	0.013	0.096	0.082	0.297	0.488	0.039							
		THQ	0.00023	0.098	0.03	4.23	0.059	0.178							
	Neotropiusatherinoides	BSAF	0.007	0.034	0.033	0.123	0.068	0.034							
		THQ	0.00009	0.029	0.014	1.13	0.008	0.131							
	Colisafascinata	BSAF	0.007	0.061	0.038	0.124	0.082	0.014							
		THQ	0.00012	0.057	0.015	1.49	0.01	0.058							
	Channastriata	BSAF	0.003	0.048	0.028	0.099	0.042	0.014							
		THQ	0.00006	0.045	0.013	1.08	0.005	0.053							
	Notopterusnotopterus	BSAF	0.005	0.036	0.028	0.095	0.171	0.011							
		THQ	0.00008	0.033	0.011	1.19	0.025	0.036							
	Batasiobatasio	BSAF	0.005	0.074	0.044	0.12	0.394	0.017							
		THQ	0.00007	0.06	0.019	1.39	0.05	0.064							
	Coricasoborna	BSAF	0.004	0.07	0.039	0.141	0.259	0.009							
		THQ	0.00005	0.058	0.017	1.54	0.027	0.033							
	Puntius chola	BSAF	0.006	0.048	0.049	0.119	0.094	0.019							
		THQ	0.00009	0.044	0.02	1.49	0.011	0.075							
Zuari Goa	Neritina (Dostia) violaceae	BSAF		0.08	0.48				0.03	0.14	0.64	0.1	135.9		Dias and Nayak (2016)
	Telescopium telescopium	BSAF		0.25	1.24				0.06	0.22	0.6	0.03	153.24		
	Cerathideacingulata	BSAF		0.17	0.65				0.02	0.21	0.7	0.19	94.77		
	Katelysiaopima	BSAF		0.05	0.15				0.01	0.07	0.36	0.02	39.48		
	Katelysiaopima	BSAF		0.13	0.47				0.01	0.42	0.53	0.08	87.99		
	Polymesoda	BSAF		0.19	0.4				0.07	0.2	1.17	0.12	58.81		

(continued)

Table 15.3 (continued)

	Species	Index	Cr	Ni	Cu	As	Cd	Pb	Fe	Mn	Zn	Co	MPI	Hg	Reference
Sundarban mangrove, India	Hermit crab	BC			7.55		0.462	4.66	375		95.5				Ahmed et al. (2011)
		BSF			0.06		1	0.02	6.08		0.13				
	Horseshoe crab	BC			5.66		0.46	5.8	480.66		76.8				
		BSF			0.04		1	0.02	7.8		0.11				
	Mud crab	BC			6.28		0.551	6.77	595.29		85.5				
		BSF			0.05		1.2	0.02	9.65		0.12				
	Gastropod	BC			3.66		0.859	6.5	235		110				
		BSF			0.03		1.87	0.02	3.81		0.15				
	Mud skipper	BC			7.23		0.725	4.99	1051		98.5				
		BSF			0.06		1.58	0.02	17.05		0.13				
Vellar estuary	Neritacrepidularia	BC	0.2319915	9.3780488	1.5909091		0.1650273	1.6326531	0.1610491	1.0871663	0.3823829				Palpandi and Kesavan (2012)

The value of Pb in few fish species shows greater than the safe limit of 0.5 mg/Kg set by WHO (2004) revealing that these fish species are polluted by Pb contamination and may pose risks to human health. In the above table, Target hazard quotients (THQs) of six heavy metals from consuming fish species are enlisted. For all the heavy metals in fish species (except As), the THQ values were less than 1, which indicates the safer consumption limits for humans. The values of Target Cancer Risk (TR) were more than the acceptable risk limit (0.000001) (USEPA 2010a, b), for As and Pb, showing that the humans eating these fish species are exposed to the contamination of As and Pb which may cause lifetime cancer risk.

In Vellar estuary, the study done by Palpandi and Kesavan (2012), on *Nerita crepidularia* shows that heavy metals occur in high amount than the ambient environment and Cu and Mn, in particular, therefore, it may be used as a biomonitoring of certain heavy metals.

In the study of Ahmed et al. (2011), on macrobenthos of Sundarbans (Bangladesh) it was found that there is accumulation of Fe, Cu, Pb due to its contamination from water and sediment through breaking and repairing activities of ship in Khulna shipyard and building, due to the degradation of organic matter by aerobic processes (Das and Nolting 1993), effluent discharge which brings the pollutants like metal rust, lubricants and various disposable solids from the ship scrapping industry.

The presence of these heavy trace metals, even if present in small amount, is injurious to the health of aquatic biota. This aquatic biota takes these metals into their tissues by the passive process (Simkiss and Taylor 1989). In some cases, the uptake may also take place by ion pumps in which the energy dependence exists (Rainbow 1995). The value of BCF<1 is expected for most of the trace metals Vassiliki and Konstantina (1984), or else the organism starts bioaccumulating the metals (Falusi and Olanipekun 2007). The high values of BCF for Cd and Fe reveals that these trace metals are highly bioaccumulated and biomagnified in the macrobenthic fauna of Bangladesh Sundarban mangrove ecosystem.

The accumulations of heavy metals in these above mentioned studies from published literature on macrobenthos, fishes, bivalves, mollusks, etc. indicate the possible pollution of heavy metals in the fauna of mangroves. This piece of knowledge may be useful for the development of new and improvement of existing policies related to the use and dumping of the toxic heavy metal in the aquatic environment.

15.5 Management of Mangroves

Acknowledging the value of mangroves, the Government of India created the committee namely 'National Mangrove Committee' in 1976 in the Ministry of Environment and Forests to advise the government about mangrove management and development. In its first meeting, the committee laid stress on the need to perform an assessment of the coverage of existing mangrove areas of the country. The government then establishes a scheme for the protection, conservation and management of mangroves with the following objectives:

- To identify the area of mangroves requiring protection.
- To prepare the management plan;
- To promote the mangrove research;
- Involve universities, state governments, local organization and research institutions to adopt a multidisciplinary approach to manage mangrove ecosystem.

15.5.1 Legislative Framework

The Indian Forest Act, since 1972, has been applied to the management of mangrove forest of Sundarban which comes under the declaration of reserved forest (Naskar and Mandal 1999). On the other hand, according to The Forest Conservation Act 1980, no area of the forest shall be used for non-forestry purposes without the permission of Government of India. The most crucial role in the management and protection of mangroves forests has been played by Environment Protection Act 1986. To conserve the coastal ecosystem, this act declares a Coastal Regulation Zone where all the activities like untreated wastewater discharge, dumping of waste, industrial activities are restricted. It also put mangrove ecosystems as the most ecologically sensitive category.

15.5.2 Mangrove Management and Conservation in Different Mangroves of India

Sundarbans mangroves, India – (Joint Forest Management (JFM) and Eco Development) - Under JFM, Government notifications were issued by the Government of West Bengal to form Forest Protection Committees (FPCs) and Eco Development Committees (EDCs). Since the issuance of the resolutions, a total of 54 FPCs and 11 EDCs have been formed in the Sundarbans covering a total of 87,287 ha and 47,325 families. As a follow up of the concept of JFM in the Reserved and Protected Forests of South Bengal and North Bengal, eco development activities have been taken up in the Sundarbans including the protected areas.

Management of mangroves in Goa – From the total land area of 370,000 ha of Goa, mangroves occupy 20,000 ha which declined sharply to 500 ha in 1987. To restore these degraded mangrove areas in 1985–1986, afforestation work has been started, and by the end of 1996–1997, 876 ha of mangroves were reported to recover with the help of this restoration program (Forest Department of Goa statistics). To govern the development of mangrove ecosystems, in 1988, a State Level Steering Committee was formed by the Government of Goa. A Multidisciplinary Project Formulation Team was created by the state government in 1990 to help in the preparation of a Comprehensive Action Plan for the development of the mangrove ecosystem. In 1991–1992, a 5-year Mangrove Management Plan was prepared by the Government of India and apply with financial support, with the plantation of around 100 ha of mangroves each year as designed.

Management of mangroves in Andaman and Nicobar Islands – Due to growing awareness regarding the protection and conservation of mangroves, since 1987, the Andaman and Nicobar Administration has resticted the extraction of mangrove wood. Empahis has been laid on the use of diesel by the plywood industries, power station, and government steam vessels.

Mangrove Management in Andhra Pradesh – The Forest Department formed the Eco Development committees (EDCs) and Vana Samarakhshana Samitis (VSSs) under the Joint Forest Management principles to protect mangrove areas. The complete Coringa and Krishna Wildlife Sanctuary are being protected by 20 and 22 EDCs respectively.

Conservation and Management of Mangroves in Karnataka – (Joint Forest Management and stakeholders' participation) – In the present time, it can be seen that the local communities and stakeholders involve in the management approach by the Department from direct implementation through the process of Joint Forest Participatory Management (JFPM).

Policy Issues

1. **Transfer of mangrove areas to Forest Department:** A policy decision is necessary to get them transfer to the Forest Department to notify them as "Conservation Reserve" under Wildlife Protection Act 1972.
2. **Integrated approach:** This approach is needed for the healthy development of mangrove forests as these ecosystems require a good combination of brackish and fresh water.
3. **Incentives for private area planting:** A scheme needs to propose to provide financial help to private farmers to encourage mangrove plantations.
4. **Addressing livelihood issues:** The private owners of mangroves should be encouraged to arrange themselves into Eco-Development Committees with the provision of financial support.
5. **Publicity and awareness:** Coastal area community, particularly the school children, should be aimed for all the awareness campaigns.

Management Practices for Mangroves in Kerala

1. Improvement of habitat for mangrove forests to eliminate non-degradable waste and toxic pollutants.
2. To remove obstructions and artificial bunds to ensure free flow of water
3. Incentives should be given to the land owners who maintain mangroves without changing land use patterns.
4. Providing approach roads, sanitation and drainage, family health care and promotion of pisciculture.

The MAP implements restoration of mangroves which are degraded with components like visits to mangrove areas, Community Center for extension activities, habitat improvement, regeneration, inventory survey, and mapping.

Bangladesh Mangrove Management

The Forest Department is one of the responsible agency who is managing the forests. At the same time the various non-government organizations like NGOs, private organizations and the group of local people now occupied in the activities of forest management. NGOs play an important role in mobilizing the community, motivating them and forms group to take care of mangrove ecosystems by providing training on the plantation and management techniques. The government also gives legal, institutional and infrastructural support to participate in the management practices. Apart from it some international organizations like JSPS, CIDA, USAID, AIDAB, UNESCO, UNDP, ADB, FAO and World Bank support various research programs and developmental activities for the management of mangrove ecosystems.

15.6 Summary

Distribution and accumulation of trace metals can be due to many factors like soil texture and organic matter as fine particle with great surface area and higher ionic attraction makes complex of inorganic and organic compound like precipitation of metals as hydroxide coating (mainly Fe and Mn) which have tendency to further accumulate other trace metals like Cd, Cu, Pb and Zn as they pass through the water en route to sediment. Trace metal has a many point and non-point anthropogenic sources like agricultural runoff, waste from sewage and sludge, industrial discharge, etc. Tsunamigenic sediment shows higher enrichment of the most of the trace metal which can be attributed to high waste carried tsunamic waves. In the vertical profile of trace metals, most of the trace metals show peak value in the subsurface layer which can be attributed to leaching, uptake by vegetation and post-depositional remineralization. Most of the trace metals show similar trend in vertical profile as Fe, Mn, and OC due to diagenetic processes. In speciation, most of the trace metals show dominance in residual fraction and are not bio-available under a prevalent condition which reduces the potential risk. Mn show potential risk with the higher value in an exchangeable fraction. BSAF in fishes follows the order as Cd>As>Ni>Cu>Pb>Cr which indicates that the rate of accumulation of Cd is higher in fish species even more than MAC (0.10 mg/kg) revealing that their consumption may lead to potential health hazards. Trace metals concentration in different fish species of mangrove and BSAF value shows that the pelagic fishes exhibit lower values of heavy metals than the bottom dwelling fishes. Bioaccumulation of Cu and Zn in both gastropods and bivalves was higher than bioavailability which revealed their preference in metal accumulation over their life span period. In the study on macrobenthos of Sundarbans (Bangladesh), it was found there is the accumulation of Fe, Cu, Pb due to its contamination from water and sediment. Thus this study shows that mangrove ecosystem act as a trap or sink for trace metals which adversely affects mangrove biodiversity, species inhabiting mangrove ecosystem and food chain. To understand the affect of trace metals on mangrove forest their relation with complex physio-chemical processes, its impact on flora and fauna a periodical study of trace metal contamination is needed; in long run it will also help in understanding changes implemented by different mangrove management and conservation plans time to time.

References

Abdallah MAM, Abdallah AMA (2008) Biomonitoring study of heavy metals in biota and sediments on the South Eastern coast of Mediterranean sea, Egypt. Environmental

Achyuthan H, Richardmohan D (2002) Trace metals concentrations in the sediment cores of the estuary and tidal zones between Chennai and Pondicherry, along with the east coast of India

Ahmed K, Mehedi Y, Haque R, Mondol P (2011) Heavy metal concentrations in some macrobenthic fauna of the Sundarbans mangrove forest, south west coast of Bangladesh. Environ Monit Assess 177(1):505–514

Ahmed MK, Shaheen N, Islam MS, Al-Mamun MH, Islam S, Banu CP (2015) Trace elements in two staple cereals (rice and wheat) and associated health risk implications in Bangladesh. Environ Monit Assess 187:326–336

Antizar-Ladislao B, Mondal P, Mitra S, Sarkar SK (2015) Assessment of trace metal contamination level and toxicity in sediments from coastal regions of West Bengal, the eastern part of India. Mar Pollut Bull 101(2):886–894

Anu G, Nair SM, Kumar NC, Jayalakshmi KV, Pamala D (2009) A baseline study of trace metals in a coral reef sedimentary environment, Lakshadweep Archipelago. Environ Earth Sci 59:1245–1266. Available at: <http://www.epa.gov/reg3hwmd/risk/human/index.htm

Attri K, Kerkar S (2011) Seasonal Assessment of Heavy Metal Pollution in Tropical Mangrove Sediments (Goa, India). J Ecobiotechnol 3:9–15

Badarudeen A, Damodaran KT, Sajan K, Padmalal D (1996) Texture and geochemistry of the sediments of a tropical mangrove ecosystem, southwest coast of India. Environ Geol 27(3):164–169

Banerjee K, Senthilkumar B, Purvaja R, Ramesh R (2012) Sedimentation and trace metal distribution in selected locations of Sundarbans mangroves and Hooghly estuary, Northeast coast of India. Environ Geochem Health 34(1):27–42

Bellucci LG, Frignani M, Paolucci D, Ravanelli M (2002) Distribution of heavy metals in sediments of the Venice Lagoon: the role of the industrial area. Sci Total Environ 295:35–49

Chakraborty P, Ramteke D, Chakraborty S (2015) Geochemical partitioning of Cu and Ni in mangrove sediments: relationships with their bioavailability. Mar Pollut Bull 93(1):194–201

Chatterjee M, Massolo S, Sarkar SK, Bhattacharya AK, Bhattacharya BD, Satpathy KK, Saha S (2009) An assessment of trace element contamination in intertidal sediment cores of Sunderban mangrove wetland, India for evaluating sediment quality guidelines. Environ Monit Assess 150(1):307–322

Chowdhury R, Favas PJ, Pratas J, Jonathan MP, Ganesh PS, Sarkar SK (2015) Accumulation of trace metals by mangrove plants in Indian Sundarban Wetland: prospects for Phytoremediation. Int J Phytoremediation 17(9):885–894

Clark MW, McConchie D, Saenger P, Pillsworth M (1997) Hydrological controls on copper, cadmium, lead and zinc concentrations in an anthropogenically polluted mangrove ecosystem, Wynnum, Brisbane, Australia. J Coast Res 13:1150–1158

Das JD, Nolting RF (1993) Distribution of trace metals in sediments and pore waters in the N. W. Mediterranean Sea. NIOZ. EROS-200 Project, p 10

Dias HQ, Nayak GN (2016) Geochemistry and bioavailability of mudflats and mangrove sediments and their effect on bioaccumulation in selected organisms within a tropical (Zuari) estuary, Goa, India. Mar Pollut Bull 105(1):227–236

Donato DC, Kauffman JB, Kurnianto S, Stidham M, Murdiyarso D (2011) Mangroves among the most carbon-rich forests in the tropics. Nat Geosci 4:293–297

Dowling CB, Poreda RJ, Basu AR, Peters SL, Aggarwal PK (2002) Geochemical study of arsenic release mechanisms in the Bengal Basin groundwater. Water Resour Res 38(9):12–18

Falusi BA, Olanipekun EO (2007) Bioconcentration factors of heavy metals in tropical crab (Carcinussp) from River Aponwe, Ado-Ekiti, Nigeria. J Appl Sci Environ Manag 11(4):51–54

Fernandes MC, Nayak GN (2016) Role of sediment size in the distribution and abundance of metals in a tropical (Sharavati) estuary, west coast of India. Arab J Geosci 9(1):33

Froelich PN, Klinkhammer GP, Bender ML, Luedtke NA, Heath GR, Cullen D, Dauphin P, Hammond D, Hartman B, Maynard V (1979) Early oxidation of organic matter in pelagic sediments of the eastern equatorial Atlantic: suboxic diagenesis. Geochim Cosmochim Acta 43:1075e1090

Kumar G, Kumar M, Ramanathan AL (2015) Assessment of heavy metal contamination in the surface sediments in the mangrove ecosystem of Gulf of Kachchh, West Coast of India. Environ Earth Sci 74:545–556. https://doi.org/10.1007/s12665-015-4062-y

Gupta A, Rai DK, Pandey RS, Sharma B (2009) Analysis of some heavy metals in the riverine water, sediments, and fish from river Ganges at Allahabad. Environ Monit Assess 157:449–458

Huang P, Li TG, Li AC, Yu XK, Hu NJ (2014) Distribution, enrichment and sources of heavy metals in surface sediments of the North Yellow Sea. Cont Shelf Res 73:1–13

Huerta-Díaz MA, Morse JW (1992) Pyritization of trace metals in anoxic marine sediments. Geochim Cosmochim Acta 56:2681e2702

Islam MS, Ahmed MK, Raknuzzaman M, Al-Mamun MH, Masunaga S (2014) Chemical speciation of trace metals in sediment and their bioaccumulation in fish of three urban rivers around Dhaka City, Bangladesh. Arch Environ Contam Toxicol 68:92–106

Islam MS, Ahmed MK, Habibullah-Al-Mamun M (2017) Heavy metals in sediment and their accumulation in commonly consumed fish species in Bangladesh. Arch Environ Occup Health 72 (1):26–38
Jonathan MP, Sarkar SK, Roy PD, Alam MA, Chatterjee M, Bhattacharya BD, Bhattacharya A, Satpathy KK (2010) Acid leachable trace metals in sediment cores from Sunderban Mangrove Wetland, India: an approach towards regular monitoring. Ecotoxicology 19(2):405–418
Kumar A, Ramanathan AL (2015) Speciation of selected trace metals (Fe, Mn, Cu, and Zn) with depth in the sediments of Sundarban mangroves: India and Bangladesh. J Soils Sediments 15 (12):2476–2486
Kumar A, Ramanathan AL, Prasad MBK, Datta D, Kumar M, Sappal SM (2016) Distribution, enrichment, and potential toxicity of trace metals in the surface sediments of Sundarban mangrove ecosystem, Bangladesh: a baseline study before Sundarban oil spill of December 2014. Environ Sci Pollut Res 23(9):8985–8999
Lambert CE, Veron A, Buatmenard P, Heyraud M, Grousset F, Simpson W (1991) The role of large biogenic particles in the transport of atmospheric pollutant Pb down to north-Atlantic sediments. Oceanol Acta 14(1):67–76
Lee SV, Cundy AB (2001) Heavy metal contamination and mixing processes in sediments from the Humber Estuary, Eastern England. Estuar Coast Shelf Sci 53(5):619–636
Marchand C, Allenbach M, Lallier-Verges E (2011) Relationships between heavy metals distribution and organic matter cycling in mangrove sediments (Conception Bay, New Caledonia). Geoderma 160(3):444–456
Massolo S, Bignasca A, Sarkar SK, Chatterjee M, Bhattacharya BD, Alam A (2012) Geochemical fractionation of trace elements in sediments of Hugli River (Ganges) and Sundarban wetland (West Bengal, India). Environ Monit Assess 184(12):7561–7577
Massoudieh A, Bombardelli FA, Ginn TR (2010) A biogeochemical model of contaminant fate and transport in river waters and sediments. J Contam Hydrol 112:103–117
Mirlean N, Andrus VE, Baisch P, Griep G, Casartelli MR (2003) Arsenic pollution in Patos Lagoon estuarine sediments, Brazil. Mar Pollut Bull 46:1480–1484
Mukherjee D, Mukherjee A, Kumar B (2009) Chemical fractionation of metals in freshly deposited marine, estuarine sediments of Sundarbans ecosystem, India. Environ Geol 58(8):1757–1767
Nair RR, Hashimi NH, Rao VP (1982) On the possibility of high-velocity tidal streams as dynamic barriers to longshore sediment transport: evidence from the continental shelf off the Gulf of Kutch, India. Mar Geol 47:77–86
Naskar KR, Mandal RN (1999) Ecology and biodiversity of Indian mangroves. 1: 3–348. Dehradun, India, Milton Book Company
Natesan U, Madan Kumar M, Deepthi K (2014) Mangrove sediments a sink for heavy metals? An assessment of Muthupet mangroves of Tamil Nadu, southeast coast of India. Environ Earth Sci 72:1255–1270. https://doi.org/10.1007/s12665-014-3043-x
Nobi EP, Dilipan E, Thangaradjou T, Sivakumar K, Kannan L (2010) Geochemical and geo-statistical assessment of heavy metal concentration in the sediments of different coastal ecosystems of Andaman Islands, India. Estuar Coast Shelf Sci 87:253–264
Noronha-D'Mello CA, Nayak GN (2015) Geochemical characterization of mangrove sediments of the Zuari estuarine system, West coast of India. Estuar Coast Shelf Sci 167:313–325
Nriagu JO, Pacyna JM (1988) Quantitative assessment of worldwide contamination of air, water and soils by trace metals. Nature 333 (6169):134–139
Olias M, Canovas CR, Nieto JM, Sarmiento AM (2006) Evaluation of the dissolved contaminant load transported by the Tinto and Odiel rivers. Appl Geochem 21:1733–1749
Palpandi C, Kesavan K (2012) Heavy metal monitoring using Neritacrepidularia-mangrove mollusk from the Vellar estuary, Southeast coast of India. Asian Pac J Trop Biomed 2(1):S358–S367
Pernetta J (1993) Mangrove forests, climate change, and sea level rise: hydrological influences on community structure and survival, with examples from the Indo-West Pacific. IUCN
Peters EC, Gassman NJ, Firman JC, Richmond RH, Power EA (1997) Ecotoxicology of tropical marine ecosystems. Environ Toxicol Chem 16(1):12–40
Purvaja R, Ramesh R (2000) Human impacts on methane emission from mangrove ecosystems in India. Reg Environ Chang 1(2):86–97
Rahman MT, Rahman MS, Quraishi SB, Ahmad JU, Choudhury TR, Mottaleb MA (2011) Distribution of heavy metals in water and sediments in Passur River, Sundarbans Mangrove Forest, Bangladesh. J Int Environ Appl Sci 6(4):537
Rahman MS, Molla AH, Saha N, Rahman A (2012) Study on heavy metals levels and its risk assessment in some edible fishes from Bangshi River, Savar, Dhaka, Bangladesh. Food Chem 134:1847–1854
Rainbow PS (1995) Biomonitoring of heavy metal availability in the marine environment. Mar Pollut Bull 31:183–192. https://doi.org/10.1016/0025-326X(95)00116-5
Raknuzzaman M, Ahmed MK, Islam MS, Habibullah-Al-Mamun M, Tokumura M, Sekine M, Masunaga S (2016) Assessment of trace metals in surface water and sediment collected from polluted coastal areas of Bangladesh. J Water Environ Technol 14(4):247–259
Raman DJ, Jonathan MP, Srinivasalu S, Armstrong AJS, Mohan SP, Rammohan V (2007) Trace metal enrichment in core sediments in Muthupet mangroves, SE coast of India: application of acid leachable technique. Environ Pollut 145:245–257
Ramanathan AL, Subramanian V, Ramesh R, Chidambaram S, James A (1999) Environmental geochemistry of the Pichavaram mangrove ecosystem (tropical), southeast coast of India. Environ Geol 37 (3):223–233
Ranjan RK, Ramanathan AL, Singh G, Chidambaram S (2008) Assessment of metal enrichments in tsunamigenic sediments of Pichavaram mangroves, southeast coast of India. Environ Monit Assess 147 (1):389–411
Ranjan RK, Singh G, Routh J, Ramanathan A (2013) Trace metal fractionation in the Pichavaram mangrove–estuarine sediments in southeast India after the tsunami of 2004. Environ Monit Assess 185(10):8197–8213
Ray AK, Tripathy SC, Patra S, Sarma VV (2006) Assessment of Godavari estuarine mangrove ecosystem through trace metal studies. Environ Int 32(2):219–223
Rogers KG, Goodbred SL Jr, Mondal DR (2013) Monsoon sedimentation on the 'abandoned' tide-influenced Ganges-Brahmaputra delta plain. Estuar Coast Shelf Sci 131:297–309
Santschi P, Höhener P, Benoit G, Buchholtz-ten Brink M (1990) Chemical processes at the sediment-water interface. Mar Chem 30:269–315
Sarangi RK, Kathiresan K, Subramanian AN (2002) Metal concentrations in five mangrove species of the Bhitarkanika, Orissa, east coast of India. Indian. J Mar Sci 31:251–253
Sarkar SK, Frančišković-Bilinski S, Bhattacharya A, Saha M, Bilinski H (2004) Levels of elements in the surficial estuarine sediments of the Hugli River, northeast India, and their environmental implications. Environ Int 30(8):1089–1098
Sarkar SK, Cabral H, Chatterjee M et al (2008) Biomonitoring of heavy metals using the bivalve molluscs in Sunderban mangrove wetland, northeast coast of Bay of Bengal (India): Possible risks to human health. Clean Soil Air Water 36:187–194. https://doi.org/10.1002/clen.200700027
Sharma VK, Rhudy KB, Koening R, Vazquez FG (1999) Metals in sediments of the Upper Laguna Madre. Mar Pollut Bull 38 (12):1221–1226

Siddiqi NA (2001) Mangrove forestry in Bangladesh. University of Chittagong, Institute of Forestry & Environmental Sciences

Silva CAR, Lacerda LD, Rezende CE (1990) Metals reservoir in a red mangrove forest. Biotropica 22:339–345

Simkiss K, Taylor MG (1989) Metal fluxes across the membranes of aquatic organisms. Rev Aquat Sci 1:173–188

Soto DX, Roig R, Gacia E, Catalan J (2011) Differential accumulation of mercury and other trace metals in the food web components of a reservoir impacted by a chlor-alkali plant (Flix, Ebro River, Spain): implications for biomonitoring. Environ Pollut 159:1481–1489

Subramanian AN (2004) Status of Indian mangrove; pollution status of the Pichavaram mangrove area, southeast coast of India. In: Vannucci M (ed) Mangrove management and conservation. United Nations University Press, Tokyo, pp 59–75

Subramanian V, Mohanachandran G (1990) Heavy metals distribution and enrichment in the sediments of the southern east coast of India. Mar Pollut Bull 21(7):324–330

Swaminathan MS (1991) Genesis of workshop (fore word). In: Sanjay V, Deshmugh, Rajeswari Mahalingarn (eds) Proceedings of the formulation workshop for establishing a global net work of mangrove genetic resource centres for adaptation to sea level rise. January 15–19, Madras, India. Proceedings No.2, CRSARD. Madras, India

Tam NF, Wong YS (1993) Retention of nutrients and heavy metals in mangrove sediment receiving wastewater of different strengths. Environ Technol 14(8):719–729

Tam NFY, Wong YS (1996) Retention and distribution of heavy metals in mangrove soils receiving wastewater. Environ Pollut 94 (3):283–291

Tao Y, Yuan Z, Xiaona H, Wei M (2012) Distribution and bioaccumulation of heavy metals in aquatic organisms of different trophic levels and potential health risk assessment from Taihu Lake, China. Ecotoxicol Environ Saf 81:55–64

US Environmental Protection Agency (USEPA) (2010a) Risk-based concentration table. Available at: <http://www.epa.gov/reg3hwmd/risk/human/index.htm

US Environmental Protection Agency (USEPA) (2010b) Risk-bsed concentration table

Vassiliki K, Konstantina AD (1984). Transfer factors of heavy metals in aquatic organisms of different trophic levels. Bulletin of Environmental Contamination and Toxicology

Veerasingam S, Vethamony P, Murali RM, Fernandes B (2015) Depositional record of trace metals and degree of contamination in core sediments from the Mandovi estuarine mangrove ecosystem, west coast of India. Mar Pollut Bull 91(1):362–367

Volkman JK, Rohjans J, Rullkotter J, Scholz-Bottcher BM, Liebezeit G (2000) Sources and diagenesis of organic matter in tidal flat sediments from the German Wadden Sea. Cont Shelf Res:1139–1158

Waldichuk M (1985) Biological availability of metals to marine organisms. Mar Pollut Bull 16:7–11

Wasserman JC, Figueiredo AMG, Pelegatti F, Silva-Filho EV (2001) Elemental composition of sediment cores from a mangrove environment using neutron activation analysis. J Geochem Explor 72:129–146

WHO (World Health Organization) (2004) Guidelines for drinking water quality, 3rd edn. World Health Organization, Geneva

Wong K, Wong PPK, Chu LM (2001) Heavy metal concentrations in marine fishes collected from fish culture sites in Hong Kong. Arch Environ Contam Toxicol 40:60–69

Zhang J, Cai L, Yuan D, Chen M (2004) Distribution and sources of polynuclear aromatic hydrocarbons in Mangrove surficial sediments of Deep Bay, China. Mar Pollut Bull 49(5):479–486

Zhang W, Feng H, Chang J, Qu J, Xie H, Yu L (2009) Heavy metal contamination in surface sediments of Yangtze River intertidal zone: an assessment from different indexes. Environ Pollut 157 (5):1533–1543

Zhang WF, Liu XP, Cheng HF, Zeng EY, Hu YN (2012) Heavy metal pollution in sediments of a typical mariculture zone in South China. Mar Pollut Bull 64:712–720

Part IV

Assessment Techniques, Ecosystem Design, and Management Strategies

Ecosystem Design: When Mangrove Ecology Meets Human Needs

16

Martin Zimmer

Abstract

At least two thirds of all ecosystems worldwide have been impacted and changed severely by human activity (MEA Millennium ecosystem assessment – ecosystems and human well-being: biodiversity synthesis. World Resources Institute, Washington, DC, 2005), mostly without considering consequences for the structure, functioning or service-provisioning of these ecosystems. The societal challenges arising from this are twofold: conserving natural heritage and resources, and at the same time providing and sustaining valuable livelihood and well-being for mankind. Once we missed the chance of preserving an ecosystem from degradation through conservation, restoration is the attempt to repair (i.e., bringing back to a past state) or otherwise enhance (i.e., promoting remaining components and structures) the function of an ecosystem that has been impacted (Suding KN, Annu Rev Ecol Evol Syst 42:465–87, 2011) into a state that warrants historical continuity (Murcia C et al., Trends Ecol Evol 29:548–553, 2014) and closely resembles natural conditions. Nevertheless, most restoration efforts lack a clear aim, and monitoring is rarely considered. Hence, an evaluation of restoration success is difficult, if not impossible. As an alternative to restoration, a new five-step concept of directed design for novel ecosystems (sensu Hobbs RJ, Arico S, Aronson J, Baron JS, Bridgewater P, Cramer VA, Epstein PR, Ewel JJ, Klink CA, Lugo AE, Norton D, Ojima D, Richardson DM, Sanderson EW, Valladares F, Vilà M, Zamora R, Zobel M et al., Glob Ecol Biogeogr 15:1–7, 2006; Morse NB, Pellissier PA, Cianciola EN, Brereton RL, Sullivan MM, Shonka NK, Wheeler TB, McDowell WH et al., Ecol Soc 19:12–21, 2014) with defined functions and services is presented in this chapter. Recent advances in restoration ecology pledge for accepting unintended novel ecosystems as valuable providers of ecosystem services in restoration efforts (Perring MP, Standish RJ, Hobbs RJ et al., Ecol Process 2:18–25, 2013; Abelson A, Halpern B, Reed DC, Orth RJ, Kendrick GA, Beck MW, Belmaker J, Krause G, Edgar GJ, Airoldi L, Brokovich E, France R, Shashar N, De Blaeij A, Stambler N, Salameh P, Shechter M, Nelson PA et al., Bio Sci 66:156–163, 2016). Ecosystem Design develops this idea further to intendedly designing novel ecosystems with the aim of providing particular services that are locally or regionally required for the well-being of mankind. Thus, in contrast to conventional restoration, Ecosystem Design places humans and their needs in the center of action. For this, Ecosystem Design first assesses local and regional needs for ecosystem services to be provided. Second, Ecosystem Design defines a set of these services as goals for the establishment of a functioning ecosystem in a degraded area. In a third step, a toolbox of information on species characteristics and requirements, as well as on the species specific contributions to service-provisioning, including interspecific interactions under the given environmental conditions, recommends a set of suitable species from the regionally available species pool. Such a toolbox requires trait-based models to determine which species assemblages are most effective (Laughlin DC, Ecol Lett 17:771–784, 2014) in providing the desired ecosystem services, and the choice of suitable and appropriate species would be facilitated by knowledge of previous community composition. The set of initial species will, in a fifth step, be installed in the degraded area, and subsequent natural succession will shape and fine-tune this novel designed ecosystem (unless this semi-natural development deviates from the aim of providing particular ecosystem services,

M. Zimmer (✉)
Department of Ecology, Leibniz Centre for Tropical Marine Research, Bremen, Germany

University of Bremen, Faculty 02 Biology/Chemistry, Bremen, Germany

IUCN-SSC Mangrove Specialist Group, Bremen, Germany
e-mail: martin.zimmer@leibniz-zmt.de; martin.zimmer@uni-bremen.de

C. Makowski, C. W. Finkl (eds.), *Threats to Mangrove Forests*, Coastal Research Library 25,
https://doi.org/10.1007/978-3-319-73016-5_16

when counteraction to semi-natural succession will be required). Upon installation and subsequent development of the designed ecosystem, long-term monitoring in the sixth step will allow for evaluating the success of the design and intervention if needed, since clear aims and goals had been defined in the second step of Ecosystem Design. Whereas this approach may in cases contrast efforts to conserve or restore biodiversity on its own sake, Ecosystem Design aligns with the Sustainable Development Goals of the United Nations in warranting human well-being in times of increasing demands for ecosystem services, especially in tropical coastal areas with ever-growing population sizes.

Keywords
Ecosystem Services · Service-Providing Unit · Novel Ecosystems · Directed Ecosystem Restoration

16.1 Introduction

16.1.1 Mangrove Restoration

Once we missed the chance of preserving an ecosystem from degradation, rehabilitation is considered a valuable alternative to conservation. Whereas rehabilitation aims at the replacement of the initial structure and function of an ecosystem, restoration – a special case of rehabilitation – is the process of assisting the recovery of an ecosystem back to its original condition as nearly as possible (Field 1998; Ellison 2000; Alexander et al. 2011). To this end, restoration is the process of repairing damage caused by humans to the diversity and dynamics of indigenous systems (Fig. 16.1). Thus, the aim of restoration is a functional ecosystem providing similar services as the original ecosystem. It is common sense that, apart from biological considerations, restoration ecology should include historical, social, cultural, political, aesthetical and ethical aspects.

Mangroves, growing on soft-sediment shores along many tropical and subtropical coasts, are among the most productive ecosystems and provide numerous ecosystem services both to local human populations and to mankind worldwide. Mangrove area loss proceeds at a rate that exceeds area loss of most other ecosystems (Bradshaw et al. 2009; Wabnitz et al. 2010; Giri et al. 2011). However, recent attempts of mangrove reforestation or restoration largely failed, because most of them were not based on scientific knowledge about the ecology and (intertidal) distribution of regionally occurring mangrove species (e.g., Elster 2000; Primavera and Esteban 2008; Lewis 2009; Alexander et al. 2011): owing to practical problems of site-selection (Field 1998), many attempts of mangrove reforestation used the "wrong species" in the "wrong environment", without taking into account species-specific requirements for habitat characteristics and location along the intertidal gradient (see also: Elster 2000; Lewis 2005; Matsui et al. 2010, with respect to habitat hydrology). Another important environmental factor that drives mangrove species distribution, both regionally and along the intertidal gradient, that should be taken into account for restoration efforts is the geomorphological settings (Balke and Friess 2015). More generally speaking, successful active restoration that is based on plantation has to consider species-specific response traits (Hedberg et al. 2013; Laughlin 2014) that determine whether a particular species can cope with, and dwell under, given environmental conditions. Thus, the framework of response- and effect-traits (Lavorel and Garnier 2002) is essential for selecting appropriate sets of species with respect to the prevailing environmental conditions of the degraded and to-be-rehabilitated or-restored ecosystem. Alternatively, providing environmental conditions –e.g., elevation above sea level and inundation regime, or hydrology and hydrodynamics – that will support the settlement success of particular species seems pivotal in cases where the abiotic environment had been changed beyond thresholds (Lewis 2005; Matsui et al. 2010) – this, in turn, requires at least basic knowledge on species-specific response traits.

As comparably little information is available on successful versus failed mangrove restoration attempts, many of these programs seem to have been carried out without any reference to lessons that might be learnt from other similar programs, contributing to the overall little success worldwide. Notwithstanding the potential of rehabilitation or even restoration of degraded ecosystem, the long-lasting

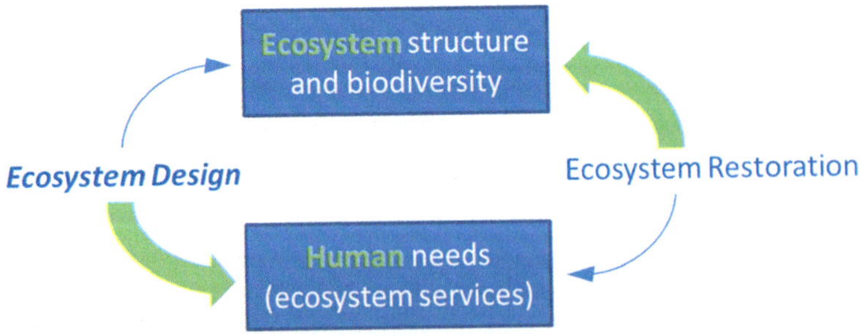

Fig. 16.1 Conceptual difference between ecosystem restoration, focusing on the ecosystem, and *Ecosystem Design*, focusing on human needs

aftereffects of mangrove degradation underscore the importance of eliminating its causes, since once sites are cleared of mangroves, it is difficult for them to recover without knowledge-driven intervention. Hence, planning mangrove reforestation or restoration should be based on scientifically sound knowledge, and any rehabilitation effort must be accompanied and monitored in the long term accordingly (c.f., Bosire et al. 2008; Alexander et al. 2011).

Unsuccessful restoration is a waste of time, money and human resources (c.f. Primavera and Esteban 2008). Hundreds of volunteers that invested time and energy in large-scale restoration campaigns must have been frustrated to see their efforts being washed away by tides and storm surges within months after having been planted. However, successful or promising examples have been implemented (c.f. Primavera and Esteban 2008; Alexander et al. 2011), e.g., through the community-based ecological mangrove restoration concept (CBEMR) of MAP (http://mangroveactionproject.org/).

Not surprisingly, naturally recovering mangroves (c.f., Kamali and Hashim 2011) might be more diverse than those restored through human action, and planting of mixed-species forests is recommended to maximize biodiversity (Alongi 2011). However, plant and crab biomass or density seem to be higher upon assisted than through natural restoration (Ferreira et al. 2015) and in reforested than in natural stands (Bosire et al. 2008; but see: Walton et al. 2007). Along the same line, protected mangroves exhibit much higher crab diversity than reforested mangroves in the Philippines, mostly owing to differences in sediment characteristics among the studied sites (Bandibas and Hilomen 2016). This leads to the question as to when rehabilitation is successful (c.f., McKee and Faulkner 2000; Ruiz-Jaen and Aide 2005). According to the aim of rehabilitation (as defined above), the answer lies in the successful re-establishment of ecosystem service-provision, and the performance of ecological processes has been suggested as one measure of restoration success (Ruiz-Jaen and Aide 2005; Walton et al. 2007; Vovides et al. 2010): regrown plantations might dwell, but a desired service might not be provided (i.e., restoration has failed: McKee and Faulkner 2000), possibly because most ecosystem services are not provided by a single species but by a community or at least a set of species, including microbial key players (c.f., Holguin et al. 2001; Berry and Widder 2014) that interact to drive underlying ecosystem processes. Interestingly, the success of mangrove restoration has rarely been estimated based on components other than the vegetation itself (but see: Macintosh et al. 2002; Bosire et al. 2008; Bandibas and Hilomen 2016), but if we include ecological processes driven by the fauna or microbiota, we will be able to better judge whether restoration was successful or not.

Along this line of how to best achieve a "functioning ecosystem" (whatever that is) upon restoration, the controversial discussion on whether coral reef restoration should be performed through artificial reef structures as hard substrate for settlement or by transplanting stocks from another, healthy reef is ongoing (Abelson 2006). The former is well-perceived in the public but has only weak scientific background, whereas the latter proves efficient and is based on a strong scientific background but is controversially perceived by the public. Obolski et al. (2016) suggest the re-establishment of grazing fish along with corals to support coral reef restoration, and Halpern et al. (2007), more generally pledge for considering (positive) ecological interactions in restoration-planning and -activities.

16.1.2 Novel Ecosystems

Human action inadvertently, or even deliberately, changes the ecosystems that surround us and are used by us. Hence, "novel ecosystems" (sensu Hobbs et al. 2006; Perring et al. 2013; Morse et al. 2014; Miller and Bestelmeyer 2016) have been established since humans started actively changing their environment by cutting forests and creating arable land. Forestry and agriculture act as designers of novel ecosystems with clear definition of goals. Similarly, fisheries management has since long ago adopted approaches of manipulating community composition in desired directions. Most human activities, however, intervene with natural systems without predicted or even predictable direction and result in undirected and unsupervised creation of novel ecosystems without clear goals and with uncertain outcome. It is interesting to note that these are widely accepted, whereas the **active design of a novel ecosystem** commonly encounters resistance, and the designers of such ecosystems have to face the reproach of acting as "playing god". Novel ecosystems bear the chance of new species combinations that will result in the potential for changes in ecosystem processes and services (Hobbs et al. 2006). If a clear goal (with respect to ecosystem processes and services) is defined, it will be possible to actively use this potential and design communities and ecosystems that drive those processes that underlie the desired services (Perring et al. 2013), particularly where it is difficult –or even impossible– or costly to return to previous ecosystem states which is the aim of classical restoration (Hobbs et al. 2006; Fig. 16.1). As recently outlined by Miller and Bestelmeyer (2016), some recommendations even include that such considerations should be made regardless of the origin of the species that drive a given process, meaning that non-native species might also contribute to ecosystem service-provisioning. Contrarily, immigration of non-native species into degraded ecosystems is considered one of the reasons for unsuccessful restoration efforts

(Suding 2011). The primary aim of **ecosystem design**, however, should clearly be to rely on the native regional species pool and to (re-)establish those species that naturally occur and would naturally colonize the to-be-designed area, if supply was guaranteed. This, actually, is also one of the most promising approaches to undirected ("community-based") rehabilitation of mangroves (Primavera and Agbayani 1997; Lewis 2009), proposing that planting mangrove seedlings or treelings will be unnecessary, if suitable conditions for natural recruitment (e.g., in terms of hydrology and shore topography) are re-implemented (Kamali and Hashim 2011).

16.2 Service-Providing Units as Target of Ecosystem-Manipulation

16.2.1 The Rationale

Forestry and Agriculture are illustrative examples of novel ecosystems that have been created in a directed approach for centuries, aiming at growing plants that provide clearly defined ecosystem services (food). Aquaculture ponds are built (also on cost of previously existing ecosystems, such as mangroves) to produce fish or shrimp for human consumption and commerce.

A less essential need of many people is covered by gardening. Several, partly contrasting, services are expected from private gardens, such as small-scale production of fruit and vegetables, provision of a quite space for leisure and recreation, and/or provision of a small-scale habitat for birds and insects within an urban area. Depending on which service the individual gardener desires or requires, they will choose the species of garden plants and the design of the garden. Wild herbs or pest species will be removed and fought from vegetable plantations but maybe not from intended insect- or bird-habitats; flowering plants with either beautiful flowers or rich in nectar for insects and nutritious fruits for birds will be chosen for the insect- and bird-habitat, whereas breeds that promise high yield will be preferred for fruit- and vegetable-production. Contrasting services can be achieved by compromising designs of the garden or spatial partitioning and compartmentalization.

16.2.2 Plants

It becomes obvious from the above that many ecosystems are already shaped by human activity to support those plant species (and breeds) that best provide services of different qualities. Thus, natural forests, including mangroves, had been (and still are being) replaced by plantations of fast-growing producers of wood for construction, furniture or charcoal, or other products. Along this line, mangroves have been clear-cut for giving space to, e.g., oil palm-plantations, even though the particular conditions under which mangroves grow are suboptimal for oil palms, and their yield is way below the value local people might gain from sustainable use of mangroves. Notwithstanding that most of the new designation of previous mangrove areas will provide money and income to either local people or companies, most corresponding management plans do not seem to take into account the related loss of valuable and irreplaceable services provided by intact and functioning mangrove forests.

Green land (and forests) had been (are still being) turned into arable land and plantations, resulting in artificial ecosystems with extremely low species diversity that, however, highly efficiently provide the desired ecosystem service of feed- and food-production. For this aim, human society has since long accepted to loose local and regional biodiversity and the provisioning of services of natural ecosystems. As for gardening, we even accept the use of highly toxic pesticides and the introduction of non-native, potentially invasive, species; the long-term consequences are only recently begun being understood. Interestingly, and in contrast to the concept of novel ecosystems or ecosystem design, society is willing to invest time, energy and money into maintaining these artificial ecosystems in a stage that guarantees (close-to-)optimal productivity and production, without even considering –potentially irreversible– ecological side-effects of their intervention.

Along this line, it seems only logical to base restoration efforts of degraded ecosystems on those plant species that most reliably and effectively provide those ecosystem services that are needed under particular circumstances. This concept is being practiced when it comes to constructed artificial wetlands for treating municipal or industrial wastewater. If, for instance, the aim of such wetland, be it designed as freshwater swamp, or in coastal areas as saltmarsh or mangrove, is the extraction of excessive nutrients from aquaculture, the best results (i.e. the cleanest water) will be obtained by using plant species that are most efficient in nutrient-uptake (and -storage).

16.2.3 Animals

The management of fish stocks is clear intervention in a natural community with the aim of (ideally sustainably) providing income and food to local populations and mankind worldwide. The concept of understanding fisheries in the light of directly changing ecosystems becomes even clearer, when looking at aquaculture rather than wild catches of fish. Here, a single species (or, if "integrated multitrophic aquaculture", IMTA, is implemented, few species) that optimally provide(s) a service is/are introduced into a novel ecosystem and its service (food production) is used to gain income.

Huge areas of mangroves have been lost through clear-cutting for the implementation of land-based aquaculture, in particular for shrimp ponds. This has happened almost all around the world, but a strong regional focus lies on South-East Asia. When shrimp ponds have to be abandoned due to decreasing productivity (and income for the owner and workers) or disease, new mangrove areas will be clear-cut to give space to new ponds. Restoring ponds into functional mangroves is possible but tedious and time-consuming.

Alternatively, we could imagine to (re-)implement those abiotic and biotic environmental conditions that will support those animal species (e.g., crabs, shrimps or fish) in a degraded mangrove area that are desired by local societies to be used for subsistence fisheries or sustainable commercial fisheries. The targeted species would, in this case, not be the implemented one, but restoration of a mangrove ecosystem would aim at those mangrove species that best provide habitat for the target species of subsequent sustainable fisheries.

16.2.4 Microbes

Microbes are being used for producing numerous medical products and food (supplements), since microbiology has provided insight into the plethora of capabilities of bacterial or fungal strains. For instance, building a fermenter for the production of food is designing an artificial single- (or few-) species ecosystem.

Much less is known about specific contributions of microbes to ecosystem processes or even the provision of ecosystem services. The knowledge of microbial physiology has proven helpful in utilizing them for habitat-amelioration and -enhancement. For instance, introducing certain strains of bacteria into seawater that has been polluted by oil spills might be the most efficient way to fight long-term aftereffects of such spills (Dombrowski and Baker 2016). Doing so is clearly aim-oriented in that the service of oil-breakdown provided by these bacteria drives the decision on which species to introduce into a degrading ecosystem.

In the case of mangrove, the pivotal role of the sediment microbiota as driver of plant-nutrient interactions has been stressed with respect to mangrove conservation (Bashan and Holguin 2002). Along the same line, mangrove restoration could be assisted by ameliorating and improving habitat quality through establishing suitable microbial communities (Holguin et al. 2001; Gomes et al. 2010). Doing so in degraded previous mangrove areas would create an inhabitable environment for recolonization by mangroves, be it aided or natural. Assuming species-specific microbial communities of the rhizosphere of (c.f. Ramírez-Elías et al. 2014), or the sediment around (Selvam and Kathiresan 2010), mangrove trees, designing a sediment microbiota by selecting certain microbial species and strains might even be used to promote mangrove species that provide particular services better than others during re-colonization processes.

16.2.5 So What?

What humans have done for centuries is undirected and unsupervised alteration of ecosystems with unpredictable, and thus unforeseen, consequences for our natural environment. It is about time to take responsibility for what we are doing and supervise changes we pose on ecosystems in a desired direction. Why not take a step further and design communities of naturally occurring plant and animal species to provide particular ecosystem services that are required (or desired) in a particular area by re-implementing regionally natural ecosystems?

16.3 Ecosystem Design

16.3.1 The Concept

"Enhancing ecosystem services (...) are exciting new directions" in restoration practice (Suding 2011), but the role of ecosystem services is often ignored in management decisions which may cause continued degradation and destruction of mangroves (Barbier et al. 2011). As any effort of improving ecosystem service-provisioning upon restoration of degraded ecosystems has to encompass social-ecological as well as socio-economic considerations (Abelson et al. 2016), clearly defined services should be included as restoration goals and be measured as criterion of success. Meeting the goals of REDD+ of reducing greenhouse gas emissions and storing carbon in forest ecosystems (including mangroves!), for instance, requires science-based restoration-planning (Alexander et al. 2011), and choosing explicitly those species that are best at driving or providing any of these services might be promising.

One step beyond trying to restore previously degraded ecosystems with the aim of possibly re-gaining a handful of previously provided ecosystem services, I propose the even more promising approach to design novel functioning ecosystems in degraded areas from scratch, according to the services locally or regionally required (Fig. 16.1), taking into account local habitat peculiarities and other environmental conditions (be they natural or man-made) as well as the pool of regionally available native species as service-providing units (SPUs: see above). From this pool, we should re-establish a minimum set of species that, according to our understanding of ecosystem service-provisioning, are necessary to drive the designed ecosystem to providing the required service(s). Upon the initial establishment of foundation species, we can expect further species to immigrate from a regional pool and trigger the development of a more natural community (for review, see Ellison 2000; Matsui et al. 2010), and taking interspecific interactions (Halpern et al. 2007) into account, might even facilitate ecosystem design and planning which species to implement. Depending on such interactions of immigrating species with initially (re-)implemented species and their interference with the desired ecosystem process (es), such natural succession may, however, be counterproductive at some stage until further species establishment will stabilize ecosystem performance and functioning (with respect to service-provision), or "gardening intervention" (see above) will become necessary.

Thus, the choice of mangrove species to be (re-) implemented in a degraded coastal area, serving as foundation species (sensu Dayton 1971) and ecosystem engineers (sensu Jones et al. 1994), must not only be based on the regional pool of native species but should be made according to the ecosystem service(s) that are sought to be provided. For instance, not all mangroves accumulate carbon, and rates of forest floor accretion are directly linked to the frequency of tidal inundation, and thus, to the composition of the mangrove community (Alongi 2011). Even though this has never been actually measured in the field, there seems to be common agreement that *Rhizophora* prop roots are more effective in capturing sediment than *Avicennia* pencil roots, as it has recently been demonstrated for particular detritus being trapped amongst mangrove roots (Gillis et al. 2016). Similarly, the sequestration of detritus-derived organic matter in anoxic mangrove sediments upon decomposition of mangrove detritus results in climate-change mitigation through huge amounts of carbon and nitrogen being stored in stable compounds in a stable environment. The structure of the sediment organic matter, however, differs among mangrove species (V. Helfer and M. Zimmer, unpublished data), as well as does the composition of the microbial community that thrives on mangrove-derived organic matter in the sediment (Holguin et al. 2001). Thus, not all mangrove species might be equally suited for restoring sedimentary C- or N-stores (c.f. Matsui et al. 2010; Alexander et al. 2011; Vovides et al. 2011).

Whereas relatively few mangrove species have been used in rehabilitation projects (Field 1998; Primavera and Esteban 2008), selecting other species than the most commonly used ones might proof a better choice, both in terms of success and benefit. However, local and regional geomorphic settings, hydrology, currents and sedimentation patterns must be taken into account (Elster 2000; Lewis 2005; Primavera and Esteban 2008; Matsui et al. 2010; Kamali and Hashim 2011; Balke and Friess 2015), when selecting the species to be used for mangrove restoration. Pre-conditioning the newly designed habitat by introducing sediment microbes with particular traits and specific interactions with mangroves might be necessary, or at least helpful and advantageous, too. Along the same line, different tree species may be associated with different mangrove crab communities that, in turn, seem to depend on the presence of habitat-mediating trees (Dahdouh-Guebas et al. 2002). Some burrowing crab species (*Ucides* spp. and *Uca* spp.), but probably not all, act in reducing sediment salinity (Pülmanns et al. 2015), particularly during dry seasons (Pestana et al. 2017), or have an effect on other sediment characteristics such as organic matter content or redox conditions.

Ecosystem design, as proposed above, requires (Fig. 16.2)

- Basic knowledge of previous and current presence of species (flora, fauna, microbiota) with potential relevance for mangrove performance: e.g., metabarcoding of environmental DNA from sediments provides a useful tool for the rapid assessment of the composition of past and present communities (Taberlet et al. 2012; Thomsen and Willerslev 2015).
- Sound knowledge of environmental requirements of species: e.g., niche-modelling and knowledge of geophysical processes of coastal environments and climatic conditions can predict mangrove recovery under given environmental conditions (Twilley et al. 1998; Balke and Friess 2015).
- Detailed understanding of interspecific interactions and mutual dependencies: e.g., the re-establishment of mangrove trees might be promoted by inoculation of seedlings or saplings with appropriate growth-promoting microbes (Holguin et al. 2001) potentially serving as initiator of microbe-based interactions of ecosystem relevance (c.f., Berry and Widder 2014).
- Reliable predictability of how consortia of species and their interactions will drive service-relevant ecosystem processes, and how environmental conditions may act mediating: e.g., successful recovery of mangrove forests may be accompanied by reduced organic carbon content of the sediment due to reduced water content (Matsui et al. 2010); along this line, high-throughput assessment of sediment organic matter structure through, e.g., pyrolysis-GC/MS (py-)GC/MS) or Near Infrared Reflectance Spectrometry (NIRS) (Fuentes et al. 2012; Gerber et al. 2012; Kleinebecker et al. 2013; Tolu et al. 2015) may help predicting the spatial distribution of ecosystem processes that relate to C- and N-sequestration as it depends on the community composition and environmental conditions.

Provided that we have this basic information at hand, the five steps of Ecosystem Design can be implemented:

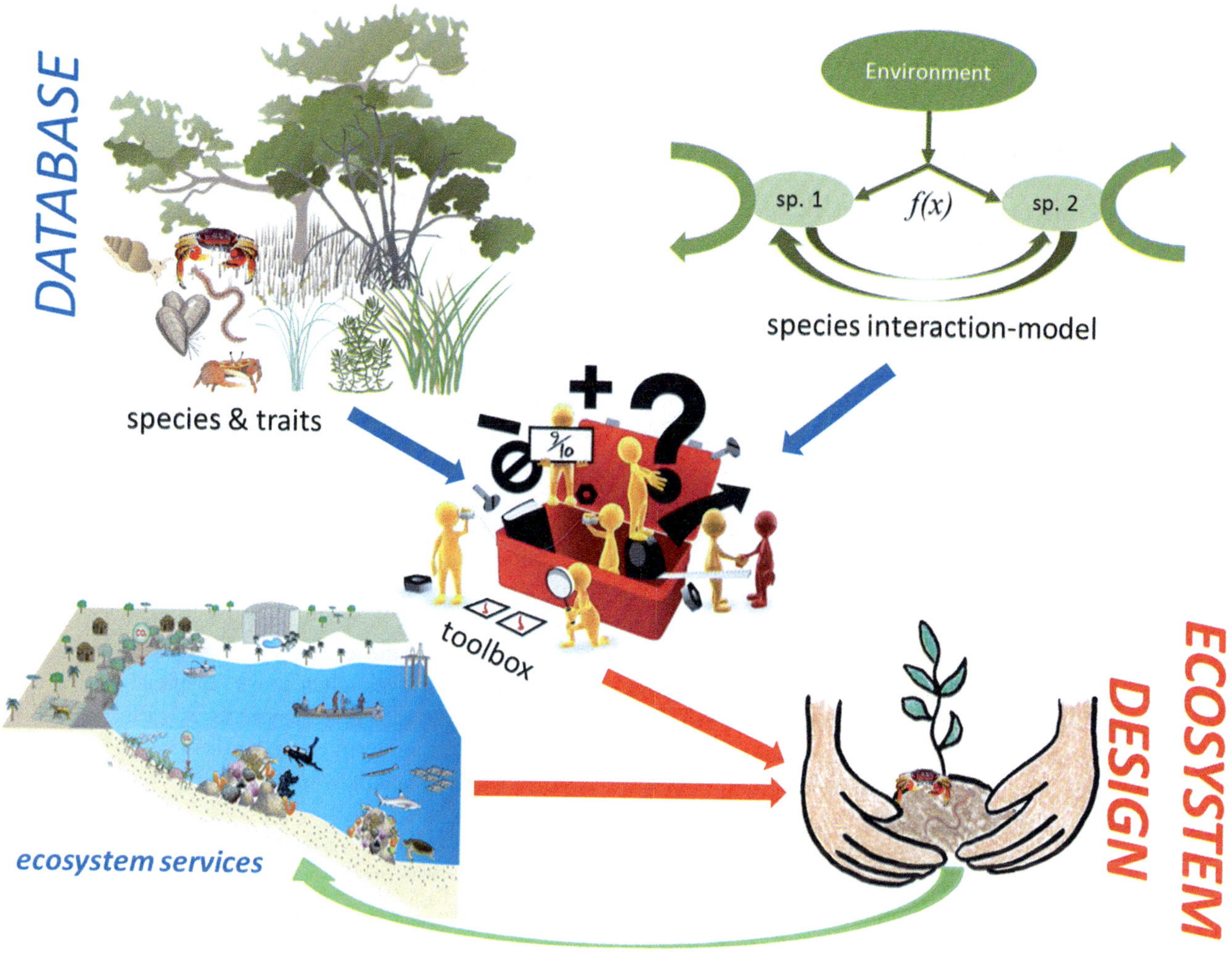

Fig. 16.2 The database of information on species and their traits and the species interaction-model jointly feed the toolbox that assists the design of an ecosystem to provide particular ecosystem services to society (Cartoons downloaded from www.clipartbest.com and ian.umces.edu)

1. Assess local and regional needs for ecosystem services to be provided.
   More specifically,
   Which ecosystem services should be provided?
   Which species and combinations of species drive those ecosystem processes that underlie these services?
   What are the specific characteristics of the target habitat?
   Which regionally occurring species are capable of thriving under these conditions?
   How will the habitat have to be modified (e.g., with respect to currents or hydrology) to support the establishment of target species and the provision of the desired services?
2. Define a set of these services as goals for the establishment of a functioning ecosystem in a degraded area.
3. A toolbox (Fig. 16.2) of information on species characteristics and requirements, as well as on the species-specific contributions to service-provisioning, including interspecific interactions under the given environmental conditions, recommends a set of suitable species from the regionally available species pool. Such a toolbox requires trait-based models to determine which species assemblages are most effective (Laughlin 2014) in providing the desired ecosystem services, and the choice of suitable and appropriate species would be facilitated by knowledge of previous community composition.
4. Install a set of initial species in the degraded area.
   Subsequent natural succession will shape and fine-tune this novel designed ecosystem (unless this semi-natural development deviates from the aim of providing particular ecosystem services, and counteraction to semi-natural succession will be required ("gardening intervention")).
5. Upon installation and subsequent development of the designed ecosystem, long-term monitoring will allow for evaluating the success of the design and intervention if needed, since clear aims and goals had been defined.

16.3.2 Toolbox for Ecosystem Design

As a toolbox for this process, we will need a database of species characteristics that will provide the information necessary to decide which species from a regional pool to implement into the to-be-designed ecosystem. A framework of response- and effect-traits (Hedberg et al. 2013; Laughlin 2014) will be essential for selecting appropriate sets of species with respect to being able to cope with the prevailing environmental conditions (response traits) and the provision of the desired ecosystem service(s) (effect traits). This requires a sound understanding of the species' environmental requirements, being reflected by their small-scale distribution along environmental gradients (Nobbs et al. 2015). On the other hand, if the geomorphological (Balke and Friess 2015) or hydrological (Lewis 2005) environmental settings of the degraded area do not meet the requirements needed as basis for restoration of mangroves, suitable environmental conditions have to be restored (Lewis 2005; Matsui et al. 2010). This, together with selecting species from the naturally occurring regional species pool is reflected in the conceptual approach of community-based ecological mangrove (mangroveactionproject.org/) and Ecological Mangrove Rehabilitation (www.mangroverestoration.com).

In many cases, of course, our understanding of species and their interactions with, and dependencies on, other species is still rudimentary and far from being deep enough to allow for such a toolbox (c.f. Ellison 2000). In some other cases of foundation or key species we might know enough to design simple functioning ecosystems and develop and implement those basal communities made up by minimum sets of species required to provide a particular ecosystem service. A simple, albeit relevant, example might be coastal protection through supporting sedimentation and preventing erosion. Mangrove species with extensive aerial root systems, such as *Rhizophora* spp., will be more effective in this regard, whereas species with dense sub-surface root systems, such as *Avicennia* spp., will better stabilize the existing sediment. A combination of both might be the best solution in terms of service-provisioning, but might not be realizable because of environmental requirements and conditions, or combining several species might conflict with the optimal provision of other services. Restoring mixed-species forests is recommended to maximize biodiversity, food web connectivity and net ecosystem production (Alongi 2011) but might counteract optimal ecosystem design, if a particular service was best provided by a monospecific stand.

The simultaneous re-establishment of edible crab species would additionally provide the basis for extracting food (and producing income) for local human populations. If, however, their burrowing activity counteracts sediment stabilization, we should refrain from co-establishing these crabs – they will, with a certain probability, establish themselves naturally with time upon assisted establishment of the basal ecosystem engineers. In some cases, co-establishment of unwanted species might even be actively prevented (see above: "gardening intervention"). Competing ecosystem services that would require different community compositions might then be handled by spatial mosaics of these different communities and compartmentalizing different sets of service-providing units, actually resulting in semi-natural situations, to simultaneously ensure the provisioning of contrasting services.

16.4 Outlook

It is still a long way to go, before we will be able to design ecosystems for more complex services, such as C- or N-sequestration, but our increasing knowledge about the ecophysiology of particular species and the dynamics of communities is paving the road – and in a time of ever-increasing ecosystem degradation and loss, it seems worthwhile taking this road, once ecosystems and their services are lost locally and their re-establishment is desired. Transferring findings derived from relatively species-poor, albeit highly productive, mangroves might prove beneficial for the restoration and design of other ecosystems (c.f. Ellison 2000) to sustain the provisioning of ecosystem services to local human populations and mankind worldwide.

Acknowledgements I am grateful to Véronique Helfer (ZMT) and the ZMT Research Focus Area *Ecosystem Design* (in alphabetical order: Werner Ekau, Sebastian Ferse, Nils Moosdorf, Hauke Reuter) for fruitful discussions on ecosystem restoration and design.

References

Abelson A (2006) Artificial reefs versus coral transplantation as restoration tools for mitigating coral reef deterioration: benefits, concerns and proposed guidelines. Bull Mar Sci 78:151–159

Abelson A, Halpern B, Reed DC, Orth RJ, Kendrick GA, Beck MW, Belmaker J, Krause G, Edgar GJ, Airoldi L, Brokovich E, France R, Shashar N, De Blaeij A, Stambler N, Salameh P, Shechter M, Nelson PA (2016) Upgrading marine ecosystem restoration using ecological-social concepts. Bioscience 66:156–163

Alexander S, Nelson CR, Aronson J, Lamb D, Cliquet A, Erwin KL, Finlayson CM, de Groot RS, Harris JA, Higgs ES, Hobbs RJ, Lewis RRR III, Martinez D, Murcia C (2011) Opportunities and Challenges for Ecological Restoration within REDD+. Restoration Ecol 19:683–689

Alongi DM (2011) Carbon payments for mangrove conservation: ecosystem constraints and uncertainties of sequestration potential. Environ Sci Pol 14:462–470

Balke T, Friess DA (2015) Geomorphic knowledge for mangrove restoration: a pan-tropical categorization. Earth Surf Process Landf. https://doi.org/10.1002/esp.3841

Bandibas MB, Hilomen VV (2016) Crab biodiversity under different management schemes of mangrove ecosystems. Glob J Environ Sci Manag 2:19–30

Barbier EB, Hacker SD, Kennedy C, Koch EW, Stier AC, Silliman BR (2011) The Value of Estuarine and Coastal Ecosystem Services. Ecol Monogr 81:169–183

Bashan Y, Holguin G (2002) Plant growth-promoting bacteria: a potential tool for arid mangrove restoration. Trees 16:159–166

Berry D, Widder S (2014) Deciphering microbial interactions and detecting keystone species with co-occurrence networks. Front Microbiol 5:1–14

Bosire JO, Dahdouh-Guebas F, Walton M, Crona BI, Lewis RR III, Field C, Kairo JG, Koedam N (2008) Functionality of restored mangroves: a review. Aquat Bot 89:251–259

Bradshaw CJ, Sodhi ANS, Brook BW (2009) Tropical turmoil: a biodiversity tragedy in progress. Front Ecol Environ 7:79–87

Dahdouh-Guebas F, Verneirt M, Cannicci S, Kairo JG, Tack JF, Koedam N (2002) An exploratory study on grapsid crab zonation in Kenyan mangroves. Wetl Ecol Manag 10:179–187

Dayton PK (1971) Competition, disturbance and community organization: the provision and subsequent utilization of space in a rocky intertidal community. Ecol Monogr 41:351–389

Dombrowski N, Baker B (2016) Can we harness bacteria to help clean up future oil spills? The Conversation June 22

Ellison AM (2000) Mangrove restoration: do we know enough? Restor Ecol 8:219–229

Elster C (2000) Reasons for reforestation success and failure with three mangrove species in Colombia. For Ecol Manag 131:201–214

Ferreira AC, Ganade G, de Attayde JL (2015) Restoration versus natural regeneration in a neotropical mangrove: effects on plant biomass and crab communities. Ocean Coast Manag 110:38–45

Field CD (1998) Rehabilitation of mangrove ecosystems: an overview. Mar Pollut Bull 37:383–392

Fuentes M, Hidalgo C, González-Martín I, Hernández-Hierro JM, Govaerts B, Sayrre KD, Etchevers J (2012) NIR Spectroscopy: an alternative for soil analysis. Commun Soil Sci Plant Anal 43:346–356

Gerber L, Eliasson M, Trygg J, Moritz T, Sunberg B (2012) Multivariate curve resolution provides a high-throughput data processing pipeline for pyrolysis-gas chromatography/mass spectrometry. J Anal Appl Pyrol 95:95–100

Gillis LG, Zimmer M, Bouma TJ (2016) Mangrove leaf transportation: do mimic *Avicennia* and *Rhizophora* roots retain or donate leaves? Mar Ecol Prog Ser 551:107–115

Giri C, Ochieng E, Tieszen LL, Zhu Z, Singh A, Loveland T, Masek J, Duke N (2011) Status and distribution of mangrove forests of the world using earth observation satellite data. Glob Ecol Biogeogr 20:154–159

Gomes NCM, Cleary DFR, Pinto FN, Egas S, Almeida A, Cunha A, Mendonça-Hagler LCS, Smalla K (2010) Taking root: enduring effect of rhizosphere bacterial colonization in mangroves. PLoS One 5:e14065. https://doi.org/10.1371/journal.pone.0014065

Halpern BS, Silliman BR, Olden JD, Bruno JP, Bertness MD (2007) Incorporating positive interactions in aquatic restoration and conservation. Front Ecol Environ 5:153–160

Hedberg P, Saetre P, Sundberg S, Rydin H, Kotowski W (2013) A functional trait approach to fen restoration analysis. Appl Veg Sci 16:658–666

Hobbs RJ, Arico S, Aronson J, Baron JS, Bridgewater P, Cramer VA, Epstein PR, Ewel JJ, Klink CA, Lugo AE, Norton D, Ojima D, Richardson DM, Sanderson EW, Valladares F, Vilà M, Zamora R, Zobel M (2006) Novel ecosystems: theoretical and management aspects of the new ecological world order. Glob Ecol Biogeogr 15:1–7

Holguin G, Vazquez P, Bashan Y (2001) The role of sediment microorganisms in the productivity, conservation, and rehabilitation of mangrove ecosystems: an overview. Biol Fertil Soils 33:265–278

Jones CG, Lawton JH, Shachak M (1994) Organisms as ecosystem engineers. Oikos 69:373–386

Kamali B, Hashim R (2011) Mangrove restoration without planting. Ecol Eng 37:387–391

Kleinebecker T, Poelen MDM, Smolders AJP, Lamers LPM, Hölzel N (2013) Fast and inexpensive detection of total and extractable element concentrations in aquatic sediments using Near-Infrared Reflectance Spectroscopy (NIRS). PLoS One 8(7):e70517

Lavorel S, Garnier E (2002) Predicting changes in community composition and ecosystem functioning from plant traits: revisiting the Holy Grail. Funct Ecol 16:545–556

Laughlin DC (2014) Applying trait-based models to achieve functional targets for theory-driven ecological restoration. Ecol Lett 17:771–784

Lewis RR III (2005) Ecological engineering for successful management and restoration of mangrove forests. Ecol Eng 24:403–418

Lewis RR III (2009) Methods and criteria for successful mangrove forest restoration. In: Perillo GME, Wolanski E, Cahoon DR, Brinson MM (eds) Coastal wetlands: an integrated ecosystem approach. Elsevier Press, Oxford, pp 787–800

Macintosh DJ, Ashton EC, Havanon S (2002) Mangrove rehabilitation and intertidal biodiversity: a study in the Ranong mangrove ecosystem, Thailand. Estuar Coast Shelf Sci 55:331–345

Matsui N, Suekuni J, Nogami M, Havanond S, Salikul P (2010) Mangrove rehabilitation dynamics and soil organic carbon changes as a result of full hydraulic restoration and re-grading of a previously intensively managed shrimp pond. Wetl Ecol Manag 18:233–242

McKee KL, Faulkner PL (2000) Restoration of biogeochemical function in mangrove forests. Restor Ecol 8:247–259

Miller JA, Bestelmeyer BT (2016) What's wrong with novel ecosystems, really? Restor Ecol 24:577–582

Morse NB, Pellissier PA, Cianciola EN, Brereton RL, Sullivan MM, Shonka NK, Wheeler TB, McDowell WH (2014) Novel ecosystems in the Anthropocene: a revision of the novel ecosystem concept for pragmatic applications. Ecol Soc 19:12–21

Obolski U, Hadany L, Abelson A (2016) Potential contribution of fish restocking to the recovery of deteriorated coral reefs: an alternative restoration method? Peer J 4:e1732. https://doi.org/10.7717/peerj.1732

Perring MP, Standish RJ, Hobbs RJ (2013) Incorporating novelty and novel ecosystems into restoration planning and practice in the 21st century. Ecol Process 2:18–25

Pestana DF, Pülmanns N, Nordhaus I, Diele K, Zimmer M (2017) The influence of crab burrows on sediment salinity in a Rhizophora-dominated mangrove forest in North Brazil during the dry season. Hydrobiologia 803:295–305

Primavera JH, Esteban JMA (2008) A review of mangrove rehabilitation in the Philippines: successes, failures and future prospects. Wetl Ecol Manag 16:345–358

Primavera JH, Agbayani RF (1997) Comparative strategies in community-based mangrove rehabilitation programmes in the Philippines. In: Hong PN, Ishwaran N, San HT, Tri NH, Tuan MS (eds). Proceedings of Ecotone V, Community Participation in Conservation, Sustainable Use and Rehabilitation of Mangroves in Southeast Asia. UNESCO, Japanese Man and the Biosphere National Committee and Mangrove Ecosystem Research Centre, Vietnam, pp. 229–243.

Pülmanns N, Nordhaus I, Diele K, Mehlig U (2015) Artificial crab burrows facilitate desalting of rooted mangrove sediment in a microcosm study. J Mar Sci Eng 3:539–559

Ramírez-Elías MA, Ferrera-Cerrato R, Alarcón A, Almaraz JJ, Ramírez-Valverde G, de-Bashan LE, Esparza-García FJ, García-Barradas O (2014) Identification of culturable microbial functional groups isolated from the rhizosphere of four species of mangroves and their biotechnological potential. Appl Soil Ecol 82:1–10

Ruiz-Jaen MC, Aide TM (2005) Restoration success: how is it being measured? Restor Ecol 13:569–577

Selvam MM, Kathiresan K (2010) Beneficial bacteria from soil of a tropical mangrove. Asian J Microbiol Biotechnol Environ Sci 12:1–8

Suding KN (2011) Toward an era of restoration in ecology: successes, failures, and opportunities ahead. Annu Rev Ecol Evol Syst 42:465–487

Taberlet P, Coissac E, Hajibabaei M, Rieberg LH (2012) Environmental DNA. Mol Ecol 21:1789–1793

Thomsen PF, Willerslev E (2015) Environmental DNA – an emerging tool in conservation for monitoring past and present biodiversity. Biol Conserv 183:4–18

Tolu J, Gerber L, Boily J-F, Bindler R (2015) High-throughput characterization of sediment organic matter by pyrolysis–gas chromatography/mass spectrometry and multivariate curve resolution: a promising analytical tool in (paleo)limnology. Anal Chim Acta 880:93–102

Twilley RR, Rivera-Monroy VH, Chen R, Botero L (1998) Adapting an ecological mangrove model to simulate trajectories in restoration ecology. Mar Pollut Bull 37:404–419

Vovides AG, Bashan Y, López-Portillo JA, Guevara R (2010) Nitrogen Fixation in Preserved, Reforested, Naturally Regenerated and Impaired Mangroves as an Indicator of Functional Restoration in Mangroves in an Arid Region of Mexico. Restoration Ecol 19:236–244

Vovides AG, Bashan Y, López-Portillo JA, Guevara R (2011) Nitrogen fixation in preserved, reforested, naturally regenerated and impaired mangroves as an indicator of functional restoration in mangroves in an arid region of Mexico. Restor Ecol 19:236–244

Wabnitz CCC, Andréfouët S, Muller-Karger FE (2010) Measuring progress toward global marine conservation targets. Front Ecol Environ 8:124–129

Walton ME, Le Vay L, Lebata JH, Binas J, Primavera JH (2007) Assessment of the effectiveness of mangrove rehabilitation using exploited and non-exploited indicator species. Biol Conserv 138:180–188

Towards Sustainability and Protection of Threatened Coastal Ecosystems: Management Strategies for a Rare Stone Mangrove in Gorda Beach, Armação dos Búzios, Brazil

17

Marcelo Obraczka, Kátia Leite Mansur, and Gerson Cardoso da Silva, Jr.

Abstract

Brazil ranks third in total mangrove area around the world, following Indonesia and Australia. Most Brazilian mangroves are poorly studied regarding hydrologic equilibrium mechanisms. Along Mangue de Pedra of Gorda beach, in Armação dos Búzios, Rio de Janeiro State, Brazil, there is a rare mangrove, which has thrived over a rocky substratum. Of particular interest is the socioenvironmental and historical background of this mangrove amid potential and emerging land use conflicts. A combination of distinct rock units bound together by a geological fault in a semiarid microclimate led to a unique setting in which fresh groundwater from hanging sedimentary rock cliffs mix with seawater creating an unusual small mangrove before a sandy beach. This attractive landscape led to intense property speculation. Sustainable land use by 'quilombolas', descendants of ancient black slaves is being threatened by that speculation, leading to conflicts. Considering the factors and aspects reported above, Mangue de Pedra has high environmental, historical-cultural and scientific relevance and, consequently, should be preserved. Public management measures, however, have not being effective to protect it, leading to a gradual loss of local bio and geodiversity. Solution for this issue includes a set of measures such as creation of a comprehensive environmental protection unit, education and environmental awareness, dissemination of knowledge about the importance of the Mangue, social control and mobilization of the population, based on participation in the decision-making forums, the paradigm shift for UC administration, among others.

Keywords

Mangrove hydrology · Groundwater flow · Búzios – Brazil · Rock substratum

17.1 Introduction

Based on a worldwide survey by remote sensing, which used satellite images obtained between 1997 and 2000, Giri et al. (2011) identified a total of 137,760 km^2 of mangroves in 118 countries and territories (Fig. 17.1). It corresponds to 0.7% of the total area of existing tropical forests. Those authors estimated that 75% of the world's mangroves stretch across 15 countries, but they have not investigated the integrity of these ecosystems. Brazil ranks third in mangrove areas in their work, following Indonesia and Australia. Brazilian mangrove area spans for 9626 km^2, representing 7% of the world distribution of the mangrove ecosystem.

The estimate was approximately 12% down from a FAO (2007) survey, using images from the period 1980–2000. In the article, situation of mangrove area distribution is described as follows: 'Five countries (Indonesia, Australia, Brazil, Nigeria and Mexico) together account for 48 percent of the total global mangrove area and 65 percent of the total is found in just ten countries [...]. The remaining 35 percent are spread over 114 countries and areas, of which 60 have less than 10,000 ha of mangroves each.'

Saenger et al. (1983) authored a pioneer study of mangrove distribution and total area evaluation in Brazil. Based on data collected by the Scientific Committee on Oceanic Research (SCOR), they report the occurrence of 25,000 km^2 of mangroves area, along a 7408 km coastline.

Spalding et al. (2010), on the other hand, estimated Brazil ranks second in the world regarding mangrove area

M. Obraczka
Departamento de Engenharia Sanitária e do Meio Ambiente, Universidade do Estado do Rio de Janeiro, Rio de Janeiro, Brazil
e-mail: obraczka@ppe.ufrj.br

K. L. Mansur · G. C. da Silva, Jr. (✉)
Instituto de Geociências, Universidade Federal do Rio de Janeiro, Rio de Janeiro, Brazil
e-mail: katia@geologia.ufrj.br; gerson@acd.ufrj.br

C. Makowski, C. W. Finkl (eds.), *Threats to Mangrove Forests*, Coastal Research Library 25,
https://doi.org/10.1007/978-3-319-73016-5_17

Fig. 17.1 World mangrove vegetation area distribution in year 2000. In addition, limits of Brazilian territory and the Brazilian mangrove forests distribution along the coastal area are presented (Modified from Giri et al. (2011)

distribution, after Indonesia, with 13,000 km^2 or 8.5% of the world total. Prates et al. (2010) presented a similar figure, with a different method, calculating the Brazilian mangrove area by means of the overlap between the ecosystems mapped area and the Conservation Units National Register area (Ministry of the Environment – MMA 2017), getting to a value of 12,254 km^2. Both Prates et al. (2010) and Spalding et al. (2010) indicate that approximately 25% of Brazilian mangrove areas have already been destroyed, with aquaculture and real estate speculation as the main causes. Table 17.1 presents a summary of surveys according to the extension of mangroves in Brazil.

Brazil also presents the largest continuous mangrove area in the world, totaling 6516 km^2, located in the north of the country, between the states of Maranhão and Pará (Webber et al. 2016).

Schaeffer-Novelli et al. (2000) report that Brazilian mangroves present a wide structural and physiognomic diversity, due to environmental factors and anthropic influence. Only one coastal state of Brazil, Rio Grande do Sul, in the extreme south of the country, does not hold mangroves, due to climate constraints, since it is farther south from mangrove climax zone.

17.2 Mangroves on State of Rio de Janeiro and Brazilian Coastline

The continental margin of Brazil, from the plate tectonics perspective, is a passive margin. Considering geodiversity it is highly diversified, since it involves varied geology, environment, climate (equatorial, tropical, tropical Atlantic and subtropical), tidal variation (macro, meso and microtides), circulation of currents, erosive and depositional processes linked to large rivers and estuaries and biodiversity, among other factors.

Brazilian mangrove forests are classified as basin, fringe and riverine types (Schaeffer-Novelli et al. 2000). They present mainly the following tree types with diverse dimension, distribution and environmental relations (Alves 2001): (a) *Rhizophora mangle*, *Rhizophora harrisonii* and *Rhizophora racemosa* = mangue-bravo or red mangrove; (b) *Avicennia schaueriana* and *Avicennia germinans* = mangue-preto (black mangrove), mangue-seriba or seriúba; (c) *Laguncularia racemosa* = mangue branco (white mangrove); and (d) *Conocarpus erecta* = button mangrove (non-exclusive species of mangrove ecosystems).

It is worth noting that the coastal zone is home to 26.6% of Brazilian population, equivalent to approximately 50.7 million inhabitants (IBGE 2011). Mangroves are part of the Atlantic Forest biome, recognized by UNESCO as a World Heritage Site. This biome originally occupied 1,300,000 km^2 in 17 States (Alves 2001). Today, only 8.5% of the original vegetation cover remains (MMA 2017).

According to Carvalho (2011), 16 of the 28 Brazilian metropolitan areas are located on the coast, where approximately 43 million inhabitants live, corresponding to about 18% of the country's population and 60% of the GDP (MMA 2008). The coastal zone has thus important economic, commercial and service activities, such as oil platforms, refineries, petrochemical centers, basic industries, ports and airports, among others. The main risks affecting the Brazilian Coastal Zone are related to the socioeconomic development and population pressure that affect large urban centers, real estate speculation, pollution and rapid expansion of some activities, such as oil and natural gas exploration and production, tourism and sea farming (Carvalho 2011).

Mangroves are some of the most impacted Brazilian ecosystems, due to human action. However, legislation considers mangroves as APP's (Areas of Permanent Preservation). Many mangroves are located in Conservation Units (Protected Areas – PA) of different administrative levels

Table 17.1 Situation of Brazil in mangrove world distribution area

Authors	Surface extension of mangrove areas around the world (km^2)	Surface extension of mangrove areas in Brazil	
		km^2	World (%)
Saenger et al. (1983)		25,000	
FAO (2007)	152.310	10,000	7
Spalding et al. (2010)		13,000	8,5
Prates et al. (2010)		12,254	
Giri et al. (2011)	137.760	9626	7

Table 17.2 Coastal ecosystems by Integral Protection Conservation Unit (federal and state jurisdiction) in ha (*)

	Wetland	Rocky shore	Dune	Estuary	Lagoon	Mangrove	Marsh	Beach	Sandbank
PAs integral protection	252,590	45,895	117,998	12,436	33,834	160,634	77	2200	95,783
Ecosystem area	4,849,671	144,475	318,312	6,696,787	1,518,426	1,225,444	12,149	82,778	469,183
% protected	5.2	31.8	37.1	0.2	2.2	13.1	0.6	2.7	20.4

Source: Prates et al. 2010
(*) Includes PAs superimposed to sustainable use units

Table 17.3 Coastal ecosystems by sustainable use PAs (federal and state jurisdiction) in ha (*)

	Wetland	Rocky shore	Dune	Estuary	Lagoon	Mangrove	Marsh	Beach	Sandbank
PAs sustainable use	2,614,665	68,274	18,119	1,375,758	8295	759,049	0	17,811	228,298
Ecosystem área	4,849,671	144,475	318,312	6,696,787	1,518,426	1,225,444	12,149	82,778	469,183
% protected	53.9	47.3	5.7	20.5	0.5	61.9	0	21.5	48.7

Source: Prates et al. 2010
(*) Excluded superposition with integral protection PA and APP and other sustainable use categories

(federal, state or municipal) and typologies (parks, environmental protection areas, ecological reserves, among others). Therefore, they are enabled to insertion, indistinctly, in integral and sustainable use protection units.

Prates et al. (2010) present the situation of Brazilian coastal ecosystems regarding their legal protection according to the National System of Conservation Units (NSCU) (see Tables 17.2, 17.3, and 17.4). Most of the mangroves are inserted, as seen in Tables 17.2, 17.3, and 17.4, in conservation units of the 'sustainable group' (about 62%) which, definitely, have less efficiency in natural patrimony preservation than those of 'integral protection'.

Mangroves in integral protection units represent only around 13% of the total Integral Protection coastal areas. Considering the estimated 12,000 km^2 mangrove area in Brazilian coasts (Prates et al. 2010), about 75% of the total are – at least theoretically – protected. This is a much higher figure than the other ecosystems, which seems to imply a bigger emphasis by environmental authorities to the protection of mangroves.

The State of Rio de Janeiro is the most densely populated in the country. It holds a population of more than 16.5 million inhabitants and a population density of 656 inhabitants/km^2. The coastline has the highest degree of anthropization in Brazil (IBGE 2011).

Rio de Janeiro State coastline extends over 1160 km, with 33 municipalities and 40.1% of the territory, with about 80% of the population of the entire state (www.inea.rj. Gov.br/Portal/Agendas/GESTAODEAGUAS/...Costeiro/index.htm&lang=). It is an area of significant economic relevance, accounting for over 90% of Brazilian oil production and 77% of the gas production, all extracted offshore. Three large bays (Guanabara, Sepetiba and Ilha Grande), 614 islands and an extensive complex of coastal lagoons make it one of the more jagged coastlines in Brazil. In these large ecosystems, there are coastal habitats of recognized ecological relevance, such as sandbars, marshes, prairies of marine phanerogams and mangroves. In recent years, the marine and coastal environments of Rio de Janeiro State have undergone a notorious and continuous process of degradation, associated with diverse activities that promote significant changes in the physical structures of habitats, biogeochemical cycles and in the natural balance of nutrients. There are also changes in geological transport and sedimentation coastal processes, causing severe material damages to numerous families that inhabit and depend on the coastal zone, as well as the overexploitation of fishing resources, introduction of exotic species and persistent pollutants in these ecosystems.

According to Polette (2008), there is a number of reasons for the serious environmental problems and impacts on the coast: (a) political (political policies incompatible with environmental policies); (b) administrative (e.g. lack of infrastructure, resources and monitoring capabilities, as well as public policies affecting the coastal zone are not integrated

Table 17.4 Costal ecosystems by integral protection and sustainable use PAs (federal and state jurisdiction) in ha

	Wetland	Rocky shore	Dune	Estuary	Lagoon	Mangrove	Marsh	Beach	Sandbank
Ecosystem área	4,849,671	144,475	318,312	6,696,787	1,518,426	1,225,444	12,149	82,778	469,183
Integral Protection PAs (IPPA)	252,590	45,895	117,998	12,436	33,834	160,634	77	2200	95,783
Sustainable Use PAs (SUPA)	2,614,665	68,274	18,119	1,375,758	8295	759,049	0	17,811	228,298
Total IPPA + SUPA	2,867,255	114,169	136,117	1,388,194	42,129	919,683	77	20,011	324,081
% protected area	59.1	79.0	42.8	20.7	2.8	75.1	0.63	24.2	69.0

Source: Prates et al. 2010

and do not have long term reach); (c) institutional (e.g. lack of vertical and horizontal integration and civil society distant from power relations between governments and private initiative); (d) ecological (e.g. lack of understanding of the structure, functioning and resilience of Coastal ecosystems); and (e) economic (e.g. excessive value of the coast and the numerous interests for its occupation).

Among the consequences of the predatory coast occupation in Rio de Janeiro, the mangroves disappearance stands out as the most serious. As stated by Alves (2001), the main impacts are deforestation, inappropriately built and managed landfills, domestic and industrial sewage and garbage, predatory fishing, impact on traditional fishing communities, impact of oil spills and its derivatives.

According to Prates et al. (2010), mangroves in State of Rio de Janeiro comprise 13,358 ha, 3107 ha of which are in protected areas and 4284 ha in APA (Sustainable Use PA). 55.3% of the state mangroves are protected by environmental legislation.

The larger mangrove terrains in Rio de Janeiro State are in Guanabara Bay region, near the city of Rio de Janeiro; Sepetiba Bay, west of Rio de Janeiro; in the north, in the Paraíba do Sul river mouth and, to the south, in Ilha Grande Bay.

Mangue de Pedra (Stone Mangrove), the case study discussed in detail in this chapter, is located in the municipality of Armação dos Búzios (Fig. 17.2), an internationally renowned resort, located about 200 km east of the state capital.

17.3 Mangue de Pedra: General Description

The Mangue de Pedra, or 'Gorda beach mangrove', is a mangrove located in the municipality of Armação dos Búzios (22°74′S and 41°88′W), or simply Búzios, in the Lakes Region of Rio de Janeiro state. It is a small fringe-type mangrove. Dominant mangrove species are *A. schaueriana* and *L. Racemosa* (Oliveira 2007). It is an ecological rarity, due to its rocky substratum (coarse sand, granules, pebbles, blocks and boulders) and fresh water fed by a groundwater discharge, hence its designation 'Stone Mangrove'. The Gorda beach mangrove is fixed on a sandy-rock substrate with structural vegetation patterns indicating it is in a frank developmental process (Oliveira 2007). This fact reinforces the relevance of its conservation, especially considering the wide variety of goods and services provided by this ecosystem, such as artisanal fishing, boat shelter, shellfish harvesting, educational and tourist visits, nutrient provision for local life development, among others. At this site, geodiversity clearly demonstrates its ecosystem service in support of biodiversity (Gray 2013).

Despite the presence of solid waste (plastic containers and packages, glass, oil waste, etc.) which is brought in by the sea tide (Fig. 17.3) or, to a lesser extent, by visitors, and organic remains (mainly fish and other marine animals), the Gorda beach mangrove does not exhibit a landscape excessively altered by anthropic tensors. The partial isolation, due to the precarious access, contributes to the good degree of conservation. Also, the fact that, since the nineteenth century, the area is a fishing and shellfish collecting area, mainly by the remaining population of 'quilombolas' (descendants of black slaves whose families fled the plantations during slavery times in Brazil) living in the neighboring area of Rasa beach. Some anthropic degradation, such as deforestation, irregular human settlements and road openings, occur in surrounding areas, reinforcing the need for conservation actions (Rebelo 2013). Nevertheless, a significant portion of its drainage area is an ancient urban allotment, and some of these lots have been gradually occupied by residences, in spite of the most recent municipal legislation (Fig. 17.3).

The area ranks third as one of the more important geosites of Cabo Frio Tectonic Domain – DTCF, geologically one of the most relevant in the whole country and, therefore, integrates the 'Costões e Lagunas Geopark' project Rio de Janeiro State (Mansur 2010).

The area of Mangue de Pedra presents seven types of geological patrimony regarding content (García-Cortés 1996): Tectonic, Geoenvironmental, Sedimentary, Paleoenvironmental, Hydrogeological, Stratigraphic and Geomorphological, as well as Historical/Cultural interest, with a very high density of typologies (Mansur 2010).

In 2012 a development/estate project (Gran Riserva 95) was licensed in the area surrounding Mangue de Pedra. The social and NGOs mobilization led to the interdiction of the project by the State Prosecutors' Office, which demanded a hydrogeological study to characterize the mangrove seawater-continental groundwater interactions. In 2013, the license was canceled by the municipal administration and the

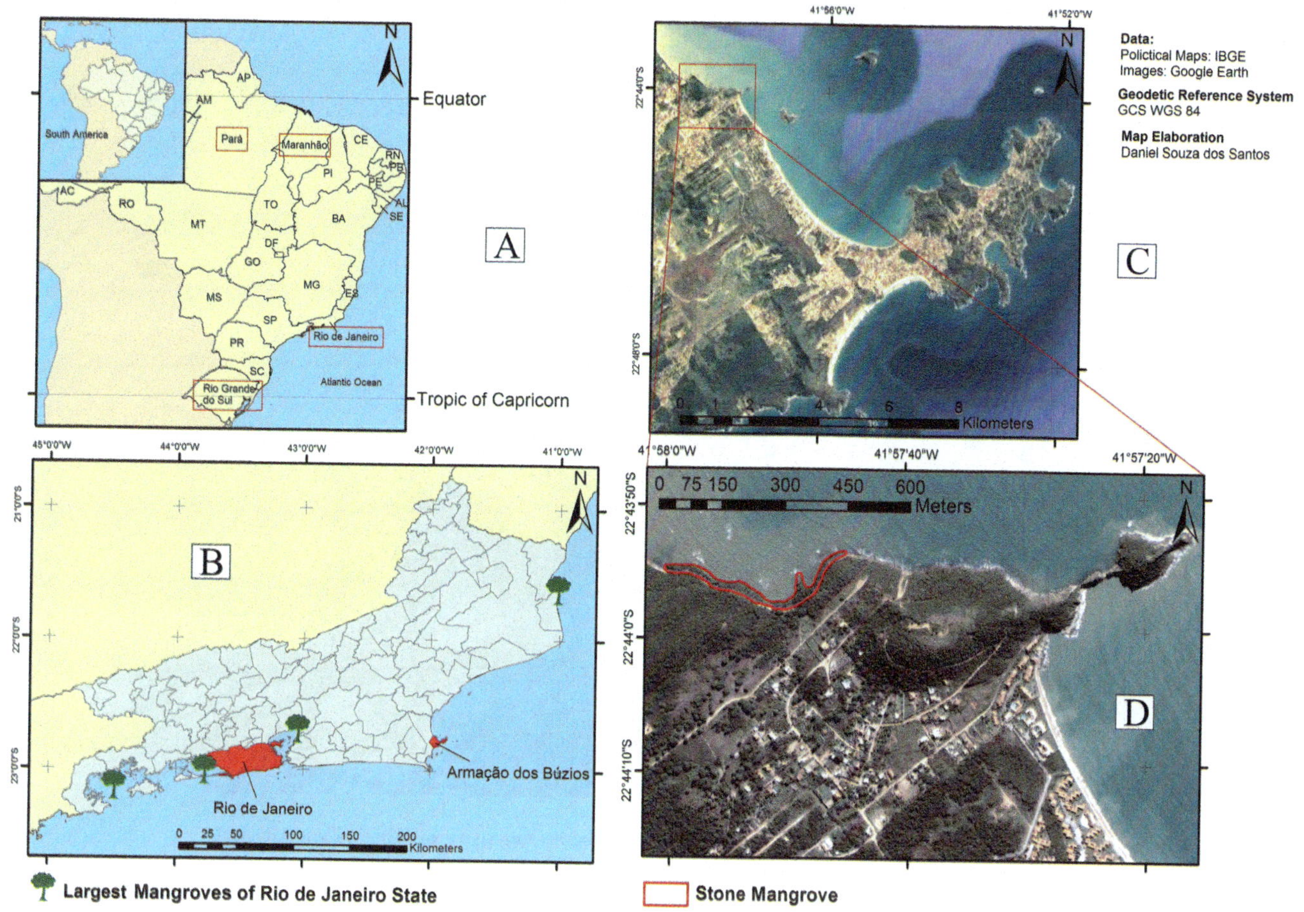

Fig. 17.2 Mangue de Pedra location map. (**a**) Brazilian map with highlights for states cited in the text; (**b**) Rio de Janeiro State where are located the largest mangroves of the state and Búzios; (**c**) Búzios; and (**d**) Mangue de Pedra localization

construction was abandoned. On site, evidence of unfinished buildings can still be observed (Fig. 17.4).

There were previous attempts to implement an urban development in the area, and all of them met fierce resistance from civil society. All these integrated factors reinforce the urgent need for the area protection. It can be achieved with the creation of an Integral Protection Conservation Unit (IPCU), which has received a favorable decision from the City Parliament in 2009, but not implemented, despite numerous requests from the population, environmental entities and researchers. The need to alienate private areas with a great development potential represents obstacles to the PA implementation to date, despite being mentioned as desirable in legislation and municipal planning (Environmental Code, Organic Law and Master Plan, for example).

17.3.1 Physiography and Climate

Búzios is located on the east coast of the state of Rio de Janeiro (Brazil), in the *Baixadas Litorâneas* (coastal lowlands) region. It comprises a very jagged peninsula and a strip of beach (Fig. 17.2). The area is 70.278 km^2, with a population of 27,560 permanent inhabitants in 2010 (https://cidades.ibge.gov.br/brasil/rj/armacao-dos-buzios/panorama). It has an eminently tourist vocation, as the fifth most sought after destination by foreign tourists in Brazil (www.dadosefatos.turismo.gov.br/2016-02-04-11-53-05.html). The tourist population in summer can reach, depending on the estimate, between 150 and 300,000 people. The renowned US magazine *Leasure and Travel* describes Búzios as 'one of the ten paradises of the world', 'a paradise with personality' (Xavier 2006). Originally, however, Búzios was a fishing village, an activity still present in the local economy.

Southeastern Brazil remains for most of the year under the Tropical Atlantic Air Mass influence, originating from the South Atlantic Anticyclone. It maintains climatic stability although, throughout the year, regional atmospheric circulation is subject to interference from polar masses and tropical instability lines (Davis and Naghettini 2001). However, Búzios region has a distinct microclimate: it is drier with respect to the rest of Rio de Janeiro State and the Southeast

Fig. 17.3 Solid waste and debris deposited on the vegetation of Mangue de Pedra mangrove area (Photo: Kátia Mansur)

Region as a whole, although the local peninsular geography favors a strong sea influence, keeping climate with little thermal amplitude and mild throughout the year. This is due to a number of factors, such as the distance from the Serra do Mar mountain range to the coastline, the peninsular condition, the presence of the great Araruama Lagoon, the abrupt change in direction of the Brazilian coastline from N-S to E-W, and especially the phenomenon of resurgence, which strongly affects this part of the coast. Thus, the climate is hot and semiarid, with rainfall around 800 mm/year and 5 months of drought, resulting in a water deficit of 317 mm/year. Winds are constant, predominantly from the northeast, with average velocity around 6 m/s and only 12% of calm winds year round.

17.3.2 Local Geological Aspects

The Province of Mantiqueira, in southeastern Brazil, comprises a Neoproterozoic orogenic system that extends from the south of the State of Bahia to the State of Rio Grande do Sul (Heilbron et al. 2004). According to the authors, the State of Rio de Janeiro is part of the Central Segment of Mantiqueira Province, in the portion named Ribeira Belt.

The state of Rio de Janeiro has a complex geological structure, subordinated to the formation and deformation event of the Ribeira Belt (Almeida and Carneiro 1998), during the Neoproterozoic orogeny, when the Gondwana paleocontinent was formed (Cordani et al. 2000; Heilbron et al. 2000). However, for Schmitt et al. (2008a), the Cabo Frio Tectonic Domain – DTCF, where Búzios is located, has an affinity with African lands and is actually exotic to the Ribeira Belt, belatedly attached to the Mantiqueira orogenic system in the Cambrian period, about 530 Ma. B.P. Schmitt et al. (2004, 2008a) consider, therefore, the DTCF as part of the African continent geological framework kept bonded to the South American territory after the Mesozoic rifting. This particular event is the 'Búzios Orogeny' (Schmitt 2001) (Fig. 17.5).

DTCF bedrock is composed mainly of felsic orthogneisses cut by aplites and pegmatites. There are still tabular bodies of (garnet) amphibolites cutting orthogneisses, interpreted as dikes associated with a late extensional Paleoproterozoic event.

Also as part of the crystalline bedrock, but more recent than the previously described units, there are volcanic and supracrustal Neoproterozoic to Cambrian rocks of Búzios, Palmital and Forte São Mateus units. Those are parts of the so-called Búzios-Palmital Basin, constituted by mafic

Fig. 17.4 Deforestation and construction in progress at Gran Riserva 95 estate project in 2012 (Mehdi and Souza 2012)

gneisses and paragnaisses, which are intercalated with the Paleoproterozoic basement, and are at least 500 m thick (Schmitt et al. 2008b). Dating, based on U-Pb analyzes, show this tectonometamorphic event occurred during the Cambro-Ordovician interface (525–490 Ma) and is part of Búzios Orogeny (Schmitt 2001). Intruding the bedrock there are Mesozoic toleitic dikes (130 million years B.P.) and Meso-Cenozoic alkali dikes.

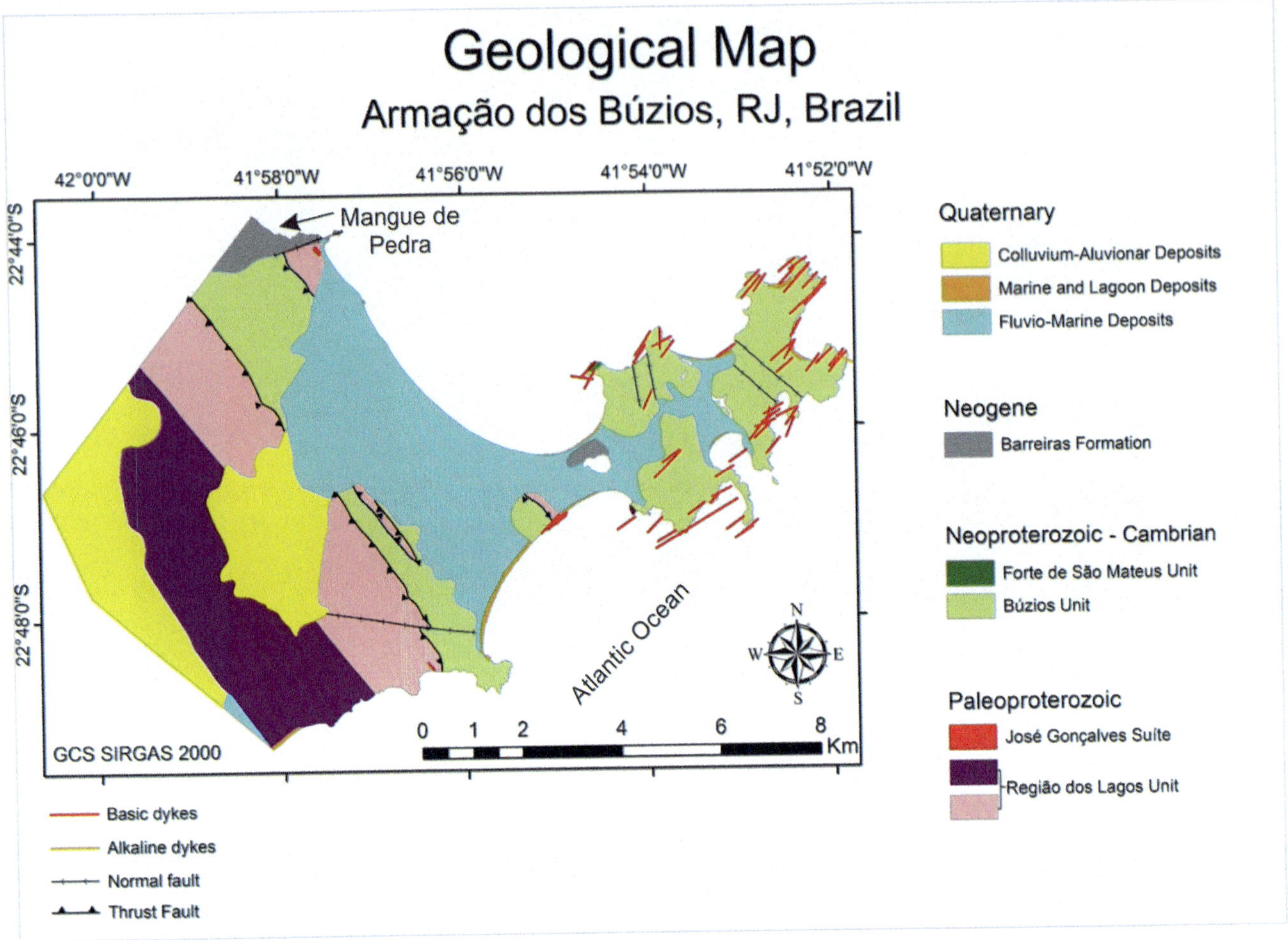

Fig. 17.5 Geological map of Armação dos Búzios (Souza et al. 2017, adapted from Schmitt et al. 2012; Tetzner 2002)

Overlying the metamorphic or igneous crystalline rocks, sedimentary materials occur, particularly the Plio-Miocene sediments of Barreiras Formation. The Barreiras Formation occurs along the Brazilian coast, from the State of Amapa to Rio de Janeiro, characterized by the nearly continuous occurrence and geomorphological regularity, with conspicuous sea bound plateau landforms. Previously considered to be of continental origin, recent work has shown paleontological and sedimentological evidence of marine influence and composition. Its origin seems to be related to global eustatic oscillations between middle Miocene and Pliocene periods, generating two depositional cycles interspersed by an erosive cycle. The erosion and reworking of the Barreiras Formation in the Quaternary must have been responsible, in part, for the current configuration of the Brazilian continental shelf (Arai 2006). Deposits of the Barreiras Formation are discontinuous and interchanging levels of friable and poorly selected sediments, ranging from conglomerates to sandy-clayey materials, sometimes with horizons of lateritic concretions near the surface. Conglomeratic layers present pebbles of various sizes, angular and/or rounded, in a sandy or at times sandy-clayey matrix. It may also contain clayey beds (Mansur 2010). Palaeocliffs of Barreiras Formation in Praia Rasa and in Gorda beach represent the moment when the sea level was lower than in the present time, and rivers laid their sediments where it is now under the sea (Morais et al. 2006).

Pleistocene and Holocene deposits occur also in the study area, in the form of sandy and clayey deposits in sandbars, current beaches, palaeobeaches, marshes, swamps, present-day lagoons, palaeolagoons, dunes, fluvial deposits, and other coastal environments (Mansur 2010).

The lithological setting is disturbed by expressive geological structures, notably faults. An important geological fault system is represented by tectonic breccias, such as the ones present in Pai Vitório fault (Rubin and Almeida 2003) and others. The Pai Vitório fault is located in the southern boundary of the Graben of Barra de São João (Mohriak and Barros 1990), and plays an important role in the formation of the Mangue de Pedra substrate: Barreiras Formation sediments in the area are interpreted as alluvial fan deposits conditioned by Pai Vitório fault scarp (Morais et al. 2006). It is a regional normal fault with N70E direction. It gathers side-by-side rocks of the Paleoproterozoic basement with conglomeratic sediments of Barreiras Formation. These sediments constitute palaeocliffs, with an age of about 5100 years B.P., when sea level was about 2.8 m above the current level (Dias 2009).

17.3.3 Hydrogeology

Groundwater is a scarce natural resource in the study area. Geological settings and intense maritimity make local aquifers of lesser relevance to human activities, although small hand dug and shallow wells are common in houses and small rural properties. It does not present voluminous and high transmissivity regional aquifers. Two main types of aquifers occur in the study area:

(a) Intergranular porosity aquifers, located in marine, fluvial and aeolic sedimentary deposits in the coastal plain, including weathered soils and rocks; and
(b) Fractured aquifers, constituted by the igneous and metamorphic crystalline bedrock previously described, barely exploited in the study area, since it is frequently affected by intense salinization (Mansur 2010).

Near the sea, (a) type aquifers constitute plateaus over the crystalline bedrock, ending in cliffs in sea bound directions, due to the coastline erosion, especially in Barreiras Formation.

As described by Silva Jr (2003), sedimentary deposits, composed predominantly by quartz sands and, secondarily, by silty and clayey deposits, have generally a shallow water table, sometimes less than 2 m deep in the lower portions. Groundwater levels may fluctuate seasonally. In sandbar areas, depth to the water level is higher near the central axis, and shallower as it approaches the borders, especially in the lagoons. These characteristics give water table aquifers a high susceptibility to bacteriological contamination. In sandy areas, the water is generally lighter and less brackish than the water of the finer sediment low-lying wetlands, although closer to the beach aquifers are usually salinized.

Sediments of Barreiras Formation, with a continental and shallow marine nature, present a typical continental characteristic in the studied region, making aquifers with medium to high favorability to the groundwater exploitation, and a high susceptibility to contamination when constituted by conglomerates, such as occur in Praia Rasa area, between Cabo Frio and Armação dos Búzios (Arai 2006).

Within the layers related to the formation of the coastal lowland, during the transgression and marine regression movements in the Holocene, between 5000 and 7000 years BP, granular type aquifers are thus present (Turcq et al. 1999; Silva Jr 2003).

Figure 17.6 shows the vulnerability map to coastal aquifer contamination presented by Seabra et al. (2009). The region around the Mangue de Pedra, including the aquifer recharge area, is classified as ranging from low to high vulnerability, reinforcing the need to implement measures to adequately manage water resources to preserve the ecosystem.

In Búzios and Arraial do Cabo, municipalities located in the semi-arid microclimate zone, typical of the Cabo Frio region, salinized aquifers are common, and even dominant, due to the significant flow of seawater landwards caused by the intensely fractured substrate and strong maritimity, wedging saline water to the inner parts of the peninsula. Therefore, water salinity increases towards the coast and in depth. Human occupation and groundwater abstraction make the problem worse, since aquifer exploitation intensifies the tendency to salinization.

17.3.4 Natural, Landscape and Environmental Aspects

Biodiversity of Búzios and Mangue de Pedra, often exclusive and endemic, is a consequence of local geodiversity, resulting in different ecosystems. The presence of unique geographic features, such as the great inflection of the Brazilian coast (see Fig. 17.1), the marine resurgence and the effect of various marine regressions and transgressions, resulted in diverse environments of lagoons, cliffs and coastlines (Mansur 2010). The rocky shores represent transitional environments between terrestrial and marine settings, with high diversity of organisms and favorable conditions to species proliferation. In the sea, banks and coral reefs shelter rich biodiversity and contribute to high fishing productivity (Mansur 2010).

The cold-water resurgence near the coast alters marine life in the region and has a great biogeographic relevance for several groups of marine organisms, including benthic algae, and represent the limit of geographical distribution of several taxa, from the typically tropical to the temperate regions, with affinity for cooler waters (Brasileiro et al. 2009).

The vegetation cover reflects the geodiversity of the region, as well as the variability of the present and past climate. It is 'a remnant of existing vegetation during the Pleistocene glacial periods, drier and colder' (Ab'Saber 1974). Between Arraial do Cabo and Armação dos Búzios there is a special vegetation, typified as Open Arboreal Steppe (Semi-Decidual Seasonal Forest). These are witnesses

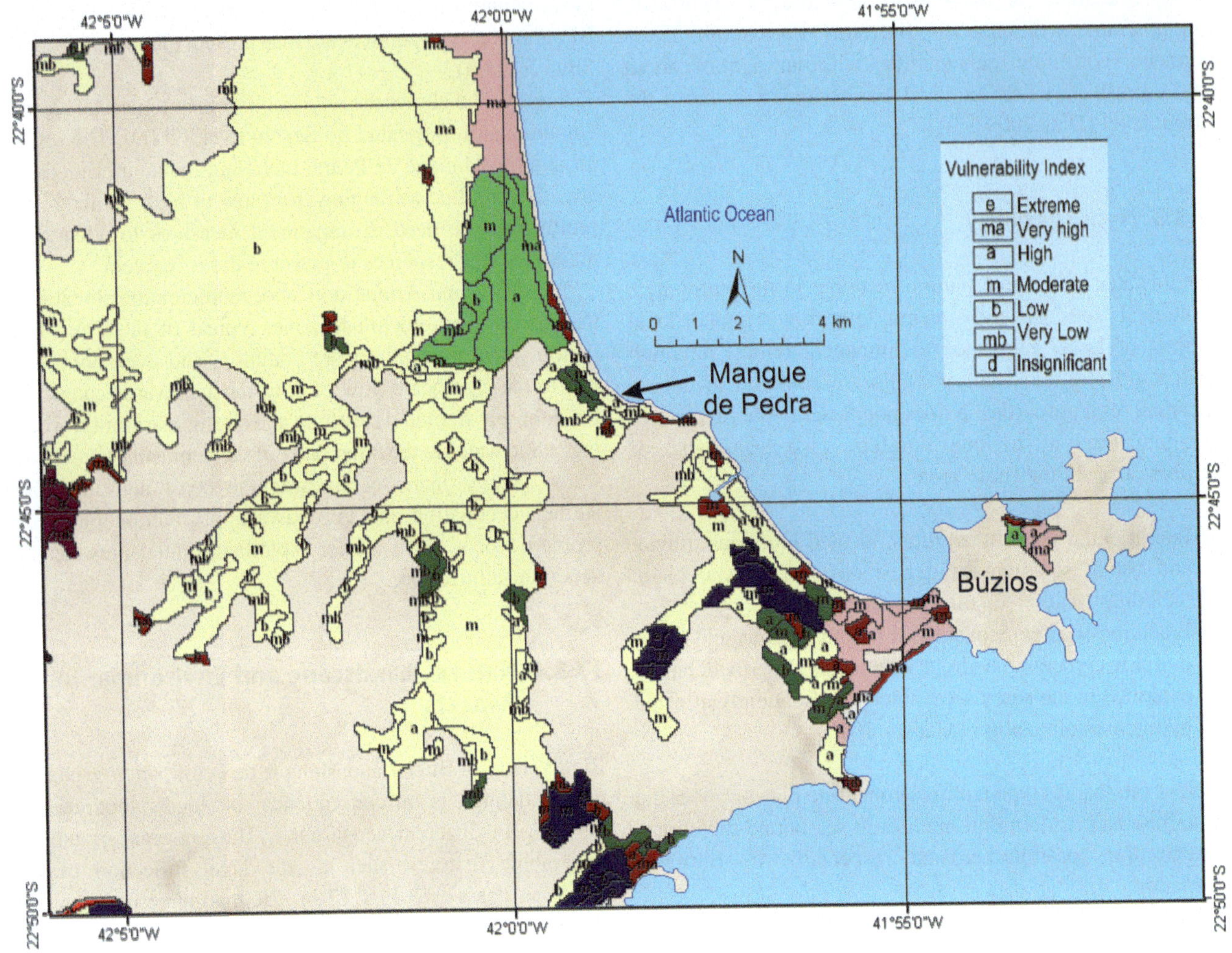

Fig. 17.6 Vulnerability map to coastal aquifers contamination in Búzios (After Seabra et al. 2009)

of sea level variations in the last 12,000–13,000 years in the state of Rio de Janeiro (Oliveira 2003). This highlights the importance of the remaining fragments of native vegetation in this region.

The vegetation is inserted in the Atlantic Forest biome, comprising forest formations and ecosystems, which include the aforementioned open steppe forest, the dense ombrophilous forest in the humid areas and the sandbar vegetation, palaeo-sandbars and dunes, as well as mangroves (such as Manguinhos and Gorda Beach/Mangue de Pedra), swamps and margins of lagoons (Rebelo 2013). Once classified as belonging to the Atlantic Forest biome, this vegetation is protected by a Federal law. Oliveira (2007) studied the structure of the mangrove forest of Gorda Beach and concluded it has a good structural development pattern. Measurements on the tree population distribution, according to the diameter of the stems, indicate a relatively young, shrubby forest, leading to conclude it is a developing mangrove. The Gorda Beach mangrove is formed by red and white mangrove. Cintrón and Schaeffer-Novelli (1983) warn 'Mangroves on stony and sandy substrates only support small trees.' However, in this mangrove, the trees reach up to four meters in height. Some of them, due to lack of substrate, end up falling but, notwithstanding, the roots continue feeding the plant, which produces propagules to colonize the environment. More importantly, beyond the mangrove there is still a significant area of native vegetation, little anthropicized. This combination of the two environments provides a unique character to the Mangue de Pedra (Fig. 17.7).

17.3.5 Importance of Sociocultural and Historical Attributes

The Gorda Beach rocky mangrove is often used in environmental education activities of schools and NGOs in the region, whether by environmental, socio-cultural or historical character. Pai Vitorio Point/Gorda Beach was an area for the

Fig. 17.7 Significant area of almost intact native vegetation around Mangue de Pedra in Gorda beach and aspects of different types of vegetation side-by-side: Mangrove and Open Arboreal Steppe (Photos: (**a**) Kátia Mansur; and (**b**) Anna Roberta Mehdi)

Fig. 17.8 *"Quilombola"* community monument (**a**) and shellfish collection activity at Gorda beach (**b**) (Photos: Kátia Mansur)

illegal arrival of enslaved African people in the nineteenth century, even before the abolition of slavery, when the traffic was already prohibited (Accioli 2013). In the region, there is a *quilombola* community living near the mangrove, collecting shellfish for feeding and, thereby, keeping alive part of their heritage (Fig.17.8) (see https://www.youtube.com/watch?v=nIKzxDtlQ_E).

17.4 Institutional Governance and Management Aspects: Legislation and Environmental Management

Brazil is a signatory of international conventions directly or indirectly related to environmental issues such as Climate, Biodiversity, RAMSAR and MARPOL areas, and commitments made at RIO 92 (Earth Summit), among others. Brazilian environmental legislation is present in the three spheres of govern, from federal (more general) to the local (municipal) sphere. The more particular (local) legislation may be more restrictive than the preceding one. Table 17.5 shows the most relevant aspects regarding legislation and legal regulations in force for mangrove management in the three spheres of power.

The obligation to create municipal councils, with guarantee of diverse social sectors' participation, including non-governmental segments, is a tendency in Brazilian legal framework. Although there is little participation and representation of civil society entities in these councils, many of which are merely consultative, there are already cases of PAs management effectively transferred by the public power to nonprofit public organizations (OSCIP's) and NGOs, going beyond the role as a mere participation in discussion forums. In addition to their efforts to raise funds

Table 17.5 Federal, state and municipal (Búzios) legal instruments and regulations related to mangrove ecosystems

Identification	Theme/concepts	Observations (only items related to Mangue de Pedra ecosystem):
Federal regulations		
Federal Law 12.651/2012	Brazilian Forest Code – **Chapter II**, Permanent Preservation Areas, **Articles 4 and 6**	Areas of permanent preservation (APPs) are natural areas whose function is to preserve water resources, landscape, geological diversity and biodiversity, facilitate the genetic flow of fauna and flora, protect the soil and ensure the well-being of human populations. An APP is to established, in rural or urban areas: (a) around springs; (B) in slopes with declivity greater than 45°; (B) in mangrove stabilizing vegetation areas; (C) in mangrove areas; (D) in the top of hills with a minimum height of 100 m and an average slope greater than 25°; (D) in areas covered with vegetation containing specimens of endangered fauna or flora; (E) in sites of exceptional beauty or of scientific, cultural or historical value; and in areas which promote public welfare; among others
CONAMA n° 303 resolution, 20/3/2002	Parameters, definitions and limits of Areas of Permanent Protection (APPs) **Article 3**:	APPs constitute areas located: (a) around springs, even if intermittent; (B) on top of hills; (C) in slopes or part thereof, with a declivity greater than 45° on the axis of biggest slope; (D) on escarpments and edges of hill plateaus; (E) in any place covered by mangrove stabilizing vegetation; (F) in mangrove areas; (G) in refuges or reproduction fauna areas containing species threatened with extinction; and (h) on beaches, in nesting and reproduction sites of wildlife
State regulations		
Rio de Janeiro State Constitution	Regulates APPs in Rio de Janeiro State. **Articles 268, 269**	APP areas are: (A) mangroves; (B) beaches and rocky coastlines; (C) springs and surface water protection zones; (D) areas containing rare, threatened, vulnerable or lesser known species; (E) areas of archaeological, historical, scientific, landscape and cultural interest. Areas of relevant ecological interest, defined as native forest cover and coastal zones
Municipal regulations		
Environmental Code: Complementary Law n° 19/2007	Defense, recovery, improvement environmental control. **Articles 3,10,24;25,31**	The objectives of Búzios Municipal Environmental Policy, among others, are to stimulate the creation of new protected areas. Specially protected territorial areas are: APPs; areas with relevant or forested vegetation; hills; beaches, seafront, rocky outcrops and the islands; and rocky shores. APPs are also mangroves and other areas declared in state and federal legislation. The beaches, islands, waterfront and rocky outcrops are areas of landscape protection
Soil Use and Occupation Law Compl. Law 27/2010	Alters Compl. Law14/ 2006 – Soil Use and Occupation. **Article 2, 29, 38**	The Municipal Master Plan must create instruments to protect the natural landscape and preserve the traditional city profile; protect water resources; promote and encourage cultural, historical and archaeological heritage protection, among others. Establishes that ***Mangue de Pedra*** is a Wildlife Conservation Zone, with limitations for human use and occupation. It is forbidden to build retaining walls on the front lots for several beaches, among them Gorda Beach
Master Plan for Sustainable Development – Compl. Law n° 13/2006	Establishes Armação dos Búzios municipal Master Plan. **Articles 3, 4, 15, 76, 78, 80, 82, 83**	Urban development policy shall observe principles for protection and preservation of the territorial sea (. . . .), swamps and mangroves, (. . . .). Environmental and cultural preservation strategies are the APPs and PAs; (. . . .); valorization of geological identity as a patrimony of humanity; archaeological and cultural rescue of prehistoric, indigenous and '*quilombo*' communities. Guidelines are the creation and implementation of the Gorda beach – Ponta do Padre Vitorio Municipal Park. The government will implement such projects to favor its conservation, preservation, or common use
Ammendment to Armação dos Búzios Organic Law n° 10/2015	New wording for Organic Law of Armação dos Búzios 11/11/1997. **Articles 199, 200, 233 e 235**	Búzios cultural heritage are public or private movable and immovable assets of historical, architectural, archaeological, environmental, (. . . .) nature. Environmental control instruments are the maintenance and defense of APPs which, due to their physiographic, geological, hydrological, biological or climatological conditions, form an important ecosystem, especially: mangroves, estuarine areas; springs; areas with endangered species of flora and fauna, as well as those that serve as places for landing, sheltering or breeding species. Government will stimulate creation and maintenance of private conservation units

for environmental projects, some of these entities – such as *FUNBIO* and *SOS Mata Atlântica* – are currently project managers and responsible for the application of environmental compensation resources. It is also worth mentioning the role of the *Mico Leão Dourado Association* (AMLD), which operates in several forums, with the objective of influencing regional territorial planning (http://www.micoleao.org.br/) for the conservation of the golden-tamarin and its ecosystem.

There is, though, little transparency in governmental acts and lack of incentive to popular participation (and even by local public managers), which is worsened by generally restrict access to information (Obraczka et al. 2017). In Rio de Janeiro, a single agency, Environmental Institute of Rio de Janeiro – INEA, concentrates all functions related to planning, management and monitoring of water, forestry, coastal and environmental resources. Regarding environmental licensing, it includes the definition of the type of licensing, scope, and project environmental analysis, issuance of licenses, supervision and monitoring.

17.4.1 Economic and Ecological Zoning and Coastal Areas' Management

In addition to the legal aspects, there are governmental programs for conservation of coastal areas as part of planning and territorial management instruments, available in the vast institutional framework of the country, with federal, state and municipal scope. Some of these mechanisms also have an interface with others, such as Coastal Management and Economic and Ecological Zoning.

Coastal Economic and Ecological Zoning in Rio de Janeiro State is an attribute of INEA. (http://200.20.53.3:8081/Portal/Agendas/GESTAODEAGUAS/Gerenciamento_Costeiro/index.htm&lang=). However, due to scale issues, the scope of this typology does not consider particularities of smaller areas, such as Mangue de Pedra. It is necessary, therefore, the elaboration of a much more specific and detailed zoning, in the local scale.

The Coastal Economic Ecological Zoning of Rio de Janeiro State (ZEEC-RJ) is a management tool provided for in the National Plan of Coastal Management, created to guide the process of land and marine land planning, in accordance with the guidelines of the Economic Ecological Zoning. It was intended as a mechanism to support monitoring, licensing, inspection and management actions, with the participation and permanent consultation of local populations, in compliance with Federal Law 7661/1988, which establishes the National Plan for Coastal Management (PNGC). Decree 5.300 / 04 regulated the PNGC and established the limits, principles, objectives, instruments and competences for the management, as well as the rules of use and occupation of the coastal zone (http://200.20.53.3:8081/Portal/Agendas/GESTAODEAGUAS/Gerenciamento_Costeiro/index.htm&lang=).

The coastal zone of Rio de Janeiro State, according to Federal Decree No. 4.983/2004, includes the legal limits of the territorial sea. The terrestrial strip comprises all municipalities covered by six coastal sectors, with Búzios located in Hydrographic Region -VI (Lagos São João).

ZEEC-RJ will advance progressively, according to a strategy defined by INEA. Among the main action lines for Coastal Management at INEA, projects and strategic actions for fragile coastal ecosystems stand out, including, e.g., the Integrated Management of Ilha Grande Bay Ecosystem Project, the first one to be addressed.

According to Obraczka et al. (2017), despite the threats to the fragile Brazilian coastal ecosystems, coastal management is still incipient in Brazil, especially on a microscale, where it barely influences land use and occupation, urban planning or environmental licensing.

On the other hand, in Búzios a coastal land occupation plan exists, the ORLA Project Intervention Plan. This approach, with a local and more specific method, was based on the General Municipal Intervention Plan of the Ministry of the Environment (MMA 2003). In this instrument, the Gorda Beach was characterized as 'non urbanized, exposed, linear, and with the presence of a rock mangrove' (MMA 2003) and inserted in a landscape unit of the type 'Rocky coastline, rustic, (although there are approved lots) interspersed by small, sheltered, rustic beaches'.

According to the MMA (2003), the Intervention Plan aims to 'promote the sustainable management within municipal limits, through a gradual approach beginning in priority areas', seeking to develop and implement actions aimed at ensuring the 'long-term environment preservation and the coastal strip use planning in Búzios'. Short-term goals comprise the inclusion, of the Report proposals in the Municipal Master Plan (MMA 2003). It would allow future interventions in the waterway possible, as well as the ordering of their use, guaranteeing the basic principle of democratization in Use of the border and guarantee access to these public spaces. This is because 'the need to reconcile the two processes is clear, since the Master Plan is nothing more than the search for a sustainable development of the entire Buzian territory and the creation of the legal mechanisms necessary for its implementation'. However, this proposal has not yet been implemented, even though some coastal aspects have been mentioned in more recent instruments such as the Municipal Environmental Code.

Table 17.6 Regulatory instruments that regulate direct or restrict land use in Búzios and more specifically the area of Mangue de Pedra

Tipology	Normative instrument (management/planning)	Sphere of power	Restrictive character for use/occupation
Planning and urban restrictions	Soil Use and Occupation Law	Municipal	Yes
	Master Plan for Sustainable Development	Municipal	Directives
	City Statute	Federal	No
Coastal management	Coastal Management National Plan (PNGC)	Federal	No
	Coastal Ecological Economic Zoning (ZEEC-RJ)	Federal and state	Directives
	ORLA Project Intervention Plan (plan for the shoreline intervention in Búzios)	Municipal	Directives
Environmental	Atlantic Rainforest Plan	Federal	Directives
	CONAMA 303/02 Resolution	Federal	Yes
	Forest Code	Federal	Yes
	Búzios Environmental Code	Municipal	Directives
	National Plan for Protected Areas	Federal	Directives
Water resources	Watershed Plan: Cabo Frio-Búzios Region	State	No
	National Water Resources Council Resolution (CNRH 51/2005)	Federal	No
	Rio de Janeiro State Water Resources Plan (PERHI)	State	No
General	Economic Ecological Zoning	State	No

Source: Adapted from Obraczka (2014)

17.5 Aspects Dealing with Territorial Integrated Management

There is a wide range of normative, management and planning instruments, directly or indirectly affecting the soil use in general in Búzios, and in Gorda beach's Mangue de Pedra in particular. According to Obraczka (2014), taking Búzios as an example, depending on the location of a certain area, there may be several federal, state or municipal laws and regulations affecting the same part of the territory that influence occupation and soil use (Table 17.6).

Multiple legal and normative mechanisms result in overlapping and often even incompatibilities on the same aspect or area (Obraczka 2014). This is the case of protected areas such as hilltops by the Forest Code (CONAMA Resolution 303/02) not considered as such in the relevant municipal legislation (Soil Use and Occupation Law or Sustainable Use Master Plan).

17.6 Proposals for Land Use: Fate of the Subject Area

Over the last two decades, there have been several proposals for occupation of the study area, almost all related somehow to large real estate projects, comprising large-scale land occupation, as well as infrastructure works and other interventions that severely modify the area (Fig. 17.9). Based on an allotment approved at the time Búzios belonged to the municipality of Cabo Frio, over the years some of these lots were occupied by large buildings since public administrators interpret the right to build is guaranteed by the legislation.

Several attempts to protect he area have been presented in the last two decades. The Búzios APP mapping, carried out by the Department of Mineral Resources of Rio de Janeiro State (DRM-RJ) in 2006, at the request of the City Hall and the Chamber of Municipal Councilors was one of them. The importance of Mangue de Pedra was highlighted and the area was classified as a rocky coast, mangrove, springs and plateau borders, in addition to nutrient convergence area from the mainland to the sea. Recommendation to preserve natural conditions in the area followed suit, resulting in areas prohibited to human occupation.

17.6.1 'Municipal Natural Monument' Proposal (Secretary of Environment and Fisheries: 2012)

In 2012, Búzios Secretary of Environment and Fisheries proposed the creation of a Municipal Conservation Unit in Mangue de Pedra area, with about 50 ha. The first area delimited to the PA included vegetation areas of sandbars, mangrove vegetation and deciduous forest vegetation, as well as (vegetation in) hilltops. The PA proposal represented a Natural Monument type, belonging to the integral protection group, according to the National System of Conservation Units (SNUC) classification. Subsequent studies showed the

a b

Fig. 17.9 Environmental impacts in Mangue de Pedra recharge basin, such as erosion (**a**) and suppression of vegetation (**b**) (Photos: Nicole D'Allincourt)

area proposed for the PA did not comprehend all protectable areas according to opinions of municipal technicians and experts. Another important finding by Obraczka (2012) is that the proposal to create this unit was not unanimous among the local population, since it would perceivably affect several apparently conflicting interests, including entrepreneurs, environmentalists, and even 'quilombolas' who claimed land rights over much of the area.

The discussion was interrupted by the licensing of a real estate project in the mangrove surroundings, named 'Gran Riserva 95'. The project aimed at the construction of five villages, with 221 twinned houses, in a condominium fashion enterprise. The works began with a series of direct environmental damages (removal of vegetation, construction of foundations on springs, cuts and landfills) and even collateral impacts, such as the removal of rolled pebbles from the beach (Fig. 17.10).

In 2012, after receiving a complaint of popular initiative, the State Public Prosecutor's Office distributed a precautionary action requesting the suspension of the works and the prohibition of sales of the Gran Riserva 95 condominium units and the judge granted an injunction. The ruling was based on the arguments that no hydrogeological study was presented, nor any geological survey supporting the slope use to guarantee the integrity of Mangue de Pedra.

17.6.2 Geopark Project 'Costões e Lagunas do RJ'

According to Mansur et al. (2012), the proposed Geopark 'Costões e Lagunas' of Rio de Janeiro State has an area of approximately 10,900 km^2 with a population of 1,585,000 inhabitants. Its scenic beauty, special cultural importance (historic and prehistoric sites, such as 'sambaquis', ancient indian settlements), endemic flora and fauna, as well as relevant geosites make it crucial to natural heritage conservation. Institutions at the federal, state and municipal levels, as well as NGOs, have been working since 2011 on the organization of information to submit an application dossier to UNESCO Global Geoparks Network.

In this proposal, the Mangue de Pedra and palaeocliffs of Gorda beach stand out as geosites of high scientific relevance. The municipality of Búzios presented in 2014 a support letter to the Geopark. In 2016, as part of the Geopark project, teachers from Búzios education network were given courses on concepts of geology and their application in Búzios, including fieldwork in Mangue de Pedra (Silveira 2017; Nascimento et al. 2016).

17.6.3 Project of the Gorda Beach State Park (NGO)

In 2012, the proposal for the creation of the Gorda beach State Park was presented through a parliamentary initiative based on a request from a municipal NGO. It was motivated by the innumerable findings of environmental damage that occurred in the area of Mangue de Pedra, especially since the beginning of the implementation of the Gran Riserva project and the failure of the municipal construction project that had been initiated (Mehdi and Souza 2012).

Fig. 17.10 Removal of quartz pebbles on a large stretch of sand beach near Ponta do Pai Vitório (Source: Mehdi and Souza 2012)

17.6.4 Inclusion in Costa do Sol State Park (PECS)

Another proposal aiming at protection and preservation of the Mangue de Pedra and surroundings was the claim by civil society and researchers to inclusion in the PECS protected areas mosaic, created by State Decree n° 42.929/2011 (https://www.jusbrasil.com.br/diarios/26496887/pg-1-parte-i-poder-executivo-diario-oficial-do-estado-do-rio-de-janeiro-doerj-de-19-04-2011). This PA has a total area of 9841 ha and covers a set of areas in coastal municipalities from Araruama to Búzios (http: //www.inepac.rj.gov.br/index.php/bens_tombados/detalhar/53). This inclusion project, although still possible, did not advance, even though there is a close proximity to the Mangue in relation to the area delimited by the Park (Fig. 17.11).

Fig. 17.11 Examples of popular initiatives aiming at the protection of areas such as Mangue de Pedra such as inclusion in PECS mosaic (Photo: https://www.facebook.com/MangueDePedraDeBuzios)

17.7 Public and Shared Management

Based on the successful experience of California (USA) on the cities and counties shared coastal management, according to a decentralized, transverse and shared institutional arrangement, Obraczka (2014) and Obraczka et al. (2017), carried

Table 17.7 Environmental and coastal management in Rio de Janeiro (Búzios): a comparative approach in light of the coastal management experience in California (USA)

Evaluated aspect/ criteria	Rio de Janeiro (Búzios)	Califórnia (USA)
General	Generally no management of coastal zone. Only urban and, eventually, environmental licensing applies	Environmental, urban and coastal licensing is not fragmented, evaluated from the coastal standpoint, including issues such as public fruition, access to the coast, and protection of ecologically important sites, among others
	Coastal strip not treated as a municipal territory, especially when managed by the Navy, with a conflict of jurisdiction between the union and local managers	Licensing is decentralized and shared between various entities and agencies. The lead agency for the process may vary depending on the type and location of the venture
	INEA is the water, forestry, environmental and coastal resources manager, responsible for licensing and monitoring	
Regulatory frameworks and support legislation	Many instances overlap	Coastal counties plan, legislate, license and oversee the use and occupation of coastal territories, provided they have their Local Coastal Management Plans (LCMPs) approved by the California Coastal Commission (CCC)
	There is no legislation on the coast, only a soil zoning carried out by the central power (Union)	Regulatory and legal framework by the California Environmental Quality Act (CEQA), the Coastal Law and the California Coastal Zone Plan, as well as other pertinent federal and state laws such as the ESHA (Endangered Species Habitat Act)
Participation of society/access to information	Society participation restricted to a single event (Public Hearing) in environmental licensing. It only occurs in licensing processes of projects considered as of big impact by the environmental body. Licenses are frequently questioned in court	There is a broad incentive for society to participate throughout the licensing process, including making it available and facilitating access to information and data on it
Management/ protection resources	Depend exclusively on investments and public resources	Coastal nonprofit agencies (such as CSCC) empowered by legislation have the autonomy and resources to invest primarily in preservation/protection/recovery of coastal sites, aiming to guarantee their enjoyment and public use
	Municipalities in general do not have autonomy and/or resources for monitoring the coast	
New tendencies	Unsuccessful attempts of inclusion and participation of coastal municipalities in coastal planning and management, such as the ORLA project	Latest aspects and concepts such as EBM (Ecosystem Based Management) and those related to climate change and sea level rise incorporated into coastal planning and management

Source: adapted from Obraczka et al. (2017) and Obraczka (2014)

out an extensive comparative analysis between that region and Búzios, particularly Mangue de Pedra. The result is shown in Table 17.7. The objective was to identify some actions successfully used in California that could be applied in Búzios, aiming at more effective planning and management of the coastal zone, especially to protect Mangue de Pedra's natural heritage, of great scenic and environmental relevance.

California central coast presents a series of similarities with Rio de Janeiro State coast, which justifies the use of this comparative methodology. The great importance of tourism in the local and regional economy; existence of a strong oil industry as well as real estate activities; an increasing trend of population concentration along the coast with a high seasonal variation due to vacation periods. On the other hand, the existence of relevant areas and sites from geo and biodiversity standpoint and of fragile and threatened ecosystems by intense economic activity (Obraczka et al. 2017).

Hence, it is a challenge to implement a true sustainable development in the two evaluated cases, considering the existence of activities that, at first sight, seem to be conflicting (or even excluding). This would be the case with, e.g., tourism, construction works and the oil industry, in one hand, and, on the other, the need for environmental preservation.

It is possible to disclose the presence of vital elements in California reality – aspects that are still incipient or even absent in the Rio scenery – that somehow explain the differences in terms of results obtained by the US state in the protection of much of its coast, especially when compared to Brazil. Regarding applicability of some of these actions in Rio de Janeiro, the effective decentralization of the licensing process, including making it more transparent and participatory, is possible to implement. In addition, it is feasible to promote incorporation of the Coastal Management's tools (ZEEC-RJ, ORLA Project and others) into municipal legal framework, especially regarding the planning and process of urban and environmental licensing.

17.8 Management Strategies and Actions for the Preservation of the Study Area

Even with the advent of a formal legal protection system, initiatives to preserve heritage areas are still eventually unsuccessful in Rio de Janeiro State. Many of those areas are privately owned, with a high market value. Economic pressures to their conversion into so-called 'productive' areas are difficult to contain. When it comes to areas owned by the government, lack of resources for land regulation, supervision and monitoring may sometimes lead to misuse. Such illegal activities often lead to the eradication and destruction of the ecosystems.

The main activities that boost Búzios municipal economy are construction and real estate market, fueled by tourism. Due to a high demand, in some places of the city land price is comparable to some of the most expensive in the country. Thus, preservation of Búzios land, despite the intrinsic importance of its rare and fragile ecosystems, and the very survival of its tourist attractiveness, that ultimately drives economy, is expensive. Preservation initiatives fell to the State government since some of them were included in the State preservation areas and others run by Rio de Janeiro's State Institute of Artistic and Cultural Heritage (INEPAC).

Locally, few and small areas were established as municipal conservation units (Lagoinha Park, Marine Corals Park, Geribá Lagoon Park, Búzios Environmental Protection Area and Artisanal Fisheries) and were never effectively implemented. Other important areas such as the dunes of Tucuns, Emerências Hills, sandbars and remnants of endemic and threatened vegetation, besides Gorda beach's Mangue de Pedra, have already been destroyed or remain unprotected.

In a scenario where public investments to expropriate relevant and endangered private natural areas is not expected in the short and even medium terms, the main challenge is to find means to achieve preservation without buying ownership of land.

17.8.1 Possible Scenarios for the Study Area

In order to make viable a better evaluation of the current and future situation, five possible scenarios taken from the present baseline are presented (see Table 17.8):

1. Inertial scenario: no initiatives in the area, both from environmental initiatives and implementation of real estate development standpoint;
2. Total conversion scenario: total or near total modification of the area with implementation of one or more real estate, development projects;
3. Partial Protection Scenario (type A) area: a smallest part of the area destined to the development and occupation, while the rest is destined for preservation/environmental initiatives;
4. Partial Protection Scenario (type B) area: most of the area is destined to a controlled occupation, and a minor part intended for preservation;
5. Integral Protection: the whole area is destined to environmental protection through a decree of Integral Protection Preservation Area (IP-PA).

In view of the previous discussion, providing minimally necessary environmental protection for the study area can only be achieved through two scenarios: (1) total protection through the creation of an integral type PA (IP-PA); or, at least, (2) implementation of a single enterprise, with low territorial demand, pending a detailed study to define environmental impact. However, Mangue de Pedra aquifer recharge areas have already been occupied, as well as being classified as APPs (DRM 2006). Second hypothesis thus could not be implemented according to environmental legislation. Considering the scenarios described above, some of the ongoing and other possible actions to preserving the area are presented.

17.8.2 Popularization of Knowledge; Environmental Awareness/Education

Environmental education and appreciation of local history and tradition are basic strategic guidelines mentioned repeatedly in municipal planning legislation. In the case of Mangue de Pedra and Gorda beach, one of the most important strategies used to preserve it is promoting public awareness of its unique environmental and natural heritage, and urgency of preservation. Relevant socio-cultural and historical aspects related to the traditional population living and/or using the area, such as "quilombolas" and artisanal fishermen are also roused. This approach is used in dissemination of scientific knowledge, through a simple language, adapted to the reality of the population, with graphs and diagrams, in some cases posted on the site of interest. As an example, in 2016 a panel was installed in the mangrove area, informing about the geological origin of Mangue de Pedra, in accessible language (Fig. 17.12), as part of the 'Geological Paths' Project (Mansur and Erthal 2004), aiming to disseminate Rio de Janeiro geology, promoting geotourism and geoconservation of geological monuments.

Diffusion and public awareness is also pursued through several meetings and lectures held over the last few years, for various types of public, some meeting demands of both the

Table 17.8 Advantages and shortcomings related to environmental, economic and social aspects of different scenarios, destinations for the study area

Possible scenarios; land use	Economic aspects: resource investments; generation of income	Environmental and natural resource and landscape protection aspects	Legal and institutional aspects: surveillance, enforcement	Other observations
(1) Inertial: current situation, no further actions	No public or private investment- low or nonexistent income generation	Continuity of *degradation tendency*; mischaracterizing and loss of heritage; disordered occupation, environmental impacts	Rising public *enforcement demands* as a consequence/ function of illegal occupation attempts, with minimum chances of success	Gradual loss of *socio-cultural*, historical and natural heritage
	Area losses value			
(2) Total conversion: approval and implementation of allotment project	No public investment needed;	Loss of natural heritage;	Illegal or informal occupation ceases	Socio-cultural, historical heritage may be completely lost
	Rising income from taxes to municipality;	Minimum preservation of some remaining natural areas considering occupation restrictions of the urban planning legislation to this area	Occupation, in this scenario, tends to restrict public use of the area, coastal sites	
	Job generation in construction and domestic jobs (gardening, housekeeping)		Occupation rate remains stable, must be enforced by government in non-private areas	
(3) Partial protection (type A)[a] partial transformation into a PA	Partial public investment (enforcement and surveillance of the protected area)	Preservation of some remaining natural areas considering low occupation rate	Occupation rate remains stable, must be enforced by government on the remaining public/ natural areas	Depends on size and location of PA
	Partial rising income from taxes	Partial loss of natural heritage;		Weak point: possibility to establish inadequate zoning and to build in inappropriate areas
	Partial job generation in construction and domestic works			Socio-cultural, historical heritage may be partially lost
(4) Partial protection (type B)[a]: transformation into a minor housing project (compare to scenario (2)	None or small public investment (enforcement and surveillance	Partial loss of natural heritage;	Occupation, in this scenario, tends to restrict public use of the area	Socio-cultural, historical heritage may be completely lost
		Increasing edge effect	Occupation rate remains stable, must be	

(continued)

Table 17.8 (continued)

Possible scenarios; land use	Economic aspects: resource investments; generation of income	Environmental and natural resource and landscape protection aspects	Legal and institutional aspects: surveillance, enforcement	Other observations
	of the protected area)		enforced by government	
	Rising public incomes from taxes			
	Job generation/ opportunities in construction and domestic jobs (gardening, housekeeping)			
(5) Integral protection: all area turned into a fully protected area (IP-PA)	Big public investment (expropriation)	Maintenance and recovery of natural heritage	Need to rise public investment to monitoring, surveillance and restoring	Better scenario form environmental standpoint
	Natural monument: no need to expropriate	Protection of natural areas through projects	Part of resources can be obtained through partnerships with companies, NGOs, etc.	Worse scenario regarding public expenditure
	Income generation from tourism			

[a]Type A – most of the area of interest is preserved/ protected; Type B – most of the area of interest is built and used by the development project

a

b

Fig. 17.12 Detail of interpretative panel from Mangue de Pedra 'Geological Interest Point' (**a**). Field classes for Búzios public school teachers (**b**) (Photo: Kátia Mansur)

government and local civil society entities. In these meetings, in addition to the main objective of increasing knowledge about the relevance of the area and its preservation, a debate on alternatives for its sustainable use, based on low impact activities such as ecological, scientific and adventure tourism, has been encouraged. Other materials such as leaflets have been drafted for the area.

The law enforcement system, both regarding protected areas and compliance on building and urban restrictions, would itself result in the protection of several remnants and important sites. Also, in coastal areas, management and planning actions can be useful. It must therefore be incentivized and valued. Current municipal coastal management, when occurs, is basically used to meet the tourist and commercial demands at the beach, with implementation of services such as kiosks, boardwalks and other similar structures.

In some U.S. PAs, such as national parks, guidelines for the use allow tourist visitation in predefined areas, based on their support capacity. Parks have a large public service infrastructure, and visitors pay a fee to use it. In order to contribute to their own sustainability, an adapted version of this model could be used in many UC's in Brazil.

Other possibilities include crowdfunding operation, like Awaroa Beach, located in Abel Tasman National Park on South Island (one of the two largest in New Zealand). The site was purchased after a crowdfunding campaign raised $ 1.7 million and integrated to a local National Park. (http: //g1.Globo.com/mundo/noticia/2016/07/apos-vaquinha-neozelandeses-compram-praia-paradisiaca-de-empresario-ea-entregam-ao-publico.html).

California coastal agencies, such as the CCSC, are examples of successful non-governmental and nonprofit activities in coastal management and the protection of environmental and scenic heritage on its central coast. In recent decades, through various initiatives and actions, including the direct acquisition of priority areas for conservation, they have managed to safeguard important portions of the California coast. Unlike Brazil, they are 'empowered' not only by their support of civil society, but also by coastal and environmental legislation itself, being formally part of the governance system.

In order to maximize the protection of threatened sites such as Mangue de Pedra, some of those experiences could be successfully implemented in Búzios, as well as in many others endangered coastal ecosystems in Rio de Janeiro and Brazil.

17.9 Concluding Remarks

Considering all the factors discussed throughout the chapter, Mangue de Pedra area has a high environmental and scientific relevance and, with these attributes, must be preserved. The current conditions for the management of this invaluable geoheritage area have not allowed its appropriate protection and, thus, a gradual loss of bio and local geodiversity has been observed.

The area, by federal, state and municipal laws and regulations, has a list of characteristics that should lead to its protection, regardless of the creation of a PA or other form of legal protection. In practice, there is no systematic and definitive path to prevent invasions, construction of urban infrastructure and homes around the ecosystem, and its degradation. NGOs and the population in general have been demanding concrete actions from the public power to protect the area, with no success.

Possibilities for solution include the creation of a comprehensive protection unit, environmental education, society mobilization and participation in decision forums, the shift of the current administration paradigm of public PAs, among others.

References

Ab'Saber AN (1974) O domínio morfoclimático semi-árido das caatingas brasileiras. Geomorfologia 43:1–39

Accioli NT (2013) José Gonçalves da Silva à Nação Brasileira: O Tráfico ilegal de escravos no antigo Cabo Frio. Ministério da Cultura. Programa Nacional de Apoio à Pesquisa – FBN/MinC. 124p

Almeida FFM, Carneiro CDR (1998) Origem e Evolução da Serra do Mar. Rev. Bras. Geociências. São Paulo. SBG, 28, 2, 135–150

Alves JRP (2001) Manguezais: Educar para proteger. Projeto PLANÁGUA SEMADS – GTZ. ISBN 85-85966-21-1, 96p

Arai M (2006) A Grande Elevação Eustática do Mioceno e Sua Influência na Origem do Grupo Barreiras. Geologia USP. São Paulo – SP. Série Científica 6(2):1–6

Brasileiro PS, Yoneshigue-Valentin Y, Bahia RG, Reis RP, Amado Filho GM (2009) Algas Marinhas Bentônicas da Região de Cabo Frio e Arredores: Síntese do Conhecimento. Rodriguésia 60 (1):039–066

Carvalho DW (2011) A zona costeira brasileira e o gerenciamento dos danos ambientais futuros. Revista Lusíada: Direito e Ambiente, Portugal, v. esp, pp 97–111

Cintrón G, Schaeffer-Novelli Y (1983) Introdución a la ecología del manglar. UNESCO/ROSTLAC, Montevideo. 109p

Cordani UG, Milani EJ, Thomaz-Filho A, Campos DA (2000) Tectonic evolution of South America, Rio de Janeiro. 31st International Geological Congress, pp 287–310

Davis EG, Naghettini MC (2001) Estudo de chuvas intensas no Estado do Rio de Janeiro. In: CPRM. Projeto Rio de Janeiro. CPRM/ Embrapa Solos/DRM-RJ, 2001, CD-ROM, Brasília

Dias FF (2009) Variações do Nível relativo do Mar na Planície Costeira de Cabo Frio e Armação dos Búzios – RJ: Reconstrução Paleoambiental Holocênica e Cenários Futuros. Tese de Dourado, 164p. Programa de Pós-graduação em Geologia, IGEO, UFRJ

DRM-RJ – Departamento de Recursos Minerais (2006) Parecer sobre delimitação de Áreas de Preservação Permanente no município de Armação dos Búzios – RJ. Relatório. 26p

Estado do Rio de Janeiro (2011) Decreto n° 42.929 de 18 de abril de 2011 – Cria o Parque Estadual da Costa do Sol e dá outras providências. https://www.jusbrasil.com.br/diarios/26496887/pg-1-parte-i-poder-executivo-diario-oficial-do-estado-do-rio-de-janeiro-doerj-de-19-04-2011

FAO – Food and Agriculture Organization of United Nations (2007) The world's mangroves 1980–2005. FAO Forestry Paper 153, Rome. ISBN 978-92-5-105856-5. 77p

García-Cortés AG (1996) Inventario del Patrimonio Geológico. In: Ministerio de Obras Públicas, Transportes y Medio Ambiente, Sociedad Española de Geologia Ambiental y Ordenación del Territorio, Comissión de Patrimonio Geológico de la Sociedad Geológica de España (eds) El patrimonio geológico. Bases para su valoración, protección, conservación y utilización. Serie monografías, Madrid, pp 53–60

Giri C, Ochieng E, Tieszen LL, Zhu Z, Singh A, Loveland T, Masek J, Duke N (2011) Status and distribution of mangrove forests of the world using earth observation satellite data. Glob Ecol Biogeogr 20(1): 154–159, 17 https://doi.org/10.1111/j.1466-8238.2010.00584.x

Gray M (2013) Geodiversity: valuing and conserving abiotic nature, 2nd edn. Wiley-Blackwell, Chichester. 508p. ISBN: 978-0-470-74215-0

Heilbron M, Mohriak W, Valeriano CM, Milani E, Almeida JCH, Tupinambá M (2000) From collision to extension: the roots of the south-eastern continental margin of Brazil. In: Talwani, Mohriak (eds) Atlantic rifts and continental margins, Geophysical Monograph Series, vol 115. American Geophysical Union, Washington, DC, pp 1–34

Heilbron M, Pedrosa-Soares, AC, Campos-Neto M, Silva LC, Trouw RAJ, Janasi VC (2004) A Província Mantiqueira. In: Mantesso-Neto V, Bartorelli A, Carneiro CDR, Brito-Neves BB (eds) O Desvendar de um Continente: a Moderna Geologia da América do Sul e o Legado da Obra de Fernando Flávio Marques de Almeida, Cap XIII. Beca Produções Culturais Ltda, São Paulo, pp 203–234

IBGE – Instituto Brasileiro de Geografia e Estatística (2011) Atlas geográfico das zonas costeiras e oceânicas do Brasil/IBGE, Diretoria de Geociências. IBGE, Rio de Janeiro. 176p. ISBN 978-85-240-4219-5

Mansur KL (2010) Diretrizes para Geoconservação do Patrimônio Geológico do Estado do Rio de Janeiro. O Caso do Domínio Tectônico Cabo Frio. Tese de Doutorado em Geologia. Universidade Federal do Rio de Janeiro/UFRJ, Brasil. 240p

Mansur KL, Erthal FLC (2004) Projeto Caminhos Geológicos e seus Desdobramentos no Estado do Rio de Janeiro. Anais do 42° Congresso Brasileiro de Geologia (CD ROM). SBG, Araxá

Mansur KL, Guedes E, Alves M, Nascimento V, Pressi L, Costa N, Pessanha A, Nascimento L, Vasconcelos G (2012) Geoparques Costões e Lagunas do Estado do Rio de Janeiro (RJ). In: Schobbenhaus C, Silva CR (eds) Geoparques do Brasil: propostas, vol 1. CPRM, Rio de Janeiro, pp 687–745

Mehdi AR, Souza R (2012) Projeto de Lei Parque Estadual do Mangue de Pedra – Praia da Gorda. Armação dos Búzios, Brasil. http://docslide.com.br/download/link/mangue-de-pedra-de-buzios

MMA – Ministério do Meio Ambiente (2003) Projeto de Gestão Integrada da Orla Marítima – PROJETO ORLA. Plano de Intervenção na Orla do Município de Armação dos Búzios – RJ março/2003

MMA – Ministério do Meio Ambiente (2008) Macro diagnóstico da Zona Costeira e Marinha do Brasil. Brasília. 242p. ISBN978-85-7738-112-8

MMA – Ministério do Meio Ambiente (2017) Mata Atlântica. http://www.mma.gov.br/biomas/mata-atlantica

Mohriak WU, Barros AZ (1990) Novas evidências de tectonismo cenozóico na região sudeste do Brasil: o gráben de Barra de São João na plataforma continental de Cabo Frio, Rio de Janeiro. Revista Brasileira de Geociências 20(1–4):187–196

Morais RMO, Mello CL, Costa FO, Santos PF (2006) Fácies Sedimentares e Ambientes Deposicionais Associados aos Depósitos da Formação Barreiras no Estado do Rio de Janeiro. Geologia USP, São Paulo, Sér. Cient 6(2):19–30

Nascimento VMR, Almeida CN, Mansur KL, Souza MM, Rosa DAR (2016) Capacitação geológica de docentes no território do Geoparque Costões e Lagunas do Estado do Rio de Janeiro – RJ: metodologia e resultados preliminares. In: Congresso Brasileiro de Geologia, 48, 2016, Porto Alegre – RS. Anais. Sociedade Brasileira de Geologia, São Paulo. 2016. http://sbgeo.org.br

Obraczka M (2012) Parecer ambiental sobre consulta prévia quanto a viabilidade de se implantar um empreendimento através de parcelamento e construção em uma área especifica situada no Município de Armação dos Búzios

Obraczka M (2014) Licenciamento ambiental e Gerenciamento Costeiro no Estado do Rio de Janeiro: Propostas de Aperfeiçoamento do Modelo Vigente à Luz da Experiência da Califórnia (EUA) Tese de Doutorado. UFRJ/COPPE, 170p

Obraczka M, Beyeler M, Magrini A, Legey LF (2017) Analysis of coastal environmental management practices in subregions of California and Brazil. J Coast Res 33:1315–1332. https://doi.org/10.2112/JCOASTRES-D-15-00239.1

Oliveira RF (2003) Vegetação de Búzios: O Passado no Presente. Parecer 3: Aspectos Biológicos. Processo E-18/1337/2003. Tombamento de duas áreas naturais em Armação dos Búzios. INEPAC, pp 74–77

Oliveira AES (2007) Caracterização do Mangue da Praia Gorda, Armação dos Búzios, Estado do Rio de Janeiro – RJ. Revista Cient. Cent. Univ. Barra Mansa – UBM. Barra Mansa – RJ 9(17):51–56

Polette M (2008) A Zona Costeira em Crise: Opções de Gestão Integrada e Participativa. UNIVALI – CTT Mar Florianópolis – SC. http://nmd.ufsc.br/files/2011/05/1a_zona_costeira_em_crise_1_polette.pdf

Prates APL, Gonçalves MA, Rosa MR (2010) Panorama da conservação dos ecossistemas costeiros e marinhos no Brasil. Secretaria de Biodiversidade e Florestas/Gerência de Biodiversidade Aquática e Recursos Pesqueiros. MMA/SBF/GBA, Brasília. ISBN 978-85-7738-142-5. 148 p

Rebelo VALC (2013) Caracterização Hidrogeológica e Hidrogeoquímica da região do Mangue de Pedra (Armação dos Búzios, RJ). Trabalho Final de Curso (Bacharel em Geologia). Universidade Federal do Rio de Janeiro, Instituto de Geociências, Departamento de Geologia, 171p

Rubin IN, Almeida JCH (2003) Petrografia de Brechas Tectônicas – Um Modelo de Descrição e Classificação. In: 3°International Symposium on Tectonics. Armação dos Búzios. Boletim de Resumos. SBG. v.1, pp 149–152

Saenger PE, Hegerl EJ, JDS D (eds) (1983) Global status of mangrove ecosystems, Commission on Ecology papers, No. 3. I.U.C.N, Gland. The Environmentalist 3(Suppl. 3):1–88

Schaeffer-Novelli Y, Cintrón-Molero G, Soares MLG, De-Rosa T (2000) Brazilian mangroves. Aquat Ecosyst Health Manag 3 (2000):561–570

Schmitt RS (2001) A orogenia Búzios – caracterização de um evento tectono-metamórfico no Domínio Tectônico Cabo Frio – sudeste da Faixa Ribeira. Tese de Doutorado. Curso de Pós-Graduação em Geologia, Departamento de Geologia, Universidade Federal do Rio de Janeiro. 271p

Schmitt RS, Trouw RAJ, Van-Schumus WR, Pimentel MM (2004) Late amalgamation in the central part of Western Gondwana: new geochronological date and the characterization of a Cambrian collision orogeny in the Ribeira Belt (SE, Brazil). Precambrian Res 133:29–61

Schmitt RS, Trouw RAJ, Van Schmus WR, Passchier CW (2008a) Cambrian orogeny in the Ribeira Belt (SE Brazil) and correlations within West Gondwana: ties that bind underwater. In: Pankhurst RJ, RAJ T, Brito Neves BB, De Wit JJ (eds) West Gondwana: pre-cenozoic correlations across the South Atlantic region, Special Publications, vol 294. Geological Society of London, London, pp 279–296

Schmitt RS, Trouw RAJ, Medeiros SR, Dantas EL (2008b) Age and geotectonic setting of a late-neoproterozoic amphibolithe and paragneiss association from southeastern Brazil based on geochemistry and Sm-Nd data. Gondwana Res 13:502–515

Seabra VS, Silva GC Jr, Cruz CBM (2009) The use of geoprocessing to assess vulnerability on the east coast aquifers of Rio de Janeiro State, Brazil. Environ Geol, Berlin 57(3):665–674

Silva Jr GC (2003) Relatório Final do Projeto Estudo de aqüíferos costeiros no leste do Estado do Rio de Janeiro. (Proc. Número 478975/2001-5). http://www.ana.gov.br

Silveira SMT (2017) Utilização dos Super Feras, mascotes do Geoparque Costões e Lagunas, no ensino e aprendizagem da Geologia em Armação dos Búzios – RJ. Monografia. Programa de Especialização em Geologia do Quaternário, Museu Nacional, Universidade Federal do Rio de Janeiro. UFRJ. 88p

Souza DS, Mansur KL, Gonçalves JB, Arruda AR Jr, Manosso FC (2017) Quantitative assessment of geodiversity and urban growth impacts in Armação dos Búzios, Rio de Janeiro, Brazil. Appl Geogr 85:184–195

Spalding M, Kainuma M, Collins L (2010) World atlas of mangroves. A collaborative project of ITTO, ISME, FAO, UNEP-WCMC, UNESCO-MAB, UNU-INWEH and TNC. Earthscan, London. 319 p

Tetzner W (2002) Tectônica, petrografia e geoquímica dos diques toleíticos do Cabo de Búzios (RJ). Dissertação de Mestrado. UERJ, 88p

Toledo PP, Motoki A, Sichel S, Guimarães PV, Silva FL, Palermo N, Pereira RM, Meneses PT (2012) Geologia e recursos minerais da folha Rio das Ostras SF.24-Y-A-IV e Cabo Frio SF.23-Z-B-VI, estado do Rio de Janeiro escala 1:100.000. CPRM, Belo Horizonte. 202p

Turc, B, Martin L, Flexor JM, Suguio K, Pierre C, Tasayco Ortega L (1999) Origin and evolution of the quaternary coastal plain between Guaratiba and Cabo Frio, State of Rio de Janeiro, Brazil. In: Knoopers B, Bidone ED, Abraão JJ (eds) Environmental geochemistry of coastal lagoon systems of Rio de Janeiro, Brazil, Geoquímica Ambiental. UFF, Rio de Janeiro, pp 25–46

Webber M, Calumpong H, Ferreira B, Granek E, Green S, Ruwa R, Soares M (2016) Mangroves. First global integrated marine assessment. Chapter 48. United Nations, pp 1–18. Disponível em http://www.un.org/depts/los/global_reporting/WOA_RPROC/Chapter_48.pdf

Xavier MAP (2006) Búzios: Estética, poder e território. Dissertação de Mestrado do Programa de Pós-Graduação em Planejamento Urbano e Regional da Universidade Federal do Rio de Janeiro – UFRJ. 243p

Assessment and Management Strategies of Mangrove Forests Alongside the Mangsalut River Basin (Brunei Darussalam, on the Island of Borneo)

18

Shafi Noor Islam, Nurin Hidayati Hj Abd Rahman, Sandra Reinstädtler, and Mohd Nur Azri Bin Aladin

Abstract

The benefits derived from mangrove forest wetland ecosystems are garnering increasing attention in coastal ecological research and mangrove forest management planning. However, because of their location, climate change issues, land use, and landscapes, coastal mangroves are vulnerable and suffer varying levels of stress and disturbance. There is often a variable and uncertain relationship between vulnerability and ecosystem structure and functional services. Degradation and vulnerability assessment and analysis can provide strategic planning initiatives with valuable insight into the processes of functional change resulting from management intervention. The study was conducted on the mangrove management and degradation situation in Brunei Darussalam. The mangrove forest covers about 4% of the country land area and has been recognized as one of the virgin management zones in the Asia Pacific region, but the present status and climate change impacts are degrading the coastal mangrove ecosystem. In addition, urbanization, settlement development, and industrial development contribute to the degradation. This study seeks the strategic management plan for the protection of coastal mangrove wetlands ecosystem in Brunei Darussalam.

Keywords

Mangrove · Degradation · Assessment · Coastal ecology · Strategic planning and management

18.1 Introduction

Brunei Darussalam is one of a small country on an island as Borneo. The northern part of Brunei Darussalam is mostly coastal as Brunei located on the north of Borneo Island facing the South China Sea. Thiessen stated that the state of mangrove forest in Brunei Bay especially on the northern of Temburong district makes Brunei Darussalam known as the largest mangroves in Southeast Asia with approximately 4% of its surface area covered by intact mangrove forest. The mangrove forest wetlands ecosystem in Brunei Darussalam has been recognized as one of the virgin management in the Asia Pacific region, but the present status and climate change impacts are forcing to degrade the coastal mangrove ecosystem (Steenberg et al. 2017). In addition the urbanization, settlement development and industrial development are due for the degradation. The benefits derived from mangrove forest wetland ecosystems are garnering increasing attention in coastal ecological research and mangrove forest management planning (JPH 2003d). However of their location, climate change issues, land use and landscapes of coastal mangroves are shifting as a vulnerable vegetation and community suffer disproportionately and varying levels of stress and disturbance (Steenberg et al. 2017).

There was often a variable and uncertain relationship between vulnerability and ecosystem structure and functional services (Steenberg et al. 2017; Thiessen 2017). Hamid (2000) stated that the most part of the mangrove forest in Brunei Darussalam are significantly dense primary forest which scholar Zamora (1992) have appointed that mangrove in Brunei Darussalam are exceedingly high regenerate capacity. Statistically, 78% of the plants are replacement pools in some areas according to the studies that have been held.

S. N. Islam (✉)
Department of Geography, Development and Environmental Studies, University of Brunei Darussalam (UBD), Bandar Seri Begawan, Brunei Darussalam
e-mail: shafi.islam@ubd.edu.bn

N. H. H. A. Rahman · M. N. A. B. Aladin
Department of Geography, Development and Environmental Studies, University of Brunei Darussalam (UBD), Jalan Tungku Link, Gadong, Bandar Seri Begawan, BE 1410, Brunei Darussalam

S. Reinstädtler
Department of Environmental Planning, Brandenburg University of Technology Cottbus-Senftenberg, Cottbus, Germany

C. Makowski, C. W. Finkl (eds.), *Threats to Mangrove Forests*, Coastal Research Library 25,
https://doi.org/10.1007/978-3-319-73016-5_18

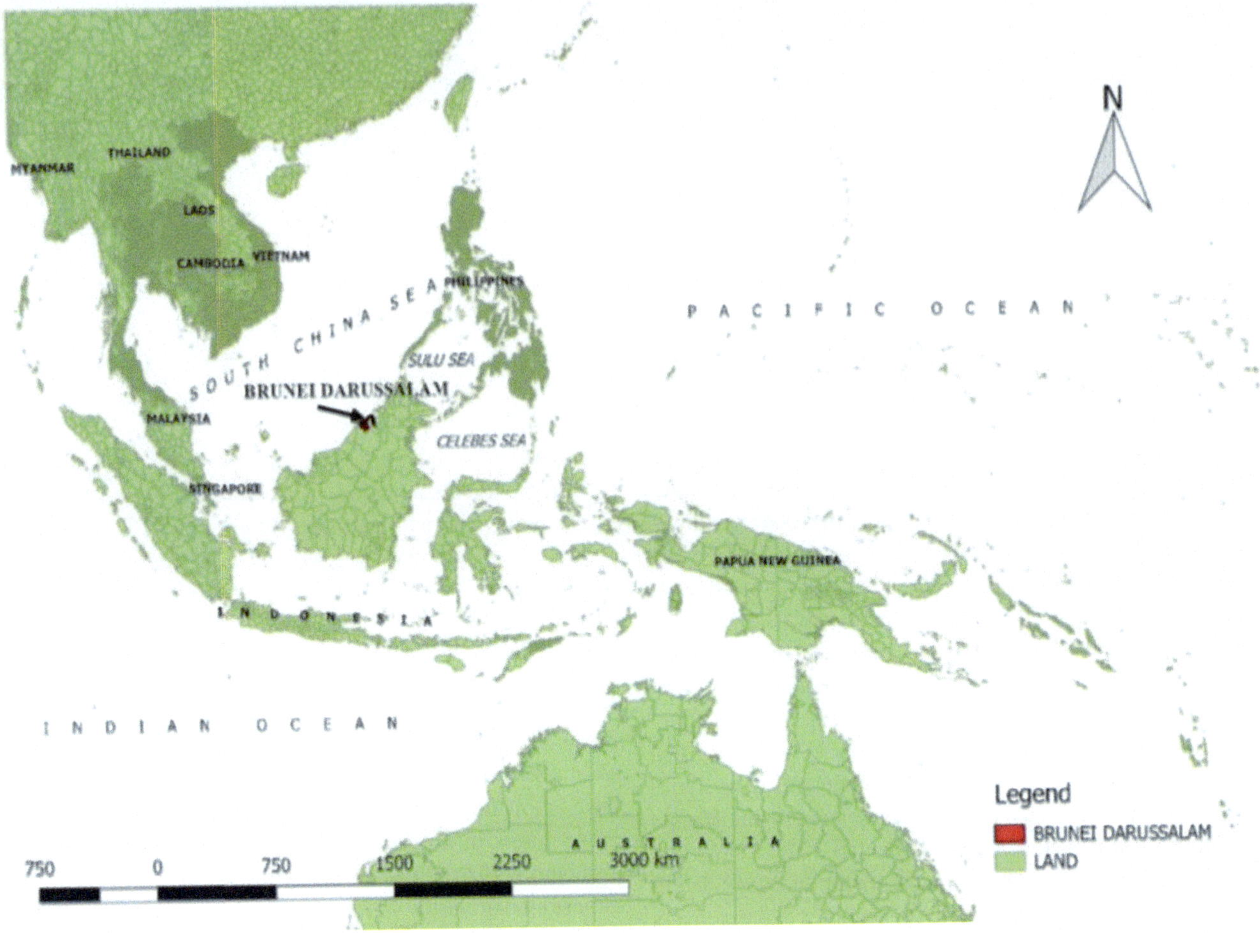

Fig. 18.1 Geographical Location of Brunei Darussalam and the Island of Borneo in Asia Pacific Region

Degradation and vulnerability assessment and analysis can provide strategic planning initiatives with valuable insight into the processes of functional change resulting from management intervention (Steenberg et al. 2017).

According to Department of Forestry (2016), about 78% of Brunei Darussalam's land area of 5765 sq. km is covered by forest, giving rise to forestry as one of the major activities in the country (Fig. 18.1 and Table 18.1). The Government of Brunei Darussalam has acted to conserve some 322,000 *ha* hectares as forest reserves. At present, about 235,520 *ha* (40% of total land area) has already been gazetted as Forest Reserve and out of this, about 50,000 *ha* has been allocated as National Park (Department of Forestry 2016) (Fig. 18.1 and Table 18.1).

At present, about 235,520 *ha* (40% of total land area) has already been gazetted as Forest Reserve and out of this, about 50,000 *ha* has been allocated as National Park (Department of Forestry 2016). Table 18.1 shows the mangrove areas and land cover area of Brunei Darussalam.

According to Chapman (1976), mangroves exist in a specific conditions dominantly brackish water, in which the condition is responsible on the growth and the degradation of the mangrove forest. Mangrove dominant in Brunei Darussalam due to the frontier ecosystem which depending on the brackish water as well it formulized by the mixing of fresh and salt water (JPH 2003b). Mangrove are special as it is able to survive on both high and low tides mangroves absorb the oxygen as it expose on the air during low tide and absorb the nutrient in the water during high tide (Thiessen 2017).

The forest of Brunei Darussalam are categorized according to the primary function they are intended to serve as follows (Department of Forestry 2016):

(i) ***Protection Forests***

Preserved forests which are intended primarily to protect critical soil and water resources; keep the country green and beautiful and the climate invigorating; prevent or minimize the occurrence of floods, droughts, erosion, pollution, and similar environmental problems; and contribute to the general ecological stability of the country.

(ii) ***Production Forests***

Natural and man-made forests, including non-timber plantations, for sustained supply of forest products for the country.

Table 18.1 Mangrove areas relative to total land area

District	Total Land area (ha)	Mangrove area (ha)	% of total mangrove area
Temburong	116.600	12,164	66.0
Brunei Muara	57,000	3937	21.4
Tutong	130,300	1784	9.7
Belait	272,500	533	2.9
Total	576,400	18,418	100.00

Source: Zamora, 1992

(iii) ***Recreational Forests***

Forested areas with natural features that are developed for outdoor recreation in order to promote social, psychological, physical, and economic well-being of the people.

(iv) ***Conservation Forests***

Undisturbed forests set aside to preserve and conserve biodiversity for scientific, educational, and related purposes.

(v) ***National Parks***

Areas with distinctive geologic, topographic and other formations and features of special interest, which are reserved to maintain biologically diverse plant and animal communities for the benefit of the present as well as the coming generations (Table 18.2).

As shown from the table below, most of the forests of the country are located in the Belait district, followed by Temburong, Tutong, and Brunei-Muara. This comparative proportion is consistent with the ratio of total land area occupied by the districts, respectively. However, the majority of the old growth, virgin forests is found in Temburong (Department of Forestry 2016) (Table 18.3).

In this study, mangrove forest will be the main forest type studied. About 4% of all forests in the country are mangrove (Department of Forestry 2015).

The mangrove forest is further subdivided into (Department of Forestry 2016):

- Bakau – *Rhizophora apiculata*
- Nyireh Bunga – *Xylocarpus granatum*
- Linggadai – *Bruguiera gymorrhiza*
- Nipah – *Nypa fruticans*
- Nipah Dungun – *Heritiera globisa*
- Pedada – *Sonneratia caseolaris*
- Nibong – *Oncosperma*

(i) ***Bakau* Forest**

This mangrove forest virtually consists entirely of a single species – *bakau minyak* (*Rhizophora apiculate*). It represents almost 50% of the total mangrove forest in the Temburong Distric. The second species, *bakau kurap* (*Rhizophora mucronata*) occurs but only very occasionally, especially along the margins of deltaic channels and creeks (Department of Forestry 2015, 2017).

(ii) ***NyirehBunga* Forest**

Nyireh bunga (*Xylocarpus granatum*) can occur in association with *bakau minyak,* or in largely stands especially on less, frequently inundated land. Large lobster mounds usually covered by *piai* fern (*Acrostichum aureum*) abound. Almost pure stands of *nyireh bunga* can be found in Selirong and Labu reserves.

(iii) ***Linggadai* Forest**

Linggadai (*Bruguiera gymnorrhiza*) is the only species of the genus of any significance in the Temburong mangrove forests. In Sarawak, abundant or pure stands of the three other species (*B. caryophylloides, B. parviflora*, and *B. sexangula*) can be found, but in Brunei these are either absent or occur only occasionally. Pure stands of *linggadai* can be found in small patches in the Labu Forest reserve, especially along the drainage of Sungai Sipuyut (Department of Forestry 2015, 2017).

(iv) ***Nipah* Forest**

Pure concentration of the *nipah* palm (*Nypa fruitcans)* occurs extensively particularly in river banks and riparian zones in the coastal areas. These are found along the lower reaches of the Pendaruan, Temburong, and Labu River in the Temburong District, and along downstream banks of the Tutong and Belait rivers (Department of Forestry 2015, 2017).

(v) ***Nipah-Dungun* Forest**

Nipah forest with an overstorey of *dungun* (*Heritiera globisa*), which can attain massive size. This occurs at the most upper limits of saline influence particularly along the Temburong and Labu rivers. Towards the down-river limits of its distribution, *buta buta* (*Excoecaria agallocha*),

Table 18.2 Functional forest classification in Brunei Darussalam

Forest category	Gazetted area (ha)	Proposed area (ha)	Total area (ha)	Forest	% of total land area
Production Forest	*13,562*	*0*	*18,562*	*3.96*	*3.22*
Conservation Forest	*28,562*	*3173*	*31,684*	*6.75*	*5.50*
National Park	*46,210*	*2644*	*48,385*	*10.41*	*8.47*
Protection Forest	*138,026*	*80,624*	*213,650*	*46.62*	*37.92*
Recreational Forest	*4211*	*234*	*4445*	*0.95*	*0.77*
Total	*235,520*	*36,675*	*322,195*	*68.69*	*55.88*

Source: Forestry Department, Ministry of Industry and Primary Resources/2015
Accessed: http://www.forestry.gov.bn/frst_forest.htm on 20th May 2017

Table 18.3 The Forest classification in Brunei Darussalam

District (ha)	Primary Forest %	Forest Reserves	Swamp Forest	PSF	MDF	Secondary Forest %	Total Forest %
Belait (278,216)	*68.1*	*45.5*	*33.5*	*19.5*	*30.2*	*2.4*	*88.3*
Tutong (121,667)	*36.8*	*29.1*	*11.0*	*4.5*	*27.7*	*11.2*	*70.7*
Brunei Muara (54,795)	*7.6*	*0.6*	*7.7*			*39.9*	*49.6*
Temburong (127,270)	*80.7*	*49.9*	*18.9*	*0.6*	*55.2*	*0.7*	*86.4*
Total	*58.6*	*40.8*	*23.2*	*10.2*	*32.2*	*7.4*	*80.6*

Source: Forestry Department, Ministry of Industry and Primary Resources
Accessed: http://www.forestry.gov.bn/frst_distribution.htm *on 20th May, 2017*

linggadai, and some *bakau* may be present (Department of Forestry 2015, 2017).

(vi) ***Pedada* Forest**

Small gregarious stands of *pedada* (*Sonneratia caseolaris*) occur on recently deposited silt along the fringes of the Temburong and Labu rivers. This species is most prominent in Kuala Labu and Temburong (Department of Forestry 2015, 2017).

(vii) ***Nibong* Forest**

The tall spiny palm *nibong* (*Oncosperma tigillarium*) is a marginal mangrove species, occurring locally in small to medium-sized clumps at the land-ward edge of mangrove forest especially in more upriver localities.

Mangrove forests grow on the coastal swamps and are flooded with salt water at every tide (Department of Forestry 2015, 2017). Only specially adapted plants are able to cope with such an environment. Some of the mangrove trees (bakau) have adapted to life on the soft and little oxygenated tidal mud by developing an extensive system of specialized roots for support (stilt roots), other species have developed breathing roots (pneumatophores) that poke above the mud (Jacqualine n.d.).

The seeds of the bakau tree do not fall in the tidal water on which they would float and be carried to sea, but they germinate while still attached to the tree. The young plant separates itself from the mother plant only when it has produced a sturdy spear like root (hypocotyl) that enables it to plant itself firmly in the mud (Jacqualine n.d).

The mangrove forest is very productive. The fallen leaves are decomposed by microorganisms that are food for many other species. Fish, prawns, crabs, clams, oysters, and snails are abundant. These in turn, provide food to a variety of animals like otters, eagles, egrets and kingfishers.

Famous mangrove inhabitants include the pot-bellied, big-nosed proboscis monkeys (*Nasalis larvatus*), unique to Borneo. They feed extensively on mangrove tree leaves and are commonly seen in large troops at the water edge around dusk, when tens of thousands of giant fruit bats (flying foxes) leave the mangrove, where they roosted during the day, in search of food (Department of Forestry 2015).

The mangrove forest is a valuable ecosystem: not only does it support unique animal life and maintain the highly productive coastal fisheries, but it also stabilizes the mudflats and protects the coastline from storms and erosion.

In this study, the main focus area will be at Sungai Mangsalut Basin (Fig. 18.2).

Hence, mangroves is important as it plays a significant role towards the ecosystems. According to Rath (2017) mangrove forest is important towards the fisheries, timber and plant production and costal ecology sustainability and coastal protection. Oswell (2016) stated that mentioned, that more than 35% of world mangrove forest are already gone (Oswell 2016).

Trigger the interest of this study as to understand the factor of the status and degradation of the mangrove forest especially for this chapter writing. Emphasizing on Mangsalut River estuary and catchment area.

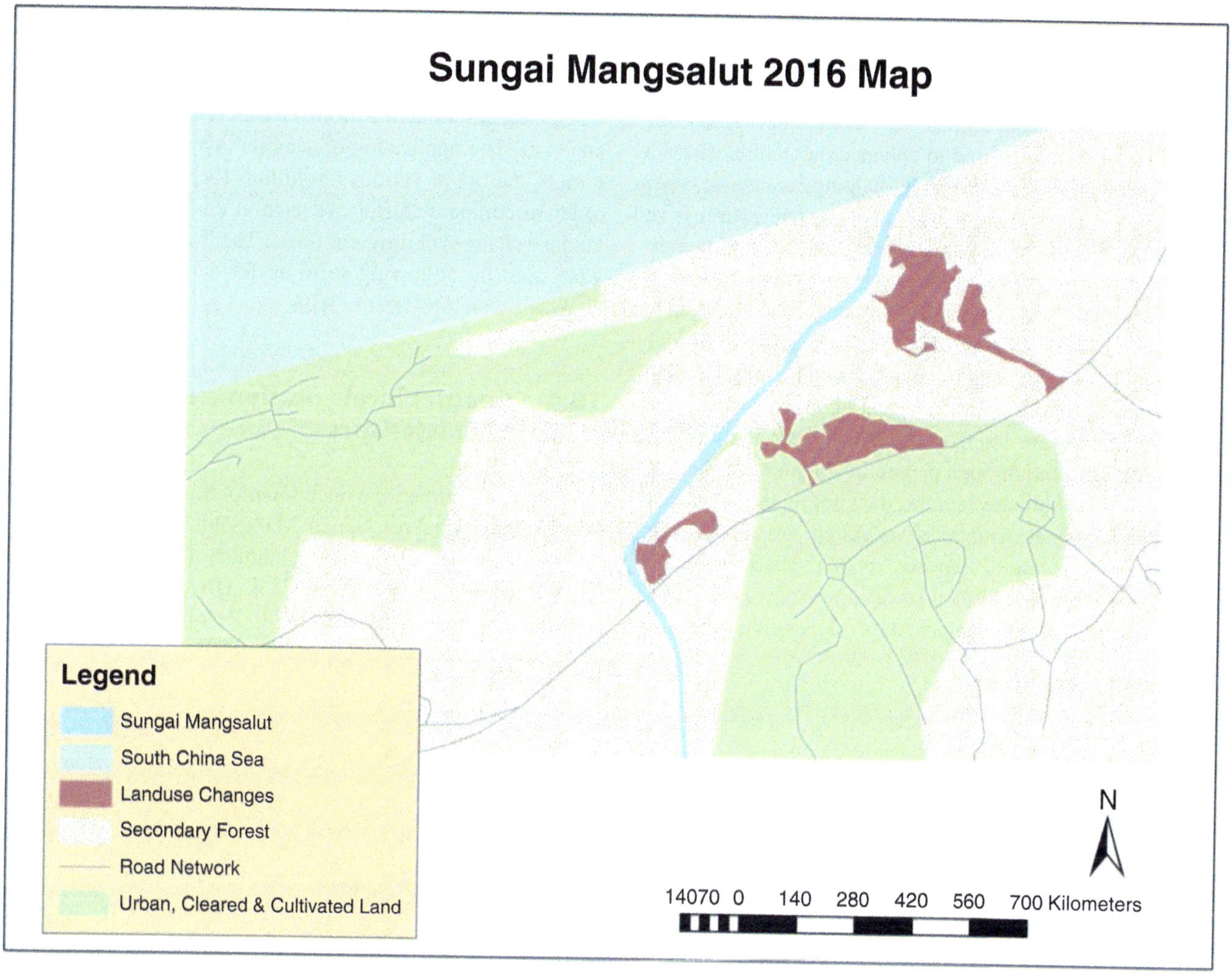

Fig. 18.2 Sungai Mangsalut in 2016 without residential areas and buildings mapped

18.2 Aim and Objectives of the Study

The aim of this chapter is to study the mangrove status and the factors of degradation of the mangrove forest especially for this book chapter and basin area in Brunei Darussalam.

The reason of doing this study especially on Mangsalut River, is due to the degradation state of the mangrove on that area which might be due to the anthropogenic and natural factors. It is important to understand the actual situation to find out the factors as the initiatives to minimize or avoiding the degradation.

To achieve the aim of this study, some specific objectives have been considered which are given bellow:

(a) To identify the coastal landuse changes and changes in vegetation cover around Mangsalut River basin area in Brunei Muara District.
(b) To use Remote Sensing Image through ArcGIS and QGIS software to compare the mangrove vegetation analysis and landuse changing pattern in the Mangsalut River Basin area (Past and present status comparison).
(c) To perform special analyses to determine the causes that influence the vulnerability of mangrove forest along the Mangsalut River Basin.
(d) To perform spatial analysis to determine the causes that influences the degradation of mangrove forest wetlands along with the Mangsalut River basin in Brunei Muara district in Brunei Darussalam.
(e) Make some applied recommendations for future development and management sustainability of Mangrove forest wetlands in Brunei Darussalam.

18.3 Data and Methodology

In this research, data collection method will be used. Data collection is the process of gathering and measuring information on variables of interest, in an established systematic fashion that enables one to answer stated research questions, test hypotheses, and evaluate outcomes (Northern Illinois University Faculty Development and Instructional Design Center n.d.).

Data collection is also an imperative part in effect assessment by giving data valuable to comprehend the courses of action behind watched results and survey changes in individuals' impression of their well-being. Furthermore subjective routines can be used to enhance the nature of review based quantitative assessments by helping create assessment theory; reinforcing the configuration of overview surveys and growing or elucidating quantitative assessment discoveries.

Data were collected from two sources, essential (primary) sources and optional (secondary) sources. Essential data are otherwise called crude (raw) data. Data are gathered from the first source in a controlled or an uncontrolled situation. Sample of a controlled situation are trial examination where certain variables are being controlled by the analyst. Then again, data gathered through perception or survey review in a characteristic setting are samples data got in an uncontrolled situation. Optional data were acquired from supporting sources, for example, reports, books, diaries, records, magazines, the web and then some other sources. There are numerous routines to gather data, depending upon our examination outline and the procedures utilized. A percentage of the basic systems are surveys, meeting and perception.

In this research, the use of remote sensing software is used where images obtained from Google Earth were used for analyzes. The application of satellite imagery from different remote sensors of various resolutions for spectral analysis in order to compare shifting vegetation cover communities or landuse changes at different times. The images obtained then were used the following software for data construction and analyzed with ArcGIS or QGIS.

18.4 Geographical Location and Physical Characteristics

As being said earlier, the main focus of this study is at Sungai Mangsalut Basin Area. Sungai Mangsalut is located at Brunei Muara district. It is located at a latitude of 5° 1′11.81 ″N and at a longitude of 114°59′20.72″E (Brown 1970; Hassan 1988; CDD 1993; Ibrahim 1988; Khatib and Sirat 2005) (Fig. 18.3).

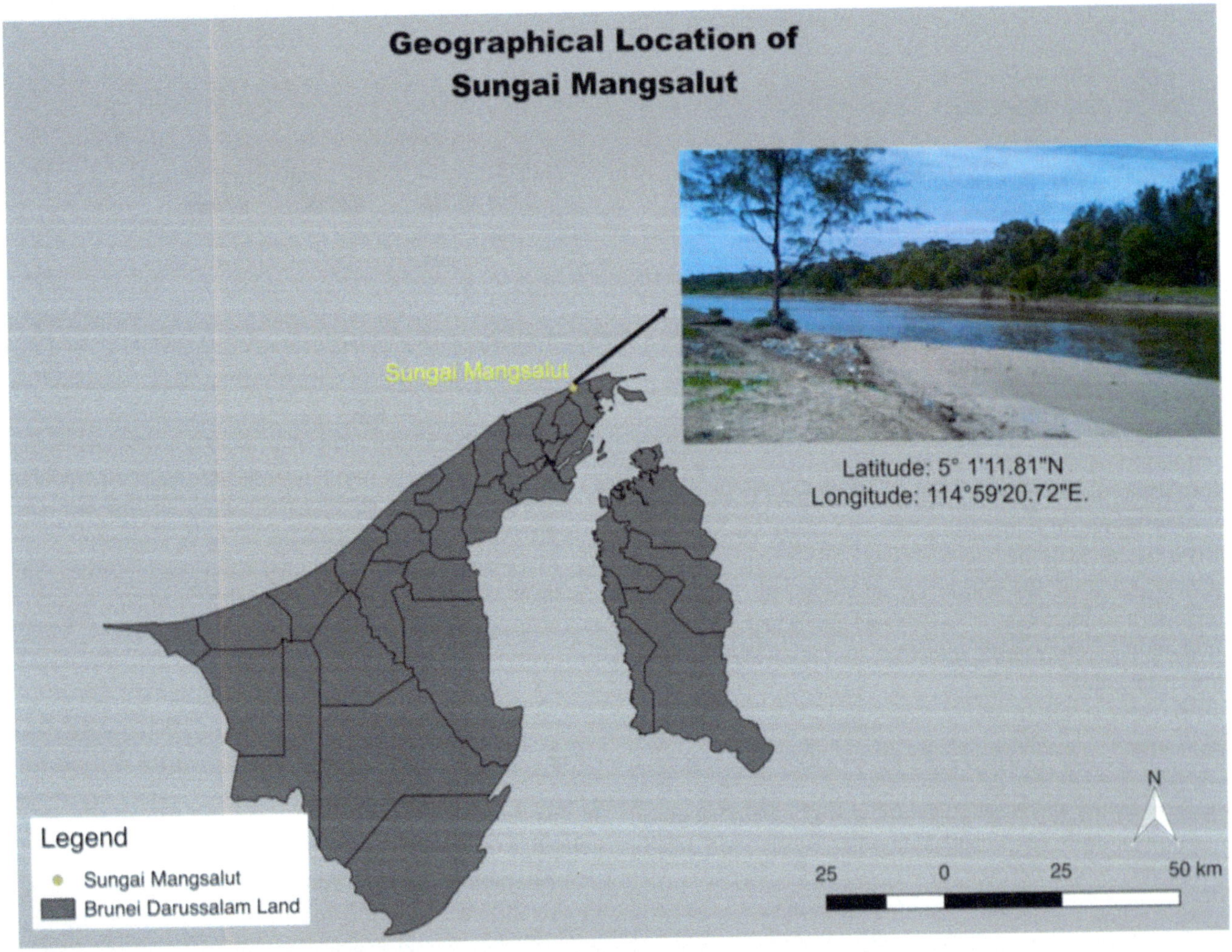

Fig. 18.3 Geographical location of Mangsalut River (Sungai Mangsalut) Case Area

Fig. 18.4 Map of head and mouth of Sungai Mangsalut (Mangsalut River)

Tracing down the Sungai Mangsalut's head with the aid of Google Earth, it can be concluded that Sungai Mangsalut's head consists of several rivers as listed below (Fig. 18.4) (JPH 2003c; Department of Forestry 2015; Thiessen 2017; Islam and Umar 2017);

1. Sungai Hanching, – (*Hannching River*)
2. Sungai Orok, – (*Orok River*)
3. Sungai Tilong – (*Tilong River*)
4. Sungai Salambigar – (*Salambigar River*)
5. Sungai Antong – (*Antong River*)
6. Sungai in Lambak – (*Lambak River*) and
7. Sungai in Tanah Jambu – (*Tanah Jembu River*)

Figure 18.4 displayed the mouth of Mangsalut River is Mangsalut River itself in the coastal region of Brunei Muara district in the direction of South China Sea.

The Mangsalut River and other six rivers are interconnected and they made a network within the rivers catchment areas (JPH 2003e). For various reasons these seven rivers network catchment area is important for urban growth, natural river networking system and communication networking, socio-economic development and urbanization, industrial development as well as settlement and housing projects initiatives and infrastructure development initiatives. All these developmental activities within the Mangsalut River catchment area the vegetation and specially the mangrove vegetation is cleanup and shrinking the mangrove forest area (JPH 2003b) (Figs. 18.4 and 18.5), the regional and geographical location can be seen in the following maps and pictures.

18.5 Results and Analysis

Map (Fig. 18.5) below shows vegetation cover around Sungai Mangsalut (Mangsalut River). Mangrove was seen to be the main vegetation cover along Sungai Mangsalut, at South of Sungai Mangsalut, swamp forest was seen to be the dominant vegetation cover as there was a swamp nearby. To the West and East of Sungai Mangsalut, the dominant vegetation cover was primary forest (Department of Forestry 2015, 2017).

However, in the present, mangrove forest cannot be seen at the river mouth and along the riverbank. Thus, indicates the shrinking of mangrove size in the coastal and peat swamp areas (JPH 2003b). The reasons as to why mangrove can be seen degrading will be explain below.

Figure 18.5 displays the mangrove vegetation locational pattern along with the Mangsalut River basin and the

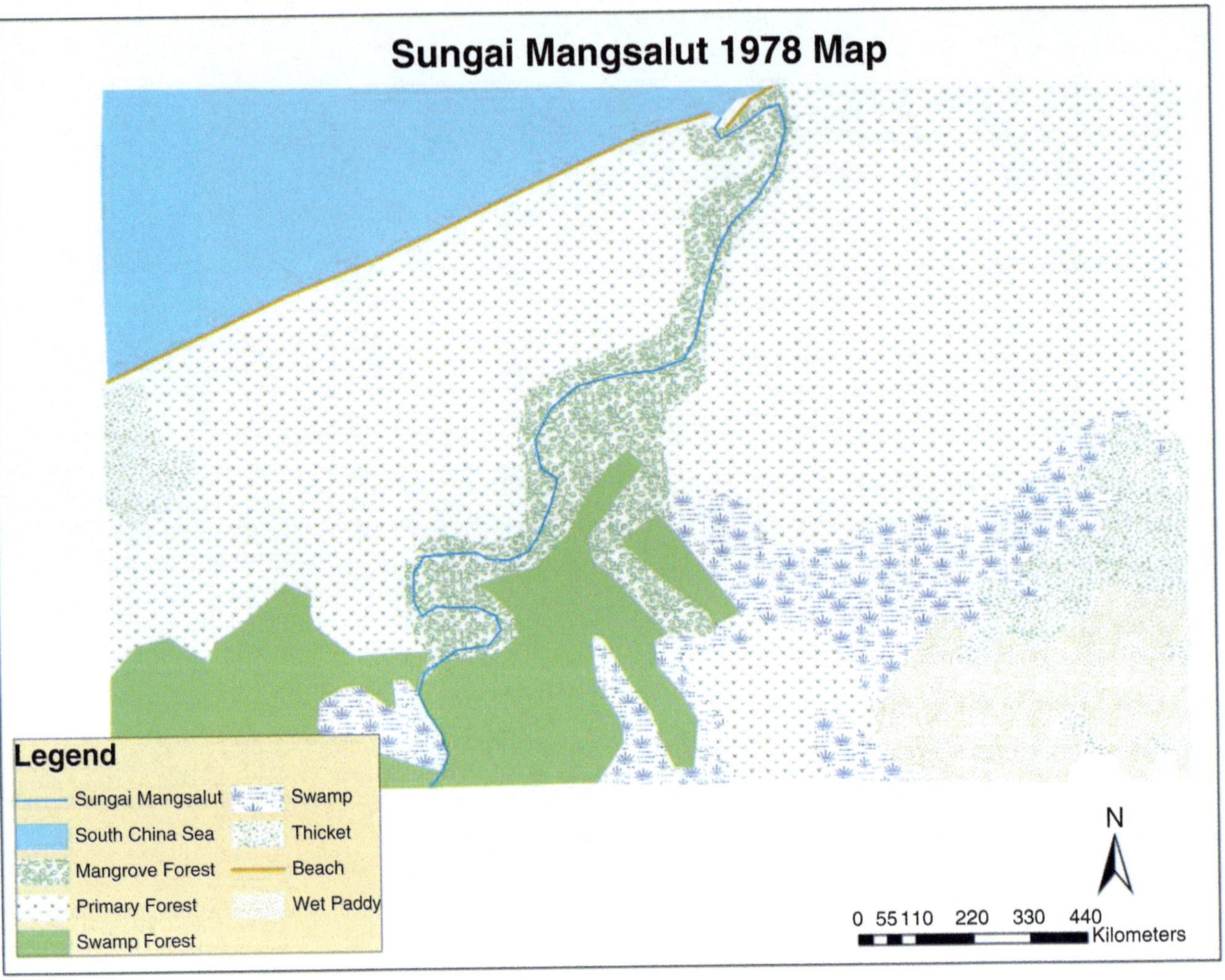

Fig. 18.5 Vegetation cover in Sungai Mangsalut (Mangsalut River) 1978

catchment area. The mangrove in the Mangsalut basin area was very rich and hotspots of biodiversity in 1978 (Fig.18.5) (Department of Forestry 2015; Islam and Umar 2017). Whereas at present the major portion of mangrove vegetation is reducing, the real scenario of mangrove forest wetland ecosystem on the Mangsalut river basin area will be seen and more detail explanation has been written in Fig. 18.11 (Sungai Mangsalut in 2016 map).

18.5.1 Reasons for the Disappearance of Mangrove Along Sungai Mangsalut (Mangsalut River)

One of the reasons as to why mangrove degrades is because of forest fire. A forest fire is an uncontrolled fire occurring in nature (JPH 2003c, d). Sometimes, the forest fire is so large that it takes a long time for the firefighting crews to gain control over the situation. This could result in massive destruction, and in this situation mangrove along Sungai Mangsalut (Mangsalut River) (Thiessen 2017; Islam and Umar 2017).

The number of forest fires varies from year to year, and quite a long time may elapse between forest fires that are considered to be large. Climatic conditions are the factor that has greatest impact on the extent of forest fires. Weather conditions such as precipitation and wind, as well as the layout of the terrain, are important factors in determining the size of the forest fire (JPH 2003a, b, c). It is estimated that as many as nine out of ten forest fires are caused by humans, although the causes of a significant number of forest fires remain unknown (Thiessen 2017). The most common cause of such fires is the use of open flames and disposable barbecue grills. Even a cigarette that is not properly extinguished can cause a forest fire. Natural forest fires are due to lightning strikes (Kriseinfo 2012).

The prove of forest fire can be seen in image below (refer to Figs. 18.5 and 18.6) obtained from Google Earth at a year

Fig. 18.6 Image obtained from Google Earth showing burn scar in 2007

of 2007 and 2009. In the image, burn scar can be clearly seen. Burn scar is a charred, barren strip of land annihilated by the fire. Devoid of vegetation, nothing remains to hold the land in place when rainfall comes (JPH 2003b, c, e). Figs. 18.6 and 18.7 displaying the disturbance and disappearing process of mangrove forest within the catchment area of Mangsalut River in Brunei Muara district in Brunei Darussalam. The natural calamities are the first step threat for the land use change and land cover changes in the catchment (Islam and Umar 2017; Thiessen 2017).

In 2009 (Fig. 18.7), burn scar can be seen intensified indicating forest fire has happened recently at the area boxed in 2009. The forest and bush fire in the catchment area is also an additional threat for the mangrove forest ecosystem in the catchment area (Islam and Umar 2017).

To strengthen prove given; image below (refer to Fig. 18.8) shows the location of mangrove in 1978 where the burn scar is currently located. The image below has been overlaid with Brunei's vegetation cover map in 1978. It is therefore, can be definitely said forest fire may be the reason as to why mangrove forest degraded alongside the Sungai Mangsalut (Department of Forestry 2015) (Fig. 18.8).

Another reason as to why mangrove has degraded along Sungai Mangsalut is because the changing course of river in Sungai Mangsalut. Blue line in image below (refer to Fig. 18.9) indicates Sungai Mangsalut in 1978, it can be seen the river to be meandering. However, presently Sungai Mangsalut course can be seen as straight.

In 1978 (refer to Fig. 18.10), mangrove vegetation was seen to be located at the meandering river. However, the river now left an Ox Bow lake. This happened because of erosional processes.

The meander is now abandoned and once high storm; that is high power flow of water cut off the meander it eventually creates an Ox-bow lake. This diversion can lead to mangroves dying out (Fig. 18.10 and Fig. 18.11).

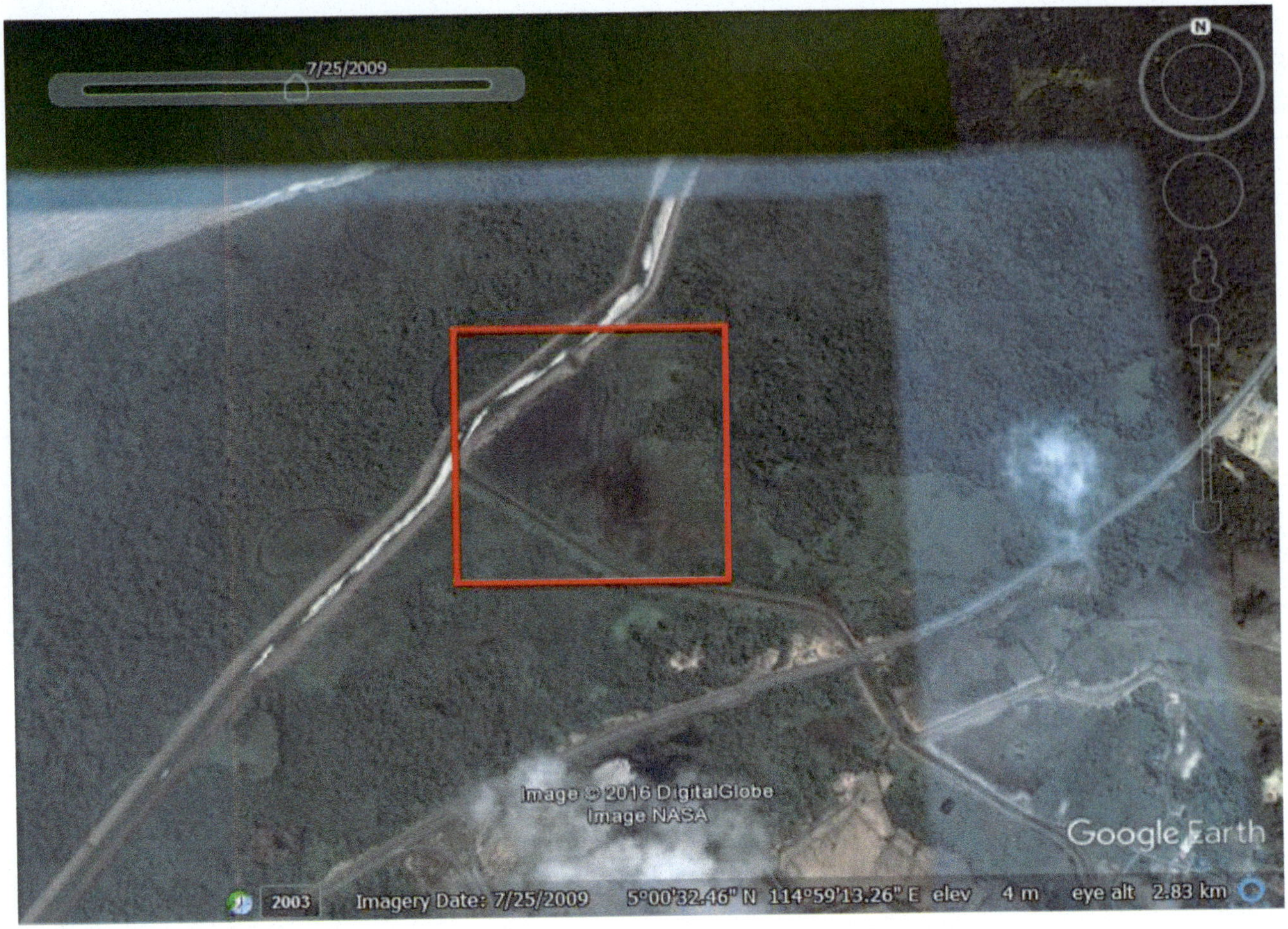

Fig. 18.7 Image obtained from Google Earth showing burn scar in 2009

Adding more to the reason as to why mangrove has degraded is because of human activities. Any major activities or development nearby can affect the mangrove ecosystem. From the map below (refer to Fig. 18.11), major landuse changes can be observed near to Sungai Mangsalut. The development involves changing natural land for human development that includes building drainage, road network system, residential area, water supply, electricity and commercial areas (Fig. 18.11).

18.5.2 Mangrove Forest Wetlands Ecosystem Degradation in Brunei Muara District

At the location of the Mangsalut River is close towards the sea, it suggest that the status of the river is at the 4th order of all the rivers and tributaries and distributaries at the catchment areas. The river stages which will determinate the weak velocity of the river current in 2016 and 2017 (Islam and Umar 2017). The river basin and river bank is also shifting due to time and natural causes as well as anthropogenic activities win the catchment areas (Fig.18.12).

The shape of river in depth, suggest the state condition of the soil along with the river either it is hard base rock or soft. The map that have been prepared and displayed below (Fig. 18.12) shows the actual and current shape of the Mangsalut river basin.

Hence, follow-up the two previous maps (Figs. 18.10 and 18.11) on the Mangsalut River and its mangrove forest, it is clearly shown that on the east side of the Mangsalut River, there are developed area where there are buildings and settlement development process is running (Fig. 18.13).

There are also, cleared area located on the South of the development area as to be shown on the map below (Fig. 18.14).

In 1967, the river Mangsalut was in sinuous or also known as meandering type of river and characters were seen. The type of river alone, suggested the possibility of the vast status of mangrove forest along the river. The data that been gathered are very limited as the old map just shows the main river channel (Fig. 18.15) in 1967.

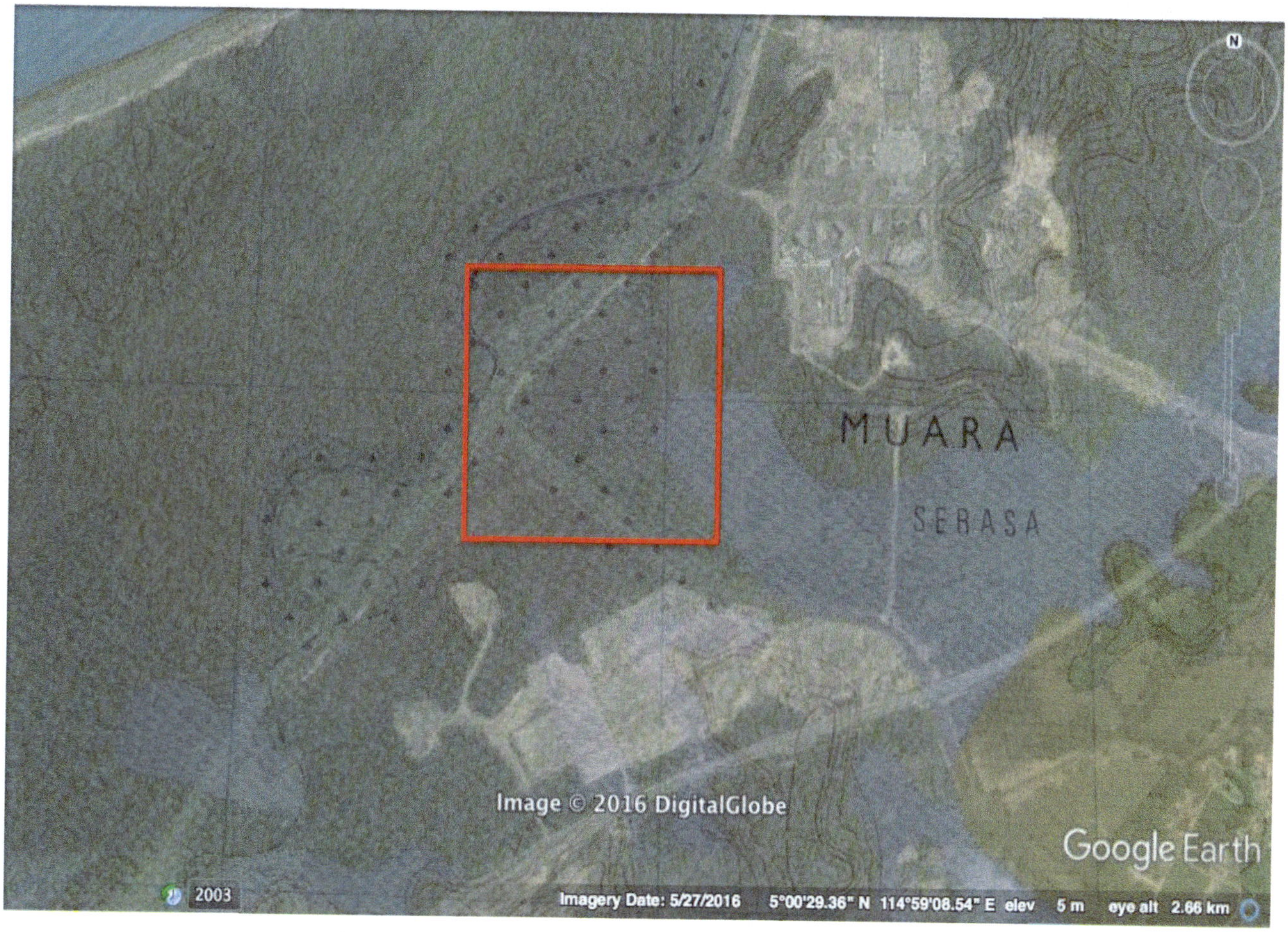

Fig. 18.8 Image showing burn scar at once to be Mangrove covered

However, since there is limitation on finding data for the mangrove forest specifically on Mangsalut River, the current status of the river will be compared by the digitized data of 2003. In 2003, the status of the river is significantly same as the present river (Fig. 18.15). Nevertheless based on the obviously still untouched. The only thing that are the same base on the landscape perspective is the post fire scar area which can be seen in the next. As a whole the mangrove vegetation and forest wetland ecosystem in the Mangsalut River basin and catchment area is ecologically vulnerable.

18.6 Discussion and Approach for Mangrove Forest Wetland Sustainability

According to Oswell (2016), there are several significant factor leads to the degradation of mangrove forest. These include the clearing overharvesting, river changes, river channel migration, over fishing, destruction of coral reefs, pollution, and climate change impacts (JPH 2003a; Department of Forestry 2015, 2016; Islam and Umar 2017). On the case of the degradation of mangrove forest along with the river Mangsalut the factors are applicable to both natural and anthropogenic factors. Through the observation focusing on the mangrove forest along the Mangsalut River (Islam and Umar 2017). The natural factors might be El-Nino and / or climate change which may leads to forest fire, and river changing as the changing of the river flow pattern leads to the changing velocity of the river, depth and also the level of water salinity. On the other hand there are variety of human intervene especially towards the development as there are land use changes occurs along with the Mansalut River from the mangrove forest to the deforestation and building construction for the governmental buildings of utilization specially on the east (Islam and Umar 2017).

By observing the current status of the mangrove forest alongside the Mangsalut River, it is significantly suggested that the invention of human activities play a tremendous rate on the degradation of the mangroves which forcing the coastal mangrove ecosystem as vulnerable condition (Breslin 2013; Islam and Umar 2017). The present management of mangrove forest wetland ecosystem is leading under the Department of Forestry of the Government of Brunei Darussalam. The forestry department following the

Fig. 18.9 Image obtained from Google Earth showing the change of course of Sungai Mangsalut

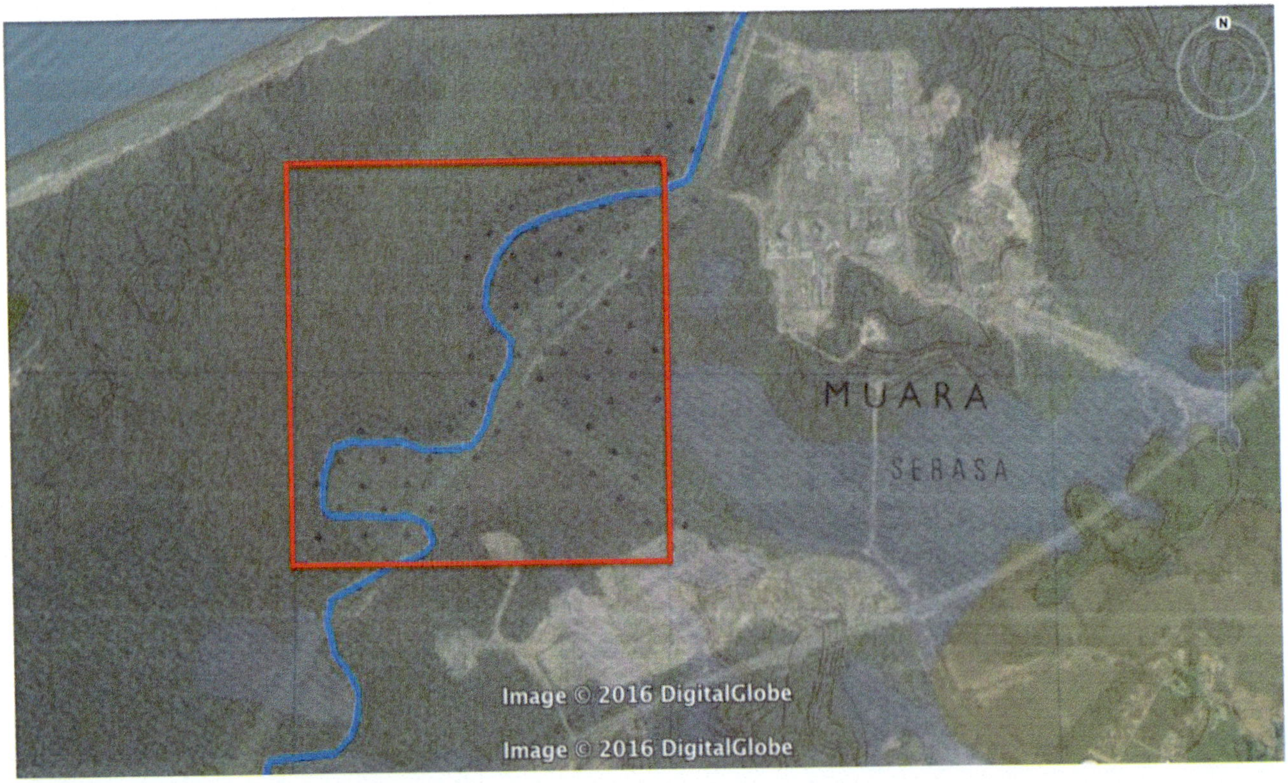

Fig. 18.10 Image obtained from Google Earth showing location of mangrove in 1978

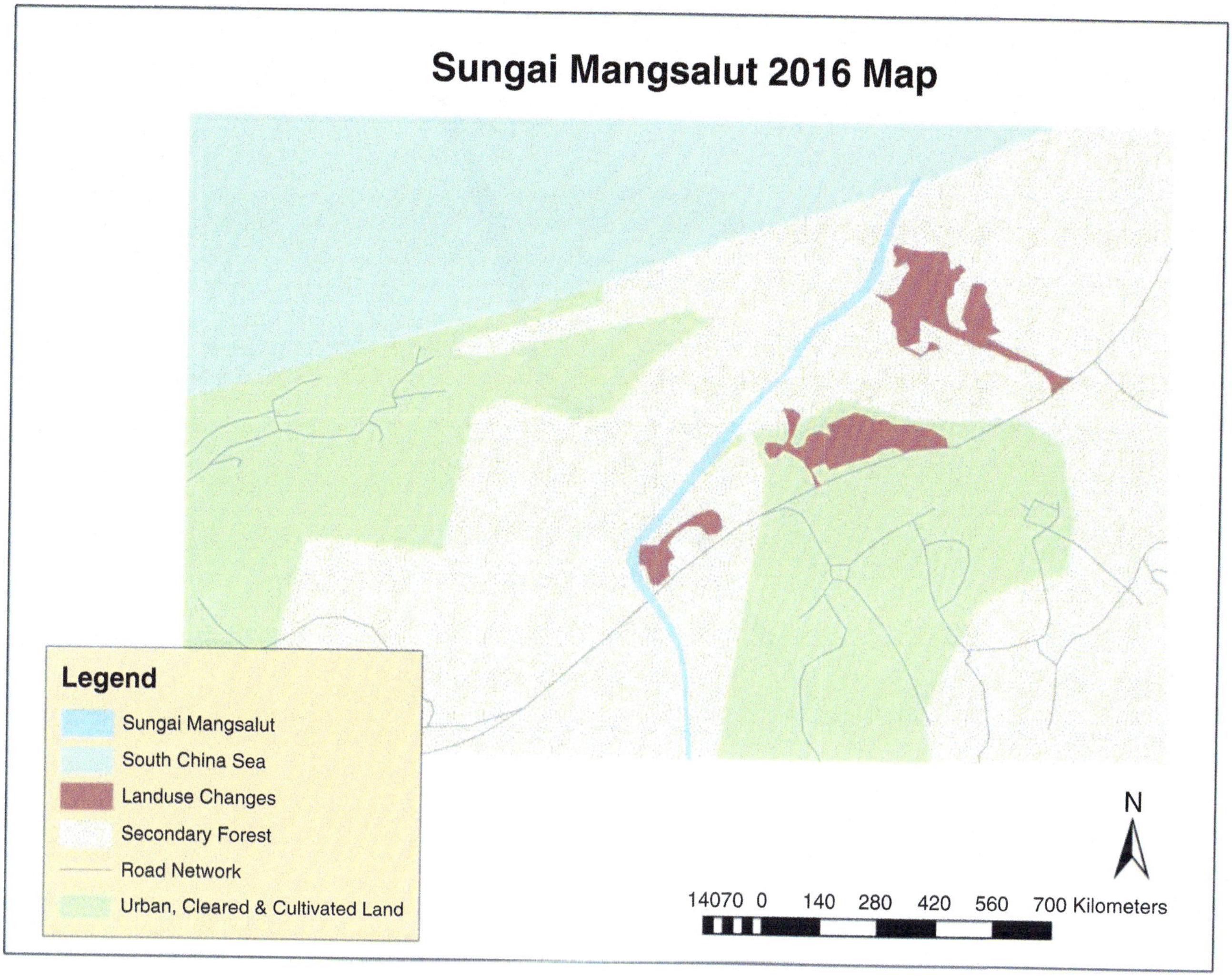

Fig. 18.11 Sungai Mangsalut (Mangsalut River) in 2016 without housing plan an buildings mapped

administrative structure of forest management in Brunei Darussalam (JPH 2003b, c; Department of Forestry 2015, 2016).

The mangrove forest is not a separate department yet even there is no separate initiatives for the coastal mangrove forest wetlands ecosystem in Brunei Darussalam. Mangrove forest in Brunei is gradually reducing for multipurpose reasons and the quality of the wetlands ecosystem is having degraded (JPH 2003b, c). For the protection of mangrove forest wetlands ecosystem the proposed model (Fig. 18.16) of mangrove forest wetland protection and sustainability could be implemented in the coastal region of Brunei Darussalam which could be ensured the management and protection of mangrove forest wetlands in the coastal region in the Brunei Darussalam (Islam and Umar 2017).

Furthermore, the cleared area/deforested area on the east side of the river are most probably due to the future plan of the governmental or industrial land use area as well as it could be used and developed for housing or settlement scheme within the river catchment area in Brunei Muara District in Brunei Darussalam.

In, nutshell the Mangsalut River system was playing a potential role towards protection of natural mangrove landscapes and micro scale ecosystem and ecosystem benefits for the local community (Stewart 1986; Islam and Umar 2017). The coastal mangrove ecosystem is gradually degrading due to anthropogenic influences and development construction, housing and settlement and climate change impacts in the coastal region in Brunei Darussalam. As a whole the ecological status of the coastal mangrove wetlands ecosystem is important community livelihoods dependency as well as for the protection of coastal ecological balance and sustainability (Islam and Umar 2017).

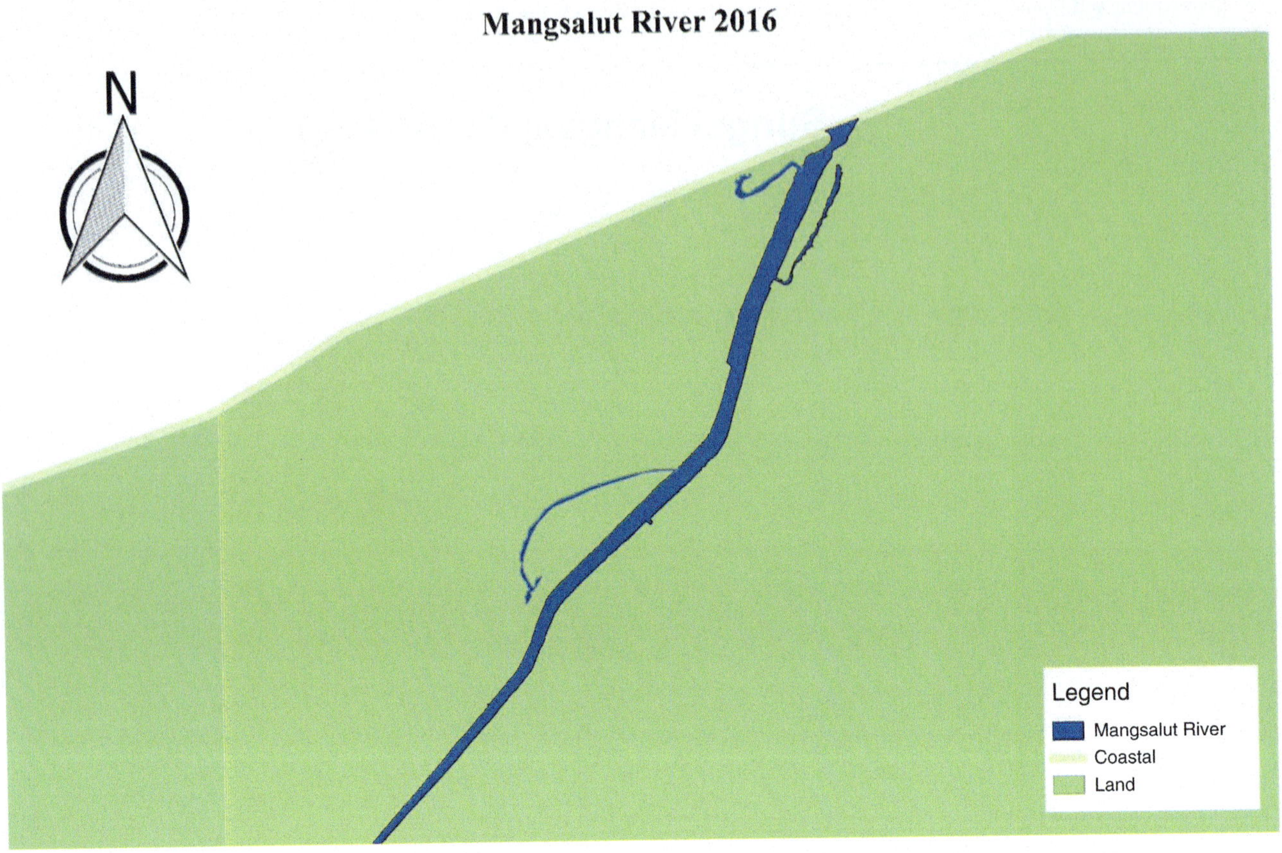

Fig. 18.12 The Features of Mangsalut River basin in 2016–2017

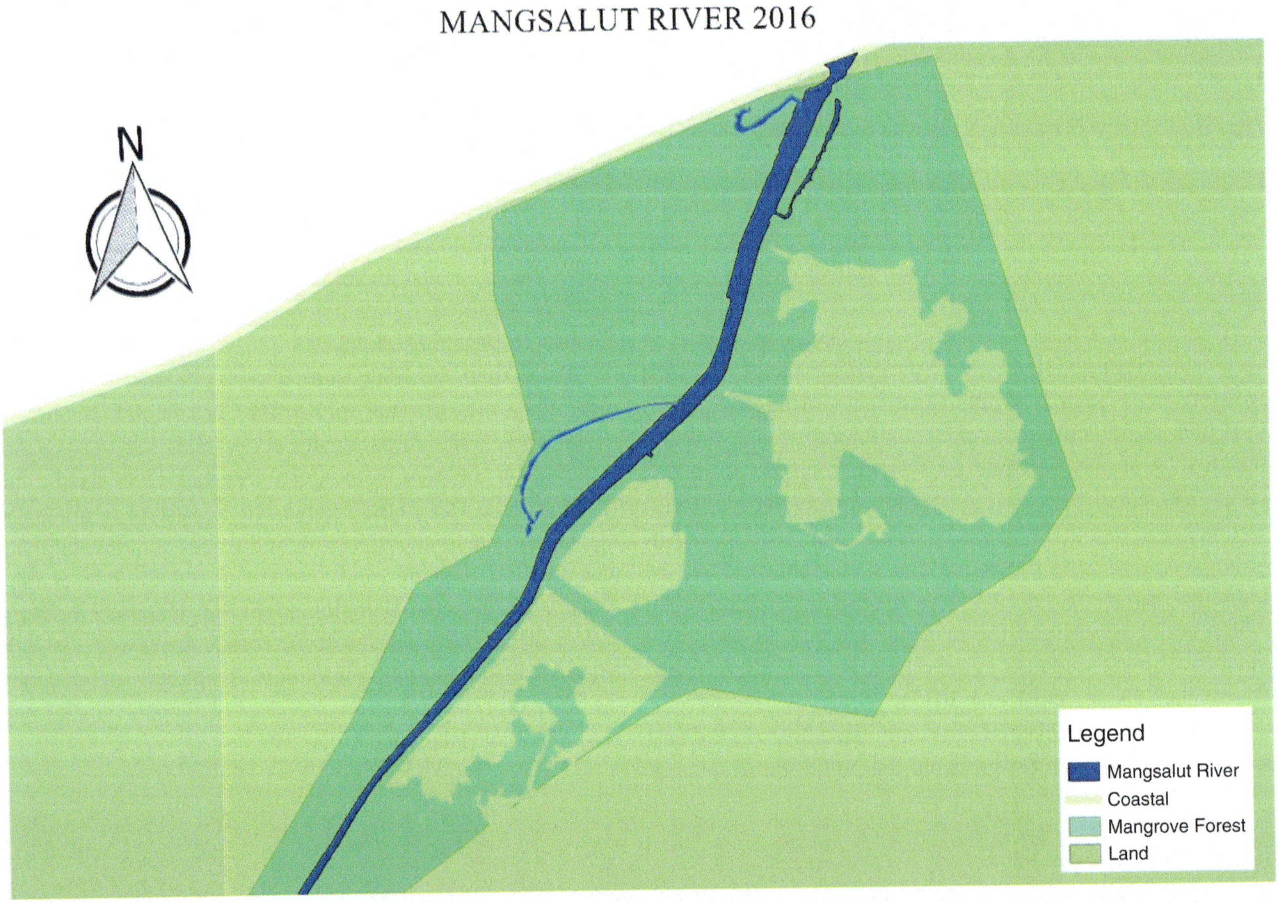

Fig. 18.13 The map of Mangsalut River with it mangroves forest in the catchment area in 2016–2017

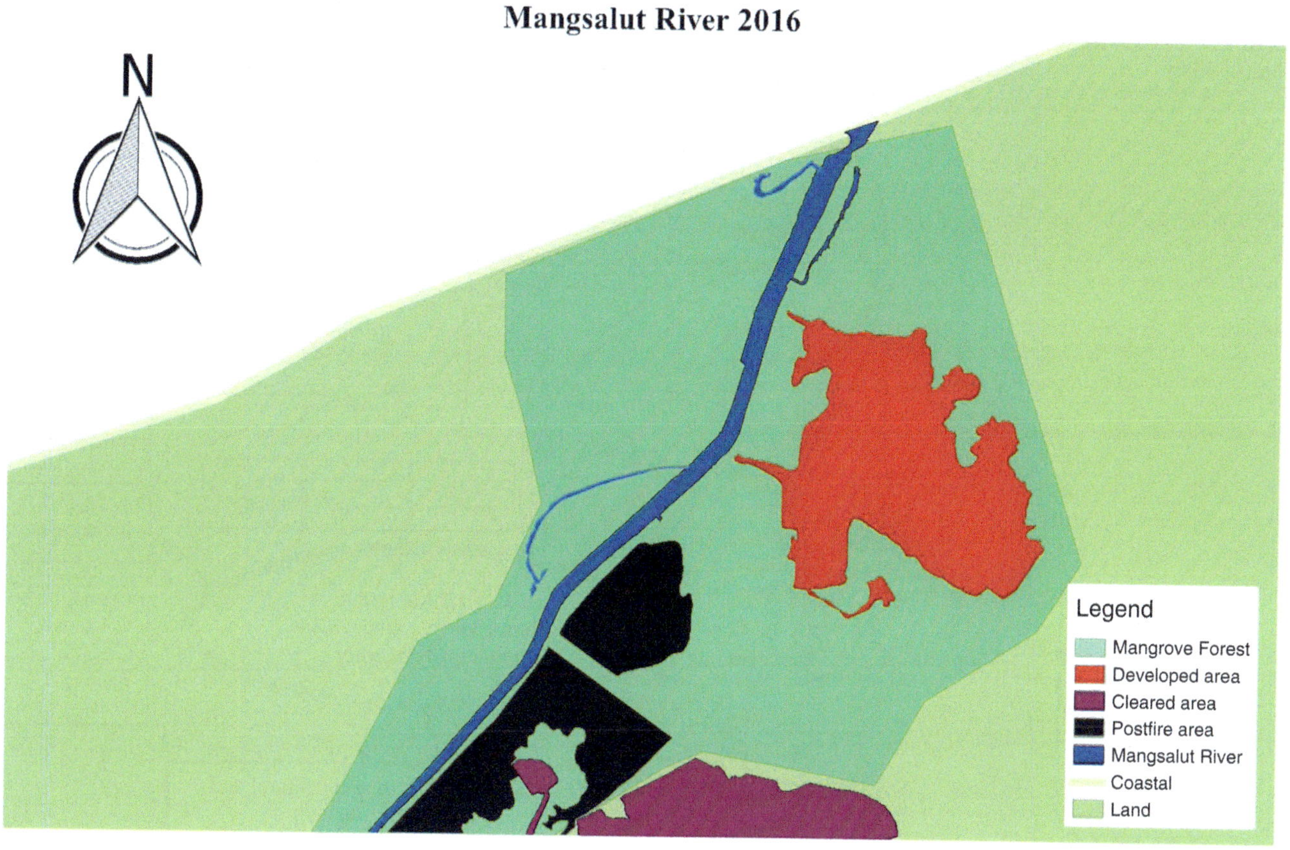

Fig. 18.14 Map of the land cover changing pattern in the Mangsalut River catchment rea in 2016-2017

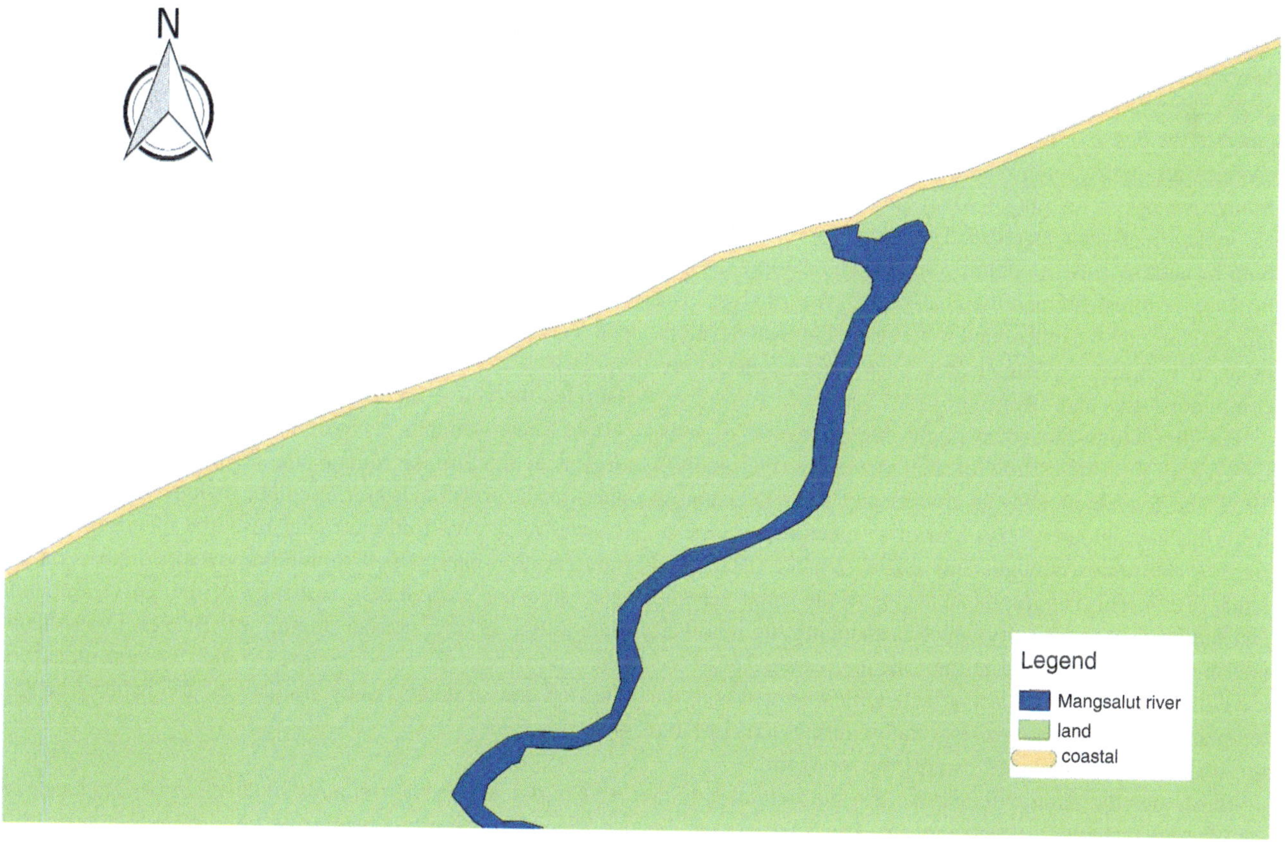

Fig. 18.15 Map of the Mangsalut River characters in 1967 (Focused)

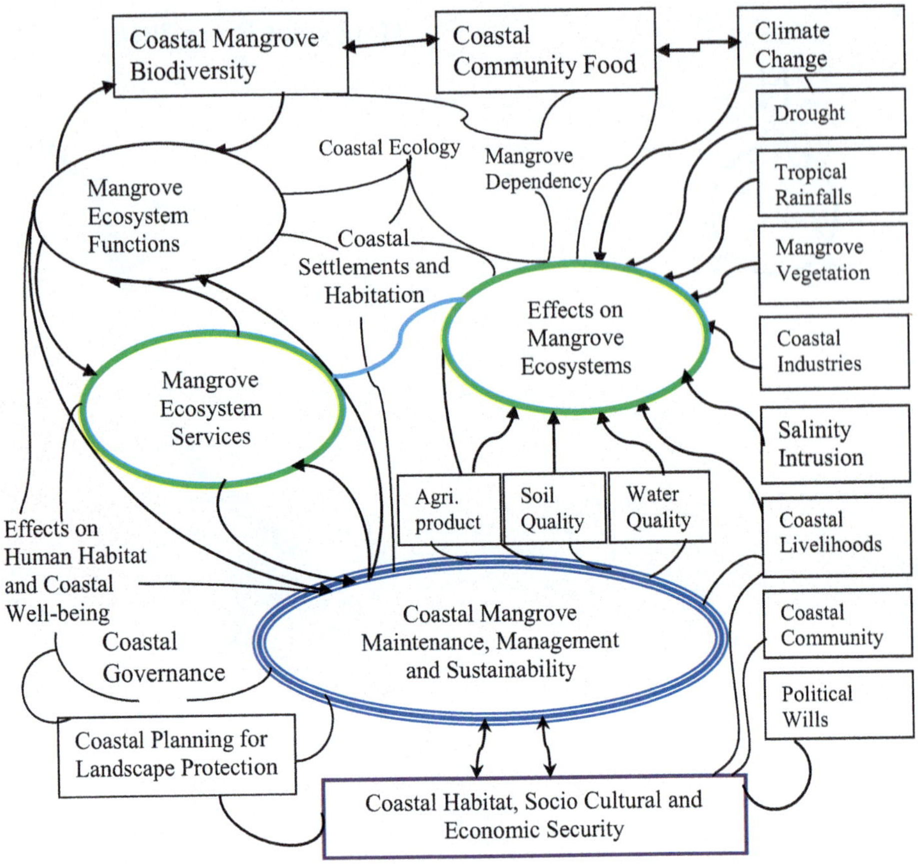

Fig. 18.16 Model of Mangrove Wetland Protection and Sustainability

18.7 Conclusions and Recommendations

There is difference, specifically on the decrease of the mangrove forest along with the Mangsalut River nowadays compared to the dense of the mangroves in the year of 2003. The changes are due to both natural and anthropogenic factors. Mangrove cover in Sungai Mangsalut as being discussed earlier has been seen degraded. The reason for this is because both of natural cause and human cause. The natural cause is because of forest fire and the changing course of river. The human cause on the other hand is because of human development in building residential area, road system, electricity, commercial area and other more.

For the natural factors, it might due to several reasons. Comparing the current flow of the river to the river bank in 1967, the river bank and the character of river is changing from the in Sinuous or also known as meandering type of river to the almost straight and single channel (Islam and Umar 2017). This suggest the changing of the water velocity and the level of the salinity which is changing the biodiversity in the basin areas and in the catchment area. Therefore, the climate change might also affecting the mangroves as the temperature are increasing especially during the El-Nino season which might trigger towards the forest fire.

Hence for the anthropogenic factors, the overall significant factors is due to the development. The deforestation are due to the process of the building construction. The change of caused use is majority be the reason of the deforestation. Sum of all, the related governmental development or private agency should be responsible on planning a better way to minimize the deforestation of the forest of Brunei Darussalam. Besides to increase the mangroves initiatives such as replanting the mangroves is one of the best policy and recommendation to conserve the qualify and the reputation of Brunei Darussalam as a green coastal space.

18.7.1 Some Important Recommendations

Besides the above mentioned and proposed model of mangrove forest wetlands ecosystem management in the coastal region in the Brunei Muara district in Brunei Darussalam the specific potential issues should be included in the national coastal mangrove forest wetland resource management and conservation plan, which should be developed based on the changing and degradation tendency of mangrove wetlands ecosystem in the coastal region of Brunei Darussalam; the following recommendations could be considered in making the strategic policy framework for better management such as;

- Coastal mangrove wetlands biodiversity maintenance in the coastal region; mangrove ecosystem functions and services should maintenance in the coastal region for the better interest of coastal communities and within the participation of coastal communities.

- Increase coastal community socio-economic enrichment and social welfare and ensure sustainable livelihoods as alternative mangrove dependency in the vulnerable coastal region.
- Provide alternative income generation opportunities to the coastal communities that should ecologically friendly and profitable to local communities.
- Increase public awareness concerning coastal ecology and environmental training and education, particularly related to the importance of the sustainable use of coastal mangrove and other natural resources.
- Provide information to the coastal communities concerning mangrove natural resources collection, properly uses and long-term protection measures.
- Ensure community involvement in maintaining and protecting of mangrove resources for sustainable coastal resource development and management planning.
- Technology transfer to the local community for sustainable use of mangrove resources and desalination measures. These seven elements are integrated to each other and could make a comprehensive coastal mangrove management plan for sustainability in the coastal region in Brunei Darussalam.

References

Breslin (2013) Wildfire burn scars: How long until they heal? Retrieved October 5 from https://weather.com/news/news/how-long-does-wildfires-burn-scar-need-heal-20130812

Brown DE (1970) Brunei the structure and history of a Borneoan Malay Sultanate Brunei Meuseum

CDD- (Curriculum Development Department) (1993) History for Brunei Darusslam (Secondary 3). EPB Pan Pacific, Singapore, pp 1–144

Chapman VJ (1976) Mangrove vegetation. Strass & Cramer GmbH, Leutershausen

Das K (2014) Estuarine morphodynamics of the Sunderbans. Retrieved September 15 from https://books.google.com.bn/books?id=ZEEBBQAAQBAJ&pg=PA202&lpg=PA202&dq=river+course+change+%2B+mangrove&source=bl&ots=D9r01_O2KH&sig=oxDCvm_I7VCBHUkb_IdSVRpV9U&hl=en&sa=X&ved=0ahUKEwioyvi6zqrQAhUGTY8KHV4yD8gQ6AEIGzAA#v=onepage&q=river%20course%20change%20%2B%20mangrove&f=false

Department of Forestry (2015) Inventory and documentation of flora & fauna in mangrove forest, Ministry of Industry & Primary Resources, Government of Brunei Darussalam, Final Report

Department of Forestry (2017) Report of Brunei Forestry, Ministry of Industry and primary resources, http://www.forestry.gov.bn/frst_distribution.htm on 20th May 2017

Hamid LA (2000) Brunei Darussalam. Mangrove- friendly aquaculture. Retrieved in March 1, 2017. File:///C:/users/USER/desktop/hamid2000-brunei-mangrove-friendly-aquaculture.Pdf

Hassan AH (1988) Elementary geography for Brunei Darussalam. Dewan Bahasa dan Pustaka Brunei, Brunei Darussalam

Henrot Jacqualine (n.d.) Forests of Brunei Darussalam. Retrieved November 1 from https://www.bsp.com.bn/panagaclub/pnhs_old/vegetati/forests/forests.htm

Ibrahim HBH (1988) culture and counter cultural forests in contemporary Brunei Darussalam. In: Thumboo E (ed) Culture in ASEAN and the 21st centuary. Uni Press, Singapore

Islam SN, Umar AABY (2017) Impacts of coastal landuse changes on mangrove wetlands at Sungai Mangsalut Basin in Brunei Darussalam. In: Finkl CW, Makowski C (eds) Coastal wetlands: alteration and remidiation, Coastal Research Library 21. Springer, Berlin, pp 133–157. https://doi.org/10.1007/978-3-319-56179-0-4

JPH- (Jabatan Perhutanan) (2003) A Unique Treasure-house of forest biological ecosystems. Retrieved September 16 from http://www.forestry.gov.bn/frst_u_intro.htm, Accessed on 10th May 2017

JPH- (Jabatan Perhutanan) (2003a) Mangrove. Retrieved September 17 from http://www.forestry.gov.bn/frst_u_mangroove.htm, Accessded on 12th May 2017

JPH- (Jabatan Perhutanan) (2003b) Distribution of forest by district. Retrieved September 18 from http://www.forestry.gov.bn/frst_distribution.htm. Accessed on 16th May 2017

JPH- (Jabatan Perhutanan) (2003c) Playing vital environmental, development, social and economic roles. Retrieved September 20 from http://www.forestry.gov.bn/frst_forest.htm, Accessed on 20th May 2017

JPH- (Jabatan Perhutanan) (2003) Geography. Retrieved October 8 from http://www.forestry.gov.bn/about_geo.htm, Accessed on 20th May 2017

Khatib MABP, Sirat DPHA (2005) Some aspect of the cultural trends for tomorrow's Brunei-A strategy. In: Thumboo E (ed) Cultures in ASEAN and the 21st country. Uni Press, Singapore, pp 33–49

Kriseinfo (2012) What is a forest fire? Retrieved September 27 from http://www.kriseinfo.no/en/Fire-and-explosions/Forest-fires/What-is-a-forest-fire/

Northern Illinois University Faculty Development and Instructional Design Center (n.d.) Responsible conduct of research. Retrieved December 14 from https://ori.hhs.gov/education/products/n_illinois_u/datamanagement/dctopic.html

Oswell A (2016) Mangrove forest: threats. Retrieved October 2 from http://wwf.panda.org/about_our_earth/blue_planet/coasts/mangroves/mangrove_threats/

Rath AB (2017) Mangrove importance. WWF, Gland

Steenberg JWN, Millward AA, Nowak DJ, Robinson PJ, Ellis A (2017) Forecasting urban forest ecosystem structure, functions and vulnerability. Environmental Management 59(3):373–392

Stewart MB (1986) A geography of Negara Brunei Darussalam. Dewan Basha dan Pustaka, Brunei Darussalam, pp 1–63

Thiessen T (2017) Borneo: natural history and conservation. Bradt, Travel Guides Ltd., Buckinghamshire

Zamora PM (1992) Mangrove resources of Brunei Darussalam: status and management. In: Silvestre G, Matdam HGH, Sharif Uddin PHY, de Silva MWRN, Chua TE (eds). The coastal resource of Brunei Darussalam. Status, utilization and management. ICLARM conference proceedings 34, 214 p. Department of Fisheries, Ministry of Industry and Primary Resources, Bandar Seri Begawan, Brunei Darussalam. And International Centre for Living Aquatic Resources Management, Manila, Philippines, pp 39–58

Mangrove Establishment in an Artificially Constructed Estuarine Channel, Sungei Api-Api, Singapore

19

Suzanna Ramos and Patrick A. Hesp

Abstract

The artificial establishment and natural regeneration of mangroves at Sungei Api-Api, a man-made estuarine channel on the north eastern coast of Singapore, is examined. Several environmental factors affecting mangrove growth are briefly detailed, including: (i) slope gradients, (ii) salinity and tidal inundation levels, (iii) substrate type, and other factors such as tidal currents and propagule establishment. An analysis of the environmental factors affecting Sungei Api-Api mangrove growth indicates that successful growth is attained where: (i) fully saline, tidal inundation occurs over low slopes and (ii) the substrate comprises fine sediments. The artificial establishment of mangroves in a man-made channel is shown to be moderately successful in this area. The Sungei Api-Api trial project provides an excellent example of a potentially significant method by which mangroves can be conserved and reintroduced in urban and near urban tropical environments.

Keywords

Mangrove re-establishment · Conservation · Singapore · Urban canals/channels

S. Ramos
Department of Educational Psychology, College of Education and Human Development, Texas A&M University, College Station, TX, USA

P. A. Hesp (✉)
Beach and Dune Systems (BEADS) Laboratory, College of Science and Engineering, Bedford Park, Flinders University, Adelaide 5042, SA, Australia
e-mail: Patrick.Hesp@flinders.edu.au

19.1 Introduction

The area of mangroves worldwide has been declining for probably several hundred years but most rapidly in the past 50–70 years as coastal urbanisation has increased, and the development of industries such as aquaculture has expanded. According to Duke et al. (2007), mangroves once covered over 200,000 km^2 of coastline globally, but in the last half-century, 30–50% of mangroves have disappeared. This rate of loss is steady and in certain locations, even increasing (Godoy and Lacerda 2015). Future climate change is expected to have a significant impact on mangroves due to increased storminess, temperature, rainfall, and CO_2 changes (Ward et al. 2016). Sea level rise is predicted to have the most significant physical, ecological and socio-economic impacts on mangrove ecosystems (Alongi 2015).

In SE Asia mangrove destruction and loss has occurred on a significant scale due to clearing for urbanization, industrial infrastructure, agriculture, palm plantations, rice production and shrimp farming (Valiela et al. 2001; Friess and Webb 2014; Richards and Friess 2016). Richards and Friess (2016) estimate a loss of 2% or ~100,000 ha between 2000 and 2012 in the SE Asian region. Reviews of mangrove conservation in Singapore indicate significant loss of mangroves from an original 13% to less than 3% by 1994 (Turner et al. 1994; Hilton and Manning 1995; Hesp 1995). By 2000, there was approximately 583 ha of remaining mangroves in Singapore (Richards and Friess 2016). Thus, there was, and still is a pressing need to conserve the last refuges of extensive mangroves which were mainly concentrated in Pulau Tekong, the eastern shore of Pulau Semokau and Pulau Ubin (Chou and Lee 1994). While mangrove loss fell to virtually zero in the 2000–2012 period according to Richards and Friess (2016), mangrove areas will continue to be under threat both in Singapore and elsewhere in SE Asia, and especially as sea level rise continues and the space for landwards translation is unavailable, unless more concerted plans for conservation and reintroduction are implemented.

C. Makowski, C. W. Finkl (eds.), *Threats to Mangrove Forests*, Coastal Research Library 25,
https://doi.org/10.1007/978-3-319-73016-5_19

With a continued demand for land for development, an alternative, or joint option is to regenerate mangroves in viable niches within the urban landscape (Lee and Chua 1985; Lee et al. 1993). This paper outlines an example of mangrove establishment and conservation in an urban environmental setting, describes an initial attempt to ascertain the environmental conditions necessary for successful mangrove establishment, and examine the success or otherwise of the project.

19.2 Study Area

Sungei Api-Api is situated in the north-eastern part of Singapore (Fig. 19.1). The catchment consists primarily of semi-compacted "Old Alluvium" of Pleistocene age (Gupta and Pitts 1992). Holocene sediments form a thin layer in isolated patches over much of the coastal area, and in river valleys and lowlands (Keng 1986). Formerly, the coastline was relatively flat and consisted of sandy beaches, with mangroves dominating the mouth of Sungei Tampines as well as along both sides of Sungei Api-Api (Fig. 19.1; Chew et al. 1987).

The low gradient beaches were utilised for reclamation projects in 1978–1980 with land being created for the development of a park to cater to the recreational needs of the residents in the area. The fill material for the reclamation was excavated from nearby Tampines hills, some 6 km away from the reclamation site. The fill consisted of semi-consolidated sand and gravel (75%) and silt and clay (25%) of Quaternary age (Chew et al. 1987). After the reclamation, the Singapore Housing and Development Board constructed an earth drain to divert Sungei Api-Api from its original course (Fig. 19.2) so that instead of flowing into Sungei Tampines, the mouth of Sungei Api-Api was relocated about 0.45 km to the north.

In May 1991, engineering works undertaken by the Drainage Department of The Ministry of the Environment, involved the upgrading of Sungei Api-Api for more effective drainage by installing 1 m diameter deep-seated contiguous reinforced piles below the low tide level for the section of the river downstream of Pasir Ris Drive 1 to the Sea (Figs. 19.2 and 19.3). Furthermore, the riverbed between the bored piles was deepened by 2 m to provide additional flow capacity for the river (Drainage Dept. Annual Report, 1993).

To enhance the natural setting of Sungei Api-Api, mangrove saplings obtained from Malaysian estuaries were planted along the banks by The Parks and Recreation Department. The saplings comprised *Avicennia marina* and *Avicennia alba,* native to the area (hence, the local term "api-api" meaning *Avicennia*). The saplings, 0.3 in height, were tied to wooden stakes and planted along the lower banks of the river from the mouth to Pasir Ris Drive 1 (Figs. 19.4a, b and 19.5a, b). The planting process began in stages in December 1992, and was completed by March 1993.

19.3 Methodologies

A range of methods were utilised to assess local geomorphic factors, environmental conditions and mangrove growth in 1994–1995.

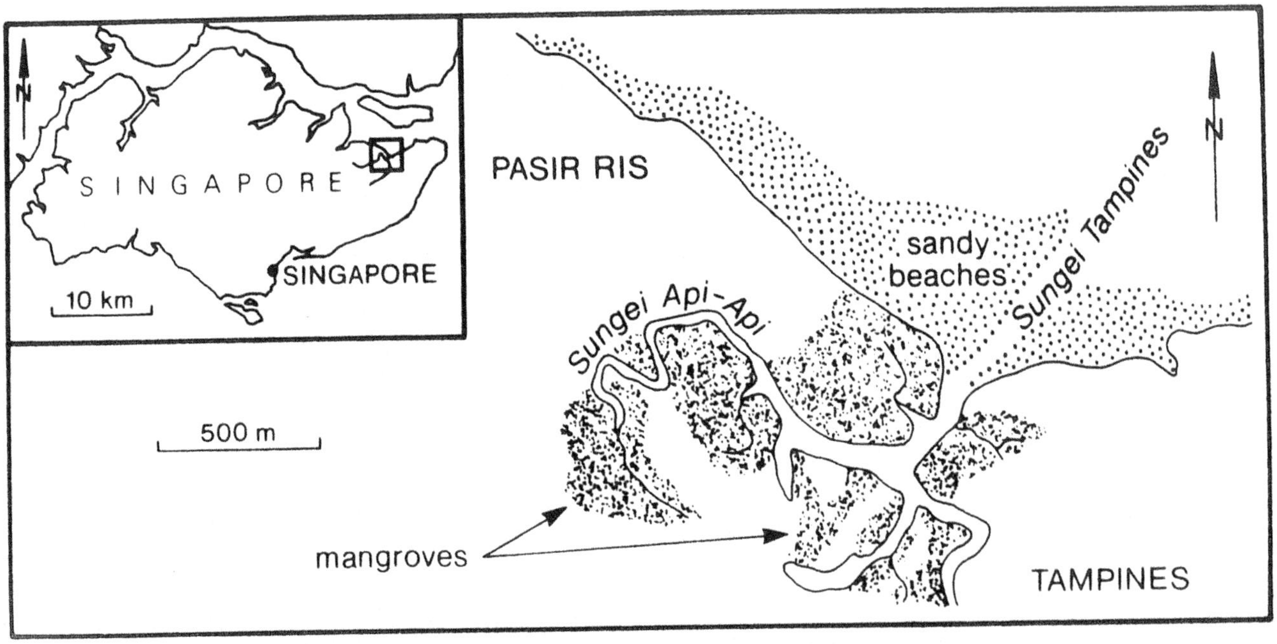

Fig. 19.1 Location map of the Sungei Api-Api area in Singapore illustrating the original estuarine morphology, extent of mangroves and sandy beaches in 1983. (Source: Topographical Map of Singapore, 1953)

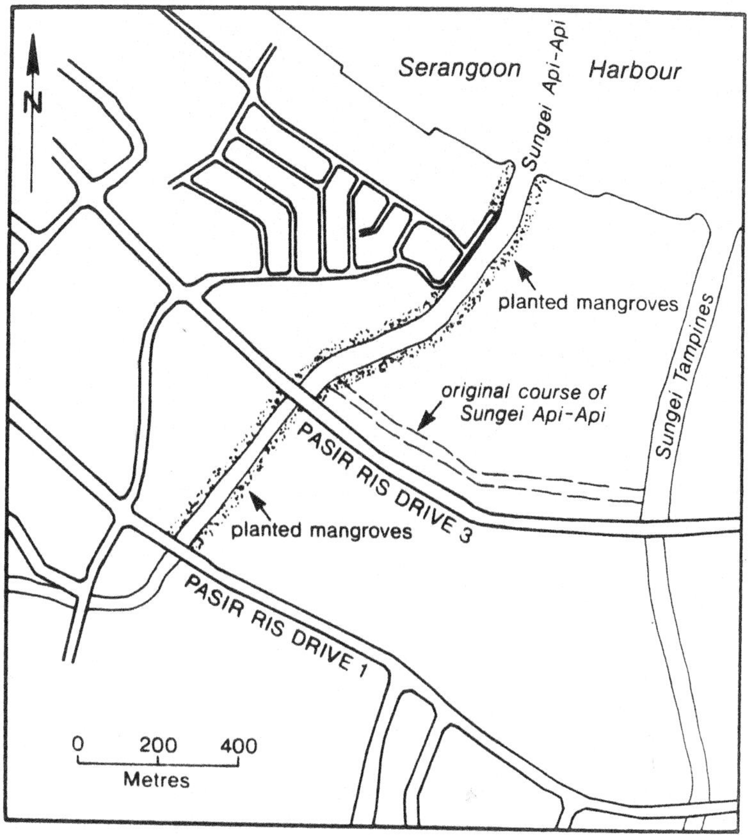

Fig. 19.2 The present constructed course and former course of Sungei Api-Api showing the extent of mangrove plantings along the channel

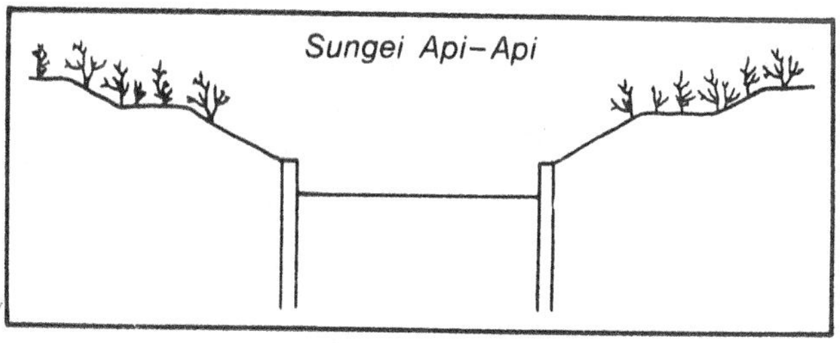

Fig. 19.3 A schematic drawing of the Sungei Api-Api constructed channel as portrayed by the Drainage Department, SingaporeMinistry of Environment

Topographic cross-sections of the river were measured by automatic level and tripod to examine links between the morphology of the man-made channel and mangrove growth, and examine channel changes since the construction of the original channel. The three sites selected for this survey were chosen on the basis that they represented steep, moderate and low gradient slopes. These sites are indicated as lines 1–3 in Fig. 19.6. Concrete drainage reserve markers on the banks were utilised as permanent benchmarks providing a fixed line for repeat surveys. All levels were tied to a standard benchmark. The cross-sections of the river are presented in Fig. 19.7.

Since mangrove communities are said to develop best in sheltered depositional environments, where there is a steady accretion of sediments (Karim et al. 1984; Woodroffe 1992), the gradients of the channel slopes were measured by inclinometer and tape survey. These sites were selected to cover all slope/soil types present and all areas of luxuriant and sparse mangrove growth (Fig. 19.7, lines A–F).

Along the transects described above, contiguous one square metre plots were placed on the ground and a vegetation survey was carried out. In each 1 m^2 plot, the number of mangrove trees, saplings and dead mangroves were noted and their heights taken.

Since sediment characteristics are one of the environmental factors directly affecting mangrove productivity and sediment structure (English et al. 1994), a description of the Munsell soil colour and texture was recorded at each plot. Sediment samples were collected at mid tide level from about 20 cm depth at each transect. This was to establish whether there was a relation between mangrove growth (e.g. as expressed by plant height) and the amount of fine sand and mud deposits.

A salinity survey was conducted at the first cross-profile of the Sungei Api-Api channel utilising a conductivity meter on site and by an analysis of water samples in the laboratory. This particular profile was selected as it was furthest inland that mangroves were found in the study area. Since salinity levels decrease inland, the objective of this was to determine

Fig. 19.4 (**a**): Mangrove plantings on the west bank of Sungei Api-Api immediately after planting in January, 1993; (**b**): The same mangrove plantings in June 1993

Fig. 19.5 (**a**) The S. Api-Api channel during the construction phase. Note the vertical pile walls of the lower channel margins. (**b**) The S. Api-Api estuary in March, 1995. A comparison with Figure 4 shows that the lower mangrove saplings did not survive, while the upper mangroves display rapid growth in the period June 1993 - March 1995. (**c**) - Bottom- Looking upstream from the mouth of Sungei Api Api in November, 2016. Mangrove growth is sustained and significant compared to 1993 and 1995

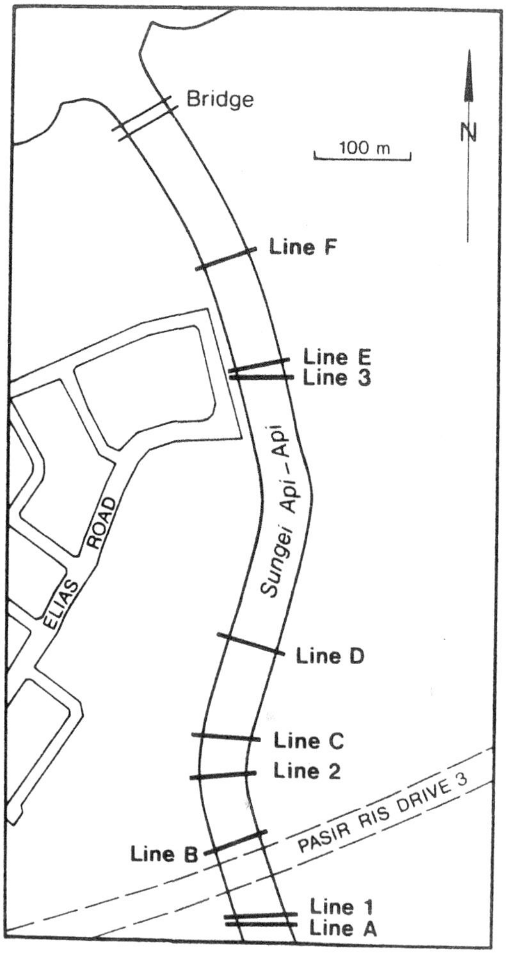

Fig. 19.6 Location of all survey transect lines along the Sungei Api-Api channel

if the salinity levels were adequate for mangrove growth in the more landward sections of the channel.

Current velocities were also measured at 0.2 and 1 m depths respectively during high, mid and low tides across the same transect (Line 1) utilising a Braystoke current metre. Both salinity and current velocity surveys are limited as both were severely hampered by instrument failure.

19.4 Environmental Conditions and Mangrove Establishment

19.4.1 Channel Changes

Schematic diagrams (Fig. 19.3) and photographs of the S. - Api-Api channel (Drainage Dept., 1991) show that the initial channel had a box-shaped profile, with vertical pile sides. By 1995, channel morphology had changed considerably (Fig. 19.7). Although quantitative comparisons cannot be made, the side slopes describe lower gradients indicating sediment accretion and smoothing of the original channel profile. The presence of steep banks in some places (e.g. Line 1, east: Line 3, west) are likely to coincide with the original canal, which was installed below the low tide level (Figs. 19.3 and 19.5a).

Therefore, it appears that the constructed channel banks of Sungei Api-Api were attaining a more natural morphology. This is due to processes such as upper bank erosion and slumping in some areas, probable deposition of sediments transported from the middle and upper catchment (where considerable earthworks were taking place), and probable transport and deposition of sediments into the estuary from the Straits of Johore.

19.4.2 Tidal Limits

Figure 19.7 illustrates three channel cross-sections on which the highest and the lowest limits of mangrove growth are indicated. Mangrove growth here refers to the presence of mangrove trees including saplings. The highest limit reached is generally at, or around mean high water springs at 2.8 m above MSL, except for the west bank of Line 1 (Fig. 19.7), where the highest limit of mangrove growth is at mean high water neap of 2.1 m. The lowest limit of mangrove growth is at, or around, mean sea level at 1.7 m. However, the only exception is Line 1 (Fig. 19.7), where both the east and west banks have lower limits of mangrove growth, which are 0.5 m below the mean sea level. Since the elevation of the slopes and surface morphology above mean sea level determines duration and extent of tidal inundation and exposure (Lear and Turner 1977), the mangrove limits shown in Fig. 19.7 correspond relatively well with the known tidal limits of mangroves. That is between the level of high water springs and a level close to, but generally above mean sea level. This highlights the fact that regardless of channel type (natural or man made), tidal levels, and the frequency of tidal inundation, play an important part in controlling the extent of mangrove establishment as commonly observed elsewhere.

However, in the case of Line 1 (Fig. 19.7), where the lowest limit of mature mangroves is a little below mean sea level, it can be seen that saplings were able to establish themselves below this level. These were saplings which have regenerated naturally. Both the banks of Line 1 exhibited generally steeper slopes than the other cross-sections. Thus, only a small number of the mangroves were able to establish themselves below mean sea level over a limited horizontal distance. The implication here is that the frequency of tidal inundation is adequate for mangrove growth along the channel.

Along the whole channel, the upper limit of mangrove growth was either at mean high water springs or a little below that. The lower limits varied from about mean high water

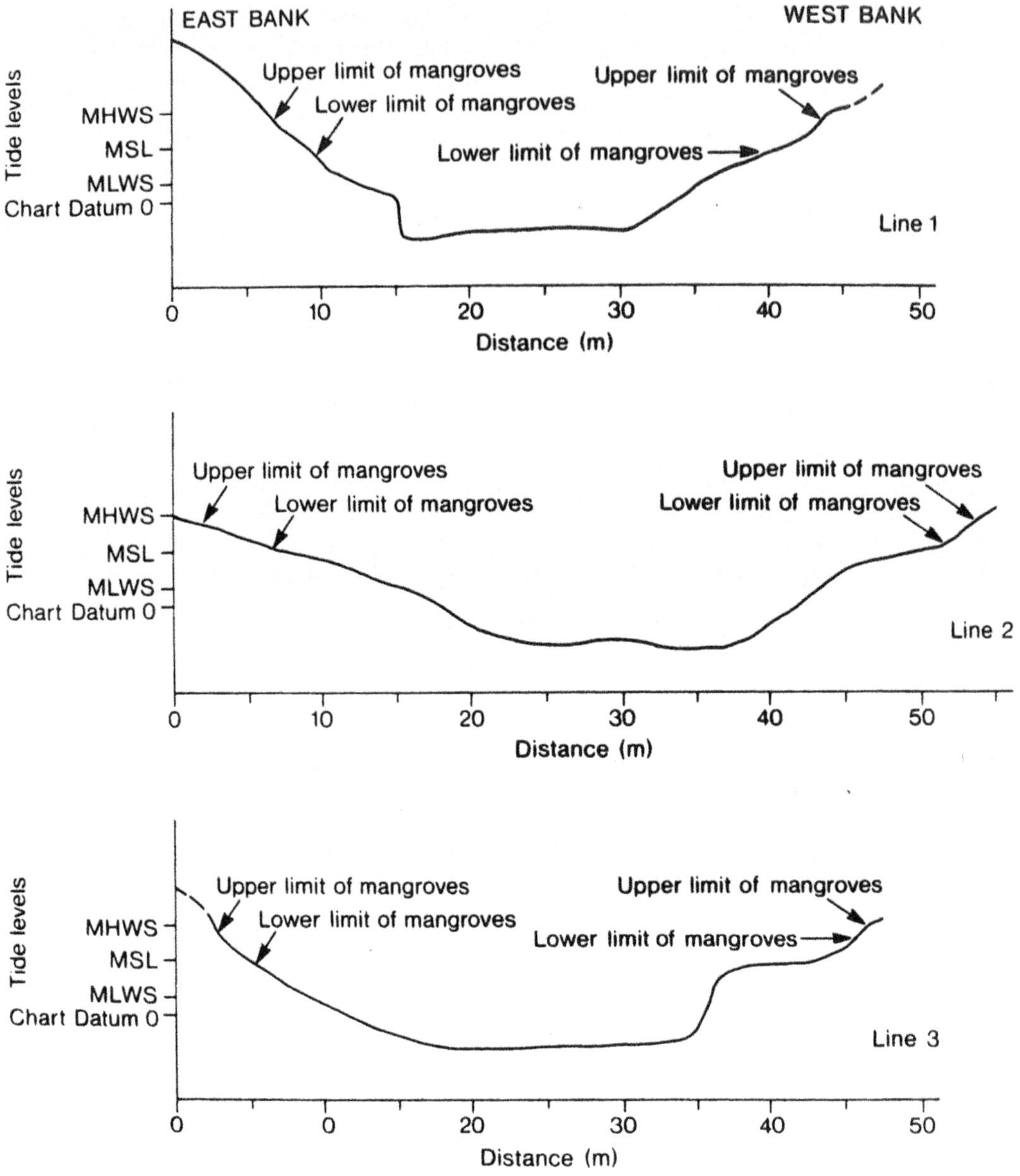

Fig. 19.7 Morphology, tidal levels and extent of mangroves on survey transect lines 1 (uppermost), 2 (middle) and 3 (lowest). Locations indicated on Figure 6

neaps, mean sea level or a little above the mean water neap level. However, in the case of mangrove trees over one metre in height, all surveys found that the mature mangrove trees were thriving at, or above mean sea level, whilst the saplings may occur below this level.

It is not clear whether the saplings will continue to grow fully over a period of time, but no mature trees can be found below mean sea level. This is clearly seen in Fig. 19.4b, where the saplings were growing right at the water's edge below mean sea level about 6 months after they were planted. However, nearly 2 years later, Fig. 19.5b shows that the mature mangroves are thriving some way above the water's edge at mid tide level. This concurs with the known natural range of mangroves (Lear and Turner 1977). Saplings which were closer to the water's edge presumably died because the levels they were planted at did not provide the most suitable conditions for further growth.

19.4.3 Slope Gradients

Surveys of the channel bank gradients (Fig. 19.6, Lines 1–3 and lines A–F) and the numbers of mature mangroves and saplings (defined as < lm in height) on those lines indicate that, with a few exceptions, mangrove trees were commonly found along gentle slopes with gradients from 4 to 16°. Saplings were found along a greater range of gradients, from 4 to 25°. This could imply that the establishment of

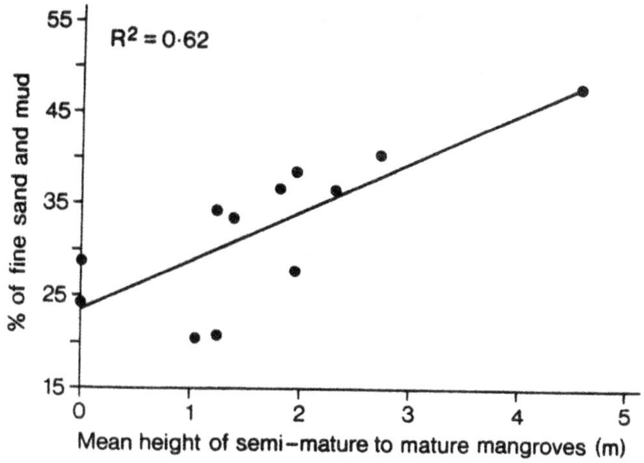

Fig. 19.8 Regression analysis of the percentage of fine sand and silt against the mean height of semi-mature to mature mangrove trees of the same age

mangrove saplings may be influenced more by tidal inundation levels rather than slope gradients, or alternatively that they can propagate across a wide slope range but not necessarily reach maturity on the relatively steeper slopes. A regression analysis of the average number of mature mangroves (excluding saplings) and the mean gradients of all survey lines (1–3; A–F) gave an R^2 value of 0.80. The greatest number of mangroves (average of 30+) was found growing on the lowest slopes (around 4°) (Raj 1995).

19.4.4 Sediments

Sediments were sampled along all bank transects. A regression analysis of the total number of plants (saplings included) on each transect line and the percentage of fine sand and mud present gave a very poor R^2 correlation of 0.23. This is principally because mangrove saplings were found growing across a wide range of substrates in which the fine sand/silt percentages ranged from 20% to 50%.

A reasonable relationship ($R^2 = 0.62$) between the mean height of semi-mature (around lm in height) to mature mangroves (heights ranging from +1 m to 6.5 m) of similar age (i.e. all planted at the same time) and the percentage of fine sand and mud was obtained by regression analysis (Fig. 19.8). The percentage of fine sand and mud increases with the heights of the mangroves. Although the soils were sampled at a depth of 20 cm, it can be reasonably assumed that the mangrove trees do assist in the accretion of fine sediments, and to a greater degree than mangrove saplings. This is probably due to their more intricate pneumatophore networks, slowing current velocities and trapping sediments. Therefore, it may be that this data only indicates that fine sediments have been trapped by the mangroves since they became established. Alternatively, it may also indicate that longer-term mature mangrove establishment is dependent on a higher percentage of fine sand and mud in the substrate as found, for example, by Karim et al. (1984) and Siddiq and Khan (1990), who demonstrated an apparent relationship between the growth of mangroves and the amount of silt deposited.

19.4.5 Salinity Levels

Salinity levels for one tidal cycle are illustrated in Fig. 19.9. At high tide, the conductance values were similar to seawater collected from the Straits of Johore (34%). Despite this, mangrove growth is not luxuriant at this cross-section (cf Fig. 19.7, Line 1). As for the lowest tide levels, the readings on the surface show that the water is predominantly brackish to freshwater, but at a depth of one metre, higher conductivities are recorded. Among all the mangrove species, *Avicennia marina* appears to have the widest range of salinity tolerance, and it can grow in almost fresh water (MacNae 1968). In the case of Sungei Api-Api, although the very limited measurements indicate a marked salinity gradient at low tide, salinity does not appear to be a limiting factor in mangrove establishment.

19.4.6 Current Velocities

Current velocities were very low in the three surveys conducted. The highest mean velocity recorded was only 9 cm s^{-1} during high water. Mean velocities were around 4–5 cm s^{-1}. This very limited data may not be at all representative of general conditions; alternatively if indicative, current velocities would certainly not appear to be a limiting factor in mangrove establishment in this channel as mangroves can survive in estuaries experiencing significant current speeds (Woodroffe 1992).

19.5 Discussion and Conclusion

The data obtained revealed that generally, mangrove growth was greater on gently sloping banks and that slope gradient in relation to tidal inundation plays an important role in sustained mangrove growth. When the mangroves were established, they were probably able to stabilise the banks further through accretionary processes. However, it was found that several parts of the riverbanks were too steep,

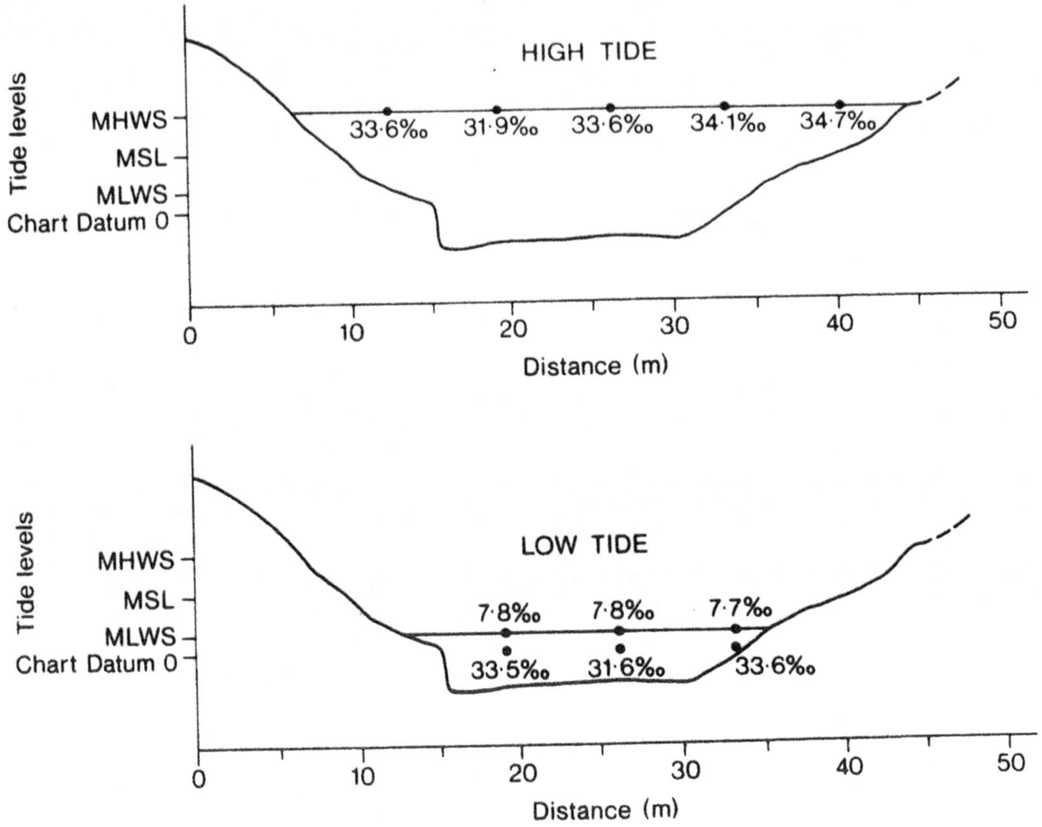

Fig. 19.9 Salinities measured at the water surface and at one metre water depth at high and low tide on the 13th Jan., 1995. A thin brackish layer overlies sea water during low tide at this location

resulting in slope instability. Although measures were undertaken to prevent slumping, such measures were proven to be ineffective, even over a relatively short term.

Therefore, rather than constructing relatively steep banks for landscaping purposes, gentler gradients should be created to allow greater numbers of mangroves to be established. In relation to this, since tidal inundation also dictates the sustained growth of mangroves, the planting of mangrove saplings should not be too far below the mid tide level. Although saplings are able to grow below this level initially, many saplings later died at S. Api-Api, apparently because of regular inundation.

On the upper reaches of the river, salinity levels have been known to decrease (Lee 1976; Chou and Lee 1994). One possible solution here could be the planting of another mangrove species that can thrive in less saline waters such as the *Bruguiera* species. One advantage of planting different species along the same channel is that any change in the quantity of fresh water inflow, or in the rate of sediment supply, leading to variations in soil and water salinity, is less likely to result in the partial or total destruction of the mangrove environment.

On the whole, the trial project at Sungei Api-Api to regenerate mangroves has been relatively successful. Many stretches of the channel exhibited luxuriant mangrove growth, and recent photographs taken in November 2016 indicate continued growth and success. It seems reasonable to conclude that as long as the gradients are relatively gentle for initial mangrove establishment, and planted mangroves are initially placed above mean sea level and up to mean high water springs, mangroves will be able to survive. A nearby source of propagules may hasten further mangrove growth, but even if this is not present, the propagules could come later from the mature mangroves themselves. The establishment and growth of mangroves at Sungei Api-Api provides valuable information on how mangrove growth is affected in an urban environmental setting. Although only limited information on the dynamics of a mangrove "ecosystem" in an artificial canal environment exists to date, the regeneration of mangroves can be extended to other parts of the island (and elsewhere in SE Asian urban environments).

The regeneration of mangroves in an urban environment can be seen as a complement to the protection of existing mangrove forests. While more research needs to be done on how the urban environment can be used as a habitat for mangroves, this initial study indicates that successful establishment is possible and that the establishment and regeneration of mangroves will most likely be an increasingly important option as urbanisation continues at a rapid rate in SE Asian coastal cities.

Acknowledgements This research was conducted while the authors were at the Dept. of Geography at the National University of Singapore. Thanks to Dr. Mike Hilton (Otago University) who provided the original idea and impetus (including the provision of the magnificent research vessel, the RV Hiltonia) for the study. The authors wish to thank the Dept. of Geography and NUS, the Singapore Housing and Development Board, and the Drainage Dept. within the Ministry of Environment for their support. We also sincerely thank Dr. Chang Chew Hung (Nanyang Technological University) for his invaluable assistance, and Karen Puklowski (Massey University) for cartography.

References

Alongi DM (2015) The impact of climate change on mangrove forests. Curr Clim Chang Rep 1(1):30–39

Chew SY; Ho SK; Wong PP, Leong YY (1987) Beach development between headland breakwaters in a low wave energy environment, Pasir Ris, Singapore. In: Proceedings of the 20th coastal engineering conference:1016–1031

Chou LM, Lee SK (1994) Conservation of marine coastal ecosystems in Singapore, ASAIHL seminar on Sustainable Development of Tropical Marine and Coastal Ecosystems, Penang

Duke NC, Meynecke JO, Dittmann S et al (2007) A world without mangroves? Science 317(5834):41–42

English S, Wilkinson C, Baker V (eds) (1994) Survey manual for tropical marine resources. ASEAN-Australia marine Science Project, Townsville

Friess DA, Webb EL (2014) Variability in mangrove change estimates and implications for the assessment of ecosystem service provision. Glob Ecol Biogeogr 23(7):715–725

Godoy MD, Lacerda LDD (2015) Mangroves response to climate change: a review of recent findings on mangrove extension and distribution. An Acad Bras Cienc 87(2):651–667

Gupta A, Pitts J (eds) (1992) Physical adjustments in a changing landscape: the Singapore story. Singapore University Press, Singapore

Hesp PA (1995) The environmental impact assessment process in Singapore with particular respect to coastal environments and the role of NGO's. J Coast Conserv 1:135–144

Hilton MJ, Manning S (1995) Conversion of coastal habitats in Singapore: indications of unsustainable development. Environ Conserv 22(4):307–322

Karim A, Hossain Z, White KJ (1984) Study of the growth of mangrove plants in relation to edaphic factors in coastal afforestation plantation of Chittagong. In: Soepadmo E, Rao AN, Macintosh DJ (eds) Proceedings of the Asian symposium on mangrove environment research and management. University of Malaya, Malaysia, pp 195–198

Keng G (1986) Estuarine systems: a systems approach, vol 1. CRC Press, Boca Raton

Lear R, Turner T (1977) Mangroves of Australia. University of Queensland Press, St. Lucia

Lee SK (1976) Problems of tree growth on reclaimed land in Singapore. In: Proceedings of symposium on our environment, Singapore, Institute of Natural Science, Nayang University, pp 76–80

Lee SK, Chua SE (1985) Creation of parks on reclaimed land in Singapore. In: Proceedings of 3rd symposium on our environment, Faculty of Science, National University of Singapore: 43–51

Lee SK, Goh KC, Yee SO (1993) Afforestation as an interim measure to improve reclaimed lands in Singapore. Land Degrad Rehabil 4:1–10

McNae W (1968) A general account of the flora and fauna of mangrove swamps and forests in the Indo-West Pacific Region. Adv Mar Biol 6:73–270

Raj SJ (1995) The management of mangroves in an urban environment: a case study of Sungei Api-Api. Unpublished honours thesis, Department of Geography, National University of Singapore

Richards DR, Friess DA (2016) Rates and drivers of mangrove deforestation in Southeast Asia, 2000–2012. PNAS 113(2):364–349

Siddiq NZ, Khan MAS (1990) Growth performance of mangrove trees along the coastal belt of Bangladesh (Part I). In: Field CD (ed) Mangrove ecosystems occasional papers, No. 8. United Educational, Scientific and Cultural Organisation, New Delhi, pp 5–14

Turner IM, Tan HTW, Wee YC et al (1994) A study of plant species extinction in Singapore: lessions for the conservation of tropical biodiversity. Conserv Biol 8(3):705–712

Valiela I, Bowen JL, York JK (2001) Mangrove forests: one of the world's threatened major tropical environments. Bioscience 51 (10):807–815

Ward RD, Friess DA, Day RH, MacKenzie RA (2016) Impacts of climate change on mangrove ecosystems: a region by region overview. Ecosyst Health Sustain 2(4)

Woodroffe C (1992) Mangrove sediments and geomorphology. In: Robertson AI, Alongi DM (eds) Coastal and estuarine studies: tropical mangrove ecosystems. American Geophysical Union, Washington, DC, pp 7–41

Interaction of Mangroves, Coastal Hydrodynamics, and Morphodynamics Along the Coastal Fringes of the Guianas

20

Erik A. Toorman, Edward Anthony, Pieter G. E. F. Augustinus, Antoine Gardel, Nicolas Gratiot, Oudho Homenauth, Nicolas Huybrechts, Jaak Monbaliu, Kene Moseley, and Sieuwnath Naipal

Abstract

The mangrove belt along the coast of the three Guianas, South America, forms a unique ecosystem and acts efficiently as a natural soft coastal defence structure. The general mechanisms have been studied for over four decades and the processes governing the coastal morphodynamics are now understood, at least qualitatively. They consist of an interaction between mangroves, hydrodynamics and sediment mechanics. Twenty percent of the mud discharged by the Amazon in the Atlantic Ocean is transported to the west along the coast by waves and currents in discrete mud banks of a few 10 s of km length, which travel at a speed of the order of 2 km/year. During the presence of a mud bank waves are damped, mud is trapped and colonized by mangroves. Once a mud bank has passed, the waves can attack the shore again. This results in a cycle of land accretion and erosion, with an average net gain of 1 m coast per cycle of roughly 30 years. However, in locations where too many mangroves have been removed, the coast has lost its natural resilience and the settlements and fields are prone to flooding, a concern that increases with climate change and predicted sea-level rise (SLR). Hard coastal defence structures, such as those in Guyana, are expensive and not sustainable. Based on many lessons learnt, pilot projects on mangrove rehabilitation have started. At the same time research efforts are undertaken to allow making quantitative estimates of the potential risks for the coastal communities. For this purpose, numerical prediction models are developed which can compute currents, wave action and sediment transport along the coast of the Guianas. Different climate change scenarios can be investigated. These models can serve in the near future as decision support tool for the local authorities for the management of the coastal zone.

Keywords

Black mangroves · Amazon river mud · Sea level rise · Morphodynamic response · Mangrove rehabilitation · Guianas coast

20.1 Introduction

Most mangroves are found in coastal lagoons and in estuaries, sheltered from waves. There are very few places around the world where they form a protecting belt along the coast itself, where they are subject to direct wave action.

E. A. Toorman (✉) · J. Monbaliu
Hydraulics Division, Department of Civil Engineering, KU Leuven, Leuven, Belgium
e-mail: erik.toorman@kuleuven.be

E. Anthony
Université Aix-Marseille, CEREGE UMR CNRS, Aix-en-Provence, France

P. G. E. F. Augustinus
Faculty of Geosciences, Utrecht University, Utrecht, The Netherlands

A. Gardel
CNRS LEEISA USR 3456, Cayenne, Guyane Française, France

N. Gratiot
Centre Asiatique de Recherche sur l'Eau CARE, Bach Khoa University, VNU-HCM, Ho Chi Minh City, Vietnam

Université Grenoble Alpes, CNRS, IRD, Grenoble, France

O. Homenauth
National Agricultural Research and Extension Institute (NAREI), Mon Repos, East Coast Demerara, Guyana

N. Huybrechts
Direction Technique Eau, Mer et Fleuves, CEREMA, Margny Lès Compiègne, France

Sorbonne Universités, Université de Technologie de Compiègne, CNRS, UMR 7337 Roberval, Centre de Recherche Royallieu, Compiègne, France

K. Moseley
Mangrove Restoration and Management Department, NAREI, Mon Repos, East Coast Demerara, Guyana

S. Naipal
Land and Water Management, Department of Infrastructure, Anton de Kom University of Suriname (AdeKUS), Paramaribo, Suriname

C. Makowski, C. W. Finkl (eds.), *Threats to Mangrove Forests*, Coastal Research Library 25,
https://doi.org/10.1007/978-3-319-73016-5_20

20.1.1 The Guianas Coast

Probably the most remarkable mangrove coastal fringe is found along the Guianas coast, stretching from the mouth of the Amazon River in northern Brazil all the way west to the mouth of the Orinoco River in Venezuela, a stretch of roughly 1600 km (Fig. 20.1). Unfortunately, the mangrove belt has been reduced or removed in many places, making place for other land use (mainly agriculture, aquaculture and urban expansion), creating many vulnerable spots where hard coastal protection structures had to be built to protect the hinterland against flooding.

The investment and subsequent maintenance costs of these dikes are extremely high. This is especially problematic for Guyana, where mangroves have been replaced by a sea defence wall along about 40% of the coastline in order to protect the polders which are used for agriculture. Pelling (1999) presents an analysis of the vulnerability of the Guyana coast from the perspective of political ecology, starting with a historical review, which traces its roots. The estimated yearly maintenance cost for Guyana is very high. A cost/benefit analysis by Odle and Arjoon (1971) indicates that too little has been invested in the maintenance of the sea wall. The Guyana Ministry of Finance (1996) estimated the approximate annual cost to maintain and rehabilitate sea defence works over the short term to about 7.5 million USD annually, 5 million USD over the medium term and 2.5 million USD over the long term. Eventually, under the tenth European Development Fund framework from 2007 to 2013, the EU–Guyana cooperation focussed on sea defences. This resulted in a Coastal Engineering Design Manual for the Guyana Sea and River Defences, compiled by four consultants, which

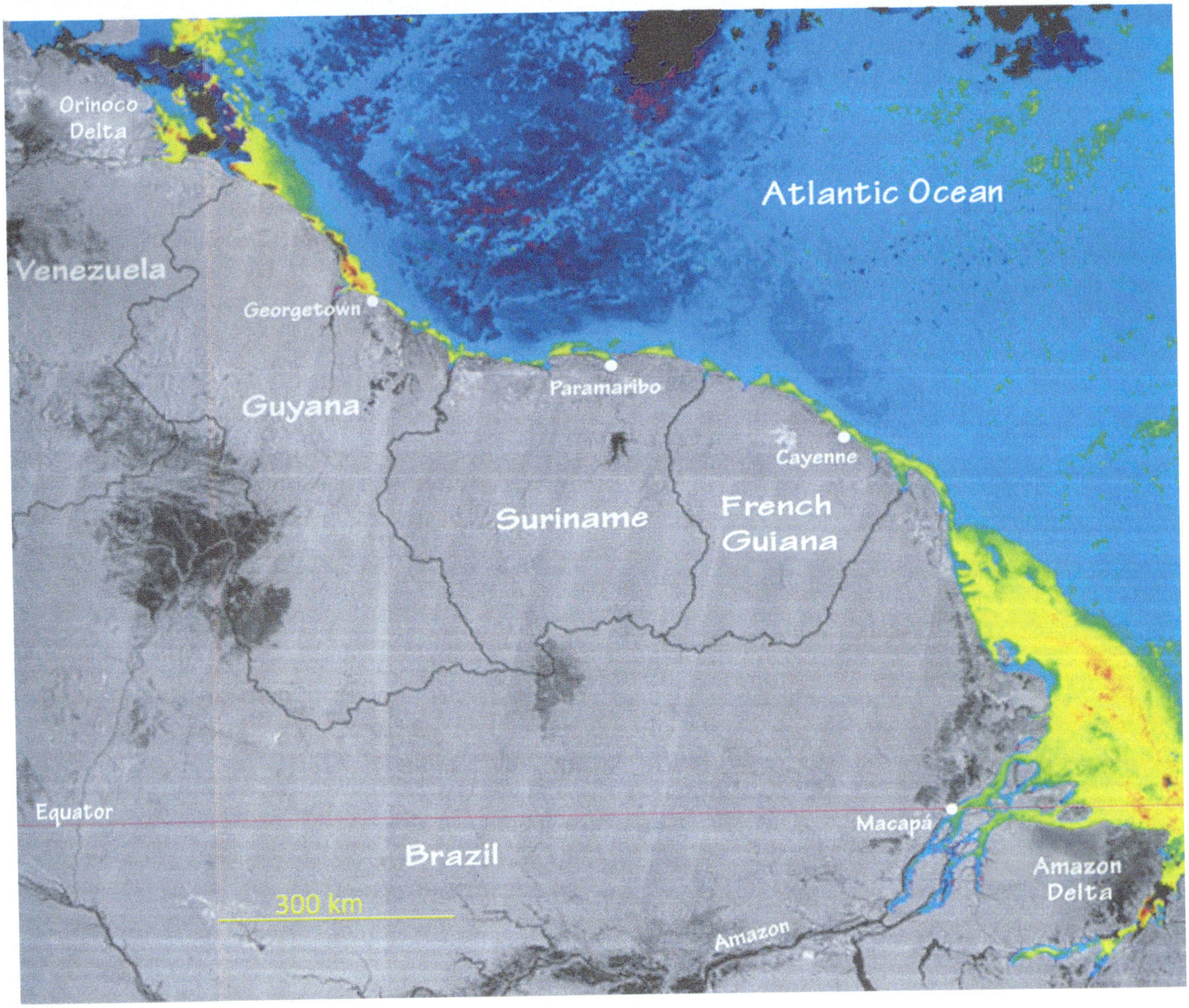

Fig. 20.1 Situation of the study area. Satellite image (processed), showing the turbidity along the north-eastern coastline of South-America (NOAA images, October 1999; Laboratoire Regional de Télédétection, CNRS – after Gratiot 2011)

was handed over in July 2016 by the EU, together with funding of 10 million USD to get the necessary actions started (Guyana Government Information Agency 2016).

Moreover, hard sea defence walls too close to the high-water line, as built in Guyana, prove to increase the problem of mangrove establishment, because the reflection of waves on the hard structure enhances the resuspension of sediments (Anthony and Gratiot 2012). By fluidizing the mud flat, the waves weaken the erosion resistance. Fluidized mud on the bottom may start to flow down the intertidal flat to deeper water offshore. By generating more turbulence, cohesive sediment flocs break up into even finer particles that more easily wash away.

The damage is much less in Suriname and French Guiana (Guyane Française). In Suriname, only a few kilometres of dike are constructed in Nickerie and recently in Coronie. By building the Coronie dike a bit further from the high-water line, behind a remnant of mangroves, waves will not directly impact with the hard structure and reflection of waves on the wall is reduced. In the meantime, the construction has been temporized because during the operations a new mud bank arrived, which has rapidly been colonized by mangroves. This decreased the necessity and changed the priorities.

In French Guiana only dozens meters of riprap have been constructed in urban places in Cayenne and Montjoly. In the Anse district of Kourou city a 900 m long sandy dune has been erected in 2016 to protect houses from erosion and submersion. Local authorities opted for this temporary solution awaiting to find a more long-lasting soft solution.

Over the past years, a growing awareness has risen among policy makers and the population that integrated coastal zone management requires a more sustainable coastal defence strategy, certainly in the light of climate change and associated sea-level rise (SLR) in particular. Thanks to increasing understanding over the past years of the natural processes controlling the interaction between morphodynamics and vegetation and their potential to create natural barriers to stabilize beaches and shores, the concept of soft shore protection has become an interesting design strategy for shore management. This chapter gives an overview of how this concept is currently under investigation for the Guianas mangrove coast.

20.1.2 Mangroves Along the Guianas Coast

20.1.2.1 Species

Mangroves are trees with unique adaptations, allowing them to survive on anaerobic, soft muddy soils in relatively high saline environments. This explains why they are found in coastal areas, where they are not competed away by other species.

Unlike other coasts, the Atlantic coast of the Americas knows only a few species of mangroves. The coastal fringe in the Guyanas is dominated by the black mangrove (*Avicennia germinans*) with prop roots (Fig. 20.2). Red mangroves (predominantly *Rhizophora mangle*), with stilt roots, occur along the borders of tidal rivers and creeks. They are rarely found along the coast itself (Fig. 20.3).

White mangroves (*Laguncularia racemosa*) are less common; they occur on somewhat higher, more sandy soils, e.g. creek levees. They also have prop roots.

20.1.2.2 Mangrove Degradation

Mangroves have adapted to grow in saline environments. Under hypersaline conditions however, even *Avicennia germinans* cannot survive. In the Guianas, these conditions may develop under natural circumstances, e.g. when a mangrove forest is closed from the sea by a new forming chenier (Fig. 20.4), impeding the natural in- and outflow of water. Especially in the dry season, the salinity of the stagnant water can rise quickly, due to evaporation, causing the sudden death of the mangroves. During the next wet season, the salinity will be lowered, and through a transition phase with saline herbs, the mangrove vegetation can restore itself again in the course of time (Fig. 20.5).

Sometimes however, the degradation of mangroves is due to human impact, such as in Coronie (Suriname), where the freshwater inflow from the coastal swamps is cut off by the main coastal road. The seawater evaporates and leaves behind extreme high-saline pools, in which the mangroves cannot survive (Fig. 20.6). In these cases, the degradation usually has a more permanent character.

20.2 Morphodynamics of the Guianas Coast

20.2.1 Mud Bank Migration

The Guianas coast has been studied for a long time, with the work by Augustinus (1978) as the main pioneering physical research. These studies have revealed the major processes that explain the formation of the Holocene coastal plain over the past 6000 years, which has gradually accreted at a mean rate of the order of 2 m per century (Rine 1980). This can be summarized as follows (Allison 2000; Augustinus 2004).

About 20% of the continuous flux of sediments by the Amazon River into the Atlantic Ocean is transported westward along the North coast of South America. Some 150×10^6 tons is transported yearly as suspension load, another 100×10^6 tons per year in the form of huge mud banks, which travel at an average speed of roughly 2 km/year in westerly direction. Waves are the driving force behind this migration (Nedeco 1972; Allison and Lee 2004). The

Fig. 20.2 Black mangrove (*Avicennia germinans*), with its well-developed pneumatophores (prop roots). (Credit: P. Augustinus)

question why an important part of the mud transport takes place in discrete banks of roughly 30 km length is still not well understood.

In the classic model, mud banks are shore face attached (e.g. Delft Hydraulics Laboratory 1962; Allersma 1971; Augustinus 1978; Wells and Coleman 1981; Rine and Ginsburg 1985). This is shown in Fig. 20.7a. The migration of the mud banks is caused by erosion of their trailing edge (East side) and the deposition, mainly of fluid mud, at their leading edge (West side).

Figure 20.7b shows the mud bank dynamics according to Allison and Lee (2004). In their hypothesis the mud bank is disconnected from the coast. During flood, fluid muds are driven shoreward over the mud bank, filling the upper intertidal area between the mud bank and the coast. The fluid mud originates from the transported suspension load as well as from erosion of the trailing edge of the mud bank. If the concentration of silt particles in the water exceeds the point of hindered settling, fluid mud is formed. It is so soft, that it absorbs the energy of surpassing waves by visco-plastic deformation (i.e. friction between the sediment particles). This means that waves running into an area with fluid mud are strongly attenuated and even may disappear (Fig. 20.8). As a result, deposition is further enhanced.

In the course of time, the mud consolidates and becomes more compact. A hydrographic pattern develops and mangroves start to grow (Fig. 20.9). Mangrove roots give rigidity to the soil and further its compaction by extracting water (Fig. 20.10).

Moreover, the complex system of above-ground prop roots, branches and stems act as an increased surface roughness for incoming waves and currents, which therefore lose energy. Vital mangroves capture sediments, resulting in coastal accretion in the quiet zone between the mud bank and the shore. New mangrove propagules will colonize the new land and stabilize its soil enhancing further accretion. However, with ongoing consolidation (i.e. self-weight compaction; Toorman 1996) towards the east, the mud matures into young clay, which can be eroded.

Therefore, when the mud bank has passed (after roughly 30 years), the waves will break without significant attenuation on the shore, washing away sediments between the mangrove roots and eventually uprooting the trees: the shore will erode. The uprooted trees may pile up against the living trees, forming extensive accumulations of stems and branches (Fig. 20.11). These act as an increased bottom roughness, slowing down erosion.

Fig. 20.3 A remarkable sight: a lonely, persisting Red mangrove (*Rhizophora mangle*) that refuses to be washed away along the eroded coast in front of Weg naar Zee, Suriname (Credit: E. Toorman)

When, after some 30 years, in which a mud bank and the adjacent interbank area has been completely migrated westward, a new mud bank settles in front of the mangrove forest under erosion, a new cycle begins.

If the width of the vital mangrove forest has been sufficient large, the coast may have accreted. However, the opposite is also possible. In the period 1947–1966 the coast of Suriname shows net erosion over its full length. From 1966 onwards, net coastal accretion takes place (Table 20.1). This net accretion can (at least partially) be explained by the observed increase of the sediment discharge of the Amazon River. A recent study by Martinez et al. (2009), based on the analysis of an extensive time series of field data (10-day sampling) and satellite data show an increase of the yearly-averaged sediment discharge from 1995 to 2007 by 20%, which is not correlated to the nearly unchanged annual river discharge. The increase may be attributed to stronger erosion caused either by climate change (intensification of rainfall) or by land cover changes resulting from deforestation (Callede et al. 2008). Since another study (Filizola and Guyot 2009) indicates that 90% of the sediment load originates from the Andes, the deforestation of the lowland rainforest of the Brazilian shield most likely is not the main contributor.

This large-scale coastal behaviour is probably caused by a (cyclic) shifting of wind frequencies. In the sixties and seventies of the former century, winds from an east-north-eastern direction steadily increase and so does the wind strength. Waves generated by winds from that direction approach the shore under a smaller angle. Therefore, the component of the wave energy flux parallel to the coast (P_2a) becomes more important, compared to the component of the wave energy flux perpendicular to the coast (P_2o) (Fig. 20.12). So sediment transport is furthered and erosion is diminished. This must lead to a growth in length of the mud banks, which is indeed the case (Fig. 20.13).

Since 2001 the average length of the mud banks is slightly decreasing. This could indicate the beginning of an increase in the frequency of more northern (or north-north-eastern) wind directions. In addition, the El Niño and the La Niña events also have impact on wind and rainfall. Sometimes even hurricane influences are noticed, even though the Guianas coast lies outside the hurricane area. To our

Fig. 20.4 Dead mangroves (*Avicennia germinans*) in a hypersaline environment behind an actual chenier in East Suriname (Credit: P. Augustinus)

knowledge, these possible effects on the mud bank migration have rarely been studied (Van Ledden et al. 2009).

20.2.2 Coastal Erosion Analysis

Erosion of the muddy coast of the Guianas begins with the development of small pits in the bare mud surface in front of the mangrove vegetation. The slightly consolidated mud in front of the *Avicennia germinans* vegetation has a brownish colour due to the presence of diatoms. The small erosion pits are filled in afterwards with lighter coloured fresh mud, which makes them easily recognizable (Fig. 20.14).

With increasing degree of consolidation, the erosion depressions become larger and more elongated. They develop in the direction of wave attack (Fig. 20.15). During ebbtide, when exposed to the sun, the walls desiccate, the clay shrinks and cracks appear parallel to the walls. With the next rising tide, rotational sliding along the sheer faces occur, resulting in a widening of the depressions. In areas where many of these depressions are concentrated the original clay surface is more lowered than elsewhere. In this way, an indented erosion coasts develops (Fig. 20.16).

Erosion is dominant in the interbank area. The incoming waves erode the clay surface and cheniers are formed. If sand is available from a local source, e.g. the Marowijne River or the Corentyne River, it is transported in the coastal zone by longshore drift and beach drift. The cheniers develop at or just above mean high tide level. During higher water levels, sand is washed over the top of the chenier to its backside. This causes the chenier to migrate landward (Fig. 20.17). In the mean time they may grow in height and therefore progressively counteract coastal erosion.

If there is no sand available from local sources, sand can be winnowed from the mud. The muddy sediment contains approximately 2% very fine sand, which originates from the Amazon River. Due to the continuous alternation of compression and decompression caused by the waves, the clay can be stirred up. The fine particles are transported to the west by weak currents, the sand grains settle down again. In this manner sand is concentrated at the surface. Chenier formation under these circumstances begins at about low water level. Due to the very fine grainsize of the sand, the cheniers are much flatter and they show ripples on the foreshore (Fig. 20.18).

In periods of severe erosion, coastal retreat can go so far that an old chenier is reached, which starts to function as a

Fig. 20.5 Recovering saltmarsh vegetation in a former hypersaline environment behind an actual chenier in East Suriname (Credit: P. Augustinus)

local sand source. Under these circumstances a sand accretionary coast will develop. A number of chenier ridges is formed, one after another, in seaward direction. Beyond backward bending parts of the coast, spits develop, e.g. at Braamspunt. These beaches often show rhythmic features like beach cusps (Fig. 20.19).

20.3 Mangroves as Sustainable Coastal Defence

It has been claimed that mangroves provide protection against waves (Mazda et al. 1997; McIvor et al. 2012), storm surges (Zhang et al. 2012) and tsunamis (Danielsen et al. 2005). This is correct in the sense that they absorb energy from these floods. But they will be uprooted and washed away nonetheless when subjected to repeatedly breaking waves (Fig. 20.20) or to a single violent flood, leaving a more vulnerable coast for the next decades. Of course, the piles of uprooted trees on the shore themselves also form a hindering for waves and currents and contribute to coastal protection (Figs. 20.11 and 20.20b).

The 2004 Indian Ocean tsunami in particular raised a discussion on the effectiveness of mangroves, mainly resulting from interpretations of the data which did not account for all factors. Eventually this debate was settled with a common publication (Feagin et al. 2010). However, the important role of the sediments has not been considered.

20.3.1 The Physical Functioning

To understand how mangroves can act as coastal defence, it is necessary to look at the interaction between three key players: the trees, the sediment and the hydrodynamic forcing. The functioning can only be understood by considering the interaction between each of these three and their combined effects.

20.3.1.1 Interaction Between Mangroves and Hydrodynamics

Many studies exist on how vegetation reduces the energy from currents and waves (Mazda et al. 1997; McIvor et al. 2012). In the case of mangroves, the vegetation can be

Fig. 20.6 Dead trees in hypersaline water in Coronie, Suriname, cut off from the inland fresh-water marshes by the main east-west road (Credit: E. Toorman)

considered rigid. Energy is dissipated by friction with the stems and, in the case of *Avicennia*, the pneumatophores, which implies conversion of hydrodynamic energy into turbulence, which is eventually dissipated by viscosity into heat. It is then evident that the degree of dissipation depends on the density of the trees and their size.

20.3.1.2 Interaction Between Mangroves and Sediment

Like all trees, mangroves find their stability by developing an appropriate root system in the soil in which they grow and from which they extract their nutrients. Mangroves typically grow on mud deposits, to which they are adapted.

Mud usually refers to a dense sediment-water mixture dominated by clay particles. Such sediment is referred to as cohesive sediment (Mehta 2014). Clay particles in water hardly occur as individual primary particles, since they tend to aggregate and form flocs. When these flocs settle they form a so-called soil skeleton where the "particles" (i.e. flocs) are stacked and support each other, whereas the open spaces between the particles (the pores) are filled with water.

A mud deposit is sometimes compared to a card house structure or a sponge structure: a low volume of solids and a large volume of voids filled with water. But due to accumulation of depositing sediment flocs, the skeleton is subject to an increasing load of particles and the network may deform and bonds between particles broken, such that the card house collapses. The latter goes very slowly, because the pore water needs to escape, but due to the small size of the pores and the hydrophilic nature of the clay particles the pore water experiences a lot of capillary flow resistance. This compaction is called self-weight consolidation (Toorman 1996; Mehta 2014).

Black mangrove propagules can develop their roots very easily in the soft top soil of these mud banks. With the growth of the tree, the tree becomes heavier, but it manages to divide its weight over a larger area by spreading the roots in the horizontal direction. The roots do not grow deeply, because

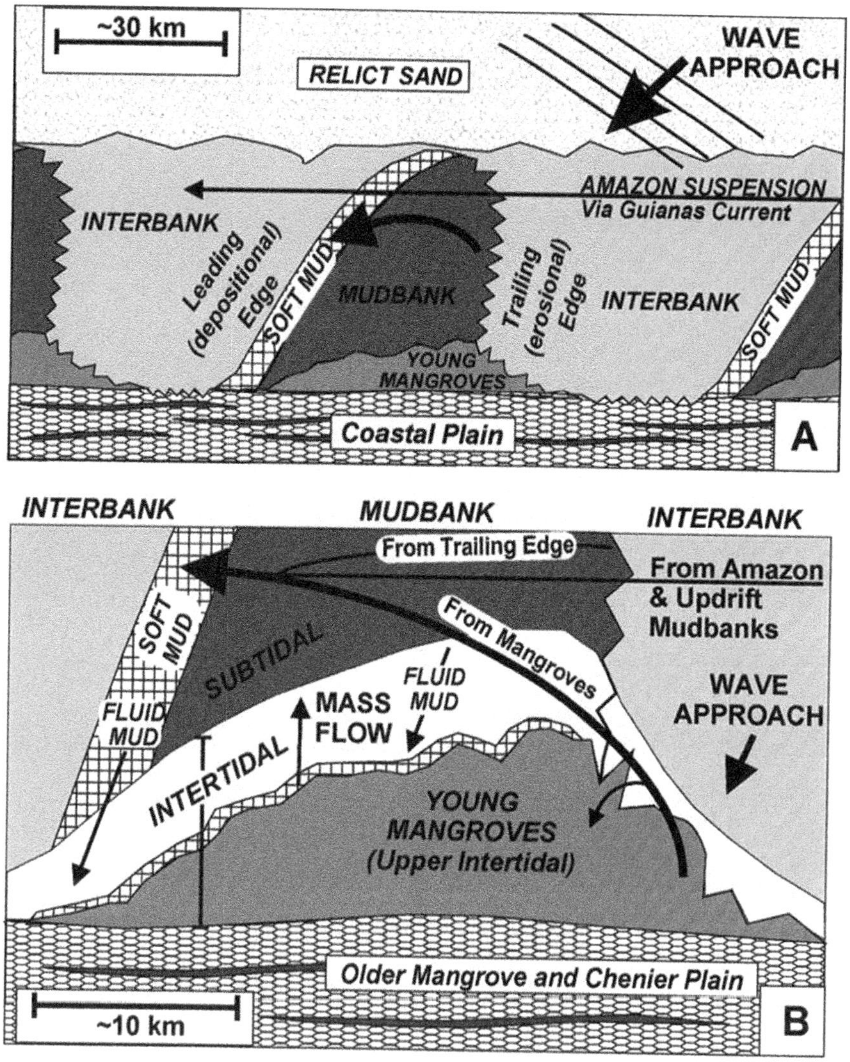

Fig. 20.7 The classic model for mud bank migration (**a**) and mud bank dynamics (**b**) (Allison and Lee 2004)

due to the low permeability of the mud, the compacting mud becomes deoxidized very quickly at 5–10 cm depths. This explains the more the adaptation of the root system.

Furthermore, the mangroves-sediment entity forms an ecosystem of high biodiversity. The growth of plants and the activity of animals alter the composition, by adding organic material to the sediments, and subsequently the structure of the sediments. From studies elsewhere, it is known that organic matter increases the strength and thus the erosion resistance of sediments (Grabowski et al. 2011).

20.3.1.3 Interaction Between Hydrodynamics and Sediment

Realizing the cohesive nature of the sediments, one can look at the interaction between marine hydrodynamics and the sediments. The sediment deposits are fed by sediments transported by tidal and wave-induced currents. But when the wave movement is more dynamic the water can also remobilize sediments that lie between the roots. When inundated and subjected to waves, the erosion resistance of the sediment may reduce due to pore pressure build-up (i.e. partial fluidization), making the sediment even more prone to erosion. Hence, mangrove trees at the seaside edge of the fringe that are most subject to waves in interbank areas are most prone to sediments weakened and eventually washed away between the roots (Fig. 20.21), which may destabilize the tree. Indeed, it is seen that many mangroves simply tip over on the eroded shore (Fig. 20.22).

Where there is persistent erosion, as seen for instance at the outlet of some drainage channels in Coronie (Suriname), cliff erosion can occur, which washes away even the older sediments beneath the root system (Figs. 20.10 and 20.23).

When waves move over a mud bank, the pores are subjected to an oscillating pressure field. This tends to induce pore water flow within the soil skeleton, but due to the capillary resistance, it cannot follow the frequency of the wave. This lag leads to pore pressure build-up. The resulting normal stresses counteract the effective stress due to self-

Fig. 20.8 To the left one notices the sudden disappearance of waves running over fluid mud (Credit: P. Augustinus)

weight consolidation. If the excess pore pressure becomes higher than the submerged weight of the overlying sediment (i.e. the effective stress), the bed is fluidized: the particles are pushed aside, bonds are broken and they can move relative to each other (similar to quick-sand).

At the other hand the cycle of expansion and compression of the pore pressure results in a net elasto-plastic deformation of the skeleton. The mechanical friction between the particles as well as the pore water flow induce energy loss. Since this energy originates from the wave motion transferred to the pore water, it therefore results in damping of the waves, when they travel over a mud bank.

Fluid mud on a slope may tend to move downhill as a visco-plastic fluid. Gratiot et al. (2007) hypothesize that this is counteracted by wave-induced drag in the wave direction (usually towards the shore). This balance may explain the cross-shore profile of a mud bank. The highest thickness is expected to occur where the waves have lost their energy.

Thus far, wave-mud interaction had been studied with an elastic description of the deformation of the mud layer (reviewed by Mehta et al. 1994). But this is only valid for very small deformations. In this case one can compute the motion of the fluid mud layer even analytically for idealized homogeneous mud layers. These results were successfully compared to laboratory experiments (Sakakiyama and Bijker 1989).

Over a soft fluidized mud layer, the deformation results in visco-plastic flow and its non-Newtonian behaviour can be represented by a (non-ideal) Bingham rheological model (Toorman 1997). The rheological parameters change in time due to the break-up and restoration of interparticle bonds. This thixotropic behaviour can be modelled with an additional kinetic equation, which describes the break-up under shear and recovery of the soil skeleton structure at rest. The corresponding rheological closure (Toorman 1997), implemented into a CFD software, has been demonstrated to be able to simulate the damping by waves (Toorman 2008; Villarroel 2009). It proved to be necessary to account for an increase in resistance with self-weight compaction, i.e. the effective viscosity increasing with depth and density, in order to obtain realistic velocity profiles, as observed in laboratory experiments (Villarroel 2009).

Therefore, mangroves need shelter, which is provided along the Guianas coast by the mud banks. Waves are damped and do not reach the shore. Moreover, the mud deposit forms a low berm which canalizes the runoff from

Fig. 20.9 A vital mangrove vegetation growing on relatively young clay (Credit: P. Augustinus)

land into a creek running in front of the mangroves (Fig. 20.24). The berm itself is also slowly colonized by propagules that get stuck on these slightly higher elevations, where optimal conditions for seedling trapping occurs (Proisy et al. 2009).

20.4 Ecological Engineering: Mangrove Restoration

Efforts have been made in Guyana and Suriname to restore mangrove woods along the coast at places where the natural belt has disappeared or became too narrow to be sufficiently resilient to survive a cycle of mud bank migration. Initially, mangroves have been planted only in the swamps behind the former coastline, where the vegetation had been degraded (Augustinus, pers. comm.). Rehabilitation on the intertidal mud flats, where the plants are subject to the forces of waves and currents has only recently been tried, with mixed success.

Worldwide, a lot of mangrove restoration projects have failed, often by lack of understanding the proper conditions for development and growth (Ellison 2000; Winterwerp et al. 2013).

Local ecological experts expressed their low confidence in the success of mangrove rehabilitation along the Guianas coast, simply because there are sufficient propagules and the necessary dispersion by currents and waves is favourable, distributing them along the entire coast (Erftemeijer and Teunissen 2009). They concluded that if natural regeneration does not occur in certain places, it implies that the local conditions are unfavourable and one should not expect that planted mangroves will survive. Therefore, the first key to mangrove restoration is to understand the local conditions. Once the unfavourable conditions are understood, one can try to design a methodology to change the local conditions to favourable ones.

20.4.1 General Principles

Consultation of various international experts and lessons learnt from other restoration projects elsewhere in the world

Fig. 20.10 Eroded coast with protruding mud, protected by roots of *Avicennia germinans* (Credit: P. Augustinus)

eventually helped to develop a general strategy (Field 1999; Lewis III 2009; Winterwerp et al. 2013).

Prior to the design of a suitable rehabilitation strategy, a study has to be made in order to understanding the ecology of the local mangrove species, i.e. reproduction patterns, propagule distribution and conditions for seedling establishment and growth. A model for the dynamics for the Guianese mangroves has been proposed by Fromard et al. (2004) (Fig. 20.25). Next, one has to understand which conditions prevent the natural regeneration of mangroves in the problematic areas.

To rehabilitate a target area, first the optimal natural conditions have to be restored. This involves measures to improve the hydrological conditions (i.e. ensuring fresh water runoff and restoring tidal creeks) and the onshore sediment fluxes. Secondly, it requires a large enough buffer zone to give sufficient flexibility to overcome periods of degradation during storms or of absence of soft mud banks (interbank periods). Since sufficient propagules are available along the Guianas coast, actual planting then should no longer be necessary.

The main challenge along the Guianas coast is to provide conditions that allow mangrove regeneration during interbank periods in areas where the minimal width of the mangrove fringe is below the critical size, which is estimated to be of the order of 1 km, while the width of a resilient coastal fringe should be of the order 5 km (Anthony and Gratiot 2012). Since there is no shelter provided by an onshore mud bank during this period, artificial shelter has to be provided which allows entrapment of sediments and reduces wave action (by reducing wave heights by damping and reducing wave reflection).

Sediment entrapment areas are well known from saltmarsh works (Winterwerp et al. 2013). The latter type is based on its successful use centuries ago for the creations of polders in the Netherlands. They consist of areas enclosed at both cross-shore boundaries, reducing longshore currents within the unit, and seaside boundary, reducing wave action and tidal currents. Fine sediments can settle and deposit more easily under these quieter conditions. Moreover, cohesive sediments can form larger flocs, which settle faster (Mehta 2014).

20.4.2 Guyana Mangrove Restoration Project

The Guyana Mangrove Restoration Project (GMRP – www.mangrovesgy.org) commenced in 2010 and was co-financed

Fig. 20.11 Accumulation of uprooted mangrove trees against the living trees in Coronie, Suriname (Credit: P. Augustinus)

Table 20.1 The net erosion (−) or accretion (+) in tons per year, for the coast of Suriname, deduced from a comparison of aerial photographs and satellite images from different years

Period		
Interval	Number of years	Net amount of sediment $\times 10^6$ tons/year
2001–2007	6	+59.44
1992–2001	9	+29.07
1981–1992	11	+48.43
1970–1981	11	+7.61
1966–1970	4	+2.18
1957–1966	9	−1.00
1947–1957	10	−0.82

Remarks: The coastline in 1981 has been drawn using aerial photographs. For 1992, satellite images with a lower resolution are used. The amount of sediment in the intermediate period is therefore less accurate (probably too high), compared to the other intervals

by the Government of Guyana and the European Union under the Global Climate Change Alliance. The Project was implemented by the National Agriculture Research and Extension Institute (NAREI) with support from a multi-agency oversight committee known as the Mangrove Action Committee. Following the completion of the project phase in 2013, the Government of Guyana committed to continued mangrove restoration and management, and the Project Unit was integrated into the NAREI structure (NAREI 2014).

GMRP adopted a multidisciplinary approach to mangrove restoration, protection and management. This included restoration through the use of coastal engineering structures and

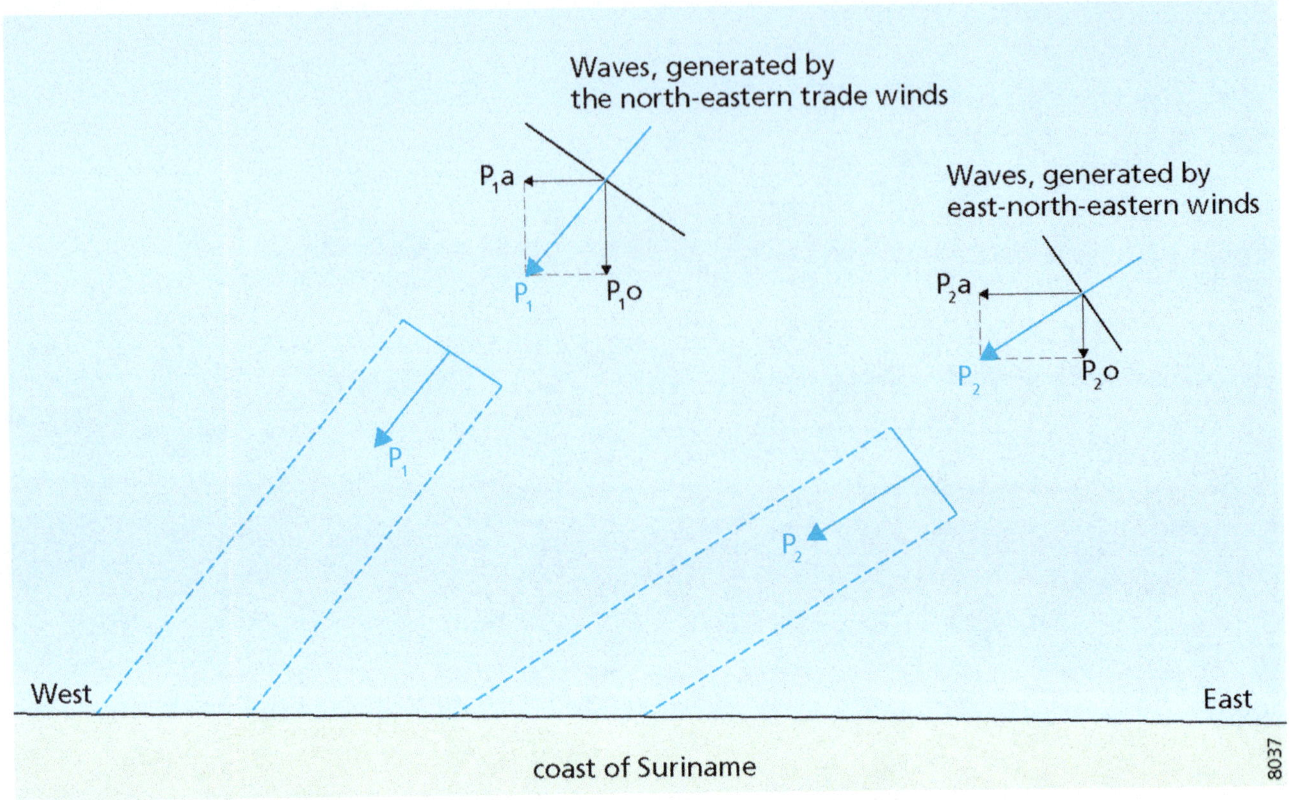

Fig. 20.12 The influence of the difference in the angle of approach of waves, generated by the north-eastern trade winds (**P$_1$**) and waves generated by east-north-eastern winds (**P$_2$**). Energy flux **P** is defined as the transport of energy through a vertical plane, parallel to the wave crests, per unit crest length. **Po** is the landward directed component of the wave energy flux, **Pa** is the component parallel to the coast (After Wong et al. 2017)

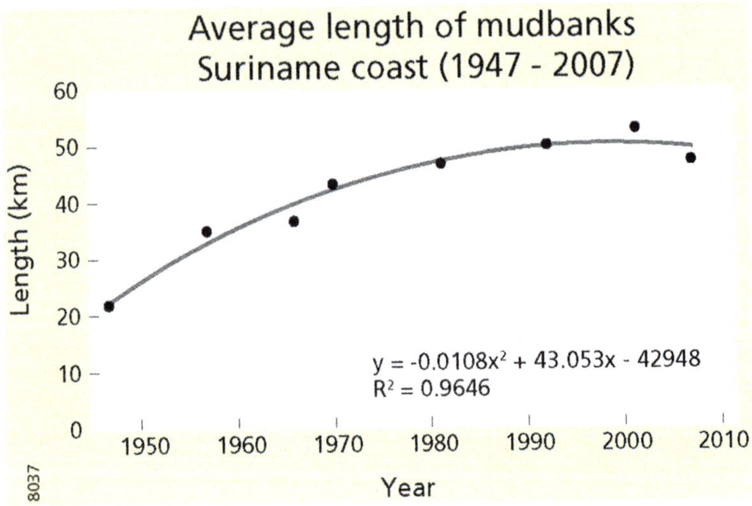

Fig. 20.13 Average length of the mud banks in Suriname, deduced from aerial photographs and satellite images (After Wong et al. 2017)

Fig. 20.14 Erosion pits in the mud surface in front of vital black mangroves (*Avicennia germinans*) in Suriname. They are filled in afterwards by lighter coloured fresh mud, which contrasts with the brownish colour of the mud surface, due to the presence of diatoms (Credit: P. Augustinus)

seedlings, enforcement by rangers, community development and alternative livelihoods, research and development and public awareness and education.

20.4.2.1 Mangrove Restoration Interventions Implemented Under GMRP

Guyana adopted the principles of Ecological Mangrove Restoration (EMR) (Lewis III 2004) to design and implement its mangrove restoration program following failed attempts at restoration primarily through seedling plantations at the beginning of the project.

A range of interventions aimed at restoring degraded coastal mangroves and creating a suitable environment for mangrove to colonize the coastline were implemented as part of the mangrove restoration programme. These interventions included:

- Mangrove seedling planting
- Construction of coastal infrastructure such as groins and breakwaters
- Spartina grass planting
- Construction of restrictive gates and fences to reduce the impact of anthropogenic activities

Mangrove Seedling Planting

Seedling planting is only possible in areas or sites that have a suitable mud elevation. Based on the most successful sites and comparison of natural forest, NAREI has established a suggested guideline range for planting *Avicennia germinans* based at 2.3–2.7 m above chart datum (Landell Mills Limited 2013). Only the *Avicennia germinans* species is planted based on the natural zonation pattern in Guyana.

From 2010 to 2016 over 500,000 mangrove seedlings were produced in community nurseries and planted in 16 villages along the coastline (Fig. 20.26).

Initial efforts of restoration were unsuccessful due to limited local expertise. The success rate increased during the second year of project implementation following the fielding of technical experts and training of local project team. Detailed site assessments where introduced and pre-planting surveys and surveys of the topography of both successful and unsuccessful planting areas showed that

Fig. 20.15 Coastal erosion in Coronie, Suriname, in 1972 (Credit: P. Augustinus)

surface elevations of unvegetated mudflats needed to be in the range of +2.0 to +2.5 m MSL in order to support planted mangroves (Landell Mills Limited 2013). Following the guiding principles of EMR, seedling planting was only utilized to increase recovery time of a site that met the necessary criteria to support restoration.

Field monitoring of planted restoration sites (Figs. 20.27, 20.28, 20.29, and 20.30) facilitated a comprehensive statistical analysis for survival and growth rates. Assessments completed (Machin and Lewis III 2013; Adams 2014) indicated that survival rates and growth varied among restoration sites and can be grouped into three scenarios:

1. Fast growing and high survival (Wellington Park, Village #6–8, Chateau Margot, Success, Lima, Better Hope). Mud elevation 2.13–2.49 cm above CD;
2. Slower growing and high survival (Greenfield, Hope Beach, Le Ressouvenir, Felicity). Mud elevation 1.61–2.37 cm above CD;
3. Low or no survival (Hope, Greenfield, Victoria, Section C Enterprise, Mon Repos, LBI/Triumph, Buxton, Nooten Zuil, Lusignan). Mud elevation 1.9–2.4 m above CD.

Research completed by Robertson (2014) on the protective capacity of the restored Chateau Margot/Success restoration site indicated that a 3 year old black mangrove forest with a bandwidth of 50 m can reduce a 0.43 m wave at open sea to 0.001 m at the coast. Modelled wave heights prove that a 50 m mangrove forest can reduce a 0.43 m wave to approximately 0.2 m (Robertson 2014).

Coastal Engineering Interventions

Coastal engineering interventions were implemented at sites along the coast to encourage accretion. These interventions aim to create a suitable environment that will encourage mangrove restoration by increasing shoreline elevation.

Following collaboration with the Guyana Sea and River Defence Division, three options for construction of low-cost coastal structures was developed using the following technologies: geotextile tubes, brushwood dams and rubble mounts.

Geotextile tube technology was implemented at two sites to form a detached offshore breakwater and a groin. Two geotextile tube projects were completed at Devonshire Castle in Regions No. 2 and Victoria in Region No. 4. *Spartina* grass was introduced at both sites following completion of the structures to support soil consolidation.

Brushwood dams were constructed using locally available building material i.e. bamboos (Fig. 20.31). The brushwood dams were designed to mimic the mangrove roots by trapping sediments thus increasing shoreline elevation. Brushwood dams were constructed at three locations in Region #2 and Region #4 in 2013, 2015 and 2016. Iron bamboos were used as piles ranging from 10 to 15 cm in diameter and having lengths between 7 and 9 m. Brushwood dams were filled with bundles of bamboos of the same size to that of the bamboo piles to complete the system.

The result of the impact of coastal structures on mangrove regeneration is expected to be a continuous long term study. Data collected on structures are varied. Results indicate that structures placed at a site with a visible mud bank offshore and experiencing accretion, have greater potential to trap sediments and increase elevation. Critical to the success of coastal structures is the completion of geotechnical surveys to ensure design parameters are suitable for the foreshore characteristics. Geotextile groin constructed at Devonshire (Fig. 20.32) showed the best results in relation to changes to the shoreline line conditions and rapid colonization of planted *Spartina* and mangrove seedlings. While there has been significant consolidation of the soil at this site, elevation taken at two periods showed a marginal decrease in elevation within a 6 month period from 2.39 to 2.35 m above chart datum (CD). This was due to consolidation of the soil.

Spartina Grass Planting

In 2013 the GMRP implemented a new restoration strategy for sites along the coastline where mangrove planting was not suitable due to poor mud consolidation and low elevation. The *Spartina brasiliensis* (*Raddi*) is one of the lesser known

Fig. 20.16 Indented erosion coast in East Suriname in 1972 (Credit: J. Schulz)

coastal salt marsh species along Guyana's coastline. The presence of *Spartina brasiliensis* beds in Guyana has been described as occurring on raised areas on nearshore mud banks where the grass quickly establishes, develops into large clumps and raises the level of the mud to form large stands. These stands are then typically colonised by *Avicennia* and *Laguncularia* mangrove species, after which they eventually disappears beneath these forests (Machin 2012).

Spartina was planted to stabilize sediments and support accretion. The use of the *Spartina* was introduced late in the project implementation as it was not a part of the original project plan and was a relatively new concept to the local project team. Following a technical mission from restoration expert Robin Lewis III this intervention was introduced and has been incorporated into the restoration program. *Spartina,* which is native to the Guyana coastline, was transplanted to nine sites along the coast which did not meet the criteria to support mangrove restoration due to poor consolidation.

The grasses planted have had mixed results. Grass planted at BV/Triumph, Kilmarnock and Walton Hall was affected by heavy waves, *Sargassum* and livestock grazing. However *Spartina* planted at Victoria, Village #6–8, Devonshire Castle and La Belle Alliance have been more resilient and have extended the area coverage significantly (Fig. 20.33). Combining *Spartina brasiliensis* with structures has shown tremendous effects at sites where implemented (Figure 20.34). The soil consolidation at these sites also improved significantly in a short period.

Challenges

While the mangrove seedling planting program was deemed relatively successful (Topper 2012), assessments conducted revealed that there are limited sites available that are suitable for mangroves to colonize. Site assessments indicated that the elevation was either too low to support mangroves or the sediments on the shoreline was too fluid and not consolidated enough to support young seedlings.

The importance of mud bank movement along the coast of Guyana and its influence on mangrove management and restoration has been recognized and documented (Gratiot 2011; Welage 2005). However due to limited research on the movement and mapping of mud banks along coastal Guyana, the project continues to be challenged to effectively plan restoration activities based on definitive data on the current and future location of mud banks.

Fig. 20.17 A landward migrating chenier along the coast of Commewijne, Suriname (Credit: P. Augustinus)

20.4.3 Mangrove Rehabilitation in Suriname

20.4.3.1 Experimental Pilot Project in Moy (Coronie District, Suriname)

From 2010 to 2013, an original pilot project by AdeKUS, mainly financed by the Suriname Conservation Foundation, was set up in Moy, along the Coronie coast to plant mangroves on a bare mudflat, the front of a newly arriving mud bank, and study the conditions. The planting and monitoring was done using mud sledges, as used by the local crab hunters. It was noticed that the small plants, grown in a nursery, where not resistant to the force of inundating water. But they were also too heavy to remain vertically upright in the upper mud layer. They tend to sink to a depth where oxygen is depleted which implies the death of the mangrove. Therefore, the young plants where fastened on a stick, which was put deep enough in the mud (Fig. 20.35). This allows the root system to develop and establish a natural equilibrium between the extend of the root system, the weight of the plant and the condition (i.e. compaction) of the sediment. This also protected the plant against being washed away.

This project continued until the natural colonization of the new mud bank overtook the experimental site. It could not be determined whether the plantation experiment had any influence on the colonization speed of the natural process. From the point of view of the abundance of propagules, one would not expect so. But the existing plantation may have speeded up the sedimentation of the migrating mud bank, allowing a faster accumulation of fresh mud at the optimal elevation for propagules to settle. It has indeed been observed that the first seedlings settled in the cracks in the by sun desiccated mud of the highest elevations of the mudflat, between the planted mangroves.

West of this pilot site, the Coronie sea defence wall is under construction. This wall blocks the free runoff of the inland fresh water, carrier of nutrients to potential new mangrove regeneration at the seaside. This water is collected in a ditch at the landside of the dike and can be flushed to the sea through a few gates in the wall. Therefore, Dr. Naipal from AdeKUS proposed to design a water distribution system to disperse this water more equally through a permeable pipeline collector at the seaside along the entire dike (since a

Fig. 20.18 Fine sandy chenier in Coronie, Suriname, in 1972. The development of this type of chenier starts at low water level. Note the ripples on the foreshore (Credit: P. Augustinus)

trench would silt up). This system has not yet been implemented and tested.

20.4.3.2 Sediment Trapping Unit Pilot Project in Suriname

A new pilot project (funded by Conservation International Suriname), based on the principles of sediment trapping units, inspired by the successful application in Indonesia (Tonneijck 2013; Tol 2016), has been implemented in 2015 at Weg naar Zee (Wanica district, near Paramaribo, Suriname), one of the most threatened coastal areas of Suriname. Here the choice went to brushwood walls forming an enclosed area over a seaward width of about 200 m on the mud flat (Fig. 20.36). The dam structure consists of wallaba (*Eperua falcata*) wooden poles (10 cm diameter), driven into the ground 2 m deep (2/3 of their length), spaced 0.75 m apart in parallel lines 0.5 m apart, filled with brushwood and tightened with plasticised wire. The advantage of this type of permeable walls is the fact that the sediment-rich water can enter the enclosed area where the sediments can deposit due to the much quieter conditions. It would actually be better to reduce the permeability close to the bottom, making sure that trapped sediment will not so easily escape with ebb currents or due to gravity currents of soft liquid mud.

The results of the first 16 months of monitoring of the sediment deposits in nine locations in this unit shows a period of deposition (5–10 cm from June to December 2015, most of which during October–December), followed by a shorter period of erosion (most of which in January 2016: rapidly down to +1 cm relative to the initial level). Subsequently gradual deposition was again observed. It is hypothesized that this alternation of erosion and deposition can be related to the on average much stronger north-eastern winds in the periods December–May, compared to the weaker, more eastern directed winds in the period June–November (NOAA Global Model wind data, averaged over the period 1997–2003; Winterwerp et al. 2005). The rapid erosion in January 2016 may indicate extreme conditions.

The following months the net deposition increased to higher levels, generating mud to the high-high waterline, reached at spring tide. Since at this elevation the fresh mud is much more exposed to drying by the sun (until the next spring tide), the material becomes much more compact and cracks are formed (Figure 20.37). These are the ideal

Fig. 20.19 Seaward extending beach near Diane Creek, Commewijne, Suriname, with rows of beach cusps (Credit: P. Augustinus)

Fig. 20.20 Left: Waves impacting on the Suriname coast. (Credit: S. Naipal) – Right: Uprooted black mangroves along the Suriname coast (Credit: E. Toorman)

Fig. 20.21 *Avicennia* roots exposed due to hydrodynamic forcing which washes away the sediments at high water (Credit: E. Toorman)

conditions for mangrove propagules to develop. Natural colonization of these higher elevations has indeed started.

The success of this STU is also explained partially by the available large sediment supply during this period, which is attributed to the approximation of a mud bank, which is crossing the Suriname river. Increased mud deposits and spontaneous mangrove colonization has been observed also a bit further to the west, but this area seems already eroding away again, indicating that the slope in front is too steep (Fig. 20.38). Further analysis and monitoring needs to be done in order to gain better insights.

Care has to be taken that the sediment entrapment would not lead to disturbing the sediment balance at the alongshore downstream side of the entrapment unit. By trapping the sediments in a specific area, there may be shortage of sediment supply behind the unit. Furthermore, the diverted longshore currents may increase in magnitude and may cause increased erosion in front of the unit. The latter will lead to a concave profile of the tidal flat which will hinder the onshore sediment flux beyond.

20.5 Research

Many studies have been undertaken along the Guianas coast in order to understand its dynamics. The coastal erosion problems in Guyana (Delft Hydraulics 1962; Nedeco 1972) and the Surinam Transportation Study (Nedeco 1968), investigating improvement of the conditions for navigation of ocean-going vessels, triggered the need to study the physical conditions in order to design the sea defence wall and for finding solutions for optimal access to the ports in between the mud banks. The corresponding engineering study reports remain an important source of information and data for the region. Major scientific contributions were subsequently realized by Augustinus (1978), Rine (1980), Wells et al. (1981) and Wells and Coleman (1981). These studies have resulted in a good qualitative understanding of the processes. However, in view of management, it is necessary to be able to make quantitative predictions of the expected evolution of the coast, particularly in view of different climate change

Fig. 20.22 *Avicennia* tree about to tip over (Credit: S. Naipal)

scenarios. Research in the twenty-first century therefore has been focusing on collecting more field data, remote sensing data and on the development of numerical models for morphodynamic predictions.

20.5.1 Research in French Guiana

The French Guiana coast constitutes an ideal laboratory for studying the coastal margin as a highly active interface between intense physical processes and induced ecological changes ranging from mud bank colonization by pioneer mangroves to adult mangrove forest destruction. The coast of French Guiana may be considered the most pristine of these coasts in terms of mangroves and their conservation, since no human impact on the mangrove system has been observed, thus far.

Research in French Guiana has been conducted on mud bank morphology, hydrodynamics and sediment characteristics, with emphasis on the processes involved in bank migration and destruction, both driven by waves, but also in the consolidation of the surfaces of mud banks that is a pre-requisite for colonization by mangroves. The aim of this research, the overarching aspects of which are summarized here, has been to gain a better understanding of overall mud bank and coastal dynamics at various spatial and temporal scales, and how these interact with mangrove ecology, especially how mangroves become established, function, and are destroyed in this highly changing setting.

Although there is corpus of old and generally descriptive unpublished studies on these aspects, much of the recent research on mud banks and seafront mangroves commenced in the late 1990s using an approach combining field monitoring and remote sensing. Given the size of a typical mud bank (several 10–100 s of km^2) and the problems of accessibility in this dynamic mud setting, remote sensing, calibrated by ground truth studies, has clearly been an area where research has led to breakthroughs in understanding sediment dispersal and accumulation patterns along what is considered as the world's longest and most dynamic muddy coast.

Fig. 20.23 Cliff erosion by waves (Credit: S. Naipal)

20.5.1.1 Field Work

Field studies on mud banks and mangroves involve considerable logistical difficulties, including the challenge of accessing experimental sites. In spite of these constraints, field work conducted in French Guiana has been fundamental in revealing processes of mudflat accretion and consolidation and mangrove colonization. Lefebvre et al. (2004) were the first to conduct high-resolution field topographic surveys that highlighted the marked surface variability of mud banks. Earlier, Fromard et al. (1998) identified, from field reconnaissance surveys, six stages of mangrove ecology, ranging from development through adulthood to destruction, that are strongly influenced by mud bank activity: (a) pioneer; (b) young; (c) adult; (d) mature; (e) mixed; (f) cemetery. The onshore-alongshore transport of mud by waves can create accreting intertidal mudflats of several km^2 in days to weeks, with very dense mangrove development in just 2–3 years.

Several other studies based on field observations and measurements, complemented by Lidar data and satellite images have further highlighted the topographic variations over the surfaces of mud banks (Gratiot et al. 2007; Anthony et al. 2008; Gensac et al. 2015). In the leading edge of a mud bank, accreted bars form a dynamic 'suture' zone with the muddy (or locally sandy) intertidal terrestrial shoreline (Anthony et al. 2008; Gardel et al. 2011). These characteristic features of the low-energy inner leading part of banks form the precursor substrate conditions for mangrove colonisation. Bars in the upper intertidal zone can become immobilized over fairly long periods of low wave energy, and, thus, progressively dry out via evaporation and dewatering (Fiot and Gratiot 2006; Gardel et al. 2009), leading to the development of mud cracks, typically during neap tides, that provide opportunities for mangrove colonization (Proisy et al. 2009). Wetting and drying cycles have been shown to vary considerably with elevation (Fiot and Gratiot 2006), while field studies have shown that very subtle elevation changes in the upper intertidal zone (order of a few centimetres) can have a strong influence on successful mangrove colonization (Anthony et al. 2008; Gardel et al. 2009; Proisy et al. 2009; Gensac et al. 2011). These studies have shown that mangrove colonization can be successfully predicted based on the definition of the geographic limits of a carefully determined elevation threshold of 2.45 m above the local datum (Fiot and Gratiot 2006; Proisy et al. 2009; Gardel et al. 2009; Gensac et al. 2011). Above this threshold, the duration of tidal emersion is sufficient to allow for mangrove propagules to take a hold on the mud bank surface. The progressive development of mud cracks, and the corresponding colonization by pioneer and then young mangroves, over a period of several months has been monitored by an innovative system of in situ time-lapse photography aimed at highlighting changes in the surface properties of a mud bank (Gardel et al. 2009). Under favourable conditions, mangrove

Fig. 20.24 Head of a mud bank at low water, deflecting the runoff from land and providing shelter for the mangroves (Credit: S. Naipal)

colonization can be extremely rapid, with plant densities exceeding 30 per m^2. Rates of mangrove colonization vary considerably, however, as a function of available intertidal area. Following colonization, extremely rapid mangrove growth (individual plants grow by up to 2 m/year) leads to the establishment of a dense fringe of young mangroves and mud stabilization. Mangroves at all stages of establishment, from young pioneers to mature forests, can however, be destroyed by mud reworking by high-energy waves, sometimes simply through burial and asphyxia of mangrove pneumatophores (Fromard et al. 1998, 2004). In some cases (i.e. Marais Sarcelle, Sinnamary) mangrove trees (*Avicennia*) can die over large areas due to hypersalinization when mangrove is cut off from the sea (formation of a new chenier in front of mangrove forest), as reported above for Suriname. The ensuing mangrove pattern may, therefore, be one of coexistence of young opportunistic rapid-growth juveniles adapted to the new substrate topography and dying and dead mangroves (cemetery stage) that are asphyxiated as mud accretion occurs. The coexistence of dead, dying, and thriving pioneer and young mangroves thus reflects active mangrove renewal that is also unique feature of the mangroves of the Guianas coast.

Sedimentation rates associated with wave reworking of mud have been monitored in the field (Gratiot et al. 2007; Gensac et al. 2015). The latter study carefully documented processes of wave overwash over mud bars in the course of the rising tide. A schematic representation of the sediment dynamics on an intertidal mudflat is proposed by authors (Fig. 20.39). Mud cracking plays a determining role in the rapid and extensive colonization, by pioneer mangroves, of a mud bank undergoing consolidation (Fiot and Gratiot 2006; Gardel et al. 2009). Gratiot and Anthony (2016) have recently shown from experimental work conducted in French Guiana that flocculation and differential settling enhance sedimentation during slack water and under low wave conditions. It can be deduced from this that enhanced settling in the leading edges of banks during such conditions is

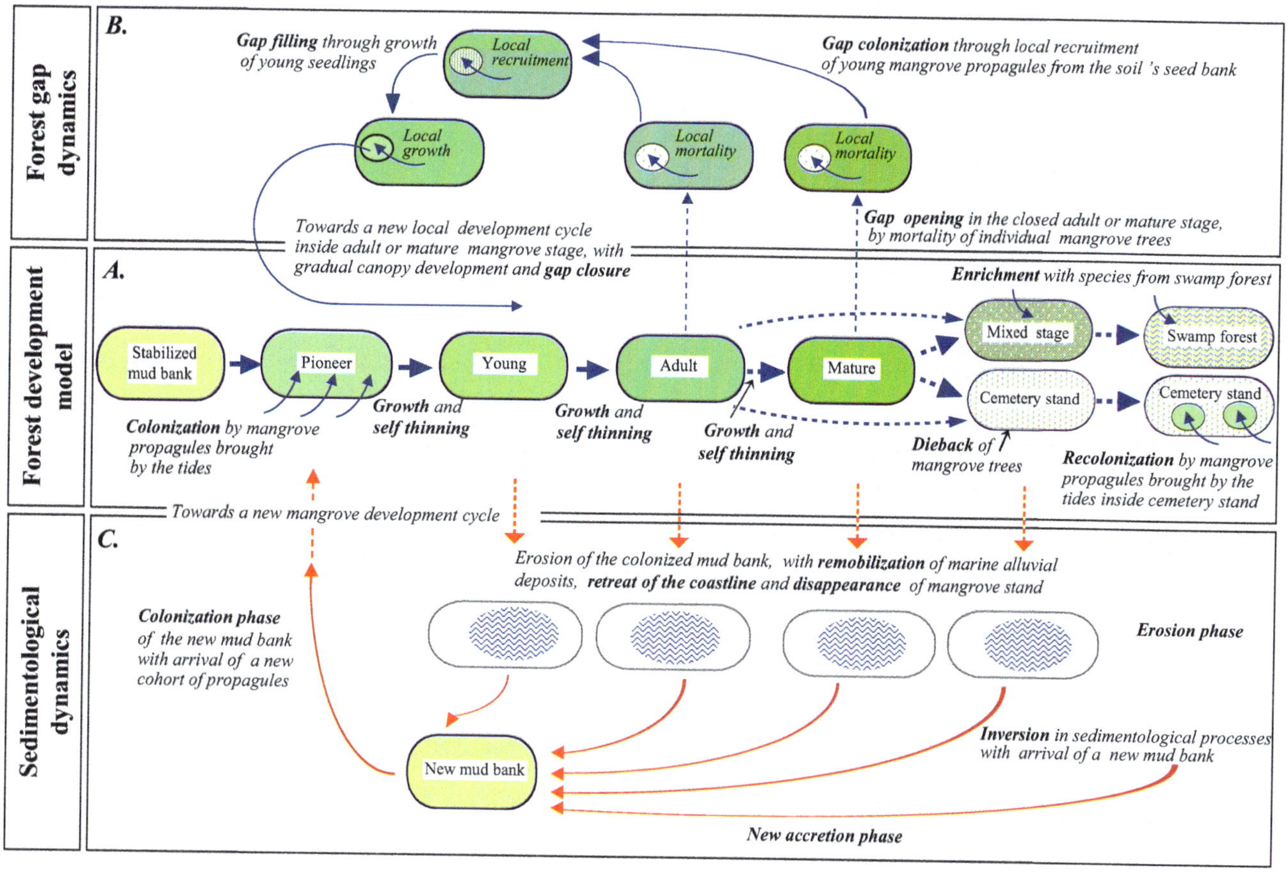

Fig. 20.25 A combined model of Guianese mangrove dynamics. (**a**) Forest development model, mainly based on growth and self-thinning processes. (**b**) Forest gaps dynamics, brought about by local decaying and death of individual mangrove trees (Adapted from Duke 2001). (**c**) Sedimentological dynamics, the major driving force along the coastal area under Amazonian influence (Fromard et al. 2004)

Fig. 20.26 *Avicennia germinans* seedlings planted along the Better Hope coastline, East Coast Demerara (Guyana), August 2016 (Credit: R. Adams, NAREI)

Fig. 20.27 Planted black mangrove seedlings along the foreshore of Village #6–8, West Coast Berbice (Guyana), Region 5. The site was planted in 2011 and 2012 (Credit: K. Moseley, NAREI)

important in the temporary sedimentation that generates under-consolidated mud bars and gel-like fluid mud patches. These characteristic features of the low-energy inner leading part of banks form the precursor substrate conditions for mangrove colonization. Flocculation, hindered settling and settling by mass are all variably favoured by mangroves. In the trailing edge, old mangrove stands are uprooted and their root and trunk system favour turbulence dissipation under wave attack and the formation of mud pebbles from reworking of the bank. This reworking promotes the formation of dilute suspensions that enhance suspended concentrations. In the leading edge, young colonising mangrove stands favour wave energy dissipation, the formation of fluid mud patches, and settling by mass.

20.5.1.2 Remote Sensing

Mud Bank Migration and SPM Dynamics in Coastal Waters

Following the example of earlier work in Suriname Augustinus (1978), Froidefond et al. (1988) used aerial photographs acquired during the late 1970s and 1980s to determine rates of mud bank migration. A mean velocity of mud bank migration of 1 km a year was determined from the interpretation of these aerial photographs.

The advent of moderate-to-high resolution satellite images (LANDSAT TM, SPOT) in the 1990s offered a new opportunity for the monitoring of the highly dynamic coast of French Guiana. Several studies have explored the capacity of these new sensors, initially to quantify mud bank migration (Froidefond et al. 2002, 2004; Baghdadi et al. 2004; Gardel and Gratiot 2005; Gratiot et al. 2007), and, more recently, suspended particulate matter in coastal waters and in mangrove dynamics (Vantrepotte et al. 2013; Gensac et al. 2016). Froidefond et al. (2002) characterized the spectral remote sensing reflectance of coastal waters. Later, Froidefond et al. (2004) established a correspondence function between in situ optical data and concentrations of suspended particulate matter obtained from coastal surveys. This function was then applied to a set of SPOT satellite images to estimate suspended particulate matter concentrations near the sea surface at different dates.

Radar (ERS and RADARSAT) and optical (ASTER) satellite images collected from 1997 to 2001 have been used to

Fig. 20.28 2016 aerial of Village #6–8, West Coast Berbice (Guyana), planted in 2011 and 2012. Successful planting of black mangrove seedlings resulted in extensive natural regeneration (Credit: C. Gittens, WSG)

Fig. 20.29 2012 aerial of planted black mangroves along the coastline Felicity to Chateau Margot (East Coast Demerara, Guyana), Region #4. Site was planted in 2011 and 2012 (Credit: I. Brierley, CATS)

Fig. 20.30 2015 aerial of restored black mangrove with extensive natural regeneration along the coastline Felicity to Chateau Margot (East Coast Demerara, Guyana), Region #4 (Credit: I. Brierley, CATS)

evaluate the capacity of these types of sensors in detecting mud banks and in monitoring coastline change (Baghdadi et al. 2004). This work showed that low-angle radar is more efficient in detecting mud banks, whereas high-angle radar is more appropriate for monitoring coastline change. These results also showed that, depending on the water level on the mud bank, optical or radar satellite images were more or less accurate.

Gardel and Gratiot (2005) attempted to extract rates of mud bank migration by applying image-analysis algorithms on SPOT images covering the period 1986–2003. They showed that rates of migration have been significantly more important over the last decade (up to 2 km/year) covered by these images, possibly as a result of enhanced wind-wave activity, as suggested by wave models of the Atlantic. Gratiot et al. (2007) highlighted an increase in wave forcing over a 44-year period (1960–2004) related to an increase in trade-wind velocities.

In more recent works using remote sensing methods of sea colour comparison of SPOT images and MODIS suspended particulate matter (SPM) maps between 2002 and 2010, Vantrepotte et al. (2013) revealed the strong spatiotemporal coupling between SPM and the dynamics of local mud banks. This study suggested that the highest MODIS SPM values (>13 g/m^3 approximately) can be significantly associated with the subtidal part of the banks as well as to the related turbid plume (Fig. 20.40). Mud bank migration rates derived from MODIS SPM data are, on average, higher than 2 km/year, in agreement with previous studies.

Gensac et al. (2016) used MODIS satellite data over the period 2000–2013 to assess the sediment dynamics in the shallow coastal waters from the Amazon River mouth to the Capes region (northern part of the Amapa region of Brazil and eastern part of French Guiana). They determined the role of continental and oceanic forcing and the implications for coastal geomorphology and mud bank formation. This work

Fig. 20.31 Brushwood dam constructed at Walton Hall (Essequibo Coast, Guyana), Region No. 2 in 2015 (Credit: L. Jack, NAREI)

Fig. 20.32 Site conditions following completion of a geotextile tube groyne at Devonshire Castle (Guyana), Region 22014 (Credit: L. Jack, NAREI)

also confirmed an average rate of mud bank migration of up to 2 km/year.

Decadal-Scale Mud Bank Dynamics and Mangroves

The dynamics of mangroves on the Guianas coast are extremely variable as their spatial and temporal extension and stages of development are determined essentially by the waxing and waning of mud banks. In this system, mangroves play an important role by stabilizing the muddy substrate and ensuring plant 'continuity' with the older muddy shoreline, from which mangrove regeneration is best assured by propagule dispersal. Each erosive inter-bank phase can result in the partial, or rarely, total erosion of a previously welded bank. Such total erosion can occur during a subsequent inter-bank

Fig. 20.33 Top: Devonshire Castle shoreline (Guyana) was planted with *Spartina* in May 2015 following construction of geotextile tube. Bottom: Rapid colonization of *Spartina* within 4 months (Credit: R. Adams, NAREI)

phase characterised by particularly high wave-energy seasons such as El Niño years or phasing of the NAO. Gardel and Gratiot (2006) showed, however, that such waves do not necessarily always have a destructive impact on mangroves. Their analysis of shoreline changes in French Guiana over the period 1995–2000, characterized by high wave energy, showed that mangroves in inter-bank areas underwent very active retreat (150–200 m/year), but at the same time the mud bank areas experienced mangrove colonization. Gratiot et al. (2007) also showed that notable phases of increased wave energy tended to be accompanied by higher annual rates of alongshore mud bank migration. Fromard et al. (2004)

Fig. 20.34 Natural colonization of black mangrove seedlings within *Spartina* grass at Devonshire Castle (Guyana), December 2016 (Credit: L. Jack, NAREI)

proposed a global scenario of mangrove forest dynamics, including a model of forest development, forest gap processes and sedimentological dynamics.

Gratiot et al. (2008) calculated a mass mud bank budget along the French Guiana coast from 1988 to 2004 based on satellite images, and suggested a link between budget fluctuations and the 18.6-year nodal tidal cycle, which they considered as an important factor in modulating coastal erosion and accretion in the Guianas. A detailed 64-year inventory (1950–2014) of mangroves and their multi-decadal area variations as a function of the spatial extent of mud bank and inter-bank zones on the French Guiana was conducted by Walcker et al. (2015). This inventory showed that the area covered by mangroves fluctuates significantly at this multi-decadal timescale depending on the global domination of bank or inter-bank phases at any time. The mangrove cover in French Guiana in 2014, for instance, has been estimated at 45,000 ha by Walcker et al. (2015). The authors showed that fluctuations in mangrove area were not due to current sea-level rise nor to the 18.6-year nodal tidal cycle identified by Gratiot et al. (2008) but to the pervasive role of NAO-driven multi-decadal fluctuations in wave energy (Fig. 20.41). Proisy et al. (2016) proposed a model of the French Guiana coastal system where interaction functions between ecological and physical processes are written to specify how ocean/mangrove shoreline/mud bank change when they interact. This approach has been applied to study mangrove shoreline variations from 1986 to 2009. A time series of remote sensing images was used during the initialization and validation phases.

All of the foregoing studies have been significant in understanding the development, morphology and sediment dynamics of the coast and their relationship with mangrove ecology. Much still remains to be done, however, regarding these aspects. Areas for forthcoming research concern changing rates of mud bank migration, how these rates will be affected by climate change, and the corresponding impacts on mangroves.

20.5.2 Coastal Engineering Research in Suriname

The first important engineering study for the Suriname coast was performed by the Dutch engineering consultants group Nedeco (1968) in the framework of improving the navigation

Fig. 20.35 Young black mangrove planted and stabilized by wooden sticks along the Coronie coast, Suriname, in 2010 (Credit: S. Naipal)

Fig. 20.36 Construction of the sediment trapping unit at Weg naar Zee, Suriname, in September 2015. Left: top view of the dam structure (see text for details) (Credit: C. Fung-A-Loi and S. Naipal, AdeKUS)

access to the ports. In-depth geomorphological research in Suriname has then been pioneered by Prof. Augustinus from the University of Utrecht (the Netherlands). Despite efforts, Suriname remained dependent for a long time on external expertise with regard to coastal engineering. This changed early twenty-first century when the Anton de Kom University of Suriname (AdeKUS) in Paramaribo appointed Surinamese scientists with the proper background, who studied abroad, as academic staff.

Within the framework of interuniversity development cooperation, funded by the Flemish Interuniversity Council (VLIR-UOS), the Hydraulics Division of the University of Leuven (KU Leuven, Belgium) has first established laboratory facilities for hydraulic engineering research. Subsequently, they ran a so-called "own initiative" project (2005–2010) to establish the necessary research capacity at the AdeKUS to study the coastal processes in more detail by carrying out field campaigns (www.kuleuven.be/hydr/SurinamCoast/SurinamCoast.htm).

The Hydraulics Laboratory, Department of Infrastructure, of AdeKUS now disposes of a small sea-borne motorboat equipped with a (by software) coupled system of GPS, dual-frequency echo-sounder and acoustic Doppler current profiler (ADCP) for transect monitoring of bathymetry, fluid mud layer thickness, velocity profiles and turbulence. Furthermore equipment is available for frame deployment to collect time series data on sediment concentrations with optical backscatter (OBS) probes, flow velocity, turbulence and water depth with acoustic Doppler velocity meters (ADV) and the same ADCP can also be used bottom-mounted, upside down to measure velocity profiles and directional waves. This allows local data collection and research to support the proposed sustainable management of the coastal zone.

A major goal of the research is the set-up of a complete morphodynamic model for the Guianas coast where all the physical processes, described above, are incorporated and local data used. This model is discussed in more detail in the following section.

20.5.3 Research in Guyana

Coastal research in Guyana is scarce and generally little known. Nevertheless, the continuous threats to the western part of the Guianas coast invited several geological studies (Nota 1958; Hawkes 1962; Morelock 1972; Daniel 1989; Lakhan et al. 1997, 2002, 2006). Important engineering studies related to the design of the coastal defence system have been carried out by Delft Hydraulics (1962) and Nedeco (1972). The "Institutional Capacity Building Activities on Guyana Sea Defences" project (2004–2005), carried out by the Dutch consortium of Delft Hydraulics and Royal Haskoning, constitutes the most prominent research for the Guyana coast, reviewing existing knowledge and including the gathering of much new field data (Les and Westra 2004; Westra 2004) and the development of a powerful modelling tool (de Graaff and Bijlsma 2005; de Graaff et al. 2005; Winterwerp et al. 2005), briefly discussed in the next section.

20.5.4 Morphodynamic Modelling of the Guianas Coast

Even though different studies have been looking at the actual time scales for the recent evolution of the Guianas coastline, there is no consensus. The main reason is that dedicated observations have only started around the 1940s of the twentieth century, which constitutes a period of about 80 years of information which is too short relative to the estimated cycle of roughly 30 years for the passing of a mud bank. Looking at the sequence of all the mud banks over the entire 1000 km stretch of the Guianas coastline may give an impression of how the mud banks evolve during their migration from east to west, which takes them about 500 years. However, the effects of climate change and sea level rise may make the interpretation back in time uncertain. Moreover, the coast has been under increasing anthropogenic pressure during the last decades, and it becomes increasingly difficult to distinguish natural from anthropogenic effects.

Furthermore, it is dangerous to extrapolate the future based on the past, certainly in view of the large uncertainty on sea level rise (certainly for South America, where little SLR data are available). There have been various studies on how a mangrove coast responds to sea level rise (Alongi 2008; McIvor et al. 2013). Based on the study of the evolution of the Guianas coast over the Holocene period (Rine 1980; Rine and Ginsburg 1985), there is some evidence that a healthy and wide enough mangrove belt along may be resilient enough and may follow sea level rise, provided there is enough sediment supply. This may explain why the Guianas mangrove coast is not considered vulnerable by IPCC (Alongi 2008; Wong et al. 2014). Much depends on how the sediment supply from the Amazon basin will evolve and respond to climate change. Nevertheless, the local authorities have worries because of the reality of increasing flood events in vulnerable spots. Numerical models can be used to assess the impact for different scenarios. Such a model would then serve as a decision support tool to study potential threats to the coastline. This model should allow to answer (a.o.) the question if the coastline will be able to follow sea level rise by sediment accumulation, depending on the expected evolution of sediment supply from the Amazon. The latest data analysis indeed indicates that the sediment supply is increasing (see section 20.2 "Morphodynamics of the Guianas Coast").

These are some reasons why numerical models for the study of morphodynamic processes are believed to be a very

Fig. 20.37 Evolution of the mud deposit of the sediment trapping unit at Weg naar Zee: situation in April 2016 (left) and June 2017 (right) (Credit: S. Naipal)

Fig. 20.38 Spontaneous colonization by mangroves of a short-lived mud deposit at the high-high-water elevation (Wanica district, Suriname). Notice the cracked surface, as well as the cliff erosion. Photo taken from a small boat, indicating that the water depth rapidly increases, too much for a stable mud deposit (Credit: S. Naipal)

useful tool. These models simulate the movement of water and sediments over periods of months to decades, depending on the specific research question, by solving the conservation equations of mass and momentum for water and the sediments. Several modelling initiatives have been undertaken over the past decade and will briefly be presented in the following sections.

20.5.4.1 French Guiana Coast Model

A 3D numerical model of the French Guiana coast was built by Chevalier et al. (2004) and further developed by Bourret et al. (2005, 2008). The model covers most of the French Guiana coastline until the 100 m isobath line (Fig. 20.42 left). This finite difference model was built to analyse the influences of river discharges and wind on the offshore current, salinity and temperature distribution. Special attention was paid to the treatment of the offshore boundary condition (passive or active boundary).

Numerical results (Bourret et al. 2008) point out the tidal currents are about 0.15–0.45 m/s on the inner shelf along the neap-spring cycle. They reach 0.55 m/s near Sinnamary and near the mouth of Oyapock river. Offshore tidal currents are

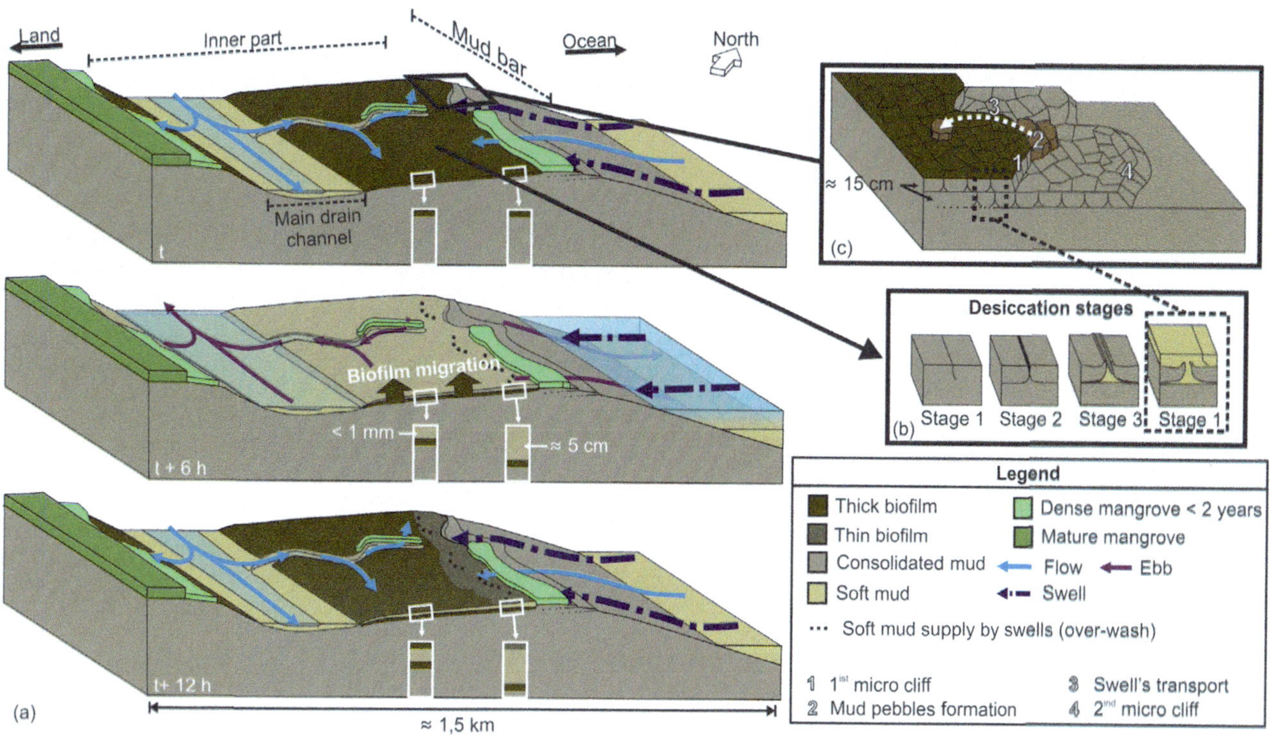

Fig. 20.39 Schematic representation of the sediment dynamics on an intertidal mudflat in French Guiana during a tidal cycle (**a**). Desiccation (**b**) and erosion (**c**) of the consolidated mudflat are also illustrated. The time scales involved in the desiccation (**b**) and erosion (**c**) stages are longer than those in (**a**) (Gensac et al. 2015)

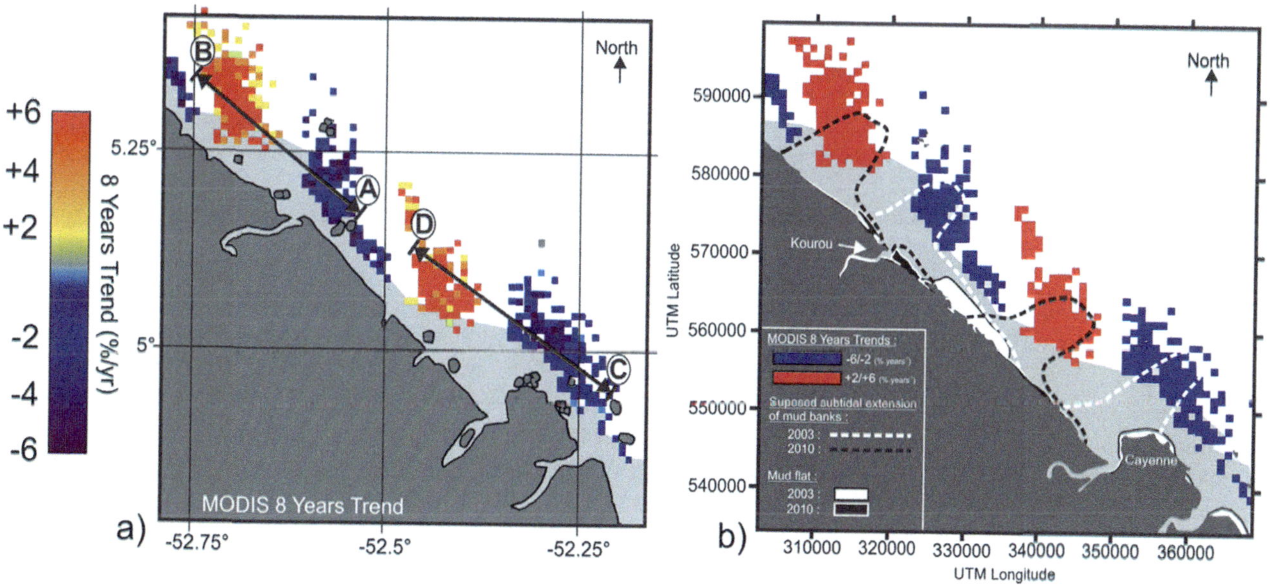

Fig. 20.40 Significant trends in SPM detected over the French Guiana coast from the 8 year MODIS time series (**a**) (in % year^{-1}) and comparison with the mud banks estimated position and morphology (**b**). SPOT estimates of the location of the mudflat and subtidal extension of the mud bank at the beginning and end of the MODIS time series considered are represented by *white* and *dark grey surfaces* and *dashed lines*, respectively. The subtidal mud bank extension has been determined from swell damping characteristics (SPOT images) (Vantrepotte et al. 2013)

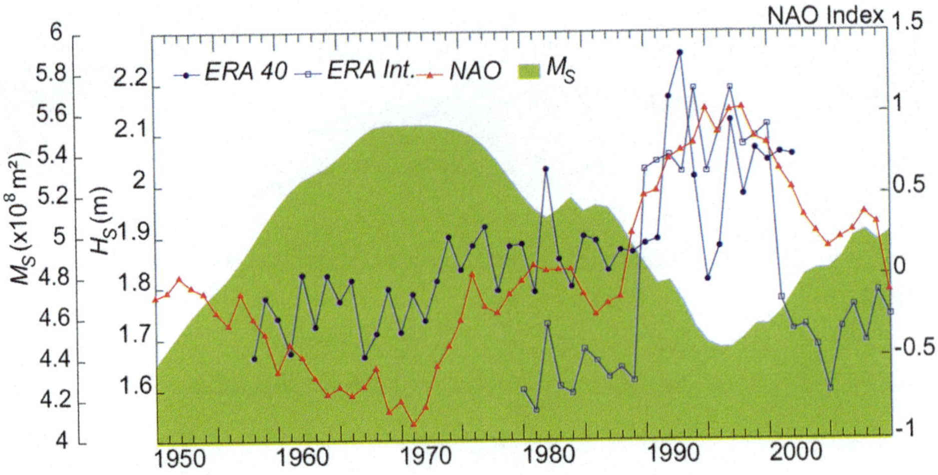

Fig. 20.41 Relationship between the mangrove surface area (MS) and significant wave heights (HS), and their link with the North Atlantic Oscillation (NAO). The ERA-40 product was used to assess the winter HS from 1958 to 2002, whereas the ERA-Interim product was used for the period of 1980–2010. The 10-year moving average of the NAO winter index is also plotted on the graph; it exhibits a significant temporal correlation with MS and HS (Walcker et al. 2015)

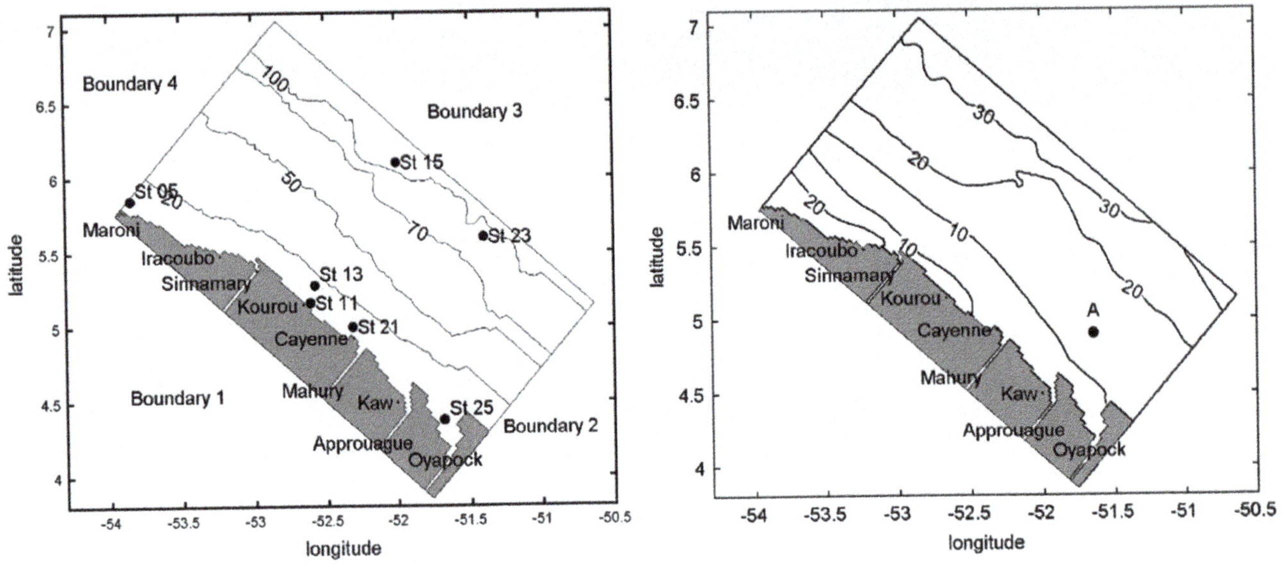

Fig. 20.42 Numerical model of the French Guiana coast. Left: model domain and bathymetry (depth in meters); right: modelled field of sea surface salinity (in psu) (Bourret et al. 2008)

weaker, 0.05–0.1 m/s, and the currents are then dominated by the oceanic circulation. The Guiana current can reach 1.1 m/s offshore and 0.3 m/s at 30 m depth. The surface salinity field (Bourret et al. 2008) describes a weak salinity tongue linked to two influences. The first one in the area between 10 and 20 psu (Fig. 20.42 right) may be attributed to the Amazon influence. The second one in the area lower than 10 psu near the coast is due to the local river discharges.

A smaller 3D model (Fig. 20.43) was also built by Chevalier et al. (2008) to analyse the combined effect of the current and wave on the mud transport around mud bank. This model extends 60 km length alongshore and up to 25 m depth. Between 20 m depth and the coastline, the bathymetry is continuously changing due to banks mobility. As no regular survey of the bathymetry is performed, Chevalier et al. (2008) rather propose to work with an idealized bathymetry (Fig. 20.44).

For this study, only hydrodynamics and suspended sediment concentrations were calculated. The influences of salinity and temperature have been neglected. Waves are modelled using Fudaa-Vag (CEREMA 2011) and the wave damping is integrated through the model of Rodriguez and Mehta (2001).

The models (Chevalier et al. 2004, 2008; Bourret et al. 2005, 2008) give general insight on the influence of the different forcing. However, full validation of these models was not feasible because of the lack of physical in situ data.

20.5.4.2 Mahury Estuary (French Guiana) Model

A 3D numerical model of the Mahury Estuary (Fig. 20.45) has been built in the framework of research on the accessibility of the port of Cayenne during a period of passage of a mud bank over the approach channel (Orseaux 2016). The model is built with the TELEMAC software (www.opentelemac.

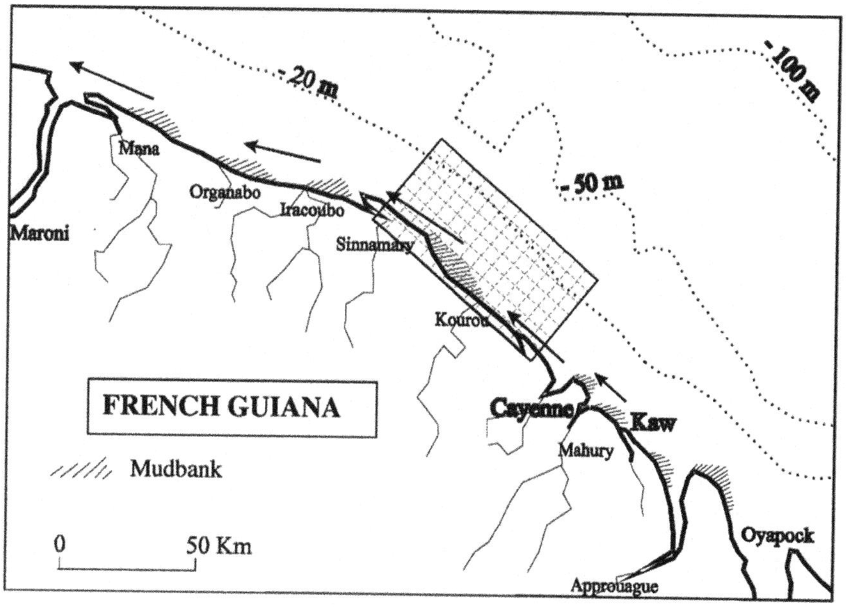

Fig. 20.43 Situation of the mud bank model of Chevalier et al. (2008)

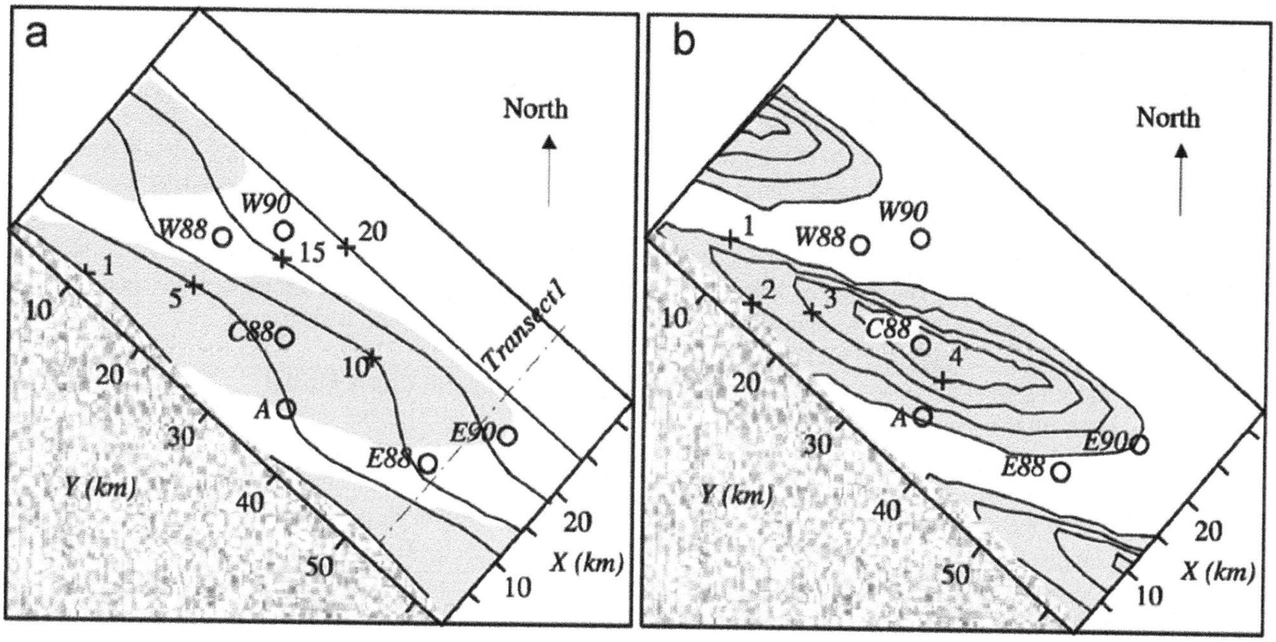

Fig. 20.44 Mud bank model: (**a**) idealized bathymetry; (**b**) mud layer thickness contours (Chevalier et al. 2008)

org). The work here focusses more on the estuarine dynamics. In situ data sets collected by Orseaux (2016) are exploited to perform calibration and validation of the model. The model extends 20 km alongshore 20 km offshore and 40 km inland. The extension has been chosen according to the annual bathymetric surveys occurring around the navigation channel. The horizontal mesh (Fig. 20.45b) comprises about 21,000 nodes. Simulations are run for 6 months with eight vertical layers. For the turbulence, a mixing length model is selected with damping function to account for stratification.

Up to now, only the 3D hydrodynamic model has been validated in regards of salinity and current. Fig. 20.46 illustrates comparison of the modelled salinity with measurement (Orseaux 2016) at the bottom on station S2 and S4 during a neap-spring cycle in the wet season. The 3D sediment transport model is still under development. In addition, it is also foreseen to build a larger scale model covering the whole French Guiana coastline, which includes the Maroni and Oyapock estuaries. This model would serve to study the Maroni dynamics and to provide offshore boundary

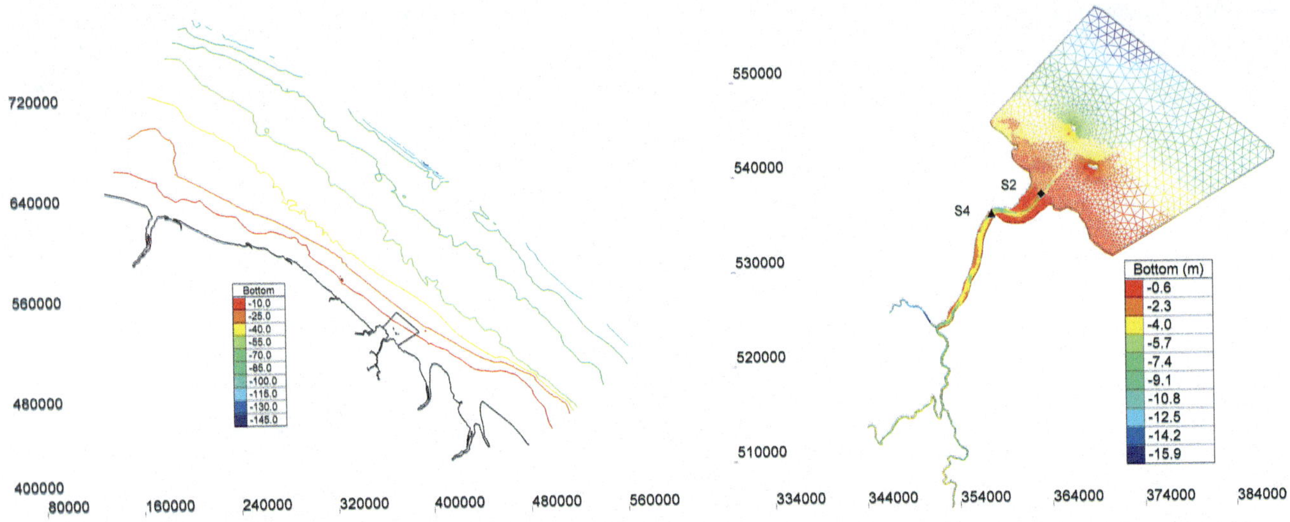

Fig. 20.45 Numerical model of the Mahury Estuary: (left) situation and model extend, (right) the horizontal mesh (Credit: S. Orseau, ULCO/CNRS/CEREMA)

conditions to local estuarine model in term of salinity and SPM plume or wave forcing.

20.5.4.3 Suriname Coast Model

The KU Leuven morphodynamic model for the Suriname coast has been under development since 2005 with the help of several MSc thesis projects (Gyssels and Van der Zype 2006; Dewaelheyns and Etneo 2007; Dirkx and Fockedey 2008; Hermans and Kroeders 2012; Pelckmans 2014) and an internship (Wongsoredjo 2017). The model is based on the open source TELEMAC software (www.opentelemac.org), using the hydrodynamic module TELEMAC (Hervouet 2007), the spectral wave module TOMAWAC (Benoit et al. 1996) and the sediment transport module SISYPHE. Several modifications to the code are developed and tested.

The model domain (Fig. 20.47) comprises the entire Suriname coast, from the Maroni River in the east, which is the border with French Guiana, to the Corentyne river in the west, which is the border with Guyana, and a part of each neighbour's coastline. The land boundary has been taken more inland, allowing inundation of the mangrove forests. The northern boundary is taken along the edge of the continental shelf.

Mud banks have been given different erosion properties than the interbank areas, since the former consist of soft mud, which is much more erodible than the overconsolidated mud layer exposed in the interbank areas. For this purpose, the mud banks had to be identified, which was done based on comparison of bathymetric data with satellite images (Fig. 20.48).

Like for the other models, by lack of sufficient in-situ data for calibration and validation, only qualitative results have been obtained thus far. Future efforts will be dedicated to including wave damping by the mud banks. Subsequently, the model can be used for scenario studies to include sea level rise and subsequent inundation of the mangroves (cf. Lovelock et al. 2015). The mangrove field can be given a higher resistance by modification of the friction coefficients for currents and waves (Escobar Ramos 2017). The latter can be developed more accurately based on the recent work of Maza et al. (2015) where wave damping by regularly and arbitrarily distributed stems has been simulated in detail at the scale of laboratory flume studies.

20.5.4.4 Guyana Coast Model

In the framework of the *Institutional Capacity Building Activities on Guyana Sea Defences* project, carried out by the Dutch engineering consultants WL Delft Hydraulics (now Deltares) and Royal Haskoning, a modelling framework was developed to study 3D currents, waves and morphodynamics along the Guyana coast (de Graaff and Bijlsma 2005; de Graaff et al. 2005; Winterwerp et al. 2005) with the software Delft3D and SWAN (Booij et al. 1999). For this purpose the SWAN model was extended to allow simulation of wave damping by fluid mud (Kranenburg et al. 2011). Necessary data have been collected during various campaigns (Les and Westra 2004; Westra 2004). The model was aimed as research tool and as assessment tool for defining operational and extreme hydraulic conditions for the design of sea defence structures. Application of the model (Fig. 20.49) has demonstrated that wave damping has to be accounted for in order to simulate observed conditions (Winterwerp et al. 2007). The model has subsequently been used to study the coastal evolution of the Guyana coast since 1950 (Welage 2005).

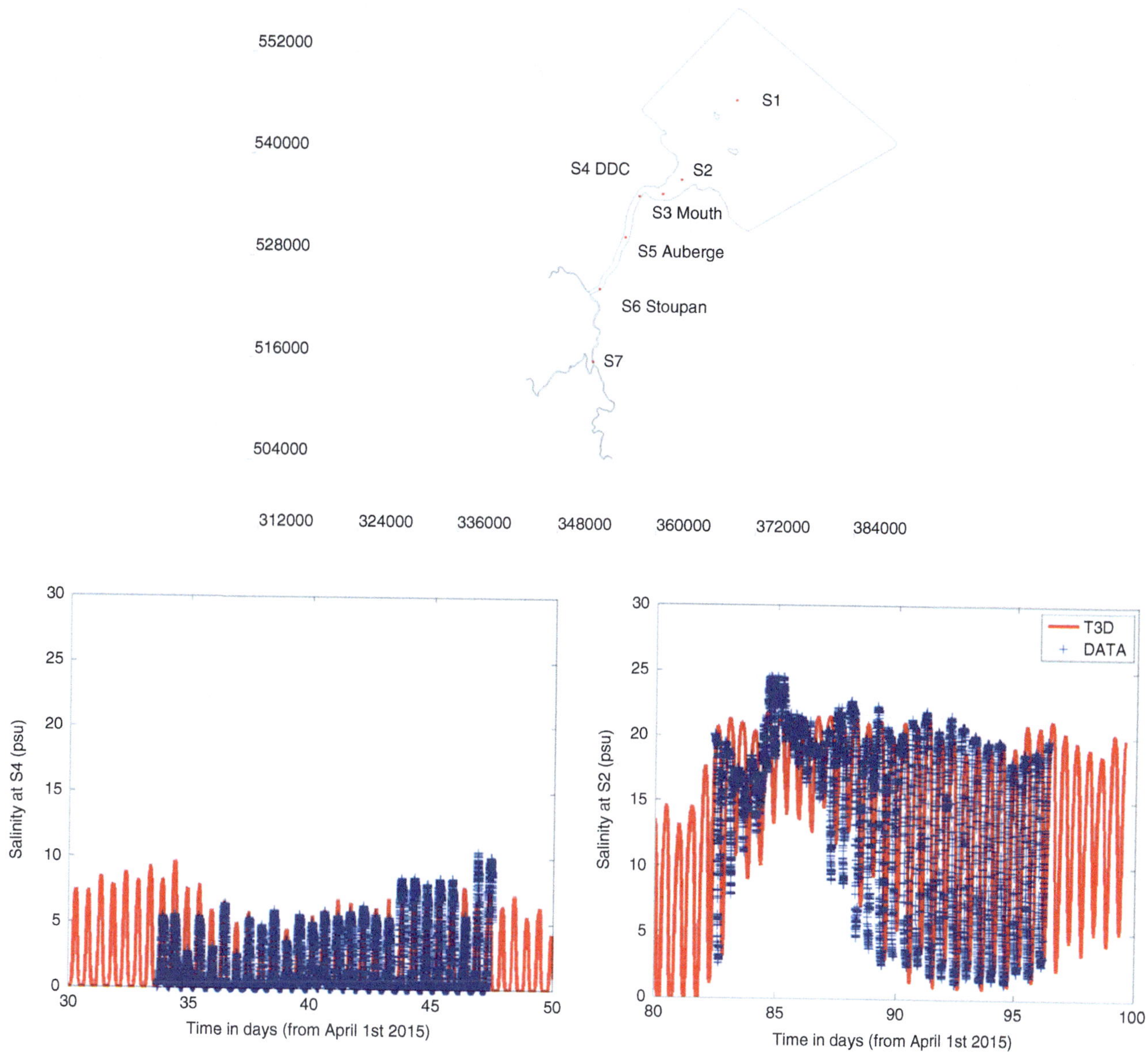

Fig. 20.46 Location of the measuring stations (top). Time evolution of the salinity at the bottom: (left) at station S4 near the only harbour of French Guiana (left) and at the mouth (right) (Credit: S. Orseau, ULCO/CNRS/CEREMA)

20.6 Conclusions and Perspectives

The potential of mangroves to protect coastal areas is very site specific. In many studies only the interaction between vegetation and hydrodynamics is considered, forgetting that the local sediment dynamics should be taken into account in order to understand how the natural system works.

In the case of the Guianas coast, the interaction between these three key players, mangroves, hydrodynamics and sediment dynamics, is now quite well understood. The mangroves require (relatively) quiet hydrodynamic conditions in order to capture and stabilize sediments. This shelter is provided when a mud bank shields the mangroves from direct wave action. During interbank periods (when there is no protecting mud bank in front of the mangroves), the mangroves are slowly destabilized by wave action and washed away, and the coast erodes; i.e. mangroves do not provide sustainable protection. In the Guianas there is an alternation of mud bank and interbank periods, a cycle of the order of 30 years. For an undisturbed mangrove belt, this results into a net accretion of the order of 1 m per cycle, which is supported by geological evidence. Furthermore, there are indications that the sediment supply form the Amazon basin has been increasing significantly over the past decades, which raises hope that flood risks by potential sea level rise can be mitigated by nature itself.

The knowledge and understanding of these processes provide the necessary arguments to protect the mangrove belt as

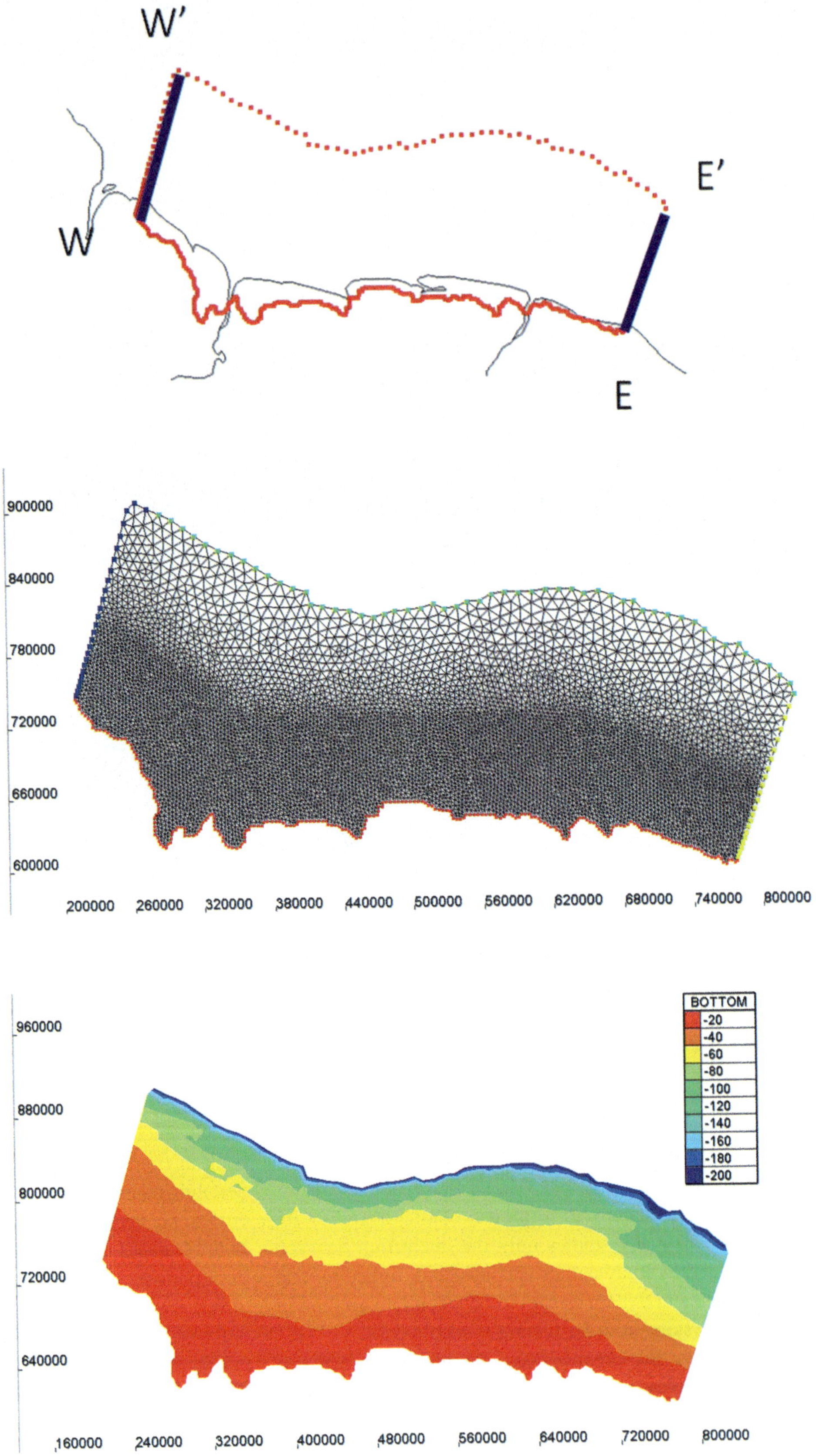

Fig. 20.47 Suriname coast model: domain boundaries, relative to the actual coastline (top); numerical model mesh (middle) and bathymetry (bottom) (Credit: Hydraulics Division, KU Leuven)

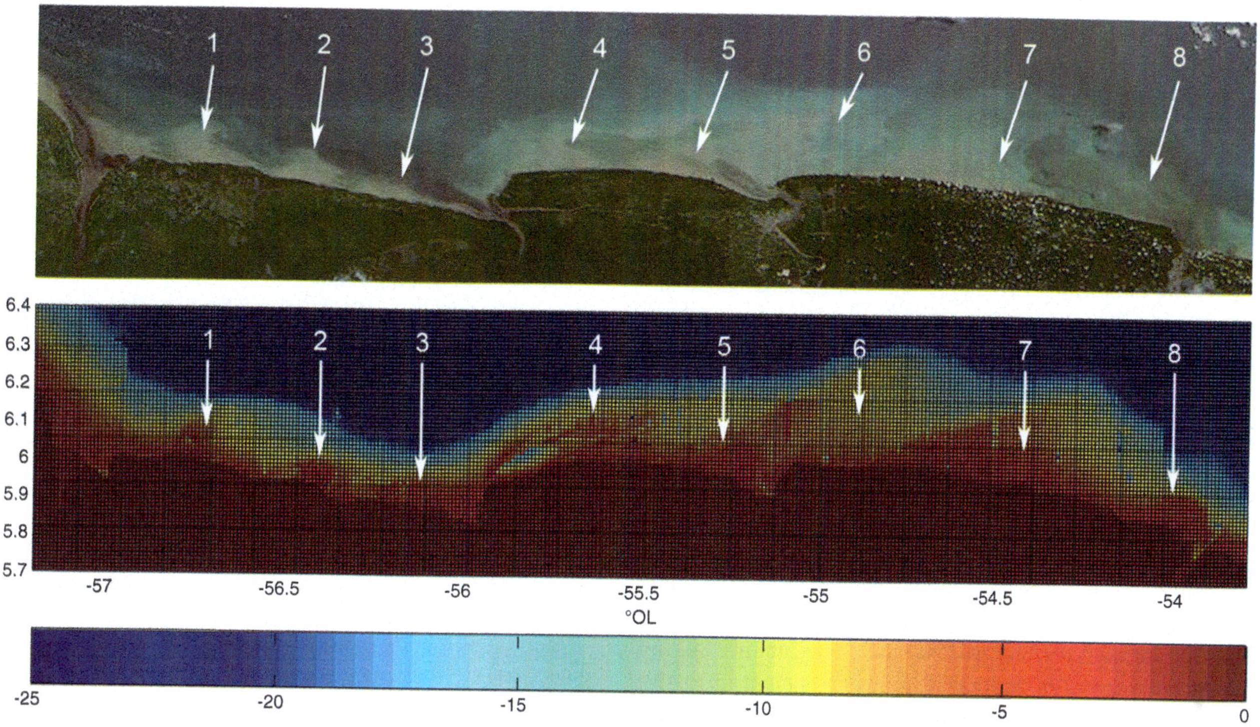

Fig. 20.48 Upper image: MERIS satellite image of 1 August 2006 in visible spectrum (ESA Earth Online), lower image: bathymetry along the coast of Suriname in 2005 (the shift in location of the indivual mud banks corresponds with the shift in time between both images) (Dewaelheyns and Etneo 2007)

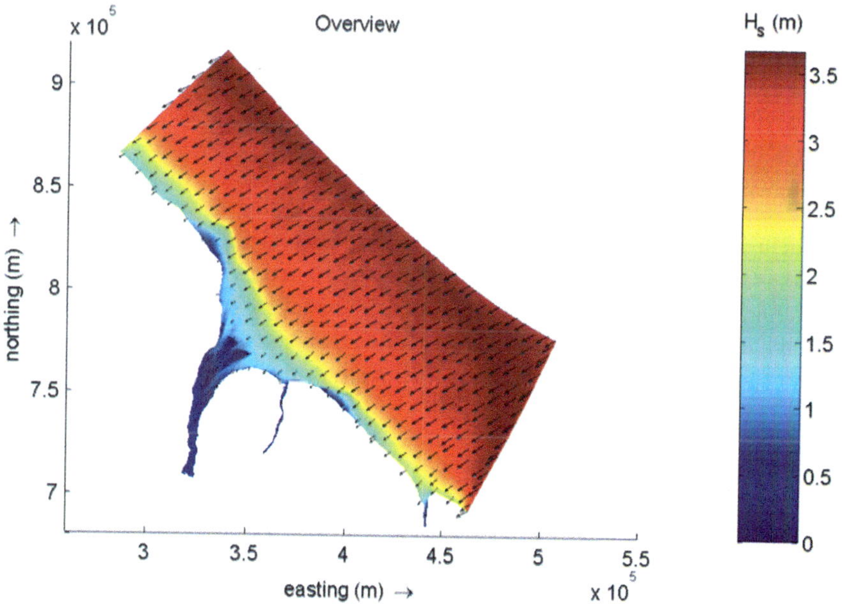

Fig. 20.49 Modelled significant wave heights (m) in the detailed Delft3D Guyana model for a return period of 1 year and the dominant north-eastern wind and wave direction of 60°, including wave damping by fluid mud (de Graaff et al. 2005)

the most efficient, cost-effective and sustainable shore protection. Action needs to be taken to prevent further deterioration of the mangrove belt. Vulnerable places where the mangrove fringe has become too narrow or has been removed are now considered for mangrove rehabilitation projects. Several pilot restoration projects are currently ongoing in Guyana and Suriname. Both mangrove rehabilitation projects do many efforts to realize community based mangrove management. They are working in the areas of administrative capacity development, monitoring and research, community development and capacity building, awareness and education.

The presented overview of recent research initiatives shows that much work still remains. Collection of field data and long-term monitoring remain too limited because of insufficient resources. Moreover, many places are difficult to access either from land or from the sea. Remote sensing plays an important role for filling up data gaps. They have already been used to analyse coastal erosion, mud bank migration and mangrove dynamics. They allow also the generation of instantaneous surface suspended sediment concentration maps for the entire coastal area (even though the conversion algorithms are still subject of research since they need improvement).

Development of numerical morphodynamic models for each of the Guianas is ongoing. Several interacting complex processes (i.e. interaction of waves with non-homogeneous fluid mud and of currents and waves with mangroves) are still difficult to be modelled correctly and require more fundamental research. Eventually, it is intended that operational models will become available to local authorities as decision support tools for coastal zone management.

Furthermore, scientific capacity building is locally realized through the international MSc program in "Sustainable Management of Natural Resources" , taught in English at the Anton de Kom University of Suriname (http://vlir-iuc.uvs.edu/smnr/), established in 2007 with support of the University Development Cooperation division of the Flemish Inter-University Council (www.vliruos.be). In this program, a specialized training course on "Coastal erosion and protection" is taught in collaboration with hydrodynamic and morphodynamic experts from the Hydraulics Division of the Department of Civil Engineering from the KU Leuven (Belgium). This course provides the general theory on the interaction between hydrodynamics, morphodynamics and mangroves and is then applied specifically to the Guianas coast. Furthermore, it provides an introduction to field monitoring techniques, numerical modelling and field excursions.

The Guianas mangrove coast in many aspects is unique. Nevertheless, the ongoing research efforts and mangrove rehabilitation projects may provide important insights and lessons for other mangrove coastlines around the world.

References

Adams R (2014) Time 0 monitoring lima, Region No.2. Report. NAREI, Mon Repos

Allersma E (1971) Mud on the oceanic shelf off Guiana. Symposium on investigation and research of the Caribbean and adjacent regions (Curaçao). UNESCO-FAO, Paris, pp 193–203

Allison MA (2000) Origin of Amazon mud banks along the northeastern coast of South America. Mar Geol 163:241–256

Allison MA, Lee MT (2004) Sediment exchange between Amazon mud banks and shore-fringing mangroves in French Guiana. Mar Geol 208:169–190

Alongi DM (2008) Mangrove forests: resilience, protection from tsunamis, and response to global climate change. Estuar Coast Shelf Sci 76(1):1–13

Anthony EJ, Gratiot N (2012) Coastal engineering and large-scale mangrove destruction in Guyana, South America: averting an environmental catastrophe in the making. Ecol Eng 47:268–273

Anthony EJ, Dolique F, Gardel A, Gratiot N, Proisy C, Polidori L (2008) Nearshore intertidal topography and topographic-forcing mechanisms of an Amazon-derived mud bank in French Guiana. Cont Shelf Res 28:813–822

Augustinus PGEF (1978) The changing shoreline of Surinam (South America). PhD dissertation, Universiteit Utrecht, Uitgave No.95, Natuurwetenschappelijke Studiekring voor Suriname en de Nederlandse Antillen, Utrecht

Augustinus PGEF (2004) The influence of the trade winds on the coastal development of the Guianas at various scale levels: a synthesis. In: F. Baltzer, M.A. Allison and F. Fromard, (eds) Material exchange between the upper continental shelf and mangrove fringed coasts with special reference to the N. Amazon-Guianas Coast. Mar Geol 208(SI):145–151

Baghdadi N, Gratiot N, Lefebvre JP, Oliveros C, Bourguignon A (2004) Coastline and mud bank monitoring in French Guiana: contributions of radar and optical satellite imagery. Can J Remote Sens 30:109–122

Benoit M, Marcos F, Becq F (1996) Development of a third generation shallow water wave model with unstructured spatial meshing. Proc. 25th Int. Conf. on Coastal Eng. (ICCE 1996, 2–6 September 1996, Orlando, Florida, USA), pp 465–478

Booij N, Ris RC, Holthuijsen LH (1999) A third-generation wave model for coastal regions, Part I, Model description and validation. J Geophys Res 104(C4):7649–7666

Bourret A, Devenon J-L, Chevalier C (2005) Investigation on passive open boundary conditions adapted to the conjunction of strong currents, standing tidal wave and high stratification: application to the French Guiana Continental Shelf. Cont Shelf Res 25 (11):1353–1373

Bourret A, Devenon J-L, Chevalier C (2008) Tidal influence on the hydrodynamics of the French Guiana continental shelf. Cont Shelf Res 28(7):951–961

Callede J, Ronchail J, Guyot JL, De Oliveira E (2008) Amazonian deforestation: its influence on the Amazon discharge at Óbidos (Brazil). Revue des Sciences de l'Eau 21(1):59–72

CEREMA (2011) VAG: code de propagation de la houle (in French). http://www.eau-mer-fleuves.cerema.fr/vag-code-de-propagation-de-la-houle-a106.html. Accessed Apr 2017

Chevalier C, Baklouti L, Ramanonjiarisoa A (2004) Modelling the influence of wind and rivers on current, salinity and temperature over the French Guiana continental shelf during the rainy season. J Coast Res 20(4):1183–1197

Chevalier C, Froidefond JM, Devenon JL (2008) Numerical analysis of the combined action of littoral current, tide and waves on the suspended mud transport and on turbid plumes around French Guiana mud banks. Cont Shelf Res 28:545–560

Daniel JRK (1989) The chenier plain coastal system of Guyana. Mar Geol 90:283–287

Danielsen F, Sørensen MK, Olwig MF, Selvam V, Parish F, Burgess ND, Hiraishi T, Karunagaran VM, Rasmussen MS, Hansen LB, Quarto A, Suryadiputra N (2005) The Asian tsunami: a protective role for coastal vegetation. Science 310:643. https://doi.org/10.1126/science.1118387

De Graaff RF, Bijlsma AC (2005) Institutional capacity building activities on Guyana sea defences: modelling report, volume 2: flow modelling. WL Delft Hydraulics

De Graaff RF, Winterwerp JC, Luijendijk AP (2005) Institutional capacity building activities on Guyana sea defences: modelling report, volume 3: wave modelling. WL Delft Hydraulics

Delft Hydraulics Laboratory (1962) Demerara coastal investigation. Report on siltation of Demerara bar channel and coastal erosion in British Guiana, Delft

Dewaelheyns Ph, Etneo L (2007) Modellering van de stromingen voor de kust van Suriname. MSc thesis, Dept. of Civil Engineering, KU Leuven (in Dutch)
Dirkx B, Fockedey B (2008) Modellering van de stromingen voor de kust van Suriname. MSc thesis, Dept. of Civil Engineering, KU Leuven (in Dutch)
Duke NC (2001) Gap creation and regenerative processes driving diversity and structure of mangrove ecosystems. Wetl Ecol Manag 9:257–269
Ellison AM (2000) Mangrove restoration: do we know enough? Restor Ecol 8(3):219–229
Erftemeijer P, Teunissen P (2009) ICZM plan Suriname – mangrove report. Analysis of problems and solutions for the management of mangrove forests along Suriname's wild coast. Final Report, Paramaribo
Escobar Ramos JS (2017) Hydrodynamic and cohesive sediments 2D modelling in Buenaventura Bay, Colombia. MSc thesis, IUPWARE program, KU Leuven
Feagin RA, Mukherjee N, Shanker K, Baird AH, Cinner J, Kerr AM, Koedam N, Sridhar A, Arthur R, Jayatissa LP, Lo Seen D, Menon M, Rodriguez S, Shamsuddoha M, Dahdouh-Guebas F (2010) Shelter from the storm? Use and misuse of coastal vegetation bioshields for managing natural disasters. Conserv Lett 3:1–11
Field CD (1999) Rehabilitation of mangrove ecosystems: an overview. Mar Pollut Bull 37(8–12):383–392
Filizola N, Guyot JL (2009) Suspended sediment yields in the Amazon basin: an assessment using the Brazilian national data set. Hydrol Process 23:3207–3215
Fiot J, Gratiot N (2006) Structural effects of tidal exposures on mudflats along the French Guiana coast. Mar Geol 228(1–4):25–37
Froidefond JM, Pujos M, Andre X (1988) Migration of mud banks and changing coastline in French Guiana. Mar Geol 84:19–30
Froidefond J, Gardel L, Guiral D, Parra M, Ternon J-F (2002) Spectral remote sensing reflectances of coastal waters in French Guiana under the Amazon influence. Remote Sens Environ 80(2):225–235
Froidefond J, Lahet F, Hu C, Doxaran D, Guiral D, Prost M, Ternon J-F (2004) Mudflats and mud suspension observed from satellite data in French Guiana. Mar Geol 208(2–4):153–168
Fromard F, Puig H, Mougin E, Marty G, Betoull JL, Cadamuro L (1998) Structure, above-ground biomass and dynamics of mangrove ecosystems: new data from French Guiana. Oecologia 115:39–53
Fromard F, Vega C, Proisy C (2004) Half a century of dynamic coastal change affecting mangrove shorelines of French Guiana. A case study based on remote sensing data analyses and field surveys. Mar Geol 208:265–280
Gardel A, Gratiot N (2005) A satellite image-based method for estimating rates of mud bank migration, French Guiana, South America. J Coast Res 21:720–728
Gardel A, Gratiot N (2006) Monitoring of coastal dynamics in French Guiana from 16 years of SPOT images. J Coast Res 39 (SI):1502–1505
Gardel A, Proisy C, Lesourd S, Philippe S, Caillaud J, Gontharet S, Anthony EJ, Brutier L (2009) A better understanding of mud cracking processes gained from in situ measurements on an intertidal mudflat in French Guiana. J Coast Res 56(SI):424–428
Gardel A, Gensac E, Anthony EJ, Lesourd S, Loisel H (2011) Wave-formed mud bars: their morphodynamics and role in opportunistic mangrove colonization. J Coast Res 64(SI):384–387
Gensac E, Gardel A, Lesourd S, Anthony EJ, Proisy C, Loisel H (2011) Short-term prediction of the evolution of mangrove surface areas: the example of the mud banks of Kourou and Sinnamary, French Guiana. J Coast Res 64(SI):388–392
Gensac E, Gardel A, Lesourd S, Brutier L (2015) Morphodynamic evolution of an intertidal mudflat under the influence of Amazon supply – Kourou mud bank, French Guiana, South America. Estuar Coast Shelf Sci 158:53–62
Gensac E, Martinez JM, Vantrepotte V, Anthony EJ (2016) Seasonal and inter-annual dynamics of suspended sediment at the mouth of the Amazon River: the role of continental and oceanic forcing, and implications for coastal geomorphology and mud bank formation. Cont Shelf Res 118:49–62
Grabowski RC, Droppo IG, Wharton G (2011) Erodibility of cohesive sediment: the importance of sediment properties. Earth Sci Rev 105 (3–4):101–120
Gratiot N (2011) Coastal erosion along the coast of Guiana final report, EuropeAid/124705/D/SER/GY project: technical assistance for capacity building and institutional strengthening of the Sea Defences, Guyana. Works Services Group, Georgetown
Gratiot N, Anthony EJ (2016) The role of flocculation and settling processes in geological development of the mangrove-colonized, Amazon-influenced mud bank coast of South America. Mar Geol 373:1–10.91
Gratiot N, Gardel A, Anthony EJ (2007) Trade-wind waves and mud dynamics on the French Guiana coast, South America: input from ERA-40 wave data and field investigations. Mar Geol 236:15–26
Gratiot N, Anthony EJ, Gardel A, Gaucherel C, Proisy C, Wells JT (2008) Significant contribution of the 18.6 year tidal cycle to regional coastal changes. Nat Geosci 1:169–172
Guyana Government Information Agency (2016) EU presents coastal engineering design manual to MPI. http://gina.gov.gy/eu-presents-coastal-engineering-design-manual-to-mpi/. Retrieved May 2017
Guyana Ministry of Finance (1996) Water management and flood control policies. National Development Strategy, Chapter 40. http://www.guyana.org/NDS/chap40.htm. Retrieved May 2017
Gyssels T, Van der Zype N (2006) Hydrodynamisch model voor Suriname: voorlopige versie. MSc thesis, Dept. of Civil Engineering, KU Leuven (in Dutch)
Hawkes DD (1962) Erosion of tidal flats near Georgetown, British Guiana. Nature 196:128–130
Hermans P, Kroeders M (2012) Modellering van sedimenttransport langs de kust van Suriname. MSc thesis, Dept. of Civil Engineering, KU Leuven (in Dutch)
Hervouet JM (2007) Hydrodynamics of free surface flows: modelling with the finite element method. Wiley, London
Kranenburg WM, Winterwerp JC, de Boer GJ, Cornelisse JM, Zijlema M (2011) SWAN-mud: engineering model for mud-induced wave damping. J Hydraul Eng 137:959–975
Lakhan VC, Pepper DA (1997) Relationship between concavity and convexity of a coast and erosion accretion patterns. J Coast Res 13 (1):226–232
Lakhan VC, Cabana K, LaValle PD (2002) Heavy metal concentrations in surficial sediments from accreting and eroding areas along the coast of Guyana. Environ Ecol 42:73–80
Lakhan VC, Ahmad SR, Pariznganeh (2006) A Investigating shifting mud banks along a coast subject to cycles of accretion and erosion. Proc Littoral Conf 2006:90–97
Landell Mills Limited (2013) Technical assistance for mangrove rehabilitation: final report. Government of Guyana, Georgetown
Lefebvre JP, Dolique F, Gratiot N (2004) Geomorphic evolution of a coastal mudflat under oceanic influences: an example from the dynamic shoreline of French Guiana. Mar Geol 208:191–205
Les BAJ, Westra MR (2004) Institutional capacity building activities on Guyana sea defences: bathymetric survey report. Royal Haskoning, Nijmegen
Lewis IIIRR (2004) Ecological engineering for successful management and restoration of mangrove forests. Ecol Eng 24(4):403–418
Lewis IIIRR (2009) Methods and criteria for successful mangrove forest restoration. Chapter 28. In: Perillo GME, Wolanski E, Cahoon DR, Brinson MM (eds) Coastal wetlands: an integrated ecosystem approach. Elsevier, Oxford, pp 787–800
Lovelock CE, Cahoon DR, Friess DA, Guntenspergen GR, Krauss KW, Reef R, Rogers K, Saunders ML, Sidik F, Swales A, Saintilan N,

Thuyen LX, Triet T (2015) The vulnerability of Indo-Pacific mangrove forests to sea-level rise. Nature. https://doi.org/10.1038/nature15538
Machin J (2012) Technical note: trial of planting coastal grasses to stabilise mud banks and facilitate natural recruitment of mangroves in Guyana. Guyana Mangrove Restoration Project, Mon Repos
Machin J, Lewis III RR (2013) Mangrove forest management and restoration in Guyana. Report. Guyana Mangrove Restoration Project, Mon Repos
Martinez JM, Guyot JL, Filizola N, Sondag F (2009) Increase in suspended sediment discharge of the Amazon River assessed by monitoring network and satellite data. Catena 79:257–264
Maza M, Lara JL, Losada IJ (2015) Tsunami wave interaction with mangrove forests: a 3-D numerical approach. Coast Eng 98:33–54
Mazda Y, Magi M, Kogo M, Hong PN (1997) Mangroves as a coastal protection from waves in the Tong King delta, Vietnam. Mangrove Salt Marshes 1:127–135
McIvor AL, Möller I, Spencer T, Spalding M (2012) Reduction of wind and swell waves by mangroves. Natural coastal protection series: Report 1. Cambridge coastal research unit working paper 40. The nature conservancy and wetlands international
McIvor AL, Spencer T, Möller I, Spalding M (2013) The response of mangrove soil surface elevation to sea level rise. Natural coastal protection series: report 3. Cambridge coastal research unit working paper 42. The Nature Conservancy and Wetlands International
Mehta AJ (2014) An introduction to hydraulics of fine sediment transport. Advanced series on ocean engineering, vol 38. World Scientific Publishing Co., Singapore
Mehta AJ, Lee SC, Li Y (1994) Fluid mud and water waves: a brief review of interactive processes and simple modeling approaches. Dredging Research Program Report DRP-94-4, US Army Corps of Engineers, Vicksburg (MS)
Morelock J (1972) Guiana-Orinoco continental shelf sediments. Bol Inst Oceanogr Univ Oriente 11(1):57–61
NAREI (2014) Mangrove department annual report 2013. National Agricultural Research and Extension Institute, Mon Repos
Nedeco (1968) Surinam transportation study: report on hydraulic investigation. Netherlands Engineering Consultants, The Hague, p 293
Nedeco (1972) Report on sea defence studies. Netherlands Engineering Consultants, The Hague
Nota DJG (1958) Sediments of the Western Guiana Shelf, reports of the Orinoco Shelf Expedition, vol II. Mededelingen Landbouwhogeschool Wageningen/Nederland 58(2):1–98
Odle MA, Arjoon PR (1971) A benefit/cost analysis of sea defence expenditure in Guyana. Soc Econ Stud 20(1):83–98
Orseaux S (2016) Dynamique sédimentaire d'un fleuve tropical sous influence amazonienne: le cas du Mahury, Guyane Française. PhD dissertation, Université du Littoral Côte D'Opale (in French)
Pelckmans A (2014) Modellering van golven en sedimenttransport langs de kust van Suriname. MSc thesis, Dept. of Civil Engineering, KU Leuven (in Dutch)
Pelling M (1999) The political ecology of flood hazard in urban Guyana. Geoforum 30:249–261
Proisy C, Gratiot N, Anthony EJ, Gardel A, Fromard F, Heuret P (2009) Mud bank colonization by opportunistic mangroves: a case study from French Guiana using lidar data. Cont Shelf Res 29:632–641
Proisy C, Degenne P, Anthony EJ, Berger U, Blanchard E, Fromard F, Gardel A, Olagoke A, Santos VF, Walcker R, Lo Seen D (2016) A multiscale simulation approach for linking mangrove dynamics to coastal processes using remote sensing observations. In: Vila-Concejo A, Bruce E, Kennedy DM, McCarroll RJ (eds), Proc 14th Int Coastal Symp (Sydney, Australia). J Coastal Research, SI75:810–814
Rine JM (1980) Depositional environments and holocene reconstruction of an argillaceous mud belt – Suriname, South America. PhD dissertation, University of Miami, Coral Gables (FL)
Rine JM, Ginsburg RN (1985) Depositional facies of a mud shoreface in Suriname, South America. A mud analogue to sandy shallow-marine deposits. J Sediment Petrol 55:633–652
Robertson R (2014) Importance of Avicennia Germinans (Black mangroves) at Chateau Margot, East Coast Demerara, Georgetown, Guyana. MSc Thesis, University of West Indies
Rodriguez HN, Mehta AJ (2001) Modeling muddy coast response to waves. J Coast Res SI 27:137–148
Sakakiyama T, Bijker EW (1989) Mass transport velocity in mud layer due to progressive waves. J Waterw Port Coast Ocean Eng 115 (5):614–633
Tol S (2016) Construction of permeable dam structures as sediment traps and basis for mangrove rehabilitation. Wetlands International. http://panorama.solutions/en/building-block/construction-permeable-dam-structures-sediment-traps-and-basis-mangrove. Retrieved May 2017
Tonneijck F (2013) Building with nature Indonesia. Opportunities for CC adaptation and Mitigation. Wetlands International/Mangrove Capital. UNFCCC Bonn 2013
Toorman EA (1996) Sedimentation and self-weight consolidation: general unifying theory. Géotechnique 46(1):103–113
Toorman EA (1997) Modelling the thixotropic behaviour of dense cohesive sediment suspensions. Rheol Acta 36(1):56–65
Toorman EA (2008) An investigation into the thixotropic wave dissipation potential of fluid mud. AGU Chapman Conference on Physics of Wave-Mud Interaction (Amelia Island, Florida, November 2008). Book of Abstracts, p 24
Topper E (2012) Assessment of the sustainable coastal zone protection through mangrove management project. Ministry of public works and communications. Works Services Group, Georgetown
Van Ledden M, Vaughn G, Lansen J, Wiersma F, Amsterdam M (2009) Extreme wave event along the Guyana coastline in October 2005. Cont Shelf Res 29(1):352–361
Vantrepotte V, Gensac E, Loisel H, Gardel A, Dessailly D, Mériaux X (2013) Satellite assessment of the coupling between in water suspended particulate matter and mud banks dynamics over the French Guiana coastal domain. J S Am Earth Sci 44:25–34
Villarroel A (2009) Stroming van vloeibaar slib. MSc thesis, Dept. of Civil Engineering, KU Leuven (in Dutch)
Walcker R, Anthony EJ, Cassou C, Aller RC, Gardel A, Proisy C, Martinez JM, Fromard F (2015) Fluctuations in the extent of mangroves driven by multi-decadal changes in North Atlantic waves. J Biogeogr 42:2209–2219
Welage SJ (2005) Coastal evolution Guyana: modelling and a historical investigation. MSc. Thesis, Delft University of Technology
Wells JT, Coleman JM (1981) Physical processes and fine-grained sediment dynamics, coast of Surinam, South America. J Sediment Petrol 31(4):1053–1068
Wells JT, Coleman JM, Wiseman WJ (1981) Suspension and transportation of fluid mud by solitary-like waves. Coastal Studies Institute, Louisiana State University, Baton Rouge
Westra MR (2004) Institutional capacity building activities on Guyana sea defences: morphological survey and oceanographic monitoring report. Royal Haskoning, Nijmegen
Winterwerp JC, de Graaff RF, Luijendijk AP (2005) Institutional capacity building activities on Guyana sea defences: modelling report, Volume 4: Morphology modelling. WL Delft Hydraulics
Winterwerp JC, de Graaff RF, Groeneweg J, Luijendijk AP (2007) Modelling of wave damping at Guyana mud coast. Coast Eng 54 (3):249–261

Winterwerp JC, Erftemeijer PLA, Suryadiputra N, van Eijk P, Zhang L (2013) Defining eco-morphodynamic requirements for rehabilitating eroding mangrove-mud coasts. Wetlands 33(3):515–526

Wong PP, Losada IJ, Gattuso J-P, Hinkel J, Khattabi A, KL MI, Saito Y, Sallenger A (2014) Coastal systems and low-lying areas. In: Field CB, Barros VR, Dokken DJ, Mach KJ, Mastrandrea MD, Bilir TE, Chatterjee M, Ebi KL, Estrada YO, Genova RC, Girma B, Kissel ES, Levy AS, MacCracken S, Mastrandrea PR, White LL (eds) Climate change 2014: impacts, adaptation, and vulnerability. Part A: global and sectoral aspects. Contribution of working group II to the fifth assessment report of the intergovernmental panel on climate change. Cambridge University Press, Cambridge, pp 361–409

Wong T, Kroonenberg S, Augustinus P (2017). Geologie en landschap van Suriname. Stichting LM Publishers, Volendam, the Netherlands, 198p (in Dutch)

Wongsoredjo S (2017) Numerical modelling of sediment transport along a mud dominated coast. MSc in Earth Sciences Internship Report, Utrecht University (the Netherlands) and KU Leuven (Belgium)

Zhang K, Liu H, Li Y, Xu H, Shen J, Rhome J, Smith TJ III (2012) The role of mangroves in attenuating storm surges. Estuar Coast Shelf Sci 102–103:11–23

The Beneficial Effects of Mangrove Forest to Sea Defence Structures

21

Henk Jan Verhagen

Abstract

At many locations, especially in deltaic areas, dikes and other flood defence structures are needed to protect the hinterland against flooding by surges (typhoons, hurricanes) and/or tsunamis. Dikes are quite costly, and mangroves in front of the dike will lower the costs of the dike. Mangroves in general allow dikes to be lower and narrower. In this chapter the hydrodynamic effects of mangroves on dikes are elaborated as well as the financial impact on the costs of dikes. The main advantage of a mangrove forest in front of a dike is that the forest decreases the wave height just in front of the dike, and therefore the freeboard needed to cope with these waves can be considerably less. Because the waves in front of the dike are small due to the mangrove forest, often no costly revetment structure is needed any more. In this chapter computational methods are presented to determine the direct benefit for dike construction.

Keywords

Sea defence structure · Levee · Dike · Overtopping · Storm surge · Cyclone surge

21.1 Sea Defence Structures

In this chapter sea defence structures are defined as structures to defend the coastal zone against the action of floods, swell and wind waves and tsunamis. The main function of these structures is to prevent overflow or overtopping of water. Overflow occurs when the water level outside the dike (levee) is higher than the crest of the structure and a continuous flow of water over the structure will occur. Overtopping occurs when the (still) water level at sea is below the crest of the dike, but the run-up of the waves is higher than the height of the dike.

In general overflow over coastal defence structures is not allowed. If overflow occurs, the area behind the dike will be flooded, and the inner slope of the dike will be damaged and usually significant repair works will be needed. Overtopping, on the other hand, is generally accepted to some degree; often a limit on the order of 1–5 l/s per running meter of dike is accepted as average overtopping discharge.

Sea defence structures are usually earthen structures. These earthen structures are called dikes, levees or embankments. The selected word depends on local habits. In the USA usually the word levee is used.

Sometimes other structures are needed in a dike trajectory, for example a discharge sluice, a lock, etc. Basically the load reduction effect from mangroves for these other structures is similar to the load reduction on dikes. In this chapter the focus will be on the earthen structures, and less on other structures in a dike trajectory. Figure 21.1 shows a stretch of dike in Vietnam where partly the mangroves disappeared and the (grass) dike had to be protected with a revetment of concrete blocks.

There is a difference in effect of mangroves on sloping structures versus on vertical structures (like concrete walls, recurved walls and sheetpiles). Attention will be paid to this difference.

21.2 Types of Loads

In general one can distinguish the following types of hydraulic loads on sea defence structures:

- Water level due to the tide
- Increased water level due to wind set-up (caused by storms, cyclones, etc.)
- Increased water level due to high river run-off
- Tsunami loads
- Attack by wind waves and swell

H. J. Verhagen (✉)
Delft University of Technology, Delft, The Netherlands
e-mail: H.J.Verhagen@TUDelft.nl

C. Makowski, C. W. Finkl (eds.), *Threats to Mangrove Forests*, Coastal Research Library 25,
https://doi.org/10.1007/978-3-319-73016-5_21

Fig. 21.1 Sea dike in Vietnam where mangroves disappeared and a block revetment had to be constructed

Each of these loads will be discussed in more detail. Other types of loads on sea defence structures (like earthquake loads, settlement of subsoil, etc.) will not be discussed in this chapter. The abovementioned hydraulic loads vary in duration, occurrence and effect on the sea defence structure.

21.2.1 Water Level Due to the Tide

Along most coasts there is a tidal variation, usually with a period of 12 h and 25 min (semi-diurnal tide). But this may vary quite a lot per location. Even in some locations close to each other tides may vary. In some locations the tide is diurnal (which means one high water and one low water per day). However, one should realise that such a diurnal tide is in most cases not a simple sinusoidal curve. Usually it is a short high water period, a short low water period and an intermediate period in-between. Especially for mangroves this is very relevant, because the various mangrove species each have their own requirements for duration of submergence per day.

In Fig. 21.2 the predicted tide is given for four stations in tropical areas. The graph for Natal, Brazil, shows a typical ("normal") semi-diurnal tide with a small variation between spring tide and neap tide. The lowest spring tide (LLWS) is usually the reference level (Chart Datum). For mangroves, the number of hours submerged and emerged is relevant. For example at level LLWS+1 m (which is approximately mean sea level) the submergence is in the order of 7 h, while the dry period is in the order of 5 h. For some mangrove species this is just tolerable. Note that the exceedance lines (right figures) are all nearly vertical lines. This means that all tides give more or less the same emergence.

At a location with a typical diurnal tide like Corpus Christi, Texas, the exceedance graphs have a different shape. Mean sea level is also here approx. LLWS+1 m. But at this level during 50% of the time submergence is more than 15 h. On a little bit higher level (LLWS+1.5 m), the 50% submergence time is already reduced to 7 h.

In other locations in the world very peculiar tides exist. Examples are given from Merauke in Irian Jaya and Duchateau Island in Papua New Guinea. Because of this shape at some level very long submergences are possible. For example, in Mearauke, at level LLWS+2.25 m during 50% of the tides the submergence is 8 h. However, 35% of all tides (100–65) have a submergence of more than 17 h. This shows that at locations with peculiar tides one should not look only to the 50% submergence tides, but have a look at the full exceedance graph of the water levels. A species which on average can survive submergence of 8 h (followed by a 6 h dry period) may not be able to survive at level LLWS +2.25 m at Merauke because during many tides this species will be submerged for more than 17 h.

Planning (and planting) a mangrove belt in front of a sea dike therefore requires always a thorough investigation of the local tide. Not only the values of high and low water have to be known, but also the spring/neap variations and the duration of the tide on each level. Tidal graphs are usually easily available. The exceedance graphs as presented in Fig. 21.2 are usually not standard available, but can easily be generated from tidal predictions. However, for this one needs not only the daily levels of high and low water, but level values some 15 min apart. The lines of Fig. 21.2 were determined using the public domain website http://tbone.biol.sc.edu/tide and a simple spreadsheet.

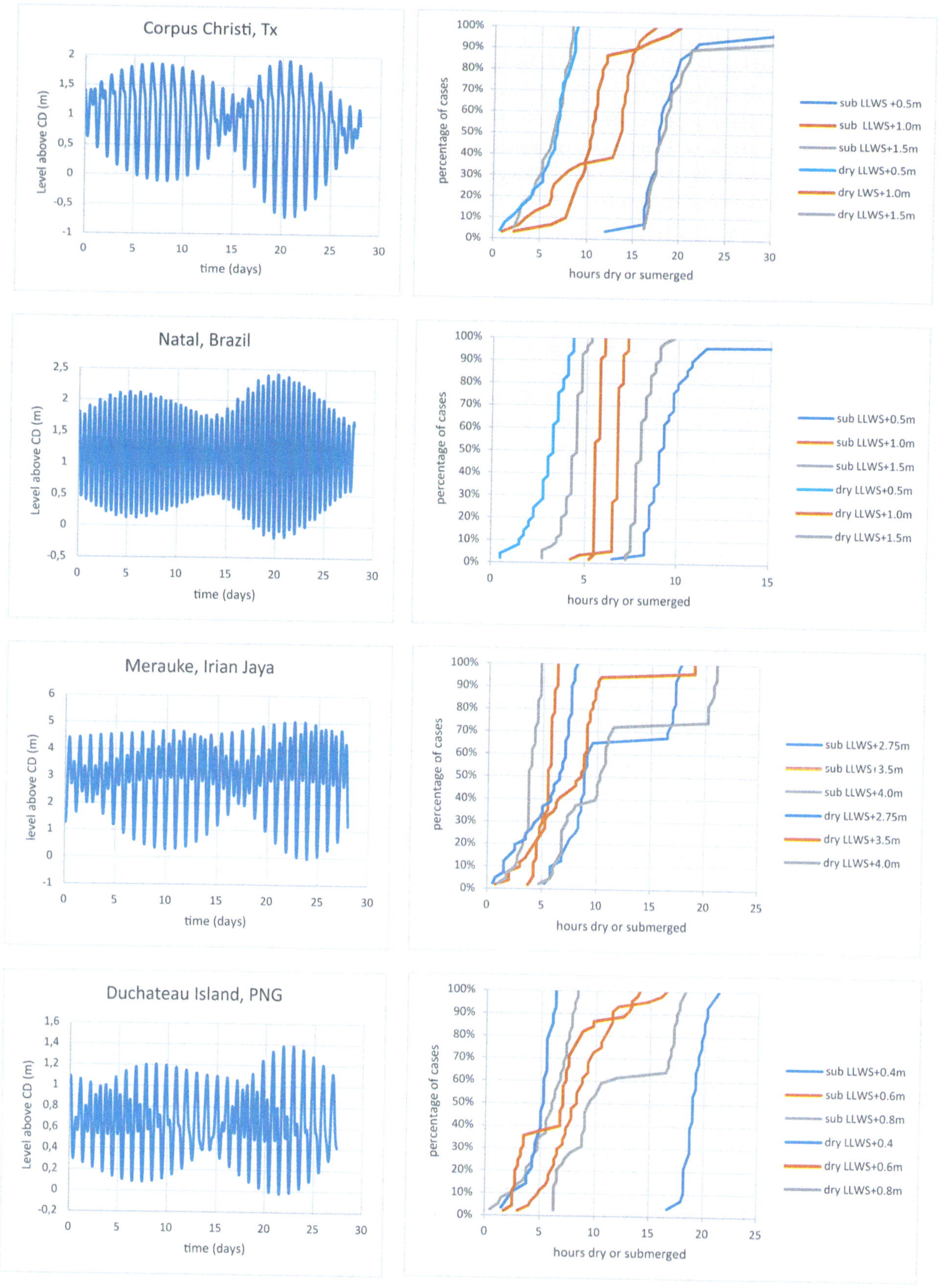

Fig. 21.2 Tidal curves and exceedance lines for four stations with different tidal regimes (waterlevel data from http://tbone.biol.sc.edu/tide)

21.2.2 Storm Surges

During storms, cyclones, hurricanes and typhoons wind blows over large water surfaces, causes a rise in the average water level at the coast. The rise of the water level (wind set-up) depends on wind speed, duration of the wind, fetch and water depth. The effect of the water depth is considerable. Especially in shallow water wind set-up is considerable.

In mangrove forests there is no interaction any more between the wind and the water surface. This implies that in such an area the wind set-up is not increasing any more. In a very simplified case this can be illustrated. Suppose we have a sea with a depth of 25 m, fetch of 200 km, and wind speed of 35 m/s. Wind duration is long enough to obtain maximum wind set-up. When there is no shallow area near the coast, the wind set-up in this case will be 2.35 m at the coastline. When there is a shallow zone near the coast of 5 km wide and 2 m deep, the total set-up will be 2.98 m. In case half of this zone (2500 m), is covered with mangroves the total set-up will be 2.69. This is a reduction of approximately 30 cm. Although the results of this calculation will be different for every location, depending on the local water depth, wind speed, etc. in general one can conclude that the effect of a mangrove forest on the wind set-up is rather marginal. The calculations for this example were made with the public domain computer program Cress (www.cress.nl, water movement → water levels).

Another aspect is that a storm surge behaves like a long wave. The resistance of the mangroves dampens the long wave and in this way mangroves will lower the amplitude of the long wave. Unfortunately this effect is very minor. As already proven by Schiereck and Booij (1995) this is effect can only become relevant with extremely wide mangrove areas. In their paper they elaborate a case where the effect of a surge in the Sundarbans (India/Bangladesh) is simulated. They simulated a surge with surge height at sea of 1.5 m on top of the normal tide. The maximum surge height 60 km inland was in the order of 50 cm less (Fig. 21.3). This shows that the effect of mangroves on waterlevels of surges is very minimal, and becomes only effective for extremely wide mangrove areas (many km wide).

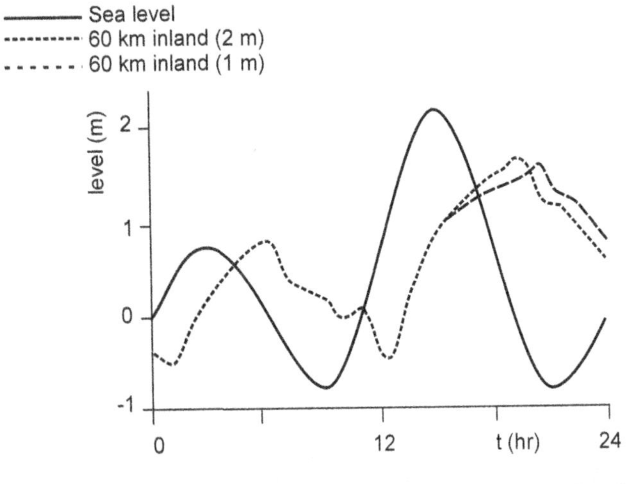

Fig. 21.3 Calculated effect of a storm surge in the Sundarbans (Based on data from Schiereck and Booij 1995)

On basis of the above, one can conclude that the beneficial effect of mangroves on the safety of dikes with respect to the storm surge in general can be neglected.

21.2.3 River Discharges

In many rivers discharge may vary a lot over the year. In estuaries one may also observe the influence of an increased river discharge. Usually this effect is only clear at some distance from the river mouth, because at the mouth usually the water levels are completely governed by the sea (tides and storm surges). For most river systems the duration of such a flood wave is in the order of days. This means a relatively slow rise of the water level. Mangrove forests along the riverbanks have hardly any effect on this water level rise. During periods of high water there is a wide floodplain and on this wide, inundated area waves may develop. The design height of a river dike depends on the water level, where a freeboard has to be added for wave action. The mangroves hardly have any effect on the water level, but the freeboard due to wave action may be decreased by the presence of mangroves. This is elaborated in section "Wave Action" below. However, in many cases along rivers, it is not possible to have mangrove forests along the dike, because during the dry period the water level at that location is too low to maintain a vital mangrove belt at that specific location.

21.2.4 Tsunamis

A tsunami is hydraulically a single, but very long wave. Such a wave behaves somewhat similar to a surge. But a normal surge has a rather slow rise and fall of the water level. Contradictory to this, a tsunami wave has a very steep front. It is therefore more similar to a translation wave (a wave very common in navigational canals due to the locking process). The speed of a translation wave depends only on the water depth. This means that the speed outside and inside the mangrove forest is identical. Mangroves do not slow down a tsunami wave.

When the cross-section of the flow profile suddenly changes (narrower or shallower) the translation wave is (partially) reflected. This also happens when a tsunami reaches the outer border of the mangrove forest. At that point a part of the energy is reflected, and a part continues to flow into the forest.

When a translation wave approaches a vertical wall, the wave is fully reflected. Because of the long character of such a wave a dike, even with a slope of 1:3, behaves like a vertical wall. Suppose at a certain nearshore point we have a tsunami approaching the coast. At that given location the tsunami has a height of 3 m above a still water level of 1 m. This tsunami approaches our dike and is fully reflected. That means that at the dike the water level will become 6 m. Suppose we have a 5 m high dike. Water level reaches then 1 m above the crest of the dike. Because behind the dike there is no water, flow over the dike will be critical. The water depth over the crest of the dike will be

$$a = \frac{2}{3}\delta h \qquad (21.1)$$

And the velocity over the dike will be

$$u = \sqrt{\frac{2}{3}g \cdot \delta h} \qquad (21.2)$$

Where:

a flow depth over the dike crest
δh water level outside the dike with respect to the crest height
u velocity water over the crest
q discharge over the dike $q = a \cdot u$

In our example this lead to a flow velocity of 2.5 m/s and a discharge of 1.7 m^3/s per meter of dike. When we have a 1 km long dike section this is 2 million m^3 in 20 min. When the area to be flooded by the tsunami is 2 km wide, the water depth will become 1 m due to the tsunami by overflow.

But it is very probable that the dike cannot survive an overflow velocity of 2.5 m/s, so probably the inner slope will fail and the dike will breach. Then, the water level inside will become much more.

However, when a mangrove forest is present, this forest will act as a partial reflector for the wave. Suppose we have a forest with a density of 1600 trees/ha, and the trees have a diameter of 20 cm (this means the trees are somewhat more than 2 m from each other). Only the first two rows act as a reflector, this means a blockage of approximately 15% ($r = 0.15$). The wave height entering the forest is given by:

$$\frac{H_1}{H_0} = \frac{1-r}{1+r} = \frac{0.85}{1.15} = 0.74 \qquad (21.3)$$

When the original wave was 3 m, the wave going into the forest is only 2.25 m high. When the forest is very regular, this wave will not decrease. But because a mature forest is not regular, there will also inside the forest some reflections. One may assume that the wave leaving the forest will become somewhat smaller, in the order of 2 m. This wave will reflect against the dike and double. So the water level at the dike will become 4 m.

Because the dike height was 5 m, the dike will not be overwashed, and not fail. In this way the mangroves have protected the area against flooding by a tsunami.

21.2.5 Wave Action

Wind waves are considerably reduced by mangroves. Basically there are two ways to calculate the decrease in wave height due to mangroves. It can be done using an overall method using a transmission coefficient or by calculating the energy dissipation step-by-step when the wave propagates through the mangrove forest. The first method is much easier, but also less reliable in detail. The second method is in principle more reliable, but requires some resistance data, which are difficult to obtain.

21.2.5.1 Overall Method

In this method the reduction of wave height in a mangrove forest is expressed with the transmission coefficient K_t, where: $K_t = H_d/H_0$, or K_t = (wave height at toe of the dike)/(wave height on the seaward side of the mangrove belt). In international literature (e.g. Quartel et al. 2007) also the value *R* is used, which is defined as:

$$R = \frac{H_0 - H_d}{H_0} = 1 - K_t \qquad (21.4)$$

Both reduction coefficients have no dimension, but depend on the width of the mangrove belt *x*. To make plots independent of the width of the mangrove belt, sometimes the value *r* is used. This is the wave height reduction per meter.

$$r = \frac{\Delta H}{H}\frac{1}{\Delta x} \qquad (21.5)$$

Where *H* is the wave height and ΔH the change of wave height over a distance Δx. This definition is for example used by Mazda et al. (2006) and Horstman et al. (2012, 2014).

The relation between *r* and K_t is given by:

$$K_t(x) - e^{-rx} \qquad (21.6)$$

Research by Quartel et al. (2007) showed that the value of *r* varies between 0.004 and 0.012 for a mangrove area in the Red river delta with water depth of 0.75–1.5 m. Quartel et al. indicate the relation between *r* and the water depth as $r = 0.004 + 0.003d$, in which *d* is the water depth. Physically there is no explanation of a decrease of *r* with a decrease of water depth. One would expect an increase. The reliability of this dependence of the water depth is rather low. Areas with observations are indicated in Fig. 21.4. Most probably this variation in the value of *r* is more due to variations in foliage, stem diameter, etc. For the beach plane and the mud bed Quartel et al. did find an increase in *r* when decreasing the depth.

Also datasets are available from Mazda et al. (2006) based on measurements along the Vinh Quang coast in Vietnam (*Sonneratia* sp.). Their result give values of r of 0.006 and smaller. Data from Horstman et al. (2014) are based on measurements in Kantang and Palian (Thailand) in areas with *Avicennia* and *Rizophora*.

The observation of the wave height decrease on mud flats by Mazda et al. (2006) and by Quartel et al. (2007) show the same tendency. As mentioned before the data from Quartel et al. (2007) in the mangrove area show strange values and a strange tendency.

Schiereck and Booij (1995) calculated the wave height reduction in terms of K_t with the mathematical model Hiswa and some laboratory data. The results from Schiereck and Booij can be converted to r-values. In Fig. 21.4 their results are compared with the observations. It seems that the dependence of the water depth as suggested by Schiereck and Booij is probably too strong. The calculated values for sparse forests coincide quite well with the observations of Mazda et al. (2006) and Horstman et al. (2014).

The variation in data is relatively large, which is problematic for making a good design. The definitions of "sparse" and "average" from Schiereck and Booij (1995) are well defined for small sticks, but absolutely not for a real mangrove forest. Therefore Horstman et al. (2014) suggest to use the vegetation density as a volume percentage in the prediction formulas. They suggest the volumetric vegetation density as a predictor instead of the water depth. First they determine the vegetation cover ρ_c (this is the total cross-sectional area of the vegetation per m^2). This means that when on a certain level over the bed there is one tree with a diameter of 20 cm every m^2, the vegetation cover is 3.141%. Very near to the bed this vegetation cover increases significantly because of the roots of the mangrove trees. In the lower 10 cm this may increase to some 15–25%. The vegetation density is now defined as the integration of the vegetation cover over the height

$$\rho_v = \frac{1}{h}\int_0^h \rho_c dh \tag{21.7}$$

where h is the water depth, but in Horstman et al. (2014) this is set to a fixed value of 1 m. This leads to vegetation densities in the order of 17–23% for *Avicennia* and 15–40% for *Rhizophora*. The reduction coefficient can be calculated with:

$$r = 5.1 \cdot 10^{-4}\rho_v + 1.5 \cdot 10^{-3} \tag{21.8}$$

Horstman et al. (2014) found a correlation of 98% for their own data (note that in Horstman et al. 2012 different

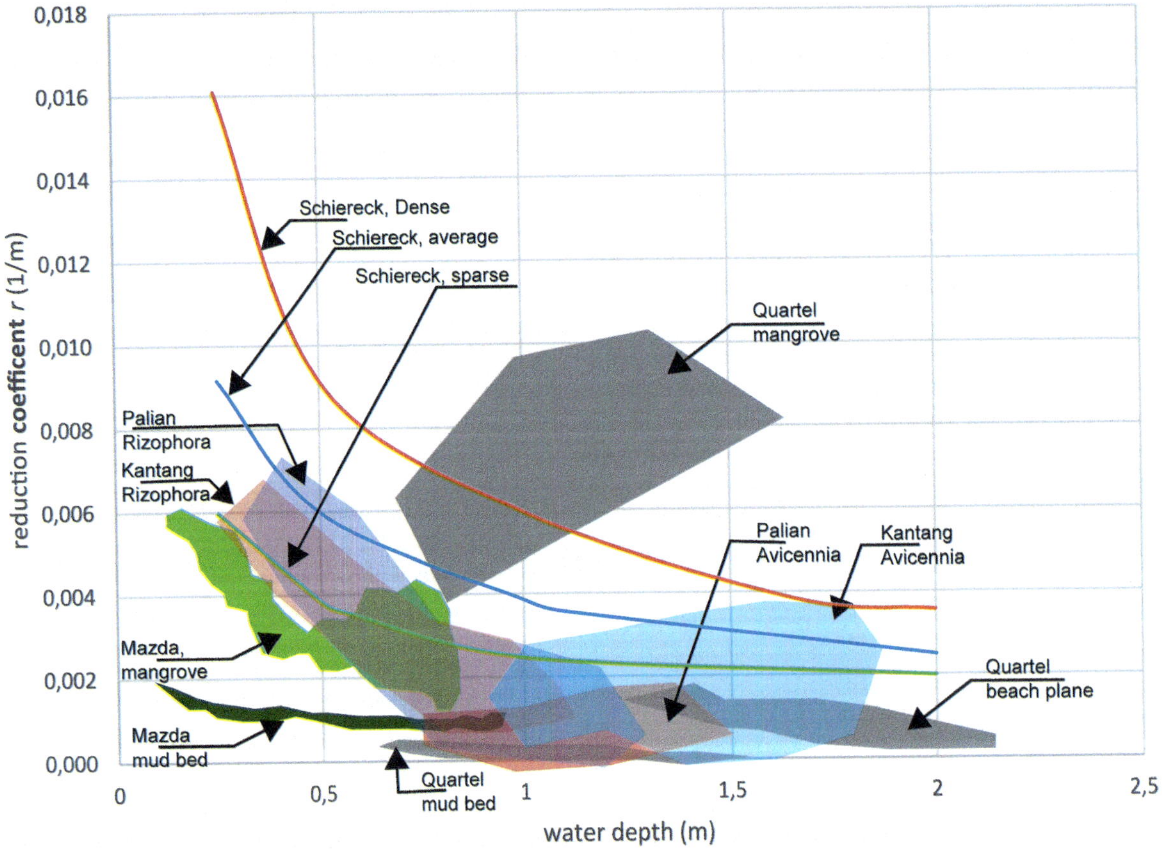

Fig. 21.4 Wave height reduction coefficient (r) as function of water depth as calculated by Schiereck and Booij (1995) compared with the data from Quartel et al. (2007), Mazda et al. (2006) and Horstman et al. (2014) (Kantang and Palian)

regression coefficients are given). The problem with this approach is that the volumetric density is difficult to measure, but even more difficult to predict.

Quynh (2010) has done some measurements in the Can Gio district. On basis of these measurements the following formula are suggested:

$$K_t = \frac{a}{H_0} e^{-bx} \quad (21.9)$$

The two coefficients are defined as:

$$a = 1.0249 H_0 - 0.00977 \ (m) \quad (21.10)$$

and average

$$b = -0.0481 + 0.016 H_{vn} + 0.00177 \ln(N) + 0.0077 \ln(T_c) \quad (21.11)$$

where:

H_{vn} average total height of the trees in the forest (m)
N density of mangrove trees taller than 1 m (trees/ha)
T_c Forest canopy cover (%)

Note: in the original publication of Quynh, b is defined as b = +0.0481–0.016.... For consistency reasons the signs have been adapted here.

The advantage of the method of Quynh (2010) above Horstman et al. (2014) is that the accuracy is not much less, but the parameters are much easier to collect. This makes that the Quynh approach is much more practical for prediction and engineering purposes.

Because the variation in field data is quite large, one should limit the accuracy of the given formulas to only one significant digit (for example $H = 1.1$ m, and not $H = 1.0867$ m). Doing so, Eq. (21.10) reduces to a = 1.0 H_0. Therefore one may approach the K_t formula of Quynh with:

$$K_t = e^{-bx} \quad (21.12)$$

This equation is identical to Eq. (21.6). This means that physically the value of b in Quynh (2010) should be identical to r. This means that also the value of b should be between 0.004 and 0.015. Because the variation in the basic data on which Quynh has derived his data, one has to be very careful with extrapolation of Eq. (21.11) outside the range of application. This range is not given by Quynh, but one may assume:

$3 < H_{vn} < 5$(m)
$1000 < N < 2000$(trees/ha)
$80 < T_c < 99\%$

Formula (21.11) is very sensitive to the tree height H_{vn}. In the field test all trees had heights between 3.2 and 3.8 m. Therefore one has to be quite careful with the use of the formula for trees with a different height. In the paper of Quynh his Table 3 gives a relation between the wave dissipation, the canopy diameter and mangrove density. This table is based on wave height measurements of reduction around an individual tree. It seems that this table gives a somewhat too optimistic required width of mangroves zones.

It is therefore suggested (Verhagen and Loi 2012) to use for practical design calculations the values of Table 21.1.

In Fig. 21.5 the computed values of K_t for various widths of mangrove belts are indicated. For dense mangroves one may use the blue line, for very sparse mangroves the pink

Table 21.1 Reduction coefficients

Density of the mangroves	Reduction coefficient r
dense	0.010
average	0.007
sparse	0.004

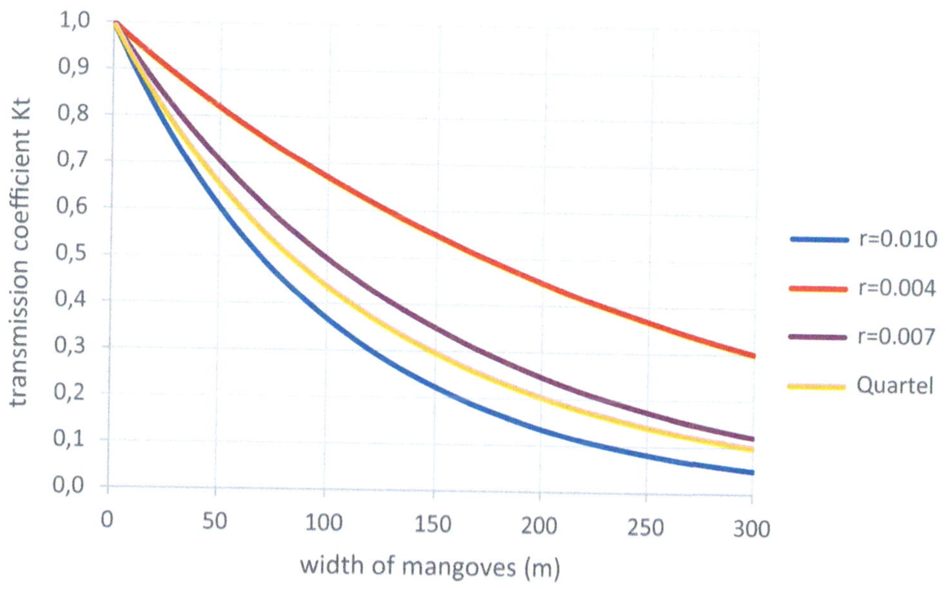

Fig. 21.5 Dissipation coefficient Kt as a function of mangrove width

line. The yellow line is the line as suggested by Quartel et al. As can be seen the difference between this line and the use of an average value of $r = 0.007$ is not very large.

21.2.5.2 Simple Design Graphs

When more details are available, the Quynh method can be used. One may use the formulas given above, but one can also use the design graphs in Fig. 21.6. From the graphs follow very clearly that the percentage of coverage is not very sensitive to the final answer, provided the coverage is above 80%.

21.2.5.3 Energy Dissipation Method

This can be illustrated by many types of mathematical modelling. One of the earliest trials to show this is the study by Schiereck and Booij (1995). They modelled a mangrove area with the mathematical model HISWA, and showed the effect. To calibrate the model, physical model tests were done, simulating the mangroves with small vertical sticks in a wave flume. Somewhat later Sumanusajonkul (1995) made similar calculations with the wave model ENDEC. Also here calibration was done with data from physical model tests. The problem with getting calibration data using physical models is that in physical scale model tests, scale effects become important. It is unclear how to translate the resistance of a real mangrove tree to a scale mangrove, because normal Froude-scaling (geometrical scaling) is not valid for resistance parameters (Fig. 21.7).

In these models mangroves are modelled as cylinders. The energy dissipation follows from:

$$S_c = \frac{2}{3\pi} n C_D D \frac{a^3 \omega^3}{gk} \left[\frac{1}{3} + \frac{1}{\sinh^2 kd} \right] \tag{21.13}$$

So, when the number of stems and stem-diameter are known, energy dissipation can be calculated, assuming a certain C_d-coefficient.

In reality a mangrove tree has a clear root system, a trunk and a canopy. These three elements have different resistance values. In order to model mangroves in a better way, one should therefore not describe mangrove trees by a single drag factor, but distinguish root, stem and canopy (Fig. 21.8).

This idea has been elaborated upon and implemented in the wave model SWAN by Burger (2005).

Later this methodology was compared to an approach using CADMAS/SURF by Suzuki (2011). This approach provided more understanding of the process and Suzuki used this knowledge to improve the vegetation module in SWAN. In SWAN the mangroves are modelled with three layers, a root layer, a stem layer and a canopy layer. Each of these layers has its own resistance coefficient. In his thesis, he elaborates a case study of the wave shielding due to mangroves on Kanika Sands (Orissa, India), see also Narayan et al. (2010). This island is oval in shape, roughly 4 km along the N-S axis and 1.5 km along the E-W axis at a maximum. The island is located around 3.5 km directly eastward of the upcoming Dhamra Port. The morphology in the region is highly unstable though showing a tendency for progradation. The island has been in existence for more than 30 years and is currently heavily inhabited throughout by mangroves of species like *Avicennia marina*, *Avicennia alba*, *Sonneratia alba*, *Rhizophora mucronata* and *Phoenix paludosa* that were found to attain a height of 10 m or more (Fig. 21.9).

The control parameters for the mangrove types were partly taken from a study by Narayan (2009). Table 21.2 shows the parameters used for the calculation.

A remaining point of uncertainty is the lack of prototype data. Only very few reliable datasets of wave dissipation are available. And these data are all collected during rather mild wave conditions. Measuring wave dissipation during design conditions is still a challenge. Some of these problems are discussed in more detail in De Vos (2004).

21.2.5.4 Practical Formulas

As seen above, the benefit of a mangrove depends very much on the local circumstances, like surge level, slope of the foreshore, wave heights, etc. With the use of mathematical models this can be calculated quite well, but for straightforward design work application of models can be too complicated. In Tusinski and Verhagen (2014) this has been elaborated for the coast of Vietnam. There the computations were carried out using latest version of wave penetration model SWAN, in which the method of Schiereck and Booij (1995) to calculate the effect of wave energy dissipation in mangrove forest was implemented by Narayan et al. (2010). The quality of modelling roots in numerical models like SWAN has been improved by a more detailed mathematical description of the root system on the basis of a photogrammetric method Bo (2012), see Fig. 21.10. Combining this with data from laboratory tests in Hannover, Strusińska-Correia et al. (2013) leads to a good model description within SWAN.

A total number of 135 simulation runs with varying parameters was carried out in order to investigate impact of them on wave height decay in the mangrove zone. The result was then a subject of statistical analysis. Presented below formulas is the result of non-linear curve fitting of equations into the data sets. The equations approximate simulated results with the accuracy of 0.02–0.15 m (standard error of deviation).

Fig. 21.6 Results of the Quynh formula for different canopy densities

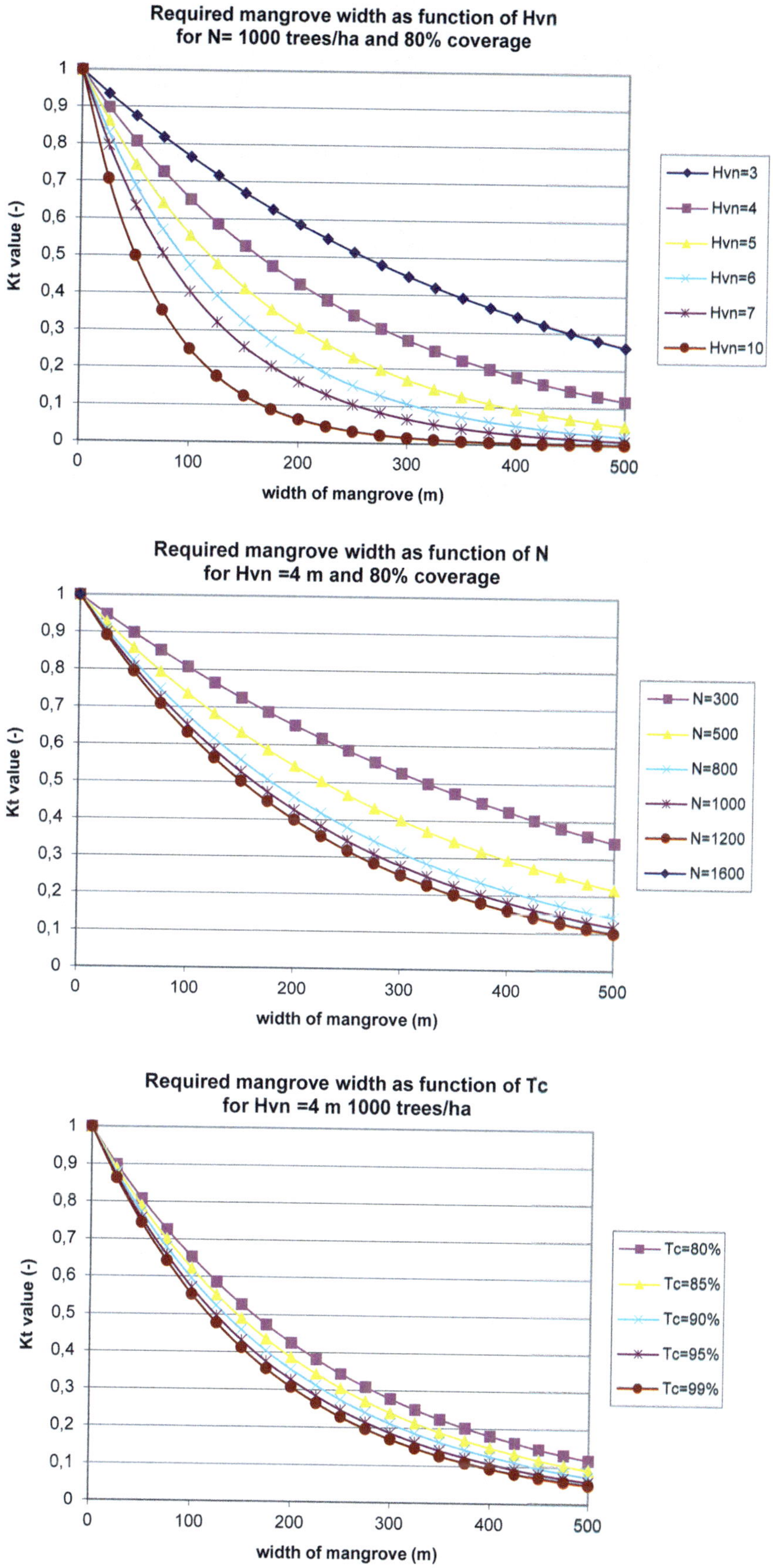

Fig. 21.7 Flume test by Schiereck and Booij (1995) to determine the dissipation parameters (Photo TU Delft)

Canopy

For the reduction effect of canopies one has to distinguish between emergent canopies and submerged canopies.

Emergent canopy:

$$K_{ce50} = \exp\left[-2.014 + 0.03C_c^3 + 1.252\ln(H_i)\right] \quad (21.14)$$

Submerged canopy:

$$K_{cs50} = \exp\left[-6.874 + \frac{0.635\left(h/h_t\right)}{\ln\left(h/h_t\right)} + 1.783\ln(H_i)\right] \quad (21.15)$$

Resulting in:

$$H_{att} = H_i - \frac{K_{c50}}{50}V_l \quad (21.16)$$

where:

K_{ce50} wave height reduction factor for 50 m of emergent canopy [m]
K_{cs50} wave height reduction factor for 50 m of submerged canopy [m]
H_{att} attenuated wave height [m]
H_i initial wave height [m] – (range of validity: $H_i = 0.5$–3.0 m)
V_l vegetation length [m] – (range of validity: $V_l = 1$–200 m)
h/h_t ratio of submergence [–] – (range of validity: $h/h_t = 1.25$–2)
h water depth [m]
h_t height of a tree [m]
C_c canopy coefficient [–]
$C_c = 1$, for lower part of a canopy
$C_c = 2$, for upper part of a canopy
$C_c = 4$, for fully submerged canopy (H/h = 1)

Trunk

The dependence of the trunks proved to be a function of the initial wave height H_i. Therefore separate formulas are given for the trunk-related wave height reduction for an initial wave height of 3, 2 and 1 m:

$$\begin{aligned} K_{t50(3)} &= 0.571 + 0.118\ln(T_d) + 0.136\ln(D) \\ K_{t50(2)} &= 0.365 + 0.074\ln(T_d) + 0.096\ln(D) \\ K_{t50(1)} &= \exp[-1.78 + 0.5051\ln(T_d) + 0.539\ln(D)] \end{aligned} \quad (21.17)$$

Fig. 21.8 Modelling mangroves in several layers (Burger 2005)

Resulting in:

$$H_{att} = H_i - \frac{K_{t50}}{50} V_l \qquad (21.18)$$

where:

$K_{t50(x)}$ wave height reduction factor of x m wave, for 50 m of trunks [m]
H_{att} attenuated wave height [m]
H_i initial wave height [m] – (range of validity: $H_i = 0.5$–3.0 m)
V_l vegetation length [m] – (range of validity: $V_l = 1$–200 m)
T_d Spatial density of trunks [trunk/m^2] – (range of validity: $T_d = 0.2$–1.8 trunk/m^2)
D Diameter of trunk [m] – (range of validity: $D = 0.1$–0.9 m)

Prop Roots

The effect of prop roots depends mainly on root density:

$$K_{pr50} = \exp\left[-1.788 - \frac{2.359}{\sqrt{R_d}} + 1.286 \ln (H_i)\right] \qquad (21.19)$$

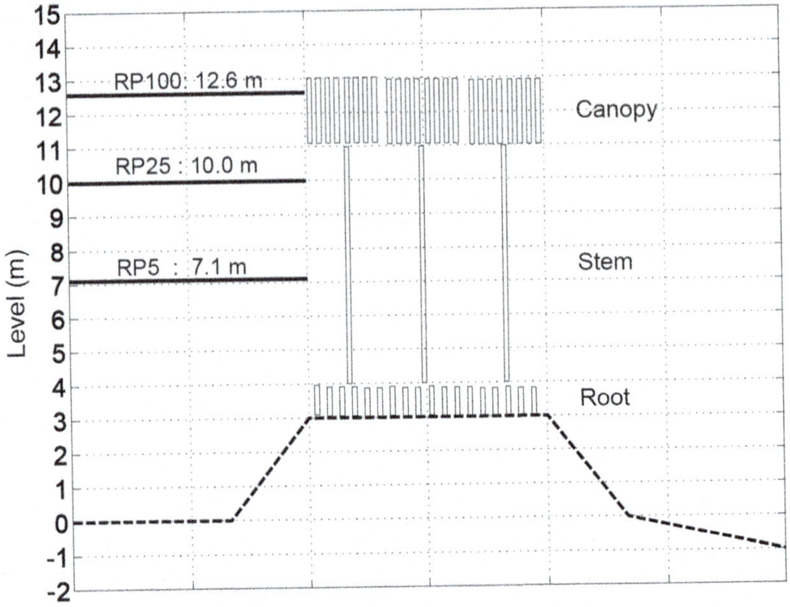

Fig. 21.9 Schematisation of a mangrove in Swan (Suzuki 2011)

Table 21.2 Coefficients for *R. Muconata*, based on Suzuki (2011) and Narayan (2009)

Element	Parameter	Value range	Control value
Root	Diameter	0.05–0.1 m	0.05 m
	Density	1–130 m^{-2}	100 m^{-2}
	Height	0–1 m	1.0 m
	Bulk drag coefficient	0.01–1	–
Stem	Diameter	0.15–0.4 m	0.2 m
	Density	0.5–1.7 m^{-2}	1 m^{-2}
	Height	5–8 m	7.0 m
	Bulk drag coefficient	0.1–1	–
Canopy	Diameter	0.02–0.1 m	0.02 m
	Density	1–100 m^{-2}	100 m^{-2}
	Height	0.2–3 m	2.0 m
	Bulk drag coefficient	0.01–1	–

Resulting in:

$$H_{att} = H_i - \frac{K_{pr50}}{50} V_l \tag{21.20}$$

where:

K_{pr50} wave height reduction for 50 m of prop roots [m]
H_{att} attenuated wave height [m]
H_i initial wave height [m] – (range of validity: $H_i = 0.5–3.0$ m)
V_l vegetation length [m] – (range of validity: $V_l = 1–200$ m)
R_d Spatial density of prop roots [root/m^2] – (range of validity: $R_d = 2–10$ trunk/m^2)

Fig. 21.10 Photogrammetric triangulation to a prop-roots (left) and pneumatophores (right) (Bo 2012)

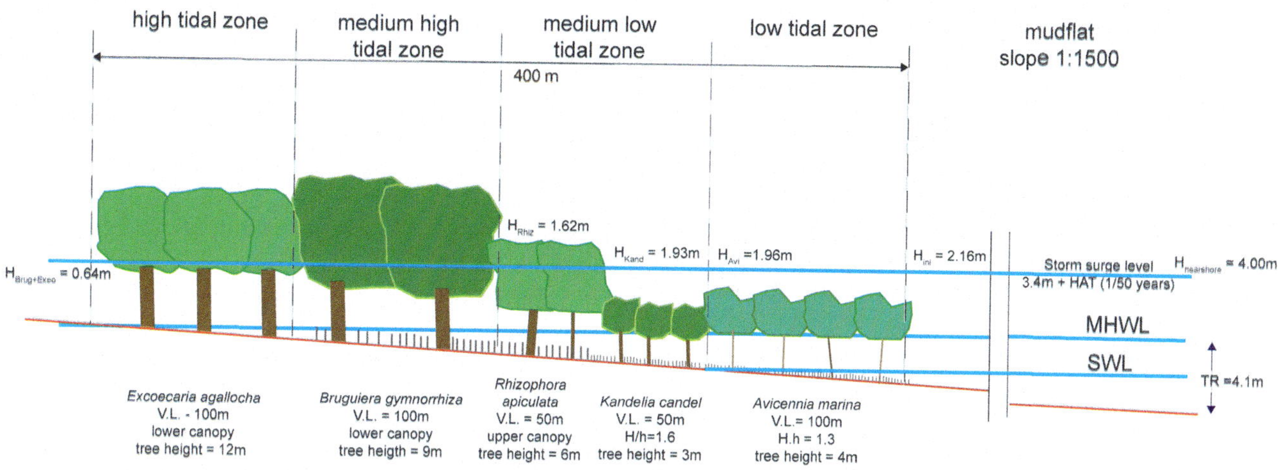

Fig. 21.11 Calculation of the wave height decrease in a mature forest in Quang Ninh province, Vietnam (Tusinski 2012)

As an example with the above formulas the wave dissipation has been calculated in the Quang Ninh mature mangrove forest in Vietnam. See Fig. 21.11.

21.3 Design of a Sea Dike

21.3.1 Design Basics

Sea dikes are earth structures with slopes on both sides, sometimes protected at the seaside with a revetment to avoid damage by waves. This revetment can be a riprap layer, placed blocks, asphalt, etc. In this chapter the design of a revetment is not discussed, because it is not relevant for the relation between mangroves and sea dikes. In the examples a block revetment is used, but this can easily be replaced by any other type of revetment (Fig. 21.12).

Dikes have a trapezoidal shape. The inner slope is usually 1:3–1:4 (vertical: horizontal). Steeper slopes are not recommended because of geotechnical instability, difficult maintenance and vulnerability to damage in case of wave overtopping. Also for the outer slope usually a 1:3 slope is used, because steeper slopes cause more overtopping. More gentle slopes are both for inner and outer slopes not recommended because of increase of the total width of the dike (and consequently the cost of the dike). In this chapter a standard inner and outer slope of 1:3 is used (Fig. 21.13).

What remains as a design parameter is the crest height of the dike. We assume for the time being that no constructions are added to the crest (like small walls, buildings, parapets, etc.).

A dike is designed for a given failure frequency. This means that the probability of failure of the dike should be lower than a given, predefined value. Usually this is a value in the order of 1/100–1/500 per year. When a value of 1/100 per year is selected, it means that every year there is a probability of 1/100 that the dike fails and that there will be a flood. Sometimes this is also expressed as a return period; on average the dike fails once in every hundred years. This might be misleading to the public, because this failure may be next year.

Dikes may fail because of overflow, overtopping, geotechnical failure, damage by the public, terrorist actions, etc. In principle one should add up all the partial failure probabilities in order to obtain the overall failure probability. However, because in most cases the probability of geotechnical failure, etc. is very small in respect to the failure due to overtopping/overflow, there contribution is neglected. This means the failure probability of the dike should be equal to the probability of just acceptable overtopping (Fig. 21.14).

21.3.2 Overtopping and Crest Height

For a normal, good quality earth dike with a solid grass cover on the inner slope no failure will occur when the average overtopping over the dike is less than 5 l/s per running meter (or 0.005 m^2/s). The amount of overtopping can be calculated with the overtopping manual (EurOtop 2016). The overtopping formula has basically the following shape:

$$\frac{q}{\sqrt{gH_{m0}^3}} = a\exp\left[-\left(b\frac{R_c}{H_{m0}}\right)^c\right] \tag{21.21}$$

in which:

q overtopping (m^2/s)

H_{m0} wave height (calculated from the energy spectrum) (m)

R_c crest height above design water level

The coefficients a, b and c depend on a number of other aspects. They are partly based on curve fitting from laboratory experiments and partly a function of other parameters of the dike, like roughness, slope, approach angle and period of the waves. For the discussion on the benefits of mangroves the variation in these coefficients is not that relevant. For details, the reader is referred to the EurOtop manual (Fig. 21.15).

However, one parameter hidden in the coefficients should be mentioned at this point. Overtopping depends also on the wave period. When the wave period changes, also the overtopping changes. This will be relevant when analysing the effects of mangroves.

So, the crest height consist of two elements:

- Design water level
- Freeboard

The design water level is the datum + tide + surge; the freeboard is the crest height above design water level to limit the overtopping discharge to the design value (usually 5 l/s per meter, for very good quality inner slopes this may be 10 l/s).

The design water level should have a predefined probability. In principle one should determine the joint probability of tides and surges. This is rather complicated. In practice surge heights are usually not determined by calculations from wind speeds, etc. but on basis of measured data. When taking the measured highest water level during storms one can put these values on a log-scale on paper and extrapolate to the desired design frequency. When following this method automatically both the probability of high tides and storm surges are taken into account.

Subsequently the wave height during this event has to be determined. Because in case of mangrove coasts, where the coastal profile is extremely gentle, one may always consider the coast as a shallow coast. This means that the wave height is limited by wave breaking. A practical rule for these cases is that the wave height is in the order of 40–50% of the water depth.

21.3.3 Revetment

When a dike is subject to serious wave attack, a revetment is needed. For low waves, a grass cover is sufficient. In general one can use a the upper limit of wave attack on a grass slope a wave height in the order of 0.5 m. Although a perfect grass cover on a good clay can withstand much higher waves during a short period, it is usually not recommended to design dikes in this way.

In case of higher waves the revetment may consist of riprap, placed blocks or other material like asphalt. On dikes placed blocks are quite common, and in this chapter the placed block solution will be used. For good quality placed blocks the stability can be calculated with (Klein Breteler 2016):

$$
\begin{aligned}
\frac{H_s}{\Delta d} &= F \cdot \xi^{-1/3} && \mathit{for}\ \xi < 2 \\
\frac{H_s}{\Delta d} &= F \cdot 2^{-1/3} + 0.5(\xi - 2) && \mathit{for}\ 2 < \xi < 5 \\
\frac{H_s}{\Delta d} &= F \cdot 2^{-1/3} + 1.5 && \mathit{for}\ \xi > 5
\end{aligned}
\qquad (21.22)
$$

in which:

ξ The Iribarren number ($\xi = \tan \alpha / \sqrt{(H_s/L_0)}$)
L_0 The deep water wave length
Δ The relative density of the blocks $(\rho_s - \rho_w)/\rho_w$
d Thickness of the blocks
F Constant, for good quality revetment blocks a value $F = 7$ is recommended

There is no need to extend the block revetment until the crest of the dike. In general it is recommended to extend the revetment to a level of H_s above design level.

In Tusinski (2012) an overview is given of revetment costs in Vietnam. For blocks the approximation formula $D = 0.3d + 600$, where d is the block thickness in m and D the cost in 1000 VND per m^2 (in 2016: 1,000,000 VND ≈40 EUR).

21.3.4 Benefit of Mangroves

When a mangrove belt is present in front of a sea dike, the design water level will remain the same, but the wave height will be reduced. The preferred reduction is such that the wave height at the toe of the dike becomes less than 0.5 m, because in that case no revetment is needed any more.

Additionally, the run-up will be less, so the required height of the dike crest can be lower. A lower dike means also a narrower dike. When assuming a slope of 1:3 on both sides, the lowering of a dike by one meter means that the footprint of the dike becomes 6 m less. Also the volume of the dike becomes much less. Because of uncertainties in the determination of the water level it is wise to make the dike crest always at least 0.5 m above design water level (Fig. 21.16).

In order to evaluate the gain in monetary terms one has to determine the construction costs of a dike. This depends on the cost of the material plus the cost of placing the material. Usually the material itself is rather cheap. This can be local material extracted from the neighbourhood of the dike. Mostly sand is the cheapest material. In case of a sand dike, one should cover the dike with a clay layer of at least 1 m thick to allow grass to grow on the dike. But it is also possible to make the dike fully out of clay, when clay is cheaper than sand at the given location.

The main cost item for the core material is therefore the execution of the work, including the finishing, making the grass cover and usually a road on top of the dike. For Vietnam an overall value of 300,000 VND/m^3 can be used (Tusinski 2012).

The coastal land value is another aspect rather difficult to establish and unify in one number. The function of the land, whether it is agricultural, residential or commercial is the governing factor here. In many cases, the land belongs to the government and its purchase is thus not a part of the cost estimate. However, it is worth noticing the value that the given land represents when not used for a flood defence structure. Such a study was carried out for agricultural land utilized for rice cultivation, the most common crop in Vietnam. The value of the land was estimated based on the average annual profit made during three harvesting periods per one hectare of land. This value was discounted over a period of 30 years with an average and estimated inflation rate of 5%. Value of the land calculated in this way allows assessing its real actual price, excluding the speculation which results from negotiations between government and the farmers, during which farmers usually ask for a price exceeding many times the land's real value (Tusinski 2012).

In order to make a balance the cost of the mangroves has to be determined. When assuming that no mangroves are present a full planting program is needed. The required minimum zone width follows from the calculation of the required K_t-value. Of course this is only possible when the tidal regime allows sufficient area for planting of mangroves. It is of no use to plant mangroves deeper or shallower that the requirements of that specific species. The cost of mangrove plantations are of course the cost of the initial planting, but also the cost of protection works to protect the seedlings from being washed away by minor waves. Usually these protection works are simple wooden or bamboo dams. Because of the simple character of these dams, there is a high probability that they are washed away by typhoons and cyclones. When this happens before the mangroves are sufficiently mature, one has to start again.

Because mangrove plantations are mainly manual work, the cost of mangrove plantations depend completely on the wage levels. From the data gathered on the mangrove restoration projects in Southern Vietnam (Ca Mau province) it appears that these costs also vary per location. In this case however the specific site and soil conditions play an important role. As it is widely known, the mangrove forests are rather difficult to establish artificially, due to their sensitivity and vulnerability to the varying environmental conditions. In the extreme cases when the trees are in their early growth stage, they need to be protected from the extreme water levels that reach their canopy and decrease their area of oxygen collection. Therefore small breakwaters need to be constructed offshore; this construction is included in the planting costs. In such cases, these costs can reach up to 80 million VND/ha (3100 Euro/ha) of mangrove vegetation. On the other hand, when no extra protection is necessary, the average costs oscillate around 13 million VND/ha (540 Euro/ha).

In order to include the risk of typhoon damage, one may assume that each plantation project has to be done twice before mangroves reach state of maturity. This means that one has to count for approx. 160 million VND/ha to obtain a functional mangrove protection.

As can be seen from the computational example below mangroves are financially much more attractive than creating a exposed dike. The main advantage is the fact that no hard revetment is needed. But even in cases where no revetment at all is feasible, there will be still a considerable advantage.

21.3.5 Ecosystem Services

Not taken into consideration are ecosystem services of a mangrove area. When these services are monetised and taken into consideration, the advantages of a mangrove belt are extremely high. As mentioned by Tas (2016) a healthy, densely vegetated mangrove forest in Vietnam may provide an ecosystem service value of 50,000 VND/m^2 per year. In the example below, 0.035 ha of mangrove was needed in front of the dike. So this provides 17.5 million VND/year per running meter of dike. Comparing this value with the creation and maintenance cost of the mangroves (5.6 million VND initially) shows that in case the value of ecosystem services are taken into consideration, creation of mangroves is by definition positive. The problem however is that the costs of creation of a mangrove forest are expenditures in real currency, while the benefits are a "virtual" currency. The income from the ecosystem services is not put into your bank account, and therefore cannot be used for, for example, paying the wages of the mangrove workers.

21.3.6 Drawbacks

Of course there are also some drawbacks to a mangrove system as part of the sea defence. An important issue is that the mangroves have to be there when the storm hits the coast. So in case of a disease or other problems with the trees, which result in the trees dying, the protection is lost and the dikes have to be strengthened in a short period (and usually at high cost). Also there is the risk that the mangrove area is misused; for example, illegal wood harvesting or the creation of illegal fishponds.

Therefore it is needed that the mangrove forest is managed well. When the mangrove forest is artificially planted, during the planting period quite some maintenance activities are needed. So there has to be an organisation in charge of the management of the mangroves. It is vital that such an organisation will continue to exist, even when the mangroves are mature. The mangroves need to have a protected status.

This requirement is fully in line with the fact that also from environmental point of view protection of the mangroves is needed. Mangroves should be part of the flood management system in a country. Usually flood defences are subject to special regulations, and a regulatory body has been established to assess on a regular basis the state of flood defences. The mangroves should be part of this assessment.

21.3.7 Computational Example

21.3.7.1 Water Levels

Wave height in front of the mangrove field (at the sea side) is $H_{m0} = 2$ m, $T = 6$ s. Allowable overtopping over the dike is 10 l/s (which implies a very good quality grass on the inner slope of the dike). Allowable wave height on the outer grass slope is 0.5 m. In case of higher waves grass is not sufficient and a revetment is needed. We need to lower the wave height.

In order to obtain a wave height reduction from 2 to 0.5 m a reduction coefficient $K_t = 0.25$ is needed. The planting process is expected to lead to an average density mangrove forest ($r = 0.007$). For obtaining a $K_t = 0.25$ one needs a mangrove belt of 200 m width.

Using the method of Quynh, and assuming that 80% canopy coverage will be achieved at the end, and that the average tree height will become 4 m, a belt of 350 m is needed, planted with 800 trees per ha.

When one assumes that mangroves may grow from 0.5 m below Mean Sea Level until the toe of the dike, a zone with a width of 750 m is available. The conclusion is that a sufficiently wide mangrove belt can be created (Fig. 21.12).

21.3.7.2 Required Dike Height Without Mangroves

Wave height in front of the mangrove field (at the sea side) is given as $H_s = 2$ m, $T = 6$ s. Allowable overtopping over the dike is 10 l/s. Allowable wave height on the outer grass slope is 0.5 m. For a dike with a 1:3 slope this means a dike height of 0.7 m above surge level, or 4.2 m above MSL.

In order to obtain a wave height reduction from 2 to 0.5 m a reduction coefficient $K_t = 0.25$ is needed. As mentioned earlier, this requires a mangrove belt on the order of 200–350 m wide.

Without mangrove reduction, the dike height has to be 3.5 m above surge level (or 7 m above MSL or 8.5 m above LLWS)., and a protection with concrete blocks is needed (Fig. 21.13).

21.3.7.3 Need of a Revetment

When no mangroves are present, the wave height is 2 m, and the wave period 6 s. The slope of the dike is 1:3. This gives a value of $\xi = 1.77$. Then $H_s/\Delta d = 4{,}1$, and thus $d = 32$ cm. The revetment should extend to H_s above surge level, so until 7 m above LLWS. This means it must it extend from LLWS +2 to LLWS+7. As the vertical height of the revetment is 5 m, the total height of the revetment along the slope is 3*5=15 m.

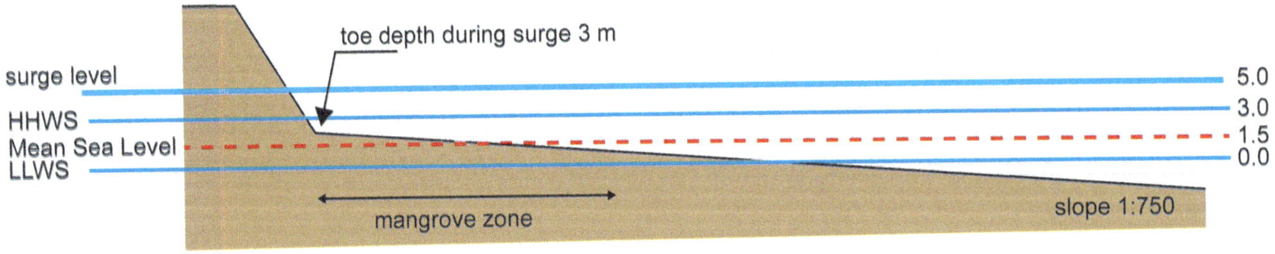

Fig. 21.12 Waterlevels in this computational example

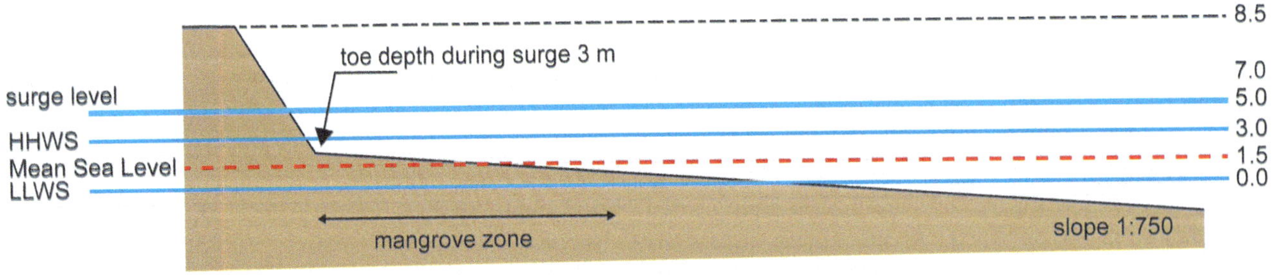

Fig. 21.13 Required dike height

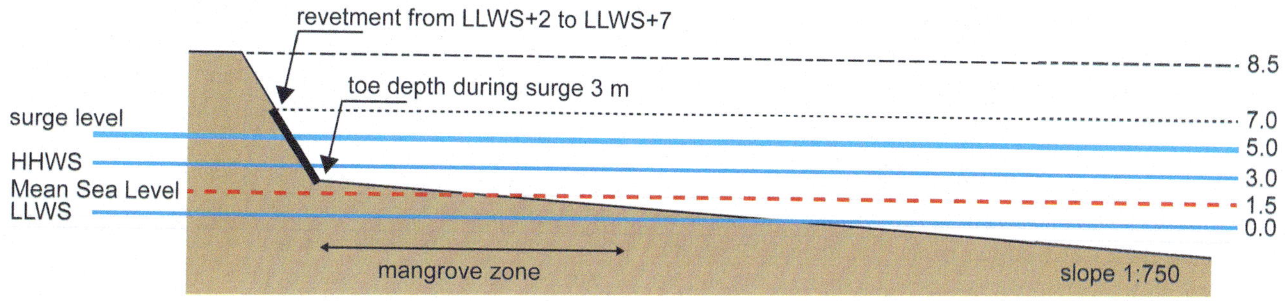

Fig. 21.14 Revetment needed

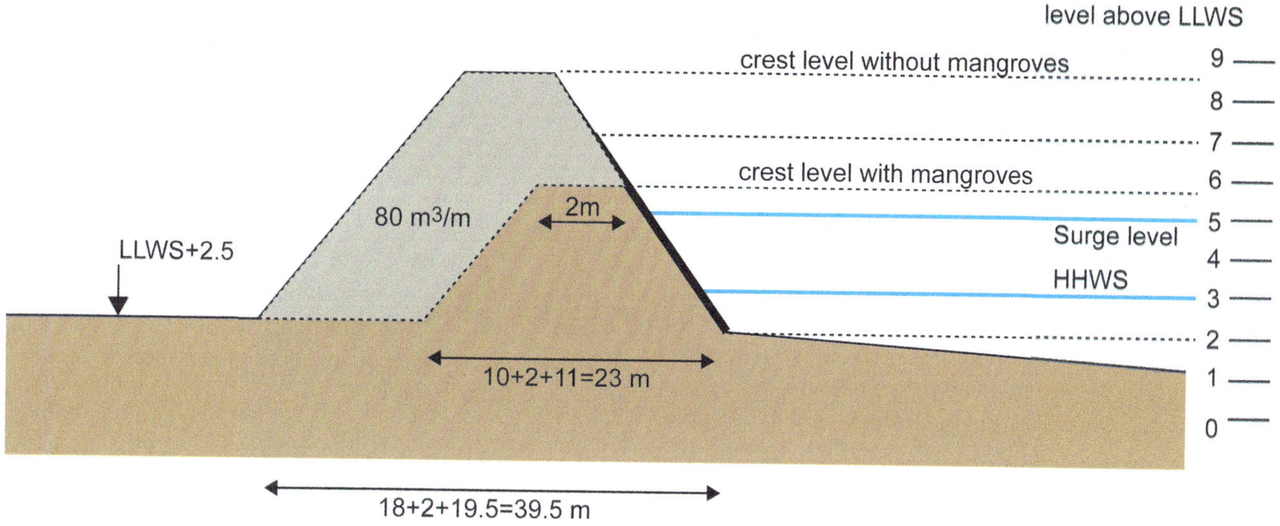

Fig. 21.15 Effect of mangroves

The volume of the revetment is 0.32*15=5 m^3 per running meter of dike (Fig. 21.14).

This means that the revetment in the example in Vietnam would cost approximately 25 million VND per running m (≈1000 EUR/m).

21.3.7.4 Effect of Mangroves

When mangroves are present, the wave height is 0.5 m, the wave period is assumed to remain 6 s. The slope of the dike is 1:3. No revetment is needed. The required height of the dike is 0.7 m above storm surge level, thus 5.7 m above LLWS (Fig. 21.15).

The benefit of having mangroves in front of the dike is for this example:

- No revetment
- 80 m^3/m less dike material

16 m^2/m less purchase of land needed

21.3.7.5 Cost Benefit

- The benefit of having mangroves in front of the dike is for this example:

No revetment	25 million VND/m
80 m^3/m less dike material	2.5 million VND/m
16 m^2/m less purchase of land needed	0.5 million VND/m
Total gain	28 million VND/m

- The additional cost for the Mangroves

A belt of 350 m wide is needed, some 350 m^2 per running meter of dike, or 0.035 ha/m. This will cost 0.035*160 = 5.6 million VND/m.

Conclusion: For this example-location mangrove planting is **five times** cheaper than making a full exposed dike.

21.4 Some Other Aspects

21.4.1 Siltation in the Mangrove Area

In the mangrove forest the siltation is much more than in a tidal flat without mangroves. On the long run this siltation is beneficial for the sea defence. It decreases the water depth in front of the dike, which decreases on its turn the wave height. However, this siltation process takes a long time. Therefore one should not take into consideration the degree of siltation in front of the dike when calculating the present day design height of a dike.

Because the siltation speed is in the same order as the sea level rise (climate effect), one could make a choice not to include the effects of sea level rise in the present day design height of a dike, provided a mangrove forest with siltation is present in front of the dike. In this way the mangrove forest contributes to the sustainability of the sea defence.

21.4.2 Changes in The Wave Spectrum

Recent research has shown that in very shallow water with extremely gentle coastlines there is a shift in wave energy to longer periods. For overtopping over a dike the period is an important parameter. Long waves cause more overtopping than short waves. In modern overtopping formulas this effect is included by not using the mean period or peak period of the spectrum, but the period based on the first negative moment of the spectrum ($Tm-1,0$). For a rather standard coastline one may use $Tm-1,0 = 0.89\ Tp$, but for very gentle coastlines the coefficient may become much more than 1. Because of this phenomenon, many dikes on very gentle coasts are under-designed.

This increase is discussed in detail in Hofland et al. (2017). This effect can often be neglected, but becomes more important for foreshores with a very gentle slope. It seems that the effect depends not only on the bed slope, but also on the short-crestedness of the waves. For short-crested waves the effect is less extreme than for long-crested waves (Fig. 21.16).

The change in spectral wave period can be described by:

$$\frac{T_{m-1,0(t)}}{T_{m-1,0(0)}} - 1 = 6\exp\left(-A\tilde{h}\right) + B\exp\left(-C\tilde{h}\right)$$
$$\tilde{h} = \frac{h_t}{H_{m0}}\left(\frac{\cot\theta}{100}\right)^{0.2} \qquad (21.23)$$

Where:

$T_{m-1,0}$ spectral wave period ($=m_{-1}/m_0$, where m_i is the ith moment of the energy spectrum)
h water depth
H_{m0} deep water wave height, based on the spectral energy
θ slope of the foreshore

the index 0 means deep water, while the index t means just in front of the toe of the dike.

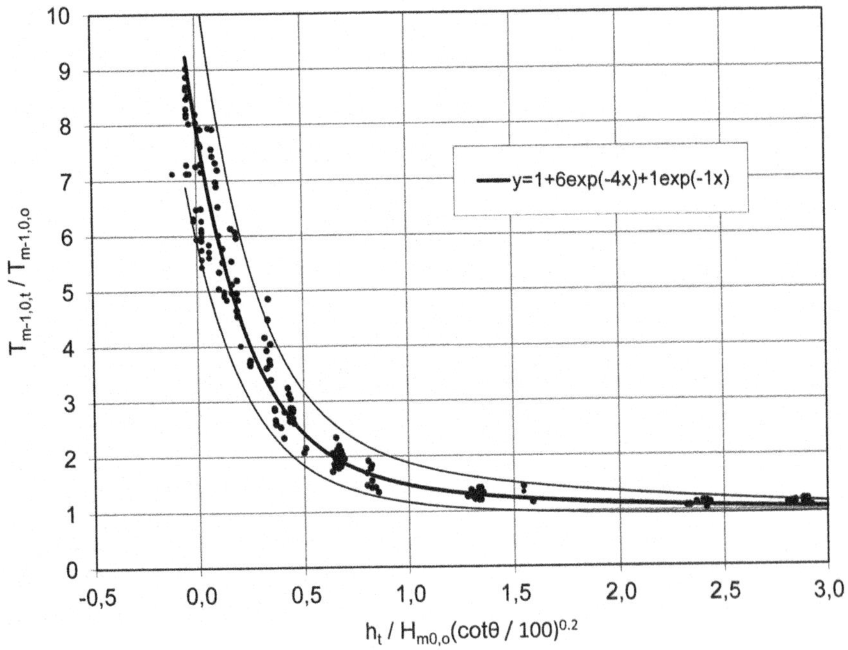

Fig. 21.16 Multiplier for the period as function of the dimensionless toe depth for a dike (Hofland et al. 2017)

The coefficients are:

	A	B	C
Short-crested waves	6	0.25	0.75
Long-crested waves	4	1	1

Example:

In deep water we have a wave generated by a typhoon with a wave height of $H_{m0} = 4$ m and a period of $T_p = 8$ s. For "young" waves one can use the relation $T_{m-1,0} = 0.9\ T_p$. So $T_{m-1,0}$ is in deep water approximately 7 s. When we have a very gentle slope (order 1/500) the multiplier is 1.38. Suppose we have at the toe a water depth of 3 m (as in the previous example) the value of dimensionless depth becomes:

$$\tilde{h} = \frac{h_t}{H_{m0}}\left(\frac{\cot\theta}{100}\right)^{0.2} = \frac{3}{4}\left(\frac{500}{100}\right)^{0.2} = \frac{3}{4}1.38 = 1.43 \tag{21.24}$$

and

$$\frac{T_{m-1,0(t)}}{T_{m-1,0(0)}} - 1 = 6\exp(-6\tilde{h}) + 0.25\exp(-0.75\tilde{h})$$
$$= 6 \cdot \exp(-6 \cdot 1.03) + 0.25\exp(-0.75 \cdot 1.03) = 0.127 \tag{21.25}$$

So in case of short-crested waves the spectral period becomes 1.127*7 = 7.9 s.

In case of long-crested waves the result is:

$$\frac{T_{m-1,0(t)}}{T_{m-1,0(0)}} - 1 = 6\exp(-4\tilde{h}) + \exp(-\tilde{h})$$
$$= 6 \cdot \exp(-4 \cdot 0.52) + \exp(-0.52) = 0.45 \tag{21.26}$$

So in case of long-crested waves the spectral period become 1.45*7 = 10.2 s.

To illustrate this effect we assume a dike with a relatively gentle slope (1:4). At the toe of the dike we have a set-up due to storm surge, and a water level of 3 m above the dike toe. Because of breaking the wave height at the toe will be not more than 2 m. The overtopping criterion we keep at 5 l/s.

When we enter this in the overtopping formula, we get a dike height of 3.85 m above design water level. When we use a period of 8 s, the result is 4.55 m. And for a period of 10 s it becomes 4.7 m.

This example shows that because of the shift to longer waves at very gentle slopes, dikes may become higher. In this example, the dike height has to increase by 70–85 cm.

21.4.3 Some Wave Height Aspects

When waves travel from offshore to onshore not only the wave height changes, but also the wave period. The change of wave period is a rather complicated process, due to wave-wave interactions and other interactions (triads, quadruplets, bed friction, surf breaking, white-capping, wind effect). Taking this all into consideration on a very gentle slope changes from deep water to an nearshore depth of for example 10 m the energy spectrum changes considerably. Tas (2016) has evaluated this for the Mekong delta, see Fig. 21.17.

Especially in case of gentle foreshores at shallow water nonlinear processes become important. Therefore near to the coastline the program SWAN is less reliable. Calculations with both SWAN and SWASH show that in SWASH the decrease of wave height starts later than predicted by SWAN. This is illustrated in Fig. 21.18 for a very gentle profile.

But when on such a gentle coastline a mangrove forest is present, this forest will have an effect of the period. It is expected that because of the very irregular structure of the bed the damping effect will be more effective on the low frequency component of the waves than on the higher frequencies. For example when a wave of 25 cm in very shallow water (1 m water depth) has a period of 4 sec, the maximum velocity at the bottom under the crest is 1.6 m/s. When under the same conditions the period is 12 s, the maximum velocity at the bottom under the cress is 11 m/s (using second order wave theory). On a smooth bottom (no vegetation, small particles) the resistance is small, and also the energy dissipation due to bed friction is low. But in a mangrove forest, especially with species with pneumatophores, this friction is considerable and consequently quite some low frequency energy is dissipated.

This effect has not yet been researched in detail. There are hardly any field data on this effect available, and including this effect in mathematical models is rather difficult when lacking good calibration data. Because the effect of friction is highly non-linear also physical scale models are problematic.

21.5 Discussion

Mangroves contribute to lowering the investment costs of flood defences considerably. As demonstrated, the costs of maintaining and/or creating mangroves in front of dikes are considerably less than the costs of dike improvement works. The exact quantification of the amount of wave attenuation due to mangroves is problematic, mainly due to the lack of sufficient prototype data. The various clouds in Fig. 21.4 illustrate this clearly. Especially data measured during extreme conditions (i.e. when the water level is much higher than normal and the canopies of the mangrove trees are partly inundated) are missing. The existing prototype data are all measured under conditions when the waves can pass under the canopies, and thus are only subjected to limited friction. It is therefore expected that the transmission coefficient K_t will be lower during surges, but because no good calibration data exist for this situation, it is quite risky to include such extra reduction in design calculations.

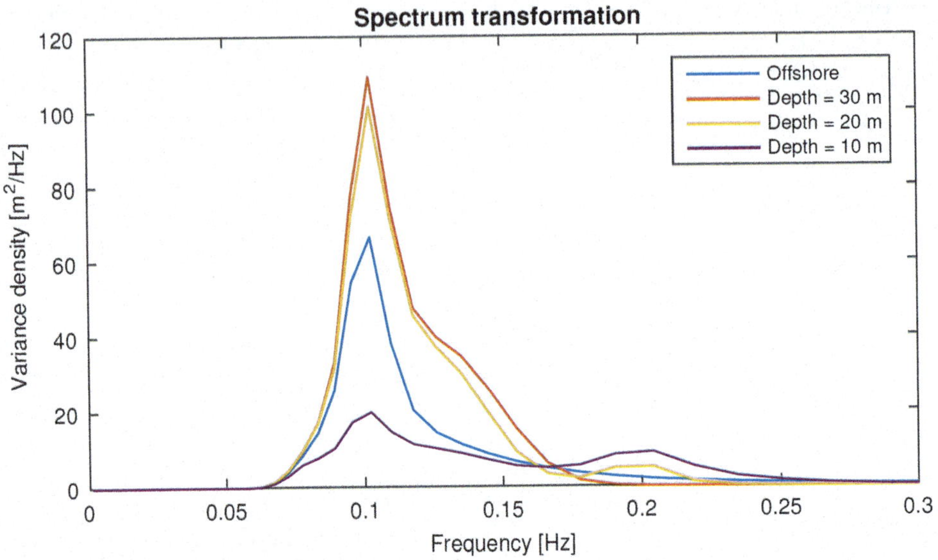

Fig. 21.17 Evolution of the wave spectrum, from offshore (JONSWAP spectrum with Hs = 6 m and Tp = 10 s, at a depth of 65 m), to nearshore (at respective depths of 30 m, 20 m and 10 m), with an onshore directed wind of 25 m/s, including white-capping (Tas 2016)

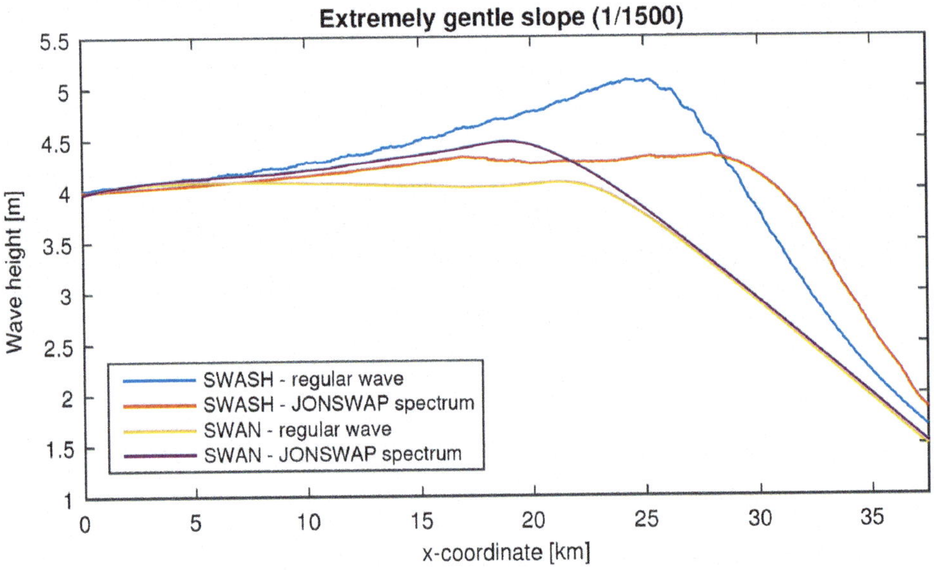

Fig. 21.18 Comparison of the wave transformation of a regular wave and a Jonswap spectrum on a slope of 1/1500 (Tas 2016)

Young mangrove plantations need protections from waves. It will take some years before the plants are mature enough to withstand serious waves. This implies that there is a considerable risk that a cyclone will destroy the plantation, as well as the temporary protection structures. Both have to be reconstructed completely after such a devastating cyclone. Nevertheless, if one capitalises the extra cost of rebuilding the plantation and protection structure, one still gets a lower cost than that of building sea dikes able to withstand the cyclones directly. A detailed risk analysis is absolutely required. In practice such risk analyses are seldom made. The temporary protection structures (often timber and bamboo walls) are easily destroyed by a cyclone. Often the designers try in such a case to lower this risk by designing more robust (and thus more expensive) constructions. However, the capitalised cost of simple bamboo walls, even if destroyed a few times by cyclones, are much lower than the capitalised cost of a heavy rock protection.

For the design of new mangrove areas (as well as for the reconstruction of previously cut mangroves) it is important to not look only to the tidal range, but also to the other aspects of the tide. Mangroves are sensitive to the number of hours submerged and emerged. One parameter to describe this is the percentage of time the mangroves are submerged/emerged, but this parameter is not sufficient. In some tidal regimes the submergence-time can vary considerably over the full neap-spring cycle.

21.6 Conclusions

The costs of sea defence structures are much lower when a healthy mangrove forest is located in front of the dike. The biggest advantage is that in many cases the cost of a hard revetment (concrete blocks or riprap) can be avoided. Comparing the construction costs of a hard revetment with mangroves shows that the cost of planting and maintaining mangroves are much lower. For a case study in Vietnam the result was a price difference of a factor five. In this analysis the value of the ecosystem services of the mangroves were not taken into consideration.

References[1]

Bo J (2012) A study on geometric and material properties of the mangrove root system in Singapore. M.Sc.-thesis TU Delft and NUS

Burger B (2005) Wave attenuation in mangrove forests. M.Sc.-thesis, TU Delft

De Vos WJ (2004) Wave attenuation in mangrove wetlands, Red river delta, Vietnam. M.Sc.-thesis TU Delft

EurOtop (2016) Manual on wave overtopping of sea defences and related structures. In: Van der Meer JW, Allsop NWH, Bruce T, De Rouck J, Kortenhaus A, Pullen T, Schüttrumpf H, Troch P, Zanuttigh B (eds). www.overtopping-manual.com

Hofland B, Chen X, Altomare C, Oosterlo P (2017) Prediction formula for the spectral wave pe-riod Tm-1,0 on mild sloping shallow foreshores. Coastal Eng., vol 123, issue May. https://doi.org/10.1016/j.coastaleng.2017.02.005

Horstman EM, Dohmen-Janssen CM, Narra PMF, Van den Berg NJF, Siemerink M, Balke T, Bouma T, Hulscher SJMH (2012) Wave attenuation in mangroves: field data obtained in Trang, Thailand. In: Proceedings of 33rd ICCE 2012, Santander, Spain

Horstman EM, Dohmen-Janssen CM, Narra PMF, Van den Berg NJF, Siemerink M, Hulscher SJMH (2014) Wave attenuation in mangroves: a quantitative approach to field observations. Coast Eng 94:47–62

Klein Breteler M (2016) Vergelijkend onderzoek zetstenen voor dijken (comparison of design methods for placed block revetments, Deltares report, 1208618–006-HYE-0022

Mazda Y, Magi M, Ikeda Y, Kurokawa T, Asano T (2006) Wave reduction in a mangrove forest dominated by Sonneratia sp. Wetl Ecol Manag 14:365–378

Narayan S (2009) The effectiveness of mangroves in attenuating cyclone-induced waves. M.Sc.-thesis TU Delft

Narayan S, Suzuki T, Stive MJF, Verhagen HJ, Ursem WNJ, Ranasinghe R (2010) On the effectiveness of mangroves in attenuating cyclone induced waves. In: Proceedings of 32nd ICCE 2010, Shanghai, China

Quartel S, Kroon A, Augustinus P, Van Santen P, Tri NH (2007) Wave attenuation in coastal mangroves in the Red River Delta, Vietnam. J Asian Earth Sci 29(4):576–584

Quynh VV (2010) Effects on wave break, sea dike protection of the tree planting formula in the coastal salinity inundated areas. In: Proceedings of the workshop on the effect of vegetation on dike design, Ho Chi Minh city, Vietnam

Schiereck GJ, Booij N (1995) Wave transmission in mangrove forests. In: Proceedings of the COPEDEC 1995, Rio de Janeiro, Brasil

Strusińska-Correia A, Husrin S, Oumeraci H (2013) Tsunami damping by mangrove forests: a laboratory study using parameterized trees. Nat Hazards Earth Syst Sci 13:483–503

Sumanusajonkul S (1995) Environment-friendly bank protection in Thailand. M.Sc.-thesis, Unesco-IHE, Delft

Suzuki T (2011) Wave dissipation over vegetation fields. Dissertation, TU Delft

Tas S (2016) Coastal protection in the Mekong Delta. M.Sc.-thesis, TU Delft

Tusinski A (2012) The role of mangroves in the design of coastal dikes: Hydrodynamic and cost related aspects. M.Sc.-thesis, TU Delft

Tusinski A, Verhagen HJ (2014) The Use of Mangroves in Coastal Protection. In: Proceedings of 34th ICCE 2014, Seoul, Korea

Verhagen HJ, Loi TT (2012) The use of mangroves in coastal protection. In: Proceedings of the COPEDEC 2012, Chennai, India

[1] All M.Sc.-theses and Dissertations mentioned below are as full-text available via http:repository.tudelft.nl, as well as most other references.

Remote Sensing of Mangrove Forests: Current Techniques and Existing Databases

22

Stuart E. Hamilton, Gustavo A. Castellanos-Galindo, Marco Millones-Mayer, and Mara Chen

Abstract

This chapter examines the major global remotely sensed mangrove databases that have become accessible since the year 2000. By doing so, we summarize the significant methodological differences between each product and provide a best estimate of post-2000 mangrove cover at the global level. We then review remotely sensed mangrove area findings in-depth for all nations in the Western Hemisphere and Oceania with mangrove holdings in the top 20 nations globally and then summarize the findings for all nations in the Western Hemisphere and Oceania with mangrove holding in the top fifty nations globally. In addition to reporting the mangrove area quantified by each national remotely sensed database, we assess the temporal domain, the spatial resolution, the instruments used, the techniques applied, the validation performed, and the error statistics reported for each national mangrove estimate. We then compare the national remotely sensed mangrove area estimates provided with each other and with the global estimates for each nation and then arrive at a post-2000 best estimate of mangrove cover for each country. Next, we review the common remote sensing techniques and instruments used to map and monitor mangrove forests throughout this chapter. Finally, we assess the future requirements of the mangrove community considering what the remote sensing community can realistically deliver. We find that national remote sensing estimates of mangrove forest area align well with the global remotely sensed measures of mangrove forest area and can, in general, be used with confidence to manage and monitor mangrove forests.

Keywords

Remote sensing · Mangrove forests · Mangrove forest area · Landsat · Mangrove

22.1 Introduction

To fully assess the threats to mangroves forests at the global, regional, national, and local scales reliable data on the presence and condition of mangrove forests are required. Beyond highly localized study sites that often utilize field-collected data, mangrove forest coverage data are most often compiled using remote sensing instruments and methodologies. Many mangrove specialists then use these remotely sensed mangrove information and applied it to their area of interest such as hazards (e.g. Blankespoor et al. 2016; Regional Centre for Mapping of Resources for Development 2015), vulnerability (e.g. Dahdouh-Guebas et al. 2004), fisheries (e.g. Hutchison et al. 2014; Paul et al. 2016), management (e.g. Kairo et al. 2002; Wang et al. 2003), ecology (e.g. Pettorelli et al. 2014), biology (e.g. Hogarth 2015; Saintilan et al. 2014), and climate change (e.g. Siikamäki et al. 2012; Pendleton et al. 2012). Despite the widespread use of mangrove databases generated by remote sensing techniques, no singular agreed upon approach to the remote sensing of mangrove forests exists. Indeed, as many approaches to remote sensing of mangroves exist as there are remotely sensed mangrove studies with each nation or academic team using a slightly differing approach. Such issues leave those researching the vulnerability and threats to mangrove forest, and those who manage the mangrove forest resource, with a difficult decision on which mangrove databases to utilize and trust and

S. E. Hamilton (✉) · M. Chen
Department of Geography and Geosciences, Henson Science Hall, Salisbury University, Salisbury, MD, USA
e-mail: sehamilton@salisbury.edu

G. A. Castellanos-Galindo
Leibniz Centre for Tropical Marine Research (ZMT), Bremen, Germany

M. Millones-Mayer
Department of Geography, University of Mary Washington, Fredericksburg, VA, USA

C. Makowski, C. W. Finkl (eds.), *Threats to Mangrove Forests*, Coastal Research Library 25,
https://doi.org/10.1007/978-3-319-73016-5_22

how to assess mangrove area estimates, that may differ from database to database.

The remotely sensed mangrove data currently available are generated from a wide range of both active and passive sensors. Active sensors used for mangrove estimates in this chapter include; light detection and ranging (lidar), synthetic aperture radar (SAR) systems including Phased Array Type L-band Synthetic Aperture Radar (PALSAR), and even use of recalibrated radar data from the historic Shuttle Radar Topography Mission (SRTM). Passive sensors used for mangrove estimates in this chapter include; Satellite Pour l'Observation de la Terre (SPOT), Advanced Spaceborne Thermal Emission and Reflection Radiometer (ASTER), Landsat Thematic Mapper (TM), Landsat Enhanced Thematic Mapper (ETM+), MODIS (Moderate Resolution Imaging Spectroradiometer), Hyperion, Ikonos, Rapid Eye, Advanced Land Observing Satellite (ALOS), and Quickbird. Remote Sensing techniques used in mangrove mapping vary even more than sensor use, ranging from simple heads-up manual digitizing of polygons using color band combinations that depict mangroves to complex heuristic machine learning algorithms that incorporate remotely sensed data from numerous differing remote sensing instruments and field control sites.

Within the mangrove remote sensing literature; no agreed upon ground verification method exist to validate the remotely sensed mangrove area estimates, and no standard error reporting statistics are used across the many differing studies. Furthermore, no agreement even exists of the level of mangrove that must be present in a remotely-sensed pixel before it is coded as having mangrove present. For these reasons, local and national mangrove estimates generated from remote sensing methods should be used with caution, particularly when being used to compare mangrove area across time or space when an entirely differing suite of instruments and techniques may have been used. In addition to these methodological issues, many of the national remotely sensed mangrove estimates are created by in-country institutes such as ministries of environment, departments of forestry, water management districts, remote sensing institutes, or similar and are difficult to obtain and often in compiled a local language.

This chapter is intended for those who are interested in mangrove monitoring and desire to use the primary remotely sensed mangrove databases at the global and national scale while gaining an insight into the instruments and methodologies behind the mangrove database construction and the reliability of their mangrove area estimates. The first section of this chapter reviews the global post-2000 high-resolution remotely sensed mangrove databases available. The second section deals with in-depth national post-2000 remotely sensed databases of mangrove forest cover for eight countries in the Western Hemisphere and two in Oceania. Next, we report the mangrove area results globally and for each country. Then we expand on the remote sensing systems and methodologies that are used to compile such estimates. Finally, we discuss our findings and assess the current trends and future needs of the mangrove community regarding what the remote sensing community can realistically achieve.

22.2 Global Remotely Sensed Mangrove Databases

Numerous remote sensing analyses have been conducted since 2000 with the goal of extracting global forest coverage across differing forest types (e.g. Bontemps et al. 2011a, b; Hansen et al. 2003, 2010, 2013; Zhu and Waller 2003), but only two remotely sensed global forest products appear dedicated to delineating the global area of mangrove forest. These two products are Mangrove Forests of the World (MFW) by Giri et al. (2011) which estimates mangrove forest area for 2000, and Continuous Global Mangrove Forest Cover for the 21st Century (CGMFC-21) by Hamilton and Casey (2016) that tracks global mangrove canopy cover from 2000 to 2012. Both products are Landsat-based, with MFW being a primary source and CGMFC-21 being a secondary source derived from other remotely sensed databases including MFW and Global Forest Cover (GFC) which itself was developed by Hansen et al. (2013).

22.2.1 Mangrove Forests of the World (MFW)

The first high-resolution remotely sensed product that attempted to map all global mangrove forests, at or below 30 m resolution, was undertaken by Giri et al. (2011). This product is often referred to as MFW. Although published in 2011, MFW represents global mangrove forest during the period 1997–2000. The pixel resolution is listed as 30 m. The polygon version of MFW can be downloaded from the Ocean Data Viewer,[1] and a raster data has recently been made available as a series of tiles from the Socioeconomic Data Applications Center.[2]

The MFW project used 1000 cloud-free Landsat scenes obtained from the Landsat Global Land Survey (GLS) product.[3] The GLS itself is a collaborative effort between NASA and the USGS; it was created to ensure scientists' have access to a consistent and terrain corrected series of Landsat data across multiple epochs. The major benefit of GLS Landsat data is that much of the pre-processing of the data is already completed and the data is provided as what is known as

[1] http://data.unep-wcmc.org/datasets/4

[2] https://doi.org/10.7927/H4J67DW8/

[3] https://lta.cr.usgs.gov/GLS/

Level-1 status. For example, GLS provides cloud-free images, therefore, avoiding time-consuming searching for suitable scenes. Additionally, the GLS Landsat scenes arrive pre-corrected for terrain using the globally consistent 90 m SRTM DEM,[4] avoiding the computationally time-consuming process of georectification when conducted at the global scale. Finally, GLS is available for download as GIS-ready GeoTiffs with pre-packaged calibration parameter files available. The currently available epochs available for download are 1975, 1990, 2000, 2005 and 2010. Individual epochs of GLS can be downloaded from the GLCF at the University of Maryland, whereas all epochs are available via the USGS Earth Explorer interface. Figure 22.1 represents the GLS 1990 and GLS 2000 representations of the mangrove forest in-and-around the Sundarbans UNESCO World Heritage Site.

As with many of the studies discussed, MFW uses a combination of the traditional supervised and unsupervised image classification techniques, but on a global scale. Supervised classification was used to mask the water bodies, and then the standard unsupervised ISODATA algorithm within Erdas Imagine was used to generate clusters (Giri et al. 2011). Once clusters were created then field data and other higher-resolution sensors, such as IKONOS and Quickbird, were used to delineate the mangrove class. Although no vigorous global mangrove classification validation is reported, the database was evaluated using other regionally available GIS datasets, and then local experts validated the dataset (Giri et al. 2011).

MFW estimates global mangrove cover in the year 2000 at 137,760 km^2 (Giri et al. 2011). No error bars, confidence intervals, or significance values are provided in the main paper to accompany this estimate, and no sensitivity analysis appears to have been conducted. The map projection used to calculate the area is not provided, and the data is presented in a non-projected coordinate system. Positional accuracy is reported as half a pixel. The actual timeframe of the GLS Landsat data used in MFW is 1997–2000. Hence, the MFW estimate is better thought of as a 4-year average at the end of the twentieth century than as an estimate for the year 2000. Although MFW is temporally limited, has only presence or absence measures of mangrove cover, has some omitted and erroneous sites in the distributed file, lacks some important methodological information, and is missing a robust error metric; it remains the preeminent remotely sensed global mangrove database. Since its creation, the MFW map has been used by numerous scientists across many disciplines in their mangrove analysis. The importance of MFW to the scientific community is evidenced by the 2017 Google Scholar citation count totaling almost eight hundred citations. Despite its many caveats, its significance in advancing mangrove is widely recognized across the mangrove research community. MFW will likely remain the base-level mangrove database used for ongoing change analysis unless an earlier Landsat epoch is additionally processed to extract global mangrove cover from the 1990s or 1980s.

When available, MFW data are included in this analysis as part of the remote sensing derived mangrove comparisons. For consistency, we use the recalibrated Giri et al. (2011) national mangrove area calculations provided by Siikamäki et al. (2012) in their supporting information file. When not available we calculated the mangrove area ourselves, using the distributed vector polygonal shapefile in equal area projection ESRI:53009. Figure 22.2 is a global representation of the MFW database.

22.2.2 Continuous Global Mangrove Forest Cover for the 21st Century (CGMFC-21)

As opposed to MFW, and many of the earlier supervised or unsupervised approaches to delineating mangrove cover, CGMFC-21[5] (Hamilton and Casey 2016; Hamilton 2016) rely on a big-data synthesis approach that incorporates many remotely sensed data products. That is, CGMFC-21 combines the remotely sensed GFC database (Hansen et al. 2013),[6] the remotely sensed MFW of the world database (Giri et al. 2011) and the expert compiled TEOW database (Olson et al. 2001). The dataset represents global mangrove forest annually starting in 2000 and ending in 2012. The pixel resolution is listed as one arc-second, or approximately 30 m at the equator and the data is only distributed only in raster format. The CGMFC-21 data can be downloaded from the Harvard Dataverse with a jumping-off page explaining and linking to the differing products also available.[7]

The CGMFC-21 database uses the high-resolution GFC database produced by Hansen et al. (2013) to add percentage tree cover to each pixel defined by mangrove in the MFW database provided by Giri et al. (2011) for the year 2000. That is, as opposed to each pixel having a value of one for mangrove presence, or a value of zero for absence, each pixel can now have a mangrove value from 0% to 100% mangrove canopy cover. These percentage tree cover values are then adjusted for latitude and converted to mangrove area calculations in square meters. The year 2001 mangrove calculation is merely the year 2000 mangrove calculation with either 2001 forest loss or forest gain applied.

[4] http://glcf.umd.edu/data/glsdem/

[5] The coauthor of this database is the author of this chapter.

[6] https://earthenginepartners.appspot.com/science-2013-global-forest/download_v1.2.html

[7] http://bit.ly/1lMJ9zj

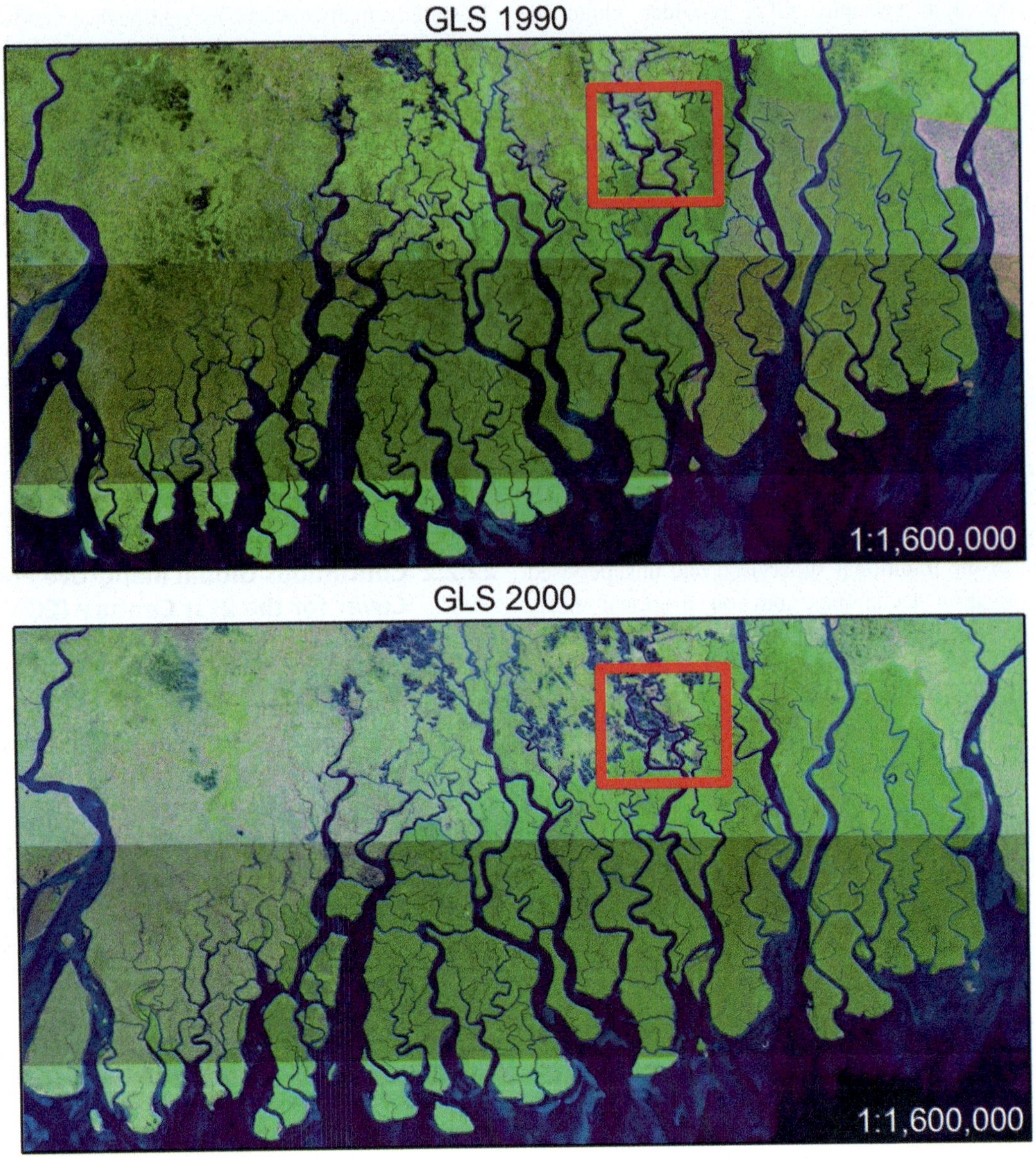

Fig. 22.1 The Sundarbans, 1990 and 2000, GLS GLS Landsat 1990 and 2010 epochs. The brighter green colors in the south of each scene represent mangrove forests. The data rendering is known as 5, 4, and 3. The red box depicts the expansion of aquaculture ponds and salt ponds between 1990 and 2000. ERSI, USGS, and NASA provided the data (Map created by Stuart E. Hamilton)

This process is the ran iteratively through 2012 with annual adjustments made to each prior year mangrove area calculation.

As with MFW, GFC was a Landsat archive based mapping process that handled approximately 650,000 Landsat scenes in a Google cloud cluster parallel computing environment (Hansen et al. 2013). The method used employed a system similar to a supervised classification that used other sensors to create training datasets and then process the Landsat scenes with the training data and a bagged decision tree methodology. They then used change to NDVI at peak growing season to help determine the global forest level and measures of global forest change (Hansen et al. 2013).

In addition to utilizing the Giri et al. (2011) mangrove forest pixels for the year 2000, the CGMFC-21 database additionally tracks forest cover in the mangrove biome as defined by the expert compiled vector (TEOW) Terrestrial Ecosystems of the World database (Olson et al. 2001).[8] Again, GFC is applied to the TEOW mangrove biome delineation. A biome is most often defined as a major ecological community, extending over a large area often

[8] http://maps.tnc.org/files/metadata/TerrEcos.xml

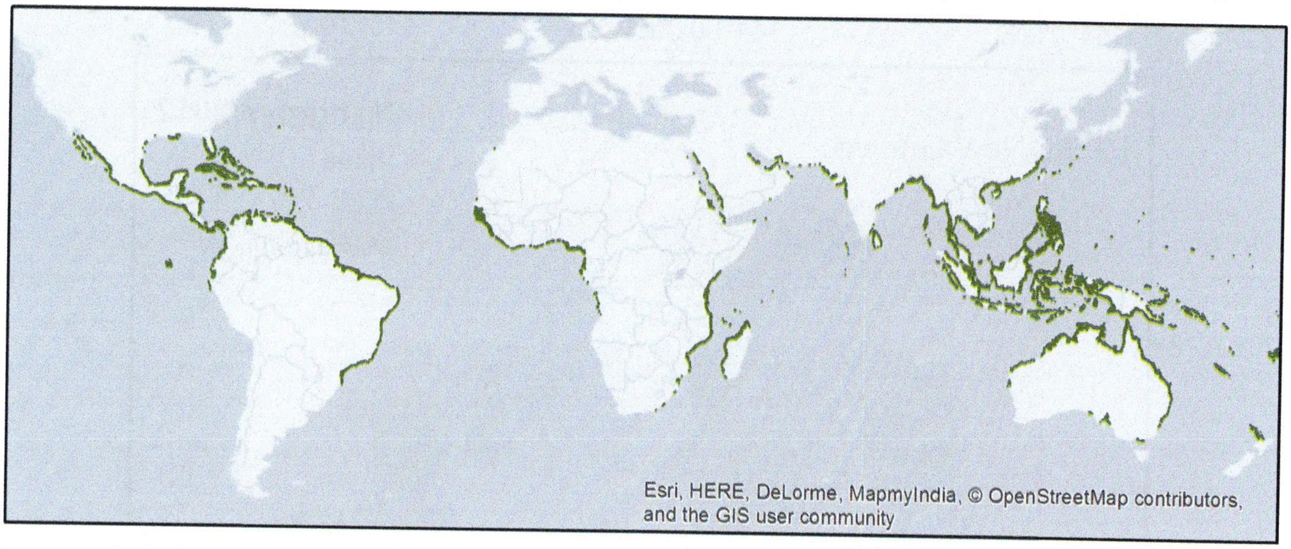

Fig. 22.2 MFW global mangroves, 2000
The MFW polygon global dataset using the v1.3 polygon shapefile provided by Giri et al. (2011) and a background map compiled by the GIS user community, ESRI, HERE, DeLorme, MapmyIndia, and OSM contributors. The green represents the mangrove forests as represented in the MFW (Map created by Stuart E. Hamilton)

characterized by a singular dominant vegetation, in this case, a mangrove-dominated region. Although not all forest within the mangrove region delineated in TEOW will be mangrove, the biome will be mangrove dominated. The TEOW database was developed over 10-years with expert local input in each region. CGMFC-21 then applies adjusted GFC measures to likely mangrove area over time using MFW and TEOW. Mangrove area estimates derived from CGMFC-21: TEOW are liable to be on the high-end of other estimates as they likely include some none non-mangrove forests by design.

CGMFC-21 estimates 2012 global mangrove forest canopy cover at 81,848.87 km^2 using the strict definition of mangrove forest in areas that are known to have mangrove in the year 2000. It additionally estimates 2012 global mangrove forest canopy cover at 167,387.10 km^2 in the mangrove biome as delineated by TEOW (Olson et al. 2001). The major caveats of CGMFC-21 are that juvenile and dwarf species are likely not captured by GFC and hence are omitted from both of the above measures and some limited mangrove stands are omitted around certain Pacific islands and atolls. Despite this fact, the CGMFC-21 supplementary material delineates the missing areas as less than <0.01% of the global mangrove total. The GFC database (Hansen et al. 2013) provides all remotely sensed measures within CGMFC-21, and hence the error metrics with GFC can are likely relevant to CGMFC-21. Hansen et al. (2013) report a tropical forest overall accuracy measure of .995 (SE .1, n = 628) and a value for the for the forest loss measure of .997 (SE .1, n = 628), and tropical forest is the most relevant category for mangrove forests. Additonally, a comparison of results produced by the 2011 National Land Cover Database (NLCD) analytical forest canopy product (Ruefenacht et al. 2015; Homer et al. 2015) was produced as part of the CGMFC-21 uncertainty analysis and the discrepancies between the two continuous forest cover datasets for mangrove forested areas of Florida was shown to be less than 4% (Hamilton and Casey 2016).

Although MFW reports the year 2000 mangrove forest area at 137,760 km^2 (Giri et al. 2011) and CGMFC-21 reports the year 2000 mangrove forest canopy cover at 83,495 km^2, these numbers are measuring two differing mangrove parameters. MFW is the global mangrove area with each Landsat pixel with mangrove present coded as having entire mangrove. That is, the entire approximately 900 m^2 of each pixel is coded as mangrove when mangrove is detected within a pixel. CGMFC-21 reports mangrove canopy cover, and in this example, the pixel could have 100 measures of mangrove such as 450 m^2, 9 m^2, or 900 m^2 as opposed to only having a value of 900 m^2 in MFW. The actual area covered by mangrove in the MFW and the CGMFC-21 year 2000 datasets is almost identical, but MFW codes the entire area as mangrove whereas CGMFC-21 has the same area delineated as mangrove but canopy cover is only approximately 60.6% of the total area. Hence the CGMFC-21 mangrove canopy area is approximately 40% lower than the MFW mangrove area estimate. This 40% reduction when moving from forest data that is compiled using presence and absence data as opposed to using continuous data is quite consistent across forest types due to edge pixels, forest degradation, rivers, creeks, and other factors that result in the entire pixel not having 100% canopy cover.

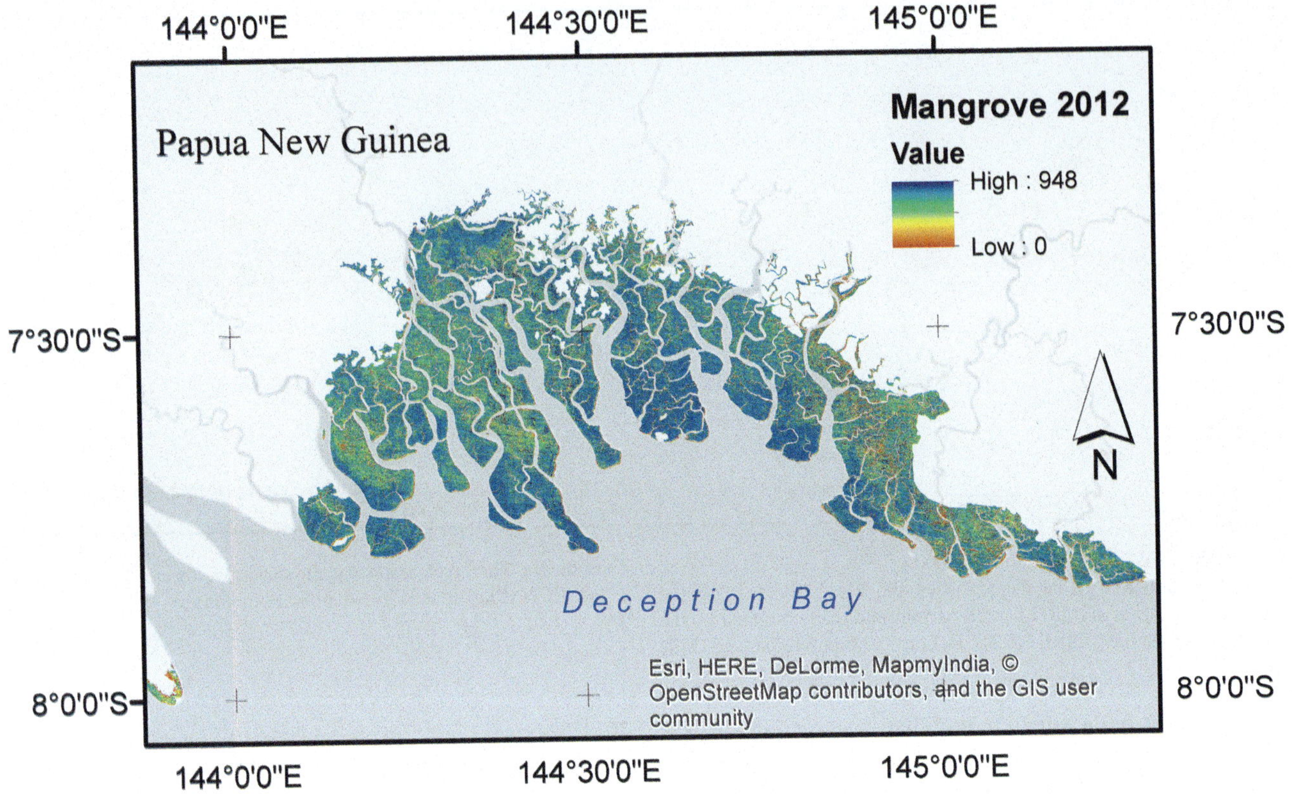

Fig. 22.3 CGMFC-21: TEOW – mangrove cover in Deception Bay, Papua New Guniea, 2012
Mangrove Cover in Deception Bay Papua New Guinea in 2012 as mapped CGMFC-21: TEOW (Hamilton and Casey 2016) (Data downloaded from doi: https://doi.org/10.13016/M2ZT44). The background map is compiled by the GIS user community, ESRI, HERE, DeLorme, MapmyIndia, and OSM contributors (Map created by Stuart E. Hamilton)

The CGMFC-21 mangrove area estimates are included in this analysis as part of the remote sensing derived mangrove comparisons. The mangrove areas are calculated by converting unprotection map reference EPSG:4326[9] pixel data into square meters using a latitudinally adjusted spherical law of cosines model. Within the results section, data are reported in the continuous system in the top row of each table and then adjusted for presence using in the lower row of each table using the 40% adjust for continuous pixel treccover data. The CGMFC-21: TEOW presence mangrove area will typically be the highest of all reported mangrove area values as this is an estimate of all forest in the wider mangrove dominated biome and not merely mangrove forests. The CGMFC-21: MFW continuous mangrove area will typically be the lowest of all reported mangrove area values as this is the strictest of all mangrove definitions presented.

The most recent remotely sensed mangrove area estimate globally that can be obtained is from CGMFC-21 and is for the year 2012. The 2012 area estimate for all forests in the mangrove biome is 167,387 km^2 and for mangrove forests alone is 81,848 km^2 (Hamilton and Casey 2016). Both these estimates use the continuous measure approach at the pixel level. These CGMFC-21 2012 estimates bound the year 2000 Giri et al. (2011) estimate of global mangrove forest area estimate of 137,760 km^2 that uses a presence or absence pixel approach that typically results in estimates approximately 40% higher than continuous measures. CGMFC-21 can be used to extrapolate mangrove forest area estimates through to the year 2014 with the global area estimate for all forests in the mangrove biome now at 163,925 km^2 and the global area estimate for mangrove forests alone at 81,626 km^2 (Hamilton and Casey 2016). Figure 22.3 represent the CGMFC-21 – 2012 mangrove cover data. Note the differing pixel values that are not present in Figs. 22.2 or 22.4 as CGMFC-21 allows for differing mangrove area values at each pixel as opposed to mere presence or absence data at the pixel level.

[9] http://www.spatialreference.org/ref/epsg/4326/html/

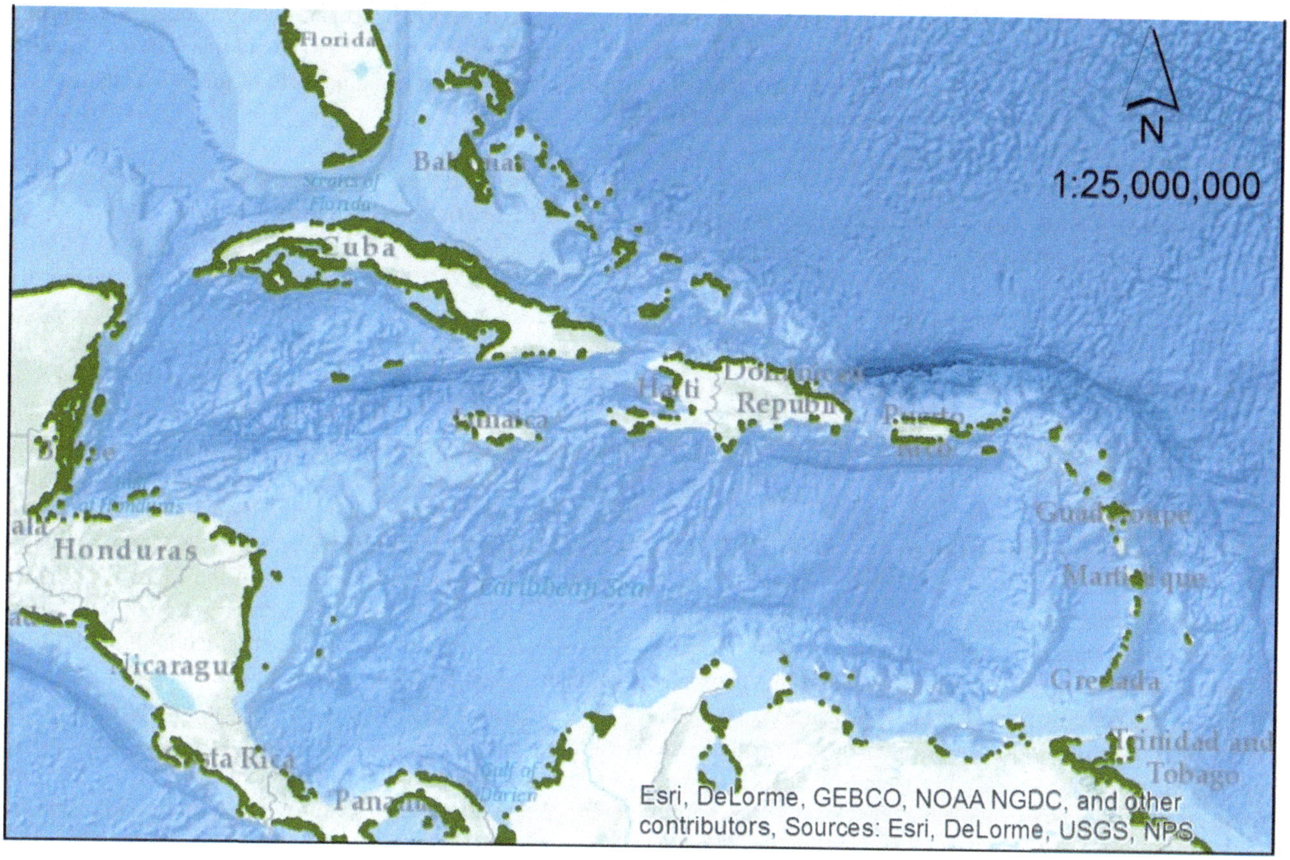

Fig. 22.4 WAM representation of the Caribbean, post-2000
The WAM polygon global dataset using the polygon shapefile provided by Spalding et al. (2010) and a background map compiled by the GIS user community, ESRI, HERE, DeLorme, MapmyIndia, and OSM contributors. The green represents the mangrove forests as represented in the WAM. The borders have been enlarged for easier viewing (Map created by Stuart E. Hamilton)

22.2.3 World Atlas of Mangroves (WAM)

Although global remotely sensed estimates of mangrove forests are important in driving the mangrove research agenda, the majority of remotely sensed mangrove products occur at either the regional, national or local scales. At the trans-regional scale Corcoran et al. (2007) created 30 m resolution mangrove maps that are then used in the World Atlas of Mangroves (Spalding et al. 2010). They used TM and ETM+ data circa 2000 at 30 m resolution and preprocessed the imagery using the COST (Chavez 1996) method (Corcoran et al. 2007). Once pre-processed, the ETM + scenes were then clipped to areas of potential mangrove and an unsupervised classification combined with a visual interpretation on band combination 5,4,3 was conducted (Corcoran et al. 2007). Verification of mangrove delineation included the use of SRTM data and local feedback. No error metrics are included, and no uncertainty analysis appears to have been conducted. The 19-countries analyzed by Corcoran et al. (2007) are all in Africa and hence are omitted from this chapter. The methodology and mangrove area for each country can be found in the UNEP Report number DEW/0913/CA[10] and in Spalding et al. (2010). To the list of 19 African countries, Spalding et al. (2010) added, El Salvador, Honduras (Pacific Coast), Nicaragua, Papua New Guinea, and Vietnam, although these are not referenced in the original report (Corcoran et al. 2007) but they do appear to use the same methodology. These Spalding et al. (2010) additions, aside from the Pacific-only estimate for Honduras and Vietnam, are included in this analysis as part of the remote sensing derived mangrove comparisons for the Western Hemisphere and Oceania.

The United Nations (UN) Food and Agricultural Organization (FAO) conducted a second major trans-regional analysis of mangroves for 31-countries globally. The methodology can be found in Rosati et al. (2008), and the results can again be found in Spalding et al. (2010). Egypt, Eritrea, French Guiana, Guadeloupe, Guam, Guyana, India, Madagascar, Malaysia, Martinique, Mauritania, Mayotte, Micronesia, Myanmar, Oman, Pakistan, Palau, Peru, Philippines, Qatar, Saint Kitts

[10]Report available at https://archive.org/details/mangrovesofweste07corc

and Nevis, Saint Lucia, Seychelles, Singapore, Solomon Islands, Somalia, Sudan, Suriname, Thailand, UAE, USA, Vanuatu, Venezuela, and Yemen were included in the analysis. Again the year is circa 2000 and the instrument in ETM+ (Rosati et al. 2008). Band composites 4,3,2 and 5,4,3 were used to visually interpret and digitize the ETM+ scenes and limited field validation was conducted in certain regions (Rosati et al. 2008; Spalding et al. 2010). Again, no error measure is reported, and no uncertainty analyses are conducted. These data for countries only in the Western Hemisphere and Oceania are included in this analysis as part of the remote sensing derived mangrove comparisons. All WAM, compiled from all sources, including non-remotely sensed data, can be download at the Ocean Data Viewer.[11]

22.3 Western Hemisphere

Within the Western Hemisphere, we examine in-depth the eight countries with mangrove holding in the top-20 globally (Hamilton and Casey 2016). These are Brazil, Venezuela, Colombia, Panama, Ecuador, Cuba, USA, and Mexico. Then for all other countries with mangrove holdings globally in the top-50, we summarize their information in tabular format. These countries are Nicaragua, Honduras, Dominican Republic, Costa Rica, Guatemala, El Salvador, Belize, and Trinidad and Tobago.

22.3.1 Brazilian Post-2000 Remote Sensing of Mangrove Studies

Brazil is ranked by CGMFC-21 (Hamilton and Casey 2016) as the second largest mangrove holding nation globally and by far the largest in the Western Hemisphere. Magris and Barreto (2010) estimate mangrove forest cover for Brazil at 1,114,399 ha, for the period of 2007–2009. Their estimate was compiled using TM. TM bands 5,4,3 were applied to the green, red, and blue color guns respectively and mangrove delineation was conducted primarily via a visual delineation based on color, texture, shape, size, context, geometry, and drainage system configuration (Magris and Barreto 2010). Validation was then conducted of the delineated mangrove regions using various ancillary sources including, published literature, research reports, aerial photographs, and private communication (Magris and Barreto 2010). No error bars or confidence intervals are reported. In addition to the mangrove area delineation, three mangrove sectors were defined along the Brazilian coast: the macrotidal between 4°N and 3°S, the mesotidal between 3°S and 15°S, and the microtidal between 15°S and 34°S. The northern macrotidal region contained the largest mangrove area in Brazil with 921,627 ha. Additionally, Magris and Barreto (2010) found that 77% of Brazilian mangroves are afforded some form of protected status.

22.3.2 Venezuelan Post-2000 Remote Sensing of Mangrove Studies

CGMFC-21 ranks Venezuela as the ninth largest mangrove holding nation globally, and the third-largest in the Americas (Hamilton and Casey 2016). Within Venezuela, Oliveira-Miranda et al. (2010), calculate mangrove cover at 853,000 ha. This estimate is derived from the visual and automated analyses of different TM, ETM+, SPOT, Aster, ALOS, PALSAR, and SAR sensor data collected between 2000 and 2009 (Huber and Oliveira-Miranda 2010). This estimation is part of a wider ecosystem delineation conducted for all of Venezuela in the context of the Libro Rojo de los Ecosistemas Terrestres de Venezuela (Rodríguez et al. 2010). No validation or error metrics are provided in this study, and the data are not available for download.

The Rosati et al. (2008) remotely sensed mangrove forest estimate that can be found in Spalding et al. (2010) places Venezuelan mangrove forest cover at 356,911 ha. The Giri et al. (2011) estimate of mangrove area in the year 2000 for Venezuela appears to have discrepancies. Within Giri et al. (2011, p. 157), Venezuela is not listed in their top-15 mangrove holding nations in their Table 1, but using the downloadable data Venezuela appears to be ranked thirteenth with 336,000 ha of mangrove forest when plotted by Siikamäki et al. (2012) in their supplemental Table SI.1. Giri himself stated that this study depicted 5,698 km^2 of mangrove forest (Carrasquel 2013), which would place it sixth. For this analysis, we use the actual downloaded Giri et al. (2011) data and estimate Venezuelan remotely sensed mangrove area as 336,000 ha as of 2000 (Siikamäki et al. 2012). This estimate aligns well with Rosati et al. (2008) and Hamilton and Casey (2016).

A comprehensive attempt to map mangrove areas in Venezuela used 50 ASTER satellite images, unsupervised classification, and field validation has appeared on the internet (Lazo 2017). The results of this project indicate that Venezuela only has 181,027 ha of mangrove but details are limited. The project was listed as complete in 2009 using data from 2000 to 2007. The data are not included in our reporting due to a lack of detailed information.

22.3.3 Colombian Post-2000 Remote Sensing of Mangrove Studies

CGMFC-21 ranks Colombia as the thirteenth largest mangrove holding nation globally, and the fourth largest in the Americas, with between 167,186 ha and 278,643 ha of

[11] http://data.unep-wcmc.org/datasets/5

mangrove forest area in 2012 (Hamilton and Casey 2016). The formal national mapping of mangrove areas in Colombia is conducted by the regional environmental authorities known as the Coorporacion Autonomas Regionales (CARs), of the 12 provinces adjacent to the Pacific and Caribbean coasts of the country. The time window for the most recent estimations at the provincial level is 2010–2011 and involves the visual interpretation of TM and ETM+ images. The Colombian Ministry of Environment then compiles the individual regional information into a national map generating a circa 2010 mangrove area estimate of 285,050 ha. Of this 285,050 ha, 90,170 ha of mangrove is on the Caribbean Coast, and 194,880 ha of mangrove is the Pacific Coast (Ministerio de Ambiente y Desarrollo Sostenible de la República de Colombia 2017). The Colombian Marine Science Institute (INVEMAR) warns against the use of this information due to the lack of a consistent remote sensing methodology employed by each region in assessing their mangrove levels (INVEMAR 2016).

In 2007, a combination of six environmental institutes (SixIns) across Colombia produced a series of remotely sensed land cover maps that included a mangrove classification (Instituto de Hidrologia et al. 2007). Mangrove area along the Pacific coast was estimated at 230,545 ha, the Caribbean coast was estimated at as containing 76,816 ha of mangrove, and the archipelagos of San Andrés, Providencia and Santa Catalina in the Caribbean, off the coast of Nicaragua, was shown to contain 176 ha of mangrove. The total estimated for the country was 307,537 ha (Instituto de Hidrologia et al. 2007). These estimates were calculated using TM and ETM+ scenes taken between 2000 and 2003 and were processed using an unsupervised ISOSeg classification algorithm. No error statistics are reports.

22.3.4 Panamanian Post-2000 Remote Sensing of Mangrove Studies

CGMFC-21 ranks Panama as the sixteenth largest mangrove holding nation globally, and the seventh largest in the Americas, with between 132,394 ha and 220,656 ha of mangrove forest area in 2012 (Hamilton and Casey 2016). Within the framework of the UN – REDD Programme in Panama, a national estimate of forest's cover, including mangrove, was produced for 2012 (ONU-REDD 2015). High-resolution 5 m Rapid Eye images were used to delineate mangrove forest based on an object-based image analysis approach as defined by MacLean and Congalton (2012). The outputs were then field verified and compared against higher resolution aerial photographs. The area of mangrove was found to be 168,677 ± 7,965 ha (CI .9) (ONU-REDD 2015). These data outputs are available in various map sources.[12]

22.3.5 Ecuadorian Post-2000 Remote Sensing of Mangrove Studies

CGMFC-21 ranks Ecuador as the 20th largest mangrove holding nation globally, and the eighth largest in the Americas, with between 93,574 ha and 155,957 ha of mangrove forest area in 2012 (Hamilton and Casey 2016). El Centro de Levantamientos Integrados de Recursos Naturales por Sensores Remotos (CLIRSEN), the remote sensing Institute of the Ecuadorian government, estimated the national mangrove holdings of mainland Ecuador in 2006 as totaling 148,230 ha (CLIRSEN 2007). The estimate was compiled using a combination of sensors that included SPOT at 5 m, ASTER at 15 m, TM and ETM+ at both 15 m and 30 m, and Ikonos at 4 m. The CLIRSEN method employed was a visual interpretation of satellite imagery and manual digitization of mangrove stands. This mangrove area estimate does not include mangroves located in the Galapagos Islands which are located approximately 1000 km away from the Ecuadorian mainland. CLIRSEN reports no error statistics or confidence intervals for this 2006 mangrove area estimate.

The Galapagos Islands mangroves' were mapped using manual digitization from Google Earth Imagery combined with field verification by the Charles Darwin Foundation (CDF) of Ecuador. CDF found the islands contain 3,690 ha of mangrove and that the data is 90% accurate with a resolution of approximately 25 m (Martin et al. 2016). The imagery dates vary across 2004 to 2015 (Martin 2017, personal communication). Combining the CLIRSEN 2006 mainland measure of 148,230 ha with the CDF Galapagos 2004–2015 measure od 3,690 ha gives a combined remotely sensed area of 151,920 ha for all of Ecuador from local remotely sensed sources.

The 2006 mainland mangrove data are available from the CLIRSEN offices in Quito in GIS format, but users are not allowed to redistribute the data, and the data is difficult to obtain due to security clearance required to enter the building on a military site. The mangrove data for Ecuador is now incorporated into the protected areas database (SNAP) and is available from the MAE offices in Quito, but again is difficult to obtain. Remotely sensed mangrove coverage data for various dates post-2000 are available for the majority of Ecuadorian mangrove forests as supplemental submissions within Hamilton et al. (2017) and Hamilton and Lovette (2015), again though the coverage is not national and excludes the Galapagos Islands. The Galapagos Islands data compiled by the Charles Darwin Institute is available from the Charles Darwin Foundation Website.

[12] https://www.gacetaoficial.gob.pa/pdfTemp/28229_A/GacetaNo_28229a_20170303.pdf and http://www.oteima.ac.pa/nueva/biblioteca/Manglares%20de%20Panama%20web.pdf

22.3.6 Cuban Post-2000 Remote Sensing of Mangrove Studies

According to Hamilton and Casey (2016), Cuba is the fifth largest mangrove holding country in the Americas and the 14th largest globally. Giri et al. (2011) rank it as the tenth largest globally and the third largest in the Americas. Part of the difference may be that some authors count the mangroves of Guantanamo Bay as under the USA, and others count them as under the Cuban counts. Menéndez (2013) estimated that mangroves occupy 70% of the Cuban coastline. Using TM and Quickbird images from 1999, 2001 and 2002, Menéndez (2013) characterized the mangrove ecosystems at the country level. Landsat images were processed using the ENVI 3.5 software. To identify Cuban mangrove areas, a classified supervision method was employed using the band combinations 4,5,2. Quickbird satellite images and aerial photographs were also used to increase the resolution of the cartographic output. Field verification of the classes occurred, but unfortunately, no error or accuracy estimates are presented. Menéndez and Guzmán (2010) combined with Menéndez (2013) estimate the mangrove area of Cuba circa 2000 as 532,400 ha.

22.3.7 USA Post-2000 Remote Sensing of Mangrove Studies

Remote sensing estimates of mangrove cover in the USA are often difficult to compare due to the inclusion or not of the designated USA Minor Outlying Islands. Thes islands include Baker Island, Howland Island, Jarvis Island, Johnston Atoll, Kingman Reef, Midway Atoll, Palmyra Atoll, and Wake Island in the Pacific Ocean and Navassa Island in the Caribbean Sea. Also, the inhabited territories of Puerto Rico, Guam, Northern Mariana Islands, United States Virgin Islands and American Samoa are sometimes included in the USA mangrove total and sometimes excluded. Finally, the Guantanamo Bay situation is discussed under Cuba. Within the 50 states, mangroves are found in Florida, Louisiana, Texas, and Hawaii but greater than 99% of these are in Florida. For this analysis, we consider the Florida mangrove area and USA mangrove area as interchangeable.

The five Floridian water management districts that cover the state routinely map landcover within their jurisdiction and all five make mangrove data available in various formats as part of their regional land cover delineations. The data compiled by the author of this chapter for this estimate are from 1999, 2004, 2005, 2006, 2008, 2009, 2010 and 2011 (Florida Water Management Districts 2012). Although compiled by five differing groups the methodology appears to be consistent using high-resolution aerial photography to digitize mangrove at the 1:12,000 scale. No error metrics can be obtained, and only limited information is available on the source of the aerial photography although it appears to be 1 m in most cases. The water management districts calculate the area of mangrove in Florida to be 245,257 ha with the area calculation conducted in map projection EPSG:3086 by this author. These data are available from the author upon request.[13]

A more systematic remotely sensed mangrove area calculation was conducted by the Florida Fish and Wildlife Commission in 2003. In this study, they used ETM+ to delineate all landcover in Florida including classifications for mangrove (Kautz et al. 2007; Gilbert and Stys 2004). All ETM+ data is from 2003, and unsupervised classification underpins the process with the resultant clusters then identified by high-resolution DOQQ photography (Kautz et al. 2007). Areas that could not be grouped in the original unsupervised process were then extracted and a second unsupervised classification was run in these regions. This process was repeated iteratively until all regions fell into a cluster (Kautz et al. 2007). The Florida Fish and Wildlife Commission find the area of mangrove in Florida to be 240,722 ha with the area calculation conducted in map projection EPSG:3086 by this author. These data are available from the Florida Geographic Data Library by submitting a keyword search for 29764.[14] Positional accuracy of within 15 m is stated and was determined using known ground control points. No classification error is reported. The methodology does state that landcover classifications in error were assessed by a supervisor and verified for correctness.

22.3.8 Mexican Post-2000 Remote Sensing of Mangrove Studies

Mexico has the second largest mangrove holdings in the Western Hemisphere and the sixth largest globally (Hamilton and Casey 2016). Within Mexico, The National Commission for Knowledge and Use of Biodiversity (CONABIO) present an estimate of 770,057 ha of mangrove cover for the entire country (Acosta-Velázquez et al. 2007, 2009). This estimate is calculated from SPOT 5 images using a variety of techniques including unsupervised ISODATA classification, image segmentation techniques, ancillary data, and visual interpretation using the band combinations 3,4,2. The temporal domain is given as 2005 and 2006. Mangrove classification accuracy is determined through field validation, aerial photography, and expert opinions. The overall accuracy of the national estimate was 90.7% with a Kappa coefficient of 0.809 ± 0.028 (Acosta-Velázquez et al. 2007, 2009).

[13] http://www.spatialreference.org/ref/epsg/3086/

[14] http://www.fgdl.org/metadataexplorer/explorer.jsp

The most recent national estimate for mangrove cover in Mexico was produced in 2015 by CONABIO using 198 scenes consisting of 182 SPOT 5, four SPOT 6, two SPOT 7, and ten Rapid Eye images. Mangroves were distributed in 17 Mexican states and covered an area of 775,555 ha as of 2015 (Valderrama-Landeros et al. 2017). Of the 775,555 ha, 266,581 ha are located along the Pacific Coast, and 508,974 ha are located along the Atlantic Coast. These estimates were validated with aerial photographs taken in 2015 and 2016, 7,662 photographs were examined with 3,300 corresponding to mangroves. Classification accuracy was determined to be between 90% and 93% (Valderrama-Landeros et al. 2017). Both sets of CONABIO data are available for download from the CONABIO GIS website by clicking on the Biodiversidad link in the table of contents and then clicking on the Monitoreo de Manglares subheading.[15]

22.3.9 Other Western Hemisphere Post-2000 Remote Sensing of Mangrove Studies

Countries outside of the top-20 global mangrove holdings nations in the Western Hemisphere are covered in detail above, the countries of Nicaragua (Meyrat 2009), Honduras (Duarte et al. 2014), Dominican Republic (Meyer-Arendt et al. 2013), Costa Rica (Calvo 2015), Guatemala (Por el Ministerio de Ambiente y Recursos Naturales 2013), El Salvador (Gallo 2005), Belize (Chang et al. 2015; Cherrington et al. 2010), and Trinidad and Tobago (Juman and Ramsewak 2013) have their remotely sensed post-2000 mangrove area reported in the results section in tabular format. Additionally, we note the primary sensor used, the acquisition dates, the main method used, the error rate reported, and the other comparable estimates for the listed country from the suite of global sources when available.

22.4 Oceania

Within Oceania, we examine in-depth the two countries with mangrove holding in the top-20 globally (Hamilton and Casey 2016). These two countries are Papua New Guinea and Australia, with the fourth and fifth largest mangrove holdings respectively (Hamilton and Casey 2016). Then for all other Oceanic nations within the top-50 mangrove holdings nations globally, we summarize the findings in tabular format. These are the four countries of New Caledonia New Zealand, Solomon Islands, and Fiji.

22.4.1 Australia Post-2000 Remote Sensing of Mangrove Studies

In Australia, the Bureau of Agricultural and Resource Economics and Sciences (ABARES) produce 5-year reports on the state of the countries forests. The last report was issued in 2013 and includes a mangrove designation (Montreal Process Implementation Group for Australia and National Forest Inventory Steering Committee 2013). The methodology is described as a Multiple Lines of Evidence Approach (Mutendeudzi et al. 2013) and is fundamentally a data synthesis approach bases on 30 differing sources that include remote sensing data, aerial photography, and ground survey. Remote sensing systems include TM, ETM +, MODIS, SPOT, and lidar (Montreal Process Implementation Group for Australia and National Forest Inventory Steering Committee 2013). Although progress is being made in assessing error, and SPOT 5 imagery in conjunction with lidar are used to evaluate the forest classes that are produced, the current 2013 v2 forest database is still unable to report error metrics (Mutendeudzi et al. 2013). The SPOT data used to assess forest classes is stated as 2.5–3 m.

The national forest data is distributed at 100 m resolution, and the temporal domain should be considered as a post-2000 synthesis (Mutendeudzi et al. 2013). The raw data are available for download in GIS format from the Australian Bureau of Agricultural and Resource Economics and Sciences website,[16] although further processing is required to extract and map the mangrove forest. The mangrove area of Australia, using the v2., 2013 data (Montreal Process Implementation Group for Australia and National Forest Inventory Steering Committee 2013), was calculated by the author of this paper to be 912,606 ha using map projection EPSG:3577. The extracted and processed national mangrove data are available upon request from the author of this chapter.

22.4.2 Papua New Guinea Post-2000 Remote Sensing of Mangrove Studies

Within Papua New Guinea the University of Papua New Guinea Remote Sensing Centre, in conjunction with the Australian Nation University produced a national forest classification map for all off Papua New Guinea for the period 2001–2007, with 62% of the data from 2002 (Shearman et al. 2009). The primary instrument was ETM+ at 30 m resolution but is supplemented in areas of high cloud cover with SPOT 4 and SPOT 5 data at 20 m and 10 m spatial resolution respectively (Shearman et al. 2009). Each image was orthorectified using

[15] http://www.conabio.gob.mx/informacion/gis/

[16] http://www.agriculture.gov.au/abares

10–15 pre-existing ground points and from SRTM data. The RMSE for the orthorectified scenes was 25–20 m for ETM+ and 15–20 m for SPOT.

The University of Papua New Guinea Remote Sensing Centre used dynamic segmentation to automatically divide the imagery into spatially continuous and spectrally homogeneous regions before vectorizing them into discrete polygons (Shearman et al. 2009). Each polygon was then classified using expert visual interpretation, with mangrove having a unique class. Shearman et al. (2009) report the mangrove area in circa 2002 for the entirety of Papua New Guinea as 574,876 ha. The mismatch error rate of land cover classification is reported as 2.2%.

22.4.3 Other Oceanic Post-2000 Remote Sensing of Mangrove Studies

Outside of the top-20 global mangrove holdings nations in Oceania covered in detail above, the countries of New Caledonia (Virly 2008), New Zealand (LINZ 2017), Solomon Islands (Bhattarai and Giri 2011), and Fiji (Bhattarai and Giri 2011) are summarized in tabular format in the results section. Additionally, we note the primary sensor used, the acquisition dates, the main method used, the error rate reported, and the other comparable estimates for the listed country from the suite of global sources when available.

22.5 Results

The results are presented globally and then nationally for the Western Hemisphere and followed finally by Oceania.

22.5.1 Global, Post-2000 Remotely Sensed Mangrove Area Estimates

Table 22.1 depicts the global remotely sensed mangrove estimates from Hamilton and Casey (2016) and Giri et al. (2011), as well as the national estimates from the trans-regional analyses noted above (Rosati et al. 2008; Spalding et al. 2010) which are upscaled to global findings. It should be pointed out that the Spalding et al. (2010) estimate may include some smaller mangrove holding national estimates that are not remotely sensed. To allow for a general comparison of presence mangrove forest data to continuous mangrove forest data an effective rule-of-thumb is to divide the continuous data by 0.6 to make it comparable. Mangroves are still generally monitored using presence or absence remote sensing techniques although other global forests measures have now converted to continuous measures at the pixel level such as Hansen et al. (2013) which is used for mangrove by analysis in the CGMFC-21 database Hamilton and Casey (2016). Using a synthesis of all globally remotely sensed mangrove area estimates, the most applicable post-2000 estimate of global mangrove cover is 141,333 km^2 $\pm$ 6% (CI 0.9, n = 4), which is calculated using G1 presence, G2 presence, G5, and G6 from Table 22.1. Aside from the CGMFC-21: TEOW, which is expected to be a high outlier due to what it measures, the remaining remotely sensed global mangrove area values are remarkably consistent.

22.5.2 Brazilian, Post-2000 Remotely Sensed Mangrove Area

Table 22.2 summarizes all remotely sensed post-2000 mangrove area estimates used in this chapter for Brazil. The Magris and Barreto (2010) remote sensing mangrove estimate of 1,114,399 ha for 2007–2009 falls close to the midpoint of the CGMFC-21 biome estimate for 2009 of 1,761,216 km^2 and the CGMFC-21 MFW 2009 estimate of 768,429 km^2. Indeed when using the 40% average difference between continuous tree cover data and presence and absence data, the CGMFC-21 MFW 2009 estimate and the Magris and Barreto (2010) estimate are similar recording 1,114,399 ha and 1,280,715 ha. Additionally, the trans-regional remotely sensed estimate of mangrove cover produced for the year 2000 by Rosati et al. (2008) that can be found in Spalding et al. (2010), places Brazilian mangrove forest cover at 1,299,947 ha. Finally, the Giri et al. (2011) of mangrove area in 2000 for Brazil is 1,063,000 ha taken from (Siikamäki et al. 2012), and approximates the other estimates but at the lower end. Aside from the CGMFC-21: TEOW, which is expected to be a high outlier due to what it measures, the remaining remotely sensed

Table 22.1 Global mangrove forest cover

	CGMFC-21: MFW		CGMFC-21: TEOW		MFW	WAM
	G1	G2	G3	G4	G5	G6
Year	*2000*	*2012*	*2000*	*2012*	*2000*	*Various post-2000*
Method	*Continuous*	*Continuous*	*Continuous*	*Continuous*	Presence	Presence
	Presence	Presence	Presence	Presence		
Mangrove area (km^2)	*83,495*	*81,849*	*173,067*	*167,387*	137,760	152,000
	139,158	136,415	288,455	278,978		

Table 22.2 Brazilian mangrove forest cover

	CGMFC-21: MFW		CGMFC-21: TEOW		MFW	Rosati WAM	Magris local
	B1	B2	B3	B4	B5	B6	B7
Year	*2000*	*2012*	*2000*	*2012*	*2000*	*2000*	*2007–2009*
Method	*Continuous*	*Continuous*	*Continuous*	*Continuous*	Presence	Presence	Presence
	Presence	Presence	Presence	Presence			
Mangrove area (ha)	*772,131*	*767, 494*	*1,816,759*	*1,741,651*	1,063,000	1,299,947	1,114,399
	1,286,886	1,279,156	3,027,931	2,902,751			

mangrove area values are all remarkably consistent. A reasonable remotely sensed post-2000 estimate of mangrove cover in Brazil is 1,208,677 ha ± 9% (CI 0.9, n = 5) using B1 presence, B2 presence, B5, B6, and B7 in Table 22.2. The major locally remotely sensed estimates of mangrove cover in Brazil appears to align quite well with the global and regional estimates.

Nascimento et al. (2013) conducted a more detailed study of a section of the macrotidal mangrove region of Brazil named the Amazon Macrotidal Mangrove Coast (AMME) They calculated 742,360 ha of mangroves in this region during 2008. This mangrove area estimate is lower than the AMME area estimate of 921,627 ha as reported by Magris and Barreto (2010) but this is not surprising as the Magris and Barreto (2010) AMME covers a longer area of coastline. This study used a combination of SAR, ALOS, and PALSAR data at 20 m resolution and TM at 30 m resolution. The object-based classification method employed in this study included rules based on relative height, SAR backscatter, and the TM reflectance values. Validation field data was collected in 2010 and used standard error matrices such as Kappa and Tau. Overall accuracy is reported as 83%, and the Kappa and Tau coefficients were 80% and 66% respectively (Nascimento et al. 2013). Comprehensive studies like this do not exist yet for the entire Brazilian coast. Neither the Magris and Barreto (2010) nor the Nascimento et al. (2013) mangrove data appear available to download publically, and users wishing to obtain the data should contact the authors as indicated in the references section of this chapter.

22.5.3 Venezuelan, Post-2000 Remotely Sensed Mangrove Area

Table 22.3 summarizes all remotely sensed mangrove area estimates used in this chapter for Venezuela. The local estimate produced by Oliveira-Miranda et al. (2010) of 853,000 ha of mangrove would make Venezuela somewhere between the second and fourth largest mangrove holding nation globally. No other literature supports such a large mangrove area for Venezuela. It is possible that Oliveira-Miranda et al. (2010) are measuring all forest in a mangrove-dominated ecological zone in a similar manner to the CGMFC-21: TEOW measure. In this scenario, the 853,000 ha would still be high but at least comparable with the 2000 finding of 753,912 ha of forest in the wider mangrove dominated biome (Hamilton and Casey 2016). The Oliveira-Miranda et al. (2010) data do not appear to be available for download to verify their findings. Due to the Oliveira-Miranda et al. (2010) data not being verifiable (we contacted all authors and the agency) and the fact the mangrove area reported appears a significant outlier (Table 22.3), as well as the unclear nature of exactly what is being mapped this mangrove area estimate is not included in the post-2000 best estimate of remotely sensed mangrove cover for Venezuela. Therefore, the most feasible remotely sensed post-2000 estimate of mangrove cover in Venezuela is 374,086 ha ± 10% (CI 0.9, n = 4) using V1 presence, V2 presence, V5, and V6 from Table 22.3. The major locally remotely sensed estimates of mangrove cover in Venezuela does not align with the global and regional estimates produced, and additionally, one of the global estimates is unclear. For these reasons, the remote-sensing mangrove area estimates of Venezuela need further clarification and estimates produced thus far should be treated with caution.

22.5.4 Colombian, Post-2000 Remotely Sensed Mangrove Area

Table 22.4 summarizes all remotely sensed mangrove area estimates used in this chapter for Colombia. The Rosati et al. (2008) remotely sensed mangrove forest estimate for Colombia, estimates Colombian mangrove forest cover at 407,926 ha in Spalding et al. (2010), again this is at the higher end of the estimates (Table 22.4). Giri et al. (2011), estimate Colombian mangrove cover at 214,700 ha, again this is at the lower end of the estimates (Table 22.4). The average remotely sensed post-2000 estimate of mangrove cover in Columbia is 295,480 ha ± 18% (CI 0.9, n = 6)

Table 22.3 Venezuelan mangrove forest cover

	CGMFC-21: MFW		CGMFC-21: TEOW		MFW	Rosati WAM	Oliveira local
	V1	V2	V3	V4	V5	V6	V7
Year	*2000*	*2012*	*2000*	*2012*	2000	2000	2000–2009
Method	*Continuous*	*Continuous*	*Continuous*	*Continuous*	Presence	Presence	Presence
	Presence	Presence	Presence	Presence			
Mangrove area (ha)	*241,584*	*240,483*	*757,908*	*753,912*	336,000	356,900	853,000
	402,640	400,805	1,263,180	1,256,520			

Table 22.4 Colombian mangrove forest cover

	CGMFC-21: MFW		CGMFC-21: TEOW		MFW	Rosati WAM	CARs local 1	Sin ins local 2
	C1	C2	C3	C4	C5	C6	C7	C8
Year	*2000*	*2012*	*2000*	*2012*	2000	2000	2010–2011	2000–2003
Method	*Continuous*	*Continuous*	*Continuous*	*Continuous*	Presence	Presence	Presence	Presence
	Presence	Presence	Presence	Presence				
Mangrove area (ha)	*167,415*	*167,186*	*631,293*	*627,470*	214,700	407,926	285,050	307,537
	279,025	278,643	1,052,155	1,045,783				

Table 22.5 Panamanian mangrove forest cover

	CGMFC-21: MFW		CGMFC-21: TEOW		MFW	ONU-REDD local
	P1	P2	P3	P4	P5	P6
Year	*2000*	*2012*	*2000*	*2012*	2000	2012
Method	*Continuous*	*Continuous*	*Continuous*	*Continuous*	Presence	Presence
	Presence	Presence	Presence	Presence		
Mangrove area (ha)	*132,786*	*132,394*	*276,844*	*270,821*	154,304	168,677
	221,310	220,656	461,406	451,368		

using C1 presence, C2 presence, C5, C6, C7, and C8 from Table 22.4. The major locally remotely sensed estimates of mangrove cover in Colombia fit in-between the extremes of the global data estimates (Table 22.4), thus giving increased confidence in the locally produced mangrove forest cover area generated by the institutes within Columbia.

22.5.5 Panamanian, Post-2000 Remotely Sensed Mangrove Area

Table 22.5 summarizes all remotely sensed mangrove area estimates used in this chapter for Panama. The Giri et al. (2011), mangrove area estimate for Panama is 154,304 ha and was calculated by this author using the downloadable Giri et al. (2011) vector polygonal version of the database available from the Ocean Data Viewer. The area calculation was conducted in map projection ESRI:53009. The Panama ONU-REDD (2015), and the Giri et al. (2011), mangrove area estimate fall between the Hamilton and Casey (2016) presence ranges, giving confidence in all estimates (Table 22.5). Again, the Giri et al. (2011) estimate is the lowest, the Hamilton and Casey (2016) the highest, and the local estimate in-between these ranges (Table 22.5). The average remotely sensed post-2000 estimate of mangrove cover in Panama is 191236.8 ha ± 21% (CI 0.9, n = 4) using P1 presence, P2 presence, P5, and P6 in Table 22.5.

22.5.6 Ecuadorian, Post-2000 Remotely Sensed Mangrove Area

Table 22.6 summarizes all remotely sensed mangrove area estimates used in this chapter for Ecuador. The Giri et al. (2011) mangrove area estimate for Ecuador in the year 2000 is 136,986 ha and again was calculated by this author using the downloadable Giri et al. (2011) vector polygonal version of the database available from the Ocean Data Viewer. Map projection ESRI:53009 was used for the area calculation. It is again the low end of the range of estimates (Table 22.6). The Rosati et al. (2008) estimate is 158,261 ha. Again, the Ecuadorian mangrove area data is remarkably consistent across all in-country and out-of-country remotely sensed sources (Table 22.6). Thus again giving confidence to the remotely sensed area estimates of mangrove cover in Ecuador. The average remotely sensed post-2000 estimate

Table 22.6 Ecuadorian mangrove forest cover

	CGMFC-21: MFW		CGMFC-21: TEOW		Giri MFW	Rosati WAM	CLIRSEN local
	E1	E2	E3	E4	E5	E6	E7
Year	*2000*	*2012*	*2000*	*2012*	*2006*	*2000*	*2006*
Method	*Continuous*	*Continuous*	*Continuous*	*Continuous*	Presence	Presence	Presence
	Presence	Presence	Presence	Presence			
Mangrove area (ha)	*93,756*	*93,574*	*197,054*	*194,527*	136,986	158,261	151,920
	156,310	155,957	328,423	324,211			

Table 22.7 Cuban mangrove forest cover

	CGMFC-21: MFW		CGMFC-21: TEOW		MFW	Rosati WAM	Menéndez local
	C1	C2	C3	C4	C5	C6	C7
Year	*2000*	*2012*	*2000*	*2012*	*2000*	*2000*	*Circa 2000*
Method	*Continuous*	*Continuous*	*Continuous*	*Continuous*	Presence	Presence	Presence
	Presence	Presence	Presence	Presence			
Mangrove area (ha)	*166,036*	*163,346*	*247,097*	*245,410*	421,538	494,405	532,400
	276,726	272,243	411,828	409,016			

of mangrove cover in Ecuador is 151,887 ha ± 5% (CI 0.9, n = 5) using E1 presence, E2 presence, E5, E6, and C7 from Table 22.6. All remotely sensed mangrove area estimates are within proximity in Ecuador giving a high-level of confidence in the locally produced mangrove forest cover area generated by CLIRSEN and other authors.

The CLIRSEN national mangrove estimate can be calibrated by using other remotely sensed provincial mangrove estimates from this period. For example, Hamilton (2012) using found that in 2006 El Oro Province had 18,807 ha of mangroves present as opposed to the CLIRSEN estimate of 16,158 ha of mangroves present in the 2006 country-wide estimate noted above (CLIRSEN 2007). The Hamilton (2012) definition of El Oro province includes approximately 15 linear km of southern Guayas Province (to avoid splitting of an estuary), and this likely explains the small amount of difference between the two estimates. The Hamilton estimate used unsupervised classification using ASTER data enhanced by manual digitization and field data collection. The similarity of the Hamilton (2012) and the CLIRSEN (2007) mangrove estimates for an entire province likely increases confidence in the 2006 mangrove area estimate produced by CLIRSEN.

22.5.7 Cuban, Post-2000 Remotely Sensed Mangrove Area

Table 22.7 summarizes all remotely sensed mangrove area estimates used in this chapter for Cuba. The Giri et al. (2011), Rosati et al. (2008), and Menéndez and Guzmán (2010) Cuban mangrove estimate are all in proximity to each other with the in-country estimate at the high end of the range (Table 22.7). Even when adjusted for continuous cover the Hamilton and Casey (2016) mangrove cover estimates are all low-end estimates. This underestimation may be partially explained if they attributed Guantanamo Bay's mangrove holdings to the USA although further verification is required as to why the estimates are so much lower when compared to other remote sensing estimates. The selected remotely sensed post-2000 average estimate of mangrove cover in Cuba is 448,454 ha ± 10% (CI 0.9, n = 5) using C3 presence, C4 presence, C5, C6, and C7 in Table 22.7 below. Aside from Hamilton and Casey (2016) MFW measure, the remaining remotely sensed mangrove area estimates are within proximity to each in Cuba, giving a high-level of confidence in the locally produced mangrove forest cover area generated by Menéndez and Guzmán (2010) as well as the estimates produced by other authors.

22.5.8 USA, Post-2000 Remotely Sensed Mangrove Area

Table 22.8 summarizes all remotely sensed mangrove area estimates used in this chapter for the USA. According to Hamilton and Casey (2016), USA is the sixth largest mangrove holding country in the Americas and the 15th largest globally. Giri et al. (2011) rank it as the 19th largest globally and the 5th largest in the Americas (Siikamäki et al. 2012). Within Spalding et al. (2010), the Rosati et al. (2008) of USA mangrove forest cover area is 302,955 ha. Table 22.8 reflects the different USA remotely sensed mangrove area estimates of the USA, again, they are in quite close general agreement. The average remotely sensed post-2000 estimate of mangrove cover in the USA is tightly clustered around 261,355 ha ± 5% (CI 0.9, n = 8) using all the presence data in Table 22.8. Such close clustering of mangrove area

Table 22.8 USA mangrove forest cover

	CGMFC-21: MFW		CGMFC-21: TEOW		MFW	Rosati WAM	WMD local 1	FWS local 1
	U1	U2	U3	U4	U5	U6	U7	U8
Year	*2000*	*2012*	*2000*	*2012*	*2000*	*2000*	*1999–2011*	*2003*
Method	*Continuous*	*Continuous*	*Continuous*	*Continuous*	Presence	Presence	Presence	Presence
	Presence	Presence	Presence	Presence				
Mangrove area (ha)	*161,204*	*156,860*	*161,574*	*158,506*	236,000	302,955	245,257	240,722
	267,673	264,766	269,290	264,176				

Table 22.9 Mexican mangrove forest cover

	CGMFC-21: MFW		CGMFC-21: TEOW		MFW	CON1 local	CON2 local
	M1	M2	M3	M4	M5	M6	M7
Year	*2000*	*2012*	*2000*	*2012*	*2000*	*2005–2006*	*2015*
Method	*Continuous*	*Continuous*	*Continuous*	*Continuous*	Presence	Presence	Presence
	Presence	Presence	Presence	Presence			
Mangrove area (ha)	*302,103*	*299,183*	*624,038*	*620,392*	730,200	770,057	775,555
	503,505	498,638	1,040,063	1,033,986			

calculations increases confidence in all estimates provided including national estimates.

The data-rich environment of the USA allows for the only remotely sensed high-resolution continuous pixel value verification of an entire nation's mangrove forest globally. The analytical forest cover product developed as part of the NLCD produces a continuous measure of forest canopy cover at 30 m for 2011 (Ruefenacht et al. 2015; Homer et al. 2015). Although the NLCD itself does not have a mangrove variable, it does provide a percentage canopy cover measure for each forest pixel that can be compared to Hamilton and Casey (2016). Comparing the entire state of Florida in 2011, which constitute greater than 99% of USA mangrove forests, the difference between the Hamilton and Casey (2016) mangrove estimate and the NLCD (Ruefenacht et al. 2015; Homer et al. 2015) analytical forest cover estimate for all pixels that contain mangrove was found to be 3.6%.

22.5.9 Mexican, Post-2000 Remotely Sensed Mangrove Area

Table 22.9 summarizes all remotely sensed mangrove area estimates used in this chapter for Mexico. The Hamilton and Casey (2016) MFW presence mangrove cover estimates are at the low-end of the Mexico mangrove cover range and as would be anticipated and the TEOW are at the high-end with all other estimates falling between these extremes (Table 22.9). The Giri et al. (2011) estimate, taken from the recalculated data generated by Siikamäki et al. (2012), is again lower than both the in-country estimates but in close proximity to them. Using M1 presence, M2 presence, M5, M6 and M7; the consensus post-2000 remotely sensed mangrove area estimate for Mexico is 655,591 ha ± 20% (CI 0.9, n = 5).

22.5.10 Other Western Hemisphere, Post-2000 Remotely Sensed Mangrove Area

Of countries in the Western Hemisphere within the top-50 mangrove holding nations globally, only Suriname and Guyana are excluded from this analysis due to the lack of a suitable remotely sensed post-2000 mangrove database. Table 22.10 depicts the remaining Western Hemisphere countries within the top-50 mangrove holding nations globally, but not within the top-20, which are covered in detail above. The CGMFC-21 variable in Table 22.10 is the average adjusted presence value of 2000 thru 2012 in the CGMFC-21 MFW database.

22.5.11 Australian Post-2000 Remote Sensing of Mangrove Studies

Table 22.11 reflects the mangrove estimates from the various remote sensing sources used for Australia. The in country ABARES (Montreal Process Implementation Group for Australia and National Forest Inventory Steering Committee 2013) and Giri et al. (2011) are in very close approximation, but the Hamilton and Casey (2016) data provide a far lower estimate even when adjusted for presence data vs. continuous data. This discrepancy needs further investigation. The range of mangrove cover in Australia post-2000 is best extracted using A5 and A6 in Table 22.11, with a low level of 912,606 ha and a high level of 952,500, and a mean of 932,553 ha. Due to the lack of reliability of the Hamilton

Table 22.10 Other Western hemisphere mangrove holding nations

Country	Area (ha)	Sensor(s)	Dates	Method	Error	CGMFC-21	WAM
Nicaragua	66,406	SPOT	2005–2006	Unsupervised	Not reported	92,637	67,068
Honduras	51,578	Rapid eye	2013–2014	Not reported	100%	88,655	Not remotely sensed
Dominican Republic	18,441	ASTER	2009–2012	Unsupervised CLUSTER	Not reported	16,848	Not remotely sensed
Costa Rica	36,153	Rapid eye	2011–2012	Not reported	89%	55,981	Not remotely sensed
Guatemala	18,840	TM, ETM+ ALOS PALSAR hyperion	2001–2012	Supervised, unsupervised	Not reported	44,537	Not remotely sensed
El Salvador	39,160	TM, ETM+	2002	Corine LC	Not reported	39,434	25,200
Belize	74,684 72,622	TM, ASTER	2010 2014	Spectral mixture analysis	Not reported	50,779	Not remotely sensed
Trinidad and Tobago	9,369	Ikonos	2007–2008	Heads up digitizing	Not reported	8,730	Not remotely sensed

Table 22.11 Australian mangrove forest cover

	CGMFC-21: MFW		CGMFC-21: TEOW		MFW	ABARES local
	A1	A2	A3	A4	A5	A6
Year	*2000*	*2012*	*2000*	*2012*	*2000*	*2000–2012*
Method	*Continuous* Presence	*Continuous* Presence	*Continuous* Presence	*Continuous* Presence	Presence	Presence
Mangrove area (ha)	*332,651* 554,418	*331,621* 552,701	*335,896* 559,826	*325,124* 541,873	952,500	912,606

Table 22.12 Papua New Guinea mangrove forest cover

	CGMFC-21: MFW		CGMFC-21: TEOW		MFW	WAM	UPNGRSC local
	PNG-1	PNG-2	PNG-3	PNG-4	PNG-5	PNG-6	PNG-7
Year	*2000*	*2012*	*2000*	*2012*	*2000*	*2000*	*Circa 2003*
Method	*Continuous* Presence	*Continuous* Presence	*Continuous* Presence	*Continuous* Presence	Presence	Presence	Presence
Mangrove area (ha)	*418,992* 698,320	*417,229* 695,381	*598,206* 997,010	*567,716* 946,193	485,000	426,500	574,876

and Casey (2016) Australia estimates, no confidence intervals can be constructed for Australia.

22.5.12 Papua New Guinean Post-2000 Remote Sensing of Mangrove Studies

Table 22.12 reflects the Papua New Guinea estimates used in this chapter. The Giri et al. (2011) and Spalding et al. (2010) are in close approximation with the local estimate (Shearman et al. 2009) falling at the high end. Again, all estimates fit between the Hamilton and Casey (2016) extreme range but this time approximate the continuous CGMFC-21 measure. This continuous data matching other estimates is an exception to the other countries analyzed but can occur when forests are generally-in-tact, spatially homogeneous, and do not exhibit fragmentation. Using PNG-1 continuous, PNG-2 continuous, PNG-3, PNG-4, PNG-5, PNG-6, and PNG-7 the best estimate for remotely sensed mangrove cover in Papua New Guinea post-2000 is 498,359 ha $\pm$ 12% (CI 0.9, n = 7). The tight clustering of mangrove area estimates in Table 22.12, again gives increased confidence in the remotely sensed mangrove area estimates.

22.5.13 Other Oceanic Post-2000 Remote Sensing of Mangrove Studies

Of countries in the Oceania within the top-50 mangrove holding nations globally, none are excluded from this analysis due to the lack of a suitable remotely sensed post-2000 mangrove database. Table 22.13 depicts the remaining Oceanic countries within the top-50 mangrove holding nations globally, but not within the top-20, which are covered in detail above. The CGMFC-21 variable in Table 22.10 is the average adjusted presence value of 2000 thru 2012 in the CGMFC-21 MFW database.

Table 22.13 Other oceanic mangrove holding nations

Country	Area (ha)	Sensor	Dates	Method	Error	CGMFC-21 presence'	WAM
New Caledonia	35,921	Aerial photography	2007	Hybrid	Not reported	16,392	Not remotely sensed
				Visual			
New Zealand	26,348	Unknown	Circa 2008	Manual	Not reported	12,999	Not remotely sensed
		Remote		Digitization			
Fiji	52,503	Ikonos	2000	Unsupervised	92.5% accurate	66,896	Not remotely sensed
		Quickbird		IsoData			
				Visual			
Solomon Islands	47,100	Ikonos	2000	Unsupervised	92.5% accurate	65,683	60,252
		Quickbird		IsoData			
				Visual			

22.6 Remote Sensing Methods and Instruments for Mangrove Mapping and Monitoring

In this section, we present a brief overview of the use of remote sensing data and methods to map the extent and condition of mangroves both locally and globally in the last 25 years. For a full and comprehensive review see Kuenzer et al. (2011) and Heumann (2011). We rely heavily on the former.

Mangrove mapping presents many unique challenges, and even some advantages when compared to traditional remote sensing of forests. On the one hand, the restricted geographic distribution of mangrove habitats between latitudinal boundaries and adjacent to salt water bodies and coasts, as well as being heavily reliant on warm water, makes it relatively easy to predict where mangroves may occur across the globe. Despite these advantages, mapping mangrove forests is still considered one of the most demanding tasks in remote sensing (Kuenzer et al. 2011; Heumann 2011). The remotely sensed spectral signal from mangrove ecosystems is composed of fractions of soil, vegetation, and water, which results in mixed pixels in the imagery. The remotely sensed signal in mangrove areas is influenced by many environmental parameters such as the seasonal variations, diurnal intertidal interactions, as well as the local landscape and species mosaic in which mangroves are embedded. This complexity makes single traditional optical remote sensing land cover mapping and change analysis techniques insufficient to, by themselves, differentiate between mangrove species, to characterize changes in plant condition, and in some cases to distinguish mangroves from other vegetation covers. This difficulty has resulted in a continued heavy reliance on intensive fieldwork and visual interpretation of high spatial resolution data. It has also fostered the development of many hybrid delineation methods, as well as the exploration methods, of object oriented methods, hyperspectral data, and the use of active sensors data such as radar and lidar.

Given that methods are very dependent on and restricted by the primary data source characteristics, we organize the following summary according to the type of data.

22.6.1 Photography Data and Methods

In this category, we include black and white, color, color infrared, video and stereo imagery captured via aircraft (e.g. Everitt et al. 1991; Binh et al. 2005; Verheyden et al. 2002). This category includes pre-satellite era aerial photography such as the archives used to produce topographic maps from the 1950s and onward, as well as more modern color and infrared photos and video. These data have been the backbone of mangrove mapping given its temporal scope and its very high sub-meter spatial resolution. This level of detail allows for visual interpretation of attributes, such as tone, texture and permits the manual onscreen delineation mangrove areas. It is particularly useful for local mapping given the tendency of mangroves to exist in narrow coastal strips. Aerial photographs are also used as a primary source of accuracy assessment for maps produced with other, typically coarser, resolution satellite data (Heumann 2011).

Common map-making methods using aerial photos include visual interpretation and manual delineation of areas of interest using interpretation keys generated from fieldwork, as well and simple unsupervised classification techniques such as ISOData, ISOCluster, and ISOseg (Chauvaud et al. 1998). Some automated techniques such as unsupervised classifications are also applied to aerial photographs, which require the manual relabeling of automatically generated spectral classes into information classes such as mangrove, other vegetation, and water.

Limitations of this category of data and methods are the lack of availability of homogeneous imagery, the various types of distortion due to flight and atmospheric conditions, and the high costs of data. Such problems are particularly problematic in areas of the world where cloud cover is prevalent (Giri et al. 2011). Additionally, these aerial photographs have a poor spectral resolution, therefore, rendering such automated methods as unsupervised classification non-optimal. While infrared photography is not uncommon, the standard continues to be the visible range within the electromagnetic spectrum (Kuenzer et al. 2011).

22.6.2 High and Medium Resolution Satellite-Based Optical Imagery (M&HR)

This category includes satellite-based sensors designed for earth observation with medium resolution (10–100 m) products, such as those produced by the SPOT and the Landsat programs, but also include more recent ones like Indian Remote Sensing System and ASTER. These data constitute the largest source of what is considered current traditional satellite-based multispectral remote sensing at various scales (Giri 2016). Such data often have an archive that spans up to three decades of relatively standardized products, and a temporal resolution of at least twice a month, which allows for the consistent monitoring of abrupt and long-term mangrove changes. These data are often available for free, at low cost to researchers and mangrove managers in the developing world. However, the most important feature of these products over aerial photography is their multispectral capability. This feature enables richer spectral differentiation between land targets according to their absorption and reflectance properties across the visible, near, to mid-infrared and thermal segments of the electromagnetic spectrum (Heumann 2011). Spatio-temporal metrics such as fragmentation indices (e.g. Seto and Fragkias 2007) and textural measures are also possible in these data, especially when a panchromatic band is added to the spectral bands in the analysis such as when pan-sharpening.

Medium resolution multi-spectral data such as the one described are commonly used for regional mapping of mangrove using traditional pixel-based classification methods such as unsupervised classification and supervised classification. Image classifications based solely on spectral differences can map mangroves up to some degree of detail: mangrove vs. non-mangrove vegetation classes, mangrove density differences, condition status, and sometimes even community dominating species. However, these methods usually need ancillary data such as DEM(s) or habitat models to produce higher accuracies. Sometimes, they can even differentiate species associations. However rarely can they distinguish between species. The increased spatial resolution has not demonstrated to be useful as more redundant pixels can overwhelm the spectral differentiation (Gao 1999).

Spectral indices and spectral band ratios, such as the normalized vegetation index (NDVI) or the Brightness soil index (BI) are also commonly used multispectral methods used in mangrove mapping workflows (e.g. Ramsey and Jensen 1996; Green et al. 1997). These image products result from mathematically manipulating spectral bands to exploit absorption and reflectance differences across bands and highlight the abundance or scarcity of certain surface type or property (e.g., red and infrared bands for vegetation vigor). These ratios and indices are typically used in combination of classification techniques as additional input, or instead of individual 'raw' spectral bands. NDVI and other vegetation indices (e.g., EVI) are also important because of their strong correlation with field measurements of biomass and leaf area index (LAI), which can help distinguish mangrove from other species or among certain mangrove species associations (e.g. Green et al. 1997; Krause et al. 2004). Other types of image spectral transformations such as those resulting from Principal Component Analysis (PCA) or Tesselcap type transformations are also commonly used as input of classification (Green et al. 1998).

Despite, and in addition to, the automatic methods described above, visual interpretation, and field data still play a major role in mapping complex fine-grained systems like mangroves. Simple false color image composites, pan-sharpening, and other visual color transformations are applied to image scenes to enhance differences visually and then manually delineate mangrove classes on the image. Visual and field analysis are also still key to perform validation and accuracy assessment (Heumann 2011).

Two relatively recent methodological developments have enhanced the mapping capability of medium and high-resolution data: are the inclusion of non-parametric and machine learning algorithms, including neural networks and decision tree classifiers, and the use of object-oriented classification methods (OBIA). Neural Networks and OBIA methods are classification techniques that often outperformed traditional classifiers such as Maximum Likelihood in map accuracy performance. They have the advantage of not relying on traditional parametric statistical assumptions such as normality of the data distribution with is often violated in remotely sensed datasets. However, these methods required considerable quantities of calibration and validation data as well as computing power that is often not available (Rogan et al. 2008). OBIA groups pixels based on neighborhood properties as opposed to spectral properties which work best with very high resolution data (see next section), it has also been used in combination with traditional spectral classification techniques using M&HR data (e.g. Wang et al. 2004).

Finally it is important to mention that, global mangrove loss numbers, such as those utilized in the global component of this chapter, have been derived solely from the analysis of medium-resolution data using mostly hybrid methods that rely on unsupervised, some supervised classification methods, and the use of higher resolution imagery and aerial photograph visual interpretation and accuracy assessment.

22.6.3 Very High-Resolution Satellite-Based Optical Imagery (VHR)

These type of satellite-based sensor data distinguish themselves for having spatial resolutions of under 10 m, and comparatively low spectral resolution which is typically four or five spectral bands, including near infrared and 'red edge' channels. Until recently, these VHR data products were only available on demand, which meant they were costly and that regular revisit frequency of observations was not guaranteed. Sensors included in this data type include products such as IKONOS, Quickbird, Rapid Eye, GeoEye, and the Worldview sensor series (Kuenzer et al. 2011).

Multispectral classification methods such as the ones described for M&HR data are also commonly used with VHR imagery. However, given the limited spectral capabilities of VHR data, methods commonly utilized for these datasets rely more on spatial resolution characteristics to achieve better mapping outcomes. Much like aerial photographs, considerable work is visual and semi-manual. One way in which VHR's spatial capabilities is utilized is by generating, texture and context-based feature images derived from second order statistics such as contrast, correlation, and entropy. Texture and context based images are then used as input to aid classification algorithms to identify mangrove classes. A more recent development that exploits VHR spatial potential is the use of OBIA methods in VHR data (e.g. Wang et al. 2004).

Spectral unmixing analysis (SMA) is a subpixel decomposition technique is often used to solve for surface type 'endmembers' such as water, greenness and soil/brightness within a mixed pixel. Which as mentioned above are prevalent and unavoidable in mangrove ecosystem landscape matrices (e.g. Kanniah et al. 2007). SMA creates single endmember decompositions; such as greenness, brightness, and wetness. These decompositions are then used as input in classifications or are thresholded to then masked out vegetation from non-vegetation areas, which will then be classified.

22.6.4 Hyperspectral Sensors (HS)

This type of data is characterized by its detail and sensitivity of the spectral information contained within it. Whereas multispectral covers roughly the same range of the electromagnetic spectrum grouped into seven bands as *M&HR* data such as Landsat and SPOT. HS data slices the same spectrum in more than 300 bands. Examples of sensors that produce HS data are Compact Airborne Spectrographic Imager (CASI) data, Airborne Imaging Spectrometer for Application (AISA+), Hymap, the Airborne Visible/Infrared Imaging Spectrometer (AVRIS) which are all airborne, and EO-1 Hyperion which is space-borne (e.g. Demuro and Chisholm 2003; Green et al. 1998; Jusoff 2006). The HS approach to spectral data promises to be able to account for much more nuanced spectral differences than multispectral sensors like SPOT or Landsat, and therefore potentially it could be able to distinguish between more mangrove cover types. In the environmental remote sensing community, this promise has encouraged work seeking to map species, including mangroves species in different landscape contexts. However, at this point, HS data and methods have yet to deliver. Much of the attempts have been based on a theory developed in labs using radio spectrometer or spectral libraries, but have yet to translate well to real-world landscape observations, which, like in the case of mangroves are impacted by background and contextual noise at the pixel level. A related problem is one of noise reduction and redundancies. With hundreds of bands, many of they contain redundant information or simply noise that make the signal to noise ratio difficult to tease out. Methods such as Principal Component Analysis and others are required to take advantage of the data (Heumann 2011; Kuenzer et al. 2011).

22.6.5 Active Sensors

In contrast to passive optical sensors such as Landsat and SPOT, active sensors like SAR, do not rely on solar energy reflectance as their radiometric source, they emit and receive their radiation located at the longer wave (microwave) end of the electromagnetic spectrum. Microwaves can penetrate clouds, water, and certain surfaces. Data produced by microwaves is backscattered according to different frequencies (C, L and P bands) and polarization modes (HH, VV, HV). This backscatter provides data about the geometric properties of the surface such as elevation, roughness, and smoothness. Elevation data produced can give insights about the architecture and structure of a forested surface. Examples of SAR sensors and applications that have been used for mangrove research and mapping include RADARSAT-1, ENVISAT ASAR, JERS-1, ERS-1, SIR-C (e.g. Kovacs et al. 2008; Lucas et al. 2007; Proisy et al. 2000; Wang and Imhoff 1993).

For mangrove systems, the backscatter signal varies according to the geometric characteristics of the mangrove stand. Properties such as canopy closure and geometry as well as the structures of the leaf, stem, cell and stem can be described. The overall roughness of the underlying structure, including the components of soil, mudflats, and water, are also 'readable' with this type of data. SRTM has allowed for global topography mapping including three-dimensional molding of mangroves at a 90 m resolution (Heumann 2011). However, several conditions can impact the backscatter response, including the polarization of the radar beams and the incidence angle of the radar waves, which makes the interpretation of radar data over mangrove ecosystems very complex.

Overall, radar has been an efficient and fertile source for mangrove mapping and analysis that complements the optical sensor datasets as demonstrated by various examples (Kuenzer et al. 2011). However, the use and interpretation of SAR data remain complex and contained within highly specialized communities, which limits the availability of research we count on at present and its transferability to other scientific communities.

Lidar technology is also an active sensor in a similar fashion as SAR based sensors. However, lidar relies on the laser, a much more focused type of radiation than a microwave. Lidar measures the time return of a light pulse reflected off a target and this measure the relative distance. It can produce elevation at different levels (returns), including canopy and bare earth levels. Therefore, lidar datasets make it possible to analyze the forest structure in at various heights. It also allows for the extraction of biophysical parameters in many forest types, including mangroves. Additionally, it has also been used to measure to mangrove biophysical parameters, mangrove spatial distributions gap identification, and assessment of hurricane disturbance (Wannasiri et al. 2013). In this sense, lidar data has mostly been used in combination of other data and methods including M&HR, VHR, and HS to better and more accurately map mangrove spatial distributions and biophysical parameters.

Until recently, most lidar data has been generated by airborne sensors, but the Geoscience Laser Altimeter System (GLAS) on ICESat, is the first spaceborne lidar instrument for global observations and has been collecting data since 2003 (Galvincio and Popescu 2016).

22.6.6 Remote Sensing Summary

Specific methods for remote sensing mapping of mangrove are first, highly depended on the primary data source utilized, and its spectral, spatial, temporal and radiometric resolutions. Some methods apply only to certain data types. Secondly, methods have evolved from purely visual interpretation to increasingly automated. However, they have remained very interactive and researcher guided. A feature that makes comparability across methods and results difficult. Thirdly, methods are evolving as the technology evolves with increased spectral capacities, radiation types and observation frequency, which at times requires an ever-increasing level of computing capabilities. Finally, different methods get at various aspects of what makes mangroves, mangroves and therefore it is very likely that methods will remain hybrid in the broad sense of the word. This is especially true given that scientists have not agreed on a precise definition of mangrove of classification schemes.

22.7 Discussion

The results indicate some potential bias with the remotely sensed mangrove estimates. Of the ten nations analyzed in-depth, the MFW database provides the lowest mangrove forest cover in 60% of them and is next to lowest in another 30%. This is in addition to providing the lowest level at the global level of the three global remotely sensed databases. This is surprising as the MFW estimate is a composite of remotely sensed data from the late 1990s, and the global mangrove trend has been decreasing since the end of the 1990s, and many of the other estimates are from the mid-2000s. Conversely, the WMA estimate provides the highest mangrove area 67% of the time it is included in the estimates, in addition to presenting the highest figure globally. Again, this may indicate some bias in this measure. The CGMFC-21: MFW database is rarely at the high-end (1) or low-end (2) of the national range when adjusted for the presence and falls in the middle ground on the global estimate as well. CGMFC-21 has an issue in Australia that requires further examination, and MFW appears to omit Venezuela in it reporting within its Table 1.

Of the ten 'local' remotely sensed mangrove estimates, nine of them fall into the reasonable range presented by the global estimates for that country. The only extreme outlier is Venezuela which the report indicates may be measuring an ecological zone as opposed to an actual land cover class. Indeed, the average of averages at the 90% CI, when estimating the mean across each suite of mangrove estimates as the country is only 11.6%. This indicates that locally remotely sensed products are a potential high-value asset when examining the hazards, vulnerability, and management of mangrove forests.

Most of the local, and some of the global, remote sensing studies analyzed in this chapter lack any form of error or uncertainty analysis. Even when included they often only refer to the pixel location, such as within half a pixel, as opposed ascertaining error or uncertainty in the original mangrove classification. Indeed, not only do they lack error and uncertainty metrics they lack comparison against other pre-existing remotely sensed estimates of their country of interest. An exception to this pattern is the recently released Hamilton et al. (2017) estimate of mangrove carbon holdings in Ecuador.[17] A full 25% of the paper is dedicated to model comparisons, accuracy assessments, estimating error, and measuring uncertainty as well as producing a suite of error metrics to measure cumulative error throughout the methodological process. Such an approach could be used going forward to assist in the development of a standard suite of error measures applicable to the remote sensing of mangrove forests.

[17] Two of the authors of this paper are also authors of the cited paper.

It is not only the measurement of error, but the actual description of what is being measured and how the measure is being calculated that is lacking in these estimates. Three concerns are noted. Firstly, the map projection information from which the area calculation is made is almost always lacking. This projection issue may explain why when this author and Siikamäki et al. (2012) in their supplemental submission obtain different area estimates for countries such as Brazil and Venezuela from the Giri et al. (2011) data that Giri's team do themselves. Indeed, the differences can be drastic, Siikamäki et al. (2012) find that Brazil is the second largest mangrove holding nation globally now matching (Hamilton and Casey 2016). Authors need to explain their geospatial method of calculating the area of mangrove once the remote sensing is finalized. Secondly, many studies do not explain just how much of their minimum mapping unit must be covered in mangrove for it to be considered mangrove. For example, at the equator Landsat has a pixel that is approximately 30 m on the x-axis and the y-axis, so the minimum area mapped is 900 m^2. Does the whole 900 m^2 require mangrove for it to get the mangrove presence flag? Is it just any mangrove in this 900 m^2? Is it 50% mangrove within the pixel? This pixel cover issue is particularly problematic in mangroves that are known to exist in fringe situations with many edge pixels and monitored almost exclusively using a presence or absence mapping system. Finally, the definition of the country needs to be fully delineated. This issue of what islands constitute a country is covered in the USA example above, but this also applies to French tropical regions, the Dutch tropical regions, and even to mangroves that exist outside of traditional country boundaries.

Interestingly, one of the major future needs is not of improved remote sensing monitoring of mangrove forests but to utilize the GLC data to compile a similar product to MFW but for the 1990s and 1980s. This would allow those examining the hazards, vulnerability, and management of mangrove forests to establish global, yet high-resolution, long-term trends in mangrove cover that are more comparable to current remote sensing area measures.

References

Acosta-Velázquez J, Rodríguez-Zuñiga T, Díaz-Gallegos JR, Cerdeira-Estrada S, Troche-Souza C, Cruz I, Ressl R, Jiménez R (2009) Assessing a nationwide spatial distribution of mangrove forest for Mexico: an analysis with high resolution images. In: 33rd international symposium on remote sensing of environment. Stressa, pp 1–4

Acosta-Velázquez J, Rodríguez-Zúñiga M, Cerdeira-Estrada S, Cruz I, Ressl R, Ascensión M (2007) Los manglares de México: estado actual y establecimiento de un programa de monitoreo a largo plazo: 1a etapa, Informe del proyecto DQ056. CONABIO, Mexico City

Bhattarai B, Giri C (2011) Assessment of mangrove forests in the Pacific region using Landsat imagery. J Appl Remote Sens 5(1):053509-053509-053511

Binh TNKD, Vromant N, Hung N, Hens L, Boon EK (2005) Land cover changes between 1968 and 2003 in Cai Nuoc, Ca Mau Peninsula, Vietnam. Environ Dev Sustain 7(4):519–536. https://doi.org/10.1007/s10668-004-6001-z

Blankespoor B, Dasgupta S, Lange G-M (2016) Mangroves as a protection from storm surges in a changing climate. Ambio 1–14

Bontemps S, Defourny P, Bogaert EV, Arino O, Kalogirou V, Perez JR (2011a) GLOBCOVER 2009-products description and validation report

Bontemps S, Defourny P, Bogaert EV, Arino O, Kalogirou V, Perez JR (2011b) GLOBCOVER 2009: products description and validation report, vol 2.2. European Space Agency, Louvain

Calvo MC (2015) Area of forests susceptible to forest management in Costa Rica and estimation of its productive potential. Fondo Nacional de Financiamiento Forestal Forest Monitoring System for REDD+ Costa Rica, Costa Rica

Carrasquel G (2013) Venezuelan coastal mangroves are shrinking. Environmental, Ecologists and Conservation News from the Americas. 06/23/2013

Chang S, Green A, Kelley E (2015) A preliminary assessment of the blue carbon capacity of Belizean mangroves with ecological, economic, and policy perspectives. Duke University, Durham

Chauvaud S, Bouchon C, Maniere R (1998) Remote sensing techniques adapted to high resolution mapping of tropical coastal marine ecosystems (coral reefs, seagrass beds and mangrove). Int J Remote Sens 19(18):3625–3639

Chavez PS (1996) Image-based atmospheric corrections-revisited and improved. Photogramm Eng Remote Sens 62(9):1025–1035

Cherrington EA, Hernandez BE, Trejos NA, Smith OA, Anderson ER, Flores AI, Garcia BC (2010) Identification of threatened and resilient mangroves in the Belize barrier reef system, 1st edn. Water Center for the Humid Tropics of Latin America and the Caribbean (CATHALAC), Panama City

CLIRSEN (2007) Actualizacion Del Estudio Multitemporal de Manglares, Camaroneras Y Areas Salinas En Las Costa Continental Ecuatoriana Al Ano 2006, vol 1. Centro De Levantamientos Integrados De Recursos Naturales Por Sensores Remotos, Quito

Corcoran E, Ravilious C, Skuja M (2007) Mangroves of Western and Central Africa, vol 26. Biodiversity related projects in Africa, vol 26. UNEP/Earthprint, Cambridge

Dahdouh-Guebas F, Van Pottelbergh I, Kairo JG, Cannicci S, Koedam N (2004) Human-impacted mangroves in Gazi (Kenya): predicting future vegetation based on retrospective remote sensing, social surveys, and tree distribution. Mar Ecol Prog Ser 272:77–92

Demuro M, Chisholm L (2003) Assessment of Hyperion for characterizing mangrove communities. In: Proceedings of the 12th JPL AVIRIS airborne earth science workshop, Pasadena, CA, USA

Duarte E, Díaz OO, Maradiaga I, Casco F, Fuentes D, Jiménez A, Emanuelli P, Milla F (2014) Mapa Forestal y de Cobertura de la Tierra de Honduras: Análisis de Cifras Nacionales. Integrando Esfuerzos Para Un Buen Manejo De Los Bosques. Deutsche Gesellschaft für Internationale Zusammenarbeit (GIZ), Bonn. https://doi.org/10.13140/RG.2.1.1553.3046

Everitt J, Escobar D, Judd F (1991) Evaluation of airborne video imagery for distinguishing black mangrove (Avicennia germinans) on the lower Texas Gulf Coast. J Coast Res 7:1169–1173

Florida Water Management Districts (2012) Mangroves Florida. Fish and Wildlife Research Institute (FWRI), St. Petersburg

Gallo M (2005) Estado del Conocimiento de la Biodiversidad en El Salvador. Project Developing Capacities and Sharing Technology for the management of biodiversity in Central America. Ministerio de Medio Ambiente y Recursos Naturales, Costa Rica

Galvincio JD, Popescu SC (2016) Measuring individual tree height and crown diameter for mangrove trees with airborne lidar data. Int J Adv Eng Manag Sci (IJAEMS) 2(5):431–443

Gao J (1999) A comparative study on spatial and spectral resolutions of satellite data in mapping mangrove forests. Int J Remote Sens 20 (14):2823–2833

Gilbert T, Stys B (2004) Descriptions of vegetation and land cover types mapped using Landsat imagery. Florida Fish and Wildlife Conservation Commission, Tallahassee, p 16

Giri C (2016) Observation and monitoring of mangrove forests using remote sensing: opportunities and challenges. Multidisciplinary Digital Publishing Institute

Giri C, Ochieng E, Tieszen LL, Zhu Z, Singh A, Loveland T, Masek J, Duke N (2011) Status and distribution of mangrove forests of the world using earth observation satellite data. Glob Ecol Biogeogr 20 (1):154–159. https://doi.org/10.1111/j.1466-8238.2010.00584.x

Green EP, Mumby PJ, Edwards AJ, Clark CD, Ellis AC (1997) Estimating leaf area index of mangroves from satellite data. Aquat Bot 58(1):11–19

Green EP, Clark CD, Mumby PJ, Edwards AJ, Ellis A (1998) Remote sensing techniques for mangrove mapping. Int J Remote Sens 19 (5):935–956

Hamilton SE (2012) The impact of shrimp farming on mangrove ecosystems and local livelihoods along the Pacific Coast of Ecuador, vol 1. Department of Geography and Geology, Doctoral dissertation, vol 3477166, 1st edn. ProQuest, UMI Dissertation Publishing, Hattiesburg

Hamilton SE, Lovette J (2015) Ecuador's mangrove forest carbon stocks: a spatiotemporal analysis of living carbon holdings and their depletion since the advent of commercial aquaculture. PLoS One 10:e0118880

Hamilton SE (2016) Introducing CGMFC-21 (continuous global mangrove forest cover for the 21st century). GLOMIS/ISME 14 (3):11–14

Hamilton SE, Casey D (2016) Creation of a high spatio-temporal resolution global database of continuous mangrove forest cover for the 21st century (CGMFC-21). Glob Ecol Biogeogr 25(6):729–738. https://doi.org/10.1111/geb.12449

Hamilton SE, Lovette JP, Borbor-Cordova MJ, Millones M (2017) The carbon holdings of northern Ecuador's mangrove forests. Ann Am Assoc Geogr 107(1):54–71. https://doi.org/10.1080/24694452.2016.1226160

Hansen M, DeFries R, Townshend J, Carroll M, Dimiceli C, Sohlberg R (2003) Global percent tree cover at a spatial resolution of 500 meters: first results of the MODIS vegetation continuous fields algorithm. Earth Interact 7(10):1–15

Hansen MC, Stehman SV, Potapov PV (2010) Quantification of global gross forest cover loss. Proc Natl Acad Sci 107(19):8650–8655. https://doi.org/10.1073/pnas.0912668107

Hansen MC, Potapov PV, Moore R, Hancher M, Turubanova SA, Tyukavina A, Thau D, Stehman SV, Goetz SJ, Loveland TR, Kommareddy A, Egorov A, Chini L, Justice CO, Townshend JRG (2013) High-resolution global maps of 21st-century forest cover change. Science 342(6160):850–853. https://doi.org/10.1126/science.1244693

Heumann BW (2011) Satellite remote sensing of mangrove forests: recent advances and future opportunities. Prog Phys Geogr 35 (1):87–108

Hogarth PJ (2015) The biology of mangroves and seagrasses. Oxford University Press

Homer CG, Dewitz JA, Yang L, Jin S, Danielson P, Xian G, Coulston J, Herold ND, Wickham J, Megown K (2015) Completion of the 2011 National Land Cover Database for the conterminous United States-Representing a decade of land cover change information. Photogramm Eng Remote Sens 81(5):345–354

Huber O, Oliveira-Miranda MA (2010) Ambientes terrestres de Venezuela. In: Rodríguez JP, Rojas-Suarez F, Giraldo-Hernández D (eds) Libro rojo de los ecosistemas terrestres de Venezuela, (Provita, Shell Venezuela y Lenovo Venezuela, Caracas). Provita, Shell Venezuela, Lenovo, Caracas, pp 29–89

Hutchison J, Spalding M, zu Ermgassen P (2014) The role of mangroves in fisheries enhancement. The Nature Conservancy and Wetlands International, Cambridge

Instituto de Hidrologia MyEA, (IDEAM), Instituto Geográfico Agustín Codazzi (IGAC), Instituto de Investigación de Recursos Biológicos Alexander von Humboldt (IAvH), Instituto de Investigaciones Marinas y Costeras José Benito Vives De Andréis (Invemar), Instituto Amazónico de Investigaciones Científicas Sinchi (I. Sinchi), (IIAP) IdIAdPJvN (2007) Ecosistemas Continentales, Costeros y Marinos de Colombia. Imprenta Nacional de Colombia, Bogotá

INVEMAR (2016) Informe del Estado de los Ambientes y Recursos Marinos y Costeros de Colombia Año 2015, Serie de Publicaciones Periódicas, vol 3, 2nd edn. INVEMAR, Santa Marta

Juman R, Ramsewak D (2013) Status of mangrove forests in Trinidad and Tobago, West Indies. Caribb J Sci 47(2–3):291–304

Jusoff K (2006) Individual mangrove species identification and mapping in Port Klang using airborne hyperspectral imaging. J Sustain Sci Manag 1(2):27–36

Kairo J, Kivyatu B, Koedam N (2002) Application of remote sensing and GIS in the management of mangrove forests within and adjacent to Kiunga Marine Protected Area, Lamu, Kenya. Environ Dev Sustain 4(2):153–166

Kanniah KD, Wai NS, Shin A, Rasib AW (2007) Per-pixel and sub-pixel classifications of high-resolution satellite data for mangrove species mapping. Appl GIS 3(8):1–22

Kautz R, Stys B, Kawula R (2007) Florida vegetation 2003 and land use change between 1985–89 and 2003. Fla Sci 70(1):12

Kovacs JM, Vandenberg CV, Wang J, Flores-Verdugo F (2008) The use of multipolarized spaceborne SAR backscatter for monitoring the health of a degraded mangrove forest. J Coast Res 24:248–254

Krause G, Bock M, Weiers S, Braun G (2004) Mapping land-cover and mangrove structures with remote sensing techniques: a contribution to a synoptic GIS in support of coastal management in North Brazil. Environ Manag 34(3):429–440

Kuenzer C, Bluemel A, Gebhardt S, Quoc TV, Dech S (2011) Remote sensing of mangrove ecosystems: a review. Remote Sens 3(5):878–928

Lazo R (2017) GIS and remote sensing: diagnosis and evaluation of mangrove areas in Venezuela. Rodrigo Lazo. http://www.rodrigolazo.com/projects/diagnosis-and-evaluation-of-mangrove-areas-in-venezuela/. Accessed 31 May 2017

LINZ (2017) LINZ data service. Land Information New Zealand. https://data.linz.govt.nz/. Accessed 31 May 2017

Lucas RM, Mitchell AL, Rosenqvist A, Proisy C, Melius A, Ticehurst C (2007) The potential of L-band SAR for quantifying mangrove characteristics and change: case studies from the tropics. Aquat Conserv Mar Freshwat Ecosyst 17(3):245–264

MacLean MG, Congalton RG (2012) Map accuracy assessment issues when using an object-oriented approach. In: Proceedings of the American Society for Photogrammetry and Remote Sensing 2012 annual conference, pp 1–5

Magris RA, Barreto R (2010) Mapping and assessment of protection of mangrove habitats in Brazil. Pan Am J Aquat Sci 5(4):546–556

Martin NM, Maldonado BD, Salinas-de-León P (2016) Inexpensive method to assess Galapagos' mangrove forests through the use of Google Earth and Open Source GIS. Paper presented at the IUCN World Conservation Congress, Hawaii, USA, 09/03/2016

Menéndez L (2013) El ecosistema de manglar en el archipiélago cubano: bases para su gestión. Universitat d'Alacant, Alicante

Menéndez L, Guzmán JM (2010) Los bosques de mangles del archipiélago cubano, caracterización, distribución y relación con el Cambio Climático. In: Hernández-Zanuy AC, Alcolado PM (eds) La biodiversidad en Ecosistemas Marinos y Costeros del Litoral de Iberoamérica y el Cambio Climático: I, vol 1. RED CYTED BIODIVMAR, Havanna, pp 90–107

Meyer-Arendt KJ, Byrd S, Hamilton SE (2013) Mangrove deforestation in the Dominican Republic, 1969 to 2012. ISME/GLOMIS Electron J 11(1):1–4

Meyrat A (2009) Ubicación, Estado y Valor de las Especies, Formaciones Vegetales y Ecosistemas de Nicaragua: Estudio de Ecosistemas y Formaciones Vegetales de Nicaragua, 1st edn. Ministerio del Ambiente y los Recursos Naturales (MARENA), Managua

Ministerio de Ambiente y Desarrollo Sostenible de la República de Colombia (2017) Manglares: Los Manglares de Colombia. Ministerio de Ambiente y Desarrollo Sostenible. http://www.minambiente.gov.co/index.php/bosques-biodiversidad-y-servicios-ecosistematicos/ecosistemas-estrategicos/manglares. Accessed 1 May 2017

Montreal Process Implementation Group for Australia, National Forest Inventory Steering Committee (2013) Australia's state of the forests report: 2013. ABARES, Canberra

Mutendeudzi M, Read S, Howell C, Davey S, Clancy T (2013) Improving Australia's forest area estimate using a 'Multiple Lines of Evidence' approach. Australian Bureau of Agricultural and Resource Economics and Sciences, Canberra

Nascimento WR, Souza-Filho PWM, Proisy C, Lucas RM, Rosenqvist A (2013) Mapping changes in the largest continuous Amazonian mangrove belt using object-based classification of multisensor satellite imagery. Estuar Coast Shelf Sci 117:83–93

Oliveira-Miranda MA, Huber O, Rodríguez JP, Rojas-Suarez F, De Oliveira-Miranda R, Hernández-Montilla M, Zambrano-Martínez S, Giraldo-Hernández D (2010) Riesgo de eliminación de los ecosistemas terrestres de Venezuela. In: Rodríguez JP, Rojas-Suarez-F, Giraldo-Hernández D (eds) Libro rojo de los ecosistemas terrestres de Venezuela, (Provita, Shell Venezuela y Lenovo Venezuela, Caracas). Provita, Shell Venezuela, Lenovo, Caracas, pp 107–235

Olson DM, Dinerstein E, Wikramanayake ED, Burgess ND, Powell GV, Underwood EC, D'amico JA, Itoua I, Strand HE, Morrison JC (2001) Terrestrial ecoregions of the world: a new map of life on earth a new global map of terrestrial ecoregions provides an innovative tool for conserving biodiversity. Bioscience 51(11):933–938

ONU-REDD (2015) La superficie boscosa y la tasa de deforestación en Panamá: Insumos para establecer datos oficiales a ser utilizados en las estadísticas nacionales, y para informar a convenciones y procesos internacionales, vol 1. Programa conjunto de las Naciones Unidas para la reducción de emisiones provenientes de deforestación y de degradación de los bosques en Panamá, Panama City

Paul TT, Dennis A, George G (2016) A review of remote sensing techniques for the visualization of mangroves, reefs, fishing grounds, and molluscan settling areas in tropical waters. In: Seafloor mapping along continental shelves. Springer, pp 105–123

Pendleton L, Donato DC, Murray BC, Crooks S, Jenkins WA, Sifleet S, Craft C, Fourqurean JW, Kauffman JB, Marbà N, Megonigal P, Pidgeon E, Herr D, Gordon D, Baldera A (2012) Estimating global "Blue Carbon" emissions from conversion and degradation of vegetated coastal ecosystems. PLoS One 7(9):e43542. https://doi.org/10.1371/journal.pone.0043542

Pettorelli N, Laurance WF, O'Brien TG, Wegmann M, Nagendra H, Turner W (2014) Satellite remote sensing for applied ecologists: opportunities and challenges. J Appl Ecol 51(4):839–848

Por el Ministerio de Ambiente y Recursos Naturales (2013) Estudio de la cobertura de mangle en la República de Guatemala. MARN, Guatemala

Proisy C, Mougin E, Fromard F, Karam M (2000) Interpretation of polarimetric radar signatures of mangrove forests. Remote Sens Environ 71(1):56–66

Ramsey EW, Jensen JR (1996) Remote sensing of mangrove wetlands relating canopy spectra to site-specific data. Photogramm Eng Remote Sens 62(8):939–948

Regional Centre for Mapping of Resources for Development (2015) Mitigating impacts of coastal hazards in mangrove. Ardhi University, Tanzania

Rodríguez JP, Rojas-Suarez F, Giraldo-Hernández D (2010) Libro rojo de los ecosistemas terrestres de Venezuela. Provita, Shell Venezuela, Lenovo, Caracas

Rogan J, Franklin J, Stow D, Miller J, Woodcock C, Roberts D (2008) Mapping land-cover modifications over large areas: a comparison of machine learning algorithms. Remote Sens Environ 112(5):2272–2283

Rosati I, Prosperi P, Latham J, Kainuma M (2008) World atlas of mangroves. In: Sessa R (ed) Terrestrial observations of our planet, GTOS 50, vol 1. Food and Agricultural Organization of the United Nations, Rome, pp 30–31

Ruefenacht B, Benton R, Johnson V, Biswas T, Baker C, Finco M, Megown K, Coulston J, Winterberger K, Riley M (2015) Forest service contributions to The National Land Cover Database (NLCD). In: Stanton SM, Christensen GA (eds) Pushing boundaries: new directions in inventory techniques & applications. Forest Inventory and Analysis (FIA) Symposium 2015, Portland, OR, 12/8/2015. U. S. Department of Agriculture, pp 241–243

Saintilan N, Wilson NC, Rogers K, Rajkaran A, Krauss KW (2014) Mangrove expansion and salt marsh decline at mangrove poleward limits. Glob Chang Biol 20(1):147–157

Seto KC, Fragkias M (2007) Mangrove conversion and aquaculture development in Vietnam: a remote sensing-based approach for evaluating the Ramsar Convention on Wetlands. Glob Environ Chang 17(3):486–500

Shearman PL, Ash J, Mackey B, Bryan JE, Lokes B (2009) Forest conversion and degradation in Papua New Guinea 1972–2002. Biotropica 41(3):379–390

Siikamäki J, Sanchirico JN, Jardine SL (2012) Global economic potential for reducing carbon dioxide emissions from mangrove loss. Proc Natl Acad Sci (Early Ed). https://doi.org/10.1073/pnas.1200519109

Spalding M, Kainuma M, Collins L (2010) World atlas of mangroves. Earthscan, London

Valderrama-Landeros LH, Rodríguez-Zúñiga MT, Troche-Souza C, Velázquez-Salazar S, Villeda-Chávez E, Alcántara-Maya JA, Vázquez-Balderas B, Cruz-López MI, Ressl R (2017) Manglares de México: actualización y exploración de los datos del sistema de monitoreo 1970/1980–2015. Comisión Nacional para el Conocimiento y Uso de la Biodiversidad, Mexico City

Verheyden A, Dahdouh-Guebas F, Thomaes K, De Genst W, Hettiarachchi S, Koedam N (2002) High-resolution vegetation data for mangrove research as obtained from aerial photography. Environ Dev Sustain 4(2):113–133

Virly S (2008) Typologies et biodiversité des mangroves de Nouvelle-Calédonie, Cartographie des mangroves. In: Paper presented at the IFRECOR – French Initiative for Coral Reefs, Nouméa, Nouvelle Calédonie

Wang Y, Imhoff M (1993) Simulated and observed L-HH radar backscatter from tropical mangrove forests. Int J Remote Sens 14 (15):2819–2828

Wang Y, Bonynge G, Nugranad J, Traber M, Ngusaru A, Tobey J, Hale L, Bowen R, Makota V (2003) Remote sensing of mangrove change along the Tanzania coast. Mar Geod 26(1–2):35–48

Wang L, Sousa W, Gong P (2004) Integration of object-based and pixel-based classification for mapping mangroves with IKONOS imagery. Int J Remote Sens 25(24):5655–5668

Wannasiri W, Nagai M, Honda K, Santitamnont P, Miphokasap P (2013) Extraction of mangrove biophysical parameters using airborne LiDAR. Remote Sens 5(4):1787–1808

Zhu Z, Waller E (2003) Global forest cover mapping for the United Nations Food and Agriculture Organization forest resources assessment 2000 program. For Sci 49(3):369–380

Urban Mangrove Biology and Ecology: Emergent Patterns and Management Implications 23

Benjamin Branoff

Abstract

Tropical coastlines will see some of the greatest urbanization rates of the twenty first century, presenting a particular threat to the management of mangrove ecosystems. This is compounded by a significant void in our understanding of how urban mangroves function. Although mangrove responses to urban environments around the world have been recorded for roughly 35 years, there remains no model describing these systems. Collectively, these studies suggest urban mangroves are characterized by patchy and mixed vegetation forests within expanding suburban fringes, but more stable and expanding forests in older city centers. Where municipal sewage is disposed, a significant body of evidence suggests enrichment of anthropogenically derived nitrogen in mangrove plant tissues. Further, benthic faunal communities in these systems are influenced by sewage effluent. Much of this is likely due to changes in nutrient inputs in these systems; however, it is also due to toxicity from heavy metals and hydrocarbons. These contaminants may also be influencing mangrove physiology and growth, although this has yet to be shown in situ. Urban faunal communities of fin-fish and birds also respond to differences in habitat and resource availability of urban mangroves. This chapter reviews the literature on urban mangroves, revealing emergent patterns that are at times confirmed by empirical evidence. These patterns are then synthesized in an urban mangrove model. Through continuing studies and the development of this model, future mangrove management along urbanizing coasts will be much more effective at optimizing ecosystem services and creating more sustainable social-ecological mangrove systems.

B. Branoff (✉)
Department of Biology, University of Puerto Rico-Río Piedras, San Juan, Puerto Rico
e-mail: benjamin.branoff@upr.edu

Keywords

Urban mangroves · Anthropocene · Social-ecological systems · Novel mangroves

23.1 Introduction

Urbanization has been identified as an increasingly ubiquitous driver of terrestrial ecology and a challenge to global sustainability (McKinney 2006; Grimm et al. 2008; McDonald 2008). This threat has sparked a growing commitment to the study of urban ecology, resulting in an expanding literature on the biology and ecology of urban ecosystems (Young and Wolf 2006). But a majority of this new research is taking place in terrestrial systems, with relatively little known of urban coastal ecology. Yet, tropical islands and coastal areas will experience some of the greatest urbanization rates over the next century, including mangrove biodiversity hotspots of the Caribbean islands and along the coasts of West and East Africa (Seto et al. 2012). With only 6.9% of global mangrove coverage protected in 2000 (Giri et al. 2011), urban expansion thus presents a particular complication in the management of mangrove ecosystems and their services. These challenges may include loss of habitat and sequestered carbon, although more nuanced changes are likely to develop, as has been observed in terrestrial systems.

Vegetated spaces in urban areas are typically characterized by lower stem densities, greater patchiness, more diversity, and more novelty, and individual plants or species are often larger, with unique adaptations to pollution, such as tolerances to heavy metals and changes in stomatal densities (Pickett et al. 2008). Animal assemblages also respond to urban landscapes, usually by being representative of more generalist species, which form less diverse but often more abundant communities composed of more non-native species. The presence of these synanthropic species and the ecosystems they form is largely due to the built environment

C. Makowski, C. W. Finkl (eds.), *Threats to Mangrove Forests*, Coastal Research Library 25,
https://doi.org/10.1007/978-3-319-73016-5_23

and an abundance of certain resources and pollutants associated with dense human populations. As a result, many of the services provided by urban ecosystems are associated with mitigating the detrimental effects of the urban landscape, and include: air filtering, micro-climate regulation, noise reduction, rainwater drainage, and sewage treatment (Bolund and Hunhammar 1999; Gómez-Baggethun et al. 2013). With this understanding, urban landscapes could be designed and managed for a better integration between the natural and developed landscapes and between the ecological and social systems.

As coastal forests, mangroves might be expected to respond similarly to urbanization as do terrestrial forests or highly vegetated spaces. Yet mangroves are unique from their terrestrial counterparts in many ways. The requirement of tidal influence and the accompanying salinity is perhaps the most notable, restricting mangrove floral diversity to a fraction of surrounding terrestrial systems, but making them critical habitat for both marine and terrestrial fauna. Their high productivity combined with anoxic conditions of the mangrove benthos result in substantially more carbon storage than the world's other major forest domains (Donato et al. 2011; Alongi 2012). This service of carbon sequestration, as well as the other services of mangroves, help value coastal wetlands above all other ecosystems, except coral reefs (de Groot et al. 2012). Thus, urban mangroves are especially unique in their potential to provide extremely valuable services to a greater number of people. Therefore, just as in terrestrial systems, understanding the ecology and function of urban mangroves will aid in managing them towards a greater synergy between their ecological and social components.

This chapter presents the current literature on urban mangroves, and the challenges these systems face under expanding urbanization. Almost all of the studies are from isolated systems, although one has examined the global influence of urban land-use on mangrove ecology (Branoff 2017). These studies represent 74 cities from 29 countries, and along coastlines of all mangrove occupied oceans (Fig. 23.1). They represent mangroves of various social and environmental backgrounds, although distinct patterns associated with the urban landscape are emergent. These patterns fall within four general themes of forest structure and composition, nutrient dynamics, sediment contamination, and community assemblages. Together, they identify several potential threats and vulnerabilities, and are finally synthesized to develop a model of urban mangroves. This initial model will aid in urban mangrove management and in the development of more effective research on urban mangroves, which will in turn serve to further refine our understanding and management capacity of these systems.

23.1.1 Deforestation and Fragmentation

Globally, mangroves experienced a loss of 0.16% annually from 2000 to 2012, with the greatest losses occurring in Indonesia (750 km^2), Malaysia (240 km^2), and Myanmar (240 km^2) (Hamilton and Casey 2014). During the same

Fig. 23.1 Locations of most of the urban mangrove studies form around the world. Studies are from nearly all parts of the world where mangroves are found and cover four basic themes of forest structure and composition, nutrient dynamics, sediment contamination, and community assemblages. Much of the contamination and toxicity studies have taken place throughout Brazil, and all of the molluscan studies are from the eastern hemisphere

period, no gains in mangrove coverage were recorded, although several countries experienced no change. The mangrove losses recorded through remote sensing in less economically developed nations of Asia, Africa, and Latin America reflect individual reports from the same regions as described below. Indeed, a large portion of twenty first century urbanization is occurring along coastlines in developing countries, where government regulation is weak and ecological vulnerability is high (United Nations Environment Program 2006; McGranahan et al. 2007; Baird 2009). Yet, cities in the more economically developed countries experienced greater losses than did their respective countries, suggesting cities are particularly vulnerable to mangrove degradation and deforestation, regardless of economic development status (Branoff 2017).These differences in mangrove responses to urbanization can be attributed to the various characteristics of the urban environment, in which the cultural, physical, chemical, and biological components have influence over the quantity and quality of mangrove forests.

Reports of urban mangroves in less economically developed nations are associated with decreasing forest coverage and stand densities, as well as increasing canopy gaps and fragmentation. In Douala, Cameroon, 53% of mangroves were lost from 1974 to 2009, concurrent with increases in roads and settlements as well as in open water and replacement vegetation (Nfotabong-Atheull et al. 2013). In nearby Lagos, Nigeria, a 77% loss of mangroves between 1984 and 2006 was accompanied by a 477% increase in urban area, as well as increases in erosion, flooding, pollution, and temperature (Obiefuna et al. 2012). In some cases, net urban mangrove losses are alleviated by small gains in mangrove area, resulting in spatial heterogeneities in mangrove deforestation and reforestation. This was the case in Panama City, Panama, where coastal development removed 100% of mangroves in some areas but in others, expansion rates of 61% were observed and attributed to the resiliency of *Laguncularia racemosa*, which dominated post disturbance urban shoreline (Benfield et al. 2005). In Mumbai, India, 36% of mangrove area has been lost since 1974, much of this occurring since 2000 (Vaz 2014). However, this loss was slightly mitigated by small increases in mangrove area, particularly in denser and older urban zones, whereas loss occurred in newer and less dense urban zones. This pattern is attributed to a replacement of rural communities dependent upon mangrove wood for fuel with more urban communities utilizing alternative fuel sources (Nagendra et al. 2013). This is a common occurrence in other urban mangrove systems, and is supported by observations in Indonesia and Tanzania, where mangrove loss was inversely correlated to distance from settlement edges (Msangameno et al. 2017; Yusuf et al. 2017).

In Mombasa, Kenya, mangroves exhibit symptoms of unsustainable harvesting, such as enlarged canopy gaps and longer closure times, as well as lower stand density and basal area when compared to average Kenyan mangrove stands (Mohamed et al. 2009). Just as in Mumbai, these symptoms tend to occur in more rural parts of the system, where inhabitants are more dependent on mangrove wood for fuel. Similar results were reported for existing mangroves of Douala, Cameroon, resulting in what the authors call more degraded peri-urban mangroves at settlement edges (Nfotabong-Atheull et al. 2013). In this case, harvesting was not uniform across species and tended to select larger individuals of *Rhizophora racemosa* more than *Avicennia germinans*. This is important when considering that the long-term regeneration of urban mangroves in Mombasa, Kenya was found to be species-specific and dependent upon both the original disturbance pattern and conspecific interactions (Olagoke et al. 2013). The repercussions of selective harvesting can therefore continue to influence forest structure long after harvest.

Despite these losses in the developing world, there is evidence that recently enacted regulations have helped mangrove recovery in India, Vietnam, and the Arabian Gulf (DasGupta and Shaw 2013; Burt 2014; Pham and Yoshino 2016), a scenario that has already been played out in more developed countries. In the city-state of Singapore, mangrove forests have increased 33% from 1993 to 2011 (Lai et al. 2015). The increase in coverage is attributed to tight restrictions, but the fragmented and restored forests there contain less biomass and carbon than older, more continuous stands (Friess et al. 2015). Mangroves in Hong Kong, and Shenzhen, China, have also shown increasing coverage, which may or may not be due to the implementation of coastal engineering projects (Tian-Hong et al. 2008). Urban mangroves of Puerto Rico show mixed results. Although mangrove coverage increased in response to strict regulations enacted in the twentieth century, a general negative correlation was reported between the distance to an urban center on the island and the probability of decreasing mangrove coverage there (Martinuzzi et al. 2009). One exception to this was in the city of San Juan, the island's largest metropolitan area and the location of the largest continuous mangrove stand on the island. From 2001 to 2010, mangrove forests of San Juan increased 36% in basal area and 48% in density, with similar growth rates to those measured in more pristine mangroves in 1938 (Lima et al. 2013).

Although low economic development is associated with poor environmental regulation and high vulnerability to habitat loss, empirical evidence from a meta-analysis has suggested no strong relationship between urban mangrove forest loss and a country's economic development or environmental performance score (Branoff 2017). Further, although patterns of urban mangrove patch characteristics are described above to be spatially dependent on distance to urban centers, the same meta-analysis found only weak correlations. Still, mangrove patch dynamics in some individual cities do seem to change with distance to the city (Fig. 23.2), but the trends both support and contradict the above reported observations. Mean mangrove patch area in the cities of Shenzhen, China and Fortaleza, Brazil, for example, are highest closer to the city centers, not farther away as would be expected from the observations in other cities. Instead, mangrove deforestation and fragmentation are likely dependent more on the dynamic interplay between the cultural and political characteristics of each individual city, which are themselves highly dependent on spatial scale, making urban habitat loss particularly difficult to model (Syphard et al. 2011; Gong et al. 2013).

Further structural studies of urban mangroves in the developed world are limited to Australia and are equally mixed in their results. In Sydney, more disturbed urban mangroves were recorded as having a higher litterfall biomass and pneumatophore density (Skilleter 1996). While another study in Sydney also showed higher pneumatophore density in structurally constrained mangroves, it reported less leaf litter and fewer saplings (Heatherington and Bishop 2012). In Brisbane, Australia, 65–80% fewer pneumatophores were counted in mangroves disturbed by path and boardwalk construction, and leaf litter biomass was very similar to control sites (Skilleter and Warren 2000). The authors attribute these differences to different types of urban disturbance. Where soil compaction associated with urban development should increase pneumatophore density, the construction of boardwalks may not necessarily compact soil and thus influence pneumatophore densities. Similarly, differences in litterfall are likely due to differences in nutrient availability

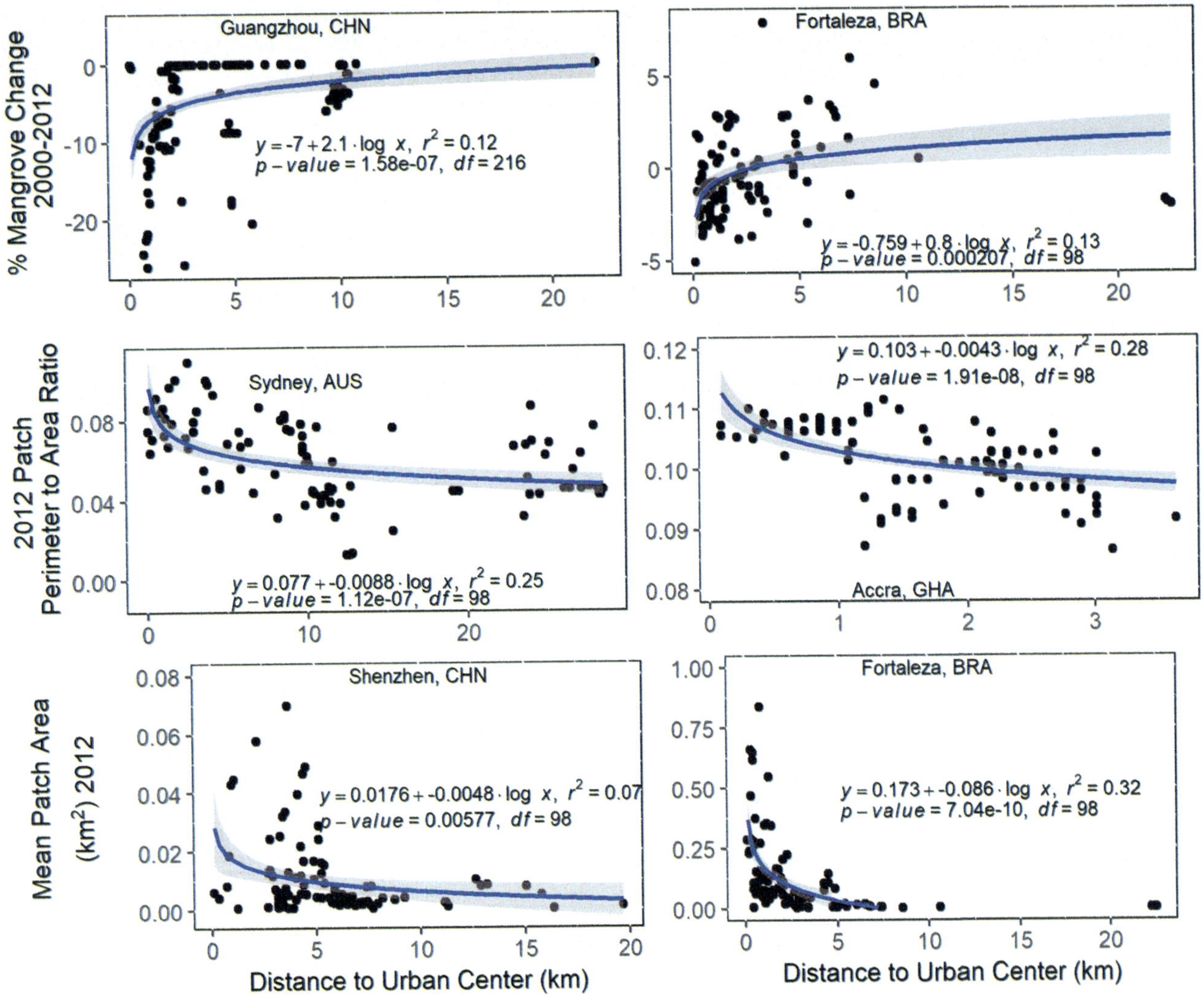

Fig. 23.2 Change in mangrove coverage for Guangzhou, China and Fortaleza, Brazil from 2000 to 2012 (top), mangrove patch perimeter to area ratio for Sydney, Australia and Accra, Ghana (middle) and mean patch size for Shenzhen, China and Fortaleza, Brazil (bottom) as a function of distance to the city center. Notice the differences in the range in y values between cities. Trends both support and contradict reports from the literature, highlighting the variability in mangrove forest responses to urbanization. Yet, both variables have been found to be important drivers of faunal diversity (Mazerolle and Villard 1999), emphasizing the potential influence of urban mangrove fragmentation on ecology (Data from Branoff 2017)

between various study sites, which has primarily been studied in the context of urban mangroves receiving sewage effluent as discussed in the next section.

One potential long-term effect of mangrove structural modification and forest fragmentation due to urbanization is a perturbation in their collective genomes. Along an anthropogenic stress gradient in Florida, USA, *Rhizophora mangle* was found to increase both outcrossing and mutation, but decrease reproductive output with increasing human population (Proffitt and Travis 2014). Further evidence suggests pollinator isolation of small patch urban mangroves in both Singapore and Australia (Hermansen et al. 2015; Wee et al. 2015). The true effect of such results is unclear, and while they may represent short-term constraints in genomic turnover, floral isolation on its own is rarely sufficient in permanently affecting genomic changes (Widmer et al. 2009).

23.1.2 Sewage Effluent

Coastal areas are particularly prone to eutrophication and altered nutrient dynamics, and these alterations are largely dependent upon surrounding land use. In urban areas, nutrient availability is altered through the effluent of municipal sewage and through an increase in impervious surfaces, which increases runoff and decreases contact time between terrestrial autotrophs and anthropogenically derived nutrients. Most of the studies on urban mangrove nutrient influences pertain to those found in wetlands receiving municipal sewage. Although the application of sewage to mangrove systems has been studied in depth and with mixed results (Clough et al. 1983), studies in urban systems report changes in mangrove biology and ecology associated with the altered nutrient dynamics in these systems. The following studies show that generally, mangroves receiving municipal sewage are nutrient enriched as evident in lower carbon to nitrogen (C:N) ratios and elevated $\delta^{15}N$ values in leaf tissues, but not necessarily in productivity. This enrichment, although in some cases also due to agricultural activity, can be measured in higher trophic levels at the regional scale (Kwon 2015). The influence of municipal sewage disposal on mangrove community ecology can also be measured in benthic faunal abundance, richness, behavior, and life-history traits.

Municipal sewage effluent typically contains elevated organic carbon and nitrogen, as well as unique isotopic concentrations of anthropogenic origin that are assimilated by lower trophic levels and persist throughout surrounding food webs (Wayland and Hobson 2001). The most commonly reported change for mangroves receiving sewage is elevated $\delta\,^{15}N$ values. This was reported for urban mangroves in Brisbane, Australia (Jones et al. 2001), Kakinada city, India (Dehairs et al. 2000), Guayaquil, Ecuador (Cifuentes et al. 1996), Mombasa, Kenya (Mohamed et al. 2008), Honolulu, Hawaii (Fry and Cormier 2011), and throughout Thailand (Wolters et al. 2016), and is reported in some cases to result in lower C:N ratios in mangrove leaf tissues (Mohamed et al. 2008; Wolters et al. 2016). These findings are confirmed in a meta-analysis of reported isotopic signatures and C:N ratios of mangroves from around the world (Branoff 2017) (Fig. 23.3). This

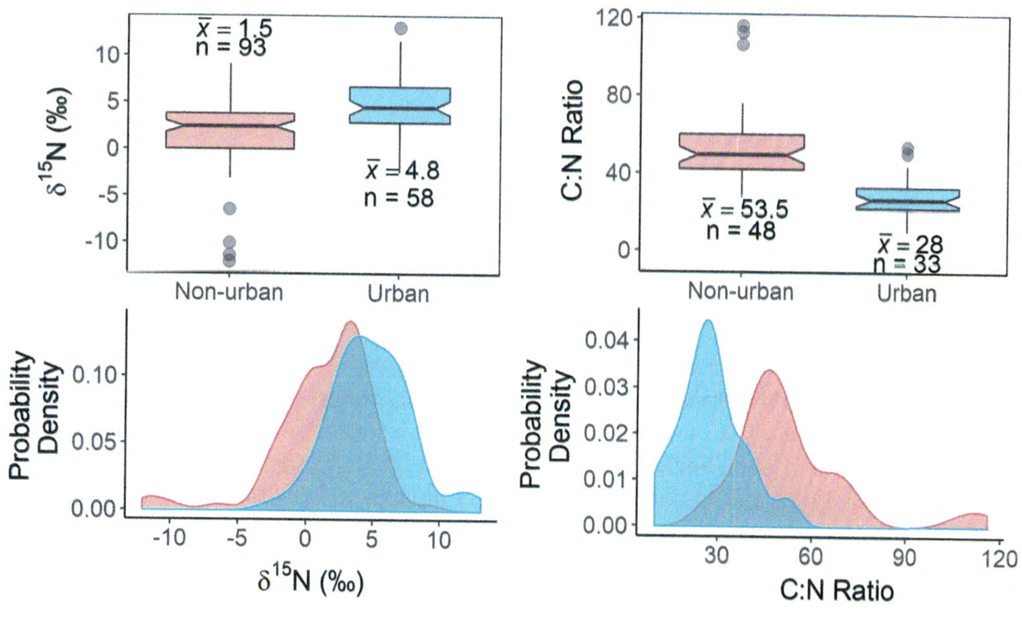

Fig. 23.3 Mangrove green leaf $\delta\,^{15}N$ content (left) and carbon to nitrogen ratio (right) in mangroves within 1 km of urban land-use (urban) and those farther than 1 km from urban land-use (non-urban). Data from (Branoff 2017). Differences are significant for both $\delta\,^{15}N$ (p-value = 4.8e-9) and C:N (p-value = 3.3e-11). Results show urban mangroves are nitrogen enriched, likely from human wastes, which carry elevated $\delta\,^{15}N$ content

analysis shows significant difference in δ ^{15}N values and C:N ratios between urban and non-urban mangrove leaves, as determined by the existence of any urban area within one kilometer of the study sites. Results confirm the above stated observations that urban mangroves are nitrogen enriched, and that the extra nitrogen is sourced from human waste and sewage effluent, as evident in elevated δ ^{15}N.

Despite the reported nutrient enrichment that accompanies sewage effluent, urban mangrove systems receiving municipal sewage are not necessarily more productive. This was the case for Darwin, Australia (Burford et al. 2008), Mombasa, Kenya (Mohamed et al. 2008), and Shenzhen, China (Wong et al. 1997), which reported no significant changes in mangrove productivity despite sewage discharge in the system. One study in Honolulu, Hawaii did report two times higher productivity for urban versus rural mangroves (Fry and Cormier 2011), resulting from differences in N and P availability, although it was not clear if these differences were due to sewage or other anthropogenic sources. Contrastingly, the sewage-fed mangroves of Jeddah City, Saudi Arabia are reported to be stunted with abnormal and sparse pneumatophore growth and increased mortality, likely due to anoxia, compared to non-sewage fed mangroves in the Red Sea (Mandura 1997). These differences in the effects of sewage discharge on urban mangrove systems reflect the inherent differences within each system, and the amount and treatment level of the effluent sewage. Although some studies have shown an ability of mangroves to effectively treat sewage, others have shown detrimental effects, depending largely on the properties of the system and the discharged sewage (Reef et al. 2010). Effects on benthic fauna are also mixed, although distinct patterns seem to be associated with crustacean communities.

Fiddler crabs of the genus *Uca* are common and in some cases integral components of mangroves throughout the world, important in the bioturbation and aeration of otherwise anaerobic soils. Many studies reported changes in crustacean community size and richness, as well as in individual organism behavior and life history traits in mangroves receiving urban sewage. In Mombasa, Kenya and Maputo, Mozambique, mangroves receiving municipal sewage contain higher crustacean biomass and richness when compared to unaffected mangroves (Penha-Lopes et al. 2009; Penha-Lopes et al. 2010; Bartolini et al. 2011), likely resulting from an increased amount of sustenance in the form of organic matter from the sewage. Further, both fiddler crabs and shrimp in these systems enjoy an extended reproductive season, increased fecundity, and improved embryo quality (Penha-Lopes et al. 2009; Penha-Lopes et al. 2011). While these may be regarded as benefits to the system, fiddler crabs in the urban mangroves foraged less, resulting in less bioturbation, and shrimp harbored higher parasite densities, reflecting the potentially detrimental effects of sewage effluent on mangrove ecology (Bartolini et al. 2009; Bartolini et al. 2011; Penha-Lopes et al. 2011).

Molluscs and Annelids are also common components of mangrove benthos, and while sewage-fed urban communities of these phyla show mixed patterns, some studies have reported significant differences when compared to more pristine benthic assemblages. In Kenya and Mozambique, sewage-fed urban benthic communities expressed decreased oligochaete, polychaete, and molluscan density and richness when compared to nearby non-urban and non-contaminated mangrove assemblages (Penha-Lopes et al. 2009; Penha-Lopes et al. 2010). In contrast, benthic communities near sewage outflow points from Bombay, India expressed higher biomass associated with more pollution tolerant species of gastropods and oligochaetes (Dwivedi and Padmakumar 1983). However, no significant differences were found in mangrove benthic macrofauna receiving municipal sewage of Shenzhen, China (Yu et al. 1997), again reflecting the different intensities of sewage influence on mangrove ecology depending upon both the system and properties of the wastewater.

23.1.3 Contamination and Toxicity

In addition to sewage, urban effluents may bring other pollutants to adjacent estuaries, likely influencing the microbial and macrofaunal benthic communities, and possibly mangrove growth in these systems. Heavy metals and hydrocarbons make up most of the studied contaminants, all likely accumulating from continuous input in more hydrologically closed systems, with relatively long turnover times, and the presence of an adsorption sink in the form of fine organic matter. These contaminants are likely sourced from animal and human wastes, as well as fuel combustion and industrial byproducts. They are concentrated in estuarine waters through rivers and sewage systems, as well as by impervious surface coverage and storm water discharge. In general, heavy metal and hydrocarbon pollution in mangrove systems are correlated with increased seedling mutation rates, stand die-back, and reduced species richness (Ellison and Farnsworth 1996). While a majority of studies found significantly higher contaminant concentrations in urban mangrove sediments, studies in Natal, Brazil (Sindern et al. 2007), French Guiana (Marchand et al. 2006), and La Paz, Mexico (Holguin et al. 2006) found no severe pollution or no significant differences between urban and non-urban areas, likely due to the complexity of contaminant transport and retention under varying estuarine conditions.

Although coastal and estuarine systems throughout the world are susceptible to anthropogenic heavy metal contamination (Fowler 1990), mangrove sediments are particularly prone to accumulating these compounds, due to their mostly

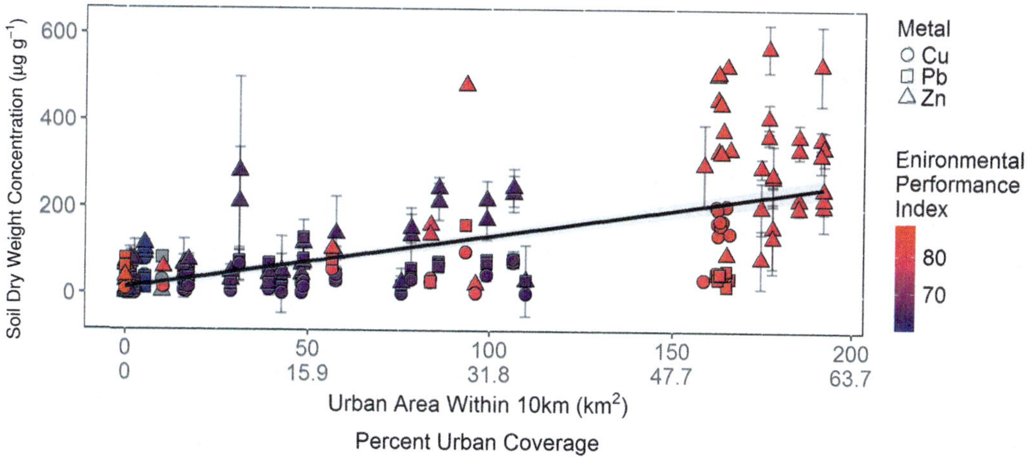

Fig. 23.4 Mangrove surficial benthic concentrations of Copper (Cu), Lead (Pb), and Zinc (Zn) as a function of the urban coverage within 10 km and the corresponding country's environmental performance index (EPI). The EPI reflects a country's policies towards environmental protection, with higher index values representing the strongest policies and enforcement. Higher scoring countries have not necessarily reduced heavy metal concentrations in urban mangrove systems, suggesting regulation is not sufficient in controlling urban mangrove metal contamination (Data from Branoff 2017)

sheltered state, the presence of fine organic particles, and high pH values, all of which encourage metal retention (Harbison 1986). Anthropogenically derived heavy metals in urban mangrove sediments were found to be significantly elevated throughout Brazil (Wasserman et al. 2000; Pagliosa and Barbosa 2006; Oliveira et al. 2014), in Taipei City, Taiwan (Lai et al. 2010), throughout Hong Kong and Shenzhen, China (Tam and Wong 2000; Li et al. 2016), Panama City, Panama (Defew et al. 2005), the United Arab Emirates (Shriadah 1999), and Townsville, Australia (Inglis and Kross 2000). The most commonly studied metals associated with these urban systems are lead (Pb), Copper (Cu), and Zinc (Zn).

These observations are backed by global analyses, which have shown a consistent trend in increasing concentrations of these metals with increasing urban coverage surrounding mangroves (Branoff 2017). The analyses also show a strong positive correlation with surrounding mangrove coverage and the combination between mangrove and urban coverage, confirming the affinity of the mangrove benthos for these contaminants and the vulnerability of urban mangroves in accumulating them. Further, although stronger environmental policies and enforcement might be expected to result in lower benthic metal concentrations, there seems to be no relationship between a country's Environmental Performance Index and the corresponding accumulation of metals in its urban mangroves (Fig. 23.4). While this accumulation represents a valuable ecosystem service in the filtration of toxic pollutants from tropical urban estuaries, these metals are taken up and partitioned in mangrove tissues (MacFarlane et al. 2007), and it is unclear how the pollutants may affect mangrove biology and ecology.

Our understanding of the influence of heavy metals on mangrove plants is limited to controlled laboratory experiments, which suggest numerous physiological responses and potential functional consequences (Table 23.1). Elevated metal concentrations in mangrove soils are associated with high peroxidase activity (Zhang et al. 2007), lower pigment concentrations (MacFarlane and Burchett 2001), and decreased photosystem two (PSII) efficiency (MacFarlane 2003) in mangrove leaves, and with the expression of genes associated with the synthesis of defensive compounds (Huang and Wang 2009). Although it is unknown how this influences mangrove growth and function, individual trees **grown in elevated metal concentrations were found to decrease in biomass and in soil microbial**

Table 23.1 Mangrove sediment concentrations of Copper (Cu), Zinc (Zn), and Lead (Pb) reported in some urban settings throughout the world and as experimental stress thresholds

		Sediment metal concentrations (µg g^{-1})		
		Cu	Zn	Pb
Urban mangroves	Abu Dhabi, UAE[a]	5–29	5–22	13–50
	Hong Kong, China[b]	80	240	80
	Panama City, Panama[c]	56	105	78
	Recife, Brazil[d]	–	84–570	–
	Florianópolis, Brazil[e]	16–32	80–130	28–40
	Townsville, Australia[f]	10	17	4
Stress limits	PSII Efficiency Decline[g]	–	500	–
	Visible Toxicity[h]	400	1000	–
	Biomass Decline[i]	50	100	50

[a]Shriadah (1999), [b]Tam and Wong (2000), [c]Defew et al. (2005), [d]Oliveira et al. (2014), [e]Pagliosa and Barbosa (2006), [f]Inglis and Kross (2000), [g]MacFarlane (2003), [h]MacFarlane and Burchett (2001), and [i]Liu et al. (2009)

activity compared to controls (Lim and Sodhi 2004; Liu et al. 2009; Yan et al. 2010).

These effects have not been corroborated with field observations, but metal toxicity in mangrove ecological communities has been reported in both laboratory and field results. Under laboratory conditions, elevated heavy metal concentrations in simulated industrial wastewater applications were found to significantly reduce the microbial activity of mangrove soils (Yim and Tam 1999), and the mangrove sediments of Recife, Brazil were used in copepod toxicology tests to determine the presence of lethal and sublethal concentrations of Zn and Pb (Oliveira et al. 2014). In Townsville, Australia, benthic macrofaunal assemblages are strongly correlated with concentrations of Zn, Pb, Cu, and hydrocarbons, despite their concentrations being well below those noted to cause ecological effects (Inglis and Kross 2000). Similar differences in benthic communities were also observed between urban and protected mangrove sediments of Brazil, with the most explicative variables being concentrations of Pb and Cu (Pagliosa and Barbosa 2006). It is therefore likely that metal toxicity is driving the composition of benthic community assemblages in some urban mangrove systems. This composition, along with the recorded changes in plant physiology, may also influence the growth, production, and forest structure of contaminated urban mangroves, although this has yet to be shown.

Hydrocarbons have also been observed in high and potentially lethal concentrations within urban mangrove sediments with reported effects on microbial communities. Because of the abundance of fine organic matter, which readily adsorbs the hydrocarbons, mangrove areas are prone to accumulate these compounds coming from petroleum use in urban and industrial areas. Mangrove areas in Fortaleza, Brazil, for instance, were found to contain higher hydrocarbon concentrations than non-mangrove areas or other cities with greater industrial development (Cavalcante et al. 2009). Although these concentrations did not reach probable effect levels (PEL), the authors note sub-PEL levels may still cause adverse biological effects. Toxic levels have been reported in organisms associated with urban mangrove sediments of Cilacap City, Indonesia, where levels sometimes reached ten times their reported toxicity limits (Dsikowitzky et al. 2011). In Rio de Janeiro, Brazil, hydrocarbons are also recorded to be accumulating in mangrove sediments (Fontana et al. 2010, 2012), and one study suggests the microbial community is not degrading them, likely because there is an abundance of more easily digested organic carbon from the mangroves (Brito et al. 2009). This was supported by other studies in Rio de Janeiro that found little difference in the microbial communities of mangrove sediments when compared to more pristine sites, although they did note an increase in microbes associated with hydrocarbons (Gomes et al. 2007, 2008). Yet another study in Brazil contradicts these findings by reporting more homogenized microbial communities in sediments of urban versus more pristine mangroves, although functional efficacy of the community remained unchanged (Nogueira et al. 2015).

Air pollutants may present an additional source of contamination in urban mangroves. Although this has yet to be shown directly in mangrove systems, urban air pollution is known to affect plant physiology (Honour et al. 2009; Horasan et al. 2015). Further, there is evidence for higher exposure to fossil fuel combustion exhausts in urban mangroves, as indicated by lower leaf $\delta^{13}C$ content (Fig. 23.5). Lower $\delta^{13}C$ in plant tissues could be indicative

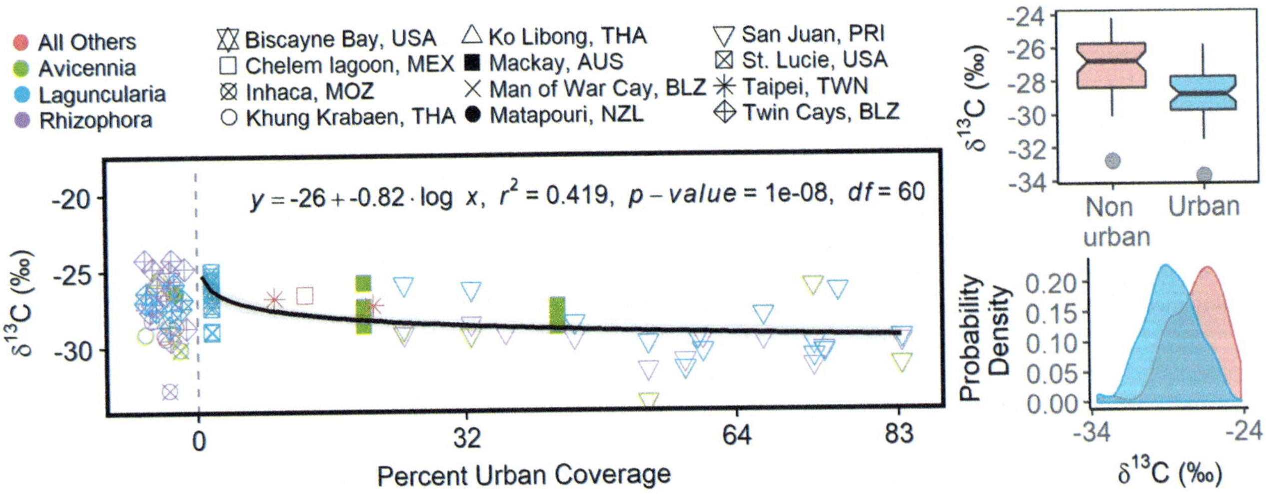

Fig. 23.5 Mangrove green leaf $\delta^{13}C$ content for trees along an urban gradient (*left*) and for trees within 1 km of urban land-use (urban) and those farther than 1 km from urban land-use (non-urban) (*right*). Differences are significant (p-value = 1.2e-6) and lower the $\delta^{13}C$ of urban leaves suggest urban mangroves may be exposed to elevated exhaust from the combustion of fossil fuels, or to environment stress that raises their water-use efficiency (Data from Branoff 2017)

of the isotopic content of source carbon in the atmosphere, and has been linked to automobile exhausts (Lichtfouse et al. 2003). But it may also be a result of water use efficiency rather than the source of carbon (Farquhar et al. 1989). In this later case, the lower $\delta^{13}C$ values associated with urban mangroves would suggest higher water use efficiency, which in turn may also indicate environmental stress (Clough and Sim 1989).

23.1.4 Community Assemblages

Mangroves serve as important faunal habitat for endemic and migratory species, as both nursery and feeding grounds, and the importance of connectivity between mangroves and off-shore reefs in coastal conservation has been well documented (Nagelkerken et al. 2002; Mumby et al. 2004; Olds et al. 2013). Along mangrove occupied shorelines, faunal abundance, richness, physiology, and life-history traits are all shown to be altered in the presence of urban land use, which may influence entire coastal systems because of their connectivity.

The most thorough study yet of the influence of urban land use on mangrove fish communities was carried out in the Tampa Bay Estuary in Florida, USA. This study monitored fish communities along a gradient of land development and found that urban areas were generally poor in species richness and in densities of economically important taxa (Krebs et al. 2014a, b). Surprisingly, this study also reported that urban land use positively influenced body condition and energetic output, and while fecundity decreased in urban areas, offspring size increased. More drastic morphological changes were observed in mangrove mudskippers of Dar es Salaam, Tanzania, in which the number of organisms with missing eyes was correlated with pollution effluent in the system (Kruitwagen et al. 2006). These fish were also limited in their size range and age classes when compared to mangroves outside of the polluted urban area. On the Atlantic coast of Florida, in the Biscayne Bay estuary of Miami, fish richness is higher along shorelines containing both mangroves and rip-rap, when compared to mangrove only shorelines, although abundance is higher in the later (Peters et al. 2015). Similar declines in species richness were also observed in an urbanized Amazon river estuary of northern Brazil (Viana et al. 2010), and homogenized fish richness of mangroves in Sydney, Australia suggests these urban mangroves hold diminished value as a habitat resource when compared to less-urban mangrove systems (Clynick and Chapman 2002).

Benthic mangrove community assemblages have also been reported to reflect influences of urban land use. Although one study has reported no significant differences in benthic assemblages of urban and non-urban sediments of a mangrove fringed estuary near Syndey, Australia (Lindegarth and Hoskin 2001), the majority of similar studies do find differences. Two additional studies in Sydney found the level of damage in urban mangroves to be significantly correlated with molluscan species richness, abundance, and the presence of particular species (Skilleter 1996), and that differences in gastropod, polychaete, and amphipod assemblages could be observed within six meters of mangrove boardwalk construction (Kelaher et al. 1998). In Darwin Harbor, disturbed urban sites were found to harbor significantly more worm taxa associated with surface deposit feeders than undisturbed sites (Metcalfe and Glasby 2008). In Brisbane, diminished molluscan richness and abundance, and increased crab abundance in disturbed mangroves was attributed to differences in pneumatophore density as a result of boardwalk construction (Skilleter and Warren 2000). Other studies attribute recorded differences in urban benthic macroinvertebrate assemblages to sediment contamination. In Townsville, Australia, urban estuaries with high benthic metal and hydrocarbon concentrations are more associated with polychaetes and lack filter feeding bivalves (Inglis and Kross 2000), although contaminant concentrations were well below those noted to affect ecology. Similar differences between pollution sensitive and pollutant tolerant benthic assemblages were observed in urban and protected mangrove sediments of Brazil, with the most explicative predictors being heavy metal and nutrient concentrations (Pagliosa and Barbosa 2006). Another explanation may lie in the physical properties of benthic sediments, which may be different in urban areas as a result of hydrological engineering throughout the catchment and the introduction of organic wastes (Dellapenna et al. 2015).

Changes in urban mangrove faunal communities hold true for bird assemblages, which have reported similar but at times mixed results. Avifauna communities of urban mangroves tend to reflect differences in species richness and abundance, notably in specialist species, although these are not always consistent across systems. In the mangroves of Paranguá Bay in Brazil, bird surveys report one of the richest avifauna communities in the country, and researchers attribute this diversity to expansive Atlantic Forests surrounding the mangroves (Mestre et al. 2007). Yet this community changes with proximity to a nearby urban area, specifically in abundances of certain specialist species associated with higher disturbance. Such observations are common in urban bird communities, where increased resources associated with human settlements support greater abundances of certain specialist species (Chace and Walsh 2006). The survey in Brazil did not mention mangrove specialists as one of the beneficiaries of urban land-use, but a separate study in peninsular Malaysia suggests that mangrove specialists have been successful in colonizing anthropogenic habitats (Noske 1995). The author attributes this success to habitat similarities between mangroves and anthropogenic land use, as well as the lower degree of specialization and higher adaptability of mangrove dependent species. Yet this same study notes an apparent lack of some mangrove specialists in areas of expanding urbanization and decreasing landscape connectivity. The idea of successful colonization of urban habitats by mangrove bird species is supported by a study in Singapore, where many of the species recorded in urban areas are typically associated with mangrove and coastal scrub ecosystems, and again attributes this observation to the similarities in mangrove and urban habitats (Lim and Sodhi 2004). A separate study in Singapore found similar results relating the structure of mangrove bird communities to the type of urban habitat (Sodhi et al. 1999). Yet, in Darwin, Australia, one survey notes that except for opportunistic nectarivores, mangrove specialists bird species have yet to colonize urban areas, while rainforest associated species have (Noske 1996). The author attributes this contradiction to a lack of ecotones between the stable and fire-resistant mangrove habitats and the seasonal and extremely fire prone suburban and urban habitats next to them. Aside from specialists, another survey in Darwin found that species richness of bird communities in mangrove patches throughout the urban area was independent of mangrove patch size and more dependent upon the diversity of habitat types surrounding the patch (Mohd-Azlan and Lawes 2011).

One study has explored these patterns in mangrove faunal communities across multiple mangrove systems and found urban intensity to be globally correlated with mangrove fauna richness for worm and bird taxa (Branoff 2017). Fish richness was more strongly correlated with mangrove coverage and patch metrics, and crustacean and molluscan richness were more strongly correlated with site distance from the meridian of maximum mangrove richness at 100°E. This indicates that local land-use may not be as influential in regional diversity in determining mangrove faunal richness for some taxa. The study also found certain sub groups, such as subsistence fisheries species to be more strongly responsive to urban coverage, suggesting the existence of synanthropic and misanthropic species within the mangrove community. Using the data from the same study, mangrove synanthropic tendencies were further explored at the family level, revealing many potential candidate families to be regarded as particularly vulnerable to urbanization (Fig. 23.6).

Figure 23.6 shows the results of a principal component analysis (PCA) performed on the data from (Branoff 2017). The ellipses in the figure represent 95% normal probability ellipses for a given family, and are plotted around the eigenscores of the study sites in which each family was found. The placement of each point represents a mix of the four land-use variables used for the PCA, as indicated by the eigenvectors (arrows), which indicate the direction of positive increase for each land-use variable. Those ellipses falling on the left, therefore, indicate families that were primarily found in relatively non-urban sites compared to those on the right side of the figure.

The fish families of Ariidae, Batrachoididae, Belonidae, Hemiramphidae, Lutjanidae, and Pristigasteridae were all observed in multiple studies from around the world but were absent in the most urban study sites. Ariidae (catfish) have been noted before to be absent in anthropogenically disturbed habitats (Oribhabor and Ogbeibu 2010; Pereira et al. 2015), and Batrachoididae (toadfish) is a vocal family and has been observed to be especially vulnerable to the high noise levels associated with boat traffic (Vasconcelos et al. 2007; Cullis-Suzuki 2016). Belonidae (needle fish) have been found in urban sites (De Angelo 2010), as has Hemiramphidae (Waltham and Connolly 2006), Lutjanidae (snappers) (Burt et al. 2013; Peters et al. 2015), and Pristigasteridae (Souza et al. 2016), suggesting their absence in the sites from Fig. 23.6 may be a result of more intense urbanization there or incomplete representation among different sampling methods.

Bird families of Falconidae (falcons) and Psittaculidae (parrots) are identified as potentially misanthropic families, and the family of Phalacrocoracidae (cormorants) are identified as potentially synanthropic. Numerous falcon species have been categorized as urban avoiders (Conole and Kirkpatrick 2011), and psittaculidae has been noted to be absent from urban landscapes (Bhattacharya and Roy 2013), further emphasizing their potential misanthropy. Cormorants have a rich history of conflict with humans (Taylor et al. 2011), suggesting they have long coinhabited with people, and continue to do so in urban wetlands (Kushwaha Shubhda and Kulkarni Neelima 2013; Traut and Hostetler 2004; Guruprasad and Basavarajappa 2016).

23.2 Urban Mangrove Hazards, Vulnerabilities, and Management

Managing urban mangroves means recognizing how and to what extent mangrove ecology and human well-being will be influenced across the range of cities and mangrove ecosystems. Studies over the past 30 years have identified several recurrent and novel patterns in these systems. From these findings, a model of urban mangroves is developed to aid in understanding their ecology and management in a future of expanding urbanization (Fig. 23.7). The model is based upon others of mangrove systems (Cintrón et al. 1978; Lugo 1990; Odum and Arding 1991), in that it attempts to incorporate both the ecological and social demands of these highly social-ecological systems. The result is an understanding of how urban areas influence mangrove function and the corresponding output of ecosystem services. These include services that have been valued around $200,000/ha, second

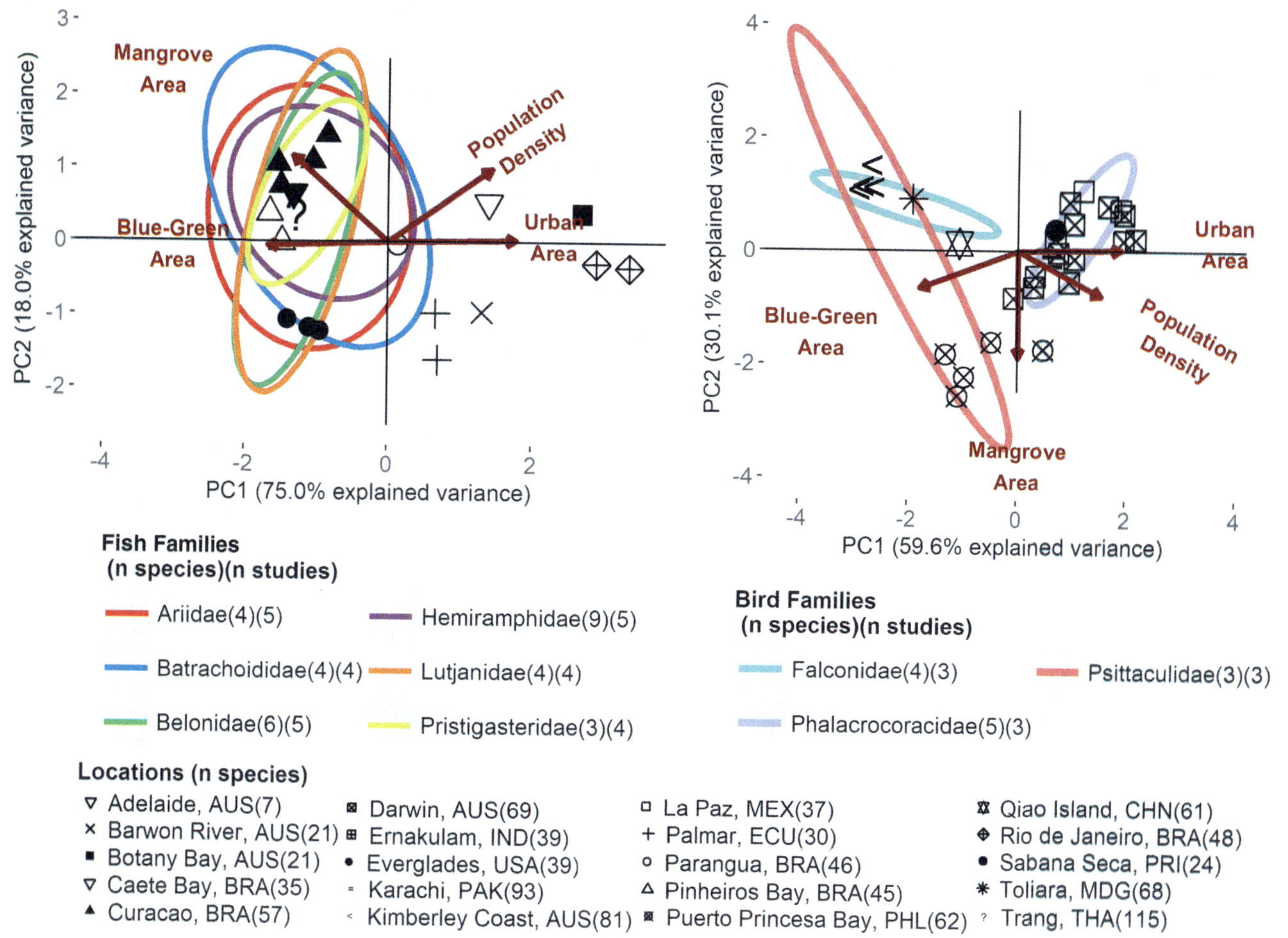

Fig. 23.6 Some mangrove faunal families are more or less likely to be found in urban areas, as indicated by a PCA analysis of data from Branoff (2017). Ellipses surround the 95% normal probability area of the landscape variables describing individual sites where each family was observed. Points to the left of the graphs represent relatively less urban mangroves, with higher mangrove and open water area, lower population density, and lower urban land-use. Ellipses surrounding these points represent potential misanthropic families that tend to avoid urban areas. Sites to the right are thus more urban mangroves, and ellipses surrounding these points may represent syntrophic species that thrive in the presence of humans

only to coral reefs (de Groot et al. 2012), making them essential to maintaining the health, resiliency, and sustainability of coastal communities around the world (Millenium Ecosystem Assessment 2005).

People are integral components of urban mangroves, living around and at times within these systems. Through their daily lives and through regulation, infrastructure, and the market, people affect how the physical, chemical, and biological components function in these systems. These anthropogenic influences, along with natural environmental conditions, impart stress on the system, which in turn contributes to the system's capacity in providing ecosystem services (Odum and Johannes 1975; Lee et al. 2014; Lugo et al. 2014). As a result, people are not only the beneficiaries of urban mangrove ecosystem services, but impart a heightened force on the quantity and quality of their provisioning. For this reason, and because they will benefit more people in urban landscapes, management of urban mangroves should focus on maximizing those ecosystem services that provide the greatest benefits to surrounding human communities. In some cases, it is unclear how prolonged and intensifying urbanization will influence mangrove ecology and potentially diminish ecosystem service provisioning. Thus, there is a need for more studies on the influence of urbanization on mangrove ecology and function. In many cases, however, optimizing ecosystem services simply means conserving or expanding existing mangroves and allowing them to function normally by maintaining sufficient tidal influence and water-quality. This is evident in the model presented in Fig. 23.7, where mangrove biomass feeds all the ecosystem services represented to the right.

Urban mangrove deforestation and fragmentation is driven by the demand for competing resources of higher economic value, as determined by local markets. This may

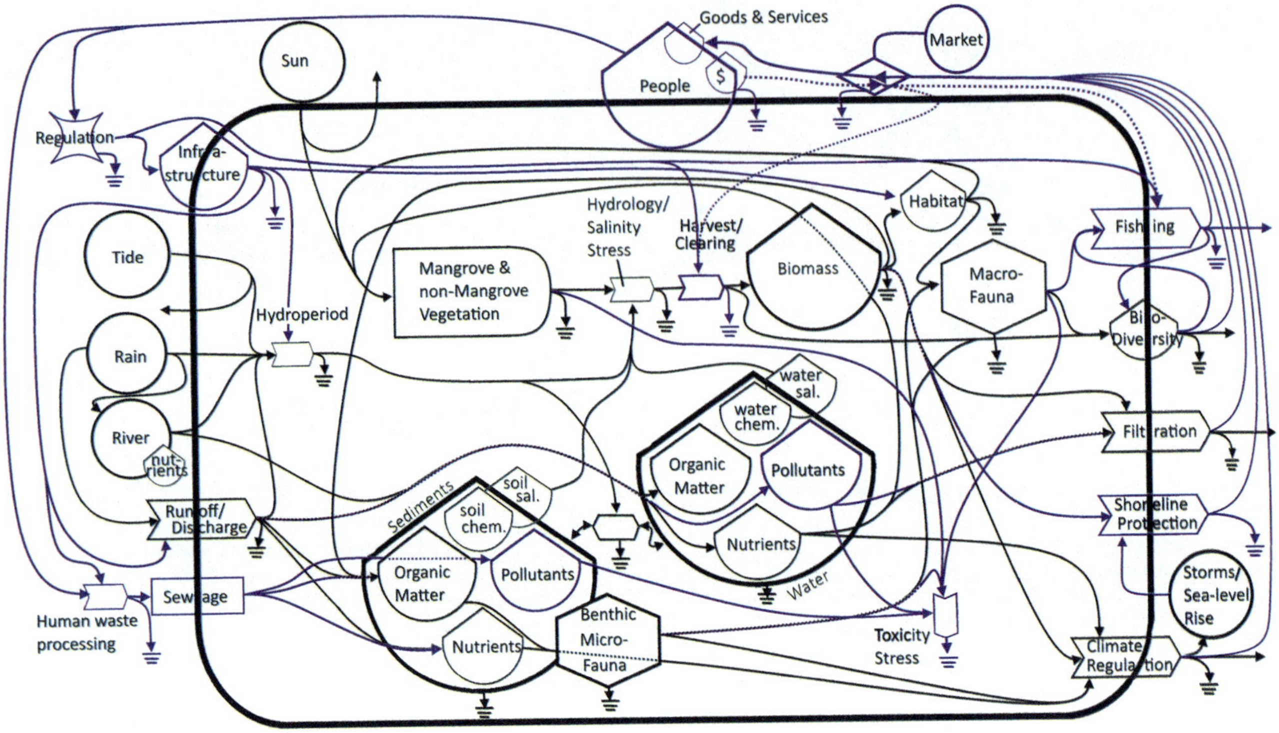

Fig. 23.7 Mangroves in urban environments are subject to several unique interactions from the social system of the city, which are colored in blue. People implement regulations that control coastal development and infrastructure, which in turn influence hydrology of surrounding landscapes through the blanketing of impervious surfaces, channelization, and storm and sewage water routing. These have downstream effects involving hydro-period and contaminant transport that likely affect mangrove biology. Additionally, through harvesting of mangrove biomass and especially nearby fisheries, people both benefit from and influence the provisioning of ecosystem services by the mangroves. These social forcings all interact with the in place ecological forcings to create novel mangrove systems

be a demand for coastal real estate, agriculture and aquaculture, or wood for fuel, depending on the region. In any case, the resulting habitat loss drives changes in biodiversity, which extends to regional scales due to the high connectivity between mangroves and surrounding ecosystems (Nagelkerken et al. 2002; Mumby et al. 2004; Olds et al. 2013). This could be potentially influential on surrounding fisheries, notably subsistence fisheries, which are highly correlated with surrounding mangrove coverage and especially important for the provisioning of ecosystem services. Additionally, the loss of mangroves in urban landscapes diminishes the filtration service these ecosystems perform, which is especially important considering the high contaminant load in urban estuaries and the high capacity mangroves have in retaining and thus sequestering or filtering them.

These contaminants derive from urban infrastructure, either in impervious surfaces runoff or sewage and industrial effluents. Although tight regulation would in theory diminish contaminants, such as heavy metals, from accumulating in mangrove benthos, it has been shown here that that may not be the case in practice (Fig. 23.4). This underscores the importance of urban mangrove conservation, as they perform a valuable filtration service for tropical urban estuaries when environmental policies and enforcement may not necessarily be effective. Still, most of the studies on urban mangrove metal and nutrient contamination point to sewage as their source, implying wastewater management might be the most direct target in reducing mangrove pollution.

The most valuable ecosystem service attributed to mangroves is that of shoreline protection (Barbier et al. 2011). This has been demonstrated during numerous tsunamis and cyclones around the world, where human communities of greater mangrove extent between them and the ocean suffered minimal loss of life and property (Dahdouh-Guebas et al. 2005; Kathiresan and Rajendran 2005; Das and Vincent 2009). This service holds greater value in more densely populated communities, such as cities, which stand to lose more under periodic natural disasters or accelerating sea level rise. Management of urban mangrove systems should therefore seek to maximize mangrove coverage surrounding city centers. This goal is often in conflict with other interests, such as that of increasing coastal development and aesthetic value, or minimizing mosquito-borne pathogen transmission. Again, such conflicts are heightened in urban areas with greater populations and diversity of interests in mangrove habitats.

One component of urban mangroves that has not been sufficiently investigated is that of carbon storage and the subsequent service of climate regulation through the sequestration or emissions of greenhouse gases. Mangroves are among the most prolific carbon storing ecosystems, primarily due to belowground storage (Alongi 2012). Maintaining urban mangroves therefore maintains their corresponding sequestration rates, thus partially mitigating the increased atmospheric carbon in urban areas, as may be indicated by stable isotopes (Fig. 23.5). However, nitrous oxide is another greenhouse gas sourced from mangrove soils and its emission rate may be influenced by the excess nitrogen associated with urban mangroves (Fig. 23.3). Further, it is unclear how toxicity from benthic metal contamination may influence mangrove production or the microbial consumption of carbon. Thus, our understanding of the capacity of urban mangroves in providing climate regulation is poor.

23.3 Future Research

As mangroves continue to be pressured by urbanization, especially in the developing world, optimal management will require science that identifies the most influential components of the urban landscape on mangrove biology, ecology, and ecosystem services. From the collective literature, it is clear that mangrove ecosystems in these urban environments reflect unique characteristics, although a coherent urban mangrove paradigm has not yet emerged. Much of the knowledge-gap in urban mangroves is likely due to the heterogeneity of these systems across climates, life zones, and coastal geography. Yet another primary explanation lies in the lack of more descriptive and definitive explanations of the urban environment, which inhibits comparisons across systems. This lack of well quantified urban intensity in ecological research has been noted in the urban ecology literature, and the use of calculated urban gradients has been proposed as a working solution. Some of the most ideal cities in the world for studying mangroves along urban gradients are identified in (Branoff 2017). Lagos, Nigeria; Recife and Rio de Janeiro, Brazil; San Juan, Puerto Rico; Klang, Malaysia; and Miami, USA all contain mangroves in densely populated and developed areas, as well as in relatively pristine areas. Establishing study sites in these cities will therefore allow for direct comparisons of mangrove biological and ecological metrics along varying degrees of urbanization. Using such gradients in future urban mangrove studies, it will be possible to make more effective decisions regarding mangrove management along rapidly urbanizing coasts, where mangrove ecosystem services are in high demand and under increasing pressure. The valuation of these services, however, are heavily dependent upon local markets, which may work against urban mangrove preservation.

Although relatively high monetary values have been placed on mangrove ecosystems, they are often dependent upon the assumption of long-term preservation. Shoreline protection and stabilization, fisheries support, and climate regulation all assume a long-term mangrove presence. Yet, the demand for these services is often outweighed in local markets by that for more short-term resources like wood fuel, fish, or coastal real-estate, resulting in a net pressure against urban mangrove preservation. A complete urban mangrove model, as well as urban mangrove management, therefore, must fully account for the social system, including the relative demand and value of ecosystem services. This will require much more emphasis on social research in these systems, perhaps through greater collaboration with social-scientists to understand how mangroves are valued and used in respective societies around the world. Under future scenarios of rapidly increasing urbanization and its corresponding pressure on mangrove ecosystems, effective management cannot ignore the social components of mangrove ecosystems. Future research should therefore not only pinpoint specific causes for the correlations emphasized in this chapter, but also how people and mangroves interact to influence the biology, ecology, and functional value of urban mangroves.

References

Alongi DM (2012) Carbon sequestration in mangrove forests. Carbon Manag 3:313–322

Baird RC (2009) Coastal urbanization: the challenge of management lag. Manag Environ Qual Int J 20:371–382. https://doi.org/10.1108/14777830910963726

Barbier EB, Hacker SD, Kennedy C, Koch EW, Stier AC, Silliman BR (2011) The value of estuarine and coastal ecosystem services. Ecol Monogr 81:169–193

Bartolini F, Penha-Lopes G, Limbu S, Paula J, Cannicci S (2009) Behavioural responses of the mangrove fiddler crabs (Ucaannulipes and U. inversa) to urban sewage loadings: results of a mesocosm approach. Mar Pollut Bull 58:1860–1867

Bartolini F, Cimò F, Fusi M, Dahdouh-Guebas F, Lopes GP, Cannicci S (2011) The effect of sewage discharge on the ecosystem engineering activities of two East African fiddler crab species: consequences for mangrove ecosystem functioning. Mar Environ Res 71:53–61

Benfield SL, Guzman HM, Mair JM (2005) Temporal mangrove dynamics in relation to coastal development in Pacific Panama. J Environ Manag 76:263–276. https://doi.org/10.1016/j.jenvman.2005.02.004

Bhattacharya R, Roy R (2013) Habitat wise distributions of local birds in the vicinity of Krishnanagar. In: Proceeding of the National Confrence on Challenges in Biodiversity and Resource Management, pp 179–187

Bolund P, Hunhammar S (1999) Ecosystem services in urban areas. Ecol Econ 29:293–301

Branoff BL (2017) Quantifying the influence of urban land-use on mangrove biology and ecology: a meta-analysis. Glob Ecol Biogeogr 26:1339–1356

Brito EMS, Duran R, Guyoneaud R, Goni-Urriza M, de Oteyza TG, Crapez MAC, Aleluia I, Wasserman JCA (2009) A case study of in situ oil contamination in a mangrove swamp (Rio De Janeiro,

Brazil). Mar Pollut Bull 58:418–423. https://doi.org/10.1016/j.marpolbul.2008.12.008
Burford MA, Alongi DM, McKinnon AD, Trott LA (2008) Primary production and nutrients in a tropical macrotidal estuary, Darwin Harbour, Australia. Estuar Coast Shelf Sci 79:440–448. https://doi.org/10.1016/j.ecss.2008.04.018
Burt JA (2014) The environmental costs of coastal urbanization in the Arabian Gulf. City 18:760–770
Burt JA, Feary DA, Cavalcante G, Bauman AG, Usseglio P (2013) Urban breakwaters as reef fish habitat in the Persian Gulf. Mar Pollut Bull 72:342–350
Cavalcante RM, Sousa FW, Nascimento RF, Silveira ER, Freire GSS (2009) The impact of urbanization on tropical mangroves (Fortaleza, Brazil): evidence from PAH distribution in sediments. J Environ Manag 91:328–335. https://doi.org/10.1016/j.jenvman.2009.08.020
Chace JF, Walsh JJ (2006) Urban effects on native avifauna: a review. Landsc Urban Plan 74:46–69
Cifuentes LA, Coffin RB, Solorzano L, Cardenas W, Espinoza J, Twilley RR (1996) Isotopic and elemental variations of carbon and nitrogen in a mangrove estuary. Estuar Coast Shelf Sci 43:781–800
Cintrón G, Lugo AE, Pool DJ, Morris G (1978) Mangroves of arid environments in Puerto Rico and adjacent islands. Biotropica:110–121
Clough BF, Sim RG (1989) Changes in gas exchange characteristics and water use efficiency of mangroves in response to salinity and vapour pressure deficit. Oecologia 79:38–44
Clough BF, Boto KG, Attiwill PM (1983) Mangroves and sewage: a re-evaluation. In: Teas H (ed) Biology and ecology of mangroves. Springer, Dordrecht, pp 151–161
Clynick B, Chapman MG (2002) Assemblages of small fish in patchy mangrove forests in Sydney Harbour. Mar Freshw Res 53:669–677. https://doi.org/10.1071/MF00147
Conole LE, Kirkpatrick JB (2011) Functional and spatial differentiation of urban bird assemblages at the landscape scale. Landsc Urban Plan 100:11–23. https://doi.org/10.1016/j.landurbplan.2010.11.007
Cullis-Suzuki S (2016) Singing fish in an ocean of noise: effects of boat noise on the plainfin midshipman (porichthys notatus) in a natural ecosystem. In: The effects of noise on aquatic life II. Springer, pp 171–179
Dahdouh-Guebas F, Jayatissa LP, Di Nitto D, Bosire JO, Lo SD, Koedam N (2005) How effective were mangroves as a defence against the recent tsunami? Curr Biol 15:R443–R447
Das S, Vincent JR (2009) Mangroves protected villages and reduced death toll during Indian super cyclone. Proc Natl Acad Sci 106:7357–7360
DasGupta R, Shaw R (2013) Changing perspectives of mangrove management in India–an analytical overview. Ocean Coast Manag 80:107–118
De Angelo JA (2010) Mangrove shoreline fish assemblages of Oleta River State Park: Baseline Conditions in an Urban System. Masters thesis, Department of Marine Affairs and Policy, University of Miami
de Groot R, Brander L, van der Ploeg S, Costanza R, Bernard F, Braat L, Christie M, Crossman N, Ghermandi A, Hein L, Hussain S, Kumar P, McVittie A, Portela R, Rodriguez LC, ten Brink P, van Beukering P (2012) Global estimates of the value of ecosystems and their services in monetary units. Ecosyst Serv 1:50–61. https://doi.org/10.1016/j.ecoser.2012.07.005
de Souza CLR, da Costa VB, Pereira S de FP, da Silva DCM, Sarpedonti V (2016) Impacts of urban life on water quality and fish larvae communities in two creeks of the Brazilian Amazon. Rev Ambient Água 11:13–23
Defew LH, Mair JM, Guzman HM (2005) An assessment of metal contamination in mangrove sediments and leaves from Punta Mala Bay, Pacific Panama. Mar Pollut Bull 50:547–552
Dehairs F, Rao RG, Mohan PC, Raman AV, Marguillier S, Hellings L (2000) Tracing mangrove carbon in suspended matter and aquatic fauna of the Gautami-Godavari Delta, Bay of Bengal (India). Hydrobiologia 431:225–241. https://doi.org/10.1023/a:1004072310525
Dellapenna T, Fielder B, Noll CIV, Savarese M (2015) Geological responses to urbanization of the Naples Bay estuarine system, Southwestern Florida, USA. Estuar Coasts 38:81–96. https://doi.org/10.1007/s12237-013-9704-2
Donato DC, Kauffman JB, Murdiyarso D, Kurnianto S, Stidham M, Kanninen M (2011) Mangroves among the most carbon-rich forests in the tropics. Nat Geosci 4:293–297
Dsikowitzky L, Nordhaus I, Jennerjahn TC, Khrycheva P, Sivatharshan Y, Yuwono E, Schwarzbauer J (2011) Anthropogenic organic contaminants in water, sediments and benthic organisms of the mangrove-fringed Segara Anakan Lagoon, Java, Indonesia. Mar Pollut Bull 62:851–862. https://doi.org/10.1016/j.marpolbul.2011.02.023
Dwivedi SN, Padmakumar KG (1983) Ecology of a mangrove swamp near Juhu Beach, Bombay with reference to sewage pollution. In: Biology and ecology of mangroves. Springer, pp 163–170
Ellison AM, Farnsworth EJ (1996) Anthropogenic disturbance of Caribbean mangrove ecosystems: past impacts, present trends, and future predictions. Biotropica 28:549–565. https://doi.org/10.2307/2389096
Farquhar GD, Hubick KT, Condon AG, Richards RA (1989) Carbon isotope fractionation and plant water-use efficiency. In: Stable isotopes in ecological research. Springer, pp 21–40
Fontana LF, da Silva FS, de Figueiredo NG, Brum DM, Netto ADP, de Figueiredo AG, Crapez MAC (2010) Superficial distribution of aromatic compounds and geomicrobiology of sediments from Surui Mangrove, Guanabara Bay, RJ, Brazil. An Acad Bras Cienc 82:1013–1030. https://doi.org/10.1590/s0001-37652010000400022
Fontana LF, Crapez MAC, de Figueiredo AG, Santos ES, da Silva FS, Ribeiro AM, da Rocha CCM, Netto ADP (2012) Characterization and distribution of polycyclic aromatic hydrocarbons in sediments from Surui Mangrove, Guanabara Bay, Rio de Janeiro, Brazil. J Coast Res 28:156–162. https://doi.org/10.2112/jcoastres-d-10-00105.1
Fowler SW (1990) Critical review of selected heavy metal and chlorinated hydrocarbon concentrations in the marine environment. Mar Environ Res 29:1–64
Friess D, Richards D, Phang VH (2015) Mangrove forests store high densities of carbon across the tropical urban landscape of Singapore. Urban Ecosyst:1–16. https://doi.org/10.1007/s11252-015-0511-3
Fry B, Cormier N (2011) Chemical Ecology of Red Mangroves, Rhizophora mangle, in the Hawaiian Islands 1. Pac Sci 65:219–234
Giri C, Ochieng E, Tieszen LL, Zhu Z, Singh A, Loveland T, Masek J, Duke N (2011) Status and distribution of mangrove forests of the world using earth observation satellite data. Glob Ecol Biogeogr 20:154–159
Gomes NCM, Borges LR, Paranhos R, Pinto FN, Krögerrecklenfort E, Mendonça-Hagler LCS, Smalla K (2007) Diversity of ndo genes in mangrove sediments exposed to different sources of polycyclic aromatic hydrocarbon pollution. Appl Environ Microbiol 73:7392–7399
Gomes NC, Borges LR, Paranhos R, Pinto FN, Mendonça-Hagler LCS, Smalla K (2008) Exploring the diversity of bacterial communities in sediments of urban mangrove forests. Fed Eur Microbiol Soc 66:96–109. https://doi.org/10.1111/j.1574-6941.2008.00519.x
Gómez-Baggethun E, Gren Å, Barton DN, Langemeyer J, McPhearson T, O'Farrell P, Andersson E, Hamstead Z, Kremer P (2013) Urban ecosystem services. In: Urbanization, biodiversity and ecosystem services: challenges and opportunities. Springer, p 175–251

Gong C, Yu S, Joesting H, Chen J (2013) Determining socioeconomic drivers of urban forest fragmentation with historical remote sensing images. Landsc Urban Plan 117:57–65

Grimm NB, Faeth SH, Golubiewski NE, Redman CL, Wu J, Bai X, Briggs JM (2008) Global change and the ecology of cities. Science 319:756–760

Guruprasad RPM, Basavarajappa S (2016) Checklist of avifauna of rourkela urban area of Sundargarh District, Odisha, India. Res J Life Sci Bioinform Pharm Chem Sci 2:34–43

Hamilton S, Casey D (2014) Creation of a high spatiotemporal resolution global database of continuous mangrove forest cover for the 21st Century (CGMFC-21). Glob Ecol Biogeogr 25:729–738

Harbison P (1986) Mangrove muds – a sink and a source for trace metals. Mar Pollut Bull 17:246–250. https://doi.org/10.1016/0025-326X(86)90057-3

Heatherington C, Bishop MJ (2012) Spatial variation in the structure of mangrove forests with respect to seawalls. Mar Freshw Res 63:926–933

Hermansen TD, Roberts DG, Toben M, Minchinton TE, Ayre DJ (2015) Small urban stands of the mangrove Avicennia marina are genetically diverse but experience elevated inbreeding. Estuar Coasts 1–10

Holguin G, Gonzalez-Zamorano P, de Bashan LE, Mendoza R, Amador E, Bashan Y (2006) Mangrove health in an arid environment encroached by urban development – a case study. Sci Total Environ 363:260–274. https://doi.org/10.1016/j.scitotenv.2005.05.026

Honour SL, Bell JNB, Ashenden TW, Cape JN, Power SA (2009) Responses of herbaceous plants to urban air pollution: effects on growth, phenology and leaf surface characteristics. Environ Pollut 157:1279–1286. https://doi.org/10.1016/j.envpol.2008.11.049

Horasan Ö, Güner ED, Güllü G, Tipirdamaz R (2015) Effects of air pollution on urban plants: Nezahat Gökyiğit Botanical Garden. In: Öztürk M, Ashraf M, Aksoy A, Ahmad MSA, Hakeem KR (eds) Plants, pollutants and remediation. Springer, Dordrecht, pp 391–401

Huang G-Y, Wang Y-S (2009) Expression analysis of type 2 metallothionein gene in mangrove species (Bruguiera gymnorrhiza) under heavy metal stress. Chemosphere 77:1026–1029

Inglis GJ, Kross JE (2000) Evidence for systemic changes in the Benthic Fauna of tropical estuaries as a result of urbanization. Mar Pollut Bull 41:367–376. https://doi.org/10.1016/S0025-326X(00)00093-X

Jones AB, O'Donohue MJ, Udy J, Dennison WC (2001) Assessing ecological impacts of shrimp and sewage effluent: biological indicators with standard water quality analyses. Estuar Coast Shelf Sci 52:91–109. https://doi.org/10.1006/ecss.2000.0729

Kathiresan K, Rajendran N (2005) Coastal mangrove forests mitigated tsunami. Estuar Coast Shelf Sci 65:601–606

Kelaher BP, Chapman MG, Underwood AJ (1998) Changes in benthic assemblages near boardwalks in temperate urban mangrove forests. J Exp Mar Bio Ecol 228:291–307. https://doi.org/10.1016/S0022-0981(98)00036-7

Krebs JM, Bell SS, McIvor CC (2014a) Assessing the link between coastal urbanization and the quality of nekton habitat in mangrove tidal tributaries. Estuar Coasts 37:832–846

Krebs JM, McIvor CC, Bell SS (2014b) Nekton community structure varies in response to coastal urbanization near mangrove tidal tributaries. Estuar Coasts 37:815–831

Kruitwagen G, Hecht T, Pratap HB, Bonga SEW (2006) Changes in morphology and growth of the mudskipper (Periophthalmus argentilineatus) associated with coastal pollution. Mar Biol 149:201–211. https://doi.org/10.1007/s00227-005-0178-z

Kushwaha Shubhda C, Kulkarni Neelima S (2013) Bird diversity of Betawade, Thane, a natural urban habitat. National Conference on Biodiversity : Status and Challenges in Conservation

Kwon M (2015) Examination of anthropogenic nutrient additions to coastal mangrove communities: a multiple isotope approach. Masters thesis, Department of Environmental Science, American University

Lai D-S, Lin J-B, Liu W-S, Pan L-K, Chu K-H, Chen C-Y, Lin D-B (2010) Metal concentrations in sediments of the Tamsui River, flows through central metropolitan Taipei. Bull Environ Contam Toxicol 84:628–634

Lai S, Loke LHL, Hilton MJ, Bouma TJ, Todd PA (2015) The effects of urbanisation on coastal habitats and the potential for ecological engineering: a Singapore case study. Ocean Coast Manag 103:78–85

Lee SY, Primavera JH, Dahdouh-Guebas F, McKee K, Bosire JO, Cannicci S, Diele K, Fromard F, Koedam N, Marchand C (2014) Ecological role and services of tropical mangrove ecosystems: a reassessment. Glob Ecol Biogeogr 23:726–743

Li R, Xu H, Chai M, Qiu GY (2016) Distribution and accumulation of mercury and copper in mangrove sediments in Shenzhen, the world's most rapid urbanized city. Environ Monit Assess 188:1–11. https://doi.org/10.1007/s10661-016-5103-z

Lichtfouse É, Lichtfouse M, Jaffrezic A (2003) δ 13 C values of grasses as a novel indicator of pollution by fossil-fuel-derived greenhouse gas CO2 in urban areas. Environ Sci Technol 37:87–89

Lim HC, Sodhi NS (2004) Responses of avian guilds to urbanisation in a tropical city. Landsc Urban Plan 66:199–215. https://doi.org/10.1016/s0169-2046(03)00111-7

Lima JMT, Staudhammer CL, Brandeis TJ, Escobedo FJ, Zipperer W (2013) Temporal dynamics of a subtropical urban forest in San Juan, Puerto Rico, 2001–2010. Landsc Urban Plan 120:96–106

Lindegarth M, Hoskin M (2001) Patterns of distribution of macro-fauna in different types of estuarine, soft sediment habitats adjacent to urban and non-urban areas. Estuar Coast Shelf Sci 52:237–247

Liu Y, Tam NFY, Yang JX, Pi N, Wong MH, Ye ZH (2009) Mixed heavy metals tolerance and radial oxygen loss in mangrove seedlings. Mar Pollut Bull 58:1843–1849

Lugo AE (1990) Fringe wetlands. In: Lugo AE, Brown S, Brinson MM (eds) Forested wetlands. Elsevier, New York

Lugo AE, Medina E, McGinley K (2014) Issues and challenges of mangrove conservation in the anthropocene. Madera y Bosques 20:11–38

MacFarlane GR (2003) Chlorophyll a fluorescence as a potential biomarker of zinc stress in the grey mangrove, Avicennia marina (Forsk.) Vierh. Bull Environ Contam Toxicol 70:90–96

MacFarlane GR, Burchett MD (2001) Photosynthetic pigments and peroxidase activity as indicators of heavy metal stress in the grey mangrove, Avicennia marina (Forsk.) Vierh. Mar Pollut Bull 42:233–240

MacFarlane GR, Koller CE, Blomberg SP (2007) Accumulation and partitioning of heavy metals in mangroves: a synthesis of field-based studies. Chemosphere 69:1454–1464. https://doi.org/10.1016/j.chemosphere.2007.04.059

Mandura AS (1997) A mangrove stand under sewage pollution stress: Red Sea. Mangroves Salt Marshes 1:255–262

Marchand C, Lallier-Vergès E, Baltzer F, Albéric P, Cossa D, Baillif P (2006) Heavy metals distribution in mangrove sediments along the mobile coastline of French Guiana. Mar Chem 98:1–17. https://doi.org/10.1016/j.marchem.2005.06.001

Martinuzzi S, Gould WA, Lugo AE, Medina E (2009) Conversion and recovery of Puerto Rican mangroves: 200 years of change. For Ecol Manag 257:75–84

Mazerolle MJ, Villard M-A (1999) Patch characteristics and landscape context as predictors of species presence and abundance: a review. Ecoscience:117–124

McDonald RI (2008) Global urbanization: can ecologists identify a sustainable way forward? Front Ecol Environ 6:99–104

McGranahan G, Balk D, Anderson B (2007) The rising tide: assessing the risks of climate change and human settlements in low elevation coastal zones. Environ Urban 19:17–37

McKinney ML (2006) Urbanization as a major cause of biotic homogenization. Biol Conserv 127:247–260

Mestre LAM, Krul R, Moraes V dos S (2007) Mangrove bird community of Paranaguá Bay – Paraná, Brazil. Braz Arch Biol Technol 50:75–83

Metcalfe KN, Glasby CJ (2008) Diversity of Polychaeta (Annelida) and other worm taxa in mangrove habitats of Darwin Harbour, northern Australia. J Sea Res 59:70–82

Millenium Ecosystem Assessment (2005) Ecosystems and human well-being: wetlands and water synthesis. World Resources Institute, Washington, DC

Mohamed MOS, Mangion P, Mwangi S, Kairo J, Dahdouh-Guebas F, Koedam N (2008) Productivity in a peri-urban mangrove: does sewage exposure have an impact? Doctoral thesis, Free University Brussels

Mohamed MOS, Neukermans G, Kairo JG, Dahdouh-Guebas F, Koedam N (2009) Mangrove forests in a peri-urban setting: the case of Mombasa (Kenya). Wetl Ecol Manag 17:243–255

Mohd-Azlan J, Lawes MJ (2011) The effect of the surrounding landscape matrix on mangrove bird community assembly in north Australia. Biol Conserv 144:2134–2141

Msangameno DJ, Jiddawi NS, Yahya SA (2017) An update on the status of mangrove forests in the western coast of Unguja Island, Tanzania: a rural vs peri-urban comparison. Trop Ecol 58:57–69

Mumby PJ, Edwards AJ, Ernesto Arias-Gonzalez J, Lindeman KC, Blackwell PG, Gall A, Gorczynska MI, Harborne AR, Pescod CL, Renken H, Wabnitz CCC, Llewellyn G (2004) Mangroves enhance the biomass of coral reef fish communities in the Caribbean. Nature 427:533–536. doi:http://www.nature.com/nature/journal/v427/n6974/suppinfo/nature02286_S1.html

Nagelkerken I, Roberts CM, Van Der Velde G, Dorenbosch M, Van Riel MC, De La Moriniere EC, Nienhuis PH (2002) How important are mangroves and seagrass beds for coral-reef fish? The nursery hypothesis tested on an island scale. Mar Ecol Prog Ser 244:299–305

Nagendra H, Sudhira HS, Katti M, Schewenius M (2013) Sub-regional assessment of India: effects of urbanization on land use, biodiversity and ecosystem services. In: Urbanization, biodiversity and ecosystem services: challenges and opportunities. Springer, pp 65–74

Nfotabong-Atheull A, Din N, Dahdouh-Guebas F (2013) Qualitative and quantitative characterization of mangrove vegetation structure and dynamics in a peri-urban setting of Douala (Cameroon): an approach using air-borne imagery. Estuar Coasts 36:1181–1192

Nogueira VLR, Rocha LL, Colares GB, Angelim AL, Normando LRO, Cantão ME, Agnez-Lima LF, Andreote FD, Melo VMM (2015) Microbiomes and potential metabolic pathways of pristine and anthropized Brazilian mangroves. Reg Stud Mar Sci 2:56–64. https://doi.org/10.1016/j.rsma.2015.08.008

Noske RA (1995) The ecology of mangrove forest birds in Peninsular Malaysia. Ibis (Lond 1859) 137:250–263

Noske RA (1996) Abundance, zonation and foraging ecology of birds in mangroves of Darwin Harbour, Northern Territory. Wildl Res 23:443–474

Obiefuna JN, Nwilo PC, Atagbaza AO, Okolie CJ (2012) Land cover dynamics associated with the spatial changes in the wetlands of Lagos/Lekki Lagoon System of Lagos, Nigeria. J Coast Res 29:671–679

Odum HT, Arding JE (1991) Emergy analysis of shrimp mariculture in Ecuador. The center

Odum WE, Johannes RE (1975) The response of mangroves to man-induced environmental stress. Elsevier Oceanogr Ser 12:52–62

Olagoke A, Mwihaki L, Bosire J, Berger U (2013) Stand structure and spatial pattern of mangrove regeneration in a degraded peri-urban coastal forest

Olds AD, Albert S, Maxwell PS, Pitt KA, Connolly RM (2013) Mangrove-reef connectivity promotes the effectiveness of marine reserves across the western Pacific. Glob Ecol Biogeogr 22:1040–1049. https://doi.org/10.1111/geb.12072

Oliveira DD, Souza-Santos LP, Silva HKP, Macedo SJ (2014) Toxicity of sediments from a mangrove forest patch in an urban area in Pernambuco (Brazil). Ecotoxicol Environ Saf 104:373–378

Oribhabor BJ, Ogbeibu AE (2010) The ecological impact of anthropogenic activities on the predatory fish assemblage of a tidal creek in the Niger Delta, Nigeria. Res J Environ Sci 4:271–279

Pagliosa PR, Barbosa FAR (2006) Assessing the environment–benthic fauna coupling in protected and urban areas of southern Brazil. Biol Conserv 129:408–417

Penha-Lopes G, Torres P, Narciso L, Cannicci S, Paula J (2009) Comparison of fecundity, embryo loss and fatty acid composition of mangrove crab species in sewage contaminated and pristine mangrove habitats in Mozambique. J Exp Mar Bio Ecol 381:25–32

Penha-Lopes G, Xavier S, Okondo J, Cannicci S, Fondo E, Ferreira S, Macamo C, Macia A, Mwangi S, Paula J (2010) Effects of urban wastewater loading on macro-and meioinfauna assemblages in subtropical and equatorial East African mangroves. West Indian Ocean J Mar Sci 9:195–212

Penha-Lopes G, Torres P, Cannicci S, Narciso L, Paula J (2011) Monitoring anthropogenic sewage pollution on mangrove creeks in southern Mozambique: a test of Palaemon concinnus Dana, 1852 (Palaemonidae) as a biological indicator. Environ Pollut 159:636–645

Pereira HH, Neves LM, Costa MR, Araújo FG (2015) Fish assemblage structure on sandy beaches with different anthropogenic influences and proximity of spawning grounds. Mar Ecol 36:16–27

Peters JR, Yeager LA, Layman CA (2015) Comparison of fish assemblages in restored and natural mangrove habitats along an urban shoreline. Bull Mar Sci 91:125–139

Pham TD, Yoshino K (2016) Impacts of mangrove management systems on mangrove changes in the Northern Coast of Vietnam. Tropics 24:141–151

Pickett STA, Cadenasso ML, Grove JM, Nilon CH, Pouyat R V, Zipperer WC, Costanza R (2008) Urban ecological systems: linking terrestrial ecological, physical, and socioeconomic components of metropolitan areas. In: Urban ecology. Springer, pp 99–122

Proffitt CE, Travis S (2014) Red mangrove life history variables along latitudinal and anthropogenic stress gradients. Ecol Evol 4:2352–2359

Reef R, Feller IC, Lovelock CE (2010) Nutrition of mangroves. Tree Physiol 30:1148–1160. https://doi.org/10.1093/treephys/tpq048

Seto KC, Güneralp B, Hutyra LR (2012) Global forecasts of urban expansion to 2030 and direct impacts on biodiversity and carbon pools. Proc Natl Acad Sci 109:16083–16088. https://doi.org/10.1073/pnas.1211658109

Shriadah MMA (1999) Heavy metals in mangrove sediments of the United Arab Emirates shoreline (Arabian Gulf). Water Air Soil Pollut 116:523–534

Sindern S, Lima RFS, Schwarzbauer J, Petta RA (2007) Anthropogenic heavy metal signatures for the fast growing urban area of Natal (NE-Brazil). Environ Geol 52:731–737. https://doi.org/10.1007/s00254-006-0510-z

Skilleter GA (1996) Validation of rapid assessment of damage in urban mangrove forests and relationships with molluscan assemblages. J Mar Biol Assoc UK 76:701–716

Skilleter GA, Warren S (2000) Effects of habitat modification in mangroves on the structure of mollusc and crab assemblages. J Exp Mar Bio Ecol 244:107–129. https://doi.org/10.1016/S0022-0981(99)00133-1

Sodhi NS, Briffett C, Kong L, Yuen B (1999) Bird use of linear areas of a tropical city: implications for park connector design and management. Landsc Urban Plan 45:123–130

Syphard AD, Clarke KC, Franklin J, Regan HM, Mcginnis M (2011) Forecasts of habitat loss and fragmentation due to urban growth are sensitive to source of input data. J Environ Manag 92:1882–1893. https://doi.org/10.1016/j.jenvman.2011.03.014

Tam NFY, Wong YS (2000) Spatial variation of heavy metals in surface sediments of Hong Kong mangrove swamps. Environ Pollut 110:195–205

Taylor B, Andrews D, Fraser GS (2011) Double-crested cormorants and urban wilderness: conflicts and management. Urban Ecosyst 14:377–394. https://doi.org/10.1007/s11252-011-0165-8

Tian-Hong L, Peng H, Zhi-Jie Z (2008) Impact analysis of coastal engineering projects on mangrove wetland area change with remote sensing. China Ocean Eng 22:347–358

Traut AH, Hostetler ME (2004) Urban lakes and waterbirds: effects of shoreline development on avian distribution. Landsc Urban Plan 69:69–85

United Nations Environment Program U (2006) World urbanization prospects: the 2005 revision

Vasconcelos RO, Amorim MCP, Ladich F (2007) Effects of ship noise on the detectability of communication signals in the Lusitanian toadfish. J Exp Biol 210:2104–2112

Vaz E (2014) Managing urban coastal areas through landscape metrics: an assessment of Mumbai's mangrove system. Ocean Coast Manag 98:27–37

Viana AP, Lucena Frédou F, Frédou T, Torres MF, Bordalo AO (2010) Fish fauna as an indicator of environmental quality in an urbanised region of the Amazon estuary. J Fish Biol 76:467–486

Waltham NJ, Connolly RM (2006) Trophic strategies of garfish, Arrhamphus sclerolepis, in natural coastal wetlands and artificial urban waterways. Mar Biol 148:1135–1141

Wasserman JC, Freitas-Pinto AAP, Amouroux D (2000) Mercury concentrations in sediment profiles of a degraded tropical coastal environment. Environ Technol 21:297–305. https://doi.org/10.1080/09593332108618117

Wayland M, Hobson KA (2001) Stable carbon, nitrogen, and sulfur isotope ratios in riparian food webs on rivers receiving sewage and pulp-mill effluents. Can J Zool 79:5–15. https://doi.org/10.1139/z00-169

Wee AKS, Low SY, Webb EL (2015) Pollen limitation affects reproductive outcome in the bird-pollinated mangrove Bruguiera gymnorrhiza (Lam.) in a highly urbanized environment. Aquat Bot 120, Part:240–243. https://doi.org/10.1016/j.aquabot.2014.09.001

Widmer A, Lexer C, Cozzolino S (2009) Evolution of reproductive isolation in plants. Heredity (Edinb) 102:31–38

Wolters J, Gillis LG, Bouma TJ, Katwijk MM, Ziegler AD (2016) Land use effects on mangrove nutrient status in Phang Nga Bay, Thailand. L Degrad Dev 27:68–76

Wong YS, Tam NFY, Lan CY (1997) Mangrove wetlands as wastewater treatment facility: a field trial. In: Wong Y-S, Tam N-Y (eds) Asia-Pacific conference on science and management of coastal environment. Springer, Dordrecht, pp 49–59

Yan ZZ, Ke L, Tam NFY (2010) Lead stress in seedlings of Avicennia marina, a common mangrove species in South China, with and without cotyledons. Aquat Bot 92:112–118. https://doi.org/10.1016/j.aquabot.2009.10.014

Yim MW, Tam NFY (1999) Effects of wastewater-borne heavy metals on mangrove plants and soil microbial activities. Mar Pollut Bull 39:179–186. https://doi.org/10.1016/S0025-326X(99)00067-3

Young RF, Wolf SA (2006) Goal attainment in urban ecology research: a bibliometric review 1975–2004. Urban Ecosyst 9:179–193

Yu R-Q, Chen GZ, Wong YS, Tam NFY, Lan CY (1997) Benthic macrofauna of the mangrove swamp treated with municipal wastewater. Hydrobiologia 347:127–137. https://doi.org/10.1023/A:1003027520750

Yusuf DN, Prasetyo LB, Kusmana C (2017) Geospatial approach in determining anthropogenic factors contributed to deforestation of mangrove: a case study in Konawe Selatan, Southeast Sulawesi. In: IOP conference series: Earth and Environmental Science. IOP Publishing, p 12049

Zhang F-Q, Wang Y-S, Lou Z-P, Dong J-D (2007) Effect of heavy metal stress on antioxidative enzymes and lipid peroxidation in leaves and roots of two mangrove plant seedlings (Kandelia candel and Bruguiera gymnorrhiza). Chemosphere 67:44–50

High-Throughput Techniques As Support for Knowledge-Based Spatial Conservation Prioritization in Mangrove Ecosystems

24

Véronique Helfer and Martin Zimmer

Abstract

The conservation of functioning ecosystems worldwide is warranted by the need for reliable and sustainable provision of ecosystem services locally, regionally and globally. Mangroves provide numerous ecosystem services both to local human communities, e.g., coastal protection or food security, and to mankind worldwide, e.g., climate change-mitigation. Nonetheless they still lack protection in many places of occurrence. Here we base spatial prioritization and planning of mangrove conservation on functional biodiversity and service-relevant ecosystem processes, being studied through cutting-edge genetic and chemical analyses of sediments to unravel the links between biodiversity, biotic interactions, ecosystem processes and ecosystem services. We nonetheless recommend multidisciplinary approaches when planning protected area networks for the sustainable use and provision of ecosystem services and pledge for (*i*) considering and prioritizing societal, biological and economic values of mangroves, (*ii*) integrating adjacent ecosystems to maintain connectivity, and (*iii*) taking into account the spatial and temporal dynamics of mangrove ecosystems and their community composition under global change, i.e. changes in the spatial distribution of species and services over time. Beyond the example of mangroves and the turnover of organic matter in mangrove sediments described herein, our approach to spatial conservation prioritization and planning is applicable to any other ecosystems and their services.

Keywords

Environmental DNA · Metabarcoding · Metabolic fingerprinting · Environmental metabolomics · Biodiversity · Ecosystem processes and services

24.1 Introduction

24.1.1 The Biodiversity-Procession: From Biodiversity to Ecosystem Services

Besides the intrinsic value of biodiversity (BD) and our moral responsibility to protect it (Cimon-Morin et al. 2014; Deliège and Neuteleers 2014), efficient strategies for its conservation should aim at sustainable provision and use of ecosystem services (ES) which, in turn, will also support biodiversity conservation, if and when ecosystem services' value or richness coincide with high biodiversity (Cimon-Morin et al. 2013). From a functional perspective, species are relevant, when their functional traits and their interactions support the internal functioning of ecosystems (EFg; De Groot et al. 2002) and/or the provision of goods and services, referred to as ecosystem functions (EF; De Groot 1992). Ecosystem services are provided by the community of living organisms (flora, fauna and microbiota) and their interactions, under given abiotic conditions (e.g., climate, characteristics of soil/sediment, air and water). Biodiversity drives ecosystem processes (EP) through biotic interactions (BI) –be they antagonistic or synergistic, competitive or facilitative– that depend upon community structure: upon interaction, groups of organisms act as ecosystem service-providing units (SPU) that control ecosystem processes. Ecosystem processes warrant ecosystem functioning in terms of performing ecosystem

V. Helfer (✉)
Department of Ecology, Leibniz Centre for Tropical Marine Research, Bremen, Germany
e-mail: veronique.helfer@leibniz-zmt.de

M. Zimmer
Department of Ecology, Leibniz Centre for Tropical Marine Research, Bremen, Germany

University of Bremen, Faculty 02 Biology/Chemistry, Bremen, Germany

IUCN-SSC Mangrove Specialist Group, Bremen, Germany
e-mail: martin.zimmer@leibniz-zmt.de

C. Makowski, C. W. Finkl (eds.), *Threats to Mangrove Forests*, Coastal Research Library 25,
https://doi.org/10.1007/978-3-319-73016-5_24

functions that, in turn, provides numerous ecosystem services (ES). Henceforth, this chain of relationships and dependencies (BD – BI – SPU –EP – EFg & EF – ES) will be referred to as biodiversity-procession (see Appendix I).

Accordingly, to support sustainable ecosystem service-provision, we need to know which species and/or communities (service-providing units) drive ecosystem processes that underlie ecosystem services under particular abiotic conditions. This requires knowledge on *who is there*, *under which environmental conditions*, and *how this correlates with ecosystem processes*. Biodiversity-procession and the resulting ecosystem services provided under particular conditions may vary regionally, depending on species-occurrence, and may change over time, owing to environmental (and land-use) change (see Brose and Hillebrand 2016).

Elucidating biodiversity-procession as sound basis for prioritization, planning and management of ecosystems requires targeting a wide range of taxa (BD) under various environmental conditions that potentially drive the ecosystem processes (EP) in focus. Along the same line, these ecosystem processes need to be studied in detail as they are linked to, and based on, biodiversity. Both requirements can nowadays be met through using cutting-edge high-resolution and high-throughput techniques of molecular biology (e.g., environmental DNA metabarcoding; see Appendix II) and (organic) chemistry (e.g., environmental metabolomics or metabolic fingerprinting; see Appendix II).

24.1.2 The Biodiversity-Procession in Mangrove Ecosystems

Studying biodiversity-procession in mangroves must encompass the assessment of local biodiversity (*who is there*) and relevant measures of the ecosystem process under focus (*how this correlates with ecosystem processes*), such as chemical properties of the leaves, roots and sediments for organic matter (OM) turnover or nutrient cycling, under different environmental settings (*under which environmental conditions*).

24.1.2.1 Who Is There?

The term mangrove refers both to the trees or the whole community and/or ecosystem, the latter encompassing the biotic community and its abiotic environment (Whitham et al. 2006). The tree component in mangrove ecosystems is essential; mangrove trees are recognized as **foundation species** or **ecosystem engineers** (see Appendix II) of mangrove ecosystems, the loss or decline of which affects adjacent ecosystem beyond boundaries of mangroves themselves (Ellison and Farnsworth 2001; Ellison et al. 2005). For instance, morphological traits of mangrove trees, such as their roots, influence sediment accretion or aeration differently, thus shaping the habitat and creating different environmental conditions for associated species (infauna, microbial communities and megafauna). Thus, mangrove trees shape their soft-bottom intertidal environment into something entirely different (i.e, they are **ecosystem engineers**) and provide a hard substratum for numerous sessile organisms and habitat for a plethora of motile marine and terrestrial animals. Similar effects can be expected from physiological traits, such as leaf litter chemistry and decomposability, which will influence sediment chemistry, particularly the structure and composition of sediment organic matter (SOM). As such, mangrove trees are expected to strongly influence ecosystem processes. However, ecosystem processes, such as organic matter (OM) turnover, are also influenced by other drivers, be they biotic (mega-, micro- and infauna; microbiota) or abiotic (salinity, temperature, inundation patterns). Crab communities are considered key players in mangrove systems; burrowing crabs contribute to habitat-shaping in many mangrove systems (Pülmanns et al. 2015, 2016; Pestana et al. 2017), and are therefore also referred to as **ecosystem engineers**, but their effects on ecosystem characteristics is more local and less far-reaching than those of mangrove trees. Some (apex) predators might act as **keystone species** (see Appendix II), but too little is known about mangrove community structure and interactions to draw straightforward conclusions in this regard. Similarly, essentially nothing is known about the occurrence, not even to mention the role, of other mangrove inhabitants (but see Cannicci et al. 2008), such as annelids or nematodes, and microbial diversity (encompassing fungi, archaea, bacteria and protists) is still usually treated as a black box when it comes to organic matter dynamics in mangroves (but see, e.g., Holguin et al. 2001; Kristensen 2007; Kristensen et al. 2008; Das et al. 2015; Luo and Gu 2016).

When it comes to the description of mangrove ecosystem processes worldwide, the global distribution of mangrove forests has to be considered. The global mangrove species distribution allows for a clear distinction of species-rich Indo-West Pacific (IWP) *versus* species-poor Atlantic Eastern Pacific (AEP) mangrove forests. A remarkable pattern is the almost complete lack of species overlap between the IWP and the AEP floras (with the exception of the fern *Acrostichum aureum* and the still discussed *Rhizophora mangle*/*R. samoensis* case). Despite the little species overlap among these biogeographical regions, morphological (and physiological) traits can be distinguished that allow for clustering species (mostly on a genus basis) into functional groups, e.g., a "*Rhizophora*-type" (prop- or stilt-roots, bowing from the trunk into the sediment), an "*Avicennia*-type" (pencil-roots, popping out the sediment like fingers), a "*Bruguiera*-type" (knee-roots, popping out the sediment and bowing back downwards), or a "*Kandelia*-type" (essentially no

Fig. 24.1 Mangrove root architecture: pneumatophores ("finger-roots" or "pencil-roots") of *Avicennia germinans* in Brazil (upper left; © V. Helfer); prop roots of *Rhizophora mangle* in Brazil (upper right; © V. Helfer); "knee-roots" of *Bruguiera gymnorhiza* in Fiji (lower left; © M. Zimmer); buttress- or plank-roots of *Xylocarpus granatum* in Fiji (lower right; © M. Zimmer)

aboveground roots, but buttress-enlarged stem basis; c.f. Tomlinson 2016). These root-types (see Fig. 24.1) differ from each other in, e.g., how they trap and stabilize sediment or particulate organic matter, aerate the belowground habitat, and serve as hard-substratum for colonization by sessile organisms in an otherwise soft-bottom environment. We expect similar biogeographically independent overlap in physiological traits, such as leaf litter chemistry and decomposability that will translate into sediment chemistry, particularly the structure and stability of sediment organic matter. Similarly, communities of burrowing crabs –often considered **ecosystem engineers** in mangrove systems– differ among biogeographical areas, but functional groups (e.g., of the genus *Uca*, feeding on microphytobenthos and bacteria and constructing shallow burrows *versus* deep- or shallow-burrowing detritus-feeders, such as, e.g., *Ucides cordatus* in Eastern American mangroves or sesarmids worldwide) occur in essentially each type of mangrove ecosystems.

24.1.2.2 Under Which Conditions?

Mangrove communities inhabit sheltered areas under tidal influence along tropical and subtropical coasts. Most probably constrained by winter seawater temperatures (Alongi 2009), their global distribution is centred on the tropical belt, reaching the 30° N and 30°S (Giri et al. 2011; but note the 40° S in Australia/New Zealand) latitudes, while their regional distribution is shaped by major oceanic currents. Thus, the Canary and Benguela cold currents restrict their extent to 20°N and c. 15°S, respectively, on the African west coast, and the Peru/Humboldt current confine mangrove distribution to latitudes not lower than c. 5°S on the American West Coast. Two important ecological descriptors of mangrove communities, tree species diversity and aboveground biomass (AGB), follow similar distribution patterns, showing a decrease from the equator towards the poles, but more importantly, a huge longitudinal variation, with the maximum richness and biomass being localized in Southeast

Asia, a pattern referred to as "the mangrove biodiversity anomaly" (see Ellison et al. 1999). While on the global scale the similar distribution patterns observed for species diversity and aboveground biomass -a potential proxy for productivity- might suggest a strong link between biodiversity and ecosystem processes linked to production, more in-depth studies are needed to understand how mangrove plant distribution shapes, or depends on, the distribution of other organisms (such as microbial communities, infauna and macrofauna), and how species interactions drive ecosystem processes and subsequent services in mangrove ecosystems.

Alike other coastal ecosystems, mangroves are subject to strong fluctuations in salinity and inundation patterns (duration and frequency). Crase et al. (2013) suggested that these environmental conditions (foremost inundation patterns, followed by sediment salinity) drive the spatial patterns of species dominance ("zonation") generally observed in mangrove communities on small spatial scales. While this may hold true for mangroves of the IWP, the zonation-like distribution of the few mangrove species dwelling in the AEP may, alternatively, be driven by species-succession along the spatio-temporal gradient of newly established and subsequently accreting and developing soft-bottom shores and banks. Hence, understanding how abiotic conditions, in combination with biotic interactions, might affect ecosystem processes, requires considering environmental conditions, such as salinity, tidal regime and temperature.

24.1.2.3 Mangroves Ecosystems Services

Mangroves are among the most productive ecosystems, and they provide manifold ecosystem services to mankind worldwide and to local societies, among them (as reviewed in Spalding et al. (2010) and Vo et al. (2012)) many ***production*** services (e.g., provision of fuel wood, construction material, food and feed), ***regulation*** services (e.g., biofiltration, coastal protection, C-storage), ***habitat*** functions (providing nursery and refuge for fish, turtles, etc.), ***information*** services (offering recreation, spiritual and historical information). More specifically, mangroves provide various goods to humans: food, tannins, fiber and dyes, fuel wood and charcoal, construction materials (wood, roofing) and even medicines; they offer protection of the coastline, and of local residents, from hurricanes and storms, and reduce coastal erosion and siltation; mangrove trees can take up nitrate and phosphate (as well as pollutants), cleaning up land-derived surface run-off and preventing the nearshore water body from eutrophication (and pollution). By this, they improve light conditions for sublittoral macrophytes, phytoplankton and coral reefs as well as for the visual orientation of juvenile fish in mangrove systems. Mangrove forests form the basis of a detrital food chain, where leaf litter provides coastal waters with much of their productivity (Mcleod et al. 2011). This high level of productivity contributes to the great biomass and wide array of species that inhabit mangroves or use them as refuges, migration sites, and nurseries: about 75% of all tropical fish and crustaceans spend part of their lives in mangrove swamps as juveniles, and adjacent mangroves improve fish abundance and diversity in coral reefs (Harborne et al. 2006; Serafy et al. 2015).

At large and even global scales, the provision of nursery and habitat to a plethora of sealife of commercial interest by mangroves is currently becoming of interest to economies worldwide. Even more important, mangroves are amongst the most efficient sinks for greenhouse gases (GHG), sequestering more carbon and nitrogen than many of their terrestrial, coastal or marine counterparts due to their high productivity (Mcleod et al. 2011) and to the vast amounts of carbon stored in sediments, and may thus contribute to counteracting GHG-driven climate change. Most of the above processes depend –directly or indirectly– on (microbial and faunal) sediment inhabitants and their interactions, and the processes of element-cycles and fluxes driven by them.

The protection of wildlife from extinction and the sustainable use of fisheries and timber resources are considered as important as the promise for natural pharmaceutical products based on traditional ethno-medical knowledge. As popular touristic sites, mangroves provide services to tourists and, at the same time, an important income for local societies to sustain their lives and improve their standard of living. The recent IPCC report stresses the worldwide importance of the provision of water, food and income. This is generally assumed the basis for preventing poverty-driven migration, societal conflicts and warfare.

24.1.2.4 Unravelling the Biodiversity-Procession in Mangroves

Along this line, our knowledge of whether and how different mangrove systems with high versus low biodiversity, and distinct floras and faunas, exhibit different ecosystem processes or provide different ecosystem services is in its infancy. A first step towards a better understanding of these relationships would be a comprehensive analysis of the complete set of species that make up the mangrove community rather than quantifying only easily detectable or enigmatic species and disregarding cryptic species (e.g., in the sediment). In a next step, this in-depth comprehension of the community would have to be complemented by acquaintance of the procession from the presence of single species or groups of species to ecosystem processes. As different mangroves host different species (e.g., IWP vs. AEP), taking into account critical effect traits of different species rather than the species themselves will foster our understanding of the link between biodiversity and ecosystem processes and services. This approach requires targeting a wide range of taxa from distinct kingdoms or domains (e.g., Bacteria,

Archaea, Protista, Fungi, Animalia, Plantae) in various environmental conditions, for instance through complex study design and sampling effort. The latter can be facilitated by targeting sediments samples from which we can assess both biodiversity and processes using eDNA metabarcoding and environmental metabolomics approaches, respectively (see § 24.3 and Appendix II for definitions).

24.2 Spatial Conservation Prioritization for Sustainable Use and Provision of Ecosystems Services

24.2.1 Rationale of Spatial Conservation Prioritization

Effective conservation of threatened ecosystems requires both acceptance by local communities and careful selection of the areas to protect (Appendix III), accounting for (*i*) present and future settings and (*ii*) connectivity among ecosystems. This is facilitated by spatial conservation prioritization approaches, such as those implemented in the Marxan (Ball et al. 2009) or Zonation (Moilanen et al. 2005) frameworks, where both ecological values (such as biodiversity) of the system and their use by human societies (related to the ecosystem processes and services) can simultaneously be accounted for. These quantitative approaches rely on spatially explicit information about e.g., biodiversity and/or ecosystem service patterns, and potential costs (to human societies, in terms of, e.g., livelihood provision). The aim is to provide scenarios for the establishment of protected area networks that are driven by evidence-based transparent rules, in order to find solutions to potentially antagonist economic or ecological demands on the considered region. Importantly, human preferences can be accounted for by attributing different weights to the different targets of the scenario, such considering not only the ecological values of the ecosystems, but also its societal and economic values. The outcome of such approaches is recommendations with respect to which areas should be prioritized for conservation (see Appendix III).

24.2.2 Information Needed to Support Spatial Conservation Prioritization

Data used for spatial conservation prioritization mostly relate to the spatial distribution of species and/or ecosystem services (e.g., Chan et al. 2006; Leathwick et al. 2008). Whereas simple analyses only include information about species distribution, additional data (see Moilanen 2012) can be profitably used in spatial conservation planning; among them (*i*) the distribution of threats to biodiversity or service provisioning (that can –deforestation– or cannot –climate change– be halted), (*ii*) the costs to human societies, not only in terms of land acquisition and maintenance, but also due to opportunity costs to various stakeholders (e.g., restriction in the use of resources).

Information on the spatial distribution of BD or EP can relate to present, past or future settings, as detailed below:

- Present spatial patterns of biodiversity (and ecosystem processes), derived from species distribution modeling (Elith and Leathwick 2009). Sarker et al. (2016) perfectly showcased the use of species distribution modeling for evaluating our ability to protect threatened mangrove species of the Sundarbans world heritage ecosystem. Their habitat suitability models show that the actual protected area network does not include the most suitable habitats of the four most prominent species in the area.
- Historical spatial patterns of biodiversity and ecosystem processes. Combining information derived from paleoclimatic reconstructions and historical occurrences of species (derived from palynology, fossils or aDNA (see Appendix II) analysis as described hereafter) allows fitting climatic niche models over different time steps using hindcasting methods (see Nogués-Bravo 2009), and thus reconstructing the evolution of spatio-temporal patterns of biodiversity. More recent changes in the distribution, species composition and/or health status of mangroves can be deduced from airborne and spaceborne remote sensing technologies that advanced drastically over the last years (see e.g., Adam et al. 2010; Vidhya et al. 2014; Asner et al. 2017). Understanding changes from past to present communities in the light of environmental change can help evaluating the changes in spatial distribution that have already taken place, thus giving us some information about what we can expect upon future climate change (Helmuth et al. 2006; Pearman et al. 2008). In the context of spatial conservation prioritization, such information can help assessing and mapping the threats, identified by a decline in BD or ES provision.
- Future spatial patterns of biodiversity and ecosystem processes. Species distributions will be affected by climate change, which will reshape communities and the distribution of biotic interactions and the entire chain of biodiversity-procession. Therefore, it is important to be able to project the future distribution of species and communities and the ecosystem processes and services they provide. Several studies already tackled the effect of climate change on mangrove distribution (e.g., Record et al. 2013) and provide forecasts of community distributions. However, their focus was on the floral community distribution, and they did not account for other units of

mangrove communities, such as the fauna or the microbiota. We hold that taking advantage of the eDNA metabarcoding approach for multi-taxa detection will help refining predictions about future mangrove communities as a whole. It will also provide valuable information about "ecological surprises" (coined by Williams and Jackson 2007) that result from non-analogue communities that might occur under novel climatic conditions. This can serve as a basis for ex-situ experimental design to test how new assemblages will affect ecosystem processes and services. In addition, future distribution scenarios should be taken into consideration for spatial conservation prioritization, in order to ensure the right choice of areas and systems to be protected for sustainable ecosystem service-provisioning at present, but also in the future.

Another aspect that has to be considered for spatial conservation prioritization is the connectivity of mangroves and adjacent coastal, marine and terrestrial ecosystems. Occupying the intertidal zone along tropical and subtropical coasts, riverine estuaries and deltas, mangrove ecosystems are at the very interface of marine, freshwater and terrestrial influences. They are ecologically linked to complexes of ecosystems (from coral reefs and seagrass beds offshore to flooded wood- and grasslands -including saltmarshes and tidal freshwater forests, as well as mudflats and salt pans of the intertidal- landwards) via fluxes of nutrients and/or organic matter and migrating organisms (Spalding et al. 2010; c.f. Hübner et al. 2015). To efficiently protect mangroves and their services, we need to understand how they connect and exchange with adjacent ecosystems in order to maintain those links that are of importance for service-provision by mangroves. Spatial subsidies (sensu Polis et al. 1997) have thus far mostly been studied through comparison of isotopic signatures, but recent developments in molecular and chemical approaches to analysing the chemical structure of organic matter and its biotic origin (c.f. Bakkar et al. 2017) may provide more detailed information than isotopic signatures (see e.g., Reef et al. 2017).

24.2.3 Reaching Consensus and Acceptance Using Spatial Conservation Prioritization

Conservation might not always result in maintaining all services provided by local mangrove systems, if they compete or are antagonistic. For instance, many wood products are being used by local societies; allowing for wood extraction for fuel production might be antagonistic to preserving intact ecological communities for climate mitigation or coastal protection. Another example is food provisioning: due to their high productivity, mangroves are targeted for aquaculture or shrimp-farming, while both activities require wood extraction and sediment excavation, which results in GHG-release into the environment. Going further, conversion of mangrove forests into agricultural land (e.g., for rice production in Myanmar, or palm oil plantation in Malaysia and Indonesia; Richards and Friess 2016) could be argued as being a good initiative, providing this is done in a sustainable way, towards the UN SDG2 (Sustainable Development Goals; SDSN 2015), i.e. "end hunger and achieve food security and improved nutrition", while constituting a severe threat regarding climate change mitigation, and thus counteracting SDG 13 ("Take urgent action to combat climate change and its impacts"). A single area can obviously not support all ecosystem services simultaneously. The solution to this resides in assigning distinct priorities to distinct areas (Appendix III) and finding spatial arrangements that allow for the maximisation of all targets (or goals) in a protected area network. A seminal demonstration of the potential of spatial conservation prioritization in reconciling different needs during the design and planning of marine protected areas was delivered by Leathwick et al. (2008). Using the Zonation framework, they were able to provide a marine protected area (MPA) scenario 2.5 times more efficient in terms of conservation benefits (biodiversity protection) and, most strikingly, at a lower cost to fishers, than the equally-sized benthic protected areas that had been implemented at the request of fishers, as the proposed network of protected areas completely avoided fishing grounds. This study showcased how consensus can be found for the benefits of all parties, resolving conflicting objectives in a transparent and scientific way and by offering alternatives to established patterns of resource-use. The need for evidence-based (or knowledge-driven) decision-making (recognized by the UN in their resolution adopted by the General Assembly on 25th September 2015[1]) has been since long advocated in the field of conservation biology (Pullin and Knight 2003; Sutherland et al. 2004; Svancara et al. 2005), but is still rarely applied in mangrove management.

To our knowledge, spatial conservation prioritization approaches have been rarely (if at all) applied for the design of protected area networks in mangrove ecosystems, while they are urgently needed to allow concealing the many various demands and counteract the threats on mangrove ecosystems. The type and amount of information needed, i.e. spatially explicit information relative to BD and EP with a high spatial resolution can be facilitated by the use of high-throughput techniques as described hereafter.

[1] Transforming Our World: The 2030 Agenda for Sustainable Development (UNGA Resolution A/RES/70/1, 25 September 2015) ('2030 Agenda').

24.3 High-Throughput Techniques for Mangrove Conservation

24.3.1 Environmental DNA Metabarcoding to Assess Biodiversity

Assessing biodiversity composition at the multi-taxa level and high spatial resolution was highly cost- and time-demanding before the development and recent spread (due to improved cost-efficiency) of high-throughput DNA-sequencing technologies, also known as next generation-sequencing (NGS; for a description of properties, pros and cons of the different available technologies, see Shokralla et al. 2012; van Dijk et al. 2014). First applied for species identification and discovery, DNA barcoding of tissues evolved into metabarcoding of environmental samples (see Appendix II) potentially revolutionizing various fields of ecology (Joly et al. 2014; Pedersen et al. 2015), notably ecotoxicology (Gardham et al. 2014), spatial ecology (e.g., Groom et al. 2016), studies of food webs and trophic interactions (Clare 2014 and references therein), species invasion (Ficetola et al. 2008), and landscape community genomics (Hand et al. 2015). In particular, the role of environmental DNA (eDNA; see Appendix II) for biodiversity-monitoring has been emphasized recently (Bohmann et al. 2014; Thomsen and Willerslev 2015; Creer et al. 2016), highlighting its potential to even track BD- and community-changes back in time (see Pedersen et al. 2015) and assess past biodiversity under past environmental conditions through the analysis of ancient eDNA (aDNA; see Appendix II). Owing to public DNA-sequence databases (such as GenBank, EMBL, SILVA) and the development of bioinformatics tools, taxon-specific metabarcodes can be targeted using newly designed primers (Riaz et al. 2011) evaluated in silico (Ficetola et al. 2010). This allows for great flexibility in fitting the markers to the specific needs of metabarcoding, i.e., the amplification of short fragments (short fragments can be recovered over longer time periods) and a given variability, allowing for targeting specific taxonomic groups with the desired taxonomic resolution. The efficiency of the metabarcode in describing community composition at the desired taxonomic resolution will depend on the availability of sequences (taxonomical coverage) in verified databases, and the presence of conserved regions allowing efficient primer-binding during DNA amplification, to ensure unbiased representation of each taxa (see Deagle et al. 2014).

Many studies used this approach to monitor biodiversity patterns in terrestrial, coastal and marine ecosystems (see Deiner et al. 2017 for a review), based on water (e.g., Foote et al. 2012; Thomsen et al. 2012), ice (e.g., Willerslev et al. 2003) and soil or sediment samples (e.g., Guardiola et al. 2015). Assessing biodiversity using high-throughput DNA-sequencing presents many advantages (Thomsen and Willerslev 2015) relative to traditional approaches: standardized sampling, non-invasiness, sensitivity for cryptic species, cost-effectiveness, detection independent of weather conditions. There are, of course, several limitations and challenges linked to this approach, due to the properties of environmental DNA (Barnes and Turner 2016) and of sequencing technologies. For example, the problem of false presences due to cross-contaminations, or false absences due to low detection probability (also a big issue in traditional monitoring of animals; see e.g., MacKenzie et al. 2002), has been discussed and statistical approaches have been developed to cope with these issues (see Ficetola et al. 2015; Ficetola et al. 2016; Lahoz-Monfort et al. 2016).

In the case of mangrove ecosystems, there are several challenges to the application of eDNA metabarcoding for present and past biodiversity assessment. One important aspect is the DNA preservation under tropical conditions; while few studies investigated eDNA in tropical ecosystems (as reviewed in Pedersen et al. 2015), characteristics of mangrove sediments (saline, anoxic) nonetheless presuppose a good DNA preservation (Parducci et al. 2012; Giguet-Covex et al. 2014; Willerslev et al. 2014), enabling present biodiversity assessment. Nonetheless, how coastal sediment translocation or water movement can affect the spatial resolution of the information obtained from eDNA analysis has still to be evaluated and the resolution of the ensuing analyses adapted.

When it comes to unravelling past biodiversity patterns (as a supporting information to spatial conservation prioritization), other challenges have to be considered. Despite the anoxic conditions of sediments that favour the preservation of ancient eDNA, warm temperatures in tropical regions might limit its persistence, thus reducing the detection probability of species present in the past. Considering approaches that increase the probability of detecting ancient DNA, such as ultrafiltration to increase DNA concentration, and multiple amplification of the same samples (8 PCR replicates for ancient DNA; Ficetola et al. 2015) to maximize true detection while ensuring minimal false presence, is needed. In addition, bioturbation (notably by burrowing crabs; see Fig. 24.2) is another aspect to be considered when reconstructing biodiversity over time using sediment cores as this might affect the reliability of the DNA chronosequences in sediment layers. Mangroves harbour a high diversity of crabs (see Fig. 24.3), among them species that dig burrows as deep as 2 m (Wunderlich et al. 2008). Depending on local sedimentation rates, this can affect the archive of more than a millennium. Therefore care should be taken, when selecting samples to compare present and past communities so that these uncertainties become less problematic. Radio-isotope-based dating should also be used to verify the chronosequence as a basis for for sample selection.

Fig. 24.2 Assemblage of large (*Ucides cordatus*) and small (*Uca* spp.) crab burrows in a Brazilian mangrove sediment underneath *Laguncularia racemosa*. The ring of excavated sediment around the large burrow in the lower center is a clear indicator of bioturbation (© M. Zimmer)

24.3.2 Environmental Metabolomics to Assess Ecosystem Processes

Ecosystem services are provided by service-providing units (SPU), species or sets of species that interact, driving those ecosystem processes that underlie these ecosystem services (see Appendix I). For instance, sedimentation and coastal protection are driven by the extensive aerial roots of some (but not all) mangrove trees that trap and hold sediment. Species, however, differ in sedimentation rates, depending on density and structure (architecture) of their roots. Hence, different communities of mangrove trees will differ in their effectiveness in trapping and holding sediment and protecting the coastline from erosion.

Another ecosystem process that underlies a globally relevant ecosystem service, namely the mitigation of GHG-driven climate change, is the turnover (through decomposition) of organic matter. As a result of this process –in addition to the above- and belowground biomass of mangrove trees– these forests store huge amounts of carbon and nitrogen in their anoxic saline sediments. The recalcitrance towards

Fig. 24.3 Large, deep-burrowing mangrove crab (*Goniopsis* cf. *pulchra*) in arid Mexico (upper left; © M. Zimmer); high density of shallow-burrowing fiddler crabs (*Uca* spp.) on a shore bank of the same mangrove (lower; © M. Zimmer); tree-climbing crab (*Episesarma versicolor*) in Singapore (upper right; © V. Helfer)

mineralization of this organic matter –originating from root exudates, detrital leachates or decay and decomposition of their detritus– and hence its stability over time, however, depends on its chemical structure that, in turn, is driven by the species composition of the mangrove (and the environmental conditions: for a general review on detrital processes, see Zimmer 2017). The species composition of the detritivorous and decomposing community of microbes and animals, on the other hand, drives decay and decomposition and the resulting structure of the organic matter stored in the sediment. This, in turn, will translate into whether and how fast SOM is either further processed into CO_2, CH_4 or N_2O that will be released into the atmosphere upon sediment turbation and aeration owing to mangrove degradation, or remains stable in anoxic sediments of functioning mangroves.

The chemical structure and composition of SOM can be determined at high resolution using technologies such as pyrolysis-gas chromatography/mass spectrometry ((py-)GC/MS; e.g., Tolu et al. 2015) and near-infrared reflectance spectroscopy (NIRS; e.g., Fuentes et al. 2012) respectively. These techniques proved cost-efficient and allow for high-throughput analysis of samples, as required under complex sampling schemes (Gerber et al. 2012; Kleinebecker et al. 2013; Tolu et al. 2015).

NIRS analyzes the near-infrared spectrum (ca 800–2500 nm) of the light reflected by solid materials or molecules in solution or suspension (via transmission spectroscopy). Owing to complex spectra emitted from organic material, it can be difficult to assign specific features to specific chemical components. However, upon calibration and subsequent multivariate statistics, particular compounds can be screened in a non-invasive, fast manner with high accuracy and at high throughput. Besides total C and N contents, the contents of specific compounds, such as proteins (N-dynamics), cellulose or lignins (C-dynamics), will provide information on decomposition processes and SOM turnover that can be related to mangrove community composition and abiotic environmental conditions.

(py-)GC/MS volatizes solids through thermal treatment (pyrolysis), to make gas chromatography possible without the need of chemical derivatization. Applied to soils, this technique has been used to analyse the chemical composition of organic matter (Buurman and Roscoe 2011) and infer changes in climatic conditions and vegetation composition over time (Vancampenhout et al. 2009; Schellekens et al. 2011). Hence, differences in the OM-signature of mangrove sediments, as obtained from (py-)GC/MS, may indicate differences in the species composition of the vegetation, and thus, can be used to track changes in mangrove communities over time. Along the same line, py-chromatograms of sediments may indicate different routes of decay or decomposition, hinting at different sediment community compositions (Buurman et al. 2009; Vancampenhout et al. 2009; Schellekens and Buurman 2011) that can be correlated with ecosystem processes and environmental conditions in time and space.

The most straightforward application of performing environmental metabolomics (Viant and Sommer 2013) is metabolic fingerprinting (sensu Du and Zeisel 2013), a multivariate comparison of the entity of detectable compounds (or groups of compounds) without detailed and individual identification of these compounds. Statistical analyses of the resulting fingerprints pinpoint those compounds that drive differences among samples, and these compounds can then be identified afterwards (e.g., Bakkar et al. 2017).

24.4 C-Sequestration and Organic Matter (OM) Turnover as Showcase

24.4.1 The Role of Mangrove Ecosystems

Originally ignored due to their little overall extent (merely 0.1% of the earth continental surface: Giri et al. 2011 and 0.6% of the tropical forest area: Hutchison et al. 2014), mangrove forests are now recognized as standing among the most important ecosystems in the context of global change. They sequester more carbon on a unit-area basis than any other terrestrial or marine ecosystem, which is of prime importance for the mitigation of GHG-driven climate change (e.g., Donato et al. 2011; Mcleod et al. 2011; Hutchison et al. 2014; Alongi 2014). Mangrove ecosystems not only store huge amounts of carbon in their aboveground biomass (modelled average of 184.8 t ha^{-1}, but with high spatial variation from less than 80 to 280 t ha^{-1}; Hutchison et al. 2014) and even more in their belowground biomass, due to a skewed belowground to aboveground biomass ratio (Komiyama et al. 2008; Lovelock 2008), but also in the sediments, where methane production (and subsequent release) by bacterial communities is generally prevented by the saline conditions. In addition, mangroves are estimated to be responsible for 11% of the total terrestrial carbon input into the ocean (Jennerjahn and Ittekot 2002) and 10% of the total oceanic dissolved organic carbon (DOC) that equals atmospheric CO_2 (Dittmar et al. 2006; Kristensen et al. 2008).

Despite their indisputable role in carbon-sequestration, mangroves still suffer from serious area loss or degradation (e.g., Richards and Friess 2016). Knowing that deforestation is the second largest source of anthropogenic CO_2 emissions (12–20% of the total; IPCC 2007; van der Werf et al. 2009) and that mangroves are evaluated to be responsible of 10% of these CO_2 emissions due to deforestation (Donato et al. 2011), it is essential for prioritizing mangrove conservation and restoration to understand who are the key players in carbon-sequestration and -storage processes in mangrove ecosystems, under different environmental settings.

24.4.2 What Are the Main Drivers of Carbon-Storage in Mangrove Ecosystems?

Thus far, mangrove research focused on productivity and ABG of the floral components of mangrove ecosystems, but little attention has been given to their microbial counterparts, while it is known since long that they play a key role in geochemical nutrient cycling in other ecosystems and as such have a huge influence on ecosystem functioning (e.g., Baskin 1994). Many of the ecosystem services provided by mangroves are driven by the microbial communities of the sediment, be it directly or indirectly. The role of microbial communities for ecosystem processes, such as soil carbon-sequestration and/or -storage, has been stressed recently (e.g., Trivedi et al. 2013). Nonetheless, little is known about microbial communities in mangroves (but see Holguin et al. 2001; Liang et al. 2007; Gomes et al. 2011; Andreote et al. 2012; Ghizelini et al. 2012; Bouchez et al. 2013; Chambers et al. 2016; Ismail et al. 2017; Ullah et al. 2017). Understanding the responses of microbial communities to changes in environmental conditions is pivotal to predicting and coping with changes in service-provisioning. For instance, several recent studies shed light on microbial community compositions of surface sediments in the Sundarbans mangroves (Basak et al. 2015, 2016; Bhattacharyya et al. 2015; Sengupta et al. 2015). Nevertheless, our knowledge on how these communities, their activities and their contributions to ecosystem processes (such as decomposition) are driven and affected by (*i*) environmental parameters (such as sediment salinity) and (*ii*) the (floral and animal) species composition of the mangrove ecosystems is, however, scarce and has to be evaluated in the context of the ongoing salinization of the area, that is threatening several species of high biological and economical value (such as *Heritiera fomes*; Sarker et al. 2016).

24.4.3 Proposed Approach

There is a lack of scientific knowledge about species-specific C- (and N-) storage capacities and involved key players (bacteria, fungi, fauna, flora) in mangrove ecosystems. This understanding, however, is essential for efficient conservation planning and implementation that need to be based on scientific evidence. Hence, studying the above processes and relationships in various mangrove ecosystems under different environmental conditions should have high priority in research activities in tropical coastal areas.

The proposed approach, showcased herein in the context of carbon-storage in sediments, consists of performing comparative studies, assessing in situ floral, faunal and microbial community composition (eDNA or cellular DNA metabarcoding depending on the taxonomic groups) and sediment organic matter content and structure (combination of py-GC/MS and NIRS). Using the same methodologies under different geographical, environmental, socio-economic and socio-ecological settings will help unravel the key drivers of carbon-sequestration and -storage, be they biological, chemical or physical. Nine different groups are supposedly playing a major role in mangrove ecosystems and/or decomposition processes: vascular plants, crustaceans, molluscs, annelids, nematodes, archaea, bacteria, fungi (Thatoi et al. 2013) and protists (esp. Thraustochytridae; Raghukumar 2002) and should be targeted for eDNA metabarcoding. With such a framework, we will be able to provide evidence-based information about C-sequestration capacity of distinct mangrove ecosystems, and gain scientific knowledge of key processes. This would then enable us to provide evidence-driven recommendations for spatial conservation planning, depending on which ES (e.g., climate change-mitigation) are targeted. Their assessment might not be possible through the approach proposed herein alone. We therefore pledge for a multi-disciplinary approach, combining the described cutting-edge technologies from ecology (in concert with traditional ecological approaches when necessary) and the knowledge derived from socio-economy and socio-ecology, to ensure the consideration of societal and economic values of mangroves in addition to its important ecological role when designing protected areas networks. Combining environmental metabolomics, (amplicon-based or shotgun) metagenomics and economics for promoting conservation, may open the venue for "conservation-omics".

24.5 Outlook

A major innovation of our approach to protecting and sustainably using biodiversity and ecosystem services is the parallel use of eDNA-metabarcoding and metabolic fingerprinting, allowing for simultaneous targeting of multiple taxonomic groups which play distinct functional roles and the analysis of the OM in the sediments as a result of specific ecosystem processes.

This approach does not go without its challenges. Among them, unravelling key biotic interactions and detecting keystone species, considering microbial, faunal and floral communities altogether, might be tedious. New statistical approaches are being developed, especially focusing on microbial communities, and should be transferable to mangrove communities as a whole (e.g., keystone species detection: Berry and Widder 2014; reconstruction of ecological networks: Bohan et al. 2017; inferring interactions: Shang et al. 2017). We reckon additional experimental studies to be performed in parallel necessary to verify inferred findings.

We emphasized here the use of high-throughput techniques in both molecular and chemical ecology as very promising tools for the documentation of biodiversity and ecosystem processes on large scales and at high resolution for conservation-omics. For microbial communities, ongoing developments encompass the description of not solely community composition, also coined potential activity (via DNA/amplicon-based metabarcoding or shotgun metagenomic DNA sequencing), but also actual activity (via metatranscriptomics) through high-throughput sequencing of cDNA or RNA (Abram 2015).

We would also explicitly stress the potential of remote sensing, when it comes to mangrove vegetation and health mapping. With the new advances in airborne and spaceborne hyperspectral remote sensing technologies, the mapping of mangrove species (Vaiphasa et al. 2005; Adam et al. 2010) and health status and its potential threats can be timely assessed on large spatial scales (see e.g., Vidhya et al. 2014; Zhang et al. 2014; Asner et al. 2017).

Acknowledgements We thank Pierre Taberlet, Eric Coissac and Francesco Ficetola for constructive discussions and insights regarding the application of eDNA metabarcoding to mangrove research and conservation. Many thanks also to Christiane Hassenrück for insights into the application of up-to-date molecular techniques on microbial communities. We are grateful to Stefano Cannicci, Joe (Shing Yip) Lee, Peter K. L. Ng and Ingo Wehrtmann for crab identification.

Appendices

Appendix I: The Biodiversity Procession

Biodiversity (BD) – degree of variation of the living world, including genetic variation, species variation (*community structure*) and ecosystem variation.

Biotic Interactions (BI) – organisms interact with each other, both intraspecifically and interspecifically, e.g., through competition, facilitation, predation, mutualism, parasitism.

Service-Providing Units (SPU) – units of interacting organisms (individuals, populations, communities) providing the functional basis for services delivered by a functioning ecosystem.

Ecosystem Processes (EP) – physical, chemical and biological activities or reactions that link organisms and their environment, such as production or decomposition.

Ecosystem Function (EF) – *sensu* De Groot (1992), EF is defined as "the capacity of natural processes and components to provide goods and services that satisfy human needs directly or indirectly". This denomination relies strongly on the human perspective, as it will depend on what humans expect ecosystems to deliver, being considered their function.

Ecosystem-Functioning (EFg) – referring to De Groot et al. (2002), internal ecosystem functioning refers to the maintenance of energy fluxes, nutrient cycling, including food-web interactions. As such, the aggregate of EP warrants the functioning of an ecosystem as entity in delivering functions and ecosystem services.

Ecosystem Services (ES) – According to the Millennium Ecosystem Assessment (MEA 2005), "ecosystem services are the benefits people obtain from ecosystems". They can be categorized into: *provisioning services* such as food- and water-supply; *regulating services* such as the regulation of floods, droughts, land degradation and diseases; *supporting services* such as soil formation, and nutrient cycling; and *cultural services*, such as the offer of recreation, spiritual, religious, and other nonmaterial benefits (see e.g., Vo et al. 2012).

Appendix II: Glossary

Ancient DNA (aDNA) – *sensu* Bohmann et al. (2014), it refers to the DNA extracted from samples that have not been intentionally taken for genetic analysis (such as archaeological finds for example). Here we also refer to eDNA that is found in historical samples, such as deep cores of ice or soil/sediment.

Barcoding – *sensu stricto* it refers to the identification of organisms to the species level using standardized DNA fragments (barcodes). *Sensu lato*, it refers to the identification of organisms at any taxonomical level using any DNA fragment (Valentini et al. 2009). Barcoding is generally applied to tissue or any DNA trace originating from a single organism while metabarcoding (see below) is applied to DNA samples of mixed composition.

Ecosystem Engineers – species that are able to create, modify or maintain physical habitat for themselves and other species by significantly changing abiotic environmental conditions (Jones et al. 1994).

Environmental DNA (eDNA) – DNA that can be recovered from environmental samples (e.g. soil/sediments, water, permafrost, faeces), without prior isolation of specific organisms; as such, eDNA is composed of a mixture of genomic DNA from various organisms, and composed of cellular DNA (from living cells) and degraded extracellular DNA (from dead cells) released upon cell structure destruction (see Taberlet et al. 2012a and references therein).

Foundation Species – dominant species that, through their considerable abundance and biomass, have a strong effect on other species, by creating conditions and stabilizing fundamental ecosystem processes, and thus community composition and biodiversity (Dayton 1972; Whitham et al. 2006).

Keystone Species – species that have a strong effect on other species, and thus community composition and biodiversity, by virtue of strong interactions with, or other indirect effects on, other species rather than of their abundance or biomass; keystone species can be rare (Power et al. 1996).

Metabarcoding simultaneous identification of multiple species or taxa (depending on the taxonomic resolution) from bulk (containing several organisms) or environmental (containing degraded DNA) samples (Taberlet et al. 2012b).

Metabolic Fingerprinting – MS-based metabolomics approach where "a global snapshot of the metabolism is acquired and compared without performing quantification and chemical identification" (Du and Zeisel 2013).

Appendix III: Traffic Light for Conservation-Planning

To increase acceptance for strictly protected areas in societies that depend on, and make use of, natural resources, alternative uses and/or alternative sources of resources must be offered. Based on the knowledge gained from the combination of traditional knowledge, social-ecological and socio-economic analyses, ecological studies (including predictive models of future distributions of species and their interactions (SPUs)) and the conservation-omics approach described herein, forecasting the fate and value of particular areas is possible. The outcome of these approaches can be translated into recommendations for spatial conservation prioritization with respect to which areas should be strictly protected (because they are valuable and delicate or sensitive to disturbance and stressors), which areas can bear sustainable use (because they are resistant or resilient to disturbance and stressors), and which areas might be declared available for open use, such as the development of infrastructure or agri- and aquaculture (because they are of little value and/or have low chances of remaining valuable over time).

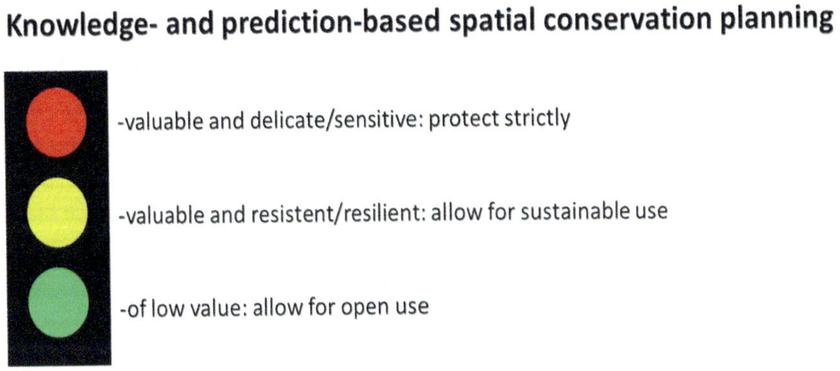

References

Abram F (2015) Systems-based approaches to unravel multi-species microbial community functioning. Comput Struct Biotechnol J 13:24–32

Adam E, Mutanga O, Rugege D (2010) Multispectral and hyperspectral remote sensing for identification and mapping of wetland vegetation: a review. Wetl Ecol Manag 18:281–296

Alongi DM (2009) The energetics of mangrove forests. Springer, New York

Alongi DM (2014) Carbon cycling and storage in mangrove forests. Annu Rev Mar Sci 6:195–219

Andreote FD, Jiménez DJ, Chaves D, Dias ACF, Luvizotto DM, Dini-Andreote F et al (2012) The microbiome of Brazilian mangrove sediments as revealed by metagenomics. PLoS ONE 7:e38600. https://doi.org/10.1371/journal.pone.0038600

Asner GP, Martin RE, Knapp DE, Tupayachi R, Anderson CB, Sinca F, Vaughn NR, Llactayo W (2017) Airborne laser-guided imaging spectroscopy to map forest trait diversity and guide conservation. Science 355:385–389

Bakkar T, Helfer V, Himmelsbach R, Zimmer M (2017) Chemical changes in detrital matter upon digestive processes in a sesarmid crab feeding on mangrove leaf litter. Hydrobiologia. https://doi.org/10.1007/s10750-017-3319-8

Ball IR, Possingham HP, Watts M (2009) Marxan and relatives: software for spatial conservation prioritisation. In: Moilanen A, Wilson KA, Possingham HP (eds) Spatial conservation prioritization: quantitative methods and computational tools. Oxford University Press, Oxford, pp 185–195

Barnes MA, Turner CR (2016) The ecology of environmental DNA and implications for conservation genetics. Conserv Genet 17:1–17

Basak P, Majumder NS, Nag S, Bhattacharyya A, Roy D, Chakraborty A, Sengupta S, Roy A, Mukherjee A, Pattanayak R, Ghosh A, Chattopadhyay D, Bhattacharyya M (2015) Spatiotemporal analysis of bacterial diversity in sediments of Sundarbans using parallel 16S rRNA gene tag sequencing. Microb Ecol 69:500–511

Basak P, Pramanik A, Sengupta S, Nag S, Bhattacharyya A, Roy D, Pattanayak R, Ghosh A, Chattopadhyay D, Bhattacharyya M (2016) Bacterial diversity assessment of pristine mangrove microbial community from Dhulibhashani, Sundarbans using 16S rRNA gene tag sequencing. Genomics Data 7:76–78

Baskin Y (1994) Ecosystem function of biodiversity. Bioscience 44:657–660

Berry D, Widder S (2014) Deciphering microbial interactions and detecting keystone species with co-occurrence networks. Front Microbiol 5:1–14

Bhattacharyya A, Majumder N, Basak P, Roy D, Nag S, Haldar A, Mukherjee S, Chattopadhyay D, Mitra S, Bhattacharyya M, Ghosh A (2015) Diversity and distribution of archaea in the mangrove sediment of Sundarbans. Archaea 2015, Article ID 968582, 14 p. https://doi.org/10.1155/2015/968582

Bohan DA, Vacher C, Tamaddoni-Nezhad A, Raybould A, Dumbrell AJ, Woodward G (2017) Next-generation global biomonitoring: large-scale, automated reconstruction of ecological networks. Trends Ecol Evol 32:477–487

Bohmann K, Evans A, Gilbert MTP, Carvalho GR, Creer S, Knapp M, Yu DW, de Bruyn M (2014) Environmental DNA for wildlife biology and biodiversity monitoring. Trends Ecol Evol 29:358–367

Bouchez A, Pascault N, Chardon C, Bouvy M, Cecchi P, Lambs L, Herteman M, Fromard F, Got P, Leboulanger C (2013) Mangrove microbial diversity and the impact of trophic contamination. Mar Pollut Bull 66:39–46

Brose U, Hillebrand H (2016) Biodiversity and ecosystem functioning in dynamic landscapes. Philos Trans Royal Soc B Biol Sci 371 (1694):20150267. https://doi.org/10.1098/rstb.2015.0267

Buurman P, Roscoe R (2011) Different chemical composition of free light, occluded light and extractable SOM fractions in soils of Cerrado and tilled and untilled fields, Minas Gerais, Brazil: a pyrolysis–GC/MS study. Eur J Soil Sci 62:253–266. https://doi.org/10.1111/j.1365-2389.2010.01327.x

Buurman P, Nierop KGJ, Kaal J, Senesi N (2009) Analytical pyrolysis and thermally assisted hydrolysis and methylation of EUROSOIL humic acid samples – a key to their source. Geoderma 150:10–22

Cannicci S, Burrows D, Fratini S, Smith UHI, Offenberg J, Dahdouh-Guebas F (2008) Faunal impact on vegetation structure and ecosystem function in mangrove forests: a review. Aquat Bot 89:186–200

Chambers LG, Guevara R, Boyer JN, Troxler TG, Davis SE (2016) Effects of salinity and inundation on microbial community structure and function in a mangrove peat soil. Wetlands 36:361–371. https://doi.org/10.1007/s13157-016-0745-8

Chan KMA, Shaw MR, Cameron DR, Underwood EC, Daily GC (2006) Conserv Plan Ecosyst Serv 4:e379. https://doi.org/10.1371/journal.pbio.0040379

Cimon-Morin J, Darveau M, Poulin M (2013) Fostering synergies between ecosystem services and biodiversity in conservation planning: a review. Biol Conserv 166:144–154

Cimon-Morin J, Darveau M, Poulin M (2014) Ecosystem services expand the biodiversity conservation toolbox – a response to deliège and neuteleers. Biol Conserv 172:219–220

Clare EL (2014) Molecular detection of trophic interactions: emerging trends, distinct advantages, significant considerations and conservation applications. Evol Appl 7:1144–1157

Crase B, Liedloff A, Vesk PA, Burgman MA, Wintle BA (2013) Hydroperiod is the main driver of the spatial pattern of dominance in mangrove communities. Glob Ecol Biogeogr 22:806–817

Creer S, Deiner K, Frey S, Porazinska D, Taberlet P, Thomas WK, Potter C, Bik HM (2016) The ecologist's field guide to sequence-based identification of biodiversity. Methods Ecol Evol 7:1008–1018

Das S, De TK, Mukherjee A, De M, Jana TK (2015) Influence of microbial composition on the carbon storage capacity of the mangrove soil at the land-ocean boundary in the Sundarban mangrove ecosystem. India Geomicrobiol J. https://doi.org/10.1080/01490451.2015.1093567

Dayton PK (1972) Toward an understanding of community resilience and the potential effects of enrichments to the benthos at McMurdo Sound, Antarctica. In: Parker BC (ed) Proceedings of the colloquium on conservation problems in Antarctica. Allen Press, Lawrence, pp 81–95

De Groot RS (1992) Functions of nature: evaluation of nature in environmental planning, management and decision making. Wolters-Noordhoff, Groningen

De Groot RS, Wilson MA, Boumans RMJ (2002) A typology for the classification, description and valuation of ecosystem functions, goods and services. Ecol Econ 41:393–408

Deagle BE, Jarman SN, Coissac E, Pompanon F, Taberlet P (2014) DNA metabarcoding and the cytochrome *c* oxidase subunit I marker: not a perfect match. Biol Lett 10:20140562

Deiner K, Bik HM, Mächler E et al (2017) Environmental DNA metabarcoding: Transforming how we survey animal and plantcommunities. Mol Ecol 26:5872–5895. https://doi.org/10.1111/mec.14350

Deliège G, Neuteleers S (2014) Ecosystem services as an argument for biodiversity preservation: why its strength is its problem – reply to Cimon-Morin et al. Biol Conserv 172:218–218

Dittmar T, Hertkorn N, Kattner G, Lara RJ (2006) Mangroves, a major source of dissolved organic carbon to the oceans. Glob Biogeochem Cycles 20:GB1012. https://doi.org/10.1029/2005GB002570

Donato D, Kauffman JB, Murdiyarso D, Kurnianto S, Stidham M, Kanninen M (2011) Mangroves among the most carbon-rich forests in the tropics. Nat Geosci 4:293–297

Du X, Zeisel SH (2013) Spectral deconvolution for gas chromatography mass spectrometry-based metabolomics: current status and future perspectives. Comput Struct Biotechnol J 4:e201201013
Elith J, Leathwick JR (2009) Conservation prioritization using species distribution models. In: Moilanen A, Wilson KA, Possingham HP (eds) Spatial conservation prioritization: quantitative methods and computational tools. Oxford University Press, Oxford, pp 70–93
Ellison AM, Farnsworth EJ (2001) Mangrove communities. In: Bertness MD, Gaines SD, Hay ME (eds) Marine community ecology. Sinauer Associates, Sunderland
Ellison AM, Farnsworth EJ, Merkt RE (1999) Origins of mangrove ecosystems and the mangrove biodiversity anomaly. Glob Ecol Biogeogr 8:95–115
Ellison AM, Bank MS, Clinton BD, Colburn EA, Elliott K, Ford CR, Foster DR et al (2005) Loss of foundation species: consequences for the structure and dynamics of forested ecosystems. Front Ecol Environ 3:479–486
Ficetola GF, Miaud C, Pompanon F, Taberlet P (2008) Species detection using environmental DNA from water samples. Biol Lett 4:423–425
Ficetola GF, Coissac E, Zundel S, Riaz T, Shehzad W, Bessière J, Taberlet P, Pompanon F (2010) An in silico approach for the evaluation of DNA barcodes. BMC Genomics 11:434
Ficetola GF, Pansu J, Bonin A, Coissac E, Giguet-Covex C, De Barba M, Gielly L, Lopes CM, Boyer F, Pompanon F, Rayé G, Taberlet P (2015) Replication levels, false presences and the estimation of the presence/absence from eDNA metabarcoding data. Mol Ecol Resour 15:543–556
Ficetola GF, Taberlet P, Coissac E (2016) How to limit false positives in environmental DNA and metabarcoding? Mol Ecol Resour 16:604–607
Foote AD, Thomsen PF, Sveegaard S, Wahlberg M, Kielgast J, Kyhn LA et al (2012) Investigating the potential use of environmental DNA (eDNA) for genetic monitoring of marine mammals. PLoS ONE 7:e41781. https://doi.org/10.1371/journal.pone.0041781
Fuentes M, Hidalgo C, González-Martín I, Hernández-Hierro JM, Govaerts B, Sayrre KD, Etchevers J (2012) NIR spectroscopy: an alternative for soil analysis. Commun Soil Sci Plant Anal 43:346–356
Gardham S, Hose GC, Stephenson S, Chariton AA (2014) DNA metabarcoding meets experimental ecotoxicology: advancing knowledge on the ecological effects of copper in freshwater ecosystems. Adv Ecol Res 51:79–104
Gerber L, Eliasson M, Trygg J, Moritz T, Sundberg B (2012) Multivariate curve resolution provides a high-throughput data processing pipeline for pyrolysis-gas chromatography/mass spectrometry. J Anal Appl Pyrolysis 95:95–100
Ghizelini AM, Mendonça-Hagler LCS, Macrae A (2012) Microbial diversity in Brazilian mangrove sediments – a mini review. Braz J Microbiol 43:1242–1254
Giguet-Covex C, Pansu J, Arnaud F, Rey P-J, Griggo C, Gielly L, Domaizon I, Coissac E, David F, Choler P, Poulenard J, Taberlet P (2014) Long livestock farming history and human landscape shaping revealed by lake sediment DNA. Nat Commun 5:3211
Giri C, Ochieng E, Tieszen LL, Zhu Z, Singh A, Loveland T, Masek J, Duke N (2011) Status and distribution of mangrove forests of the world using earth observation satellite data. Glob Ecol Biogeogr 20:154–159
Gomes NC, Cleary DF, Calado R, Costa R (2011) Mangrove bacterial richness. Commun Integr Biol 4:419–423
Groom C, White NE, Mitchell N, Roberts JD, Mawson P (2016) Assessing the spatial ecology and resource use of a mobile and endangered species in an urbanized landscape using satellite telemetry and DNA faecal metabarcoding. Ibis 159:390–405
Guardiola M, Uriz MJ, Taberlet P, Coissac E, Wangensteen OS, Turon X (2015) Deep-sea, deep-sequencing: metabarcoding extracellular DNA from sediments of marine canyons. PLoS ONE 10:e0139633. https://doi.org/10.1371/journal.pone.0139633
Hand BK, Lowe WH, Kovach RP, Muhlfeld CC, Luikart G (2015) Landscape community genomics: understanding eco-evolutionary processes in complex environments. Trends Ecol Evol 30:161–168
Harborne AR, Mumby PJ, Micheli F, Perry CT, Dahlgren CP, Holmes KE, Brumbaugh DR (2006) The functional value of caribbean coral reef, seagrass and mangrove habitats to ecosystem processes. Adv Mar Biol 50:57–189
Helmuth B, Mieszkowska N, Moore P, Hawkins SJ (2006) Living on the edge of two changingworlds: forecasting the responses of rocky intertidal ecosystems to climate change. Annu Rev Ecol Evol Syst 37:373–404
Holguin G, Vazquez P, Bashan Y (2001) The role of sediment microorganisms in the productivity, conservation, and rehabilitation of mangrove ecosystems: an overview. Biol Fertil Soils 33:265–278
Hübner L, Pennings SC, Zimmer M (2015) Sex- and habitat-specific movement of an omnivorous semi-terrestrial crab controls habitat connectivity and subsidies: a multi-parameter approach. Oecologia 178:999–1015
Hutchison J, Manica A, Swetnam R, Balmford A, Spalding M (2014) Predicting global patterns in mangrove forest biomass. Conserv Lett 7:233–240
IPCC (2007) Climate change 2007: the physical science basis. contribution of working group I to the fourth assessment report of the intergovernmental panel on climate change. Cambridge University Press, Cambridge
Ismail Z, Sam CK, Yin WF, Chan KG (2017) Tropical mangrove swamp metagenome reveals unusual abundance of ecologically important microbes. Curr Sci 112:1698–1703
Jennerjahn TC, Ittekot V (2002) Relevance of mangroves for the production and deposition of organic matter along tropical continental margins. Naturwissenschaften 89:23–30
Joly S, Jonathan Davies T, Archambault A, Bruneau A, Derry A, Kembel SW, Peres-Neto P, Vamosi J, Wheeler TA (2014) Ecology in the age of DNA barcoding: the resource, the promise and the challenges ahead. Mol Ecol Resour 14:221–232
Jones CG, Lawton JH, Shachak M (1994) Organisms as ecosystem engineers. Oikos 69:373–386
Kleinebecker T, Poelen MDM, Smolders AJP, Lamers LPM, Hölzel N (2013) Fast and inexpensive detection of total and extractable element concentrations in aquatic sediments using Near-Infrared Reflectance Spectroscopy (NIRS). PLoS ONE 8(7):e70517
Komiyama A, Ong JE, Poungparn S (2008) Allometry, biomass, and productivity of mangrove forests: a review. Aquat Bot 89:128–137
Kristensen E (2007) Carbon balance in mangrove sediments: the driving processes and their controls. In: Tateda Y, Upstill-Goddard R, Goreau T, Alongi D, Nose A, Kristensen E, Wattayakorn G (eds) Greenhouse gas and carbon balances in mangrove coastal ecosystems. Maruzen, Tokyo, pp 61–78
Kristensen E, Bouillon S, Dittmar T, Marchand C (2008) Organic carbon dynamics in mangrove ecosystems: a review. Aquat Bot 89:201–219
Lahoz-Monfort JJ, Guillera-Arroita G, Tingley R (2016) Statistical approaches to account for false-positive errors in environmental DNA samples. Mol Ecol Resour 16:673–685
Leathwick J, Moilanen A, Francis M, Elith J, Taylor P, Julian K, Hastie T, Duffy C (2008) Novel methods for the design and evaluation of marine protected areas in offshore waters. Conserv Lett 1:91–102
Liang J-B, Chen Y-Q, Lan C-Y, Tam NFY, Zan Q-J, Huang L-N (2007) Recovery of novel bacterial diversity from mangrove sediment. Mar Biol 150:739–747
Lovelock CE (2008) Soil respiration and belowground carbon allocation in mangrove forests. Ecosystems 11:342–354

Luo L, Gu JD (2016) Nutrient limitation status in a subtropical mangrove ecosystem revealed by analysis of enzymatic stoichiometry and microbial abundance for sediment carbon cycling. Int Biodeterior Biodegrad https://doi.org/10.1016/j.ibiod.2016.04.023
MacKenzie DJ, Nichols JD, Lachman GB, Droege S, Royle JA, Langtimm CA (2002) Estimating site occupancy rates when detection probabilities are less than one. Ecology 83:2248–2255
Mcleod E, Chmura GL, Bouillon S, Salm R, Björk M et al (2011) A blueprint for blue carbon: toward an improved understanding of the role of vegetated coastal habitats in sequestering CO_2. Front Ecol Environ 9:552–560
Millennium Ecosystem Assessment (2005) Millennium ecosystem assessment – ecosystems and human well-being: biodiversity synthesis. World Resources Institute, Washington, DC. 86 p
Moilanen A (2012) Spatial conservation prioritization in data-poor areas of the world. Natureza Conservação 10:12–19
Moilanen A, Franco AMA, Early R, Fox R, Wintle B, Thomas CD (2005) Prioritising multiple-use landscapes for conservation: methods for large multi-species planning problems. Proc Royal Soc Lond B Biol Sci 272:1885–1891
Nogués-Bravo D (2009) Predicting the past distribution of species climatic niches. Glob Ecol Biogeogr 18:521–531
Parducci L, Jørgensen T, Tollefsrud MM, Elverland E, Alm T, Fontana SL et al (2012) Glacial survival of boreal trees in northern Scandinavia. Science 335:1083–1086
Pearman PB, Randin CF, Broennimann O, Vittoz P, van der Knaap WO, Engler R, Lay L, Zimmermann NE, Guisan A (2008) Prediction of plant species distributions across six millennia. Ecol Lett 11:357–369
Pedersen MW, Overballe-Petersen S, Ermini L, Sarkissian CD, Haile J, Hellstrom M et al (2015) Ancient and modern environmental DNA. Philos Trans Royal Soc B Biol Sci 370:20130383. https://doi.org/10.1098/rstb.2013.0383
Pestana D, Pülmanns N, Nordhaus I, Diele K, Zimmer M (2017) The influence of crab burrows on sediment salinity in a Rhizophora-dominated mangrove forest in North Brazil during the dry season. Hydrobiologia. https://doi.org/10.1007/s10750-017-3282-4
Polis GA, Anderson WB, Holt RD (1997) Toward an integration of landscape and food web ecology: the dynamics of spatially subsidized food webs. Annu Rev Ecol Syst 28:289–316
Power ME, Tilman D, Estes JA, Menge BA, Bond WJ, Mills LS, Gretchen D, Castilla JC, Lubchenco J, Paine RT (1996) Challenges in the quest for keystones. Bioscience 46:609–620
Pullin AS, Knight TM (2003) Support for decision making in conservation practice: an evidence-based approach. J Nat Conserv 11:83–90
Pülmanns N, Nordhaus I, Diele K, Mehlig U (2015) Artificial crab burrows facilitate desalting of rooted mangrove sediment in a microcosm study. J Mar Sci Eng 3:539–559
Pülmanns N, Mehlig U, Nordhaus I, Saint-Paul U, Diele K (2016) Mangrove crab **Ucides cordatus** removal does not affect sediment parameters and stipule production in a one year experiment in Northern Brazil. PLoS ONE 11:e0167375. https://doi.org/10.1371/journal.pone.0167375
Raghukumar S (2002) Ecology of the marine protists, the Labyrinthulomycetes (Thraustochytrids and Labyrinthulids). Eur J Protistol 38:127–145
Record S, Charney ND, Zakaria RM, Ellison AM (2013) Projecting global mangrove species and community distributions under climate change. Ecosphere 4:1–23
Reef R, Atwood TB, Samper-Villarreal J, Adame MF, Sampayo EM, Lovelock CE (2017) Using eDNA todetermine the source of organic carbon in seagrass meadows. Limnol Oceanogr 62:1254–1265. https://doi.org/10.1002/lno.1049
Riaz T, Shehzad W, Viari A, Pompanon F, Taberlet P, Coissac E (2011) ecoPrimers: inference of new DNA barcode markers from whole genome sequence analysis. Nucleic Acids Res 39:e145. https://doi.org/10.1093/nar/gkr732
Richards DR, Friess DA (2016) Rates and drivers of mangrove deforestation in Southeast Asia, 2000–2012. Proc Natl Acad Sci U S A 113:344–349
Sarker SK, Reeve R, Thompson J, Paul NK, Matthiopoulos J (2016) Are we failing to protect threatened mangroves in the Sundarbans world heritage ecosystem? Sci Rep 6:21234. https://doi.org/10.1038/srep21234
Schellekens J, Buurman P (2011) n-Alkane distributions as palaeoclimatic proxies in ombrotrophic peat: the role of decomposition and dominant vegetation. Geoderma 164:112–121
Schellekens J, Buurman P, Fraga I, Martínez-Cortizas A (2011) Holocene vegetation and hydrologic changes inferred from molecular vegetation markers in peat, Penido Vello (Galicia Spain). Palaeogeogr Palaeoclimatol Palaeoecol 299:56–69
Sengupta S, Pramanik A, Ghosh A, Bhattacharyya M (2015) Antimicrobial activities of actinomycetes isolated from unexplored regions of Sundarbans mangrove ecosystem. BMC Microbiol 15:1–16
Serafy JE, Shideler GS, Araújo RJ, Nagelkerken I (2015) Mangroves enhance reef fish abundance at the Caribbean regional scale. PLoS ONE 10:e0142022. https://doi.org/10.1371/journal.pone.0142022
Shang Y, Sikorski J, Bonkowski M, Fiore-Donno A-M, Kandeler E, Marhan S et al (2017) Inferring interactions in complex microbial communities from nucleotide sequence data and environmental parameters. PLoS ONE 12:e0173765. https://doi.org/10.1371/journal.pone.0173765
Shokralla S, Spall JL, Gibson JF, Hajibabaei M (2012) Next-generation sequencing technologies for environmental DNA research. Mol Ecol 21:1794–1805
Spalding M, Kainuma M, Collins L (2010) World atlas of mangroves. Earthscan, London/Washington DC
Sustainable Development Solutions Network (2015) Indicators and a monitoring framework for the sustainable development goals launching a data revolution for the SDGs
Sutherland WJ, Pullin AS, Dolman PM, Knight TM (2004) The need for evidence-based conservation. Trends Ecol Evol 19:305–308
Svancara LK, Brannon R, Scott JM, Groves CR, Noss RF, Pressey RL (2005) Policy-driven versus evidence-based conservation: a review of political targets and biological needs. Bioscience 55:989–995
Taberlet P, Coissac E, Hajibabaei M, Rieberg LH (2012a) Environmental DNA. Mol Ecol 21:1789–1793
Taberlet P, Coissac E, Pompanon F, Brochmann C, Willerslev E (2012b) Towards next-generation biodiversity assessment using DNA metabarcoding. Mol Ecol 21:2045–2050. https://doi.org/10.1111/j.1365-294X.2012.05470.x
Thatoi H, Behera, Mishra RR (2013) Ecological role and biotechnological potential of mangrove fungi: a review. Mycology 4:54–71
Thomsen PF, Willerslev E (2015) Environmental DNA – an emerging tool in conservation for monitoring past and present biodiversity. Biol Conserv 183:4–18
Thomsen PF, Kielgast J, Iversen LL, Wiuf C, Rasmussen M, Gilbert MTP, Orlando L, Willerslev E (2012) Monitoring endangered freshwater biodiversity using environmental DNA. Mol Ecol 21:2565–2573
Tolu J, Gerber L, Boily J-F, Bindler R (2015) High-throughput characterization of sediment organic matter by pyrolysis–gas chromatography/mass spectrometry and multivariate curve resolution: a promising analytical tool in (paleo)limnology. Anal Chim Acta 880:93–102

Tomlinson P (2016) The botany of mangroves. Cambridge University Press, Cambridge. https://doi.org/10.1017/CBO9781139946575

Trivedi P, Anderson IC, Singh BK (2013) Microbial modulators of soil carbon storage: integrating genomic and metabolic knowledge for global prediction. Trends Microbiol 21:641–651

Ullah R, Yasir M, Khan I, Bibi F, Sohrab SS, Al-Ansari A, Al-Abbasi F, Al-Sofyani AA, Daur I, Lee SW, Azhar EI (2017) Comparative bacterial community analysis in relatively pristine and anthropogenically influenced mangrove ecosystems on the Red Sea. Can J Microbiol 63:649–660. https://doi.org/10.1139/cjm-2016-0587

Vaiphasa C, Ongsomwang S, Vaiphasa T, Skidmore AK (2005) Tropical mangrove species discrimination using hyperspectral data: a laboratory study. Estuar Coast Shelf Sci 65:371–379

Valentini A, Pompanon F, Taberlet P (2009) DNA barcoding for ecologists. Trends Ecol Evol 24:110–117

van der Werf GR, Morton DC, DeFries RS et al (2009) CO_2 emissions from forest loss. Nat Geosci 2:737–738

van Dijk AH, Jaszczyszyn Y, Thermes C (2014) Ten years of next-generation sequencing technology. Trends Genet 30:418–426

Vancampenhout K, Wouters K, De Vos B, Buurman P, Swennen R, Deckers J (2009) Differences in chemical composition of soil organic matter in natural ecosystems from different climatic regions – a pyrolysis–GC/MS study. Soil Biol Biochem 41:568–579

Viant MR, Sommer U (2013) Mass spectrometry based environmental metabolomics: a primer and review. Metabolomics 9:S144–S158

Vidhya R, Vijayasekaran D, Ahamed Farook M, Jai S, Rohini M, Sinduja A (2014) Improved classification of mangroves health status using hyperspectral remote sensing data. ISPRS Int Archiv Photogramm Remote Sens Spat Inf Sci XL-8:667–670

Vo QT, Kuenzer C, Vo QM, Moder F, Oppelt N (2012) Review of valuation methods for mangrove ecosystem services. Ecol Indic 23:431–446

Whitham TG, Bailey JK, Schweitzer JA, Shuster SM, Bangert RK, LeRoy CJ, Lonsdorf EV, Allan GJ, DiFazio SP, Potts BM et al (2006) A framework for community and ecosystem genetics: from genes to ecosystems. Nat Rev Genet 7:510–523

Willerslev E, Hansen AJ, Binladen J, Brand TB, Gilbert MTP, Shapiro B, Bunce M, Wiuf C, Gilichinsky DA, Alan Cooper A (2003) Diverse plant and animal genetic records from holocene and pleistocene sediments. Science 300:791–795

Willerslev E, Davison J, Moora M et al (2014) Fifty thousand years of Arctic vegetation and megafaunal diet. Nature 506:47–51

Williams JW, Jackson ST (2007) Novel climates, no-analog communities, and ecological surprises. Front Ecol Environ 5:475–482

Wunderlich AC, Pinheiro MAA, Rodrigues AMT (2008) Biologia do caranguejo-uçá, Ucides cordatus (Crustacea: Decapoda: Brachyura), na Baía da Babitonga, Santa Catar Santa Catarina, Brasil. Rev Bras Zool 25:188–198

Zhang C, Kovacs JM, Liu Y, Flores-Verdugo F, Flores-de-Santiago F (2014) Separating mangrove species and conditions using laboratory hyperspectral data: a case study of a degraded mangrove forest of the Mexican Pacific. Remote Sens 6:11673–11688

Zimmer M (2017) Detritus. In: Jorgensen SE, Fath BD (eds) Encyclopedia of ecology, 2nd edn. Elsevier, Oxford. (in press)

Part V

Conservation, Rehabilitation, and Governance

Mangrove Concessions: An Innovative Strategy for Community Mangrove Conservation in Ecuador

25

Fausto Vinicio López Rodríguez

Abstract

The sustainable use and custody agreements of mangrove forests (Mangrove forest concessions) emerged in 1999 as a complementary strategy to the national protected area policy strategy. This innovative strategy supports the conservation of mangrove forests by providing legal security to traditional mangrove users, conceding them land rights concessions and thereby promote the participation of local communities in the conservation of mangroves, allowing the sustainable capture of biodiversity mainly of sea shells and crabs. This chapter includes a case study of the association of artisanal fishermen and related activities named "Costa Rica", located in the province of El Oro. This association obtained in 2000 a mangrove forest concession, being one of the first in the country. After 17 years of signing the agreement with the National Government, in spite of anthropogenic pressure, mainly from the shrimp industry, the area maintained its mangrove forest cover and the association has been able to sustainable use the mangrove resources, especially the black ark or blood cockle *(Anadara tuberculosa).*

Keywords

Mangroves · Mangrove concessions · Participatory management of mangroves · Mangrove deforestation · Shrimp industry · Black arch production

25.1 Introduction

Mangrove forests are a type of tropical forest of great importance for the livelihoods of millions of people around the world. Globally, mangroves cover about 152,360 km^2 (15′236.000 ha) in 123 countries (Spalding et al. 2010). More than 100 million people worldwide live within 10 km of mangroves, benefiting on a permanent basis from a wide variety of ecosystem goods and services provided by the mangrove forests (UNEP 2014).

Communities directly benefit from goods such as wood (used for construction, furniture, firewood, etc.) and aquatic fauna (mollusks, fish, crustaceans, etc.), which is crucial for their food security (UNEP 2014). Mangroves are among the most productive source of renewable resources on the planet. The primary production levels of mangroves surpass those of many agricultural systems, producing about 10 g/m^2 per day of organic material which amounts up to 37 metric tons per hectare per year (Center for Integrated Natural Resource Mapping by Remote Sensing et al. 2007). These ecosystem goods and services are estimated to amount up to US $33.000–57.000 annually per hectare for the national economies of the developing countries that possess mangroves (UNEP 2014).

The economic potential of mangroves comes from three key sources: forestry products, estuary and littoral fishing and ecotourism. Mangroves also play a key role in the protection of coasts and the management of habitats for species that are threatened and critically endangered (FAO 1994). This means that in addition to the important ecosystem services provided to human beings, mangroves are key for the conservation of species that are threatened due to anthropogenic activities.

In addition to the benefits mentioned above, mangrove ecosystems are capable of storing about 1.000 tons of CO_2 per hectare in their biomass and subsoil. This fact makes of mangroves one of the richest ecosystems in terms of carbon

F. V. L. Rodríguez (✉)
Departamento de Ciencias Biológicas, Research Group: Gobernanza, Biodiversidad y Áreas Protegidas, Universidad Técnica Particular de Loja, Loja, Ecuador
e-mail: fvlopezx@utpl.edu.ec

C. Makowski, C. W. Finkl (eds.), *Threats to Mangrove Forests*, Coastal Research Library 25,
https://doi.org/10.1007/978-3-319-73016-5_25

in the planet. Consequently, the greenhouse gas emissions from mangrove deforestation are among the highest produced by land use changes in the tropics. Greenhouse gas emissions due to mangrove deforestation represent up to 1/5 of the global emissions produced by deforestation (UNEP 2014).

Despite its importance, mangroves are disappearing worldwide by 1–2% per year, a deforestation rate greater than tropical rainforests. It is estimated that nearly 35% of global mangrove forests no longer exist (World Wildlife Fund 2015). Losses are occurring in almost every country that has mangroves, especially in developing countries, where more than 90% of the world's mangroves are located. The economic cost of mangrove deforestation is US $42 billion per year (UNEP 2014). Fragmentation and deforestation put the long-term survival and essential ecosystem services of mangroves at great risk (Polidoro et al. 2010). Main drivers of mangrove loss are aquaculture (shrimp farming), coastal development and logging for timber and fuel production (Kavanagh 2014; Polidoro et al. 2010).

World catch of shrimp is about 3.4 million tons per year. China, together with four other Asian countries, accounts for 55% of the world catch. Globally, 40% of shrimp production comes from shrimp farming. Annual export of shrimps is currently worth more than US$10 billion, or 16% of all fishery exports (FAO 2015_1). Although it's a very important industry, shrimp fisheries are the major source of fisheries conflict and problems in the mangrove forests (Gillett 2008).

Mangroves are critically endangered and approaching extinction in 26 out of the countries having mangroves (Polidoro et al. 2010). Eleven of the 70 mangrove species (16%) are at elevated threat of extinction, particularly in the Atlantic and Pacific coasts of Central America, where as many as 40% of mangroves species present are threatened with extinction (Polidoro et al. 2010; Kavanagh 2014).

Ecuador is the smallest of the mega diverse countries in the world, and perhaps the most diverse of the planet according to the number of species per hectare. This amazing biodiversity is represented at the genetic, species and ecosystem level and is due to factors such as: its equatorial location, presence of the Andes and Coastal mountains, multiple climates and the circulation of ocean currents: the cold Humboldt current and the warmer El Niño current (García et al. 2015). Thus, in Ecuador exist 24 of the 27 globally recognized marine and coastal ecosystems.

Of the 13 globally recognized coastal ecosystems, 12 can be found in Ecuador, amongst this is the mangrove ecosystem (Hurtado-Domínguez et al. 2012; Ministerio del Ambiente and GIZ 2012; FAO 2012).

Of the South Pacific countries, Colombia has the largest area of mangrove forests with 370.000 ha (more than two-thirds are located in the Pacific) (CPPS et al. 2013). Ecuador has 161.835 ha of mangrove forests (Ministerio del Ambiente de Ecuador 2017_1) while Peru has around 6.000 ha of mangroves (CPPS et al. 2013). The mangrove surface related to country surface in Ecuador is 0.60%, followed by Colombia with 0.31% and Peru with 0.01% (Yáñez–Arancibia; Lara–Domínguez 1999).

Approximately, 39% of Ecuador's territory lies within 100 km of the coast. 58% of the Ecuadorian population lives in this area, and depends on the services provided by marine and coastal ecosystems, such as mangroves (FAO 2012; Terán et al. 2006).

In spite of the importance of this ecosystem, in Ecuador, in the late 1970s, mangroves were considered unproductive areas, leading to aggressive deforestation causing the loss of 27.7% of its coverage. In 2006, the mangrove land coverage was 147228.6 ha (Bravo 2013), an extension that increased by 14,067 ha in 2016, recovery attributed to reforestation and natural regeneration of mangrove forests (Ministerio del Ambiente de Ecuador 2017_1).

This chapter presents the historical and current situation of mangroves in Ecuador, their importance and main threats, and the conservation efforts taken to combat mangrove deforestation, including policies and regulations, the creation of protected natural areas and, upon request of ancestral communities[1] and traditional mangrove users,[2] the creation of the innovative conservation mechanism called: Sustainable Use and Custody Agreements of Mangrove Forests also called Mangrove Forest Concessions.

25.2 Mangrove Forests in Ecuador

25.2.1 Distribution and Coverage of Mangrove Forests

Located in the Eastern South Pacific, the coast of continental Ecuador has a length of about 2500 km (excluding the Galapagos Islands). Within 15 km from the coast there are 345 human settlements with a total population of 3,943,912 people, equivalent to 32.44% of Ecuador's population (Ministerio del Ambiente S.f.). Many of the coastal communities depend on marine and coastal resources for their livelihood (Ministerio del Ambiente and Conservación Internacional-Ecuador 2017). Fishing, aquaculture and coastal tourism are some of the main economic activities. (See Fig. 25.1)

[1] Ancestral communities are human groups that historically lived in a specific geographical area in which they have coexisted with their ecosystems, obtaining benefits from the natural resources for their subsistence (Ministerio del Ambiente de Ecuador 2017_1).

[2] Traditional users are human groups that over a long period of time have benefited of the natural resources of a specific geographical area for their subsistence (Ministerio del Ambiente de Ecuador 2017_1).

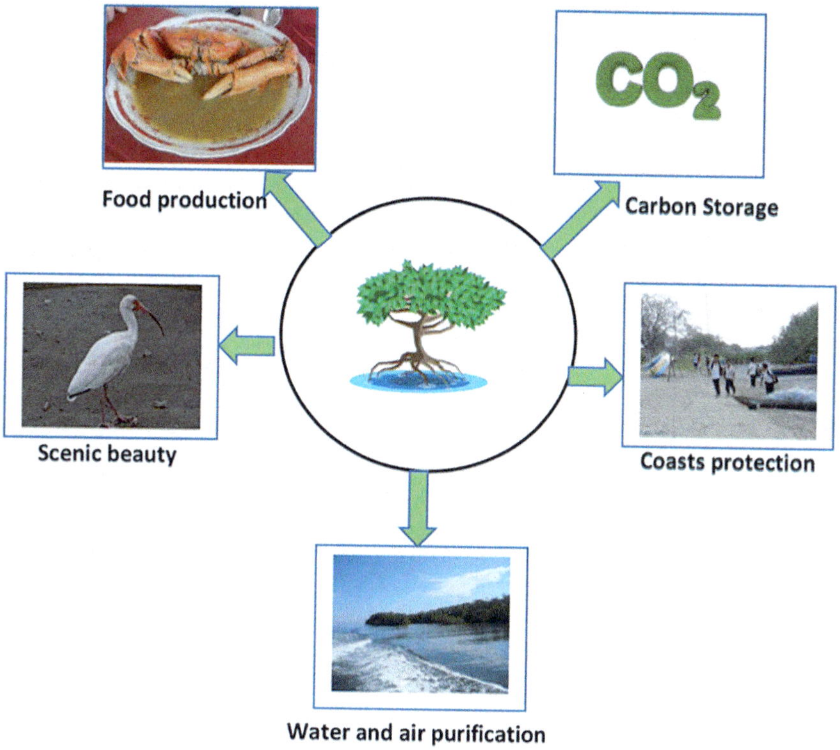

Fig. 25.1 Five key environmental services provided by mangrove forests. Mangroves help build and maintain the integrity of coastal areas, are highly productive and maintain a complex network of ecological interactions. They are the breeding grounds for a wide variety of fish, mollusks and crustaceans and a food source for many vertebrate and invertebrate species. It is estimated that at least two-thirds of edible species of fish need the mangrove to survive. In Ecuador, thousands of artisanal fishers depended on mangroves for generations onwards, for subsistence fishing, as well as a principal source of income. Ecuador determined the economic cost of the loss of environmental goods and services due to mangrove destruction and its restoration cost per hectare on USD 89,273.01 (Gobierno de la República del Ecuador 2011a, b; CPPS et al. 2015; Beitl 2010)

In Ecuador, mangrove forests are distributed along the coast from the province of Esmeraldas in the north of Ecuador to the province of El Oro in the south of the country (Quizhpe 2008).

In 1969, the surface covered by mangroves in Ecuador was 203.624 ha. According to the Center for Integrated Natural Resource Mapping by Remote Sensing et al. (2007), by 2006, this area had been reduced to 147.228,60 ha, that is a reduction of 56.395,9 ha or 27.7% of the mangrove present in 1969. Ecuador's Ministry of Environment, states that currently there are 161.835 ha of mangroves in the country. This increase in mangrove cover is likely to be a result of reforestation programs and natural regeneration (Ministerio del Ambiente de Ecuador 2017_1).

According to a study carried out by Center for Integrated Natural Resource Mapping by Remote Sensing et al. (2007), 99.8% of remaining mangroves are found in six key estuaries: Cayapas Mataje (21.400 ha), Muisne (1.558 ha), Cojimies (2.742 ha) and Chone (933 ha), all in the north of the country; and the Gulf of Guayaquil (105.130 ha) and the Jambelí Archipelago (15.208 ha) both in the south of the country. (See Fig. 25.2)

The Cojimies estuary (79.1%), the Chone estuary (76.5%) and the Jambelí Archipelago estuary (56.2%) have lost most of their mangrove coverage, caused mainly due to land use change related to shrimp farming and urban development (Bravo 2013). (See Table 25.1)

25.2.2 Flora of Mangrove Forests

Mangrove forests are mainly composed by the following mangrove species: *Rhizophora mangle, R. racemosa, R. x harrisonii, Laguncularia racemosa var. racemosa, L. racemosa var. Glabriflora y Avicennia germinans* (MAE and FAO 2014). A recent study by Cornejo (2014) recorded 179 species of vascular plants in the country's mangroves (Cornejo ed. 2014). The red mangrove (*Rhizophora mangle L.*) is the most abundant species found in Ecuador (Bravo 2013). (See Fig. 25.3)

25.2.3 Fauna of Mangrove Forests

Although the diversity of plant species in mangrove forests is relatively low compared to other ecosystems, the diversity of animals is much higher. In Ecuadorian mangroves there are 52 species of birds, 15 species of reptiles, 19 species of mammals, approximately 100 species of fish, 20 of crustaceans and 70 of mollusks. These animals can be found in this ecosystem primarily due to its importance as an area for shelter, food and / or reproduction (MAE and FAO, 2014).

Many of the animal species found in the mangroves, whether visitors or residents, are of food and commercial importance for nearby communities because of their nutritional value and abundance, among which we can

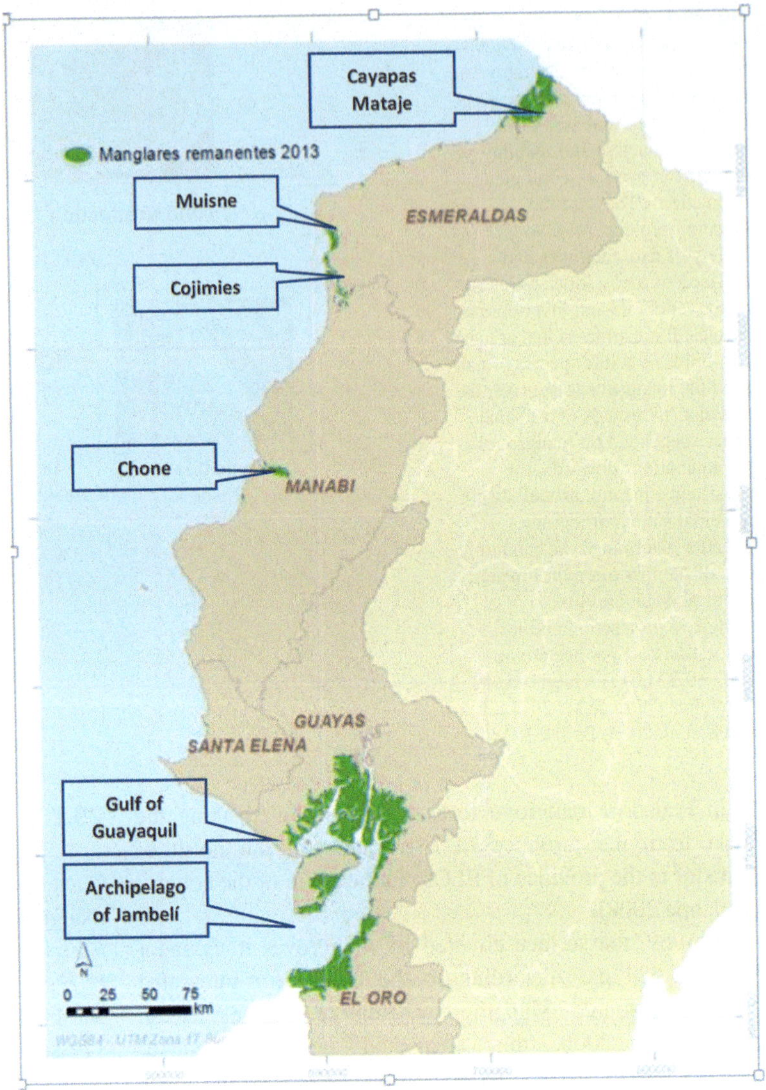

Fig. 25.2 Location of main mangrove areas of the continental territory of Ecuador. The Gulf of Guayaquil estuary is the largest area of mangrove forest (105,130 ha) followed by the Cayapas Mataje estuary with 21,400 ha and the Jambelí Archipelago estuary with 15,208 ha. The Chone estuary (933 ha) is the smallest of all (Center for Integrated Natural Resource Mapping by Remote Sensing et al. 2007). (Map: Conservacion Internacional-Ecuador 2016)

Table 25.1 Mangrove coverage in the five main estuaries of Ecuador

Estuary	1969	1984	1987	1991	1995	1999	2006	% of coverage loss (using data from 1969 and 2006)
Cayapas Mataje	23.677	23.653	23.507	22.863	21.947	22.057	21.400	9,6
Muisne	3.282	2.701	2.445	1.340	830	1.187	1.187	52,5
Cojimíes	13.123	9.917	8.466	6.028	3.651	1.921	2.742	79,1
Chone	3.973	1.673	1.040	784	391	705	932	76,5
Golf of Guayaquil	124.320	119.277	115.784	109.608	102.108	104.715	105.130	15,4
Archipelago of Jambelí	34.712	24.592	23.570	21.092	17.697	19.111	19.111	56,2
Total	203.624	181.815	174.815	161.718	146.628	149.699	146.971	27,6

The highest absolute annual deforestation rate occurred in the period from 1991 to 1995 (2.35% per year), and in the period from 1995 to1999 a mangrove recovery was observed. For the period from 1999 to 2006, the deforestation rate is 0.13% per year (Center for Integrated Natural Resource Mapping by Remote Sensing et al. 2007). On the contrary, in the decade from 2006 to 2016 there is again a mangrove recovery of 15,835 ha (Ministerio del Ambiente de Ecuador 2017$_1$)

mention the crabs *(Ucides occidentalis, Callinectes arcuatus, C. toxins)*, shrimp *(Litopenaeus stylirostris, L. vannamei)*, black ark *(Anadara tuberculosa)*, mussels *(Mytella guyanensis)*, oysters *(Crassostrea columbiensis)*, clams *(Protothaca sp., Chione subrugosa)* and dozens of fish species like Lily *(Mugil cephalus)*, corvina *(Cynoscion albus)*, sea bass *(Centropomus spp.)*, Snapper *(Lutjanus sp.)* etc. (Castellanos-Galindo et al. 2012).

Fig. 25.3 Roots of red mangrove (*Rhizophora mangle* L.) during low tide in the Costa Rica Island, Archipelago of Jambelí. This specie is characterized by its large stilt-like roots that accumulate sediments, making up the habitat of mollusks and crustaceans that play a key role in the local economy of nearby communities. This mangrove species developed physiological and anatomical adaptation strategies as a marked tolerance to high concentrations of salt and adaptations to deal with unstable soils. Due to its high resistance in saline environments, the red mangrove wood has been used in the construction sector as columns on docks and in the construction of houses. It is also used to make hammocks and crafts. The red mangrove is currently the most used mangrove species in reforestation programs in Ecuador (MAE and FAO 2014). Photo credit: Adrián Orihuela

Mangrove productivity, which is reflected by the presence, abundance and status of the populations of its species, is directly related to the extent, state and quality of conservation efforts of mangrove forests (Castellanos-Galindo et al. 2012; Bravo 2013).

Amongst the mangrove fauna in Ecuador, there is also a tiny species of mite *(Hattena rhizophorae)*, a visitor to the flowers of the red mangrove. This species has been discovered in the country's mangroves in 2006, being its genus unknown to the American continent (Cornejo and González 2015).

The most relevant bird species found are the Magnificent Frigatebird (*Fregata magnificens*), Blue-footed booby (*Sula nebouxii*), Neotropic Cormorant (*Phalacrocorax brasilianus*), Brown Pelican (*Pelecanus occidentalis*), Black-crowned Night-Heron (*Nycticorax nycticorax*), Roseate Spoonbill (*Platalea ajaja*) amongst others. These bird species are attracted to mangroves due to the abundance of prey and nesting places. Only in the Jambeli Archipelago (El Oro province), 115 bird species have been recorded (A. Orijuela, personal communication, May 22, 2017). This number is relatively high given the previous high rates of mangrove deforestation of the 70s and the 80s. Due to this fact, Birdlife International has included the remaining mangrove forest of the Jambelí Archipelago among its 108 important bird areas (IBAs) in Ecuador (BirdLife International and Conservation International 2005) (Fig. 25.4).

Many of these bird species are abundant and can be easily seen perching and nesting on the mangroves or flying overhead making them an important ecotourism attraction (Bravo 2013).

In terms of reptiles, the once abundant American crocodile (*Crocodylus acutus*) is rarely seen in the mangroves of Ecuador. Green marine turtles (*Chelonia mydas*) are common in the Jambeli Archipelago. Similarly, researchers have been able to track Hawksbill (*Erytmochelys imbricata*) nesters and juveniles in the estuaries of the archipelago (García et al. 2015). This is very important reason for mangrove conservation given the endangered status of these marine turtle species.

Fig. 25.4 The Roseate Spoonbil (*Platalea ajaja*) is a very large pink colored bird with an unmistakable long flattened and spatulate beak. It measures from 71 to 79 cm and feeds on small fish and crustaceans. Photo credit: Adrián Orihuela

25.2.4 Threats to Mangrove Forest in Ecuador

Anthropogenic activities are the main drivers of mangrove destruction around the world. The destruction of mangroves is related to habitat destruction, over exploitation of resources, lack of urban planning, shrimp farming, cattle ranching, agriculture, aquaculture, urban solid waste, industrial pollution, pesticide and fertilizer pollution, oil spills, and alteration of hydrological conditions (Spalding et al. 2010). Anthropogenic activities substantially alter the composition, structure and function of mangroves reducing the ecosystem services these provide.

In Ecuador, changes in land use from mangroves to shrimp farms and urban development are the main factors of mangrove forest loss. Mangroves are increasingly threated by aquaculture activities and especially by shrimp farming. (See Fig. 25.5)

Although the shrimp industry is of great importance for the country, this industry is the principal cause of mangrove destruction in Ecuador. This activity encourages mangrove forest fragmentation and results in a reduction of goods and services provided by the mangrove ecosystem (Polidoro et al. 2010; Aschenbroich et al. 2015). In Ecuador, before the development of shrimp farms, the subsistence economy of coastal inhabitants was based on a traditional use of the mangrove ecosystem that included the extraction of mollusks and crustaceans and the production of vegetable coal (Latorre 2014).

After the startup of the first shrimp farms in the 1960's, the Ecuadorian government started to see mangroves as a source of economic development and promoted shrimp farming. This is how Ecuador became one of the major producers of farmed shrimp in the world, which caused a consequent loss of a significant area of mangrove forest.

According to Ecuador's National Chamber of Aquaculture, in 2015 there were 213,000 ha allocated to the production of shrimp, out of which 181,000 ha were originally mangrove ecosystems. There are around 3000 shrimp farms in Ecuador (Revista Líderes 2015). The Guayas province had the most shrimp farming activity with about 140,000 ha of shrimp pools and generating 66% of Ecuador's total shrimp production. The provinces of El Oro, Manabí, Esmeraldas and Santa Elena followed with 18%, 9%, 6% and 1% of Ecuador shrimp production respectively (Cámara Nacional de Acuacultura 2017).

In addition to the direct loss of mangrove forests during the construction of a shrimp farm, these also liberate effluents rich in organic and inorganic particles causing destruction to nearby mangroves due to contamination (Hurtado-Domínguez et al. 2012; FAO 2015_2).

The deforestation of mangroves in Ecuador coincides with the growth of the shrimp farming industry. Recent studies have shown that 80% of the loss of mangrove carbon in Ecuador is due to the direct displacement of mangroves by shrimp farms (Hamilton and Lovette 2015).

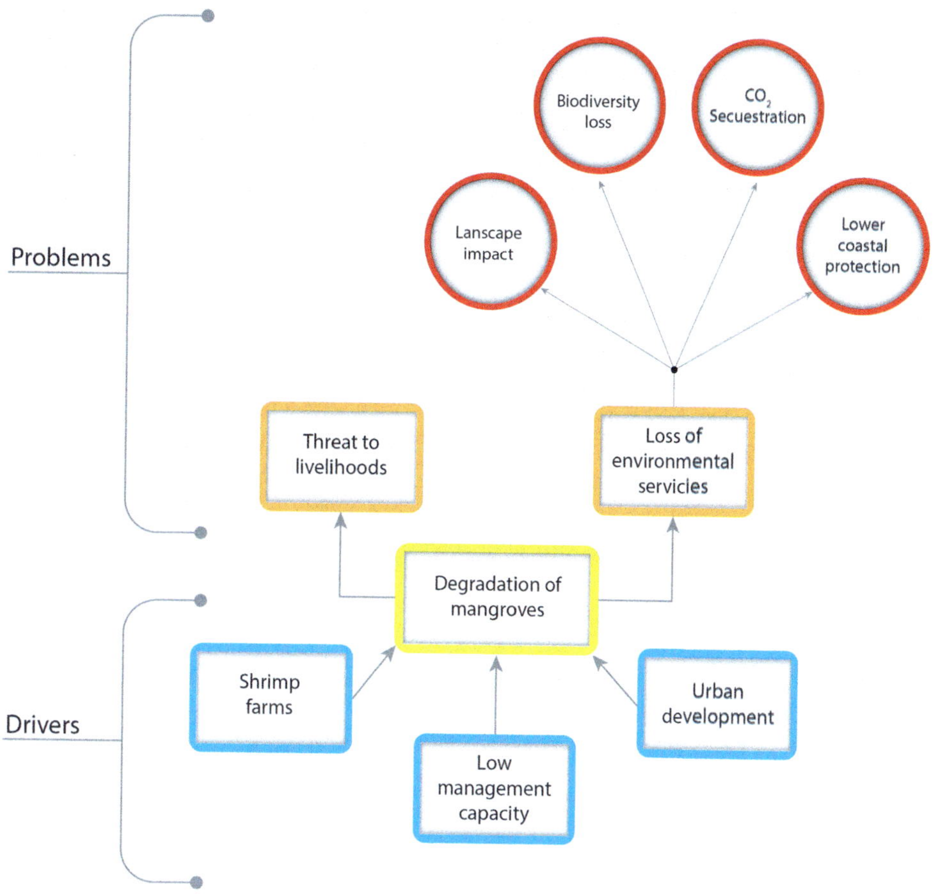

Fig. 25.5 Drivers of mangrove degradation in Ecuador. This generates a series of problems for the human being: the loss of food that is extracted directly from the mangrove s and the loss of environmental services provided by this ecosystem. Many cities on the Ecuadorian coast are located in areas that were mangrove areas in the past. In 1994 it was estimated that approximately 3000–5000 ha of mangrove forests have been destroyed by uncontrolled growth of cities (Guayaquil, Machala and Esmeraldas). Guayaquil, the country's largest city, is located in what was used to be a mangrove area (Spurrier 2012)

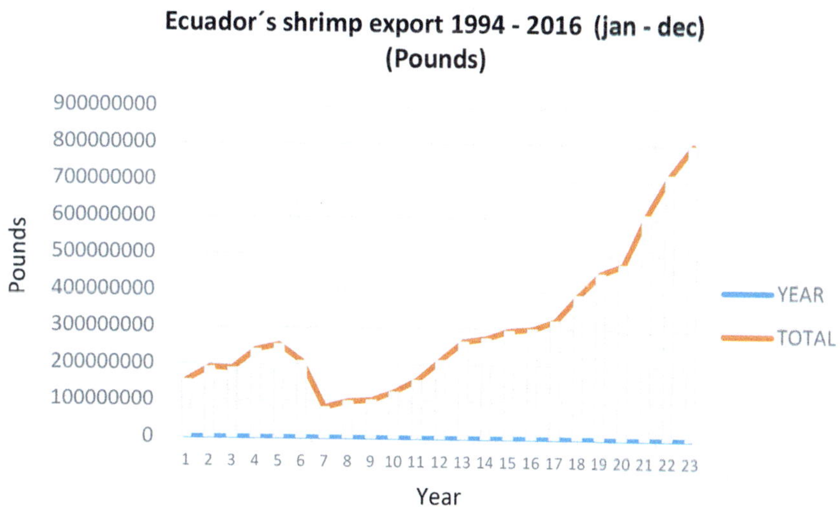

Fig. 25.6 Ecuador's shrimp export during the period 1994–2016. While in 1994, Ecuador exported 156 million pounds of shrimp equivalent to more than US $514 million, in 2016 the country exported close to 800 million pounds (US $2.5 billion), the highest amount of income since 1994 (Cámara Nacional de Acuacultura 2017)

25.2.5 Shrimp Farming in Ecuador

Ecuador is one of the major producers of shrimp around the world. The provinces with the most shrimp farming activity are Guayas, El Oro and Esmeraldas (IUCN and CI Ecuador 2016). In 2016, Ecuador sold close to US $2.5 billion of shrimps, 13% more than in 2015, making shrimp the second major non-petroleum export product according to Ecuador's Central Bank. According to Ecuador's National Chamber of Aquaculture, 20% of non-petroleum exports are generated by the shrimp industry. The growth of the shrimp industry has been progressive over the last 20 years (Cámara Nacional de Acuacultura 2017). (See Fig. 25.6)

The principal export market for Ecuadorian shrimps is the Asian continent, receiving 58% of its export volume in 2016 (and 61% in 2015). The United States and Africa were second with 19% each in 2016. It is estimated that the shrimp industry generates 180.000 direct and indirect jobs and that together with the fishing industry it makes up 5% of Ecuador's employment (Bernabé-Argandona 2016).

It is important to mention that the growth of the shrimp sector in Ecuador has been related to some illegitimacy issues. Even though, article 54 of Ecuador's Secondary Environmental Law (Texto Unificado de Legislación Ambiental Secundaria-TULAS), (Gobierno de la República del Ecuador 2003) states that: "*no government institution or entity will authorize the construction of new shrimp farm pools or the extension of existing shrimp farms located in the mangrove ecosystem and its buffer zone.*", many shrimp farms were built since this legislation entered into force in 2003. Ecuador's Aquaculture Sub-Secretariat (Subsecretaria de Acuacultura) estimated that in 2013, 66% (6.192 ha) of the shrimp farms in Esmeraldas province were illegal. Similarly, in El Oro, Manabí and Guayas provinces, illegal shrimp farms made up to 59% (8434 ha), 39% (12,576 ha), and 18% (17,437 ha) of the shrimp farming land respectively (Government of the Republic of Ecuador 2008).

The executive decrees No. 1391 (Government of the Republic of Ecuador 2008) and No. 852 (Government of the Republic of Ecuador 2016) describe the regulation of shrimp enterprises and establish that shrimp enterprises created after 1999 are illegal when they occupy areas that belong to the State or are located within protected areas. It also defines the maximum extension of shrimp concessions: for natural persons up to 250 ha and for legal persons up to 1000 ha. In these executive decrees the period of concessions was increased from 10 to 20 years, period

Fig. 25.7 Shrimp pond. The construction of shrimp farms has been directly linked to mangrove destruction, the main cause of the decline in the reproduction of mollusks and crustaceans (shell and crab). In addition, a study carried out by the UnderSecretariat of Marine and Coastal Management of the Ministry of Environment (MAE-SGMC) determined that in 4 out of 6 analyzed protected areas that belong to the National Protected Area System there were new shrimp farmers installed or old ones expanded their area after the date of the declaration of the protected area. On the other hand, and in compliance with Decree 1391, the MAE-SGMC, until October 2012, had approved 745 reforestation plans submitted by shrimp farmers. Most of these plans belonged to the province of El Oro (263) and Guayas (221) (Ministerio del Ambiente 2017_2). The Ministry of Environment estimate that between 2008 and 2017, 2614 ha of deforested mangroves where reforested. Photo credit: Javier Vásquez

Table 25.2 Evolution of the shrimp areas in the period of 1984–2006

	Year							
	1969	1984	1987	1991	1995	1999	2006	2016
Shrimp farms (ha)	0	89.368	117.728	145.998	178.071	175.253	175.748	210.000

The main expansion of the surface of shrimp pools in Ecuador occurred during the first two decades after its first introduction to the country. Since 1995 the expansion of this industry stabilized (Center for Integrated Natural Resource Mapping by Remote Sensing et al. 2007)

Table 25.3 Evolution of the mangrove area in the period 1969–2016

	Year								
	1969	1984	1987	1991	1995	1999	2000	2006	2016
Mangrove (ha)	203.695	182.157	175.157	162.186	146.938	149.556	127.690	148.230	161.835

Since the period 1995–1999 the negative deforestation rate changed into a situation of mangrove recovery of 0.23, becoming again negative for the period 1999–2006, with 0.025% per year (Fig. 25.8)

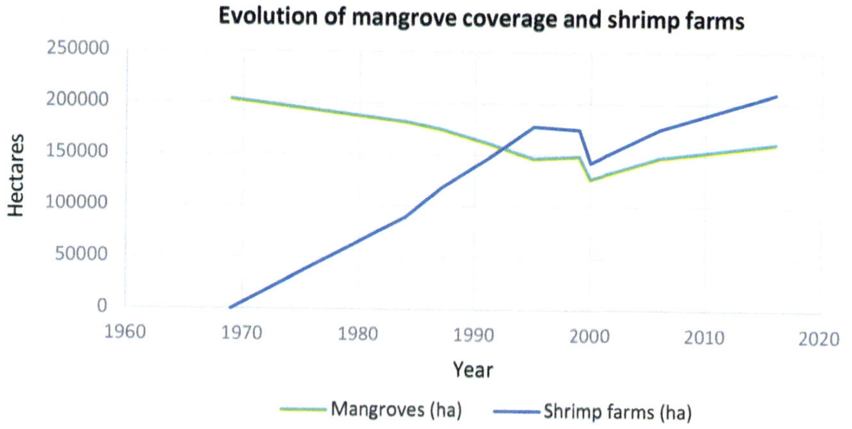

Fig. 25.8 Comparison of mangrove coverage (ha) and extension of shrimp farms (ha) between 1969 and 2016. Since 2006 the coverage of mangroves has remained relatively stable, while the area covered by shrimp farming has increased to 210,000 has. The main loss of mangrove coverage occurred between 1969 and 1995 (Center for Integrated Natural Resource Mapping by Remote Sensing et al. 2007)

that can be renewed, but obliges the concession holders to reforest mangrove forests being cut until 2015. In compliance with this regulation, between 2010 and 2012, 40 shrimp farms have been evicted adding up to 2495.86 ha of shrimp farms (Government of the Republic of Ecuador 2016) (Fig. 25.7).

In addition, the Ministry of Environment's resolution 056 establishes fines up to US$89,273 per hectare for cutting down mangroves (Gobierno de la República del Ecuador 2011a, b).

According to the Center for Integrated Natural Resource Mapping by Remote Sensing et (2007) the evolution of the surface of shrimp pools from 1984 to 2006 is as follows. (see Table 25.2):

25.2.6 Environmental and Socio-economic Impacts of Shrimp Farming in Ecuador

Despite the economic importance of the shrimp industry, it is the main cause of the destruction of the mangrove forest in Ecuador, and related to the reduction of ecosystem services of mangroves (Latorre 2014; Revista Líderes 2015) (Table 25.3).

Latorre (2014) indicate that in Ecuador, before the development of the shrimp farming industry, traditional uses (mainly collection of crab and shellfish, along with coal production) played an important role in the subsistence economy of the inhabitants living along the Ecuadorian coast. With the expansion of shrimp farming at the end of the 1960's, the situation changed, and mangrove areas were seen by the Ecuadorian State as a potential source of economic development through the promotion of the shrimp industry.

The destruction of mangroves and degradation of the associated biodiversity had multiple negative impacts on the communities and traditional users who benefited from resources and ecosystem services provided by mangroves. Increased effort is required to collect biodiversity species that provide food security and income to the local population. Unemployment rates increased amongst traditional mangrove users, losing their traditional work, as shell and crab collection (Bravo 2013). Mangrove conversion has caused the overexploitation of the mangrove cockle (*Anadara tuberculosa* and *A. similis*), a bivalve mollusk harvested from the roots of mangrove trees, locally known as concha prieta (Beitl 2010). Another social impact caused by mangrove deforestation are community displacement, the decrease of resource rights, the reorganization of local economies, increase in economic

Table 25.4 State managed protected areas with mangrove forests coverage

	Name of the protected areas	Month/year of creation	Total extension (ha)
1	Mangroves of Churute ecological reserve	July 1979	49,389
2	Cayapas Mataje ecological reserve	October 1995	51.300
3	Esmeraldas River wildlife refuge	June 2008	242
4	Mangroves Muisne River wildlife refuge	March 2003	3173
5	Islas Corazón y Fragatas wildlife refuge	October 2002	2811
6	Mangroves El Morro wildlife refuge	September 2007	10.030
7	Mangroves of El Salado Reserve for Fauna	November 2002	10,635
8	Isla Santay national recreation area	February 2010	2215
9	Arenillas ecological reserve	May 2001	13,170

Source: http://areasprotegidas.ambiente.gob.ec

These nine protected areas all belong to the *State managed natural protected areas system (PANE in Spanish)*, one of the four subsystems that make up the *National System of Protected Areas* (SNAP). Mangrove areas under the Sustainable Use and Custody agreement have an important conservation role due to its close distance to natural protected areas of the SNAP, some are even located within the protected areas

Table 25.5 Distribution of mangroves according to their legal status

Mangrove distribution	Surface coverage (ha)	%	Source
Protected areas (SNAP) (2012)	73.071	46,5	Astudillo et al. (2014)
Mangrove concessions (2014)	65.933	42,0	SGMC
Mangrove forests without conservation status (to 2014)	18.020	11,5	Astudillo et al. (2014)
TOTAL	157.024	100	

Protected areas contain 73.071 ha of mangrove forests or 46.5% of the total existing mangrove forests in Ecuador. 42% of the mangrove forest is protected through sustainable use and custody agreements or mangrove concessions. The remaining 18.020 ha of mangroves are protected according to law, but are not covered by the national protected area system or mangrove concessions (see Table 25.5)

disparity and social conflict (Beitl 2010; C-CONDEM 2007; Martinez-Alier 2001; Stonich and Vandergeest 2001).

In addition to the direct loss of mangroves due to shrimp pool construction; shrimp farms also impact neighboring mangroves by releasing large quantities of effluents rich inorganic and inorganic particles and nutrients (Páez-Osuna 2001). Shrimp production in pools requires the use of antibiotics, in addition to fertilizers, pesticides and other chemicals for the treatment of water and soil, causing the indirect destruction of nearby mangroves due to water pollution by the above mentioned chemicals.

After leaving the shrimp farming activity the soil is unsuitable for any other productive activity due to soil salinization and soil contamination (Bravo 2013).

25.3 Mangrove Forest Conservation Efforts in Ecuador

25.3.1 Protected Areas

The first strategy to stop rapid loss of mangroves in Ecuador was the creation of protected areas. In 1979 the Manglares Churute Ecological Reserve was created, located north of the equator in the province of Esmeraldas, followed by the Cayapas Mataje Ecological Reserve in 1995.

Nowadays, Ecuador has 14 marine and coastal protected areas, together accounting for about 8% of all total coverage of protected areas in the country. Nine of these areas contain (total or partially) mangroves forest. (See Table 25.4)

25.3.2 The Mangrove Sustainable Use and Custody Agreements (Mangrove Forest Concessions)

The term “traditional or ancestral communities” is enshrined in the Constitution of the Republic of Ecuador. It recognizes the existence of indigenous peoples and Afro-Ecuadorians who define themselves as nationalities with ancestral roots, which are part of the Ecuadorian State. These are groups of people who live according to organizational traditions established prior to the emergence of the modern state (National Constituent Assembly 2008). The Unified Text of Secondary Environmental Legislation (TULAS), establishes that “. . . the ancestral communities may request to be granted the sustainable use of mangroves for subsistence, development and marketing of fish, mollusks and crustaceans, among other species, which grow in this habitat.” Such requests of ancestral users and communities is agreed upon by a “Sustainable Use and Custody Agreements of the mangrove forest (Gobierno de la República del Ecuador 2003).

The idea of these “Sustainable Use and Custody Agreements for the of the mangrove forest” already emerged in 1994, due to recognizing the need to defend mangrove forests from logging and the consequent loss of biodiversity associated with mangrove (Bravo 2013).

Formally, the "*Acuerdos para el Uso Sustentable y Custodia del Manglar* ¨ (In English: *Sustainable Use and Custody Agreements for mangrove forests),* also known as "Mangrove Forest Concessions", were introduced in 1999 in order to prevent the clearing of mangrove forests and to provide legal land use rights to ancestral communities and traditional users of the mangrove forests. They aim to promote community participation in the defense and conservation of the ecosystem and to maintain the traditional use of fish, mollusks and crustaceans by local communities. These agreements give ancestral communities and traditional users the exclusive use of mangrove resources for 10 years, in turn, they are requested to protect the ecosystem. Thereby, these agreements complement the national conservation strategies of the nine protected areas of Ecuador containing mangrove forests (Gobierno de la República del Ecuador 1999).

Already before various efforts had been made to stop the felling of mangrove forests, such as the modification of the *Forestry and Conservation of Natural Areas and Wildlife Law* in 1990, through which it was legislated that: "mangroves, even those existing in particular properties, are considered State property and cannot be, they are not susceptible to possession or any other means of appropriation and may only be exploited by concession.

To establish the regulatory framework for the *Sustainable Use and Custody Agreements* for mangrove forests the following executive decrees and ministerial agreements were emitted.

Executive decrees and ministerial agreements

Executive Decree No. 1102.

Released in 1999, gave ancestral communities the right to apply for concessions for the sustainable use and custody of mangrove areas. This Executive Decree was revoked in 2002, and included as a chapter on mangroves in the Unified Text of Secondary Environmental Legislation of Ecuador (Coello et al. 2008).

Ministerial Agreement No. 172.

Issued in 2000, this agreement established the instructions for the granting of agreements of sustainable use and custody of mangroves, and started the issuing of concessions of mangrove forests to the ancestral communities and traditional users settled along the coastal line (Coello et al. 2008).

Ministerial Agreement No. 024.

Through this agreement in 2009, the Sub-Secretary for Marine and Coastal Management (of the Ministry of Environment), was given the responsibility to issue and monitor the Agreements of mangroves use to ancestral communities and traditional users (Coello et al. 2008).

Ministerial Agreement No. 129.

This agreement issued in 2010, specified and detailed the instructions for conceding the agreements of sustainable use and custody of mangroves in favor of ancestral communities and traditional users (Coello et al. 2008).

Ministerial Agreement No. 144.

This Agreement, issued in 2011 reformed the Ministerial agreement No. 129. It particularly addresses the minimum content of the Management Plans of the Mangrove Concessions (Coello et al. 2008).

Ministerial Agreement No. 198.

This agreement, issued in 2014, specified instructions for a governmental incentive to ancestral communities and traditional users for the conservation and sustainable use of mangroves, named Mangrove Partner (Socio Manglar), which seeks to contribute to the consolidation of mangrove areas to which concessions have been issued (Coello et al. 2008).

Socio Manglar (Mangrove Partner)

This program provides a direct economic incentive to organizations that hold agreements for sustainable use and custody of mangroves. The amount of the incentive depends on the area of the concession and is aimed at strengthening local organizations through supporting the implementation of their mangrove management plans. The incentive is set up for 10 years, it supports the development of the following activities:

- Control and Monitoring
- Administrative Management
- Technical Assistance for the implementation, monitoring and/or evaluation of management plans.
- Organizational strengthening.
- Financing of productive and social projects, that benefit partners.

The amount of the economic incentive depends on the area of the concession and is transferred in two installments per year. The amount is calculated according to the following table:

Categoría	Rango de Has en Acuerdo	Incentivo Fijo Anual
1	100 a 500 has	$7.000 USD
2	501 a 1.000 has	$10.000 USD
3	Mayor a 1.000 has	$15.000 USD
Más un incentivo variable de $3USD/Ha		

By 2017, 24 associations with more than 23,000 ha of mangrove are receiving funds from Socio Manglar.

Outline of Requirements for Sustainable Use and Custody Agreements

The concession of a certain area of mangroves can be issued to ancestral communities and traditional users, organized in legally constituted cooperatives or association registered by the Sub-Secretary for Marine and Coastal Management, to whom they have to present a Management Plan. The Ministerial Agreement No. 144 (Gobierno de la República del Ecuador 2011a, b) defines the minimum content of these

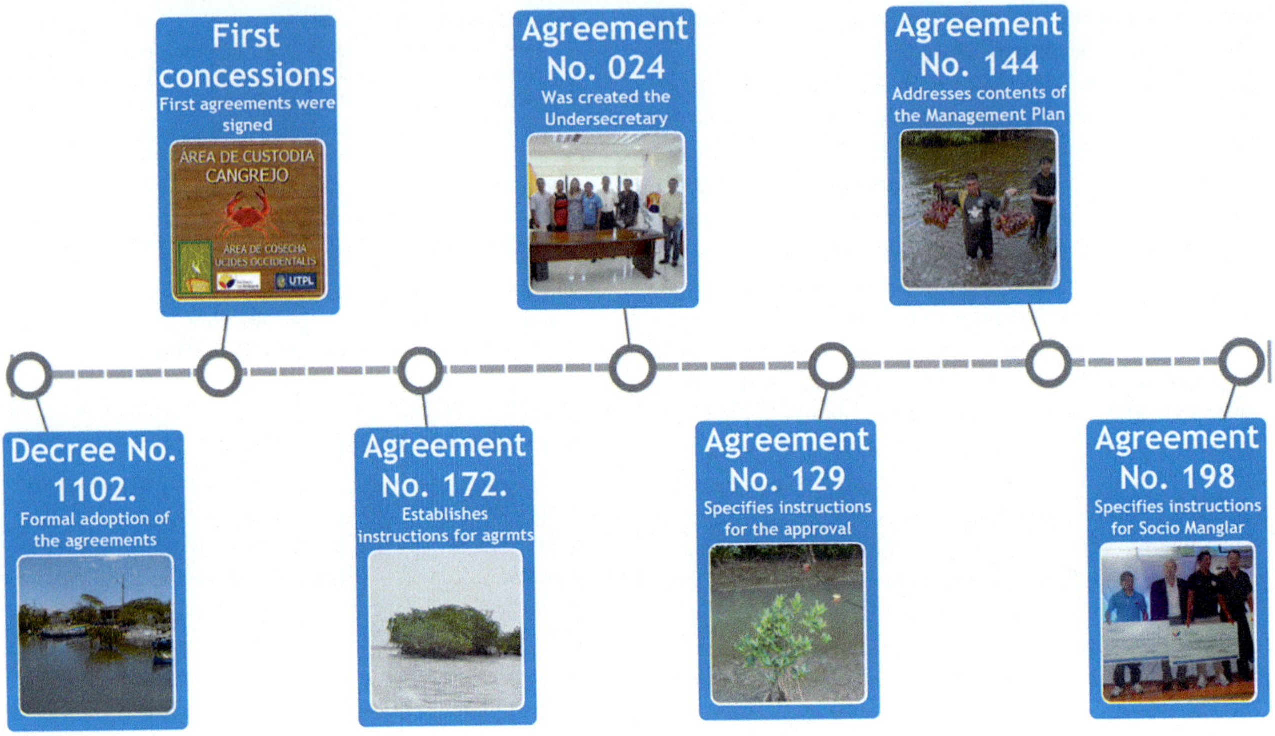

Fig. 25.9 Even though, sustainable use and custody agreements of mangrove forests were established from 1999, it was from the beginning of the decade of the 90s that social organizations and NGOs began promoting the issuance of mangrove areas in favor of ancestral and traditional users. The last stage of the policy framework for community mangrove conservation consisting in the Mangrove Partner Incentive sets the beginning of the consolidation of mangrove concessions

Management Plan. The Plan should contain three programs: Sustainable Use or Resources, Control and Surveillance and Monitoring and Evaluation. These programs have to contain activities and well defined indicators.

The application must be accompanied by a technical assistance agreement with a university, research institute, NGO or governmental institution.

The activities permitted in the area of the custody agreement correspond only to sustainable activities and non-destructive use of the mangrove areas: Catching of fishes and invertebrates; Aquaculture of fish, invertebrates, mammals, reptiles or other species of native wildlife through practices that do not affect the coverage of mangroves or the dynamics of the water bodies; mangrove reforestation; eco-tourism and non-destructive mangrove recreation activities, education and scientific research.

From 2000 to 2015, 66 agreements for sustainable use and custody of mangroves had been issued to fishermen's organizations allocated in 5 coastal provinces: El Oro, Guayas, Santa Elena, Manabí and Esmeraldas. By 2017, 47 Associations have mangrove concessions with a total of 59.208 ha (Ministerio del Ambiente de Ecuador 2017_1). (See Fig. 25.9)

Evaluation of Use and Custody Agreements

In 2008, the Ministry of Environment, CI and IUCN, performed an evaluation of 27 of the existing sustainable use and custody agreements of mangrove forest, during which 27 management plans were assessed. This evaluation indicated that in general terms, the planned activities of the management plans were implemented successfully. While there were organizations that achieved 100% compliance with the implementation of the management plans there were others that have failed to comply all activities. The most relevant conclusions of this evaluation were: mangrove agreements are a positive tool for the conservation of mangroves and have served as a legal instrument to guarantee ancestral mangrove users exclusive access to these areas, generated social and economic benefits and improved living conditions; contributed significantly to stop the felling of mangrove. On the other hand, control and monitoring have been the most difficult tasks as well as the administration of areas. Technical assistance provided by NGOs, universities, among others was a very important support (Coello et al. 2008).

25.3.3 Other Legislation for Related to Mangrove Conservation

Until the end of the 70s, mangroves were considered unproductive areas by the Ecuadorian government, without representing economic benefit for the country. As a result, concessions for using the beach and mangrove forests to establish shrimp farms were emitted, due to the production value generated by this activity, causing the destruction and felling of large areas of mangrove. From the 1980s onwards, the recognition of the importance of mangrove forests to society led to the development of mechanisms for conservation and protection of mangroves, being included in the Constitution of Ecuador.

Several laws, executive decrees and ministerial agreements have incorporated elements related to the protection and conservation of the mangrove. Among the most important are:

Constitution of the Republic of Ecuador (2008)
Article 405 states that in the areas of the National System of Protected Areas (SNAP), the State will promote the "Participation of communities, peoples and nationalities that have traditionally inhabited areas protected in their administration and management". While article 406, mentions that the State shall regulate the conservation, management and sustainable use, recovery, and domain limitations of the fragile and threatened ecosystems; among which is the mangrove.

National Plan for Good Living (2013)
This plan refers to the mangrove in different sections. It is commanded to strengthen the conservation and sustainable use of biodiversity, it will recognize, respect and promote knowledge and ancestral knowledge, innovations, and sustainable traditional practices of communities, peoples and nationalities with their full and effective participation (Gobierno de la República del Ecuador 2013).

Environmental Organic Code (2017)
Several articles of this Code mention the importance of the knowledge of the communities of their ancestors and the priority to conserve fragile ecosystems such as mangroves. Furthermore, it emphasizes that the conservation of mangroves and other fragile ecosystems it in the public interest as well as the establishment of areas of custody and sustainable use of mangrove and the creation of economic incentives for the conservation of fragile ecosystems (Government of the Republic of Ecuador. 2017).

Ramsar Convention
The Ramsar Convention is an intergovernmental Treaty adopted on 2 February 1971, signed in Ramsar, Iran; it serves as the framework for national action and international cooperation for the conservation and wise use of wetlands and their resources. Ecuador is one of the 168 Contracting Parties to this Convention, and entered into force in Ecuador on January 7, 1991. Ecuador currently has 18 sites designated as wetlands of international importance (Ramsar sites), with an area of 290,815 hectares. Five of these Ramsar Sites have some percentage of mangrove cover (Echeverría 2008).

25.4 Case Study: Mangrove Concessions at Costa Rica Island

25.4.1 Archipelago of Jambelí

The archipelago of Jambelí is located at the southern end of the Ecuadorian coast, in waters of the Gulf of Guayaquil. It belongs to the canton of Santa Rosa, province of El Oro. It has an approximate area of 41,706 ha. The archipelago of Jambelí is comprised of five island communities: Bellavista, Costa Rica, Las Huacas, Las Casitas and Pongalillo. (See Fig. 25.10)

In the province of El Oro, until 2016, there were 26 mangrove agreements for sustainable use and custody of the mangroves, which cover an area of approximately 8328 ha, equivalent to more than 55% of the extent of mangrove in the province of El Oro. By 2017, there are 21 agreements with an extension of 11.318 ha (75,45%). From the point of view of conservation, the

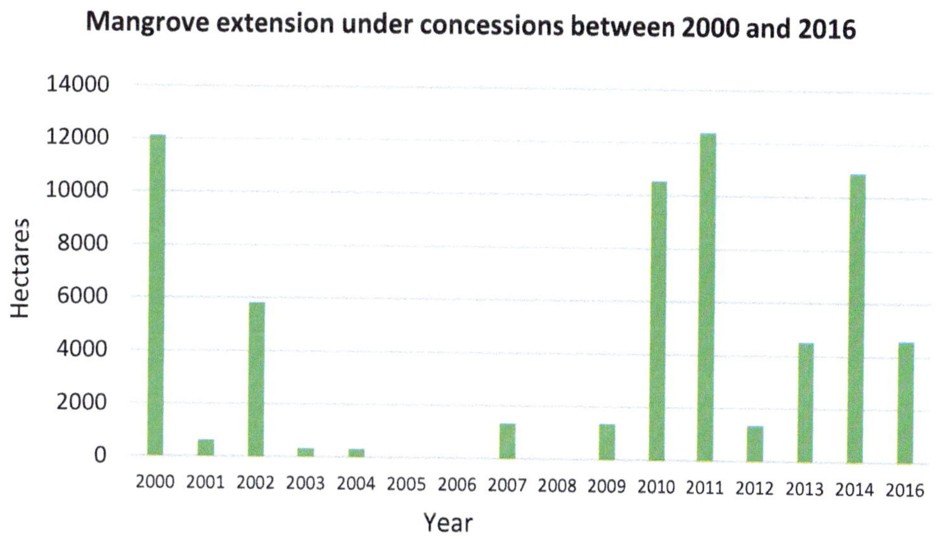

Fig. 25.10 Mangrove extension under concessions between 2000 and 2016. The periods between 2000–2002 and 2010–2014 recorded the highest number of agreements signed with beneficiaries mainly from El Oro and Guayas provinces (Ministerio del Ambiente de Ecuador. 2017l)

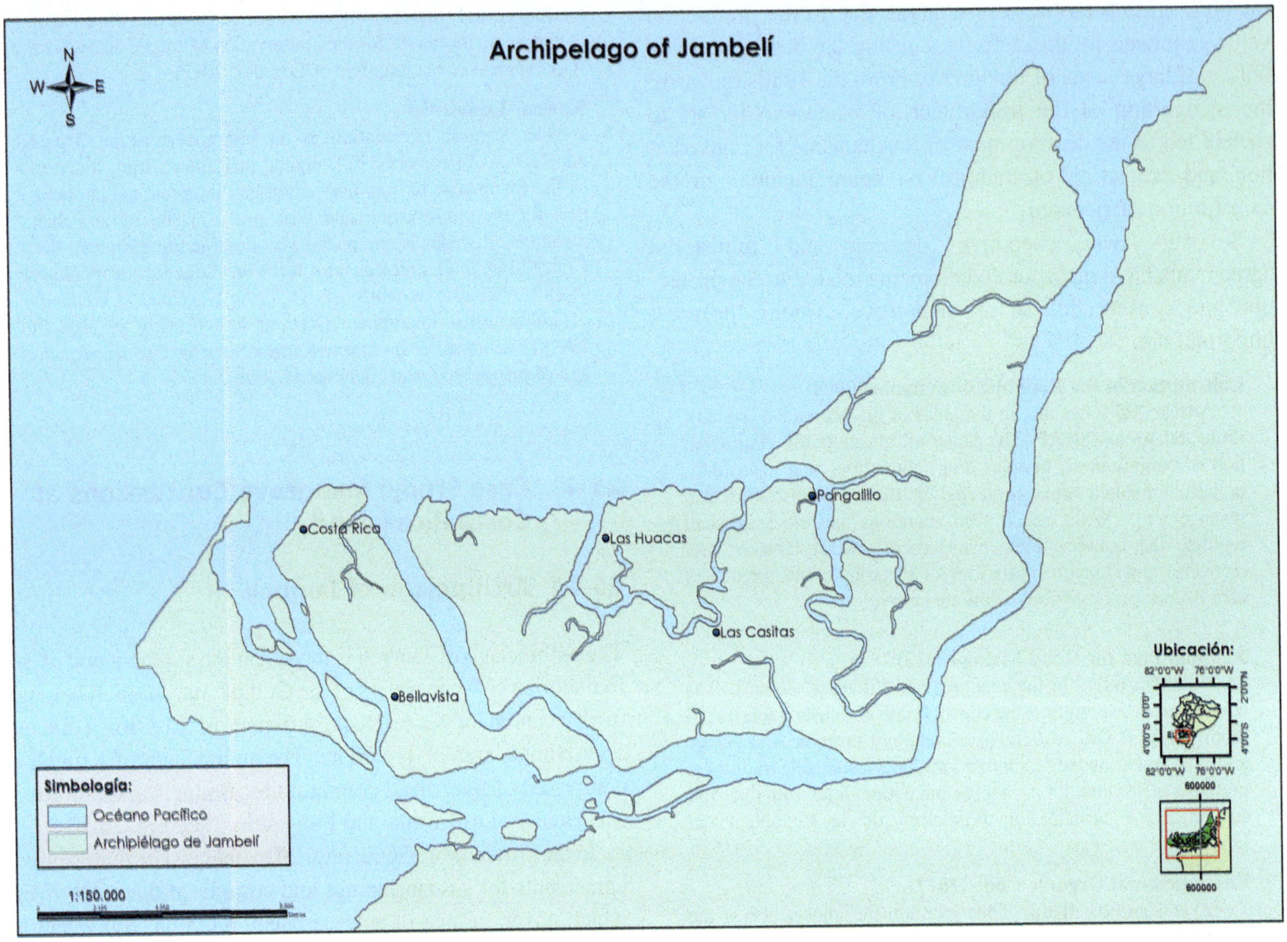

Fig. 25.11 The five islands of the archipelago of Jambelí, through their grassroots organizations made up of fishermen, black ark collectors ("concheros") and red crab collectors ("cangrejeros"), have been awarded mangrove concessions, since the same year that these were put into effect (2000). The concessions include more than 2.300 ha of mangrove forest (Gobierno Provincial Autónomo El Oro and Fundación Ecológica Arcoiris 2013)

area of mangroves protected through these agreements in this province is very important as the Arenillas Ecological Reserve, a State managed protected area, contains only 1.266 ha of mangrove forest (7.41% of the total area of the reserve).

The mangroves at the Archipelago of Jambelí are one of the mangrove ecosystems most impacted by human activity in Ecuador. The current landscape that can be seen in this area is a mosaic of shrimp farms, patches of mangrove forest, networks of canals and estuaries, settlements of small towns and the scene of traditional activities such as fishing and collecting of shells and crabs (Bravo 2010). (See Fig. 25.11)

The Island Costa Rica

Costa Rica is the urban capital of the rural Jambelí Parish, belonging to the Santa Rosa municipality, and the province of El Oro. Costa Rica has an approximate population of 86 families, with 165 men and 175 women. Its inhabitants collect shells and crabs, or work in the fishing and tourism sector.

25.4.2 The Management of the Mangrove Concession

The mangrove concession was granted to the Costa Rica Artisanal Fishermen Association (In Spanish: Asociación de Pescadores Artesanales y Afines "Costa Rica") on August 16, 2000 with an extension of 519 ha (Gobierno Provincial Autónomo El Oro and Fundación Ecológica Arcoiris 2013). The agreement of the Association was renewed in July, 2016, with the same extension. (See Figs. 25.12 and 25.13)

This Association has developed a successful management model which is based on four key activities (Fig. 25.14):

(a) *Sustainable harvesting of back ark*

The Costa Rica Association, as part of its management plan, promotes the sustainable use of the black ark (*Anadara tuberculosa*). They respect the minimum size for extraction

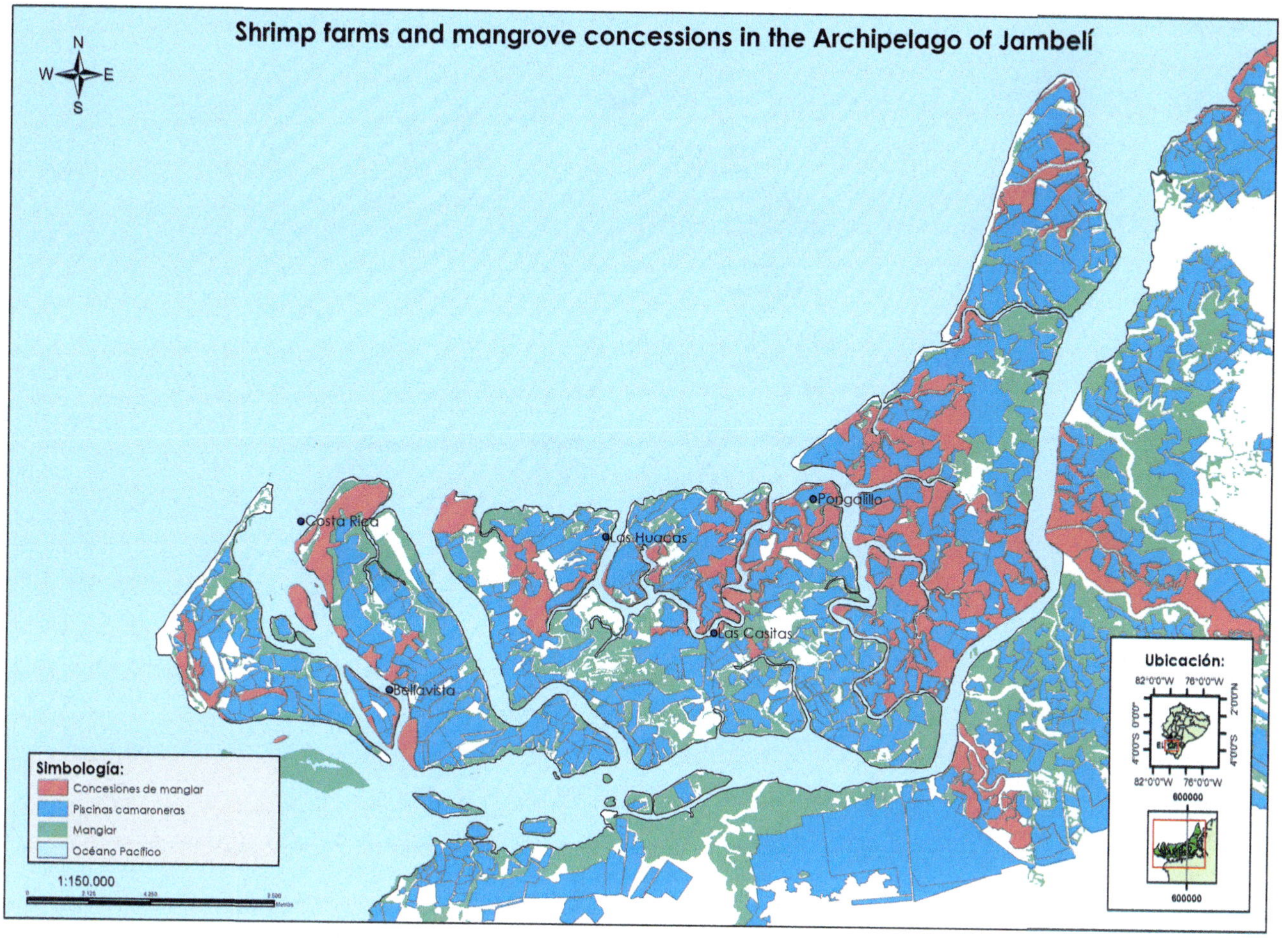

Fig. 25.12 Shrimp pools and concessions of mangrove forest in the Archipelago of Jambelí. The shrimp industry occupies most of the mangrove forest territory of this area. According to the Cámara Nacional de Acuacultura-National Chamber of Aquaculture (2017), El Oro is the second province dedicated to the production of shrimp, with more than 30,000 ha, i.e. 15% of the 210.000 ha of shrimp farms that currently exist in Ecuador. This activity was developed even in protected areas such as the Arenillas Ecological Reserve, where 500 ha of shrimp farms were identified

and commercialization, being 4,5 cm as regulated by the Ministry of Agriculture, cattle raising, aquaculture and fisheries (Gobierno de la República del Ecuador 2008). The Costa Rica Association gathers approximately 3000 shells per day, at an average of 100 shells by each partner of the association (Total of 30 currently active partners). There are five main collection sites mentioned below, which are harvested in a period of 10 days:

- **Encenada or Chupador**. A day of collection
- **Chalet**. Two days of collection
- **Los Chanchos**. Four days of collection
- **Diluvio**. Four days of collection
- **Costa Rica (in front of the town)**. Two days of collection

During the last two decades, the black ark has begun to show signs of overexploitation throughout the Ecuadorian coast. The smaller sizes and fewer catches indicate signs of stress (Beitl 2010) (Fig. 25.15).

(b) *Reforestation*

Although there is little area available for reforestation, the association reforested all spaces without vegetation within their area, which accounts to 55 ha approximately since the beginning of the concession.

(c) *Custody of the area*

The area is patrolled almost every day. The association organized patrol groups formed by two members of the association. The treat that limits the continuity of this activity is the lack of resources for fuel and the piracy danger.

(d) *Ecotourism*

The Costa Rica Association has a Committee of Ecotourism, which currently is promoting this activity on the island. A pristine beach, known as San Gregorio, is located very close to

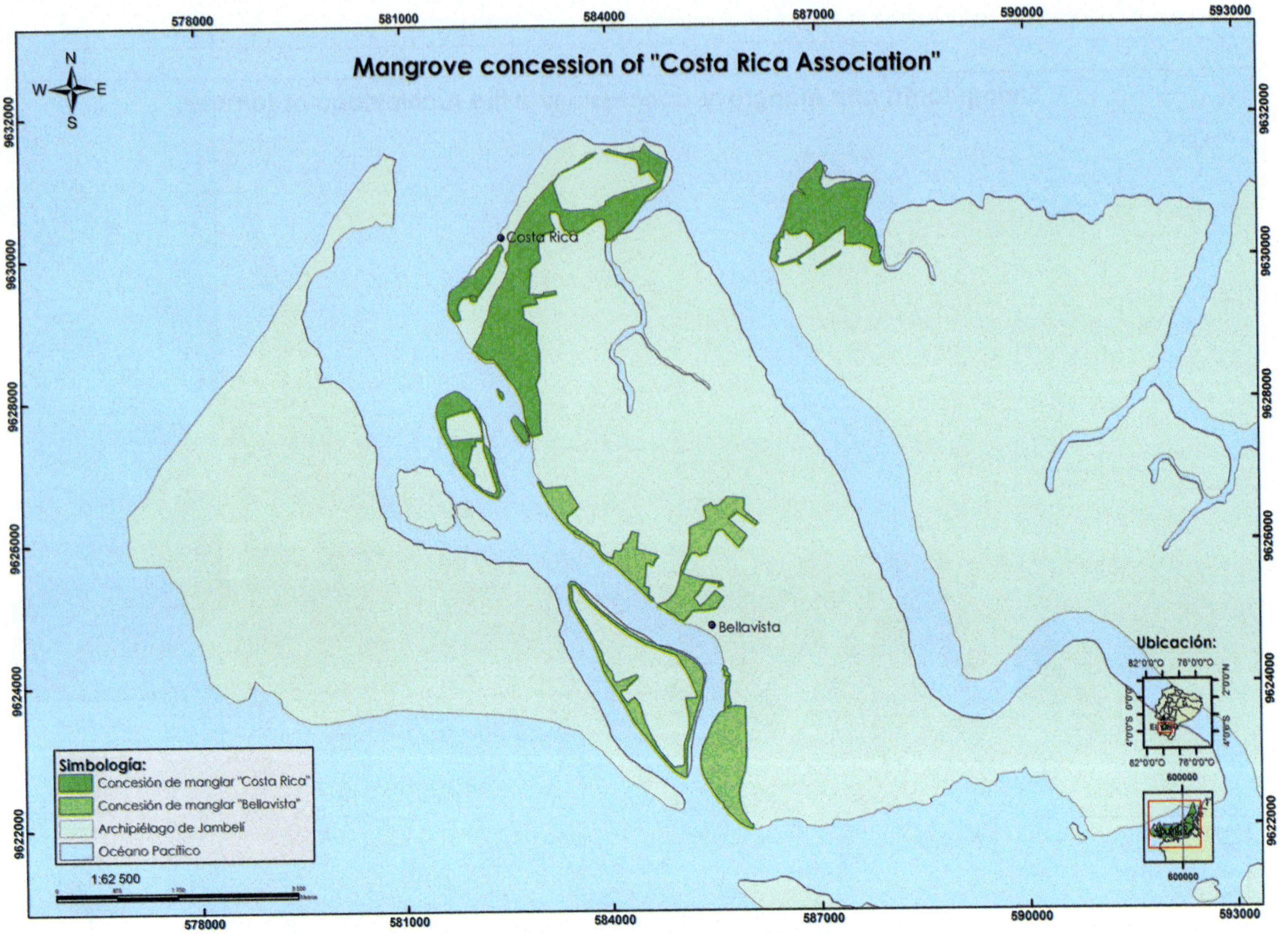

Fig. 25.13 Map of the mangrove concession "Costa Rica". The agreement expired in 2010 but it was renewed in July 2016 by the Sub-Secretary of Marine and Coastal Management after updating the management plan. The area of the Costa Rica Association is close to the area of the Bellavista Association a very important aspect for the biological connectivity

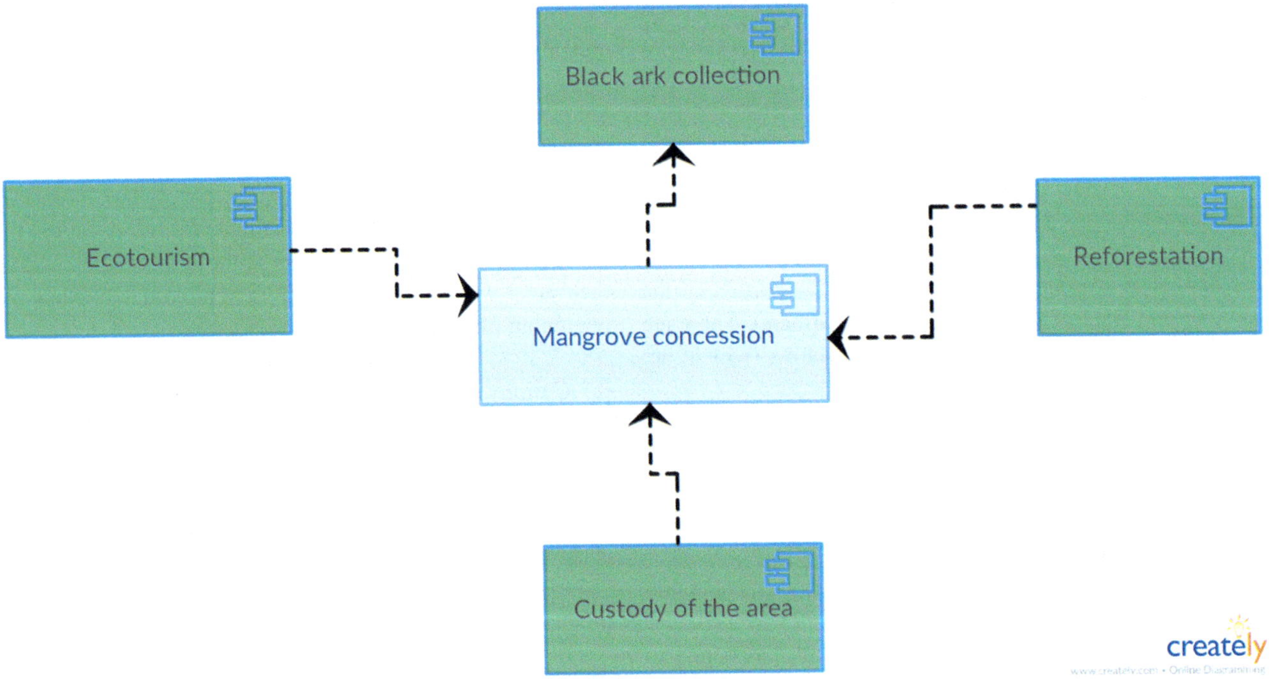

Fig. 25.14 The management model of the area mangrove concession of the Costa Rica Association. The Association of Costa Rica is one of the first organizations that received a mangrove concession. Originally it received technical and economic support from the Fundación Ecológica Arcoiris (Rainbow Foundation) and the UNDP – GEF Small Grants Program to implement activities such as reforestation, reproduction of black ark in captivity, ecotourism, among others

Fig. 25.15 **Black Ark (*Anadara tuberculosa*).** This species lives in the mangrove sludge (associated with the roots of *Rhizophora mangle*), buried from the root zone up to about 5 cm deep, or on soft bottoms between 15 and 50 cm deep (IMARPE n.d.). This mangrove resource has suffered a significant decrease in numbers due to the deterioration of the mangrove forests and overexploitation. The National Fisheries Institute-INP (2017) estimates that around 25'000.000 shells are extracted annually by about 3.000 fishermen, although at present they collect 50% less than previous years. C-CONDEM estimates that at present each fisherman is able to collect 100 shells per day unlike previous years when they collected up to 1000. There is a permanent prohibition to collect shells up to 4.5 cm, established by law. To reach this size, the shell requires between 12 and 18 months, period in which it has reproduced at least once. At this rate of collection it is estimated that in 10 years this species might disappear since more than 50% of the collected shells do not meet the minimus size of 4.5 cm (in the province of El Oro is estimated at 59%) (Instituto Nacional de Pesca 2017). Photo credit: Javier Vásquez

the urban population. Besides the beach, bird watching and local gastronomy of the island are the main tourism attractions, receiving more and more visitors day by day.

25.4.3 State of Mangrove Forest Conservation of the Island Costa Rica

In order to determine the state of conservation of mangrove forest under the concession, multi-temporal analysis were implemented in 1991, 2000, 2013 and 2014, which took into consideration a 1 kilomenter buffer zone.

The main land use coverage's identified by this study include: mangrove, deforested areas and shrimp ponds; the coverage related to sea and sand banks were not analyzed as they were not considered relevant according to the mangrove concession system. (See Fig. 25.16)

Mangrove deforestation had a peak during the first year of the mangrove concession (2000) due to the deforestation caused mainly by shrimp farms, accounting up to 394.19 ha which represents a 13% of the analyzed extension; the deforestation rate reduced to 176.53 ha (5%) in 2013, but the following year it raised to 6% 182.19 ha.

The area covered by shrimp pools in the zone increased from 1991 to 2013, from 176 ha to 467 ha which accounts a 15,4% of the analyzed area. During 2014 a small reduction of shrimp pools was observed 463 ha (15,2%). See Tables 25.6 and 25.7

The data presented suggests that the main cause of mangroves forest loss during the last few decades is a result of deforestation due to the construction of shrimp-culture pools. Aschenbroich et al. (2015) states that mangroves are increasingly threatened by aquaculture development especially shrimp-culture fostering through this practice the fragmentation of the ecosystem, and consequently the reduction of goods and services it offers.

25.5 The Future of Mangrove Concessions in Ecuador

The economic, social and environmental importance of the areas under Sustainable use and custody agreements of mangrove forests is receiving recognition by governmental authorities, which is reflected in the political, technical and financial support. The Mangrove Partner program (Socio Manglar) for instance allowed the direct assignation of economic resources to beneficiaries of mangrove concessions during 10 years to support the management of areas, allowing a long term management model. At the moment, not all concession holders are participating in the programme, as funding is limited, hopefully in coming years all communities with concessions can enter in the programme, so that by having financial means for control activities their mangrove management can be improved.

Another fundamental aspect in mangrove long term conservation is put all mangrove forest under the direct involvement of local associations by providing new concessions or increase the areas of existing concessions, since it represent to them a legal framework to safeguard their food security and a source of stable economic income.

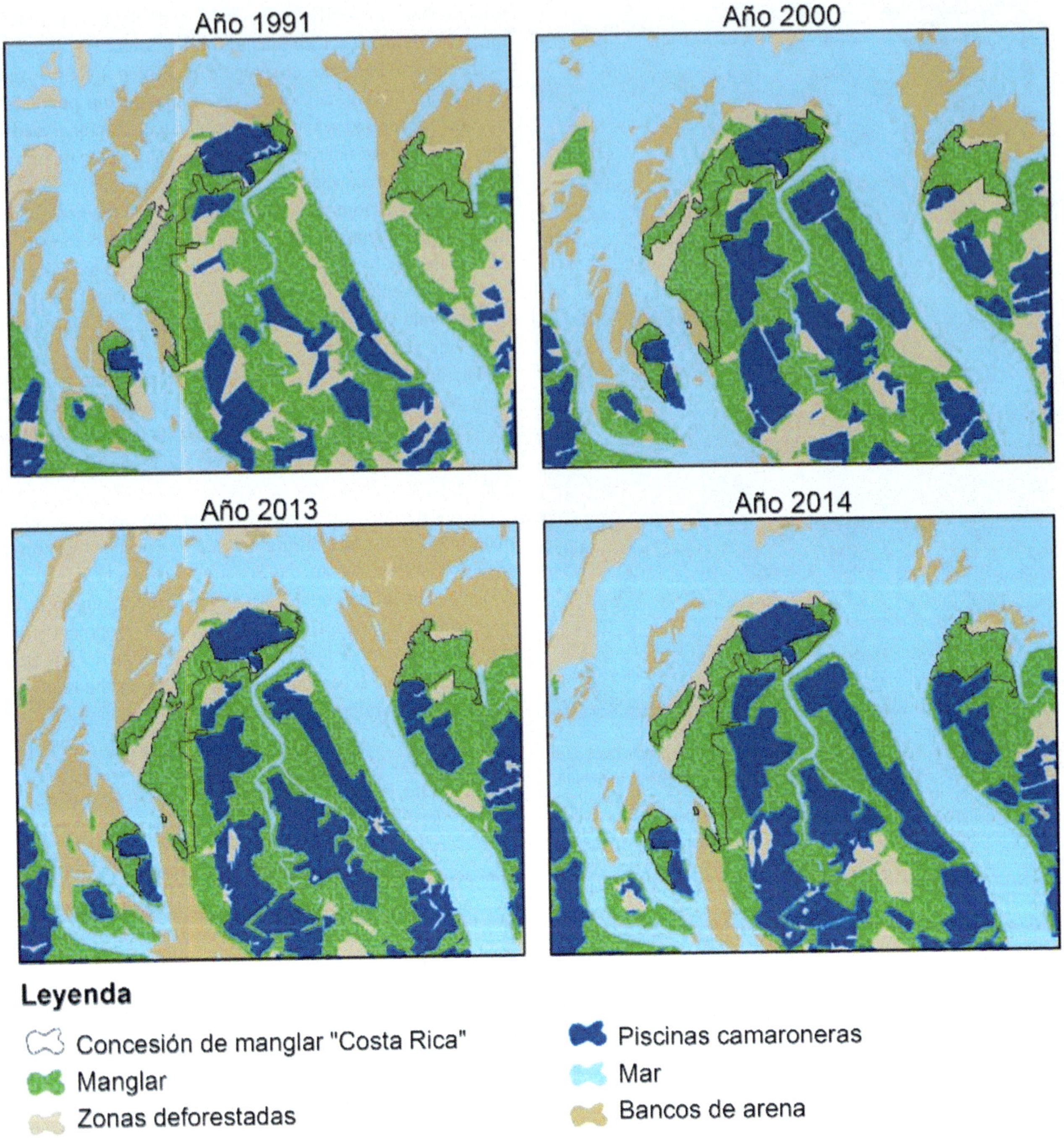

Fig. 25.16 Land use change in the Costa Rica Island. Images show the beginning of land use change due to mangrove deforestation by shrimp farms construction between 1991 to 2014

Even though there are persistent conflicts with shrimp farmers, there are recent examples of collaboration between shrimp farmers businesses and community associations, like joint control and surveillance activities, reforestation of degraded areas amongst other, which bring up important success factors for long term conservation of community mangrove areas. Include mangrove contamination into this dialogues, and search together for innovative measures to reduce water pollution is desired to guarantee the conservation of flora and fauna of the mangrove forests.

On the other hand, some associations or concessionaries show weaknesses related to the management of their mangrove areas, for instance, related to a good organizational structure and competences related to technical aspects, particularly related to monitoring and evaluation of the area under concession. Therefore, the agreements that are signed

Table 25.6 Surface in hectares of the different land uses on the Costa Rica Island considering 1 kilometer buffer zone

	1991		2000		2013		2014	
Land use extensions	Ha	%	Ha	%	ha	%	Ha	%
Mangrove	880.07	29.02	798.39	26.33	823.19	27.15	808.84	26.67
Deforested areas	394.19	13.00	242.76	8.00	176.53	5.82	182.19	6.00
Shrimp pools	176.21	5.81	413.82	13.64	467.02	15.40	463.02	15.27
Sea	1025.52	33.82	1208.49	39.85	785.77	25.91	1416.58	46.72
Sand banks	555.85	18.33	368.38	12.15	779.33	25.70	161.21	5.31
TOTAL	3031.86	100	3031.86	100	3031.86	100	3031.86	100

The mangrove coverage in 1991, of approximately 880 ha was reduced to 798,39 ha (26%) by 2000; later during 2013 and 2014 mangrove coverage changed from 823.19 ha (27%) to 808,84 (26%), respectively

Table 25.7 Analysis of the mangrove cover change rate

Period	Surface (ha) Year 1	%	Surface(ha) Year 2	%	Change Rate
1991–2000	880.07	29.02	798.39	26.33	−1.08
2000–2013	798.39	26.33	823.19	27.15	0.23
2013–2014	823.19	27.15	808.84	26.67	−1.75

The quantitative analysis of the data above, shows the surface occupied by mangrove coverage in periods: 1991–2000, 2000–2013 and 2013–2014. The deforestation and reforestation values are represented by the negative/positive sign respectively.The data shows a loss of mangroves in the period 1991–2000, a period in which there was a high increase in deforestation, corresponding to 81,68 has; whereas in the period 2000–2013 there was an increase in the mangrove coverage. The mangrove coverage increased partly due to reforestation of some areas, carried out by the Rainbow Foundation (Fundación Arcoiris in Spanish) jointly with community members of Costa Rica Island, which corresponds to 24.8 has. Moreover, in the last period a loss of mangroves was identified corresponding to 14.34 ha. This information might be related to having insufficient data available in comparison with earlier periods

should be supported by technical cooperation agreements with a university or non-governmental organization (NGO), which contributes to overcome these weaknesses. In this aspect, it is important to mention the support at organizational level that the Union Network of Organizations of Artisanal Fishery Production (Red Unión de Organizaciones de Producción Pesquera Artesanal de El Oro – UOPPAO) provides to most of concessionaries that are affiliated to this network.

Income diversification is another important aspect in which associations are working. It includes activities such as ecotourism and direct commercialization of red crab (*Ucides occidentalis*) and black ark (*Anadara similis* y *Anadara tuberculosa)* pulps, without intermediaries, as well as give added value to their product like crab pulp.

Recently (2017) conflicts arised between independent collectors (no legalized) pirates and mangrove concessionaries, however conflict solving meetings are taking place to find solutions to these problems.

25.6 Conclusions

In Ecuador, mangroves are protected by different strategies: national protected areas, mangrove concessions and by environmental law all mangroves, being fragile ecosystems are declared protected forests (by executive decree) although the latter have not been formally created (they do not contain management plans for example).

By the mangrove concessions strategy, through providing land rights, communities are involved in protecting 40% of the Ecuadorian mangroves against deforestation. Their control activities have proven to be effective. Therefore the Mangrove Agreements strategy is considered of great importance for the future of this ecosystem in Ecuador.

In the province of El Oro, 11,300 ha of mangroves are in the hands of ancestral communities and traditional users, which is equivalent to 75% of the mangrove that exists in this province. In contrast, only 08% of this ecosystem is within the Arenillas Ecological Reserve, an area that belongs to the National System of Protected Areas, therefore the Concessions strategy being of great importance in the El Oro Province.

Multi-temporal analysis of the forest cover in the mangroves of the Costa Rica Association and other areas showed that the main changes over time have been mainly due to the development of the shrimp industry. There is however evidence of mangrove recovery as an indicator of the effective functioning of the areas under community custody.

In the El Oro Province, mangrove deforestation has been reduced almost entirely in the areas covered by concessions, but contamination from shrimp farming has not been tackled by the concession strategy. Water pollution affects the flora and fauna of this ecosystem and alternative strategies should be implemented to reduce mangrove degradation due to contamination.

Although traditionally the relationship between shrimp farmers and mangrove concessionaires in the province of El Oro has been marked by conflicts, in recent years there has been an improvement in relations between these two groups, therefor government institutions, communities with mangrove concessions, shrimp farmers and other supporting actors like Universities might, through dialogue find solutions to water pollution.

Although all mangrove concessions have important biological information, it is necessary to define basic indicators to design and implement a monitoring program that ensures scientific evidence for the positive impacts of mangrove conservation on flora and fauna, and define whether contamination impacts mangrove biodiversity.

The incentive of the Mangrove Partner Programme, has a positive impact on the management of the Mangrove concessions, providing financial means for buying equipment (boats, motor, etc.), as well as fuel for the control operations in the Mangrove to prevent invasive fisherman, detect mangrove deforestation and contamination activities etc. The Mangrove Partner Programme, also has a positive impact on the management of the Mangrove as it provides financial means for organizational strengthening. The mangrove concession of Costa Rica Association, for being a community protected area without funding from the Mangrove Partner Programme, has much less funds for its management, causing that the financial sustainability is one of the main weaknesses to exercise good management of the area.

The mangrove forests have several resources and attractions that could be used to develop ecotourism projects. However technical resources, transportation basic infrastructure, and improvement of the client service are still needed. Greater involvement of public and private institutions is required, to help strengthen social cohesion and promote community-based tourism, also the community requires inputs and resources required to strengthen them.

Acknowledgements Irma Jurrius (Deutsche Gesellschaft für Internationale Zusammenarbeit-GIZ); Gabriela Costa (Universidad Técnica Particular de Loja)

References

Asamblea Nacional Constituyente (2008) Constitución de la República del Ecuador

Aschenbroich A, Marchand C, Molnar N, Deborde J, Hubas C, Rybarczyk H, Meziane T (2015) Spatio-temporal variations in the composition of organic matter in surface sediments of a mangrove receiving shrimp farm effluents (New Caledonia). Sci Total Environ 512–513:296–307. https://doi.org/10.1016/j.scitotenv.2014.12.082. http://www.sciencedirect.com/science/article/pii/S0048969714017938

Astudillo RF, Pérez J, Proaño F, Coello S (2014) Determinación de suelos con potencial de reforestación de manglar por debajo de la línea de influencia de costa escala: 1:50 000. Informe para la Subsecretaría de Gestión Marina y Costera, Ministerio del Ambiente. Ecobiotec del Ecuador, Quito, Ecuador, 54 pp + 3 anexos + 2 mapas

Beitl C (2010) The role of collective action in the social-ecological resilience of mangroves and artisanal fisheries on the ecuadorian coast. Prepared for the Regional Meetings of the International Association for the Study of the Commons, Tempe AZ. September 29 – October 3, 2010. Panel: Coastal Commons. https://dlc.dlib.indiana.edu/dlc/bitstream/handle/10535/6556/Beitl.pdf?sequence=1

Bernabé-Argandona L (2016) Sector Camaronero: Evolución y proyección a corto plazo. Escuela Superior Politécnica del Litoral. http://www.revistas.espol.edu.ec/index.php/fenopina/article/view/100/107

BirdLife International and Conservation International (2005) Áreas Importantes para la Conservación de las Aves en los Andes Tropicales: sitios prioritarios para la conservación de la biodiversidad. Quito, Ecuador: BirdLife International (Serie de Conservación de BirdLife No. 14)

Bravo M (2010) Interpretación del estudio multitemporal (CLIRSEN 1969–2006) de las coberturas de manglar, camaroneras y áreas salinas en la franja costera del Ecuador Continental. Tesis de grado para la obtención de título de magister en ciencias con énfasis en manejo sustentable de recursos bioacuáticos y medio ambiente, Universidad de Guayaquil, Facultad de Ciencias Naturales, Ecuador, p 56

Bravo M (2013) Alianza público-privada para la gestión de los manglares del Ecuador: Los acuerdos para el uso sustentable y custodia. Ecuador. USAID Costas y Bosques Sostenibles. http://pdf.usaid.gov/pdf_docs/PA00JVG6.pdf

Cámara Nacional de Acuacultura (2017) Análisis de las Exportaciones de Camarón Diciembre – 2016. http://www.cna-ecuador.com/

Castellanos-Galindo GA, Krumme U, Rubio EA, Saint-Paul (2012) Spatial variability of mangrove fish assemblage composition in the tropical Eastern Pacific Ocean. Reviews in Fish Biology and Fisheries. https://link.springer.com/article/10.1007/s11160-012-9276-4

C-CONDEM (2007) Certificando la Destrucción: Análisis integral de la certificación orgánica a la acuacultura industrial de camarón en Ecuador. Corporación Coordinadora Nacional para la Defensa del Ecosistema Manglar, Quito

Centro de Levantamientos Integrados de Recursos Naturales por Sensores Remotos -CLIRSEN and the Plan de Manejo de Recursos Costeros- PMRC (2007) Actualización del estudio multitemporal de manglares, camaroneras y áreas salinas en la costa continental ecuatoriana al año 2006. (s.n, ed) Ecuador. http://cpps.dyndns.info/cpps-docs-web/planaccion/docs2013/manglares/Informe%20Final%20PMRC-CLIRSEN.PDF

Coello S, Vinueza D, Alemán R (2008) Evaluación del desempeño de los acuerdos de uso sustentable y custodia de manglar de la zona costera del Ecuador. Ministerio del Ambiente del Ecuador – Conservación Internacional – Unión Mundial para la Naturaleza (UICN) – Comisión Mundial de Áreas Protegidas de UICN – Programa de apoyo a la gestión descentralizada de los recursos naturales en las tres provincias del norte del Ecuador (PRODERENA) – Ecobiotec. Julio de 2008: 52 pp + 4 Figuras + 17 Tablas + 5 Apéndices + 29 mapas

Cornejo (ed) (2014) Plants of the South American Pacific mangrove swamps (Colombia, Ecuador, Peru). Universidad de Guayaquil

Cornejo X, González G (2015) Contribución al conocimiento de la fauna entomológica de los manglares: *Olla roatanensis* Vandenberg y *Cheilomenes sexmaculata* Fabricius, dos nuevos registros de Coleoptera: Coccinellidae para Ecuador y Perú. https://www.researchgate.net/publication/306106516_Contribucion_al_conocimiento_de_la_fauna_entomologica_de_los_manglares_Olla_roatanensis_Vandenberg_y_Cheilomenes_sexmaculata_Fabricius_dos_nuevos_registros_de_Coleoptera_Coccinellidae_para_Ecuador_y_P

CPPS/UNESCO/CI/HIVOS (2013) CPPS, Conservación Internacional & UNESCO. Memorias. Taller Internacional sobre manglares y

Desarrollo Sostenible. 23–26 de julio de 2013. Santa Marta. Colombia http://cpps.dyndns.info/cpps-docs-web/planaccion/docs2014/enero/III-EXTRA-AG/referencias/CPPS_SE_AG_III_REF_01.pdf

CPPS/UNESCO/CI/HIVOS (2015) Plan de acción regional para la conservación de los manglares en el Pacífico Sudeste. CPPS, UNESCO, Conservación Internacional e Hivos. Guayaquil. http://unesdoc.unesco.org/images/0024/002464/246476s.pdf

Echeverría H (2008) La Convención Ramsar en Ecuador. Guía sobre la conservación y uso racional de los humedales. Centro Ecuatoriano de Derecho Ambiental, Quito

FAO Food and Agriculture Organization (1994) Directrices para la ordenación de manglares. Estudio FAO, Montes

FAO Food and Agriculture Organization (2012) Estado de las áreas marinas y costeras protegidas en América Latina. Elaborado por Aylem Hernández Avila. REDPARQUES Cuba, Santiago de Chile, 620 pp. http://www.fao.org/3/a-as176s.pdf

FAO Food and Agriculture Organization (2015_2) Shrimp – March 2015. Market Report. http://www.fao.org/in-action/globefish/market-reports/resource-detail/en/c/338028/. Accessed 4 May 2017

FAO GlobeFish (2015_1) Analysis and information on world fish trade. Accessed 4 May 2017 http://www.fao.org/in-action/globefish/market-reports/resource-detail/en/c/338028/

García M, Parra D, Mena P (2015) El país de la biodiversidad: Ecuador. Ministerio del Ambiente, Ecofondo & Jardín Botánico del Ecuador

Gillett R (2008) Global study of shrimp fisheries. FAO fisheries technical paper. No. 475. Rome, FAO, 331p. http://www.fao.org/docrep/011/i0300e/i0300e00.htm

Gobierno de la República del Ecuador (1999) Decreto Presidencial No. 1102. 21 de julio de 1999 Registro oficial N°243 del 28 de julio de 1999

Gobierno de la República del Ecuador (2003) Texto Unificado de Legislación Ambiental Secundaria. Decreto Ejecutivo 3516

Gobierno de la República del Ecuador (2008) Acuerdo Ministerial 149, RO No 412, del 27 de agosto de 2008. Ministerio de Agricultura, Ganadería, Acuacultura y Pesca

Gobierno de la República del Ecuador (2011a) Resolución No. 056. 28 de enero de 2011. Ministerio del Ambiente. Registro Oficial No. 496. 21 de julio de 2011. http://www.derechoecuador.com/productos/producto/catalogo/registros-oficiales/2011/julio/code/19983/registro-oficial-no-496---jueves-21-de-julio-de-2011

Gobierno de la República del Ecuador (2011b) Acuerdo Ministerial No 144. 9 de agosto de 2011. Ministerio del Ambiente. http://www.ambiente.gob.ec/wp-content/uploads/downloads/2014/09/43b5f146d2ad0fb8c3a3679aa5b26959c57111f2.pdf

Gobierno de la República del Ecuador (2013) Plan Nacional para el Buen Vivir 2013–2017. Secretaría Nacional de Planificación y Desarrollo-SENPLADES, Quito

Gobierno Provincial Autónomo El Oro, Fundación Ecológica Arcoiris (2013) Plan de Desarrollo Estratégico de la Parroquia Jambelí 2010–2025. Gobierno Provincial Autónomo El Oro. El Oro: s.n

Government of the Republic of Ecuador (2008) Presidential Decree No 1391.15 October 15 2008. Official Registry No. 454 october 27 2008. http://www2.congreso.gob.pe/sicr/cendocbib/con2_uibd.nsf/916E4710EDDF265E0525781D0057B26E/$FILE/4_Decreto_N%C2%B0_1391.pdf

Government of the Republic of Ecuador (2016) Presidential decree no. 852 28 January 2016 Official Registry No 694 19 February 2016. https://www.registroficial.gob.ec/index.php/registro-oficial-web/publicaciones/suplementos/item/7571-suplemento-no-694.html

Government of the Republic of Ecuador (2017) Código Orgánico Ambiental. Asamblea Nacional. Registro Oficial No 983. Abril 12 de 2017. http://www.asambleanacional.gob.ec/es/multimedios-legislativos/38792-codigo-organico-del-ambiente

Hamilton SE, Lovette J (2015) Ecuador's mangrove forest carbon stocks: a spatiotemporal analysis of living carbon holdings and their depletion since the advent of commercial aquaculture. PLoS One 10(3):e0118880. http://journals.plos.org/plosone/article?id=10.1371/journal.pone.0118880

Hurtado-Domínguez M, Hurtado-Gualán M, Hurtado-Domínguez L (2012) Estado de la Contaminación Marina en Ecuador. Comisión Permanente del Pacífico Sur. Plan de Acción del Pacífico Sudeste. http://cpps.dyndns.info/cpps-docs-web/planaccion/docs2013/mar/xix_ag/025.%20INFORME_FINAL_ECU_CONTAMINACI%C3%93N_MARINA.pdf

IMARPE (n.d.) Concha negra. http://www.imarpe.gob.pe/tumbes/especies_comerciales/invertebrados/concha_negra.pdf. Accessed 25 July 2017

Instituto Nacional de Pesca (2017) Concha. http://www.institutopesca.gob.ec/programas-y-servicios/concha/. Accessed 24 July 2017

IUCN, CI Ecuador (2016) National Blue Carbon Policy Assessment. IUCN, Conservation International Ecuador, Ecuador. 25 pp

Kavanagh K (2014) A world without mangroves? Letters. Science, vol 317. 6 July 2007. https://www.researchgate.net/publication/6223233_A_World_Without_Mangroves

Latorre S (2014) Resisting environmental dispossession in Ecuador: whom does the political category of 'ancestral peoples of the mangrove ecosystem' include and aim to empower? J Agrar Chang 14:541–563. https://doi.org/10.1111/joac.12052?. http://onlinelibrary.wiley.com/doi/10.1111/joac.12052/abstract

MAE (Ministerio del Ambiente del Ecuador) and FAO (Organización de las Naciones Unidas para la Alimentación y la Agricultura, IT) (2014) Árboles y Arbustos de los Manglares del Ecuador. Quito, 48 p. http://www.flacsoandes.edu.ec/libros/digital/55818.pdf

MAE. Ministerio del Ambiente (2017_2) http://www.ambiente.gob.ec/proyecto-regularizacion-de-camaroneras/. Accessed 23 May 2017

Martinez-Alier J (2001) Ecological conflicts and valuation: mangroves versus shrimps in the late 1990s. Environ Plann C-Gov Policy 19:713–728

Ministerio del Ambiente and GIZ (2012) Situación actual de los ecosistemas marinos – costeros y sus posibilidades de conectividad en el Ecuador. Ministerio del Ambiente del Ecuador / Dirección Nacional de Biodiversidad, Programa Gestión Sostenible de los Recursos Naturales (GESOREN)-GIZ. Quito. 243 pp.

Ministerio del Ambiente and Conservación Internacional-Ecuador (2017) El manglar, un ecosistema altamente productivo. Cinco servicios clave que nos provee el manglar

Ministerio del Ambiente de Ecuador (2017_1) Guía de derechos y deberes de las organizaciones custodios del manglar. Ministerio del Ambiente de Ecuador, Conservación Internacional Ecuador, Instituto Humanista para la Cooperación con los Países en Desarrollo, Organizaciones de las Naciones Unidas para la Alimentación y la Agricultura y Fondo para el Medio Ambiente Mundial. Guayaquil, Ecuador

Páez-Osuna F (2001) The environmental impact of shrimp aquaculture: causes, effects and mitigating alternatives. Environ Manag 28:131–140

Polidoro B, Carpenter K, Collins L, Duke N, Ellison A, Ellison J, Elizabeth Jean Farnsworth E, Fernando E, Kathiresan K, Koedam N, Livingstone S, Miyagi T, Moore G, Ngoc-Nam V, Eong Ong J, Primavera J, Salmo S, Sanciangco J, Sukardjo S, Wang Y, Wan Hong J (2010) The loss of species: mangrove extinction risk and geographic areas of global concern. PLoS One 5(4):e10095. http://journals.plos.org/plosone/article?id=10.1371/journal.pone.0010095

Quizhpe (2008) Áreas de ecosistemas de manglar concesionadas en la provincia de El Oro–Ecuador. Revista DELOS (Desarrollo Local Sostenible) vol 1. http://www.eumed.net/rev/delos/02/pqc.pdf costa

Revista Líderes (06 de 07 de 2015) José Camposano: 'A la industria camaronera le tomó 13 años recuperarse'. http://www.revistalideres.ec/lideres/jose-camposano-industria-camaronera-le.html

Spalding M, Kainuma M, Collins L (2010) World atlas of mangroves. ITTO, ISME, FAO, UNESCO-MAB, UNEP-WCMC and UNU-INWEH

Spurrier G (2012) Estudio del impacto de la acuicultura camaronera en el Ecuador. Guayaquil

Stonich S, Vandergeest P (2001) Violence, environment, and industrial shrimp farming. In: Peluso NL, Watts M (eds) Violent environments. Cornell University Press, Ithaca, pp 261–286

Terán, M.C., Clark, K., Suárez, C., Campos, F., Denkinger, J., Ruiz, D. y Jiménez, P. 2006. Análisis de Vacíos e Identificación de Áreas Prioritarias para la Conservación de la Biodiversidad Marino-Costera en el Ecuador Continental. Resumen Ejecutivo. Ministerio del Ambiente. Quito

UNEP (2014) The importance of mangroves to people: a call to action. In: van Bochove J, Sullivan E, Nakamura T (eds) United Nations environment programme world conservation monitoring centre, 128 pp. http://newsroom.unfccc.int/es/el-papel-de-la-naturaleza/la-onu-alerta-de-la-rapida-destruccion-d-los-manglares/

World Wildlife Fund (2015) Mangrove forests: threats. Accessed 10 June 2017, from WWF Global: http://wwf.panda.org/about_our_earth/blue_planet/coasts/mangroves/mangrove_threats/

Yáñez–Arancibia A, Lara–Domínguez AL (1999) Los manglares de América Latina en la encrucijada, p 9–16. In: A. Yáñez–Arancibia y A. L. Lara–Domínquez (eds) Ecosistemas de Manglar en América Tropical. Instituto de Ecología A.C/UICN/ORMA/NOAA/NMFS Silver Spring, México/Costa Rica/MD USA, 380 p. http://www.observatorioirsb.org/cmsAdmin/uploads/manglares-en-am-rica-latina_y-ez_lara_(1999).pdf

Conserving Mangroves for Their Blue Carbon: Insights and Prospects for Community-Based Mangrove Management in Southeast Asia

26

Dixon T. Gevaña, Leni D. Camacho, and Juan M. Pulhin

Abstract

Mangrove forests of Southeast Asia have significantly decreased to give way to more favored economic activities such as aquaculture production, rice farming, and very recently oil palm plantation development. The severe cover loss implies the serious reduction of essential ecosystem services, particularly climate change mitigation. This chapter therefore sought to synthesize and describe the blue carbon stock potential of the region's mangrove forests, in the context of developing community-based payments for ecosystem services (PES) schemes that could offset the declining trend. Estimates showed that Southeast Asian mangroves could store as much as a kiloton carbon stock per hectare, an ecosystem value that is worth sustaining. Lessons learned from pilot PES projects has revealed some governance concerns that need to be addressed in order to ensure the benefits and commitment of local communities in forest conservation. These include: (1) clarification of tenure rights; (2) provision of equitable financial incentives to offset mangrove-degrading livelihoods; (3) development of appropriate and acceptable methodologies to account and trade blue carbon credits; (4) inclusion of local needs and interests in PES program and other coastal resource management plans; and (5) ecological considerations in plantation development. Some of the helpful policy and institutional recommendations to overcome these challenges include: (1) stronger incorporation of mangroves into marine protected areas; (2) full adoption of community-based mangrove management approach to further improve local capacities to manage mangroves; and (3) forging private sector partnership to support the blue carbon project.

Keywords

Carbon stock · Climate change · Local community · Mangroves · PES

26.1 Introduction

Coastal ecosystem plays an essential role in mitigating climate change and its impacts. Collectively termed as blue carbon ecosystem, mangrove forest, salt marsh and seagrass meadows have been regarded as powerhouse of carbon sequestration. The role of blue carbon ecosystems, particularly mangroves, are now being incorporated in the climate change mitigation modalities particularly in the Reducing Emission from Deforestation and Forest Degradation or REDD (IUCN 2012). Interest on their rehabilitation has also increased.

Mangroves provide countless ecosystem benefits and services. They enrich coastal waters, produce commercial forest and marine products, stabilize coastlines, regulate tidal actions, and sequester vast amount of carbon to help mitigate climate change, to name a few. The term *mangrove* denotes two different concepts according to Lugo and Snedaker (1974) and Alongi (2010). Biologically, it is a group of salt-tolerant plants that can be classified into nine orders, 20 families, 27 genera, and roughly 70 species. These plants grow well in saline coastal sediments of the tropical and subtropical (32°N to 32°S), delimited by major ocean currents and 20 °C isotherm of seawater in winter. Mangrove plants are also resilient to natural stressors such as high temperature, high salinity, anaerobic sediments, strong winds and extreme tides. However, their numbers are continuously in decline as they are being disturbed to favor coastal development such as for aquaculture production, settlements and industries (Kathiresan and Bingham 2001; FAO 2007; Gevaña et al. 2015).

D. T. Gevaña (✉) · L. D. Camacho · J. M. Pulhin
Department of Social Forestry and Forest Governance, College of Forestry and Natural Resources, University of the Philippines Los Banos, Los Baños, Laguna, Philippines
e-mail: dtgevana@up.edu.ph

C. Makowski, C. W. Finkl (eds.), *Threats to Mangrove Forests*, Coastal Research Library 25,
https://doi.org/10.1007/978-3-319-73016-5_26

The world's forests are estimated at 4 billion ha. Of these, mangrove forests cover only 0.4% or 15.6 million ha. Mangroves are present in 112 countries, with the largest in five countries namely: Indonesia, Brazil, Nigeria, Australia, and Mexico (FAO 2010). They are one of the world's most threatened ecosystems since about 50% have been lost over the past century. This change reflects a substantial damage to the vital economic and ecological values such as carbon sequestration.

By estimate, global mangroves contain about 4–20 billion tC (Donato et al. 2011). Bulk of this stock is stored in sediments amounting to at least 5 billion tC. They sequester carbon at a rate of 1.15–1.39 t/ha annually (Bouillon et al. 2008). Ironically, continuous deforestation emits 35 million t Cyear^{-1} globally (Siikamäki et al. 2012). If mangroves are converted to aquaculture ponds, the emission could reach as much as 112–392 tC ha^{-1} year^{-1}. According to Kauffman et al. (2014), the carbon emission from converting mangrove areas to shrimp ponds could reach as much as 1135 tC ha^{-1}, a value that is almost equal to their maximum storage capacity.

Local communities have pivotal role in ensuring the health and well-functioning of mangroves. They can be tapped as co-stewards of the government for forest conservation since they live closely and often directly dependent on mangrove resources. Several countries in Southeast Asia had adopted community-based approach to manage coastal forest areas. This chapter therefore describes the achievements of community-managed mangroves of Southeast Asia in view of the potential blue carbon storage and sequestration of the mangroves they co-manage with the government. Some issues and challenges on the sustainable implementation of community-based mangrove management and blue carbon offset projects were also described.

26.2 Mangrove Value

According to Costanza et al. (1998), global mangrove forest is worth US$ 180,900,000,000 and their average monetary value is estimated at US$ 10,000 ha^{-1} year^{-1}. These values can be expressed in various ecosystem goods and services such as (FAO 2007; Miththapala 2008; Garcia et al. 2014):

1. Provisioning service
   - Mangrove timber is used to build and make houses, furniture, fences, and boats.
   - Mangrove wood is also used as fuelwood and still provides ninety percent of the fuel used in Viet Nam.
   - The leaves of species such as the Mangrove Palm (*Nypa*) and Screw Pine (*Pandanus* sp.) are used for thatching and weaving.
   - The breathing roots of various *Sonneratia* spp. – with their aerenchyma tissue – are used to make corks and fish floats.
   - Mangrove plants are a source of sodium and the ash of some species such as *Avicennia* is used as soap.
   - Bark of many species produces gums and tannins.
   - In Bangladesh and India, honey from mangroves is an important local industry, producing 20 ton of honey every year from 200,000 ha of mangroves.
   - Mangrove leaves, fruits, shoots, and roots serve as vegetables and edible fruits.
   - About 70 different mangrove plants have medicinal value.
   - Roughly eighty percent of global fish catches are directly or indirectly dependent on mangroves. Annual commercial fish harvests from mangroves have been valued at 6200 USD per km^2 in the United States to 60,000 USD per km^2 in Indonesia.
2. Regulating Service
   - Trees protect shoreline from tidal surges. A good example of this is the 2004 Tsunami in the Indian Ocean where mangroves in Sri Lanka helped minimize the horrendous impact of tidal waves.
   - Mangrove reduces the effects of flooding and sea water intrusion.
   - Mangrove filters pollutants from both land and sea.
3. Supporting Services
   - Mangrove trees and sediments are rich blue carbon sink.
   - They help enhance organic accumulation, and land accretion therefore providing food and shelter to diverse marine flora and fauna.
4. Cultural services
   - Mangrove forests have aesthetic or recreation values.
   - Mangrove sustains traditional fishing practices in many regions.

Albeit countless benefits, they are often undervalued and viewed as wastelands, wilderness and unhealthy environments (Clough 2013). The high population pressures in coastal zones had cleared mangrove areas to favor urban development. In order to augment food security and boost national economies, many governments have also encouraged the agricultural development through shrimp and fish farming, and salt and rice production in mangrove areas. These have therefore resulted to massive fragmentation, degradation, and pollution of coastal forests.

26.3 Mangroves of Southeast Asia

Southeast Asia covers about 4.5 million km^2 of land, which harbors 39% of the world's mangrove biodiversity (FAO 2007). Mangroves in this region are well-distributed along its coasts covering about 4 million ha (Fig. 26.1). According to the latest records of FAO (2015), Indonesia has the largest cover with 2.24 million ha (Table 26.1). It was followed by

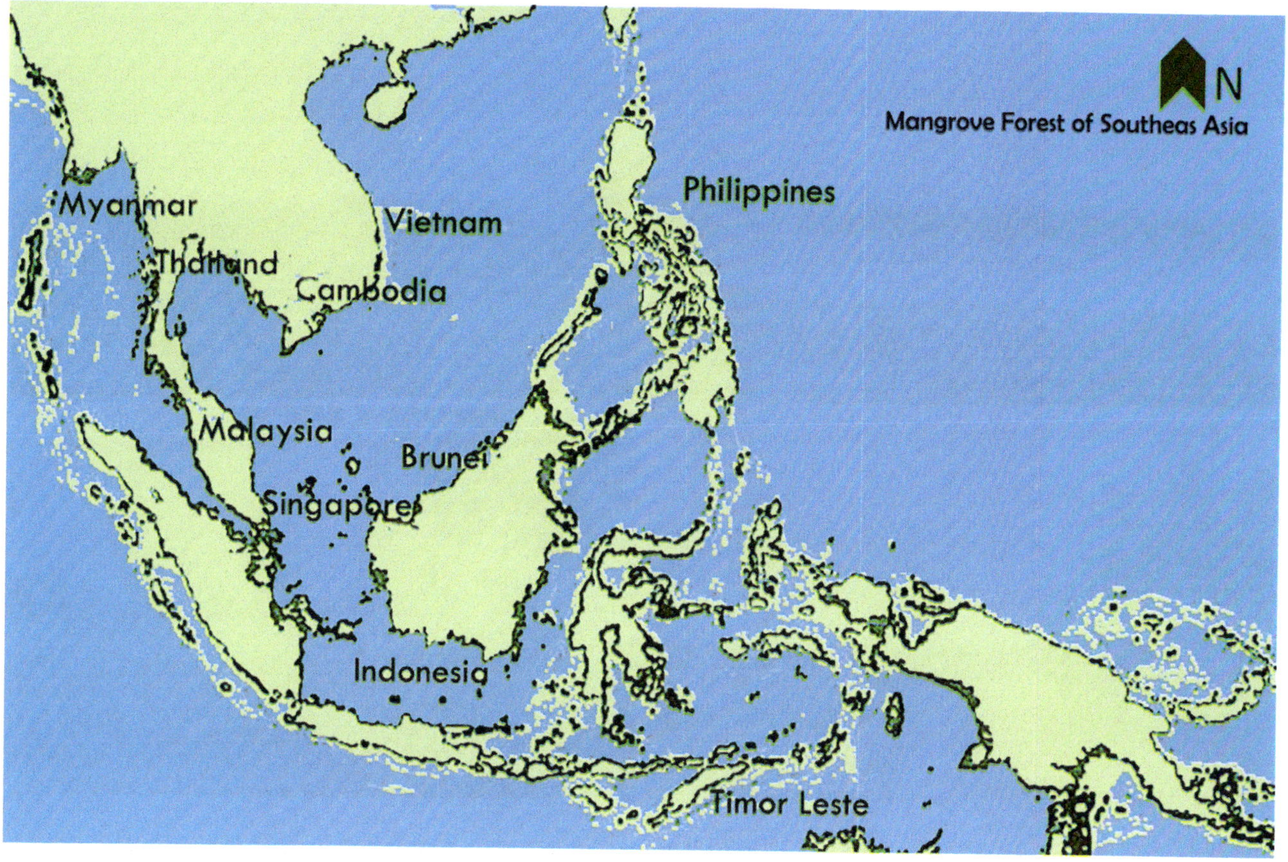

Fig. 26.1 Map of Southeast Asian mangrove forests as shown in dark green lines

Table 26.1 Mangrove cover and rate of cover change in Southeast Asia

Country	Mangrove cover in year 2015 ('000 ha)[a]	Mangrove loss between 2000 and 2012 (%)[b]
Indonesia	2244	1.7
Myanmar	299	5.5
Malaysia	521	2.8
Thailand	240	1.4
Philippines	356	0.5
Cambodia	50	2.3
Vietnam	270	0.3
Brunei	18	0.4
Timor-Leste	2	0.2
Singapore	1	0.0
Southeast Asia	4001	2.1

[a]Based on FAO (2015)
[b]Based on Richards and Friess (2016)

Malaysia (521,000 ha) and Philippines (356,000 ha). Timor-Leste and Singapore has the smallest cover with only about 3000 ha combined.

Mangrove deforestation and degradation remains as a major challenge in this region (Fig. 26.2). Among the countries listed, deforestation was highest in Myanmar with 5.5% loss from the period of 2000 to 2012. This was followed by Malaysia with almost 3%. Richards and Friess (2016) reported that the region has lost 2.1% of its mangrove cover over the past decade. The major driver of mangrove loss was aquaculture development, which contributed 30% loss in forest cover. This was followed by rice farming 21.7% loss. They also found out that mangrove conversion to oil palm plantation has also emerges as a significant threat.

26.4 Community-Based Mangrove Management

Previous mangrove management approaches can be generally described to be government-centered. This centralized governance was viewed to be inadequate in dealing with deforestation and rural poverty problems. A number of studies suggested that poor governance leads to limited local commitment and participation in mangrove conservation because the needs and interests of local stakeholders are often overlooked (Walters 2004; Locatelli et al. 2014; Friess et al. 2015; Wylie et al. 2015). Community forestry or community-based forest management therefore offers a better approach to effectively manage mangrove forests (Gilmour

Fig. 26.2 Major drivers of mangrove deforestation in Southeast Asia

Fig. 26.3 Satellite image of Banacon Island, Bohol, Philippines (Google Earth Ver. 7.1.2.2019. Typical mangrove vegetation types are: (**a**) *Rhizophora stylosa* plantation; (**b**) *Avicennia & Sonneratia*-dominated natural stand (Gevaña et al. 2015)

and Fisher 1991; Pulhin et al. 2017). Community-based forest management decentralizes the management functions especially to resource-dependent local communities since they have better stakes to their protection and conservation. It is enthusiastically being promoted in Southeast Asia today because of its benefits in ensuring tenure security and equitable local access to mangrove goods and services.

One of Southeast Asia's successful community-based mangrove management cases is the Banacon Island in Bohol, Philippines (Fig. 26.3). It is lauded as Asia's largest community-initiated mangrove plantation and a recipient of the FAO's Outstanding Tree Farmer Award in 1991. It is home to at least 300 households and has an area of about 1775 ha. Around 80% of this island is covered by mangrove forest (Gevaña and Im 2016). It is located at 10° 03′ 30″–10° 15′ 30″ north and 124° 03′ 30″–124° 14′ 30″ east, forming part of the unique and eco-diverse protected marine sanctuary of Danajon Double Barrier Reef (Pichon 1977).

The island used to be devoid of mangrove cover due to decades of illegal cutting for fuelwood and charcoal production. This dismal condition was reversed when the government pursued Community-Based Forest Management Agreement or CBFMA which transferred the forest management rights and responsibilities to local community. This program paved the way for the organizing and strengthening of People's organization known as *Banacon Fisherfolks and Mangrove Planters' Association* or BAFMAPA. BAFMAPA envisioned itself to be a mangrove champion with their strong commitment to pursue mangrove planting (Camacho et al. 2011; Carandang et al. 2013). Their success in plantation development received several recognitions which have further stimulated interests of private organizations to partner with them in mangrove plantation projects. Among these was the Kanepackage Philippines Inc., an international corrugated box production company, which has recently funded a 200-ha plantation as part of their corporate social responsibility (CSR) program. The whole plantation amassed to 485 ha, a potential source of carbon credits in the future (Gevaña and Im 2016). According to Camacho et al. (2011), the carbon stocks of mangrove plantations is around 370 tC ha^{-1}.

26.5 Carbon Stock of Community-Managed Mangroves

A greater interest to account for blue carbon stocks has emerged over the past two decades. Studies on mangrove blue carbon were likely driven by four major factors, namely: (1) recognition of huge carbon stock capacity of mangroves; (2) pursuing mangrove protection enhances other essential ecosystem services particularly coastal protection against tsunami and tidal waves; (3) interests on carbon offset project as alternative income source to local livelihoods that utilize and degrade mangrove resources; and (4) empowering coastal communities leads to effective mangrove rehabilitation and conservation hence improvement of carbon stocks (Bouillon et al. 2008; Mcleod et al. 2011; Ullman et al. 2013; Siikamäki et al. 2012; Thomas 2014). In view of the two latter reasons, Table 26.2 provides a summary of the potential carbon stocks of community-managed mangroves in Southeast Asia. At a glance, blue carbon stocks vary across countries, area, vegetation type and carbon pool (vegetation and sediment) assessed.

In view of natural mangrove sites, values range from 22.6 tC ha^{-1} (in Tha Po Village, Surat Thani Province) to 939.3 tC ha^{-1} (in Sulawesi, Indonesia). The apparent disparities among these estimates are reflective of the carbon pool assessed. This means that those studies that accounted both vegetation and sediment carbon pools have relatively larger values than those that only included vegetation carbon stocks. Further, stand density (trees per unit area) is likely to have contributed to varying estimates but unfortunately no information was obtained from the reports that were reviewed.

In view of community-managed mangrove plantations, estimates range from 40.1 to 370.0 tC ha^{-1}. The low value recorded for the 12-year old *Rhizophora apiculata* in Samut Songkram, Thailand is indicative of the stand's young age and small stem diameter (3.41 cm) and height (11.4 m). Stand density of this site is around 22,089 trees ha^{-1}. As compared to Banacon Island, Philippines, larger carbon stock observed from 55-year old *Rhizophora stylosa* plantation is reflective of relatively larger stem diameter and height i.e. 7.3 cm and 12.7 m, respectively. Stand density of this site was estimated at 11,580 trees ha^{-1}.

Studies that accounted for both vegetation and sediment carbon pools have values ranging from 115.5 to 939.3 tC ha^{-1}. This underscored the sediment's significant contribution to the total blue carbon stock. Murdiyarso et al. (2009) reported that 87.5% of Sulawesi mangrove carbon stocks is stored in sediment. This site has a sediment layer of 1.2 m depth. In general, sediment carbon stock increases as sediment layer becomes thicker. Protecting the aboveground vegetation is therefore crucial to avoid this huge blue carbon stock from washing away.

Figure 26.4 provides some of the major activities involved in mangrove blue carbon stock assessment. These activities are usually done in a participatory manner which involves the local community in tree measurements and resource mapping.

26.6 Prospects and Challenges of Community-Based Blue Carbon Project

26.6.1 Payments for Mangrove Ecosystem Services

Developing payment for ecosystem services (PES) schemes for mangroves of Southeast Asia could provide a sustainable income source for local communities to engage in mangrove restoration and protection (Locatelli et al. 2014; Warner et al. 2016). This will greatly benefit landless and impoverished communities whose survival is intimately dependent on mangroves. Vietnam is among the countries that took the

Table 26.2 Blue carbon stock estimates from community-managed mangrove forests in Southeast Asia

Country	Site	Vegetation type	Area (ha)	Est C-stock (t C ha^{-1})	Year	Carbon pool	Method used	References
Indonesia	Karimunjawa, Java Sea	Natural – mixed spp.	446.66	269.9	2010	Vegetation	Remote sensing and GIS	Wicaksono et al. (2015)
	Nusa Penida, Bali	Natural – mixed spp.	164.6	90.7	2014	Vegetation	Non-destructive plot sampling	Kusumaningtyas et al. (2014)
	Segara Anakan, Java	Natural – mixed spp.	Not determined	586.0	2009	Vegetation + sediment	Non-destructive plot sampling	Murdiyarso et al. (2009)
	Bunaken National Park, Sulawesi	Natural – mixed spp.	Not determined	939.3	9009	Vegetation + sediment	Non-destructive plot sampling	Murdiyarso et al. (2009)
Myanmar	Bogalay, Ayeyarwady Delta	Plantation (6 years old) – *Avicennia* spp. and *Sonneratia* spp.	122,770	240.0	2012	Vegetation + sediment	Non-destructive plot sampling	Thant et al. (2012)
Malaysia	Delta Kelantan in Kelantan	Natural – mixed spp.	1200	305.34	2017	Vegetation + sediment	Non-destructive plot sampling	Nazri (2017)
Thailand	Yeesarn, Samut Songkram	Plantation (12 year-old) – *Rhizophora apiculata*	1276	40.1	2012	Vegetation	Non-destructive plot sampling	Kridiborworn et al. (2012)
	Tha Po Village, Surat Thani Province	Natural – mixed spp.	400	22.6	1998	Vegetation	Non-destructive plot sampling (trees only)	Sathirathai (1998)
	Pak Phanang, Nakhon Si Thammarat Province	Plantation (4,10,14,20 and 25 years old) - *Rhizophora* spp.	18448.5	121.7	2009	Vegetation	Non-destructive plot sampling	Sriladda and Puangchit (2009)
	Suksamran Subdistrict, Ranong Province	Natural – mixed spp.	151	155.0	2013	Vegetation	Non-destructive plot sampling	Jachowski et al. (2013)
Philippines	Banacon Island, Bohol	Plantation (55 years old) – *Rhizophora stylosa*	1775	370.0	2011	Vegetation	Non-destructive plot sampling	Camacho et al. (2011)
	San Juan, Batangas	Natural-*Rhizophora* spp. dominated	15	115.5	2009	Vegetation + sediment	Non-destructive plot sampling	Gevaña and Pampolina (2009) Gevaña et al. (2008)
	Ibajay, Akalan	Natural – mixed spp.	75	562.5	2013	Vegetation + sediment	Non-destructive plot sampling	Thompson et al. (2014)
Vietnam	Quang Ninh	Natural – mixed spp.	17,354	26.6	2010	Vegetation	Remote sensing and GIS	Vu et al. (2014)
	Mui Ca Mau National Park	Natural – mixed spp.	13,400	768.7	2012	Vegetation	Non-destructive transect sampling	Tue et al. (2014)
Brunei	Brunei Bay	Natural – mixed spp.	18,418	169.4	1961	Vegetation	Not determined (AGB only)	Verwer and van der Meer (2010)
Timor-Leste	East Dili	Natural – mixed spp.	Not determined	221.5	2007	Vegetation	Non-destructive transect sampling	Alongi and Carvalho (2008)

(continued)

Table 26.2 (continued)

Country	Site	Vegetation type	Area (ha)	Est C-stock (t C ha^{-1})	Year	Carbon pool	Method used	References
Singapore	Palau Semakau	Natural – mixed spp.	38.5	36.6	2014	Vegetation	Remote sensing, GIS and non-destructive e plot sampling	Friess et al. (2015)
	Pasir Ris	Natural – mixed spp.	7.3	105.4	2014	Vegetation	Remote sensing, GIS and non-destructive e plot sampling	Friess et al. (2015)
	Palau Ubin	Natural – mixed spp.	249.3	54.8	2014	Vegetation	Remote sensing, GIS and non-destructive e plot sampling	Friess et al. (2015)

Fig. 26.4 Participatory activities in mangrove blue carbon stock assessment

first initiative to develop PES for mangroves. Pham et al. (2013) examined the PES pilot project in Red River in Northern Vietnam and noted a number of concerns. Among these was the high cost (e.g. money, time and effort) of rehabilitating mangrove areas but with limited or low compensation. They suggested that the government should explore additional means to incentivize local communities aside from the payment for labor during mangrove planting.

Further, they underlined the need to intimately involve local communities in land use and mangrove management planning so that their interests on what crops to plant and what livelihoods to be developed under PES program will be sufficiently captured. PES on blue carbon was also identified as a potential commodity that must be considered. In a related review conducted by Nam (2017) for the Payment for Forest Ecosystem Services (PFES) pilot project in Ngoc Hien District, Ca Mau Province, Vietnam, a similar concern on the difficult requirement to qualify as project beneficiary was reported. This project targeted small shrimp farming households with the aim of promoting internationally certified and mangrove-friendly aquaculture i.e. by keeping 50% forest cover in their aquasilviculture farm. Such requirement was perceived to be difficult to implement since there is limited support for mangrove restoration and unclear income benefits after having their shrimp ponds certified. Payment for mangrove blue carbon is therefore being explored as potential additional incentive for this scheme.

The United Nations Reducing Emissions from Deforestation and Forest Degradation or REDD is a well-sought scheme for forest carbon projects. Unfortunately, the inclusion of mangroves in this framework is still being debated despite the increasing evidences of their huge contribution to carbon sequestration. The scope of REDD is also under negotiations whether it should remain exclusively for carbon emission reduction or whether it should include co-benefits such as livelihood development and biodiversity conservation (Visseren-Hamakers et al. 2012). With such uncertainty, it remains difficult for the government and local communities to develop community-based carbon offset program under this scheme. Locatelli et al. (2014) therefore stressed the importance of developing appropriate methodologies and approaches for accounting and trading mangrove blue carbon. This should also include the improvement of carbon price to make it more sustainable and feasible for the community to pursue. Based on the latest carbon transactions reported by Ecosystem Marketplace (2017), voluntary forest carbon price is around USD 5.1 per ton of carbon dioxide equivalent (tCO_2e). This indicates the mangroves huge economic potential in REDD negotiations particularly in the voluntary market schemes. In 2011, 29% of forest carbon credits were transacted in the voluntary carbon market, a significant increase from the 7% during the previous year (Peters-Stanley et al. 2011).

26.6.2 Improving Mangrove Governance

Fisher (2014) suggested that empowerment of local communities and livelihood improvement are crucial elements for PES schemes to succeed. Notwithstanding the growing appreciation of the effectiveness of community-based approach in mangrove management, there is still a number of challenges that most countries of Southeast Asia need to overcome. These include: (1) unclear tenure rights; (2) poor ecological consideration in mangrove reforestation; and (3) poor coastal management planning (Primavera 2000, Walters 2004, Locatelli et al. 2014; Friess et al. 2015; Wylie et al. 2015; Pulhin et al. 2017). Friess et al. (2016) listed some recommendations for establishing effective community-based mangrove PES platform. First, there should be a stronger incorporation of mangroves into marine protected areas as this could help resolve some policy and resource management conflicts in declaring mangroves exclusive for protection purpose. Second, community-based or co-management practice should be employed to effectively reconcile varying interests of local stakeholders and address conflicting policy objectives on mangrove use. Third, private-sector initiatives should be encouraged to protect and restore mangroves, and develop suitable local livelihoods.

26.7 Summary and Recommendations

Mangrove forests of Southeast Asia have significantly declined to give way to more favored economic activities such as aquaculture production, rice farming, and very recently oil palm plantation development. The severe loss translates to reduction of ecosystem services, particularly climate change mitigation. A review of the potential blue carbon stock estimates of community-managed mangroves in this region provides the rationale of putting forward conservation efforts. Mangroves are indeed exceptional carbon sinks. Based on conservative estimates, community-managed plantations could store around 370.0 tC ha^{-1}. If blue carbon stock assessment will fully account the contributions of vegetation and sediment, values could reach as much as a kiloton carbon stock per hectare. Many studies confirmed that carbon is richer at the sediment layer hence serious protection of the vegetation cover is critical to avoid this carbon pool from washing away.

The pursuit of effective and efficient ways to bring forward mangrove conservation has underscored the potential of developing community-based PES schemes. Lessons from pilot PES projects revealed some governance concerns that need to be addressed in order to fully realize and sustain the benefits of mangrove conservation in favor of the local communities. These include: (1) clarification of tenure rights; (2) provision of equitable financial incentives to offset mangrove-degrading livelihoods; (3) development of appropriate and acceptable methodologies to account and trade blue carbon credits; (4) inclusion of local needs and interests in PES program and other coastal resource management plans; (5) ecological considerations in plantation development; to name a few. Some of the helpful policy and institutional recommendations to overcome these challenges were also identified, namely: (1) stronger incorporation of mangroves into marine protected areas; (2) full adoption of community-based mangrove management approach where local communities are empowered (organized and equipped) to manage mangroves and PES project; and (3) forging private sector partnership to support the blue carbon projects of local communities.

References

Alongi D, Carvalho (2008) The effect of small-scale logging on stand characteristics and soil biogeochemistry in mangrove forests of Timor Leste. For Ecol Manag 255:1359–1366

Alongi D (2010) The Energetics of Mangrove Forest. Springer, New York, 216pp

Bouillon S, Borges A, Castaneda-Moya E, Diele K, Dittmar T, Duke N, Kristenses E, Lee S, Merchand C, Middleburg J, Rivera-Monroy V, Smith T, Twilley R (2008) Mangrove production and carbon sinks: a revision of global budget estimates. Glob Biogeochem Cycles 22:12

Camacho L, Gevaña D, Carandang A, Camacho S, Combalicer E, Rebugio L, Youn Y (2011) Tree biomass and carbon stock of a community-managed mangrove forest in Bohol, Philippines. For Sci Tech 7(4):161–167

Carandang A, Camacho L, Gevaña D, Dizon J, Camacho S, de Luna C, Pulhin F, Paras F, Peras R, Rebugio L (2013) Economic valuation for sustainable mangrove ecosystems management in Bohol and Palawan, Philippines. For Sci Tech. https://doi.org/10.1080/21580103.2013.801149

Clough B (2013) Continuing the journey amongst mangroves. ISME Mangrove Educational Book Series No. 1. International Society for Mangrove Ecosystems (ISME), Okinawa, Japan, and International Tropical Timber Organization (ITTO), Yokohama, Japan

Costanza R, d' Arge R, deGroot R, Farber S, Grasso M, Hannon B (1998) The value of the world's ecosystem services and natural capital. Ecol Econ 25:3–15

Donato D, Kauffman J, Kurnianto S, Stidham M, Murdiyarso D (2011) Mangroves among the most carbon-rich forests in the tropics. Nat Geosci 4:293–297

Ecosystem Marketplace (2017) Unlocking potential: state of the voluntary carbon markets 2017. In: Hamrick K, Gallant M (eds) Forest trends' ecosystem marketplace. Washington, DC, USA, 52 pp

FAO (2007) The world's mangrove: 1980–2005: a thematic study prepared in the network of the Global Forest Resource Assessment 2005, FAO Paper No. 153. Food and Agriculture Organization, Rome

FAO (2010) Global forest resources assessment 2010, FAO Forestry Paper No. 163. Food and Agriculture Organization, Rome, p 378

FAO (2015) Global Forest Resources Assessment 2015: Desk reference. Food and Agriculture Organization, Rome, p 64

Fisher RJ (2014) Lessons learned from community forestry in Asia and their relevance for REDD? USAID-supported Forest Carbon, Markets and Communities (FCMC) Program, Washington DC, 45pp

Friess D, Richards D, Phangl V (2015) Mangrove forests store high densities of carbon across the tropical urban landscape of Singapore. Urban Ecosyst 19(2):795–810

Friess D, Thompson B, Brown B, Amir A, Cameron C, Koldewey H, Sasmito S, Sidik F (2016) Policy challenges and approaches for the conservation of mangrove forests in Southeast Asia. Conserv Biol 30 (5):933–949

Garcia K, Malabrigo P, Gevaña D (2014) Philippines' mangrove ecosystem: status, threats and conservation. In: Hakeem et al (eds) Mangrove ecosystem in Asia: current status, challenges and management strategies. Springer, pp 81–94

Gevaña D, Pampolina N (2009) Plant diversity and carbon storage of a rhizopora stand in Verde Passage, San Juan, Batangas, Philippines. J Envi Sci and Mgt 12(2):1–10

Gevaña D, Pulhin F, Pampolina N (2008) Carbon stock assessment of a mangrove ecosystem in San Juan, Batangas. J Environ Sci Manag 11 (1):15–25

Gevaña D, Camacho L, Carandang A, Camacho S, Im S (2015) Landuse characterization and change detection of a small mangrove area in Banacon Island, Bohol, Philippines using maximum likelihood classification method. For Sci Technol 11(4):197–205

Gevana D, Im S (2016) Allometric models for Rhizophora stylosa Griff. in dense monoculture plantation in the Philippines. Malays Forester 79(1/2):39–53

Gilmour D, Fisher R (1991) Evolution in community forestry: contesting forest resources. Community forestry at crossroads: reflections and future directions in the development of community forestry. In: Victor M, Lang C, Bornemeir J (eds) Proceedings of an International Seminar. 17–19 July, vol 16. RECOFTC Report, Bangkok, pp 27–44

IUCN (2012) Blue Carbon Policy Framework: based on the discussion of the International Blue Carbon Policy Working Group. In: Herr D, Pidgeon E, Laffoley D (eds) IUCN,Gland/Arlington, 39pp

Jachowski N, MQuak M, Friess D, Duangnamon D, Weeb E, Ziegler A (2013) Mangrove biomass estimation in Southwest Thailand using machine learning. Appl Geogr 45:311–321

Kathiresan K, Bingham B (2001) Biology of mangroves and mangrove ecosystems. Adv Marine Bio 40:81–251

Kauffman J, Heider C, Norfolk J, Payton F (2014) Carbon stocks of intact man- groves and carbon emissions arising from their conversion in the Dominican Republic. Ecol Appl 24:518–527

Kridiborworn P, Chidthaisong A, Yuttitham M, Tripetchkul S (2012) Carbon sequestration by mangrove forest planted specifically for charcoal production in Yeesarn Samut songkram. J Sustain Energy Environ 3:87–92

Kusumaningtyas M, Daulat A, Suryono D, Ati R, Sudirman N, Hutahaean A (2014) Blue carbon stock of mangrove ecosystem in Nusa Penida, Bali. In: Paper presented 12th biennial conference of Pan Ocean Remote Sensing conference (PORSEC 2014). November 4–7, 2014. Bali, 8pp

Locatelli T, Binet T, Kairo JG, King L, Madden S, Patenaude G, Upton C, Huxham M (2014) How blue carbon and payments for ecosystem services (PES) might help save mangrove forests. Ambio 43:981–995

Lugo A, Snedaker S (1974) The ecology of mangroves. Annu Rev Ecol Syst 5:39–65

Mcleod E, Chmura G, Bouillon S, Salm R, Björk M, Duarte C, Lovelock C, Schlesinger W, Silliman B (2011) A blueprint for blue carbon: toward an improved understanding of the role of vegetated coastal habitats in sequestering CO_2. Front Ecol Environ 9(10):552–560

Miththapala S (2008) Mangroves. IUCN Coastal Ecosystems Series 2. Ecosystems and Livelihoods Group Asia, Sri Lanka, 28pp

Murdiyarso D, Alongi D, Kauffman J, Kurnianto S, Stidham M, Kanninen M (2009) Carbon storage in mangrove and peatland ecosystems: a preliminary account from plots in Indonesia. CIFOR Working Paper no. 48. Center for International Forestry Research (CIFOR), Indonesia. 40pp

Nam V (2017) Payment for forest environmental services in mangrove of Ca Mau Province, Viet Nam. In: Paper presented in the international conference on Sustainable Mangrove Ecosystems in Bali, Indonesia, April 18–21, 2017. Accessed at: www.itto.int/mangrove2017/presentations/

Nazri N (2017) Blue carbon stock assessment of mangrove vegetation and sediment in two selected mangrove sites in Penninsular Malaysia. Masters thesis, University of Malaya, Kuala Lumpur, pp 1–20

Peters-Stanley M, Hamilton K, Marcello T, Sjardin M (2011) Back to the future: State of the voluntary carbon markets 2011. Ecosystem Marketplace and Bloomberg New Energy Finance Report, New York/Washington, DC, 77pp

Pham TT, Bennet K, Vu TP, Brunner J, Le ND, Nguyen DT (2013) Payments for forest environmental services in Vietnam: from policy to practice. Occasional Paper 93. CIFOR, Bogor

Pichon M (1977) Physiology, morphology and ecology of the Double Barrier reef of North Bohol (Philippines). Proceedings of the thirds international coral reef symposium, Miami, pp 261–267

Primavera J (2000) Development and conservation of the Philippine mangroves: institutional issues. Ecol Econ 35:91–106

Pulhin J, Gevana D, Pulhin F (2017) Community-based mangrove management in the Philippines: experience and challenges in the context of changing climate. In: Gupta, Shaw (eds) Participatory mangrove management in a changing climate. Springer, pp 247–262

Richards D, Friess D (2016) Rates and drivers of mangrove deforestation in Southeast Asia, 2000–2012. Proc Natl Acad Sci 113(2):1–2

Sathirathai S (1998) Economic valuation of mangroves and the roles of local communities in the conservation of natural resources: case study of Surat Thani, South of Thailand. Economy and Environment Program for Southeast Asia (EEPSEA), 60pp

Siikamäki J, Sanchirico J, Jardine S, McLaughlin D, Morris D (2012) Blue carbon: global options for reducing emissions from the degradation and development of coastal ecosystems. Resources for the Future, Washington, DC

Sriladda C, Puangchit L (2009) Carbon sequestration of mangrove plantations at pak phanang, nakhon si thammarat province. In: The conference on mangrove ecology, 12–14 September, 2008, Cha-um, Petchburi, pp 379–389

Thant Y, Kanzaki M, Ohta S, Than M (2012) Cabon sequestration by mangrove plantations and a natural regeneration stand in the Ayeyarwady Delta, Myanmar. Tropics 12(1):1–10

Thomas S (2014) Blue carbon: knowledge gaps, critical issues, and novel approaches. Ecol Econ 107:22–38

Thompson B, Clubbe C, Primavera J, Curnick D, Koldewey H (2014) Locally assessing the economic viability of blue carbon: a case study from PanayIsland, the Philippines. Ecosyst Serv 8:128–140

Tue N, Dung L, Nhuan M, Omori K (2014) Carbon storage of a tropical mangrove forest in Mui Ca Mau National Park, Vietnam. Catena 121:119–126

Ullman R, Bilbao-Bastida V, Grimsditch G (2013) Including blue carbon in climate market mechanisms. Ocean Coast Manag 83:15–18

Verwer C, van der Meer P (2010) Carbon pools in tropical peat forest – towards a reference value for forest biomass carbon in relatively undisturbed swamp forest in Southeast Asia. Altera, Wageningen. Altera-report, Netherlands, 67pp

Visseren-Hamakers IJ, Gupta A, Herold M, Peña-Claros M, Vijge MJ (2012) Will REDD+ work? The need for interdisciplinary research to address key challenges. Curr Opin Environ Sustain 4:590–596

Vu T, Takeuchi W, Van N (2014) Carbon Stock Calculating and Forest Change Assessment Toward REDD+ Activities for The Mangrove Forest in Vietnam. Trans. JSASS Aerospace Tech. Japan 12 (29):23–31

Walters B (2004) Local management of mangrove forests in the Philippines: successful conservation or efficient resource exploitation. Hum Ecol 32(2):177–195

Wicaksono P, Danoedoro P, Nehren H, Nehren U (2015) Mangrove biomass and carbon stock mapping of the Karimunjawa Islands using multispectral remote sensing. Int J Remote Sens 37(1):26–52

Wylie L, Sutton-Grier A, Moore A (2015) Keys to successful blue carbon projects: lessons learned from global case studies. Mar Policy 65:76–84

Mangrove Habitats in São Tomé and Príncipe (Gulf of Guinea, Africa): Conservation and Management Status

27

R. Haroun, A. Herrero Barrencua, and A. D. Abreu

Abstract

São Tomé and Príncipe is a small island state located in the Gulf of Guinea (West Equatorial Africa) with fragile mangrove habitats along its coastlines. These habitats are threatened by historical conversion to land for agriculture uses, overharvesting for firewood and charcoal, changing hydrology and coastal erosion; the last impacts increasing their vulnerability to sea-level rise. In the case of São Tomé mangroves, road construction (sometimes very close or even crossing through the habitat itself) is considered as a major factor leading to its transformation.

In this contribution, we identified the major botanical and faunal (vertebrate and invertebrates) species encountered in this unique forest ecosystems. We also described some ecosystem services such as food supply and nursery habitat for diverse fish and invertebrates, with a short mention of its carbon sequestration role.

Most of the mangrove hábitats in São Tomé are located inside the Parque Natural Obô of São Tomé, thus, they are under some degree of protection. In the case of Príncipe Island, there are three main remnants of mangrove forests: Praia Salgada, Praia Caixão and Praia Grande, all of them outside the Parque Natural Obô of Principe.

The potential development of ecotourism activities (such as birdwatching, trails, etc.) similar to those already in place at Malanza mangrove (São Tomé) and the restoration of the remnants of mangrove habitats combined with capacity building actions could support community development and job opportunities, especially for women and young people, with high relevance at local level.

27.1 Introduction

The Democratic Republic of São Tomé and Príncipe, one of the youngest and smallest countries in Africa, is made up of two main islands and several islets located in the Gulf of Guinea about 440 km to the south and 225 km to the west, respectively, of the coasts of Nigeria and Gabon, very near to the equatorial line.

In the recent worldwide review of mangrove forests by Spalding et al. (2010) these habitats were cited only for a number of estuaries in the bigger island of São Tomé, whereas in Príncipe island their presence was not confirmed, probably as the major regional environmental information sources, such as Corcoran et al. (2007) and FAO (2007) did not have distributional data on these keystone ecosystem from this equatorial insular country.

Mangroves are essential ecosystems, providing multiple ecological services including fisheries, shoreline stabilization, nutrient and sediment trapping and high biodiversity. For example, mangroves in Western Central Africa are among the most carbon-rich ecosystems in the world, with estimates that 1299 tonnes of carbon dioxide storaged in pristine mangrove habitats (Ajonina et al. 2014). Although mangroves forests are internationally recognized for their importance to the well-being of human populations (FAO 2007; Spalding et al. 2010; UNEP 2016a), there is scarcely any information on these ecosystems for São Tomé and Príncipe State. In recent years, the economic development of São Tomé and Príncipe is highly dependent on blue economy sectors, and this represents both challenges and great potential for biodiversity conservation issues (UNEP 2016b).

R. Haroun (✉) · A. Herrero Barrencua
Biodiversity and Conservation Research Group, Institute ECOAQUA, University of Las Palmas de Gran Canaria, Scientific and Technological Marine Park, 35214 Telde, Spain
e-mail: Ricardo.haroun@ulpgc.es

A. D. Abreu
Príncipe Island UNESCO Biosphere Reserve, Cidade de Santo António do Príncipe, Região Autónoma do Príncipe, Santo António, Democratic Republic of São Tomé e Príncipe

C. Makowski, C. W. Finkl (eds.), *Threats to Mangrove Forests*, Coastal Research Library 25,
https://doi.org/10.1007/978-3-319-73016-5_27

This chapter provides updated information about environmental, conservation and management issues related to mangrove habitats present along the coasts of São Tomé and Príncipe islands. We described the botanical and faunal composition of this unique forest ecosystem, which hosts many species coming from both the marine and the terrestrial realms. Then, its current distributional pattern along the shorelines of both islands are shown. After that, the historical and current uses are presented. Subsequently, we characterized the major threats affecting the remnants of these unique habitats in the islands. Finally, the current management issues are presented together with the most relevant conservation initiatives to promote the sustainable use of diverse ecosystem services provided by these transitional habitats.

27.1.1 Geographical Features

São Tomé and Príncipe islands are aligned in a volcanic island chain in the Gulf of Guinea (Equatorial Atlantic), surrounded by deep-sea trenches and host one of the world's biodiversity hotspots (Fig. 27.1), with several terrestrial endemisms (Gillespie and Clague 2009), whereas its marine biota start to be investigated. Both islands are within the wet tropical belt, with annual rainfall ranges from 1000 to over 4000 mm with a raining season running from October to May; average annual temperatures range from 18–21 °C minimums to 30–35 °C maximums. In general, the islands are very steep, covered with lush tropical forests and with lowlands along the coastlines as the only flat areas.

Mangroves forest thrive in saline coastal sediments in the Western Central Africa (UNEP 2016a). They may occur in intertidal zones of sheltered or low exposed coasts, gulfs and inlets, coastal lagoons, marshes and estuaries or river mouths, covered with soft bottoms (of sands, silts or clays, never rocky). Mangroves habitats are found only along the coasts of the islands of São Tomé and Príncipe, not in the islets, in estuarine and scattered sheltered areas.

27.2 Biological Characteristics

The current knowledge about the main biological characteristics of the mangroves habitats encountered in this insular country will be described in the following sections,

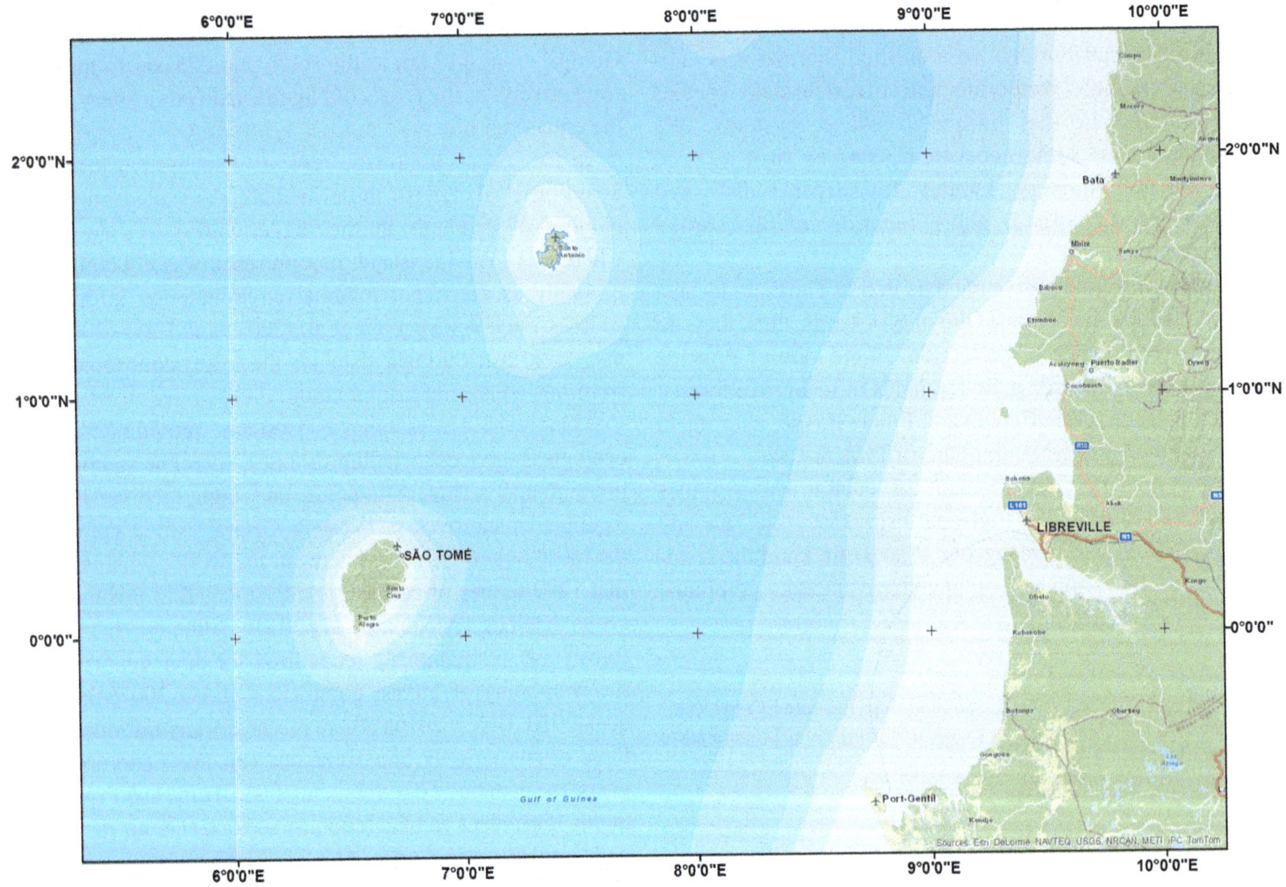

Fig. 27.1 Location of São Tome and Príncipe Islands in the Gulf of Guinea, Equatorial Africa (Modified from ESRI, 9 June 2017)

covering both the botanical and faunal aspects. Part of these descriptions comes from specialized works done, for example with birds or marine turtles, whereas other scientific data were gathered by a concerted effort of the co-authors partially supported by the UNESCO Reserve of Biosphere of Principe Island. It is important to mention that an exhaustive account of the biodiversity associated to mangroves are not yet available and many taxonomic groups are still unknown to the science.

27.2.1 Flora

27.2.1.1 Species Composition

In Table 27.1 is included the catalogue of botanical species encountered in the mangrove habitats, both in São Tomé and Príncipe islands. Inside that table are included several useful plants, such crop, fruit or timber species originated from historical human interventions in those mangrove areas.

The most characteristic botanical species of the mangroves hábitats in São Tomé and Príncipe are the following ones:

- Red Mangrove: *Rhizhopora harrisonii Leechman/R. racemosa* G. Mey/*R. mangle* Linnaeus. Aside from its characteristic arched prop reddish aerial roots and shinning spatulate leaves, the species of red mangrove trees have tiny yellow flowers. The main taxonomic differences among the reported species in the area are related to some morphological characters in the flowers. After fertilization, red mangroves develops distinguished pencil-shaped long fruits, which turns into propagules (= embryonic root structures). When ripe, the young seedlings detach from the parent tree and float until reaching a suitable substratum to root.
- Black Mangrove: *Avicennia germinans* (L.) Linnaeus. This species is widespread in coastal and estuarine areas of the tropics and subtropics to about 28° in both the northern and southern hemispheres. The tree is characterized by long horizontal roots and root-likc projections known as pneumatophores, protruding from the soil and providing oxygen to the underground and underwater root systems (Figs. 27.2 and 27.3).
- Golden Leather Fern: *Acrostichum aureum*. This species is a large understory fern that occurs worldwide in mangrove forests and other wetlands.

Table 27.1 Catalogue of major botanical species observed in São Tomé and Príncipe mangroves

Common local name	Common English name	Scientific name
Amoreira	Iroko	*Milicia excelsa*
Bambú	Bamboo	*Bambusa vulgaris*
Bananeira	Banana	*Musa* sp.
Bolinha	–	*Xylopia staudtii*
Cacaouero	Cocoa tree	*Theobroma cacao*
Café	Coffee	*Coffea* spp.
Cajamangueira	Golden apple	*Spondias cytherea,*
Candeia	–	*Hernandia beninensis*
Carozeiro	Indian almond tree	*Terminalia catappa*
Colma	–	*Milettia thonningii*
Coqueiro	Coconut palm	*Cocos nucifera*
Eritrineira	Coral tree	*Erytrina poeppigiana*
Figueira	Common fig	*Ficus carica*
Fruteira	Breadfruit	*Artocarpus altilis*
Gofe	Trumpet tree	*Cecropia peltata*
Goiabeira	Common guava	*Psidium guajava*
Guêguê	Yellow mombin tree	*Spondias mombin*
Grigô	Brimstone tree	*Morinda lucida*
Isaquinteiro	African breadfruit	*Treculia africana*
Jaqueira	Jackfruit	*Artocarpus heterophyllus*
Lemba-lemba	–	*Turraea thonningii*
Lima	Lime	*Citrus* spp.
Mamoeiro	Papaya	*Carica papaya*
Mangue	Red mangrove	*Rhizophora racemosa/R. hardisonii/R. mangle*
Mangue	Black mangrove	*Avicennia germinans*
Mangueira	Mangoo tree	*Mangifera indica*
Marapinhao	Satinwood	*Fagara macrophylla*
Moandí	African oil bean	*Pentaclethra macrophylla*
Mucumbli	–	*Lannea welwitschii*
Oká pequeño	Kapok	*Ceiba pentandra*
Palmeira	African oil palm	*Elaeis guineensis*
Pau-caixão	African nutmeg	*Pycnanthus angolensis*
Pau-esteira	Endemic screwpine	*Pandanus thomensis*
Pau-ferro	Pheasant-berry tree	*Margaritaria discoidea*
Pau-leite	False rubber tree	*Funtumia africana*
Pau-lisha	Sandpaper fig	*Ficus exasperata*
Safu d'Óbô	African grape	*Pseudospondias microcarpa*
Sap-sap	Soursop	*Annona muricata*
–	Golden leather fern	*Acrostichum aureum*

The White Mangrove: *Laguncularia racemosa* (L.) Gaertn., which is reported in West Africa from Senegal to Angola, appears cited in some national reports, but its presence needs to be confirmed in the mangrove habitats of these islands. The same situation applies to the Buttonwood Mangrove (*Conocarpus erectus* L.) that is a tree or shruh, somehow similar to the White mangrove, but with dense flower heads that grow in branched clusters, forming cone-like fruit. A potential explanation of their absence in the nature is either both species have never been found in these

Fig. 27.2 Black Mangrove (*Avicennia germinans*) and Red Mangrove (*Rhizophora*) trees at the water edge in Malanza mangrove

Fig. 27.3 Flowers of Red Mangrove (*Rhizophora*) and Black Mangrove (*Avicennia germinans*) trees

islands or historically they have been extirpated from the corresponding mangrove forest during the agro-transformation process of the mangrove habitat.

27.2.1.2 Reproductive Anual Cycle

Preliminary observations on the reproductive stages of mangroves trees, allow us to define a broad reproductive anual cycle inside the mangrove forest, at least for the Red Mangrove species and Black Mangrove tree (Fig. 27.3). The buds of the flowers start to develop from late September and October, whereas the fruits are seen in late November, becoming ripe in February–March, almost at the end of the raining season. The propagules detaches from the parent trees mainly in April and may act as long-range and short-range dispersion mechanism for the species. In the case of the *Avicennia germinans* tree, the flowers are mainly observed between March and May, a bit latter than the previous species, while the fruits could be found during a longer period.

27.2.2 Fauna

27.2.2.1 Marine Invertebrates

In Table 27.2 are listed a preliminary number of marine mollusks and crustaceans species that could be found in the soft bottoms as well as in the arched roots of red mangroves trees. These species were mainly identified between September–November 2016 in Principe Island (Fig. 27.4).

Table 27.2 List of invertebrates observed in the mangrove habitats of Príncipe

Taxa	Family	Species	Local common name
Mollusca	Archatinidae	*Archachatina marginata*	Buzio de terra
	Neritidae	*Clypeolum owenianum* (*W. Wood, 1828*)	–
		Gastropoda sp.	–
	Hipponicidae	*Hipponix antiquatus* (L.1767)	–
	Mactridae	*Leptospisula nívea* (Gmelin, 1791)	–
	Littorininae	*Littorina* sp.	–
	Muricidae	*Tympanotonus fuscatus* (Linnaeus, 1758)	–
	Ostreidae	*Saccostrea cucullata* (Born, 1778)	Ostra
	Arcidae	*Senilia senilis* (Linnaeus, 1758)	–
	Solecurtidae	*Tagelus adansonii* (Bosc, 1801)	Almexa
Crustacea	Ocypodidae	*Afruca tangeri* (Eydoux, 1835)	Caranguejo violinista
	Portunidae	*Callinectes* sp.	Caranguejo tartaruga
		Crab sp. 1	–
		Crab sp. 2	–
	Gecarcinidae	*Cardisoma armatum* Herklots, 1851	Caranguejo de mangle
	Coenobitidae	*Coenobita rubescens* Greeff, 1884	Eremita terrestre
	Grapsidae	*Goniopsis pelii* (Herklots, 1851)	Caranguejo negro
	Grapsidae	*Grapsus adscensionis* (Osbeck, 1765)	Caranguejo de roca
	Gecarcinidae	*Johngarthia weileri* (Sendler, 1912)	Cracke
	Ocypodidae	*Ocypode cursor* (Linnaeus, 1758)	Caranguejo fantasma
	Palaemonidae	*Palaemon concinnus* Dana 1852	Camaron de mangle

Few records, such the blue crab *Callinectis* or the arch shell *Senilia senilis*, have been reported by other authors for Malanza mangrove in São Tomé (Pisoni et al. 2015). Besides, in one ocasion a large crustacean specimen resembling *Atya gabonensis*, commonly known as Giant African Fan Shrimp, was collected in Praia Grande mangrove, but this record still needs some extra sampling and more research to confirm its taxonomic identification (Cumberlidge 1999; Rossignon 1999).

This list represents the first account of marine invertebrates species for mangrove habitats in São Tomé and Príncipe, although some mollusks and crustacean specimens as well as other marine invertebrates taxonomic groups were not yet fully identified.

27.2.2.2 Fishes

In the Table 27.3 is presented the 22 fish species reported for the mangrove habitats of São Tomé and Príncipe. Species were taxonomically ranked according to Nelson (2006) and fish identification was performed using the relevant fish guides and available checklists (Afonso et al. 1999; Lévêque et al. 1990; Quéro et al. 1990; Schneider 1990; Stiassny et al. 2007; Wirtz et al. 2007; Vasco-Rodriguez et al. 2016). This catalogue of extant species have been compiled by two different approaches: review of the pubished data for this geographical area as well as the samples obtained during recent fisheries surveys done in the mangrove hábitats of Príncipe island (September–December 2016), using diverse fishing gears. One of the common species is the mudskipper (Fig. 27.5) that could be found easily moving outside the water in soft bottoms at the mangrove edges.

In a recent survey of fish population of mangroves in São Tomé, Félix and collaborators (2016) finds out that the highest number of both juveniles and adults (inshore and residents) was observed at Malanza, the largest mangrove forest in both islands and presenting the highest hábitat heterogeneity. These authors concluded that the species richness found in São Tomé mangroves, particularly in Malanza, is similar to that of other insular mangroves of equivalent size or of continental mangroves in the Gulf of Guinea. Félix and collaborators (2016) also mentioned the presence of introduced Nile Tilapia specimens in the Malanza mangrove (*Oreochromis mossambicus*), with some question marks about its potential effects on authoctonous fish species.

27.2.2.3 Birds

The avifauna is another important vertebrate group that can be encountered inside the mangrove habitats (Fig. 27.6). During the surveys done by one of the authors between September and December 2016 in the mangrove habitats of Príncipe island, a total number of 1533 birds were observed, belonging to 26 species, 10 of them endemics (Christy 2001; Birlife International 2017). In the case of São Tome, Pisoni

Fig. 27.4 Bivalve shells of hooded oyster (*Saccostrea cucullata*) and clams (*Tagelus adansonii*) in the soft bottom of Praia Salgada mangrove, Príncipe Island

Table 27.3 List of fish species reported in the mangrove habitats of São Tomé and Príncipe

Family	Species	Local common name
Elopidae	*Elops senegalensis* Regan, 1909	Colepinha balabo
Megalopidae	*Megalops atlanticus* Valenciennes, 1847	Tainha congo
Ophicththidae	*Dalophis cephalopeltis* (Bleeker, 1863)	–
Mugilidae	*Liza grandisquamis* (Valenciennes, 1836)	Tainha tarrafa
	Liza dumerili Steindachner, 1870	Tainha tarrafa
Lutjanidae	*Lutjanus agennes* Bleeker, 1863	Corvina preta
	Lutjanus endecacanthus Bleeker, 1863	Corvina
	Lutjanus goreensis (Valenciennes, 1830)	Corvina vermelha
Gerreidae	*Eucinostomus melanopterus* (Bleeker, 1863)	Parente
Haemulidae	*Plectorhinchus macrolepis* (Boulenger, 1899)	Peixe-porco
	Pomadasys jubelini (Cuvier, 1830)	Roncador
Polynemidae	*Galeoides decadactylus* (Bloch, 1795)	Barbudo
Monodactylidae	*Monodactylus sebae* (Cuvier, 1829)	Cozinheiro
Kyphosidae	*Kyphosus incisor* (Cuvier, 1831)	Sopa
Cichlidae	*Oreochromis mossambicus* (Peters, 1852)	Papé
Eleotridae	*Bostrychus africanus* (Steindachner, 1879)	Xarroco-cherne
Gobiidae	*Awaous lateristriga* (Duméril, 1861)	Xarroco
	Awaous bustamantei (Greeff, 1882)	Xarroco
	Bathygobius burtoni (O'Shaughnessy, 1875)	Xarroco
	Periophthalmus barbarus (L., 1766)	Cocumba
	Porogobius schlegelii (Günther, 1861)	Xarroco-blabo
Poeciliidae	*Aplocheilichthys spilauchen* (Duméril, 1861)	Tose tose

and collaborators (2015) did some bird census in Praia das Conchas counting 239 birds, belonging to 26 species, whereas in the case of Malanza mangrove the number of birds counted raised towards 1942 individuals belonging to 29 species. Moreover, the only species of endangered birds identified during that field work were the São Tomé Green Pigeon (*Treron sanctithomae*) and São Tomé White Eye (*Zosterops feae*), both with vulnerable status and observed only in the Malanza mangrove.

Fig. 27.5 An adult mudskipper *Periophthalmus barbarus* (Linnaeus, 1766), locally know as cocumba (Praia Salgada, Príncipe Island), a conspicuous species in Central Africa mangroves

Fig. 27.6 Common bird species: Water duck (*Phalocrocorax africanus*) and White heron (*Egretta intermedia*) associated with mangrove habitats in São Tomé and Príncipe

27.2.3 Mangrove Habitat Distribution

27.2.3.1 São Tomé

The current distribution of main mangrove habitats along the coastline of São Tomé island are represented in Fig. 27.7. Specially mangroves sites located in the northen and western side of the islands have a more reduced surface coverage compared to historical records (Corcoran et al. 2007; Spalding 2010), probably as these areas were partially transformed into agriculture lands (Pisoni et al. 2015) and affected by coastal road construction.

27.2.3.2 Malanza

The mangrove of Malanza is located in the south (0.8°02′04.6″N, 0.6°83′10.48″E), occupies an area of 0.7 km^2 and the maximm depth is 3.5 m. This site has the most extensive mangrove forest in the all country (Fig. 27.8). Of the total

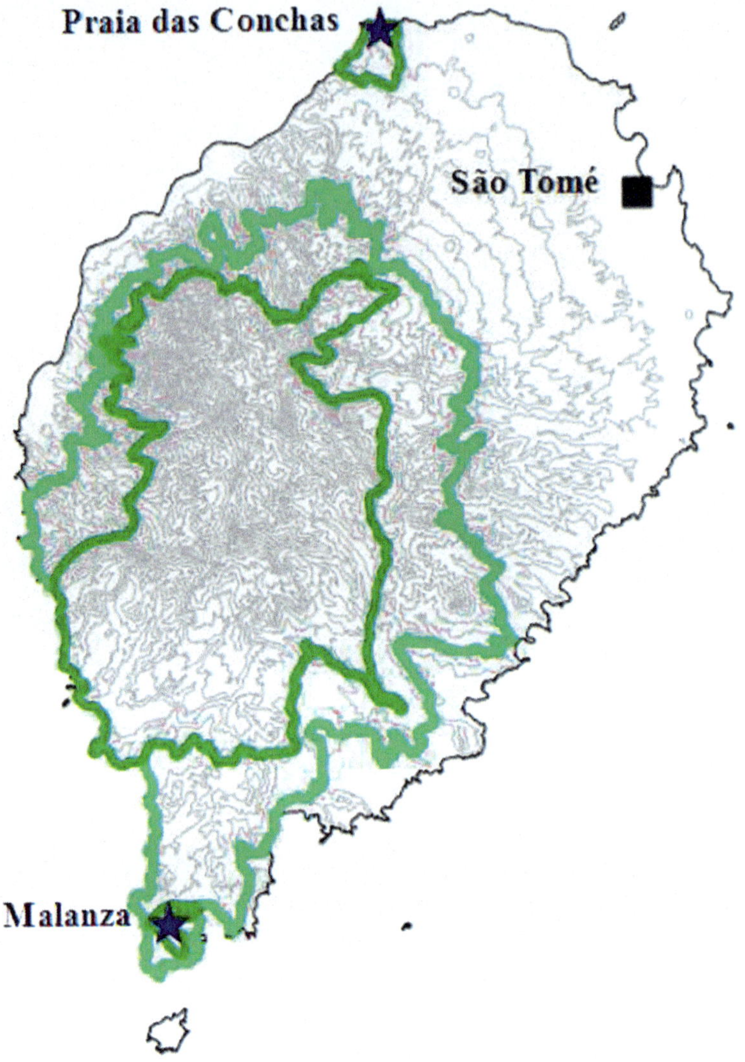

Fig. 27.7 Locations of the two main mangrove hábitats identified in the coast of São Tomé Island: Praia das Conchas in the Northwest and Malanza in the South (Modified from Pisoni et al. 2015)

339 ha of Malanza area is estimated to exist 68.6 ha of mangrove.

27.2.3.3 Praia das Conchas

The mangrove of Praia das Conchas is located in the north of the island (0.8° 24′02.1″N, 0.6°83′70.2″E), covers an area of 0.01 km^2 and is a shallow system (maximum depth 0.5 m). There are about 0.8 ha of mangrove, divided between the beach of the same name (0.5 ha) and Praia Quinze (0.3). Its historical distribution have been reduced as consequence of the construction of the coastal road.

27.2.4 Príncipe

In the case of the island of Principe, the Fig. 27.9 pointed out the three locations of mangrove habitats that can be recognized at present time along its coasts: Praia Caixão, Praia Grande and Praia Salgada. There are not historical records about the former extension of these unique forests (Corcoran 2007; FAO 2007; Spalding et al. 2010), but some local people are familiar with uses and products extracted from them, therefore, it is presumed that these three areas represent the remnants of more extensive mangrove forests.

27.2.4.1 Praia Grande

The mangrove habitat of Praia Grande, in the Northeast side of the island (1°40′01.5″N 7°26′43.1″E), occupies the borderline of a coastal lagoon. The mangrove trees develops fringing the inner side of the lagoon whereas the sandbar that separates the freshwater body from the sea is still holding many specimens of a historical coconut plantation (Fig. 27.10). This location is an important nesting site for marine turtles, with nests of different endangered species.

27.2.4.2 Praia Salgada

This mangrove hábitat is located in the East side of the island (1°37′57.1″N 7°27′14.1″E), inside the Bay of Abades. The mangrove hábitat was mainly transformed in a coconut plantation, which is at present time abandoned (Fig. 27.11).

Fig. 27.8 General view of main water course inside the Malanza mangrove, São Tomé

Nevertheless, the main botanical species encountered in the area is still the Coconut palm (47%), whereas the Red Mangrove is the second most observed species occupying 15% of the area, with the Golden Leather Fern also quite common as understore species. Aside from the coconut palms, there are still several botanical useful species planted in the sandy bottoms occurring in the seafront.

27.2.4.3 Praia Caixão

In this case, the mangrove habitat is located in the estuary of the Caixão river, in the West side of the island (1°37′08.3″N 7°22′09.6″E). This river mouth has a permanent flow of freshwater and its mouth is open directly to the sea. The transition of tree zonation from the mangrove riverine habitat to the secondary tropical forest is very quick and only the Golden Leather Fern is recognized few meters inside the forest from the water edge.

27.3 Uses

At international level, mangrove forests are mainly recognized for their ecological role in coastal protection and nursery habitat for commercially valuable marine species (either fish or shrimps) (Rönnbäck et al. 1999; Blasco et al. 2005; Nagelkerken et al. 2008). They also play important functions in terms of wood and food supply (both fishing and hunting). These habitats have been used as indicators of environmental risk, to avoid or reduce coastal erosion, and as a keystone carbon sink areas (FAO 2007; Spalding et al. 2010; UNEP 2016b).

27.3.1 Tradicional Use

Mangroves provide various traditional products and services to the population of São Tomé and Príncipe; more recently, new and innovative uses of the mangrove are been developed. The large Malanza mangrove, in the South of São Tomé island, has important ecosystem services at the level of food supply (fish, crabs and shrimps) and coastal protection; more recently, its aesthetic value couple with the possibility to encounter some bird species is underpinning the development of ecotourism activities by some local companies (Pisoni et al. 2015). Its role as nursery area for commercial demersal fish species have recently been evaluated (Félix et al. 2016). Other ecological services, such as aquatic bioremediation and pollution regulation as

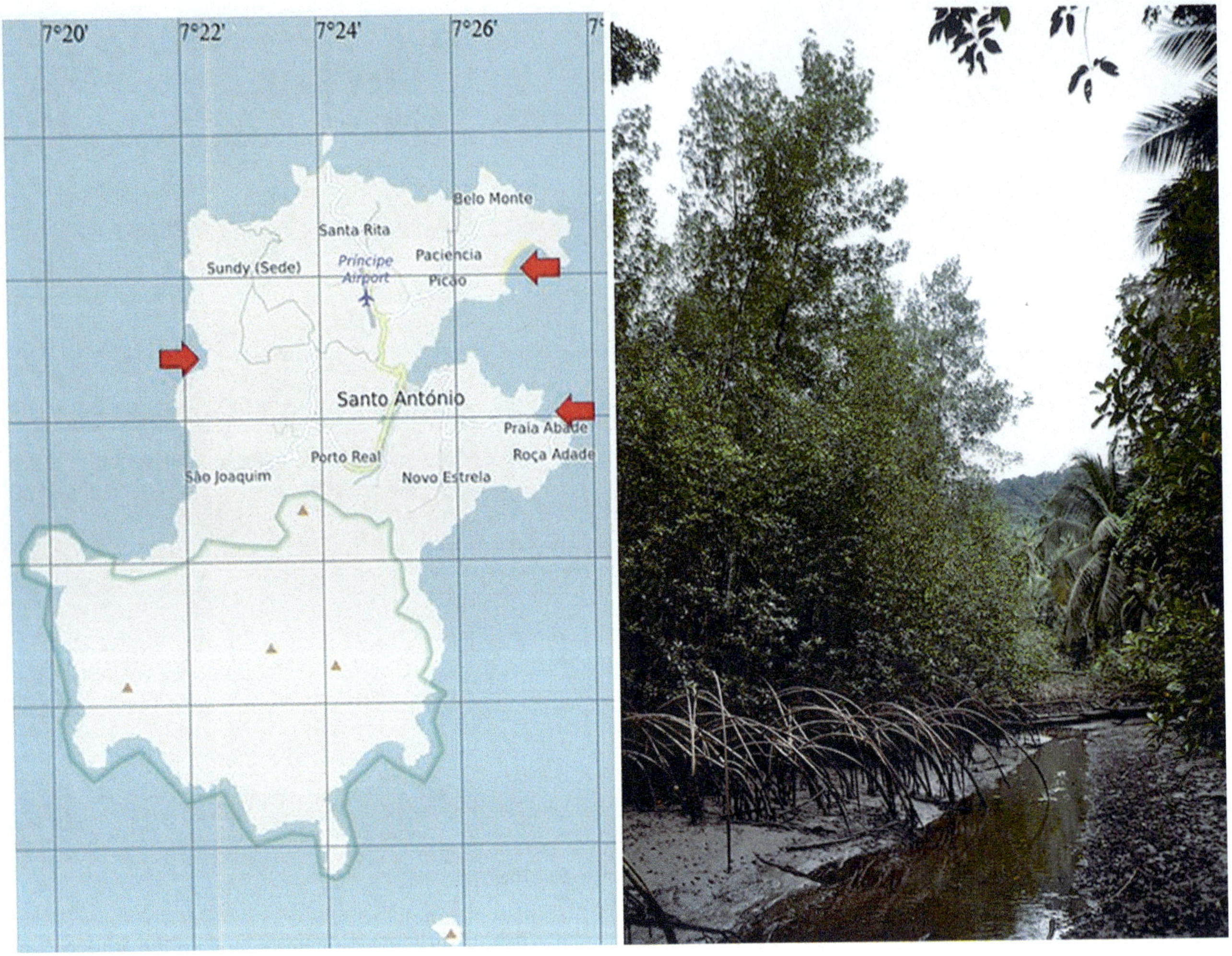

Fig. 27.9 Locations of the mangrove hábitats identified in the coast of Principe Island. Praia Caixão in the West, Praia Grande in the Northeast and Praia Salgada in the East

well as carbon sequestration are not yet evaluated. In the same island, the main services provided by the mangrove located at Praia das Conchas are the supply of wood, sand and charcoal (energy source) and has potential to improve its ecosystem service delivery in coastal protection and carbon sequestration.

Another traditional product delivered by mangroves, which is still operational in Príncipe's mangroves, is the application of the pigments (tannins) extracted from the leaves as staining product to dye natural fiber fishnets of a purple colour (Fig. 27.12). Nowadays, this activity is becoming more and more rare, due to the import of synthetic coloured fishnets.

27.4 Threats

The islands of São Tomé and Príncipe were uninhabited prior to their discovery in 1470–1471. Since then, they were quickly populated by the Portuguese and subsequent transformation of the authoctonous vegetation and landscapes have been reported mainly in the lowlands and coastal areas. As already pointed out by Corcoran and collaborators (2007), there are indications that much of the coastal mangrove forests were destroyed just before 1990 mainly for agriculture land.

27.4.1 Deforestation

Before the arrival of human beings, both islands were covered by lush tropical forests and along the coastline mangrove habitats were probably more common and widespread in estuarine areas that at present time. Sugar cane was the first agricultural activity and very soon became the dominant crop. Coffee and cocoa were introduced later, in the nineteenth, and extensive shade plantations of these crops were developed until nowadays. In recent decades, deforestation process concentrated in the coastal habitats,

Fig. 27.10 Aerial view of the coastal lagoon, coconut plantation and mangrove habitat at Praía Grande, Príncipe Island

Fig. 27.11 Red mangrove, coconut palms and other plants in Praia Salgada area

Fig. 27.12 Cuttings of Red Mangrove leaves used to stain traditional fishing nets

Fig. 27.13 Mangrove trees and Coconut palms mixed at Praia Salgada mangrove, Príncipe Island

and oil palm and coconut plantations are dominant crops in the lowlands areas.

After the independence from Portugal in 1975, the agricultural activities were reduced and many plantations were abandoned, thus stimulating some forms of regeneration to secondary forests. De Lima et al. (2012, 2014) have analyzed the distributional presence and abundance of endemic birds in primary forests and shade plantations (cocoa and coffee), showing a reduction of endemisms in shade plantations.

Particular areas, such as the lowland forests, have been widely cleared for agriculture (sugar cane, African oil palm and coconuts plantations), sometimes affecting the mangrove forests, such in the case of Praia Salgada and Praia Grande mangrove areas in Príncipe (Fig. 27.13). Coconut plantations and diverse tropical trees, such as Indian almond tree, African breadfruit, Jackfruit, Common guava, Golden apple, African nutmeg, Soursop, etc. together with some medicinal shrubs were cultivated. Nowadays, some of these botanical useful species still remains in the mangrove habitat range, specially in the case of the mangrove habitats in the northern side of São Tomé and in those currently identified in Príncipe Island.

27.4.2 Habitat Transformation

In the case of São Tomé mangroves, road construction (sometimes very close or even crossing through the habitat itself) is considered as a major factor leading to the transformation of the habitat. For example, in Praia das Conchas mangrove the water flow is limited by a road that crosses the downstream area, limiting superficial intrusion of seawater; thus, the development of the road resulted in a slight increase in height in the area's topography and restricted seawater intrusion inside this mangrove mainly during spring tides. For the Malanza mangrove area, that mangrove have a constrained southeastern opening towards the sea associated to the foundations of the Malanza bridge, which halt the tidal influence; therefore, sedimentation inside the mangrove habitat is an increasing threat.

Another factor leading to habitat transformation is the introduction of exotic species. Terrestrial introduced species is probably the chief conservation concern in these tropical islands and may play a critical role in the management of primary vegetated habitats (Brown 1991; Dutton 1994). The islands have the usual mammals introductions associated with human translocations, such as rats, cats, dogs and pigs, whereas civets, weasels, and mona monkeys have been introduced on purpose to São Tomé and more recently, to Príncipe. The negative effects of these introductions may range from endemic bird reproduction disruption to reduction in primary forest seedlings. It it not posible to assess now the damage caused by agricultural activities and the introduction of various mammals (domestic and wild) and even by terrestrial gastropods, on the authoctonous vegetation distribution and structure, including mangroves.

Another threat affecting the hábitat structure and composition of these mangroves is the extraction of timber, wood and charcoal. In both islands, mangrove trees have been commonly used as a regular source of wood and charcoal for daily life for the nearby local communities. In these sense, large trees are almost absent in any of the current mangrove forests.

27.4.3 Global Change, Sea-Level Rise and Coastal Erosion

As in many other regional seas, equatorial african coasts are subject to natural erosion and sedimentation processes, including high wave energy and strong littoral transport, which are intensified by human activities such as sand mining, port construction, and mangrove deforestation (UNEP 2016a). Moreover, acording to Penhor et al. (2011), São Tomé and Príncipe will have during 2040–2060 a reduction in precipitation combined with an increase of mean temperatures, which may increase the stress of remaining mangrove forests.

Moreover, the sea-level rise projections along Africa's coastal zones by 2100 are approximately 10% higher than the global mean (World Bank 2013) meaning that the coastal lowlands including mangroves of 37 countries, many of them in the Gulf of Guinea, will be vulnerable at various spatial and temporal scales (UNEP 2016a).

Already, some coastal communities and villages both in São Tomé and in Príncipe are increasingly and more frequently submitted to flooding and erosion damage apparently due to increased storm intensity and significant shifts in the duration and intensity of the rainy seasons that have begun to overlap with the storm season (Fig. 27.14). Due in part to this overlap, a number of communities are experiencing record flooding and loss of property since 2008. Sand mining, mangrove deforestation and other human interventions are sigificantly contributing to exarcebate the impacts of extreme events. Loss of access to the sea, damage and even destruction in houses and public facilities (schools) are some of the most significant impacts leading to loss of income and fear as some communities are also forced to run from their houses (e.g. Praia das Burras in Príncipe Island and Iô Grande in São Tomé). Several adaptation projects are ongoing including the identification and adaptation interventions for reducing the vulnerability of coastal community infrastructure and productive assets related with climate change impacts.

In summary, the mangrove habitats in these islands are threatened by historical conversion to land for agriculture uses, overharvesting for firewood and charcoal, changing hydrology and coastal erosion.

Fig. 27.14 Coastal erosion at Praia Salgada mangrove (Abades Bay, Principe Island) (© Pedro Gómez)

27.5 Conservation and Sustainable Management Issues

Aside from the above mentioned traditional usage of mangroves forests, there are very few studies on the ecosystems services they may provide to local communities. These keystone coastal and estuarine ecosystems have been recognized recently as nursery and feeding grounds for resident fish species (Félix et al. 2016). Mangroves in Western Central Africa are among the most carbon-rich ecosystems in the world, with estimates that 1299 tonnes of carbon dioxide storaged in pristine mangrove habitats (Ajonina et al. 2014). Mangroves has higher primary production levels that most tropical forests, due to to very high standing and below-ground biomass, with large storage capacity of blue carbon. Therefore, degradation of the mangrove ecosystem compromises its resilience and erodes its natural capacity for carbon sequestration. A recent study in Guinea-Bissau by Vasconcelos and collaborators (2015) showed that reversing mangrove deforestation could contribute to mitigation of climate change effects plus preserving the many other ecosystem services.

The ecosystem services and products derived from mangroves in São Tomé and Príncipe start to be evaluated and in near future addtional research may bring new quantitative data to valorize these habitats and contribute to its conservation at national scale.

Most of the mangrove hábitats in São Tomé are located inside the Parque Natural Obô of São Tomé; therefore they are subjected to some degree of protection. In the case of Príncipe Island, there are three main remnants of mangrove forests: Praia Salgada, close to the community of Praia Abade, Praia Caixão and Praia Grande, all of them outside the Parque Natural Obô of Principe. A preliminary assessment of these three habitats was developed in 2016–2017, by the UNESCO Biosphere Reserve of Príncipe Island aiming to identify the main existing species, to confirm their current extension and assess the potential for expansion and potential development of ecotourism activities (such as birdwatching, trails, etc.) similar to those already developed for Malanza mangrove forest in São Tomé with partial support of European funds (Fig. 27.15).

In the case of Príncipe Island, the preliminary results indicate that Praia Salgada mangrove, close to the community of Praia Abade, shows a great potential for conservation and

Fig. 27.15 Advertisement of local ecotourism activities at Malanza Mangrove, South of São Tomé Island

Fig. 27.16 New mangrove trees (*Rhizophora*) sprouting in the edge of an abandoned coconut plantation in the historical range of the mangrove habitat (Praia Salgada, Príncipe Island)

expansion (Fig. 27.16). The restoration of the mangrove forest combined with the development of ecotourism and educational activities could support community development and job opportunities, especially for women and young people, with high relevance at local level. Recovery and expansion of this mangrove will also contribute to prevent coastal erosion and prevent potential damages to infrastructures (road, bridges) and access to natural resources such as water, beach and coastal fisheries. Praia Caixão and Praia Grande mangroves, although not having established communities, also show potential for ecotourism and educational activities in combination with other conservation (marine turtles and bird watching) and responsible fisheries activities.

Another ecosystem service that may play an important role for the conservation strategy of mangrove hábitats at a national level is related to the blue carbon financing (Herr et al. 2012), in this case under the Abidjan Convention Region. Here there are opportunities, as well as some constraints, and issues of uncertainty associated with payments for maintaining blue carbon stocks, which needs to be overcome on improved measurement, reporting and verification. Despite the constraints, blue carbon payments have been advancing in a number of developing countries (Ajonina et al. 2014; UNEP 2016a). UNEP and CIFOR (2014) have produced a detailed guidance on planning a blue carbon project, from concept development to regulatory compliance.

In this sense, the Blue Forests Project is a powerfull conservation tool implemented by the United Nations Environment Programme (UNEP) and executed by GRID-Arendal through funds from the Global Environment Facility (GEF) and partner co-finance. The project was initiated in January 2015 and will run over the course of 4 years. The Blue Forests Project represents a milestone opportunity for sustainable financing of coastal ecosystem management through economic values associated with carbon and wider ecosystem services.

São Tomé and Príncipe mangrove habitats have suffered a tremendous impacts on historical times, reducing its potential surface coverage and modifying its structure and composition. Nevertheless, this type of ecosystem has shown its resilience in the scattered mangrove areas that can be observed along the coastlines of both islands. Nowadays, there are some research efforts, national policies and international financing tools that may contribute to foster under novel, innovative approaches the sustainable use and conservation of this magnificent habitats in the middle of the Gulf of Guinea.

References

Afonso P, Porteiro FM, Santos RS, Barreiros JP, Worms J, Wirtz P (1999) Coastal marine fishes of São Tomé Island (Gulf of Guinea). Arquipelago Life Mar Sci 17A:65–92

Ajonina G, Kairo GG, Sembres T, Chuyong G, Mibog D, Nyambane A, FitzGerald C (2014) Carbon pools and multiple benefits of mangroves in Central Africa: Assessment for REDD+. 72pp. Available online at: http://staging.unep.org/pdf/REDDcarbon_lowres.pdf

BirdLife International (2017) Country profile: São Tomé e Príncipe. Available online at: http://www.birdlife.org/datazone/countrysao-tome. Checked: 2017-06-12

Blasco F, Aizpuru M, Besnehard J (2005) Mangroves, ecology. Encyclopedia of Earth Science Series, pp 606–611

Brown DS (1991) Freshwater snails of São Tomé, with special reference to *Bulinus forskalii* (Ehrenberg), host of *Schistosoma intercalatum*. Hydrobiologia 209:141–153

Christy P (2001) São Tomé and Príncipe. In: Fishpool LDC, Evans MI (eds) Important bird areas in Africa and associated islands: priority sites for conservation. Birdlife Conservation Series No. 11. Pisces Publications and BirdLife International, Newbury and Cambridge, UK, pp 727–731

Corcoran E, Ravilious C, Skuja M (2007) Mangrove of Western and Central Africa. UNEP – Regional Seas Program, UNEP/WCMC. 92 pp. Available online at: http://www.unepwcmc.org/resources/publications/UNEP_WCMC_bio_series/26.htm

Cumberlidge N (1999) The freshwater crabs of West Africa: family Potamonautidae. IRD, Paris

Dutton J (1994) Introduction of mammals in São Tomé and Príncipe: possible threats to biodiversity. Biodivers Conserv 3:927–938

FAO (2007) The world's mangroves 1980–2005. FAO Forestry Paper 153. Rome, Italy. 89 pp. ISBN 978-92-5-105856-5. Available online at: ftp://ftp.fao.org/docrep/fao/010/a1427e/a1427e00.pdf

Faustino de Lima R, Dallimer M, Atkinson PW, Barlow J (2012) Biodiversity and land-use change: understanding the complex responses of an endemic-rich bird assemblage. Divers Distrib 19 (4):411–422

Faustino de Lima R, Viegas L, Solé N, Soares E, Dallimer M, Atkinson PW, Barlow J (2014) Can management improve the value of shade plantations for the endemic species of São Tomé Island? Biotropica 46(2):238–247

Félix PM, Chainho P, Lima GRF, Costa JL, Almeida AJ, Domingos I, Brito AC (2016) Mangrove fish of São Tomé Island (Gulf of Guinea): new occurrences and habitat usage. Mar Freshw Res 68:123–130. https://doi.org/10.1071/MF15392

Gillespie RG, Clague DA (2009) Encyclopedia of Islands. University of California Press, Berkeley/Los Angeles, 1110 pp

Herr D, Pidgeon E, Laffoley D (2012) Blue Carbon Policy Framework: based on the discussion of the International Blue Carbon Policy Working Group. Gland, Switzerland: IUCN and Arlington, USA: CI. vi+39pp

Lévêque C, Paugy D, Teugels GG (1990) The fresh and brackish water fishes of West Africa. Institut Françiais de recherche scientifíque pour le Developpement en cooperation. Paris, France

Nagelkerken I, Blaberb SJM, Boullon S, Green P, Haywood M, Kirton LG, Meynecke J-O, Pawlik J, Penrose HM, Sasekumar A, Somerfield PJ (2008) The habitat function of mangroves for terrestrial and marine fauna: a review. Aquat Bot 89(2):155–185

Nelson JS (ed) (2006) Fishes of the world, 4th edn. Wiley, Hoboken

Pisoni T, De Lima RF, Brito AC, Chainho P, Félix PM, Caçador I, Cavalho A (2015) Planos de gestão participativa para dois sítios de mangal na Ilha de S. Tomé: Praia das Conchas e Malanza. Relatório Final. Caracterização biofísica e socioeconómica. ALISEI, São Tomé. 79 pp

Quéro, J-C, Hureau J-C, Karrer C, Post A, Saldanha L (1990) Checklist of the fishes of the eastern tropical Atlantic (CLOFETA), vol 2. Lisbon (JNICT; UNESCO). 1492 pp

Rönnbäck P, Troell M, Kautsky N, Primavera JH (1999) Distribution pattern of shrimps and fish among *Avicennia* and *Rhizophora* microhabitats in the Pagbilao Mangroves, Philippines. Estuar Coast Shelf Sci 48(2):223–234

Rossignon O (1999) Contribution à l'ecologie des crevettes dulcaquicoles de São Tomé: du cadre limnologique à l'elevage. Faculté Universitaire des Sciences Agronomiques de Gembloux, France

Schneider W (1990) FAO species identification sheets for fishery purposes. Field guide to the commercial marine resources of the Gulf of Guinea. FAO, Rome

Spalding, M., Kainuma, M., Collins, N. 2010. World atlas of mangroves. Earthscan, London/Washington, DC. ISBN: 978-1-84407-657-4

Stiassny M, Teugels GG, Hopkins CD (2007) The fresh and brackish water fishes of lower Guinea, West-Central Africa. Vols I and II. IRD, Publications scientifiques du Muséum MRAC. Paris, France

UNEP (2016) GEO-6 regional assessment for Africa. United Nations Environment Programme, Nairobi, 215 pp

UNEP (2016a) Blue carbon financing of mangrove conservation in the Abidjan Convention Region: a feasibility study. United Nations Environment Programme, Abidjan Convention Secretariat and GRID-Arendal, Nairobi, Abidjan and Arendal. ISBN: 978-82-7701-163-9

UNEP and CIFOR (2014) Guiding principles for delivering coastal wetland carbon projects. United Nations Environment Programme, Nairobi, Kenya and Center for International Forestry Research, Bogor, p 57

Vasconcelos MJ, Cabral AI, Melo JB, Pearson TR, Pereira HdA, Cassamá V, Yudelman T (2015) Can blue carbon contribute to clean development in West-Africa? The case of Guinea-Bissau. Mitig Adapt Strateg Glob Chang 20(8):1361–1383. https://doi.org/10.1007/s11027-014-9551-x

Vasco-Rodriguez N, Fontes J, Andrade BA (2016) Ten new records of marine fishes for São Tomé, West Africa. Acta Ichthyol Piscat 46 (2):123–129

Wirtz P, Ferreira CEL, Floeter SR, Fricke R, Gasparini JL, Iwamoto T, Rocha LA, Sampaio CLS, Schliewen UK (2007) Coastal fishes of São Tomé and Príncipe islands, Gulf of Guinea (Eastern Atlantic Ocean) – an update. Zootaxa 1523:1–48

World Bank (2013) Turn down the heat: climate extremes, regional impacts, and the case for resilience, A report for the World Bank by the Potsdam Institute for Climate Impact Research and Climate Analytics. World Bank, Washington, DC. 252 pp. License: Creative Commons Attribution—Non Commercial–No Derivatives 3.0 Unported license (CC BY-NC-ND 3.0)

The Success of Hydrological Rehabilitation in Mangrove Wetlands Using Box Culverts Across Coastal Roads in Northern Yucatán (SE, México) 28

Claudia Teutli-Hernández and Jorge A. Herrera-Silveira

Abstract

The construction of roads, shelter ports and artificial mouths has altered the balance between freshwater and saltwater intakes, flow, tidal levels and water quality. This is reflected in the deterioration and death of the mangrove. Restoration, rehabilitation and reforestation methods have been carried out, but their success has not been determined, and sometimes perhaps do not meet the proposed objectives. For example, the opening of sewers as a method of restoration did not meet the objectives proposed in all localities, with Chuburnà, Progreso site east, Vía Antigua, Chicxulub, and Chabihau, being the most impacted. While San Benito, Dzemul, La Línea, and Telchac, had better results with mangroves being preserved. Chabihau, for maintaining a permanent connection with the sea, represents a real salinization. In the north, the concentration of inorganic nutrients in the water column is mainly phosphorus, by the removal of sediments during rain events, which facilitates the flow of nutrients between the sediment and the water column. The high concentrations of sediments suggest that these are basins of accumulation, and therefore, more susceptible to impacts. Among these impacts, organic pollution can be highlighted, resulting in the deterioration of flora and fauna. This condition implies that management measures, such as using box culverts across roads, should be carried out according to the characteristics of each locality.

Keywords

Mangrove restoration · Forest structure · Hydrology · Roads impacts

C. Teutli-Hernández (✉)
CINVESTAV-IPN, Unidad Merida, Merida, Mexico

University of Barcelona, Barcelona, Spain

J. A. Herrera-Silveira
CINVESTAV-IPN, Unidad Merida, Merida, Mexico
e-mail: jorge.herrera@cinvestav.mx

28.1 Introduction

Human activities in the coastal zone of tropical and subtropical regions are increasingly expanding through the construction of roads, tourism infrastructure, urban developments, ports and marine construction, ponds for aquaculture practices, and artificial hydraulic infrastructure for navigation that directly connect freshwater and brackish wetlands to coastal waters (Rivera-Monroy et al. 2006; Twilley and Rivera-Monroy 2005; Twilley et al. 1999). All of these activities are altering salinity and hydrological gradients and the water quality in estuaries and coastal wetlands. In addition, as result of hydroperiod changes, hypersalinity condition are more frequent in coastal regions where evaporation exceeds precipitation as result of salt accumulation (Rivera-Monroy et al. 2006; Castaneda-Moya et al. 2006). These perturbations are reflected on the degradation and increasing mortality of coastal wetlands in tropical regions where mangroves are the dominant vegetation (Giri et al. 2011).

In addition to high ecosystem productivity, mangrove forests are critical habitats for a variety of species, act as buffers against flooding, improve water quality, and dissipate storm energy reducing potential damage to coastal zones (Alongi 2011). Moreover, mangrove forests are identified as critical habitats due to their capacity to provide goods and services to the local and regional human populations (Costanza et al. 1997; Polidoro et al. 2010). Continued coastal development along the tropical and subtropical coastal regions requires the construction of roads, which in many cases have been built over extensive wetland areas, modifying the spatial and temporal hydrological patterns. Indeed, road construction has been identified as one of the

C. Makowski, C. W. Finkl (eds.), *Threats to Mangrove Forests*, Coastal Research Library 25,
https://doi.org/10.1007/978-3-319-73016-5_28

principal causes of mangrove mortality (Batllori-Sampedro and Febles-Patrón 1999; Valiela et al. 2001; Alongi 2008), and a threat to the ecological integrity of subtropical and tropical coastal-marine ecosystems.

Multiple efforts have been implemented to avoid, remedy, mitigate or reverse the impacts of hydrological modifications in mangrove forests throughout restoration or rehabilitation programs (Field 1996; Bosire et al. 2003). However, despite ambitious and expensive efforts, few reports are available describing actual outcomes (i.e., success, partial or total failure). An understanding of the appropriate environmental conditions of the area selected for rehabilitation and a clear definition of rehabilitation goals/objectives, are key components to increase the success of these programs, particularly since mangrove rehabilitation trajectories depend on: (a) the magnitude and type of disturbance; (b) the environmental setting; and (c) the geophysical and ecological processes regulating ecosystem structural and productive attributes (Twilley and Rivera-Monroy 2005). Successful mangrove rehabilitation programs require systematic monitoring of indicator variables such as pore-water salinity, changes in vegetation coverage, productivity and biological diversity, among other variables (Boesch and Paul 2001; Hyman and Leibowits 2001; Twilley and Rivera-Monroy 2005; Lewis 2009). Monitoring of these variables certainly help to define the success of rehabilitation actions, particularly if a set of structural and functional variables are analysed under an adaptive management scheme and a transdisciplinary approach (e.g., hydrology, ecology, engineering, sociology, economy) over a wide variety of spatial and temporal scales (Twilley 1998; Lewis III 2005).

Despite the multiple descriptions concerning the impact caused by modification of the levels, frequency and time of flooding of wetlands (particularly in mangrove ecosystems) (Cahoon et al. 1995; Long and Nestler 1996; Valentine-Rose and Layman 2011; Powell et al. 2011), in Mexico there is still a lack of basic ecological knowledge which permits propose and implement rehabilitation programs and specific actions in order to revert the impacts in different environmental settings in coastal wetlands.

The mangroves of the north coast of the Yucatan Peninsula (SE, México) have been impacted by changes in land use since 1948 and are related to the land use changes, coastal highways construction and railway lines. For example, the construction of seven harbours between the 1970s and 1980s (POETCY 2007), triggering widespread urbanization, promoting land use changes, and changes on the hydrological pattern in the mangroves of these locations. Changes in pore-water salinity and sediments lead to a process of sediment salinization causing a gradual loss of mangrove forest up to 60% of the original coverage (Euan-Avila 1998; Batllori-Sampedro and Febles-Patrón 1999; Zhao et al. 2016). Other land use changes affecting mangrove forest coverage in Yucatán are land reclamation for human settlements, suburban dumps (e.g. rubber), construction of large scale aquaculture infrastructure (semi-intensive shrimp ponds), mangrove tree cutting for fishing arts, and large scale climatic events such as storms and hurricanes (Euán-Avila and Witter 2002).

Land use change assessments in the Yucatan coastal landscape region between 1950 and 1990 indicated that roads and highway construction around urban centres (e.g., Mérida, Motul) facilitating access to beaches and ports was one of the major causes in the gradual increase in human population densities along the coast (Batllori-Sampedro and Febles-Patrón 1999; Euán-Avila and Witter 2002). For example, when these roads were constructed over wetlands the water surface flow was restricted and impacted the hydraulic connectivity between east and west sites of each road and creating two artificial micro-basins (one of each site of the road) and favouring extensive mangrove mortality (Fig. 28.1).

Due to the extension of the mangrove area impacted by the roads (>5000 ha), a rehabilitation program was implemented with the major goal of improving hydrological conditions. The main actions in this program included: (1) construction of box culverts to reconnect the surface water flow between both sides of the roads; and (2) the removal of sediments accumulated in water springs to increase fresh water discharge to reduce excess salt accumulation in affected mangrove areas (Perry and Berkeley 2009; Vyas and Sengupta 2012; Rovai et al. 2013).

However, this program did not consider that all actions of ecological restoration should be accompanied by a program that includes a diagnostic and verification of the degree of success of the actions undertaken (SER 2004). In other words, there is no control or possibility of integrating such actions through adaptive management in order to reach the objective of recovering the functions of the degraded ecosystem.

The objective of this chapter was to evaluate if constructed box culverts improved soil physicochemical properties in areas undergoing mangrove mortality along the Yucatan coastal region. Selected soil properties were considered as performance measures (*sensu* Twilley and Rivera-Monroy 2005) and measured 1 year after project implementation to assess potential changes in forest structural variables (e.g. natural regeneration) and representing rehabilitation trajectories based on "in situ" measurement of non-impacted mangrove forest adjacent to the study area. By selecting areas impacted by roads and undergoing potential hydrological changes as well as evaluating performance of project rehabilitation tasks we wanted to initiate a monitoring baseline to determine long term project success. We also aimed to establish an adaptative monitoring program that could be applied to other disturbed mangrove areas in the unique karstic geomorphic setting of the Yucatan Peninsula.

Fig. 28.1 Road that crosses the mangrove and the impact caused by the interruption of water flows

In the restoration program of the mangroves of Yucatan implemented by the local government the aim was to improve the hydrological conditions and thereby favour the natural regeneration of the mangroves (POETCY 2007). Therefore, in this study physical and chemical variable of the sediment were evaluated as signs of short term recovery (1 year) and structural characteristics of the mangroves were evaluated as a baseline for ecological conditions, identifying whether or not differences exist between both sides (east and west) of the roads along the north coast of Yucatan. The results of this are intended to contribute to the establishment of an ecological baseline for the rehabilitation of mangrove ecosystems in semiarid and karstic areas, as well as to initiate a monitoring program to determine whether or not the rehabilitation actions undertaken were successful in at least reducing the differences in the environmental variables of the water, sediments and structure of the mangroves between both sides of the roads of the mangrove flood zone in the north of Yucatán.

28.2 Materials and Methods

28.2.1 Study Area

The study area is located in the northern region of the state of Yucatán (SE, México). Average annual precipitation and temperature is <500 mm year^{-1} and the average temperature is 25.5 °C (García 1988). Climate warm and dry with the greatest rainfall occurring between July and September and scarce the rest of the year. Evaporation exceeds precipitation particularly March to June with an average value of 1460 mm year^{-1} (Dasilva SMD94 1999). The area target for rehabilitation is the Chuburnà-Chabihau basin which runs parallel to the coastline and is limited to the west by the Chuburnà-Sierra Papacal road and to the east by the Chabihau-Yobain road, approximately 170 km^2. The rehabilitation project was performed by NGO's, state and federal agencies (SECOL, SEMARNAT, SCT) and international institutions (NORTH AMERICAN WETLANDS COUNCIL). The Chuburnà-Chabihau basin is subdivided into micro-basins (Table 28.1) delimited by perpendicular roads to the coastline. Box culvert construction was performed in 1998–1999, with estimated water exchange volume of 10,000 m^3/day.

Due to the soil karstic origins, rainwater infiltrates quickly and freely to the subterranean aquifer, hence there is no surface water flow and the fresh water inputs to the coastal wetlands into the temporal swamp called locally "ciénagas" (swamps), are through point (spring) and non-point underground water discharges. Freshwater from these sources reduces salinity and primarily supplies inorganic dissolved nutrients mainly silicates and nitrates to coastal ecosystems (Herrera-Silveira and Ramírez-Ramírez 1997; Gonneea et al. 2014). A hydrological connection between wetlands and coastal waters in Chuburna, Yucalpeten, Telchac and Chabihau study sites is via man made harbours; whereas in

Table 28.1 Micro-basins on the Chuburna-Chabihau

Sampling sites	Latitude	Longitude
Chuburna- (CHB–W)	21° 14′ 16.8″	89° 49′ 42.0″
Chiburna-E (CHB–E)	21° 14′ 16.8″	89° 49′ 40.4″
Progreso–W (PRG–W)	21° 16′ 31.8″	89° 39′ 39.3″
Progreso-E (PRG–E)	21° 16′ 31.8″	89° 39′ 38.03″
Vía Antigua–W (VAT–W)	21° 16′ 57.3″	89° 38′ 10.5″
Vía Antigua– E(VAT–E)	21° 16′ 57.3″	89° 38′ 8.7″
Chixchulub–W (CHX–W)	21° 17′ 05.7″	89° 36′ 02.1″
Chicxulub–E (CHX–E)	21° 17′ 05.7″	89° 36′ 01.1″
San Benito–W (SBN–W)	21° 19′ 09.0″	89° 25′ 34.2″
San Benito-E (SBN–E)	21° 19′ 09.0″	89° 25′ 32.2″
Dzemul-W (DZE–W)	21° 19′ 34.0″	89° 20′ 56.1″
Dzemul-E (DZE–E)	21° 19′ 34.0″	89° 20′ 55.0″
La Línea–W (LLN–W)	21° 20′ 06.9″	89° 16′ 58.8″
La Línea–E (LLN–E)	21° 20′ 06.9″	89° 16′ 56.3″
Telchac– W (TLH–W)	21° 20′ 17.1″	89° 15′ 44.1″
Telchac– E (TLH–E)	21° 20′ 17.1″	89° 15′ 41.9″
Chabihau– W (CHH–W)	21° 21′ 15.7″	89° 13′ 33.7″
Chabihau– E (CHH–E)	21° 21′ 15.7″	89° 13′ 31.2″

the Carbonera zone area connectivity it is via a natural channel. As result of marine water inputs, microtidal regime tide (<0.7 m) and restricted water circulation in mangrove wetlands due to roads construction, there is a significant salt intrusion into the swamps. Salt accumulation as results of seawater exchange and high evaporation is reflected on high salinity vanes in pore-water and swamps water column (Herrera-Silveira 1996; Zaldivar 1999).

According to Euan-Avila and Witter (2002) and the Zoning Coastal Program of Yucatan (POETCY 2007), the roads and the artificial harbours to connect the sea and wetlands are the origin of the hydrological changes, and the consequences was the loss of 12,807 ha of mangroves (Valderrama-Landeros et al. 2017). Therefore, after the aerial photographs analysis and field trips, nine sampling sites (Table 28.1) were stablished on both sides (west-east) of the coastal roads which go from Chuburna to Chabihau (Fig. 28.2).

Monthly sampling trips were conducted between November 2000 and June 2001 at both sides of each road. Interstitial and water surface salinity were measured "in situ", as well as temperature (°C) and dissolved oxygen (mg l^{-1}) with a YSI-85 multiparametric probe. Surface water samples were collected and analyzed by dissolved inorganic nutrients as dissolved inorganic nitrogen (DIN), soluble reactive phosphorus (SRP) and silicates (Si) measured using standard spectrophotometric techniques described by Strickland and Parsons (1972).

In order to identify the effect of the roads on the mangrove community of north Yucatan, two vegetation sample sites were established in each location (east and west of each road). The vegetation structure of the mangrove forest was determined in two 100 m^2 plots in each road side, using the methods described by Lugo and Snedaker 1974. The structural attributes selected for this study were: diameter at breast (DAP); basal area (m^2 ha^{-1}); density (No. stems ha^{-1}); and height (m) (Shaefeer-Novelli and Cintrón 1990). The complexity index (dimensionless), which is the sum of the forest structural characteristics such as number of species, total density, total basal area and height, was calculated as a quantitative proxy descriptor of the tropical vegetation's structural complexity (Holdridge et al. 1971).

28.2.2 Data Analysis

Physical and chemical variables of the sediment taken from the study sites were analysed through a three-way nested ANOVA, taking into account location, site and the location*season interaction as a source of variation. In the cases where significant differences were observed, an "*a posteriori*" LSD (least significant difference) ($P < 0.05$) analysis was conducted. These analyses were performed using the statistics program JMP® 4.02 (SAS Institute 2000).

In order to determinate significant differences of water quality variables (temperature, dissolved oxygen, salinity and nutrients) between sites (west and east) of each sampling location, a non-parametric ANOVA was conducted using the Kruskal-Wallis significance test and represented with "box-whisker" plots using the STATGRAPHICS 1995 program.

Multidimensional non-metric scaling (MDS) was used to evaluate whether or not the rehabilitation actions had been successful. The ANOSIM test was applied to determine whether or not there were significant differences among sampling sites (Warwick and Clarke 1993), according to their vegetation structural characteristics. This test evaluates the significant differences between groups of replicates of the vegetation structural variables of the plots against a series of random simulations, resulting in a statistic (R). If this R statistic is close to 0 there are no significant differences between groups and increases to a maximum of 1 when the data of both matrices are totally different (Warwick and Clarke 1993).

28.3 Results and Discussion

28.3.1 Physicochemical Variables (Water and Sediments)

The results of the three-way nested ANOVA (Table 28.2) indicate that there are significant differences between both sides of the roads (east and west) of each sampling site with regards to the physical and chemical variables of the water. These suggest that rehabilitation actions 2 years after completing them had not achieved the hydrological rehabilitation.

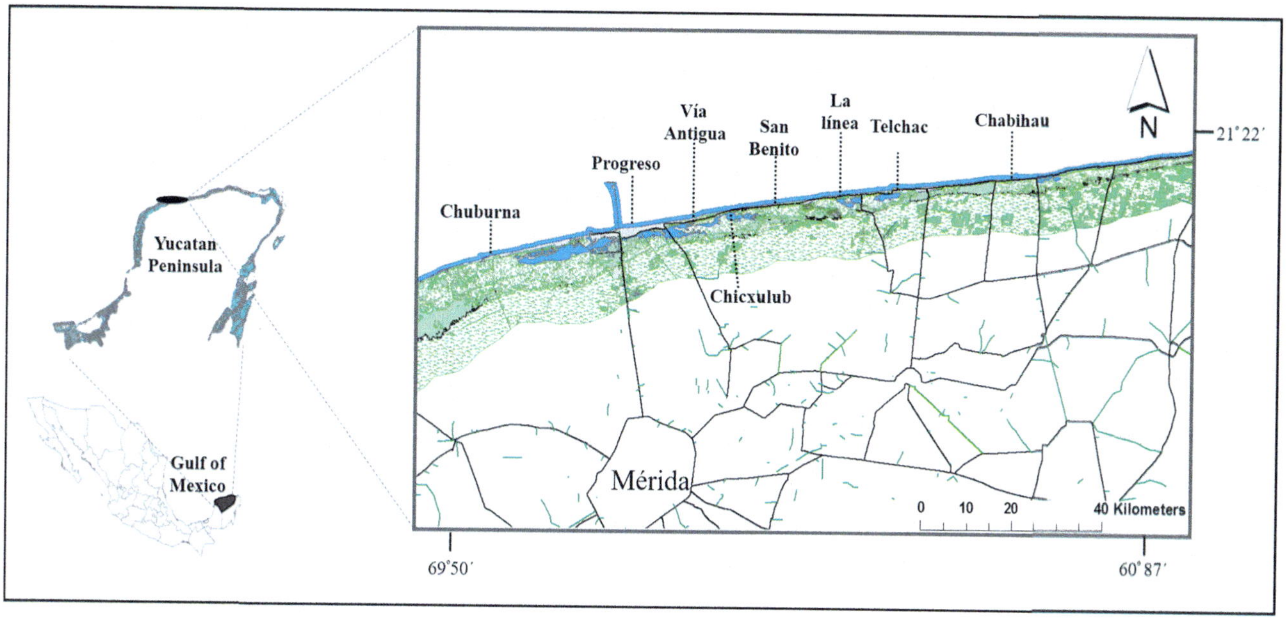

Fig. 28.2 Location of the nine sampling sites

Table 28.2 Chemical characteristics of water among micro-basins

Variables	Oxygen mg l^{-1}	Temperature °C	Salinity psu	DIN µmol l^{-1}	PO_4 µmol l^{-1}	SiO_4 µmol l^{-1}
Station (locations)	.8354	.9969	.4750	.2492	.7719	.9623
Locations	1.72	*.0001	*.0001	*.0018	*.0257	*.0175
Season	.1259	.0604	*.0001	.9700	*.0001	*.0001
Season *locations	.1035	*.0001	*.0001	.5845	.2234	.0278

Similarly, were observed significant differences between locations ($P < 0.05$), indicating variability of the physical and chemical characteristics of water among micro-basins, suggesting that rehabilitation actions should be site-specific (Bashan et al. 2013).

The surface temperature of the water showed variations with an average of 20 °C (Telchac-W) and 30.4 °C (Chuburna-W) (Fig. 28.3a). Higher temperatures (26.9–30.4 °C) were observed toward the west of the study area (Chicxulub W and E and Chuburná-W Chuburna-Chicxulub), while lower temperatures (20 °C in Telchac-W and 24.75 °C in Dzemul-W) were registered between San Benito and Chabihau. The lower temperatures could be associated with a greater influence of groundwater discharges since these freshwater sources have been characterized by low temperatures as well as high concentrations of nitrates and silicates (Herrera-Silveira 1994). In contrast, the zone with higher temperatures probably receives lower groundwater supplies and the residence time of the water is greater, favouring the warming and evaporation of the water column in the swamp. High temperatures have negative effects on the metabolic reactions of the seedlings, limiting the recruitment of new individuals and the growth of the trees (Harry and Nyle 1982). The dissolved oxygen concentrations varied from 0.4 mg l^{-1} (San Benito) to 7.15 mg l^{-1} (Vía Antigua) (Fig. 28.3b). In San Benito and Telchac low values were consistently registered through sampling period. Even though the dissolved oxygen concentrations did not present significant differences ($P < 0.05$) between sites of the roads (west, east) at each location, the differences were observed among locations. Water oxygen balance is mainly related with production and decomposition process in wetlands and the last one is dominant (Reddy and Delaune 2008), so that, in places with high production and decomposition of the leaf litter could stimulate the oxygen consumption in the water column.

In mangroves ecosystems has been demonstrated a direct relationship between the vegetation structure and the productivity, this last measured as litterfall (Wang'Ondu et al. 2014). In San Benito and Telchac structural variables as basal areas and heights show high values (Table 28.3) suggesting that the organic matter in the sediments are high, then this could explain the low dissolved oxygen registered in these sampling sites (Fig. 28.3b).

Surface water salinity showed spatial variations with lowest concentrations in Telchac (1.45 g/kg) and highest in Chabihau (50.6 g/kg). While there were no significant differences in surface water salinity between both sites of the roads, greater seasonal variations were observed among

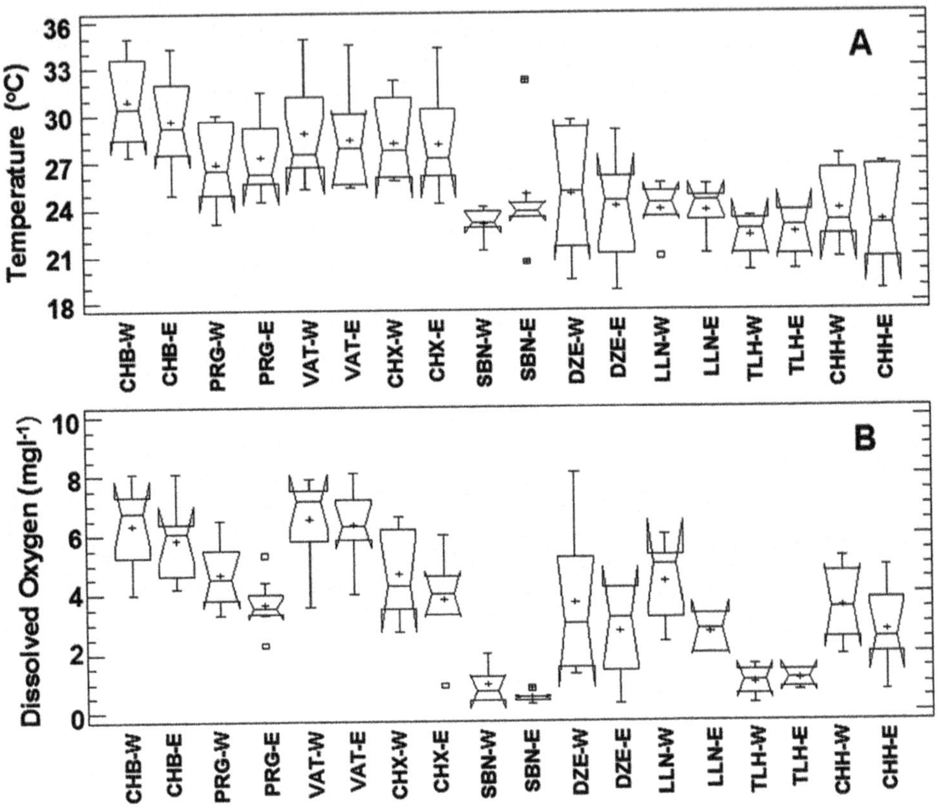

Fig. 28.3 Temperature and dissolved oxygen

locations (Fig. 28.4a), with the greatest variation in Chicxulub (<5 a>80 g/kg) in contrast with Telchac (5–10 g/kg). The hyperhaline conditions in the micro-basins of Chicxulub and Chabihau can be explained by the effect of roads which limit the ebb and flow of surface water, favouring a greater water residence time, evaporation and salt concentration. On the other hand, micro-basins as the Chuburná-Progreso has direct connection (natural or artificial) with the sea, and together with the arid climate favour soil salinization, whereas the sites that show lower surface and pore-water salinity such as San Benito and Telchac (FIG) could be influenced by fresh groundwater inputs and surface water floods.

Spatial differences in the water characteristics between the sub-basins of Chuburna-Chicxulub and San Benito-Telchac, probably are related with another important feature of the karstic system found in Yucatan, known as the "ring of cenotes". This is characterized by large discharges of freshwater on land and at sea, mainly in areas where the "ring of cenotes" meets the coastal zone (Pacheco Martinez and Alonzo Salomon 2003), that in the case of the study area is in the sub-basin San Benito-Telchac, explaining the water characteristics of these. Progreso (west site), San Benito, Dzemul, La Línea and Telchac showed mangrove trees of greater basal areas and heights values (Table 28.3). The hydrochemical characteristics in San Benito, Dzemul, La Línea and Telchac suggest that these locations are influenced by freshwater discharges supplying nutrients and regulating their interstitial salinity (<13 g/kg) (Fig. 28.4b).

The dissolved inorganic nitrogen (DIN = $NO_3 + NO_2 + NH_4$) showed average concentrations less than 10 μmol l^{-1} and no significant differences were registered between sites of roads in each location (Fig. 28.5). Nevertheless, high average concentrations (>20 μmol l^{-1}) were registered in Progreso-W, Via Antigua-E, and in both sites of the road in Chabihau. In the case of Progreso and Vía Antigua, high DIN concentrations might be related with the influence of the leachates from the municipal rubbish dump located in the swamp and influenced by flood during high tides and rainy season. However, the DIN concentrations registered in Chabihau could be explained through decomposition and remineralization processes associated to greater water residence time, favouring the increasing concentrations of reduced forms of nitrogen (Teutli 2003).

The average phosphate concentration was generally low (0.60 μmol l^{-1} and 1.22 μmol l^{-1}) (Fig. 28.6a), and because it is present in calcareous substrates soil fertility is low (Koch and Snedaker 1997). As a result of the karstic characteristics and high water alkalinity of the groundwater in Yucatan region, the supplies of phosphorus via groundwater discharges are low, since the calcium carbonate act as a phosphorus sink through geochemical processes of adsorption by carbonates to form apatite (Carpenter et al. 1998). Nevertheless, high concentrations can be seen in

Table 28.3 Structural characteristics of the mangrove

Locality	Basal area (m^2 ha^{-1})	Density (Ind. ha^{-1})	Height (m)	No. Sps	Species	Complexity index	Type of mangrove
CHB-W	0.75	187	2.83	1	*A. germinans*	0.0039	Basin
CHB-E	225	620	1.90	3	*A. germinans, C. erectus, L. racemosa*	0.0796	Basin
PRG-W	18.89	8182	3.35	3	*R. mangl, A. germinans, L. racemosa*	15.517	Fringe
PRG-E	0.09	100	1.00	1	*R. mangle*	0.0001	Fringe
VAT -W	4.17	3286	2.42	3	*A. germinans, R. mangle, L. racemosa*	0.996	Basin
VAT -E	0.12	102	1.73	3	*A. germinans, R. mangle, L. racemosa*	0.0006	Basin
CHX -W	0.14	77	1.28	2	*A. germinans, L. racemosa*	0.0003	Basin
CHX -E	0.12	74	1.14	1	*A. germinans*	0.0001	Basin
SBN-W	2.46	2222	3.28	2	*R. mangle, A. germinans*	0.3591	Basin
SBN-E	23.00	3333	3.83	1	*R. mangle*	2.9377	Basin
DZE-W	22.75	8663	4.20	1	*R. mangle*	8.275	Basin
DZE-E	24.10	5500	5.39	3	*R. mangle, L. racemosa, A. germinans*	21.427	Basin
LLN -W	25.23	3100	6.14	2	*R. mangle, L. racemosa*	9.517	Basin
LLN -E	20.84	4300	4.47	2	*R. mangle, A. germinans*	8.014	Basin
TLH -W	12.76	9600	4.36	2	*L. racemosa, A. germinans*	10.670	Fringe
TLH -E	11.19	4067	4.85	1	*A. germinans*	2.2065	Fringe
CHH -W	0.40	360	1.89	4	*R. mangle, L. racemosa, A. germinans, C. erectus*	0.0082	Basin
CB - E CHH -E	3.83	2400	3.16	4	*R. mangle, L. racemosa, A. germinans, C. erectus*	0.5806	Basin
*	17 ± 3.7	1760 ± 350	41.3 ± 8.8	3 ± 4		36.4 ± 10.3	Riverine
	9.0 ± 0.7	3580 ± 394	18.5 ± 1.6	2.3 ± 0.2		15.4 ± 3.2	Cuenca
	8.2 ± 1.1	5930 ± 3005	17.9 ± 2.7	2.0 ± 0.3		145.6 ± 4.7	Fringe
	1.0	25,030	0.6	1		1.5	Dwarf

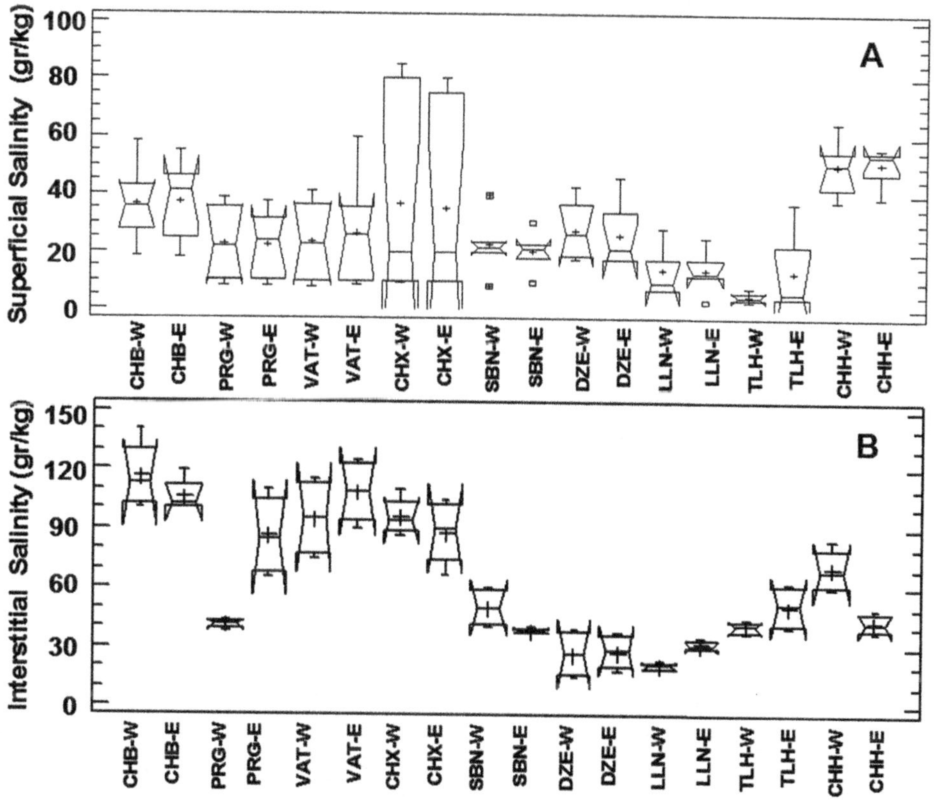

Fig. 28.4 Superficial and interstitial salinity

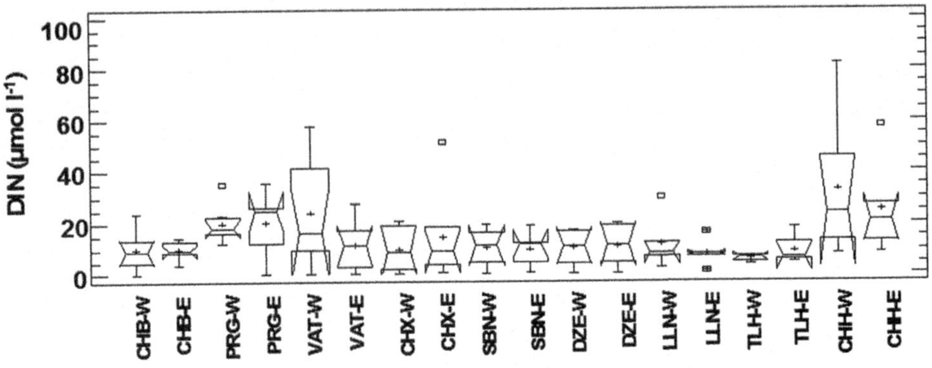

Fig. 28.5 Dissolved inorganic nitrogen

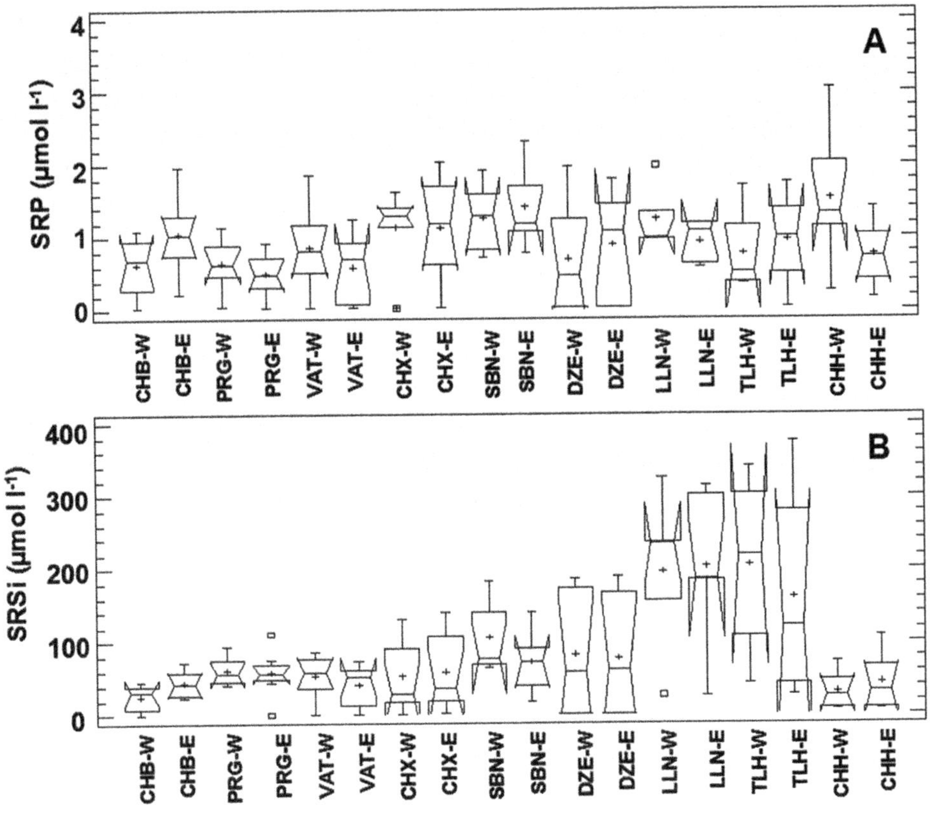

Fig. 28.6 Soluble reactive Phosphorous (SRP) and soluble reactive silicates (SRSi)

Chuburná-E, Vía Antigua-W, San Benito-E and Chabihau-W (Fig. 28.6), which might be associated in the case of Vía Antigua to the leachates of the municipal rubbish dump of the Puerto Progreso. The differences in the phosphorus concentrations between locations are probably, on the one hand, associated with the presence of abundant birdlife, with higher concentrations where they are present, since this zone represents an extensive wetland used at least seasonally by migratory birds (SECOL 2004). The fertilization role of birds to mangrove has been registered in a close area of the study sites (Adame et al. 2015). On the other hand, they are probably associated with inputs from the sea in the locations, which have an artificial connection. Nutrient inputs via bioturbation and faeces from aquatic birds are considered to be autochthonous and haloctonous respectively, contributing to increasing the aquatic productivity of the aquatic ecosystem (Comín and Herrera-Silveira 2000). However, it has also been registered that the haloctonous inputs of phosphorus from sea connections to mangroves zones can proceed from the coastal waters, as occurs in the west zone of Florida (Chen and Twilley 1998).

Silicate concentrations showed significant differences ($p < 0.05$) between locations, with average concentrations varying between 20.8 μmol l^{-1} in Telchac-W and 105 μmol l^{-1} in La Línea-E (Fig. 28.6b). Spatially, a zone of low concentrations and lower variability is observed between

Chuburná and Chicxulub, while between San Benito and Telchac greater value concentrations and greater variability were observed. This indicates that the latter locations are influenced by groundwater inputs via springs, since it has been determined that this nutrient shows high concentrations in groundwater discharges (Herrera-Silveira 1994).

According to the results of the water and sediment variables, the aims of favouring the recovery of the impacted zones of mangroves were not achieved in 2 years. High spatial variability of the hydrology in the study area is maintained and no significant changes in surface and pore-water salinity are observed 2 years after the opening of drains in the coastal roads of Yucatán.

28.4 Mangrove Forest Characteristics

The results of the structural characteristics of the mangroves in the study area reflect high variability (Table 28.3), indicating differences in the environmental variables among locations, since it is recognized that the development of mangrove vegetation is a reflection of the magnitude of local hydrological stressors and resources available in the sediments (Twilley 1995; Twilley and Rivera-Monroy 2005).

Basal area is an indicator of the development achieved by the mangrove community with regards to forest biomass; this characteristic varied between 0.09 m^{-2} ha^{-1} in Progreso-E and 25.23 m^{-2} ha^{-1} in La Línea-W. On the other hand, the density of trees in a mangrove forest is a reflection of its state of development; therefore, it acts as an indicator of vulnerability and response capacity in face of anthropogenic and natural impacts (Shaeffer-Novelli and Cintrón 1990). The density of mangroves in the area of study varied between 74 and 9600 trees ha^{-1}, indicating that there are locations with a low level of impact (Dzemul-E), and others with a high impact level (Progreso-E, Chicxulub-E). The average height of trees varied between 1 m (Progreso- E) and >6 m (La Línea-W). It is recognized that the variability in the tree height responds to hydrological characteristics such as salinity; therefore, at lower salinities values of structural characteristics such as height are high due to the freshwater has been related with nutrient inputs (Twilley 1995). On the other hand, the frequency of natural events such as hurricanes have an impact on the average height of the mangrove community, since these have a "pruning" effect, and the magnitude of the impact is related to the degree of exposure that the vegetation has to the path of the maximum winds (Lugo et al. 1976; Cintron et al. 1978; Twilley 1998). Given that the study area is exposed to hurricanes trajectories, this could limit its average height (<5 m). The differences in height between locations may be associated with the high interstitial salinities as in Chuburná-E, Progreso-E, Via Antigua-E and Chicxulub in both sides (Fig. 28.3b) that showed the lowest average heights values (Table 28.3).

The four species of mangrove present in the region of the Gulf of Mexico were registered in the study area: *Rhizophora mangle, Avicennia germinans, Laguncularia racemosa, Conocarpus erectus*. With regards their distribution, there were locations with the four species (Chabihau), and those of monospecific forests (Chuburna-W, Chicxulub-E) (Table 28.3). The abundance and dominance of a mangrove species has been used as an indicator of the specificity of hydrological and sediment variables as salnity, flood, an nutrients among other, acknowledging that *Avicennia germinans* dominates in sites of high salinities, less waterlogged and highly reduced soils (López-Portillo and Ezcurra 1989; Mckee 1995); while *Laguncularia racemosa* is associated with zones of low salinities and relatively greater concentrations of nutrients (Mckee 1995), and *Rhizophora mangle* establishes mainly in sites directly exposed to the flood and tide, where the sediment is less consolidated, salinities are variable (5–60 g/kg) and concentrations of nutrients are low (Jiménez and Lugo 2000). This coincides response of *R. mangle* with what Twilley and Chen (1998) described, in that the distribution of the species of mangrove is related to the interaction of hydrological and sediment variables, which in the case of the north coast of Yucatán are highly heterogeneous (Table 28.3).

In accordance with the complexity index (IC) of Holdridge et al. (1971), which is an integrative measure of structural development of the mangrove forest, high heterogeneity was observed in the study area (Table 28.3). The IC values of the locations between San Benito and Telchac may represent the structure of the typical mangroves for Yucatán, since their characteristics are similar to other sites of the Yucatán Península that have been defined as being in a good state of conservation (Zaldivar et al. 2004). The structural characteristics of the mangroves of Chuburná, Progreso-E, Vía Antigua, Chicxulub and Chabihau are below the values reported in the literature typifying mangrove forests (Table 28.3). These suggest that these sites reflect severe impacts where the hydrological modifications by roads have put them in poor conditions with a lower capacity of resilience in face of natural events such as hurricanes.

The surface and pore water results suggest that these locations are impacted by a process of sediments salinization induced by the seawater inputs to the wetland through the ports and artificial openings that connect both systems (Euan-Avila and Witter 2002). High pore-water salinities (>50 g/kg) in mangrove ecosystems is an stressor for the forest structure of these vegetation being reflected in a low complexity index (Cintrón et al. 1978; Twilley 1995).

The locations that showed high complexity index values, which may be associated with a good conservation status as Progreso (west site), Dzemul, La Línea and Telchac, these could be used as mangrove reference sites (Twilley and Rivera-Monroy 2005). These locations were characterized by a flooding pattern similar to tides with a delay time and subterranean water inputs via springs, which additionally represent a nutrient input (Herrera-Silveira 1996). In particular, the locations of San Benito and Telchac may characterize typical mangroves structure of the central-northern area of Yucatan, since the karstic conditions, dry climate and hurricanes frequency are environmental condition at which these mangroves has been growing. They show a maximum height of 5 m (Telchac-E) and a minimum basal area of 2.46 m^2 ha^{-1} (San Benito-W). These locations exhibit structural characteristics similar to the mangrove forests of Quintana Roo, in which an average height of 3.5 m was observed and densities were between 1520 and 25,000 trees ha^{-1}, indicating that the structure of this community is strongly controlled by the geohydrology and climate patterns including tropical storms (Lara-Dominguez et al. 2005).

As can be seen, the location of Progreso is quite unique, since the characteristics of the mangroves on both sides of the road differ significantly, making the impact by obstruction of water flows to evident (Table 28.3). The Progreso-E site showed very poor condition, where the interruption of water flow has favoured the accumulation and stagnation of sediments (Batllori-Sampedro and Febles-patrón 1999), thereby preventing mangrove seeds from establishing and developing. In contrast, the west site is in a good state of conservation (Table 28.3 and Fig. 28.6b) as a result of the free ebb-flow of the water by the action of the tide.

The differences in the characteristics of the mangrove vegetation between sites of each location suggest that in some cases the roads have affected the hydrological characteristics and have impacted the mangrove sites (Progreso, Vía antigua, San Benito and Chabihau). However, the locations of Chicxulub, Dzemul, La Línea and Telchac did not present differences in the structural characteristics of the mangrove vegetation on both sides of the respective roads, suggesting that local conditions mainly related with groundwater discharges and water flow form the swam to mangroves (Adame et al. 2013) could favour the type of vegetation observed.

According to the results of mangrove structure, the aims of rehabilitation program to favour the recovery of these areas impacted by hydrological changes were not achieved after 2 years. High variability was observed in mangrove structural characteristics between both sides of the roads. These suggest that the restoration activities should be specific according to the environmental characteristics of each mangrove zone. In locations such as Chuburná, Progreso-W, Vía Antigua and Chicxulub which are badly damaged the actions of hydrological rehabilitation should be applied, while in the sites such as Dzemul, La Línea and Telchac which showed signs of being in good condition, no specific rehabilitation action are recommended conservation actions should be the priority.

The results of the nonparametric multidimensional scaling analysis (MDS) using the physical and chemical variables of the water and sediments in addition to the structural characteristics of the mangrove forests, and separating "a priori" the sites that present characteristics of better structural development of the mangrove vegetation (Fig. 28.7 and Table 28.3), indicate that both sets of variables effectively represent different environmental and ecological conditions. Figure 28.7a and the ANOSIM test showed that even though according to the physical and chemical characteristics of the water there are sites in favourable conditions for the mangroves (San Benito-W), their structure vegetation characteristics are poor (Table 28.3). The water characteristics favour the ordination of sites less significant (global $R = 0.577$; $P < 0.01$) than the mangroves forest variables (global $R = 0.839$; $P < 0.001$). These differences suggest that in addition to the physical and chemical variables of the water, other components of the ecosystem such as the level and frequency of flooding, and the fertility of the sediments could play an important role in mangrove condition (Rivera-Monroy et al. 2006). This suggests that the restoration actions should not be directed solely at re-establishing the quality of the surface water and the surface flow of water, but should also take into account characteristics such as the variables of the sediment, hydroperiod, and seasonal changes in the flooding levels among others ecosystem variables. Reforestation should be considered as the last option if after all the other actions the natural regeneration has not been favoured (Lewis III 2005).

Documenting the success or failure of the rehabilitation actions in areas of mangroves through monitoring different types of actions, contributes to generating scientific bases and a conceptual framework on the ecology of restoration of these ecosystems. On the other hand, it helps to define improved rehabilitation actions at a regional scale, the specific plan of action according to the causes of deterioration of each site in particular, and the ecological characteristics of each location, understanding that the process of ecological rehabilitation is complex and dynamic, therefore monitoring is a tool for the definition of actions under the context of adaptive management.

Fig. 28.7 Multivariate analysis

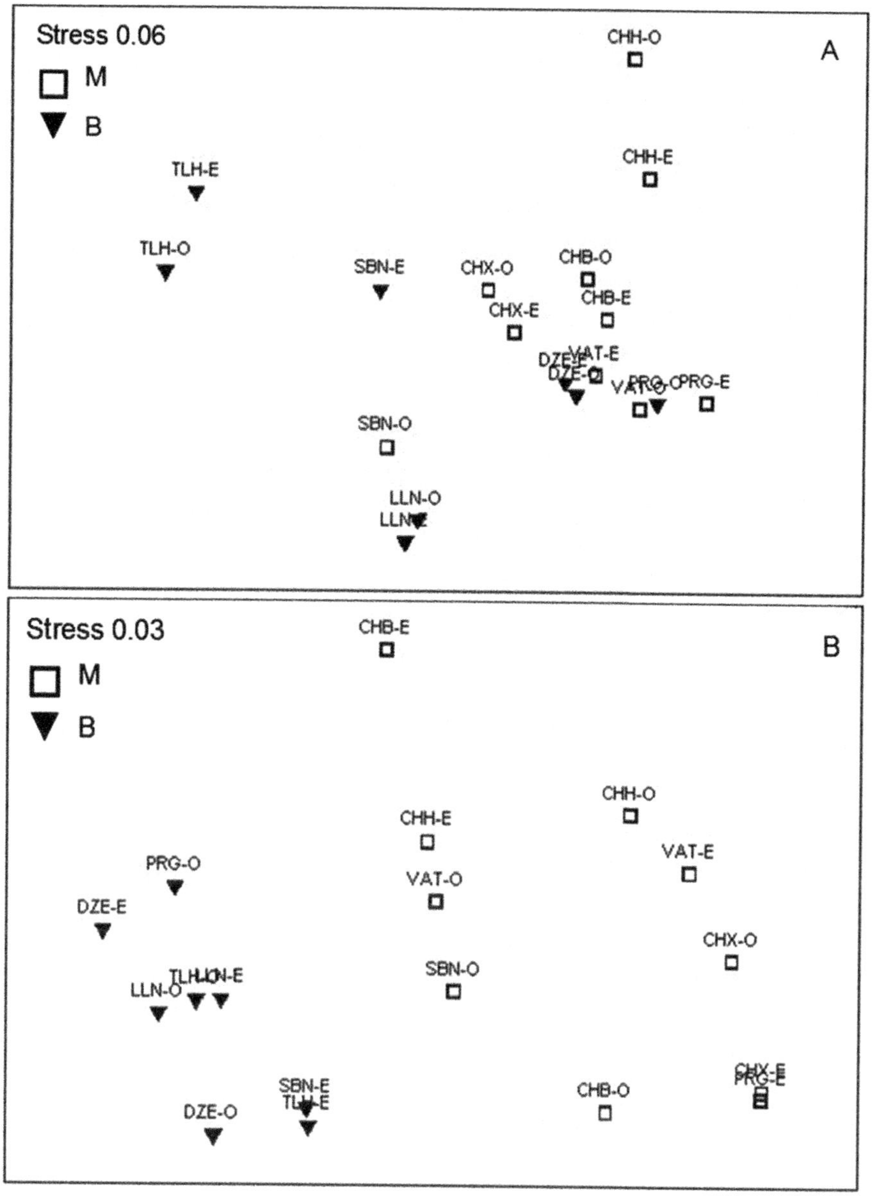

References

Adame MF, Zaldivar-Jimenez A, Teutli C, Caamal JP, Andueza MT, López-Adame A, Cano R, Hernández-Arana H, Torres-Lara R, Herrera-Silveira JA (2013) Drivers of mangrove litterfall within a karstic region affected by frequent hurricanes. Biotropica 45(2):147–154

Adame MF, Hermoso V, Perhans K, Lovelock CE, Herrera-Silveira JA (2015) Selecting cost-effective areas for restoration of ecosystem services. Conserv Biol 29(2):493–502

Alongi DM (2008) Mangrove forests: resilience, protection from tsunamis, and responses to global climate change. Estuar Coast Shelf Sci 76(1):1–13

Alongi DM (2011) Carbon payments for mangrove conservation: ecosystem constraints and uncertainties of sequestration potential. Environ Sci Policy 14(4):462–470

Bashan Y, Moreno M, Salazar B, Alvarez L (2013) Restoration and recovery of hurricane-damaged mangroves using the knickpoint retreat effect and tides as dredging tools. J Environ Manage 116 (15):196–203

Batllori-Sampedro E, Febles-Patrón JL (1999) Landscape change in Yucatan's Norwest Coastal Wetland (1948-1991). Hum Ecol 16(1):8–20

Boesch DF, Paul JF (2001) An overview of coastal environmental health indicators. Hum Ecol Risk Assess Int J 7(5):1409–1417

Bosire JO, Dahdouh-Guebas F, Kairo JG, Koedam (2003) Colonization of non-planted mangrove species into restored mangrove stands in GAzi Bay, Kenya. Aquat Bot 76(4):267–279

Cahoon DR, Reed DJ, Day JW (1995) Estimating shallow subsidence in microtidal salt marshes of the southeastern Unites States: Kaye and Barghoorn revisited. Mar Geol 128:1–9

Carpenter SR, Cole JJ, Essington TE, Hodgson JR, Houser JN, Kitchell JF, Pace ML (1998) Evaluating alternative explanations in ecosystem experiments. Ecosystems 1(4):335–344

Castañeda-Moya E, Rivera-Monroy VH, Twilley RR (2006) Mangrove zonation in the dry life zone of the Gulf of Fonseca, Honduras. Estuar Coasts 29(5):751–764

Chen R, Twilley RR (1998) A gap dynamic model of mangrove forest development along gradients of soil salinity and nutrients resource. J Ecol 86:37–51

Cintrón G, Lugo AE, Pool DJ, Moris G (1978) Mangrove of arid evironments in Puerto Rico and adyacent islands. Biotropica 10 (2):110–121

Comín FA, Herrera-Silveira JA (2000) The role of birds on the trophic structure and nutrient cycles of aquatic ecosystems: a review. In: Comin FA, Herrera-Silveira JA, Ramírez J (eds) Proceedings Limnology and Waterfowl, Monitoring, Modeling and Management, Workshop, Aquatic Birds Working Group. Societas Internationalis Limnologiae, Universidad Autónoma de Yucatán, pp 205–218

Costanza R, Darge R, de Groot R, Farber S, Grasso M, Hannon B, Limburg K, Naeem S, Oneill RV, Paruelo J, Raskin RG, Sutton P, van den Belt M (1997) The value of the world's ecosystem services and natural capital. Nature 387:253–260

DASILVA SMD 94 (1999) IRI/LDEO climate Data Library Magellan star Site. Lamont – Doherty Earth Observatory of Columbia University

Euán-Ávila JI (1998) Análisis de Detección de Cambio del Paisaje: Una Herramienta para el Manejo de las Costa Euan-Avila JI, Witter SG (2002) Promoting integrated coastal management in the Yucatan Peninsula. Mex J Policy Stud 12:1–15

Euan-Avila JI, Witter SG (2002) Promoting integrated coastal management in the Yucatan Peninsula. Mex J Policy Stud 12:1–15

Field C (1996) Restoration of mangrove ecosystems. Internacional tropical Timber Organization and Internacional Society for Mangrove Ecosystems, Okinawa

García E (1988) Modificaciones al sistema de clasificación de Koppen 4ª edición. México, D.F

Giri C, Ochieng E, Tieszen LL, Zhu Z, Singh A, Loveland T, Masek J, Duke N (2011) Status and distribution of mangrove forests of the world using earth observation satellite data. Glob Ecol Biogeogr 20:154–159

Gonneea ME, Charette MA, Liu Q, Herrera-Silveira JA, Morales-Ojeda SM (2014) Trace element geochemistry of groundwater in a karst subterranean estuary (Yucatan Peninsula, Mexico). Geochim Cosmochim Acta 132:31–49

Harry OB, Nyles CB (1982) Naturaleza y propiedades de los suelos. Editorial UTEHA, México, 388p

Herrera-Silveira JA (1994) Spatial heterogeneity and seasonal patterns in a tropical coastal lagoon. J Coast Res 10(3):738–746

Herrera-Silveira JA, Ramírez-Ramirez J (1997) Salinity and nutrients in the coastal lagoons of Yucatan. Mex Verh Internat Verein Limnol 23:1348–1351

Herrera-Silverira JA (1996) Salinity and nutrients in a tropical coastal lagoon influenced by groundwater discharges. Hydrobiologia 321:165–176

Holdridge LR, Grenke WC, Hathaway WH, Liang T, Tosi J (1971) Forest environments in tropical life zones, a pilot study. Pergamon Press, New York

Hyman JB, Leibowitz SG (2001) JSEM: a framework for identifying and evaluating indicators. Environ Monit Assess 66(3):207–232

Jiménez JA, Lugo AE (2000) Rhizophora mangle L. Red mangrove. SO-ITF-SM-2. New Orleans, LA, US, Department of Agriculture, Forest SErvice, Southern Forest Experiment Station, 450–455

Koch MS, Snedaker SC (1997) Factors influencing Rhizophora mangle L. seedling development in everglades carbonate soils. Aquat Bot 59:87–98

Lara-Dominguez A, Day J, Villalobos G, Twilley RR, Alvarez G, Yañez-Arancibia (2005) Structure of unique inland mangrove forest assemblage in fósil lagoons on the Caribbean Coast of Mexico. Wetl Ecol Manag 13:111–122

Lewis RR (2005) Ecological engineering for successful management and restoration of mangrove forests. Ecol Eng 24(4):403–418

Lewis RR III (2009) Methods and criteria for successful mangrove forest restoration. Coastal Wetlands: an integrated ecosystem approach. Elsevier, Amsterdam, pp 787–800

Long KS, Nestler JM (1996) Hydroperiod changes as clues to impacts on cache river riparian wetlands. Wetlands 16:379–396

López-Portillo J, Ezcurra E (1989) Litter fall of Avicenncia germinans L. in a one-year cycle in a mudflat at the lagun Mecoacan,Tabasco, Mexico. Biotropica 17(3):186–190

Lugo AE, Sell MA, Snedaker SC (1976) Mangrove ecosystem analysis. In: Systems analysis and simulation in ecology, vol 4. Academic, New York, pp 114–145

Mckee KL (1995) Seedling recruitment pattern in a Belizean mangrove forest: effects of establishment ability and physico-chemical factors. Oecologia 101:448–460

Pacheco Martinez JL, Alonzo Salomon LA (2003) Characterization of karst material of Carrillo Puerto formation in Yucatan Autonomous University of Yucatan, Mérida, México. Acad Mag Eng 7:7–19

Perry C, Berkely A (2009) Intertidal substrate modifications as a result of mangrove planting: impacts of introduced mangrove species on sediment microfacies characteristics. Estuar Coast Shelf Scence 81:225–237

Polidoro BA, Carpenter KE, Collins L, Duke NC, Ellison AM, Ellison JC, Livingstone SR (2010) The loss of species: mangrove extinction risk and geographic areas of global concern. PloS One 5(4)

Powell N, Osbeck M, Tan SB, Toan VC (2011) Mangrove restoration and rehabilitation for climate change adaptation in Vietnam. World Resources Report Case Study. World Resources Report, Washington, DC. URL: http://www.worldresourcesreport.org

Programa de ordenamiento ecológico del Territorio Costero del Estado de Yucatán (POETCY) (2007) Informe Final. Portal de la Secretaría de Ecología del Estado de Yucatán

Reddy KR, DeLaune RD (2008) Biogeochemistry of wetlands: science and applications. CRC Press, Boca Raton

Rivera-Monroy VH, Twilley RR, Mancera E, Alcantara-Eguren A, Castañeda-Moya E, Casas O, Reyes P, Restrepo J, Perdomo L, Campos E, Cotes G, Viloria E (2006) Aventuras y Desventuras en Macondo: Rehabilitación de la Ciénaga Grande de Santa Marta, Colombia. Ecotropicos 19:72–93

Rovai AS, Barufi JB, Pagliosa PR, Scherner F, Torres MA, Horta PA, Soriano-Sierra EJ (2013) Photosynthetic performance of restored and natural mangroves under different environmental constraints. Environ Pollut 181:233–241

SAS Institute I (2000) Statistical Discovery Software,JMP V4.02. SAS Institute Cary, North Carolina

Schaeffer-Novelli Y, Cintron G (1990) Métodos para la descripción y estudio de áreas de manglar. Ingeniería marítima, ciencias del mar. Escuela superior politécnica del litoral. Puerto Rico

SECOL (2004) Gobierno del Estado de Yucatán. Reporte técnico de aprovechamiento y monitoreo de aves acuáticas de interés cinegético del estado de Yucatán, 68p

SER (Society for Ecological Restoration International Science & Policy Working Group) (2004) The SER International Primer on Ecological Restoration. Society for Ecological Restoration International, Tucson, Arizona. Available from www.ser.org. Accessed Apr 2005

Strickland JD, Parsons TR (1972) A practical handbook of sea water analysis. Fish Res Board Can Bull 122:311

Teutli-Hernández C (2003) Estimación del éxito de la restauración hidrologica de zonas de manglar en el norte de Yucatán, México. Tesis de licenciatura. Licenciatura en Biología. Benemérita Universidad Autónoma de Puebla, 51pp

Twilley RR (1995) Properties of mangrove ecosystems relate to the energy signature of coastal environments. In: Hall C (ed) Maximun Power: the ideas and applications of H.T. University of Colorado, Colorado, pp 43–22

Twilley RR (1998) Mangrove wetlands. In: Messina MG, Conner WH (eds) Southern forested wetland, ecology and management. Lewis Publishers, Boca Raton, pp 445–473

Twilley RR, Chen R (1998) A wáter Budget and hydrology modelo f a basin mangrove forest in Rookery Bay florida. Mar Fresw Res 49(4):309–323

Twilley RR, Rivera-Monroy VH (2005) Developing performance measures of mangrove wetlands using simulation models of hydrology,nutrient biogeochemistry, and community dynamics. J Coast Res 40:79–93

Twilley RR, Rivera-Monroy VH, Chen R, Botero L (1999) Adapting an ecological mangrove model to simulate trajectories in restoration ecology. Mar Pollut Bull 37(8):404–419

Valderrama-Landeros LH, Rodríguez-Zúñiga MT, Troche-Souza C, Velázquez-Salazar S, Villeda-Chávez E, Alcántara-Maya JA, Vázquez-Balderas B, Cruz-López MI, Ressl R (2017) Manglares de México: actualización y exploración de los datos del sistema dc monitoreo 1970/1980–2015. Comisión Nacional para el Conocimiento y Uso de la Biodiversidad. Ciudad de México, 128pp

Valentine-Rose L, Layman CA (2011) Response of fish assemblage structure and function following restoration of two samll bahamian todal creeks. Restor Ecol 19(2):205–215

Valiela I, Bowen JL, York JK (2001) Mangrove Forests: one of the world's threatened major tropical environments: at least 35% of the area of mangrove forests has been lost in the past two decades, losses that exceed those for tropical rain forests and coral reefs, two other well-known threatened environments. Bioscience 51(10):807–815

Vyas P, Sengupta K (2012) Mangrove conservation and restoration in the Indian Sundarbans. Sharing Lessons on Mangrove Restoration, 93pp

Wang'ondu V, Bosire J, Kairo J, Kinyamario J, Mwaura F, Dahdouh F, Koedam N (2014) Litter fall dynamics of restored Mangroves (Rhizophora mucronata Lamk. and Sonneratia alba Sm.) in Kenya. Restor Ecol:824

Warwick RM, Clarke KR (1993) Increased variability as a symptom of stress in marine communities. J Exp Mar Biol Ecol 172(1–2):215–226

Zaldivar JA (1999) Estructura de la comunidad del manglar y su relación con las características físicas y químicas del sedimento en laciénaga Sisal – Chuburnà, Yucatán, México. Tesis de licenciatura. Licenciatura en Biología. Universidad Autónoma de Yucatán, 61pp

Zaldivar-Jiménez AZ, Silveira JH, Molina CC, Parra DA (2004) Estructura y productividad de los manglares en la reserva de biosfera Ría Celestún, Yucatán, México. Madera Bosques 10(1):25–35

Zhao Q, Bai J, Huang L, Gu B, Lu Q, Gao Z (2016) A review of methodologies and success indicators for coastal wetland restoration. Ecol Indic 60:442–452

Mangroves on the Brazilian Amazon Coast: Uses and Rehabilitation

29

Marcus E. B. Fernandes, Francisco Pereira Oliveira, and Indira A. L. Eyzaguirre

Abstract

Mangroves are enormously important for the survival of traditional communities found in coastal zones around the world. This is typical of many areas in Brazil, in particular on the Amazon coast, which encompasses the largest continuous tract of mangrove forest found anywhere in the world. A multidisciplinary approach was used to describe the patterns of appropriation and uses of the resources available in this ecosystem, through the investigation of the local ecological knowledge of the traditional mangrove dwellers. Semi-structured interviews based on a set of specially formulated questions were used to better understand the knowledge of the local populations and their practices in relation to the mangrove. These people fish, harvest shellfish, and extract timber for both domestic and productive uses (e.g. construction of fishing weirs). In this region, the natural resources harvested by estuarine-coastal extractivists are used for both subsistence and sale. The principal problems in this region are the increase in the population, overfishing, predatory fishing practices, and the degradation of the mangroves. The initiative of the local community for reforesting degraded mangroves sites stimulated projects for the rehabilitation of the mangrove ecosystem in the surroundings of the communities. The establishment of new forest stands and return of invertebrate species, as *Ucides cordatus*, marked this rehabilitation. Ultimately, it is important to understand that the participative rehabilitation of the mangrove not only guarantees its productivity and the extractive activities, but also contributes to the development of the social, economic, political, and cultural dimensions of local communities, by reinforcing their perception of the need for measures to guarantee the sustainability and conservation of this ecosystem.

Keywords

Mangrove uses and appropriation · Replantation · Rehabilitation · Brazilian Amazon Coast

29.1 Introduction

The Mangrove is an ecosystem of vital importance for many species of marine and terrestrial animals (Hogarth 1999), bringing together a diverse biota that is very useful in the extractivist social context. It is well known that for a long time the extractivism has been important, and it is still up to the present day, for the great majority of people and traditional communities (Rodrigues and Cajueiro 2003; Passos and Di Beneditto 2005). There are records that identify the mangroves as an ecological unit, of which at least two thirds of the world fishery population depend directly on this ecosystem (Spalding et al. 2010). These records provide other indicators, among which the importance of various animals from mangroves are stressed (Pannier and Pannier 1980). Consequently, this importance triggers a process of economic valuation of environmental resources incorporated in this ecosystem.

In Brazil, there are seventeen coastal capitals, where most of the population is distributed in a narrow strip of land near the coast (IBGE 2007). In addition, ecosystems located in this coastal region eventually become the target of the pressure exerted by the anthropic action,[1] resulting from an intense and disorderly occupation (Rossi and Mattos 2002). In fact,

[1] It is understood, in the present work, by anthropic action any human activity that, somehow, interferes in the natural mechanisms of operation of an ecological unit or ecosystem. The anthropic action, therefore, is an activity that causes some kind of impact on the environment or in an ecosystem, interfering in their natural functioning (Mellanby 1982). In addition to direct human action on the natural environment, there is the presence of the so-called anthropic elements, such as buildings and objects, which act as agents of degradation and modification of the environment (CETESB 1985).

M. E. B. Fernandes (✉) · F. P. Oliveira · I. A. L. Eyzaguirre
Laboratório de Ecologia de Manguezal, Instituto de Estudos Costeiros, Universidade Federal do Pará, Bragança, Pará, Brasil
e-mail: mebf@ufpa.br

C. Makowski, C. W. Finkl (eds.), *Threats to Mangrove Forests*, Coastal Research Library 25,
https://doi.org/10.1007/978-3-319-73016-5_29

coastal dwellers use and explore several ecosystems. In the case of the Brazilian Amazon coast, the natural fields, rivers and streams, the beaches and mangroves are the target of this exploitation, being that the latter stands out for having economic, cultural, social and environmental meanings. In addition, the mangroves, in most cases, represent the main income and food source in this region (Maneschy 1993, 1995).

The mangrove ecosystem receives pressure from all sides, either through natural events or through anthropogenic activities. These, in turn, have their economic, social, cultural, and environmental basis supported in natural resources that this system offers local mangrove dwellers. These people who take ownership, use and live in the mangroves and its surroundings are considered estuarine-coastal extractivist workers (Oliveira 2015), and therefore have local ecological knowledge that allow them ownership and the use of these existing resources along the coast.

Among the most exploited resources in the Amazon mangroves are the mangrove crab (*Ucides cordatus*), and the mangrove trees (*Rhizophora mangle* L., *Avicennia germinans* (L.) L. and *Laguncularia racemosa* (L.) c.f. Gaertn., shrimps, mussels, oysters, crabs, and others.

In an ecological panorama, *U. cordatus* has a keystone role in nutrients cycling in the mangroves, being responsible, for example, for the consumption and degradation of more than half of the senescent leaves in this environment (Schories et al. 2003). In addition, this crustacean represents about 80% of the epifauna biomass of mangroves, serving as food resource for many groups of animals that inhabit or regularly visit the mangroves (Koch and Wolff 2002).

The mangrove forests are related to the importance of the knowledge and practice of the peoples who developed through observation and experimentation, a thorough knowledge about the natural processes. This knowledge is considered, currently, practices of land use and management adapted to tropical forests (Meggers 1997; Anderson and Posey 1989; Descola 1990; Glaser et al. 2005).

The appropriate resources by traditional communities are diverse and used for multiple functions, such as: (i) income generation for families; (ii) operations of the local, regional and extra-regional market; (iii) social reproduction of peoples and their cultures of the environment and the surrounding ecosystems and (iv) production practices from the use of environmental resources. Thus, the profile of appropriation and use of natural resources of mangroves has been described in accordance with the local estuarine-coastal extractivists. Considering the different uses and impacts on the resources of the mangroves, the various aspects of the rehabilitation process of impacted mangrove areas were also presented. Thus, an interdisciplinary approach is used, adding ecological, social, economic, and conservationist aspects, in order to present the coexistence and the productive practices of the estuarine-coastal extractivists on the Ajuruteua Peninsula, Bragança, Brazilian Amazon coast.

29.1.1 Site Selection and Procedures

The state of Pará is part of the Brazilian Amazon coast that extends from the mouth of the Oiapoque River, in the state of Amapá, until the eastern part of the state of Maranhão. This region is located on the Bragança coastal Plain, between Ponta do Maiaú and the mouth of the Caeté River, featuring a coastline of approximately 40 km (Wolff et al. 1999).

The estuarine and coastal area of the city of Bragança has circa of 180 km^2 of mangroves (Souza Filho 2005), with its characteristic biotic and abiotic components. The area that is around the mangroves, especially that occupied by traditional communities are here called "surroundings" (Fig. 29.1). It is worth pointing out that the mangrove forests of the Ajuruteua Peninsula are part of the Caeté-Taperaçu Marine Extractive Reserve (MER), as well as two communities: Castelo and Bonifácio (Fig. 29.1).

Based on the criterion of concentration of mangrove crabbers, whose information was obtained through direct contact with mangrove dwellers on the peninsula, the main estuarine-coastal communities inhabited by these people were identified and georreferenced: Acarajó, Bacuriteua, Bonifácio, Caratateua, Castelo, Tamatateua, and Treme.

The number of estuarine-coastal extractivists interviewed was 80, who have as their main activity the extraction of mangrove crabs in the mangroves of MER, besides being residents from different communities distributed throughout the Ajuruteua Peninsula.

The criterion to define the sample number s (i) the main productive practice, (ii) knowledge about the resources extracted from mangroves, (iii) access to zones of extraction for commercial production and (iv) permission to the researcher to accompany them in their productive practice. The implicit knowledge of the users was determinant in the location of the main extraction zones. Since then, their activities started to be monitored in order to: (a) georeferentiate the extraction zones and (b) describe each resource extracted.

29.1.2 Environmental Perception of Estuarine-Coastal Extractivists

In order to describe the profile of estuarine-coastal extractivists, data collection described by means of a questionnaire was performed. It is a set of questions systematically articulated that are intended to raise written information from the respondents (Severino 2007). The questionnaire presented semi-structured questions, which allow not only

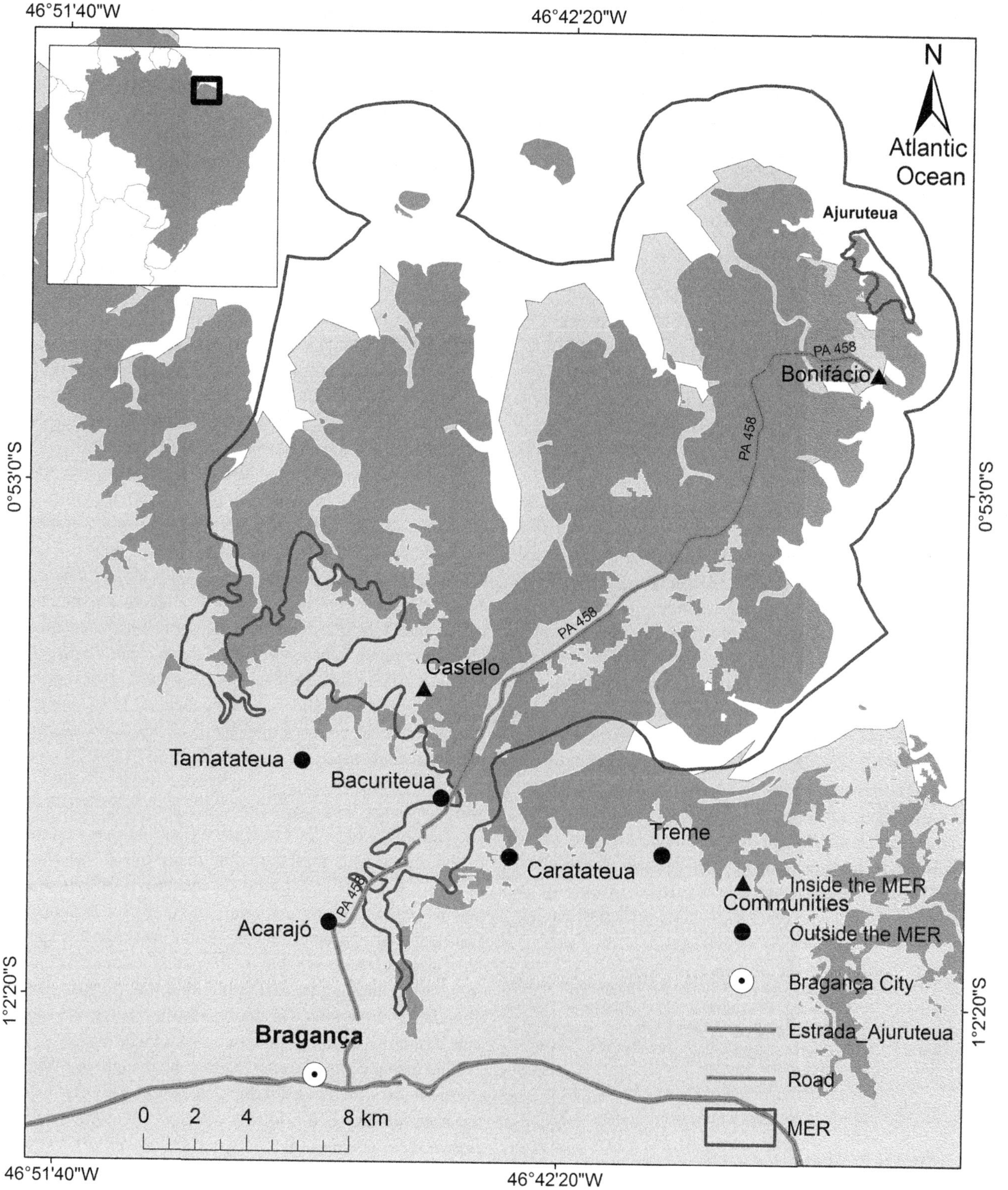

Fig. 29.1 Location of the Caeté-Taperaçu Marine Extractive Reserve on the Ajuruteua Peninsula, Bragança, State of Pará, Brazilian Amazon coast

the completion of questions that are necessary for the survey, but it may also give rise to other questions that can lead to a better understanding about their activities (Oliveira 2009). Additionally, the environmental perception of these respondents about the natural resources was used, since the planning for the extraction until the commercialization process. At first sight, one may think that the concept of environmental perception goes from physiology to semiotics, passing by the social representations or the functionalism (Del Rio and Oliveira 1996; Fernandes et al. 2004). However, the environmental perception is defined here as the operation that exposes the logic of language that organizes the expressive signs of uses and practices of a place. It is an expression of the image of a place, conveyed in signs that a community builds around itself. In this sense, the environmental perception is revealed through a semiotics reading of discursive, artistic, architectural, etc. production of the community (Ferrara 1993).

29.1.3 Appropriation and Use of Mangroves Resources

There are circa of 700 crabbers along the Ajuruteua Peninsula in Bragança. They live in communities in the surroundings of mangroves and have the mangrove crab extraction and the logging as primary activities. The communities that extract these resources the most are: Acarajó, América, Bacuriteua, Bonifácio, Caratateua, Castelo, Enseada Funda, Patalino, Porto da Mangueira, Praia de Ajuruteua, Retiro, São Benedito (Acarajozinho), São Domingos, São Pedro, Tamatateua, Tapreval, Taperaçu-Campo, Taperaçu-Porto, Treme, and Vila do Meio. Table 29.1 presents the main activities (Fig. 29.2) carried out by the estuarine-coastal extractivists on the Ajuruteua Peninsula and shows that they live on the fisheries extractivism (95.0%). Thus, it is evident that the practice of fishing, the plant extractivism and agriculture overlap with other activities, such as extraction of clay, recreation and leisure. The majority of respondents (82.5%; n = 66) reported they practice extractivism during the entire year, while 17.5% (n = 14) engage in only one of the seasons of the year (Table 29.2). Whereas the results in Table 29.3 show that the activities carried out by mangrove crabbers combined with other productive activities, namely:

Table 29.1 Description of activities practiced by estuarine-coastal extractivists (n = 80) in the mangrove forests of the peninsula of Ajuruteua, Bragança, Pará, Brazilian Amazon coast. n = number of respondents. Each respondent can engage in more than one activity

Activities	n (%)
Fish (fishing weir)	38 (47.5)
Crab (catch handling)	36 (45.0)
Shellfish (catch handling)	36 (45.0)
Agriculture	26 (32.5)
Firewood domestic uses	15 (18.8)
Medicinal plants	12 (15.0)
Wood for construction	09 (11.3)
Firewood for selling	05 (06.3)
Clay extraction	04 (05.0)
Tourism	03 (03.8)
Recreation and leisure	02 (02.5)

(i) fishing for crab and fish (95%), (ii) fishing for crab and other shellfish (85%) and (iii) fishing for crab and extract mangrove wood (55%).

In parallel to the information on the production of all the resources extracted, it was sought for the destination given to the products. The majority of fishermen (70.0%; n = 56) reported use for sale, while the remainder (30.0%, n = 24) said they use for their own consumption and both (consumption and sale) (Table 29.4).

Considering both production and destination of different resources extracted from the mangroves, it was possible to reveal the key factors that may affect their productive activities in the mangroves, and hence those that adversely affect the availability of these resources in mangroves. The results pointed out, according to the scale of importance, that: (i) predatory activities, (ii) increase of people using mangroves, (iii) logging, (iv) over-exploitation of mangrove products, are the most relevant factors (Table 29.5).

Few respondents (5%) said they were satisfied with the professionalization of fishermen in the age range from 15 to 55 years, while from 56 years old on the number of respondents satisfied was three times higher (20%) (Table 29.6). Out of the factors that led to this professionalization, the class from 18 to 55 years' points as the main factor (45%) the "*lack of opportunity for formal work*", while the older ones claim to have "*inherited the occupation from their families*" (35%). The "*low educational level*" also appears as a relevant factor to the class of younger (25%), while the group above 56 years of age "*They do not know or do not intend to perform another activity*" (25%) (Table 29.6).

Considering the typical mangrove trees, it was asked which species they used the most and how. The results showed, in an order of importance, that the majority (66%) of the respondents used the white mangrove (*L. racemosa*), circa of 20% and 14% the black mangrove (*A. germinans*) and red mangrove (*R. mangle*), respectively. Respondents justified their appropriation due to the fact that this species have size and shape suitable for the manufacture of fish weirs. In addition, they report that is the kind that presents greater abundance at the edges of the mangroves, being the species that has more durability in direct contact with water.

The respondents also showed that the mangrove wood has different uses for the estuarine-coastal extractivists. One of the main utilities is for domestic uses, such as charcoal and firewood for cooking. The firewood from mangrove wood is widely used for cooking the mangrove crab (*U. cordatus*) in the crustacean processing. This process starts with the capture of the crustacean (*catch handling*) in the mangroves, passing by cooking with water and salt (*brine*). Then, the crabmeat is removed from the crab and stored in plastic bags for sale. For the meat removal it is required some skills so that pieces of the exoskeleton are not crushed and mixed to the meat losing its quality (Machado 2007).

Fig. 29.2 Uses and products from the mangrove ecosystem on the Ajuruteua Peninsula, Bragança, Brazilian Amazon coast. Fresh fish (A) and salted fish (B), live mangrove crab (C) and cooked mangrove crab (D), swimming crab (E), shrimp (F), cassava (G), firewood (H), leaves of tobacco (I), clay sculpture (J), recreation and leisure (K), and handicraft (L)

The "wood for construction" is in the scenario of making fixed fishing traps, which can be divided into two types: weir and *fuzarca* net (Fig. 29.3). They are structures built with woods from the mangroves (*R. mangle*, *A. germinans*, and *L. racemosa*) or other stems as the bacuri tree (*Platonia insignis* Mart.). Both have entries in the form of "V", called

Table 29.2 Productive periods for crab fishing and wood extraction in the mangrove swamps of the peninsula of Ajuruteua, Brazilian Amazon coast

Period	n (%)
Rainy (December to May)	06 (07.5)
Dry (June to November)	08 (10.0)
During entire year (January to December)	66 (82.5)
Total	80 (100)

n = number of respondents

Table 29.3 Resources extracted by estuarine-coastal extractivists (n = 80) of mangrove crabs and mangrove wood in the peninsula of Ajuruteua, Bragança, Brazilian Amazon coast

Resources	n (%)
Mangrove crab and fish	76 (95.0)
Mangrove crab and Shellfish	68 (85.0)
Mangrove crab and wood	44 (55.0)
Mangrove crab and shrimp	12 (15.0)
Mangrove crab and others	04 (05.0)

n = number of respondents

Table 29.4 Destination of mangrove crab's production and wood extracted by esturarine-coastal extractivists (n = 80) on the Ajuruteua Peninsula, Bragança, Brazilian Amazon coast

Destination	n (%)
Sale	56 (70.0)
Own consumption	13 (16.3)
Both (consumption and sale)	11 (13.7)
Total	80 (100)

n = number of respondents

Table 29.5 Factors affecting productive activities of resources extracted from mangroves on the Ajuruteua Peninsula, Bragança, Brazilian Amazon coast

Factor	n (%)
Predatory activities	38 (47.5)
Increase of people using mangroves	36 (45.0)
Logging	32 (40.0)
Over-exploitation of mangrove products	30 (37.5)
Overfishing	24 (30.0)
Tides elevation	12 (15.0)
Pollution	12 (15.0)
Agriculture	05 (06.3)
Dunes advance	05 (06.3)
Oil spill (boats)	04 (05.0)

n = number of respondents

"*espias*", which directs the fish inside the trap. In the case of *fuzarca* net, the tunnel entrance leads to a conic net, and the weirs, directs to an area called sty where fish are caught. Both fisheries do not have target species, considering that they are characterized as generalist fisheries. The harvest is performed usually twice per day during the low tide.

In this same scenario, mangrove tree species (*L. racemosa* and *A. germinans*) are used by local people as a pen for rearing pigs in the communities surrounding mangroves, such as: Castelo, Porto da Mangueira, and Tamatateua (Fig. 29.4).

The "commercialization of mangrove woods" follows the parameters ruled out by the commerce demanded by brickyards, bakeries, and fishing traps, because according to the respondents, for the brickyards and the bakeries the three species of mangroves are used. However, for the fishing traps, the respondents claim to use more *L. racemosa*, followed by *A. germinans*, and as a third option, *R. mangle*. Similarly, it was possible to identify the use of mangroves wood for other activities such as production of firewood for manufacturing cassava flour (Fig. 29.5). This productive activity is common among the estuarine-coastal extractivists, because 32.5% of the respondents practice agriculture and, among the most cultured, the planting of cassava (*Manihot esculenta* Crantz) and manufacture of the cassava flour. In a scale of importance such activities are always in the first place. From this production, one part is for the families' own consumption and the other for commercialization. The main species used for firewood are *R. mangle* and *L. racemosa*. However, it is emphasized that this use is sporadic and occurs only when the wood is dry, which was reported by only 10% of respondents.

According to the respondents, the manufacturing of charcoal for domestic use and commercialization has emerged as the main goals in the production of this activity (Fig. 29.6), which in turn is common among almost all the families on the Ajuruteua Peninsula. The main mangrove wood used for this purpose is *R. mangle* and *L. racemosa* representing 31%, while *A. germinans* is used to a lesser extent (13%).

The dying of fishing nets by using stain of vegetable origin was also singled out as a common activity among the estuarine-coastal extractivists. This process takes place so that the fishing net durability is prolonged. The raw material (tannin) used in this process comes from the bark of *R. mangle*. In addition, a more detailed description of uses of mangrove products, based on the respondents' speech, showed that the use of mangrove tree species is ample and common on the Ajuruteua Peninsula (Table 29.7). *Laguncularia recemosa*, for example, was present in all the speeches (100%), followed by *A. germinans* (62%) and *R. mangle* that reached 18%.

The areas of mangroves are therefore of utmost importance for the peoples from coastal communities, since a good part of the animal origin proteins is originated therein, so essential for their livelihood. In this sense, strong traces in the inhabitants of these communities are realized when it comes to the uses of resources from the mangroves and also cultivated in the surrounding areas (Table 29.8). The process of appropriation and use are due to meeting the interests of

Table 29.6 Factors leading to the professionalization of estuarine-coastal extractivists (n = 80) on the Ajuruteua Peninsula, Bragança, Brazilian Amazon coast

Age groups (years)	Factors	n (%)
18–55	Lack of opportunity for formal work	18 (45%)
	Low educational level	10 (25%)
	Inherited the occupation from their families	06 (15%)
	They do not know or do not intend to perform	04 (10%)
	Satisfied with the profession	02 (05%)
> 56	Inherited the occupation from their families	14 (35%)
	They do not know or do not intend to perform *another activity*	10 (25%)
	Satisfied with the profession	08 (20%)
	Low educational level	06 (15%)
	Lack of opportunity for formal work	02 (05%)

n = number of respondents

Fig. 29.3 Weir (A) and *Fuzarca* net (B) used at fishing artifacts in the Caeté River, Ajuruteua Peninsula, Bragança, Brazilian Amazon coast

both the estuarine-coastal community in search of their livelihood as the interests of local merchants.

29.2 Rationale for Rehabilitation of Mangrove Forests on the Ajuruteua Peninsula

According to the Federal Decree 97.632 from 1989 of Brazilian Institute of the Environment and Renewable Natural Resources (IBAMA), the terms *Restoration*, *Recuperation*, and *Rehabilitation* are inverse paths to degradation and important to make it easier the communication when it comes to the process choice to be adopted in the degraded area. The term *Restoration* implies "mandatory return to original state of the area before the degradation", what is technically and economically questionable and an impractical process. The term *Recuperation* results in "Return of the site, reduced to a form of use, according to a pre-established plan for the use of the soil, aiming to produce a stable environment", that is, for repair of resources by establishing the composition and frequency of original species of the place. Rehabilitation, in turn, aims at the "Return of the degraded area to a suitable biological condition, which can mean the implementation of activities of (i) profitable use, but not long-term and (ii) less profitable use, such as recreation and/or aesthetic and ecological enhancement. In fact, the rehabilitation of a degraded area or a system is part of a growing approach where goals and targets should be well defined, in this way, adding the concept of environmental sustainability.

Fig. 29.4 Pens for rearing pigs in estuarine-coastal communities, with emphasis on the use of *Laguncularia racemosa* and *Avicennia germinans*

Fig. 29.5 The manufacuring process of cassava flour highlighting the place (A) and toasting, when they use firewood (B) from mangroves

The main motivation in rehabilitating the degraded mangroves on the Ajuruteua Peninsula arose from the isolated effort of the local estuarine-coastal communities in reforesting the degraded areas due to the construction of the state road PA-458, in the municipality of Bragança, State of Pará, Brazil. In 2005, it was joined to this initiative the bilateral agreement between Japan and Brazil, through the project entitled "Recuperation of degraded mangrove areas in Bragança", between the Japan International Cooperation Agency (JICA) and the Federal University of Pará (UFPA). Through the Technical Cooperation from Community Projects, which has as its purpose to contribute to the social and economic development of developing countries at the local level. The following activities were implemented: (i) establishment of partnerships between the community and the project, (ii) construction of a plant nursery with mangrove seedlings (genera: *Rhizophora, Avicennia*, and *Laguncularia*), (iii) direct performing of community monitors with the technicians and researchers, and (iv) preparation of a handbook of environmental action to explain the importance of the conservation and preservation of mangroves to the community.

In the same way, the sustainable production of mangrove resources, such as the mangrove trees that are used for a wide variety of local practice, was used as a justification for the rehabilitation of mangroves on the peninsula. *Laguncularia racemosa* is the most used mangrove tree species throughout the region, being used for the construction of large fishing

Fig. 29.6 The manufacturing process of charcoal using mangrove wood

Table 29.7 Species and uses of wood by estuarine-coastal extractivists on the Ajuruteua Peninsula, Bragança, Brazilian Amazon coast

Scientific name	Common name	Products	Uses
Avicennia germinans	*Siribeira*	Poles	Housing
	Black mangrove	Charcoal	Domestic, commercialization
		Pen	Rearing pig and cow
		Canoe for cassava	Processing cassava flour
		Firewood	Mosquito repellent, cooking
		Weir	Fishing
Laguncularia racemosa	*Tinteiro*	Weir	Fishing
	White mangrove	Fence	Housing
		Pen	Rearing pig and cattle
Rhizophora mangle	*Mangueiro*	Medicine	Fever, intestinal infections.
	Red mangrove	Firewood	Domestic, manufacturing cassava flour
		Charcoal	Domestic, commercialization
		Stain	Dying fishing nets
		Roof	Housing
		Pen	Rearing cattle
		Weir	Fishing

weir (Voigt 2011). The return of the mangrove crab (*U. cordatus*) to the rehabilitated areas is also a great benefit to local residents, who use them for consumption and commerce (Monteiro et al. 2014). In addition, it is worth mentioning that the conservation of mangrove ecosystem may imply also in the conservation of the genetic stock, the maintenance of the majority of ecological processes or simply the landscape, whose purpose can be related to recreation and the protection of coastal areas (Field 1998).

29.2.1 General Description of Replanted Mangrove Areas

The areas of replanted mangroves are located at the peninsula of Ajuruteua, in the communities of Tamatateua (00°57′12.5″S and 46°47′02.4″W) and Taperaçu-Campo (00°57′29.5″S and 46°64,5′58.7″W), about 15 km away from the city limits of the city of Bragança, State of Pará, Brazilian Amazon coast (Fig. 29.1).

According to Moraes et al. (2005), this region has a tropical, hot and humid climate, with annual rainfall of approximately 2300 mm and average temperature of 26 °C, due to the influence of the Inter-Tropical Convergence Zone. The driest months comprise the period from July to December, while the rainiest from January to July. The Ajuruteua Peninsula presents semidiurnal tidal cycles (DHN 2013), dominated by macrotides ranging between 4 and 6 m in height (Souza Filho 2005).

The texture of the soil in some replanted areas was classified as light or heavy clay, varying the content of organic matter between 16.3 and 23.9 g kg^{-1}. The percolation water

Table 29.8 Extracted resources, their uses and products from the mangroves of Bragança-PA, Brazilian Amazon coast

Resources	Products	Uses
Fishes	Dishes, fish flour	Food, commercialization
	Leather, swim bladder, scale	Handicraft, commercialization
Mangrove crab	Dishes	Food, commercialization
	Organic residuals	Handicraft, commercialization
Shimps	Dishes	Food, commercialization
Swimming crabs	Dishes	Food, commercialization
Oysters	Dishes	Food, commercialization
	Organic residuals	Handicraft: bijou, etc.
Mussels	Dishes	Food, commercialization
	Organic residuals	Handicraft, commercialization
Shipworm	Dishes	Food, commercialization
Scalet Ibis	Meat, feather	Food, handicraft, commercialization
Heron	Meat, bait	Food, fishing
Ducks	Meat, bait, feather	Food, fishing, handicraft, commercialization
Honey	Meat, Medicines	Food, health
Beans	Dishes	Food, commercialization
Rice	Dishes	Food, commercialization
Corn	Dishes	Food, commercialization
Cassava	Root	Food, commercialization
Tabacco	Cigarettes	Food, commercialization
Clay	Domestic utensil, handicraft	Commercialization
Reeds	Domestic utensil, handicraft	Handicraft, commercialization
Fruits	Pulp	Food, commercialization
Mangrove trees	Charcoal, firewood, medicines, stain, weir, pole	Domestic fuel, commercialization, medicine, dying

presents values that vary according to the period of the year. In the rainy period, the pH is close to 6, while in the dry period rises to 8. Salinity also varies between the dry and wet periods ranging between 2 and 80, defining a seasonal fluctuation.

29.2.2 Site Preparation and Planting Techniques

The place where the seedlings nursery was built flooded on a daily basis. Tidal creeks connected hydrologically through narrow channels ensured natural flood of the nursery. The area occupied by the nursery was 32 m^2 (8 × 4 m), with a depth of 40 cm, and wooden support to prevent collapse due to the tidal movements. Three rows were built, and each has the capacity of 800 seedlings, totaling circa of 2400 units (Fig. 29.7).

For the seedlings production the most dominant species were selected along the peninsula: *R. mangle*; *A. germinans*, and *L. racemosa*, considering also the dominant landscape closer to the replanted areas in an attempt to reproduce them. The species *A. schaueriana* Stapf and Leechman *ex* Moldenke also occurs on the peninsula but in shorter density (Abreu et al. 2016), and as no individuals next to the replanted areas were registered, this species was not used for reforestation. The collection of mature seeds were carried out during the rainy season (January to May), when there is the peak of production of reproductive phenophases in the coastal region of the Brazilian Amazon (Fernandes 1999; Nascimento et al. 2016).

The seeds collected went through a pre-treatment, except the seeds of *R. mangle*. *Avicennia germinans*, for example, were left about 8 h immersed in water brackish water, when the peels begin to loosen. In the case of *L. racemosa*, the seeds were immersed for 5–7 days. On the third day almost all seeds already presented the beginning of a primary root and should be planted during the sixth or seventh day, before the secondary roots appear by the ninth day, because the latter break very easily.

The mangrove seeds have peculiarities according to the species, which must be considered during the storage process. The seeds of *R. mangle*, for example, loses the viability over time, having to be planted as soon as possible, while *L. racemosa* can be kept for about a week, or more weeks in the case of *A. germinans*, if kept dry and in a cool and in the shade. However, it is important to bear in mind that seeds must be stored as short as possible.

In the case of *R. mangle*, one third of the seed was buried in mud, being the depth adjusted as the flood and hardness of the mud to prevent loss by tidal movements. The same procedure was performed for the other two species, but

Fig. 29.7 Nursery rows divided by narrow channels with wooden support to prevent collapse

Fig. 29.8 Direct sowing of *Rhizophora mangle* (A), *Avicennia germinans* (B), and *Laguncularia racemosa* (C) seeds into degraded mangrove soil

being up to five seeds for *R. mangle* and *A. germinans*, while *L. racemosa* from five to ten seeds per hole, since this last species has a high mortality rate in this first stage of planting (Fig. 29.8). Direct sowing provides great results regarding seeds survival due to the proximity between the seeds, thus stimulating the competition and accelerating the individuals' growth (Tsuji and Fernandes 2008).

The reforestation of mangrove patches by planting seeds or seedlings enable the recovering of degraded mangrove landscape. Planting seeds by using direct sowing is highly efficient. However, in addition of being dependent upon the period of greatest seed production, it offers higher risk of non-viability of seeds planted or mortality of those sprouted seeds. However, it is important to emphasize that growing seeds indoor presents some advantages in addition of creating a favorable environment for the development of seedlings. This method is also very efficient since it avoids or reduces the loss of seedlings (i) by tidal movements, (ii) cattle trampling and (iii) by pest attacks, thus ensuring greater quality, survival and success in the reforestation (Tsuji and Fernandes 2008). Results using the conventional method with nursery-grown seedlings in different parts of the world showed considerable variation in the percentage of survival of the plants, such as 40–60% in Australia (Saenger 1996), up to 85% in Indonesia (Soemodihardjo et al. 1996), and 90% in India (Untawale 1996). The quality of seedlings of *R. mangle*, *A. germinans*, and *L. racemosa* produced by nurseries in the amazon region was tested considering some variables, which may interfere in the seedlings survival rates, such as the type of substrate and percentage of shading.

Regarding the substrate, the mangrove seedlings in the first year of development depend on the energy contained in the cotyledon (Tamai and Iampa 1988). In addition, these cotyledon reserves sustain the production of epigeal and hypogeal parts of seedling, being extremely important for their establishment (Kitajima 2002). However, the seedling shoots growth is one of the main effects of the substrate on the roots (Hartmann et al. 1990). In fact, Araujo et al. (2014) showed that the type of substrate affects the seedling growth, which was evidenced by evaluating the quality of seedlings. According to Oliveira et al. (2008), the index of Dickson (IDQ) that uses the variable total dry weight (TDW; g); the shoot height (H; mm); the stem diameter (SD; mm); shoot dry matter (SDM; g), and root dry matter (RDM; g), can be used to express the quality of a seedling. The experiment with seeds/seedlings of *R. mangle*, *A. germinans*, and

L. racemosa carried out in Brazilian Amazon coast, after 270 days, showed that IQD values are larger for seedlings developed in yellow latosol substrate than in sandy or muddy substrates, and were confirmed in the study carried out in the same region by Costa et al. (2016). Thus, it is important to bear in mind that the use of new substrates for the production of mangrove seedlings in nursery seems not only an efficient procedure, but also an alternative measure to reduce the impact on the mangrove soil.

Some types of mangrove trees as those of the genus *Bruguiera, Heritiera*, and *Xylocarpus* can be less tolerant to light than others because their seedlings normally develop under the canopy of different tree species (Hanley et al. 2008). Although mangrove trees are pioneers and heliophilous plants, they also feature their spatial distribution influenced according to their tolerance to light, as is the case of *R. mangle* that is less tolerant to light (Smith and Lee 1999). Indeed, the seedlings of *L. racemosa* and *A. germinans* are more tolerant to brightness than *R. mangle* (Lopes et al. 2013). These authors showed that through monitoring over 9 months and the evaluation of biometric and morphologic parameters, seedlings of *A. germinans* develop better at 30% of shading and in full sun, while the seedlings of *L. racemosa* were more tolerant to 30 and 60% shade, but showed a better development in full sun. Additionally, the production of seedlings in nurseries enables reforestation outside the period of seeds production of local species.

The planting procedures applied to degraded mangroves areas are important for the rehabilitation and maintenance of the resources and landscape, but this process must achieve greater success in places where certain environmental characteristics have not been lost yet, mainly the flooding system. From this point of view, the reforestation carried out on the Ajuruteua Peninsula showed high mortality of seedlings planted in dry soils due to road construction (PA-458) that acted as a barrier, avoiding flooding from tidal waters. As a consequence of this impact on mangrove forest, several soil variables as organic matter, macro, and micronutrients increased reference values for carbon (C), ratio C: N, organic matter, iron (Fe), pH, sand, and clay.

29.2.3 Environmental, Social, and Economic Impacts of Mangrove Rehabilitation

From rehabilitation of impacted areas of mangroves, it is expected that the effect on the substrate be gradual and positive through greater retention of moisture, turning the dried soil to a muddier substrate. Therefore, a more adequate environment is obtained for the decomposition and retention of organic matter, which is essential in the process of nutrient cycling, as well as an environment for the recruitment of a new seedling generation. Likewise, this new substrate provides greater mobility for the recruitment of polychaetes that are important to the process of nutrient cycling and represent on the Ajuruteua Peninsula around 85% of the benthic fauna (Beasley et al. 2010). In addition, the new environment is also attractive to other invertebrate species, but especially the mangrove crab (*U. cordatus*) that has a high rate of larval recruitment in habitats favorable to the development of juveniles (Simith and Diele 2008). In fact, the study focused on the recruitment of *U. cordatus* in replanted mangrove sites showed that this species is able to settle down successfully in this new environment (Aviz 2016).

From a social point of view, the rehabilitation of this ecosystem is of great value and an essential tool for the recuperation and maintenance of natural resources inherent to the mangrove forests for the estuarine-coastal communities. Thus, the emergence of new propagules and seedlings in the areas of replanted mangroves reaffirms the fact that reforest impacted areas is a good practice. It is essential not only for the return of the forest per se, but the wood of mangrove trees for sustainable uses. The recruitment of fauna typical of invertebrates, especially of *U. cordatus*, in turn, clearly indicates the importance of rehabilitation of this system for the conservation of benthic macrofauna, specially the commercially relevant species. In fact, it is important to bear in mind that the actions for the maintenance of the structure and functioning of the mangroves are relevant to the social well-being, with the improvement of life through the generation of income for the estuarine-coastal extractivists worldwide. Through the implementation of new methods of management in the mangrove forests of Can Gio, in Vietnam, for example, many jobs have been created by generating income for a large number of fishermen through the reforestation, cutting firewood when thinning or pruning, and catching small crabs, oysters and clams on the forest floor (Hong 1996).

On the other hand, there are different types of natural threats or not, which foster the increase of mortality in the mangrove planting process. The natural threats concern the different herbivores that attack the seeds and seedlings causing some sort of damage. It can be exemplified by the action of beetles (e.g. *Poecilips Fallax* – Scolytidae) on propagules of the genus *Rhizophora* (Agaloos 1994; Hong 1994). Aquatic invertebrates (e.g. clams and oysters) also retard the growth of seedlings and damage the plumules and young leaves, respectively. Replanted mangrove sites and seedling nurseries, on the Ajuruteua Peninsula, were adversely affected by several episodes of herbivory. Records include attacks by crustaceans of the genus *Uca* (Ocypodidae) on the first leaves of seedlings of *A. germinans* (Tsuji and Fernandes 2008), as well as the presence of endoparasites (gall-inducing insects) on leaves of *A. germinans*. Considering the region and the places chosen for the implementation of planting a higher number of threats should be expected. Interference from marine algae (Rhodophyta species) can bend and break the seedlings (Cabahug et al. 1986), until the presence of vertebrates can be felt. Wild boars (*Sus scrofa*) have already been reported by Nam (1994) in the plantation of mangroves, as the presence of long tailed macaques (*Macaca fascicularis*) that forage for crabs and mollusks in newly planted areas of *R. mucronata* (Hong 1996) are both examples of negative natural effects on the replanted mangrove sites. In addition, it is worthy highlighting that during the seedlings production, when using local tidal regime for irrigation, the spring tides took away several seeds and even the bags with seedlings, mainly, of *L. racemosa* (Tsuji and Fernandes 2008).

The anthropic effects are less harmful and worrying. Nam (1994) reported that, in Vietnam, the growth of seedlings of mangrove species (*R. apiculata* BL., *R. mucronata* Lamk., and *Ceriops tagal* (Perr.) C.B. Robinson), was inhibited by competition with *Acrostichum aureum* and *Phoenis paludosa*. It resulted from the fact that seedling were planted in higher lands that had been sprayed with herbicides. Hong (1994) also reported that the seedlings were broken due to the passage of boats through the plantations, or even by dweller trampling who gather shellfish and fish in replanted mangrove sites. On the Ajuruteua Peninsula, the loss of seedlings of *R. mangle*, *A. germinans*, and *L. racemosa* is also caused by both dweller trampling during logging and/or crabbing activities and cattle herds trampling that graze freely on the peninsula and hence invade plantations for foraging.

29.3 Final Considerations

The description and understanding of the appropriation and use of resources, as well as productive activities held in the mangroves and the participative involvement in the rehabilitation process of this environment by the estuarine-coastal extractivists is always the result of involvement of a multidisciplinary approach in a socio-environmental perspective. Thus, it was possible to identify the use, extraction, and the reason for the mangroves resources. As well as their identities in the productive practices. The productive activities are not restricting only to the economical dimension, but they are also built through social relationships. In fact, at the traditional communities, the productive activities comprise multiple dimensions (e.g. social, economic, political, cultural, and environmental) (Castro 1997). It is worth noting, on the basis of the discussion presented by this author that productive activities developed in traditional communities are associated with nature, as well as they reflect on the accumulation process of knowledge over the generations. In the same way, Diegues (2004) defines traditional community, as they emphasize that it is a socio-political space dynamic and sociocultural, but highlights the economic organization as one of the central elements. The same author points out that the economic organization is based on the use of renewable natural resources with patterns of consumption, which combine various activities dependent on the nature cycles, as well as their culture is founded on symbologies, myths, and rituals associated with production activities, such as: hunting, fishing, and several extractive activities.

The resources extracted from the mangroves are for the commercialization and this unified information to those who sell and eat (13.7%) gains amplitude, aiming more than 80% of this resource to commerce. Based on the above information, there is the conception that the mangroves natural resources of mangroves on the Arujuteua Peninsula are part of the commercial dynamics and, hence, increases the pressure towards the predatory practice of intensification in the exploitation of these resources, without considering, the negative impacts resulting from this activity. Nevertheless, the analysis of users' perception brought to light the main problems that most affect the production activities of the estuarine-coastal extractivists. It is stressed the "*employment forms of predatory production*", which is directly related to the use of instruments both for crabbing and logging. The latter activity results from the use of mangrove wood for the manufacturing of fishing weirs, fences for domestic yards, load and unload fisheries ports, stowages, huts that serve for shelter and selling to brickyards and bakeries. Another factor that stands out is the "*increase of people using the mangroves*", especially during the holiday periods (January, February, and July), when mangrove crabs is notably more consumed in bars, beaches, and restaurants in the city of Bragança, also increasing the demand for other surrounding municipalities.

Being a dweller in mangrove areas in the Amazon region is not always a choice, but a result of lack of opportunities, low educational level and family heritage. There are many causal factors influencing this reality on the Brazilian Amazon coast. This region is characterized by the lack of industries or enterprises that may promote access to formal employment. This is corroborated by elder respondents (aged 56 or over), who highlighted the continuity of the family and the community's heritage, from the learning obtained with the family members and due to, therefore, having mastered the fishing activity skills (Maneschy 1995; Castro 2000).

It is evident that there are many approaches when it comes to the mangroves forests management. Pristine mangroves need fewer interventions than the impacted ones, together with the best choice regarding the restauration, recuperation or rehabilitation process. The most important is that the users, participatively, have their roles defined in this process. The gradual transformation of soil and its properties, the recruitment of flora and fauna point out the environmental gains, so that the environmental goods and services are rehabilitated. So, the income generation and the social well-being have guarantee in the system rehabilitation for the sustainable use by the estuarine-coastal extractivists. Surely, it reflects the reality of most communities on the Brazilian Amazon coast. It is important to emphasize that many forms of appropriation and use of the mangrove resources in this region is relevant to a better understanding of the interaction between users and the mangrove as an asset of common use. Finally, all participative activities for rehabilitation of this ecosystem becomes essential to sensitize mangrove dwellers towards sustainable use of mangrove resources.

References

Abreu MMO, Veiga N, Costa-Neto SV, Fernandes MEB (2016) Vegetação Arbórea: Distribuição Espacial. In: Santa Cruz (ed) Os Manguezais da Costa Norte Brasileira, vol IIII, 1st edn. Belém, pp 35–78

Agaloos BD (1994) Re-afforestation of mangrove forests in the Republic of the Philippines. In: The proceedings of the ITTO project workshop development and dissemination of reafforestation techniques of mangrove forests, Bangkok, Thailand, 18–20 April 1994

Anderson AB, Posey DA (1989) Management of a tropical scrub Savanna by the Gorotire Kayapó of Brazil. Adv Econ Bot 7:159–173

Araujo EC, Costa RS, Lopes EC, Daher RF, Fernandes MEB (2014) Qualidade das mudas de espécies arbóreas de mangue cultivadas em viveiro e diferentes substratos. Acta Ambiental Catarinense 11(1–2):21–32

Aviz BP (2016) Estrutura da vegetação e aspectos populacionais do caranguejo-uçá *Ucides cordatus* (Ucididae) em manguezais replantados. Thesis, Universidade Federal do Pará

Beasley CR, Fernandes MEB, Figueira EAG, Sampaio DS, Melo KR, Barros RS (2010) Chapter 7 Mangrove Infauna and Sessile Epifauna. In: Saint Paul U, Schneider H (eds) Mangrove dynamics and management in North Brazil, 1st edn. Bremen, pp 109–123

Cabahug DM, Ambi PM, Nisperos SO, Truzan Jr NC (1986) Impact of community-based mangrove forestation to mangrove dependent families and to nearby coastal areas in Central Visayas: a case example. In: UNDPIUNESCO Techn. Report on the research and training pilot programme on mangrove ecosystems in Asia and the Pacific (RAS/79/002) Mangroves of Asia and the Pacific: Status and management, Paris, France, p 441–466

Castro E (1997) Território, biodiversidade e saberes de populações tradicionais. In: Castro E, Pinton F (eds) Faces do trópico úmido: conceitos e questões sobre desenvolvimento e meio ambiente, 1st edn. Belém, p 445

Castro E (2000) Território, Biodiversidade e Saberes de Populações Tradicionais. In: Diegues C (ed) Etnoconservação: Novos Rumos para a Proteção da Natureza, 1st edn. São Paulo, pp 164–182

CETESB Companhia de Tecnologia de Saneamento Ambiental (1985) Relatório de avaliação dos projetos da CETESB – 1985. São Paulo

Costa RS, Araujo EC, Aguilar ECL, Fernandes MEB, Daher RF (2016) Survival and growth of mangrove tree seedlings in diferente types of substrate on the Ajuruteua Peninsula on the Amazon coast of Brazil. Open Access Libr J 3:2–9. https://doi.org/10.4236/oalib.1102777

Del Rio V, Oliveira L (1996) Percepção Ambiental: A Experiência Brasileira. São Paulo

Descola P (1990) Límites ecológicos e sociais do desenvolvimento da Amazônia. In: Bologna G (ed) Amazônia Adeus, 1st edn. Rio de Janeiro, p 288

DHN Diretoria de Hidrografia e Navegação (2013) The DHN register. https://www.marinha.mil.br/chm/dados-do-segnav/dados-demare-mapa. Accessed 10 May 2013

Diegues A (2004) Pesca construindo sociedades. São Paulo

Fernandes MEB (1999) Phenological patterns of *Rhizophora*, *Avicennia*, and *Laguncularia* Gaertn. f. in Amazonian mangrove swamps. Hydrobiologia 413:53–62

Fernandes RS, Souza VJ, Pelissari VB, Fernandes S (2004) O uso da percepção ambiental como instrumento de gestão em aplicações ligadas às áreas educacional, social e ambiental. Paper presented at the 2nd Encontro da ANPPAS, Associação Nacional de Pós-Graduação e Pesquisa em Ambiente e Sociedade, Indaiatuba, 26–29 May 2004

Ferrara L (1993) Olhar Periférico: informação, Linguagem, Percepção Ambiental. São Paulo

Field C (1998) Restoration of mangrove ecosystems. Okinawa

Glaser M, Berger U, Macedo R (2005) A sustentabilidade na gestão de floresta de manguezal sob condições de ilegalidade. In: Glaser M, Cabral N, Ribeiro A (eds) Gente, ambiente e pesquisa: manejo transdisciplinar no manguezal, 1st edn. Belém, pp 69–86

Hanley R, Mamonto D, Broadhead J (2008) Coastal forest rehabilitation. Manual for Aceh province and north Sumatra. Bangkok

Hartmann H, Schleicher M, Noegel AA (1990) Heterodimeric capping proteins constitute a highly conserved group of actin-binding proteins. Xenbase 11:396–337

Hogarth P (1999) The biology of mangroves. Oxford Univ Press, Oxford

Hong PN (1994) Reforestation of mangrove forests in Vietnam. In: The proceedings of the ITTO project workshop development and dissemination of reafforestation techniques of mangrove forests, Bangkok, Thailand, 18–20 April 1994

Hong PN (1996) Restoration of mangrove ecosystems in Vietnam. A case study of Can Gio District, Ho Chi Minh city. In: Field C (ed) Restoration of mangrove ecosystems, 2nd edn. Okinawa, pp 76–96

IBGE Instituto Brasileiro de Geografia e Estatística (2007) The IBGE register. http://www.ibge.gov.br/home/estatistica/populacao/contagem2007/defaulttab.shtm. Accessed 16 Feb 2014

Kitajima K (2002) Do shade-tolerant tropical tree seedlings depend longer on seed reserves? Functional growth analysis of three Bignoniaceae species. British Ecological Society. Funct Ecol 16:433–444. https://doi.org/10.1046/j.1365-2435.2002.00641.x

Koch V, Wolff M (2002) Energy budget and ecological role of mangrove epibenthos in the Caeté estuary, North Brazil. Mar Ecol Prog Ser 228:119–130

Lopes EC, Araujo EC, Costal RS, Daher RF, Fernandes MEB (2013) Crescimento de mudas de mangue sob diferentes níveis de sombreamento na península de Ajuruteua, Bragança, Pará. Acta Amazonica, Ciências Florestais 43:291–296. https://doi.org/10.1590/S0044-59672013000300004

Machado D (2007) Catadoras de caranguejo e saberes tradicionais na conservação de manguezais da Amazônia brasileira. Rev Estud Fem 15(2):485–490. https://doi.org/10.1590/S0104-026X2007000200016

Maneschy M (1993) Pescadores nos manguezais: estratégias técnicas e relações sociais de produção na captura de caranguejo. In: Furtado L, Leitão W, Fiúza A (eds) Povos das Águas: Realidade e Perspectivas na Amazônia, 1st edn. Belém, pp 19–62

Maneschy MC (1995) A Mulher está se afastando da pesca? Continuidade e mudança no papel da mulher na manutenção doméstica entre famílias de pescadores no litoral do Pará. Boletim do Museu Paraense Emilio Goeldi. Série Antropologia 11 (2):145–166

Meggers B (1997) Amazônia, a ilusão de um paraíso. Rio de Janeiro

Mellanby K (1982) Biologia da poluição. São Paulo

Monteiro MAB, Oliveira FP, Araujo JN, Fernandes MEB (2014) Productive chain of the mangrove crab (Ucides cordatus) in the town of

Bragança, in the Northern Brazilian State of Pará (Amazon Region). J Coast Res 70:443–447

Moraes BC, Costa JMN, Costa ACL, Costa MH (2005) Variação espacial e temporal da precipitação no Estado do Pará. Acta Amazon 35:207–214

Nam VN (1994) Plantation to rehabilitate mangroves in Can Gio, Ho Chi Minh City. In: Proceeding of the national workshop reforestation and afforestation of mangroves in Vietnam Can Gio, Ho Chi Minh City, Hanoi National Pedagogic University, 6–8 August 1994

Nascimento AAM, Gardunho DCL, Fernandes MEB (2016) Fenologia Reprodutiva. In: Fernandes MEB (ed) Os manguezais da costa norte brasileira, vol III, 1st edn. Belém, pp 80–89

Oliveira CL (2009) Um apanhado teórico conceitual sobre a pesquisa qualitativa: tipos, técnicas e características. Revista Travessias. Pesquisa em educação, cultura, linguagem e arte 2(3):1–16

Oliveira F (2015) Análise da percepção dos extrativistas estuarino-costeiros sobre o zoneamento da extração do caranguejo-uçá (*Ucides cordatus*) e da madeira nos manguezais da RESEX-Marinha Caeté-Taperaçu, Pará, costa amazônica brasileira. Dissertation, Universidade Federal do Pará

Oliveira RB, Lima JS, Souza CAM, Filho SM (2008) Produção de mudas de essências florestais em diferentes substratos e acompanhamento do desenvolvimento em campo. Revista Ciência Agrotécnia 32(1):122–128

Pannier R, Pannier F (1980) Estructura y dinamica del ecosistema de manglares: um enfoque global de la problematica. In: Memorias del Seminario sobre el estúdio científico e impacto humano en el ecosistema de manglares, Oficina Regional de Ciencia y Tecnologia de Ia UNESCO para a America Latina y el Caribe – ROSTLAC, UNESCO,Cali, 27 November-1 December 1980

Passos CA, Di Beneditto APM (2005) Captura comercial do caranguejo-uçá, Ucides cordatus (L. 1763), no manguezal de Gargaú RJ. Biotemas 18(1):223–231

Rodrigues I, Cajueiro L (2003) Caracterização preliminar dos problemas ambientais no estuário do Rio Caeté, Bracuriteua (Pará, Brasil). In: Claudino-Sales V, Tonini I, Dantas EWC (eds) Congresso de Ecologia do Brasil, Fortaleza, 2003

Rossi M, Mattos I (2002) Solos de mangue do estado de São Paulo: caracterização química e física. Revista do Departamento de Geografia 15:101–113. https://doi.org/10.7154/RDG.2002.0015.0010

Saenger P (1996) Mangrove restoration in Australia: a case study of Brisbane international airport. In: Field C (ed) Restoration of mangrove ecosystems, 2nd edn. Okinawa, pp 36–51

Schories D, Barletta-Bergan A, Barletta M, Krumme U, Mehlig U, Rademaker V (2003) The keystone role of leaf-removing crabs in mangrove forests of North Brazil. Wetl Ecol Manag 11:243–255

Severino A (2007) Metodologia do trabalho científico. São Paulo

Simith DJB, Diele K (2008) Metamorphosis of mangrove crab megalopae, Ucides cordatus (Ocypodidae): effects of interspecific versus intraspecific settlement cues. J Exp Mar Biol Ecol 362:101–107

Smith SM, Lee DW (1999) Effects of light quantity and quality on early seedling development in the red mangrove, *Rhizophora mangle*. Bull Mar Sci 65:795–806

Soemodihardjo S, Wiroatmodjo P, Mulia F, Harahap MK (1996) Mangroves in Indonesia: a case study of Tembilahan, Sumatra. In: Field C (ed) Restoration of mangrove ecosystems, 2nd edn. Okinawa, pp 97–110

Souza Filho PWM (2005) Costa de Manguezais de Macromaré da Amazônia: Cenários Morfológicos, Mapeamento e Quantificação a partir de Dados de Sensores Remotos. Revista Brasileira de Geofísica 23:427–435

Spalding M, Kainuma M, Collins L (2010) World atlas of mangroves. Earthscan, London

Tamai S, Iampa P (1988) Establishment and growth of mangrove seedlings in mangrove forests of southern Thailand. Ecol Res 3:227–238

Tsuji T, Fernandes MEB (2008) Replantando os manguezais: técnicas de reflorestamento. Bragança

Untawale AG (1996) Restoration of mangroves along the Central West Coast of India. In: Field C (ed) Restoration of mangrove ecosystems, 2nd edn. Okinawa, pp 111–125

Voigt M (2011) The co-management approach of the Marine Extractive Reserve "Caeté – Taperaçu" in northern Brazil: analysis of selected biological and socio-economic aspects. Thesis, Technische Universität Dresden

Wolff M, Koch V, Isaac V (1999) A trophic flow model of the Caeté mangrove estuary (North Brasil) with considerations for the sustainable use of its resources. Estuar Coast Shelf Sci 50:789–803. https://doi.org/10.1006/ecss.2000.0611

Mangrove Restoration and Mitigation After Oil Spills and Development Projects in East Africa and the Middle East

30

David I. Little

Abstract

This chapter describes inter-governmental and industry initiatives for spill prevention, preparedness and oil spill response (OSR) planning in East Africa and the Middle East, where mangroves are vulnerable to spills and other threats. The two main regional examples of oil spills affecting mangroves are deliberate spills by Iraq following the 1st Gulf War of 1991 (Saudi Arabia), and the 'Katina P' innocent passage incident of 1992 in Maputo Bay (Mozambique). The scale of the Gulf War spills led to the United Nations Compensation Commission (UNCC) programs on remediation of oiled shorelines, mangrove restoration, and nature reserves in Saudi Arabia. Also described are recommendations for mangrove mitigation and creative conservation in Environmental and Social Impact Assessment (ESIA) of coastal developments (Qatar). Marine and terrestrial ecological sensitivities in East Africa and the Middle East are acknowledged and widespread, but often data-deficient. There are many important coastal areas with high biological diversity and abundance, including mangroves, coral reefs, seagrass beds, saltmarsh, brackish and freshwater wetlands. The associated communities of fish, turtles, marine mammals and birds in East Africa and the Middle East are often linked by migratory fauna that are functionally dependent at certain life stages on mangroves and other, often adjacent habitats (e.g., mangroves occur in 57% of coastal important bird areas). Local economies throughout these regions rely heavily on natural capital for subsistence livelihoods and sustainable development (e.g., high value tourism). The chapter concludes that unless local communities are genuinely involved with, and also benefit more equitably from, extractive and other coastal developments, both directly and through local investments paid for by the taxes levied by their governments, there is a risk that they reject the industry's license to operate. As described by Gundlach, and by Little et al. (in this Coastal Research Library publication), this negative point is illustrated by pipeline spills of 2008 followed by numerous subsequent spills from 'hot-tapping' and 'artisanal' refining in the mangroves near Bodo, Nigeria. The diligent protection, restoration and integration of mangroves into landscape-scale conservation that is inclusive of the rural poor are key strategies to avoid this extreme precedent.

Keywords

Environmental sensitivity · ESIA follow-up · License to operate · OSR constraints · Cleanup trade-offs · Regional development · UNCC follow-up

30.1 Introduction

30.1.1 Coastal and Biological Resources in East Africa

The variations in environmental sensitivity of the coastal environment play an important role in determining oil spill response (OSR) options in East Africa, as elsewhere. The coastline is tropical and mainly low-lying, seasonally humid in the south, humid in Tanzania, and semi-arid further north. Coastal environments are characterized by steady SE. swells, augmented by easterly winds and periodic tropical cyclones in the south and by monsoons in the north and south (Orme 1996). Figure 30.1 shows the countries and oceanographic features of the Western Indian Ocean (WIO) region.

Except in the lee of Madagascar where they are muted, the swells drive shore zone dynamics and push non-cohesive sediments northwards in the littoral drift. The finer-grained sediments are seasonally discharged from major rivers such as Tugela in the Republic of South Africa (RSA), and the

D. I. Little (✉)
Swavesey, Cambridge, Cambridgeshire, UK

C. Makowski, C. W. Finkl (eds.), *Threats to Mangrove Forests*, Coastal Research Library 25,
https://doi.org/10.1007/978-3-319-73016-5_30

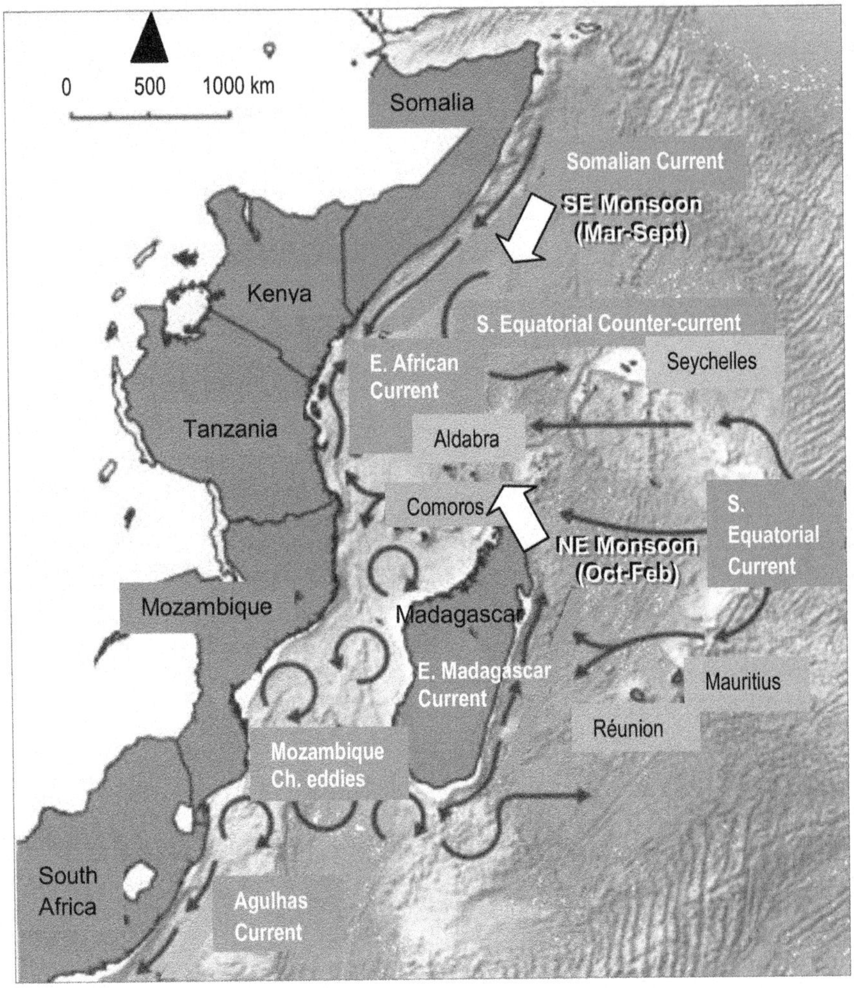

Fig. 30.1 The countries, bathymetry, main currents and monsoon winds of the WIO region (Adapted from UNEP and WIOMSA 2015)

Limpopo and Zambezi in Mozambique. These sediments are deposited in rich coastal lagoons and mangrove forests, often sheltered behind northward-trending spits and nearshore bars. Silt and clay sediments are also flushed seawards and are entrained southwards in the 6 km/h Mozambique and Agulhas currents. In the Sofala Bight, with an almost 6 m tidal range, much of the fine sediment discharge escaping from the 7150 km^2 Zambezi delta is deposited in the wide tidal flats, or else it crosses the broad continental shelf and is deposited in a massive cone beyond the shelf edge.

Madagascar coastlines are characterized by a linear eastern shore controlled by fault line interrupted by only one large embayment, the Baie d'Antongil. Persistent SE. Trade winds and westward moving cyclones (December–March), are crossed by swells originating in storms in the Southern Ocean. This high energy produces strong wave action that has created an 800 km long linear system of beach barriers with biologically-diverse shallow coastal lagoons behind. The north-western coast is a rich range of habitats produced between Cape d'Ambre and Cape St Andre by a recent marine transgression which has created drowned valleys (rias), deltas, mangrove swamps and islands. The semi-arid south western coast has dune systems, and on a narrow shelf also supports coral reefs in a wide range of formations and ages: fringing, banks and islets.

Mangroves are the dominant coastal vegetation type between the Tana and Zambezi deltas and on the sheltered west coast of Madagascar. In all these areas where climatic, edaphic and topographic conditions are favorable, typically five species of mangroves occur, with some trees up to 15–25 m high (*Sonneratia alba* is a pioneer in exposed zones, *Avicennia marina* is a primary seaward colonist, *Rhizophora mucronata* in muddy creek soils, *Ceriops tagal* in drier areas, and *Bruguiera gymnorrhiza* in wetter areas). *Avicennia marina* is sometimes the only mangrove species in the semi-arid zones in NE. Kenya, although the drier sandy shores throughout the region support an important pioneer community of *Scaevola thunbergii*, *Ipomoea biloba* (south) and *I. pes caprae* (north) that promotes fore-dune formation.

In contrast, within the Zambezi delta the more humid conditions encourage *B. gymnorrhiza* and *R. mucronata*. Fiddler crabs occupy the seaward mangroves, with snails and other invertebrates found inland among *A. marina*. Ground cover is richer in *Sesuvium portulacastrum*, *Arthrocnemum* spp., *Suaeda monoica*, algae and grasses where mangroves are less dense. The Zambezi delta also demonstrates a rich series of transitions from saline wetlands through brackish to freshwater swamps. The brackish transition is marked by ferns, the mangrove *Hibiscus tiliaceus* and the tree *Heritiera littoralis*. The freshwater transition supports extensive reed swamp communities (*Phragmites australis*, *Cyperus papyrus*, *Juncus* spp., and *Scirpus* spp.).

In addition to cetaceans offshore, the coastal wetlands support abundant waterbirds, fish, crustaceans, molluscs and insects in habitats dominated by both trees and reeds. The sea cow (*Dugong dugon*), crocodile (*Crocodylus niloticus*), hippopotamus (*Hippopotamus amphibus*) and monitor lizard (*Varanus niloticus*) are the dominant aquatic megafauna. Section 30.5 gives further information on the rich biological environment of the East Africa region, and also the southern Arabian Gulf.

30.1.2 Offshore and Onshore Industry

Oil spill damages arising from tankers are typically covered by a range of compensation schemes, as described by the International Tanker Owners Pollution Federation (ITOPF, www.itopf.co.uk/). In contrast, there are no relevant international compensation schemes or conventions covering oil spills originating from offshore or onshore installations. Operators of facilities are required to meet national legislation regarding insurance and liability, and this includes oil spill damages. The mechanism for negotiating these topics is the Production Sharing Agreements (PSAs) or Production Sharing Contracts (PSCs) for each lease awarded (Deloitte 2014). Two aspects of PSAs/PSCs are particularly relevant in African contexts: decommissioning and local content. The terms for decommissioning in the PSAs of East African countries have varied widely. Along with provision for the costs to be tax deductible (all nations), terms ranged from the setting up of abandonment reserves (Kenya) or abandonment funds (Tanzania) or decommissioning funds (Mozambique), to site restoration being required but no fund has been established (Uganda). These PSA terms could have OSR or other environmental implications in future. Local content provisions could impede procurement or mobilisation of OSR equipment, while decommissioning terms would affect environmental cleanups.

The latter risk is likely to become acute long before the decommissioning phase in the event of major spills from exploration and production (E&P) facilities in offshore, lacustrine or inland fields. This is because these platforms, pipelines, pump stations and downstream depots or refineries will not be covered by existing oil spill compensation arrangements that are applicable to tanker accidents. Because there has been little experience of cleanup in East Africa, costs and expectations for compensation are unpredictable. The various governments and the oil and gas industry will need suitable models to address the issue of funding cleanup (e.g., PSA terms, Escrow accounts, and inclusion in Regional Oil Spill Contingency Plans; ROSCPs).

Local content of goods, services and labor is one of the ways in which host nations maximise returns from the industry to the local and national economy. Although encouraged in all countries, it was found that no minimum local content is prescribed for contractors to follow. In Mozambique no requirement for local content is expected for goods and services. Although contractors are required to comply with national policy, in Tanzania this policy has not been published (Deloitte 2014). Unlikely and unpalatable though it may seem at the start of the project cycle, the failure of local content provisions in terms of health, educational and job opportunities is a major cause of the loss of the license to operate among local communities. Unless arrangements are made in advance for funding restoration costs in East Africa, cleanup will not be straight forward. This loss of confidence in Nigeria has led to illegal 'tapping' of pipelines, oil theft, vandalism or terrorist attacks on oilfield installations.

Another complication is that the oil spill compensation conventions do not cover spills resulting from acts of war and terrorism. The unprecedented deliberate sabotage by Iraq during the First Gulf War in 1990–1991 causing fires and spills led to the establishment of the UN Compensation Commission (UNCC) to administer the funds accumulated from Iraq's controlled oil sales. The UNCC was perceived as a model for solving unfunded cleanup and restorations. The UNCC succeeded in assessing and moderating 2.7 million individual claims from several nations. Ultimately U.S. $ 52.4 billion was awarded to approximately 1.5 million successful claimants. However, in Saudi Arabia the restoration did not commence until 20 years later, with a budget of about U.S. $1.14 billion. Despite good progress in restoring intertidal mudflat channels and salt marshes along with successful mangrove transplants, the project has not yet lived up to expectations especially in terms of restoring desert environments, ecological services, human and natural capital, and establishing marine protected areas. The much smaller spill during the Israel-Lebanon war in 2006 resulting from the bombing by Israel of the Al-Jiyeh power plant in southern Lebanon was suggested by the United Nations as another candidate project for the UNCC, but this was not adopted. In fact, the UNCC did not assist in any more cleanups after the 1991 Gulf War spills between 1991 and the fulfilment of its mandate in 2012. Although €50 million was given by the European Union (EU) and others via the Regional Marine Pollution Emergency Response Centre for the Mediterranean Sea (REMPEC), Lebanon was not compensated for the pollution and cleanup despite efforts by the Food and Agriculture Organisation, UN Development Programme, UN Environment Programme and World Bank (UN 2007a, b).

Article 11 of the Nairobi Convention requires that Parties shall co-operate to combat pollution in OSR emergencies. It calls for the development of contingency plans and notification procedures. Further to Article 11, a Protocol Concerning Co-operation in Combating Marine Pollution in Cases of Emergency in the Eastern African Region (the 'Emergency Protocol') entered into force in 1996. This Protocol complements the 1990 International Convention on Oil Pollution Preparedness, Response and Co-operation (OPRC), with further detail on co-operation in case of marine pollution incidents and the establishment of contingency plans in the region, based on mutual support from national systems. Implementation of the Protocol is not complete. Regional projects under the Global Environment Facility (GEF) have made a contribution to assisting the countries meet the Protocol's requirements, as described in the following sections.

30.2 Cooperation in OSR in East Africa

30.2.1 International Stakeholders

As in western and central Africa with its work on mangroves (UNEP 2007), the UN plays a significant role in East Africa, through the International Maritime Organization (IMO) working in conjunction with UNEP and UNDP. The Nairobi Convention Secretariat and the Indian Ocean Commission are key bodies. The World Bank and the GEF played major roles in financing and supporting past oil spill preparedness projects and may continue to do so in future. Overseas national programs to support development in the oil sector are on-going with a focus on oil spill preparedness. These include the U.S. AFRICOM, the Norwegian OfD program and, possibly, the IFUSE program in UK. These programs are described in Sects. 30.2.5, 30.2.6 and 30.2.7. This group of stakeholders is likely to provide opportunities for engagement, alignment of objectives and OSR resources.

30.2.2 Civil Society and Non-governmental Organizations

The potential concerns of the local and international community in relation to petroleum sector development are typically channelled through campaigning Civil Society Organizations (CSOs) and environmental Non-Governmental Organizations (NGOs). There have been numerous campaigns by high-profile organizations (e.g., World Wide Fund for Nature, WWF; and Greenpeace) that protest against petroleum developments. East Africa has many of the factors that could trigger new targeted campaigns against oil development in the region (e.g., biological diversity and abundance, reliance on artisanal fisheries and agriculture, and the other serious potential challenges to effective emergency response discussed in Sects. 30.5 and 30.8). There are opportunities for industry to engage constructively with CSOs and NGOs, specifically in relation to OSR, ESIA transparency, follow-up and reputation management.

30.2.3 WIO Islands Oil Spill Contingency Planning (1999–2004)

The GEF provided financing for the development of national and regional oil spill contingency planning activities in Comoros, Madagascar, Mauritius and Seychelles (Fig. 30.1). The project was implemented through the Indian Ocean Commission (IOC) with support from others, including the governments of RSA and France (Réunion), IMO, and IPIECA (the global oil and gas industry association for environmental and social issues). To assist regional cooperation and implementation, a regional coordination centre (RCC) was established in Madagascar in 2004 under the WIO Islands oil spill contingency planning project. However, there is no evidence that the RCC is functional. Whilst covering only one of the countries in the present study, the WIO Islands project design was partly utilized in the subsequent GEF WIO Marine Highway project (see following). The Parties to the Nairobi Convention viewed both these projects as a means of realising their obligations under the convention.

30.2.4 WIO Marine Highway Project (2008–2012)

30.2.4.1 Aims and Achievements

The Western Indian Ocean Marine Highway (WIOMH) development and coastal and marine contamination prevention project was a U.S. $ multi-million project financed through the GEF, and covered eight of the coastal countries in East Africa. In this project, national and regional OSR capacity was improved through the IOC, of which the key outputs were:

- Implementation of fully tested National Oil Spill Contingency Plans (NOSCPs)
- Implementation of a regional oil spill contingency plan
- Capacity building through training courses
- Draft 'Regional Contingency Plan for Preparedness for and Response to Major Marine Pollution Incidents in the Western Indian Ocean' (the 'Regional Plan')

- Terms of reference for a future Regional Coordination Centre (RCC)
- Detailed description of human and materials resources, work plan and RCC budget.

The project completion reports state that a website structure was developed, with the intention that the site be adopted by the RCC beyond the life of the project. This has not yet happened and there appears to be a risk that some of the project output is in danger of being either lost or not adopted within national frameworks. This presents an urgent opportunity for consideration by those preparing OSR plans. Various workshops and meetings were held, including the involvement of IMO, IOC, UNEP and World Bank. This culminated in seven of the eight WIOMH project recipient countries signing the Regional Agreement on Cooperation on the Regional Oil Spill Contingency Plan, which entered into force in November 2011. This regional agreement is linked to the Nairobi Convention and its Emergency Protocol, and its purpose is to strengthen the Regional Plan in alignment with the OPRC Convention. The Regional Draft (version 3, dated 26 July 2010) is based on the IMO content and format used in various other regional seas, with its general objective as follows: "to organize a prompt and effective response to accidental marine pollution affecting or likely to affect the area of responsibility and/or the area of interest of one or more of the countries concerned and to facilitate the cooperation in the field of marine pollution preparedness and response." The Regional Plan identified a series of specific objectives and actions:

- Implementation of operational procedures and financial conditions for cooperation
- Principles of command and control
- Areas of responsibility and types of assistance which might be provided
- Developing appropriate preparedness measures and effective systems for detecting and reporting pollution incidents
- Developing and implementing a program of training courses and practical exercises for different levels of personnel.

The Regional Plan contains the framework, terminology and recommended procedures to meet these objectives. Ten appendices support the plan with technical data and information on the arrangements to be provided by each country. In October 2010 country level data was still being sought for the technical appendices. In December 2011 all the WIOMH participating countries adopted the 'Host Country Agreement for the Regional Coordination Centre for Preparedness and Response to Oil, Hazardous and Noxious Substances Pollution in the Western Indian Ocean'. The South African Maritime Safety Authority (SAMSA) Centre for Sea Watch and Response (CSWR) would be the location for the RCC, and its ability to perform the role of host was verified by the project consultants.

30.2.4.2 Lessons from the WIOMH Project

The WIOMH project was a significant effort to develop oil spill preparedness and response in the region. The final report reviewed the project's success and challenges during implementation, with areas of concern including the following (WIOMH 2013).

Awareness and Ownership

The involvement of the countries in the preparation phase of this regional project was uneven, with some agencies being unaware of the project or without means to influence its scope. This reflected the changes in responsible personnel over the long period. The project was endorsed at the government level but in some cases without adequate consultation at a technical level, causing delay and potentially undermining ownership and sustainability.

Engagement Across Ministries

Institutional responsibilities in the participating countries were not fully considered in designing the arrangements for implementation. The delineation of responsibilities for project activities between the maritime and environmental authorities was not given sufficient attention during the preparation phase. Initially, the implementation required the nomination of a single focal point per country, generally in the maritime sector under the transport administration. Unfortunately, this led to a de-emphasis on the environmental activities in the project, in some cases causing tension between the ministries. This prompted the designation of an additional focal point during implementation, most often the environmental ministry, to advance the environmental activities and enhance communications.

Sustainability

Despite the achievements in creating or updating the NOSCPs, these plans should not be static. To sustain their usefulness requires the appropriate resources to be maintained in a state of readiness, and this involves regular (annual) training, including simulations of risk events with all stakeholders. However, there is concern that the current response capability may not be sustained in all participating countries, particularly where ownership and commitment was found lacking during implementation. Some countries took the opportunity to update parts of their maritime law or to launch a national effort on disaster preparedness, but others struggled to take the necessary actions. The development of Environmentally Sensitive Area (ESA) maps is constrained by the fact that several stakeholders are waiting for compensation for time spent on the activity or sharing the geographic data.

Regional Cooperation

At the end of the project, most stakeholders recognized the need to organize a regional response capacity to combat major oil spills; at the same time they were concerned that the agreements are mainly theoretical and should be put into practice. Unfortunately, there has been no opportunity to test the ROSCP and the evidence was inconclusive from the only event occurring during the project large enough to require regional cooperation. Furthermore, the issue of compatibility between oil spill equipment of the countries of the region was not addressed in the project; with existing incompatibilities forming a technical barrier to cooperation during major spills.

It should be noted that this is the second attempt to establish a RCC: the previous GEF project established one in Madagascar, but it is no longer functioning. Whether the lessons learned from this experience were used in the design of the new RCC is unclear, but the commitment of RSA and SAMSA will undoubtedly ensure sustainability. At the close of the project in late 2012 it was reported that to sustain the RCC's viability there was a need for the countries to revise the RCC budget and work plans. The participating countries have not been consistently responsive during implementation of the regional cooperation which led to delay. In this context, and based on the experience of a similar regional centre (REMPEC) in the Mediterranean, SAMSA decided not to ask for financial contributions from the other countries to operate the RCC. This decision reflects the commitment of RSA and SAMSA in particular towards the RCC.

30.2.5 United States AFRICOM Oil Spill Missions

The U.S. African Command (AFRICOM) of the Department of Defense has assisted in developing OSR and awareness programs for marine oil spills since 2008. This program is delivered mainly through national OSR courses attended by various environment, health and tourism ministries, military and local officials from the host venue, as well as industry representatives. These were delivered by a multi-disciplinary team (AFRICOM, U.S. Coast Guard, National Oceanic and Atmospheric Administration, U.S. Navy and other personnel). Reported missions in East Africa included Mozambique in 2008 and Kenya and Tanzania in 2012.

30.2.6 Norwegian Oil for Development Program

The Oil for Development (OfD) program was initiated by the Norwegian Government in 2005. The aim of the OfD program is to reduce poverty by promoting economically, environmentally and socially responsible management of petroleum resources. The OfD secretariat is in the Norwegian Agency for Development Cooperation (NORAD). The main approach in the OfD program is capacity development through institutional collaboration. Norwegian public institutions enter into long-term agreements with public institutions in a partner country. The key implementing parties are the Norwegian Petroleum Directorate, Norwegian Environment Agency, Petroleum Safety Authority Norway, Norwegian Oil Taxation Office and Norwegian Coastal Administration (NCA).

A Regional Seminar for East Africa 'Sharing Experiences – Safety and Oil Spill Preparedness and Response' was organized under OfD in 9–12 September 2013 in Bagamoyo, Tanzania. The list of attendees shows representatives from Uganda, Kenya, Tanzania and Mozambique. Although the group worked on OSR challenges in the region, no report on the outcome was obtained. IPIECA was represented at the meeting by a consultant and a presentation made on cooperation between governments and industry in the East Africa region. In Mozambique, Tanzania and Uganda the OfD activities included support to oil spill preparedness mainly through the NCA.

The OfD program engaged with WWF Norway to review environmental concerns and facilitate engagement of CSOs in the development of petroleum resources in East Africa (NCG 2012). The project summary stressed the contribution that WWF can make by mobilizing international expertise and conducting "meaningful dialogue with high-level stakeholders in the petroleum sector". Technical reports, policies and legal documents can be scrutinized for compliance with environmental standards. The WWF contribution also involved capacity building in the public sector dealing with petroleum, and this can be transferred to local CSOs in short or medium terms. This capacity building is still weak in those project countries lacking knowledge and technical competence in the petroleum sector. The project has strengthened networks but in future large international agencies (*e.g.*, WWF) will still be working with small local partners.

Many CSOs lack credibility when promoting the interests of the local communities in exploration areas according to NCG (2012). The local CSOs must provide legitimacy through dedication to specific districts, and this will be of more value for cooperation than the professional national organizations operating from the capital. Project follow-up should aim to select suitable candidates and provide them with the resources to work locally. Despite the need for civil society engagement in the petroleum sector, the opportunities for such engagement vary from acceptance or 'lip service' in some project countries to hostility from the public sector in others. Typically the international oil companies are more professional, recognizing the need to work with both international environmental organizations and local CSOs. Although acting from commercial motives, the companies wish to ease conflict with local populations and avoid bad publicity and accusations of neglect. With the exception of "politically backward" countries the demand for CSO participation could soon increase (NCG 2012).

The project summary recommended the project should continue and should concentrate on core objectives in a limited number of countries (Uganda, Tanzania and Mozambique). Madagascar and Kenya are no longer within the OfD program, and in Madagascar the poor achievements were because government institutions were reluctant to partner with civil society and it was considered the country will probably not be a major petroleum producer for many years (NCG 2012). In response to the report, WWF raised some concerns about conclusions drawn (e.g., they disagreed with the proposal to remove Madagascar from the project). Nevertheless, they found the overall evaluation to be "thorough, accurate, and useful in developing the project further".

30.2.7 Other Programs

Oil, gas and mining companies traditionally have engaged in a variety of social development activities in the main countries of East Africa, such as construction of community clinics, rehabilitation of roads and community buildings, and support to cyclone-affected populations (e.g., in Madagascar). Whereas this is to the credit of the companies concerned, there is a danger that governments in the study region get left behind as industry continues to grow and self-regulate. The risks this presents are related to the long-term nature of these developments, the significant legacy when industry moves on, the risk of environmental incidents, and the need for a coordinated response. Regional regulatory networks such as the East African Network for Environmental Compliance and Enforcement (EANECE) run courses for enforcement officials through the environmental management cycle. Opportunities have also been identified to link up with their work in oil, gas and mining to achieve sustainable growth.

The recent oil and gas discoveries in East Africa and the potential scale of African mineral deposits, present the challenge of raising the capacity of governments to manage the sector effectively and equitably:

- Good environmental management systems for extractive industries must be in place to protect local communities and livelihoods
- Predictable and stable regulation is needed to attract investment to the sector
- A holistic approach to the sector is needed for international safeguards to be meaningful and to ensure regulation can keep pace with fast-moving developments.

Capacity-building is needed among host government ministries with environmental and extractive industry portfolios. The UK Department for International Development (DFID) proposes to run a leadership program for East African officials responsible for regulation of the regional oil, gas and mining industries. The DFID Investment Facility for Utilising UK Specialist Expertise (IFUSE) has a potential to partner with NORAD's OfD program to offer technical assistance under three main management themes: resources, revenues and environmental protection. NORAD is interested in potential collaboration for Tanzania, Mozambique, Uganda, Ghana and South Sudan, for which the Norwegian Environment Agency would be the delivery partner.

30.3 Regional Development in East Africa

30.3.1 History

The colonial history of Africa to some extent still shapes the present trade patterns in terms of how transport routes responded to geographical barriers and corridors to the coast. Only 12% of Africa's trade takes place within the African continent, compared with 60% of the trade in Europe. This is a product of slow visa approvals and the poor implementation of free-trade agreements as a result of there being too many inefficient African trade blocs (some countries belong to three blocs or more). The tendency of producers to export to former colonial powers is reinforced by inherited national language blocs (French, Portuguese and English) since independence. The language barriers in East Africa are probably still a key factor in regional education and business, and in implementation of regional OSR planning at the local level.

30.3.2 East African Community (EAC)

The EAC is the regional intergovernmental organisation of the republics of Burundi, Kenya, Rwanda, Tanzania, Uganda, with headquarters in Arusha, Tanzania. The 'Treaty for Establishment of the East African Community' was signed in November 1999 and entered into force in July 2000. The Republic of Rwanda and the Republic of Burundi became members with effect from 1 July 2007. The EAC aims to increase co-operation in political, economic, cultural and social fields to improve quality of life. The EAC established a customs union in 2005 and a common market in 2010.

The next phase of integration will be a monetary union and ultimately a political federation of the East African states. The realisation of EAC as a regional economic bloc of 130 million people (2010), an area of 1.82 million km^2 and a GDP of U.S. $74.5 billion (2009), is of strategic significance. Working with EAC, national governments and CSOs, Trademark East Africa (TMEA) is funded by development agencies to foster prosperity and reduce poverty by trade, increased market access, product competitiveness and standards. A focus of TMEA is the effects of oil, gas and mining on patterns and volumes of trade, including the logistics, power and infrastructure demands that restrict or enable trade. TMEA has headquarters in Nairobi, and offices in Kampala, Arusha and Dar es Salaam.

30.3.3 OSR Implications

Although sustainability of the natural environment is not their main priority, regional development strategies emphasise the importance of improving the free flow of goods and services across national borders within a framework of consistent standards. The achievement of these aims will improve OSR and preparedness in East Africa, and this will reduce impacts to mangroves. In advance of a major spill, regional cooperation is needed to facilitate visas for personnel, mutual aid for OSR equipment, cross-border customs clearance and transit arrangements. OSR equipment must be compatible with equivalent items from neighboring states when the size and severity of a spill necessitate scaling up and cooperation. This should be tested during drills and desktop exercises. Any necessary improvements should be made as part of the tiered response within bilateral national mutual aid organisations and the ROSCP.

30.3.4 Fisheries

A large amount of primary protein in the human diet in East Africa is sourced from lacustrine, coastal and maritime fisheries, and these represent a strategic resource for development. Mangroves are an important nursery area for fish. Although free-swimming fish rarely suffer significant adverse impacts from oil spills, this will not necessarily apply in spills in mangrove creeks, inland or riverine areas, or where subsistence fishponds and commercial aquaculture (with seawater intakes) have been developed on the coast. Management systems such as the use of scientific observers in local fisheries and aboard fishing vessels, the control of invasive species, and the development of more rigorous fish stock assessments will need significant improvement in future. An ecosystem approach, catch per unit effort (CPU), by-catch reporting, and coastal livelihood assessments are all feasible management methods in the region (TDA 2012). It is important both before and after a significant oil spill that impacts be properly described using scientific fish stock assessment. The status of fish stocks is also affected by market and trade demands. Kenya, Madagascar and Mozambique have signed the South Indian Ocean Fisheries Agreement (SIOFA) which entered into force in 2012. SIOFA has a mandate to take legally-binding decisions over high seas fisheries, except for sedentary species on continental shelves and for migratory species covered by the Indian Ocean Tuna Commission (IOTC). Regulatory instruments and fisheries protection organisations are shown in Table 30.1.

Countries of the WIO region have benefitted from preferential trade between African, Caribbean and Pacific (ACP) countries and the EU, which receives 68% of total fish and fish products exported by ACP. This is changing with the advent of Fishery Partnership Agreements and Economic Partnership Agreements (EPAs) as follows, and the impact of such changes on fish stocks is unknown:

- East Africa Community (Kenya, Tanzania), with Mozambique and RSA
- Southern Africa Development Community (SADC)
- Eastern and southern Africa (including Madagascar)
- Interim EPAs between EU and Madagascar, Mozambique, Kenya, Tanzania (2007–9).

30.4 Oil Spill Response (OSR) in East Africa

30.4.1 Options and Constraints

A balance must always be struck in OSR planning and preparedness between available response options and receptor importance and sensitivities. Reference will need to be made to the texts of national regulations and OSCPs for the products and policies that are still in some cases to be authorized in each East African country. This will involve significant capacity- and consensus-building efforts in future for all the nations involved, particularly for the most sensitive mangrove coastlines and for oil and gas assets located inland and in freshwater environments.

No OSR response technology is a panacea, and IPIECA among others encourages national response frameworks that include access to a complete response toolkit. OSR equipment stockpiled in the region should be fit-for-purpose in the differing environments and hydro-meteorological conditions. For example, in mounting mechanical oil recovery operations even in calm weather, the Mozambique Channel has strong tidal streams that could cause failure of booms and skimmers, adding to the significant challenges posed by oil spreading and the fragmentation of oil slicks. This is because despite

Table 30.1 International, regional, sub-regional instruments in marine fisheries (TDA 2012)

	UNCLOS = United Nations Convention on Law of the Sea	Fish Stocks 1995	Port State FAO (Food and Agriculture Organisation) 2009	**IOTC** = Indian Ocean Tuna Commission	**SIOFA** = South Indian Ocean Fisheries Agreement 2012	**SWIOFC** = South West Indian Ocean Fisheries Commission	**ASCLME** = Agulhas and Somalia Current Large Marine Ecosystem	**COI-IOC** = Indian Ocean Commission	**EAC** = East African Community	**SADC** = Southern Africa Development Community
Kenya	✓	✓	✓	✓	✓	✓	✓		✓	
Tanzania	✓			✓		✓	✓		✓	✓
Mozambique	✓	✓	(a)	✓	✓	✓	✓			✓
Madagascar	✓			✓	✓	✓	✓	✓		✓
South Africa	✓	✓		(b)		✓	✓			✓

(a) = signatory only, (b) = cooperating party only

being more sheltered from tropical storms and wave action than the open WIO, the Mozambique Channel is classified as 'lower macrotidal' (tidal range 3.5 m to 5 m). The resulting current velocity through the water column is augmented by the south-flowing Mozambique and Agulhas currents and complicated by numerous large-scale eddies (Fig. 30.1).

Chemical dispersants are potentially useful where mechanical recovery is impossible due to sea state, spill size and/or availability of recovery equipment. Moreover, the spilled oil must first be amenable to chemical dispersion. Some crude and refined oils are not amenable (e.g., oils with viscosities exceeding 10,000 cSt), or have a restricted 'window of opportunity' for use of chemical dispersants. The waxy crudes oils recently discovered in Uganda and Kenya may not respond to chemical dispersants due to their high viscosities despite the relatively high sea and air temperatures that may assist in lowering oil viscosities compared to temperate climates. Excepting possible future marine export options, this need not be an issue of concern for selecting among OSR options because oils produced from Lake Albert (freshwater) in Uganda and Lake Turkana (brackish-saline) in Kenya will not be spilled in suitable habitats for the use of dispersants. Madagascar will use steam injection to produce a heavy crude oil from the Tsimiroro field, and if spilled during export, this oil is also unlikely to respond to dispersants.

Good planning and rapid mobilization maximize the dispersant window of opportunity prior to the inevitable increase in oil viscosity due to weathering. High initial viscosity and the formation of water-in-oil emulsion ('mousse') during wave action also render oil more difficult to treat. Where available, aircraft application systems can provide a rapid response time to take advantage of the window of opportunity. Finding the location of the thicker parts of oil slicks quickly is a key factor in effective responses using dispersants. With time the oil may be broken into 'spillets' of widely varying oil quantities and forms. Equally importantly, aerial application of dispersant achieves higher oil encounter rates than surface craft for the treatment of larger oil slicks; close supervision by spotter aircraft during the spraying operation is essential. For offshore well incidents, subsea dispersant injection systems are developed which greatly improve both response time and oil encounter rate.

IPIECA and IOGP (2014) strongly supports development of national lists of approved dispersant products and the situations under which their use may be pre-authorized. Pre-authorization does not imply that blanket usage is approved, merely that the products are licensed, tested and available, and could be considered among the tools for OSR on suitable oil types and in certain locations and water depths. The lists of approved chemical dispersants may differ between East African nations, or may not be developed as yet. E&P companies are interested in the approval status (if any) of tried and tested products in each nation.

After the Macondo incident in 2010 the industry invested very heavily in establishing a global stockpile of chemical dispersants (Carter-Groves 2014). Three products are currently included in this global stockpile (Corexit 9500, Slickgone NS and Finasol OSR 52), for which considerable practical and scientific experience in their use has been gained worldwide. RSA is the regional location for this industry initiative, and has an abundance of hauliers. There is a potential issue with regards to an emergency mobilization: working hours for export clearance are reduced on weekends and may result in delay. Good dialogue with the customs authorities will help, and with this information to hand in the plan, mobilization can begin with the most accessible stock and not necessarily the closest.

Even when approved product schedules for dispersants are already developed by the regulatory and technical agencies responsible for OSR in each nation, the procedures and constraints on pre-authorization may not yet be enshrined in national regulations (e.g., as was the case in 2013 in Tanzania). Having clear criteria for the authorization of dispersants allows for their immediate use without delays stemming from prolonged debate during an incident. Water depth and distance to shore are the simplest and most commonly used criteria to protect mangroves and other sensitive coastlines (Table 30.2).

In more complex situations (e.g., shallow water or where there may be specific ecological considerations; both applicable to mangroves), responders who have worked together during response training and desk-top or live drills can quickly assess relative risks between use and non-use in a real spill, and are thus more likely to reach consensus on using dispersants. The wildlife conservation agencies are responsible for the most sensitive environmental resources, including designated sites. These agencies should be involved in the development of dispersant authorization criteria and understanding of the potential net environmental benefit of dispersant use. They should be part of training and drills to the greatest extent possible, not least because otherwise they can help or hinder an agreed decision to use dispersants. An additional advantage of engagement with conservation groups is that some of their facilities and personnel are often located in remote coastal zones that are of most interest for mangroves and wildlife that is sensitive to oil. Such decentralised locations and the ability of conservation agencies and closely-allied NGOs to help attract, inform and rapidly train volunteers may be able to assist response (e.g., shore surveys and collection of injured wildlife; IPIECA 2004).

Table 30.2 Comparison of the minimum water depth and distance offshore for dispersant use adapted from a framework agreed in East Africa through the GEF project (WIOMH 2012)

	Tier 3	Tier 1 and 2	Implement/modify
Kenya	Three (3) zones: Conditional approval (<10 m) Pre-authorized (>10 m) Special consideration areas		Kenya Maritime Authority
Uganda	No anticipated use of dispersants in freshwaters or brackish inland waters		
Tanzania	>200 t spill:	<200 t spill:	Commission
	>20 m depth and >1 km from shore	>10 m depth and >0.5 km from shore	
Mozambique	>200 t spill:	<200 t spill:	Commission
	>20 m depth and >1 km from shore	>10 m depth and >0.5 km from shore	
Madagascar	>100 t spill:	<10 t spill:	Ministry of Environment
	>30 m depth and >44 km from shore	>20 m depth and >3.7 km from shore	
		10–100 t spill:	
		>30 m depth and > 23 km from shore	
South Africa	Review of draft dispersant policy at workshop in November 2014		

30.4.2 Response Support Tools

The information in Table 30.2 will help to define the sea areas in which dispersant use may be feasible. A useful next step in developing OSR documentation is the translation of those sea areas into tactical response maps that can be included in regional, national and international plans. The widespread adoption from the 1980s onwards of the Environmental Sensitivity Index (ESI) and ESI maps funded by IMO, UNEP and NOAA has produced good ESI maps in East Africa. Other examples include recent Kenyan maps with assistance from Denmark, and the IMO/UNEP response atlas for Mauritius (Gundlach and Murday 1989).

The use of Geographic Information Systems (GIS) in mapping Environmentally Sensitive Areas (ESA) and resources, ESI, and tactical dispersant plans is not standardized across East Africa. The WIOMH project produced outputs of mapping but only copies that lacked the functionality and resolution of the original GIS could be found. Unless the original WIOMH (2012) source files can be located and are viable, the standardization of maps is still needed for future OSR activities in East Africa.

30.4.3 Receptor Sensitivity and 'Trade-Offs'

The choice of response options depends on weighing up evidence in advance for the likely sensitivity of various receptors to the range of oil types likely to be spilled. If leaving the oil alone to degrade naturally causes less adverse impact than the cleanup, then a net environmental benefit analysis (NEBA) suggests natural recovery is preferable. Similarly, if there is likely only to be partial success of any response operation, the choice typically depends on where the remaining oil goes, and in what quantities and forms. In this case, the NEBA may rest on comparing the likely intolerance to oil of a variable set of receptors. For example, when deciding whether to use approved chemical dispersants, in some cases habitats nearer to the shoreline such as mangroves, salt marshes, coral reefs, lagoons, seagrass beds, as well as seabirds generally benefit from protection by the use of dispersants offshore in deeper water. This is because if successful, dispersant use offshore helps prevent the ingress of untreated oil into the more sensitive shoreline habitats. This greater dispersion may adversely affect fish offshore, but the NEBA helps decide where the balance is to be struck. However, if some of the oil is not fully dispersed before reaching habitats and species that are sensitive to dispersed oil, the chemically-dispersed oil may cause greater adverse shoreline impacts (*e.g.*, bleaching and necrosis of algae and corals, mortality of seagrass, and loss of sediment infauna due to deeper oil penetration into fine-grained sediments).

The toxicity of oil(s) and dispersant(s) is only part of the NEBA, because the analysis must also make a series of trade-offs between different volumes and physico-chemical states of the oil and the effect they may have on a different set of receptors. Heavy persistent oils are problematical in this respect because they are not responsive to dispersants and smother receptors for longer periods on oiled surfaces and in infaunal burrows. Lighter hydrocarbon compounds in crude oil or refined products may kill benthic grazers and, due to the resulting bloom of algae, alter habitat functioning such as blocking creeks draining tidal flats. Impacts may thus include habitat deterioration as well as species injuries or death.

In addition, the intolerance and vulnerability may change during different life stages or at different seasons for different marine and socio-economic resources. For example, mangroves in Kenya, Tanzania and Mozambique (at least five species) differ from those on the open eastern coast of Madagascar (usually two species), and all differ from those in the brackish wetlands of the Zambezi delta (see Sect. 30.5). One of the same mangrove species (*Avicennia marina*) also occurs at the limit of its range in the arid Middle East where it may be more sensitive to pollution stress than it is in tropical areas with a much less prolonged dry season. Recoverability is the ability of a habitat, community, species or socio-economic resource to return to a state close to that which existed before the activity or event that caused change. For example, mangrove and intertidal mudflat habitats near Maputo, Beira, Dar es Salaam and Mombasa suffer chronic pollution from urban areas, tankers and other stressors. In such areas, opportunistic infaunal communities may be somewhat pre-adapted to oil pollution stress; however, the less robust and more rare species will be among the spill casualties in a more pristine area.

The presence of biological communities of flora and fauna of particular interest or rarity, and their designation as features in protected sites are factors in planning an OSR capability. ESI mapping is essential in this respect, but wildlife is not statically-defined within conservation areas. Existence of migratory wildlife feeding, loafing or breeding sites of international significance is crucial for many species. In East Africa many migratory bird species are shared with the Middle East, central Asia and the arctic (e.g., African-Eurasian Waterbirds Agreement species rely on flyway integrity and not simplistically on the designation of selected sites).

30.5 Environmental Sensitivity and Conservation

30.5.1 Conservation Priorities and Perceptions in East Africa

A perception of wildlife and habitat conservation in East Africa is that it is a mature science-based movement that also generates valuable local income from ecotourism enterprises. This perception is reinforced by decades of wildlife documentaries on television around the world, and drives a major tourist industry that is vulnerable to oil spills. While it can often be an accurate perception, it also is one that should be tempered by the realities of big game poaching and of the stressors imposed by climate change, desertification, biodiversity loss, poverty and population growth.

The inequitable distribution of wealth and of access to water, sanitation and electricity in many parts of the region are harsh constraints on health, education, employment and security. Where such areas of rural poor coincide with wildlife reserves, there must be genuine involvement of the local communities in the management of reserves for them to be successful. This includes maintaining vigilance at all levels of civil society and governments, and zero-tolerance of poaching or incremental habitat destruction by informal settlements and inappropriate development. This can be difficult when local people resent nature reserves that stop their previously legal activities or their hunting of wildlife pests (Adams 1996; Adams et al. 2004; Getz et al. 1999). It becomes extremely difficult when, for example, corrupt regulators and the local poor are recruited by organized crime to poach elephant ivory that is then hidden among antique ivory and sold legally. Despite the gains made by elephant conservation since the 1989 ivory ban, many were lost since 2008.

The wildlife conservation movement of East and southern Africa, although far from secure, is undeniably more mature than in most other parts of Africa and many other tropical regions. Part of the reason for this is the above-mentioned premium of ecotourism, which can bring real benefits to GDP, and to the people, habitats and species of interest, be they coastal, forest, savanna or mountains. Many of these habitats are well represented in the established protected area networks managed nationally, but many are trans-boundary and globally unique (e.g., the lakes and mountains of the African Rift Valley). Such national parks and wildlife sanctuaries are likely to be already mapped in ESI and natural resource atlases, and as such should already be factored into the OSR planning now being done by industry and government. However, for OSR purposes this has historically had a primarily coastal emphasis. The differences to be expected in freshwater lakes compared to the sea, in terms of evaporation, wave action and the ecological and socio-economic resources at risk, include:

- Lakes have less potential than the sea for natural dispersion and dilution of oil
- Lakeside communities depend on the lake waters for fishing and water supply
- The lakes are isolated and ancient, with hundreds of endemic species.

The trajectory model results of a 30,000 tonne spill of crude oil in the June–August period of southerly wind showed that oil travelled 300 km from the southern to the northern end of Lake Tanganyika (Burundi) within 20 days. Oil also beached on both western (Democratic Republic of the Congo; DRC) and eastern (Tanzania) shorelines of the lake, which would represent a trans-boundary incident. IPIECA has supported the oil industry in its operations in sensitive and challenging environments including East Africa by providing guidance along the lines outlined in Baker (1991, 1992), and IPIECA-OGP (undated a, b).

Whether inland or coastal, there is more to be accomplished than the initial safeguarding of protected areas from spills and other threats. Even in protected areas there is a constant threat that implementation of management plans will be stymied by poor enforcement of environmental laws due to under-resourcing, lack of leadership and training, corruption, and collapse of security.

Since the Rio (UNCED) summit in 1992, the Convention on Biological Diversity (CBD) and the Large Marine Ecosystem (LME) concept have in theory strengthened commitment to the maintenance of biodiversity in the wider environment (IPIECA-OGP 2007, undated a, b). The CBD presented an approach whereby wild resources and ecological services are monetized to motivate their conservation through less intensive farming, forestry, fisheries, aquaculture and other development. The aspiration is for landscape-scale conservation, not just in nature reserves, and this has the potential to bring a unity of purpose to ordinary people in East African rural areas and the cities. Few wealthy urban dwellers have experienced directly the environmental risks of soil erosion, deforestation or pollution in the wider environment, or the human consequences of breakdown in ecological services. In contrast, the rural poor have often directly experienced these stressors and in response have migrated to the cities' informal settlements in search of work. The challenge in many cases is to educate the governing and business elites in the rigorous implementation of the conservation and sustainable development pledges they made at Rio and Rio+20.

The wider environment both on land and at sea may be characterized in East Africa by examination of the Ecologically or Biologically Significant Areas (EBSAs) promulgated under the CBD. The nations in the study region have ratified the CBD on the dates given, as follows: Kenya (1994), Uganda (1993), Tanzania (1996), Mozambique (1995) and Madagascar (1996). They have also all ratified the Cartagena Protocol on Bio-safety. They have recently prepared national Biodiversity Strategies and Action Plans, as follows: Kenya (2000), Uganda (2014), Tanzania (2011), Mozambique (2003) and Madagascar (date unknown). Nations signatory to CBD are committed to protecting these large sites using the biodiversity strategy and action plans they have all produced.

In addition to the marine EBSAs in the study region, further information on sensitivity may be derived from Endemic Bird Areas (EBAs) and Important Bird Areas (IBAs). These areas may or may not already enjoy some degree of protection. They represent conservationists' best estimate of where biodiversity losses will be unacceptably severe unless a science-based approach to their designation is energetically pursued. The achievement of this will improve the capability of resource managers to respond effectively to oil spills as well as other stressors in both marine and terrestrial settings.

Oil spill and other environmental impacts are not limited to biological communities. The key vulnerabilities are the protected or designated conservation sites, fisheries, shellfisheries, tourism amenities, industrial and municipal water intakes. These receptors are often identified on ESI maps and well-integrated in OSR planning around the world. East Africa, however, is a test case in that the extractive industries are comparatively recent compared to oil in Nigeria and in North Africa, and to minerals and coal in the Copper Belt and South Africa. Local communities' stability and a company's license to operate are potentially very finely-balanced. The commercial, ecological, health, cultural and social (including tribal) importance and sensitivities of resources are tightly interlinked. In extreme situations, due to infiltration of government and industry by organized crime, these links can break down and threaten security. This has been found most forcefully in onshore oilfields of SE. Nigeria, including mangrove areas (IPIECA 2008, 2012; UNEP 2007, 2011; Gundlach 2018, Chap. 13 this volume; Little and Galperin 2017). The East African implications of this scenario are examined in the context of OSR.

30.5.2 Ecologically Sensitive Areas

30.5.2.1 WIO Ecologically or Biologically Significant Marine Areas

Introduction to EBSAs

EBSAs in the WIO region have been identified under the umbrella of the UN Convention on Biological Diversity (CBD). Expert workshops were used for this purpose, providing a regional overview. The regional meeting was held in Flic en Flac, Mauritius, 31 July – 3 August 2012. Recommendations were approved at the eighteenth meeting of the Subsidiary Body on Scientific, Technical and Technological Advice (SBSTTA) in Montreal, Canada, 23–28 June 2014. In this way, proposed sites are 'moderated' by expert workshops in other regions. The EBSA findings in the study area are described below, and are summarized together with the internationally accepted criteria for their selection in Table 30.3.

The areas listed have often been highlighted as important sites for conservation, for example many of them overlap with the marine IBAs of the Southern Indian Ocean described in Section "Birds of mangroves in important bird areas (IBAs) of the Middle East and East Africa" below. Unlike IBAs, the selection of CBD sites under the SBSTTA process does oblige nations that are signatory to the CBD to designate sites in national laws. However, even if sites are designated, the implementation of monitoring, and of effective regulatory protection are by no means guaranteed in practice. The EBSA site descriptions below refer to those sub-sites that have already been designated within the areas proposed under SBSTTA.

The concept of landscape scale conservation inherent in CBD and the Large Marine Ecosystem approach is recognised as the essential first step in designing a network of effective conservation sites and the linkages between them. The challenge for the national agencies responsible for the large marine sites of the southern Indian Ocean will be the scale of necessary surveillance and enforcement action for conservation. In the case of threats from floating oil pollution, the widely-

Table 30.3 Selected areas meeting Convention on Biological Diversity (CBD) criteria for Ecologically or Biologically Significant Marine Area (EBSA) in the Southern Indian Ocean

EBSA name (In matrix, H = High, M = Medium, L = Low)	**Mangrove** habitat is a key feature in EBSA	Uniqueness or rarity	Special importance for life-history stages of species	Importance for threatened, endangered or declining species and/or habitats	Vulnerability, fragility, sensitivity, or slow recovery	Biological productivity	Biological diversity	Naturalness
1. Incomati River to Ponta do Ouro (S. Mozambique)	✓	M	M	H	M	H	H	M
2. Delagoa Shelf Edge, Canyons and Slope		H	H	M	M	M	H	H
3. Save River to San Sebastian (Mozambique)	✓	H	H	H	M	H	H	H
4. Morrumbene to Zavora Bay (S. Mozambique)	✓	H	H	H	M	H	H	L
5. Quelimane to Zuni River (Zambezi River Delta)	✓	H	H	M	L	H	–	M
6. Tanga Coelacanth Marine Park		H	M	H	M	L	M	L
7. Pemba-Shimoni-Kisite	✓	M	M	M	M	M	M	L
8. Baixo Pinda – Pebane (Primeiras & Segundas Is.)		M	M	M	M	M	H	M
9. Zanzibar (Unguja) – Saadani		M	M	M	M	M	M	M
10. Rufiji – Mafia- Kilwa	✓	M	M	M	M	M	H	M
11. Watamu Area	✓	H	M	M	M	M	M	M
12. Pemba Bay – Mtwara (Mozambique Channel)		H	M	M	H	H	H	L
13. Mozambique Channel		H	H	H	H	H	M	H
14. The Iles Éparses (part Mozambique Channel)		M	H	M	H	H	M	H
15. Lamu-Kiunga Area	✓	H	M	M	M	M	M	L
16. Northern Mozambique Channel		M	H	H	H	H	H	L
17. Moheli Marine Park		H	H	H	H	H	H	H
18. Southern Madagascar (Mozambique Channel)		H	H	H	M	H	H	H
19. Tromelin Island		H	H	H	H	H	H	H

(continued)

Table 30.3 (continued)

EBSA name (In matrix, H = High, M = Medium, L = Low)	**Mangrove** habitat is a key feature in EBSA	Uniqueness or rarity	Special importance for life-history stages of species	Importance for threatened, endangered or declining species and/or habitats	Vulnerability, fragility, sensitivity, or slow recovery	Biological productivity	Biological diversity	Naturalness
20. Mahe, Alphonse and Amirantes Plateau	✓	H	H	H	M	H	H	L
21. Atlantis Seamount		H	M	H	H	M	H	M
22. Blue Bay Marine Park		H	H	M	H	–	H	H
23. Saya de Malha Bank		H	H	–	–	H	H	H

Appendix to annex IV of the Report of the Southern Indian Ocean Regional Workshop to Facilitate the Description of EBSAs, UNEP/CBD/RW/EBSA/SIO/1/4, based on the Subsidiary Body on Scientific, Technical and Technological Advice 18, Recommendation XVIII/3 http://www.cbd.int/recommendations/sbstta/?m=sbstta-18

H high, *M* medium, and *L* low each represent scores against important criteria, and mangrove habitats are indicated (✓)

dispersed archipelagos of the Mozambique Channel and Mascarene plateau area are key examples. National ESI mapping programs and the knowledge of sensitivity to oil and dispersed oil of corals and mangrove shorelines are relatively well–known. Nine of the 23 EBSAs (39%) include rich mangrove habitats. This knowledge will need to be supplemented by an improved understanding of risks arising from sub-sea blowouts and to potential impacts from spills of HFO (*e.g.*, 'Katina P') and noxious cargoes (*e.g.*, 'Gulser Ana') from third-party carriers. The Tsimiroro field in Madagascar's Morondava basin holds a heavy crude oil being developed using steam-enhanced recovery that when exported will be a new source of risk to mangroves and other wetlands in western Madagascar. Thirty percent of the production at the major refinery in the region at Mombasa (Kenya) is fuel oil, and up to 10 Mt. per year of oil is shipped into Dar es Salaam (Tanzania) and Beira (Mozambique). The total tanker movements through the region are between 50–100 Mt. per year on each side of Madagascar (ITOPF 2011 data).

EBSA Sites with Mangroves As a Key Feature (Numbers Refer to Table 30.3)

EBSA 1 Incomati River to Ponta do Ouro (Southern Mozambique):

Location: Approximately 26°S and 33°E. This area covers Maputo Bay from the Incomati River mouth, the Lagoa Bight, and the regions of Baixo Danae. It also covers the whole coastline and offshore from Inhaca Island to Ponta do Ouro (the border between Mozambique and RSA in KwaZulu Natal).

Importance: Maputo Bay is diverse and supports critical habitats (e.g., extensive mangroves and seagrass beds, and the largest, southernmost coral reefs in sub-equatorial Africa, in addition to sandy and rocky beaches, exposed and sheltered coastlines). It has extremely high biodiversity in various taxa, including commercially significant fisheries and shrimp. The bay also has several taxa of special concern, such as dugongs, dolphins, three species of turtles (leatherback turtle, *Dermochelys coriacea*, loggerhead turtle, *Caretta caretta*, and green turtle, *Chelonia mydas*), sharks, whales, seahorses, endangered bivalves, and the vulnerable seagrass, *Zostera capensis*. Inhaca Island has recorded 33% of all bird species occurring in Southern Africa. The area is designated for the marine and terrestrial reserves of Inhaca Island and Machangulo peninsula.

EBSA 3 Save River to San Sebastian (Central Mozambique):

Location: Bazaruto Archipelago is located up to 20 km off the Mozambique coast within latitudes 21°30′–22°10′S and longitudes 35°22′–35°30′E. This area also covers the Twelve Mile Reef at 21°21′S; 35°30.2′E.

Importance: The Bazaruto Archipelago site, which is home of the most viable dugong population in East Africa is already a marine protected area. There are many megafauna, such as dugongs, turtles, dolphins and marlins, as well as seagrass meadows and mangrove forests found in the area.

EBSA 4 Morrumbene to Zavora Bay (Southern Mozambique):

Location: The area covers Inhambane Bay, the peninsula, and Tofo up to Zavora (covering regions of Pomene and Paindane).

Importance: This area has abundant megafauna, mainly the reef manta (*Manta alfredi*), giant manta ray (*Manta birostris*), and whale shark (*Rhincodon typus*), described as among the largest populations in the world. The area supports dugongs, five species of turtles as well as coral reefs (one of which is unique) and mangroves forests with extensive seagrass beds, mainly around Morrumbene and Inhambane Bay. The area has recently become a focus of research, and recent reports of new species of nudibranch around Pomene/Zavora support the value of this emerging hotspot of biodiversity in Mozambique.

EBSA 5 Quelimane to Zuni River (Zambezi River Delta):

Location: The area extends from the river dos Bons Sinais and the Zuni River in the south (mid-way from Chinde, main delta branch to Beira city). The delta gives rise to the Sofala Bank, which extends from Save River to the chain islands of Ilhas Primeiras e Segundas.

Importance: The delta and island chain is the largest and among the most productive fisheries areas in Mozambique yielding close to 50% of the entire industrial catches of Mozambique (some 50,000 tons in 2002). Sofala Bank is here represented by the Zambezi delta (Quelimane to Zuni River, about 200 km coastline). The productivity of this area for fisheries is directly related to the extensive mangroves of the Zambezi River delta, the largest mangrove forest in all of East Africa, covering some 100,000 ha.

EBSA 7 Pemba-Shimoni-Kisite:

Location: Between the latitudes 04° 50′ - 05° 30′S. The Pemba Channel has a high diversity comprising pelagic fishes, turtles, dolphins, dugongs and whales.

Importance: The Kisite-Mpunguti area, located in the Shimoni area on the southern coast of Kenya, incorporates the Kisite Marine Park, the largest no-take area in Kenya (28 km^2), and the adjacent Mpunguti Marine Reserve, Kenya's smallest reserve (11 km^2). The area supports a high diversity of marine life, including corals, reef fish and sea turtles, and is important for the life history of the coconut crab, a rare and endemic species. Kisite Island is an Important Bird Area (IBA), hosting species such as the sooty tern (*Sterna fuscata*) and large numbers (up to 1000 breeding pairs recorded) of crested tern (*Thalasseus bergii*) and roseate terns (*Sterna dougallii*), and encompasses a wide range of habitats, including mangrove forests, coral reefs, seagrass beds and offshore waters, which are considered important fish nursery grounds. The Pemba-Shimoni-Kisite area thus provides prime habitat for sea mammals and various types of corals and associated fish species.

EBSA 10 Rufiji – Mafia – Kilwa:

Location: Between latitude 7.1°S–9.0°S and longitude 39.2°E–40.6°E.

Importance: The area hosts significant populations of a variety of endangered marine species, such as dugong, sea turtles, coelacanth and other fin fish, shellfish and birds. The largest continuous mangrove areas are to be found on the coasts of Mafia, Kilwa and the delta of the Rufiji River.

EBSA 11 Watamu Area:

Location: Between 39.9°E, 3.5°S and 40.2°E, 3.3°S.

Importance: Habitats in this area include intertidal rock, sand and mud, fringing reefs and coral gardens, coral cliffs, sandy beaches and the Mida Creek mangrove forest. Biodiversity in this area includes fish, turtles, dugongs and crabs. The area is surrounded in part by the Mida Creek forest and has a high diversity of mangrove species, including *Ceriops tagal, Rhizophora mucronata, Bruguiera gymnorrhiza, Avicennia marina* and *Sonneratia alba*. These provide refuge to a variety of both resident and migrant bird species.

EBSA 15 Lamu-Kiunga Area:

Location: This area covers 40.3°E, 3.2°S and 41.9°E, 1.5°S.

Importance: The mangrove and tidal flat habitats in the area of Lamu on the Indian Ocean coast of north-eastern Kenya close to the Somali border, are known as some of the most extensive and species-rich along the entire coast of East Africa. They are highly valuable in terms of biodiversity, climate protection (blue carbon), fisheries, nature-based tourism and coastal protection.

EBSA 20 Mahe, Alphonse and Amirantes Plateau:

Location: Between 50°00 - 58°00′E and between 0°00′ - 10°00′S.

Importance: This is an area of high diversity and a breeding, feeding and nursery area for cetaceans, providing migratory paths for these species and important feeding sites for pelagic fish, especially tuna and shark species. Coral reefs and mangroves characterize this area, providing important sites for fish spawning and nurseries, while mangroves help in reducing sedimentation and runoff to coral reefs. The plateau assists in the conservation of seabirds through provision of breeding and feeding sites. Important nesting sites for green and hawksbill turtles are found here.

30.5.2.2 Endemic Bird Areas and Important Bird Areas

There are several sources of species data and distributions that are typically used by the conservation agencies and NGOs to assess and rank sites. The information is also useful in preparation of ESIAs and Environmental Risk Assessments. 'Sea Alarm' Country Wildlife Profiles are available for Kenya, Tanzania, Madagascar and Mozambique; these describe the species and habitats most vulnerable to oil pollution. Some examples of the available information follow, including the work of BirdLife International and its East African country partner organizations.

Endemic Bird Areas (EBAs)

These are areas enclosing the entire world population of two or more bird species with a restricted range (<50,000 km^2). Secondary Areas (SAs) are areas with one or more Restricted Range Species (RRS), but are not classed as EBAs because fewer than two species are entirely confined to it. There are eight SAs in the study area but none are coastal (Stattersfield et al. 1998). EBAs are clearly of world-scale conservation importance, being vital to survival of three-quarters of the planet's threatened bird species. Worldwide, most of these are forest-dwelling species. The EBAs in the study area are shown in Table 30.4. The majority of restricted range birds are confined to the EBA in question. Six of these EBAs are designated by UNESCO as World Heritage Sites (EBAs 093, 099, 100, 106, 107 and 108).

The main threats to the EBAs include forest loss and agricultural clearance (called 'tavy' in Madagascar). Clearings may be illegal and used to grow cardomoms, cloves, cocoa and coconuts. Coastal forests in N. Kenya at Boni and Dodori Nature Reserves are poorly-studied due to security concerns. The Tana River Primate Reserve is also at risk in this area, including from poaching. This valley is an unspoiled meandering channel and associated wetlands crossing a semi-arid region. EBA 105 is also poorly-studied since threats to security have precluded expeditions.

Other EBAs are of world-class importance, for example, EBAs 106 and 107 together comprise the 'Central Refugium' one of Africa's major centres of endemism and species richness for forest biota; it is one of the very few places that forest communities are known to have survived during the driest spells of the Quaternary Ice Age, and is of the highest biogeographic and evolutionary importance. EBA 106 contains UNESCO World Heritage Site Virunga National Park around Lake Edward, the Itombure mountains north of Lake Tanganyika, and the Bwindi Impenetrable

Table 30.4 Endemic Bird Areas (EBAs) in the East Africa region showing numbers of Restricted Range Species (RRS; corresponds to IBA Category A2)

EBA site number, name	Height (m)	Priority	RRS Threat	RRS Total	Habitats (coastal in bold)
092. SE. African coast	0–200	High	0	4	**Dune forest, coastal scrub** (MO, RSA)
093. W. Malagasy dry forest	0–800	High	4	8	Dry deciduous forest
094. E. Malagasy wet forest	0–2000	Critical	14	23	Lowland evergreen rain forest
095. E. Malagasy wetlands	0–2700	Critical	3	8	Lakes, marsh, **rocky shores, mangroves**
096. W. Malagasy wetlands	0–1500	Critical	5	7	Lakes, rivers, **marsh, deltas, mangroves**
097. S. Malagasy spiny forest	0–500	Urgent	2	10	Dry spiny forest, thorn scrub
098. Comoros Islands	0–2600	Critical	6	18	Forest, montane evergreen rain forest, heath
099. Aldabra	0–20	High	0	3	**Atoll, coral, mangroves, dune scrub**
100. Granitic Seychelles	0–900	Critical	8	11	Dry forest, upland rain forest
101. Réunion	0–3000	Urgent	1	7	Lowland/montane and tamarin forest, heath
102. Mauritius	0–800	Critical	7	10	Upland evergreen, dwarf and exotic forest
103. Rodrigues	0–300	Critical	2	2	Native and introduced evergreen trees
104. E. Zimbabwe Mountains	800–2400	High	1	3	Montane forest, dense edge undergrowth
105. Tanzania-Malawi Mts.	200–3000	Critical	20	37	Montane and lowland forest, grass/scrub
106. Albertine Rift Mountains	1000–4300	Urgent	11	37	Montane forest, bamboo, alpine heath
107. E. DRC lowlands	700–1800	High	3	6	Lowland rain forest (UG, Semliki forest)
108. Serengeti plains	1000–2200	High	2	6	Grassland, acacia scrub (E. shore L. Victoria)
109. Kenyan Mountains	900–4400	Urgent	3	9	Montane forest, bamboo, grass, alpine scrub
110. Pemba	0–100	High	0	4	**Coastal forest**, woodland (TA)
111. E. African coastal forests	0–500	Urgent	5	7	**Coastal forest**, acacia scrub (TA, KE, Somalia)

Stattersfield et al. (1998)
Threat Threatened

National Park. Virunga and Bwindi support most of the world's mountain gorilla population, and together are home to 31 RRS in one contiguous succession from low-elevation forest to highland forest.

Compared to the endemism, species richness and importance of the above world-class conservation areas and also to Africa's largest mangrove forests in the Niger Delta, the relatively small mangrove areas in East Africa and the Middle East are disjunct and therefore particularly vulnerable to piecemeal coastal development. The largest mangrove sites in East Africa are Zambezi Delta in Mozambique (100,000 ha) and Rufiji Delta in Tanzania (54,500 ha) compared to approximately 725,000 ha remaining in the Niger Delta (Gundlach 2018, Chap. 13 this volume).

Mangrove stands of all sizes are vulnerable to coastal 'squeeze' between the pressures of reclamation for development and aquaculture exacerbated by the rising sea levels with climate change. These losses of mangroves including the organic muddy sediment of their habitat, are all the more disastrous given their role as a carbon sink (blue carbon). Unfortunately, mangrove used as fuel wood and building materials or clear-felled for shrimp farms offers subsistence livelihoods and employment for rural poor who cannot necessarily focus on global or regional development issues. In Table 30.4 two of the EBAs regarded as 'critical' are in Madagascar coastal wetlands of which one of the main habitats is mangrove forest. Including Aldabra, three (3) EBAs in total have mangroves (15% of the EBAs). Although not an endemic species, the black-bellied starling *Notopholia corusca* breeds almost exclusively in mangroves (e.g. at Rufiji in EBA111) throughout the East African coastal flyway.

Birds of Mangroves in Important Bird Areas (IBAs) of the Middle East and East Africa

In collaboration with national agencies and experts around the world, BirdLife International maintains a global database on the IBA system. The status of IBAs confers no statutory protection, although some are already protected to a degree in national parks and nature reserves. However, each of these sites is a candidate for statutory protection by the respective national environmental agencies. For example 130,000 ha of the Tana delta (IBA KE022) have recently been designated by Nature Kenya as a community conservation area to protect endangered species and address UN Sustainable Development Goals. They are sites recognised by national and international ornithologists using accepted criteria (Table 30.5) as being key for conservation of birds whether they are in breeding, feeding, roosting, migrating or wintering concentrations. Species with the codes defined in Table 30.5 have been used to designate IBAs in the region (Table 30.6). Table 30.6 also shows the important conservation conventions across the region and, for some nations, the assessment and management of levels of threat. Marine, coastal and wetland birds are often threatened by oil and plastics pollution.

Although not as important for restricted-range species, the Arabian Peninsula is a major flyway for migrating birds, with up to 200 species passing through Saudi Arabia during both spring and autumn passage on the flyway between sub-Saharan Africa and the Palearctic. Many IBAs are designated or potentially qualify for designation, as Ramsar sites on the basis of their waterbird populations. Twenty-six (26) out of 46 IBAs (57%) and three EBAs (two in Madagascar and Aldabra) in the flyway include mangrove habitats.

Figure 30.2 shows IBAs in the Middle East. Al-Dhakira in Qatar has dense stands of mangrove *Avicennia marina*, broad mudflats (tidal amplitude is ~2 m) and saltmarsh vegetation; the mangrove is grazed by camels. The bay is protected from the open sea to the east by a spit, fed by longshore drift from north to south. There are important spawning and nursery areas for economically valuable fish and shrimp stocks, and some artisanal fishing occurs. In UAE, Abu al Abyadh Island supports mature *Avicennia* woodland and breeding crab plover *Dromas ardeola*, and the Islands off Sir Bani Yas are rich in breeding terns. Surveys on offshore islands in the Arabian

Table 30.5 Important Bird Area (IBA) criteria

A:	**Globally important**
A1	Globally threatened species
A2	Restricted-range species
A3	Biome-restricted species
A4	Congregations
A4i	Sites on a regular basis with >1% of biogeographic population of congregatory waterbird species
A4ii	Sites on a regular basis with >1% of the global population of congregatory seabird species
A4iii	Sites on a regular basis with >20,000 waterbirds or >10,000 pairs of seabirds of one or more species
B:	**Regionally important in Middle East**
B1i	Sites on a regular basis with > 1% of a flyway or other distinct population of waterbird species
B1ii	Sites on a regular basis with > 1% of a distinct population of seabird species
B2	Species with an unfavorable conservation status in the Middle East
B3	Species with favorable conservation status, but concentrated in Middle East

Table 30.6 Summary of IBAs and EBAs of selected countries of the Middle East and East Africa

	Conventions		Ramsar	Marine	Total	PSR [c]	Bird species	Globally- threat.		Endemic
	Nairobi[a]	Other[b]	sites	IBA	IBA	data	number	sp.	EBAs	species
Saudi Arabia		✓	0	11	39		391	19	1	0
Qatar		✓	0	3	5		232	8	0	0
UAE		✓	0	10	19		295	12	0	0
Kenya	✓	✓	7	6	66	✓	1060	42	8	11
Uganda		✓	0		34	✓	1000	28	6	2
Tanzania	✓	✓	7	5	77	✓	1078	50	9	29
Mozambique	✓	✓	0	2	16		674	29	4	2
Madagascar	✓	✓	4	10	84		246	35	7	115

BirdLife International http://datazone.birdlife.org/country accessed 20 May 2017
[a] *Nairobi* 1992 Convention and Protocol Concerning Protected Areas and Wild Flora and Fauna in the East African Region
[b] *Other* 1999 African-Eurasian Waterbird Agreement (AEWA) and Memorandum of Understanding on the Conservation of Migratory Birds of Prey in Africa and Eurasia; 1992 Convention on Biological Diversity (CBD); 1979 Convention on Migratory Species of Wild Animals (CMS); 1971 Convention on Wetlands of International Importance (Ramsar); 1978 Convention on the International Trade in Endangered Species of Wild Flora and Fauna (CITES); 1992 UN Framework Convention on Climate Change (UNFCCC) and Kyoto Protocol
[c] *PSR Pressure* (threats facing sites), *State* (conditions of birds and habitats), *Response* (actions being taken to conserve the site)

Fig. 30.2 Saudi Arabia, Qatar and UAE coastal and island IBAs (*green*), BirdLife International http://datazone.birdlife.org/country

Gulf are rarely carried out. Shara'awh Island in October 1973 had mummified chicks of Socotra cormorant *Phalacrocorax nigrogularis*, and nesting terns *Sterna bengalensis* and *S. repressa*. The Al-Ashat islands in 1970–1972 also had dead chicks of *P. nigrogularis*.

Coastal and offshore areas are crucial to seabird conservation in WIO international waters (Fig. 30.3). Considering the areas of ocean and archipelagos used by birds (e.g., many species of albatross, petrel and skua) and traversed by oil tankers, protection of these IBAs is a high priority.

Examples of coastal IBAs in East Africa include a chain of about 50 offshore islands and coral reefs in the Lamu Archipelago, running for 60 km parallel to the coastline off northernmost Kenya (KE015; also included in EBSA 15). The islands are ~2 km offshore, lying inshore of the fringing reef; they vary in size up to 100 ha, and the smaller islands provide nest sites for seabirds with little vegetation (e.g., *Salicornia* and the succulent *Sanseveria*). The coast has sandy beaches, and mangroves grow in the shelter of the larger islands. Over 5000 pairs of *Sterna dougallii* were recorded in August 1970, with 1195 pairs noted on islets near Kiunga in 1961. Other species nesting from June to September include *Larus hemprichii*, *Sterna repressa*, *S. anaethetus* and *Anous stolidus*. Over 800 *Dromas ardeola* were noted on southern Kiwayuu Island in November 1996 and migrant waders frequent the sheltered flats and mangrove creeks, mainly from August to April. Three globally threatened species of turtles (*Chelonia mydas*, *Eretmochelys imbricata* and *Lepidochelys olivacea*) nest on the beaches, and there are records of the globally threatened mammal *Dugong dugon*.

Mida Creek (KE016) is an important passage and wintering area for Palearctic migrant waders, the coastline supports important *Sterna saundersi* populations. Populations of *Charadrius leschenaultii*, *C. mongolus* and *Dromas ardeola* are internationally important with up to 6000 waders of many species. The creek is a feeding area for *Egretta gularis*, *Sterna bengalensis* and *S. dougallii*. Migrant shorebirds include *Calidris alba*, *C. ferruginea*, *Numenius phaeopus*, *Pluvialis squatarola*, *Charadrius leschenaultii* and *C. mongolus*. *Sterna dougallii* and *S. anaethetus* nest on Whale Island (June–October) in some years. *Sterna saundersi* occurs in internationally important numbers feeding close to shore.

The coastal zone and wetland birds of the SW. Arabian Gulf are listed in Table 30.7, taken from survey reports, Birds of the Western Palearctic (BWP 1977–1994), the IUCN Red List (2003), BirdLife International (BL 2003, 2005, 2008 and 2017), and field guides (e.g., Porter et al. 1996, 2004). Taxonomic order generally follows BWP, and conservation status is assessed using IUCN (2003), Collar et al. (1994), Tucker and Heath (1994) and Tucker and Evans (1997). The table shows the birds' usage of wetlands (48 species) and in particular mangroves (53 species), the seasonality and abundance, together with an assessment of conservation status in

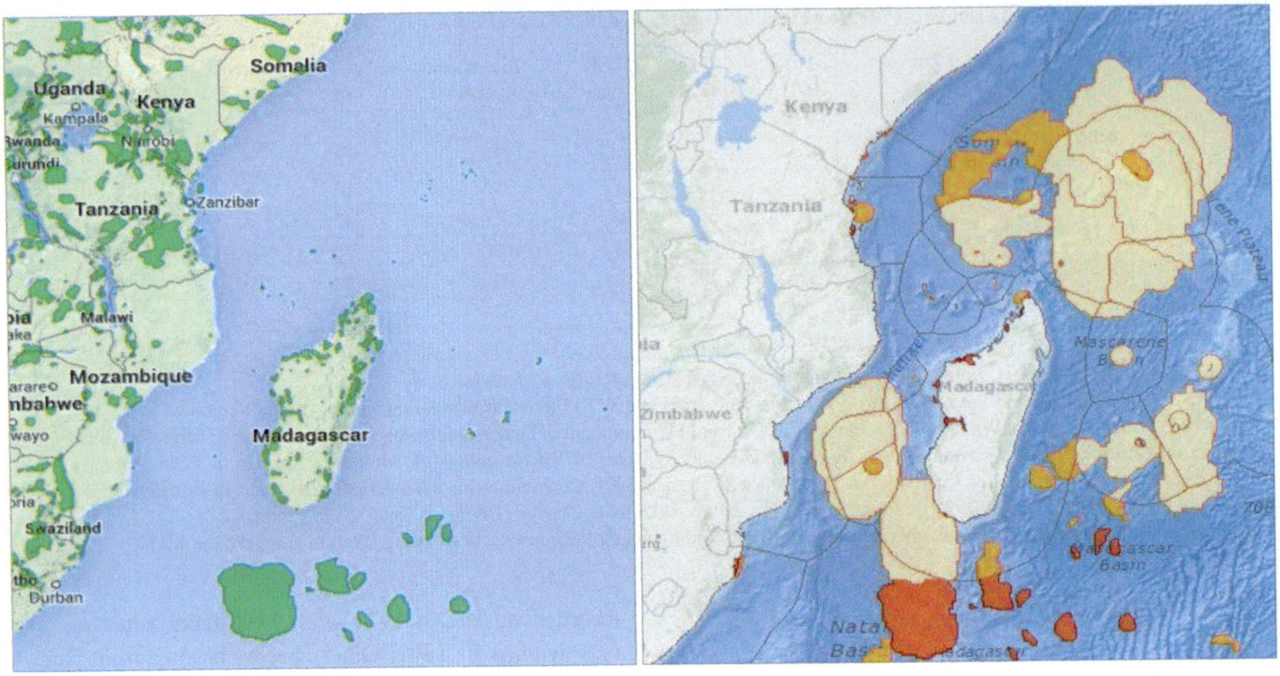

Fig. 30.3 (left) East African IBAs; (right) Marine IBAs (*red* confirmed, *orange* proposed, *yellow* candidate), BirdLife International http://datazone.birdlife.org/country

the Middle East using the codes in Table 30.5. The birds using mangroves represent more than 20% of the typical avifauna of the SW. Arabian Gulf states (about 255 species). The proportion increases to almost 40% when birds using all wetland types are included. In arid zones the conservation of IBA sites is especially important (i.e., a network of wetland habitats for conservation and management of flyways and areas of global or regional importance to distinct biogeographic populations of shorebirds). Yet the frequency of IBA assessment and recent counts are often poor, and the conservation of many IBAs is data-deficient.

The 1999 African-Eurasian Migratory Waterbird Agreement (AEWA) links nations on flyways between sub-Saharan Africa and the Palearctic. The marine IBAs on the flyway through Saudi Arabia, Qatar, UAE, Kenya, Tanzania, Mozambique and Madagascar are listed in Table 30.8 with the IBA codes. Most are being considered for designation as protected areas in each country. Also shown in Table 30.8 is the spatial coexistence between mangroves and corals, turtles and marine mammals, which is important in OSR planning because of the trade-offs often required between species and habitats (see Sect. 30.6).

30.6 Effects of Oil Spills

30.6.1 Introduction and Sensitivity Matrix

Mangrove, coral, seagrass and saltmarsh tropical ecosystems are particularly susceptible to oil spills, to discharges from shipping and also to coastal building developments in low latitudes. Although unlikely to occur, a significant oil spill may have disastrous and trans-boundary ecological, economic and reputation impacts. Oil spills are usually accidental, and the size and impacts of the incidents are unpredictable. Mangroves, coral reefs, seagrass beds and saltmarsh often co-exist in close proximity to each other, and a spill could occur at some distance away that would affect many of the most sensitive habitat(s). Table 30.9 shows how relative and seasonal sensitivities may be characterized, and this may require 'trade-offs' between communities and species during OSR.

Table 30.10 summarizes the likely range in the impacts of spills, approximate volumes and types of oil spilled, and cleanup operations, from examples of incidents occurring in tropical marine biological communities since the 1960s.

Table 30.7 Bird seasonality and abundance in wetland habitats particularly mangroves (bold) in SW Arabian Gulf, also showing species triggering IBA criteria

Common name	Scientific name	Seasonality	Habitat	Abundance	IBA	References
Little Grebe	*Tachybaptus ruficollis iraquensis*	Breeding & winter	Pond, reed	Locally common		QP 2002
Great Cormorant	*Phalacrocorax carbo*	Winter	Wetland, pelagic		B1i	BL 2003, BWP
Socotra Cormorant	*Phalacrocorax nigrogularis*	Breeding Jan/Mar & winter	Sand/gravel islands, G. Salwah	Common, but long-term decline	A1, A4i, B1i, B2	BL 2003, BWP, IUCN Red, Dome 2004 (D. 2004)
Dalmatian Pelican	*Pelecanus crispus*	Winter possible – unlikely	Lagoon	Vagrant		BWP
Squacco Heron	*Ardeola ralloides*	Winter	**Mangrove (Mangr.)**, reed	Rare		QP 2002
Night Heron	*Nycticorax nycticorax*	Winter	**Mangrove**, reed	Rare		QP 2002
Little Bittern	*Ixobrychus minutus*	Breeding, passage & winter	Pond, lagoon			Porter 1996, 2004
Striated Heron	*Butorides striatus*	Resident	**Mangrove**, mud, sand	Fairly common		D. 2004, Porter 1996, 2004
Western Reef Heron	*Egretta gularis*	Breeding & winter Jun/Aug	Lagoon, sand, pelagic	Locally common	B1i	BL 2003, BWP, QP 2002
Little Egret	*Egretta garzetta*	Passage & winter	**Mangrove**, mud, sand	Uncommon		QP 2002
Cattle Egret	*Bubulcus ibis*	Passage	**Mangrove**, farm	Uncommon		QP 2002
Great White Egret	*Egretta alba*	Passage & winter (Nov-Feb)	**Mangrove**, mud	Scarce		QP 2002
Grey Heron	*Ardea cinerea*	Passage & winter	**Mangrove**, mud, sand	Moderate	B1i	BL 2003, QP 2002
Purple Heron	*Ardea purpurea*	Winter	**Mangrove**, mud, sand	Uncommon		BWP, QP 2002
Spoonbill	*Platalea leucorodia*	Winter	**Mangrove**, mud	Rare		QP 2002
Glossy Ibis	*Plegadis falcinellus*	Passage	**Mangrove**, reed	Scarce		QP 2002
Yellow-billed Stork	*Mycteria ibis*	Winter (escape?)	**Mangrove**, mud	Rare		QP 2002
Greater Flamingo	*Phoenicopterus ruber*	Passage & winter	Mud, sand	Locally common		BWP, QP 2002
Shelduck	*Tadorna tadorna*	Winter	**Mangrove**, pond, lagoon, mud	Rare		QP 2002
Marsh Harrier	*Circus aeruginosus*	Winter	**Mangrove**, reed	Moderate		BWP, QP 2002
Spotted Eagle	*Aquila clanga*	Passage, winter	Wetland, arable	Vulnerable		IUCN Red
Imperial Eagle	*Aquila heliaca*	Passage, winter	Wetland, arable	Vulnerable		IUCN Red
Osprey	*Pandion haliaetus*	Breeding Jan/Mar & winter	**Mangrove**, lag., pel., rock, jetty	Scarce, UAE on islands		BWP, QP 2002, D. 2004
Peregrine Falcon	*Falco peregrinus*	Winter	Savannah, coasts			BWP
Water Rail	*Rallus aquaticus*	Breeding & winter	Reed, pond			Porter 1996, 2004
Little Crake	*Porzana parva*	Winter (Nov–Feb)	Reed	Rare		QP 2002
Eurasian Oystercatcher	*Haematopus ostralegus*	Passage & winter	**Mangrove**, lagoon, mud, sand	Uncommon	B1i	BL 2003, BWP, QP 2002
Black-winged Stilt	*Himantopus himantopus*	Pass. & winter, breeds wet yrs	Wetland, sand	Fairly common	B1i	BL 2003, BWP, QP 2002, D. 2004
Avocet	*Recurvirostra avosetta*	Passage & winter	Wetland	Rare		BWP, QP 2002
Crab Plover	*Dromas ardeola*	Breeding & winter (Nov–Feb)	**Mangrove**, mud, sand	Rare (Bubiyan Is. most)	A4i, B1i, B3	BL 2003, QP 2002

(continued)

Table 30.7 (continued)

Common name	Scientific name	Seasonality	Habitat	Abundance	IBA	References
Little Ringed Plover	*Charadrius dubius*	Breeding & winter	**Mangr.**, pond, lag., mud, sand	Moderate		BWP, QP 2002
Ringed Plover	*Charadrius hiaticula*	Passage & winter	**Mangr.**, pond, lag., mud, sand	Uncommon		BWP, QP 2002
Kentish Plover	*Charadrius alexandrinus*	Breeding & winter resident, opportunistic	**Mangr.**, lag., pond, mud, sand, sabkha, islands	Locally abundant	B1i	BL 2003, BWP, QP 2002
Lr. Sand (Mongolian) Plover	*Charadrius mongolus*	Passage & winter	**Mangrove**, dunes, lag., pond	Common	B1i	BL 2003, BWP, QP 2002
Greater Sand Plover	*Charadrius leschenaultii*	Passage & winter	**Mangrove**, lagoon, mud, sand	Moderate, common	B1i	BWP, QP 2002, D. 2004
Grey Plover	*Pluvialis squatarola*	Passage, winter	**Mangrove**, pond, mud, sand	Locally common	B1i	BL 2003, BWP, QP 2002
Red-wattled Plover	*Hoplopterus indicus*	Breeding & winter	Pond, lagoon	Vagrant		BWP, Porter 1996, 2004
White-tailed Plover	*Chettusia leucura*	Summer & passage	Pond, lagoon (shallow)			BWP, Porter 1996, 2004
Sociable Plover	*Chettusia gregaria*	Winters occasional UAE	**Mangr.**, grass, reed, desert	Vulnerable, rare		IUCN Red, QP 2002
Lapwing	*Vanellus vanellus*	Winter	Grassland, coasts	Rare		BWP
Great Knot	*Calidris tenuirostris*	Passage & winter	Mud	Scarce		Porter 1996, 2004
Knot	*Calidris canutus*	Passage	Sand, mud	Scarce	B1i	BL 2003
Sanderling	*Calidris alba*	Passage & winter	Sand	Uncommon		BWP, QP 2002
Little Stint	*Calidris minuta*	Passage & winter	**Mangrove**, pond, mud, sand	Locally common		BWP, QP 2002
Temminck's Stint	*Calidris temminckii*	Passage & winter	**Mangrove**, pond, lagoon, mud	Uncommon		BWP, QP 2002
Curlew Sandpiper	*Calidris ferruginea*	Passage & winter	**Mangrove**, lagoon, sand, mud	Scarce – Fairly common		BWP, QP 2002, D. 2004
Dunlin	*Calidris alpina*	Passage & winter	**Mangrove**, pond, mud, sand	Common	B1i	BL 2003, BWP, QP 2002
Broad-billed Sandpiper	*Limicola falcinellus*	Passage & winter	**Mangrove**, lagoon, pond	Scarce	B1i	BL 2003, QP 2002
Ruff	*Philomachus pugnax*	Passage & winter	**Mangrove**, pond, lagoon, reed	Uncommon		BWP, QP 2002
Jack Snipe	*Lymnocryptes minimus*	Passage & winter	Wetlands, farm			BWP
Snipe	*Gallinago gallinago*	Passage & winter	**Mangrove**, pond, lagoon, farm	Uncommon		BWP, QP 2002
Great Snipe	*Gallinago media*	Passage	Grassland, wetlands	Data deficient, Vuln.		IUCN Red, BL 2003
Pintail Snipe	*Gallinago stenura*	Passage	Grassland, wetlands			Porter 1996, 2004
Black-tailed Godwit	*Limosa limosa*	Passage & winter	**Mangrove**, lagoon, mud, sand	Uncommon		QP 2002
Bar-tailed Godwit	*Limosa lapponica*	Passage, winter	**Mangrove**, lagoon, mud, sand	Fairly common	B1i	BL 2003, BWP, QP 2002, D. 2004
Whimbrel	*Numenius phaeopus*	Passage & winter	**Mangr.**, lag., mud, sand, coral	Uncommon	B1i	BL 2003, BWP, QP 2002
Eurasian Curlew	*Numenius arquata*	Passage & winter	**Mangrove**, mud, sand	Locally common	B1i	BL 2003, BWP, QP 2002

(continued)

Table 30.7 (continued)

Common name	Scientific name	Seasonality	Habitat	Abundance	IBA	References
Common Redshank	*Tringa totanus*	Passage & winter	**Mangrove**, pond, lagoon, mud	Locally common	B1i	BL 2003, BWP, QP 2002
Spotted Redshank	*Tringa erythropus*	Passage & winter	**Mangrove**, pond, lagoon	Locally moderate		BWP, QP 2002
Marsh Sandpiper	*Tringa stagnatilis*	Passage & winter	Wetland, reed, mud	Widely scattered		BWP, Porter 1996, 2004
Greenshank	*Tringa nebularia*	Passage & winter	Pond, mud, sand	Uncommon		BWP, QP 2002
Green Sandpiper	*Tringa ochropus*	Passage & winter	Pond, lagoon, inc. man-made			BWP
Wood Sandpiper	*Tringa glareola*	Winter	Pond, lagoon, inc. man-made			BWP
Terek Sandpiper	*Xenus cinereus*	Passage & winter	**Mangrove**, lagoon, mud, sand	Moderate	B1i	BL 2003, BWP, QP 2002
Common Sandpiper	*Actitis hypoleucos*	Passage & winter	Pond, sand	Moderate		BWP, QP 2002
Red-necked Phalarope	*Phalaropus lobatus*	Passage	Pond, lagoon, pelagic (winter)	Scarce		QP 2002
Great black-headed Gull	*Larus ichthyaetus*	Winter	Pelagic, lagoon			BWP
Black-headed Gull	*Larus ridibundus*	Passage & winter	Wetlands, pelagic			BWP
Slender-billed Gull	*Larus genei*	Winter	Wetland		B1i	BL 2003, BWP, D. 2004
Lesser black-backed Gull	*Larus fuscus*	Winter	Pelagic, human sites			BWP
Herring Gull	*Larus argentatus*	Winter	Pelagic, human sites			BWP
Gull-billed Tern	*Gelochelidon nilotica*	Passage & winter	Lagoon, mud		B1i	BL 2003, BWP, D. 2004
Caspian Tern	*Sterna caspia*	Breeding June/Aug & winter	**Mangrove**, lag., mud, sand, pel.	Moderate	B1i	BWP, QP 2002
Swift Tern	*Sterna bergii*	Breeding June/Aug & winter	Lagoon, sand, pelagic	Uncommon		BWP, QP 2002
Lesser Crested Tern	*Sterna bengalensis*	Breeding June/Aug & winter	Pelagic; Coral Islands KSA largest nesting site in world	Uncommon	B1i	BL 2003, BWP, QP 2002
Sandwich Tern	*Sterna sandvicensis*	Passage & winter	Wetland, mud, sand			BWP
Roseate Tern	*Sterna dougallii*	Summer	Maritime	Vagrant		Porter 1996, 2004
Common Tern	*Sterna hirundo*	Winter	Wetlands, sand			BWP
White-cheeked Tern	*Sterna repressa*	Breeding June/Aug & winter	Lagoon, sand, pelagic	Scarce-common	A4i, B1i, B3	BL 2003, BWP, QP 2002, D. 2004
Bridled Tern	*Sterna anaethetus*	Breeding & winter	Scrub, islands, pelagic, sand	Common	B1i	BL 2003, BWP, D. 2004
Little Tern	*Sterna albifrons*	Passage	Lagoon, pelagic			BWP
Saunders' Little Tern	*Sterna saunders'i*	Breeding June/Aug	**Mangrove**, lag., mud, sand, pel.	Mod. Abu Ali KSA	B3	BL 2003, QP 2002, D. 2004
Short-eared Owl	*Asio flammeus*	Winter	**Mangrove**, farm, desert	Scarce		QP 2002
Common Swift	*Apus apus*	Passage	Farm, pond, lagoon	Scarce		QP 2002
Pallid Swift	*Apus pallidus*	Breeding (Nov)	Farm, pond, lagoon	Uncommon		BWP, QP 2002
Sand Martin	*Riparia riparia*	Passage	**Mangrove**, farm, lag., reed	Moderate		QP 2002
Pale Crag Mart./Af. Rock M.	*Ptyonoprogne fuligula*	Passage	**Mangrove**, farm, pond	Rare		BWP, QP 2002

(continued)

Table 30.7 (continued)

Common name	Scientific name	Seasonality	Habitat	Abundance	IBA	References
Crag Martin	*Ptyonoprogne rupestris*	Passage	**Mangrove**, farm, pond	Scarce		QP 2002
House Martin	*Delichon urbica*	Winter	**Mangrove**, farm, pond, reed	Uncommon		QP 2002
Barn Swallow	*Hirundo rustica*	Passage	**Mangrove**, farm, pond, sand	Common		BWP, QP 2002
Red-rumped Swallow	*Hirundo daurica*	Passage	**Mangrove**, farm, pond	Scarce		QP 2002
Pied Kingfisher	*Ceryle rudis*	Winter	**Mangrove**			BWP
Kingfisher	*Alcedo atthis*	Passage & winter	**Mangrove**, mud, pond	Uncommon		BWP, QP 2002
White-breasted Kingfisher	*Halcyon smyrnensis*	Winter	**Mangrove**	Rare		QP 2002
White-cheeked Bulbul	*Pycnonotus leucogenys*	Breeding & winter	**Mangrove**, garden, sand	Abundant		BWP, QP 2002, D. 2004
Yellow-vented Bulbul	*Pycnonotus xanthopygos*	Breeding & winter	**Mangrove**, garden	Resident		BWP
Bluethroat	*Luscinia svecica*	Winter	Mud, wetland			BWP, QP 2002
Moustached Warbler	*Acrocephalus melanopogon*	Breeding & winter	Low reed and sedge	Al-Hasa Lagoon KSA		BWP
Marsh Warbler	*Acrocephalus palustris*	Passage	Wetland	Vagrant Qatar		Porter 1996, 2004
Clamorous Reed Warbler	*Acrocephalus stentoreus*	Breeding & winter	**Mangrove**, reed, pond, sand	Scarce (250–750 pr.)		BWP, QP 2002, D. 2004
Chiffchaff	*Phylloscopus collybita*	Winter	**Mangrove**, plantation, reed	Common		BWP, QP 2002

Not included in Table 30.10 are anecdotal reports on long-term damage caused by spills in tropical habitats in World War II, because in most cases no estimate of the amount of oil spilled was made (NRC 1985). However, reports from long-term spill impact studies following the 1991 Gulf War are included, because of good documentation of spill volumes and their residues remaining after natural attenuation and cleanup (Michel et al. 2005; Getter et al. 2005).

In addition to follow-up studies conducted at the sites of marine and coastal oil spills, there are field and mesocosm experimental studies summarized at the end of Table 30.10.

Results from laboratory studies may not be representative of field conditions during an oil spill, and some laboratory studies do not include data on the water column or tissue oil concentrations actually causing the response in the organism. Laboratory studies also are often hampered by the static nature of the test, where additional stress on the subject can build up in the test system due to microbial action, oxygen depletion and waste product accumulation (Knap et al. 1985).

30.6.2 Inter-Related and Sensitive Tropical Coastal Habitats

30.6.2.1 Mangroves

Mangroves are found on 75% of tropical coastlines (25°N to 25°S) and represent a major part of tropical coastal biological productivity, comprising approximately 70 species of tree and bush, each adapted to life in salt water which they desalinate by ultra-filtration. Mangroves are under pressure from urban and industrial coastal developments, non-sustainable wood harvesting, salt lagoons and aquaculture ponds. In addition to the trees themselves, mangroves provide food and habitat to a range of associated species, and also support a detrital food web that recycles nutrients from fallen leaves. Mangrove prop roots are often encrusted with filter-feeding and grazing molluscs, algae and barnacles, and the sediment below is habitat to numerous crabs and infauna. Mangrove-fringed bays are an important nursery site for juvenile fish, and support nesting, feeding and migrating birds. They are a principal and sustainable form of coastal

Table 30.8 Important Bird Areas (IBAs) of selected countries (N to S) in Middle East and East Africa showing IBA trigger species and correspondence with mangroves, corals, turtles and marine mammals (*SA* Saudi Arabia, *QA* Qatar, *AE* United Arab Emirates, *KE* Kenya, *TZ* Tanzania, *MZ* Mozambique and *MG* Madagascar)

Site	Site name	Descriptive IBA code	**Mangroves**	Coral/ turtles	Mam-mals	IUCN Red List	Area (ha)
SA007	Gulf coral islands	A1, A4i, A4iii, B1i, B2, B3		✓		VU[e]	2000
SA008	Abu Ali	A4i, B1i, B3		✓		Terns, osprey	12,500
SA010	Tarut Bay	A4i, A4iii, B1i, B3	Formerly	✓	✓	EN[i] NT[fg] terns	41,000
SA012	Al-Hasa Lagoon (inland)	A4iii, B1i, B2				Ferruginous duck	7500
SA013	Gulf of Salwah	A1, A4i, A4iii, B1i, B2, B3			✓	VU[e] grebe, terns	62,500
QA001	Al-Dhakira Mangrove	B1i	✓<1000 ha			Greater flamingo	3000
QA002	Al-Aliyah Island	A1, B2, B3		✓		VU[e] terns	65
QA003	Shara'awh Island	A1, B2, B3		✓		VU[e] formerly, terns	22
QA004	Al-Ashat islands	A1, B2				VU[e] formerly	62
QA005	Khor al-Udeid	B1i		✓	✓	Greater flamingo	12,000
AE003	Siniyah Island	A1, A4i, B1i, B2	✓			VU[e]	1000
AE005	Khor al Beideh	A4i, B1i, B3	✓			EN[i] NT[fg]	5000
AE012	Qarnayn Island	A1, A4i, A4ii, B1i, B1ii, B2, B3		✓✓		VU[e] Red-b-tropicbird	300
AE014	Ghaghah Island	A1, A4i, B1i, B2				VU[e] Osprey	800
AE015	Islands off Sir Bani Yas	A1, A4i, B1i, B2, B3	✓	✓		VU[e] terns	190
AE016	Yasat islands	A1, A4i, B1i, B2, B3		✓✓	✓	VU[e] terns	2000
AE017	Abu al Abyadh Island	A4i, B1i, B3	✓	✓	✓	Shorebirds, terns	60,000
AE018	Umm Amin Island	B1i, B3		✓		Shorebirds, terns	50
KE014	Kisite Island	A4i		✓		Roseate tern	34,285
KE015	Kiunga Marine National Reserve	A4i	✓	✓✓	✓	Crab plov. roseate tern	25,000
KE016	Mida Creek, Whale Is., Malindi-Watamu	A4i	✓5 spp. 1600 ha	✓		Shorebirds, terns	26,100
KE022	Tana River Delta	A1, A3, A4i, A4iii	✓	✓	✓	EN[h] NT[de] shorebirds, terns	130,000
TZ012	Mafia Island	A4i		✓ fossil		Shorebirds	115,000
TZ021	Dar es Salaam Coast	A4i, A4iii	✓			VU[c] shorebirds	61,000
TZ027	Latham Island	A4i, A4ii, A4iii		✓ fossil		Noddy, terns	3
TZ028	Mnazi Bay	A4i				Shorebirds	10,000
TZ032	Rufiji Delta	A4i	✓7 spp. 54,500 ha			NT[e] shorebirds, terns	72,000
TZ036	Tanga South	A4i	✓			Shorebirds	4000
TZ044	Zanzibar Island – South Coast	A4i	✓	✓		Shorebirds, terns	4000
TZ045	Zanzibar Island – East Coast	A4i	✓			Shorebirds, terns	10,000
TZ056	Pangani coastal forest	A1, A2, A3	✓600 ha			NT[d]	4600
MZ001	Maputo special reserve	A1, A2, A3	✓			EN[fg] NT[d]	104,000
MZ007	Zambezi River Delta	A1, A3, A4i	✓			VU[d] NT[d]	500,000
MG001	Cape Anorontony Archipelago	A1, A3, A4i		✓✓		CR[a]	13,464
MG009	Ambavanankarrana wetland	A1, A2, A3, A4i	✓			CR[a] EN[ab]	51,442
MG012	Ampasindava Bay wetlands	A1, A2, A3	✓			CR[a] EN[b]	163,000
MG018	Sahamalaza – Radama Island	A1, A3, A4i	✓8 spp. 8400 ha	✓		CR[a] EN[bc] NT[a]	56,878
MG020	Loza Bay wetlands	A1, A3	✓		✓	CR[a]	57,734
MG023	Mahajamba – Anjavavy wetlands	A1, A2, A3, A4i	✓47,500 ha			CR[a] EN[b]	186,637
MG024	Bombetoka Bay – Marovoay NPA	A1, A2, A3, A4i	✓45,000 ha		✓	CR[a] EN[ab] VU[a] NT[b]	78,814
MG026	Baly Bay national park	A1, A2, A3, A4i	✓	✓	✓	CR[a] EN[abc] VU[ab] NT[ac]	396,789
MG028	Cape St Andre Forest and wetland	A1, A2, A3	✓4 spp.			CR[a] VU[a] NT[ac]	110,261

(continued)

Table 30.8 (continued)

Site	Site name	Descriptive IBA code	**Mangroves**	Coral/ turtles	Mam-mals	IUCN Red List	Area (ha)
MG035	Tambohorana wetland NPA	A1, A2, A3, A4i	✓			CR[a] EN[ab] VU[b] NT[c]	83,441
MG036	Barren Islands marine protected area	A1, A3, A4i				EN[b]	74,930
MG059	Tsiribihina Delta and river	A1, A2, A3, A4i	✓5 spp.			CR[a] EN[abc] VU[bc]	264,100
MG062	Lake Ihotry – Mangoky Delta complex NPA	A1, A2, A3, A4i	✓			CR[a] EN[cde] VU[ab]NT[abc]	176,105

BirdLife International (2017) http://datazone.birdlife.org/country
IUCN Red List codes and key to trigger species:
CR Critically endangered [a]Madagascar fish-eagle *Haliaeetus vociferoides*
EN Endangered [a]Madagascar teal *Anas bernieri*, [b]Madagascar heron *Ardea humbloti*, [c]Madagascar pond-heron *Ardeola idea* [d]Sakalava rail *Zapornia olivieri* [e]Madagascar marsh harrier *Circus macrosceles* [f]Cape vulture *Gyps coprotheres* [g]Spotted ground thrush *Geokichla guttata* [h]Basra reed warbler *Acrocephalus griseldis* [i]Great knot *Calidris tenuirostris*
VU Vulnerable [a]Madagascar grebe *Tachybaptus pelzelnii* [b]Black-banded plover *Charadrius thoracicus* [c]Madagascar Pratincole *Glareola ocularis* [d]Wattled crane *Bugeranus carunculatus* [e]Socotra cormorant *Phalacrocorax nigrogularis*
NT Near threatened [a]Madagascar crested ibis *Lophotibus cristata* [b]Lesser flamingo *Phoeniconaias minor* [c]Madagascar jacana *Actophilornis albinucha* [d]Southern banded snake-eagle *Circaetus fasciolatus* [e]Curlew sandpiper *Calidris ferruginea* [f]Eurasian curlew *Numenius arquata* [g]Bar-tailed godwit *Limosa lapponica*

Table 30.9 Generalized seasonal ecological sensitivities of mangroves (bold) and other coastal and offshore biota in south-western Arabian Gulf: in northern summer many species die back or leave the area due to very high temperatures; winter is generally richer with many benthic species in their reproductive and growth phase

Resource	J	F	M	A	M	J	J	A	S	O	N	D	Comment
Mangroves	Most	Most	Least	Least	Moderately	Moderately	Most	Most	Most	Least	Least	Most	Most sensitive to heat and salinity stress in dry season (northern summer)
Seagrass/ saltmarsh	Most	Most	Moderately	Moderately	Least	Least	Least	Least	Least	Moderately	Moderately	Most	Low biomass in summer , higher sensitivity in growth season (winter)
Plankton	Least	Least	Least	Least	Moderately	Moderately	Moderately	Moderately	Least	Least	Least	Least	Productivity highest in spring, but most populations resilient to stressors
Benthos	Most	Most	Least	Most	Moderately	Moderately	Moderately	Moderately	Most	Least	Most	Most	Spawning, larval and moulting stages sensitive to pollution
Coral	Most	Most	Most	Most	Moderately	Moderately	Moderately	Moderately	Moderately	Most	Most	Most	Temperature stress already resulting in coral bleaching; sedimentation impacts
Fish	Most	Most	Most	Least	Moderately	Moderately	Moderately	Least	Least	Most	Most	Most	Spawning sensitive period; eggs, fish larvae, and juveniles all vulnerable
Turtles nesting	Least	Least	Least	Most	Most	Most	Most	Least	Least	Least	Least	Least	Hawksbills nest May-July; green turtles nest June-Sept (both, islands)
Turtles hatching	Least	Least	Least	Least	Least	Most	Most	Most	Moderately	Least	Least	Least	Hawksbill hatchlings emerge July-Aug; green turtles emerge Jul-Oct
Turtles feeding	Most	Most	Most	Most	Most	Moderately	Moderately	Moderately	Most	Most	Most	Most	Hawksbill on coral; green turtles are sensitive if seagrass biomass is low
Shorebirds general	Moderately	Moderately	Moderately	Moderately	Moderately	Least	Least	Least	Least	Moderately	Moderately	Moderately	Moderate sensitivity due to numbers & diversity of migrants Eurasia-Africa
Shorebirds breeding	Most	Most	Most	Least	Least	Most	Most	Most	Most	Least	Least	Least	Some (*e.g.*, cormorant and osprey) breed Jan-Mar; most terns June-Aug
Dugong	Most	Most	Most	Most	Most	Moderately	Moderately	Moderately	Most	Most	Most	Most	Generally absent during lowest seagrass biomass; calving is most sensitive

Legend: Most Sensitive; Moderately Sensitive; Least Sensitive

Table 30.10 Selected examples of oil spill and cleanup impacts focusing on mangroves (bold) but often including nearby corals, saltmarsh and seagrass

Oil spill/experiment	Tonnes (t) and type of oil	Reported impacts of oil and cleanup	References
'Argea Prima', Guanica, Puerto Rico, 1962	10,000 t crude oil	Extensive damage to **mangroves** and mortalities of adult and juvenile lobster, crabs, sea urchins, sea stars, sea cucumbers, gastropods, octopus, turtles and fish; *Thalassia* sp. seagrass beds degenerated; rocks denuded	Diaz-Piferrer (1964)
'Witwater', Galeta Island, Panama, 1968	2720 t diesel oil and Bunker C	**Mangroves** (*Rhizophora mangle, Avicennia* sp.) 49 ha killed of ~490 ha **mangrove** overall; impact >23 years 80% recovered; impact on meiofauna, fiddler crabs, sessile animals on prop roots and algae eliminated for >5 years; reef corals least affected	Rützler and Sterrer (1970), Keller and Jackson (1993) and Duke (2016)
Pipeline break, Tarut Bay, Saudi Arabia, 1970	13,600 t Arabian light crude oil	**Mangroves** *Avicennia marina* defoliated at 3 months but survived and less affected than invertebrates and fish; mortalities among crabs, bivalves, gastropods, fish; no detrimental effects on corals and associated reef fauna; good subsequent recovery	Spooner (1970)
'Santa Augusta', St. Croix, US Virgin Islands, 1971	12,000 t crude oil	*Rhizophora mangle* 5 ha completely destroyed of about 50 ha **mangrove** overall; little or no recolonization after 7 years; hand-planting of propagules successful after 8 years shows that this may be simply due to lack of propagules reaching the site	Lewis and Haines (1980) and Lewis (1983)
'Zoe Colocotronis', Cabo Rojo, Puerto Rico, 1973	5030 t Venezuelan crude oil	*Rhizophora mangle* and *Avicennia germinans* death of adult **mangroves** across 1–2.7 ha within 3 years; prop root invertebrates reduced; partial recolonization by 1977; some sediment oiling still affecting *R. mangle* fringe in 1978–79; recovering in 2002	Nadeau and Bergquist (1977), Gilfillan et al. (1981) Getter and Lewis (2003) and Lewis (2005)
'Garbis', Florida keys, 1975	200–400 t crude oil	*Rhizophora mangle* and *Avicennia germinans (nitida)* young **mangroves** killed when >50% foliage was oiled; crabs eliminated for several months	Chan (1977)
'Showa Maru', Karang Banteng, Malacca Straits, Indonesia, 1975	7350 t Arabian light, Berri and Murban crude oils, some dispersant used offshore	*Sonneratia* and *Rhizophora* spp. significantly affected in sheltered bays, with *Littorina* snails and *Uca* crabs associated with oiled sediments; *Rhizophora* **mangrove** (20 ha) killed in backshore pockets where oil was able to penetrate edge of swamp	Baker (1982)
Pipeline break, Corpus Christi Bay, Texas, 1976	45 t crude oil	*Avicennia germinans* **mangroves** burned to remove oiled trees, un-cleaned trees recovered after minor defoliation	Holt et al. (1978)
'Peck Slip', barge spill, Puerto Rico, 1977	1420–1490 t bunker C oil	*Rhizophora mangle* **mangroves** significantly affected; also crabs, snails and epiphytes	Gundlach et al. (1979) and Getter et al. (1981)
'Howard Star', Tampa, Florida, 1978	130 t (20% diesel and 80% bunker C oil)	*Rhizophora mangle, Avicennia germinans, Laguncularia racemosa* **mangrove** mortalities and root abnormalities, also molluscs and polychaetes affected	Lewis (1980)
Spills in Gulf of Mexico and Caribbean Sea	Historical comparison of five oil spill sites	Responses of oiled **mangroves** across various sites were similar, including defoliation and tree/seedling mortality, and leaf deformation; however,	Getter et al. (1981)

(continued)

Table 30.10 (continued)

Oil spill/experiment	Tonnes (t) and type of oil	Reported impacts of oil and cleanup	References
		differences in the physical environment (e.g., topography, wave exposure) affected the fate and persistence of the oil	
Funiwa-5 well blowout, offshore Niger Delta, Nigeria, 1980	27,000–57,000 t light Nigerian crude oil, some dispersant used offshore	340 ha **mangroves** killed along approximately 35 km of affected shoreline (5100 ha)	Gundlach et al. (1995), Little et al. (2003) and Aghalino and Eyinla (2009)
Gulf War oil spills, 1991–present	1500,000 t Kuwait crude oil and burn residues	Over 800 km of Saudi coastline were affected by the deliberate Gulf War spills, with 8 million m^3 of oiled sediments, 45% in mudflats and 23% in marsh and **mangroves**; 400 ha **mangroves** (almost one-third of the mangrove resource) were oiled, and of these 120 ha were killed; oil is mostly found in crab burrows down to >50 cm depths; >70% of upper intertidal zones of marsh/**mangrove** have reduced species richness and were not fully recovered after 12 years; waterlogged tidal flats impede oil weathering and natural removal; cyanobacteria mats are widespread indicators of grazer mortality	MEPA (1991), Michel et al. (2005) and Getter et al. (2005), UNCC (2016)
'Katina P' Mozambique 1992	67,000 t HFO	An estimated 500 t was stranded on shorelines and approximately 1450 km shoreline was oiled, mostly sporadic tarballs on exposed sandy shores, some of it in Republic of South Africa; **mangroves** along about 3 km of Maputo Bay were oiled, but proportions killed are unknown in areas of intense fuel wood cutting and erosion	Little et al. (2003), CEDRE (2015), ITOPF, NOAA (2016)
Petrobras refinery spill, Guanabara Bay, Brazil, 2000	1145 t Bunkers and diesel	Many Guanabara Bay shorelines were affected within rocky, sand and **mangrove** habitats and including nearshore artisanal fish traps, but the proportions of trees killed are unknown	Brito et al. (2009) and NOAA (2014)
Malaysian university experiments, Pantai Acheh, 1984	Arabian light and Malaysian Tapis crude oils, some treated with dispersant Corexit 9527	Untreated crude oil was more toxic than dispersed crude oil to **mangrove** saplings; untreated oil in the surface sediments needed longer to weather and depurate than chemically dispersed oil; however, dispersed oil was more toxic than untreated oil to associated fauna	Hoi-Chaw and Meow-Chan (1985)
TROPICS experiment on two 900 m^2 sites, Laguna de Chiriqui, Atlantic Panama coast, 1984	North slope (Prudhoe Bay) crude oil 20–50 ppm for 24 h, glycol ether-based dispersant in 30 × 30 m plots comprising 50% mangrove and 50% seagrass/coral	Fresh untreated crude oil had minor effects on *Thalassia testudinum* seagrasses and corals, but >25 year effects on *Rhizophora mangle* **mangroves**' survival and associated fauna; oil that was chemically dispersed just offshore had minor impacts on **mangroves** but did affect corals in the short-term; a classic example of an environmental 'trade-off' decision	Ballou et al. (1989) and DeMicco et al. 2011
Turkey point experiment, Florida, 1985	South Louisiana crude oil, non-ionic water-based and glycol ether-based dispersants in 3 m^2 plots	*Rhizophora mangle* **mangroves** treated with pre-dispersed light crude oil simulating a slick partially dispersed offshore, showed no greater mortality than those in untreated plots; high pressure seawater washing of mangroves one day after spill was ineffective in reducing toxicity of oil, as was washing using non-ionic dispersant	Teas et al. (1989)

protection in hurricane zones, and stabilise suspended sediment regimes, acting as natural water clarifiers (see below) and desalination agents (Baker 1982).

The 1973 'Zoe Colocotronis' spill (Puerto Rico), the 1975 'Showa Maru' spill (Indonesia) and the 1980 'Funiwa-5' well blow out (8 km offshore Nigeria) were among the many spills reported to have caused widespread mangrove mortality (Gundlach et al. 1995). These incidents, especially the last, occurred in areas that also support human communities heavily dependent on mangrove ecosystem services such as fishing, hunting, gathering building materials, poles for fencing and fish traps, and firewood. In densely populated coastlines near mangroves there may be pressure to apply strenuous cleanup measures. Cleanup in mangroves is difficult due to their inaccessibility, and in soft sediment the consequent danger of increasing the severity and duration of impacts by trampling, deeper oil penetration, and prolonged physical intrusion.

The impact of crude oil on mangroves includes both direct toxicity (to roots, shoots, leaves and particularly saplings), and the effect of smothering due to air being excluded from the spongy breathing tissues by the blocking of mangrove lenticels on the prop roots and pneumatophores (breathing roots). Root cell membranes are directly sensitive to low molecular weight hydrocarbon compounds, and to secondary impacts due to the resulting impairment of the cellular salt exclusion process that protects the osmotic balance in healthy trees (Lai Hoi-Chaw et al. 1993).

The weathering of the oil by evaporation and oxidation leads to reduction in the direct toxic impacts, but heavier oils are more damaging in terms of smothering impact. Severely oiled leaves suffer defoliation due to the damage to trichomes, which regulate water losses through the stomata. Total defoliation will invariably kill the trees, although recovery can begin within months if leaf losses are only partial. Degradation of oil and reduction of toxicity can be rapid in the tropics, and allow seedling growth within one year. However, in anoxic sediments the oil does not degrade so fast, and depending on water circulation and energy level, the recovery of the mangroves may be greatly protracted, at least 10 and sometimes more than 30 years.

Attempts are sometimes made to accelerate mangrove recovery by transplantation programs. These mitigation actions have been most successful where sediment oil toxicity has declined by weathering processes, but the trees have not recovered unaided because for some reason propagules are not reaching the area. Transplantation is a long–term commitment and involves skilled labour, intensive work and adequately equipped nursery facilities.

Mangrove forests (and saltmarsh plants) effectively trap oil when the tide is high, and thereafter can act as secondary sources of oil pollution to nearby corals and seagrass communities. Their fine sediment substrates are sites of long-term entrapment of pollutants. Mangrove forests are also filters of suspended sediment particles that would otherwise be available to smother nearby corals by direct sedimentation and also to suppress their growth by the interruption of the light needed for zooxanthellae photosynthesis. Because of this functional coupling between mangrove and coral reef systems, it is possible that a mangrove forest killed by oil can create indirect impact on nearby corals. Spill impacts are complex in space and time, and thus difficult to predict from the records of real spill incidents.

The TROPICS field experiment described below for corals compared the impacts of simulated chemically-treated and untreated oil slicks on mangrove communities. In this case, the untreated oil was more damaging than dispersed oil to the trees and also to the associated epifauna on prop roots and infauna in the sediment. Observations made at this 30-year study site showed continuing presence of oil in sediments, substrate erosion and less recovery of the mangroves where dispersant had not been used (Baca et al. 2005). In contrast to coral reefs, it may be concluded that mangroves may be protected by judicious use of chemical dispersants offshore.

30.6.2.2 Coral Reefs

Corals are generally colonial animals that depend on symbiotic relationships with microscopic plants called zooxanthellae. These plants receive nutrients and shelter and in return the coral obtains energy from the photosynthetic activity of the zooxanthellae, which builds the corals' skeletal material. Because of this need for light to drive photosynthesis, corals are usually limited to shallow and clear waters.

There are still gaps in fundamental knowledge of the coral ecosystem, and little is known of the long-term effects of oil on corals in the presence of other stresses, including elevated sea surface temperatures at the global scale. Corals are prone to widespread bleaching and mortality under these circumstances. They have been locally affected by inappropriate coastal development, sediment plumes from dredging, metals from smelters, and blasting for fishing, navigation and construction. Anti-fouling paints are an additional concern for corals (Owen and Knap 2002; Negri et al. 2002).

Large coral barrier/fringing reefs and atolls support diverse fish and invertebrate communities. Although less spectacular, blue-green algae fix nitrogen on the reef and coralline algae help cement the reef structure. The ecological and human use values of these smaller patch reefs, coralgal reefs and re-worked skeletal sediments are more obscure than the larger reefs. In all coral systems the delicate balance of zooxanthellae, coralline and blue-green algae, allows corals to survive, fix CO_2 and support diverse biological communities in some of the harshest marine environments. This complexity makes it difficult to make general predictions of impact and recommendations for mitigation.

Field study results on oil spill impacts on coral reef systems are often contradictory, ranging from mass mortality of coral, fish and invertebrates to no apparent effect. To simplify and control the multitude of field variables, mesocosm experiments offer some clarity and have shown oiling effects such as decreases in coral growth rates, reproduction, colonization capacity, feeding, and behavioral impacts.

Temporary tissue swelling, rupture and polyp retraction were observed in a flow-through test on *Diploria shigosa*. Using C_{14} uptake measurements, changes in lipid metabolism and a reduction in zooxanthellae primary productivity were observed, but the corals recovered after one to a few days in clean water (Knap et al. 1985). Alizarin red staining was used to mark coral skeletons non-destructively, and assess the impacts of oil and dispersed oil (Corexit 9527) on coral growth in the field mesocosms; no significant differences were seen over 1 year. Long-term bioassays and coupled measurements of oil, dispersed oil and water-soluble fractions (WSF) were made in this program, reproducing field conditions in terms of exposure to hydrocarbons.

Oil slicks only come into direct contact with corals at extreme low tide and in particular when the oil has weathered and reached a density when it can begin to sink. Unrealistically high oil concentrations were often used in laboratory studies, for example, direct contact with oil for between minutes to 2 h caused tissue death and no re-growth. Oil droplets are formed in surf zones, and the suspended sediment loadings can cause oil to sediment out of the water column. Oil tends to stick faster to branching species such as *Acropora* because of surface area effects. The larval stages are particularly sensitive to oil pollution, as with many animal species.

Oil droplets also adhere to the mucus coating of corals, but will normally be shed along with the mucus, leading to less sensitivity to oil in the case of *Fungia* and *Symphyllia* spp. that are abundant producers of mucus. Depuration by this mechanism was estimated to take about 14 days (Neff and Anderson 1981). Impacts on zooxanthellae photosynthesis were not observed for *Madracis*, *Oculina*, *Favia* and *Montastrea* spp. that were exposed to south Louisiana crude, No. 2 fuel oil and pure phenanthrene at 1–6 ppm for up to 72 h. However, in some species calcium deposition was altered in proportion to the oil dose. These authors also reported naphthalene accumulation after acute exposure.

A range of responses from oil-impacted corals has been documented, including mortality and reduced reproductive capability (by around four times), both leading to reef degeneration. Oil can reduce numbers of breeding colonies, average numbers of ovaria per polyp, and numbers of larvae per coral head. Coral polyps exposed to WSF of oil have been shown to expel their larvae prematurely, which may be harmful at the population level, considering the spasmodic nature and synchronicity of many corals' spawning behavior (Knap 1992). *Stylopiora pistillata* only release larvae at night, for example, and so premature release (abortion) upon exposure to oil or WSF at an effective concentration of <1.0 ml/l, would decrease chances of settlement and survival (Loya and Rinkevich 1979).

Chronic impacts on corals are often more severe than those from individual acute spills, as shown by long-term observations adjacent to oil refineries and terminals in Aruba (Knap 1992) and the Gulf of Eilat, Red Sea. In addition, heat stress has weakened coral viability in some industrial outfall zones (e.g., Ruwais Industrial city in UAE). Chronic pollution or thermal stress can make the corals more susceptible to impacts from other natural or anthropogenic factors (*e.g.*, after coral mortality caused by an extreme low tide in the Gulf of Eilat, it took about 10 years for corals to recover in a clean environment but no recovery took place in the chronically-polluted reef).

The effects of oil spills have often been shown to include the mortality of the associated fauna of reefs such as fish, crustaceans and urchins rather than the corals themselves. In a reef affected by the 'Witwater' spill off Panama (Jackson et al. 1989), impact was observed on corals down to 3–6 m water depths, with branching corals more susceptible than massive corals. Effects were greater and deeper than recorded in other incidents, perhaps due to the trapping of oil in nearby mangroves for over 2 years, which then leached out back into the reef system.

The TROPICS experiment simulated a classic 'trade-off': attempting to protect shoreline biological communities by the offshore use of chemical dispersants, at the risk of the dispersed oil causing adverse impacts on near-shore seabed flora and fauna (Ballou et al. 1989). The experiment demonstrated that corals might be slightly damaged by untreated oil that was made available to them by natural WSF dispersion downward by 0.5 m into the water column. Deeper corals in this case remained unharmed. The mangroves, however, received the full brunt of the untreated oil (see above). When chemically dispersed, which effectively protected the mangroves, the oil did affect corals and seagrasses to a greater depth. The oil would have passed relatively harmlessly above the coral if the simulated slicks had been left untreated, but obviously at the expense of the mangroves.

30.6.2.3 Seagrass Beds

These are very sensitive temperate and tropical habitats not only for the seagrass itself, but also for sediment infauna and epibiota, foraging areas for marine turtles (tropics) and birds, and as nursery areas for fish. Therefore, seagrass communities are a vital habitat-forming element in many temperate and tropical coastal sedimentary environments. Seagrasses are the only truly marine examples of flowering higher plants. Seagrass tends to colonize relatively hard-packed sediment substrates that are in equilibrium with erosion and sedimentation processes. Both sand and muddy sand substrates are colonized in a range of inter-tidal and subtidal settings. As with salt marshes and mangroves, seagrass beds are vulnerable to coastal engineering, construction and reclamation projects. The temperate seagrasses have also been decimated by disease at intervals during the twentieth century.

When assessing the sources of impacts on the flora and fauna of seagrass beds, it should be remembered that many stresses may appear slight in isolation, but in combination even at different times of year they may be significant. For example, marine turtles and dugongs are both heavily dependent on seagrass in the southern Arabian Gulf. These megafauna may be affected not only by oil pollution, but also by the deposition of debris (plastics, fishing lines, litter, polythene bags), sewage from vessels and urban and industrial sites, and by noise and disturbance from marine operations. Turtles are particularly prone to ingesting marine litter (Tomás et al. 2002). Turtles and dugongs are at risk of collision with small craft, and in the case of turtle hatchlings, distraction by artificial lights. Both corals and seagrasses can be damaged by ships' hulls and anchor arrays, pipe-laying activities, scouring and dredging of sediments as well as by changes in sedimentation patterns, reducing photosynthesis and smothering benthic organisms (Allen 1992).

Although found in the intertidal zone, the main beds of tropical seagrass tend to be located in subtidal lagoons and shallow areas protected by corals, spits and bars. These locations are often those in which flood-tidal current asymmetry and shelter from wave action cause oil to be trapped. In such areas, it has been observed that long-term impacts are often more severe when oil has penetrated the sediment and is not able to flush out naturally. This is particularly likely to occur when the oil is located in animal burrows and other macropores within sheltered low energy environments (Howard and Little 1987; Kingston et al. 1999; Michel et al. 2005).

As with corals, mangroves and salt marsh habitats, field and laboratory experiments on oil and cleanup methods have been carried out in seagrass beds. In agreement with results from temperate field experiments (Howard et al. 1989), it was found by Thorhaug et al. (1986) that mixtures of oil and dispersant were most harmful to subtropical and tropical seagrasses in terms of reduced percentage cover and mortality. The epiphytic algae living on seagrass have also been shown to be very sensitive to dispersed oil. There was great variation in effects depending on which of three Caribbean test species was concerned. Two of the three seagrass species were also affected by the dispersant when applied alone. This latter result contrasted with the temperate field experiment, but it should be noted that these laboratory aquaria experiments ran for between 5 and 100 h contact time between oil, dispersant and seagrass specimens, a longer contact time than the 6 h of low tide in the temperate field experiment.

30.6.2.4 Subtropical Marsh

As with mangroves, soft sediment marshes are very sensitive habitats that support a diversity of invertebrate, bird and fish communities at different stages of the tide. They are frequently under threat of reclamation, chronic pollution and construction activities, and are vulnerable to oil spills due to the tendency of oil to accumulate at high water in the sheltered areas occupied by marshes. These process are often amplified by the complicated creek systems and the surface area (often dry and thus prone to oil adhesion) offered by the plants themselves. Halophytic plants colonize sheltered bays, estuaries and in arid zones such as the southern Arabian Gulf, hypersaline marine inlets known as khors. Such vegetation also occurs on sand and gravel beach ridges, and in all these geographic settings may be vulnerable to stranding and wind-blown oil.

Research into oil and cleanup impacts in tropical coastal marshes extended the early work from temperate and sub-tropical sites that showed the greater impact of repeated small spills and also the variations in impact between annual and perennial plant species. Some marsh halophytes are also extremely sensitive to perturbations that upset osmosis or the oxygen fluxes in the root systems of plants inhabiting anoxic fine-grained sediments. Oil on leaves or penetrating sediments can interfere with these processes, and early dispersants and crude oil in combination caused plant mortality, with little significant differences between treatments (Baker 1976). Later dispersants (Corexit 7664 and BP1100WD) applied alone had little or no impact on *Spartina* marsh (cord grass), but were not effective in removing the oil (Baker et al. 1984). When dispersants were used to treat stranded Nigerian crude and emulsified medium fuel oils they were effective in reducing sediment oil concentrations (Little and Scales 1987).

More recent experiments on subtropical salt marshes, however, show that protecting marsh communities from oil by using dispersants offshore (i.e., before oil enters the marsh) can be highly advantageous. The modern dispersant JD-2000 relieved the adverse effects of crude, diesel, and No. 2 fuel oils on *Spartina alterniflora* and *Sagittaria lancifolia* that were affected by low doses of untreated oils (Qianxin Lin and Mendelssohn 2005). Nevertheless, when chemical dispersants are used to prevent oil coming into contact with vegetation and the communities they support, including birds, there are possible impacts to water column and seabed communities to consider as part of the 'trade-off'.

Recovery times for oiled marshes are generally up to 5 years, irrespective of cleanup, with longer recovery times in the case of thick oiling, widespread light toxic oils (penetrating sediments and killing roots), or where inappropriate cleanup methods are used, such as removal or trampling of oiled sediments by the labor-intensive operations. Consequently, salt marsh is generally best left alone after being oiled because of the potential severity of such ancillary impacts during cleanup. However, some very narrow fringing marshes and halophytes on beach ridges are accessible to manual cleanup crews. Small groups of workers may be able to retrieve oiled strandline debris, if access is optimised by helicopter, hovercraft or shallow-draft vessels.

If the oil has not penetrated below the sediment litter layer, dispersants may also be effective in helping to flush oil out of marsh creek systems. If the marsh is predicted to die in any case due to the severity of the initial oiling, chemical dispersants applied aerially may speed recovery by reducing sediment oil content and preventing the formation of oil crusts and pavements. Similarly, in wetlands where the rooting zone is protected by a high water-table, it may be acceptable to employ *in situ* burning to remove heavily oiled shoots and stems, depending on the time of year with respect to nesting birds.

30.6.3 Oil Spill Case Studies

30.6.3.1 Kenya Oil Spills

The National Contingency Plan lists 33 marine spills in Kenya between 1972 and 2005. ITOPF report that several groundings have occurred on the reefs outside Mombasa harbour but with no resulting pollution. They highlight a 1988 spill of 5000 t of fuel oil at Makupa causeway, Kibarani reported to have killed 10 ha of mangrove. The Oil Spill Mutual Aid Group (OSMAG) was formed in 1992 and report three significant spills:

- The 'Rattan Shalini' oil spill at the Kipevu Oil Terminal in 2005 is the largest oil spill in Kenyan waters handled so far. Approximately 350 t of crude oil was spilled at the harbor after the ship's hull was punctured while berthing. OSMAG in conjunction with Kenya Ports Authority (KPA) mounted a fast response and effective cleanup with no serious environmental impact
- OSMAG together with KPA responded to a spill in 2009 from a truck laden with 20 m^3 of waste oil at the Makupa causeway, Kibarani
- A power producer's shore tanks released HFO into Makupa creek in January 2012.

30.6.3.2 Mozambique Oil Spills

The largest spill so far to have affected the East African coast was in Mozambique, stemming from the loss of the tanker 'Katina P', one of the top 20 tanker spills. Figures 30.4, 30.5, 30.6 and 30.7 show photographs taken in late May 1992 illustrating the 'Katina P' incident during work on behalf of the government of Mozambique by Arthur D. Little.

On 17 April 1992 the tanker was disabled by a 'freak' wave in the Mozambique Channel. It was carrying 66,700 tonnes of HFO from Venezuela to the UAE. The vessel lost hull plating amidships resulting in a release of approximately 3000 tonnes of cargo. It was intentionally grounded on a sand bar six miles offshore of Maputo Bay to prevent sinking. In an attempt to avoid further coastal

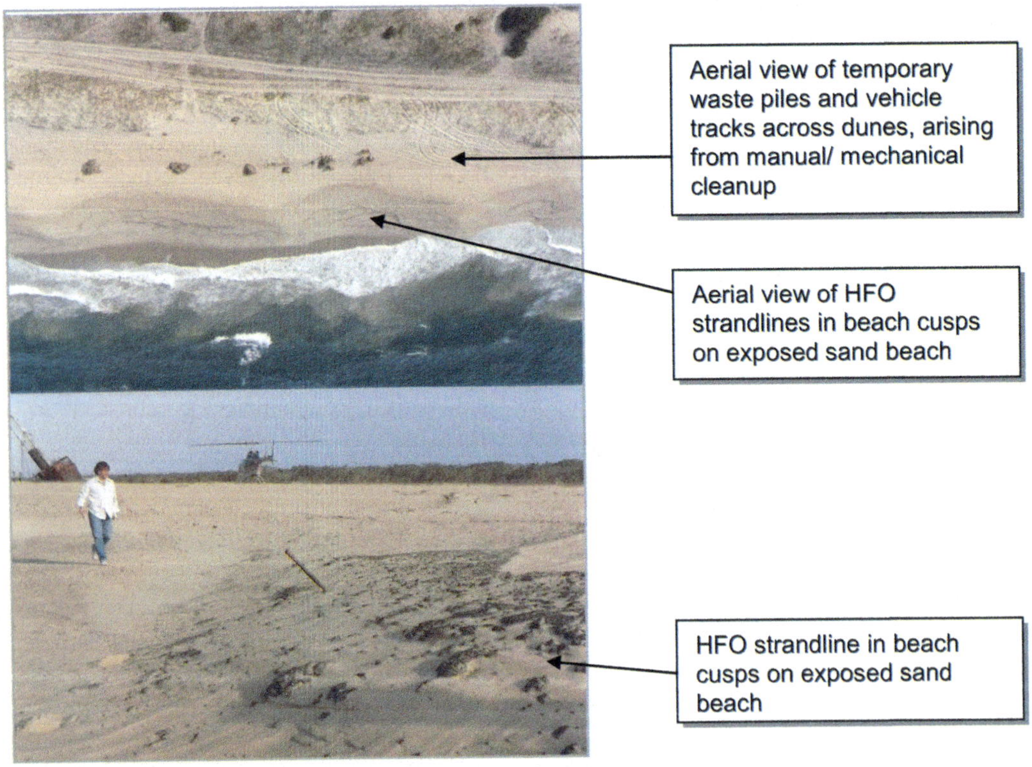

Fig. 30.4 'Katina P' HFO near Macaneta Point, Mozambique, 25 May 1992 (Credit: Little)

Fig. 30.5 'Katina P' HFO impacts on shorelines in the lower River Incomati estuary, Mozambique, May 1992: (clockwise from top left) oiled mangroves over tidal range; evidence of pre-existing mangrove cutting for fuel wood and building materials; mangrove peat substrate exposed by erosion and pre-existing damage from cutting (not significantly oiled); HFO in exposed sand beach showing 4 cm penetration depth; tar pavement in sand strandline (Credit: Little)

pollution, the vessel was towed out to sea for lightering to another tanker. However, the vessel broke in two while under tow, and sank on 26 April approximately 85 nautical miles off Mozambique, with a further release of oil. Some of the oil initially stranded on various beaches in and outside Maputo Bay. Oil that spilled when the vessel sank travelled south on the Agulhas current, with limited amounts of weathered oil stranding on the RSA coastline.

Shoreline cleanup was organised by the Ministry of Public Works with the municipalities using local labor, excavators and trucks. However, work was hampered by limited resources and the ongoing civil war. Oil entered mangrove stands in Maputo Bay. These were left to clean naturally in order to minimise further damage. A fishing ban imposed in the bay immediately after oil came ashore was soon lifted.

Fig. 30.6 'Katina P' HFO impacts on low-energy shores in south Maputo Bay, Mozambique, May 1992: (clockwise from top left) oiled mangroves; close-up of oiled pneumatophores; HFO residues in sediments being cleaned manually using hoes (as a pilot test) and piled at top of shore, but the advice given was not to continue due to destruction of pneumatophores and the low oil penetration depths and low risk of persistence in sandy sediment (Credit: Little)

Salt pans along the Matola River were closed for some weeks.

UNEP-WCMC reporting on mangroves of East Africa in 2003 stated that there had been 14 minor and two major crude oil spills in Maputo Bay that affected mangroves. Following the 'Katina P' such repeated spills in combination with sewage runoff in the area have produced a high pollution baseline in Maputo Bay (Sect. 30.8.7.2).

Although not affecting mangroves, a 30 m^3 spill of oil sludge was reported on 10 May 2014 by Anadarko in

Fig. 30.7 'Katina P' HFO showing manual/mechanical cleanup and temporary waste storage in south Maputo Bay, Mozambique, May 1992: (clockwise from top left) liquid oil in unlined pits with co-mingled solid waste; oiled sand and bags in unlined pit also being used as a latrine; vehicle tracks in mangroves, note sumps (*) used to collect free-phase oil; HFO-coated rocks on moderately-exposed shores of the bay (Credit: Little)

northern Mozambique during drilling operations about 46 km off the coast of Mocimboa da Praia district. Approximately 30,000 l of oil sludge spilled, and there was a prompt response to the spill including closure of the well that was being drilled and removal of the drilling platform. No cleanup was needed since mud or oil slicks were not detected during aerial surveys.

30.6.3.3 Madagascar Oil Spills

Also not affecting mangroves, a significant spill occurred in southern Madagascar after the loss of the bulk carrier 'Gulser Ana' which ran aground near Faux Cap in August 2009. The vessel was carrying 39,250 tonnes of rock phosphate as well as 568 tonnes of HFO. The integrity of the vessel quickly deteriorated, resulting in a significant oil spill and loss of cargo, with further smaller losses over the following months. By the end of 2009 all the phosphate cargo and HFO had been lost causing approximately 47 km of oil contaminated of sandy beaches (Laruelle 2012).

ITOPF reported that although 'Gulser Ana' cargo was observed dissipating into the sea, none was observed to have washed ashore. Faux Cap is a remote coastal settlement of a few thousand inhabitants and many outlying villages interlinked by sandy tracks. There is a lack of infrastructure, vehicles and supplies, and this was a major logistical challenge. The cleanup operations were organized by international contractors using local workers (Reynolds 2014).

The nearest airport is located in Fort Dauphin, a 7 h drive away, and the capital of Madagascar, Antananarivo, is 1200 km to the north. ITOPF agreed a cleanup plan and work began on 3 September, just over a week after the incident occurred. Subsistence fisheries using hand gathering from intertidal reefs are important to the local livelihood. There is also a limited amount of fishing by canoes and commercial gathering of lobster. This dependency on fishing was the main driver for the cleanup. The affected coastline was sand beach where manual techniques using basic tools recovered the light to moderate oiling, including buried oil and submerged oil in shallow lagoons. Spill responders trained villagers on the most effective cleanup techniques to remove only oil and oily material, producing as little waste as possible. An additional phase of cleanup was recommended for heavily oiled rocks using a combination of flushing and high pressure washing by specialist equipment mobilized from France.

The rationale for conducting the monitoring studies during the 'Gulser Ana' spill was concern over seafood safety and contamination of drinking water wells near beaches. Sampling and monitoring protocols are a topic in OSR where international standards are applicable. It is important that the results of sampling are related to the relevant standards (e.g., whether concentrations of contaminants measured in biota exceed maximum allowable concentrations given in seafood safety guidelines). Where the country does not have a national standard, ITOPF recommends use of appropriate international or national guidelines and standards (*e.g.*, WHO, EU, etc.).

In April 2014 the oil tanker 'Tromso' ran around in the approaches to Diego Suarez Bay in northern Madagascar, carrying a cargo of around 10,000 tonnes of crude oil. A lightering and salvage operation was carried out. Boom and skimming equipment was mobilized by the salvage contractors but was not deployed as no oil was spilled. If either the 'Tromso' or 'Gulser Ana' incidents had occurred on the east coast of Madagascar the likelihood of prolonged mangrove impacts would have been significantly higher.

30.6.3.4 First Gulf War Oil Spills

Early Effects

After the Gulf War oil spills of about 1.5 Mt. in early 1991, the assessment and monitoring studies revealed no serious, general, long-term impacts of the oil on seagrass and coral communities in the waters of Saudi Arabia (Kenworthy et al. 1993; Price et al. 1994; Poulsen et al. 2006). In a monitoring study by Saudi Aramco during 1981–1993, there was no significant increase in the levels of n-alkanes accumulated in the clams *Meretrix meretrix* as a consequence of the oil spills at five sites (Safaniyah Bay, Tanajib Bay, Manifa Bay, Ras Az Zawr, and Abu Ali North). Dibenzothiophenes and phenanthrenes increased significantly in the clam tissues during the war but concentrations of these aromatic hydrocarbons returned to pre-war levels within 2 years (Vazquez et al. 2000). Anomalous pneumatophores and adventitious roots in *A. marina* mangroves were observed 2 years after the spills (Böer 1993).

Longer-term Effects

Mangroves are relatively sparse and highly localized (only about 4 km^2 along the Saudi Arabian Gulf coast compared to about 200 km^2 along its Red Sea coast). About 90 % of mangroves in the Eastern Province were lost due to land reclamation since 1955, particularly in Tarut Bay, which was not oiled in the 1991 spills. The Gulf War oil spills affected seven mangrove areas, of which the most heavily affected were the mangroves on Qurmah Island in the sheltered southern portion of Dawhat ad-Dafi (MEPA 1991). The potential for long-term impact on mangroves had been shown previously by Burns et al. (1994), Garrity et al. (1994), and Levings and Garrity (1995). Mangrove mortality due to the Gulf War spills was estimated at 30%, with adverse impacts also on the associated flora and fauna. Initial cleanup using high-volume sprinklers to flush oil from the exposed mangrove root systems successfully increased mangrove density compared with untreated areas (De Clerck and Koppejans 1994). Further monitoring in Saudi Arabia revealed the slow recovery of the mangroves and associated biota, and yet presumably on the basis of avoiding additional damage during remediation, no major corrective action was taken for the affected mangrove sites. Another reason for this was that impacts on the existing mangroves on the Saudi coastline of the Arabian Gulf were also less in spatial and temporal terms than the impacts on saltmarshes and tidal flats. However, Saudi Aramco's eco-park in Tarut Bay clearly shows the potential for mangrove restoration http://www.ipieca.org/media/2781/saudi_aramco_biodiversity_protection_areas.pdf

Initially, it was found that 48 saltmarsh sites were heavily oiled, which accounted for 149 km of oiled shoreline and 677,000 m^3 of oiled sediment (MEPA 1991). Long-term impacts on marshes are common after acute, repeated and chronic oil spill impacts. The literature on saltmarsh restoration shows that assessment of the ecological condition and recovery potential of an oiled marsh, and the identification of restoration goals, is prerequisite for deciding between restoration and non-intervention (e.g., Baker et al. 1990, Clewell and Aronson 2007). A model was established for predicting saltmarsh recovery under moderate, heavy and extreme oiling conditions (Baker et al. 1996). The model predicted that 60% of marshes would be non-recovering after 11–12 years, and the surveys in late 2002-early 2003 showed that marshes were indeed between 60–80% non-recovering. Although comparison with the model's 'extreme' category was good, large areas of dead saltmarsh were still visible decades later. In some Gulf war-affected shores, a thin layer of clean sediment was deposited over the 1991 oiled surface. This allowed some annual seedlings (*e.g., Salicornia* sp.) to germinate and survive, but longer-term recovery was unlikely. This is partly due to the very low rates of sedimentation measured: averages between 1991 and 2002 were 0.8 cm in saltmarshes, 1.1 cm in mangroves, 1.6 cm on mudflats and 2–3 cm on sandflats, compared with >20 cm on exposed sand beaches.

The degree of biological recovery that has taken place in the affected areas since 1991 is very low compared to previous oil spills, such as those considered in the model proposed by Baker et al. (1996). Foremost among the reasons for this slow recovery is the huge volume of oil that was released (1.5 Mt). Also, oil was concentrated into 800 km of linear shoreline which caused oil deposits of up to 20 cm thick at the time, and penetrating deeply in animal burrows. Although about a tenth of the oil was recovered by the cleanup, the extremely wide intertidal embayments, complex channels and creeks, marshes and ridge-and-runnel geomorphology were able to absorb the oil. This is also a probable reason that the wider Arabian Gulf environment was less badly affected than widely expected (see above). Additional factors impeding natural recovery are that sediment erosion and deposition have neither removed and cleaned the contaminated sediment, nor buried it deeply under clean sediment which would have allowed recolonization of plants above the oiled layers.

The United Nations Compensation Commission (UNCC)

The UNCC was established in 1991 specifically to process claims and oversee the disbursements *via* national government focal points in several countries affected by the Gulf War spills and fires (UNCC 2016). The compensation included the environmental remediation and restoration projects moderated by the F4 Panel (Decision 258). In the Saudi Arabian shoreline project (UNCC claim number 5000451), the lack of ecological recovery required specialist work to determine when intertidal zone sediment was functioning hydrologically and was sufficiently free of oil residues to allow the natural recovery of *Halocnemon* (a dominant saltmarsh perennial), and the successful propagation and out-planting of nursery stock of *Avicennia marina*. Much of the work in the first few years of the project was investigatory rather than the application of standard civil engineering, shoreline cleanup, and agronomic methods that could lead to failure in the sensitive environment. UNCC Decision 258 anticipated this by emphasizing small- to large-scale field tests using adaptive management to identify the best restoration approaches and then to apply them at full scale.

How should the extent and rates of ecological recovery be measured? Ecological recovery was assessed in 2002 from species abundance histograms that compared impacted and non-impacted sites in terms of species richness thresholds. Sites were then classified as 'non-recovering' if they failed to reach the threshold, and as 'recovering/disturbed' if they had a richer assemblage of species than non-recovering sites, but a different community (e.g., including opportunist species) than was characterized by the reference sites. Using this model of recovery, the scale of the ongoing biological impacts in 2002 was stark. Whereas the percentage of non-recovering transects on exposed sand beaches was between 28–38% and on exposed rocky shores was 19–31%, on the more sheltered fine-grained sediments, in contrast:

- Mangrove-dominated transects were 89% non-recovering
- Saltmarsh-dominated transects were 87% non-recovering
- Sheltered muddy tidal flat transects were 71% non-recovering.

Algal mats were more than five times higher in occurrence on oiled habitats and were identified as 'probabilistic obstacles' to ecological recovery. The algae were found in much higher quantities because the grazers such as crabs and molluscs had been killed by the oil. On many exposed sand flats, mudflats and saltmarshes, habitat modification by algal mats formed a barrier to channel and creek function, to sediment burrowing fauna, and also to chemistry processes (*e.g.*, oxygenation). Physical barriers to recovery also included hardened surficial oil residues which were likely to continue to affect:

- Sediment pore moisture content, availability, redox potential and water table behavior
- Dissolved O_2 and nutrient availability, and salt exclusion and flushing processes
- Larval settlement, propagules emplacement and seed germination
- Crab burrowing and the feeding behavior of epibenthic grazers.

Figure 30.8 shows halophyte aspects of the UNCC project in Saudi Arabia.

Because the UNCC aim was to enhance ecological recovery, it needed to be shown why remediation was still required after what was then two decades since the spills. A key requirement is to remove high concentrations of oil without causing new impacts compared with shores where no cleanup was attempted, and compared with the 'stalled recovery' of the severely-impacted sites, as below:

- Select appropriate reference model(s) for restoration, and this could include upper shore and supra-littoral mangroves/marshes even where they did not exist pre-1991
- Set up non-oiled reference sites, preferably within or close to the spill zone
- 'Set-aside' sites that were oiled in 1991 and where cleanup is not proposed.

Monitoring objectives then included continual comparisons of treated sites with both reference and set-aside sites. If no progress is made with time against the reference model, then a changed goal could be considered. If no progress is made with time against the set-aside sites then changes may be needed in cleanup technology, tailoring of the method, supervision level, or reverting to 'no treatment except monitoring'.

Over 26,000 measurements of total petroleum hydrocarbons (TPH) showed the following mean concentrations, where 1% is 10,000 mg/kg TPH in dry sediment as a function of heavy (H), moderate (M) and light (L) oiled residue (OR) and oiled burrows (OB) respectively:

- Heavy (HOR 4%; HOB 1.4%; together 31% of total volume of oiled sediment)
- Medium (MOR 2%; MOB 0.8%; together 41% of total volume of oiled sediment)
- Low (LOR 0.5%; LOB 0.3%; together 25% of total volume of oiled sediment).

A schematic of the occurrence of these categories of residual oiling and the candidate remediation and restoration methods is shown in Fig. 30.9, including mangrove habitat creation to landward to accommodate rising sea level. Progress on continuing restoration of shorelines in Saudi Arabia is summarized by Cormack et al. (2011) and Langman et al. (2012). The report by UNCC (2016) was current to mid-2013 when Saudi Arabia was released from the program having assured the UN of its commitment to continue cleanup, to restore habitats including mangroves, and to create nature reserves. In contrast to the many legal and scientific publications on the spills and the role of UNCC in the 1990s, no comparable updates have appeared more recently despite this probably being the "largest and most complex environmental cleanup project in human history" (UNCC 2016). Although the promised advisory panel is due to meet in 2018, the findings of seven biannual Independent Reviewers' reports between 2008 and 2012 remain undisclosed, unusually for UN programs.

30.7 Development Projects in Oil and Gas

30.7.1 Oil Spill Contingency Plan (OSCP)

30.7.1.1 Introduction

The aim of an OSCP is to deliver a co-ordinated response to a significant oil spill arising as part of the construction and operation of a new development. The OSCP is typically developed by collaboration between industry and government. For example, Qatar Petroleum (QP) is the national responder for Qatar, Mesaieed Industrial City (MIC) Port Operations is responsible for MIC port area, and the Head of Marine Operations is the On-Scene Commander (OSC). The QP Safety, Quality and Environment departments, the Offshore Support Department, the MIC Environment Department, MIC industries, and the Supreme Council for Environment and Natural Resources (SCENR) are all involved.

Separation of vessels by Vessel Traffic Control, obligatory use of tugs, and pilotage and notification procedures are described. Spill response drills and audits of the MIC activities improve emergency alarms and response procedures, coupled with training. Emphasis is placed on the Tier 2 (<200 tonne) spill, as this size of spill is predicted to be more probable than a Tier 3 (>250 tonne) spill.

30.7.1.2 Response Strategy and Equipment

The following oil spill response (OSR) equipment is provided:

- Operational spills of <0.1 t are contained using equipment on board ships or dockside without reaching the water
- Tier 1: OSR equipment onboard vessels and placed at strategic locations in MIC by the Port Operations Department consists of containment and cleanup equipment for a spill of up to 1.5 t, and apart from notification does not require outside assistance

Fig. 30.8 UNCC project 5000451 Saudi Arabia, Gulf War spills: (clockwise from top right) oil in poorly-drained tidal flat, June 2010; waterlogged algal mats and new channels, June 2011; mangrove nursery, September 2011; replanting area in Alfrea, January 2012; oiled marsh creek sediments and non-oiled mangroves nearby, September 2011; migrating birds in marsh high tide roost (*Charadrius* sp.), September 2011; mangrove transplant, December 2010; sediment mounds created from relatively non-oiled spoil during channel construction, July 2012 (Credit: Little, except Sandrin for migrating birds in marsh)

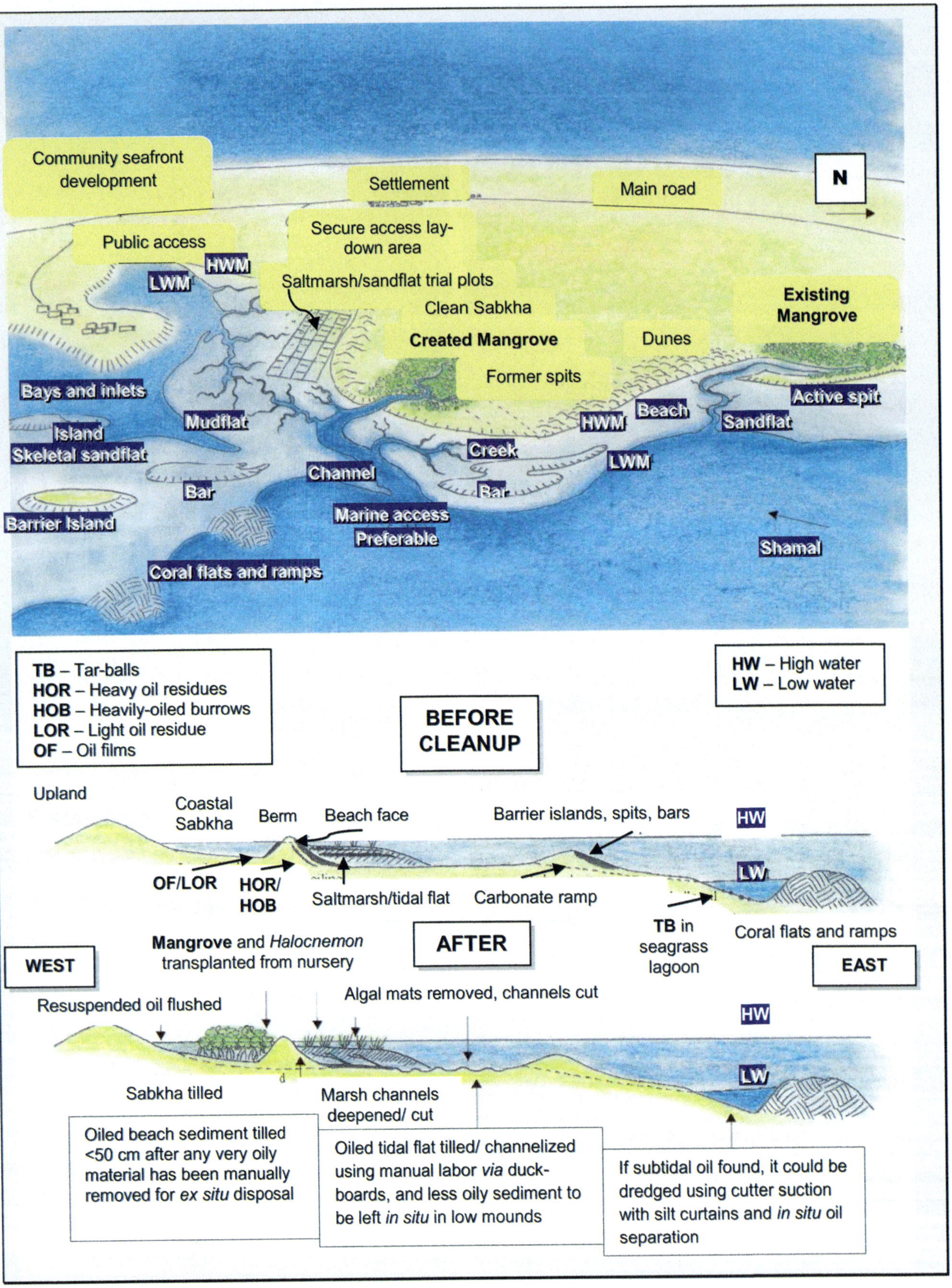

Fig. 30.9 Visualization of candidate remediation and restoration methods, including landward mangrove habitat creation in coastal sabkha to accommodate rising sea level, southern Arabian Gulf (Credits: Little and Little)

- Tier 2: OSR equipment from the main stockpile at MIC Small Boat Harbour, and supported by QP, could be mobilized by vessel and would respond to spills <200 t (*e.g.*, damaged diesel fuel tank, product handling facilities, or partial loss of pipeline integrity)
- Tier 3: OSR equipment/manpower available on national and international call-out for potential large-scale spill events >250 t. These resources would be transported by sea or air from stockpiles at Ras Laffan, and internationally (e.g., OSRL in Southampton, UK).

30.7.1.3 Response Methods

OSR faces numerous challenges in Qatar, including some remote and uninhabited shorelines and islands, and risks of oil penetrating coarse sediments or moving inland into khors with storm surges. Mangrove and saltmarsh ecosystems are characterized by sensitivity to oil pollution, long persistence of oil in fine-grained, organic sediments, and relatively slow ecological recovery. There are few onshore points of access, and a very ill-defined 'beach' in many locations, making ancillary impacts (*e.g.*, trampling) more likely in sabkha and tidal flats. It is often relatively shallow inshore, restricting access for response vessels, especially around mangroves and coral reefs most sensitive to pollution. There are few arrangements for oiled animal welfare and adequate oiled waste disposal facilities, although MIC Environmental Department determines recovered oil disposal routes according to the nature of the waste. Methods include burial, incineration, re-use in the QP refinery, temporary storage and stabilization using cement, lime, clay, asphalt, binding agents, and composting.

Water intakes are a high priority for mechanical protection by booming. In addition to all the water intakes along the MIC water front, Ras Abu Funtas power station is well-protected by booming equipment. Al Wakra has a small boat harbour, and there are artisanal fish traps both north and south of the town which would be difficult to protect due to the complexity of the coral rubble that would trap oil on the shoreline. The Sealine Resort is an important recreational area that needs protection, although stranded oil would be easily recovered from the firm sandy beach.

Modelling is used in the OSCP to predict the spread of oil from hypothetical points of spillage in the approach channels and in the case of a tanker aground on the Fasht Al-Arif. These predictive studies are adequate for the defined spill sizes, however, the loss of larger amounts of oil further offshore should be considered. Areas threatened could include mangroves at Al-Dhakira IBA (QA001) and Rawdat Sabt. Impacts on fisheries and bird life in the Fasht Al-Adaid and the Khor Al-Adaid could be severe and prolonged, especially if sediments were to be contaminated in what would be a trans-boundary incident. The Khor Al-Adaid is a tentative UNESCO World Heritage Site (Fig. 30.10 and Box 30.1).

30.7.1.4 Oil Spill Protection/Cleanup Options

Table 30.11 summarizes the main protection and mitigation options that are available to oil spill responders in the Qatar near-shore zone (after API 1985). Natural recovery and/or physical removal of the oil using booms and skimmers are generally the preferred methods. However, these operations are logistically challenging, especially in bad weather, and this has led to development of alternative methods. Of these, most of the controversy during oil spill cleanup decision-making relates to the use of chemical dispersants.

Effective use of chemical dispersants offshore may successfully protect some of the key tropical coastal communities, particularly mangrove and saltmarsh. However, caution is advised because there is evidence that incompletely dispersed oil in water emulsions can increase damage, especially to seagrass, but also to corals at deeper water depths than would have been affected if the oil had been left alone (Sect. 30.6.2). Chemical dispersants are not envisaged by QP as a first response in the OSCP in waters less than 20 m deep, but the QP plan provides an overview of dispersant options using the Regional Organisation for Protection of the Marine Environment (ROPME) approved list of dispersants.

Case-by-case approval for dispersant use is required by SCENR. Chemical dispersant use strategies and tactics are best worked out in advance of the spill as part of the OSCP. This is because there is very little time to do this properly during the spill itself before the effective window of opportunity for dispersants has closed: typically less than a few days depending on the type and weathering of the oil, and ambient temperature.

Shoreline responders should mark 'no-go' areas for cleanup crews, where mangroves, coral reefs or seagrass beds occur adjacent to other marine and coastal habitats that are in the process of cleanup. All shoreline cleanup methods used, whether chemical, mechanical or manual, must be well supervised and documented. Methods such as sediment removal or burial, indiscriminate cutting and burning of oiled dead vegetation, sinking agents, steam cleaning and high pressure flushing are all to be avoided in tropical marine habitats. Natural cleaning and low pressure flushing are the preferred responses to stranded oil in mangroves, seagrass beds and saltmarshes (API 1985). Viable methods include the use (and recovery) of sorbents in all shoreline types. Depending on accessibility, vacuum pumping and recovering the oil may also be possible in mangrove and saltmarsh creeks, and bioremediation is viable if shoreline sediment oil concentrations are low enough.

In all these habitats, effective protection or cleanup operations will remove the majority of the free-phase oil without doing more damage than leaving the oil to weather naturally. Representatives of SCENR and the NGOs play a vital conservation role in helping to judge whether any cleanup impacts that do occur are justified by the benefits

Fig. 30.10 Satellite image of (left and center) Khor Al-Adaid, Qatar and (right) Ras Abu Kamis peninsula, Saudi Arabia

gained by oil removal. Note that effective cleanup does not attempt to remove all trace of the oil. The target for cleanup should not be set by an arbitrary hydrocarbon concentration remaining in the environment. Instead, end points are set on the basis of net environmental benefit in terms of recovery of ecosystem function and biodiversity, recognizing that intrusive cleanup can do more damage than leaving some oil to degrade in the environment.

For purposes of emergency response, the shoreline is divided into linear segments on the basis of QP's aerial video over-flights conducted in 2003, to enable tactical management of any significant spill and appropriate planning of personnel and equipment staging areas. Expert local knowledge will be used to determine which options for cleanup are most suitable for the different offshore and shoreline habitats found in the area, and joint decisions documented for 'lessons learned' and possible cost recovery from the spillers' insurers.

For oil spills in the offshore zone, mechanical recovery or *in situ* burning counter-measures are likely to be focused on the leading edge of the slick, which tends to be composed of the thickest oil layers. Chemical dispersants are unlikely to be required due to the predicted short residence times of offshore oil. However, as described above, a suitable product will be identified in advance, and pre-approval for its use agreed with the SCENR.

In the near-shore zone where the water depth is less than the maximum draught of response vessels (<10 m) and where the presence of vegetation stands makes ponding of oil more likely than in deeper water, the cleanup options include mechanical oil recovery. Vacuum pumping may be feasible in these areas, assuming shallow draft craft can gain access. *In situ* burning is also viable, but less preferred due to possible smoke/SO_2 impacts on coastal settlements. A range of techniques to enhance *in situ* burning, including fire-retardant booms, ignition devices and smoke suppressants (e.g., ferrocene) will be assessed. If appropriate, the equipment will be procured and training in their implementation scheduled in advance of commissioning.

30.7.1.5 Disposal of Recovered Oil

Because of the difficulties posed by poor access to many parts of the Qatar coastline, careful planning will be needed to ensure that the temporary storage and disposal of recovered oil during cleanup is effectively co-ordinated. It is very important to manage this aspect of cleanup effectively, for if a bottleneck is created the whole spill response effort can be brought to a halt.

Table 30.12 summarizes the disposal options most suitable for different types of oily waste. Expert local knowledge will be used to establish the methods used. The most

Box 30.1 Tentative UNESCO-designated site, showing idealized transect from upland (left) to coastal habitats (right) down-wind on the 'Shamal' wind vector (SCENR 2008, personal communication, Böer (UNESCO, Qatar); (Credits: Little and Little)

Khor Al-Adaid is a large (120 km^2) tidal inlet with sandy islands, located mainly in the south-east corner of Qatar and sharing its southern margin with Saudi Arabia. Including the large landward hinterland, the area was submitted by SCENR on 8 March 2008 as a tentative UNESCO World Heritage Site, on account of its excellent range and globally unique combination of unspoiled desert and marine landforms and transitional ecosystems. It is internationally significant for geology and geomorphology, including wadis, barchans, parabolic dunes, mesas and inselbergs, and for desert soil types and aeolian-sculpted fabrics that are among the least subject to degradation in the region. The wadis are probably Pleistocene paleochannels attaining depths of up to several tens of metres, and showing a dendritic drainage network cut into a desert peneplain consisting of hamada bedrock and serir gravel.

Khor Al-Adaid is an example of a 'negative' estuary, with negligible freshwater inflow today, very high evaporation rates and thus well-developed coastal sabkha that are unusual in that quartz sand is derived from landward dunes. The sabkha are therefore relatively low in $CaCO_3$, although shells coated with $CaCO_3$ precipitates – Pisolites – are the only ones known to exist in a matrix of quartz sand. In addition, the hyper-saline ground waters in the more landward parts of the sabkha contain primary dolomite crystals, again unusual in quartz systems. Landward-oriented sediment bedform structures support shallow benthic communities. Evidence of changing sea level is seen in the form of raised beaches and salt hummocks in Khor Al-Adaid.

Breeding species include osprey *Pandion haliaetus* and Caspian tern *Sterna caspia*. In winter, large concentrations of greater flamingo *Phoenicopterus ruber* are reported, and significant numbers of other waterbirds may also occur, including cormorants *Phalacrocorax carbo*, *P. nigrogularis*, gulls, terns, and migrant waders. Arabian gazelle, dugongs and marine turtles inhabit the area, and archaeological and cultural sites occur. Although commercial fishing was banned in 1993, Khor Al-Adaid is vulnerable to other threats: marine oil and chemical pollution, construction of major marine and overland transport and communications corridors, and inappropriate tourism developments such as recreational disturbance by camping and the indiscriminate use of four-wheel drive vehicles.

http://whc.unesco.org/en/tentativelists/5317/

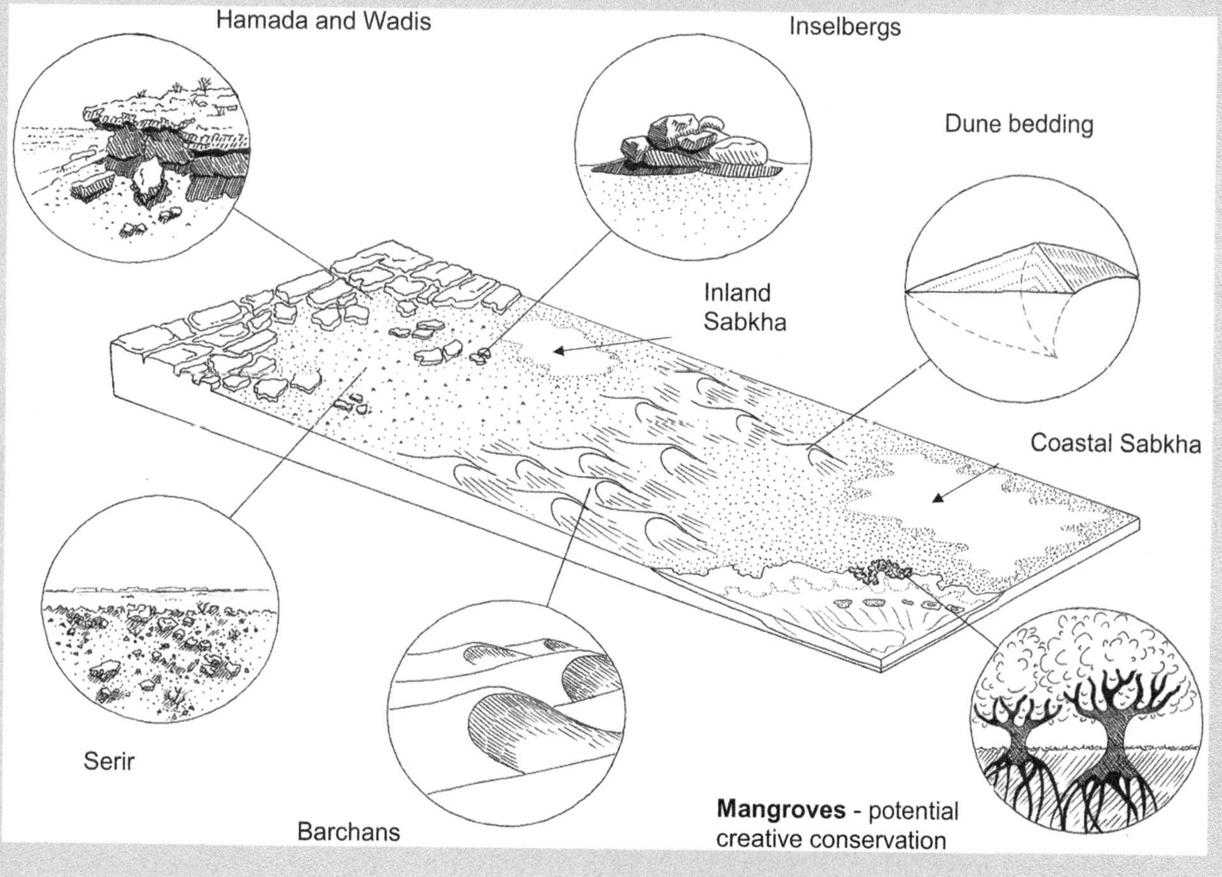

Table 30.11 Suggested protection and cleanup methods for offshore (open water), mangrove (bold) and other shoreline habitats in south-western Arabian Gulf

Protection/ Cleanup Methods	Open Water			Shoreline Habitats			
	Offshore	>10 m deep	<10 m deep Near-shore	**Mangrove**	Rocky substrates	Saltmarsh	Sediment flats & sabkha
Booms/ skimmers	P	P	V	A	A	A	A
Incineration *in situ*	V / P	V	V	A	V	A	A
Chemical dispersants	V	V	A	A	A	A	A
Herders/ elastomers	NA	V	V	A	A	V	A
Vacuum recovery	V	P / V	P	V*	V	V	V
Low-pressure flush	A	A	A	P	P	V	V
Sorbents	NA	NA	V	V	V	V	V
Manual methods	A	A	A	V	V	V	V
Substrate/ Veg. removal	A	A	A	V*	V	NA	NA
Natural cleaning	V	V	V	P / V	P / V	P / V	P / V
Bioremediation	NA	NA	NA	V	V	V	V

* In mangrove creeks vacuum recovery and vegetation cutting may be viable during high water conditions

P	**Preferred**
V	Viable
NA	Not advisable
A	Avoid (or not feasible)

Table 30.12 Suggested storage and disposal methods for oil and oiled vegetation following a significant spill in south-western Arabian Gulf

Disposal Method	Type of oil/ oiled waste material				
	Light - medium oils (fresh condensate, diesel)	Heavy oils, weathered emulsion, Bunker C	Combustible oiled debris (*e.g.*, oiled sorbents)	Organic slurry of vegetation	Non-combustible oiled debris/ sediment
Natural degradation	V	A	A	V	V
Temporary/ compost	V	V	A	V	V
Landfill	V	V	V	V	V
Combustion	P / V	V* / A	P	P	NA
Re-refining	P	V	A	A	A
Recycling	V	V	A	V	V
Well injection	V	A	A	A	A
Demulsification	V	P / V	NA	NA	NA

*Certain incinerators may be viable

P	**Preferred**
V	Viable
NA	Not advisable
A	Avoid (or not feasible)

appropriate methods will be re-processing at the QP refinery/ MIC oil slops system for light oil and condensates, and demulsification followed by combustion for heavier oils. The suitability of these waste streams for reprocessing, and the terms under which they might be acceptable at the point of disposal should be worked out in advance of a spill, and included in the OSCP. In very large incidents, it may be necessary to procure a rotary kiln incinerator (possibly barge-mounted) as a means of local destruction of wastes, if onshore incineration is deemed unacceptable due to the perceived SO_2 risk. If acceptable, then burn pit(s) could be sited at strategic locations remote from sensitive receptors.

Collected oily wastes, including some booms and sorbents, oiled strandline vegetation and rags (but not plastics, such as personal protective equipment) may also be burned using modified 45-gallon drums. Alternatively, oiled vegetation with relatively low oil concentrations (but no man-made materials) may be piled into compost heaps along the shoreline, provided a suitably impermeable base can be rapidly built. This organic material will begin to biodegrade rapidly in typical Qatar weather conditions. Nutrient (nitrogen and phosphorus) addition may also be utilized to enhance biodegradation. Alternatively, there may be opportunities to land-farm the oiled waste material in contained and supervised sites.

30.7.2 Mitigation Proposals

30.7.2.1 Introduction

The MIC proposals for construction of a major port expansion and industrial development in Qatar involved predicted coastal and offshore impacts that were largely unavoidable. Inevitably, such impacts will be amenable to mitigation mainly at the 'compensation' level of the mitigation hierarchy. From scoping sessions with MIC, SCENR, UN Education and Science Council (UNESCO) and UN Food and Agriculture Organisation (FAO) there were a number of mitigation opportunities to be explored. Some involved creative habitat conservation using indigenous halophyte species, with potential increases in local biodiversity and carbon sequestration.

These compensatory measures go beyond landscaping, and would help to mitigate impacts of offshore construction, including:

- Land take by reclamation and build-out of the quay-walls
- Dust and thermal pollution
- Sedimentation impacts
- Removal or deterioration of benthic communities such as seagrasses, and
- Climate change impacts including 'coastal squeeze'.

Before the proposed mitigation, sabkha was the main coastal habitat north of the city, and this is being degraded. With input from UNESCO, mitigation involved designing what would be one of Qatar's largest stands of mangroves in place of part of the sabkha. In such a very large proposed expansion of MIC the environmental carrying capacity had been seen as a roadblock to obtaining approvals. With the proposed mitigation, the regulator might have perceived the expansion more positively in terms of biodiversity offset, CO_2 sequestration, coastal protection and green landscaping, potentially unblocking the project's approval. These sustainable options are discussed below for the 'North' and 'South' ports.

30.7.2.2 North Port

The suggested mitigation and creative conservation measures proposed for the area to the north of the new North port were designed on the assumption that there would be no maritime traffic on its northern edge. This would depend on designing the North port to accommodate a landscaped buffer zone to protect the Fasht Al-Arif, the principal sensitive receptor in the north. Littoral sand accumulation would be encouraged, engineered and landscaped to actively develop a nature reserve, where human access would be controlled and the habitats specifically managed for wildlife.

Port lighting would be directed southwards away from the new sand beach habitat. Noise sources in the new port adjacent to the reserve would be minimised by careful zoning and screening of activities on the edge of the North port. If successful, one of the aims of this mitigation proposal would be the attraction of marine turtles, as has been accomplished at Ras Laffan Industrial City.

Depending on the final selection of the cooling water intake location for the Kahramaa power plant, it was proposed that the routing of the main intake channel be used to draw excess cool seawater into the coastal zone for conservation as well as cooling and desalination purposes. The North port mitigation proposal would encourage better flood and ebb tidal flushing of the extensive sabkha areas to the north of MIC, in order to progressively remove salts from the surficial layers of sabkha sediments. To achieve this flushing, it would be necessary to build an extensive network of channels to draw in water on both flood and ebb tides.

A preliminary design for the mangrove and seagrass dominated engineered wetland is shown in Fig. 30.11 in the form of an oblique aerial visualization, before and after wetland construction. The landscape schematic is viewed from offshore and above the Fasht Al-Arif over the conceptual North port looking west. The mangroves would be transplanted as propagules from Qatar's existing tree nurseries, along the edges of raised areas to be developed using clean dredged material and fill that is already being

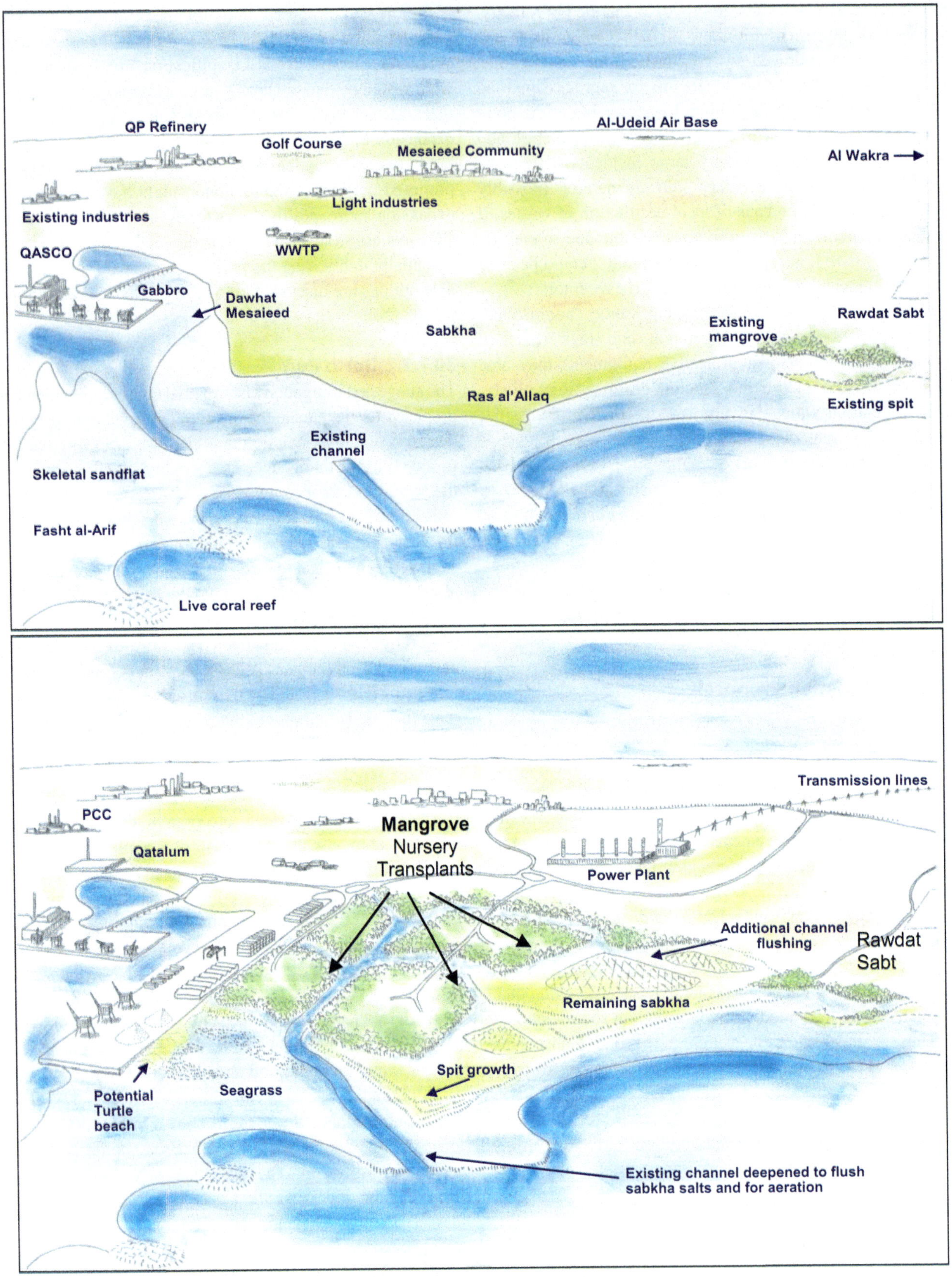

Fig. 30.11 Before (top) and after (bottom) visualization of MIC North port expansion and mangrove mitigation scheme, showing new channels and creative conservation features across sabkha to accommodate rising sea level (Credits: Little and Little)

deposited at the site as 'made-ground' at elevations up to 2 m above the existing sabkha.

Under this mitigation/compensation scheme, the difference would be that the layout of the clean fill would be less rectilinear than currently, and would involve a dendritic channel pattern lined with fine sand and silt-grade sediments suitable for the growth of *Avicennia marina*, once the initial salt concentrations had been sufficiently reduced by enhanced tidal flushing. It would be necessary to test such a mangrove mitigation proposal in a pilot scale area, and both UNESCO and FAO have indicated a strong interest in such a 'halophyte park'. Key factors would include careful selection of sediment materials and their successful emplacement and continued deposition along the margins of the shallower channels, while leaving unimpeded water flows through the deeper channels to maintain flushing.

The made ground blocks shown in Fig. 30.11 (lower) could be inter-connected by light roadways. The central portions of each block would then be available for appropriate development in the form of light service and high-technology industries that would be located in an attractive science and business park environment. Air quality and amenity would be improved over current conditions during 'shamal' winds at MIC, due to the baffling of winds and trapping of entrained dust by mangroves. Appropriate commercial development should not impair the conservation value of the created wetlands. Mitigation recommendations for the North port include mangrove re-planting along each of the margins of the above channels. This would encourage sedimentation of calcareous mud and the natural colonization by seagrass and polychaete/molluscan benthic communities.

Additionally, displaced juvenile fish from the inner harbour of Dawhat Mesaieed Bay and birds from Al Bushariya Island would be attracted. In addition to improving the attraction of the North port for invertebrate, bird and fish communities, it is possible that the engineered wetlands might also attract dugong and marine turtles. Mangroves at nearby Rawdat Sabt would be the main source of recruitment of associated mangrove species (Figs. 30.11 and 30.12).

30.7.2.3 South Port

In contrast to the North port area, the main receptor in the South port area is socio-economic, consisting of tourism and amenity use centred on the Sealine Resort and the adjacent barchan dune area. The nature of this impact is primarily visual, as perceived by the human receptors using the Sealine Resort. However, the adverse visual impact could lead to a cumulative adverse economic impact if the resort facility's revenues were to be affected. Mitigation and landscaping would focus here on developing a buffer zone including a marina, just outside the commercial area of the South port, to compensate for the potential adverse economic impact by increasing the attractiveness of the southern area to Sealine Resort users (Fig. 30.13).

The mitigation proposal after marina construction shown in Fig. 30.13 is in the form of an oblique aerial visualization, viewed from the south and above the coastline over the conceptual South port. The southern edge of the South port would not be used for heavy or potentially-polluting commercial harbour activities. Instead, it would be designed to grade into the marina facility, which in turn presents a more acceptable visual and land-use interface toward the existing recreational area at the Sealine Resort.

Potential synergies with the existing resort would also be explored, including mitigation options that include creation

Fig. 30.12 Mangroves near MIC, east of Rawdat Sabt, February 2005 (Credit: Little)

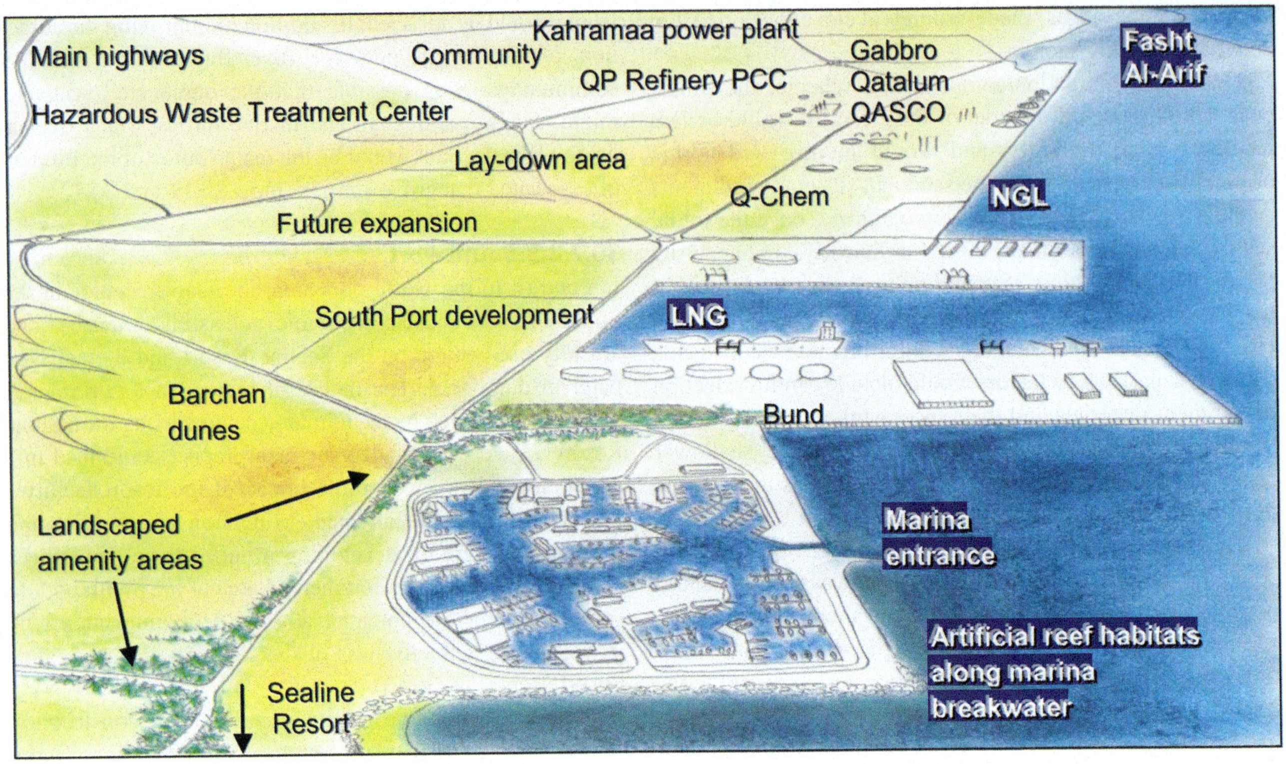

Fig. 30.13 South port development proposal showing buffer zone and marina between MIC and Sealine Resort (Credits: Little and Little)

of artificial reefs for enhancing the southern area offshore of the Sealine Resort for recreational diving. This was discussed in the Gabbro project ESIA, and the reefs would be attractive to invertebrate colonizers and divers alike. The combined effect of the mitigation proposed for the South port would be that the recreational activities to the south of MIC would be enhanced instead of being destroyed. In the latter case, an entirely new location for the Sealine Resort would be needed, and far greater disturbance would be felt both environmentally and economically.

Despite discussion with MIC and presentation of these options for mangrove habitat creation in the strategic EIA, it was decided to build the new port entirely on the north side of Mesaieed. This now occupies the area envisioned above for mangroves, with more than 15 km of commercial frontage and waterways. One of these is immediately adjacent to surviving mangroves at Rawdat Sabt. There are other coastal developments nearby (e.g. 'The Pearl', several island causeways, Hamad international airport, Samsung Hitachi Zozen power and desalination plant with two solid >2 km intake/outfall causeways). Together, these reclamations have destroyed sabkha, spits, benthos and seagrass habitats, interrupted sediment movements, and largely foreclosed the option to create extensive new mangrove and seagrass wetlands. See satellite imagery for the coastal zone south of Doha at https://www.google.com/maps/d/viewer?mid=1lboxnH6DwQkR8Zatf7sax7Sg5xE&hl=en_US&ll=25.238271490032933%2C51.38826296093748&z=10

30.7.2.4 Green Landscaping and Wilderness Areas

The mitigation should be planned in the context of MIC's water resources management plan. The existing MIC requirement for landscaping is to allocate 3.5% of the land take needed by new industry for 'greening', most of which has visual and amenity aims around MIC industries' offices and residential areas. Although worthwhile as both habitat (particularly for mobile insects and birds) and also for re-use of some of the 5000 m^3/day treated domestic waste water in an innovative manner, this planting does not currently lead to the development of natural biological communities. This should include usage for irrigation of increasing proportions of the estimated 8000 m^3/day domestic wastewater during expansion. It is also envisaged that process wastewater that will no longer be disposed of at sea after 2007 should be treated in an engineered wetland and re-used for irrigation of native species. This initiative is to be developed in partnership with QP-affiliated industries and managed as MIC's Environmental Conservation Park.

CO_2 offsets should be considered as part of the MIC Strategic Master Plan for each individual new industry's planned development, and this could include continuing with mangrove forestation elsewhere in Qatar, in conjunction with experts from UNESCO and FAO, and the Ministry of Municipal Affairs and Agriculture, Qatar. The recommended form of the new compensatory program would be halophytes such as mangrove because of their high biodiversity and coastal protection roles, particularly important in the context of sea level rise. The North port mangrove mitigation proposal outlined above could have been an invaluable pilot scheme for wider area mitigation.

It was recommended that MIC collaborate with UNESCO tentative World Heritage Site proposals for the Khor Al-Adaid to the south of MIC (Fig. 30.10 and Box 30.1), such that this wilderness area is only 'developed' in a sensitive and sustainable manner, without increasing off-road impacts in serir (stony) desert and coastal habitats in particular. This would involve MIC working with planning authorities to develop a land and traffic management plan for the additional traffic that Route 55 might encourage to encroach into the southern hinterland of MIC. The form of such collaboration could also include the establishment of a visitor and interpretive centre that would have educational benefits as well as managed ecotourism goals.

Further planning for the borderland region with Saudi Arabia is underway. This is to make improvements to transport corridors including a new rail link, and the GCC electrical grid into the south of Qatar and onwards into Saudi Arabia. It was recommended that the individual infrastructure corridors are carefully selected and if possible, harmonised into one 'macro-corridor', to minimise the adverse impacts of intrusion into a nationally designated wilderness area or Khor Al-Adaid World Heritage Site.

30.8 Discussion

30.8.1 Introduction

The factors affecting successful development of OSR capability in East Africa and the Middle East include recognizing the unusually wide range of challenges. Some of the uncertainties are caused by the diversity of countries, language groups, geographies and economies, those in East Africa being mostly at a comparatively early stage in the production of oil and gas. For example, it is clear that acquiring information and getting engagement from the authorities is challenging, driven by some of the factors described by NCG (2012). The perceived barriers encountered might be explained by a lack of priority placed by governments on the value of committing resources to oil spill preparedness. This may also explain the lack of full follow-up since the close of the WIOMH project in 2012. Given the initial desirability of face-to-face communications especially in Africa, and given fuller engagement on the ground, the challenges described below will need to be managed in terms of their impact on oil spill preparedness and response. Each has the potential to adversely affect mangroves in the event of a spill.

30.8.2 Security and Maritime Safety

Marine safety was one of the main focus points for the WIOMH project funded by the GEF until 2012. Achievements included construction of lighthouses in the Mozambique Channel for example at Punta Zavora (Mozambique). Offshore oil and gas discoveries in East Africa have since increased the need for maritime security and aerial surveillance. Coastal nations in East Africa have seen the incidence of piracy soar; offshore Somalia in particular suffered increased piracy from 2007, and piracy peaked after increasing by 11% from 2010 to 2011 (TDA 2012). At that time, during the life of the WIOMH project, the western Indian Ocean was the main area of piracy. In many cases it was far offshore, and oil tanker traffic was diverted eastwards, thus increasing the distance and cost of transit between the Arabian Gulf and Cape of Good Hope. Piracy poses a serious development challenge, with 4% declines in GDP reported (e.g., Seychelles). Asian long line and purse seine fishing fleets have been displaced eastwards, but have partly moved back due to strengthened security arrangements for the vessels (TDA 2012). The impacts of piracy on long term fish stocks are unknown, but could be beneficial due to any avoidance of the fishing grounds by vessels. In most cases, significant losses of oil do not occur as it is in the interest of both owner and pirate to maintain the integrity of the target vessel.

More recently the worst-affected states for piracy have been in West Africa, from Guinea to Angola but most notoriously Nigeria ('The Economist' 29 November 2014). The balance has shifted to areas closer to the coast rather than the high seas, and this makes it imperative that responses by national security forces improve. The most extreme piracy attacks in Nigeria are aimed at fisheries, oil installations and coastwise traffic. Success against piracy in East Africa was due to unprecedented cooperation among all five permanent members of the UN Security Council, the EU's 'Operation Atalanta' and NATO, as well as national naval and coastguard forces. Multi-national anti-piracy forces were UN, NATO or EU-mandated because Somalia was a failed state. However, these forces are stretched and they are reducing in strength in East Africa without reinforcing in West Africa. These multi-national forces would not be so effective against the inshore and Niger delta activities because in the coastal zone they are subject to national laws. The recent decline of piracy in East Africa could be reversed if successful countermeasures are relaxed. These include surveillance, high transit speeds, convoy formation, nearby escorts, and passive defenses onboard (e.g., bullet-proof rooms to delay crews being overwhelmed until rescue).

For many years organized crime in Nigeria has been involved in piracy, kidnap, narcotics and weapons trafficking; there are signs that these are also connected to the Boko Haram insurgency in the north-eastern Borno State. It may be reasonably supposed that piracy may be connected with fund-raising for mainstream politicians' election campaigns, because piracy tends to increase during the lead up to Nigerian elections, as in 2015. Sometimes the vessels targeted by pirates are themselves involved in smuggling ('The Economist' 29 November 2014). In the sense that piracy is often made possible by corruption and insider intelligence from within the security forces, it would be a natural outgrowth of the long-established crime organizations benefitting from large numbers of oil thefts from pipelines inland.

Insurgencies elsewhere such as Islamic State fuelled their activities with 3 million barrels from hijacked pipelines and simple mobile refineries on skid-mounted units ('Petroleum Review' December 2014). From this, it is sensible to assume that oil and gas facilities may also be at risk in East Africa, from South Sudan and the Elemi Triangle to Al Shabaab's area of operations, or the border zone between Uganda and the DRC with the Lord's Resistance Army. This risk became real in the September 2013 attack on Westgate Mall in Nairobi. Security risks exist elsewhere in the region, for example Renamo in Mozambique resumed operations, and also as far south as Madagascar where security has caused ExxonMobil problems.

In 2013 military spending in Africa increased in total by 8.3% according to the Stockholm International Peace Research Institute (SIPRI). Kenya and Uganda over the past decade increased military spending by up to 50%, while Tanzania more than doubled its spending. Kenya has shifted in 2014 military/security procurements away from reliance on importation to manufacturing in-country, with subsidiaries of 'Eurocopter' (Nairobi) and 'Fincantieri' ship-building (Mombasa).

Refugees and internally-displaced persons are another unfortunate effect of security problems caused by terrorism, insurgency and also by desertification, famine or disease. Northern Kenya receives refugees from the north-eastern border with Somalia due to Al-Shabaab. The Ebola outbreaks of 2014 in west and central Africa were unprecedented in scale, and some refugees from the epidemic fled to Kenya. The long-term impacts are not restricted to the health services being overwhelmed and increased numbers of malaria and HIV cases due to avoidance of clinics. There is also the risk that such outbreaks create panic and lead to security breakdown. Expatriate companies posted health and security advice and links to the World Health Organization (WHO) regarding their personnel travelling in West Africa. The WHO was criticized for its lack of central coordination, despite its major representation across the affected region. The economic impact of the Ebola epidemic was estimated as having reduced GDP of the three most affected states by between 2% and 3.5% (IEA 2014). The fear of crowds during the epidemic meant that clinics were avoided and schools closed because pupils left school to help their families survive ('The Economist' 3 January 2015).

30.8.3 Health and Safety (H&S) in Remote Locations

During OSR operations, H&S must remain the priority throughout, regardless of location. Travel, health, safety and security are matters of increasing concern for oil company personnel under normal operations in many East Africa countries. In the event of an oil spill incident, because of its unpredictable nature and the uncertain locations of any shoreline landfalls, there is a need for rapid and effective risk screening procedures to be in place and tested as part of OSR planning.

The workforce engaged in remote ORS and in particular shoreline cleanup is largely drawn from the local community, as opposed to trained spill responders. These individuals are unlikely to be familiar with international H&S best practice. Those personnel managing the response should use their experience to ensure that best practice is applied, and those carrying out the work are sufficiently trained in principles of assessing risks and taking mitigating actions. For example, if the local teams are not familiar with personal protective equipment (PPE) they are reluctant to wear it, and particularly so in hot climates.

The overall approach to H&S at a spill should not differ between accessible and remote incidents. However, in remote areas the response can be significantly affected by difficult access to shores, the presence of dangerous wildlife, and limited emergency healthcare. The hazards and limited options for mitigating their consequences may mean that working methods must be significantly modified, or that some operations are deemed too risky. This has clear implications for the cleanup activity achievable in remote areas, including cost. For example, a risk assessment might determine that it is necessary to have an on-site doctor or a helicopter available for emergencies. Shoreline topography and prevailing weather and sea conditions may mean that cleanup is impossible in some areas due to H&S constraints (Reynolds 2014). The constraints arising in exploration areas and how to mitigate them during OSR activities in East Africa are fruitful training themes.

30.8.4 Communications and Cooperation

Reliable communications are essential to an effective OSR. Satellite links are widely available but can be unreliable or inadequate for downloading large data sets. To improve this situation, East Africa is connecting up the regional fibre-optic broadband networks until now owned or leased by several entities, and this is largely being achieved by private equity investors and long term leases. An offshore cabling system installed by private consortia runs north to south from offshore Somalia with major landfalls in Mombasa, Dar es Salam, Maputo and finally to Lesotho on the coastline of RSA where it connects to Johannesburg. The main Kenyan networks connect from Mombasa *via* Uganda and Tanzania to Rwanda and Burundi. Tanzania is covered separately from landfall at Dar es Salam, and is also linked with Zambia, Zimbabwe, and RSA. This inland network has reached south-eastern DRC *via* Zambia's Copper Belt ('The Economist' 5 July 2014). In contrast, Mozambique and Madagascar are still comparatively isolated.

ITOPF personnel consider spills to be 'remote' not based solely on geographic location, but also on little or no available infrastructure and OSR preparedness. A lack of infrastructure and resources means that responders are unable to get on site as quickly as desired, that not all of the preferred equipment may be mobilized and consequently some OSR options may not be feasible. In such cases the ship owner may assume leadership of the response and call in external help as necessary. Net environmental benefit analysis (NEBA) has been described by IPIECA (2000) as the process of weighing up and comparing the advantages and disadvantages of different response options with the advantages and disadvantages of natural cleaning. However, there are certain aspects to remote spills that impact on the practicalities of actually going through the NEBA process and its outcome (Reynolds 2014).

Tullow oil in Uganda is planning to cooperate with other operators in Kenya to construct an export pipeline to the port of Lamu. Various options are being considered, and general risks and opportunities are being explored, perhaps without immediate urgency due to a combination of low oil price and security concerns in north-eastern Kenya.

30.8.5 Geopolitical and Trans-boundary Factors

A constitutional difficulty arose in April 2014 in Tanzania between the mainland (old Tanganyika) and the islands of Zanzibar and Pemba. The latter want more autonomy from the central government in Dodoma which they consider to be influenced too strongly by mainland issues. Most Zanzibaris are not satisfied by the new draft constitution despite its proposed improvements in human rights. The central government fears that greater autonomy may lead to Zanzibar independence; even if this does not occur there is a potential realignment or devolution of coastal and offshore responsibilities. The risk is then that confusing surveillance and response arrangements might ensue rather than Dar es Salaam authorities retaining maritime supervision duties. The constitutional issue may also be delaying agreement on offshore oil and gas licensing in Zanzibar.

Oil spill response is complicated by real or perceived trans-boundary pollution and the threats of spills from 'innocent passage' far offshore. Areas prone to these threats include crude and refined product oil cargos in transit along the entire WIO seaboard of Africa. The neighboring states most likely to suffer trans-boundary pollution by refined products from coastal vessel movements are between Kenya and Tanzania, and between southern Mozambique and RSA. The 1992 'Katina P' spill offshore Maputo involved HFO which was feared persistent enough to threaten RSA. Some oil did move southwards *via* the St. Lucia coastline (iSimangaliso Wetland Park), but very little shoreline oil contamination was found in RSA perhaps due to the northerly littoral drift under the prevailing SE trade winds. However, fuel oil escaping from the deep water wreckage of the 'Katina P' would have become entrained as droplets in the Agulhas current, and this could result in chronic pollution in RSA and islands of the southern Mozambique Channel.

The Anadarko spill in May 2014 was the first environmental accident reported since foreign companies began prospecting in Ruvuma in Cabo Delgado province northern Mozambique in 2006. The gas field is about 46 km from the coast of Mocimboa da Praia district, and because of its location potentially could have affected southern Tanzania. Low toxic drilling mud (30,000 l) spilled into the sea off Mozambique's northern coast at 4000 m below the ocean surface. At this depth the likelihood of surface impacts is slight and any drift would be to the south due to the Agulhas current. Legally, however, drill muds and cuttings must be disposed of in landfill sites and not at sea, and in this case it was reasonable to assume that trans-boundary impact could have occurred. The release consisted of 53% oily lubricants, 30% water, and 17% solids (barite, chlorides and calcium), deriving from the geological formations drilled. Resources at risk included the Quirimbas archipelago (EBSA site 12), coral reef islands designated under the Convention on Biological Diversity for the highest diversity of corals recorded in the south-western Indian Ocean region. It is also popular with recreational divers.

The southern part of Tanzania, from Mtwara southwards is also part of the large EBSA site 16, a centre of high marine biodiversity designated for its strong gyres and eddies connecting the islands, and also supports science field centres and ecotourism facilities. Anadarko responded to the spill by immediately closing the well, and moving the drilling platform to avoid any further spillage. They then dispatched a helicopter to survey the surrounding area, but no oil slicks or other detritus could be detected. No cleanup operations on the sea surface were conducted since no effects were visible, but overflights continued to search for possible signs of the muds or impacts on wildlife. An investigation into the causes of the spill was to be reported to the Mozambique Environmental Impact Evaluation Agency in Maputo, but it is not known whether transboundary impacts occurred.

30.8.6 Access to Energy

30.8.6.1 Primary Energy Supply

The International Energy Agency (IEA) published a study in 2014 in their World Energy Outlook series. According to their report, sub-Saharan African development is impeded by an inefficient energy sector, and the region's future economic and social stability increasingly depends on improving equitable access to a modern energy mix. Grid-based power generation in sub-Saharan Africa is dominated by RSA (>50%), and much of the power is generated around large cities and mining industries (e.g., Copper Belt). Two-thirds of the investment in sub-Saharan African energy industries since 2000 has been for export, perpetuating the colonial and post-colonial extractive model. As noted above, the unreliability of electricity supply has resulted in the widespread ownership of back-up diesel generators. In general, these localised and unilateral arrangements are restricted to better-off residential properties, factories, higher-end shops and offices, and essential services.

Only a third of the sub-Saharan population have access to electricity, meaning that over 620 million people do not. Indoor cooking with biomass fuels (wood and charcoal) is used by 730 million people and is highly detrimental to human health (gases, soot and in particular 4, 5, 6 ring polycyclic aromatic hydrocarbons – PAHs – are carcinogenic and persistent in human lipids and the wider environment). The environmental (e.g., biodiversity conservation) impact of such solid fuels is expanding due to uncontrolled wood-harvesting around rural and urban settlements where modern energy sources are unavailable. The energy resources of sub-Saharan Africa are diverse, abundant and becoming more available, with the region boasting almost 30% of the global oil and gas discoveries made since 2010. There are also high-quality and widespread renewables resources in hydro, wind, solar, and geothermal; however, all are still comparatively under-developed (IEA 2014).

The energy outlook based on UNDP demographic projections from 2013 allows IEA (2014) to model a scenario for a doubling of the African population to 1.75 billion in 2040 and an energy demand increase of 80%, with the following consequences: "The sub-Saharan power system expands rapidly, with generation capacity quadrupling to 385 GW. The power mix becomes more diverse, with coal (mainly RSA) and hydropower (all regions), being joined by greater use of gas (Nigeria, Mozambique, Tanzania), solar (notably in RSA and Nigeria) and geothermal (East Africa); the share of renewables in total capacity more than doubles to 44%. Total power sector investment averages around U.S. $46 billion per year, with just over half of it in transmission and distribution".

In 2040, the IEA predict that sub-Saharan African coal supply will grow by 50%, mainly in RSA but increasingly in Mozambique and elsewhere. The sub-Saharan demand for oil products will grow by 100% to 4 million bbl/d; more than its relative share of the above 80% increase in total energy demand. Therefore, according to IEA, despite an increase and diversification of energy supply, and despite improvement in capacity and efficiency of transmission and distribution, the region's net contribution to the global oil balance will be further squeezed by 2040. The demands placed upon the energy systems mean that many of the existing challenges will also remain, due to rapid population growth. Whereas almost a billion people are predicted to gain access to electricity by 2040, more than 500 million will still not be connected. Between 2014 and 2040 the reduction in those unconnected to electricity supply will have been less than 20% (IEA 2014).

Relative to population in 2040, according to IEA the energy capacity per capita will also remain low, with charcoal and other biomass fuels remaining in wide usage. Biodiversity around human settlements will continue to be depleted. Impacts of energy non-equitability on social cohesion and political stability are thus likely to be negative in many areas. The IEA predicts that stressors will be higher in those areas where climate change impacts are most severe, even though the region contributes little to the energy–related CO_2 emissions globally. The recognition of this paradox alone will increase tensions around the issues of fuel poverty and wealth distribution, and increase emigration and exploitation of refugees. Examples of adverse climate change impacts due to sea-level rise include the Seychelles and other south Indian Ocean islands. Examples due to desertification include the savannah grazing lands and upland forest habitats in Tanzania, Uganda and Kenya.

In the relatively new onshore exploration zones of East Africa, these stressors could eventually lead to disproportionate risks of antisocial behavior, and of crime including illegal tapping of pipelines, increased poaching, and people trafficking. The risks are likely to materialise unless electrical power supply and distribution are adequate in advance of the extractive oil and gas economy reaching the full production stage. If national and local governments and oil operators fail to recognise this risk in the onshore developments in East Africa, and fail to ensure greater equitability in the benefits arising from oil and gas, then exactly the same situation is likely to be faced in East Africa as has caused and dominated oil spill impacts and lack of cleanup in the Niger delta.

The evidence for this in East Africa is mainly circumstantial at present, but includes the acknowledged existence of corruption in all the nations of the study region, the linkage between organised crime and for example large-scale big game poaching, the legacy of colonial politics and geography clashing with the grazing cultures of tribal common lands. In the most desperate situations, the last resort is crime, terrorism, or exile as refugees (see Sect. 30.8.2 on security). In Kenya as well as Tanzania there are devolutionary tensions between national and local governments that may result in permitting and developmental issues being resolved locally or in an *ad hoc* manner. Local content provisions in PSAs and PSCs, although well-intentioned, can potentially become a source of graft in procurements ('The Economist' January 2015). The risks arising from this uncertainty in the regulatory processes increase in proportion to the lack of clarity in the national oil and gas legislation, and to the lack of effective enforcement of environmental regulations on the ground. Warning signs also arise from the connections between energy availability, economic opportunity, the rural poor, and environmental conservation (Myers 1996; Adams 1996).

From the precedent of the Niger delta, it is thus possible to see that the social and environmental problems caused by oil theft would begin to be solved if the demand-side of the market could be curtailed by creation of a reliable and efficient power supply system. The first essential steps were taken in 2013 when privatisations of generating and distribution entities were initiated in combination with matching funds within a sovereign wealth fund of U.S. $550 million (Financial Times 11 February 2014). Encouraged more recently and most urgently by the falling oil price, by the consequent rapid reduction of its foreign currency reserves, and by an election in February 2015, Nigeria is trying once again to tackle the economic wastage and environmental impacts of losing a large part of its crude oil exports and tax revenues to theft. After a 4 year impasse since the UNEP (2011) report, negotiations resumed between UNEP, the oil industry and the Federal Government of Nigeria resulting in talks in Geneva in November 2014. Plans to begin cleanups across the delta are imminently expected (Little et al. 2018, Chap. 14 this volume).

It is to be hoped that these measures will reduce demand for illegal fuels by making them less profitable, and will reduce or eliminate 'hot-tapping' of pipelines and illegal refineries in the mangroves. The safety and security in the delta will improve when criminal activities are curtailed by better enforcement. In future, pipeline inspections and maintenance can fully resume, and this will then facilitate cleanup, remediation and restoration of wetland ecosystems, communities and livelihoods. The extreme difficulty of restoring this 'virtuous circle' is a salutary lesson for any new oil and gas province, particularly in Africa.

30.8.6.2 Electricity Supply

In much of sub-Saharan Africa, electricity supply has been state-run for many years and is often interrupted by chronic shortfalls in generating capacity and acute power distribution outages. This infrastructural weakness necessitates the routine use of local diesel generators on a wide scale, and the additional costs involved are estimated to reduce the average GDP of African states by more than 2%. In an extreme case such as Nigeria the estimate of GDP loss is 4% of what is now the largest economy in Africa. The use of local diesel generators can add up to 40% to the cost of doing business in Nigeria, and so powerful vested interests have evolved to make the use of generators part of the *status quo*, with environmentally and socially disastrous consequences ('The Economist' 27 September 2014).

In terms of average per capita generating capacity, sub-Saharan Africa lags Latin America by a factor of two, and Asia by a factor of three. Electricity consumption per capita in Africa is an order of magnitude less than in Europe and almost 15 times less than in Australia and North America ('Petroleum Review' December 2014). Africa's electrical supply capacities are generally far lower than required for sustainable development: for example, a study by McKinsey quoted Angola and Nigeria as presently having annual capacities of 1800 and 3200 MW respectively, and their aim is to increase this by more than five times within the next 5–10 years, at which point they will be at roughly the present capacity of India ('The Economist' 27 September 2014).

Interruptions of power supply would also constrain the timeliness and effectiveness of a major spill response, in terms of power for command centres (e.g., health and safety, lighting, heating and A/C, IT support, etc.). Power helps ensure functional, safe and sanitary accommodation for response teams (e.g., first aid clinic, transport and medical evacuation, reverse osmosis water maker, catering, waste and sewage treatment). In OSR operations offshore the marine supply vessels and facilities can generate their own power (and many other of the above utilities), whereas in the inland and shore zones any remote or rural cleanup operations will not be so well-provided at the point of deployment. This will be especially true in the earliest phase of OSR operations.

Electrical distribution networks are another potentially serious constraint, typically being most complete in more densely-populated areas such as in metropolitan areas (usually only capital and some coastal cities). However, power supply is sometimes prone to interruption due to infrastructure maintenance problems and in some cases poor management systems. A pervasive weakness has been lack of investment for many decades, although this is now changing. Large increases in private investor activity in recent years have assisted in plugging infrastructure gaps in many East African nations, especially in Kenya and Tanzania. Usually as part of major extractive industry or agricultural investments and often led by the Chinese, such developments have included infrastructure projects from electricity, water supply and sewerage, to road building and health care facilities.

Interruption of OSR operations by power cuts could occur over a range of scales and durations dependent on the level of preparedness. At the local or provincial level it will be relatively easy to work-around power shortfalls. OSR responders will bring uninterrupted power supplies (UPS) to the work site(s) as necessary during the incident. In a large spill incident with a prolonged response mobilized out of a Regional Coordination Centre and a series of temporary local centres, UPS would be standard equipment. At the regional or national level, as mentioned above, there are often deficiencies of governance and finance in the state-owned enterprises that run electrical and other utilities. Such deficiencies represent opportunities for vested interests both in the mainstream and informal economies. In some cases this has encouraged criminal activity across the socio-economic spectrum of political and civil society. Examples of the vested interest and/or criminal activity include:

- Beneficiaries of nepotistic supply chains and corrupt procurement procedures
- Inefficient fuel subsidies that distort the economy and delay or prevent modernization
- Importation of refined fuel for diesel generators, even in oil-producing regions
- Crude oil theft from cross-country pipelines and facilities.

Nigeria is suffering from all the above problems and as such represents an extreme negative model for African oil and gas development and OSR preparedness. In the Niger delta many of the ubiquitous diesel generators are powered by illegal fuel supplies arising from stolen crude oil ('bunkering'). This stolen oil is refined in several hundred 'artisanal' refineries hidden in the mangrove creek system and sold openly in plastic bottles and other re-used containers from street stalls across the region (UNEP 2011). This situation is exploited by organised crime and protected by corrupt officials (*e.g.*, not pressing charges after any arrests that are made). Criminal activity makes it hazardous for spill responders to make OSR interventions, and has delayed mangrove cleanup and restoration for a decade (Little et al. 2018, Chap. 14 this volume). Spills onshore in Africa are between 2667 and 3152 t/year, often involving illegal third-party damage in Nigeria (IOGP http://www.iogp.org/data-series/#environment). Theft is estimated by the Extractive Industry Transparency Initiative (EITI) and by the UN as being approximately 10,000 t/day, more than 3% of Nigerian oil production.

30.8.6.3 Renewable or Off-Grid Sources of Energy

There are many very high-quality and widespread renewables resources in East Africa. The importance of this is that renewables in many cases can provide a quicker and more dispersed 'off-grid' source of local power at lower-cost than larger infrastructure. In future, therefore, with the growth in renewables the access to electricity should become more widespread and available. That would help to mitigate climate change, reduce rural poverty and improve education and jobs in trade, crafts, services and ecotourism. Because it would also help to avoid a situation where local communities begin to resent the production of oil and gas in their locality, any disenchanted citizens would not so readily resort to oil theft or other acts of piracy and perhaps in extreme cases, terrorism. Reducing risks of pipeline tapping and sabotage is clearly of direct relevance to OSR preparedness.

Renewables should not therefore be perceived as competition for the oil and gas sector. In fact, the reverse is true if the early and decentralised development of renewables reduces fuel poverty and thus allows the extractive industries to build and maintain a license to operate. Less attention locally will be focussed on the proportion of product that is for export if the local communities are well-provided with sustainable and affordable energy sources. The locals in some countries will also be less vulnerable to poor internal redistribution of oil export revenues from the urban elites and governments.

The costs of alternative energy sources are rapidly falling in East Africa because of the existence in the region of some of the best sites for renewables technologies (wind, solar, geothermal and hydroelectric power). Kenya is exploring for geothermal energy in the Rift Valley at Hell's Gate National Park, and plans to build the world's largest single geothermal plant. Lake Turkana, as well as having Tullow Oil's E&P activity, is also the construction site of Africa's largest wind turbine power plant (private investment). Reasons for this include the lower risk to investors in renewables compared to large fossil fuel or hydroelectric power plants. The development of costly infrastructure for renewables such as large hydroelectric schemes is avoided in the case of wind and solar ('The Economist' 27 September 2014). Non-fossil fuel electricity generating in East Africa includes:

- Kenya – hydroelectric, thermal, geothermal, and wind
- Uganda – nuclear
- Rwanda and Burundi – hydroelectric
- Tanzania – hydroelectric, thermal, and wind
- Mozambique – hydroelectric
- Madagascar – hydroelectric
- RSA – hydroelectric, solar, thermal, nuclear, and wind.

Another renewables advantage is the off-grid power that they can provide direct to customers without congesting the existing transmission lines. Cumbersome bureaucracy is also bypassed by building smaller renewable plant that has not been developed by state-owned enterprises. A common factor in both renewable energy development and electricity privatisations going on in Africa is the creation of investment holding entities that obviate the need for sole state-ownership or state subsidies for fossil fuels ('The Economist' 27 September 2014). If successful and transparent, these governance factors would also build confidence of the people in their leaders and hence reduce the temptation of unemployed people to interfere with oil industry infrastructure.

The USAID 'Power Africa' program announced in December 2013 that it would invest U.S. $7 billion to double the access to electricity in sub-Saharan Africa. The World Bank's 'African Electrification Initiative' has 500 projects across sub-Saharan Africa. It is estimated, however, that by 2030 still only half of the currently unconnected population will be served, due to the inflexibility and high cost of large-scale transmission grids. Combined with the aim to reduce the climate change impact of large generation schemes, the future of off-grid service is clearly vital to the rural poor of whom less than 14% are connected to electricity supplies ('Petroleum Review' April 2014).

Accenture Development Partnerships (ADP) is looking at ways to bring together energy providers, governments and CSOs to help rural communities without electricity. In Uganda ADP is piloting an electricity and internet connectivity project with NGOs and Notre Dame University. A range of obstacles must be overcome (e.g., supply chain, education and finance) and without creating new challenges unintentionally, for example fuel subsidies to support otherwise over-expensive energy sources. The UN agencies under one of the post Rio + 20 initiatives, 'Sustainable Energy for All', will provide seed funding and local resources. Off-grid power from solar or biofuels can be started up around an anchor client such as a telecom company powering their mobile network masts, and then other consumers may be attracted and more rapidly develop their own businesses; all without extending the main power grid into all the rural areas. In the meantime local health centres, schools and evening homework using electric lighting and internet connections will reap immediate rewards in the countryside especially for children ('Petroleum Review' April 2014).

Shell has been supporting the 'Global Alliance for Clean Cookstoves' for over a decade. The strategy is market-based and provides sustainable cooking and heating utensils to reduce the estimated 4 million premature deaths mainly among women and children from indoor cooking fuel pollution in Africa ('Petroleum Review' April 2014).

30.8.7 Procurement and Waste Management

30.8.7.1 Oiled Wastes Arising

Major oil spills affecting coastlines nearly always produce very large quantities of solid and liquid wastes, in particular those spills of persistent oil types that interact with shoreline sediments and mangrove or marsh vegetation over a long time period (Little et al. 2003). International best practice for OSR waste management is to consider during the contingency planning phase how any waste generated during a spill may be handled (*i.e.*, options for storage, transport, processing and disposal). Table 30.11 illustrated the plan in Qatar. Adopting the waste hierarchy is the key goal, whereby the waste generated is reduced, reused and recycled so that the volume requiring final disposal is as low as possible. However, by the nature of spills in remote areas such as East Africa, this level of preparedness does not typically exist.

In many countries, it is seldom very clear in advance what the waste management arrangements will be after an oil spill, and much will depend on the location of the beached oil, and how much is physically recovered. In the 'Gulser Ana' (Madagascar) and 'Oliva' (South Atlantic) spills reported by Laruelle (2012) it was clearly challenging to bring specialist waste handling materials and personnel to the sites, and in the case of the 'Oliva' to take the wastes off Tristan da Cunha 2816 km to South Africa. In the 'Gulser Ana' spill, wastes were handled in-country, but 112 tonnes of quicklime had to be procured from France. During the incident, waste was collected in heavy duty plastic bags which were taken from the cleanup site to intermediate storage sites *via* quad bikes, 4 × 4 pickups or manpower. Waste was then transported to a site near to the capital (approximately a 3 day drive) for either incineration or neutralisation using quicklime prior to being used as raw material for construction. Whilst the options available for final treatment/disposal of waste may be more limited in remote spills and the logistics involved with the transport of waste may result in higher than average costs, it is in the interest of the responsible parties and OSR team leaders to ensure that their solution involves no long term impacts arising from how the waste is managed (Reynolds 2014).

This pragmatic approach has led ITOPF to recommend in another tropical incident ('Asian Lily', December 2012 in a remote island of Papua New Guinea), where only relatively limited quantities of lightly contaminated natural debris were generated (*e.g.*, wood, palm fronds, coconuts), that these be collected and burned. The context of this decision was that the local population live by open fires. In many other parts of the world, this solution would have been in contravention of national legislation and thus illegal (Reynolds 2014). This ethical dilemma may also be interpreted by some stakeholders as hypocritical in terms of best international practice (*i.e.*, 'first world *vs.* developing world'). Given the last paragraph of Sect. 30.8.6.3 regarding the aims of the 'Global Alliance for Clean Cookstoves', it is indeed a major challenge, not least to the reputations of the expatriate entities and national leaders involved.

A spill in Brazil in early 2000 from the Petrobras refinery near Rio de Janeiro released 1145 t of bunker fuels into the sheltered Guanabara Bay near the Guapimiri protected area of mangroves. Much of the oil stranded on sand beaches and was recovered manually as oily waste. Like the HFO lost by 'Katina P', the highly persistent pollutant severely oiled the dry rock cliffs on a rising part of the tidal cycle. Mangroves, waterbirds and artisanal fish traps were more lightly oiled. These examples highlight the difficulties that occur during the emergency response to oil spills affecting mangroves and associated resources in more remote regions.

In contrast, E&P operators have arguably more time to prepare and at least know their work sites in advance. Nevertheless, they may not have been able to fully assess the capability and existence of local waste treatment and disposal facilities, and whether they are fit-for-purpose. These details seldom attract the attention of the negotiators of PSAs and PSCs, although decommissioning terms might provide a guide to OSR preparations in some cases. However, it is not always clear what remediation targets are enshrined in the environment regulations and some are either accidentally or conveniently confusing (*e.g.*, EGASPIN in Nigeria) in terms of trigger levels and close-out levels for remedial action on residual petroleum hydrocarbons in marine sediment and soils (UNEP 2011). UN agencies also reported that fly-tipping of wastes arising from oil spill cleanup has occurred in both Nigeria and after the Al-Jiyeh power plant spill in Lebanon in 2006, where there was also a large volume of demolition debris after the Israeli-Hizbullah conflict.

30.8.7.2 Sewerage Infrastructure

Microbiological contamination arises from land-based (domestic, industrial, agriculture and livestock) and marine (mariculture, shipping) sources. The general picture emerging for all the countries described below (TDA 2012) and others in the WIO region such as Uganda (and including impoverished parts of South Africa), is that the health and economic potential of people in most rural communities and urban informal settlements are adversely affected by the lack of sewerage and biological waste treatment infrastructure.

Additionally, there is anecdotal information that oil exploration crews (seismic surveys and exploration drilling) on land are also challenged by the difficulties of waste treatment, and have sometimes resorted to informal waste disposal in shallow pits. During the response to the 'Katina P' oil spill in southern Mozambique in 1992 it was observed that local people used beaches and mangroves for defecation (and also the temporary waste storage sites provided for the OSR operations around Maputo Bay). In contrast, during the 'Exxon Valdez' response in 1989, one of the first tasks of responders was to design and obtain permits under the U.S. National Pollutant Discharge Elimination System to provide sewage treatment facilities for the cleanup personnel. At the peak the numbers reached 10,000, all housed aboard marine vessels without any permanent structures based on deserted Alaskan shorelines.

It seems that respective governments and large scale aid programs in East Africa have so far failed to solve completely the problem of treating organic wastes. There are now available low cost systems that have been designed specifically and have a proven ability to operate in rural areas of developing countries and during emergency situations such as flooding (www.sowtech.com; www.flexigester.com). They are butyl rubber units that are resistant to UV light. Sealed anaerobic units generate methane that is being used as biogas fuel for the local community. Units are inexpensive, light enough for air transport, easily-transported in-country and assembled in one day. Several units have been installed in Malawi and Uganda in East Africa, plus Sierra Leone, and Chennai, India. In Kampala urban sewage sometimes discharges into papyrus beds; this is now being considered formally for development as tertiary treatment of wastewater. Such efficient intermediate technologies should be more widely available in the study region, not least to improve sanitation during OSR operations.

30.8.7.3 Sewage Treatment (TDA 2012)

Kenya

It was reported that 83% of the population of Mombasa were using pit latrines or septic tanks, with sludges dumped at Kibarani landfill or in other unspecified areas of mangrove (Munga and Mwaguni 1994; Williams et al. 1996). Microbial water quality studies have been completed in a number of locations (UNEP et al. 2009). In Mombasa, microbial pollution levels near urban centres were several orders of magnitude higher than in coastal waters in rural areas. Over 50% of all reported diseases have been attributed to inadequate wastewater management, although the source of the microbial contamination was not always clear; whether from drinking contaminated water or contact with contaminated seawater during coastal recreation.

Tanzania

Raw sewage is directly released into estuaries and other coastal habitats and from coastal towns. High counts of faecal coliform and total coliform bacteria of up to several thousand per ml of seawater were also recorded in Zanzibar. High counts have prompted health concerns and warnings of health risks to swimmers, and some beaches on Zanzibar and the mainland in Dar es Salaam (e.g., Ocean Road and Banda) have been closed for swimming and other recreation due to microbial contamination. WIO-LaB monitoring in 2007 showed water contamination in Stone Town and Dar es Salaam (near mangrove-rich TZ021, an IBA 'in danger').

Mozambique

There is only one sewage treatment plant in Mozambique, located in Maputo City, which treats about 50% of the city's sewage. Faecal coliform, faecal streptococci and *E. coli* were detected both in marine waters and shellfish tissues in Maputo Bay. Where the discharge of sewage takes place, such as Miramar, it is not safe for swimming. Faecal coliform counts in the water within the channel near the Infulene River were high (460,000 bacteria counts per 100 ml) and values exceeded 2400 bacteria counts per 100 ml in the river mouth. The bacteria *Vibrio parahaemolyticus* and *V. mimicus* have been found in clams at the Incomati River mouth and near Matola in the Maputo estuary. *Vibrio* spp. causes severe gastro-intestinal illnesses. High levels of microbial pollution have also been found at Beira and Nacala Bays, although the concentrations are low compared to those observed in Maputo Bay.

Madagascar

Studies conducted around Taolagnaro measured high *E. coli* counts (13,300 counts/100 ml) in coastal waters. High counts of total coliforms were reported from Mahajanga and Nosy Bé, and were attributed to defecation on the beaches as well as inappropriate treatment of municipal wastewater. Over the period 1993–1998) there were 18 cases of human illness caused by eating contaminated seafood: marine turtles (seven cases), sharks (eight cases), fish (one case) and molluscs (two cases). WIO-LaB water and sediment quality monitoring surveys confirmed that microbial pollution is an ongoing problem in coastal Madagascar (UNEP et al. 2009).

Comoros

There is no sewerage, drainage or wastewater treatment in the Comoro Islands. Households typically use pit latrines which can leak and contaminate groundwater and coastal and marine environments. More than two thirds of the population relies on rain, surface and ground water supplies. No water quality monitoring and control mechanisms exist, and only occasional salinity tests are done. On some islands there is a massive risk of groundwater pollution by seepage from septic tanks. Comoro Island (EBA098) is critically in need of priority conservation action.

Mauritius

At present, 73% of the households use cesspits or septic tanks whilst 2% use pit latrines; most of the effluents are discharged directly to the sea or carried to the sea by runoff and rivers with higher potential for microbial pollution, particularly after heavy rains. Total and faecal coliforms are monitored monthly at public beaches at Flic en Flac, Albion, Pointe aux Sables, Trou aux Biches, Mon Choisy, Le Goulet, Grand Baie and Blue Bay (UNEP et al. 2009). The Ministry of Fisheries reported that waters at most beaches were within recommended guidelines for contact recreation in 2004 (counts of total coliforms <1000 per 100 ml and faecal coliforms <200 per 100 ml). Good water quality was confirmed by quarterly monitoring of waters in 2007 through the WIO-LaB project. The highest levels of faecal and total coliforms were at Pointe aux Sables, near Port Louis, where the beach is closed to swimming (UNEP et al. 2009).

Seychelles

Coastal waters are generally good quality, except during the rainy season when areas with significant river inflows have higher microbial quantities. Effluent from wastewater treatment plants discharged directly into the ocean was found to contain between 2000 and 5000 total coliform counts per 100 ml, far above the recommended standards of 500 per 100 ml. During a monitoring survey conducted during 2007, high microbial counts were recorded at Beau Vallon Bay during the rainy season. These were mostly associated with runoff from non-point sources such as rivers and small streams. Outbreaks of water- and insect-borne diseases usually occur during the rainy season and associated with defective on-site wastewater disposal systems (UNEP et al. 2009).

30.8.8 Environmental and Social Impact Assessment (ESIA)

In many oil and gas developments the first stages of E&P activity are characterized by low risks of oil spills during the seismic survey phase. Risks increase during the drilling phases (e.g., well-kick, blow-out and well-testing) and may peak during the construction and development phases due to intensification of land and marine transport including river-crossings and over-water fuel-handling operations. The host countries typically review the development ESIAs for a description of the proposed technologies and of the local baseline of the natural and human environment, followed by risk-based predictions of potential impacts and their mitigation. OSR arrangements are usually introduced at this stage (Personal communication, Tullow Oil). Care is needed to ensure that communications and dialogue regarding OSR plans and sensitivity mapping are not left too late in the E&P development life cycle. These good practices ideally should proceed along with stakeholder engagement under the ESIA umbrella.

Predictions of impacts made during ESIAs will be tested in public hearings, by the media and by researchers. Any challenges from the host government and local community and the lessons learned from them, should be recorded. If impact predictions are not defensible, such challenges will be an early warning of the sorts of problems that would eventually materialise during OSR. Consequently, tracking the lessons learned should continue seamlessly into OSR contingency planning using a 'hazard chain of custody' that systematically binds in a succession of contractors and suppliers in the development life cycle.

In that they seek to obtain a permit to operate, ESIA predictions might appear arrogant if they are too vague or optimistic. On the other hand, when impact predictions are too precise, they can cause difficulty during later OSR incidents if the real fate and effects of spills become wider-ranging or more intense. The answer is always to ensure that baseline data and the results of OSR drills, trajectory modelling, and ESIAs are made public, allowing NGOs, CSOs and the local people to reach their own conclusions. Effective planning is central to providing a benchmark standard, training and case studies that make sense and avoid surprises to communities, industry and government.

ESIA principles should also be followed in the case of tanker accidents affecting important natural and human resources, although the location is not predictable and thus pre-spill baseline survey data are usually missing. The consequences of an incident occurring in a remote location as opposed to in a busy shipping lane are different, but are not necessarily less significant. A vessel grounded and losing oil in a busy port may cause significant disruption to shipping activity in the area, but the same vessel grounded and spilling oil in a remote location may cause significant disruption to local subsistence fishermen. The people impacted in remote locations often rely more directly and heavily on those natural resources that have been most affected, including mangroves (Reynolds 2014).

Whether from fixed installations or tankers, after an oil spill some monitoring and impact assessment is typically conducted to determine the overall area of impact and to provide information on the likelihood and timescales of natural recovery. In any monitoring or impact assessment planning is essential, and this is even more crucial when working on remote incidents (Reynolds 2014). Best practice suggests that if the need for studies is reasonable and if logistical and H&S considerations allow the work to be carried out within a meaningful time scale, then such work should be conducted and paid for by the responsible party. Although post spill monitoring is desirable, the feasibility and cost effectiveness of deploying and maintaining resources means that often it is not possible to carry out such activities successfully. This is particularly unfortunate given the lengthy timescales that are often involved in recovery of damaged mangrove ecosystems, and the human uses that are affected.

30.8.9 Reputation Management

The extent to which a responsible party during a spill can maintain its reputation with stakeholders (*e.g.*, staff, shareholders, investors, neighbors) will depend on prevention, response and mitigation of the adverse impacts of oil spills. These are principles central to the IPIECA Global Initiative (GI) approach. Some national governments have permitted their mining and oil & gas companies to seek exploration and development rights abroad without attempting to influence these companies' behavior towards the host country and to the local communities. In the event of poor behavior, nations like USA already have regulatory instruments to prosecute offences alleging corruption (e.g., the U.S. Foreign Corrupt Practices Act and further rule-making by the U.S. Securities and Exchange Commission).

Corporate Social Responsibility (CSR) and the Extractive Industry Transparency Initiative (EITI) aim to improve transparency. The EITI is reporting the revenues in taxes and fees from the developments that are paid to the host country. There can be resistance to a project-by-project reporting of these payments, thus limiting the potential value of EITI. Some host country governments are content to leave details of their arrangements unaccountable and opaque. This has occurred in a major OSR context (>U.S. $1 billion mitigation and restoration projects) when suggestions were rejected to use governance frameworks such as CSR and EITI to improve the government organisation of the cleanup projects (see Sect. 30.6.3.4). The reason given by officials was that these measures apply only to extractive industry, but this runs counter to the principle of government and industry collaboration enshrined in the OPRC Convention (see Sect. 30.1.2).

Some governments actively promote their countries' extractive industries abroad, and there are initiatives to make this support conditional on the reputations of their corporations remaining in good standing. CSR and EITI are among the accepted tools that help maintain a positive reputation. According to new rules in Canada, any company with foreign development portfolios that fails to maintain good relations with local communities must implement conflict-resolution measures or else the home government will withdraw its economic diplomacy on that company's behalf. Unfortunately, such measures are the exception. Reputation management, auditing, and Environmental, Social and Governance (ESG) methods (www.cdcgroup.com) often appear unable to improve accountability. National and international companies and wealthy citizens are not deterred from evading or legally avoiding taxes. These taxes would benefit local communities in some of the poorest host nations; for example, the private equity firm Actis (created in 2004 from privatization of part of UK's former Commonwealth Development Corporation) has used a Mauritius holding company to reduce tax liabilities and extract management fees from a Uganda power utility. Private equity can be a catalyst for investment and growth (see Sect. 30.8.4). However, tax arrangements that are opaque in terms of beneficiaries are draining money from Africa, as shown by the 'Paradise Papers', so far involving Mauritius tax havens in the Namibian fishing industry and an Angolan sovereign wealth fund. The role of Mauritius and others in African tax avoidance seems unethical; doubly so, given the lack of sewage treatment facilities for three-quarters of the population of Mauritius, and the consequent health and environmental effects of marine pollution (see Sect. 30.8.7.3). Although elites of independent nations can legally hold wealth in offshore funds and property, such tax avoidance may be seen as colonialism by other means. Accountability, transparency and good reputation are thus equally vulnerable as they are in authoritarian states, and mangroves are but one of the natural capital assets at risk.

30.9 Conclusions

The marine and terrestrial ecological sensitivities in East Africa and the Middle East are acknowledged and widespread, and include mangrove habitats. However, during coastal development projects and after oil spills, the management of these sensitivities is often data-deficient and faces serious challenges. A cyclic tendency can militate against sustained OSR planning efforts in Africa: industry is inevitably in thrall to the fluctuating oil price, while inter-governmental initiatives are often susceptible to a lack of continuity in training and implementation. During real oil spills and coastal development projects, the follow-up environmental monitoring and mitigation can be patchy or non-existent. Because mangrove habitats are vulnerable to adverse impacts due to their sensitivity and long recovery times, their conservation outlook is deeply uncertain in all phases of the economic cycle.

Although it has been 25 years since the most significant oil spills affecting mangroves in the region, there are many important mangrove areas with high biological diversity and abundance, some also including adjacent corals, seagrass beds, saltmarsh, brackish and freshwater wetlands. Oil spill response must balance the habitats' sensitivities and often 'trade-off' an adverse impact on one habitat to protect other habitats nearby. The trade-off decision must be made collaboratively with local people. Each of these habitats' associated communities of fish, turtles, marine mammals and birds are linked by migrant fauna that are functionally dependent at certain life stages on mangroves and other often adjacent habitats such as intertidal flats in East Africa and the Middle East (*e.g.*, along bird 'flyways' that also reach northern Eurasia). In the countries studied, the proportions of designated sites that include mangroves are high: 57% of coastal IBAs; 39% of all EBSAs; and 15% of all EBAs. Local economies throughout the region rely heavily on natural capital both for subsistence livelihoods and in terms of efforts to promote sustainable development (*e.g.*, high value tourism and UN Sustainable Development Goals).

Mangrove restoration and creative conservation programs need to be evaluated and, if feasible, taken more seriously in project ESIAs. Follow-up monitoring by developers and regulatory agencies should be scrupulous. Unless local communities are genuinely involved and also benefit more equitably from extractive industries and other coastal developments, both directly and through local investments paid for by the taxes levied by the communities' government, there is a risk that the community will reject industry's license to operate. As described by Gundlach, and by Little et al. (in this Coastal Research Library publication), this negative point is illustrated by pipeline spills of 2008 followed by numerous subsequent spills from 'hot-tapping' and 'artisanal' refining in the mangroves near Bodo, Nigeria. The diligent protection and restoration of mangroves and their integration into landscape-scale conservation that is inclusive of the rural poor are key strategies to avoid this extreme and at times violent precedent.

Acknowledgements The preparation of this chapter was self-funded, and the views expressed are my own. The author is grateful for the assistance of the following individuals and organizations: BirdLife International for data on IBAs, EBAs and bird habitats; Government of Mozambique, Richard Golob and Jeff Hyland for 'Katina P' study; Joseph Sandrin (CH2M) for Saudi Arabian marsh photograph; Phil Ruck (IPIECA, now of ITOPF) and Peter Taylor (Petronia Consulting Limited) for East Africa oil spill preparedness project; Melanie Little for landscape and mangrove sketches; Mesaieed Industrial City, Arthur D. Little Limited, Benno Böer (UNESCO), and John Barker and the late Phil Beall for environmental assessment and mangrove mitigation in Qatar; John Mullett for 'Flexigester' data; Martin Guard (UNCC Secretariat), Kingdom of Saudi Arabia Presidency of Meteorology and Environment, Pandion Technology Limited, Yakov Galperin, and the Independent Reviewers for coastal remediation and restoration projects in Saudi Arabia after the 1st Gulf War; and Muralee Thummarukudy (UNEP) for post-conflict environmental projects in Lebanon and Nigeria.

References

Adams WM (1996) Conservation and development. In: Adams WM, Goudie AS, Orme AR (eds) The physical geography of Africa, Oxford Regional Environments. Oxford University Press, Oxford, pp 367–382

Adams WM, Aveling R, Brockington D, Dickson B, Elliott J, Hutton J, Roe D, Vira B, Wolmer W (2004) Biodiversity conservation and the eradication of poverty. Science 306:1146–1148

Aghalino SA, Eyinla E (2009) Oil exploration and marine pollution: evidence from the Niger Delta, Nigeria. J Hum Ecol 28(3):177–181

Allen WH (1992) Increased dangers to Caribbean marine ecosystems. Bioscience 42:330–335

API (1985) Oil spill response: options for minimising adverse ecological impacts. American Petroleum Institute (API) Publication 4398

Baca B, Ward G, Lane CH, Schuler PA (2005) Net Environmental Benefit Analysis (NEBA) of dispersed oil on near shore tropical

ecosystems derived from the 20 year 'TROPICS' field study. In: Proceedings international oil spill conference. API, Washington, DC
Baker JM (1976) Marine ecology and oil pollution. Applied Science Publishers, Barking
Baker JM (1982) Mangrove swamps and the oil industry. Oil Petrochem Pollut 1(1):5–22
Baker JM (1991) Guidelines on biological impacts of oil pollution, IPIECA Report Series, vol 1, 21p
Baker JM (1992) Oil and African Great Lakes. Mitt Internat Verein Limnol 23:71–77
Baker JM, Crothers JH, Little DI, Oldham JH, Wilson CM (1984) Comparison of the fate and ecological effects of dispersed and non-dispersed oil in a variety of intertidal habitats. In: Allen TE (ed) Oil spill chemical dispersants: research, experience and recommendations, STP 840. American Society for Testing and Materials, Philadelphia, pp 239–279
Baker J, Oldham J, Wilson C, Dicks B, Little D, Levell D (1990) *Spartina anglica* and oil: spill and effluent effects, cleanup and rehabilitation. In: Gray A, Benham P (eds) *Spartina anglica* – a research review. HMSO, London, pp 52–62
Baker J, Little A, Heaps L (1996) Guidelines for assessing the ecological condition and recovery of oiled shorelines. AURIS Report to BP Shipping Ltd., Marine Pollution Control Unit (UK), Marine Spill Response Corporation (USA), Amoco Corporation, and Exxon Research and Engineering Company, 89p
Ballou TG, Hess SC, Dodge RE, Knap AH, Sleeter TD (1989) Effects of untreated and chemically dispersed oil on tropical marine communities: a long-term field experiment. In: Proceedings international oil spill conference. API, Washington, DC
BirdLife International (2003) BirdLife's Online World Bird Database: the site for bird conservation, Versions 2.0 and 2.1. BirdLife International, Cambridge. http://www.birdlife.org. Accessed 3 Jul 2005, 21 Jun 2008 and 20 May 2017
Böer B (1993) Anomalous pneumatophores and adventitious roots of *Avicennia marina* (Forssk.) Vierh. mangroves two years after the 1991 Gulf War oil spill in Saudi Arabia. Mar Pollut Bull 27:207–211
Brito EMS, Duran R, Guyoneaud R, Goñi-Urriza M, de Oteyza TG, Crapez MAC, Aleluia I, Wasserman JCA (2009) A case study of *in situ* oil contamination in a mangrove swamp (Rio de Janeiro, Brazil). Mar Pollut Bull 58:418–423
Burns K, Garrity S, Jorissen D, MacPherson J, Stoelting M, Tierney J, Yelle-Simmons L (1994) The 'Galeta' oil spill (ii). Unexpected persistence of oil trapped in mangrove sediments. Estuar Coast Shelf Sci 38(4):349–364
BWP (1977–1994) Handbook of the birds of Europe, the Middle East and North Africa; birds of the Western Palearctic. In: Cramp S et al (eds) in nine volumes. Oxford University Press, Oxford
Carter-Groves M (2014) Global dispersant stockpile: part of the industry solution to worst case scenario readiness. In: Proceedings international oil spill conference. API, Washington, DC
CEDRE (2015) Centre of Documentation, Research and Experimentation. Accidental water pollution, Database of spill incidents and threats in waters around the world
Chan EI (1977) Oil pollution and tropical littoral communities; biological effects of the 1975 Florida Keys oil spill. In: Proceedings international oil spill conference. API, Washington, DC
Clewell A, Aronson J (2007) Ecological restoration: principles, values, and structure of an emerging profession. Society for Ecological Restoration International & Island Press, Washington, DC, 216p
Collar N, Crosby M, Stattersfield A (1994) Birds to watch 2: the world list of threatened birds, BirdLife Conservation Series No. 4. BirdLife International, Cambridge
Cormack CD, Hale JA, Gabriel JJ, Langman O (2011) *Nasima* and oil – do they mix? Assessing crab survival in oiled sediment. In: Proceedings international oil spill conference. API, Washington, DC
De Clerck O, Koppejans E (1994) Status of the macroalgae and seagrass vegetation after the 1991 Gulf War oil spill. In: Abduzinada A, Krupp F (eds) The status of coastal and marine habitats two years after the Gulf War oil spill. Courier Forschungsintitut Senkenberg
Deloitte (2014) Guide to oil & gas in East Africa: uniquely structured, 2014 edition, 48p
DeMicco E, Schuler PA, Omer T, Baca B (2011) Net Environmental Benefit Analysis (NEBA) of dispersed oil on nearshore tropical ecosystems – 'TROPICS' – 25th year research visit. In: Proceedings international oil spill conference. API, Washington, DC
Diaz-Piferrer M (1964) Effects of an oil spill on the shore of Guanica, Puerto Rico. Abstr Deep Sea Res 11(5):855–856
Dome (2004) Taweelah B extension project EIA: mangrove study report, unpublished
Duke NC (2016) Oil spill impacts on mangroves: recommendations for operational planning and action based on a global review. Mar Pollut Bull 109:700–715
Garrity S, Levings S, Burns K (1994) The 'Galeta' oil spill (i). Long-term effects on the physical structure of the mangrove fringe. Estuar Coast Shelf Sci 38(4):327–348
Getter CD, Lewis RR III (2003) Spill response that benefits the long-term recovery of oiled mangroves. In: Proceedings international oil spill conference. API, Washington, DC
Getter CD, Scott GI, Michel J (1981) The effects of oil spills on mangrove forests: a comparison of five oil spill sites in the Gulf of Mexico and the Caribbean Sea. In: Proceedings international oil spill conference. API, Washington, DC
Getter C, Michel J, Hayes M (2005) Oiled subtropical salt marshes and tidal flats: biological recovery of intertidal macro-epibenthos and associated vegetation from the 1991 oil spill on the Saudi Arabian coast. In: Proceedings international oil spill conference. API, Washington, DC
Getz WM, Fortman L, Cumming DHM, Du Toit J, Hilty J, Martin RB, Murphree M, Owen-Smith N, Starfield AM, Westphal MI (1999) Sustaining natural capital and human capital: villages and scientists. Science 283:1855–1856
Gilfillan ES, Page DS, Hanson RP, Cooley JF, Hotham JR (1981) Fate of the 'Zoe Colocotroni' oil spill and its effects on infaunal communities associated with mangroves. In: Proceedings international oil spill conference. API, Washington, DC
Gundlach ER (2018) Oil-related mangrove loss east of Bonny River, Nigeria. In: Makowski C, Finkl CW (eds) Coastal Research Library (CRL): threats to mangrove forests: hazards, vulnerability and management solutions. Springer, Cham
Gundlach ER, Michel J, Scott GI, Hayes MO, Getter CD, Davis WP (1979) Ecological assessment of the 'Peck Slip' oil spill in eastern Puerto Rico. In: Proceedings ecological damage assessment conference Society of Petroleum Biology. American Petroleum Institute, Washington, DC, pp 303–318
Gundlach ER, Murday M (1989) Coastal sensitivity atlas of Mauritius for oil spill response, Ministry of Housing, Lands and the Environment of Mauritius, in cooperation with IMO and UNEP, Port Louis, Mauritius, 56 p
Gundlach ER, Little DI, Neff JM, Aurand DV (1995) Evaluation of historic spill sites for long-term recovery studies. In: Proceedings international oil spill conference. API, Washington, DC
Hoi-Chaw L, Meow-Chan F (1985) Field and laboratory studies on the toxicities of oils to mangroves. In: Proceedings international oil spill conference. API, Washington, DC
Hoi-Chaw L, Teas HJ, Pannier F, Baker JM (1993) Biological impacts of oil pollution: mangroves, IPIECA Report Series, vol 4. IPIECA, London, 20p
Holt SS, Rabalais N, Rabalais S, Cornelius S, Holland JS (1978) Effects of an oil spill on saltmarshes at Harbour Island, Texas. In: Conference on assessment of ecological impacts of oil spills. American Institute of Biological Sciences, Arlington, pp 3435–3352

Howard S, Little DI (1987) Effects of infaunal burrow structure on oil penetration into sediments. In: Proceedings international oil spill conference. API, Washington, DC
Howard S, Baker JM, Hiscock K (1989) The effects of oil and dispersants on seagrasses in Milford Haven. In: Dicks B (ed) Ecological impacts of the oil industry. Wiley, Chichester, pp 61–96
IEA (2014) Africa energy outlook: a focus on energy prospects in sub-Saharan Africa, world energy outlook special report. International Energy Agency, Paris, 242p
IPIECA (2000) Choosing spill options to minimize damage: net environmental benefit analysis, IPIECA Report Series, vol 10, 24p
IPIECA (2004) A guide to oiled wildlife response planning. IPIECA, London, 51p
IPIECA (2008) Operating in areas of conflict: an IPIECA guide for the oil and gas industry, 38p
IPIECA (2012) Preventing corruption: promoting transparent business practices, prepared for Rio+20, June 2012, 2p
IPIECA-IOGP Joint Industry Project (2014) Regulatory approval of dispersant products and authorization of their use. IPIECA and IOGP
IPIECA-OGP (2007) The convention on biological diversity: a guide for the oil and gas industry. IPIECA and International Association of Oil and Gas Producers, 24p
IPIECA-OGP (undated-a) An Ecosystem Approach to Oil and Gas Industry Biodiversity Conservation. IPIECA-OGP Biodiversity Working Group, 2p
IPIECA-OGP (undated-b) Sensitive Environments: Challenging Operations, 2p
IUCN (World Conservation Union) (2003) International Union for Conservation of Nature and Natural Resources 'Red List' of threatened species. Available at: www.redlist.org
Jackson JBC, Cubit JD, Keller BD, Batista V, Burns K, Caffey HM, Caldwell RL, Garrity SD, Getter CD, Gonzalez C, Guzman HM, Kaufman KW, Knap AH, Levings SC, Marshall MJ, Steger R, Thompson RC, Weil E (1989) Ecological effects of a major oil spill on Panamanian coastal marine communities. Science 243:37–44
Keller BD, Jackson JBC (1993) Long-term assessment of the oil spill at Bahia Las Minas, Panama, synthesis report, volume I: executive summary, OCS Study MMS 93–0047. U.S. Department of the Interior, New Orleans, 129p
Kenworthy W, Durako M, Fatemy S, Valavis H, Thayer G (1993) Ecology of seagrass in north-eastern Saudi Arabia one year after the Gulf War oil spill. Mar Pollut Bull 27:213–222
Kingston P, Little DI, Harkantra S (1999) Biological impacts of oil pollution: sedimentary shores, IPIECA Report Series, vol 9. IPIECA, London. 20p
Knap AH (1992) Biological impacts of oil pollution: coral reefs, IPIECA Report Series, vol 3. IPIECA, London. 16p
Knap AH, Wyers SC, Dodge RE, Sleeter TD, Frith HR, Smith SR (1985) The effects of chemically and physically dispersed oil on the brain coral *Diploria shigosa* (Dana) – a summary review. In: Proceedings international oil spill conference. API, Washington, DC
Langman O, Hale JA, Cormack CD, Risk MJ, Madon SP (2012) Developing multimetric indices for ecological restoration progress in saltmarshes. Mar Pollut Bull 64:820–835
Laruelle F (2012) Responding to spills in remote locations: 'Gulser Ana' (Madagascar) and 'Oliva' (South Atlantic), In: Proceedings Interspill Conference, London, 13–15 March 2012
Levings SC, Garrity SD (1995) Oiling of mangrove keys in the 1993 Tampa Bay oil spill. In: Proceedings international oil spill conference. API, Washington, DC
Lewis RR (1980) Oil and mangrove forests: observed impacts 12 months after the 'Howard Star' oil spill, Tampa, FL. Fla Sci 43(Suppl):23
Lewis RR III (1983) Chapter 19: impact of oil spills on mangrove forests. In: Teas H (ed) Biology and ecology of mangroves, tasks for vegetation science. Springer, Dordrecht, pp 171–183
Lewis RR III (2005) Ecological engineering for successful management and restoration of mangrove forests. Ecol Eng 24:403–418
Lewis RR, Haines KC (1980) Large scale mangrove restoration on St. Croix, U.S. Virgin Islands, second year. In: Proceedings 7th annual conference on restoration and creation of wetlands. Hillsborough Community College, Tampa. 294p
Lin Q, Mendelssohn IA (2005) Dispersants as countermeasures in near shore oil spills for coastal habitat protection. In: Proceedings international oil spill conference. API, Washington, DC
Little DI, Scales DL (1987) Effectiveness of a type III dispersant on low energy shorelines. In: Proceedings international oil spill conference. API, Washington, DC
Little DI, Deakin T, Fichaut B, Meech RJ (2003) Some observations on heavy fuel oil spills: trends, impacts and comparisons with crude oil spills. In: Proceedings 26th Arctic and marine oilspill program technical seminar, Ottawa, pp 971–998
Little DI, Galperin Y (2017) The assessment of hydrocarbon contamination in contrasting sedimentary environments. In: Heimann K, Karthikeyan OP, Muthu SS (eds) Biodegradation and bioconversion of hydrocarbons: research advances and recent developments. Springer Science+Business Media, Singapore. isbn:ISBN: 978-981-10-0199-4. http://link.springer.com/chapter/10.1007/978-981-10-0201-4_1
Little DI, Holtzmann K, Gundlach ER, Galperin Y (2018) Sediment hydrocarbons in former mangrove areas, southern Ogoniland, eastern Niger Delta, Nigeria. In: Makowski C, Finkl CW (eds) Coastal Research Library (CRL): threats to mangrove forests: hazards, vulnerability and management solutions. Springer, Cham
Loya Y, Rinkevich B (1979) Abortion effect in corals induced by oil pollution. Mar Ecol Prog Ser 1:77–80
MEPA (1991) Coastal survey summary report, Gulf oil spill. Unpublished report by former Meteorological and Environmental Protection Administration, Kingdom of Saudi Arabia
Michel J, Hayes MO, Getter CD (2005) The Gulf War oil spill twelve years later: consequences of eco-terrorism. In: Proceedings international oil spill conference. API, Washington, DC
Munga Y, Mwaguni O (1994) Assessment of land-based sources of marine pollution along the Kenyan Coast. Draft report to FAO by Kenya Marine and Fisheries Research Institute, Mombasa
Myers N (1996) Biodiversity and biodepletion. In: Adams WM, Goudie AS, Orme AR (eds) The physical geography of Africa, Oxford Regional Environments. Oxford University Press, Oxford, pp 356–366
Nadeau RJ, Berquist ET (1977) Effects of the March 18, 1973 oil spill near Cabo Rojo, Puerto Rico on tropical marine communities. In: Proceedings international oil spill conference. API, Washington, DC
National Research Council (1985) Oil in the sea: inputs, fates and effects. National Academy Press, Washington, DC
NCG (2012) Review of: integration of environmental concerns and civil society engagement in petroleum-related Norwegian development cooperation and the development of petroleum resources in developing countries, 49p
Neff JN, Anderson JW (1981) Response of marine animals to petroleum and specific petroleum hydrocarbons. Applied Science Publishers, London, 177p
Negri AP, Smith LD, Webster NS, Heyward AJ (2002) Understanding ship-grounding impacts on coral reefs: potential effects of anti-foulant paint contamination on coral recruitment. Mar Pollut Bull 44(2):111–117
NOAA (2014) Oil spills in Mangroves. Planning & response considerations. Seattle, Washington, U.S. Department of Commerce, National Oceanic and Atmospheric Administration (NOAA), National Ocean Service, Office of Response and Restoration, 96 p
NOAA (2016) Incident archive (complete list of all publicly released information on any oil spill incident). US Department of Commerce, National Oceanic and Atmospheric Administration

Orme AR (1996) Coastal environments. In: Adams WM, Goudie AS, Orme AR (eds) The physical geography of Africa, Oxford Regional Environments. Oxford University Press, Oxford, pp 238–266

Owen R, Knap AH (2002) Inhibition of coral photosynthesis by the anti-fouling herbicide 'Irgarol' 1051. Mar Pollut Bull 44(7): 623–632

Porter R, Christensen S, Schiermacker-Hansen P (1996) Field guide to the birds of the Middle East, 2nd edn 2004. Christopher Helm, London, 460 p

Poulsen A, Burns K, Lough J, Brinkman D, Delean S (2006) Trace analysis of hydrocarbons in coral cores from Saudi Arabia. Org Geochem 37:1913–1930

Price A, Downing N, Fowler S, Hardy J, Le Tissier M, Mathews C, McGlade J, Medley P, Oregioni B, Readman J, Roberts C, Wrathall T (1994) The 1991 Gulf War: environmental assessments of IUCN and collaborators, A Marine Conservation and Development Report. IUCN, Gland, 48p

Qatar Petroleum (2002) Distribution and abundance of coastal and marine birds in Mangrove Swamp at Ras Laffan Industrial City, Qatar. Report by Tayab M, Bernandas A, Olivenza R

Reynolds K (2014) Applying international standards in response to oil spills in remote areas. In: Proceedings international oil spill conference. API, Washington, DC

Rützler K, Sterrer W (1970) Oil pollution damage observed in tropical communities along the Atlantic seaboard coast of Panama. Bioscience 20:222–224

SCENR (2008) Supreme Council for the Environment and Natural Resources submittal of Khor Al-Adaid as candidate world heritage site to UNESCO., 18 March 2008

Spooner M (1970) Oil spill in Tarut Bay, Saudi Arabia. Mar Pollut Bull 1:166–167

Stattersfield AJ, Crosby MJ, Long AJ, Wege DC (1998) Endemic bird areas: priorities for biodiversity conservation, Biodiversity Conservation Series No. 7. BirdLife International, Cambridge, 846p

TDA (2012) Trans-boundary diagnostics analysis, Agulhas and Somalia current large marine ecosystem and South West Indian ocean fisheries program, Vol 1, Baseline

Teas HJ, Lasday AH, Luque L, Morales RA, de Diego ME, Baker JM (1989) Mangrove restoration after the 1986 Refinería Panama oil spill. In: Proceedings international oil spill conference. API, Washington, DC

Thorhaug A, Marcus J, Booker F (1986) Oil and dispersed oil effects on subtropical and tropical seagrasses. Mar Pollut Bull 17:357–361

Tomás J, Guitart R, Mateo R, Raga JA (2002) Marine debris ingestion in loggerhead turtle *Caretta caretta* from the western Mediterranean. Mar Pollut Bull 44(3):211–216

Tucker G, Evans M (1997) Habitats for birds in Europe: a conservation strategy for the wider environment, BirdLife Conservation Series No. 6. BirdLife International, Cambridge

Tucker G, Heath M (1994) Birds in Britain and Europe; their conservation status, BirdLife Conservation Series No. 3. BirdLife International, Cambridge

UN (2007a) Resolution adopted by the general assembly, oil slick on Lebanese shores, 61st Session, Agenda item 53(a), United Nations General Assembly Resolution 61/194, 2006, 2p

UN (2007b) Oil slick on Lebanese shores. Report of the secretary-general to the United Nations General Assembly 62nd Session, Agenda item 54 Sustainable Development, October 2007, 14p

UNCC (2016) Post-conflict environmental restoration: the UNCC Follow-up programme for environmental awards. United Nations Compensation Commission, 110p

UNEP (2007) Mangroves of Western and Central Africa. UNEP-Regional Seas Programme/UNEP and World Conservation Monitoring Centre, 92p

UNEP (2011) Environmental assessment of Ogoniland, United Nations environment programme, ISBN: 978-92-807-3130-9, 257p

UNEP-Nairobi Convention, CSIR, WIOMSA (2009) Regional synthesis report on the status of ratification and implementation of international environmental conventions relevant to land-based activities and sources of pollution of the coastal and marine environment of the WIO region. UNEP, Council for Scientific and Industrial Research, Western Indian Ocean Marine Science Association, Nairobi, Kenya, 53p

UNEP-Nairobi Convention, WIOMSA (2015) Chapter 36: Overall assessment of the state of the coast in the Western Indian Ocean. In: Paula J (ed) The regional state of the coast report: Western Indian Ocean. UNEP and WIOMSA, Nairobi, 546p

Vazquez M, Allen K, Kattan Y (2000) Long-term effects of the 1991 Gulf War on the hydrocarbon levels in clams at selected areas of the Saudi Arabian Gulf coastline. Mar Pollut Bull 40:440–448

Williams TM, Rees J, Kairu KK, Yobe AC (1996) Assessment of contamination by metals and selected organic compounds in coastal sediments and waters of Mombasa, Kenya. British Geological Survey report WC/96/37, 89p

WIOMH (2012) West Indian ocean marine highway project post completion report – final – December 2012

WIOMH (2013) West Indian ocean marine highway implementation completion and results report., June 2013

31 Environmental Governance As a Framework for Mangrove Sustainability on the Brazilian Amazon Coast

Indira A. L. Eyzaguirre and Marcus E. B. Fernandes

Abstract

The mangrove ecosystem is subject to the dynamics of symbiotic relationships between the social and ecological system, according to the use of the ecosystem services of this forest. However, the environmental conflicts have increased due to the social metabolism and the terms of economic development. The dwellers are environmentally responsible for the mangrove ecosystem governance on the Brazilian Amazon coast. They heavily depend on the mangrove resource, such as dewllers who live inside and in the surroundings of the Caeté-Taperaçu Marine Extractive Reserve (MER). This type of Conservation Unit (CU) aims to generate sustainable development, and users identify direct and indirect drivers of change according to the dependence of the mangroves and the handling experience of this ecosystem. Besides them, the Brazilian State, through the promulgation of laws also govern environmentally forests of mangroves in the MER, but mandatorily. The practice of environmental inspection is run by federal institutions through conservation tools ruled by the federal policy guidelines. Finally, the MER environmental governance, as a space where ecological and social dynamics occurs, is formed by networks that include some principles such as the social actors' responsibilities, as well as the representations, the transparency and accountability of the management, and the resilience as a strategic vision for the sustainable conservation of these units.

Keywords

Common resources · Ecosystem services · Mangrove dwellers · Environmental policy · Conservation strategies

I. A. L. Eyzaguirre · M. E. B. Fernandes (✉)
Laboratório de Ecologia de Manguezal, Instituto de Estudos Costeiros, Universidade Federal do Pará, Bragança, Pará, Brazil
e-mail: mebf@ufpa.br

C. Makowski, C. W. Finkl (eds.), *Threats to Mangrove Forests*, Coastal Research Library 25,
https://doi.org/10.1007/978-3-319-73016-5_31

31.1 Introduction

The ecosystems were degraded or used unaffordable in a percentage of more than 60% at global level. At the same time, more than 24% s the structure of the earth's surface has been transformed due to: land use, preferably modified to cultivable systems, the expansion of agriculture and the loss of more than 35% of the forests of mangroves and their ecosystem services (MEA 2005). The environmental degradation of more than 60% of the ecosystems worldwide were caused by the unsustainable use as the transformation of more than 24% of the structure of the earth's surface caused by the change of land use to cultivable systems (Barbier 2014). In the same way, the promotion of agricultural expansion has led to the loss of more than 35% of the forests of mangroves, along with its ecosystem services worldwide (MEA 2005). These mangrove forests are located in tropical and subtropical areas, where socio-ecological systems cover an area of approximately 200,000 km^2 around the world (Duke et al. 2007).

In Brazil, such ecosystems cover about 6800 km of coastline (Spalding et al. 1997) located within the so-called "Blue Amazon" (Wiesebron 2013; IPS 2015). Nevertheless, the mangrove forests with a greater extension are placed on the Brazilian Amazon coast, mainly in the states of Amapá, Pará, and Maranhão (Spalding et al. 1997). In the state of Pará, three species of mangrove trees dominates the landscapel: *Rhizophora mangle* L., *Avicennia germinans* (L.) L., and *Laguncularia racemosa* (L.) c.f. Gaertn. (Abreu et al. 2016). In the northeastern coast of Pará, the mangrove forests are the most distributed ecosystem at the Caeté-Taperaçu Marine Extractive Reserve (MER), located in the municipality of Bragança (0°53′ 22″ S, 46°40′08″ W) (Fig. 31.1), this reserve has an area of 42,068,086 ha (ICMBio 2012). Caeté-Taperaçu MER is considered a Federal Conservation Unit (CU) of sustainable Use (Brasil 2000), the same that fulfills ecological, social and economic functions (Glaser et al. 2010). The mangroves forests are considered the most

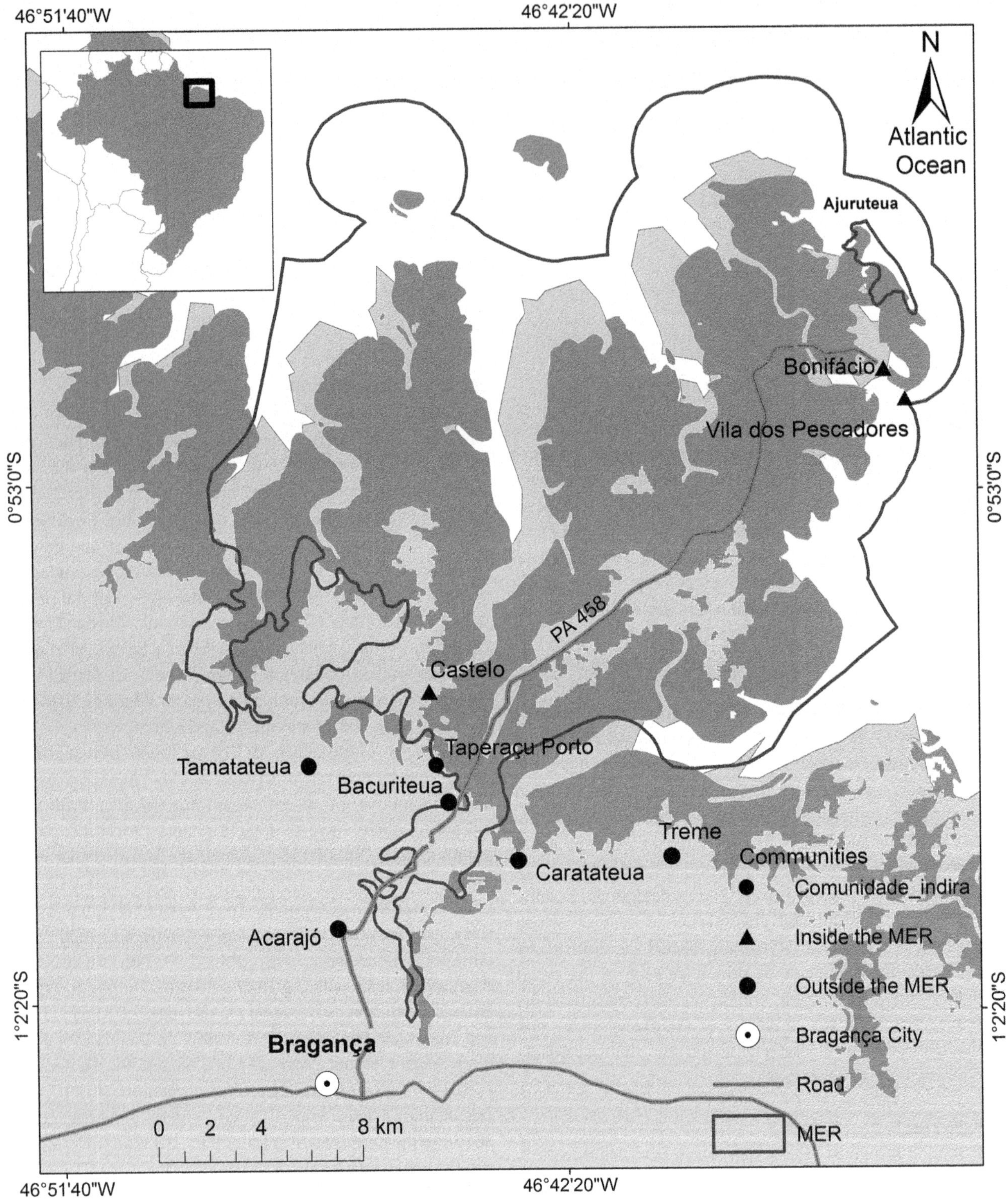

Fig. 31.1 Locations of onsite surveys in the Caeté-Taperaçu Marine Extractive Reserve (MER), Brazilian Amazon coast. Estuarine-coastal communitites inside of the MER: Vila dos Pescadores, Vila do Bonifácio, and Vila do Castelo. Estuarine-coastal communitites outside of the MER: Acarajó, Bacuriteua, Caratateua, Tamatateua, Taperaçu-Porto, and Vila do Treme

productive and vital to the livelihoods of many traditional communities due to the quality and quantity of goods and services offered (Odum and Barret 2008). However, a total of 57 communities live inside and outside MER (ICMBio 2012). Among the main extractive activities practiced by these people, are: fishing, crabbing (*Ucides cordatus*), agriculture, gather shellfish, medicinal plants, clay, and wood for firewood and charcoal, but also tourism, recreation and leisure activities (Oliveira 2015).

Currently, the human activities have accelerated natural processes such as the greenhouse effect and climate change, two global environmental problems that directly affect the stability of the mangroves (IPCC 2002). Two groups of direct and indirect drivers of change can be identified, at the analysis of mangroves ecosystems. Among the direct drivers are climatic factors change such as temperature and precipitation, the invasion of alien species, the landscape change, overexploitation of resources and pollution and change of habitat. Regarding indirect drivers of change, are scientific and technological factors, cultural, political, economic, and demographic factors as indicators of changes of ecosystems (IPCC 2002; UNEP 2014), along with the quality and quantity of their ecosystem services (MEA 2005). Furthermore, the lack of recognition of social and ecological value of services of mangroves on the Brazilian Amazon coast is evident. This fact is based on three factors: (i) the exclusion of externalities, the costs and benefits on the market, (ii) the inefficiency of economic incentives, along with the political decisions, and (iii) the lack of recognition as a common good (Groot et al. 2007). All these factors of change in the mangrove ecosystem influence on the environmental governance and in the implementation of participative policies.

Considering the fact that the political processes on international environmental decisions are neither linear nor rational, on the contrary, they depend on interaction, power networks, participation and the stakeholders' interests that make up the environmental governance (Charlie et al. 2016). This is the case of the highway (PA-458 → 00° 55′36,3″S and 046°42′12,9″W) built in Bragança in the 1970s to link the city of Bragança to the Vila dos Pescadores community. It crossed 26 km inside mangrove forest generating the main ecological impact on the Brazilian Amazon coast (Fernandes et al. 2007). In fact, this highway produced a Corridor of linear Disturbances (Odum and Barret 2008) by changing the structure of the mangroves woods and prevent the water from Caeté River from flooding an area of 90 ha, causing massive mortality of mangrove trees (Fernandes et al. 2007). Consequently, local people and the bilateral agreement (Federal University of Pará/Japan International Cooperation Agency) implanted a participative environmental project REDEMA (Recuperation of degraded mangrove areas in Bragança) in order to recover mangrove degraded areas. The project replanted more than 80,000 seeds/seedlings of the three main local species of mangroves (Tsuji and Fernandes 2008).

The model of MER environmental is represented in two ways: the obligatory and voluntary, both forms are part of the scenario of the environmental governance system. However, the management model of social groups, such as the estuarine-coastal communities, is structured by the model of social group's management, just like the estuarine-coastal communities is structured voluntarily. These social key actors use the ecosystem services of mangroves in a common way, i.e., with a system of rules of use and rules of access to places where they extract resources from the mangroves (Oliveira et al. 2016). These traditional communities can be innovative in the ways of management and in the decisions taking that benefit them with the goal of obtaining sustainable economic development (Ostrom 1990). Many of the rules of environmental management of the mangrove ecosystems stems from the political system of the State, since it is another key social actor in governance of this ecosystem. The Brazilian political structure, for example, is influenced by international agreements and treaties since the signing of the Charter of the United Nations in 1945 (UN 1945). The governance also includes the economic and social development, which is based on international cooperation and has the objective of promoting standards and guidelines for reducing inequalities in an economic degree (Ocampo 2015).

As a result, eight priority objectives at a global level are established among them to ensure environmental sustainability of the participating nations (PNUD 2017). Additionally, it is included the five P's in the Agenda 2013 for the development of Brazil, as the Planet which has the objective of protecting natural resources and the climate for future generations (UNDP 2017). Based on the maturity of the policies on an international, federal, state and municipal scale; environmental governance is the decision-making process which includes the power of stakeholders along with their interrelationships (Gravez et al. 2015). All this dynamics is part of the network of environmental governance and includes: (i) the location where this dynamics takes place; (ii) the objective of the activity, in this case the sustainable conservation of MER; (iii) the resources that are part of this network as the government benefits and even the traditional knowledge of local communities; and iv) the bodies of research developed around the mangrove ecosystem (Charlie et al. 2016). In the same way, the five principles for an effective environmental governance are proposed by the United Nations (UN), among them are: the participation, the search for consensus, accountability, transparency of information, resilience as performance, equity and social justice, and finally the strategic vision (Gravez et al. 2015). Finally, the strategies for an environmental governance in favor of the sustainability of ecosystems, such as mangroves, includes the instruments of environmental management as the designation of CUs and the RAMSAR sites according to the guidelines of the national environment (Brasil 2000). On the other hand, voluntary strategies such as the guidelines of ISO 26000 on the social responsibility are also considered in the market, but they are part of economic development (ISO 2010).

In this chapter, we will first introduce the importance of ecosystem services of mangroves around the MER, even so, we will identify the environmental conflicts in this socio-ecological system from the extraction of common features of mangrove. Then, we will describe the environmental governance in two ways: voluntarily managed by estuarine coastal extractive workers who reside in the estuarine coastal communities as the main social actors in MER management, and the obligatory through the history of environmental legislation in Brazil, placing social speeches in the representations of the extractive representations of MER. Finally, this chapter emphasizes the importance of environmental governance as a framework for the sustainability of the mangrove ecosystem from the ecological dynamics of this system of international importance.

31.2 The Importance of the Ecosystem Services of Mangroves for the Sustainability of the Brazilian Amazon Coastline

Mangroves provide common resources to many local coastal communities (Barbier 2014) by direct influence to the human well-being of these populations (UNEP 2014). Therefore, the traditional communities depend on the direct or indirect benefits of ecosystem services (Costanza 1997), the same ones that are classified into four main categories: (i) support services, (ii) currency regulation, (iii) provision services and (iv) cultural services (MEA 2005). On the northeastern coast of Pará, in Brazil, the support services were classified into two groups: Biodiversity being found different taxonomic groups of mammals, groups of marine vertebrates (e.g. shipworms, mollusks, crustaceans, and oysters) and terrestrial invertebrates (e.g. ants, spiders, and termites). Among the regulation services there are biological factors associated with the vegetation of mangroves as the regulation of trophic networks through the floristic and faunistic resources of mangroves, as well as hydrologic bio-remediation against erosion, saline intrusion for coastal protection of the region. The services of provision are the availability of ecological biomass (e.g. the biochemical cycles of mangroves) and efficiency (e.g. extraction of shellfish and fish, wood, coal, and domestication of animals such as the bee of mangroves for extraction of honey and extraction of non-mineral resources for the preparation of local handicrafts). Finally, among the cultural services is the production of scientific and cultural information from the mangroves (e.g. conservation projects with community participation and the scientific development of educational institutions). In the symbolic group is the symbiotic relationship of the social and natural system (e.g. the symbolic uses of resources of mangroves in the festivals of fish and crab of the estuarine-coastal communities), as the development of local ecological knowledge (e.g. medical use of species of mangroves used by dwellers). In addition, there is the conservation of artisanal fishing and crabbing spaces in the mangrove, based on social representations. It is the case of the apparition tales regarded as a magical force in the form of evil spirits (Galvão 1955) such as *Ataíde*, the spirit who lives in the mangrove forests and protects them, it is the site narrative classified as wonderful story (Silva et al. 2014).

The benefits of ecosystem services of mangroves are used by 57 estuarine-coastal communities (ICMBio 2012; Oliveira 2015) or mangrove civilization (Diegues 1995) that lives in the marine-coastal biome (Toledo 2001), and that directly or indirectly depend on MER, due to ecological, social, cultural and economic support (Glaser et al. 2010). These mangrove dwellers extract resources from the mangroves that are of common use (e.g. timber, fish, and shellfish) through the transmission of local ecological knowledge. The artisanal fishermen, for example, perform the role of "guardians", as they are responsible for the sustainability of the ecosystem of global importance (Cordell 2000). All the ecosystem services that supply the mangroves are managed directly or indirectly by the dwellers on an informal basis, although this ecosystem is ruled by the government entities. The same who manage this ecosystem according to the socioeconomic development of the state of Pará, protecting these services through the supervision for the use of this ecosystem for future generations (Wyman and Stevenson 2007; Saunier and Meganck 2009; Collin 2014).

In fact, mangrove ecosystem services have holistic vision for the interrelationship between ecology, society, and economy (Farley et al. 2009), and its future energetic use and commercial value (Quoc et al. 2012; Glicksman and Kaime 2013). All the services identified influence human wellbeing, because many social structures depend on the mangrove ecosystem services as units of common resources (Quoc et al. 2012), as the subsistence of the estuarine-coastal communities in the northern coast of Pará, which depend on the mangrove services. The symbiotic relationship between humans and non-humans (Latour 1993) based on the re-appropriation of ecosystem services (Leff 2004) is a relevant for sustainable development because the local users perceived the key threats to mangrove forests (Warren-Rhodes et al. 2011). All this is important for decision making of environmental management according to the models of governance of the socio-ecological system of mangroves (Warren-Rhodes et al. 2011).

31.3 Socio-environmental Conflicts in the Management of Mangroves

In recent decades, the marine ecosystems have been affected by various human activities, such as coastal development, growth of tourism without planning, change of soil, among others (IPCC 2002). On the other hand, over the last 20 years, it has been discussed the costs of economic development on the ecosystem services through the macroeconomic accounts according to the indicators of sustainability in terms of social metabolism (Martinez-Alier 2002). The environmental impact is caused by natural and anthropological factors, the latter based on economic systems. For example, climate change, as an environmental problem at a global scale, is a natural terrestrial process, but it is in process of acceleration caused by human activities leading to changes in the estuaries that are sensitive ecosystems taking into account their adaptation, such as the case of mangrove forests (IPCC 2002). Currently, it is estimated that more than 20% of the mangroves have already been lost between the years of 1980 and 2005, caused by natural and anthropogenic impacts (UNEP 2014). The mangrove ecosystem in the MER, for example, is impacted in a natural way by changes caused by oceanographic variations (Souza Filho et al. 2000) and physical processes (Thom 1984; Behling et al. 2001). Modifications on the margins of the beach on the Ajuruteua Peninsula are affecting the local infrastructures (El-Robrini et al. 2006). However, estuarine-coastal communities within the MER (Vila dos Pescadores and Vila do Bonifácio) have been adapted continuously to these changes through the houses made out of wood, which are removable (Fig. 31.2).

Among the main environmental impacts of anthropogenic origin affecting the mangroves of MER are: the deforestation of mangroves due to logging (Oliveira et al. 2016), the predatory fishing activity caused by the use of weirs and nets (IBAMA 2012). The lack of a general sanitary system is also responsible for mangrove forest loss (Pereira et al. 2006), the majority of MER communities do not rely on this basic service, however the provision of domestic sewage ends at the Caeté River and finally on the beaches of the Atlantic Ocean along this region. In the same way, the final disposal of household and industrial wastes (e.g. fuel from small and large boats, ships, etc.) are bio-accumulated in the mud of the mangroves located on the banks of the same river. Another factor of anthropogenic impact is caused by unsustainable management of stakeholders in this region, mainly the estuarine coastal communities and residents of the MER (e.g. the use of ichthyologic plants as *timbó* and *cunambi* for fishing) (IBAMA 2012). Crabbing represents another threat by the lack of organization of the communities by triggering local and regional conflicts (IBAMA 2012). In addition, estuarine-coastal extractivists recognize that pollution of the mangroves is also due to the disorganized growth of their communities caused by the increase in population. All the impacts mentioned above affect directly the quality and quantity of ecosystem services (MEA 2005), to the well-being of the estuarine and coastal waters though. Nonetheless, there are strategies to mitigate these impacts,

(A)

(B)

Fig. 31.2 Wooden houses in the estuarine-coastal communities inside the MER (**A**) Vila do Bonifácio, and (**B**) Vila dos Pescadores intervention projects that include the mangrove dwellers are still missing, those who are the main social actors in the management of the mangroves on the Brazilian Amazon coast. The social representations, for example, are part of the environmental governance and can be used as a tool of sustainability with the goal of preserving the mangrove forest. With this strategy, it is possible to know the use of ecosystem services, along with the rules of access to common resources by the estuarine-coastal communities. In the year 2016, a total of 47 estuarine-coastal activists of the MER were interviewed, it is described the representations of the direct and indirect drivers of change of the mangrove ecosystem along with the process of adaptation of these communities (Fig. 31.3).

In addition to the socio-environmental conflicts, the estuarine-coastal communities recognize other difficulties, and many of these are related to resource extraction from the mangroves. They also claim that the local inspection institutions are not efficient at fulfilling their tasks, such as crabbing and logging. For the latter it is necessary permit issued by the supervisory institutions to be licensed for logging that is an illegal activity in Brazil. The same thing applies to artisanal fisheries that present a local dispute with the fishing enterprises for the fishing resources of the MER. On the other hand, economic incentives, through government programs, make that many of extractive industries become dependent economically on government (e.g. family

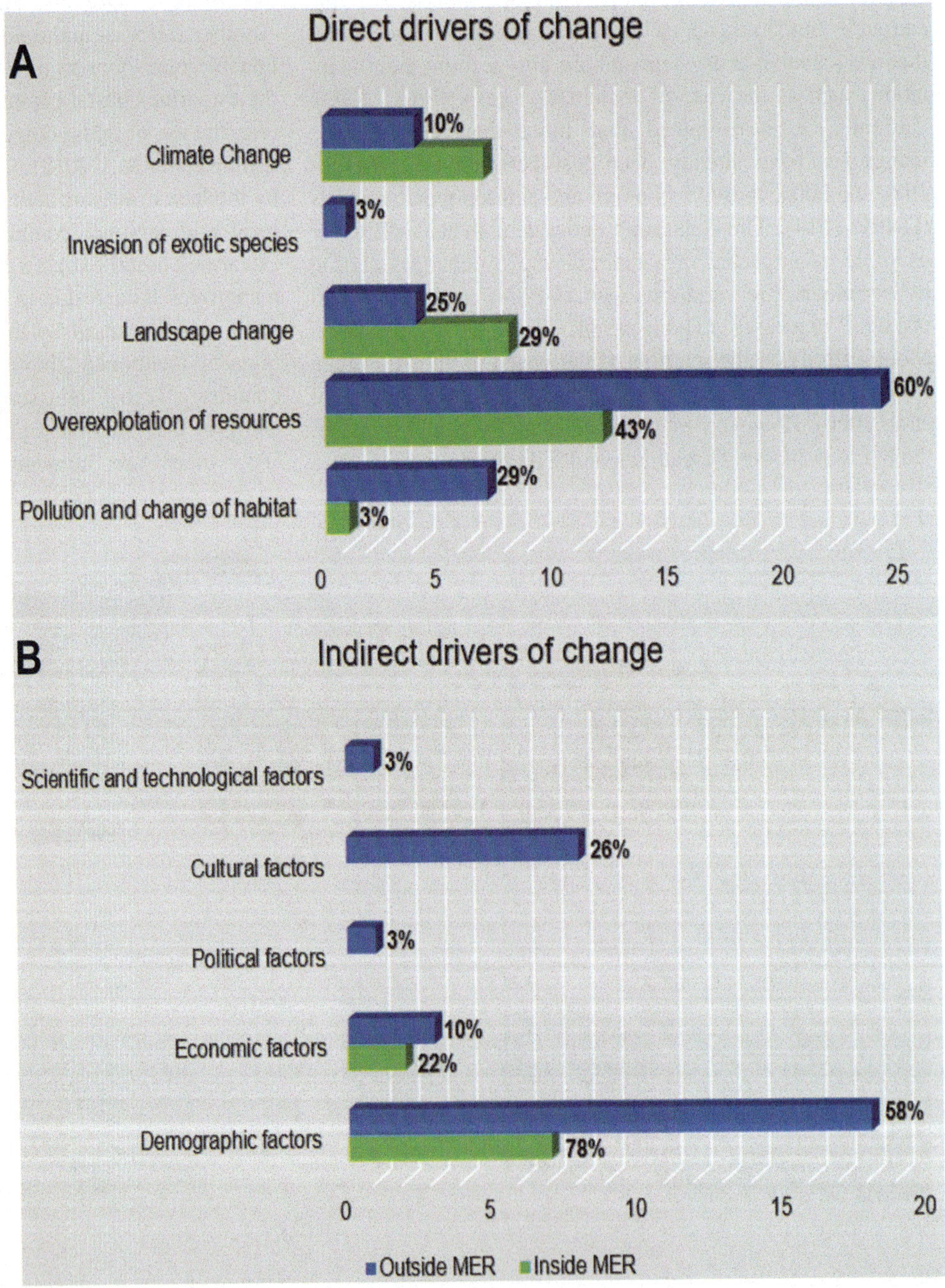

Fig. 31.3 Environmental representations of estuarine-coastal activists on the (**a**) direct and (**b**) indirect drivers of climate change in the MER

allowance, school allowance and "green" allowance), although, according to interviews, many of these extractives workers are not aware about the purpose of incentives. As in the case of the ignorance of the objective of the "green" allowance program funded by the Caixa Economical Federal bank that serves to encourage the conservation of the ecosystem of mangrove and at the same time increase the economic growth of the communities living in extreme poverty (MMA 2017). All of these conflicts are catalysts of the destruction of the common resource of the mangroves. However, the main consequence influences the efficiency of environmental governance of this important system. For example, many of the estuarine coastal activists of MER do not know how to preserve the mangrove forests, due to the lack of information of the importance of ecosystem services, which in turn causes the unsustainability of the socio-ecological system (Rahman et al. 2010).

31.4 Measures of Institutional Intervention in Environmental Management of Mangroves

The measures of environmental interventions in the environment at a global level began in the early 1970s after evaluating the impacts caused by environmental disasters, as well as in the case of MER (see Sect. 31.1.3). However, in 1974, the UN organized the conference in Cocoyoc, Mexico, where institutional obligations on the environment were defined. In addition, it was defined the responsibilities of states on environmental governance, according to article 30 of the Charter of Economic Rights and Duties of States, the protection and preservation of the environment must be through the development and strengthening of policies (Martinez-Alier et al. 2016). Later on, with the strengthening of environmental policies and agreements on global environment, the local policies in Brazil are consolidated. However, the socio-ecological system of mangroves is governed compulsorily and voluntarily by both the federal institutions and the mangrove dwellers, respectively. This environmental governance depends on the dynamics of the relationship of stakeholders (Spitzeck and Hansen (2010)), especially the social representations and the participative inclusion of estuarine coastal communities in case of MER. The main goal of governance both voluntary and mandatory is to reduce and/or mitigate the negative impacts on forests of mangroves, mainly from anthropogenic origin (Ortmann 2017). In the following subtitle, it is told how the intervention of institutions that are composed of stakeholders, that handle the mangrove ecosystem voluntarily and mandatorily.

31.4.1 Stakeholders of Mangroves: The Volunteer Social Management

Environmental governance of the mangrove is based on the symbiotic relationship of social groups with their natural environment. The stakeholders influence at different levels and at primary level they are part of the management of the mangroves directly. The estuarine-coastal communities, for example, are part of the social group that has the main role in the management of the ecosystem services of mangroves, but they are the first to be affected before any change, due to the dependence of this ecosystem (Terry and Timothy 1998). At secondary level the stakeholders are affected indirectly, i.e., people who make use of cultural services such as tourism on the beach of Ajuruteua. The stakeholders are also categorized according to the influences on decision making, often those who have more genuine power, usually are the governmental institutions, which decide the management method of the mangrove ecosystem. The stakeholders who are consulted in decision making are often the communities who manage directly the SES of mangroves (Terry and Timothy 1998). However, it is important to trace the profile of stakeholders with the aim of knowing the legitimacy in decision-making on the management of an ecosystem, such as the mangrove. In the profile of interest groups, dimensions of intervention are identified. On the other hand, the ability to use the influence and power over the decisions related to mangrove management is important, since it is possible to obtain management in a sustainable manner. Many of the estuarine-coastal communities of MER depend on the ecosystem services of mangroves, but the social representations (see Sect. 31.1.3) about the social and ecological functions of this system is an important dimension in environmental governance (Krick et al. 2006).

However, the environmental management of ecosystem services on the Brazilian Amazon coast is associated with the presence of social actors, such as the users of the estuarine-coastal communities, those that manage voluntarily this socio-ecological system. These users directly influenced the creation of the MER after conflicts regarding crabbing in some municipalities of the State of Pará (IBAMA 2012). The social actors' contributions such as educational institutions were also important in the creation of MER, through research on the ecological and social importance of the mangrove ecosystem for the livelihood of the estuarine-coastal communities (IBAMA 2012). With this knowledge, it was soon identified and classified the main social actors according to the dependence of the mangrove, i.e., according to the socio-economic activities practiced in this region. Nine social actors were categorized according to the location of the

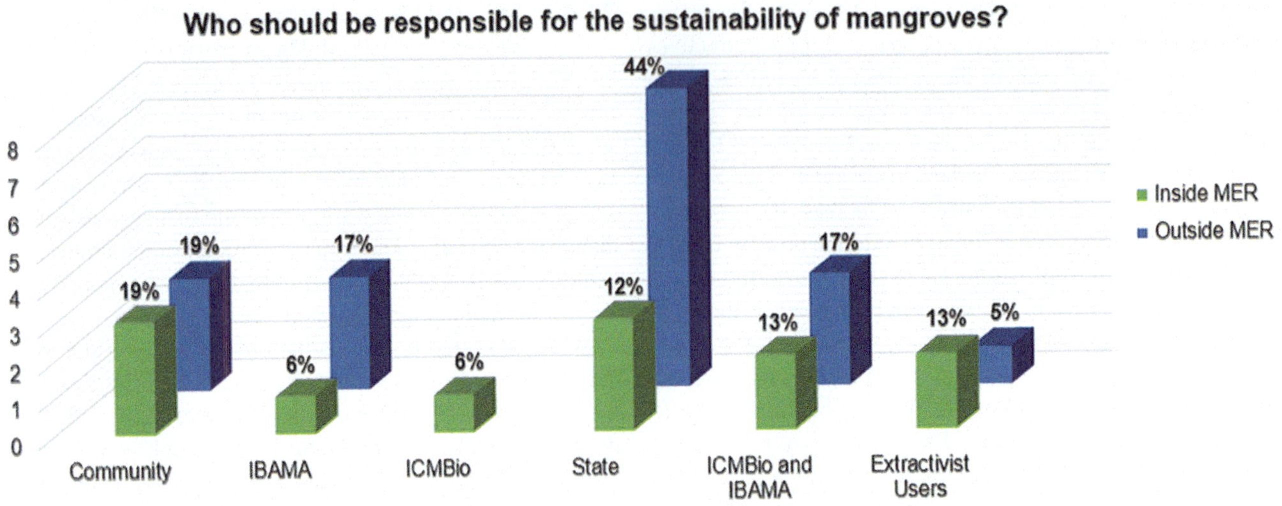

Fig. 31.4 Stakeholder dynamics management of the mangrove ecosystem (UFPA: Federal University of Para, IFPA: Federal Institute of Pará, IBAMA: Brazilian Institute of Environment and Renewable Natural Resources, ICMBio: Chico Mendes Institute for Biodiversity Conservation, SEMMA: Environmental Secretary, IDATAM: Development Institute and Amazon Technical Assistance, SCPA: Artisanal Fishermen Union)

communities inside and outside of MER. In the same way, eight actors were established according to their socio-economic activities undertaken: artisanal fishermen, crabbers, female shellfish gatherers, farmers, loggers, community in general, educational institutions and supervisory institutions. Finally, the social actors were identified according to the distribution of power and competences (legislative, executive and legal). It was identified 22 social actors (Fig. 31.4). It is evident that the social actors identified influence directly or indirectly the management of mangrove ecosystem services. However, the identification of these services (see Sect. 31.1.2) and the relevant social actors help on the dynamics of SES for the decision-making process in environmental management of the mangrove (Voinov and Bousquest 2010). Many of these social actors manage compulsorily the mangrove ecosystem (e.g. supervisory institutions), whereas other manage voluntarily as the case of the estuarine-coastal communities located inside and outside MER. The latter are the main social actors involved in environmental management of mangroves.

According to the SES theory, the volunteer management belongs to users of traditional communities (Ostrom 1990). So, the estuarine-coastal communities of the MER have their own system of self-organization related to their own dynamics of access to common features such as the extraction of fish, shellfish and wood (see Sect. 31.1.3). The sharing and reciprocity between the estuarine-coastal activists of MER, most artisanal fishermen and crabbers, are evident in four characteristics of this voluntary governance described by Oliveira et al. (2016). The first is the work effort when many of the extractive activities require teamwork. In crabbing activity, for example, most users go to the mangroves in groups of friends of family. They say that the artisanal extraction has to be divided, i.e., they are organized in a way that they form rules for collective use of this common resource. Secondly, another feature is the acquisition of fishing tools, usually the resident communities of the MER build their fishing tools (e.g. meshes, development of appropriate clothing and construction of hooks for crabbing). In the Castelo community, for example, the elders sculpt needles made of wood from red mangrove (R. mangle) the same tool they mend their meshes. In the Tamatateua community, women make appropriate garments (e.g. rubber boots) for crabbing in the mangrove. All this local ecological knowledge is transmitted to the new generations, and is part of the volunteer governance of ecosystem services for the benefit of human well-being. The definition of environments for extraction of marine resources is the third feature, MER users decide together the places for crabbing according to the council of elders, the availability of transport to the places (e.g. bus, walking, biking, boat tail). Finally, the last feature on the use of common resources of mangroves, are the agreements and the setting of rules of use. These rules are defined voluntarily by the estuarine-coastal activists themselves of MER, based on representations of the mangroves. They emphasize that the places are defined according to the availability of recourse to extract, the accessibility according to the external factors, such as the supervision of the government entities and climatic conditions.

The social actors of volunteer governance, as the estuarine coastal are considered by MER, have social representations about the mandatory governance. In a study conducted in the

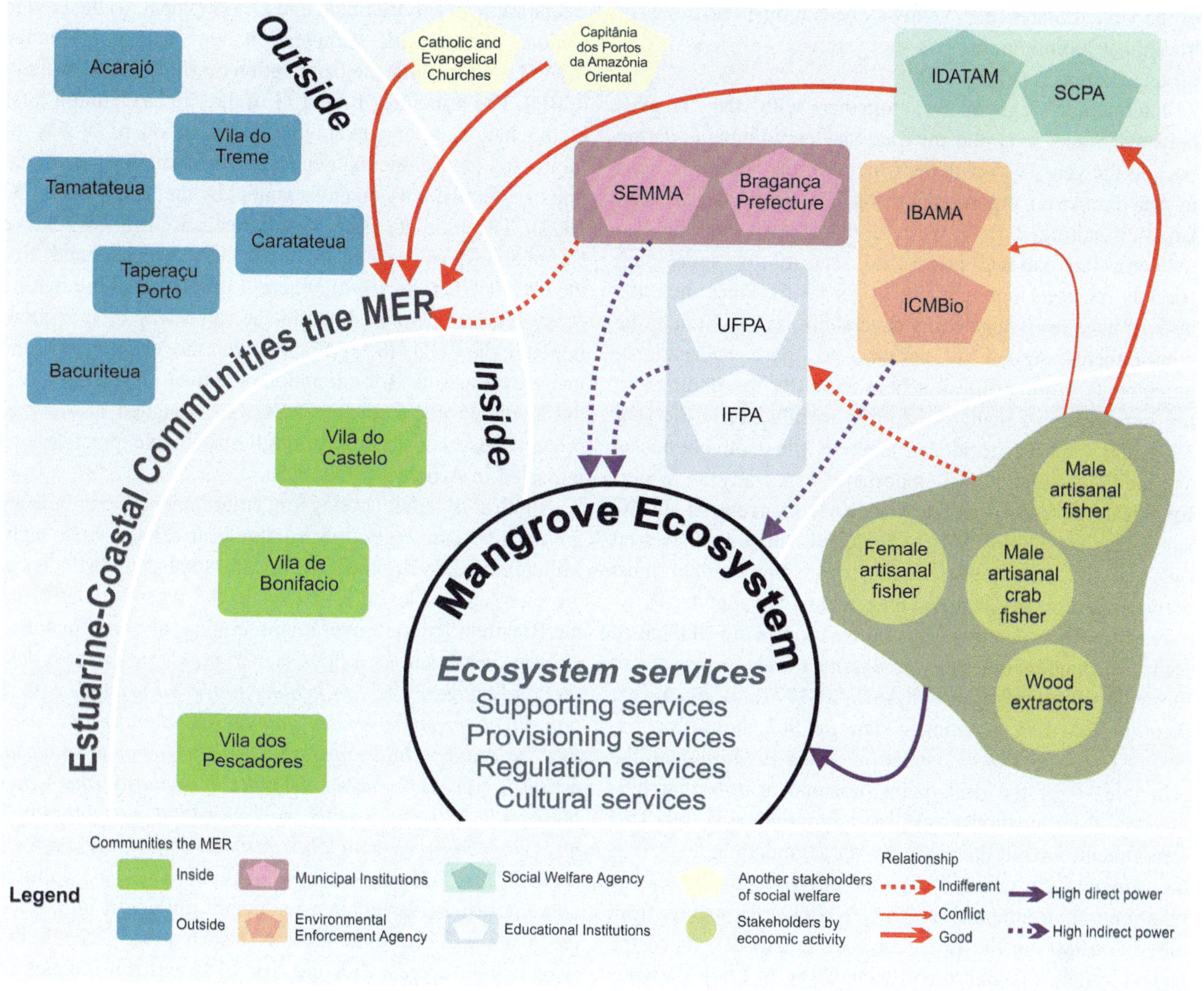

Fig. 31.5 Environmental representations of estuarine-coastal activists on the responsible for the sustainability of mangroves

year 2016, a total of 47 estuarine-coastal activists, who live in nine communities inside and outside the MER, were interviewed. They were questioned – in a hypothetical case – Who do you think is responsible for the management of mangroves in the case of some environmental disaster either natural or caused by human activities? The majority of respondents who live outside MER (44%) answered: In general, the State is the first responsible if any environmental disaster happen. The users who live outside the MER (5%) answered that the responsibilities are the estuarine coastal activists, i.e., they identified themselves as the responsible party for the management of the mangroves. On the other side, estuarine-coastal activists who live within the MER answered that the primary responsibility is the community as a whole (19%), and to a lesser extent the regulatory agencies (13%) (Fig. 31.5).

31.4.2 Institutional Political Dynamics in the Mandatory Management of Mangroves

The estuarine-coastal communities have influence on the management of mangroves due to being a resource for common use, while the State is also another social actor that manages the mangrove forests, through the promulgation of legislation applicable within the Brazilian territory. The Caeté-Taperaçu MER is a CU where the residence of traditional communities is allowed, according to the Brazilian environmental legislation. This reserve is part of the National System of Conservation Units. This classification is the result of a set of international agreements, which later influenced the Brazilian environmental legislation, as in the case of the concept of unspoilt nature such as models of conservation of the United States (e.g. Yellowstone National Park), where traditional communities are not allowed to live inside protected areas (Diegues 2008).

International agreements, together with the Treaties between States over the international commitment, dating back to the year 1945 with the Charter of the United Nations in San Francisco organized by the UN. In the same year, Brazil is admitted to be part of this Treaty that aims to maintain international peace and security by developing friendly relations among nations (UN 1945). Since then, the nations have been developing several efforts to maintain the commitments signed. In addition to these international agreements in the political sphere and after the findings on the importance of mangrove forests worldwide, in the year 1971 the Nations decide to meet in the Convention on Wetlands of International Importance especially as Waterfowl Habitat, known as the RAMSAR Convention. In the same year, Resolution VIII.14 is signed, where it is stressed out that the stakeholders (traditional communities) must actively participate in the management of wetlands. However, one of the results of this convention is the Manual of Ramsar, which includes the ecosystems of mangroves due to the wide variety of habitat (RAMSAR 1971a, b, c). A year later, the Brazilian government signs the Stockholm Declaration of United Nations Convention on the Environment, the principle 19 of the declaration hereunto affirms that it is essential the educational tasks in environmental issues. This step aimed to avoid the environment degradation, as well as the principle 20, which fosters the development of science regarding environmental issues (UN 1972). However, from the year 1980 on, the Brazilian government developed the largest system of science and technology in Latin America along with more than 15 thousand scientists and researchers in the country (Schwartzman 1993). In fact, it is evident, for example, in the MER where social actors such as the Federal University of Pará (UFPA) and the Federal Institute of Education, Science and Technology of Pará (IFPA), both educational institutions, foster the growth of research in order to increase the knowledge about mangroves on the Ajuruteua Peninsula, Brazilian Amazon coast. The aim of the research is to promote the conservation of mangrove forests distributed along the marine-coastal biome (IBGE 2017). On the other hand, in 1992, under economic development, it is signed the Non-Legally Binding Authoritative Statement of Principles for a Global Consensus on the Management, Conservation and Sustainable Development of all Types of Forests. It is in accordance with the sovereign right to exploit their own resources, along with the application of their own environmental policies without harming the territory from other states. Even so, it is also stated that the integration of land use includes sustainable productivity of forests (UN 1992). In addition, three documents are signed of worldwide importance in terms of the environment: i) the Rio Declaration on Environment and Development, ii) the United Nations Framework Convention on Climate Change (CMNUCC), and iii) the Convention on Biological Diversity (CBD). The principles 10 and 11 of the Rio Declaration refer to the inclusion of stakeholder participation as a way of improving environmental issues, together with the promulgation of effective laws on environment by the State (UNESCO 1992). The ultimate goal of this declaration is to achieve sustainable development, as stated in the article 3 and 4 of the CMNUCC (UN 1992), where it is reaffirmed the marine ecosystems as carbon sinks, in the same way as it is mentioned in the CBD (1992). Finally, after the recognition of the mangroves as sinks for mitigation of climate change, in 1997, the Kyoto Protocol to the CMNUCC is signed in order to protect these ecosystems through sustainable practices, as indicated in Article 2 (UN 1998).

All this dynamic policy on environmental compulsory governance converges toward the year 2000, when eight Millennium Development Goals are stablished (MDGs) at the international level. Recently, in the Agenda 2030 from the Brazilian federal government, eight goals were included and are distributed as follow: *(i) "to eradicate extreme poverty and hunger," (ii) "to achieve universal primary education, (iii) to promote gender equality and empower women," (iv) "to reduce child mortality," (v) "to improve maternal health," (vi) "to combat HIV/AIDS, malaria and other diseases," (vii) "to ensure environmental sustainability", and (viii) "to develop a global partnership for development"* (PNUD 2017). However, these goals influence the political dynamics of the Brazilian government. In fact, it influenced the creation of economic incentives such as the green allowance, which agrees with the first goal and the policies of creation of CUs, which together with the supervisory policies, are part of the seventh goal. The international agreements, signed by Brazil, are ideally suited to the regiments of the Federal Constitution of 1988, which currently is the instrument of highest political power in the Brazilian State. The guidelines on the environment are listed in section VI of this instrument, specifically in article 225. It mentions that everyone has the right to a balanced environment, in common use, and in favor of improving quality of life by means of promoting environmental education for preserving environments, such as the Amazon forest and the coastal zone, considered as national heritage (Brasil 1988a, b). Although some laws and federal decrees on the environment have already been enacted before the Brazilian constitution of 1988. In the case of the mangrove ecosystem within the Brazilian State, the Articles 30 and 39 of Federal Decree 3.438 mention that no one can explore the mangroves for being on land of the Navy (strip of land with 33 meters wide, counted from the high tide line), as it was also cited in the Federal Law 970/1946 (Brasil 1941, 1946). In addition, the article 29, subsection VI of the national policy on the environment, enables the creation of territorial spaces protected by public authorities, federal, state, and municipal levels as the areas of natural protection of relevance and interest such as Extractive Reserves (Brasil 1981). Later in the year 2000, article 14 of Act no. 9985, it is defined the Units of Conservation (UC) as instruments for the conservation. This is the case of an Extractivist Reserve that is defined as an area where the traditional communities coexist in the extraction activity (Brasil 2000). It is established in article 12 of Federal Decree 4.340, the management plan of the Conservation Unit proposal by the owner of reserve (Brasil 2002). In the case of the creation of the Caeté Extractive Marine Extractive Reserve (MER), the Brazilian Institute of Environment and Renewable Natural Resources (IBAMA), at that time, is granted the administrator of this reserve (Brasil 2005). In the same way, it is established the national policy for wetlands in accordance with the criteria of the RAMSAR Convention (Brasil 2006a, b). As a result, seventeen RAMSAR sites are registered according to the nine criteria for identifying wetlands (MMA 2017). However, the Caeté-Taperaçu MER is still not part of the RAMSAR sites, although it counts on a plan of environmental management.

Consequently, the mangrove ecosystem in the MER, as a dynamic system with social and ecological networks, is governed by social actors (see Sect. 31.1.4.1), such as the estuarine-coastal communities, the State, and educational institutions. This entire dynamic is governed by mandatory and voluntary rules, on an international, federal, state and municipal scale. These rules determine the obligations and responsibilities of social actors in taking decisions on the management of the mangrove forests in favor of sustainability. There are eleven principles for effective environmental governance (Gravez et al. 2015). The first is the transparency by sharing and transmission of local ecological knowledge of the local people from MER, as an indicator of the existing cooperativity in voluntary governance (Oliveira et al. 2016). Although there are internal conflicts among the MER communities, there are also rules on the use of access to common resources. This principle, together with the accountability, is part of the network relationship among the main stakeholders in the management of the mangrove forest, as the case of the users of the MER and inspection agencies. The supervision of the mangrove resources is an obligation that derives from the Federal Law 10.165, which specifies in Article 17 defining the rate of Environmental Inspection and Control rate (Brasil 2000). Therefore, the environmental surveillance of logging and fishing resources (fish and mangrove crab) it is the IBAMA's responsibility, though this institution complies with the environmental audit in the estuarine-coastal communities inside and outside of the MER. This activity is supported by article 50 from Federal Law 9.605/98 and articles 35 and 46 from Federal Decree 6.514; both mention that people that extract fish above permitted will be fined up to $100,000.00. The same is applied for intensive logging of mangrove wood for charcoal with commercial purposes without a license, whose fine reaches up to U$ 1500.00 per cubic meter of charcoal (Brasil 2008). In a study on social representations, some users of the Caeté-Taperaçu MER mentioned the conflicts surrounding the inspection performed by IBAMA when responding if they knew some kind of penalty for the extraction of resources from the mangroves.

1. Environmental Agent Volunteer from of the community Vila dos Pescadores
   Usually there is fine, but we try to monitor the staff, usually we talk with the staff, especially not to extract wood. However, on Sunday we caught on the act people deforesting, they (external estuarine-coastal extractivits) *stole wood, and I believe they will be soon arrested and fined. They extract wood to build pens; they unforested an area with about three thousand trees. I have everything recorded and I will show to IBAMA.*
2. Environmental Agent Volunteer from the Treme Community
   We inspected, "condurua" (female mangrove crab) *that could not be killed, as it was female. We also supervised the "mangueiros"* (R. mangle trees)*; it cannot be cut green because it is the food of the crab and other animals. There are people who do not want to obey, who do not like it. I was threatened to death when I was a supervisor there* (Treme community). *The people invited me to follow them, but they thought I was denouncing them, because I was an agent.*
3. Male artisanal fisher from Taperaçu Port community
   In my opinion, at least I think that there should be closing seasons in fishing, just as in crab closed season. Because when it comes to harvest (period of abundance), *there are many small fish and the people who work with weirs, mainly the "curraleiros"* (fishermen) *who work with fuzarca* (predatory fishing tools) *keep killing all fish. Big fish, small fish, it does not matter, all of them are already dead, there is no way of throwing them back to water. I called IBAMA, and they told me they had nothing to do with that, the responsibility was the "curraleiros" and there is nothing to do with the artisanal fishing. They said that they had nothing to do with it and that I should contact Brasilia* (capital of Brazil), *so I tried to contact them, but it was that difficult* (bureaucracy), *then I got upset and gave it up.*

The arguments 1 and 2, agree with the Principle of Direction in management, since the Volunteer Environmental Agents (VEA) are defined by IBAMA. The functions assigned to them are legitimate and have the objective to monitor and communicate the MER's political decisions. This principle is a system of chain of leadership on the common features of mangrove and is transmitted mandatorily (IBAMA) and voluntarily (VEA's helpers). As it can be observed in argument 3, the internal conflicts of the estuarine-coastal communities influence on the efficiency of the surveillance system. The absence of an implemented education policy in respect of duties and rights of the estuarine-coastal extractivists is evident in the management of the MER. The principle of direction adds to the principle of responsibility, since each social actor has a specific function within the management of the SES of mangroves. Besides being able to be legally attributed, this power is often exceeded and not compensated in the local reality, as evidenced in the argument 3.

The participation of stakeholders is another principle of the environmental governance (see Sect. 31.1.4.1). This fact highlights the users' speech, such as the relationship of indifference with municipal institutions as the city hall. For many of the users this institution is present in the communities eventually through the collection of household waste or only in the election stage. The participation of educational institutions such as the university are also important in environmental governance of the MER, as the case of the initiatives of bodies of research that include the active participation of communities, as in the case of REDEMA project through the Laboratory of ecology of Mangroves, Institute of Coastal Studies, Federal University of Pará. Regarding the consensus principles, it is evident since the creation of MER that was on the initiative of important actors, such as the communities and the university after results of bodies of research that broaden the scientific knowledge of mangrove forests in the region. Social justice and equity are two principles that are connected, because they are important factors in the economic development of the mangrove estuarine-coastal communities. Many of the artisanal fishermen are benefited from technical courses offered by external actors governmental and non-governmental institutions (e.g. IDATAM) for equal opportunities of societies considered in extreme poverty. This category is based on the features of social benefits as the *green allowance*, many of the users of the MER receive this federal benefit according to the guidelines of the Federal Law 12.512 and its Federal Decree 7.572 promulgated in the year 2011. Finally, the strategic vision and resilience are the last principles of environmental governance. Although to make the MER socio-economic system more resilient, it is necessary the strengthening of the adaptation capacity of these communities. Many of the estuarine-coastal activists mention continuously the shortage of marine resources (fish and mangrove crab) that they realize over time. Although, many of them identify the responsibility for the conservation of mangrove forests in case of an environmental disaster. There is an extensive database of scientific data, which covers research on interdisciplinary mangroves on the Brazilian Amazon coast. In the same way, environmental projects have been planned that are still incipient, such as the case of reuse of the crab's carapace for the preparation of crab flour, which is a raw material for plant fertilizers, in the community of Treme. These data and projects are part of the principle of strategic vision that is still emerging in environmental governance of the mangroves in this region.

31.5 Conservation Strategies for the Sustainability of the Mangroves on the Brazilian Amazon Coast

The instruments for the conservation developed in favor of sustainability, which fall within the environmental policy, are involved in the dynamics of the mangrove ecosystem and are characterized at the international, federal, state and municipal level.

The categorization of UCs in Brazil is an instrument of conservation, which retains the ecosystems on the units as in the case of MER, according to the guidelines of the Federal Law 9.985 (Brasil 2000). This categorization influences on MER environmental management. Even so, the management plan of this Reserve is also an instrument for achieving sustainability, the same that serves as a guideline for environmental management of this ecosystem (MMA Ministério do Meio Ambiente 2017). In this Plan, five programs are applied of socio-economic and environmental sustainability, each of them with their sub-programs applied to MER (IBAMA 2012). The program of natural resources and productive chains includes sustainable artisanal fishing and crabbing, shellfish, non-timber resources, as well as the management of mangroves *per se* and the development of tourism. Among the programs, environmental protection is also included through monitoring and environmental control of the use of mangrove ecosystem services, according to the Federal Law of Environmental Crimes or Federal Law of Nature 9.605 (Brasil 1998). In Caeté-Taperaçu MER there is an environmental control through the supervision of three key features of mangroves: i) logging for commercial purposes, the predatory fishing and crabbing at the time of closed season. The control stocks of timber, fishes, and crabs are based on the Article 45 of Federal Decree 6.514, where it is reported that it is not allowed the transformation of wood originating from the forest without permission from the environmental inspection agency (IBAMA), through an environmental permit (Brasil 2008). In addition, the article 35 states that it is forbidden to fish quantities greater than those established by law. Crabbing, in turn, is monitored according to the Interministerial Normative Instruction, on Article 1. This legal instrument establishes that it is prohibited the capture, transport, and marketing of any individual of the species *U. cordatus* in the states of Pará, Maranhão, Piauí, Ceará, and Rio Grande do Norte. In addition, it is established that the dates of closed season until the year 2019 are specified (Brasil 2017). The incentive to generate knowledge includes the sub-programs of research, environmental education, identification of priority areas for conservation, and development of technologies. It includes, for example, projects that have been implemented by educational institutions, the social actor responsible to perform research according to both (i) Federal Law 7.797/89 supported by the National Fund for the Environment (Brasil 1989) and (ii) Federal Law 9.795 on environmental education as mentioned in the Article 1 (Brasil 1999). The participative management and the social organization through awareness, communication, strengthening of the Deliberative Council and the associations and cooperatives is also another group of programs for sustainability of MER. Through the promotion of the *green allowance* it is included the participation of the estuarine-coastal communities in the conservation of mangroves, according to article 3 of Federal Decree 7.572 and Federal Law 12.512 (Brasil 2011). Although they lack the presence of an environmental education and policy on the objective of this federal program, this program reaches the aims of the education program as mentioned in the Management Plan of the MER. The last program of the MER is the operationalization of this reserve through the territorial consolidation, administration, infrastructure and the inter-institutional relationships (IBAMA 2012). The MER administration was defined by Federal Decree without number (2005) the creation of MER and the Federal Decree 17 with the creation of the Deliberative Council (Brasil 2007), both political tools come from the second article 12 of the Federal Decree No. 4340 (Brasil 2002).

The majority of these programs agree with the principles of environmental governance, as well as other instruments for conservation. For example, the principle of strategic vision is observed in the networks of the dynamics of the environmental governance system of MER (Fig. 31.6). After all this system aims at reaching a future sustainable management through factors of the conservation opportunities. The social representations of stakeholders are an important factor in environmental governance (Eyzaguirre 2017), as the arguments (1, 2, and 3) where it is realized the level of involvement and the relationship with the other stakeholders. The presence and the impact of other social actors influence the MER's environmental governance (Fig. 31.4). The arguments (1, 2, and 3) demonstrate the role of actors such

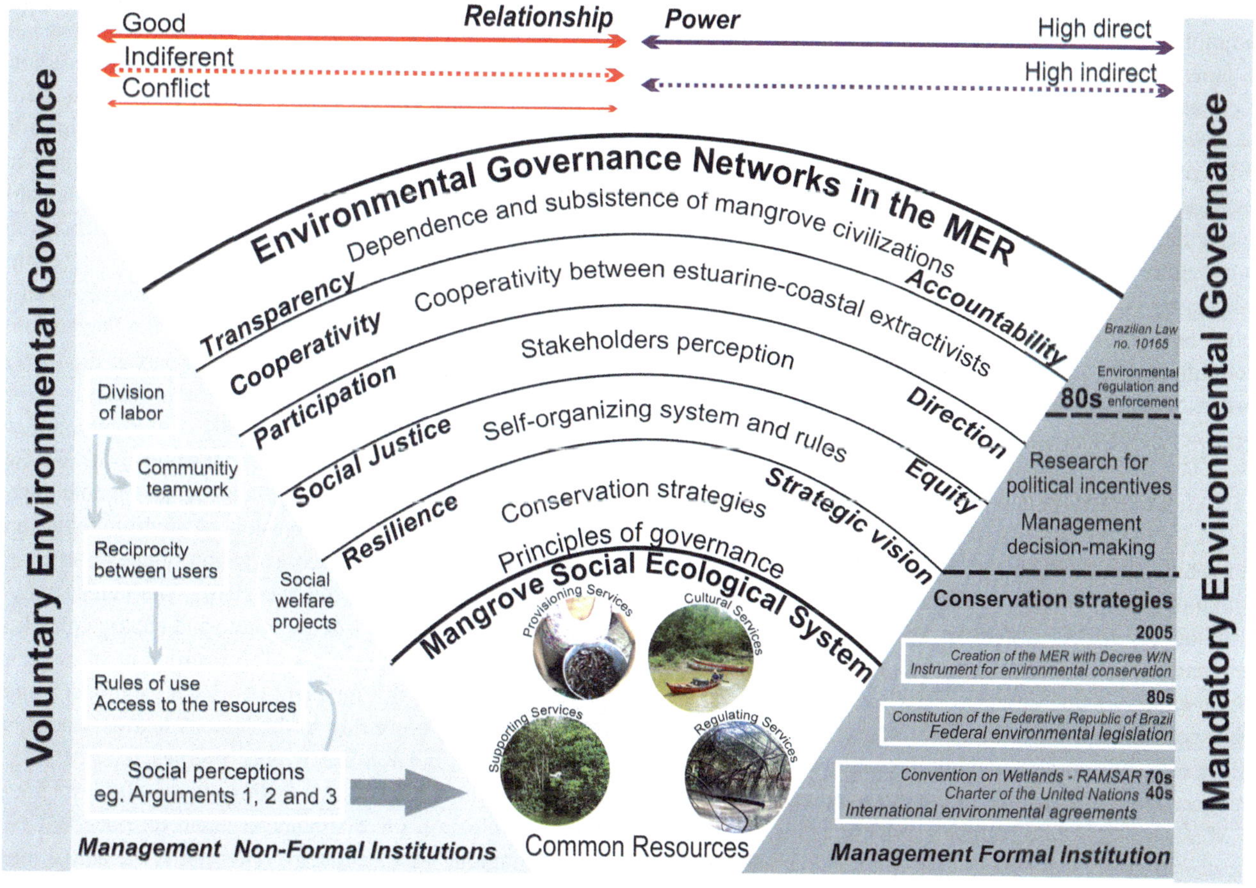

Fig. 31.6 Environmental governance networks in the MER

as VEA and the same users of MER, but also the presence of other actors such as IBAMA through inspection impacts the governance. However, the rules of access and the rules of use of the ecosystem services of mangroves in the MER are included on a voluntary basis in participatory dynamics of the estuarine-coastal communities. On the other hand, the presence of environmental policy around the management of the MER confronts these volunteer dynamics, as evidenced by the argument 3. However, the stakeholders' relationships will influence on the networks of government along with the dynamics of the principle of management, such as the case of the relationship of IBAMA with communities especially with crabbers, often generating a conflict relationship (Fig. 31.4).

Since the signing of the RAMSAR Convention, it is proposed manuals for the management of wetlands at worldwide level in order to establish the Ramsar sites according to the nine criteria for identifying the importance of wetlands. Just as the Environment Ministry (2017) has 17 RAMSAR sites, MER is not still part of this list, although it has the instrument of conservation on the management of UC on the planning and implementation of the Management Plan of this Reserve (IBAMA 2012). In addition, MER counts on the main themes of Federal Decree 5.758 on the Strategic Plan of Protected Areas, through the participation of local communities in direct management of MER (Brasil 2006a, b). However, in order to include the MER in this instrument of conservation (Ramsar sites), a review of compliance with the nine criteria must be performed. The importance of MER to be part of this instrument is related to the sustainable development, which is based on external funding, since among the benefits of international recognition of RAMSAR sites is the priority in the implementation of government policies by political positioning. In fact, it is important for being considered a priority area and having the possibility of international negotiations and access to programs such as the carbon market (MMA 2017).

31.6 Conclusions and Recommendations

This study described the MER's Environment Governance Structure, emphasizing it as a FRAMEWORK for the sustainability of the mangrove system. The environmental services offered by the mangroves in MER, is an important factor for achieving sustainable conservation of SES, since these common resources are governed on a voluntary basis by the estuarine-coastal communities. However, the services recognition is important for the understanding the dynamics of MER, which is based on the symbiotic relationship of the social and ecological system. The conflicts and the impacts from natural and anthropogenic origin of MER influence the relationship and involvement of stakeholders, although the estuarine-coastal extractivists realize the direct and indirect drivers of change. However, it is still missing the inclusion of the stakeholders' social representations regarding the MER's environment governance. On the other hand, the compulsory environmental governance is also present in the SES management of mangroves, through the impact and influence of environmental policies at international, federal, state and municipal scales. We described the eleven principles of environmental governance based on studies of identification of stakeholders, the verification of applicable legislation to MER along with environmental impacts identified by the social representations of the estuarine-coastal communities of MER. The dynamics of environmental governance of this UC depends on the interaction of principles such as transparency of information, accountability on the part of the administrating institutions of the reserve, as well as the strategic vision on the future through the promotion of conservation strategies. In addition to the strategies that come from these international agreements and Treaties, it is possible to include the Clean Development Mechanisms (IPCC 2002) as the implementation of the Strategies for Reducing Emissions from Deforestation and Forest Degradation (REDD) in Latin America, where Brazil is still in phase 2 of this strategic plan (Aguilar-Stoen et al. 2016).

However, Brazil was against this strategy due to the possible menaces to the control of its forests. However, the political changes in 2003 promoted discussion among scientists and activists, who discussed proposals to create compensation mechanisms to pay for deforestation in order to avoid this problem (Aguilar-Stoen et al. 2016). The Carbon Market is another example, a strategy of sustainable conservation practiced by many private organizations, which has the objective of joining in markets with a capital increase (added value) from the *green marketing*. For example, there are the voluntary certification from the International Organization for Standardization (ISO) as part of the market on an international level, where mainly the industries compete to obtain a Certification ISO 14001 (Environment management) or ISO 26000 (social responsibility). These two certificates contain guidelines that include sustainable development, human rights, the inclusion of environmental and social justice, as well as the conservation of ecosystems and their services (Aras and Crowther 2010). The latter strategy complements the strategy of economic and ecological valuation of the ecosystem services. The instrument of environmental valuation is still incipient (Bockstael and McConnell 2007). However, according to the lessons learned about the valuation of use and non-use in other reserves, it is possible to implement this strategy for MER. For a successful implementation of valuation there should be taken into account two main limitations of this strategy. The first is the actual measurement of support services and regulation of ecosystems, being necessary to know more about the functioning of this ecosystem along with the direct and indirect drivers of change (Bockstael and McConnell 2007). On the other hand, the analysis of the resilience is a characteristic that describes the capability of the system, but it should be included as a strategy for conservation (Bockstael and McConnell 2007).

In general, the environmental management of mangrove forests can have three kinds of failures: (i) the lack of qualified administrators, (ii) the lack of institutions that manage effectively the ecosystem services of mangroves, and (iii) the mapping of environmental conflicts present and those who can probably occur (Rahman et al. 2010). Finally, the path to a successful sustainability around the environmental management of ecosystems, and still MER as a Conservation Unit, is the promotion of strategies of sustainability by using instruments for conservation as the programs of protection, monitoring, supervision, and the partnership through various social actors (Kairo 2008). It is also relevant to consider the inclusion of political ecology and ecological economics in the system of environmental governance to prevent environmental conflicts caused by the use of common resources and the direct and indirect drivers of change (Martinez-Alier 2002).

References

Abreu MMO, Veiga N, Costa-Neto SV, Fernandes MEB (2016) Vegetação Arbórea: Distribuição Espacial. In: Cruz S (ed) Os Manguezais da Costa Norte Brasileira vol IIII, 1st edn. Belém, pp 35–78

Aguilar-Støen M, Toni F, Hirsch C (2016) Forest governance in Latin America: strategies for implementing REDD. In: Castro F, Hogenboom B, Baud M (eds) Environmental governance in Latin America. Palgrave Macmillan, Basingstoke, pp 205–233

Aras G, Crowther (eds) (2010) A handbook of corporate governance and social responsibility. Routledge, London

Barbier E (2014) Account for depreciation of natural capital. Nature 515 (6):32–33

Behling H, Cohen MCL, Lara RJ (2001) Studies on holocene mangrove ecosystem dynamics of the Bragança peninsula in north-eastern Para, Brazil. Palaeogeogr Palaeoclimatol Palaeoecol 167:225–242

Bockstael NE, McConnell KC (2007) Environmental and resource valuation with revealed preferences a theoretical guide to empirical models. In: Bateman IJ (ed) The economics of non-market goods and resources, vol 7. Springer, Dordrecht, p 375

Brasil (1941) Federal Law 3.438/1941. Available from: http://www.planalto.gov.br/ccivil_03/decreto-lei/1937-1946/Del3438.htm (in Portuguese)

Brasil (1946) Federal Law 9.760/1946. Available from: http://www.planalto.gov.br/ccivil_03/decreto-lei/Del9760.htm (in Portuguese)

Brasil (1981) Federal Law 6.938/1981. Available from: http://www.planalto.gov.br/ccivil_03/leis/L6938.htm (in Portuguese)

Brasil (1988a) Constitution of the Federative Republic of Brazil 1988. http://www.planalto.gov.br/ccivil_03/constituicao/constituicao.htm. Accessed 15 May 2017

Brasil (1988b) Federal Law 3.438/1941. Available from: http://www.planalto.gov.br/ccivil_03/decreto-lei/1937-1946/Del3438.htm (in Portuguese)

Brasil (1989) Federal Law 7.797/1989. Available from: http://www.planalto.gov.br/ccivil_03/leis/L7797.htm (in Portuguese)

Brasil (1998) Federal Law 9.605/1998. Available from: http://www.planalto.gov.br/ccivil_03/leis/L7797.htm (in Portuguese)

Brasil (1999) Federal Law 9.785/1999. Available from: http://www.planalto.gov.br/ccivil_03/leis/L9795.htm (in Portuguese)

Brasil (2000) Federal Law 10.165/2000. Available from: http://www.planalto.gov.br/ccivil_03/leis/L10165.htm (in Portuguese)

Brasil (2002) Federal Decree 4340/2002. Available from: http://www.planalto.gov.br/ccivil_03/decreto/2002/d4340.htm (in Portuguese)

Brasil (2005) Federal Decree 20/2005. http://www.planalto.gov.br/ccivil_03/_Ato2004-2006/2005/Dnn/Dnn10531.htm (in Portuguese)

Brasil (2006a) Federal Decree 5.758/2006. Available from: http://www.planalto.gov.br/ccivil_03/_ato2004-2006/2006/decreto/d5758.htm (in Portuguese)

Brasil (2006b) Federal Law 9.785/1999. Available from: http://www.planalto.gov.br/ccivil_03/leis/L9795.htm (in Portuguese)

Brasil (2007) Portaria 17. Available from: http://www.icmbio.gov.br/portal/images/stories/imgs-unidades-coservacao/portarias/RESEX%20Caete%20Tapera%C3%A7u%20Port%2016%20de%2024%2009%202007.pdf (in Portuguese)

Brasil (2008) Federal Decree 6.514/2008. Available from: http://www.planalto.gov.br/ccivil_03/_ato2007-2010/2008/decreto/d6514.htm (in Portuguese)

Brasil (2011) Federal Law 12.512/2011. http://www.planalto.gov.br/ccivil_03/_ato2011-2014/2011/lei/l12512.htm (in Portuguese)

Brasil (2017) Instrução normativa interministerial Mapa/MMA 6/2017. Available from: http://www.iobonline.com.br/Repository/ConsultaDoc?guid=I46C2774E6F9FB380E05330B5DE0A9E04&produto=iof (in Portuguese)

Charlie C, Pearlman M, King B (2016) A conceptual framework for environmental governance networks: an analysis of small island destinations in Indonesia and the Coral Triangle. Transactions on Ecology and The Environment. Sustain Tour 161(5):189–202

Collin P (2014) Dictionary of environment & ecology. Bloomsbury, London

Cordell J Remapping the Waters: The significance of sea tenure-based protected áreas. International Center for Research on Environmental Issues. http://nupaub.fflch.usp.br/sites/nupaub.fflch.usp.br/files/remapping.pdf (2000). Accessed 25 Feb 2017

Costanza R (1997) The value of the world's ecosystem services and natural capital. Nature 387(15):253–260

Diegues A (1995) Povos e mares: Leituras em sócio antropologia marítima, 1st edn. Núcleo de Apoio à Pesquisa sobre Populações Humanas e Áreas Úmidas Brasileiras, São Paulo

Diegues AC (2008) O mito moderno da natureza intocada. NUPAUB, São Paulo

Duke NC, Meynecke JO, Dittmann S, Ellison AM, Anger K, Berger U, Cannicci S, Diele K, Ewel KC, Field CD, Koedam N, Lee SY, Marchand C, Nordhaus I, Dandouh-Guebas F (2007) A world without mangroves? Science 317:41–42

El-Robrini M, Silva MMA, Souza Filho PWM, El-Robrini MHS, Silva OG, França CF (2006) Parpa. In: Muehe D (ed) Erosão e progradação no litoral brasileiro. Ministério do Meio Ambiente, Brasília

Eyzaguirre IAL (2017) Avaliação holística da dinâmica dos serviços ecossistêmicos dos manguezais e suas representações ambientais: estudo de caso na Reserva Extrativista Caeté-Taperaçu, costa amazônica brasileira. Thesis, Universidade Federal do Pará

Farley J, Batker D, De la Torre I, Hudspeth T (2009) Conserving mangrove ecosystems in the Philippines: transcending disciplinary and institutional borders. Environ Manag 45(1):39–51. Springer

Fernandes MEB, Fernandes JS, Muriel-Cunha J, Sedovim WR, Gomes IA, Santana DA, Sampaio DS, Andrade FAG, Oliveira FP, Brabo LB, Silva Júnior MG, Elias MP (2007) Efeito da construção da rodovia PA-458 sobre as florestas de mangue da Peninsula

Bragantina, Braganca, Para, Brasil. UAKARI: Revista eletrônica, Instituto de Desenvolvimento Sustentável Mamirauá 3(1):55–63
Galvão E (1955) Santos e visagens. Companhia Editora Nacional, São Paulo
Glaser M, Krause G, Oliveira R, Herazo-Fontalvo M (2010) Mangroves and people: a social-ecological system. In: Saint Paul U, Schneider H (eds) Mangrove dynamics and management in North Brazil, Ecological Studies Series 211. Springer, Bremen, pp 307–388
Glicksman RL, Kaime T (2013) A Comparative analysis of accountability mechanisms for ecosystem services markets in the United States and the European Union. Transnatl Environ Law 2:259–283. https://doi.org/10.1017/S2047102513000125
Gravez V, Rivadeneira C, Heylings P, Luna D, Cabrera P, Curi M (2015) Gobernanza para el manejo de los recursos naturales y las áreas protegidas, Quito, Ecuador
Groot R, Stuip M, Finlayson M, Davidson N (2007) Valoración de humedales lineamientos para valorar los beneficios derivados de los servicios de los ecosistemas de humedades, Gland, Suiza
IBAMA (2012) Plano de manejo da Reserva Extrativista Marinha de Caeté-Taperaçu (PA)-Volume I, 1st edn. Instituto Chico Mendes de Conservação da Biodiversidade, Brasília
IBGE Instituto Brasileiro de Geografia e Estatística (2017) Biomas. http://www.ibge.gov.br/home/presidencia/noticias/21052004biomashtml.shtm. Accessed 15 Jan 2017
ICMBIO Instituto Chico Mendes de Conservação da Biodiversidade (2012) Plano de manejo da Reserva Extrativista Marinha de Caeté-Taperaçu (PA). Diagnóstico, Brasília
IPCC Intergovernmental Panel on Climate Change (2002) Climate Change and Biodiversity, Technical paper V
IPS Inter Press Services (2015) The blue Amazon, Brazil's new natural resources frontier. http://www.ipsnews.net/2015/05/the-blue-amazon-brazils-new-natural-resources-frontier/. Accessed 15 May
ISO International Organization for Standardization (2010) ISO 2010. https://www.iso.org/home.html. Accessed 15 Apr 2017
Kairo JG (2008) Sustainable management of mangrove resources through a participatory approach – Kenya. In: Examples of successful experiences in coastal community development, vol 16. Special Unit for South-South Cooperation, United Nations Development Programme, New York, pp 97–108
Krick T, Forstater M, Monaghan P, Sillanää M (2006) El compromiso con los stakeholders. Manual para la práctica de las relaciones con los grupos de interés. Ottawa, Canadá
Latour B (1993) We have never been modern, 1st edn. Massachusetts, Cambridge
Leff E (2004) Racionaldidad ambiental la reapropiación social de la naturaleza. Ciudad de México, México
Martínez-Alier J (2002) The environmentalism of the poor: a study of ecological conflicts and valuation. Edward Elgar, Cheltenham
Martinez-Alier J, Baud M, Sejenocihv H (2016) Origins and perspectives of Latin American environmentalism. In: Castro F, Hogenboom B, Baud M (eds) Environmental governance in Latin America, 1st edn. Palgrave MacMillan, New York, pp 29–57
MEA Millennium Ecosystem Assessment (2005) Ecosystems and human well-being, synthesis. Island Press, Washington, DC
MMA Ministério do Meio Ambiente (2017). Available via DIALOG. http://www.mma.gov.br/areas-protegidas/instrumentos-de-gestao/s%C3%ADtios-ramsar. Accessed 16 Jan 2017
Ocampo J (2015) La gobernanza económica y social y el sistema de las Naciones Unidas. In: Ocampo J (ed) Gobernanza global y desarrollo, nuevos desafíos y prioridades de la cooperación internacional, 1st edn. Argentina Siglo Veintiuno Editores, Buenos Aires, pp 31–70
Odum E, Barret W (2008) Fundamentos de ecología. Ciudad de México, México
Oliveira F (2015) Análise da percepção dos extrativistas estuarino-costeiros sobre o zoneamento da extração do caranguejo-uçá (Ucides cordatus) e da madeira nos manguezais da RESEX-Marinha Caeté-Taperaçu, Pará, costa amazônica brasileira. Dissertation, Universidade Federal do Pará
Oliveira F, Maneschy A, Fernandes M (2016) O caranguejo-uçá e a civilização do mangue. In: Fernandes MEB (ed) Os manguezais da Costa Norte Brasileira Vol. III, Gráfica e Editora Santa Cruz. Laboratório de Ecologia de Manguezal LAMA, Bragança, pp 146–175
ONU (1945) Charter of the United Nations. San Francisco, USA
ONU (1972) Report of the United Nations conference on the environment. http://www.un-documents.net/aconf48-14r1.pdf. Accessed 28 May 2017
ONU (1992a) Non-legally binding authoritative statement of principles for a global consensus on the management, conservation and sustainable development of all types of forests. http://www.un.org/documents/ga/conf151/aconf15126-3annex3.htm. Accessed 25 Feb 2017
ONU (1992b) United Nations Framework Convention on climate change. https://unfccc.int/resource/docs/convkp/conveng.pdf. Accessed 15 Feb 2017
Ortamann S (2017) Environmental governance in Vietnam, institutional reforms and failures. Kowloon Tong, Hong Kong
Ostrom E (1990) Governing the commons. Political economy of institutional and decisions. Washington, DC, USA
Pereira L, Ribeiro M, Guimarães D, Souza Filho P, Costa R (2006) Formas de uso e ocupação na praia de Ajuruteua-Pará (Brasil). Desenvolvimento e Meio Ambiente 13:19–30
PNUD Programas de las Naciones Unidas para el Desarrollo (2017). Available via DIALOG. http://www.br.undp.org/content/dam/brazil/docs/agenda2030/undp-br-Agenda2030-completo-pt-br-2016.pdf. Accessed 16 Jan 2017
Quoc T, Kuenzer C, Quang M, Moder F, Oppelt N (2012) Review of valuation methods for mangrove ecosystem services. Ecol Indic 23:431–446
Rahman MM, Rahman M, Islam KS (2010) The causes of deterioration of Sundarban mangrove forest ecosystem of Bangladesh: conservation and sustainable management issues. Bioflux 3(2):77–90
RAMSAR (1971a) Convention on wetlands of international importance especially as waterfowl habitat. http://www.ramsar.org/sites/default/files/documents/library/scan_certified_e.pdf. Accessed 25 May 2017
RAMSAR (1971b) Resolution VIII.14. Teherán, Iran
RAMSAR (1971c) The Ramsar convention manual. Teherán, Iran
Saunier R, Meganck R (2009) Dictionary & introduction to global environmental governance. Earthscan, London
Schwartzman S (1993) Ciência e Tecnologia no Brasil: Uma nova politica para um mundo global. São Paulo, Brasil
Silva A, Cardoso S, Pinheiro L (2014) O lugar dos seres encantados e dos saberes acadêmicos: o realismo maravilhoso em narrativas orais da região bragantina. A Palavrada 5:10–25
Souza Filho PWM, Tozzi HAM, El-Robrini M (2000) Geomorphology, land-use and environmental hazards in Ajuruteua macrotidal sandy beach, northeastern Pará, Brazil. Paper presented at the 1st Simpósio Brasileiro de Praias Arenosas, Associação Brasileira de Estudos do Quaternário, Itajaí
Spalding M, Blasco F, Field C (1997) World mangroves atlas. International Society for Mangrove Ecosystems, Okinawa
Spitzeck H, Hansen E (2010) Stakeholder governance: how stakeholders influence corporate decision making. Corp Gov 10(4):378–391
Terry Y, Timothy H (1998) Using stakeholder processes in environmenal decision making. An evaluation of lessons learned. In: Key issues and future challenges. Available via DIALOG. https://

www.facebook.com/pg/gdrc.org/about/?ref=page_internal. Accessed 25 Sept 2016

Thom BG (1984) Coastal landforms and geomorphic processes. In: Snedacker S, Snedacker J (eds) The mangrove ecosystem: research methods. UNESCO, Paris, pp 3–17

Toledo V (2001) Povos/comunidades tradicionais e a biodiversidade. São Paulo, Brasil

Tsuji T, Fernandes M (2008) Replantando os manguezais: técnicas de reflorestamento. Bragança, Brasil

UNDP United Nations Development Programme (2017) Available via DIALOG. http://www.undp.org/content/undp/en/home/sustainable-development-goals.html. Accessed 16 Jan 2017

UNESCO (1992) The Rio declaration on environment and development. http://www.unesco.org/education/pdf/RIO_E.PDF. Accessed 30 Apr 2017

UNEP Unidade de Ensino e Pesquisa Integrada (2014) The importance of mangroves to people: a call to action. In: van Bochove J, Sullivan E, Nakamura T (eds) United Nations Environment Programme World Conservation Monitoring Centre, Cambridge. 128 pp

Voinov A, Bousquet F (2010) Modelling with stakeholders. Environ Model Softw 25(11):1268–1281

Warren-Rhodes K, Schwarz AM, Boyle L, Albert J, Agalo SS, Warren R, Bana A, Paul C, Kodosiku R, Bosma W, Yee D, Ronnback O, Crona B, Duke N (2011) Mangrove ecosystem services and the potential for carbon revenue programmes in Solomon Islands. Environ Conserv 38:485–496

Wiesebron M (2013) Blue Amazon: thinking the defense of Brazilian maritime territory. Aust Braz J Strateg Int Relat 2(3):101–124

Wyman B, Stevenson H (2007) Dictionary of environmental science. Washington, DC

Erratum to: Threats to Mangrove Forests

Christopher Makowski and Charles W. Finkl

Erratum to:
C. Makowski and C. W. Finkl (eds.), *Threats to Mangrove Forests*,
Coastal Research Library 25,
https://doi.org/10.1007/978-3-319-73016-5

Chapter 2: The Dynamics of Expanding Mangroves in New Zealand

The original version of this chapter was inadvertently published with an incorrect affiliation for Dr. Richard H. Bulmer. The correct affiliation is updated in the current version as:

National Institute of Water & Atmospheric Research Ltd., Hamilton, New Zealand

Chapter 15: Heavy Metal Distribution and Accumulation from Natural and Anthropogenic Sources in Tropical Mangroves of India and Bangladesh

The original version of this chapter was inadvertently published with incorrect citation. The correct citation is updated in the current version as:

Soto DX, Roig R, Gacia E, Catalan J (2011) Differential accumulation of mercury and other trace metals in the food web components of a reservoir impacted by a chlor-alkali plant (Flix, Ebro River, Spain): implications for biomonitoring. Environ Pollut 159:1481–1489.

The updated original online version for this chapter can be found at DOI
https://doi.org/10.1007/978-3-319-73016-5_2
https://doi.org/10.1007/978-3-319-73016-5_15
https://doi.org/10.1007/978-3-319-73016-5

C. Makowski and C. W. Finkl (eds.), *Threats to Mangrove Forests*, Coastal Research Library 25,
https://doi.org/10.1007/978-3-319-73016-5_32

Index

C. Makowski, C. W. Finkl (eds.), *Threats to Mangrove Forests*, Coastal Research Library 25,
https://doi.org/10.1007/978-3-319-73016-5

C

D

N

O

P

Q

R

S

T

U

V

W

The manufacturer's authorised representative in the EU is Springer Nature Customer Service Centre GmbH, Europaplatz 3, 69115 Heidelberg, Germany. If you have any concerns regarding our products, please contact ProductSafety@springernature.com

Printed and bound by CPI Group (UK) Ltd, Croydon, CR0 4YY
01/10/2026
02225991-0005